PROPERTIES OF VARIOUS HOMOGENEOUS SOLIDS

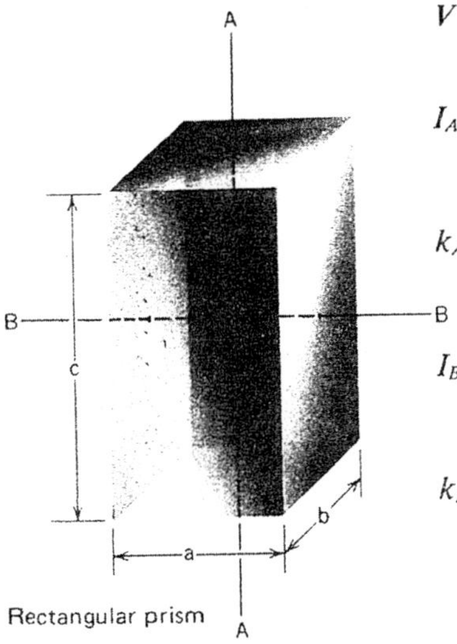

Rectangular prism

$$V = abc$$

$$I_{AA} = \frac{1}{12} M(a^2 + b^2)$$

$$k_{AA} = \sqrt{\frac{a^2 + b^2}{12}}$$

$$I_{BB} = \frac{1}{12} M(b^2 + c^2)$$

$$k_{BB} = \sqrt{\frac{b^2 + c^2}{12}}$$

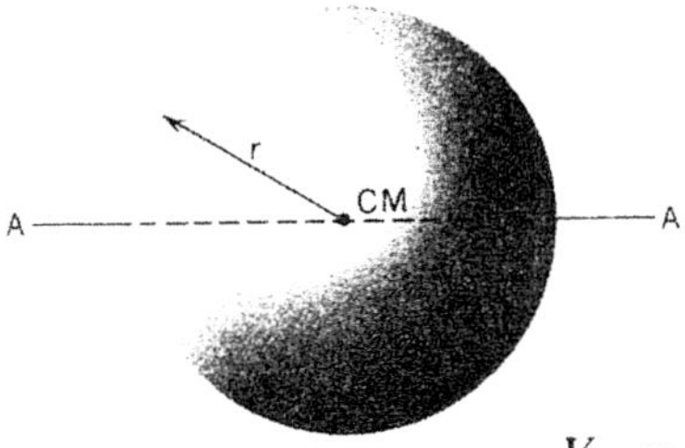

Sphere

$$V = \frac{4}{3} \pi r^3$$

$$I_{AA} = \frac{2}{5} Mr^2$$

$$k_{AA} = \frac{2r}{\sqrt{10}}$$

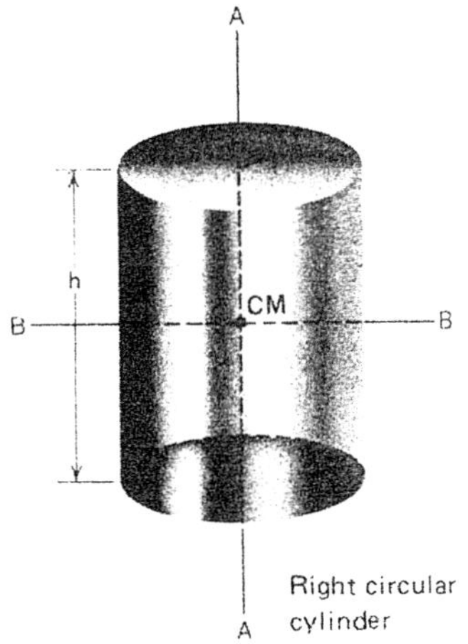

Right circular cylinder

$$V = \pi r^2 h$$

$$I_{AA} = \frac{1}{2} Mr^2$$

$$k_{AA} = \frac{r}{\sqrt{2}}$$

$$I_{BB} = \frac{1}{12} M(3r^2 + h^2)$$

$$k_{BB} = \sqrt{\frac{3r^2 + h^2}{12}}$$

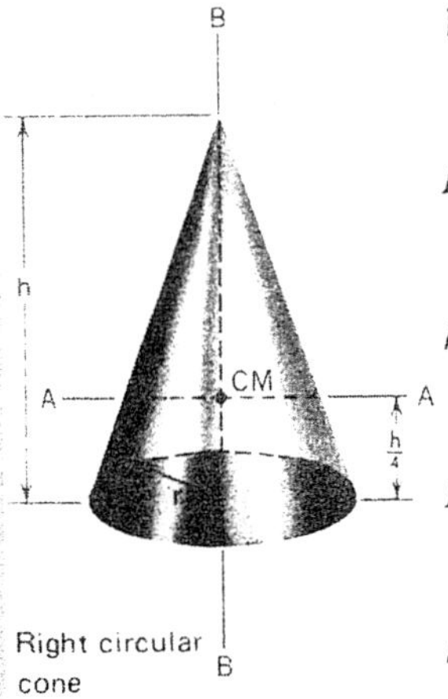

Right circular cone

$$V = \frac{1}{3} \pi r^2 h$$

$$I_{AA} = \frac{3}{20} M\left(r^2 + \frac{h^2}{4}\right)$$

$$k_{AA} = \sqrt{\frac{3}{80}(4r^2 + h^2)}$$

$$I_{BB} = \frac{3}{10} Mr^2$$

$$k_{BB} = \frac{3r}{\sqrt{30}}$$

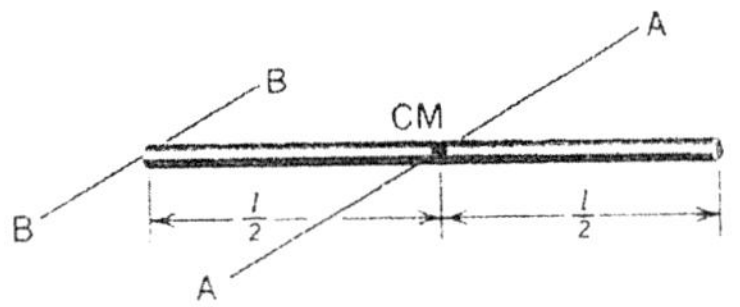

Slender rod

$$I_{AA} = \frac{1}{12} Ml^2$$

$$k_{AA} = \frac{l}{\sqrt{12}}$$

$$I_{BB} = \frac{1}{3} Ml^2$$

$$k_{BB} = \frac{l}{\sqrt{3}}$$

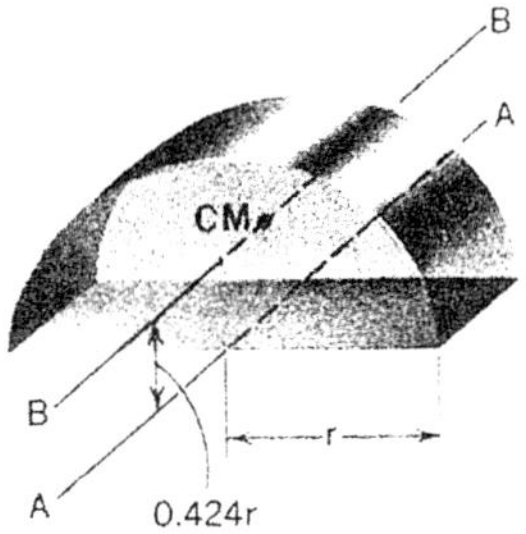

Semi-circular cylinder

$$I_{AA} = \frac{1}{2} Mr^2$$

$$I_{BB} = .320\, Mr^2$$

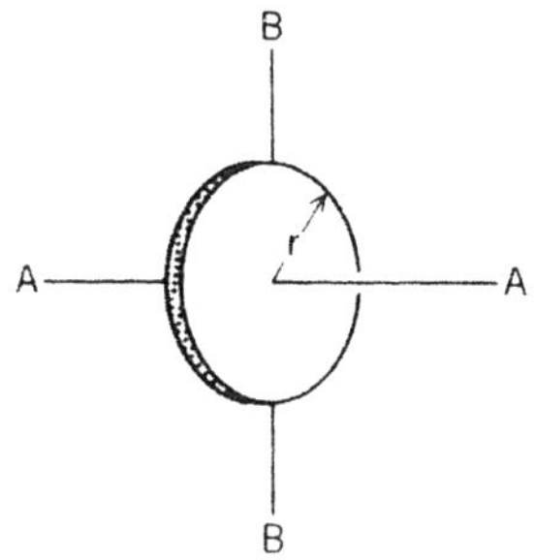

Thin disc

$$I_{AA} = \frac{1}{2} Mr^2$$

$$I_{BB} = \frac{1}{4} Mr^2$$

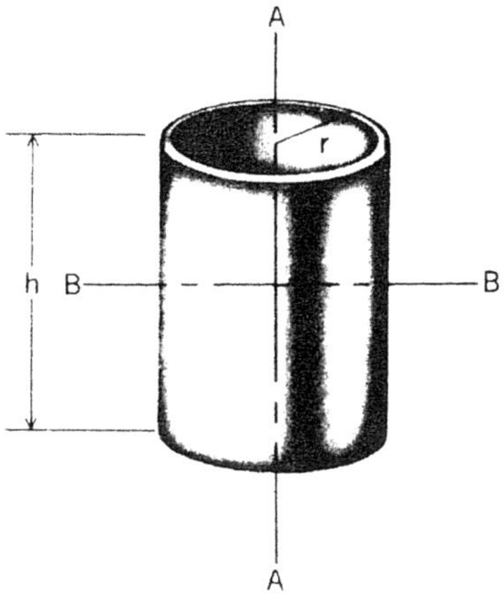

Thin-walled cylinder

$$I_{AA} = Mr^2$$

$$I_{BB} = \frac{M}{2}\left(r^2 + \frac{h^2}{6}\right)$$

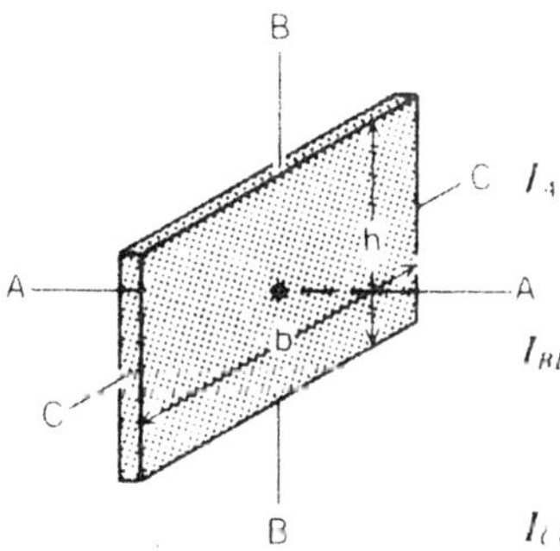

Thin rectangular plate

$$I_{AA} = \frac{1}{12} M(b^2 + h^2)$$

$$I_{BB} = \frac{1}{12} Mb^2$$

$$I_{CC} = \frac{1}{12} Mh^2$$

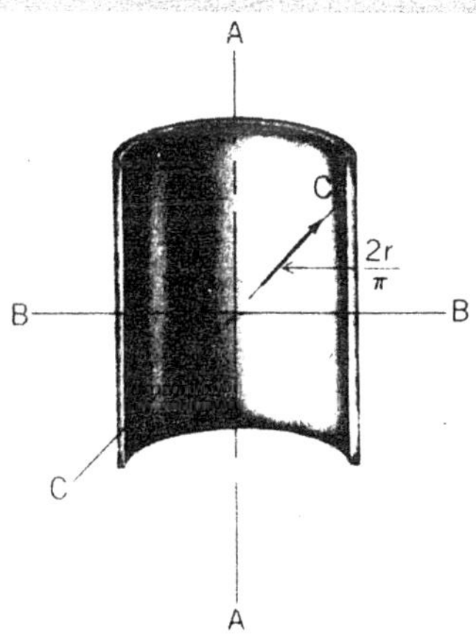

Thin half cylinder

$$I_{BB} = \frac{1}{2} M \left(\frac{r^2 + h^2}{6}\right)$$

$$I_{AA} = Mr^2$$

$$I_{CC} = \frac{1}{2} M \left(\frac{r^2 + h^2}{6}\right)$$

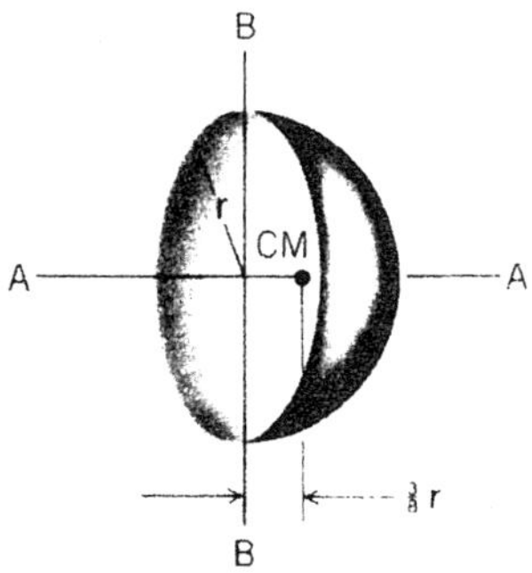

Hemisphere

$$V = \frac{2}{3} \pi r^3$$

$$I_{AA} = \frac{2}{5} Mr^2$$

$$I_{BB} = \frac{2}{5} Mr^2$$

THIRD EDITION

Introduction to Solid Mechanics

Irving H. Shames

The George Washington University

James M. Pitarresi

State University of New York at Binghamton

Prentice-Hall International, Inc.

Original edition, entitled INTRODUCTION TO SOLID MECHANICS, 3rd Edition, by SHAMES, IRVING H.; PITARRESI, JAMES M., published by Pearson Education, Inc, publishing as Prentice Hall, Copyright © 2000.

ISBN 978-89-450-0128-3
ISBN 89-450-0128-X

Contents

A deep-draft caisson vessel to be used for off-shore oil drilling.
Photo by Wilfred Kruger / Black Star / Courtesy Exxon Corporation.

Preface

With the publication of the third edition, this book enters the third decade of its existence—this time with a co-author who used the book, first as an undergraduate student and then later for many years as a professor. The main thrust has not changed from that of its predecessors. We have strived to make this treatment careful and thorough without short changing or sneeking around important but challenging fundamentals. We have presented the theory in a mature manner that we have found for many years and for varied classes to be within the reach of sophomores.* The goal has been to get to a point wherein students know the theory well enough so as to solve problems from first principles. That is, we have tried to avoid presenting in a weak matrix of discussion, lists of procedures for solving various classes of problems and for which strings of examples are presented in a way to encourage the mapping of homework problems from the examples. In short, we have tried to avoid a "black box" approach whose main thrust is on methodology. We have found over the years that students will retain the material much better when learning stems from fundamentals to applications directly with a minimum of rote learning of recipes accompanied by excessive problem mapping. What is worse, we believe that students that "plug and chug," relying on the recipes while cramming for exams, do not mature analytically in mechanics as they should. Furthermore, we believe that well grounded theory in solids will give the student a more meaningful experience and grasp in later mechanics-based courses such as fluids, structures, machine design etc.

The steps we have taken in this edition to help foster problem solving from first principles and to increase retention of basics are as follows:

1. At appropriate places, we have inserted "A Look Back" section that reviews the material covered in Statics, Dynamics, and Physics for purposes of continuity and to make for greater ease in dealing with new material that depends on these earlier studies. These sections are short and to the point.

*Later, more will be said about sophomore coverage as well as possible use of this book.

2. There are a number of starred sections entitled "A Look Ahead" which open-end the text toward future work that may interest the student. We found that the more serious students will look at this material and at a later in more advanced courses will come back to these sections for valuable linkups with the solids courses. At the least, they will see that the theory they are studying with some depth does continue on in later courses.
3. At the end of each chapter, there is a "Closure" that reminds the Student of the contents of the chapter and sets the stage for the next chapter which with an "Introduction" section picks up the subject thread.

Thus, with these three items, we are attempting to provide continuity first between courses in the curriculum and also continuity between chapters in the course. This is important since mechanics in a vertical subject and solids is but one part of a very extensive discipline. That is another reason for emphasizing the theory since it is the glue that binds the various aspects of the subject. There is yet one more new item that we have included.

4. At the end of each chapter, there is a "Highlights" section. Here, we go over the essence of the chapter without mathematical and developmental details to give a physically meaningful discussion of key items. This reading comes after the student has read relevant parts of the chapter for his/her course, done the assigned homework, participated in classroom work, and heard and participated in discussions in and out of the classroom. This possibility of an uncluttered but informed overview we find to be very valuable for the learning process.

We wish to emphasize that the developments and the examples are firmly built around what we call the three pillars of solid mechanics, namely **equilibrium, constitutive laws,** and **compatibility**.* Compatibility is carefully defined early in the book in terms of requiring the strain fields to be properly related to single-valued and continuous displacement fields. We do not use the second-order partial differential equations to satisfy compatibility but instead use geometry and trigonometry to accomplish this goal. As an example, for a truss we make sure at a joint that, when the pin is associated with one of the members, its movement is compatible with the movement of this pin when it is associated with any other member at the joint. That is, the pin must end up at the same position, independent of which member it might be associated with. Thus, in this way using geometry and trigonometry we insure a single valued, continuous deformation of the truss. Those readers that get into the energy formulations will see that compatibility plays an analogous and an equally important role as equilibrium. In energy formulations we can use the vital strain-displacement relations developed early in the book to form compatible strain fields. The three pillars are identified as they occur in the energy material and are thus highlighted throughout the entire book.

*Notice from the cover that even the publisher is supporting **"SOLID MECHANICS"** with three pillars—presumably the three mentioned above.

As is well known, there are two conventions in use for shear forces, bending moments, etc. One is highly favored by civil engineers (called the structural convention) and the other (called the stress convention) is used by a significant number of mechanical and aerospace engineers as well as by some applied mechanicians. We have introduced both conventions. However, at the request of users of the earlier editions, we have gone over to the structural convention in this text. Also, we have used the more standard notation, v, for the deflection of the neutral surface.

Next, we have presented the second Castigliano theorem carefully via the interesting Maxwell-Betti reciprocal theorem fairly early in the text and it is used for trusses at its introduction. Later, it is used in the beam and torsion chapters. In presenting this very useful theorem, we point out that it is the third derived principle in the **energy force methods** and that there is an analogous set of principles called the **energy displacement methods**. The interested student is invited to examine these beautiful and powerful systems of principles in chapter 18. Advanced work in solid mechanics will require a thorough understanding of this material. We have found that this chapter, although rigorous, is within the reach of competent sophomores should the instructor desire to use it for extra credit or honors work.

To discourage excessive mapping, the homework problems are placed at the end of the chapters. Two-thirds of the problems indicate in brackets the latest section of the chapter for which knowledge is needed for a solution; the last third is in random order. The instructors manual will give the instructor the information as to what last section is needed for a solution of these random problems. Also, the manual includes a three-level rating system of the degree of difficulty to be expected for each problem. Finally, examples are not adjusted to fit each on a single page nor are they delineated as a series of steps. Instead generous explanations are made to elaborate how the theory and the modeling have been applied without such artificial constraints of space and form. In short, the examples are meant more to be read and studied and less to be mapped.

It is the feeling of the authors that the student should be familiar with the use of modern engineering computer-based tools. These include (but are by no means limited to) such general purpose codes as *Mathematica*, MathCAD, and Maple. Such computer programs allow for both symbolic and numeric manipulations of expressions. This can be very helpful for both the derivation of equations as well as their solution. But perhaps most importantly, codes such as these permit the students to conveniently maintain certain key parameters as variables within the solution equations. By subsequently plotting the relationship between the variables, insight into the nature of the solution is gained. In this way, the students can accelerate their development of "engineering feel." These problems are double-starred and are at the very ends of the chapters. We have presented a series of problems throughout the text using this philosophy. We have made the problems independent of any particular software package. The focus is on the nature of the solution and the interaction of the variables. We want to encourage the student to try these problems—they are often challenging and fun! To help foster and maintain computer programming skills, a number of problems and/or projects are also presented whereby the student is asked to write a program using high level computer language such as FORTRAN or C.

Design is blended in throughout the text. We have not presented specialized formulas for this phase of the text. A number of regular problems and particularly computer-based problems and projects are design oriented.

We continue to include a treatment of singularities. We do not use the "disappearing brackets" approach. Instead, we use the step function, the delta function and the doublet function. This approach takes more time to learn, but once learned it can be used in important numerical methods such as the boundary element method, not to speak of heat transfer, electromagnetic theory, and many forms of continuum mechanics. Finally, we wish to point out that our singularities stem from a legitimate area of mathematics called *distribution theory*. We normally do not cover this material in our first course except to explain the gains to be realized from its use. To our delight, we have found that there are some students each year who learn the method on their own in order to spare themselves of the excessive amount of arithmetic for solving the deflection of beams problems.

Now a word about the chapter (19) on the introduction of finite elements. Our reviewers approved of its inclusion. However, our primary reason for including it stems from the following experience. Many of our seniors do design projects and use finite element codes for computing stresses. When queried, we found they were working without the slightest understanding of the finite element method and spending much time as a "black box" participant. We believe it is not wise to have students working blindly with computer software and so we continue to urge these students to study with our help the introductory chapter on finite elements when they use software for finite elements.

A quick inspection of the text will indicate that there is a fair amount of what is considered "advanced" material for sophomores and that there is more material than can be comfortably covered in a three-credit or even a four-credit course. This is deliberate on the part of the authors and we make no apologies for this. We have found that students in our curricula come back to this text in their later studies for linking up new material in solid mechanics with expositions that are extrapolations of their earlier familiar work. As examples, we have included treatments of three-dimensional stress and three-dimensional strain, the six principles of energy methods, singularity functions, a gentle introduction to second-order tensors, viscoelasticity and creep, composite materials, and a direct way to determine the sign of shear stress in thin-walled open members. Furthermore, we have endeavored to write this text to be flexible in its possible uses. Thus, for a number of years an earlier edition of this book was used as a second course at Binghamton when engineering was an upper division program there and the students were transfers from community colleges where they received the usual first course in solid mechanics. Also, by deleting starred material; by carefully choosing topics in the text: and by using problems in the lower levels of difficulty as specified in the instructors manual, the instructor can still present an excellent, rigorous course of modest proportions should that be called for. The authors employ about 65 percent of the book in their respective schools in a first 3 credit hour course.

We have set for ourselves the goal of continuing the development of a quality, time-tested text for sophomore or junior solids courses, and in addition to serve students for the remainder of his/her college work and even beyond.

The following professors were reviewers of this edition of the book:

1. Professor Joseph M. Bracci, Dept. of Civil Engineering, Texas A&M University
2. Professor Makola M. Abdullah, Dept. of Civil Engineering, Florida A&M University
3. Professor Eliot Fried, Dept. of Theoretical and Applied Mechanics, University of Illinois at Urbana-Champaign
4. Professor Michael D. Symans, Dept. of Civil and Environmental Engineering, Washington State University

We thank them for their efforts and their encouraging as well as useful comments. Professor Symans, in addition to an overall review mentioned above, gave us an in-depth line by line critique of the book including rewrites of selected short but important portions of examples and text. We thank him profusely. Professor Shahid Ahmad, a most esteemed colleague of coauthor Shames for many years, alternated teaching solid mechanics each semester with him to large sophomore classes at Buffalo using earlier editions of this book. His comments and ideas for using the book most effectively have found their way into this edition. Most importantly, his support and enthusiasm for the book is most gratefully received. Coauthor Pitarresi wishes to thank his former instructor Professor Toby Richards at Buffalo for permitting him to use a half-dozen of his excellent problems for this edition. And both coauthors wish to thank their other respective colleagues at Buffalo and at Binghamton for their continued support and approval of the contents and educational philosophy of the book. Finally, Professor Shames would like to take the liberty of extending his profound appreciation toward his new distinguished colleagues at The George Washington University for the welcome he has received from them. They have encouraged him to play a significant role in their academic program. And they have welcomed his continued writing efforts while at the same time they have insisted on using his existing books wherever possible.

About the Authors

Professor Shames spent 31 years at the State University at Buffalo where he attained the titles of Faculty Professor and Distinguished Teaching Professor. Since 1993 he has been teaching full time at The George Washington University as Visiting Distinguished Professor and as Professor. At the present time, he has written 10 books in mechanics including undergraduate and graduate texts. These books have been used world wide in English and in a half dozen translations. His books have had a number of important "firsts" that have become mainstays as to the way mechanics is now taught. Remarkably, virtually everything that Professor Shames has published during a period of over four decades is still in print. The following are his textbooks:

- *Engineering Mechanics—Statics,* Prentice-Hall, Inc.
- *Engineering Mechanics—Dynamics,* Prentice-Hall, Inc.
- *Engineering Mechanics—Statics and Dynamics,* Prentice-Hall, Inc.
- *Mechanics of Deformable Solids*, Krieger Publishing Co.
- *Introduction to Statics,* Prentice-Hall, Inc.
- *Introduction to Solid Mechanics*, Prentice-Hall, Inc.
- *Mechanics of Fluids*, McGraw-Hill, Inc.
- *Elastic and Inelastic Stress Analysis* (with F. Cozzarelli), Taylor & Francis
- *Solid Mechanics—A Variational Approach,* (with C.L. Dym), McGraw-Hill, Inc.
- *Energy and Finite Element Methods in Structural Mechanics,* (with C.L. Dym), Taylor & Francis

James M. Pitarresi is an Associate Professor in the Department of Mechanical Engineering at the State University of New York at Binghamton (Binghamton University). Dr. Pitarresi received his BS (1981), MS (1983) and PhD (1986) from the Department of Civil Engineering at the State University of New York at Buffalo. Throughout his career, he has been closely involved with the development and use of computers in solving complex engineering problems. He has worked as a stress analyst for the automotive and aerospace industry and as a postdoctoral research associate at the National Center for Earthquake Engineering Research. Dr. Pitarresi is a consultant to industry and government and is an active researcher having written over thirty-five technical papers. He is currently the director of the Opto-Mechanical Research Lab and co-director of the Vibration Research Lab at Binghamton University. In 1999, he was honored with the Chancellor's Award for Excellence in Teaching and the Peter A. Engel Memorial Teaching Award.

CHAPTER 1

A Review of Some Fundamental Notions

1.1 Introduction

One of the principle concerns of an engineer is to analyze and design safe, useful structures—that is, structures that do not fail unexpectedly or have excessive deformation making them useless. In addition, they should be built without wasted material. By the term **structure** we refer to the components needed to support loads and to keep the deformation within accepted limits. Therefore, such structures appear in a variety of engineering disciplines, including mechanical, civil, aerospace, biomedical, and more. In order to understand the limits of our designs, it is important that we be able to relate the geometry of the structure, the loads, the material response, and the deformation. To do this, we use the science of mechanics.

Mechanics is the physical science concerned with the movement and deformation of a body acted on by mechanical, thermal, or other loads. Such considerations are involved in many situations confronting the engineer; therefore, mechanics lies at the core of much engineering analysis. In fact, it is the oldest of all the physical sciences. The writings of Archimedes (287–212 B.C.) covering buoyancy and the lever were recorded before 200 B.C. Early studies of beams and rods under loads were conducted by Leonardo da Vinci (1452–1519) and Galileo Galilei (1564–1642). Our modern knowledge of gravity and motion was established by Isaac Newton (1642–1727), whose laws founded Newtonian mechanics.[1]

In your approach to the study of mechanics, it is common to first learn the basic concepts as applied to *rigid bodies*—that is, the mechan-

[1]See S.P. Timoshenko, *History of Strength of Materials*, Dover Publications, Inc., New York, NY, 1983

ics of structures in which we are *not* concerned with the internal distribution of load or deformation. These topics are commonly covered in your introductory physics, statics, and dynamics courses. This book is the next step in that learning process. Here, we are chiefly concerned with *deformable bodies;* that is, we want to know the deformation and distribution of loads *within* the body. Such knowledge is critical for the design of safe, useful structures. This branch of mechanics is referred to as Solid Mechanics. (Introductory courses covering this area of mechanics are sometimes called "Strength-of-Materials," "Deformable Body Mechanics," or "Mechanics of Materials," but they all refer to the same basic concepts of solid mechanics.) In this book we shall be concerned with the statics of deformable bodies. We shall now discuss certain general considerations that form the foundation of our future efforts.

1.2 Fundamental Concepts

A cornerstone to our approach is that of making certain assumptions and simplifications to both the structure and the analysis when necessary. These idealizations allow us to model the behavior of a structure in such a way as to make the equations governing its response easier to use. We will make good use of certain assumptions and simplifications throughout this book. We must be sure, of course, that the results of our idealizations have some reasonable correlation with reality. Also, the idealizations may limit the usefulness of the equations we develop. It is important to not only understand the simplifications and assumptions, but to also recognize the limitations they place on the analysis. All analytical physical sciences must resort to this technique of simplification, and, consequently, their computations are not cut and dried but involve a considerable amount of imagination, ingenuity, and insight into physical behavior. We shall at this time set forth the most fundamental idealizations that we shall rely on in this book.

The continuum. Even the simplification of matter into elementary bodies such as molecules, atoms, electrons, and so on is too complex a picture for most problems of engineering mechanics. In most cases we are interested only in the average measurable effects on the body. Pressure, density, and temperature are actually the gross effects of the actions of the many molecules and atoms. They can be conveniently assumed to arise from a hypothetically continuous distribution of matter, that we shall call the *continuum.* Without such a simplification, we would have to consider the interaction of each of these elementary bodies—a virtual impossibility for most engineering problems!

The rigid body. In many cases involving the action on a structure by a force, we simplify the continuum concept even further. The simplest case is that of a rigid body: a piece of continuum that undergoes no deformation whatever. Actually, every body must deform a certain amount when acted on by forces, but often the deformation is too small to affect the analysis. It is then preferable to consider the body as rigid and proceed with the simplified computations. For example, assume that we are to determine the forces transmitted by a beam to the foundation as the result of a load P (Fig. 1.1). If P is reasonably small, the beam will undergo little deflection and we can carry out a straightforward simple analysis as if the structure were rigid. That is, we can use the *undeformed geometry*. If we were to attempt a more accurate analysis, even though a slight increase in accuracy is not required, we would then need to know the exact position that the load assumes relative to the foundation *after* the beam has ceased to deform, as shown in an exaggerated manner in Fig. 1.2. To do this accurately is a difficult task, especially when we consider that the supports must also "give" a little bit. Although the alternative to a rigid body analysis here leads us to a very complex series of calculations, situations do arise in which more realistic models must be employed to yield the accuracy required. The guiding principle is to *make such simplifications as are consistent with the required accuracy of the results*. To a large extent, this is difficult to do at first. However, with experience, the student can develop a "feel" for the analysis and know when to make certain assumptions.

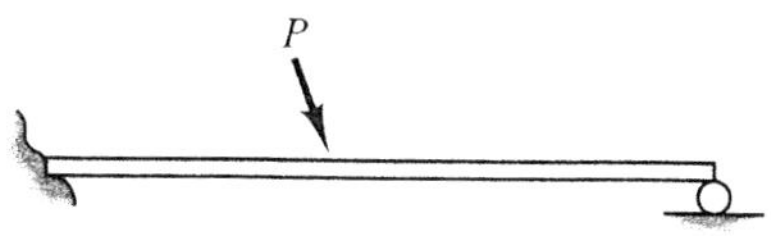

Figure 1.1. Beam showing undeformed geometry.

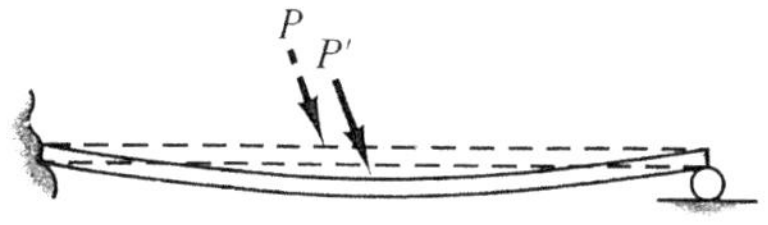

Figure 1.2. Beam showing exaggerated deformation.

We must generally abandon the rigid body model when the applied loads cause the body to deform to such an extent that the final orientation of these applied loads is not known with sufficient accuracy for the problem at hand. At other times we find that the deformation of a body, however small, must be taken into account to solve the problem. We shall encounter the latter problems often; they are called *statically indeterminate* problems, for reasons that will soon be clear.

The elastic body and other idealized deformation patterns. Another continuum that we shall find useful, when deformations have to be taken into account, is the elastic body. Such bodies resume their original unloaded shape when the loads are removed from the body. If, additionally, the measure of deformation throughout a body is directly proportional to the load, we have what is called a linearly elastic body. This concept was first put forth by Robert Hooke (1635–1703).

There are other idealizations that closely resemble many kinds of material response in solids—for instance, we have the elastic-plastic body, and the rigid-plastic body. We shall discuss these at greater length later in Chapter 4.

Point force. A finite force exerted on one body by another must cause a finite amount of local deformation and always creates a finite area of contact between the bodies through which the force is transmitted. However, since we have formulated the concept of the rigid body, we should also be able to imagine that a finite force is transmitted through an infinitesimal area or point. This simplification of a force distribution is called a *point force*. In the many cases where the actual area of contact in a problem is very small but is not known exactly, the use of the concept of the point force results in little sacrifice in accuracy. In Figs. 1.1 and 1.2 we used the graphical representation of the point force.

Many other simplifications pervade mechanics. The frictionless surface, the frictionless fluid, and so on will become quite familiar to you as you study various phases of mechanics.

1.3 Vectors and Tensors

From your previous course work, you will recall that a vector can be represented by three orthogonal rectilinear components along the axes of a Cartesian reference. Furthermore, if we rotate the reference to a new orientation, three new components will generally be associated with the new axes to represent the vector. These components take on values that depend in a definite way, through the parallelogram law, on the new orientation of the reference. In more formal language we can say that the components of a vector *transform* in a certain way under a rotation of axes. In your more advanced studies of mechanics it is this transformation law that becomes the basic definition of a vector rather than the one employed in the earlier courses.

There are more complex quantities, such as stress and strain, that we shall soon deal with in solid mechanics that have nine scalar components associated with a Cartesian reference. When the reference is rotated to a new orientation at the point, a new set of nine components associated with the new reference may be used to represent the quantity. As in the case of a vector, the new components are computed with another specific (and more complex) *transformation* equation. We shall examine such quantities in Chapters 7 and 8, and at that time we shall introduce the concept of the transformation of a *second-order tensor*.

1.4 Force Distributions

We shall deal with force distributions as well as point forces in this book. We can profitably subdivide force distributions into two categories at this time. On the one hand, force distributions that exert influence directly throughout the body are termed *body force distributions* and are given in terms of per unit of mass or per unit volume of the matter that they directly influence. Examples of body force distributions are the action of gravity on a body or the action of a magnetic field on a magnetized body. Note that these forces act throughout the entire body. On the other hand, force distributions over a *surface* are called *surface force distributions* or *surface tractions* and are given in terms of per unit area of the surface directly influenced. Remember that surface tractions have both normal components and tangential components of distributed force. A simple example is the force distribution on the surface of a body submerged in a fluid. In the case of a static fluid the force from the fluid on an area element is always *normal* to the area element and directed in *toward* the body. The force per unit area stemming from such fluid action is called *pressure* and is denoted as p. Like force components, pressure is a scalar quantity. The direction of the force resulting from a pressure on a surface is given by the orientation of the surface. [You will recall from earlier studies that an area element can be considered as a vector which is normal to the area element and which is directed outward from the enclosed body (Fig. 1.3).] The infinitesimal force on the area element is then given as

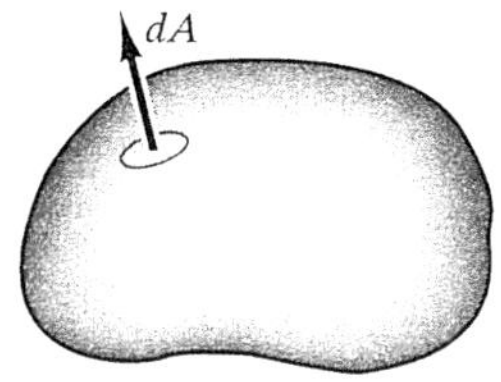

Figure 1.3. Infinitesimal area element on a body.

$$df = -p dA$$

Still a more specialized, but nevertheless common, surface force distribution that we shall see in Chapters 10, 11, and 12 is that of a continuous load on a beam. This is often a parallel loading distribution that is symmetric about the center plane of a beam, as illustrated in Fig. 1.4. We can therefore replace it by an equivalent coplanar distribution that acts at the center plane. The loading is given in terms of per unit length of the beam and is denoted as w, *the intensity of loading*. The force on an infinitesimal element of the beam, then, is $w\ dx$.

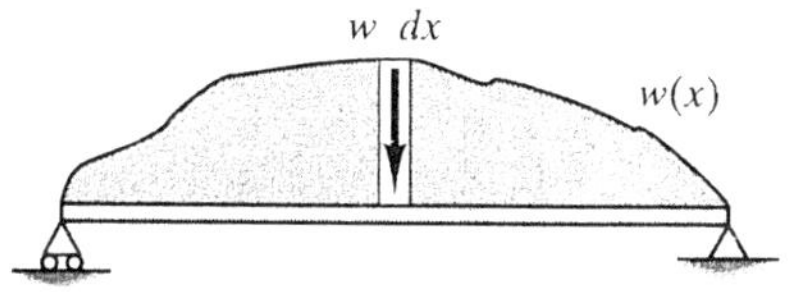

Figure 1.4. Two-dimensional loading distribution on a beam.

1.5 A Note on Force and Mass

In USCS units, we define the amount of mass that accelerates at the rate of 1 ft/s^2 under the action of 1 lbf in accordance with Newton's law as the *slug*. The pound-force could be defined in terms of the deformation of an elastic body such as a spring at prescribed conditions of temperature. Unfortunately, a unit of mass stipulated independently of Newton's law is also in common usage. This stems from the law of gravitational attraction, wherein it is posited that the force of attraction between two bodies is proportional to the masses of the bodies—the very same property of a material that enters into Newton's law. Hence, the *pound-mass* (lbm) has been defined as the amount of matter which at the earth's surface is drawn by gravity toward the earth by 1 lbf.

We have thus formulated two units of mass by two different actions, and to relate these units we must subject them to the same action. Thus, we can take the pound-mass and see what fraction or multiple of it will accelerate at 1 ft/s^2 under the action of 1 lb of force. This fraction, or multiple, will then represent the number of units of pound-mass that are physically equivalent to 1 slug. It turns out that this coefficient is g_0, where g_0 has the value corresponding to the acceleration of gravity at a position on the earth's surface where the pound-mass was standardized.The value of g_0 is approximately 32.2. We may then make the statement of equivalence that

$$1 \text{ slug} \equiv 32.2 \text{ lbm}$$

How does weight fit into this picture? *Weight is defined as the force of gravity on a body.* Its value will depend on the position of the body relative to the earth's surface. At a location on the earth's surface where the pound-mass is standardized, a mass of 1 lbm has the weight of 1 lbf, but with increasing altitude, the weight will become smaller than 1 lbf. The mass remains at all times a pound-mass, however. If the altitude is not exceedingly high, the measure of weight, in pound-force, will practically equal the measure of mass, in pound-mass. Therefore, it is an unfortunate practice in engineering to think erroneously of weight at positions other than on the earth's surface as the measure of mass and consequently to use the same symbol W to represent pound-mass and pound-force.

In USCS there are two units of *mass,* the slug and the lbm. In contrast, SI units, as used by many people, involve two units of *force,* as we shall soon see. The basic unit for mass in SI is the *kilogram,* which is the amount of mass that will accelerate 1m/s^2 under the action of 1 N (newton) force. Unfortunately, the kilogram is also used as a measure of force. That is, one often comes across such statements as "body C weighs 5 kg." A kilogram of force is the weight measured at the earth's surface of a body A having a mass of 1 kg. Note that at positions appreciably above the earth's surface, the weight of the body A will decrease, but the mass remains at all times 1 kg. Therefore, the weight in kilograms equals numerically the mass in kilograms only at the earth's surface, where the acceleration of gravity is 9.806 m/s^2. Care must be taken accordingly in using the kilogram as a measure of weight. In this book we use only the newton, the kilonewton, and so on as the unit for force. Note, a newton is about one-fifth of a pound and a kilonewton is about 200 lb.

Finally, we wish to point to a unit that will be used in much of this book, and that is the *pascal* (Pa). One pascal is equivalent to $1 \; N/m^2$ and can accordingly at times be a unit of pressure, albeit a very small unit. Hence, we will also use kilopascal, denoted kPa and equal to 1000 Pa, as well as megapascal denoted MPa and equal to 10^6 Pa.

In this book we use both USCS units and SI units.

1.6 Closure

In this chapter we have presented a number of fundamental ideas needed to develop the statics of deformable bodies. In particular we considered the concept of the continuum as well as the concepts of point forces and distributed forces. A basic concern to us will be to describe the effects of such forces when they are applied to various kinds of continua. As a first step we shall set forth means of effectively describing the manner in which external loads are transmitted through a body. Accordingly, we begin our studies with the concept of stress.

1.7 A Look Back

In your statics course, you dealt with the vital concept of the **free-body diagram.** You will recall that you isolated a rigid body and showed all the external forces acting on the body. The reason you did all of this was to be able to use Newton's law for that body. Since each such force affects the motion or the lack of motion of the body, clearly it was vital to include all the external forces. And, since internal forces cannot affect the motion of a rigid body, it was also vital not to include such forces in the free-body diagram.

In this course, we shall deal with deformable bodies. We shall continue to use the free-body diagram as in your previous course. We learned in your dynamics course that the motion of the center of mass of any body will be dependent on all the external forces and will be independent of the internal forces. But now there will generally be a difference in the manner of showing the external forces. When we mathematically cut a body in forming the free-body diagram, instead of inserting at the cut the resultant force that existed before we made the cut, we often insert what we shall call **stresses,** which are certain force **distributions,** in place of the resultants as was used in your statics course. In the next chapter, we shall carefully define stresses and will use them freely in our free-body diagrams. As you will soon see, one vital concern for us will be the evaluation of stresses.

CHAPTER 2

Stress

2.1 Introduction

In our studies of rigid-body mechanics, the deformation of bodies was of no significance in the problems we were able to solve. In such problems, Newton's law was all we needed in order to compute certain unknown forces acting on bodies in equilibrium. However, we encountered problems where the use of Newton's law alone was insufficient for the handling of the problem even though the bodies involved seemed quite "rigid" from a physical point of view. For those problems, called *statically indeterminate* problems, the deformation, however small, is significant for the determination of the desired forces. As an illustration, consider the simply supported beam shown in Fig. 2.1(a) with the free-body diagram of the beam shown in Fig. 2.1(b). You may readily solve for the supporting forces A, B_x, and B_y by the method of rigid-body mechanics, provided that there is little change of position of the external loads as they are applied to the beam. We can accomplish this because we know that the total resultant force system on the free body is of zero value and so, by setting this resultant equal to zero while using the *undeformed geometry* of the problem, we can easily solve for the three unknowns. Suppose next that there are three supports for the beam instead of two, as shown in Fig. 2.2. Clearly, the deformation of the beam can be expected to be even smaller in extent than for the two-support system, so there should be no difficulty in this problem arising from a changing geometry. The *total supporting force system from the three supports must be equivalent to the total supporting force system from the two supports in accordance with the dictates of rigid-body mechanics.* However, in the latter problem, we cannot determine the value of supporting forces A, C, B_x, and B_y uniquely since there are an infinite number of combinations of values that will give the required equivalent

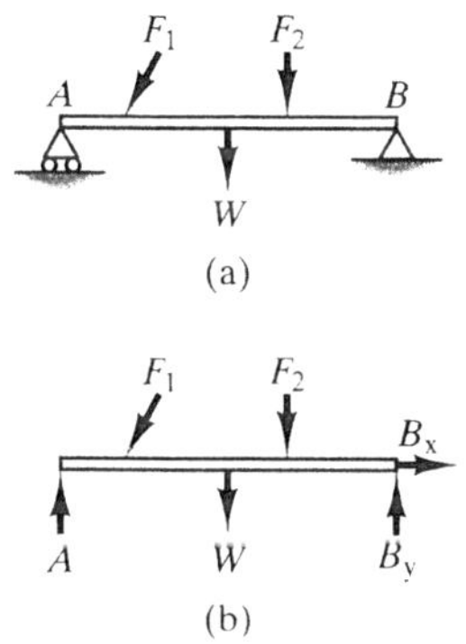

Figure 2.1. Statically determinate beam.

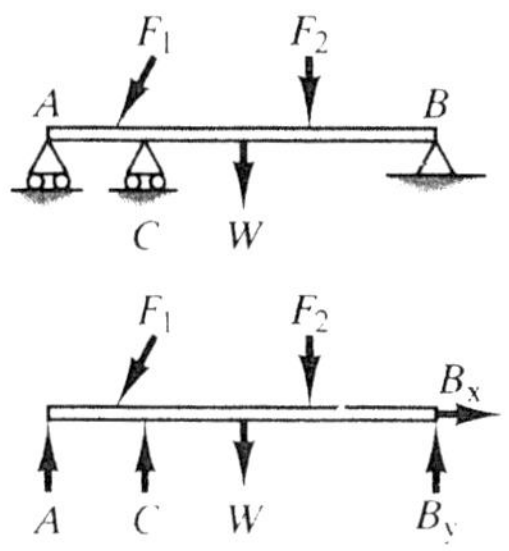

Figure 2.2. Statically indeterminate beam.

resultant supporting force system. Since the resultant supporting force system for Fig. 2.1 satisfies the rigid-body equations of equilibrium, to have the *same resultant,* the resultant supporting force system in Fig. 2.2 must *also* satisfy the rigid-body equations of equilibrium.

> To choose the particular values of the supporting forces for indeterminate beams, now requires the consideration of the deformation of the beam, small as this deformation might be.

In summary we see that the rigid-body considerations afford us a *necessary requirement* for the *resultant* of the supporting force system, whereas the deformation analysis supplies the additional information *sufficient* to determine the values of each supporting force. Hence, we must still use rigid-body equilibrium equations. Statically indeterminate problems akin to the one discussed will be a class of problems that we shall soon undertake.

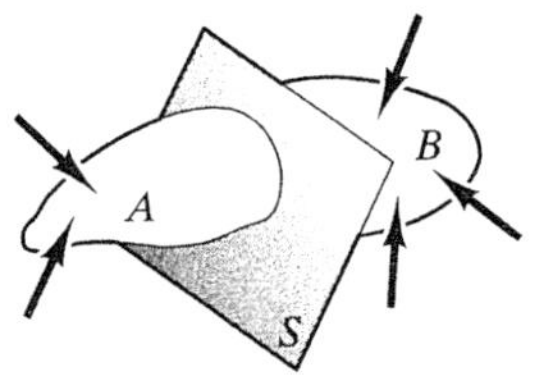

Figure 2.3. Body with a mathematical cut.

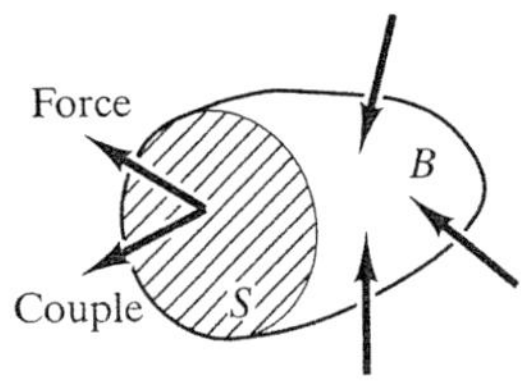

Figure 2.4. Free body showing cut surface.

Next, consider some arbitrary solid in equilibrium, as shown in Fig. 2.3. Suppose that we pass a hypothetical plane S through the body as shown in the diagram. We wish to determine the force distribution that is transmitted from one portion of the body A to the other portion B through this interface. Considering part B as a free body (Fig. 2.4), we can find by the methods of rigid-body mechanics a force and couple moment at any position in the section that is a correct resultant force system for the desired force distribution, provided, of course, that the applied forces have not appreciably changed their initial known orientation as a result of deformation. But just like the supports of the indeterminate beam problem, there are an infinite number of *force distributions* that can yield this resultant force system. As before, we must investigate the deformation of the body in order to obtain sufficient additional information for establishing a unique force distribution.

In investigating deformations in the cases above, we shall have to know something about how forces are distributed throughout the bodies—a topic we shall begin to investigate in the following section. It is to be pointed out meanwhile that the knowledge of force distribution in solids is of vital importance to design of many systems.

2.2 Stress

As pointed out in Chapter 1, we may apply external loads to a body in the form of surface tractions and body force distributions. These external loads transmit their effects throughout the body and cause the body to deform. We shall want to describe quantitatively the manner of transmission of the external forces through a body. We have already shown in Section 2.1 the resultant force $\boldsymbol{F}_R$ and resultant couple moment $\boldsymbol{C}_R$ transmitted across a surface S in a body. We shall now examine an area element δA of surface S, as shown in Fig. 2.5. There will be a resultant force transmitted across this interface given as $\delta \boldsymbol{F}_R$ with a couple moment $\delta \boldsymbol{C}_R$. Consistent with the continuum concept, we may now let δA go to zero size to allow us to generate a force *intensity* at a point on the surface S. In doing this, we shall first decompose $\delta \boldsymbol{F}_R$ into three orthogonal components—one component normal to the interface and two components tangent to the interface along some chosen directions t_1 and t_2. These components are denoted in Fig. 2.6 as δF_n, δF_{t_1}, and δF_{t_2}, respectively. We may now define the *normal stress* τ_n and the *shear stresses* τ_{t_1} and τ_{t_2} by the following limiting processes:

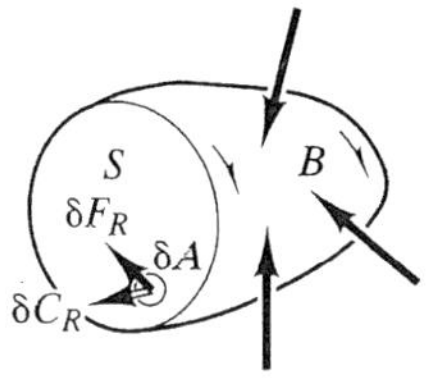

Figure 2.5. Force system on δA.

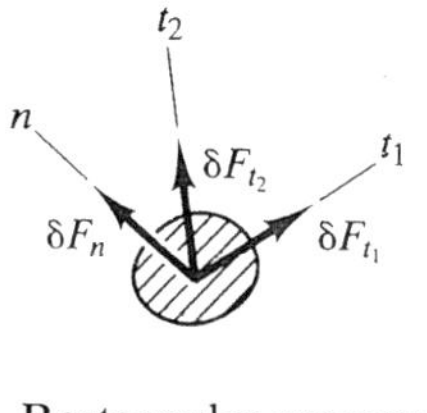

Figure 2.6. Rectangular components of force on δA.

$$
\begin{aligned}
\tau_n &= \lim_{\delta A \to 0} \frac{\delta F_n}{\delta A} = \frac{dF_n}{dA} \\
\tau_{t_1} &= \lim_{\delta A \to 0} \frac{\delta F_{t_1}}{\delta A} = \frac{dF_{t_1}}{dA} \\
\tau_{t_2} &= \lim_{\delta A \to 0} \frac{\delta F_{t_2}}{\delta A} = \frac{dF_{t_2}}{dA}
\end{aligned} \tag{2.1}
$$

It should be clear that τ_n, τ_{t_1}, and τ_{t_2} as defined are scalar quantities and are force intensities (i.e., forces per unit area). As for the couple moment $\delta \boldsymbol{C}_R$ we note that as the area δA approaches zero size, any continuous force distribution on this area approaches that of a parallel unidirectional distribution. We learned in statics of rigid bodies that the simplest resultant for this kind of distribution is a single force with a specific line of action and a zero couple moment. As the area shrinks to zero size we can say that this line of action is at the position of the area dA. We have already accounted for this loading in terms of stresses, so we can dispense with the couple moment $\delta \boldsymbol{C}_R$ in the limit. Thus, we get no contribution from $\delta \boldsymbol{C}_R$ in this process.

For a given set of directions t_1 and t_2, the three stresses will vary in value over the area S, and their specifications as functions of position on S gives us a means of describing the distribution of the force transmitted across the internal section S.

It should be clearly understood that stress is not restricted to solids. Our conclusions here, and indeed, throughout the entire chapter, apply to any continuous medium exhibiting viscosity or rigidity.

We now consider some examples which you are urged to examine carefully.

Example 2.1

Shown in Fig. 2.7(a) is a *thin-walled cylinder* ($\varepsilon \ll D$) at whose ends equal and opposite twisting torques T have been respectively

Example 2.1 (Continued)

applied. What are the stresses on cross sections of the cylinder taken normal to the centerline?

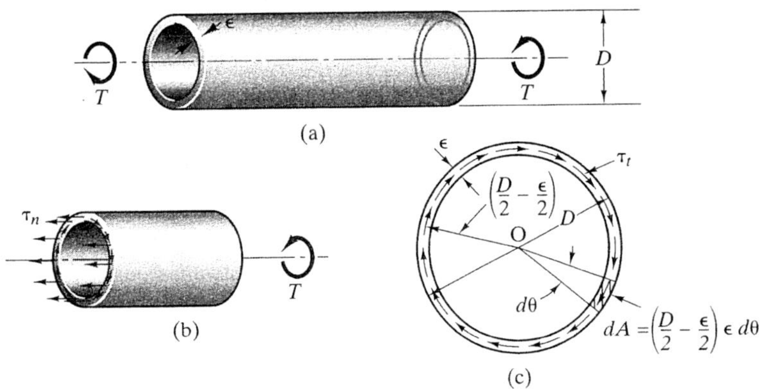

Figure 2.7. Thin-walled cylinder under torsion.

Imagine "cutting" the cylinder over a cross section and forming a free body of the portion on one side of the cut [see Fig. 2.7(b)]. Assume for a moment that a normal stress τ_n exists for the section. Because the cylinder wall is very *thin,* τ_n may be considered constant in value across the thickness of the wall, and additionally, because of *axial symmetry,* τ_n does not vary along the midcircle. Accordingly, τ_n is then constant over the entire section. When we sum forces in the axial direction **equilibrium** tells us what our instincts probably have already told us—namely, that τ_n must be zero. Furthermore, by choosing directions at each point that are *radial* and *transverse* to the section [i.e., out from the center O in Fig. 2.7(c) or tangent to the midcircle, respectively], we may assume that the shear stress at each point in the radial direction is zero[1] and that only the transverse shear stress τ_t is nonzero. Finally, because the cylinder is thin, τ_t is taken as uniform over the thickness of the cylinder, and, because of the axial symmetry, τ_t is also uniform in value along the transverse direction of the section. Thus, τ_t is constant for the entire cross section. Before proceeding further, remember that τ_t, like pressure, is a force *intensity,* so that for shear force the increment from shear stress τ_t acting over area dA is given as $\tau_t dA$. Accordingly, we can say from

[1]You may readily comprehend this assumption when you study the complementary property of shear stress in Section 2.4.

Example 2.1 (Continued)

rigid-body mechanics on summing moments about the axis of the cylinder [see Fig. 2.7(c)] that

with $\left(\frac{D}{2} - \frac{\varepsilon}{2}\right)$ as the mean radius and with $dA = \left(\frac{D}{2} - \frac{\varepsilon}{2}\right)\varepsilon\, d\theta$.

$$T = \iint_A \left(\frac{D}{2} - \frac{\varepsilon}{2}\right)\tau_t dA = \int_0^{2\pi} \tau_t \left[\left(\frac{D}{2} - \frac{\varepsilon}{2}\right)\right]^2 \varepsilon\, d\theta = 2\pi\tau_t \left(\frac{D}{2} - \frac{\varepsilon}{2}\right)^2 \varepsilon$$

$$\therefore\ \tau_t = \frac{2T}{\pi(D - \varepsilon)^2 \varepsilon}$$

Example 2.2

Consider a *thin-walled* tank containing air at a pressure of 100 psi above that of the atmosphere [see Fig. 2.8(a)]. The outside diameter D of the tank is 2 ft, and the wall thickness t is 1/4 in. We consider as a free body from the tank wall a vanishingly small element such as $ABCD$ in the diagram having the shape of a rectangular parallelepiped. What are the stresses on the cut surfaces of the element? Neglect the weight of the cylinder.

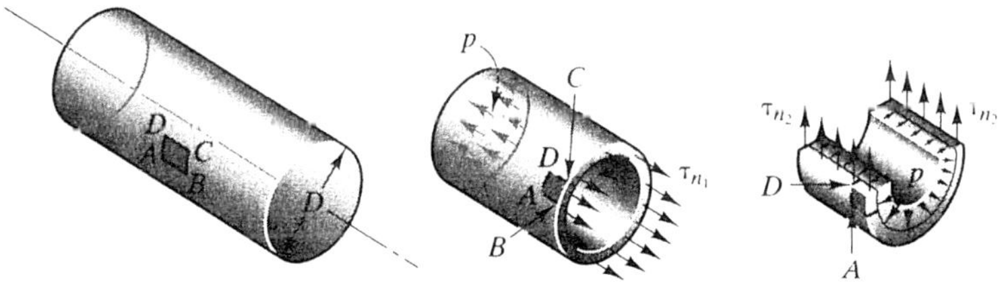

Figure 2.8. Thin-walled tank with inside gauge pressure p.

We can examine face BC by considering a free body of part of the tank, as shown in Fig. 2.8(b) exposing BC. Because there is a net force from the air pressure only in the *axial* direction of the cylinder, we can expect only normal stress over the cut section of the cylinder, as has been indicated in the diagram. Furthermore, because the wall of the tank is *thin* compared to the diameter, we can assume that the stress τ_{n_1} is *uniform* across the thickness. Finally, for reasons of *axial symmetry* of geometry and loading, we can expect this stress to be uniform around the entire cross section. Now we may say from considerations of **equilibrium** in the axial direction that

Example 2.2 (Continued)

$$\tau_{n_1}\pi\left[\frac{D^2}{4}-\frac{(D-2t)^2}{4}\right]-p\pi\frac{(D-2t)^2}{4}=0$$

$$\therefore \tau_{n_1}=\frac{p(D-2t)^2}{D^2-(D-2t)^2}=\frac{(100)\left(24-\frac{1}{2}\right)^2}{24^2-23.5^2}=2325 \text{ psi} \qquad \text{(a)}$$

Hence, on face BC we have a uniform stress of 2325 psi. Clearly, this must also be true for face AD.

To expose face DC next, we consider a half-cylinder of unit length such as is shown in Fig. 2.8(c). Because it is *far* from the ends and because the wall is *thin*, we can assume that the stress τ_{n_2} shown is uniform over the cut section.[2] A pressure p is shown acting normal to the inside wall surface. (The stress τ_{n_1}, computed earlier is not shown, to avoid cluttering the diagram.) Now we consider an end view of this body in Fig. 2.9 for **equilibrium** in the vertical direction. We have

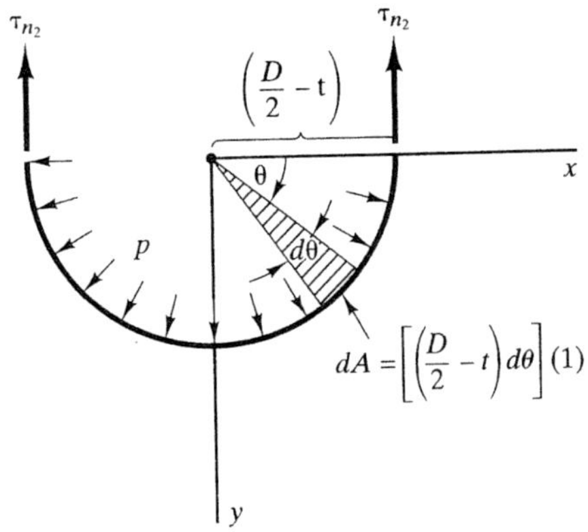

Figure 2.9. Free body of part of cylinder.

$$2[(\tau_{n_2})(t)(1)]-\int_0^{\pi}p\left(\frac{D}{2}-t\right)d\theta(1)\sin\theta=0 \qquad \text{(b)}$$

$$\therefore \tau_{n_2}=\frac{1}{2t}\left[p\left(\frac{D}{2}-t\right)\right](-\cos\theta)\Big|_0^{\pi}$$

$$\tau_{n_2}=\frac{p(D/2-t)}{t}=4700 \text{ psi}$$

We point out now that the force in a particular direction from a uniform pressure on a curved surface equals the pressure times the projected area of this surface in the direction of the desired force. (You will learn this in your studies of hydrostatics.) Thus for the case at hand the projected area is that of a rectangle $1\times(D-2t)$, so that the second expression of Eq. (b) becomes $(p)(D-2t)$. You may readily verify that this gives the same result as above.

[2]Near the ends of the tank the stress distribution *varies* in value because of the proximity of the *complicated geometry*. We shall point out several times later that we get stress *concentrations* in regions of complex geometry.

Example 2.2 (Continued)

The stress τ_{n_2} is called the *hoop stress;* it is about twice the *axial stress* τ_{n_1}.[3] We show element $ABCD$ with the stresses present in Fig. 2.10.

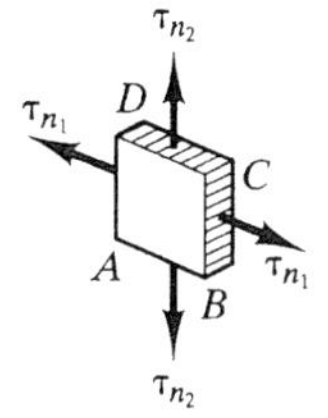

Figure 2.10. Free body of an element of the cylinder.

Example 2.3

A shaft is connected to a tube via a rubber insert that is glued to each member as shown in Fig. 2.11. On cylindrical surfaces of the rubber connector concentric with the shaft, what is the shear stress as a function of r, the radius of the cylindrical surface? The force P is 2000 N. What is the normal stress on cross sections of the tube below the rubber connector? The inside diameter D of the tube is 50 mm, and its thickness t is 3 mm.

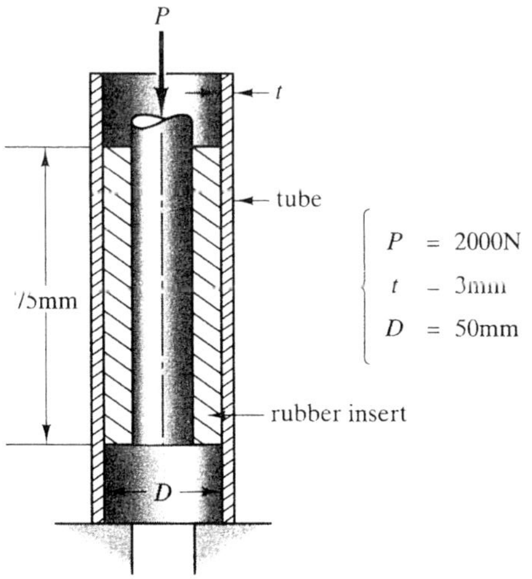

Figure 2.11. Axially loaded vertical rod supported by a tube via a rubber insert.

[3]For thin-walled cylinders where a single radius measure r is to be used rather than accounting for an inside radius and an outside radius, we suggest you show the following simple results

$$\tau_{Axial} = \frac{pr}{2t} \qquad \tau_{Hoop} = \frac{pr}{t}$$

where t is the thickness of the cylinder wall. We also shall have use of these simple formulas in this text.

Example 2.3 (Continued)

A free body is first formed by slicing downward to form a surface of rubber which is concentric with the shaft and having a radius r. This is shown in Fig. 2.12. By doing this we expose a *uniform* vertical shear stress distribution on a circular rubber surface having a radius r. By this surgery, we form a free body allowing us to make a statement of **equilibrium** involving the desired shear stress distribution. Thus we can say

$$\underline{\Sigma F_y = 0} \qquad \tau_S A_{CUT} - P = 0$$

$$\therefore\ \tau_s(2\pi r)(.075) = P = 2000$$

where τ_s is the desired shear stress. Solving for τ_s, we get the shear stress distribution as a function of r. Thus[4]

$$\tau_s = \frac{2000}{(.075)(2\pi r)} = 4244 / r \text{ Pa}$$

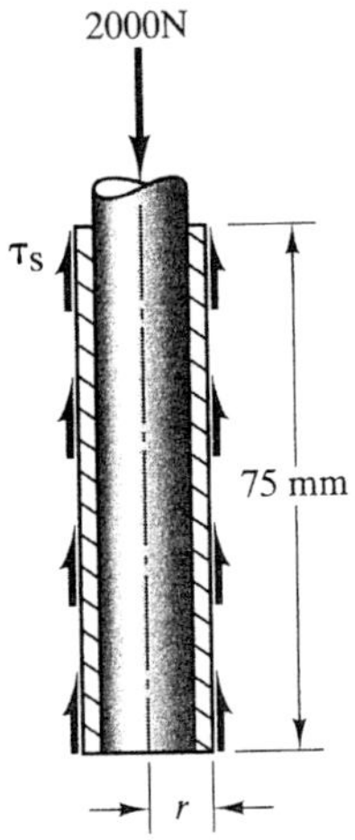

Figure 2.12. Free body exposing desired shear stress in rubber.

To get the normal stress on sections of the tube below the rubber insert, we must again perform some surgery in order to expose the normal stress in a free-body diagram so we can use a statement of **equilibrium.** This is shown in Fig. 2.13. Now summing forces in the vertical direction, we have

$$\underline{\Sigma F_y = 0} \qquad \therefore\ \tau_n\left[\left(\frac{\pi}{4}\right)[(.056)^2 - (.05)^2]\right] = P = 2000$$

$$\tau_n = 4.004 \times 10^6 \text{ Pa}$$

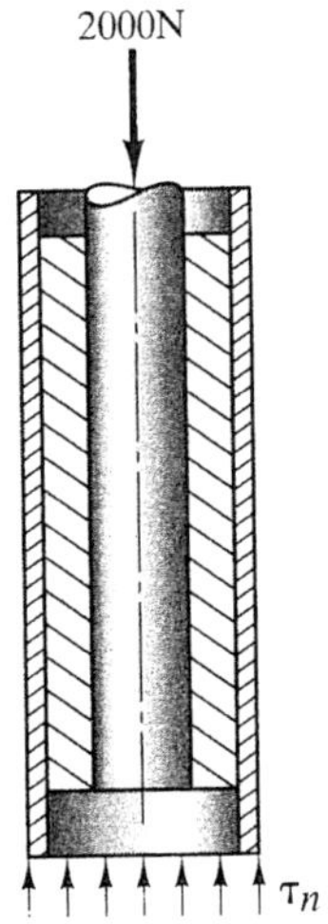

Figure 2.13. Free body exposing the desired normal stress in tube.

[4]Please note that only values of r within the rubber are considered in this problem.

Example 2.4

Solid rods of diameter 30 mm are welded to a rigid drum D (see Fig. 2.14), which is turning at a speed of 500 rpm. What is the normal stress on cross sections of the rods at the base of the rods and also at 100 mm from the base? The rods have a mass per unit length of 5 kg/m. Disregard the static weight of the rods.

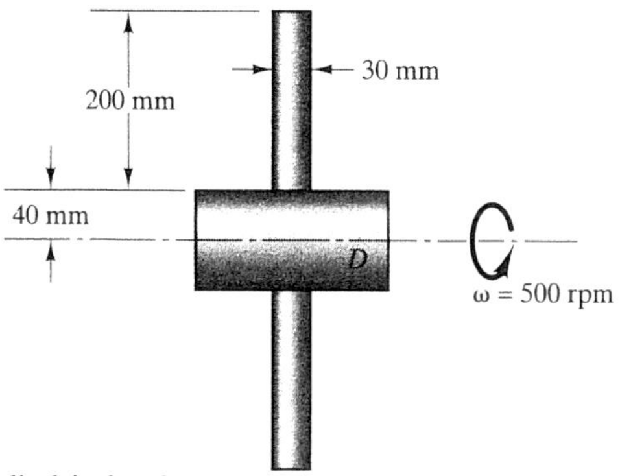

Figure 2.14. Cylindrical rods are rotating with a rigid drum.

We will use two different approaches for this example. The first will use the calculus.

Method I. **Calculus.**

a). We first show a free body in Fig. 2.15 exposing the base section of the rod where we wish to compute the normal stress. Now using **Newton's law** ($F_r = Mr\omega^2$) we can say if we neglect the weight

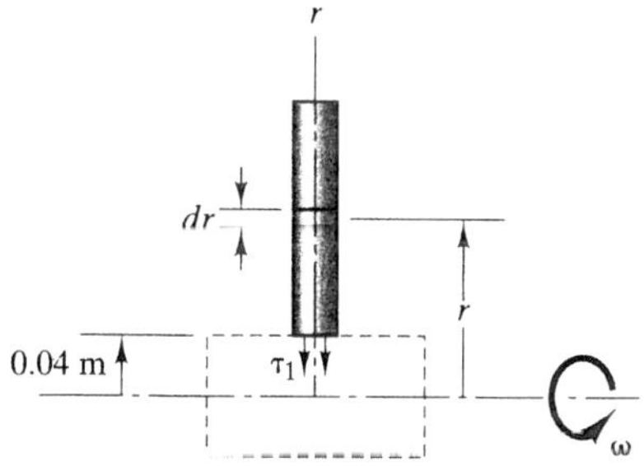

Figure 2.15. Free body exposing bottom cross section of the cylinder.

$$\tau_1 \frac{\pi(.03)^2}{4} = \int_{.04}^{.240} (5dr)\,(r)\left[\frac{(500)(2\pi)}{60}\right]^2$$

Integrating and solving for τ_1 we get

$$\tau_1 = \frac{4}{\pi(.03)^2}(5)\left[\frac{(500)(2\pi)}{60}\right]^2\left(\frac{1}{2}\right)(.240^2 - .04^2)$$

$$\tau_1 = 5.430 \times 10^5 \text{ Pa}$$

b). We next cut the rod at .1 m above the base to form a second free body for the computation of the second stress. This is shown in Fig. 2.16. Again from **Newton's law** we have

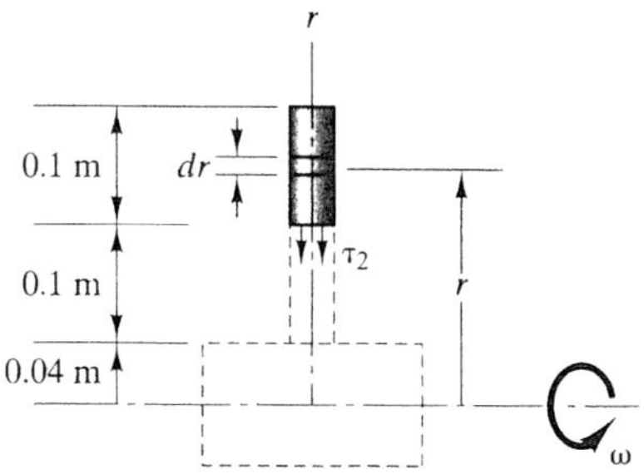

Figure 2.16. Free body exposing an interior cross section of the rod.

$$\tau_2 \frac{\pi(.03)^2}{4} = \int_{.15}^{.24} (5dr)\,(r)\left[\frac{(500)(2\pi)}{60}\right]^2$$

$$\tau_2 = 3.865 \times 10^5 \text{ Pa}$$

Example 2.4 (Continued)

Method II. **Center of Mass.**

Using **Newton's law** for the center of mass for each of the free bodies of Figs. 2.15 and 2.16 we get

a) $$\tau_1 \frac{(\pi)(.03)^2}{4} = (5)(.2)(.14)\left[\frac{(500)(2\pi)}{60}\right]^2$$

$$\tau_1 = 5.430 \times 10^5 \text{ Pa}$$

b) $$\frac{\tau_2(\pi)(.03)^2}{4} = (5)(.1)(.19)\left[\frac{(500)(2\pi)}{60}\right]^2$$

$$\tau_2 = 3.685 \times 10^5 \text{ Pa}$$

Example 2.5

A 500-N tank A is full of water and is connected to open tank B also full in Fig. 2.17 through a pipe. If the wall of tank A is 2 mm in thickness, determine certain tensile stresses τ_{xx} and τ_{yy} from air and water. The stresses desired are: *a*) in the tank wall at $y = 3$ m and *b*) in the cross section of each of the 40 bolts at the base. The diameter of the bolts is 3 mm. Note that hydrostatic pressure of the water is 9806 times the distance below the free surface of the water. Also, any pressure on the free surface extends undiminished throughout the entire fluid. Hence, the force from atmospheric pressure on the free surface will be cancelled on our free bodies by atmospheric pressure on the outside of our free bodies. Thus we will use gage pressure (pressure above atmosphere). Take the center of gravity of tank A to be at 3 m above the base.

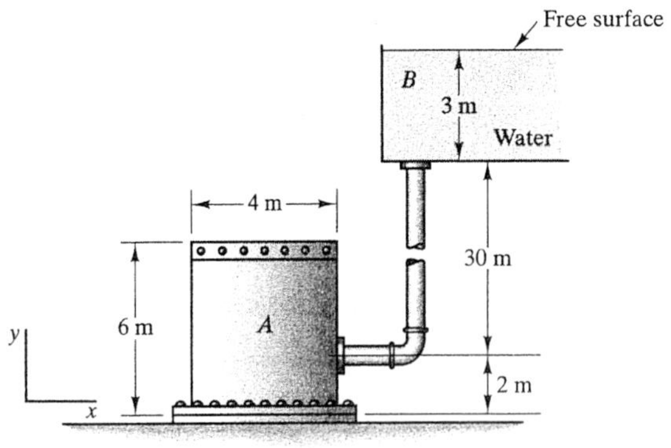

Figure 2.17. Tanks containing water whose specific weight is 9806 N/m³.

Example 2.5 (Continued)

We shall solve this problem in two parts starting with the stress in the tank wall.

Part a. **Stress in tank *A*.**

The pressure p_1 of the water at 3 m below the top of tank A is (going down from the free surface a distance of 33 m and then going up a distance of 1 m)

$$p_1 = (9{,}806)(3 + 30 - 1) = 3.14 \times 10^5 \text{ Pa gage}$$

From **equilibrium,** we have for free-body diagram (F.B.D.) I (see Fig. 2.18)

$$p_1(A)_{water} - (\tau_{yy})(A)_{tank\ wall} - (W_2)_{tank} - (W_1)_{water} = 0$$

$$(3.14 \times 10^5)(\pi)\left(\frac{4^2}{4}\right) - (\tau_{yy})[(\pi)(4)](.002) -$$

$$\left(\frac{1}{2}\right)(500) - (9{,}806)\left[(\pi)\left(\frac{4^2}{4}\right)\right](3) = 0$$

$$\tau_{yy} = 1.423 \times 10^8 \text{ Pa}$$

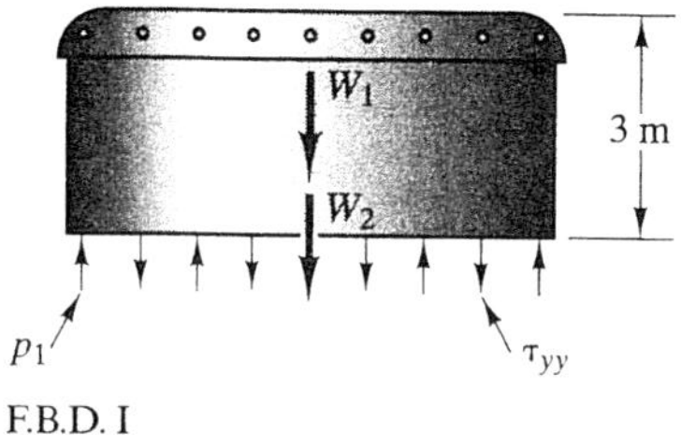

Figure 2.18. Free body exposing τ_{yy}.

Next consider the F.B.D. II for a half unit strip of the tank in Fig. 2.19. From **equilibrium** again,

$$2[\tau_{xx} A_{cut}] = p A_{projected}$$

$$2[(\tau_{xx})(.002)(.001)] = (3.14 \times 10^5)(4)(.001)$$

$$\tau_{xx} = 3.14 \times 10^8 \text{ Pa}$$

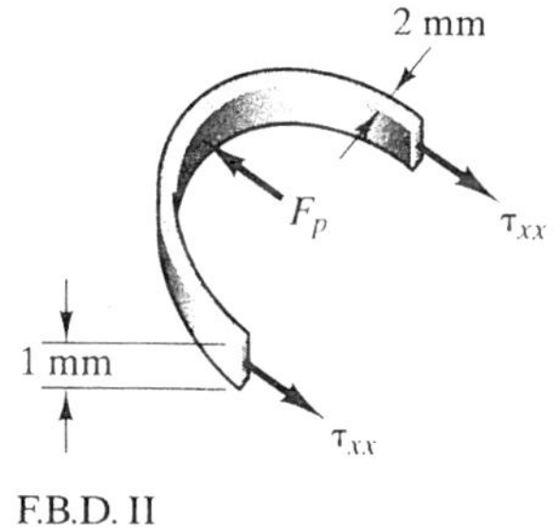

Figure 2.19. Free body exposing τ_{xx}.

Part b. **Stress in bolts at bottom of tank *A*.**

A new pressure is needed now. At the base of the tank,

$$p_2 = (9{,}806)(3 + 30 + 2) = 3.432 \times 10^5 \text{ Pa gage}$$

Consider F.B.D. III in Fig. 2.20 now. From **equilibrium,** we have

$$p_2 A_{water} - F_{bolts} - W_{tank} - W_{water} = 0$$

$$(3.432 \times 10^5)\left(\frac{\pi 4^2}{4}\right) - 40F - 500 - (9{,}806)\left(\frac{\pi 4^2}{4}\right)(6) = 0$$

$$F = 89.3 \text{ kN per bolt}$$

Hence,

$$\tau_{xx} = \frac{(89.3)(1000)}{\left(\frac{\pi}{4}\right)(.003)^2} = 1.2693 \times 10^{10} \text{ Pa}$$

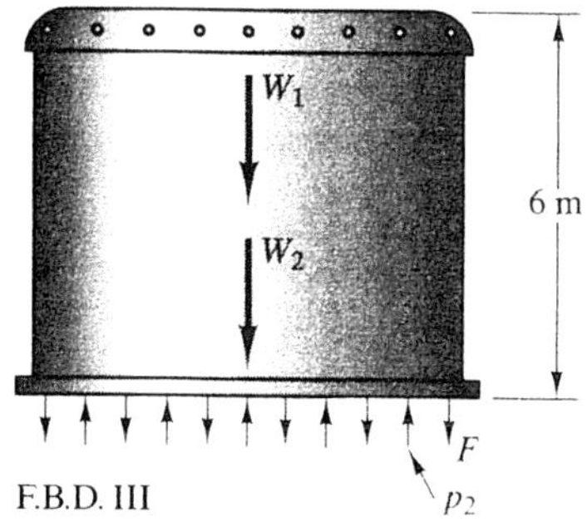

Figure 2.20. Free-body diagram cutting liquid and bolts.

These examples show the importance of forming free bodies using appropriate cuts to expose the stresses that may interest us. In your statics course, you formed free bodies by both separating and/or cutting, the latter generally to expose forces or resultants of force distributions. The skill you have acquired in doing this will be vital for success in the subject matter of this book.

It should be clear that at a given point in the body we could expose by mathematical cuts area elements having *any* arbitrary orientation whatsoever.[5] This has been illustrated in Fig. 2.21 where through a given

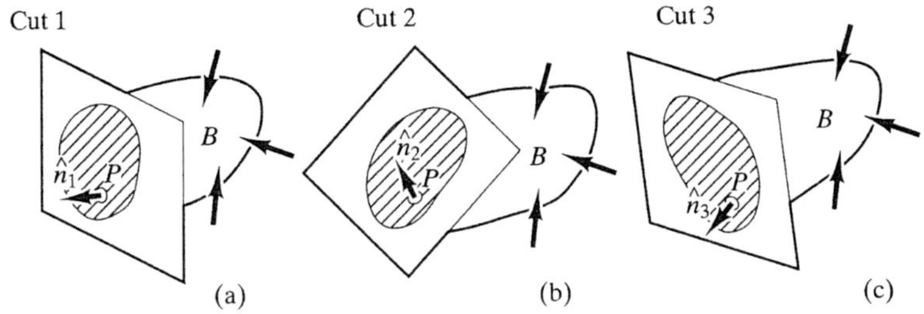

Figure 2.21. Different cuts through the point P.

point P a series of cuts has been shown exposing area elements having different unit normals **n.** The force intensity (and thus the stresses) transmitted through the element at point P will depend on the orientation of the element. This is illustrated further in the following example.

Example 2.6

Shown in Fig. 2.22(a) is a uniform bar subjected to a tensile force P whose line of action coincides with the centerline of the bar. A cross section A is shown away from the applied loads. We may assume that the force intensity transmitted through this section is *uniform* over the section. Clearly, with section A normal to P, there will be only a normal stress present on A. Neglecting the force of gravity, this stress, τ_n, is given everywhere on the section as

[5]Up to now the orientation of the interfaces used in the examples was obvious on inspection. The normals were either horizontal or vertical.

Example 2.6 (Continued)

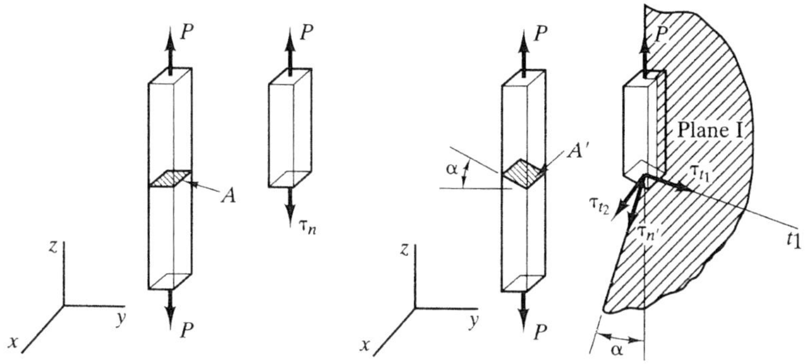

Figure 2.22. One-dimensional member with different cuts.

$$\tau_n = \frac{P}{A} \tag{a}$$

What are the stresses for other sections there?

Now suppose that we consider another section of area A', such as is shown in Fig. 2.22(b). Here we may again assume that the force intensity over the section is uniform. However, we shall have a new normal stress $\tau_{n'}$, and, for directions parallel to the edges of the section, we have shown shear stresses τ_{t_1} and τ_{t_2}. Note that the force P and the stresses $\tau_{n'}$ and τ_{t_1} are coplanar (Plane I); it is then apparent that τ_{t_2} lying normal to Plane I must be zero from considerations of **equilibrium** in that direction. We can then say that:

$\Sigma F_z = 0$:

$$P - \tau_{n'} A' \cos\alpha - \tau_{t_1} A' \sin\alpha = 0 \tag{b}$$

$\Sigma F_y = 0$:

$$-\tau_{n'} A' \sin\alpha + \tau_{t_1} A' \cos\alpha = 0 \tag{c}$$

Solving simultaneously for $\tau_{n'}$ and τ_{t_1} we get, noting that $A = A' \cos\alpha$,

$$\begin{aligned} \tau_{n'} &= \frac{P}{A}\cos^2\alpha \\ \tau_{t_1} &= \frac{1}{2}\frac{P}{A}\sin 2\alpha \end{aligned} \tag{d}$$

We see here that the normal and shear stresses *vary* with the orientation α of the section.

We shall investigate the variation of stresses at a point as the orientation of the area element at the point is changed for more general cases later.

We now make the following conclusion. The force intensity (and hence the stresses) transmitted across an area element depends on the *position* of the area element in the body and also on the *orientation* of the element. This means that to convey properly information regarding stress at a point, we must give the position of the point and also the orientation of the area element at the point. We can readily give position through the specification of coordinates. The indication of orientation of the area element will be discussed at some length in the following section.

2.3 Stress Notation

It will be shown in Chapter 15 that if stresses are known for *three orthogonal interfaces at a point,* stresses for *any* interface at the point can readily be determined by a straightforward algebraic formulation. Hence, anticipating this formulation, we shall concentrate on means of specifying stresses on an orthogonal set of interfaces at a point. Accordingly, consider interfaces at a point P parallel to reference planes of an xyz reference, as shown in Fig. 2.23(a), where faces of a *vanishingly*

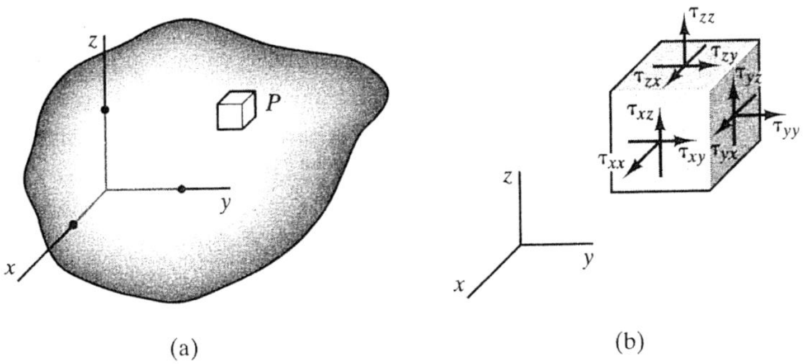

Figure 2.23. Stress components parallel to xyz.

small rectangular parallelepiped serve as the desired interfaces. To identify stresses parallel to the coordinate axes on three interfaces, we employ two subscripts for τ. The first subscript identifies the interface by giving the coordinate direction of the *normal* to the interface. Thus, τ_{xy}, τ_{xz}, and τ_{xx} indicate that each of these stresses is applied to an interface having its normal parallel to the x axis. The second subscript gives the direction of the stress itself. Thus τ_{xy} is a stress pointing in the y coordinate direction. It should be clear that stresses with repeated indices must be normal stresses, while stresses with mixed indices must be shear stresses. The stresses for the aforementioned orthogonal faces are shown on the enlarged diagram in Fig. 2.23(b).

We shall introduce the following sign convention for stresses: A stress acting on an area element whose outward normal vector points in the positive direction of any coordinate axis will be taken as positive if the stress itself also points in the positive direction of any coordinate axis. (Note that the axes for the area normal and stress need not be the same axes.) A stress is positive also if both the outward area normal and the stress point in the negative direction of the same coordinate axis or of different coordinate axes. (You will note on inspection that the stresses shown in Fig. 2.23 are all positive stresses.) If, now, the outward area normal and the stress are not directed simultaneously in either positive or negative coordinate directions, the stress is negative.

We shall show in Section 2.4 that the stresses on one side of an interface (see Fig. 2.24) are equal in value and have opposite senses as the corresponding stresses shown lighter on the other side. Clearly, the *same sign* and the *same notation* apply no matter what side you choose to work with.

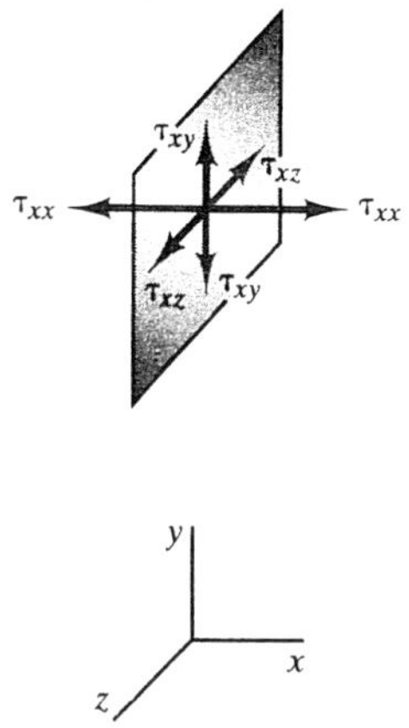

Figure 2.24. Stresses on opposite sides of an interface.

As we pointed out at the beginning of this section, nine stresses for three orthogonal interfaces at a point determine the stresses for *any interface* at the point. Using as the three interfaces at a point those parallel to the planes of the reference xyz, we may consider the set of nine stresses parallel to the coordinate axes

$$\tau_{ij} = \begin{pmatrix} \tau_{xx} & \tau_{xy} & \tau_{xz} \\ \tau_{yx} & \tau_{yy} & \tau_{yz} \\ \tau_{zx} & \tau_{zy} & \tau_{zz} \end{pmatrix}$$

as the rectangular components of a quantity which we shall call a *second-order tensor.*[6] This is analogous to vectors such as $\mathbf{A}$, where for a reference xyz we have three rectangular components parallel to the axes

$$\begin{pmatrix} A_x \\ A_y \\ A_z \end{pmatrix}$$

from which we can determine the component of $\mathbf{A}$ in any direction $\mathbf{n}$. We shall discuss the computation of stresses at some arbitrary interface at a point in terms of the nine rectangular stress components for reference xyz later in Chapter 15.

[6]Note that in this format for representing the nine rectangular stress components at a point, the first subscript identifies the row and the second subscript identifies the column. The main diagonal of the array (going from upper left to lower right) contains the normal stresses. The shear stresses are then said to be the off-diagonal terms. Discussions of the definition of second-order tensors will be taken up in Chapters 7, 8, and 15.

2.4 Complementary Property of Shear

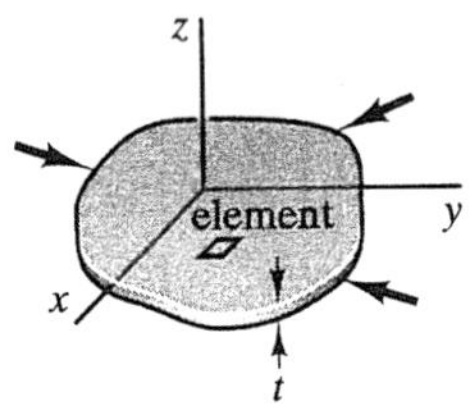

Figure 2.25. Body under plane stress.

We shall define the *plane stress* state as one where for a reference xyz the only nonzero stresses are τ_{xx}, τ_{yy}, τ_{yx}, and τ_{xy}. That is, all stresses in the z direction are zero. Such a state of stress may be found to a reasonable degree of approximation in a thin plate with external loads applied only in the midplane of the plate. Taking z to be the direction normal to the midplane of the plate, we show the desired situation in Fig. 2.25, where the xy plane coincides with the midplane of the plate. With no loads on the "faces" of the plate, it is easy to understand on inspection why stresses τ_{zz}, τ_{zx}, τ_{zy}, and so on, are small enough throughout to be considered negligible compared to stresses such as τ_{xx}, τ_{yy}, and so on.

We shall now consider a finite element of a plate of thickness t shown in Fig. 2.25. Viewing the element along the z direction (normal to the plate) we see as edges those faces having nonzero stresses. This is shown in Fig. 2.26. Corners A, B, C, and D of the element have coordinates (x, y), $(x, y + \Delta y)$, $(x + \Delta x, y + \Delta y)$, and $(x + \Delta x, y)$, respectively. Positive stresses and the body force intensity components are shown. Asterisks are used to indicate that the stresses and the body force intensity components are average values. Note also that a prime has been used on the stresses on the sides BC and DC to distinguish these stresses from the corresponding stresses on faces BA and AD, respectively.

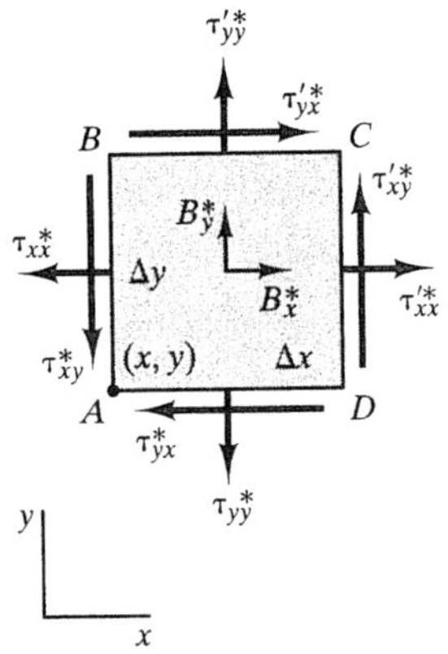

Figure 2.26. Plane element.

Now, summing forces, we have for **equilibrium** in the x direction,

$$-\tau^*_{xx}\Delta y\, t + (\tau'_{xx})^*\Delta y\, t - \tau^*_{yx}\Delta x\, t + (\tau'_{yx})^*\Delta x\, t + B^*_x\Delta y\Delta x\, t = 0 \tag{2.2}$$

Canceling t, we first take the limit as $\Delta y \to 0$. This gives us

$$-\tau^*_{yx}\Delta x + \tau'^*_{yx}\Delta x = 0$$

Cancel Δx from the equation. Now letting Δx go to zero (thus causing the rectangle to collapse to a point) permits us to drop the asterisk, so that in the limit

$$\tau_{yx} = \tau'_{yx}$$

This indicates that on *opposite* sides of an interface at a point having a normal in the y direction the corresponding shear stresses are equal in value and are oppositely directed. Since the y axis could have been chosen in any direction in the plate, the conclusions above hold for any interface that is normal to the faces of the plate.

Now we let Δx go to zero first in Eq. (2.2). We then have from the equilibrium equation, on canceling t,

$$-\tau^*_{xx}\Delta y + \tau'^*_{xx}\Delta y = 0$$

Next cancel Δy from the equation. Finally, letting Δy go to zero, thereby shrinking to a point and deleting the asterisks, we conclude that

$$\tau_{xx} = \tau'_{xx}$$

Similarly, considering equilibrium in the y direction we may arrive at the result

$$\tau_{yy} = \tau'_{yy}$$

Thus, on opposite sides of an interface the normal stresses are equal and directed in opposite senses. This justifies the assertions made in Section 2.3 (see Fig. 2.24) regarding stresses on opposite sides of an interface.

By next taking moments about an axis parallel to the z axis at corner A (Fig. 2.26), we have, canceling t,

$$-(\tau'_{yx})^*\Delta x\Delta y + (\tau'_{yy})^*\frac{\Delta x}{2}\Delta x - \tau^*_{yy}\frac{\Delta x}{2}\Delta x + \tau^*_{xx}\frac{\Delta y}{2}\Delta y + (\tau'_{xy})^*\Delta y\Delta x - \tau'^*_{xx}\Delta y\frac{\Delta y}{2} - B^*_x\Delta x\Delta y\frac{\Delta y}{2} + B^*_y\Delta x\Delta y\frac{\Delta x}{2} = 0$$

We divide through by $\Delta x\Delta y$ in the equation above and we get, on canceling terms,

$$-(\tau'_{yx})^* + [(\tau'_{yy})^* - (\tau^*_{yy})]\frac{1}{2}\left(\frac{\Delta x}{\Delta y}\right) + [(\tau^*_{xx}) - (\tau'_{xx})^*]\frac{1}{2}\left(\frac{\Delta y}{\Delta x}\right) + (\tau'_{xy})^* - B^*_x\frac{\Delta y}{2} + B^*_y\frac{\Delta x}{2} = 0$$

Now let $\Delta x \to 0$ and $\Delta y \to 0$. In the limit we may drop the asterisks and we may invoke the conclusions from the previous consideration that $\tau_{yy} = \tau'_{yy}$ and $\tau_{xx} = \tau'_{xx}$. Clearly, all terms but the shear terms then vanish and we arrive at the result on faces opposite corner A:

$$\tau'_{yx} = \tau'_{xy}$$

By next considering moments about corner C we can similarly conclude on faces opposite corner C that

$$\tau_{xy} = \tau_{yx}$$

Finally, by considering moments about corners B and D we may also conclude that on faces opposite corners B and D, respectively,

$$\tau'_{xy} = \tau_{yx}$$
$$\tau'_{yx} = \tau_{xy}$$

This indicates that the *cross-shears* (i.e., shear stresses on adjacent faces at a point which thus have reversed indices) are equal and this is the so-called *complementary* property of shear. Furthermore, shear stresses τ'_{xy} and τ'_{yx} are in a proper orientation relative to each other, as shown in Fig. 2.26. [Otherwise, we would have incurred a minus sign in Eq.

(2.3) for one of the terms.] Thus, the shears are pointing *toward C*. If we had originally oriented these shears pointing away from C, we would have arrived at the same equation (by multiplying by -1), indicating that this is also a possible permissible relative orientation of the cross-shears. Thus, the shears either must point toward or away from the corner. From the aforementioned moment considerations about other corners of the element, we may conclude at a point that the relative orientation of the shear stresses must be given as one of the two possibilities shown in Fig. 2.27. Notice that the *shears must either point toward or away from each corner* of the vanishingly small rectangle representing the neighborhood of a point.[7]

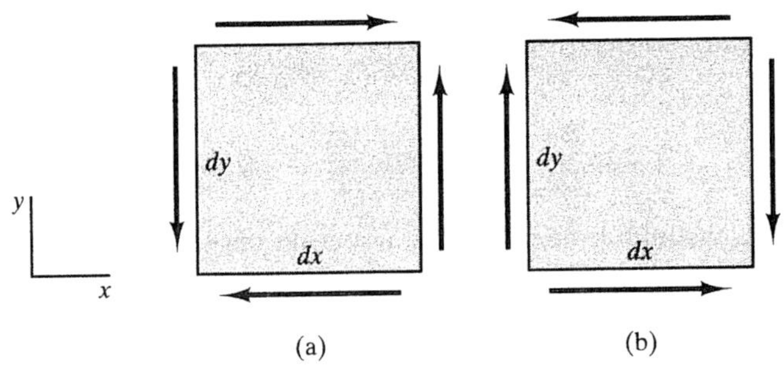

Figure 2.27. Proper shear-stress orientations.

The conclusions above may be extended to the three-dimensional state of stress. Thus, the shear stress possibilities for xz axes and yz axes on a vanishingly small rectangular parallelepiped are shown in Fig. 2.28(a) and (b), respectively. Notice that the shears either point toward or away from an edge. Furthermore, we can say that

$$\begin{aligned} \tau_{xz} &= \tau_{zx} \\ \tau_{yz} &= \tau_{zy} \end{aligned} \tag{2.4}$$

That is, the so-called cross-shears are equal at a point. In the matrix representation of the stresses we conclude further that the values of the shear stresses form mirror images of each other around the main (left to right) diagonal. For this reason we say that the stress tensor is *symmetric*. In Fig. 2.29 we show a complete set of stresses with shear stresses satisfying the complementary property of shear.

The conclusions above, although derived for statics, are valid for dynamic conditions. This is easily understood when one realizes that the inertial effects may be considered as body force contributions and hence, like the body force contribution of gravity, drop out in the limit.

[7]Note that a single sign for all shear stresses at a point for a given reference always applies. Thus, in Fig. 2.27(a) we have a positive shear on all faces, while in Fig. 2.27(b) we have a negative shear on all faces. The particular state of shear stress at a point in a body will depend on the loading on the body.

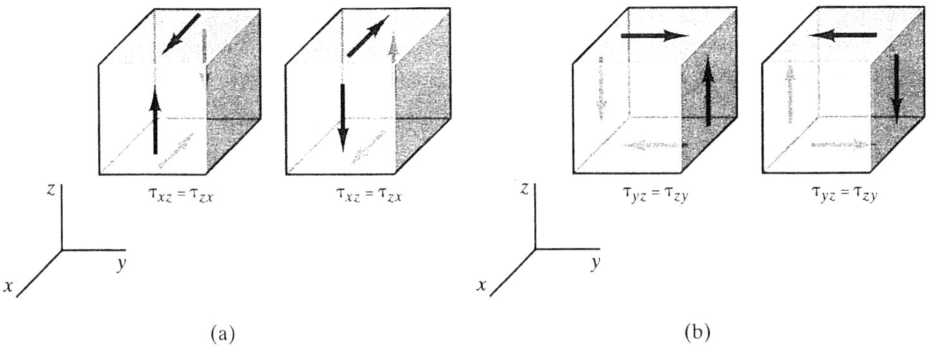

Figure 2.28. Other proper shear-stress orientations.

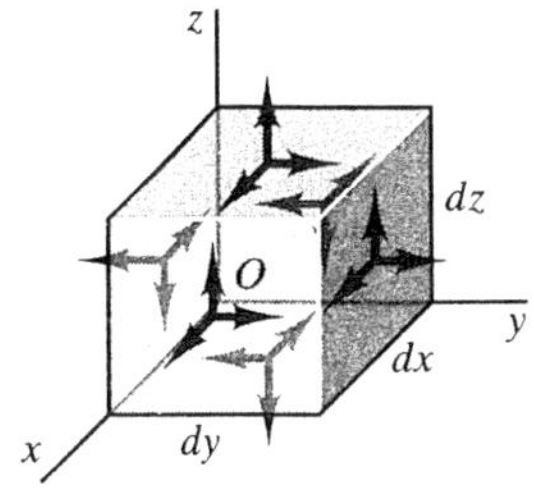

Figure 2.29. Orthogonal stress components.

2.5 A Comment on the Complementary Property of Shear

In Example 2.1 concerning the torsion of a thin-walled cylinder (see Fig. 2.7), we stated that the direction of the shear stress in a cross section of the cylinder was directed essentially parallel to the circular boundaries of the cross section. Furthermore, this shear stress was assumed uniform over the thickness of the section. We appealed at that time to your engineering judgment to accept these propositions. We can now be a bit more authoritative in these assertions by making use of the complementary property of shear.

For this purpose, examine in Fig. 2.30 an enlarged portion of the thin-walled cylinder on which we have shown two infinitesimal rectangular parallelepipeds, one having a face A on the *outside* lateral surface of the cylinder and the other having a face B on the *inside* lateral surface of the cylinder. On the surface A, clearly we have only air pressure and hence no shear stress possible. This means that a shear stress on this face directed normal to edge ab [see Fig. 2.31(a)] must be zero, and hence from the complementary property of shear, the radial shear stress on face C in the cross section must also be zero. The same conclusion can be reached [see Figs. 2.30 and 2.31(b)] for faces B and E. Thus at the edges of the cross section there can be no shear stresses directed radially. Since the wall of the cylinder is thin, we can reasonably assume that shear stress in the radial direction is zero over the entire thickness of the cross section. We are accordingly now in a better position to accept what is probably intuitively apparent to you—that the shear stress direction is parallel to the boundaries of the cross section for a thin-walled cylinder.

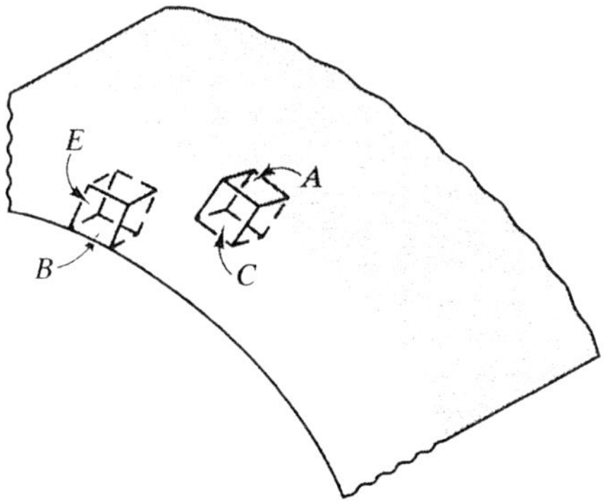

Figure 2.30. Infinitesimal elements at the edges of the thin-walled cylinder.

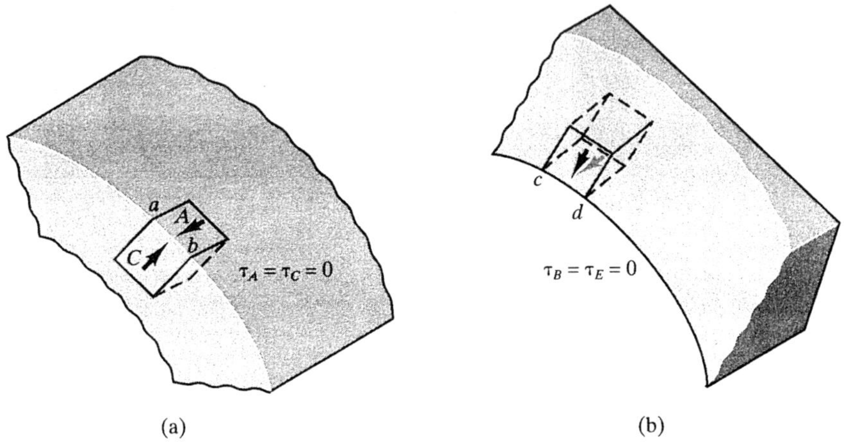

Figure 2.31. Elements showing hypothetical shear stress directed normal to boundary.

2.6 Equations of Equilibrium in Differential Form

In earlier courses in statics we set forth equations of equilibrium for discrete particles and rigid bodies. The forces and torques were finite, and the resulting equations were *algebraic* equations. At this time we shall consider a vanishingly small element of a deformable body which in its *deformed* equilibrium configuration is a rectangular parallelepiped under the action of stresses and body forces. In Fig. 2.32 we show such an element but include only those stresses which generate a force component in the x direction to avoid cluttering the diagram. We wish to establish with this element the requirements for the manner in which stresses must vary with position to guarantee equilibrium in a domain (i.e., we wish the *differential* form of the equations of equilibrium). On face 1 we have shown a stress τ_{xx}, while on the face parallel to face 1, namely face 2 in the diagram, dx apart from face 1, we have shown the same stress but with an infinitesimal change in value. That is, on face 2 we have stress $[\tau_{xx} + (\partial\tau_{xx}/\partial x)dx]$.[8] This you will notice is similarly done for other pairs of faces of the rectangular parallelepiped. Summing forces in the x direction, we then get

$$\left(\tau_{xx} + \frac{\partial\tau_{xx}}{\partial x}dx\right)dy\,dz - \tau_{xx}\,dy\,dz + \left(\tau_{yx} + \frac{\partial\tau_{yx}}{\partial y}dy\right)dx\,dz - \tau_{yx}\,dx\,dz$$
$$+ \left(\tau_{zx} + \frac{\partial\tau_{zx}}{\partial z}dz\right)dx\,dy - \tau_{zx}\,dx\,dy + B_x\,dx\,dy\,dz = 0$$

[8] We are using a two-term Taylor series expansion here.

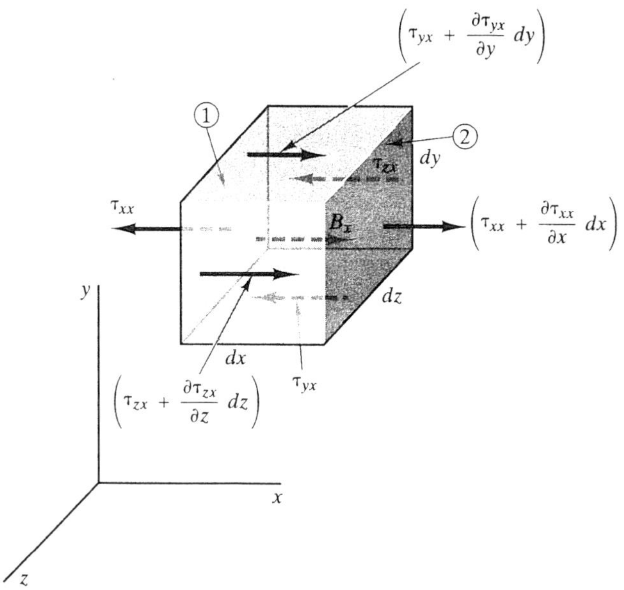

Figure 2.32. Stresses giving rise to a force in the x direction.

Canceling terms and dividing by $dxdydz$, we then get

$$\frac{\partial \tau_{xx}}{\partial x} + \frac{\partial \tau_{yx}}{\partial y} + \frac{\partial \tau_{zx}}{\partial z} + B_x = 0$$

Using the complementary property of shear stress and making similar computations in the y and z directions, we may give the equations of equilibrium at a point in the body as follows:

$$\begin{aligned} \frac{\partial \tau_{xx}}{\partial x} + \frac{\partial \tau_{xy}}{\partial y} + \frac{\partial \tau_{xz}}{\partial z} + B_x &= 0 \\ \frac{\partial \tau_{yx}}{\partial x} + \frac{\partial \tau_{yy}}{\partial y} + \frac{\partial \tau_{yz}}{\partial z} + B_y &= 0 \\ \frac{\partial \tau_{zx}}{\partial x} + \frac{\delta \tau_{zy}}{\partial y} + \frac{\delta \tau_{zz}}{\partial z} + B_z &= 0 \end{aligned} \tag{2.5}$$

Notice that the equations of equilibrium now involve quantities expressed as functions of position—that is, the stress and body force distributions, rather than as discrete quantities, namely forces and moments. These are so-called *field* variables. Also, we have here a set of partial differential equations for these field variables rather than the usual algebraic equations. What happens at the application point of a point load? Here the stresses become infinite, and the equations above cannot be used at the point. The procedure is to solve the foregoing differential equations in regions where point loads do not exist and then to adjust the solutions so as to take into account the point loads which are taken as boundary conditions for such problems. Although we have used deformed geometry for the element in this development, we wish to point out that for small deformation we can generally apply the above equations to the undeformed geometry.

2.7 Closure

We have introduced in this chapter the concept of stress as a means of describing how an external force distributes through a body. In particular we introduced notation for nine stresses parallel to reference axes *xyz* on three orthogonal interfaces at a point oriented parallel to the reference planes of reference *xyz*. We pointed out that knowing these nine stress components at a point is sufficient to determine any stress on an interface oriented arbitrarily relative to *xyz* at the point. We termed these nine quantities at a point the tensor components of the stress tensor. Not all nine components of the stress tensor are arbitrary, however. Through Newton's law we found that shear stresses with reversed indices at a point are equal—this is the complementary property of shear—so only six of the nine stress terms are independent at a point. Finally, for equilibrium these six quantities had to vary with position in a certain manner dictated by the equations of equilibrium, a set of partial differential equations.

Now that we have considered how to measure the way external forces distribute through a body, we shall consider next how we can measure the primary effects of these forces, namely the deformation of the body.

*2.8 A Look Ahead: Hydrostatics

When you study the dynamics of fluids and deformable solids, you will see that the complementary property of shear stress still holds. The reason for this is that inertial terms for a vanishingly small element of fluid or solid have the same number of differentials as the body force of gravity and, similar to our development, drop out in the limit. In a more rigorous, general study of the complementary property of shear for the three-dimensional case, one uses the moment of momentum equation $\boldsymbol{M}_A = \dot{\boldsymbol{H}}_A$ that you studied in dynamics to prove that the stress tensor must be symmetric and hence must insure the complementary property of shear stress.

From time to time, we shall present problems from hydrostatics (liquids in equilibrium). There will then be no shear stress present and the normal stress at a point will be the same in all directions

(Pascal's Law). This stress will equal minus the pressure. Equations 2.5 will then simplify to

$$\frac{\partial p}{\partial x} = 0 \quad \frac{\partial p}{\partial y} = 0 \quad \frac{\partial p}{\partial z} = -\gamma$$

where γ is the specific weight. This may be further simplified to $dp / dz = -\gamma$, to become an ordinary differential equation. From this equation, we understand the hint given in some problems such as Example 2.5 that the pressure at a point in a static liquid (where γ is constant) is the product of specific weight times the distance of the point below the free surface. Also, considering external pressure on the free surface, we may conclude from this ordinary differential equation that this pressure extends unchanged throughout the entire fluid.

Highlights (2)

A key concept in solid mechanics is to know how to measure the way external forces on a body distribute inside a body. For this purpose, we consider an **interface,** which is an infinitesimal plane surface either on the boundary of the body or inside the body. We decompose the infinitesimal force, coming from just outside the interface and acting on the interface, into a normal component and two orthogonal shear components. The ratios of each of these infinitesimal forces divided by the infinitesimal area are **intensities** (forces per unit area). We call them **stresses.** The stresses depend on the location of the interface (a point in the body) and also on the orientation of the interface (i.e., the direction of the area normal of the interface). To get stresses, it would appear at first a very complicated job because of the difficulty in identifying both the position and orientation of an interface. However, we will learn that knowing nine stresses on three orthogonal interfaces at a point, we can readily compute the stresses on any interface at the point in question. Thus, to convey quantitatively how the forces distribute through a body, we can proceed by using an orthogonal reference to measure stress fields on interfaces everywhere parallel to the faces of the reference (thereby generating the three orthogonal interfaces everywhere). We thus generate nine stress field distributions so that at each point in the body we have nine stresses permitting us to give three stresses at any interface orientation at that point. The notation that we use to specify the nine stress field distributions throughout the bodies are accordingly related to the faces of the orthogonal reference xyz. (Stress distributions referred to other coordinate systems such as cylindrical coordinate systems will be discussed later.) We use two subscripts for identifying the nine stresses at a point. The first identifies the coordinate for the normal to the interface, and the second identifies the coordinate in the direction of the stress itself. If the interface normal and the stress direction are both in positive or are in negative coordinate directions, the stress is positive. If there is a mix of signs for the aforementioned coordinate directions, the stress is negative. In this process of identification, the coordinates need not be the same. The identification process for stress is thus manageable.

As a further aid, we have the complementary property of shear stress, which says that the shear stresses at a point with reversed subscripts such as τ_{xy} and τ_{yx} are equal in value and are directed always either both toward the edge or both away from the edge of the infinitesimal cube having the three orthogonal interfaces. This is a result of **Newton's law** and is valid for any continuum exhibiting shear and is valid for both static and dynamic conditions. We reached these conclusions, however, for simplicity by considering equilibrium.

PROBLEMS

2.1. [2.2] To attach a beam to a column, a seat bracket (Fig. P.2.1) is sometimes used. Compute the allowable load P that can be transmitted to the column considering only shear stress in the bolts. The bolts have a diameter of 1/2 in. Neglect friction and use a working shear stress of 18,000 psi. Assume that each bolt carries one-half of the total load P.

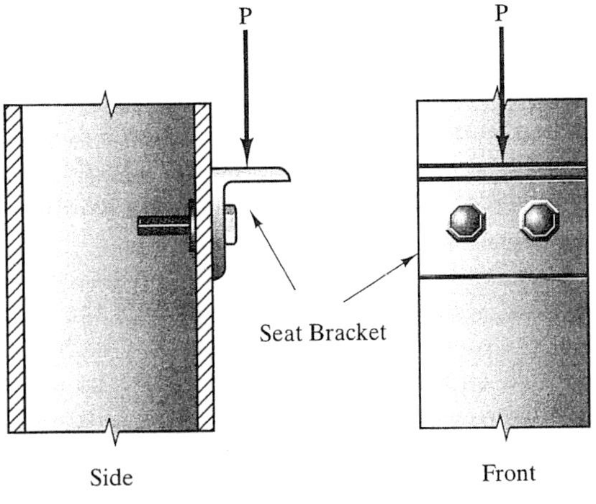

Figure P.2.1.

2.2. [2.2] What force F is needed to make the longitudinal stress equal to the hoop stress for the thin-walled pressure vessel shown in Fig. P.2.2? Use

$$\tau_{hoop} = \frac{pr}{t} \text{ and } \tau_{long} = \frac{pr}{2t}.$$

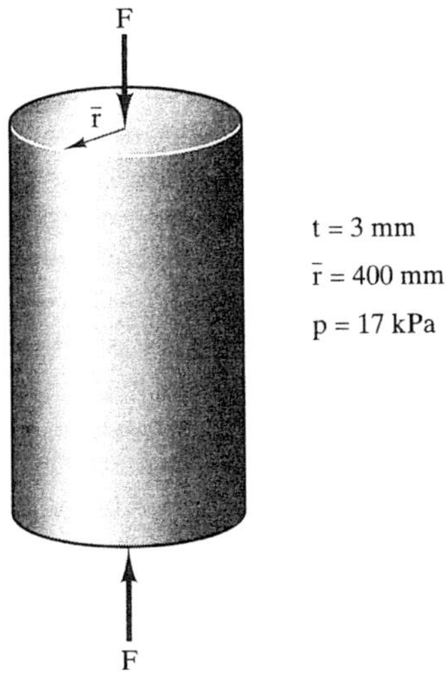

Figure P.2.2.

2.3. [2.2] Part of an apparatus for testing the shear strength of materials is shown in Fig. P.2.3. Two rigid plates A and B sandwich four identical samples of the material under test. A force P and a torque T of 10 ft-lb is applied. If the system remains stationary under these conditions, what is the average shear stress on the samples? The samples are cubes .1″ on an edge.

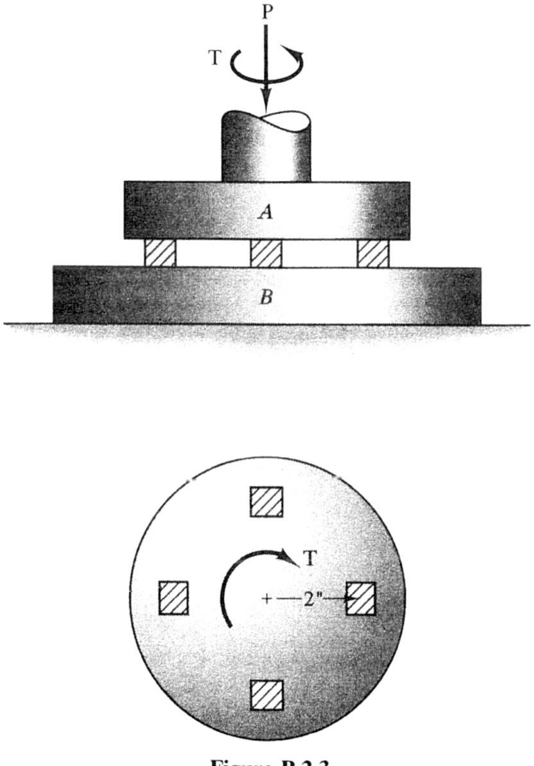

Figure P.2.3.

2.4. [2.2] A shear pin is sometimes used to connect two tubes of slightly different diameters but with the same wall thickness (see Fig. P.2.4). Determine the maximum shear stress in the shear pin for conditions given in the diagram.

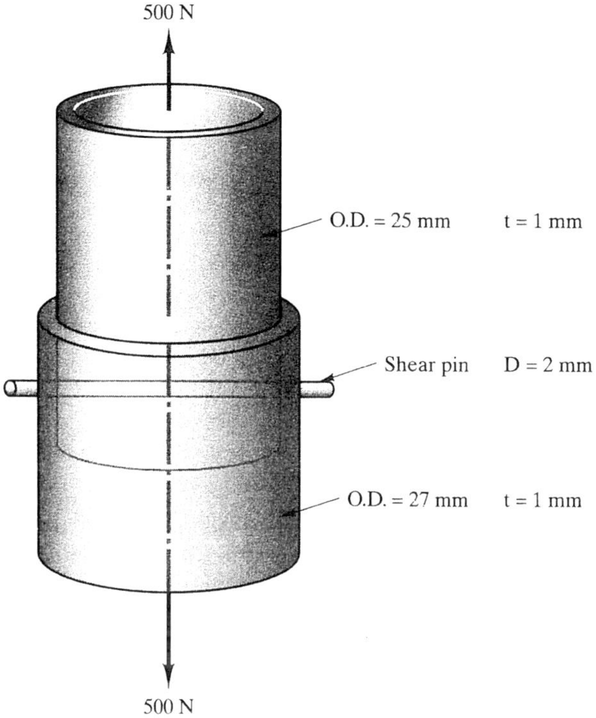

Figure P.2.4.

2.5. [2.2] Two pieces of wood in Fig. P.2.5 are glued together as shown. What is the average shear stress along the joint when a force $P = 300$ lb is applied?

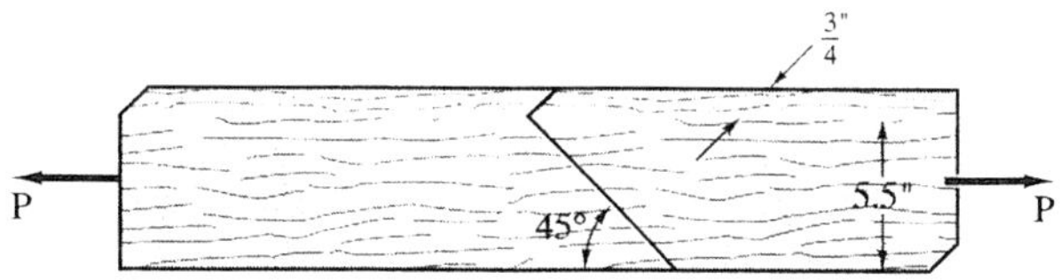

Figure P.2.5.

2.6. [2.2] A cylindrical pressure tank is shown in Fig. P.2.6. The cylinder wall is only .0125 m in thickness, so we may consider that we have a thin-walled vessel. An internal pressure of 5.3×10^5 Pa above atmosphere is maintained by the tank. What are the normal stresses $(\tau_n)_1$ and $(\tau_n)_2$ on interfaces formed by cutting out (mathematically) a vanishingly small rectangular parallelepiped, as has been shown in the diagram?

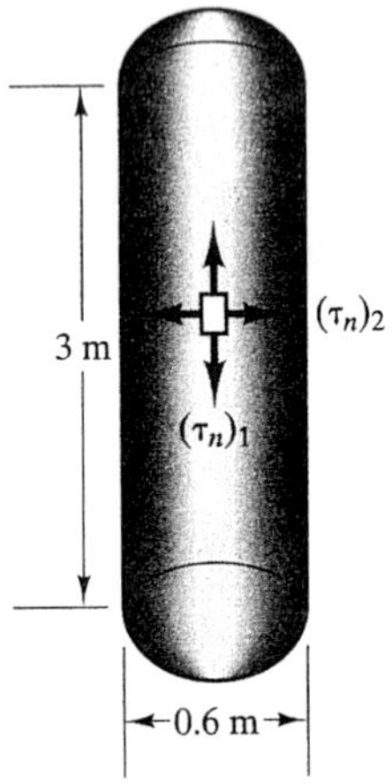

Figure P.2.6.

2.7. [2.2] A drive shaft in a device is a thin-walled cylinder having a diameter of 150 mm and a wall thickness of 6.25 mm. What is the shear stress on sections of the drive shaft if a torque of 30 N-m is transmitted by the shaft?

2.8. [2.2] A simplified model of an arm lifting a weight W is shown in Fig. P.2.8. Compute the stress in the bicep muscle at A when a weight W of 150 N is held in the position shown. The cross-sectional area at A is 600 mm^2.

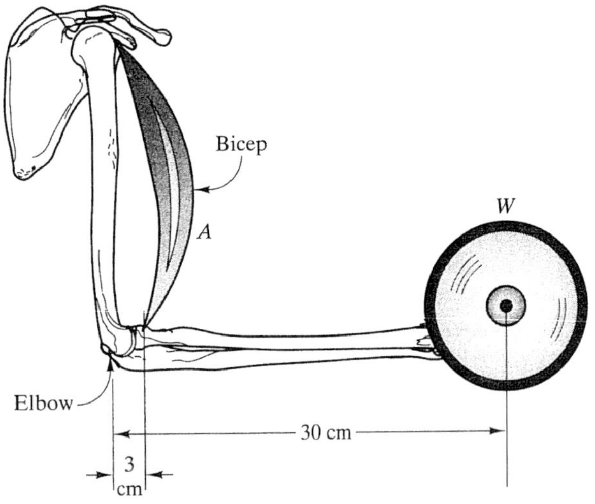

Figure P.2.8.

2.9. [2.2] Shown in Fig. P.2.9 are cross sections for two thin-walled pressurized tube testing designs. Compare the stress states in the tube wall for the two arrangements.

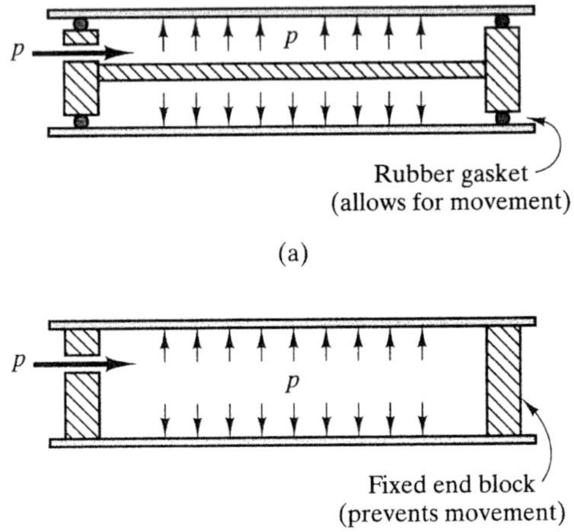

Figure P.2.9.

2.10. [2.2] A coupling between two shafts (see Fig. P.2.10) transmits a torque of 50 N-m and an axial load of 5000 N. Four bolts having a diameter each of 13 mm connect the two units and before loading of the shaft have negligible tensile forces. Assuming that each bolt undergoes the same stress, what is the average shear stress at the section in the bolt between the coupling members, and what is the normal stress there?

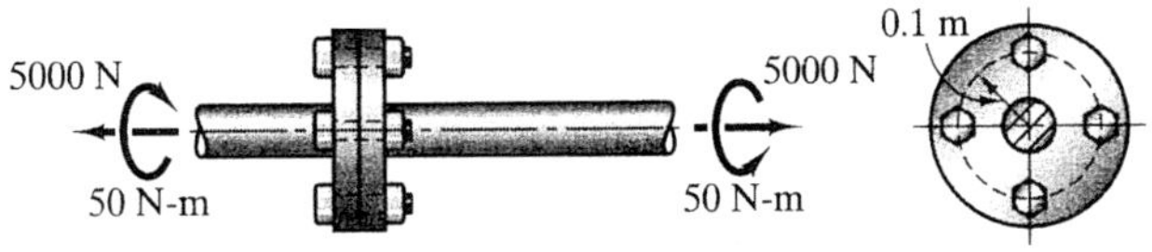

Figure P.2.10.

2.11. [2.2] In Fig. P.2.11 we show a mechanism whereby a torque T acts on bevel gear A having a mean diameter of 3 in., which in turn acts on bevel gear B having a mean diameter of 15 in. A torque is thus transmitted to the vertical tube C. This tube also supports a force P, as can be seen in the diagram. What are the shear stress and normal stress over cross sections of the tube for the following data?

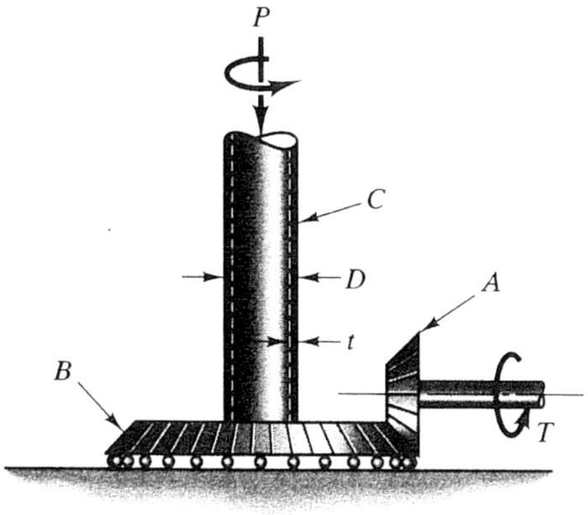

Figure P.2.11.

$P = 500$ lb
$T = 300$ in.-lb

$t = 1/8$ in.
$D = 3$ in.

2.12. [2.2] Shown in Fig. P.2.12 is a plate having the shape of a parallelogram. A force distribution of 7×10^5 Pa acts on sides *DC* and *AB*, while a force distribution of 3.5×10^5 Pa acts on sides *AD* and *BC*. What are the shear and normal stresses on these faces? The thickness of the plate is 25 mm. Is the plate in equilibrium?

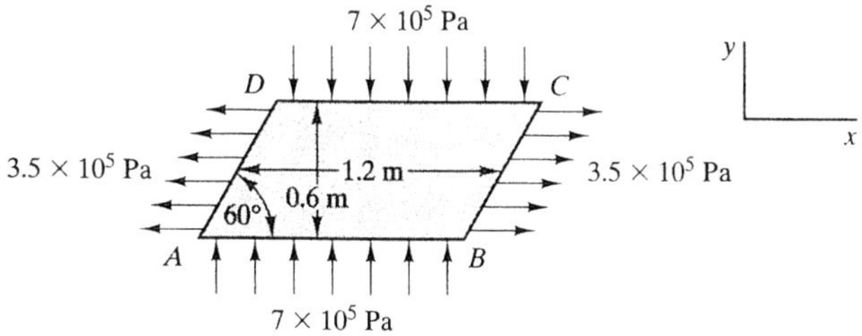

Figure P.2.12.

2.13. [2.2] A triangular plate in equilibrium is shown in Fig. P.2.13, where on surfaces *BC* and *AC* we have uniformly distributed forces of 4500 and 3600 N, respectively. What are the normal and shear stresses on face *AB* if the force F_{AB} is uniformly distributed? The thickness of the plate is 25 mm.

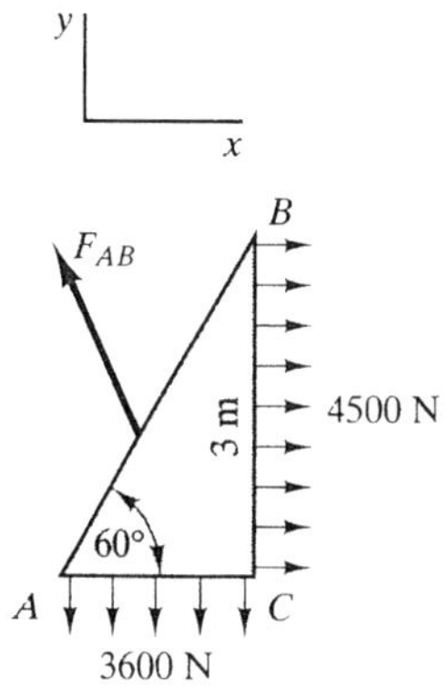

Figure P.2.13.

2.14. [2.2] In Example 2.3 give the normal stress on cross sections of the tube which are, respectively, 25 mm and 50 mm downward from the top of the rubber connector. Use solution $\tau = 4244/r$ Pa for shear stress.

2.15. [2.2] A horizontal shaft transmits a torque T of 50 N-m to a tube, as shown in Fig. P.2.15, through a rubber connector glued to both members. What is the shear stress on elements of cylindrical surfaces in the rubber connector concentric with the shaft as a function of r, the radius of the aforementioned cylindrical surface?

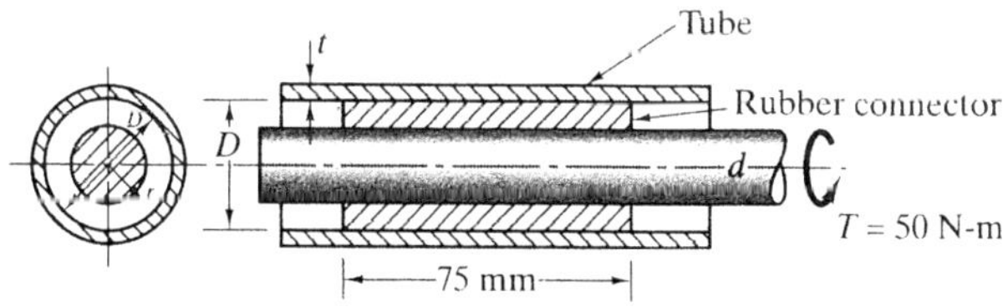

Figure P.2.15.

2.16. [2.2] In Problem 2.15, give the shear stress on a cross section of the tube at a distance of 25 mm to the right of the left end of the rubber connector. Take $t = 3$ mm and D as 50 mm. List the assumptions that you make to arrive at your answer.

2.17. [2.2] Shown in Fig. P.2.17 are two plates fastened together by three rivets in what is called a *lap joint*. What is the average shear stress over the sections of the rivets in the plane of the interface between the two plates, assuming that each rivet transmits the same load? In rivet calculations it is American practice to discount frictional forces between the plates (this gives an added measure of safety in design). If across all cross sections of the plates (i.e., along cuts across the width of the plates including those like A–A) the normal stress is assumed uniform for each such section, what section has the largest stress, and what is the value of this stress? Again, do not consider friction.

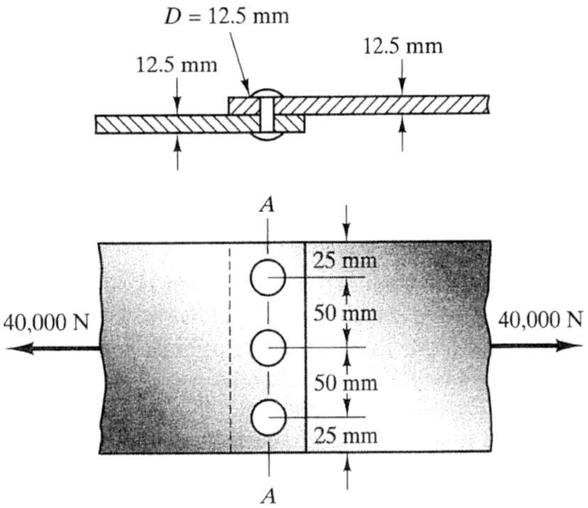

Figure P.2.17.

2.18. [2.2] Determine the diameter of the pin at C in Fig. P.2.18 if the maximum allowable stress is 70 MPa.

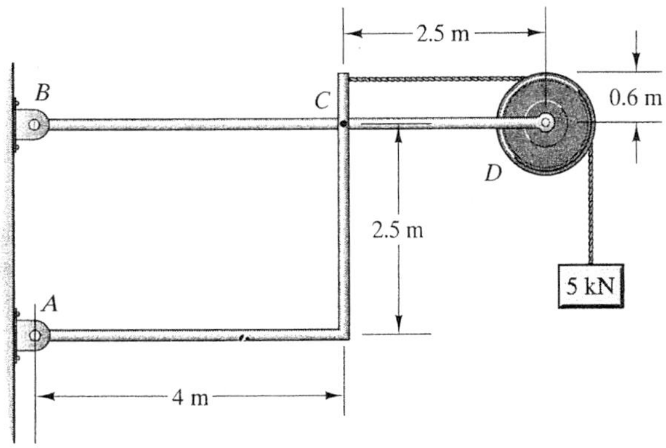

Figure P.2.18.

2.19. [2.2] Find the normal stress and maximum shear stress on a section inclined by 60° as shown in Fig. P.2.19. Neglect the weight of the member. Work from first principles.

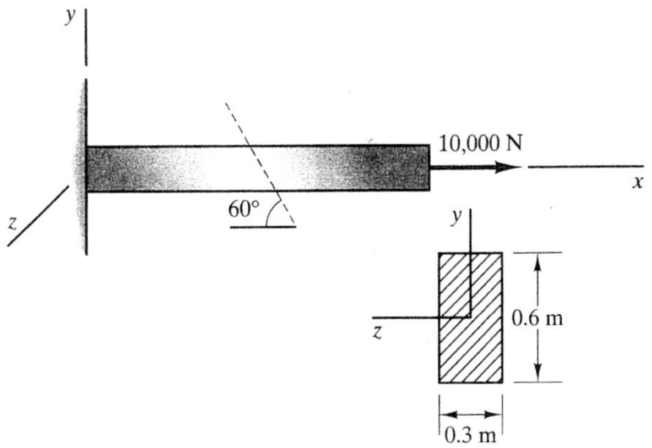

Figure P.2.19.

2.20. [2.2] A 1000-N load acts along the centerline of a circular shaft (see Fig. P.2.20). What is the *average normal* stress on a section inclined by 30° as shown? The area of an ellipse is πab, where a and b are the semimajor and semiminor diameters. Compute the *average* shear stress τ_t. Neglect the weight of the member. Work from first principles.

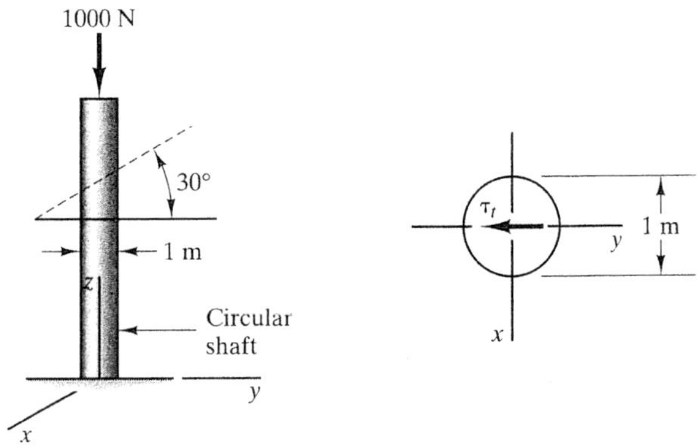

Figure P.2.20.

2.21. [2.2] Determine the maximum and minimum shear stresses on cylindrical surfaces in the rubber connector shown in Fig. P.2.21 about the centerline of the upper shaft.

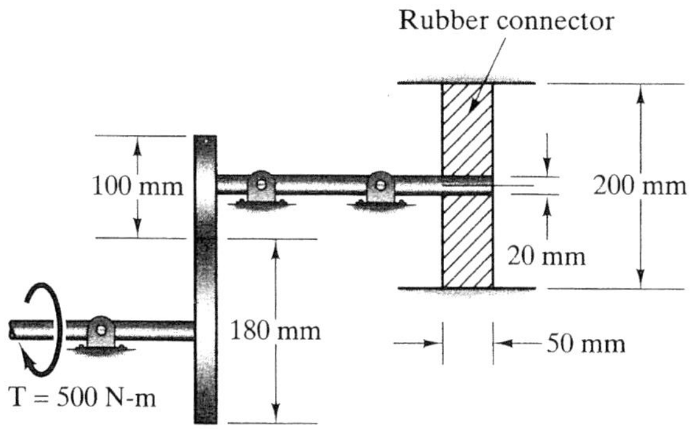

Figure P.2.21.

2.22. [2.2] A thin-walled tube transmits 20 kW of power while rotating at a speed of 100 rpm. If the outside diameter is 50 mm and the thickness is 5 mm, what is the shear stress in the section?

2.23. [2.2] Develop the relationship between the geometry of a compression fitting, the applied pressure, the coefficient of friction μ, and the loads P and T for the system shown in Fig. P.2.23.

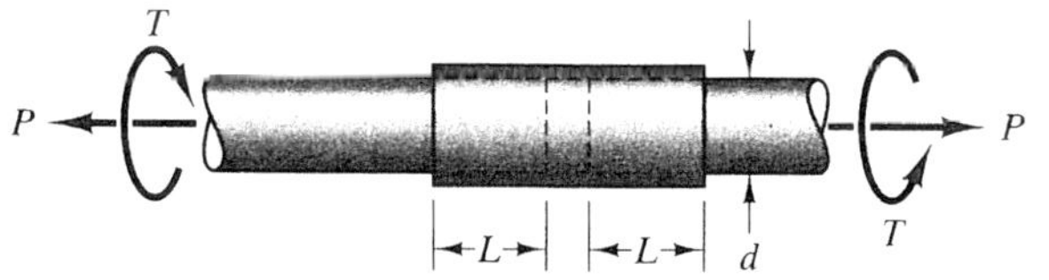

Figure P.2.23.

2.24. [2.2] In Fig. P.2.24 a tank contains water under pressure from air at the free surface having a pressure of 10 psi above the atmosphere. What is the vertical normal stress τ in the tank itself as a result of the air pressure and the water as a function of y from the free surface? The pressure just from the weight of the water alone is $p = \gamma d$ where γ is the specific weight (62.4 lb/ft^3) and d is the depth below the free surface. Any external pressure on the free surface extends unchanged throughout the water. The wall thickness of the tank is 3/4 in. and the inside diameter is 3 ft. The weight of the tank above the position of the free surface shown is 400 lb. The weight per unit length of the tank below the position of the free surface is $200y$ lb/ft with y in units of ft.

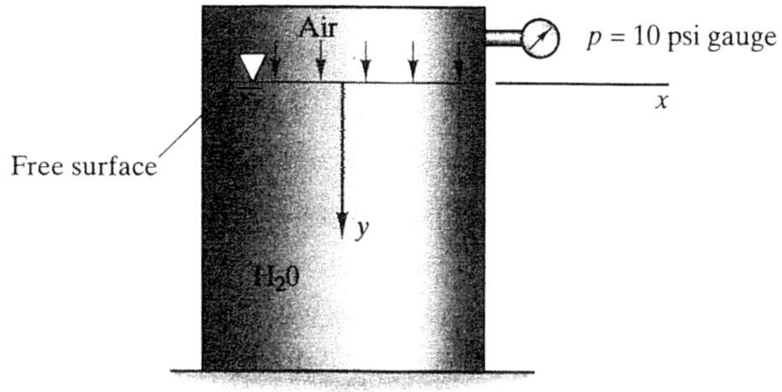

Figure P.2.24.

2.25. [2.3] Label the stresses shown in the infinitesimal rectangular parallelepiped in Fig. P.2.25.

2.26. [2.3] Label the stresses shown in Fig. P.2.26 and indicate the proper signs.

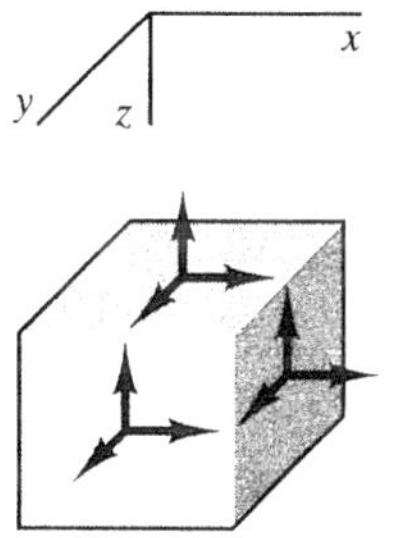

Figure P.2.25.

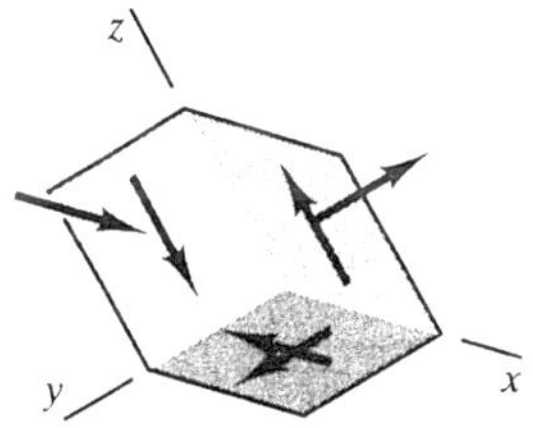

Figure P.2.26.

2.27. [2.3] Consider the stress distribution given over the circular cross section shown in Fig. P.2.27. Stresses $\tau_{z\theta}$ and τ_{zr} have been shown on a convenient element of this section. Note that this notation bears the same relation to cylindrical coordinates as does the notation of Section 2.3 to Cartesian coordinates. Suppose that we are given the following stress distribution over this section:

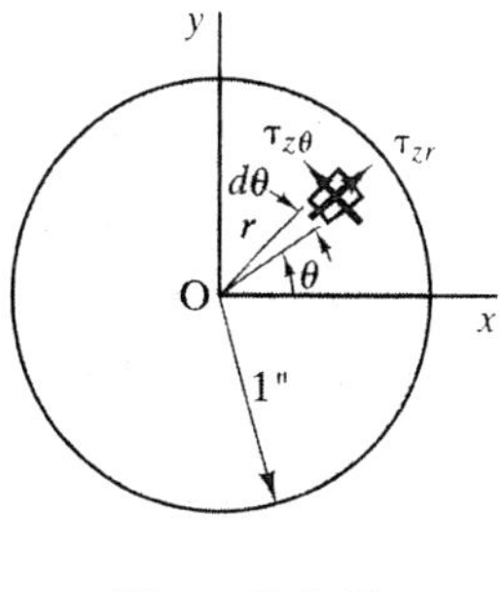

Figure P.2.27.

$$\tau_{z\theta} = (3r^2 + 2\theta) \times 10^3 \text{ psi}$$
$$\tau_{zr} = (2\theta + r\theta) \times 10^3 \text{ psi}$$
$$\tau_{zz} = (\theta^2 r + r^3) \times 10^3 \text{ psi}$$

What is the resultant force of this distribution?

2.28. [2.4] Show in Section 2.4 by considering corners B that

$$\tau'_{xy} = \tau_{yx}$$

2.29. [2.5] Explain why in Example 2.1 we can take the shear stress to be oriented tangent to the centerline of the section.

2.30. [2.6] Given the following stress distribution in a domain,

$$\tau_{xx} = 6x^2 + 3yx + 20 \text{ Pa}$$
$$\tau_{yy} = 2xz + 10 \text{ Pa}$$
$$\tau_{zz} = 4x^2 + 3y + 10yz \text{ Pa}$$

what must the body force distribution be for equilibrium? What is the body force intensity at (3, 2, 4) m?

2.31. [2.6] What are the differential equations of equilibrium for plane stress for no body forces? Show that if we give the stresses as follows,

$$\tau_{xx} = \frac{\partial^2 \phi(x,y,z)}{\partial y^2}$$
$$\tau_{yy} = \frac{\partial^2 \phi(x,y,z)}{\partial x^2}$$
$$\tau_{xy} = -\frac{\partial^2 \phi(x,y,z)}{\partial x \partial y}$$

where ϕ is a scalar function called the *Airy stress function,* we automatically satisfy the equations of equilibrium. Hence, in the theory of elasticity we often work with this function, thereby satisfying the equations of equilibrium intrinsically.

***2.32. [2.6]** By methods analogous to those of Section 2.6, develop the following equations of equilibrium for plane stress using cylindrical coordinates (see Fig. P.2.32):

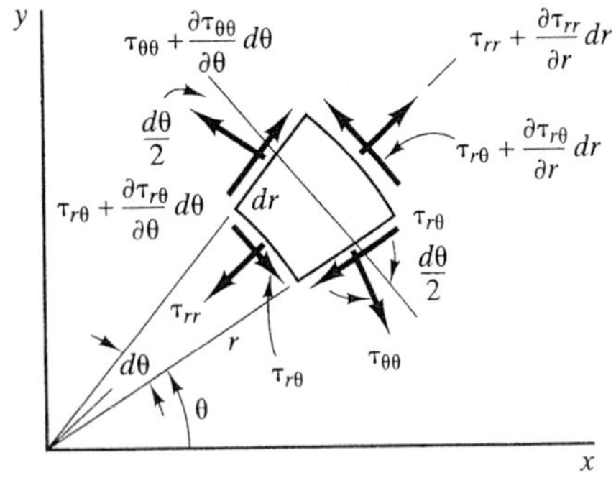

Figure P.2.32.

$$\frac{\partial \tau_{rr}}{\partial r} + \frac{1}{r}\frac{\partial \tau_{r\theta}}{\partial \theta} + \frac{\tau_{rr} - \tau_{\theta\theta}}{r} = 0$$
$$\frac{1}{r}\frac{\partial \tau_{\theta\theta}}{\partial \theta} + \frac{\partial \tau_{r\theta}}{\partial r} + \frac{2\tau_{r\theta}}{r} = 0$$

Employ Fig. P.2.32 as an aid. Assume no body forces. (*Hint:* Approximate the sine and cosine of a small angle.)

2.33. We show in Fig. P.2.33 a simple universal joint whereby power is transmitted, albeit with some loss, from one axis of rotation to another axis of rotation inclined at some angle α to the first. If the torque T in the horizontal axis is 50 N-m, what is the average shear stress over a cross section of the pin B for the case where $\alpha = 0$?

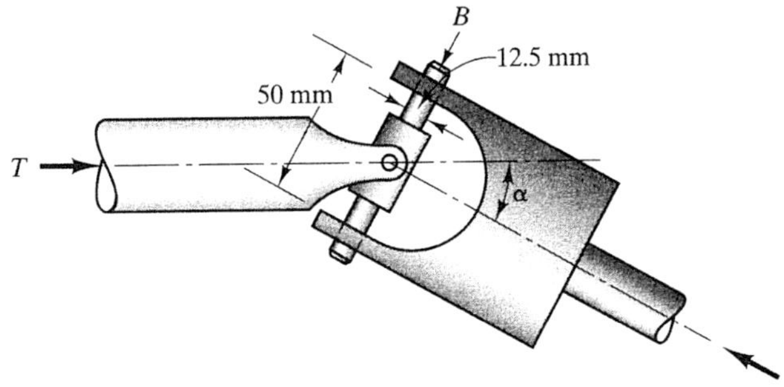

Figure P.2.33.

2.34. A thin-walled spherical pressure vessel (Fig. P.2.34) has an internal pressure p_i of 100 psi above atmosphere. The outside diameter D of the sphere is 2 ft, and the thickness t of the wall is 1/2 in. Imagine sections of the vessel formed by taking cuts that go through the center of the sphere. What is the normal stress along such sections? (*Hint:* The force in a given direction from a uniform pressure on a curved surface equals the force from the same pressure on the *projection* of the surface onto a plane normal to the direction of the force desired.)

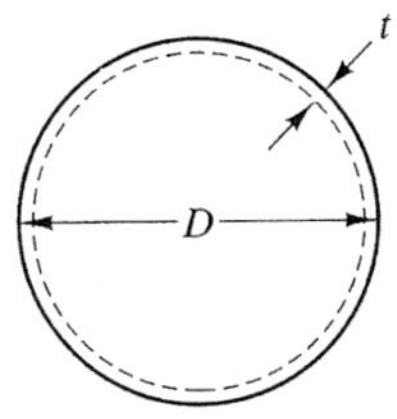

Figure P.2.34.

2.35. The camping trailer shown in Fig. P.2.35 weighs 12,000 lb and is pulled by a truck moving at 65 mph. During an emergency stop, the truck and trailer come to rest in five seconds with constant deceleration along a straight and level section of road. Assuming that there are no brakes on the trailer and ignoring friction, compute the shear stress in the "king-pin" connecting the truck and trailer. The king-pin is a circular shaft with inside diameter of 2.5 in and outside diameter of 3.0 in.

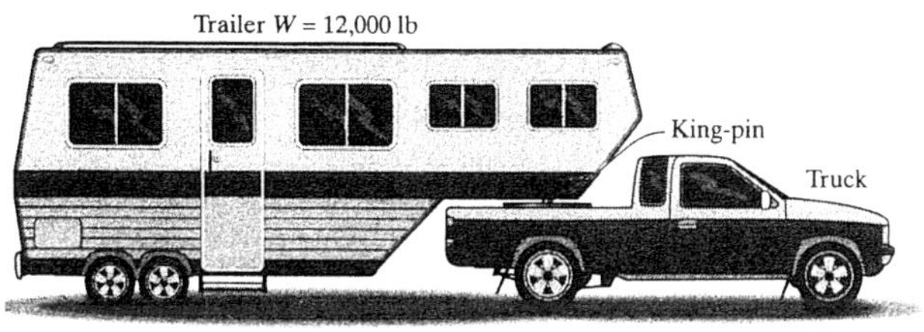

Figure P.2.35.

2.36. Determine the required bolt diameter at B (see Fig. P.2.36) for an allowable shear stress of 10,000 psi.

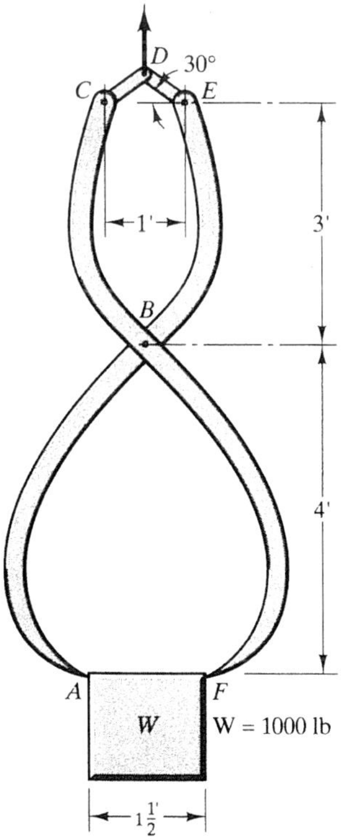

Figure P.2.36.

2.37. Determine the maximum stress in the bolts approximately for the eccentrically loaded connection in Fig. P.2.37. Neglect the weight of the plate.

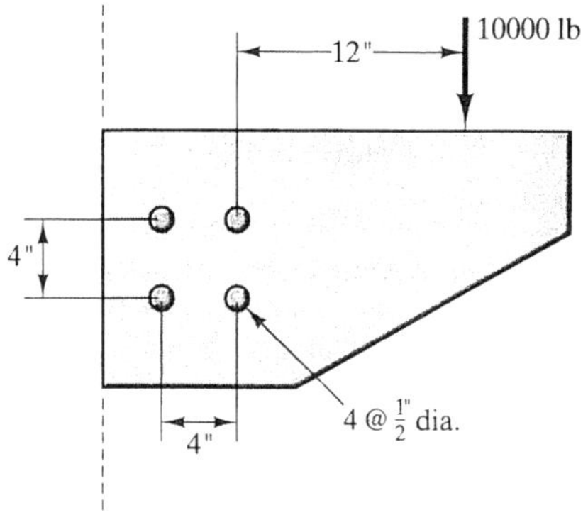

Figure P.2.37.

2.38. In sport rock climbing, a hanger (see Fig. P.2.38) is sometimes bolted into the rock so that the lead climber can clip the rope to it for protection. A typical bolt and hanger design is shown. Compute the strength of the design for 1/4-in. and for 3/8-in. bolts. List your assumptions.

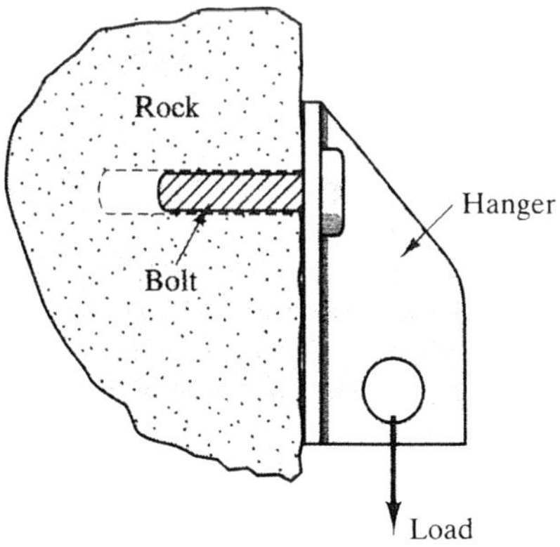

Figure P.2.38.

2.39. What is the maximum torque T transmitted by the flange connection in Fig. P.2.39? There are eight bolts each having a diameter of 5 mm. The maximum allowable stress per bolt is 300 MPa.

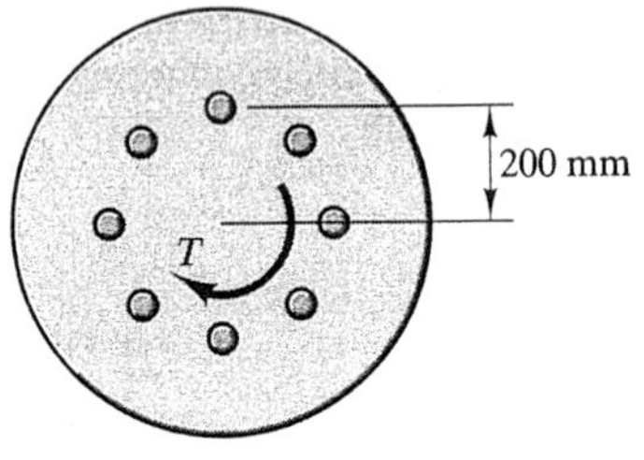

Figure P.2.39.

2.40. A thin-walled cone is shown in Fig. P.2.40 supporting a 100-lb force. Compute the normal stresses pointing toward 0 on interfaces with normals pointing toward 0 having a vertical distance from point 0 of 5 ft. Now compute normal stresses oriented in the vertical direction on interfaces formed by taking a horizontal cut of the cone 5 ft from the top. Disregard shear stresses.

2.41. A cylinder of length L and constant cross-sectional area A is held from above as shown in Fig. P.2.41. Taking the specific weight γ to be constant, what is the normal stress at the cross sections of the cylinder as a function of z and other pertinent variables?

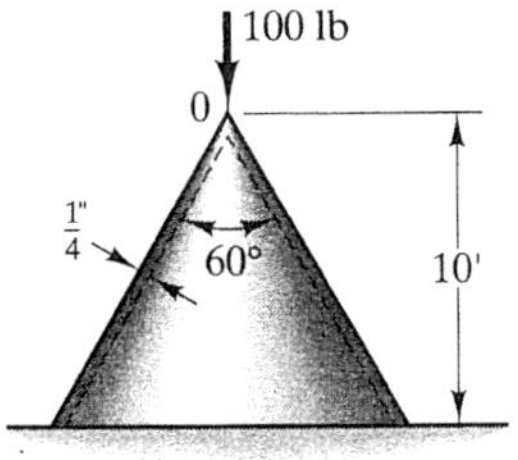

Figure P.2.40.

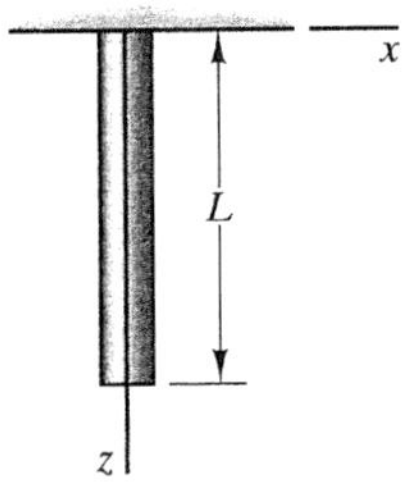

Figure P.2.41.

2.42. A cone hangs by its own weight (see Fig. P.2.42). What is the average normal stress at its cross sections as a function of z and other pertinent parameters? The specific weight of the cone material is a constant given as γ.

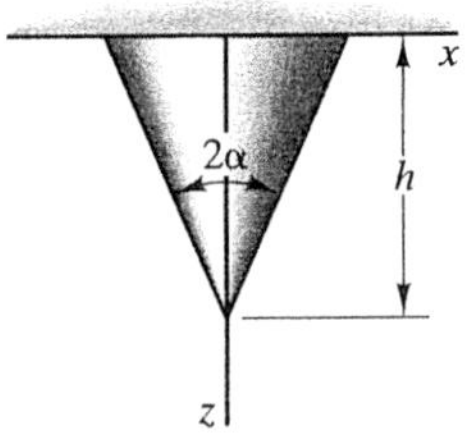

Figure P.2.42.

2.43. A compression fitting in Fig. P.2.43 is used to connect two rods under tension. What minimum pressure must the fitting provide for the connection? Take $\mu = 0.80$.

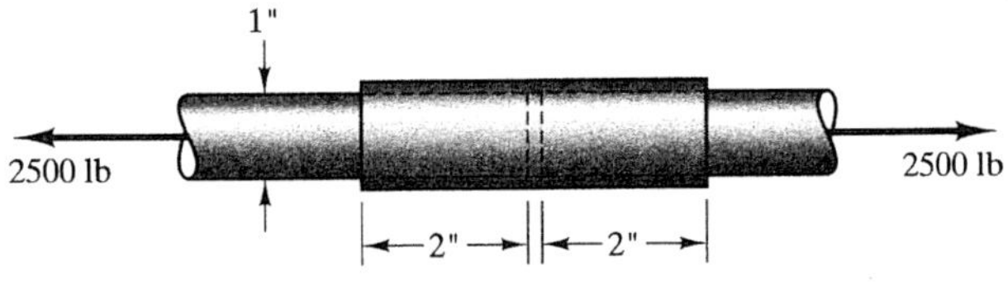

Figure P.2.43.

2.44. Consider a thin-walled shell used for a pressure vessel. A free-body diagram of a section of the vessel is shown in Fig. P.2.44. The radii of curvature are given by r_ϕ and r_θ. Derive an equation relating the wall stresses, the radii, and the wall thickness t. Specifically, show that

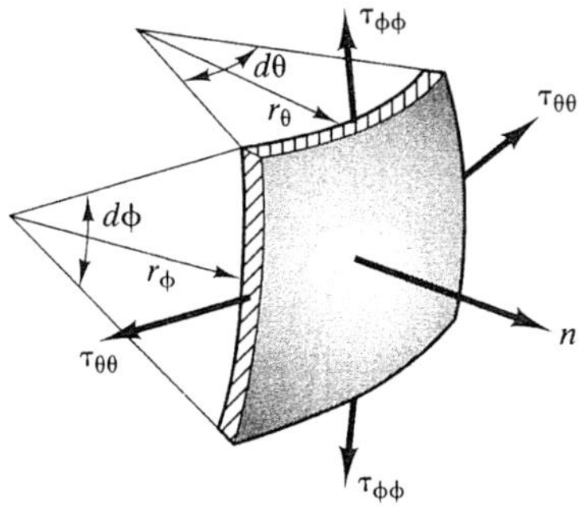

Figure P.2.44.

$$\frac{\tau_{\theta\theta}}{r_\theta} + \frac{\tau_{\phi\phi}}{r_\phi} = \frac{p}{t}$$

Take $\sin\frac{d\theta}{2} = \frac{d\theta}{2}$ and $\sin\frac{d\phi}{2} = \frac{d\phi}{2}$. Also note that the force on a curved surface from a uniform pressure equals the pressure times the projected area of the surface in the direction of the desired force.

2.45. Solid rods of diameter 30 mm are welded to a rigid drum D (see Fig. P.2.45), which is turning at a speed of 500 rpm. What is the maximum tensile stress on cross sections of the rods at the base of the rods? The rods have a mass per unit length of 5 kg/m. Include the static weight of the rods.

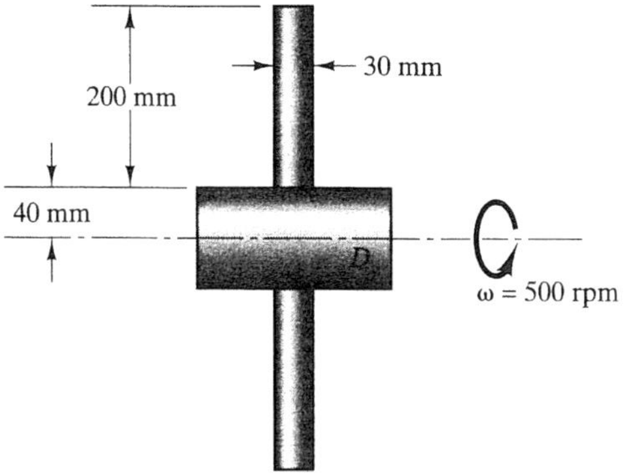

Figure P.2.45.

2.46. A wooden pile is being driven into a hole in the ground as shown in Fig. P.2.46 by a force $F = 100{,}000$ lb. A friction force of f lb/unit length opposes this force. The force intensity f is proportional to the square of the distance z from the top surface, being zero at the top. If $L = 30$ ft and $D = 12$ in., what is the normal stress as a function of z at sections of the pile away from the ends?

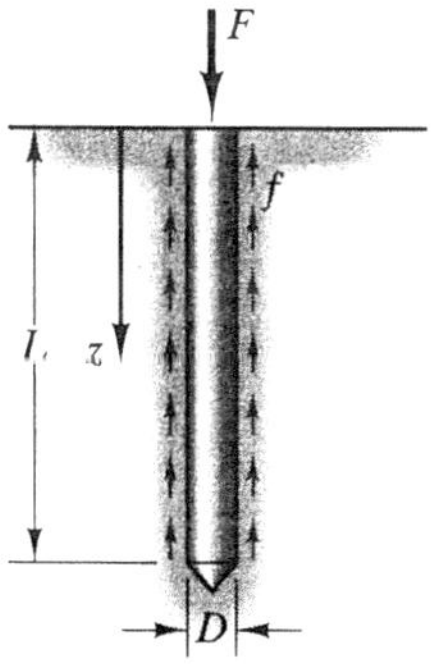

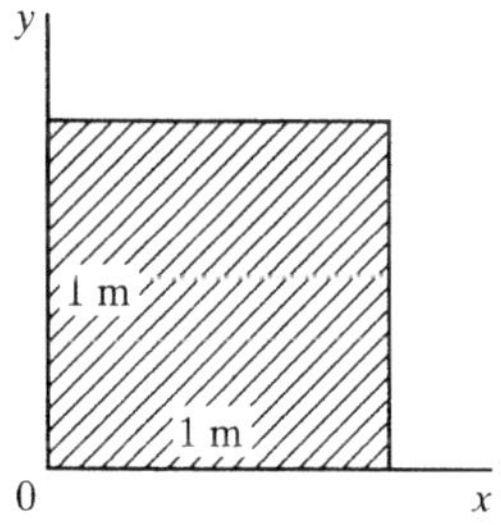

Figure P.2.46. **Figure P.2.47.**

2.47. Suppose that the following stress distribution is known over the plane rectangular cross section shown in Fig. P.2.47:

$$\tau_{zy} = (2x^2 + 3y) \times 10^3 \text{ kPa}$$
$$\tau_{zx} = 1000 + (2x) \times 10^3 \text{ kPa}$$
$$\tau_{zz} = (xy + 2x^2) \times 10^3 \text{ kPa}$$

What is the resultant force from such a distribution?

2.48. Shown in Fig. P.2.48 is a flow of water in a channel. The velocity profile is shown as parabolic. *Newton's viscosity law* gives the shear stress τ on an interface parallel to the floor of the channel as $\tau = \mu(\partial V/\partial y)$, where μ is the *coefficient of viscosity*. What is the shear stress on an element normal to y at an elevation of 2.5 m in terms of μ? What is the drag on the channel bed, in terms of μ, per unit area if we had a linear profile (shown dashed)?

2.49. In Fig. P.2.49 we show a section of a shaft A rotating concentrically with speed ω inside a journal bearing B and separated from B by a film of oil of thickness ε and viscosity μ. Newton's viscosity law $\tau = \mu(\partial V/\partial r)$ gives the shear stress on interfaces normal to radius r in the oil. If there is a linear profile of the oil and if it "sticks" to the metal surfaces (i.e., it has the same speed as the surface at the surface), what is the torque T needed to rotate the shaft? Take the length of the bearing to be L.

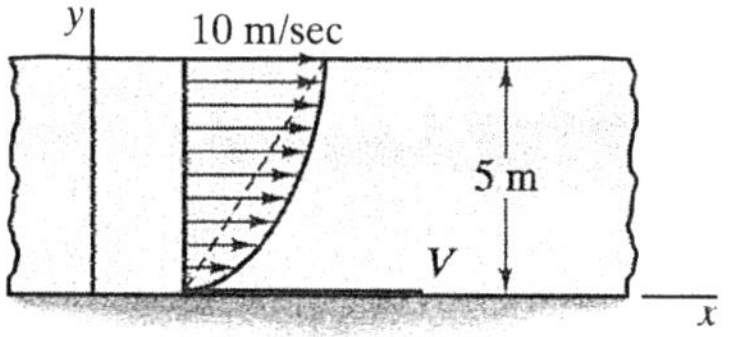

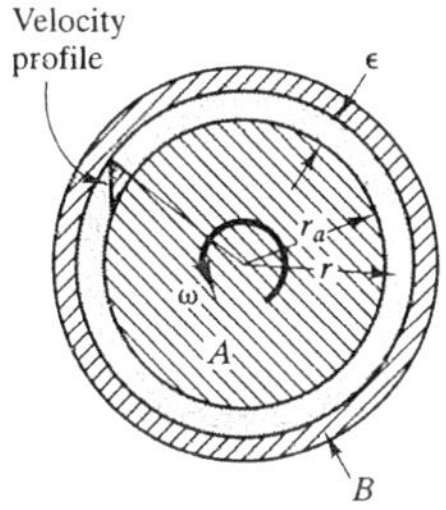

Figure P.2.48. **Figure P.2.49.**

2.50 How many g's of acceleration can we allow the space vehicle shown in Fig. P.2.50 to accelerate vertically in outer space if the normal stress in the cross sections of rods BA and CA are not to exceed 8×10^6 Pa? The rods are parallel to the ZY plane and have a diameter of 35 mm. Neglect the mass of the rods and take the mass of the body pin-connected to the rods at point A to be 100 kg.

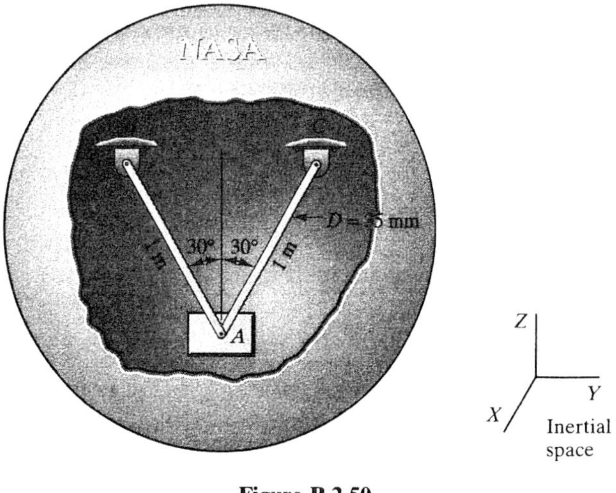

Figure P.2.50.

2.51. The *efficiency of a riveted joint* is the ratio of the maximum transmissible force permitted by the joint to the maximum force that would be permitted if the joint were not present and the material were continuous at the joint. If the maximum shear stress in the rivets is 8.7×10^7 Pa before failure and the maximum stress in the plates is 1×10^8 Pa before failure, give two efficiencies for Problem 2.17 and explain the nature of the failure. Note again that we do not include frictional effects for such rivet calculations.

2.52. We have illustrated in Fig. P.2.52 a *butt-joint* rivet connection. Read Problem 2.17 and obtain the same results for this joint as was found for the lap joint of that problem.

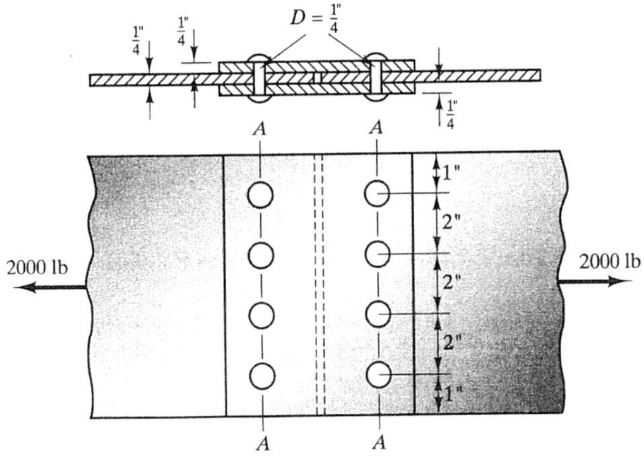

Figure P.2.52.

2.53. In Fig. P.2.53 water fills a spherical tank which is supported from below. Pressure p_1 = 300 kPa gage. Fifty bolts hold the upper half of the tank to the lower half with a force between flanges of 5000 N. What is the stress in each bolt? Each half of the sphere weighs 2000 N. The diameter of the bolts is 3.4 millimeters. *Hint:* Take the upper half of the system as a free body. Also note that the pressure in the water decreases linearly as you move upward such that $\Delta p = \gamma \Delta d$ where Δd is the distance moved upward and γ is the specific weight of the water.

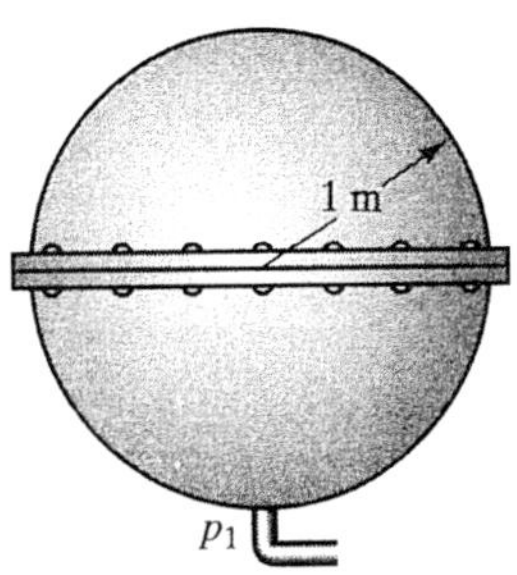

Figure P.2.53.

2.54. An open-ended 60° conical container in Fig. P.2.54 is bolted to a cylinder. The cylinder contains oil and water with oil extending into the conical container filling the latter. Find the normal stress on each of 30 bolts connecting the cone and cylinder so that there is a force of 6000 N between flanges of the two containers. The volume of a cone is $1/3(A_{(\text{Base})})$ (height). The diameter of the bolts is 30 mm. *Hint:* Consider the cone and its oil as a free body.

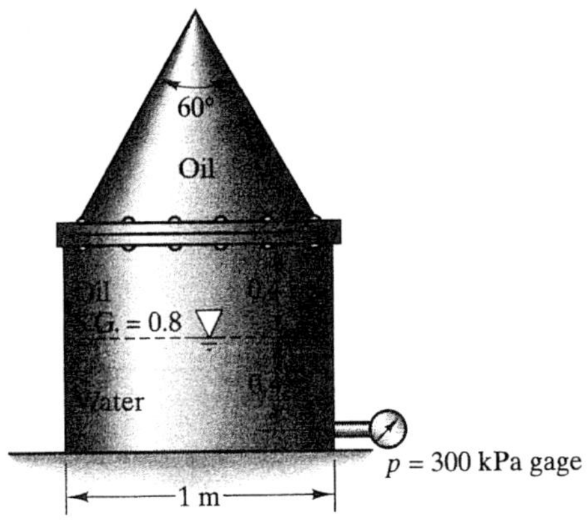

Figure P.2.54.

2.55. For the tip-loaded cantilever beam in Fig. P.2.55, the following stresses may be written (you will learn more about these stresses in Chapter 11):

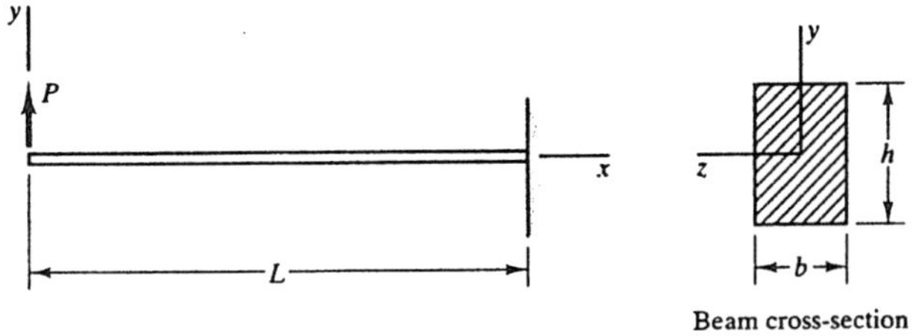

Figure P.2.55.

$$\tau_{xx} = -\frac{12My}{bh^3}$$

$$\tau_{xy} = -\frac{6V}{bh^3}\left(\frac{b^2}{4} - y^2\right)$$

Where M and V are the moment and shear force, respectively, on cross sections of the beam at any location x due to the applied load P. If we assume that all other stresses are zero, do these stresses satisfy the equations of equilibrium? Take $M = Vx$.

2.56. A wooden structure is bolted together and loaded as shown in Fig. P.2.56. If the allowable shear stress for each bolt is 9,000 psi, what is the maximum force P that can be supported? The bolts are 1/4-in. diameter and each is in single shear.

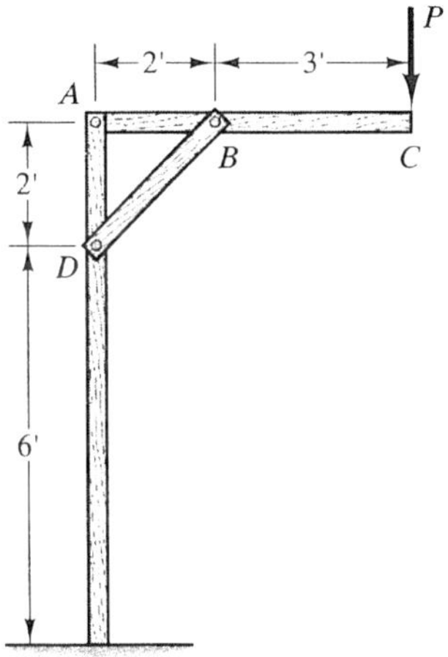

Figure P.2.56.

2.57. For the disk clutch shown in Fig. P.2.57, compute the equation for the torque T transmitted as a function of the applied force F, the coefficient of friction μ, and the clutch geometry. The clutch pad is a circular disc. Assume uniform contact pressure on the pad.

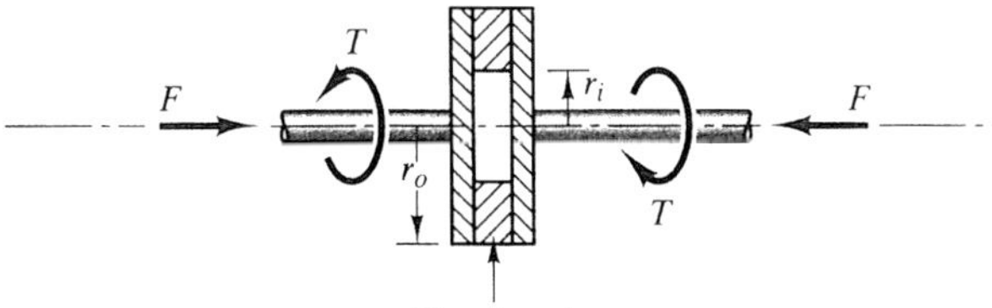

Figure P.2.57.

***2.58.** Determine the lay-up angle α for the filament-wound cylindrical pressure vessel (see Fig. P.2.58) that results in equal strength in the longitudinal and hoop directions. Assume that the vessel is thin-walled, with thickness t, radius r, and internal pressure p. Isolate a rectangular element so that there are the same number of filaments in the vertical and horizontal edges. Let P be the force per unit length along the filaments. Use equilibrium and thin-walled cylinder formulation in the longitudinal and hoop directions to find the angle α.

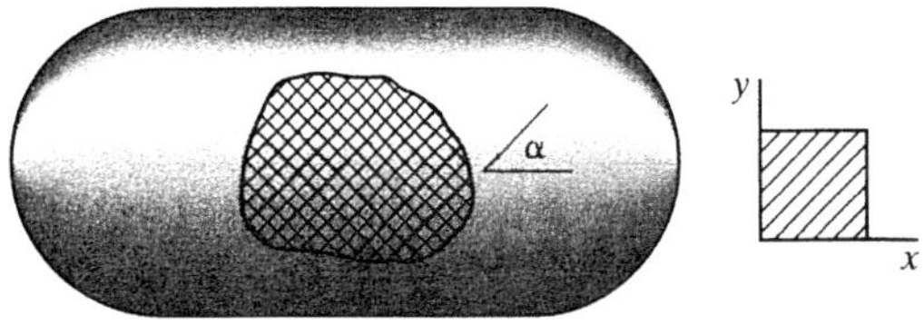

Figure P.2.58.

****2.59.** For the journal bearing considered in Problem 2.49 for which

$$T = \frac{2\pi\mu\omega L r^3}{\varepsilon},$$

plot

$$\frac{T}{\mu\omega L}$$

versus oil film thickness ε for $0.05'' \leq \varepsilon \leq 0.50''$ and shaft radii r of 0.5, 1.0, 1.5, and 2.0 in. What combination of oil film thickness and shaft radius gives the highest resisting torque? What combination is best for a journal bearing?

****2.60.** Consider the pile presented in Problem 2.46. Develop a general solution for the problem by keeping the cross-sectional area A of the pile and the depth z of the free-body cut as variables. Plot the normal stress in the z-direction versus depth for pile diameters D = 0.5, 0.75, 1.0, 1.5 and 2.0 ft. Comment on the peak normal stress in the pile as the diameter changes. From problem 2.46 we have

$$\tau_{zz} = \frac{100}{27A}(30^3 - z^3)$$

****2.61.** For the symmetric roof truss shown in Fig. P.2.61, construct a design table that lists the forces in each member, indicating tension or compression, for roof pitches (h/L) of

1:1, 1:2, 1:3, 1:4, and 1:5. Use the computer to aid in the repetitive calculations.

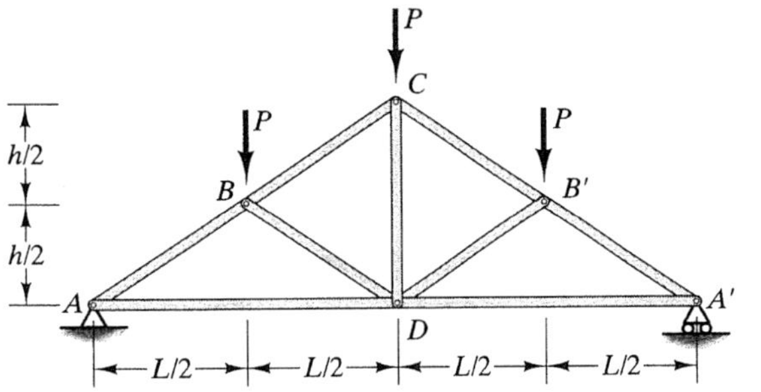

Figure P.2.61.

****2.62.** The theory of elasticity gives an "exact" solution for the hoop stress in an internally pressurized *thick-walled* cylinder is given as (see Fig. P.2.62):

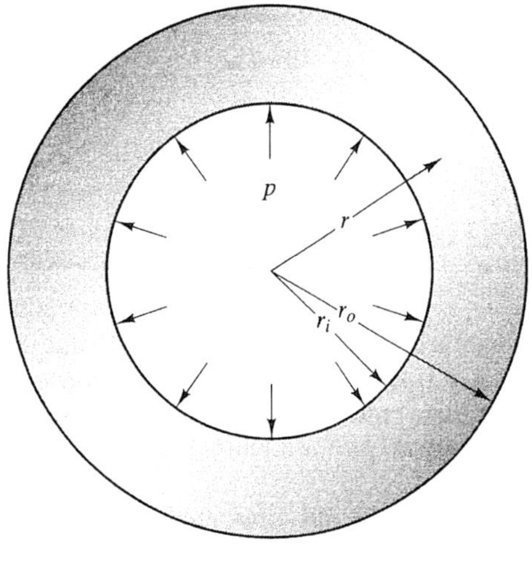

Figure P.2.62.

$$\tau_{hoop} = \frac{pr_i^2}{r_o^2 - r_i^2}\left(1 + \frac{r_o^2}{r^2}\right)$$

Compare this stress with the hoop stress obtained using the thin-walled approximation presented in this Chapter,

$$\tau_{hoop} = \frac{pr_i}{t}.$$

Plot the two solutions as τ_{hoop} / p versus t/r_i. At what ratio of t/r_i is the approximate solution in error by less than 10%? (*Hint:* evaluate the theory of elasticity solution at $r = r_i$.)

****2.63.** Consider the two pieces of wood in Fig. P.2.63, each of width w and thickness t, butt-joined at an angle α as shown. Plot the shear stress at the joint as the angle α varies from 30° to 60°.

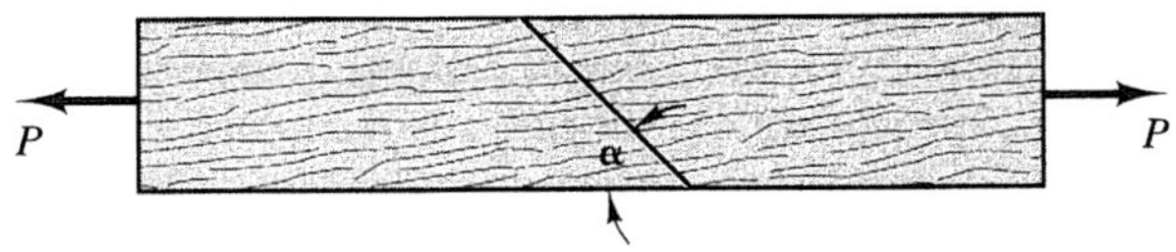

Figure P.2.63.

****2.64.** A 225-lb ball is attached to a 35-ft wire by a 3/8-in. diameter shear pin as shown in Fig. P.2.64. If the ball is then rotated about the z-axis, determine the stress in the pin as a function of the velocity V. Consider the range $0 \leq V \leq 50$ ft/s.

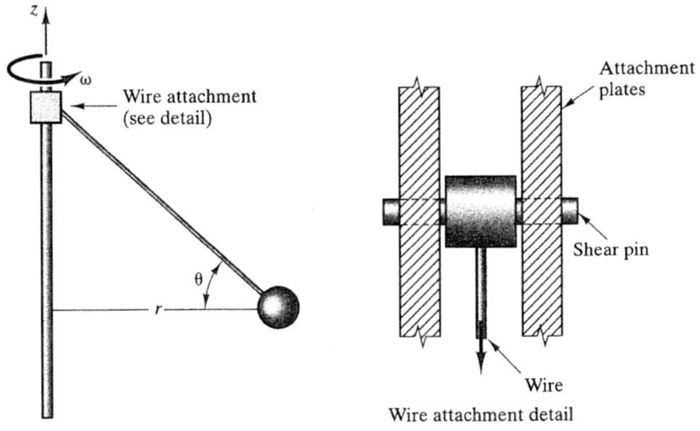

Figure P.2.64.

Project 1: Compressor Problem

Find the maximum tensile strength in the cylinder wall in Fig. P.2.65 as the piston moves from right to left compressing the air in an isothermal manner.

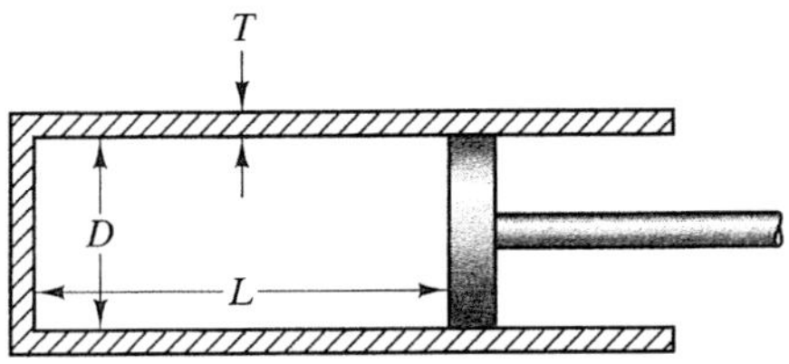

Figure P.2.65.

Ask the user what he or she wishes for

$D \equiv$ outside diameter (inches)
$L \equiv$ initial distance of piston from end plate (inches)
PRI $\equiv$ initial pressure of air above atmosphere (psi)
$T \equiv$ thickness of wall (inches)

Have the piston move $0.8L$ in 20 increments. Print out:

Pressure (psi) Distance moved (in.) Stress (psi)

for 20 increments. Run the program for the following data:

$L = 5$ in.
$D = 3$ in.
PRI $= 0$
$T = 0.125$ in.

Project 2: Stresses in a Leaking Tank

Air at some initial pressure p and water up to a certain height h_2, are contained in a tank in Fig. P.2.66. A number of successive leakages numbering no greater than 5 occur at the bottom of the tank. If the air expands isothermally, find the hoop stress and axial stress in the tank after each leakage at 10 points equally spaced and extending downward from near the top of the tank to near the bottom of the tank.

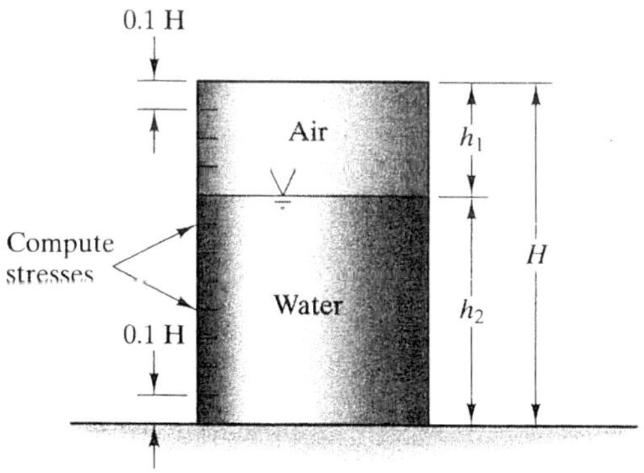

Figure P.2.66.

The user supplies the following data:

Initial pressure of air, p_A (Pa) gauge
Diameter of the tank, D (m)
Wall thickness, t (m)
Initial height of air, h_1 (m)
Initial height of water, h_2 (m)

Treat the process above as a *quasi-steady* process. After each leakage the water level drops by $0.05\ H$ (m). The stresses are computed after each successive leakage. Neglect the region $= 0.1H$ near the top and bottom of the tank (i.e., divide the remaining 80% of H into 10 sections). Neglect the weight of the tank.

The input is to be accepted in the following format:

Outer diameter and wall thickness (m)
Gauge pressure of the air initially (Pa)
Initial heights of air and water (m)

The results are to be displayed as follows:

First leakage section	*Abs. pressure of air* ______ *Axial stress*	*Hoop stress*
1	______ Pa	______ Pa
2	______ Pa	______ Pa
3	______ Pa	______ Pa
4	______ Pa	______ Pa
5	______ Pa	______ Pa
6	______ Pa	______ Pa
7	______ Pa	______ Pa
8	______ Pa	______ Pa
9	______ Pa	______ Pa
10	______ Pa	______ Pa

Second leakage section	*Abs. pressure of air* ______ *Axial stress*	*Hoop stress*
1	______ Pa	______ Pa
2	______ Pa	______ Pa
3	______ Pa	______ Pa
4	______ Pa	______ Pa
5	______ Pa	______ Pa
6	______ Pa	______ Pa
7	______ Pa	______ Pa
8	______ Pa	______ Pa
9	______ Pa	______ Pa
10	______ Pa	______ Pa

(etc. up to fifth leakage)

Run the program for the five leakages for the following data:

$D = 0.5$ m
$t = 0.01$ m
$p_A = 100{,}000$ Pa gauge
$h_1 = 0.4$ m
$h_2 = 0.8$ m

CHAPTER 3

Strain

3.1 Introduction

Now that we are able to describe quantitatively the manner in which a force distributes through a body, we shall proceed to set forth means of describing quantitatively the deformation of a body stemming from the applied external loads. First we present the *displacement field*, which permits us to describe the change of geometry, from loading action, on a point-by-point basis. Although the displacement field (a vector field) is sufficient to describe all geometric characteristics of deformation, we shall find it expeditious to develop a tensor called the strain tensor in order to relate deformation more directly to the stress tensor. Most of the chapter centers around a discussion of the strain tensor.

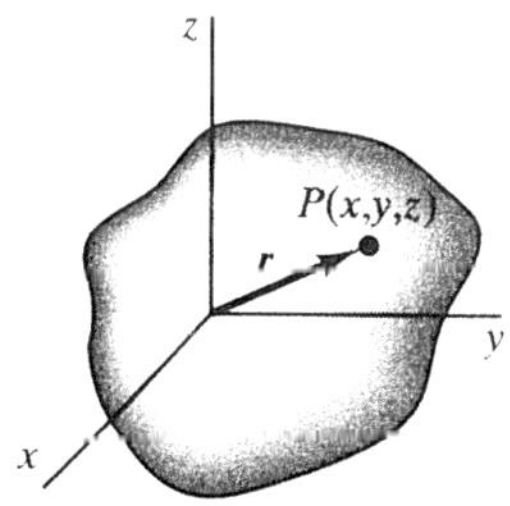

Figure 3.1. Position vector in a body.

3.2 The Displacement Field

Consider a body in its undeformed geometry. We may locate any point P by using a position vector $\boldsymbol{r}$ corresponding to point P. This is shown in Fig. 3.1. Now assume that loads have been applied and that the body has deformed to some new configuration, shown dashed in Fig. 3.2. Each point P at (x,y,z) moves to a new point P', as shown in the diagram. The continuum of vectors giving the displacement from each point P to its position P' in the deformed geometry is the *displacement field*, $\boldsymbol{u}(x,y,z)$. Thus, if we know $\boldsymbol{u}$ for a particular problem, we can determine the displacement of any particular point of the body by substituting the coordinates of that point in the vector function $\boldsymbol{u}(x,y,z)$.[1]

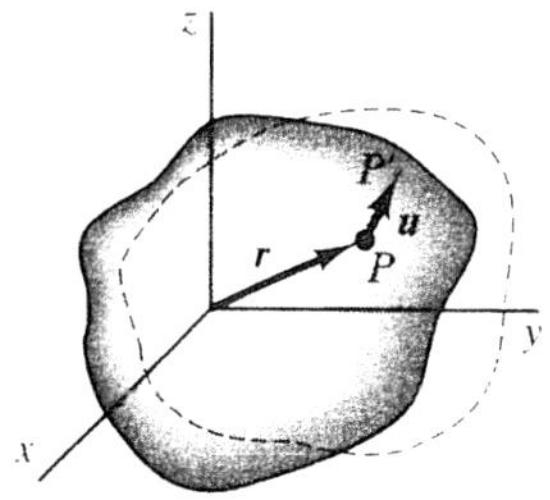

Figure 3.2. Displacement field vector.

[1]The displacement field can also be a function of time such that $\boldsymbol{u} \equiv \boldsymbol{u}(x,y,z,t)$.

The following example illustrates the use of the displacement field.

Example 3.1

The following displacement field represents the deformation of a body in a given domain:

$$\boldsymbol{u} = [x^2\mathbf{i} + (x+3)\mathbf{j} + 10\mathbf{k}] \times 10^{-2}\,\mathrm{m}$$

What is the displacement of the point originally at the position $\boldsymbol{r} = \mathbf{j} + \mathbf{k}$ m in the undeformed geometry? What is the new position vector?

To find the displacement of this point, we substitute $x = 0$, $y = 1$, and $z = 1$ in the displacement field above. We get

$$\boldsymbol{u}(0,1,1) = [0\mathbf{i} + (0+3)\mathbf{j} + 10\mathbf{k}] \times 10^{-2}$$

$$\boldsymbol{u} = (3\mathbf{j} + 10\mathbf{k}) \times 10^{-2}\,\mathrm{m}$$

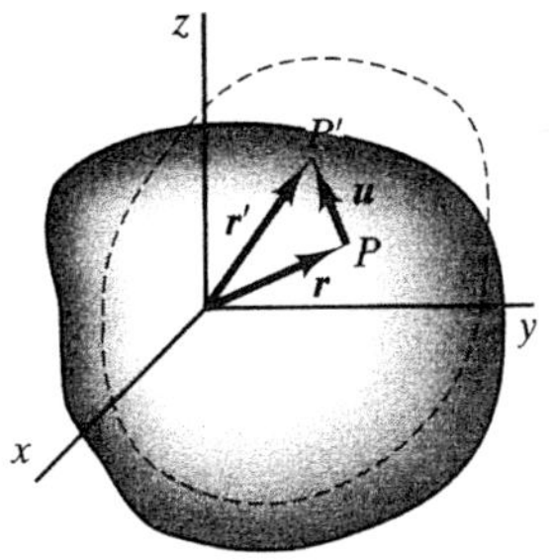

Figure 3.3. Movement of point P.

We have shown this in Fig. 3.3. The new position vector $\boldsymbol{r}'$ of the given point in the deformed geometry is then

$$\boldsymbol{r}' = \boldsymbol{r} + \boldsymbol{u}$$

$$= \mathbf{j} + \mathbf{k} + (3\mathbf{j} + 10\mathbf{k}) \times 10^{-2}$$

$$\boldsymbol{r}' = 1.03\mathbf{j} + 1.10\mathbf{k}\ \mathrm{m}$$

3.3 Strain Components

The displacement field presented in Section 3.2 can fully describe the manner in which a body moves point by point. It can accordingly yield any and all information concerning the change of geometry of a body as a result of loads. However, for relating this change in geometry most directly to the loads or, more conveniently, the stress distribution, there is a better descriptive procedure than directly using the displacement field. To understand this, recall from physics that it was not directly the position of an atom that gave rise to a bonding force but rather the *relative position* of pairs of atoms. Similarly, in solid mechanics it is not the position of a point (given by $\boldsymbol{u}(x,y,z)$) that relates to stress but certain aspects of the *relative movement* of *adjacent points* (points infinitesimally close to each other). In this section we shall set forth definitions for describing the *relative movement of adjacent points lying along three orthogonal axes at a point* (see Fig. 3.4). Once we have this information, we shall later show that the relative motion between *any pair* of adjacent points in the neighborhood can be readily found. In particular along *any* set of orthogonal axes rotated relative to the original one (see Fig. 3.5), we can then determine the relative movements of adjacent points. It will be convenient to consider the orthogonal infinitesimal line segments connecting the adjacent points, as discussed above, to be the edges of an infinitesimal rectangular parallelepiped (see Fig. 3.6). Then we can readily describe the relative movement of the adjacent points by considering the deformation and the rigid-body rotation (the rigid-body translation introduces no relative movement between adjacent points) of the infinitesimal rectangular parallelepiped.

We shall accordingly consider now a vanishingly small rectangular parallelepiped (see Fig. 3.7, solid lines) in the undeformed geometry on

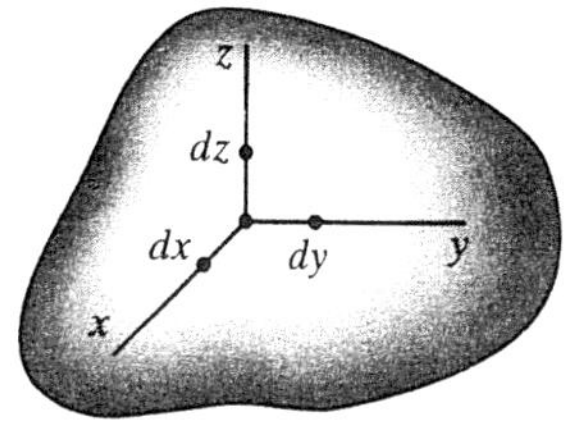

Figure 3.4. Four adjacent points.

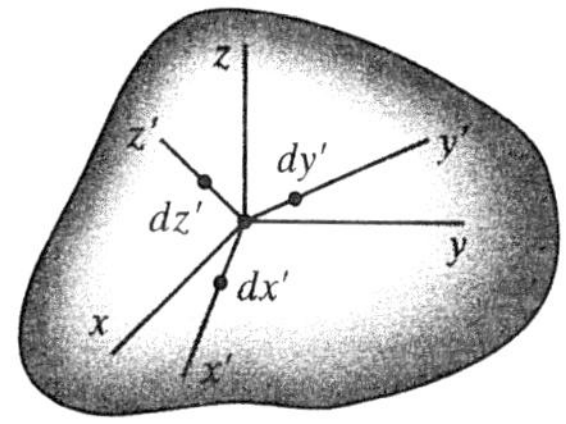

Figure 3.5. Adjacent points for $x'y'z'$.

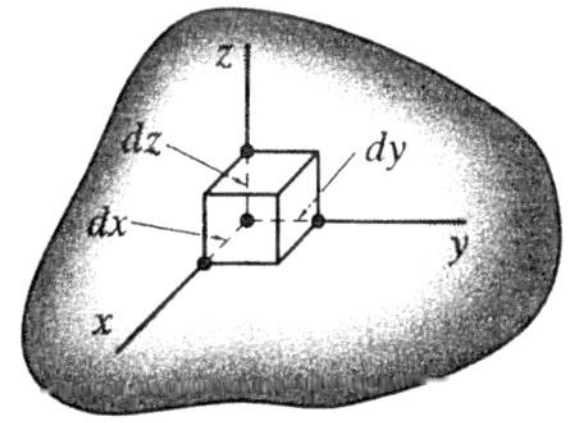

Figure 3.6. Adjacent points form a rectangular parallelepiped.

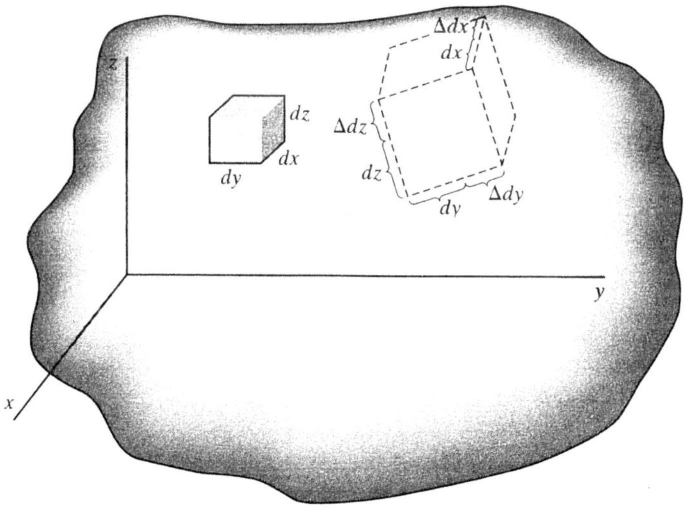

Figure 3.7. Deforming rectangular parallelepiped.

whose faces only *normal stress* is developed as a result of loading. It can then be expected as a result of loading that in addition to possible translation and rotation as a rigid body, the element will *deform* in such a manner as to remain a rectangular parallelepiped—that is, its essential shape will be retained—with the lengths of the sides possibly changing. This is shown in Fig. 3.7, where the dashed lines represent the deformed geometry (shown enlarged for clarity). The original lengths of the sides of the rectangular parallelepiped are denoted as dx, dy, and dz. In the deformed geometry the corresponding sides are shown in the diagram to have lengths $dx + \Delta dx$, $dy + \Delta dy$, and $dz + \Delta dz$, respectively. There is thus possible a change in volume of the element, or, as we say, a *dilatation.* As a measure of the deformation, we shall now define three *normal strains* as follows:

$$\begin{aligned} \varepsilon_{xx} &= \frac{\Delta dx}{dx} \\ \varepsilon_{yy} &= \frac{\Delta dy}{dy} \\ \varepsilon_{zz} &= \frac{\Delta dz}{dz} \end{aligned} \tag{3.1}$$

From this definition it is clear that we can interpret a *normal strain* ε_{xx} *as the change in length per unit original length of a vanishingly small line element connecting adjacent points originally in the x direction.* The interpretations for ε_{yy} and ε_{zz} follow directly.

Consider next an element inside the loaded body subject only to shear stresses parallel to the yz plane (Fig. 3.8). In addition to transla-

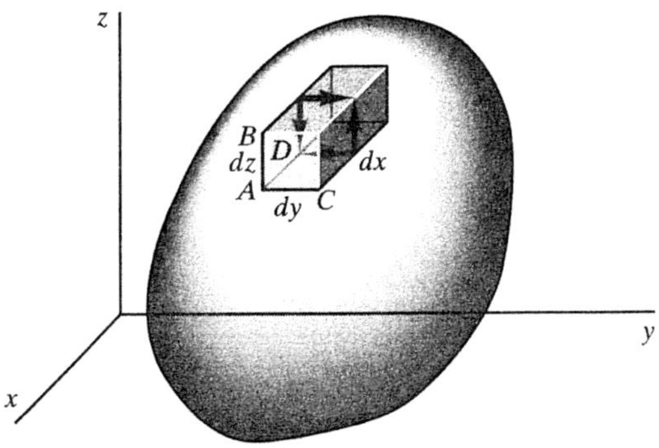

Figure 3.8. Element under pure shear.

tion and rigid-body rotation of the element, which we now ignore, we propose for small deformation that there may take place changes of the *relative orientation* of the faces of the element on which the shear stresses act involving only negligible changes in volume of the element. In short, we propose that the shear stress causes a change in shape (i.e., distortion) but no appreciable dilatation. This change in relative orientation of the faces for the case at hand may be measured by the change of angle from that of a right angle of line segments dy and dz. This angle is called the *shear angle* or *engineering shear strain* and is denoted as γ_{yz} or γ_{zy}. Thus (see Fig. 3.9 where we have shown face $ABCD$ in a deformed state)

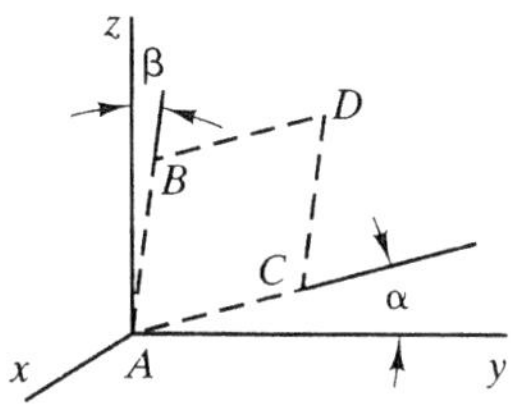

Figure 3.9. $\gamma_{yz} = \alpha + \beta$.

$$\gamma_{yz} = \gamma_{zy} = \alpha + \beta \tag{3.2}$$

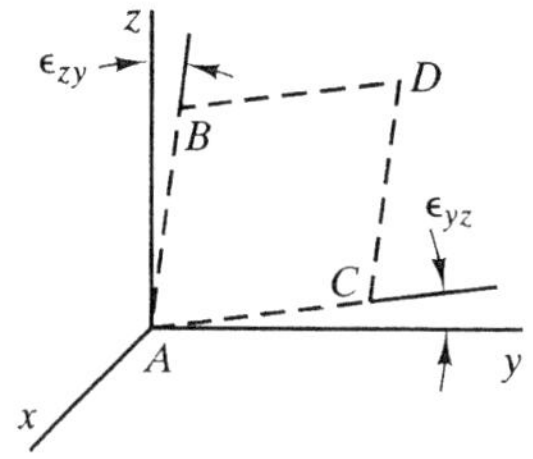

Figure 3.10. Shear strains $\varepsilon_{yz} = \varepsilon_{xy}$.

It will be convenient to imagine a rigid-body rotation of the element about the x axis so as to make α equal to β. We then denote each of these angles as ε_{yz} and ε_{zy}, respectively, as shown in Fig. 3.10. Note that the first subscript identifies the axis associated with the particular ε. We now define *shear strains* ε_{yz} or ε_{zy} as

$$\varepsilon_{zy} = \varepsilon_{yz} = \frac{1}{2}\gamma_{yz} = \frac{1}{2}\gamma_{zy} \tag{3.3}$$

In a similar way we can define four more shear strains ε_{xy}, ε_{yx}, ε_{xz}, and ε_{zx}, where terms with reversed indices are equal to each other.

As for the sign convention that we shall employ for strain components, we shall adopt the following rules. For a normal strain, an extension per unit length is a positive normal strain, while a contraction per unit length is a negative normal strain. As for shear deformation, a *decrease* in the angle from that of a right angle for a pair of line segments gives *positive* shear strains for these line segments, while an *increase* of the angle from that of a right angle gives *negative* shear strains. The shear deformation shown in Fig. 3.10 clearly yields positive shear strains ε_{yz} and ε_{zy}. A moment's thought while observing Fig. 3.8 will reveal that positive shear strain corresponds to the expected action of a positive shear stress.

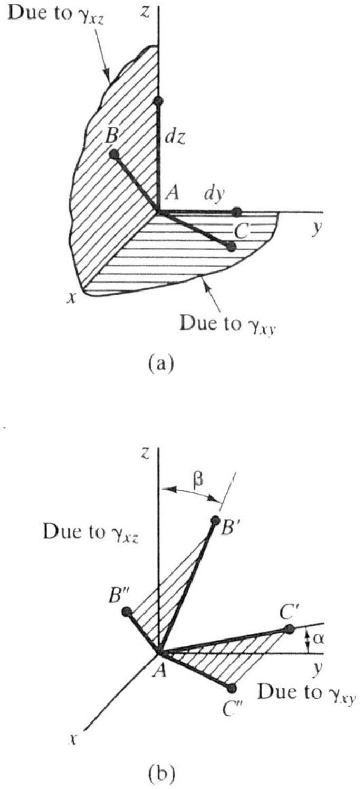

Figure 3.11. Line segments dz and dy in the presence of γ_{xz}, γ_{xy}, and γ_{yz}.

Up to now we have considered shear strains separately (i.e., one at a time). Now let us consider that these strains can exist simultaneously. To start, let us assume that a state of shear deformation γ_{xz} exists alone at point A in Fig. 3.11(a). A vanishingly small line segment connecting adjacent points in the z direction in the undeformed geometry will rotate in the xz plane to line AB, as shown in the diagram. Note that it rotates in a plane normal to the yz plane. Now consider a line segment AB' in Fig. 3.11(b) inclined to the z axis by a *small angle* β and in the yz plane for the initial state. If a shear deformation γ_{xz} is once more applied, we can assume for *small deformation* that the line rotates in a plane *again* normal to the yz plane (plane $AB'B''$ in the diagram) to position AB''. Returning to Fig. 3.11(a) we note next that a shear deformation γ_{xy} will cause a vanishingly small line segment along the y axis to rotate to position AC in the xy plane—that is, the line segment rotates in a plane normal to the zy plane. And we may assume for small deformation, as before, that a line AC' inclined by a *small angle* α to the y axis [see Fig. 3.11(b)] rotates *also* in a plane normal to the yz plane, as a result of shear deformation γ_{xy}, to position AC''. Now AB' and AC' in Fig. 3.11(b) may be considered line segments that have been deformed from the z and y directions, respectively, as a result of shear deformation γ_{yz}. The discussion then has shown the *subsequent* effects on the segments when additional shear deformations γ_{xz} and γ_{xy} are applied. Specifically, the elements AB' and AC' rotate out of the plane yz to form two planes $AB'B''$ and $AC'C''$ normal to yz and inclined to each other by an angle $C'AB'$—the same angle that the segments had as a result of γ_{yz} *before* subsequent shear deformations have been applied. The amount of shear strain γ_{xz} and γ_{xy} present determines how much AB' and AC' rotate, respectively, in planes $AB'B''$ and $AC'C''$. Thus, AB'' *and* AC'' are the line segments *originally* in the z and y directions, respectively, after *three* shear strains γ_{yz}, γ_{xz}, and γ_{xy} are imposed at the point A. How do we determine the shear strain γ_{yz} for the line segments dz and dy when we have also γ_{xz} and γ_{xy} present? Clearly, we can project AB'' and AC'' *back* onto the plane yz [see Fig. 3.11(b)]. We then add angles α and β as before to determine the shear angle γ_{yz}. In general for any two orthogonal directions n, t at a point, we can then say that γ_{nt} is the change in right angles of two vanishingly small segments connecting adjacent points originally along the n and t directions *when these segments in the deformed geometry are projected back onto the original plane of the segments.*[2]

We thus heuristically propose nine quantities to describe the relative movement of adjacent points along orthogonal axes giving rise to the deformation of the associated vanishingly small rectangular parallelepiped. The three normal strains supply a dilatation effect with no change in basic shape, while, independently, the six shear strains supply a change in shape with negligible change in volume. We shall express these nine quantities in an array as we did stress, using the first index to identify the row and the second index to identify the column. Thus,

$$\varepsilon_{ij} = \begin{pmatrix} \varepsilon_{xx} & \varepsilon_{xy} & \varepsilon_{xz} \\ \varepsilon_{yx} & \varepsilon_{yy} & \varepsilon_{yz} \\ \varepsilon_{zx} & \varepsilon_{zy} & \varepsilon_{zz} \end{pmatrix} \tag{3.4}$$

Note that the diagonal terms represent the normal strains, while the off-diagonal terms whose values are symmetric about the left to right diagonal of the array (just like shear stresses) are shear strains.

[2]We shall make use of this procedure twice in this chapter.

We now present some examples to illustrate normal and shear strains.

Example 3.2

A bar having a square cross section (Fig. 3.12) is elongated by forces P applied uniformly at the ends as shown in the diagram. If the total elongation is 2 in. and assuming the volume of the bar is unchanged, what are the normal strains ε_{xx}, ε_{yy}, and ε_{zz}? The cross section is 1 in. on an edge in the undeformed geometry.

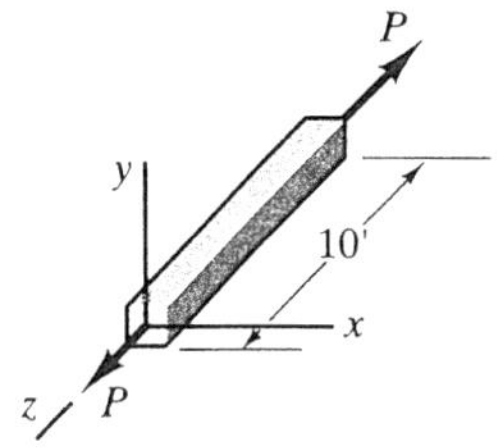

Figure 3.12. Axially loaded bar.

In this problem all equal rectangular parallelepipeds, having edges parallel to xyz, undergo the same deformation. We call such a deformation *uniform strain*. It should then be clear that the ratio of the elongation to original length is the same for corresponding sides of all elements having edges parallel to xyz regardless of the size of the element. Accordingly, we can use the entire block itself to get ε_{zz} in the following manner:

$$\varepsilon_{zz} = \frac{2}{(10)(12)} = .01667$$

Notice that normal strain, having a ratio of two lengths, is dimensionless.

To get ε_{xx} and ε_{yy} we need the change in length of the sides of the body in the x and y directions. Since the total volume is constant, we can say, equating volumes before and after deformation, that

$$(120)(1)^2 = (120 + 2)(1 + \delta)^2$$

where δ is the change in length of the sides of the cross section. Solving for δ, we get

$$\delta = -.00823 \text{ in.}$$

Hence, we can say that

$$\varepsilon_{xx} = \varepsilon_{yy} = -\frac{.00823}{1} = -.00823$$

It should be clear that the sides of any and all rectangular parallelepipeds having edges parallel to xyz remain orthogonal after deformation. Hence, the shear strains are zero for the xyz directions.

Example 3.3

A rubber block is glued between two rigid surfaces AB and CD (Fig. 3.13) as well as rigid members EF and GH, which are hinged to

Example 3.3 (Continued)

AB and *CD*. If *CD* moves upward a distance of .1 mm, as shown in Fig. 3.14, what is the shear angle γ_{xy} at any point in the rubber block? What are the corresponding shear strains ε_{xy} and ε_{yx}?

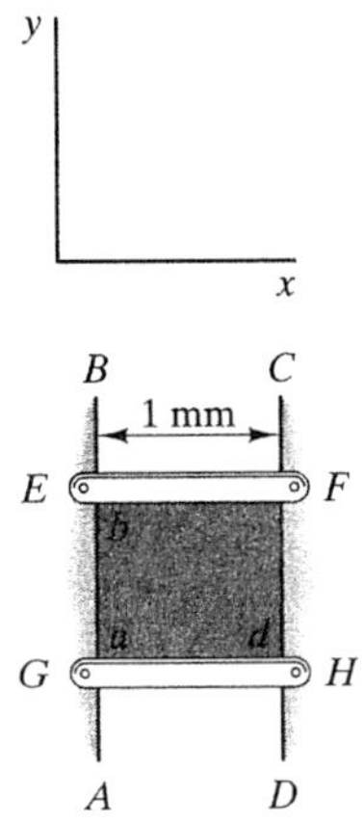

Figure 3.13. Rubber block.

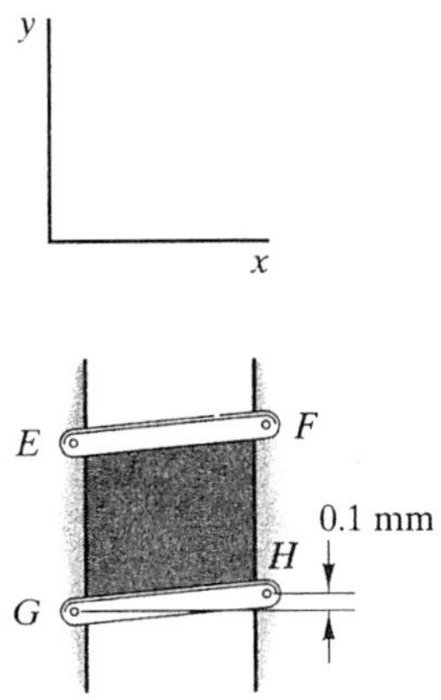

Figure 3.14. Deformed state of block.

We have shown the rubber block in its deformed geometry in Fig. 3.14. We have here again a case of *uniform strain* since all pairs of vanishingly small line segments parallel to the *x* and *y* axes undergo the same loss of perpendicularity as a result of the deformation and thus develop the same shear angle γ_{xy}. Indeed, we may consider finite edges *ad* and *ab* (Fig. 3.13) during deformation since they remain straight, undergoing the same change in right angles as any pair of line segments *dx* and *dy*. Accordingly, noting that we have positive shear strain, we conclude that

$$\gamma_{xy} = \tan^{-1}\frac{.1}{1}$$

Since for small angles the tangent may be taken equal to the angle itself in radians, we have

$$\gamma_{xy} = .1$$

Consequently, we may say for the shear strains that

$$\varepsilon_{xy} = .05 \qquad \varepsilon_{yx} = .05$$

These results are shown in Fig. 3.15. Since changes in length of line segments *dx* and *dy* are second-order, then $\varepsilon_{xx} = \varepsilon_{yy} = 0$.

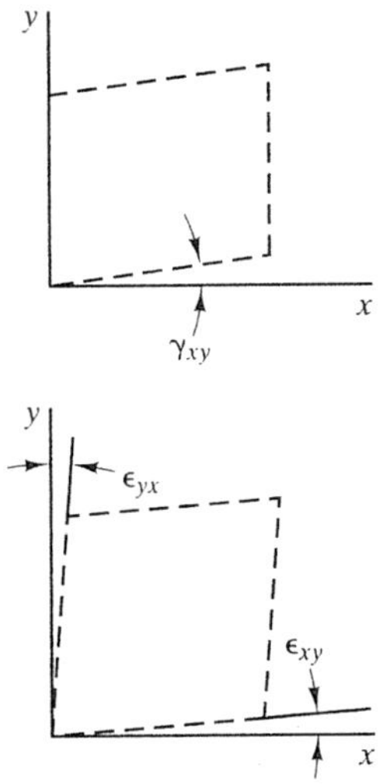

Figure 3.15. Measures of shear deformation.

Example 3.4

The thin-walled cylinder shown in Fig. 3.16(a) is twisted by torques T so that end B rotates $\Delta\phi = 15°$ relative to end A. What is the shear strain ε_{xy} at point P on the surface of the cylinder?

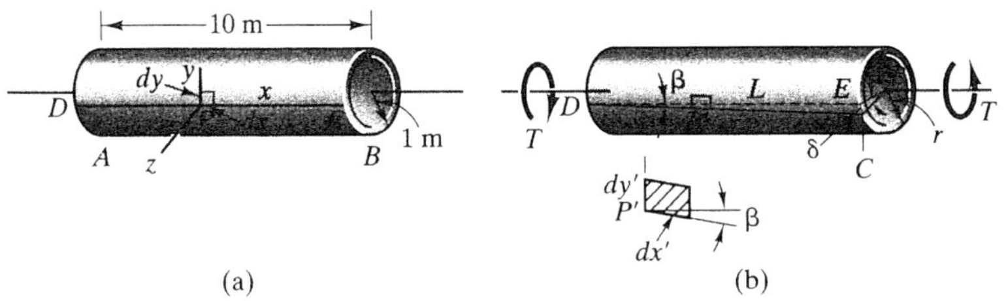

Figure 3.16. Thin-walled cylinder under torsion.

For convenience we have shown an infinitesimal rectangle at P with sides along the xy axes. Also we have shown a line DE on the cylinder surface parallel to the axis of the cylinder in the undeformed geometry. When the cylinder is twisted (it is convenient in this regard to imagine one end, say the left end, as fixed while the other end rotates the prescribed amount) straight line DE becomes the helix DC [Fig. 3.16(b)]. We can conclude that the element dx' in the deformed state when projected back to the xy plane has rotated an angle β. Observing Fig. 3.16(b), we have for angle β,

$$\beta - \tan^{-1}\frac{\delta}{L} = \tan^{-1}\frac{r\Delta\phi}{L} = \tan^{-1}\frac{(1)\left(\frac{15}{360}\times 2\pi\right)}{10} = .0262$$

On the other hand, we can expect that line element dy' in the deformed state when projected back to the xy plane has undergone *no* rotation—only translation. Consequently, for the segment dx and dy chosen we have negative shear strain, given as

$$\begin{aligned}\gamma_{xy} &= -\ .0262\\ \varepsilon_{xy} &= -\ .01310\\ \varepsilon_{yx} &= -\ .01310\end{aligned}$$

In this problem, if we consider s to be the *axial* direction of the cylinder and t to represent the *transverse* direction (i.e., tangent to the outer circumference of the section), then $\gamma_{st} = -.0261$ everywhere on the cylinder. In this sense (using directions corresponding to cylindrical coordinates) we have again a case of uniform strain. As in Example 3.3 $\varepsilon_{xx} = \varepsilon_{yy} = 0$.

In Examples 3.2 to 3.4 we have used the *overall deformation* of finite bodies in ascertaining certain strain components *at a point.* We shall now reverse this process in the following examples, where, from considerations of strain fields, we shall now deduce certain conclusions applicable to the overall deformation of finite bodies.

Example 3.5

A given body is subject to the following hypothetical strain distribution[3]:

$$\varepsilon_{ij} = \begin{pmatrix} x^2 + 2yx & 3 + 10zx & z^3 + 10xy \\ 3 + 10zx & 5y + z^2x & x^2 + 2yz^2 \\ z^3 + 10xy & x^2 + 2yz^2 & x^2 + y^2 \end{pmatrix} \times 10^{-2}$$

What is the new length of line AB, which in the undeformed geometry lies along the x axis from $x = 2$ ft to $x = 4$ ft (see Fig. 3.17)?

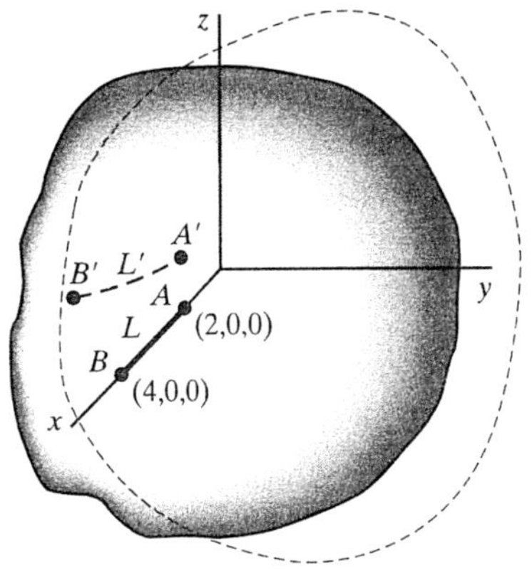

Figure 3.17. Find change in length of AB as body deforms.

From our physical interpretation of normal strain the change in length of a line element dx in the original geometry is $\varepsilon_{xx} dx$ [see also Eq. (3.1)]. The change in length ΔL of the line AB can then be given as follows:

$$\Delta L = \int_{x=2}^{x=4} \varepsilon_{xx} dx \tag{a}$$

Note that the general expression for ε_{xx} is

$$\varepsilon_{xx} = (x^2 + 2yx) \times 10^{-2} \tag{b}$$

[3]We will ask you to show in Problem 3.31 that this deformation is not physically realizable.

Example 3.5 (Continued)

Along the x axis we take $y = 0$ and for ε_{xx} we have

$$[\varepsilon_{xx}]_{x \text{ axis}} = (x^2) \times 10^{-2} \qquad \text{(c)}$$

We must use Eq. (c) to establish the appropriate strain function in Eq. (a). Thus,

$$\Delta L = \left[\int_2^4 (x^2) dx \right] \times 10^{-2} = \left. \frac{x^3}{3} \right|_2^4 \times 10^{-2} = .187 \text{ ft}$$

The new length L' of AB will now be

$$L' = L + \Delta L = 2.187 \text{ ft}$$

Note that AB in the deformed geometry is generally no longer a straight line (see Fig. 3.17).

In the following two examples, we shall cut a body to expose a stress in a free body diagram and then use equilibrium to solve for this stress. Then we shall use a given simple stress-strain law to get the strain field. Finally, just as in the preceding example, we shall go from the strain field to determine a desired change in overall geometry of a given body. This may give you insight as to how the concept of strain fields relates to the real world.

Example 3.6

A circular rod hangs by its own weight as shown in Fig. 3.18. The specific weight is constant for the body. If the normal strain in any direction is $1/E$ times the normal stress in that direction, what is the total deflection Δ_A of the end A of the member as a result of the weight?

To expose the stress τ_{zz} at any section a distance z from the support, we cut the cylinder at a position z and consider as a free body the portion of the cylinder below z. This is shown in Fig. 3.19.

Note that the weight $W(z)$ of this free body is $\gamma \dfrac{\pi D^2}{4}(L - z)$.

The equation of **equilibrium** in the vertical direction is then

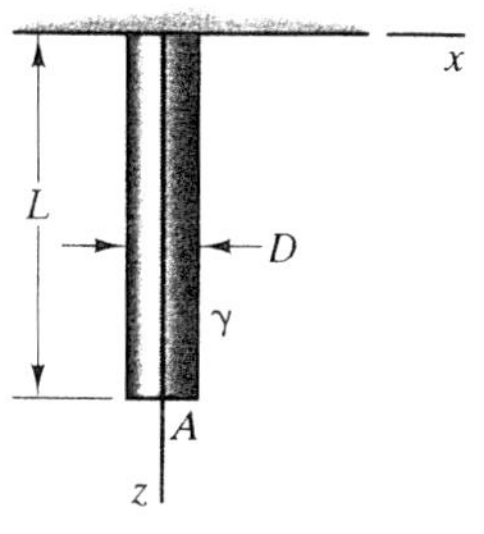

Figure 3.18. A rod suspended from above.

Example 3.6 (Continued)

$$-\tau_{zz}\frac{\pi D^2}{4}+\gamma\frac{\pi D^2}{4}(L-z)=0$$

$$\therefore\ \tau_{zz}=\gamma(L-z)$$

We shall soon point out that a relation between stress and strain is called a **constitutive law.** Here we have the simple relation $\varepsilon_{zz}=\tau_{zz}/E$ and so the strain ε_{zz} at section z is then

$$\varepsilon_{zz}=\frac{\gamma}{E}(L-z)$$

Next examine Fig. 3.20 showing an infinitesimal slice dz of the rod. The elongation $\Delta(dz)$ of this slice is $\varepsilon_{zz}dz$, using the definition of normal strain. Hence we can say

$$\Delta(dz)=\frac{\gamma}{E}(L-z)dz$$

Using the concept of **compatibility**[4] we see that all the increments of elongations can be superposed and so we integrate the above equation to get the total movement Δ_A of the end of the rod. Thus

$$\Delta_A=\int_0^L\frac{\gamma}{E}(L-z)dz=\frac{\gamma}{E}\left(L^2-\frac{L^2}{2}\right)=\frac{\gamma L^2}{2E}$$

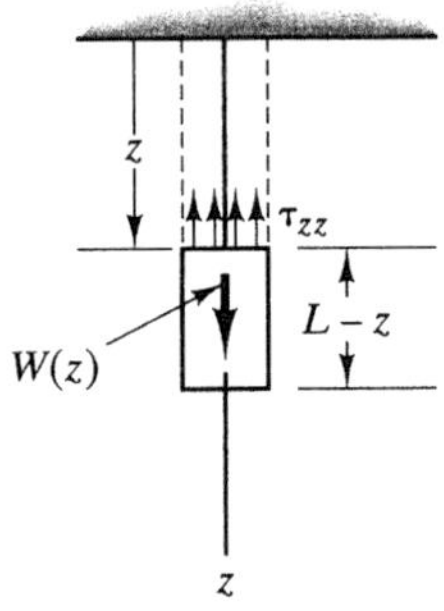

Figure 3.19. Free body is formed by cutting at z to expose τ_{zz}.

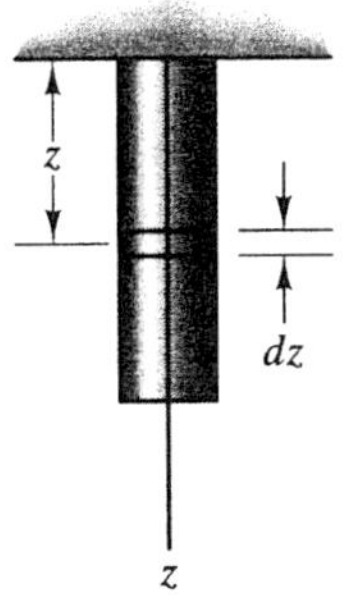

Figure 3.20. The rod with slices of length dz.

Note that in the preceding example we were able to evaluate the weight of the rod from position z to the end by a simple algebraic calculation because of the uniformity of the specific weight as well as the uniformity of the cross-section. In the next example, the specific weight varies and here we will make use of a second variable, ζ, which we call a dummy variable, to enable us to easily get the weight of the rod *below* the position denoted by the basic variable z. And in certain homework problems, we will also have nonuniformity of the cross-section making the dummy variable approach very useful. Please note carefully from the preceding example and the next example as to when it is appropriate to use the dummy variable and when it is not necessary to do so.

[4]The requirement of compatibility here means that we can superpose the elongations of the slices of the member without mathematically creating a void and without mathematically making two slices occupy the same space. Thus the deformation is assured of being physically realizable here. More will be said about compatibility in Section 3.5. You will see that it is a vital consideration in solid mechanics.

Example 3.7

A circular rod hangs by its own weight, as shown in Fig. 3.21. The specific weight[5] γ varies linearly from γ_0 at the top to γ_A at the bottom for the body. If the material behaves such that the normal strain in the vertical direction is $1/E$ times the normal stress in that direction, what is the total deflection of the end A of the member as a result of the weight?

Our method of approach will be to find the strain ε_{zz} at any position z. Then we can ascertain the elongation of the entire rod by an integration procedure. As a first step, we determine γ as a function of position z as follows:

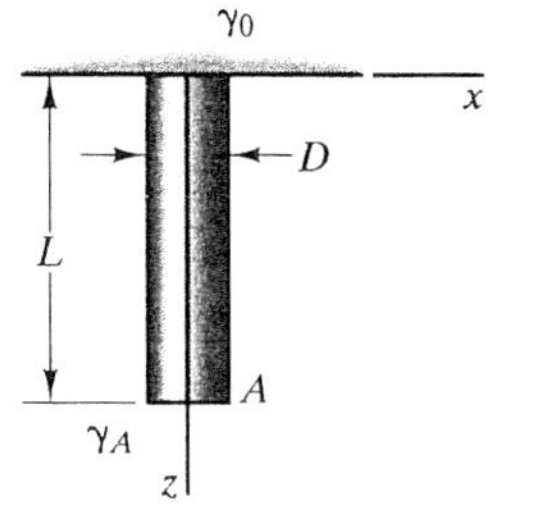

Figure 3.21. Circular rod hanging by its own weight.

$$\gamma = \gamma_0 + \frac{z}{L}(\gamma_A - \gamma_0) \tag{a}$$

We shall need the stress τ_{zz} at any section z [see Fig. 3.22(a)]. The stress at z, considering the material below (shown crosshatched) as a free body [Fig. 3.22(b)], supports the weight of this material and may be considered uniform over the section. Accordingly, we shall need the weight of that part of the cylinder below position z; we shall then consider z in the next few steps as a fixed parameter. We have shown a slice of material in the cylinder below section z and have used the "dummy" variable ζ (zeta) to locate this element [see Fig. 3.22(a)] and $d\zeta$ to give its thickness. In summing the weights of these slices, the dummy variable ζ runs from z (the top of the crosshatched region) to L. Accordingly, for the weight, which we shall denote as $W(z)$, we have

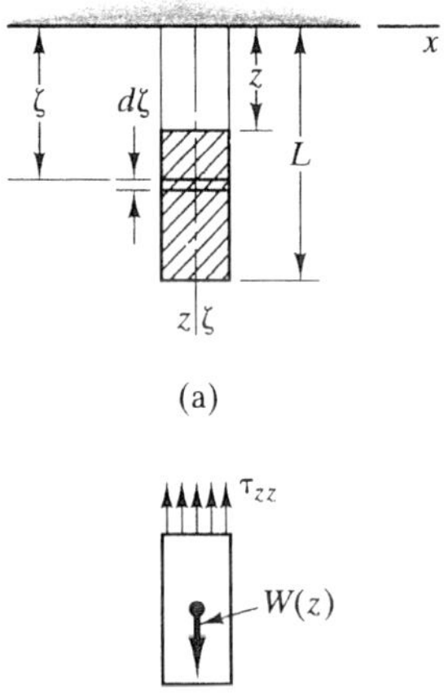

Figure 3.22. Free body exposing a section at z.

$$W(z) = \int_z^L \gamma \frac{\pi D^2}{4} d\zeta \tag{b}$$

Now γ at any position ζ below the top can be given from Eq. (a) with z replaced by ζ. Accordingly, we have

$$\begin{aligned} W(z) &= \int_z^L \left[\gamma_0 + \frac{\zeta}{L}(\gamma_A - \gamma_0) \right] \frac{\pi D^2}{4} d\zeta \\ &= \frac{\pi D^2}{4} \left[\gamma_0 (L - z) + \frac{\gamma_A - \gamma_0}{2L}(L^2 - z^2) \right] \end{aligned} \tag{c}$$

[5]The use of γ with none or one subscript denotes specific weight and should not be confused with γ with two subscripts, which denotes shear angle.

Example 3.7 (Continued)

Notice that the dummy variable ζ "integrates out" and disappears, leaving the right-hand side as a function of the parameter z. To get the stress we employ the free body shown in Fig. 3.22(b) and we conclude from **equilibrium** that

$$\tau_{zz} = \frac{1}{\pi D^2/4} W(z) = \gamma_0(L - z) + \frac{\gamma_A - \gamma_0}{2L}(L^2 - z^2) \qquad \text{(d)}$$

Also, for ε_{zz} we have (**constitutive law**)

$$\varepsilon_{zz} = \frac{1}{E}\tau_{zz} = \frac{1}{E}\left[\gamma_0(L - z) + \frac{\gamma_A - \gamma_0}{2L}(L^2 - z^2)\right] \qquad \text{(e)}$$

Now we have ε_{zz} at any position z, and we can consider z to be a variable rather than a parameter in the remainder of the discussion.

In Fig. 3.23 we have shown a slice dz of the cylinder. The increase in length in the z direction of this slice, $\Delta(dz)$, is seen from Eq. (3.1) to be

$$\Delta(dz) = \varepsilon_{zz} dz = \frac{1}{E}\left[\gamma_0(L - z) + \frac{\gamma_A - \gamma_0}{2L}(L^2 - z^2)\right] dz$$

To get the elongation of the entire rod we sum the elongations of all the elements in the rod (i.e., we integrate from $z = 0$ to $z = L$). Thus, from **compatibility** (see footnote in Example 3.6)

$$\Delta_A = \frac{1}{E}\int_0^L \left[\gamma_0(L - z) + \frac{\gamma_A - \gamma_0}{2L}(L^2 - z^2)\right] dz \qquad \text{(f)}$$

$$\Delta_A = \frac{L^2}{E}\left(\frac{1}{3}\gamma_A + \frac{1}{6}\gamma_0\right)$$

In later calculations we shall proceed with a second approach in problems of this type.

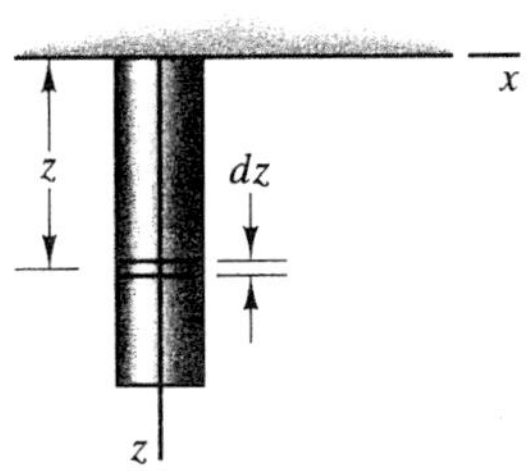

Figure 3.23. Consider slice dz.

We have thus far set forth nine strain terms at a point for axes parallel to reference xyz. And from the development, we know that shear strains with the same but reversed subscripts are equal to each other—we therefore have six independent strain terms at a point. We shall show in Chapter 16 that knowing the six independent strain terms at a point for axes parallel to reference xyz, we can then determine the strain terms for any set of axes at the point.[6] In fact the formulations for

[6]That is, we can determine the deformational relative movements of adjacent points on orthogonal axes oriented arbitrarily relative to xyz at a point. This can be accomplished by observing the deformation of an infinitesimal rectangular parallelepiped having edges oriented along the aforestated orthogonal axes at the point.

doing this are the *same* as that for stress at a point. Indeed the strains presented here are rectangular components of a second-order, symmetric tensor just as stresses are rectangular components of a second-order, symmetric tensor. Mathematically this is why stresses are most simply related to strains rather than to displacement fields. We shall have more to say about stress and strain as second-order, symmetric tensors in Chapters 7, 8, 15, and 16.

As we pointed out earlier, there is yet a third contribution to the relative movement of our adjacent points, and that is *rigid-body rotation.* We shall not be concerned with this form of relative movement in this book. However, when you study fluid mechanics, this kind of relative movement will be of vital importance.[7]

3.4 Strains in Terms of the Displacement Field

We have thus far presented two approaches to describing the geometry of a body movement. The first, the displacement field, gives the movement of *each point* in a body. The strain terms meanwhile describe the *deformational relative movement* of *adjacent points* originally along orthogonal axes. (We can now also say that the strain terms measure the stretching of originally orthogonal infinitesimal line segments at a point as well as the change in relative orientation of these line segments.) Since the displacement field describes the movement of each and every point in the body, we can expect that the strains are expressible in terms of the displacement field. In this section we shall set forth these relationships for the special circumstance of *small deformation.* These are the very important *strain-displacement relations.*

We may readily express the normal strain ε_{xx} in terms of $\boldsymbol{u}$ by consulting Fig. 3.24, where we have shown an element Δx in the undeformed geometry of a body. Displacement vectors are shown for ends A and B of the element to the positions of these points A' and B' in the deformed geometry. From our discussion in Section 3.3, we can give ε_{xx} as follows using an overbar to denote distance between points:

$$\varepsilon_{xx} = \lim_{\Delta x \to 0} \frac{\overline{A'B'} - \Delta x}{\Delta x}$$

[7]See I. H. Shames, *Mechanics of Fluids*, 3rd ed. (McGraw-Hill, New York, 1992), Chap. 4.

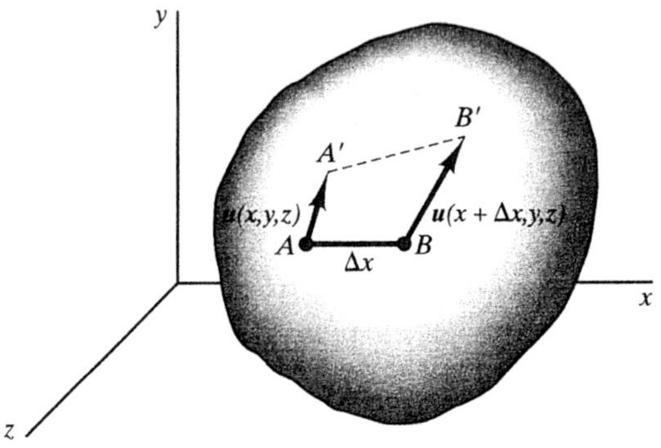

Figure 3.24. Line segment Δx in the undeformed geometry.

If the shear strains present are small, the angle between $\overline{A'B'}$ and $\overline{AB}$ will be small, and we can use the projection of $\overline{A'B'}$ in the x direction rather than $\overline{A'B'}$ itself in the formulation above.[8] That is,

$$\varepsilon_{xx} = \lim_{\Delta x \to 0} \frac{(\overrightarrow{A'B'})_x - \Delta x}{\Delta x}$$

We can easily express $[(\overrightarrow{A'B'})_x]$ as the sum of Δx plus the net movement in the x direction of the end points of the segment Δx as a result of deformation. Thus

$$(\overrightarrow{A'B'})_x = \Delta x + [u_x(x + \Delta x, y, z) - u_x(x, y, z)]$$

We can thus give ε_{xx} as follows on canceling Δx:

$$\varepsilon_{xx} = \lim_{\Delta x \to 0} \frac{u_x(x + \Delta x, y, z) - u_x(x, y, z)}{\Delta x} \tag{3.5}$$

The right-hand side you will recognize as simply the partial derivative $\partial u_x / \partial x$. Hence, we can say that:

$$\varepsilon_{xx} = \frac{\partial u_x}{\partial x} \tag{3.6}$$

[8]This means that we can interpret a normal strain ε_{nn} as the elongation *component* of a vanishingly small line segment in the *original direction of the line segment* ($\hat{\boldsymbol{n}}$) per unit original length. That is, we do not have to use the *total* elongation of the segment.

Similarly,

$$\varepsilon_{yy} = \frac{\partial u_y}{\partial y} \tag{3.7}$$

$$\varepsilon_{zz} = \frac{\partial u_z}{\partial z} \tag{3.8}$$

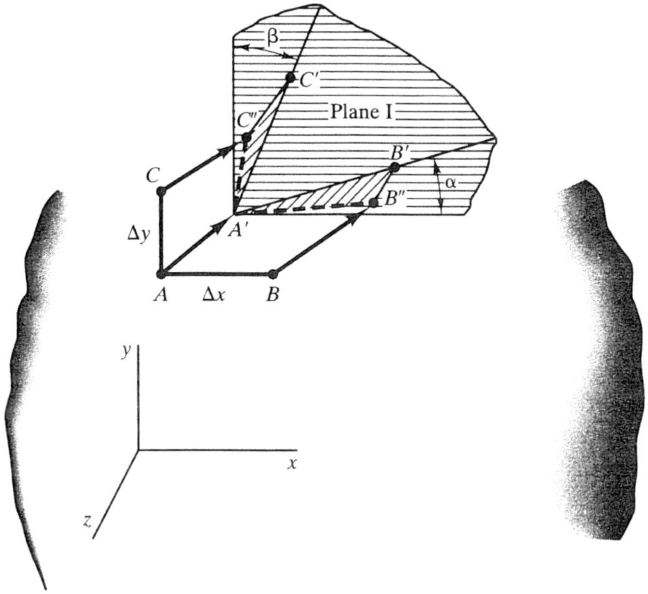

Figure 3.25. Line segments Δx and Δy in the undeformed geometry.

As for the shear strain, we consider line elements Δx and Δy in Fig. 3.25 in the undeformed geometry at a point A at position xyz. Displacement vectors have been shown connecting points A,B,C at the ends of these segments to the corresponding points A',B'',C'', respectively, in the deformed geometry. Now to measure γ_{xy} project $A'B''$ and $A'C''$ onto Plane I—that is, onto a plane parallel to the original plane of the segments Δx and Δy, in accordance with the discussion in Section 3.3. The angle that projection $A'B'$ makes with the x direction is denoted as α, and the angle that projection $A'C'$ makes with the y direction is denoted as β. Their sum equals the change in right angles of the vanishingly small segments Δx and Δy when the deformed segments are projected onto the plane parallel to the original undeformed plane of the segments. We can then say as a result of Section 3.3 that

$$\gamma_{xy} = \alpha + \beta \tag{3.9}$$

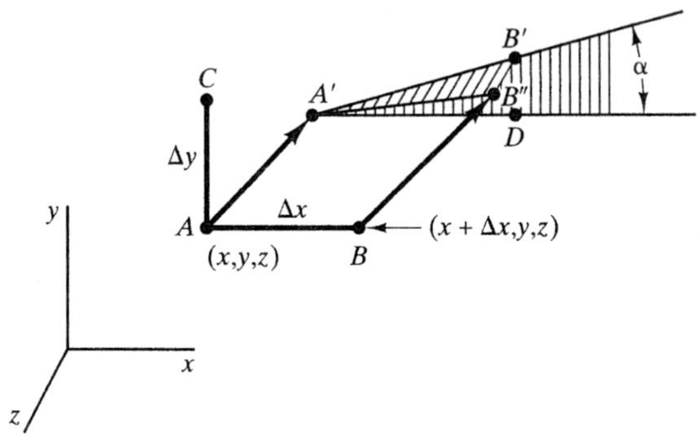

Figure 3.26. Formation of angle α.

Let us first consider the angle α (see Fig. 3.26). It is clear that

$$\tan\alpha = \lim_{\Delta x \to 0} \frac{DB'}{A'D}$$

For small deformation we can take $\tan\alpha = \alpha$, so we have

$$\alpha = \lim_{\Delta x \to 0} \frac{DB'}{A'D}$$

But DB' is the difference in the displacement vector components in the y direction of points B and A, as you can see by considering Fig. 3.26. That is,

$$DB' = u_y(x + \Delta x, y, z) - u_y(x, y, z)$$

Also, for small normal strain we can say that

$$A'D = \Delta x$$

Accordingly, for α we have

$$\alpha = \lim_{\Delta x \to 0} \frac{u_y(x + \Delta x, y, z) - u_y(x, y, z)}{\Delta x}$$

The right-hand side of the equation is simply $\partial u_y / \partial x$, so we have

$$\alpha = \frac{\partial u_y}{\partial x}$$

In a similar way we can show that

$$\beta = \frac{\partial u_x}{\partial y}$$

Thus, we can say that

$$\gamma_{xy} = \frac{\partial u_x}{\partial y} + \frac{\partial u_y}{\partial x}$$

$$\therefore \varepsilon_{xy} = \varepsilon_{yx} = \frac{1}{2}\left(\frac{\partial u_x}{\partial y} + \frac{\partial u_y}{\partial x}\right) \tag{3.10}$$

Similarly, we have for the other shear strains,

$$\varepsilon_{xz} = \frac{1}{2}\left(\frac{\partial u_x}{\partial z} + \frac{\partial u_z}{\partial x}\right) \tag{3.11a}$$

$$\varepsilon_{yz} = \frac{1}{2}\left(\frac{\partial u_y}{\partial z} + \frac{\partial u_z}{\partial y}\right) \tag{3.11b}$$

Now, rather than confining the consideration of strain components to that of infinitesimal rectangular parallelepipeds or their bounding line segments, we may also consider strain components directly in terms of the displacement fields. Thus, knowing the displacement field $\boldsymbol{u}(x, y, z)$ we can give the strain field $\varepsilon_{ij}(x, y, z)$ as follows:[9]

$$\varepsilon_{ij} = \begin{pmatrix} \frac{\partial u_x}{\partial x} & \frac{1}{2}\left(\frac{\partial u_x}{\partial y} + \frac{\partial u_y}{\partial x}\right) & \frac{1}{2}\left(\frac{\partial u_x}{\partial z} + \frac{\partial u_z}{\partial x}\right) \\ \frac{1}{2}\left(\frac{\partial u_y}{\partial x} + \frac{\partial u_x}{\partial y}\right) & \frac{\partial u_y}{\partial y} & \frac{1}{2}\left(\frac{\partial u_y}{\partial z} + \frac{\partial u_z}{\partial y}\right) \\ \frac{1}{2}\left(\frac{\partial u_z}{\partial x} + \frac{\partial u_x}{\partial z}\right) & \frac{1}{2}\left(\frac{\partial u_z}{\partial y} + \frac{\partial u_y}{\partial z}\right) & \frac{\partial u_z}{\partial z} \end{pmatrix} \tag{3.12}$$

We have used the assumption of small deformation in arriving at our relation between the strain and the displacement field. This assumption requires the strain terms to be small compared to unity. Is this a serious limitation? Actually, in most engineering problems it is not, since for safety the strains that we permit in structural members must be small; the results of this section are then valid for a wide range of engineering applications.

Example 3.8

Given the following displacement field,

$$\boldsymbol{u} = [(x^2 + 3)\mathbf{i} + (3y^2z)\mathbf{j} + (x + 3z)\mathbf{k}] \times 10^{-2}\,\text{m}$$

what are the strain components at (0, 2, 3)?

We first determine the strain components as functions of position (i.e., the strain field) as follows:

[9]These equations are the so-called *strain-displacement relations,* which play a vital role in solid mechanics.

Example 3.8 (Continued)

$$\varepsilon_{xx} = \frac{\partial u_x}{\partial x} = \frac{\partial}{\partial x}[(x^2 + 3) \times 10^{-2}] = 2x \times 10^{-2}$$

$$\varepsilon_{yy} = \frac{\partial u_y}{\partial y} = \frac{\partial}{\partial y}[(3y^2 z) \times 10^{-2}] = 6yz \times 10^{-2}$$

$$\varepsilon_{zz} = \frac{\partial u_z}{\partial z} = \frac{\partial}{\partial z}[(x + 3z) \times 10^{-2}] = 3 \times 10^{-2}$$

$$\varepsilon_{xy} = \varepsilon_{yx} = \frac{1}{2}\left(\frac{\partial u_x}{\partial y} + \frac{\partial u_y}{\partial x}\right) = \frac{1}{2}(0 + 0) \times 10^{-2} = 0$$

$$\varepsilon_{xz} = \varepsilon_{zx} = \frac{1}{2}\left(\frac{\partial u_x}{\partial z} + \frac{\partial u_z}{\partial x}\right) = \frac{1}{2}(0 + 1) \times 10^{-2} = \frac{1}{2} \times 10^{-2}$$

$$\varepsilon_{yz} = \varepsilon_{zy} = \frac{1}{2}\left(\frac{\partial u_y}{\partial z} + \frac{\partial u_z}{\partial y}\right) = \frac{1}{2}(3y^2 + 0) \times 10^{-2} = \frac{3}{2}y^2 \times 10^{-2}$$

At the position of interest we then get

$$\begin{array}{ll} \varepsilon_{xx} = 0 & \varepsilon_{xy} = \varepsilon_{yx} = 0 \\ \varepsilon_{yy} = .36 & \varepsilon_{xz} = \varepsilon_{zx} = .005 \\ \varepsilon_{zz} = .03 & \varepsilon_{yz} = \varepsilon_{zy} = .06 \end{array}$$

Example 3.9

We now go back to Example 3.7 to compute the movement of section A using a displacement field approach. We note first that

$$\frac{\partial u_z}{\partial z} = \varepsilon_{zz} \tag{a}$$

Since u_z does not depend on any variable other than z in this problem, we can rewrite the above using an ordinary derivative. Employing Eq. (e) of Example 3.7 we thus have

$$\frac{du_z}{dz} = \varepsilon_{zz} = \frac{1}{E}\left[\gamma_0(L - z) + \frac{\gamma_A - \gamma_0}{2L}(L^2 - z^2)\right]$$

Now separate the variables (bring dz to right hand side) and integrate both sides:

Example 3.9 (Continued)

$$u_z = \frac{1}{E}\left[\gamma_0\left(Lz - \frac{z^2}{2}\right) + \frac{\gamma_A - \gamma_0}{2L}\left(L^2 z - \frac{z^3}{3}\right)\right] + C \qquad \text{(b)}$$

where C is a constant of integration. We can set the constant equal to zero by noting that at the support ($z = 0$) the displacement is zero. To get the displacement of section A we need only use $z = L$ in the Eq. (b). Thus,

$$u_z(L) = \Delta_A = \frac{1}{E}\left[\gamma_0\left(L^2 - \frac{L^2}{2}\right) + \frac{\gamma_A - \gamma_0}{2L}\left(L^3 - \frac{L^3}{3}\right)\right]$$

This becomes

$$\Delta_A = \frac{L^2}{E}\left(\frac{1}{3}\gamma_A + \frac{1}{6}\gamma_0\right)$$

This is the same result as that found earlier.

3.5 Compatibility Considerations

In our efforts to describe the movement and deformation of a solid, we introduced the displacement field $\boldsymbol{u}$. For a displacement field to represent some *real* movement and deformation of a solid which develops no *cracks* and *fissures,* it is clearly necessary that the displacement field be a *single-valued, continuous* function. If it is not single-valued, it means that certain points can have more than one displacement at a time, which is physically impossible. And a discontinuity in the displacement field means that originally *infinitesimally* close points would be separated by a *finite* amount in the deformed geometry. This can only happen if a crack or a fissure is developed on deformation—a possibility that we are not considering in this text. Meanwhile, the strain field, which is a means of describing deformation of a solid for purposes of relating simply to the stress field, must *also* have certain requirements imposed on it in order that it be properly associated with a physically realizable deformation of a solid, free from cracks and fissures. That is, we cannot simply choose any set of strain component functions and expect them to automatically be associated with a physically realizable movement of a solid—that is, a movement representable by a single-valued and continuous displacement field.[10] For instance, suppose that a solid is imagined before deformation to consist of a system of infinitesimal contiguous cubes. Upon application of loads, these cubes will have dilatation as a result of normal strains, and their faces will become parallelograms from the shear strains. Now the normal and shear strains will vary with position, but they must vary in such a way that the deformed elements *remain contiguous*—that is, the elements must not come apart so as to expose holes and fissures. Nor must a pair of cubes partially or fully occupy the same space at the same time. One way to ensure that the strains are compatible is to *start with a single-valued, continuous displacement field and develop the strains from this field in accordance with the strain-displacement equations prescribed in this chapter.* There are times, however, when we wish to work initially with the strains. (Thus, we may be able to make some reasonable assumptions about strains but not about the displacement field.) To ensure that the strains are properly related in accordance with the discussion above (to ensure that the strains are "compatible"), one can derive certain equations, called *compatibility equations,* which guarantee for certain classes of bodies that the strains are proper functions. The development of these equations is beyond the level of this book.[11] These equations are of vital importance and are used considerably in more advanced studies of solid mechanics. However, in the cases we shall consider in

[10]It should be understood that we can choose the strain components arbitrarily at *one point* in a body assuming the small deformation requirement is maintained. We are concerned with the *distribution of strain,* and it is here that restrictions are needed.

[11]Note in Chapter 2 we found that to satisfy equilibrium, stresses had to vary with position in such a way as to satisfy certain partial differential equations, namely the equations of equilibrium. Similarly, strains must vary with position so as to satisfy certain partial differential equations in order to represent physically realizable deformations. They are

$$\frac{\partial^2 \varepsilon_{xx}}{\partial y^2} + \frac{\partial^2 \varepsilon_{yy}}{\partial x^2} = 2\frac{\partial^2 \varepsilon_{xy}}{\partial x \partial y} \tag{3.13a}$$

$$\frac{\partial^2 \varepsilon_{yy}}{\partial z^2} + \frac{\partial^2 \varepsilon_{zz}}{\partial y^2} = 2\frac{\partial^2 \varepsilon_{yz}}{\partial y \partial z} \tag{3.13b}$$

$$\frac{\partial^2 \varepsilon_{zz}}{\partial x^2} + \frac{\partial^2 \varepsilon_{xx}}{\partial z^2} = 2\frac{\partial^2 \varepsilon_{zx}}{\partial x \partial z} \tag{3.13c}$$

$$\frac{\partial^2 \varepsilon_{xx}}{\partial y \partial z} = \frac{\partial}{\partial x}\left(-\frac{\partial \varepsilon_{yz}}{\partial x} + \frac{\partial \varepsilon_{xz}}{\partial y} + \frac{\partial \varepsilon_{xy}}{\partial z}\right) \tag{3.13d}$$

$$\frac{\partial^2 \varepsilon_{yy}}{\partial z \partial x} = \frac{\partial}{\partial y}\left(-\frac{\partial \varepsilon_{zx}}{\partial y} + \frac{\partial \varepsilon_{yx}}{\partial z} + \frac{\partial \varepsilon_{yz}}{\partial x}\right) \tag{3.13e}$$

$$\frac{\partial^2 \varepsilon_{zz}}{\partial x \partial y} = \frac{\partial}{\partial z}\left(-\frac{\partial \varepsilon_{xy}}{\partial z} + \frac{\partial \varepsilon_{zy}}{\partial x} + \frac{\partial \varepsilon_{zx}}{\partial y}\right) \tag{3.13f}$$

See I. H. Shames and C. Dym, *Energy and Finite Element Methods in Structural Mechanics* (Hemisphere Publishing Corporation, New York, 1985), for development of these equations.

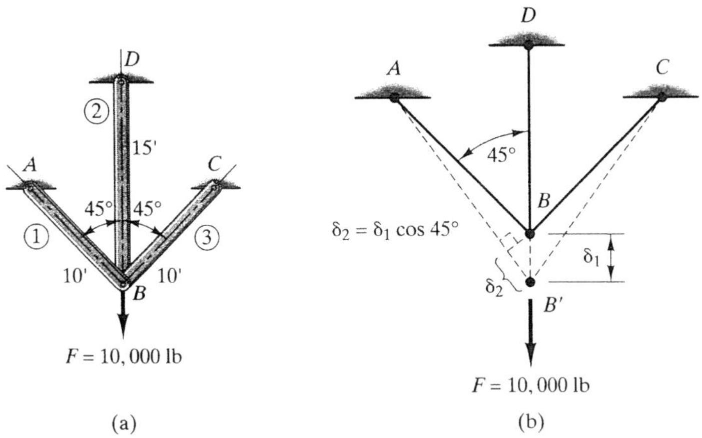

Figure 3.27. (a) Simple loaded truss. (b) Truss showing deformation.

this text, the strains will either stem from single-valued, continuous displacement fields and so be compatible, or they will be such that *physical reasoning* will be sufficient to assure us of their compatibility.

Thus, consider the truss in Fig. 3.27 (a) consisting of three tensile members supporting a load F. There will take place a deflection of the pin at B downward. We will now demonstrate the use of the compatibility concept for this simple deformation. We will first associate the pin with the vertical member (BD) and we note that this member allows the pin to move downward a distance δ_1[see Fig. 3.27 (b)]. We next associate the pin with the member AB. In order for the pin to go to the very same position as when it was associated with the vertical member and thus satisfied compatibility, the following must be true. Member AB must rotate and elongate as has been shown by the dashed line. By erecting a perpendicular from B to the dashed line AB' we can see the elongation δ_2 of the member AB. For the pin to go to its proper position, we see from the diagram that the elongation δ_2 of AB must relate to δ_1 by the equation $\delta_2 = \delta_1 \cos \angle AB'D \approx \delta_1 \cos 45°$. Thus, we have used geometric reasoning to insure compatibility.

We will use similar reasoning throughout the text to ensure compatibility in each problem and example. Those students that study energy methods later in the text (Chapter 18) will realize that compatibility is every bit as important as equilibrium in developing methodologies for solving problems.

3.6 Closure

In this chapter we have been concerned with means of quantitatively describing the manner in which a body deforms. We first presented the displacement field $\boldsymbol{u}(x, y, z)$ to give us a means of describing the movement of each point in the domain. However, as with bonding forces between atoms in physics, it is the relative deformational movement between sets of adjacent points that relates most directly to stress. In particular we indicated that relative movement of adjacent points along orthogonal axes at a point determines the relative movement between any pair of adjacent points in the neighborhood. We accordingly developed as a consequence nine strain terms. Recall that the physical interpretations of the strain terms could be effectively set forth by considering a vanishingly small rectangular parallelepiped formed from the aforementioned adjacent points. At a point there are only six independent strain terms—as with stress, the off-diagonal terms (shear strains) with reversed indices are equal. We pointed out, furthermore, that strain terms, like stress terms, form rectangular components of a second-order tensor. It is for this reason that strain rather than the displacement field is most simply related to stress. We shall begin to investigate the relation between stress and strain for a special but important case in Chapter 4.

*3.7 A Look Ahead; Fluid Mechanics I

The study of fluid mechanics has many direct relations to the study of solid mechanics. We will point out some of the more salient features as we go along in this text.

For instance, in this chapter we looked at the relative displacements of four adjacent points in order to describe the deformation of an infinitesimal rectangular parallelepiped wherein changes of length of the sides of the rectangular parallelepiped gave us normal strains and the change from right angle of these sides gave us engineering shear strains. The relative movement of the adjacent points due to rigid-body rotation of the aforementioned element was *not* needed for our purposes in this text.

In fluid mechanics, note first that an initial rectangular parallelepiped will deform continuously in a flow, unlike the solid wherein we generally go from an initial configuration to a final configuration. However, we still use the same adjacent points, as we have done in this chapter, at a point in the flow. But now, because of the continuous deformation, this rectangular parallelepiped will be in the process of deforming. We accordingly have **normal strain rates** and **shear strain rates** rather than normal strains and shear strains for the aforementioned rectangular element. One important difference must now be noted and that has to do with the rigid-body rotation of fluid elements. Unlike solid mechanics, this rotation is of vital importance in fluid mechanics. This rotation in a flow makes a world of difference in the manner of solution of fluid flows. Indeed, we have different names for flows with rotation and for flows without rotation. They are called rotational flows and irrotational flows.

In *rigid-body mechanics* we use the free body wherein we isolate a body or a portion of a body and we identify all the external forces acting on the body so that we can employ Newton's law.

In *fluid mechanics,* we may either make a free body of some chosen chunk of fluid (here it is called a system), but more likely it will be more profitable to identify some volume in space involving fluid flow through the volume. Such a volume is called a *control volume.* Here, as in the case of a free body, we must specify *all* the *external forces* such as tractions on the bounding surfaces of the control volume and, in addition, body forces on the material inside the control volume. This identification and force specification is needed to ensure, in appropriate equations, that Newton's law and other laws are satisfied for the fluid and other bodies inside the control volume at any time t.

Highlights (3)

The displacement of field $\boldsymbol{u}(x,y,z,t)$ gives the movement of every point at position x,y,z of the body at time t from the undeformed geometry to the deformed geometry. However, to relate deformation more directly to stress, we consider the **relative** movements of certain points infinitesimally close to each other. The points of interest that give rise to useful results are the points at the corners of an infinitesimal rectangular parallelepiped. What are these relative movements? First, there is the relative movement that elongates or shortens the sides of the rectangular parallelepiped and gives rise to three normal strains. For instance, the normal strain ε_{xx} is the change in length per unit original length of the line segment originally in the x direction. This is similar for line segments in the y and z directions. Note that for this deformation there is no essential change in shape of the element; it remains a rectangular parallelepiped. Next, we have the changes in angle from that of a right angle of each pair of line segments giving rise to shear strains. Note, here there is only a second-order change in volume of the element. Finally, there is rigid-body rotation which also gives relative movement. However, in solid mechanics, since there is no distortion involved we disregard this contribution. There are then nine strain terms consisting of three normal strains and six shear strains. Like stress, the nine-term array is symmetric so that shear strain ε_{xy} between the x and y axes equals shear strain ε_{yx} between the y and x axes, and so on.

The strain components can be formulated from the displacement field since the displacement field describes all kinds of deformations, and therefore can also yield the relative movement of the adjacent points we have been discussing. The result is the vital **strain displacement equations** (Eq. 3.12) consisting of various partial derivatives. We shall use these equations from time to time in our studies where they are needed for proper, careful development.

Finally, we come to one of the three pillars of solid mechanics, and that is the **compatibility** consideration. To describe deformations that do not develop cracks during deformation, we must require the displacement field to be **continuous.** This prevents in our formulations points that are infinitesimally close to each to be separated by a finite chasm of space in the deformed state as a result of a jump in the displacement field. Additionally, we want the displacement to be **single-valued.** This prevents the displacement field having an incompressible element from going to two different points at the same time and also does not allow two incompressible elements from occupying the same volume at the same time. The strain field to satisfy compatibility must be linked to a continuous, single-valued displacement field. There are partial differential equations that will ensure this for a broad class of geometries. Another approach is to use the above mentioned strain-displacement equations to form the strains from acceptable displacement fields. Generally, we will use simple geometry and elementary trigonometry to satisfy the conditions of compatibility. This step will become abundantly clear as we proceed in the text.

PROBLEMS

The Displacement Field	3.1–3.6
Strain	3.7–3.23
Strain-Displacement Equations and Compatibility	3.24–3.31
Unspecified Section Problems	3.32–3.45
Computer Problems	3.46–3.49
Programming Project 3	

3.1. [3.2] Given the following displacement field,

$$\boldsymbol{u} = [(x^2 + y)\mathbf{i} + (3 + z)\mathbf{j} + (x^2 + 2y)\mathbf{k}] \text{ m}$$

what is the deformed position of a point originally at (3, 1, −2) m?

3.2. [3.2] Two points in the undeformed geometry are originally at (0, 0, 1) m and (2, 0, −1) m. What is the distance between these points after deformation? (Assume that the displacement field given in Problem 3.1 is imposed on the body.)

3.3. [3.2] A displacement field is given as

$$\boldsymbol{u} = (.16x^2 + \sin y)\mathbf{i} + \left(.1z + \frac{x}{y^3}\right)\mathbf{j} + .004\mathbf{k} \text{ m}$$

As a result of deformation, what is the increase in distance between two points, which in the undeformed geometry have position vectors

$$\boldsymbol{r}_1 = 10\mathbf{i} + 3\mathbf{j} \text{ m}$$
$$\boldsymbol{r}_2 = 4\mathbf{k} + 3\mathbf{j} \text{ m}$$

3.4. [3.2] Given the following displacement field,

$$\boldsymbol{u} = xy\mathbf{i} + (3 + y)\mathbf{j} + (x + 2y + z)\mathbf{k} \text{ m}$$

what is the loss in perpendicularity of two line segments each of unit length originally along the x and y axes at the origin as a result of the displacement field?

3.5. [3.2] In Problem 3.4, find the loss in perpendicularity of unit line segments originally along the x and y directions at position (0, 1, 1) m.

3.6. [3.3] A block with a square section (see Fig. P.3.6) is compressed by a uniform loading P an amount equal to 9 mm. If the volume remains constant, compute the strains for reference xyz.

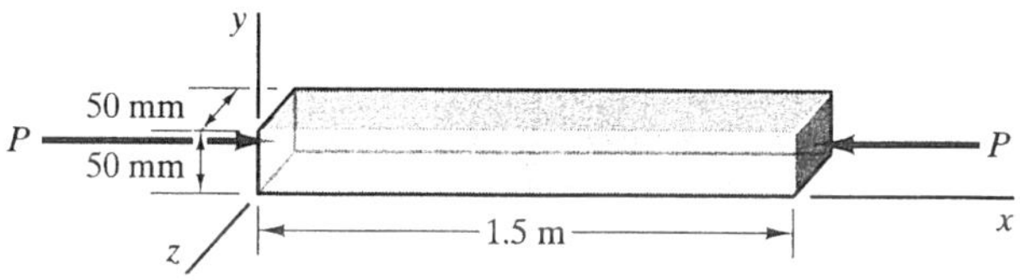

Figure P.3.6.

3.7. [3.3] Two cylinders (see Fig. P.3.7) are welded together and are held between immovable walls. A force F at the interface of the members causes this interface to move 1/4 in. to the left. What are the normal strains in each member along the axis of the cylinders?

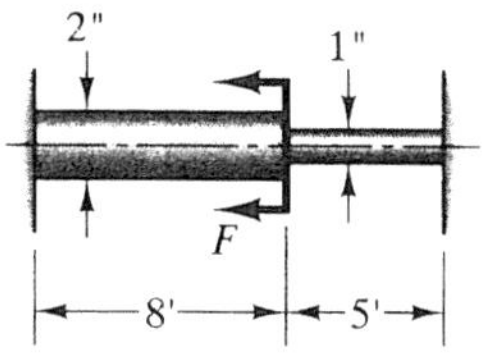

Figure P.3.7.

3.8. [3.3] What is the change in diameter of a long thin-walled tube of thickness t, mean radius r, internal gage pressure p, and elastic modulus E?

3.9. [3.3] A steel shaft 25 mm in diameter is connected to a sleeve having an inside diameter of 37.5 mm (see Fig. P.3.9) via a rubber connector firmly glued to both shaft and sleeve. If the shaft is moved 1.25 mm along its axis relative to the sleeve, what is the shear strain in the rubber assuming shaft and sleeve to be rigid in comparison to the rubber?

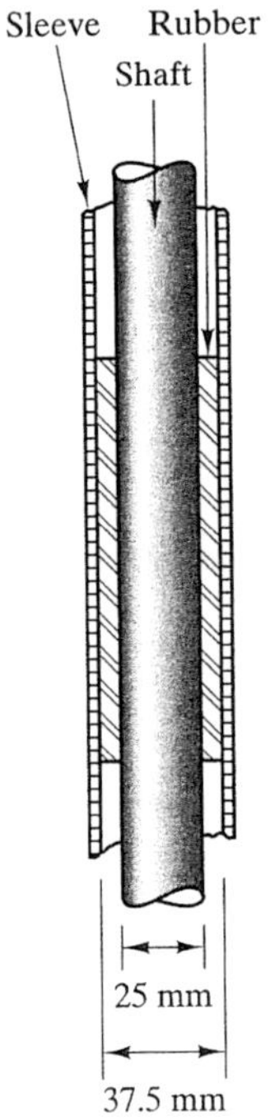

Figure P.3.9.

3.10. [3.3] A thin-walled tank is shown in Fig. P.3.10. Torques T at the ends cause the faces to change relative orientation by $\theta°$, while an internal pressure causes the ends to separate by an amount Δ and the diameter to increase by δ. What are the strains ε_{xx}, ε_{yy}, and ε_{xy} at point P away from the ends as shown in the diagram?

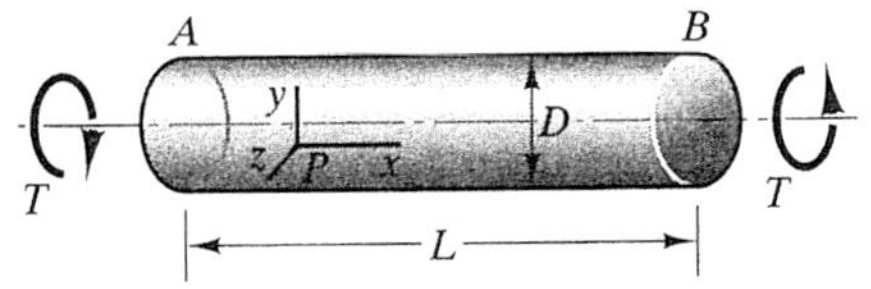

Figure P.3.10.

3.11. [3.3] Given the following strain field,

$$\varepsilon_{ij} = \begin{pmatrix} x^2 & 2x & 3y \\ 2x & 3+2y & 0 \\ 3y & 0 & z+2x \end{pmatrix} \times 10^{-2}$$

what are the strains at position (3, 2, 4)?

3.12. [3.3] In Problem 3.11 find the cubical dilatation at position (2, 1, 3). Next find the new volume of the material in a region $-1 \le x \le 1$ m, $-2 \le y \le 2$ m, $-2 \le z \le 1$ m in the undeformed geometry. See Problem 3.36 before doing this problem. What reservations do you have about the validity of results?

3.13. [3.3] A thin cylinder is subject to a twisting moment distribution along its length so that (see Fig. P.3.13) the rotation of a cross section as a result of the loading varies as the square of z from zero at $z = 0$ to ϕ_0 rad at the end. What is the relative rotation for sections Δz apart? Using a limiting process where $\Delta z \to 0$, find the strain ε_{xy} as a function of z. The radius of the cylinder is R. Show that $\varepsilon_{xy} = R\phi_0 z \,/\, L^2$.

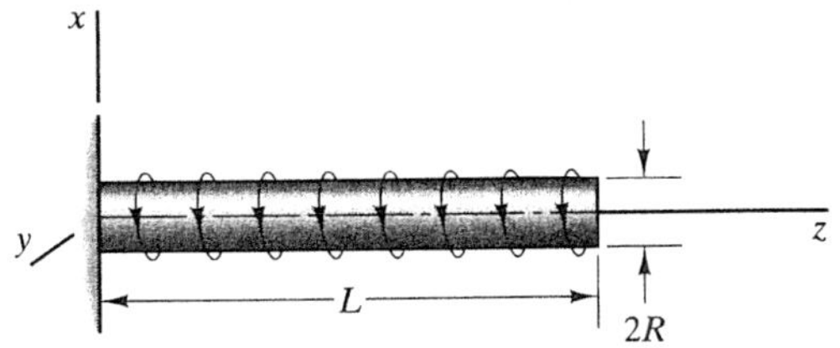

Figure P.3.13.

3.14. [3.3] A circular rod hangs by its own weight and force F as shown in Fig. P.3.14. The specific weight γ is constant for the body. If the normal strain in any direction is $1/E$ times the normal stress in that direction, what is the total deflection of the end A of the member as a result of the weight and F?

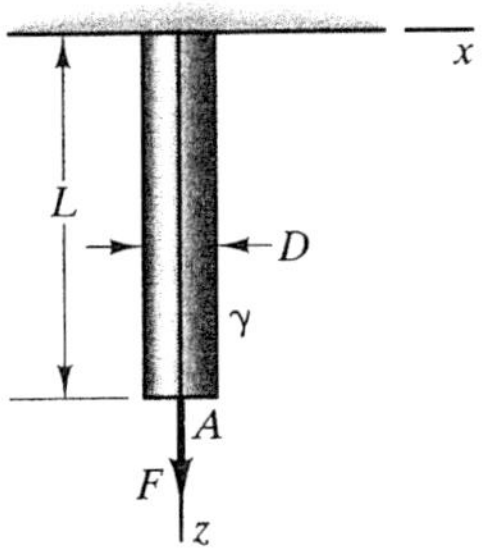

Figure P.3.14.

3.15. [3.3] A cone is suspended from a support as shown in Fig. P.3.15 and is acted on by its own weight. If, for the material, normal strain in a direction is equal to $1/E$ of the normal stress in that direction, what is the deflection of end A as a result of the action of gravity? Take γ as 200 lb/ft^3 and E as 10×10^6 psi.

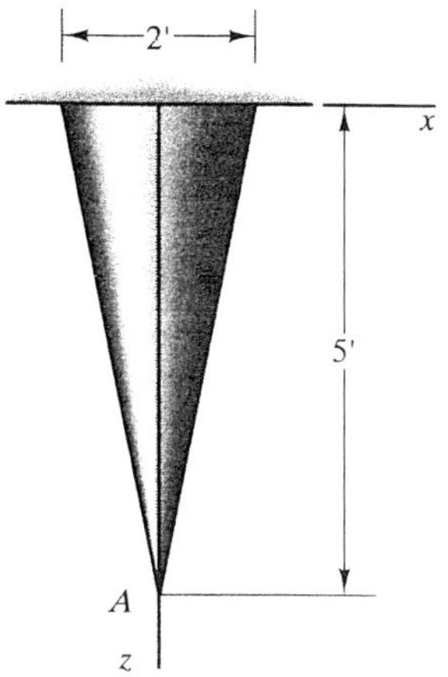

Figure P.3.15.

3.16. [3.3] A plate having rectangular cross sections is shown in Fig. P.3.16. If the specific weight is 6.28×10^4 N/m^3 and if the thickness of the plate is 25 mm, what is the deflection of point A as a result of the deformation from its own weight? Assume that the normal strain in the vertical direction is $1/E$ times the stress in this direction. Take $E = 1.380 \times 10^{11}$ Pa.

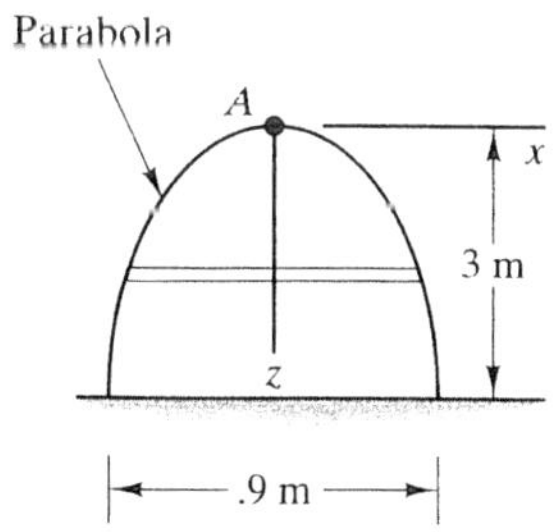

Figure P.3.16.

3.17. [3.3] A space device (Fig. P.3.17) is accelerating in the Z direction in inertial space at a rate of 9 g's far from gravity of the earth. What is the elongation on the rod AD as a result of this acceleration? The mass of the rod is 20 kg/m. Rigid disks E and B each having a mass of 50 kg are welded to the rod. Take $E = 2 \times 10^{11}$ Pa. (*Hint:* Consider two domains. The shaft goes through E and B but does not elongate at E and B because of the rigidity of the welded disks.)

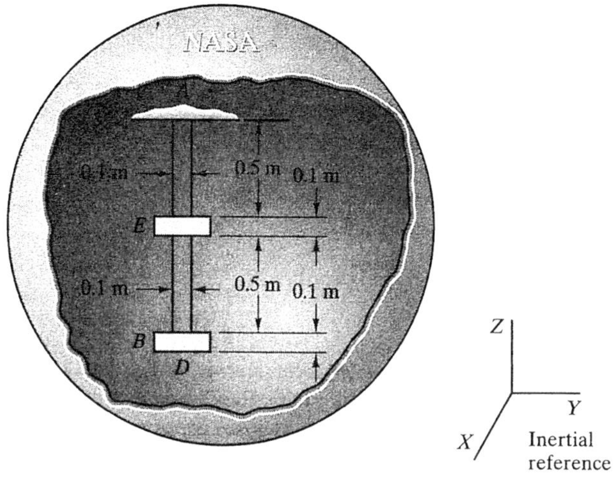

Figure P.3.17.

3.18. [3.3] Find the deflection of end A for the cylinder (Fig. P.3.18) under the action of a friction force given per unit length and varying linearly from 0 N/m at the top to 5000 N/m at the bottom. Assume linear elastic behavior. Neglect weight.

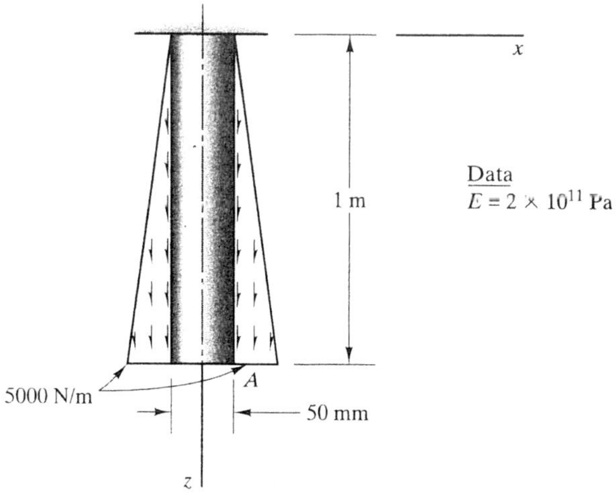

Figure P.3.18.

3.19. [3.3] A balloon having the shape of a sphere is shown in Fig. P.3.19, partially blown up. A small rectangular element has been shown along a great circle of the sphere. If when the balloon is blown up further, the diameter increases from 8 in. to 14 in., what are the strains ε_{xx}, ε_{yy}, and ε_{xy}?

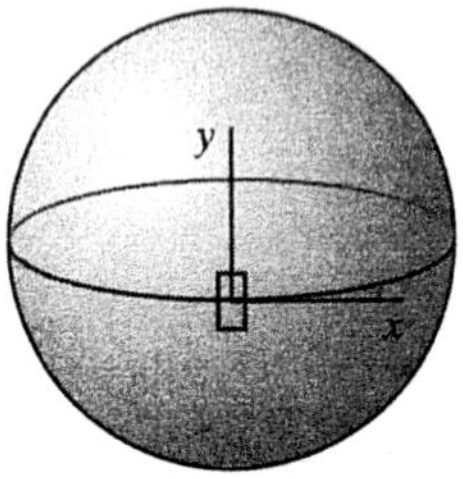

Figure P.3.19.

3.20. [3.3] Two thin-walled cylinders are welded to a rigid connecting plate (black) and are held by immovable walls, as shown in Fig. P.3.20. A torque C causes the connecting plate to rotate 8°. What are the shear strains in the cylinders?

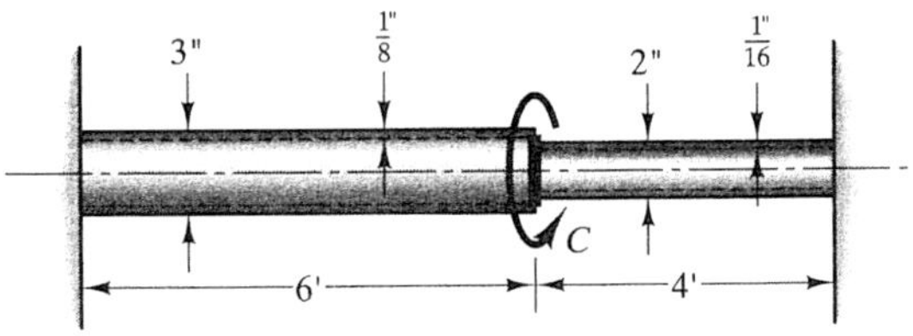

Figure P.3.20.

3.21. [3.3] Water is flowing along a canal as shown in Fig. P.3.21. If the velocity profile is a straight line, as shown in the diagram, what is the time rate of change of ε_{xy}? This is called a shear *strain rate.*

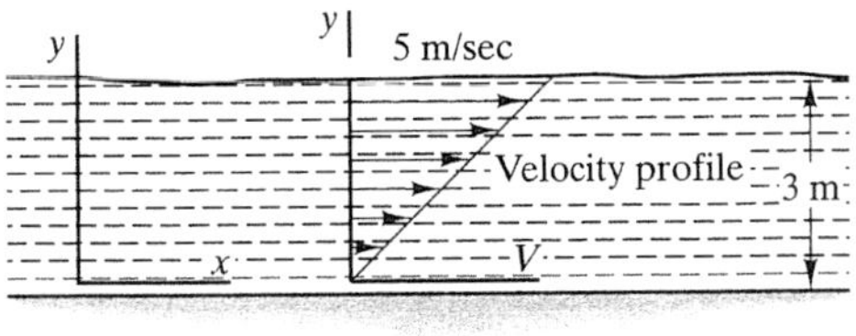

Figure P.3.21.

3.22. [3.3] The velocity profile of the canal in Problem 3.21 has been more realistically shown as a parabola in Fig. P.3.22. What is the time rate of change of shear strain ε_{xy} at $y = 2$ m?

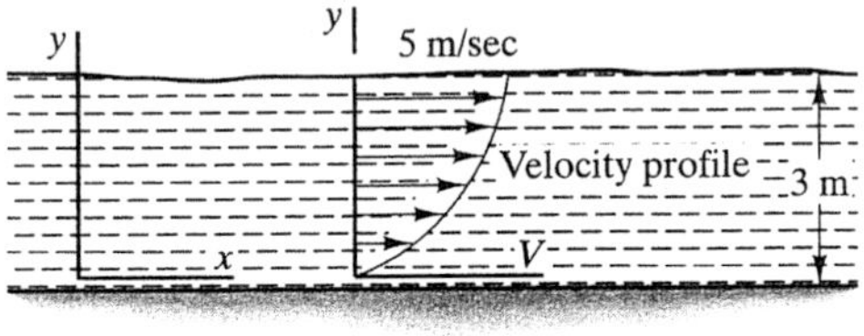

Figure P.3.22.

3.23. [3.3] Do Problem 3.14 when γ varies as the square of z from γ_0 at $z = 0$ to γ_A at $z = L$.

3.24. [3.4] For the following displacement field,

$$\boldsymbol{u} = (\sin x\mathbf{i} + yz\mathbf{j} + x^2\mathbf{k}) \times 10^{-2} \text{ m},$$

what are the strain components at $\boldsymbol{r} = 2\mathbf{i} + \mathbf{j} - \mathbf{k}$ m?

3.25. [3.4] Given the displacement field

$$\boldsymbol{u} = (x^2\mathbf{i} + 3y\mathbf{j} + 10\mathbf{k}) \times 10^{-2} \text{ m}$$

what are the strain components at position (1, 2, 0) m? What are the strain components at $\boldsymbol{r} = 2\mathbf{i} + \mathbf{j} - \mathbf{k}$ m?

3.26. [3.4] A displacement field is given as follows:

$$u_i = (\lambda_{ix})x + (\lambda_{iy})y + (\lambda_{iz})z \text{ m}$$

where i represents either x, y, or z. If the λ's are a set of constants, the deformation is said to be *affine.* Suppose for the λ's that

$$\begin{array}{lll} \lambda_{xx} = .2 & \lambda_{xy} = -.05 & \lambda_{xz} = -.1 \\ \lambda_{yx} = .03 & \lambda_{yy} = .1 & \lambda_{yz} = -.02 \\ \lambda_{zx} = .003 & \lambda_{zy} = -.2 & \lambda_{zz} = .03 \end{array}$$

What is the displacement of a point which in the undeformed geometry is at $\boldsymbol{r} = \mathbf{i} - \mathbf{j} + 3\mathbf{k}$ m? What is the strain field throughout the body?

3.27. [3.4] Show as we did in the text that

$$\varepsilon_{yy} = \frac{\partial u_y}{\partial y}$$

3.28. [3.4] Do Problem 3.14 using a displacement field approach. Delete force F.

3.29. [3.4] Do Problem 3.23 using a displacement field approach for the case where γ varies as the square of z from γ_0 at $z = 0$ to $\gamma = \gamma_A$ at $z = L$. Delete force F.

3.30. [3.4] Consider a rigid-body rotation of an infinitesimal rectangular parallelepiped. Show that a small rotation about the y axis given as ϕ_y is related to the displacement field as follows:

$$\phi_y = \frac{1}{2}\left(\frac{\partial u_x}{\partial z} - \frac{\partial u_z}{\partial x}\right)$$

Permute the coordinates now to get other rotation components ϕ_z and ϕ_x. These rotations are called *rigid-body rotations* at a point.

3.31. [3.4] Using the equations of compatibility (Eqs. 3.13), check to see if the strain field given in Example 3.5 is a compatible strain field.

3.32. A rod has cylindrical volumes cut out of it (Fig. P.3.32). It has a specific weight of 500 lb/ft^3. If the material is linear elastic with $E = 30 \times 10^6$ psi, what is the deflection downward of end A as a result of gravity?

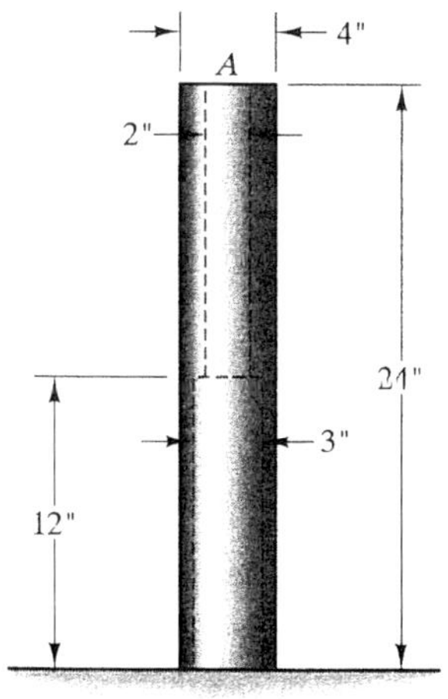

Figure P.3.32.

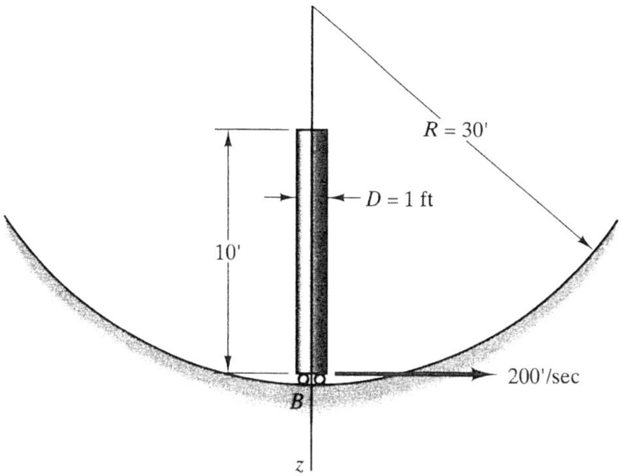

Figure P.3.33.

3.33. A rod moves in a horizontal plane so that end B has a speed of 200 ft/sec along a circular path (see Fig. P.3.33). The rod remains at all times perpendicular to the circular path. If $E = 30 \times 10^6$ psi for the rod, how much does it shorten because of the motion? The density is 450 lbm/ft^3.

3.34. A friction force f acts on the surface of the cylinder (see Fig. P.3.34). It varies with distance from position O and is given *per unit length* as $f = 10x^2$ lb/ft with x in feet. If $E = 10 \times 10^6$ psi, what is the movement of O from the loading?

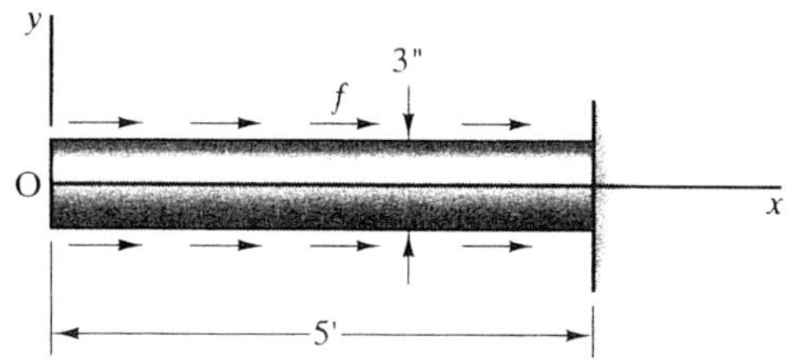

Figure P.3.34.

3.35. Two solid shafts B are welded (see Fig. P.3.35) to a rigid drum A. What is the maximum speed ω permissible if the shafts are not to touch the stationary walls (cross-hatched)? The clearance of .01 in. shown in the diagram corresponds to $\omega = 0$. Neglect the weight of the cylinders.

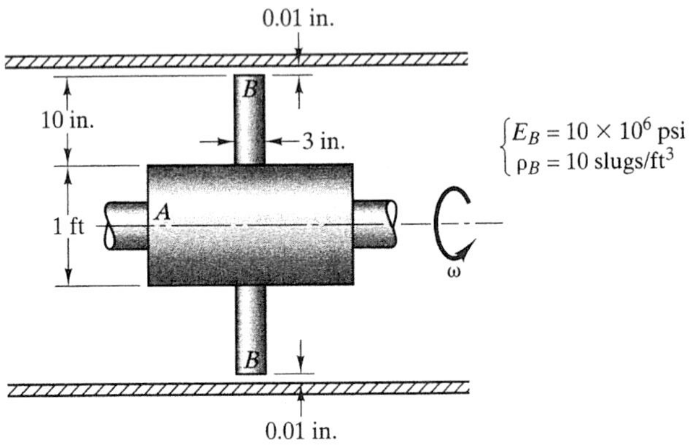

Figure P.3.35.

3.36. A vanishingly small rectangular parallelepiped at a point undergoes only normal strains ε_{xx}, ε_{yy}, and ε_{zz}. If we take as the original lengths of the sides dx, dy, dz, respectively, give an expression for the change in volume of the element in terms of the strains and dx, dy, and dz. If we assume, because strains are small, that products of strains are negligible in comparison to the strains themselves, show that the change in volume per unit volume of the element is $\varepsilon_{xx} + \varepsilon_{yy} + \varepsilon_{zz}$. This is called the *cubical dilatation.*

3.37. We will learn in Chapter 4 that for certain materials the normal stress τ_{xx} equals $E\varepsilon_{xx}$ at a point where E is the *Young's modulus.* Find per unit area the elongation or shortening of each member of the plane truss shown in Fig. P.3.37. Take $E = 2 \times 10^{11}$ Pa.

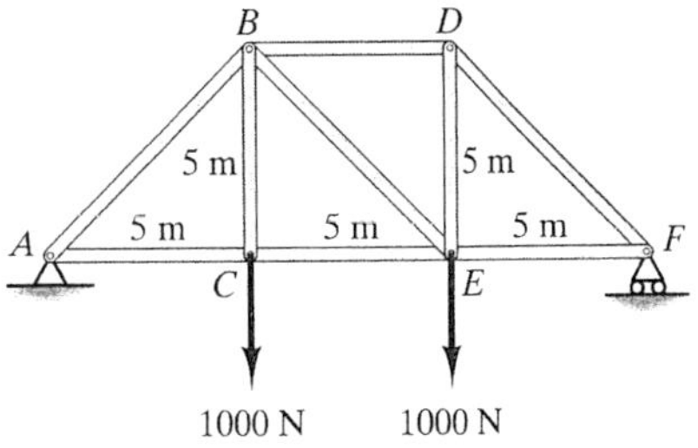

Figure P.3.37.

3.38. Given the following strain field,

$$\varepsilon_{ij} = \begin{pmatrix} 2x + y & -3y & 0 \\ -3y & x + y & x \\ 0 & x & 0 \end{pmatrix} \times 10^{-2}$$

what is the new length of a line parallel to the x axis going from (3, 0, 2) to (10, 0, 2) m in the undeformed geometry after the strain field above has been imposed?

3.39. In Problem 3.38, find the new length of a line going from $\mathbf{r}_1 = (2\mathbf{i} + 3\mathbf{j} + \mathbf{k})$ m to $\mathbf{r}_2 = (2\mathbf{i} - 8\mathbf{j} + \mathbf{k})$ m.

3.40. The thin cylinder shown in Fig. P.3.40 is twisted by torques T so that end B rotates 10° relative to end A. In addition a set of tensile forces at A and B cause an increase in separation of the end faces by 50 mm. What are the strains ε_{xx}, ε_{xy}, and ε_{yx} at P?

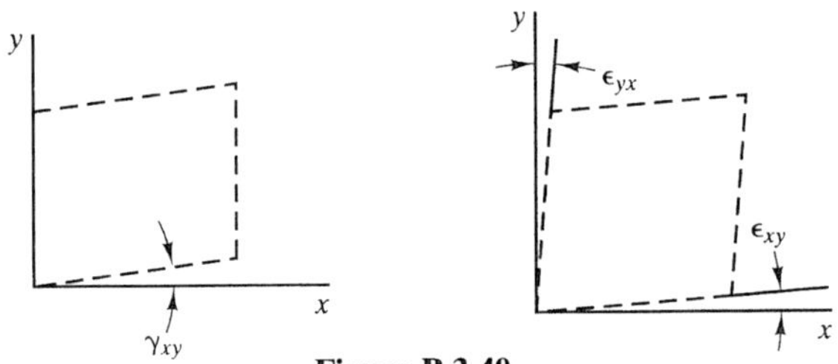

Figure P.3.40.

3.41. For the thin-walled cylinder shown in Fig. P.3.41, what is the distance Δ that a point on the rim of the wheel rotates if it is known that the engineering shear strain γ_{xy} on the surface of the cylinder is .025? Let $R_a = 100$ mm, and $R_b = 400$ mm.

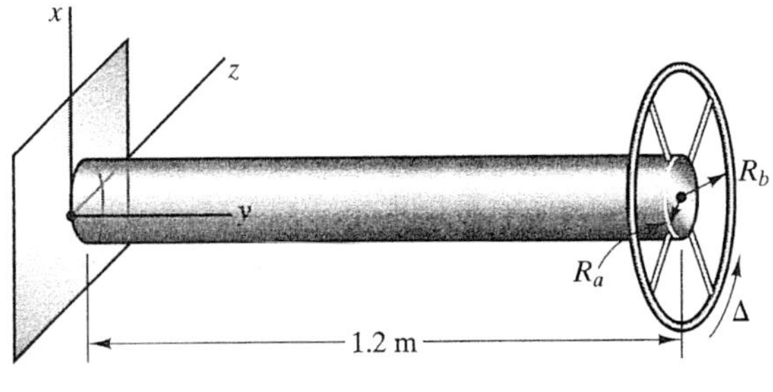

Figure P.3.41.

3.42. Shown in Fig. P.3.42 is a rigid bar AC supported by thin rods. A load P and couple M on the rigid rod cause AC to descend a distance of 25 mm and to rotate 2° about B. What are the normal strains along the axes of the rods?

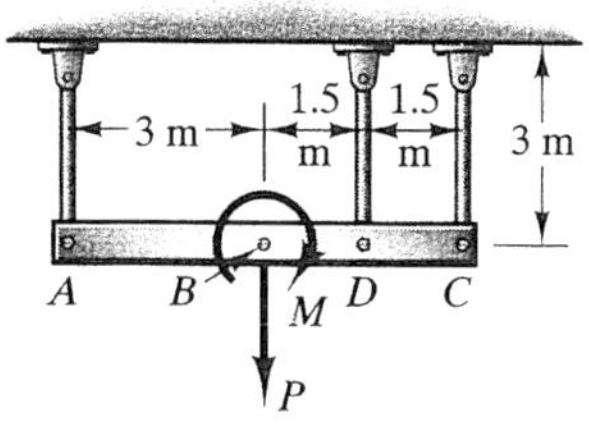

Figure P.3.42.

3.43. In Fig. P.3.43 is shown a rigid plate CD glued to identical rubber cubes A and B. This plate supports a 500-lb load. Rigid side plates GH and IK are also glued to the cubes as are members HC, CK, GD, and DI. The latter members are hinged as shown. If plate CD moves downward a distance of .1 in., what is the strain ε_{xy} inside blocks?

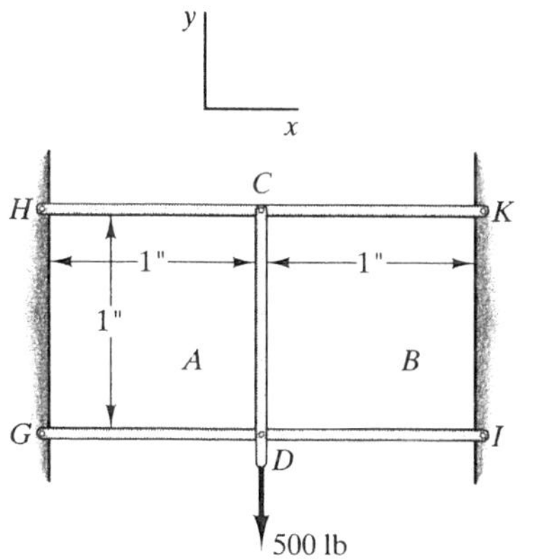

Figure P.3.43.

3.44. Shown in Fig. P.3.44 are two pin-connected members supporting a load P. Neglect the weight of the members. Show that for a small vertical deflection Δ of pin A we can approximate the normal strain in each member as $\Delta \cos \alpha / L$.

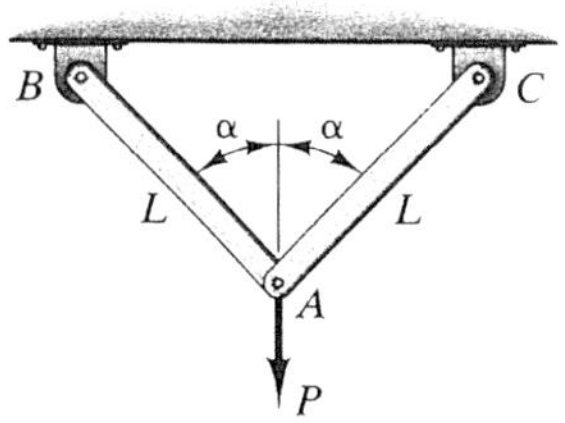

Figure P.3.44.

3.45. Determine the axial displacement $\mu(z)$ of rod A of the shock mount shown in Fig. P.3.45. Use the relation $\gamma = \tau/G$ where τ is the shear stress and G is called the *shear modulus.*

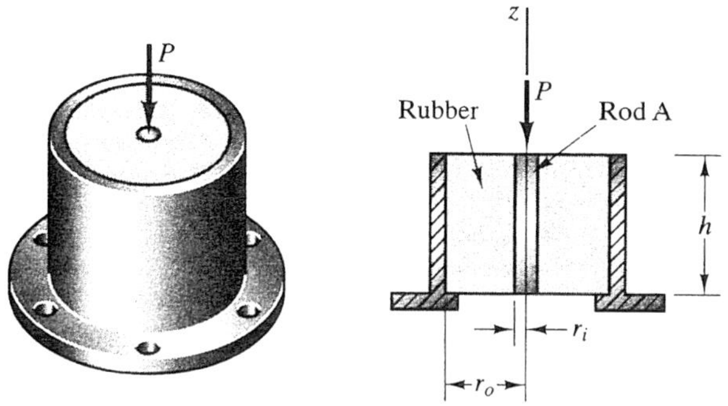

Figure P.3.45.

****3.46.** Consider a rod held fixed between rigid walls (see Fig. P.3.46). The initial temperature of the rod is 0°C. The final temperature varies from 100°C at $x = 0$ to 0°C at $x = L$. Plot the linear displacement Δ in the rod divided by αL as a function of position x. Where is the displacement a maximum? The rod has an area A, modulus E, and expansion coefficient α. Take $T_i = 0$°C and $T_f = 100$°C at $x = 0$ and 0°C at $x = L$.

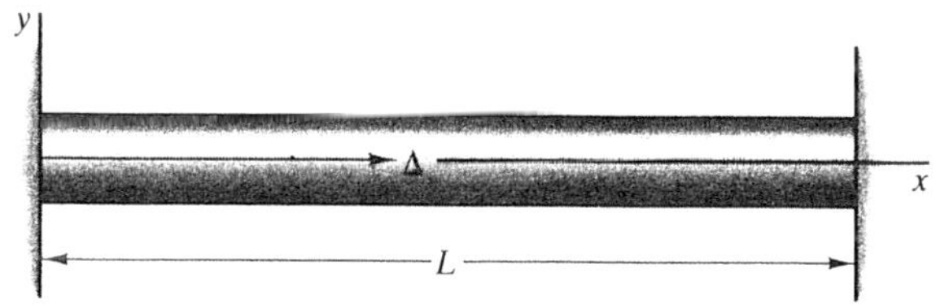

Figure P.3.46.

****3.47** For the shock mount considered in Problem 3.45, plot the mount height h vs. radius ratio r_o / r_i for a force of 1000 lb and $G = 15{,}000$ psi. Consider the range $2 \le r_o / r_i \le 10$ and displacement limits of 0.01″, 0.025″, 0.050″, 0.075″ and 0.10″. From the solution for problem 3.45 we have

$$\delta = \frac{F}{2\pi G h} \ln\left(\frac{r_o}{r_i}\right)$$

****3.48.** Consider a generalization of Problem 3.14. Now, let the weight per unit volume $\gamma_A = \alpha\gamma_o$ where α is a nondimensional parameter. Plot $\varepsilon_{zz}\, E/(\gamma_o L)$ vs. z/L for $\alpha = 5, 2, 1, 1/2, 1/5$. Is the relationship linear? Which causes a larger peak strain: when $\gamma_o > \gamma_A$ or when $\gamma_A > \gamma_o$?

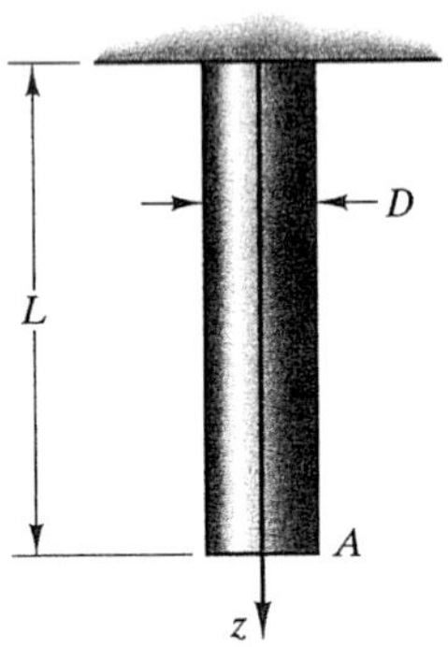

Figure P.3.48.

****3.49.** For the structure shown in Fig. P.3.42, plot the strain in each of the three members for a rotation varying between 0° and 5° and for values of the rigid bar displacement of 0, 25 mm, and 50 mm. How does the strain vary with the rotation angle? As a check, what is the strain in each member when the rotation is zero? Does this show on your plot?

Project 3 Layered Stalactite Problem

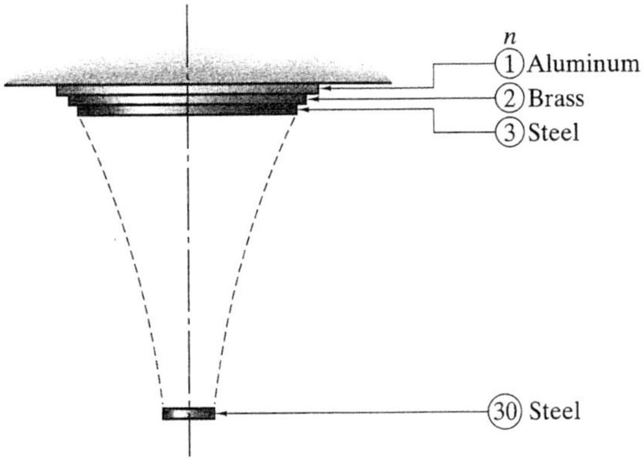

Figure P.3.49.

Find the elongation of a stalactite in Fig. P.3.49 made up of successive slices of thickness 0.01 m and of circular shape with varying diameter. The material of the slices is *aluminum* for the first slice, *brass* for the second slice, and *steel* for the third slice, with this sequence repeated over and over as you go down for 30 slices. The slices are glued together firmly. The elongation is due to two effects:

1. The elongation of a slice due to its own weight = $\gamma L^2 / 2E$
2. The stress stemming from holding up all the slices below it

The *diameter* of each slice is given as

$$D = \frac{e^{1/n}}{10}$$

where n is the number of the slice, starting with unity at the top and 30 at the bottom.

Employ the following data:

Specific weight of:	aluminum = 26.6×10^3 N/m³
	brass = 83.8×10^3 N/m³
	steel = 76.5×10^3 N/m³
Elastic modulus of:	aluminum = 71×10^9 Pa
	brass = 106×10^9 Pa
	steel = 207×10^9 Pa

CHAPTER 4

Introduction to Mechanical Properties of Solids

4.1 Introduction

In previous chapters we have considered stress and strain separately. It is intuitively clear that these quantities are related, and in this chapter we make an introductory inquiry into such relations. These relations are called *constitutive laws*.

Almost all the working knowledge we now possess for such relations stems from *macroscopic* testing of materials, and it is the results of such tests, as well as the macroscopic theories stemming from such tests, that concern us in this chapter. For some time, however, solid-state physicists and engineers have been intensively studying the *microscopic* bases for mechanical properties (i.e., actions at the atomic and molecular levels). Much progress has been made along this avenue of approach, although a thorough understanding of the mechanisms involved has not yet been reached. It is expected that in the future we shall turn more and more to this fundamental approach. Modern technology is putting our structures into more complex environments and under more complex conditions for which macroscopic laboratory tests, such as the ones we describe in this chapter, are becoming less meaningful. Needed for a better understanding of how a material is to behave under a combination of conditions, such as high temperature, dynamic loads, radiation, temperature gradients, and vibration, is a comprehension of how mechanical action relates to atomic and molecular structure. You will learn more about these theories in your courses in materials science.[1]

[1]For a discussion of microscopic properties, see I. H. Shames, *Mechanics of Deformable Solids*, R.E. Krieger, 1964, Melbourne, Fla. Appendix X.

4.2 The Tensile Test

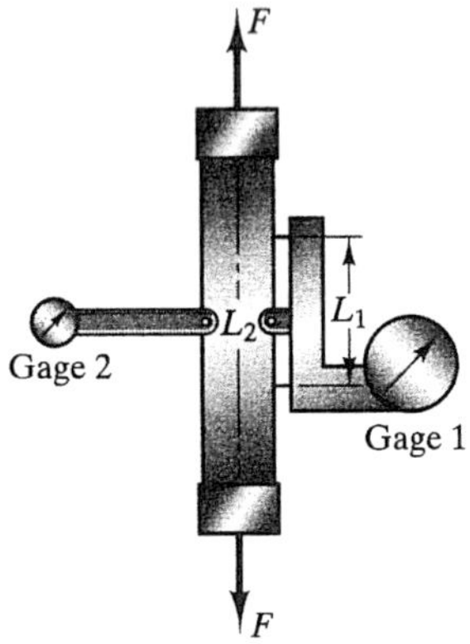

Figure 4.1. Tensile specimen with gauges.

We shall now consider a simple state of stress, the one-dimensional normal state of stress introduced in Example 2.6. You will recall that such a state of stress in a domain is defined for reference xyz as having one direction, say z, throughout the domain for which the only nonzero stress is τ_{zz}. That is, all the stresses are zero except for the normal stress in the z direction. Many practical engineering problems consist of, or may be approximated as, one-dimensional stress problems. We shall examine some of these systems later.

The most basic test in the study of stress-strain relations is the simple one-dimensional tensile test.[2] The specimen to be tested is subjected to an axial force F along its centerline. We will assume here that the specimen is a flat sheet with a rectangular cross section like the one shown in Fig. 4.1. A gage (1) is used to measure the change in length between two points, carefully aligned in the direction of the load, originally L_1 apart. Another gage (2) meanwhile measures the change in length between two points, originally L_2 apart, but at a right angle to the load direction, as shown in the diagram. As the force F is increased, we measure the change in length L_1 at gage (1) and L_2 at gage (2) for each value of F. Hence, at any value F, we have the following information:

1. *Engineering stress* $(\tau_{zz})_{\text{Eng}}$: Computed as F/A_o where A_o is the initial unstrained cross-sectional area of the specimen.
2. *Strain* (ε_{zz}): Computed by the ratio $\Delta L_1/L_1$, where ΔL_1 is the change in length found using gage 1 and L_1 is the unstrained length.
3. *Strain* (ε_{yy}): Computed by the ratio $\Delta L_2/L_2$, where ΔL_2 is the change in length found using gage 2 and L_2 is the unstrained length.

The values of stress and strain from such measurements are actually average values. However, if the test specimen is carefully made, the stresses and strains away from the ends of the specimen are close to being one-dimensional and uniform. And so these stress and strain values may be considered as the correct values at any point *away from the ends.*

Since the volume of the specimen will change slightly during the test, there will take place a contraction of the original cross-sectional area A_o as the tensile load is increased. We can define a new stress called the *actual stress* that is computed as F/A_{act}, where A_{act} is the actual cross-sectional area, for any value of F. For a tensile load, the engineering stress will be less than the actual stress at all times. For small loads, A_{act} will not be significantly smaller than A_o so that little difficulty is encountered by using the simpler engineering stress. How-

[2]Please refer to the appropriate ASTM standard for suggested test procedures. These are available in most engineering libraries or from ASTM, 1916 Race Street, Philadelphia, PA 19103.

ever, for large loads there will be a significant difference between A_{act} and A_o, with the result that the curve $(\tau_{zz})_{Eng}$ versus ε_{zz} will diverge markedly from that of $(\tau_{zz})_{act}$ versus ε_{zz}. Since $(\tau_{zz})_{Eng}$ is simply related to F through the constant $1/A_o$ and since the change in area is not easily measured, engineers are often motivated to use the engineering stress $(\tau_{zz})_{Eng}$, rather than the actual stress $(\tau_{zz})_{act}$. Unless otherwise noted, we shall use the engineering stress, which is based on the initial, unstrained geometry.

We show a plot of engineering axial stress $(\tau_{zz})_{Eng}$, versus the axial strain ε_{zz} for a simple tensile test in Fig. 4.2. Traditionally, values of

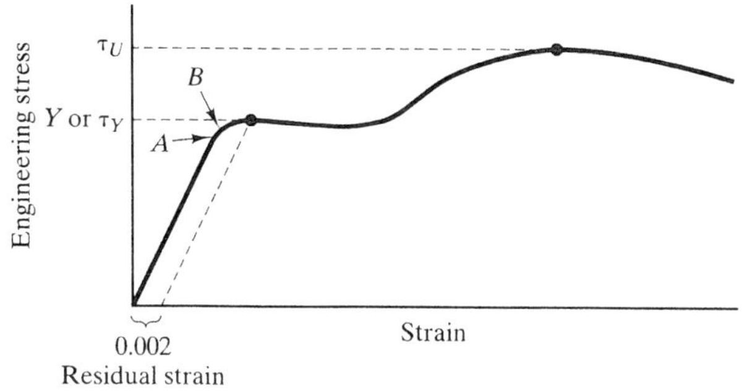

Figure 4.2. Typical stress-strain curve for mild steel.

stress are plotted vertically (ordinates) while values of strain are plotted horizontally (abscissas). This is an idealized stress-strain diagram for a steel specimen. Although stress-strain diagrams may be quite different for other materials, we shall consider this curve in some detail so as to set forth certain general definitions. Notice that the stress-strain response is a straight line at the early stages of the loading; that is, the stress is proportional to strain and we may then state that

$$\tau_{zz} = E\varepsilon_{zz} \tag{4.1}$$

where the proportionality constant E is called *Young's modulus*, or the *modulus of elasticity* having dimensions F/L^2, as you may easily verify yourself. Essentially, this result was reached about 300 years ago by Robert Hooke, who as a result of his experiments with metallic rods under axially applied tensile loads concluded that "*ut tensio sic vis*" or roughly "the extension is proportional to the force," a relation known today as *Hooke's* law. Materials that obey this relationship are called *Hookean materials.*

When using Eq. (4.1), it does not matter if you are loading or unloading the specimen; the stress-strain response is a straight line of slope E. The stress at which this linear relationship ceases is called the *proportional limit;* it is shown as point A in Fig. 4.2. Its value, however, is not easily measured. For the steel specimen considered, there is a stress level close to the proportional limit such that when the specimen is loaded beyond this level, then unloaded, it does not return to its original length; it will have a *permanent set.* That is, it has shown *inelastic* behavior. We call this stress level, which gives the limit of elastic behavior and the onset of inelastic behavior, the *elastic* limit, shown by point B on the diagram. A material having a proportional limit close to the elastic limit, such as steel, is termed a *linear, elastic material.* That is, linear, elastic materials have an appreciable straight-line portion at the outset of the stress-strain diagram and for stress values below the elastic limit, the loading and unloading response follows the same path. Thus, linear elastic material returns to its original (unloaded) shape when the loads are removed.

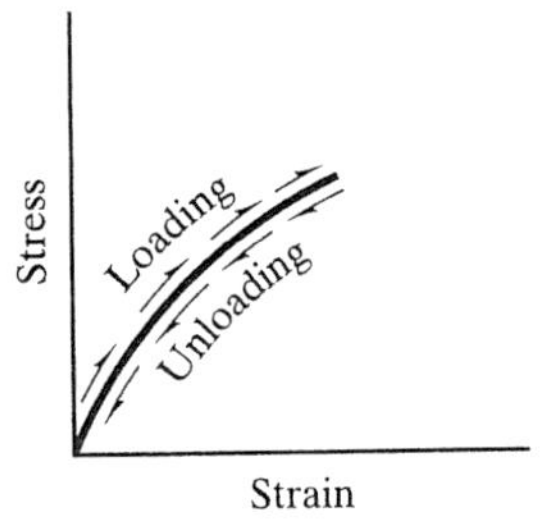

Figure 4.3. Nonlinear elastic behavior.

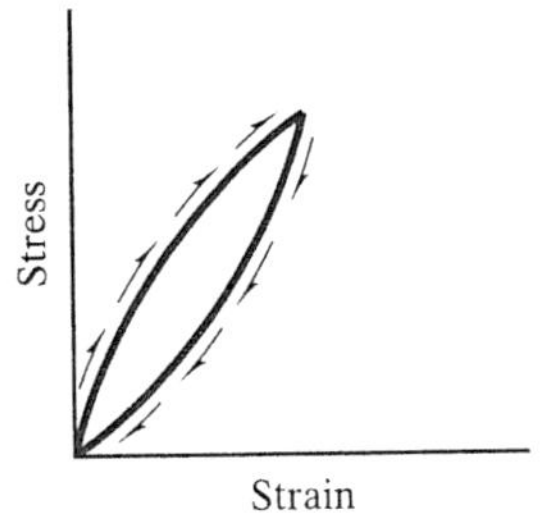

Figure 4.4. Anelastic behavior.

Not all materials have a finite straight-line portion at the outset of the stress-strain diagram. For instance, rubber is a material that generally does not, and an idealized stress-strain curve for a particular specimen of this material is shown in Fig. 4.3. Despite the apparent difference in appearance between the curves for steel and rubber in Figs. 4.2 and 4.3, there is an important similarity to be pointed out. That is, if that particular rubber specimen is unloaded to zero load, it will return to its original geometry along the loading curve, as will the steel specimen, provided that the load on the latter develops a stress below the elastic limit. For these reasons both materials are said to be elastic. Materials that have a proportional limit far below the elastic limit and that have loading and unloading curves that coincide, such as the cases of the particular rubber specimen discussed, are called *nonlinear, elastic materials.* The key is that to be termed elastic, the specimen must return to its original shape after loading.[3] The linear or nonlinear classification is simply a description of the shape of the stress-strain diagram below the elastic limit.

Returning to the stress-strain diagram for steel, it is to be pointed out that the elastic limit, like the proportional limit, can be difficult to measure accurately. Hence, we *define* the onset of inelastic behavior as the stress, called the *yield stress* (or sometimes called the *yield strength*) resulting in a small specified residual strain upon unloading for a one-dimensional test. A value of residual strain of 0.002 (or 0.2%) is commonly used as the specified residual strain; however, other values are sometimes used. The stress denoted as Y or τ_Y on the stress-strain diagram in Fig. 4.2 corresponds to the yield stress. As can be seen in the figure, the 0.2% offset yield stress is readily determined from the stress-strain diagram by considering the intersection of the measured stress-

[3]If the specimen returns to the original shape but along an unloading curve *different* than the loading curve (see Fig. 4.4), the material is called *anelastic.* Since the work put into the material is proportional to the area under the stress-strain curve, as will be discussed later, a net amount of work is done on the material during a loading and unloading cycle. This work goes into the material as thermal energy and is eventually dissipated to the surroundings. Materials having stress-strain diagrams as shown in Fig. 4.4 are very useful for damping vibrations.

strain curve with a line drawn in the diagram with a slope equal to Young's modulus and offset by a 0.2% strain.

Our discussion of the stress-strain diagram has taken us thus far only to the yield stress. In the domain up to the yield stress, the actual cross-sectional area and the original area of the specimen differ by a very small amount and so it does not matter which area one uses for computations. We have been using A_o for reasons set forth earlier. However, as pointed out, at all times during the tensile test, a continual decrease in the cross-sectional area of the specimen takes place as the load is applied. (In the case of a compression test, there is clearly a corresponding increase in the cross-sectional area as the load is applied.) After the yield point, there may be a rapid increase in the strain ε_{zz} and simultaneously there will then be appreciable change in the cross-sectional area. This will cause the values of engineering stress and actual stress to diverge appreciably from each other. To illustrate this, Fig. 4.5

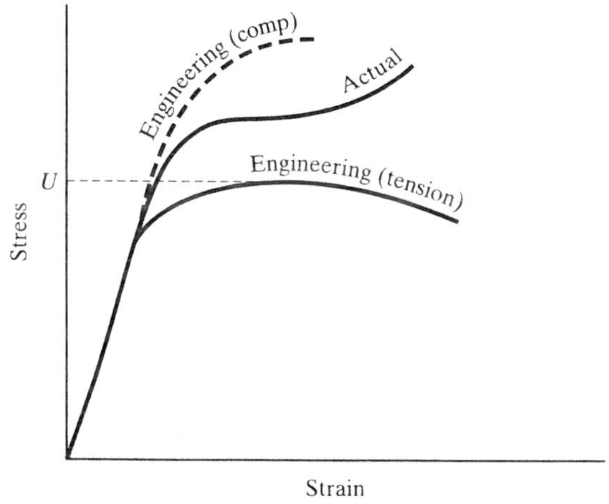

Figure 4.5. Different kinds of stress versus strain plots for a ductile metal.

shows sample stress-strain curves using both the actual stress and the engineering stress for a tensile test. (Also shown dashed is the corresponding compression test using engineering stress.) You will notice that the actual stress continually increases until the specimen breaks.

As pointed out earlier, the engineering stress is proportional to the force F and so the maximum load carrying capability of the specimen is developed at the maximum elevation of the engineering stress-strain curve (see Fig. 4.2). The value of the engineering stress at this point, denoted as U in the diagram, is termed the *ultimate stress* (it is sometimes referred to as the *tensile strength*). To better understand the ultimate stress let us follow the tensile test after passing the yield stress. The strain increases rapidly now for small increases in load. The cross-sectional area accordingly decreases measurably in contrast to the very small changes during the elastic range of loading. The *actual stress* increases because of the combined effects of increasing load and decreasing cross-sectional area. The load-carrying capacity of the specimen (and thus the engineering stress) meanwhile increases because of the increasing actual stress, but is *adversely affected* by the ever-decreasing cross-sectional area, which tends to contribute to a decrease of the specimen's loading-carrying capacity. At the ultimate (engineering) stress the effect of the decrease in cross-sectional area begins to dominate, and, even though the actual stress continues to rise as the test continues, the cross-sectional area decreases at an ever faster pace so that the load-carrying capacity of the specimen falls off after the ultimate stress has been reached. The test specimen then goes rapidly to destruction.

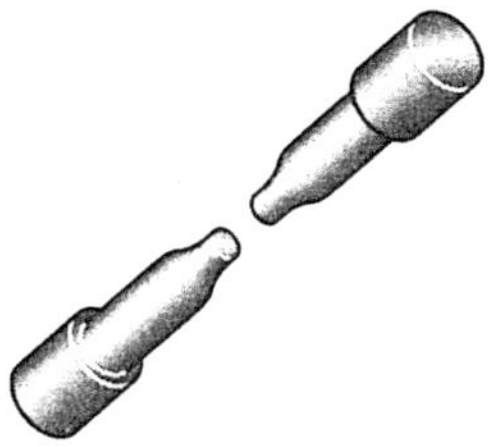

Figure 4.6. Broken cylindrical specimen showing evidence of plastic flow.

The loss in load-carrying capacity beyond the ultimate-stress point just described does not occur as a result of rapid area decrease of the entire specimen. Rather, it occurs as a result of a rapid area decrease at some *localized* portion of the specimen. We call this action "necking" of the specimen, because the specimen forms a "neck" or narrowed region. The position in the specimen where necking takes place depends primarily on local imperfections of the material. We have shown a diagram of a cylindrical specimen loaded to destruction in Fig. 4.6. The necking action can be easily seen by observing the broken portion of the specimen. When large inelastic deformation occurs rapidly in a small domain, as in the necked region of the tensile specimen, we say there is *plastic flow* in this domain.

We have thus far discussed the axial stretching of the specimen. In addition to this, a tensile load on the specimen causes a lateral contraction (i.e., shortening) *perpendicular* to the load direction, that we call the *Poisson effect.*[4] (By the same mechanism, a compression test induces a lateral extension.) In the *linear, elastic range,* we find from these tests that the lateral strain ε_{lat} due to this contraction (in our case ε_{yy}) is proportional to the longitudinal strain $\varepsilon_{\text{long}}$ due to the load (in our case ε_{zz}) and may be expressed as follows:

$$\varepsilon_{\text{lat}} = -\nu\varepsilon_{long} \tag{4.2}$$

where the constant of proportionality ν is called *Poisson's ratio.* For many materials, it ranges between 0.2 to 0.5; however, for some materials, especially composites,[5] it can vary outside this range. As a general

[4]So named after the French mathematician S. D. Poisson (1781–1840) who first reported this phenomena.

[5]Some composite materials are constructed from layers of strong thread with varied directions per layer and separated by filler materials. Thus the composite material can be designed to have desired maximum strengths in prescribed directions for a lighter, stronger design. Such materials thus have properties that are dependent on direction and are said to be *anisotropic.* As such, composites can have more than one Poisson ratio. More will be said about isotropy in Chapter 5.

rule-of-thumb, brittle materials (glass, ceramic) tend to have a low value for ν, while flexible materials (plastic, rubber) tend to have a high value for ν. Meanwhile, it is clear that simple tensile tests permit the evaluation of this constant. As defined here, Poisson's ratio is valid only below the proportional limit. In Chapter 6 we shall see that ν along with E can be considered as two of the fundamental mechanical properties characterizing the behavior of linear, elastic, homogeneous, isotropic materials.

Up until now we have considered the response of a material to a tensile load in terms of stress and strain. However, we can view the same test from a different perspective: as an energy (or work) exchange. To deform the specimen, work was done on it by the testing machine. In turn, the specimen either stores the energy or converts it to another form, such as heat. Such an energy approach has found many applications in solid mechanics including the study of fracture in materials and the analysis of structural systems.

In your study of mechanics, you learned that mechanical work can be expressed as force times the distance the force moves through. Let us now consider an *increment* of work on the specimen. We denote this increment as dW_k. If the deformation is purely elastic, this increment of work goes into an increment of strain energy denoted as dU. The elongation of the specimen is ΔL and so the increment of displacement is then $d(\Delta L)$. With force F, the increment of strain energy then becomes

$$dU = Fd(\Delta L)$$

The total strain energy done on the specimen as it undergoes any deformation $\Delta L'$, is obtained by summing the work increments as follows

$$U = \int_0^{\Delta L'} Fd(\Delta L) \tag{4.3}$$

For the strain energy per unit volume, we divide Eq. (4.3) by the original volume $A_0 L_0$ and define $\mathcal{U}$ as the *strain energy density* (dimensions F/L). Thus

$$\mathcal{U} = \frac{U}{A_o L_o} = \int_0^{\Delta L'} \left(\frac{F}{A_o L_o}\right) d(\Delta L) = \int_0^{\Delta L'/L_o} \left(\frac{F}{A_o}\right) d\left(\frac{\Delta L}{L_o}\right)$$

From the above, we recognize F/A_o as the normal stress in the specimen τ_{zz} and $\Delta L/L_o$ as the normal strain ε_{zz}. We can now express the work per unit volume needed to deform the specimen to a strain level ε'_{zz} as

$$\mathcal{U} = \int_0^{\varepsilon'_{zz}} \tau_{zz}\, d\varepsilon_{zz} \tag{4.4}$$

It is evident that Eq. (4.4) represents the area under the stress-strain diagram up to a strain level ε'_{zz}. So we can say that for loads that are below the elastic limit, the area under the stress-strain diagram represents the energy stored by the specimen to be given up when the load is released. For strains above the elastic limit, some of the energy is stored elastically and can be recovered upon unloading; however, some energy is converted into another form as the material deforms plastically. Most of this plastic work is converted into heat.

If we use the offset yield stress as the limit of elastic behavior, the area under the stress-strain diagram up to the yield stress is called the *modulus of resilience*. By substituting the stress-strain relationship from Eq. (4.1) into Eq. (4.4), we can express the modulus of resilience as a function of the yield stress, as

$$\mathcal{U}_R = \frac{\tau_Y^2}{2E} \tag{4.5}$$

This modulus can be regarded as a measure of the material's ability to store energy then release it when the load is removed, without sustaining permanent deformation.

This brings us through the tensile test of a steel specimen. We thus have examined one of the most important of structural materials. What about other materials? We have already considered some rubber materials in this section, and we see that there can be great departures from the case exemplified by mild steel. However, we can use mild steel as a basis of comparison in our discussion of other material so as to make our communication more meaningful. Furthermore, the definitions that we set up while discussing the steel case hold for general discussions. In Appendix IV-F we have listed some of the parameters we have been discussing in this section for certain important materials. (For more precise, detailed information of this type, you are urged to consult structural or materials handbooks, as well as information provided by the material manufacturer.)

4.3 Strain Hardening and Other Properties

In materials such as mild steel, aluminum, and copper, it is observed that ever-increasing actual stress is required for continued deformation beyond the yield point. This is the case for the stress-strain diagram shown in Fig. 4.7. We call this *strain hardening*.

There is another important phenomenon in the plastic range that is also referred to as strain hardening. It has to do with the unloading from the plastic range of a specimen having a linear, elastic range. You will recall that in the preceding section we discussed the unloading of a linear, elastic material when the load was in the elastic range as well as

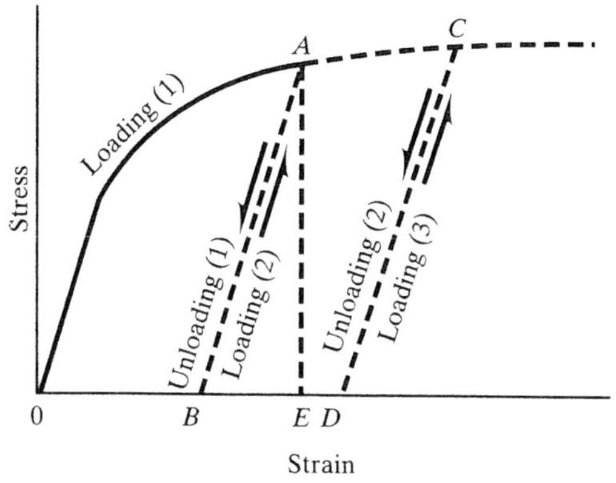

Figure 4.7. Illustration of the strain hardening process.

the unloading of a nonlinear, elastic material. In those cases, complete removal of load results in restoration of the original geometry (i.e., no permanent set). Furthermore, the unloading path must retrace the loading path in the stress-strain diagram. *In unloading a material with a linear, elastic range from a load in the plastic range, we do not retrace the loading path but instead move along a new path which is essentially parallel to the linear-elastic portion of the original loading path.* This is shown in Fig. 4.7, where the initial loading has been stopped at A and the first unloading is shown to take place along a straight line to point B on the abscissa. Thus, we have introduced a *permanent set* given by OB on the abscissa. The *elastic recovery,* on the other hand, is clearly BE. Now on the second loading we move along the path BA. A second unloading from a stress below that corresponding to point A will essentially move along path BA back to B and so we have for practical purposes a linear-elastic range from B to A. An inspection of the diagram will indicate that the yield point has been *raised* for the second loading as a result of the first loading into the plastic range. The raising of the yield point by this action is the second phenomenon referred to as *strain hardening.* Beyond the new yield point, the second loading proceeds along AC, which you will notice is along the stress curve that would be followed by an uninterrupted first loading. At C, a second unloading is shown, and the same process is repeated.

It is to be pointed out that unloading and reloading curves do not exactly overlap. Instead they form a small hysteresis loop, as shown in Fig. 4.8 in an exaggerated manner. There is an energy loss during a cycle represented by the area of the hysteresis loop. This energy, however, is very small.

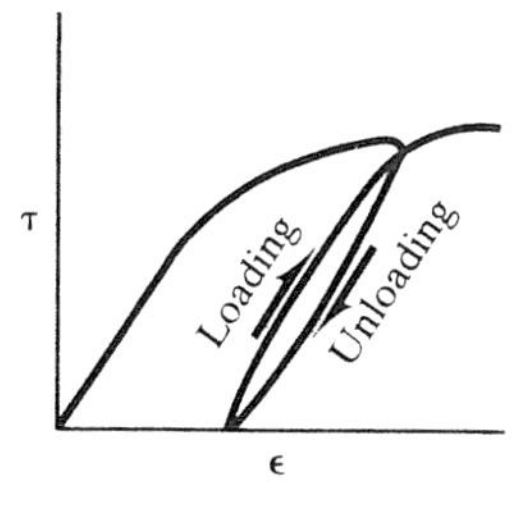

Figure 4.8. Illustration of hysteresis.

Finally, it should be pointed out that the change in yield point by strain hardening is observable only in the direction of initial loading.

That is, there is no increase in yield stress in the material at right angles to the direction of the initial loading.

On the basis of the simple test described earlier and the simple compression test, which is essentially the same except for direction, we can make additional useful classifications which are meaningful in describing mechanical behavior of materials.

First we can form two classes of materials on the basis of the behavior of a specimen in a tensile test carried out to the point where the specimen fractures. Materials exhibiting little or no plastic deformation up to fracture, such as glass, are called *brittle* materials. Materials exhibiting substantial plastic deformation up to the point of breaking, such as mild steel, are called *ductile* materials. For a brittle material, a stress-strain curve carried out in a tensile test will differ from the corresponding curve carried out as a compression test. Furthermore, brittle materials exhibit a considerable scatter in breaking points, found by a series of many tests. Ductile materials, on the other hand, give essentially the same stress-strain curve for a tensile or compression test and have yield points, breaking points, and so on, which are considerably more reproducible in a series of tests.

4.4 Idealized One-Dimensional, Time-Independent, Stress-Strain Laws

It should be apparent by now that stress-strain relations in general are of great complexity with the possibility of many ramifications. To permit analytical treatment of material behavior we employ, at times, idealizations of stress-strain relations.

The most simple stress-strain idealization is of course the rigid-body idealization shown in Fig. 4.9. We have used such a model in rigid-body mechanics courses.

In Fig. 4.10 we show the stress-strain curve for a *linear, elastic* material. This model is the one we shall employ in the major portion of the text. We must not forget that the stress-strain diagram is taken from a simple one-dimensional state of stress, and accordingly in Chapter 6 we shall generalize this model for a general state of stress. The resulting mathematical formulation is called the *three-dimensional Hooke's law*. We shall be able to use these results for the analysis of bodies composed of the usual structural materials, such as steel and aluminum, in cases where the stress has not exceeded the yield stress.[6]

[6]We shall also generalize the criterion for yielding in a general state of stress in Chapter 9.

Figure 4.9. Rigid-body behavior.

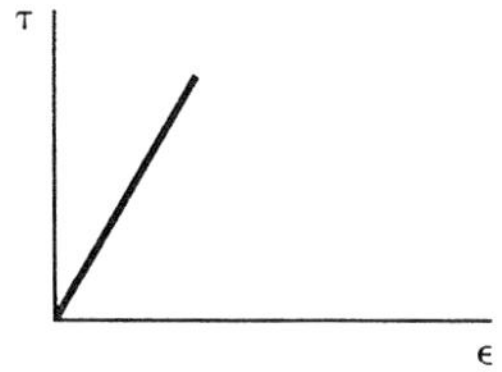

Figure 4.10. Perfectly-elastic behavior.

There are situations where there may be plastic deformations involved which far exceed the elastic deformations present, and it may be profitable to formulate the idealization of a stress-strain diagram shown in Fig. 4.11, which embodies rigid-body behavior up to a certain stress and then exhibits what we shall call *perfectly plastic* behavior. During perfectly plastic behavior there is no change in stress possible and the material "flows" at constant stress. If next we include strain hardening in the plastic range, our model become more accurate albeit more complex. We show in Fig. 4.12 the idealization *rigid, plastic behavior with strain hardening* to illustrate this case.

Figure 4.11. Rigid, perfectly-plastic behavior.

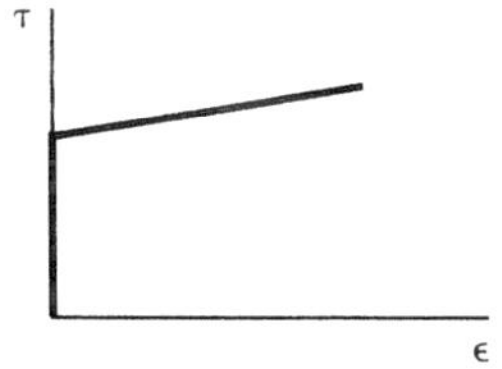

Figure 4.12. Rigid-body, plastic behavior with strain hardening.

There may be times when the elastic deformation cannot be deleted from considerations and where there is little strain hardening. For such cases, one may be able to employ the idealization shown in Fig. 4.13, called the *linearly elastic, perfectly plastic* stress-strain curve. Finally, allowing for strain hardening, we get the curve shown in Fig. 4.14, which is reasonably close to certain actual stress-strain diagrams.

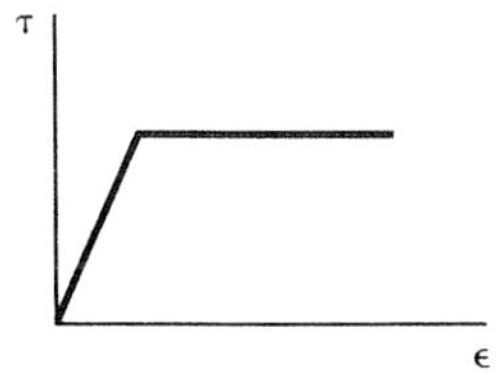

Figure 4.13. Linearly-elastic, perfectly-plastic behavior.

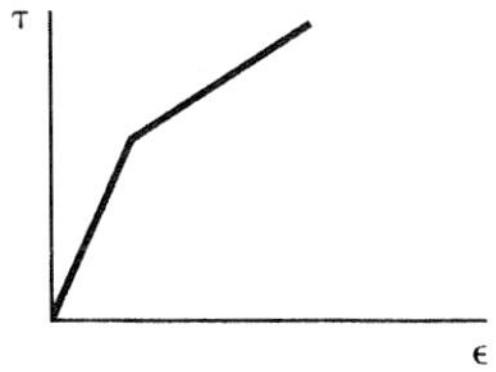

Figure 4.14. Linearly-elastic, plastic behavior with strain hardening.

The two most used idealizations in this text are the perfectly elastic modulus case (Fig. 4.10) and the linearly elastic, perfectly plastic case (Fig. 4.13). The first one needs no explanation but the second one requires some discussion. For materials like mild steel (see Fig. 4.2), there is a region around the yield stress where large strains occur with little increase in stress. The stress-strain curve is close to being horizontal for a range of strain before strain hardening begins to take place. In cases where the strain is expected to be in this range, clearly the linearly elastic, perfectly plastic idealization can be very useful despite the unrealistic notion of a metal "flowing" like chewing gum.

*4.5 A Look Ahead; Viscoelasticity and Creep

There are solids that exhibit both elastic and viscous flow behavior simultaneously. Such materials are said to be **viscoelastic.** Amorphous polymers such as plastics and synthetic rubbers may frequently behave in a viscoelastic manner. In addition, fibrous materials (e.g., silk, rayon, and cellulose), glasses, ceramics, biomaterials (skin and muscle), and nonmetals in general may frequently be considered as viscoelastic materials. Finally, composite materials can be entirely viscoelastic. Clearly, the knowledge of viscoelastic behavior is of growing importance.

As an aid in studying viscoelastic behavior, we make use of simple, familiar mechanical devices whose behavior may be considered analogous to that of the material itself. As a most simple illustration of such a device, consider a linear spring with spring constant K subjected to a force F as shown in Fig. 4.15. It should be clear that the force on the spring and the resulting extension are analogous, respectively, to the stress and the resulting strain in a tensile test specimen. Why so? Force and extension are linearly related for the spring, while stress and strain are linearly related for a Hookean material. And, for a hypothetical, entirely viscous solid, we may say for the corresponding device (a dashpot) that $F = \mu\dot{\delta}$, where μ is the so-called coefficient of viscosity stemming from the dashpot (see Fig. 4.16). Correspondingly, for the material specimen, the dashpot force and the rate of deflection are related to the material stress and the strain rate, respectively. That is, $\tau_{zz} = \eta\dot{\varepsilon}_{zz}$, where η is equivalent to μ. To aid in describing viscoelastic behavior and to help order our thinking, we employ combinations of springs, and dashpots. Thus, in Fig. 4.17(a)

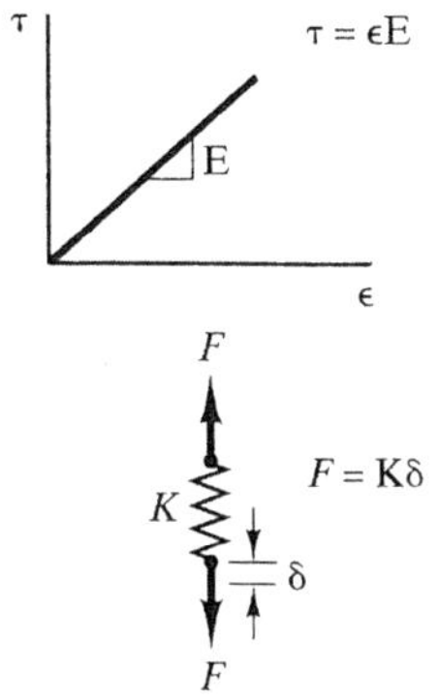

Figure 4.15. Linear elastic model.

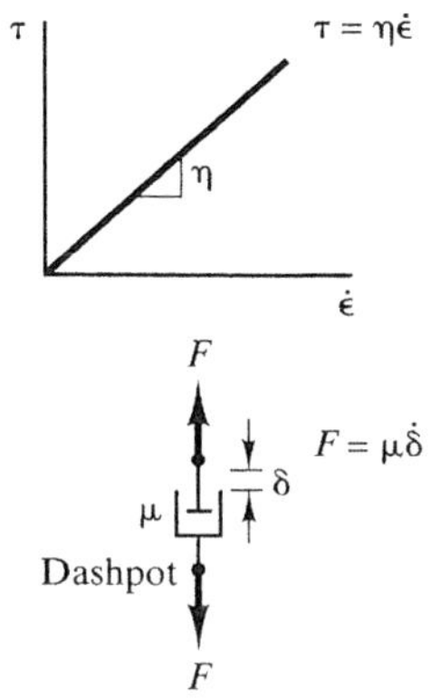

Figure 4.16. Linear viscous model.

we show what is called a *Maxwell* model which consists of a linear spring and a linear dashpot in series. We can say for this model

$$\dot{\delta} = \frac{\dot{F}}{K} + \frac{F}{\mu}$$

For the one-dimensional material specimen represented by this model, we can accordingly say

$$\dot{\varepsilon} = \frac{\dot{\tau}_{zz}}{E} + \frac{\tau_{zz}}{\eta}$$

Another basic model is that of a linear spring and dashpot in parallel as shown in Fig. 4.17(b). This model is called a *Kelvin* model. Other more complex models may be put together with springs and dashpots, or we may assemble Maxwell and Kelvin models within various systems in order to convey certain properties of viscoelastic materials.

If we cannot find suitable models for a particular material, we may perform tests to obtain experimental data for a particular material. For one such test, called a *creep test*, a constant stress level, τ_o, is maintained on a one-dimensional test specimen, while the strain is observed as a function of time. The ratio of $\varepsilon(t)$ over τ_o, is called the *compliance* and is denoted generally as $J(t)$. In another test, we maintain a constant strain, ε_o, on a one-dimensional specimen of the material and observe what we call the *relaxation modulus*, $\tau_t(t)/\varepsilon_o$, often denoted as $Y(t)$. The test is called the *relaxation test*.

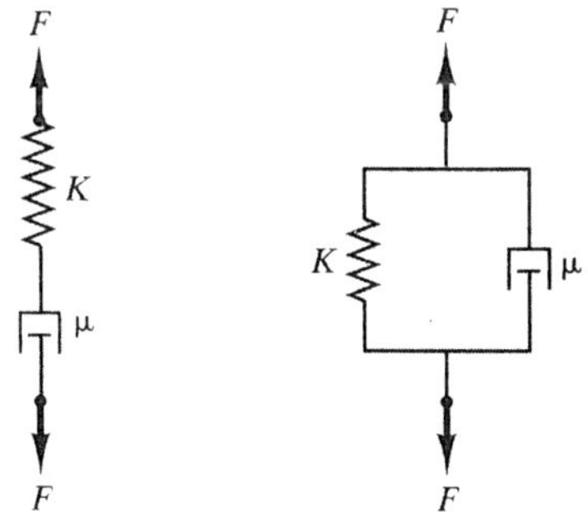

Figure 4.17. Viscoelastic models.

An example of viscoelastic behavior is the phenomenon called **creep.** For many important materials, there will first take place on loading an immediate response predicted by a stress-strain diagram. However, if the load is maintained, let us say for simplicity, at a constant level, there will take place a very small strain rate which, over a long enough time period, can significantly change the geometry of the specimen so as to result in a failure of the material. Creep increases with temperature. Accordingly, for such structures as boilers, reactor shells, and gas turbines, where elevated temperatures prevail and long life under stress must be achieved, creep must be carefully considered in design.

4.6 Fatigue

Now we shall consider a phenomenon called *fatigue,* which depends primarily on *variation of stress.* Experience indicates that many repetitions of load can cause fracture in a material even though the maximum stress is far below the ultimate stress. Such a failure is called *fatigue failure.* Physically, what takes place is that a minute crack originates at a point of high stress. This point may be at a surface scratch or at some imperfection in the material. The crack enlarges during the continued repetitive loadings until there is insufficient material left to support the load. The importance of fatigue failure was dramatized by the series of tragic accidents experienced by the early British Comet jet airplanes.

One of the simple ways to study the fatigue phenomenon is by employing a modification of the simple tensile test apparatus, as shown diagrammatically in Fig. 4.18. Clearly, a sinusoidal load at amplitude

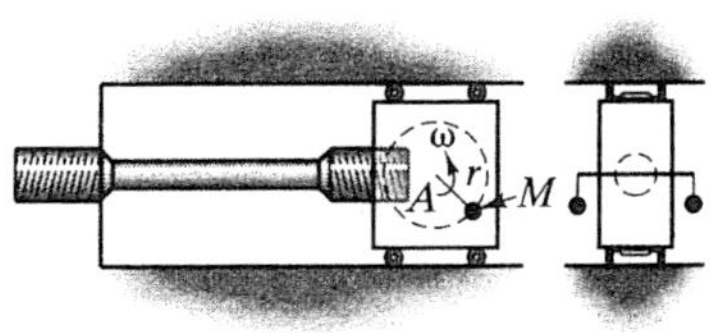

Figure 4.18. Fatigue test apparatus.

equal to $Mr\omega^2$ is developed on the cylindrical test specimen by this apparatus. We have plotted the stress amplitude from such a load in Fig. 4.19(a), and you should note that the mean stress is zero. If we plot

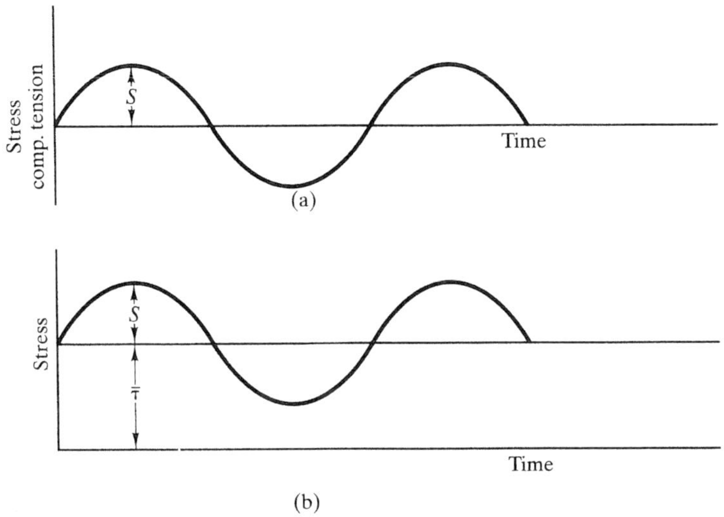

Figure 4.19. Sinusoidal stress on a specimen: (a) zero mean; (b) non-zero mean.

the number of cycles N that is required by such a test to achieve fatigue failure versus the amplitude of the sinusoidal stress variation S, we in general get a curve which for many materials is similar to the one shown in Fig. 4.20(a). This is the well-known S–N curve. Notice the amount of scatter of the experimental points about the average curve indicated. This shows that the number of cycles for a fatigue failure at a particular alternating stress depends to an appreciable extent on local conditions, such as minor surface irregularities and imperfections in the material. Nevertheless, the S–N curve gives us information for a simple alternating tensile stress on the probable number of cycles N to which the material can be subject for a given stress amplitude S before fatigue failure is to occur. This value of N for a given value S is called the *fatigue life.* Also, for a given N, the corresponding value of S is called the *fatigue strength.* Thus, for a given stress amplitude, a number of one-dimensional loading repetitions below the corresponding fatigue life N of the material will have a high probability of not causing a fatigue failure in the material. On the other hand, with a stress amplitude below the fatigue strength S, the material will probably undergo N cycles of loading (where N corresponds to S) without fatigue failure.

For some materials (e.g., ferrous metals and alloys) the S–N curve approaches a horizontal line as N gets very large, as shown in Fig. 4.20(b). The use of a log-log grid brings out the location of the "knee"

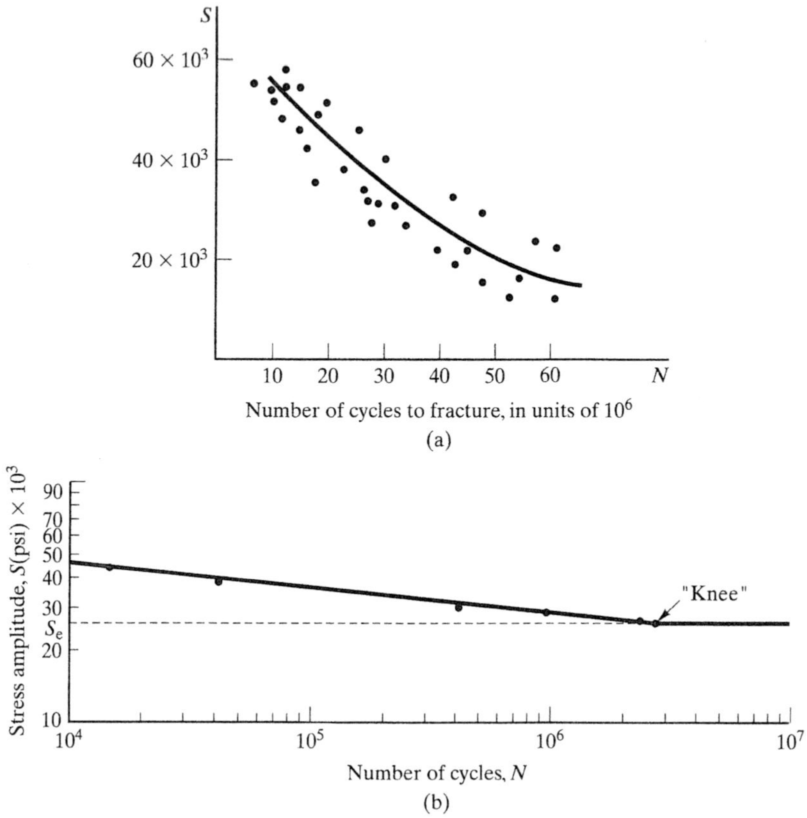

Figure 4.20. *S–N* curves: (a) general behavior; (b) material with an endurance limit S_e.

in the curve, which might not be apparent if a uniform grid were used. This limiting strength is called the *fatigue* or *endurance limit* S_e, which is thus the stress below which the material has a high probability of not failing in fatigue no matter how many cycles of stress are imposed on the specimen. For nonferrous metals and alloys the *S–N* curve usually does *not* approach a horizontal limit, and hence such materials do not have an endurance limit.

The average stress for the tests described up to now has been zero. Suppose that we consider next a sinusoidal stress superposed on a static stress such as it shown in Fig. 4.19(b). We have indicated the static or mean stress as $\bar{\tau}$ and the amplitude of the harmonic stress variation as S. Experiment indicates that increasing the mean stress $\bar{\tau}$ tends to depress the *S–N* curve, so that there is greater susceptibility to fatigue failure. That is, for a given alternating stress of amplitude S, there will be a smaller fatigue life, or for a given number of cycles N, there is a smaller fatigue strength. We have shown a plot of S versus N for different values of $\bar{\tau}$ in Fig. 4.21 where $\bar{\tau}_1 < \bar{\tau}_2 < \bar{\tau}_3$.

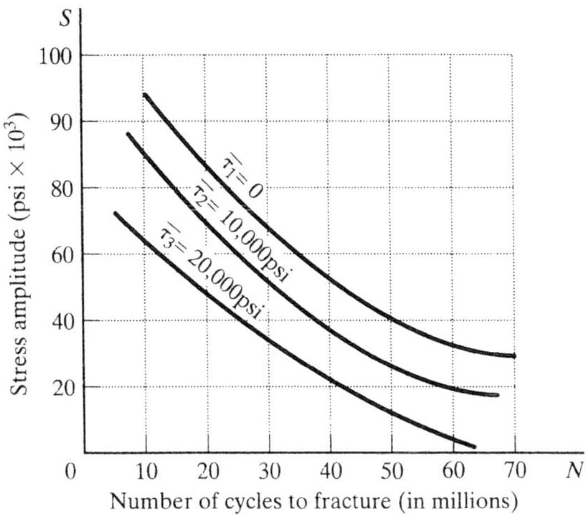

Figure 4.21. Fatigue curves with alternating stress superposed on uniform stress.

For a given value of fatigue life (i.e., N) we may plot S for each value of mean stress $\bar{\tau}$, and some typical data points are shown in Fig. 4.22. For design purposes it is convenient to pass some simple curve (not shown) through these points. Clearly, at the intercept of any such curve with the abscissa, S is equal to zero and we have the case of failure due to $\bar{\tau}$ reaching the ultimate stress U. On the other hand, the intercept at the ordinate represents the fatigue strength for the case where we have a zero mean stress (i.e., the very first case of simple sinusoidal loading discussed in this section). Knowing, then, two intercepts from a simple tension test and an alternating-stress fatigue test with zero mean stress, we can next draw a straight line (shown as a dashed line in the figure) and have a simple and reasonably good representation of the data points. This approximation is called the *modified Goodman line*, and as shown in Fig. 4.22 it gives results on the safe side. Knowing the average stress and choosing the modified Goodman line corresponding to the number of cycles of simple unidimensional normal stress that the specimen is expected to be subjected to in its lifetime as part of a machine, we can estimate the alternating stress S for which the element will probably not fail in fatigue during its lifetime of use. By restricting ourselves below this stress S, we increase the probability that the element will not fail by fatigue during its expected lifetime.

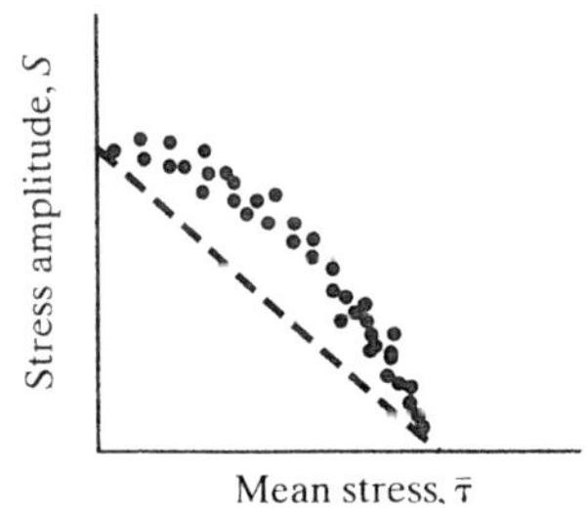

Figure 4.22. Modified Goodman diagram.

In physical problems, many devices are subject in certain domains to alternating stresses not dissimilar from that which we have been discussing in our modified tensile test. For example, rotating shafts are subject during a rotation to a cycle of stress essentially similar to a cycle shown in Fig. 4.19(b) for each rotation of the shaft. This is also true for vibrating turbine blades and propellers. It is to be pointed out that a common means of running fatigue tests for simple normal stress reversal is to observe a vibrating beam or a rotating rod. We may make good use of experimental data of the type discussed in the proper design of machines.

It should be pointed out that we have presented data for very simple normal stress reversal by having used the very simple geometry of a standard cylindrical test specimen, which has a particular size, surface finish, and temperature. The actual endurance limit S_e of a machine element may be considerably smaller than the endurance limit obtained by the standard specimen. The difference may be explained by the presence of a variety of modifying factors in the actual working condition, which are not present under experimental conditions. For example, if the test cylinder has an irregularity such as the small notch, we will have a lower fatigue strength than expected for a uniform specimen. This occurs because the actual stress at the notch is quite complicated and has a maximum normal stress that may considerably exceed the average normal stress computed for the narrowest section as F/A. Such a region of high stress is called a *stress concentration* (discussed in the next section).

Furthermore, we must understand that the foregoing procedure can be expected to give only approximate results when we extend simple one-dimensional stress results to a case of general three-dimensional stress. Actually, one should more properly relate fatigue effects in the general case to a certain combination of stresses at a point called the *octahedral shear stress.* We shall examine this in Chapter 9 when we consider general yield criteria. For now, we shall merely point out that designers should be careful of small fillets, holes, and so on. These innocent-looking elements can give rise to disastrous fatigue failures if not carefully considered from that viewpoint.

Finally, the reader should be aware that this section was devoted to the so-called simple classical method of design against fatigue failure. In the past two decades a more sophisticated method of design against fatigue and fracture has evolved which is based on *fracture mechanics.* Very briefly, the fracture mechanics approach begins with the assumption that all raw materials contain microscopic cracks. Then, by using the stress concentration factor K for a particular microscopic crack shape along with a suitable law for the rate of crack propagation, the designer is able to estimate the fatigue life of a component. For detailed discussions on this subject, the reader is referred to the literature on engineering fracture mechanics.

4.7 Stress Concentration

The one-dimensional members we have been discussing quite often have cut into them for various purposes small holes, notches, keyways, and so on. When this is the case, the stress distribution no longer remains a simple one-dimensional stress distribution in the vicinity of these cuts. There does occur a considerable increase in the maximum stress at these locations over that which would obtain were there no such cut present. We say that we have a *stress concentration* or a *stress raiser* present. It is difficult to determine analytically the exact nature of the stress distribution in the domain of a stress concentration. However, as a result of experimental methods (such as photoelasticity and strain gauge techniques) as well as some sophisticated theoretical and numerical work, we have available, for many kinds of cuts, *concentration factors K* which, when multiplied by a simply computed maximum stress found from one-dimensional analysis of the cut member, gives the actual maximum stress. The simply computed, one-dimensional maximum stress thus disregards the actual complex stress distribution around the cut which gives rise to the stress raiser; the factor K accordingly takes this stress raiser into account. For our one-dimensional cylindrical members the simply computed maximum stress is F/A_{min}, where A_{min} is the smallest cross-sectional area. The actual maximum stress is then given as

$$\tau_{max} = K\frac{F}{A_{min}} \tag{4.6}$$

Concentration factors are to be found in engineering handbooks.[7] However, we show in Fig. 4.23 the case of a tensile member with a simple

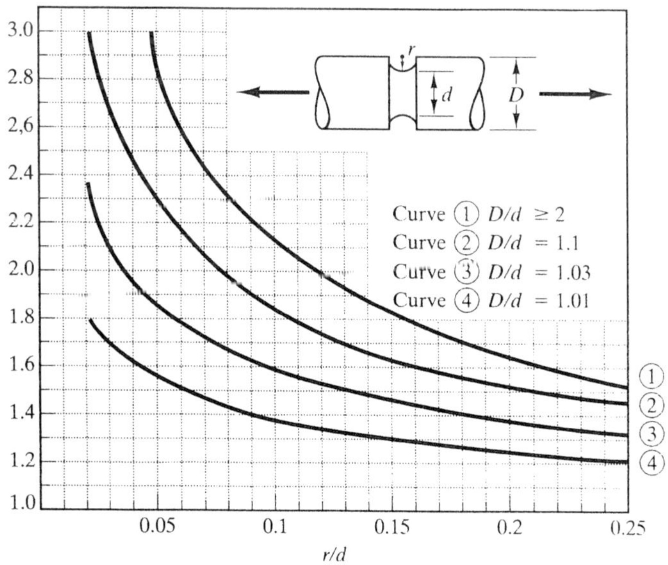

Figure 4.23. Stress concentration factor curves.

[7]See C. Lipson and R. C. Juvinall, *Handbook of Stress and Strength* (New York, 1963). Also, for a greater discussion of stress concentration, see R. C. Juvinall, *Engineering Considerations of Stress, Strain, and Strength* (McGraw-Hill, New York, 1967), Chap. 13.

notch. Notice in Fig. 4.23 that for small fillets ($r < 1$) we can get a higher than threefold concentration factor.

We wish to point out at this time that, for the question of yielding in a ductile material, stress concentrations of the kind discussed are not significant. When the stresses get inordinately high in these regions local yielding mainly at the skin surface takes place with little effect on the load-carrying capacity of the member as a whole. However, the opposite is true for *brittle materials* or when *fatigue* is a problem.[8] Here *the stress concentration must be fully taken into account to prevent failure.*

We shall have more to say about stress concentration when we discuss bending and torsion.

4.8 One-Dimensional Thermal Stress

Our considerations thus far have been based on the assumption that the body is of uniform temperature throughout (i.e., the so-called *isothermal case*). The strains induced for such a case result purely from applied external loads. We shall now release this restriction and consider first the case where there may be a nonuniform temperature field in an elastic body. However, we shall assume that the temperature variation is not great enough to cause the elastic moduli of the material to vary appreciably throughout the body.

An unconstrained, isotropic solid element will dilate (or shrink) uniformly in all directions when there is a change in temperature. Thus, a sphere will remain a sphere but will undergo a change in radius; a cubic element at any orientation at a point remains a cubic element but will undergo a change in length of edge. This means that there will be *equal normal strain in all directions but no shear strain* for the unconstrained element whose temperature has been changed. If the element is completely confined so that no change in shape or size is permitted as the temperature is changed, there is no strain whatever. But since the element tries to change length uniformly in all directions, there will be a state of stress like that found in a static fluid wherein the normal stress in all directions has the same value and where the shear stress for all interfaces is zero. This is a *hydrostatic stress.* On the other hand, if only a *partial expansion* of the element is permitted, there can be some *general state* of strain and some *general state* of stress in the element dependent on the nature of the constraint. In an isotropic body subject to a nonuniform temperature distribution, the elements attempt to undergo dilation (or shrinkage) as a result of the changes in temperature from

[8]For a discussion of fatigue in the presence of a stress concentration, see R. C. Juvinall, *Engineering Considerations of Stress, Strain, and Strength* (McGraw-Hill, New York, 1967), Chap. 14.

some initially uniform temperature. However, the elements cannot dilate (or shrink) in an unrestricted manner. Since the body must remain continuous during the change in the temperature, there will be partial constraint internally even if the body externally is unconstrained as to change in geometry.[9] We may then introduce in this way a general stress field in the body. This stress field is called *thermal stress.* Our high-speed aircraft and space vehicles are subject to considerable thermal stress from aerodynamic heating on the outside surfaces and from the heat originating in their propulsion systems. In addition, electronic systems generate considerable heat during operation. We can see from these examples that the study of these thermally induced stresses is of vital importance in present-day technology.

In addition to thermal stress from a nonuniform temperature distribution on a body, there may be a thermal stress distribution developed when a body changes temperature uniformly but is confined in some way from changes in volume by external constraints.

How do we measure the strain developed in a body when it is subject to a temperature field as well as to a system of loads? Most simply, we superpose the strain associated with free dilatation or shrinkage of the element with the strain associated with the *total actual state of stress* of the element. This stress includes the thermally induced stress as well as stress caused by the external loads. Thus, we can imagine that an element is first allowed to dilate (or shrink) freely, giving a strain ε'. The restraints resulting from nonuniform expansion and external constraints then lead to a thermally induced stress which causes a strain ε''. With no external loads, the net strain from these separate actions is the actual strain. Thus, for a completely confined element, mentioned earlier, the strain ε' and the strain ε'' from the aforementioned actions give a zero total strain when superposed. If there are external loads, there will be additional strain ε''', which we superpose. Thus, for the strain at a point we have

$$\varepsilon = \underset{\substack{\text{(from free}\\ \text{expansion)}}}{\varepsilon'} + \underset{\substack{\text{(from thermally}\\ \text{induced stress)}}}{\varepsilon''} + \underset{\substack{\text{(from stress due}\\ \text{to external load)}}}{\varepsilon'''}$$

We compute ε'' and ε''' in terms of stress using an appropriate stress-strain law. As for ε' we employ the equation for thermal expansion, which says that, for temperature changes $\Delta T(x,y,z)$, the change of length of a vanishingly small line segment L is

$$\Delta L = \alpha L \Delta T \tag{4.7}$$

[9]If this is not clear, imagine the body at a uniform initial temperature to be composed of elemental equal cubes. If the temperature is now raised nonuniformly, the cubes expand differently. If no voids are allowed in the material, it should be clear that the cubes will interact with each other, creating mutual constraint.

where α is the so-called *coefficient of expansion.* Like most material properties, this coefficient is dependent on temperature but is usually taken as constant for reasonably small temperature changes. The strain ε' in the direction of L is then given as

$$\varepsilon' = \lim_{\Delta L \to 0} \frac{\Delta L}{L} = \alpha \Delta T \tag{4.8}$$

For a *one-dimensional state of stress* in the z direction, we can then say for a Hookean material that

$$\varepsilon_{zz} = \alpha \Delta T + \frac{1}{E} \tau_{zz} \tag{4.9}$$

where τ_{zz} is the actual stress resulting from both thermal effects and loading.

In our discussion thus far, we have implied that the computation of the temperature field in a solid can be carried out separately from that of the stress distribution. It can be shown[10] that for most practical problems the effect of stress and strain on temperature is very small and can be neglected. The computation of the temperature field is then a heat transfer problem involving the first and second laws of thermodynamics, as well as such laws as the Fourier conduction law.

4.9 Closure

We have begun in this chapter an introductory study of the mechanical properties of solids. This is an important area of the engineering sciences, and you will undoubtedly pursue such studies both from macroscopic and microscopic viewpoints in later courses. Although we have restricted our attention to one-dimensional test results, you will later see (Chapter 6) that many of the measurements and definitions carry over to more general states of stress and strain.

In the following chapter we shall continue to work with one-dimensional stress problems. These problems are of interest themselves and deserve a careful study. More than that, they give us an opportunity of considering the basic laws of solid mechanics while the computations are still reasonably simple. The overview that this study will afford will be invaluable when, in later chapters, we get into the more complex studies of beams and shafts.

[10]B. Boley and J. Wiener, *Theory of Thermal Stress* (New York: Wiley, 1960), Chap. 2.

4.10 A Look Back

In the next chapter, we shall begin solving stress problems starting with one-dimensional stress problems and going on later to more complex cases. We of course shall be making free-body diagrams. It is important to know how many independent equations of equilibrium are available for each free body. You should recall from your statics course that the following rules give the maximum number of independent scalar equations of equilibrium.

a. For a three-dimensional problem, **six scalar equations.**
b. For a coplanar problem, **three scalar equations.**
c. For a problem with a general parallel force system, **three scalar equations.**
d. For a parallel, coplanar problem, **two scalar equations.**

Also, for a body acted on by forces at two points only (a two-point loaded system), recall from statics that equilibrium requires that the external forces be colinear and equal and opposite. The forces must be colinear with the connecting line between the two points of loading.

*4.11 A Look Ahead; Composite Materials

Composite materials compose a broad group of engineering materials that have applications in fields ranging from aerospace to electronics, from biomedical to construction, to name only a few. Their unique feature is that they may be custom-made to suit a particular application. That is, their material properties may be designed to give a balance of attributes required for a given application. For example, a portion of an airplane fuselage may need to have different strength and stiffness characteristics in different directions. Through the use of a custom-made composite material, these characteristics may be tailored to meet the requirements while minimizing weight and cost.

A good working definition, although by no means all-encompassing, of a composite material is given by Schwartz[11]:

> A composite material is a material brought about by combining materials differing in composition or form on a macroscale for the purpose of obtaining specific characteristics and properties. The constituents retain their identity such that they can be physically identified and they exhibit an interface between one another.

Many structural composite materials are composed of two main ingredients—a "strong" component and a "matrix" component that surrounds and binds the strong part. The main purpose of the matrix is to hold the strong component in place and provide a means for distributing the load. In addition, the matrix may also provide a certain degree of environmental protection for the strong part. Other composites may be developed not for their strength but for other desirable properties, such as electrical or thermal properties.

Composite materials can be categorized in many different ways: by matrix, by strong part, by performance, by application, and so on. Some common categories include fiber composites, flake and particulate composites, and layered composites. In fiber composites, high-strength fibers, often made from glass, are embedded in a resin matrix. An example of a fiber composite is Kelvar[12]—a popular high-performance fiber composite used in tires, sports equipment, and architectural fabrics, to name only a few applications. Flake and particulate composites are made by combining flake or particulate materials into a matrix. Composites of this type are often developed to enhance certain material characteristics such as to help match the coefficient of thermal expansion of an adhesive with its bonding materials. Another category are the so-called layered composites. These are typically made from sheets of materials combined so as to maximize the composite strength properties.

In your study of composite materials, you will extend your knowledge beyond the realm of isotropic material response. However, the foundations for studying this growing field are firmly embedded in the concepts presented in this chapter.

[11]M. Schwartz, *Composite Materials Handbook*, 2nd ed. McGraw-Hill, 1992.

[12]Trademark of E. I. du Pont de Nemours & Co.

Highlights (4)

The *tensile test* involving only one nonzero stress along the axis of the test rod is a standard method for obtaining widely used *macroscopic* physical properties. In particular, the tensile test of a steel specimen permits the establishment of a number of important definitions that often have use for other kinds of structural materials. Thus, we have:

1. The *actual stress,* $(\tau_{zz})_{\text{Act}}$. This is the tensile load divided by the actual cross-sectional area of the specimen.
2. The *engineering stress,* $(\tau_{zz})_{\text{Eng}}$. This is the tensile load divided by the initial cross-sectional area of the specimen.
3. The *proportional limit.* The stress above which stress is no longer proportional to strain.
4. The *elastic limit.* The stress above which the specimen no longer returns to its original geometry when the load is removed.
5. The *Poisson ratio,* ν. The fraction of strain lateral to the axis of load, ε_{lat}, such that $\varepsilon_{\text{lat}} = -\nu\varepsilon_{zz}$. The ratio is generally defined only in the linear part of the stress-strain diagram.
6. The *Young's modulus E.* The ratio of stress to strain for the linear portion of the stress-strain diagram.
7. The *ultimate stress.* The very maximum engineering stress attainable in the tensile test.
8. The *yield stress.* The stress that results in a permanent strain of .002 when completely unloading the specimen. This stress is used because both the stresses corresponding to the proportional limit and elastic limit are difficult to measure.

In this text, we use the *perfectly elastic model* and the *elastic, perfectly plastic model* (see Fig. 4.13) as idealizations of material behavior. The latter is useful for materials such as steel where, once the yield stress is reached, there is a notable increase in strain for very small increases in stress. At some point, however, the stress begins to increase noticeably for further increase in strain. The latter is called *strain hardening.* The elastic, perfectly plastic model is useful for strains occurring after the yield stress has been exceeded and before we begin to have appreciable strain hardening.

It is important to remember that when the geometry is made more complex by a keyway or by a sudden change in diameter or some other sudden change of geometry, there will be a maximum stress in the region of the sudden change of geometry. In fact, there may be double, triple, or even higher maximum stress over that which one would obtain by simplified calculations. This is called a *stress concentration* or a *stress raiser.* One may often find a stress concentration factor from handbooks which, when multiplied by the simplified stress computation, will then give the maximum stress. Remember that stress concentrations must be fully taken into account in the design involving *brittle* materials and when there is the possibility of *fatigue,* the latter occurring when there are many variations of stress at a point during the longtime operation of a device.

PROBLEMS

The Tensile Test	4.1–4.7
Idealized Stress-Strain Laws	4.8–4.16
Fatigue and Stress Concentration	4.17–4.19
Thermal Stress	4.21–4.24
Computer Problems	4.25–4.27
Programming Project 4	

4.1. [4.2] A tensile specimen has a diameter of 1/2 in. The increase in length recorded by the longitudinal gauge for a load of 1000 lb is .001 in. for an original length L of 2 in. The decrease in width as measured by the lateral gauge is .00005 in. What is the engineering stress? Compute the modulus of elasticity. Compute Poisson's ratio.

4.2. [4.2] In Fig. P.4.2 is shown a hypothetical stress-strain curve. What is the proportional limit? The ultimate stress? The modulus of elasticity for this material?

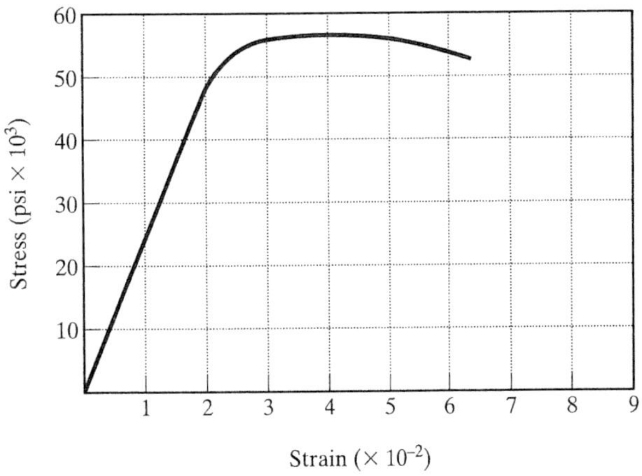

Figure P.4.2.

4.3. [4.2] A rope of length 3 m extends 25 mm when a force of 2250 N of tension is exerted on it. If the diameter when loaded is 25 mm, what is the spring constant for the rope and the modulus of elasticity assuming linear elastic behavior?

***4.4. [4.2]** Consider a nylon cord with a diameter of 1/4 in. and a length of 5 ft. If this material has a modulus of elasticity of 3×10^5 psi and a Poisson ratio of .4, what is the stress and the diameter when pulled by a force of 500 lb? (*Hint:* Since the strain is large, the diameter varies appreciably with stress and a trial-and-error calculation will be needed.) Compute the average spring constant for the range of load from 0 to 500 lb.

4.5. [4.2] A "mechanical fuse" is sometimes used to limit the load transferred between two parts of a device. Such a fuse is shown in Fig. P.4.5. What is the maximum force that it will allow to transfer if the ultimate stress for the fuse material is 36,000 psi.

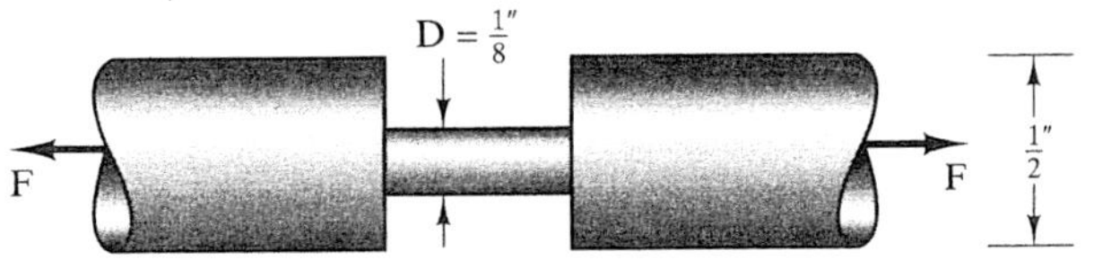

Figure P.4.5.

4.6. [4.2.] A circle of radius r is drawn so it is centered on a plate loaded in uniform tension with stress τ_o (see Fig. P.4.6.). If the circle is distorted due to the loading as shown, compute E and ν from the loading and geometry.

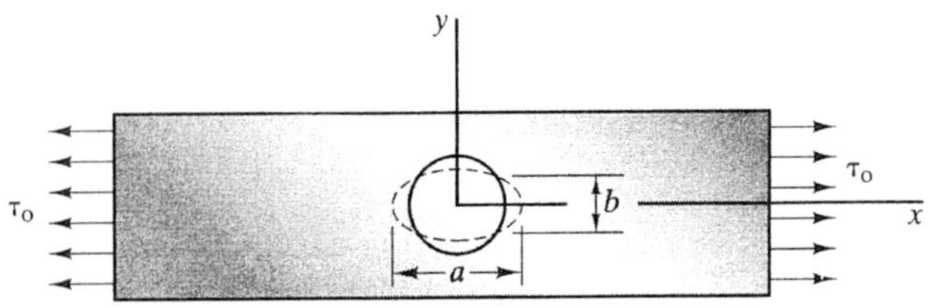

Figure P.4.6.

4.7. [4.2] A square outline is drawn centered on a plate loaded in uniform tension with stress τ_0 (see Fig. P.4.7). If the sides of the square deform such that the loaded shape is that of a rhombus of dimensions a and b as shown, determine the modulus of elasticity and Poisson's ratio.

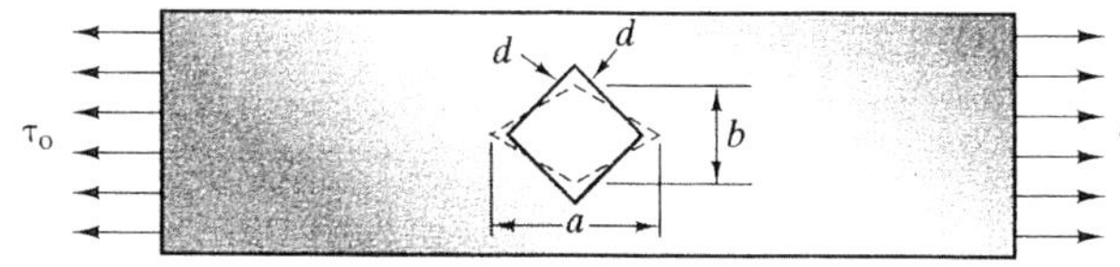

Figure P.4.7.

4.8. [4.3] If in the tensile test corresponding to the graph in Fig. P.4.2., the material is unloaded from a stress of 55,000 psi, what is the elastic recovery of strain? The permanent set? The proportional limit of the material when it is reloaded?

4.9. [4.4] Shown in Fig. P.4.9. are (1) rigid, perfectly plastic, (2) elastic, perfectly plastic, and (3) elastic, plastic with strain hardening stress-strain idealizations. What stress is needed in each case to have a strain of .001? What is the stress in each case needed for a strain of .004?

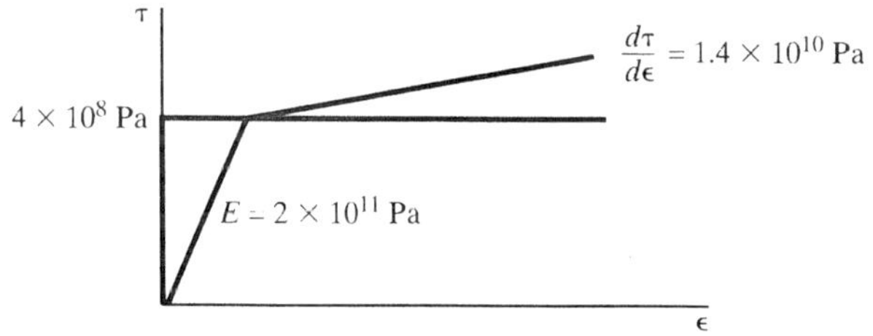

Figure P.4.9.

4.10. [4.4] In Problem 4.9, what is the permanent set in each case if the specimen is unloaded from a strain of .004?

4.11. [4.4] Cylinder A in Fig. P.4.11 may be considered an elastic, perfectly plastic material with a yield stress of 50,000 psi and an elastic modulus of 20×10^6 psi. Cylinder B, welded to cylinder A, is also to be considered elastic, perfectly plastic with a yield stress of 100,000 psi and an elastic modulus of 30×10^6 psi. What is the maximum force that can be imposed without causing permanent set? What is the maximum movement of end D without permanent set? What force is needed to double this deflection? Neglect the weight of the members.

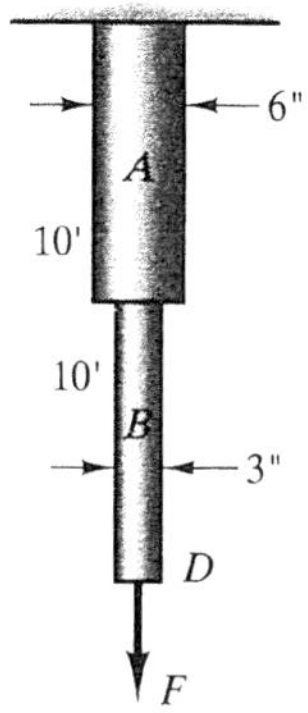

Figure P.4.11.

4.12. [4.4] If a force of 3.6×10^5 N is applied to the cylinder shown in Fig. P. 4.12(a) having a cross-sectional area of 6.25 $\times$ 10^{-4} m^2, what is the deflection of the end B as a result of this loading? The stress-strain diagram for the material is shown in Fig. P.4.12(b).

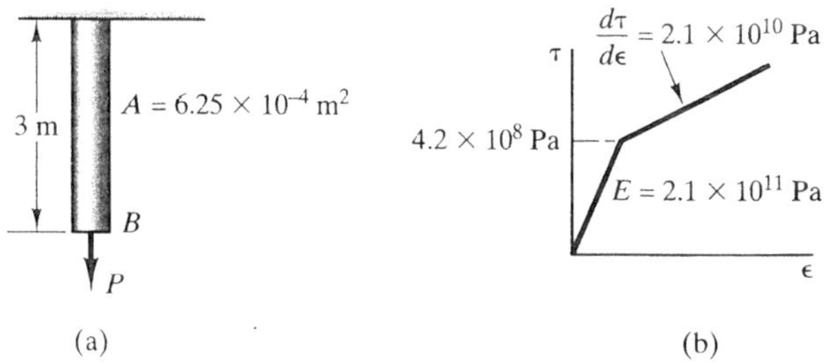

Figure P.4.12.

4.13. [4.4] A cylinder having a cross-sectional area of 1 in.^{2} is welded to a shaft which rotates with an angular speed ω of 10,000 rpm (see Fig. P.4.13). If the cylinder has a stress-strain law as given by Fig. P.4.12(b) and if it has a density of 460 lbm/ft^3, what is the total elongation of the cylinder caused by the centrifugal force developed by the rotation?

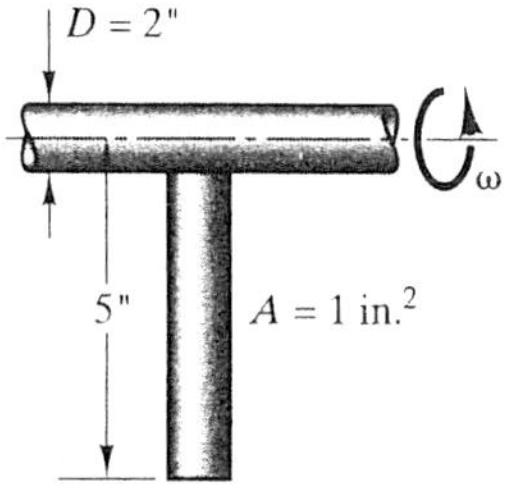

Figure P.4.13.

4.14. [4.4] An aluminum rod A and a steel tube B support a load P (see Fig. P.4.14). The following data apply to A and B:

A	B
$E = 1.5 \times 10^{11}$ Pa	$E = 2 \times 10^{11}$ Pa
$Y = 4 \times 10^{8}$ Pa	$Y = 6 \times 10^{8}$ Pa
$\left(\dfrac{d\tau}{d\varepsilon}\right)_{plas} = 1.5 \times 10^{10}$ Pa	$\left(\dfrac{d\tau}{d\varepsilon}\right)_{plas} = 1.9 \times 10^{10}$ Pa

(a) What is the maximum load for elastic behavior throughout? (b) What is the total deflection due to a load which is 1.5 times that of the load from part (a)?

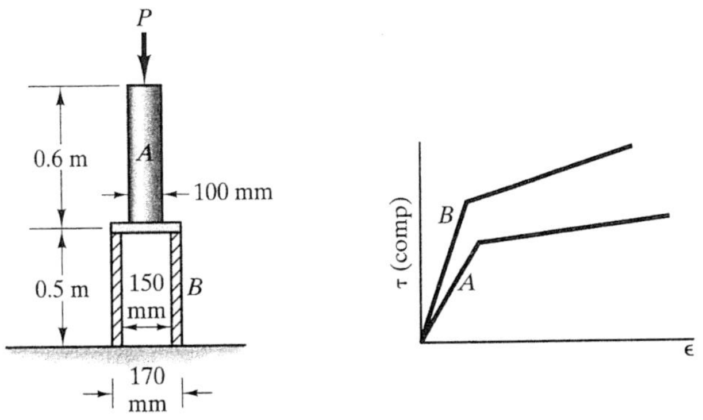

Figure P.4.14.

4.15. [4.4] A rod of diameter .02 m rotates (see Fig. P.4.15) about B at a speed ω rpm. It has a mass per unit length of 5 kg/m. It has a yield stress of 4×10^8 Pa. At the end is a 10-kg rigid mass A. What is the maximum angular speed for elastic elongation of the rod? What is this elongation? Take $E = 2 \times 10^{11}$ Pa. The system rotates on a horizontal plane sliding on a frictionless surface. Approximate action of A as that of a particle at the mass center of A.

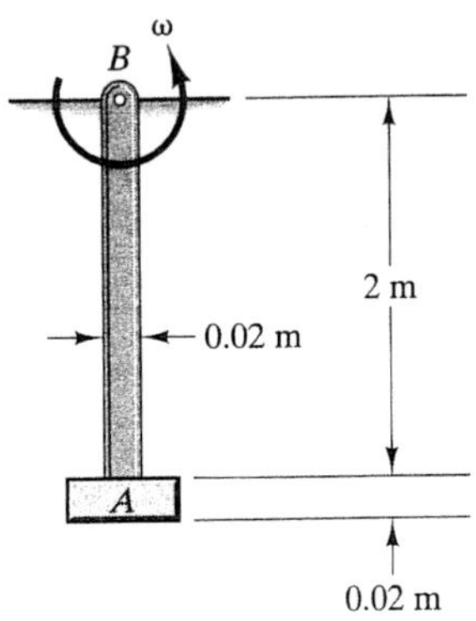

Figure P.4.15.

4.16. [4.4] In Problem 4.15 the rod has a stress-strain behavior which is elastic with strain hardening beyond the yield stress so that $(d\tau/d\varepsilon)_{plas} = 1.6 \times 10^{10}$ Pa. The maximum speed ω for elastic behavior is 64.6 rad/sec. What is the maximum stress if the speed ω is brought up to 70 rad/sec? At what position does yielding start? What is the maximum strain?

4.17. [4.6] What is the fatigue life of a material subject to an alternating stress with an amplitude of 20,000 psi? [Use Fig. 4.20(a).] What is the fatigue strength for this material when it is subject to 40 million cycles of stress?

4.18. [4.6] In Fig. 4.18, we have the following data:

$$\omega = 1000 \text{ rad/sec}$$
$$r = 50 \text{ mm}$$
$$M = .5 \text{ kg}$$

The diameter of the test specimen is 12 mm. If we do not consider the inertial effects of the nonrotating part of the apparatus, how long would you expect it to take for a fatigue failure, assuming the material behaves according to Fig. 4.20(a)?

4.19. [4.6] In Problem 4.18, assume that there is superposed a constant tensile force of 7798 N. If the test specimen behaves according to Fig. 4.21, what is the fatigue life for a fatigue failure? For Problem 4.18, S = 32,058 psi.

4.20. [4.7] A brittle one-dimensional member of diameter 50 mm has a notch of radius r = 2.27 mm giving an inner diameter of 45.45 mm. What is the maximum normal stress if the material behaves as shown in Fig. 4.23? The axial force on the specimen is 2250 N.

4.21. [4.8] Two materials A and B are welded together as shown in Fig. P.4.21. Initially, the temperature is uniform at 60°F. If the temperature is then raised to 100°, what are the strains in the members and what is the new length of the system? If now an axial tensile force of 10,000 lb is added to the system, what is the strain in each member and the change in length? Take the following data:

$$\alpha_A = 6.5 \times 10^{-6}/°F \qquad \alpha_B = 10 \times 10^{-6}/°F$$
$$E_A = 30 \times 10^6 \text{ psi} \qquad E_B = 15 \times 10^6 \text{ psi}$$

Neglect the weight of the members.

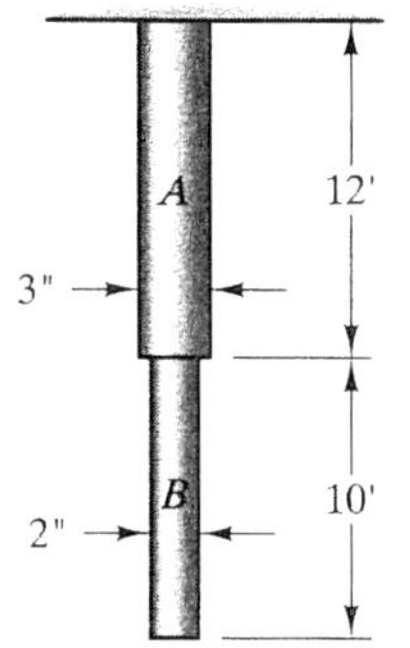

Figure P.4.21.

4.22. [4.8] In the member shown in Fig. P.4.22 the temperature varies from 60°C to 80°C as the square of the distance z after being initially at 50°C uniformly throughout. What is the thermal strain at $z = 1$ m, and what is the change in length of the whole member? Take $\alpha = 14 \times 10^{-6}/°C$.

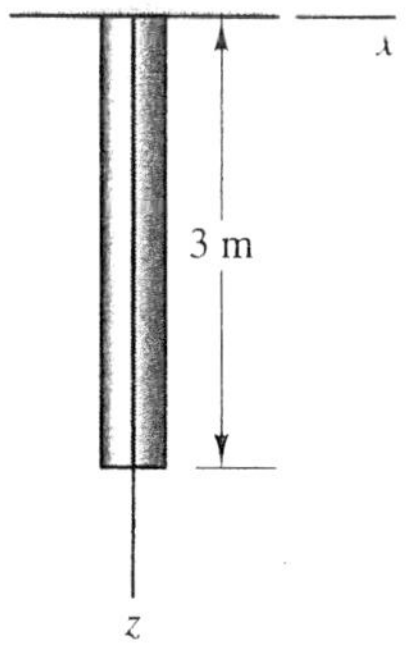

Figure P.4.22.

4.23. [4.8] Shown in Fig. P.4.23(a) is a split pulley held together by four rivets [see Fig. P.4.23(b)]. These rivets are heated to a high temperature and then installed in the slots. Finally a "head" is pounded at the end so that in the expanded condition the rivets just fit snug in the slots as shown. They are then allowed to cool to room temperature. If we assume that the pulley is rigid, what force holds the pulley together for the following data?

Initial temperature of rivet = 300°F
Final temperature of rivet = 60°F
Average α for interval = $6 \times 10^{-6}/°F$
Average E for interval = 20×10^6 psi
Average diameter for interval = 1/2 in.

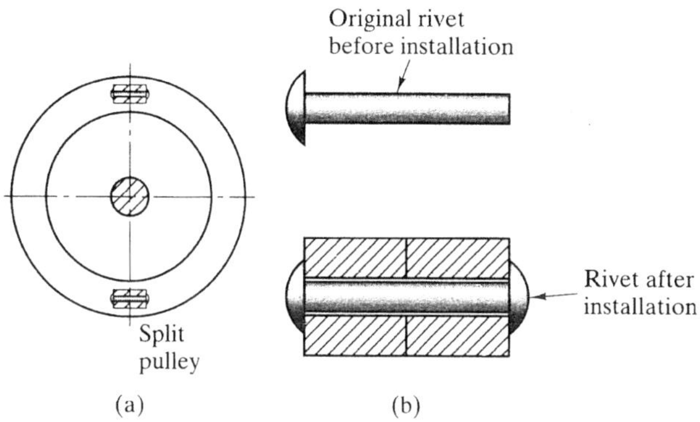

Figure P.4.23.

4.24. [4.8] A rod held between two walls (Fig. P.4.24) is heated from 30°C to 120°C. What is the stress in the rod if $\alpha = 18 \times 10^{-6}/°C$ and the walls move apart a distance of .3 mm? Take $E = 1.4 \times 10^{11}$ Pa.

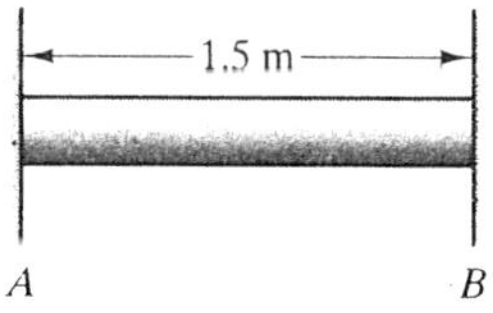

Figure P.4.24.

****4.25.** For the materials listed below, obtain a visual comparison of their response by plotting the stress-strain curve to yield for each material on the same graph.

Material	E (10^3 ksi)	Y (ksi)	Strain at yield
Aluminum	10	45	4.50×10^{-3}
Brass	13	20	1.54×10^{-3}
Copper	17	40	2.35×10^{-3}
Magnesium	6.5	23	3.54×10^{-3}
Structural Steel	29	35	1.21×10^{-3}
Stainless Steel	28	80	2.86×10^{-3}

4.26. A circular prismatic rod of length $2L$ is spinning about its center inside a cowl as shown in Fig. P.4.26. Show that the maximum temperature rise ΔT of the rod without it hitting the cowl is given by:

$$\Delta T < \frac{1}{\alpha}\left(\frac{\delta}{L} - \frac{2Y}{3E}\right)$$

In the above, δ is the gap between the cowl and the tip of the rod when $\omega = 0$, Y is the yield stress, E the elastic modulus, and α the thermal expansion coefficient of the rod. Assume that these material properties do not change significantly with temperature.

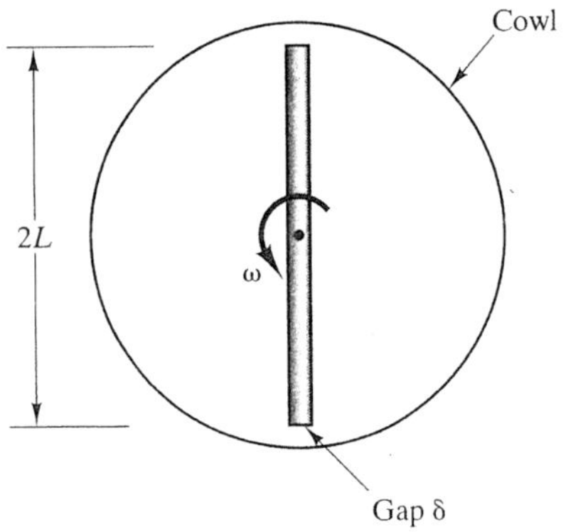

Figure P.4.26.

4.27. The following data was measured from a tension parallel to the grain test on an oak specimen. An extensometer (see Fig. P.4.27) measured the strain and a load cell measured the force on the specimen. Plot the stress-strain curve. What is Young's modulus? What is the ultimate stress? The highest recorded force was 1270 lb and the specimen's narrowest cross-section measured 3/16 in by 3/8 in.

Force (lb)	Strain (in/in)
0	0.00000
47.5	0.00038
139	0.00113
218	0.00177
326	0.00261
451	0.00351
572	0.00473
691	0.00560
803	0.00661
915	0.00740
1020	0.00824
1140	0.00915
1270	0.01030

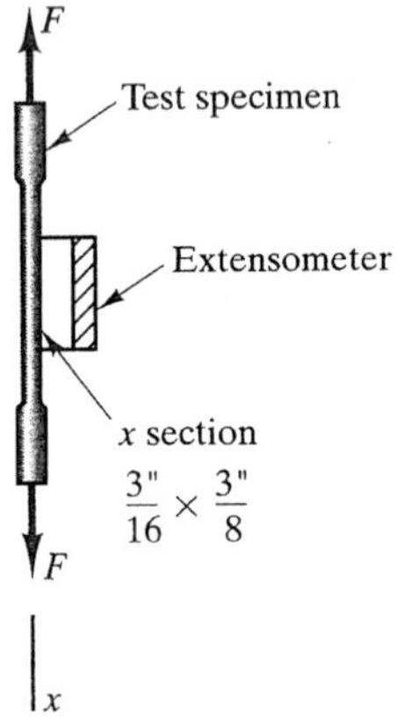

Figure P.4.27.

Project 4 Elongation of a Rod with Variable E and Variable ν

Find the elongation of a 10-foot rod of original diameter of 0.1 ft under successive loadings increasing by 100-lb increments. The modulus of elasticity is given as

$$E = 30 \times 10^6 + 0.001\tau^{1.5} + \tau \text{ psi}$$

The Poisson ratio ν is given as

$$\nu = 0.2 + 10\varepsilon^2$$

where ε is the longitudinal strain.

The results are to be displayed as follows:

Force	Elongation
100	________
200	________
⋮	⋮

Procedure: Start with a 100-lb load and set up an iterative procedure to find the elongation of the rod. For each cycle use newly computed ν and E and recompute elongation. Repeat process until new elongation and preceding elongation are within 0.5% of each other. Then increase load by 100 lb and go through the same iterative process using initially unloaded values of ν and E for the initial cycle, and so on. Neglect weight.

CHAPTER 5

One-Dimensional Problems

5.1 Introduction

There are many engineering structures of interest that are composed of members having one-dimensional stress distributions over much of their volumes. One example that we should be familiar with is the simple truss considered in statics courses. We are sometimes interested in determining stresses as well as deflections in such systems under the action of given loadings for a given geometry of the members. This is called *analysis*. There is also the possibility that we may wish to choose the cross-sectional areas of the members in order for them to withstand expected loads safely and economically. The latter is an example of what we call *design* and consists generally of a series of analyses leading in a convergent manner to the optimal system.

5.2 Basic Considerations

What laws govern us in carrying out analyses and/or designs? First, as we have already pointed out, the stresses and forces on any element of the system must satisfy the equations of **equilibrium.** Second, we must generally employ the proper **constitutive law** that best describes the way the material behaves under stress. Third, we must ensure that the strains formulated must be such that the resulting deflections of the members are compatible with each other. This is the **compatibility condition** presented in Chapter 3. Finally, if the system is subject to a temperature change, we may have to account for thermal expansions or contractions that can give rise to significant stresses and strains—so-called **thermal effects**.

In the process of design, we often require that the geometry and materials be so chosen that nowhere does the stress equal the yield stress of the material. This ensures that there is no excessive deflection of the system and further is an assurance that the material does not physically fail (i.e., does not reach the ultimate stress). In any such attempt the designer has certain nagging doubts. Some of these doubts are:

1. How accurate is the constitutive law in representing the particular materials being used?
2. How close to specifications will the properties of the materials be from the manufacturer, and how close will the tolerances be kept in the fabrication of proposed members?
3. How good are the estimations of stress concentrations used in the calculations?
4. Will there possibly be greater loads (or other loads) coming onto the system than the loads used in the calculation?

There are many other doubts that may persist in the mind of the designer. There is, in short, a certain *degree of uncertainty* in his or her effort—an unavoidable uncertainty irrespective of the skill and experience of the designer. To decrease this uncertainty the designer uses a stress lower than the yield stress in ascertaining the extreme state of stress permissible. This stress is called the *working stress,* and we may define the *factor of safety n* as

$$n = \frac{\text{yield stress}}{\text{working stress}}$$

The higher the value of n, the more probable it is that the system will not fail during a long enough interval of time. On the other hand, a safety factor that is too high may make your design uncompetitive. Thus, the assignment of a safety factor in a design depends on many factors, one of which is the danger to life and limb of people who are using or who are near the structure. Clearly, we cannot give here any prescription for assigning a safety factor except to say that factors of 2 and 3 are not uncommon for certain endeavors.

Before proceeding to the problems, we shall present the very useful *St.-Venant principle.* You will recall that when solving problems in rigid-body mechanics we found it profitable to employ the concept of a point force when we had a force distribution over a small area. At other times, we employed the rigid-body resultant force system of some load distribution in the handling of a problem. Such replacements led to reasonably accurate and direct solutions.

Akin to this procedure, we have for the study of elastic behavior of bodies the procedure stemming from *St.-Venant's principle.* This principle states that the stresses reasonably distant from an applied load on a boundary are not significantly altered if this load is changed to a second load which is equivalent to it from the viewpoint of rigid-body mechanics. We may call such a second load the *statically equivalent load.* Thus, the reasonably distant effects from surface tractions over a part of the boundary can be thought of as dependent on the rigid-body resultant of the applied loads on this surface. By this principle, we can replace the complex supporting force system exerted by the wall on the cantilever beam (Fig. 5.1) by a single force and couple as shown in the diagram for the purpose of simplifying computations of stresses and strains in the domain to the right of the support.

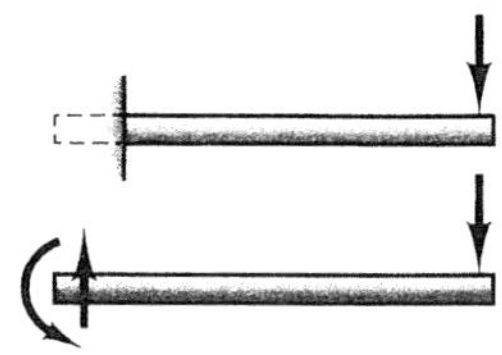

Figure 5.1. Cantilever beam.

Many of the problems to be undertaken in this text will permit us to use the *St.-Venant principle* to help simplify the computations without incurring serious error. Specifically, in the one-dimensional stress problems to be undertaken, we shall often employ only the rigid-body resultant force exerted by the supporting and loading devices on the members. The computed stresses will then have validity at positions *away* from these supporting and loading devices.

5.3 Statically Determinate Problems

In this and ensuing sections we shall present a series of one-dimensional problems illustrating the use of the laws and theorems presented in earlier sections. The following four problems are statically determinate. This means that the forces and stresses in the members can be solved completely *independently* of the deformation—that is, they can be solved by the statics of rigid bodies alone. Note this as you proceed through the examples.

Example 5.1

(a) Shown in Fig. 5.2 are two solid cylinders of different materials welded together at B. A load F is applied along the centerline of the system. If the materials are linearly elastic having moduli of $E_1 = 15 \times 10^6$ psi and $E_2 = 30 \times 10^6$ psi, respectively, what is the deflection of the endpoint A as a result of the load $F = 10{,}000$ lb?

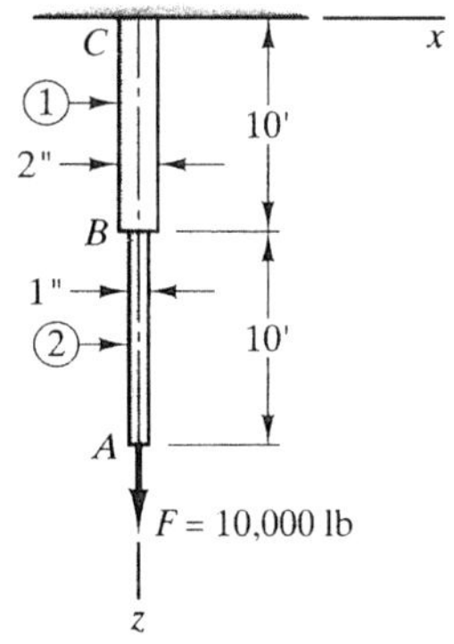

Figure 5.2. Statically determinate axial force problem.

Neglecting the weight of the members and considering sections away from C, B, and A where there are stress concentrations, the stress τ_{zz} may be obtained from **equilibrium** considerations. Thus

$$(\tau_{zz})_1 = \frac{F}{A_1} = \frac{10{,}000}{\pi(2^2/4)} = 3183 \text{ psi}$$

$$(\tau_{zz})_2 = \frac{F}{A_2} = \frac{10{,}000}{\pi(1^2/4)} = 12{,}730 \text{ psi}$$

To get the total deflection we now compute the strain ε_{zz} assuming that it is uniform throughout each rod. We now use the **constitutive law** (Hooke's law) for elastic materials as follows:

Example 5.1 (Continued)

$$(\varepsilon_{zz})_1 = \frac{(\tau_{zz})_1}{E_1} = \frac{3183}{15 \times 10^6} = .212 \times 10^{-3}$$

$$(\varepsilon_{zz})_2 = \frac{(\tau_{zz})_2}{E_2} = \frac{12{,}730}{30 \times 10^6} = .424 \times 10^{-3}$$

The elongation of the rods can easily be ascertained as follows from **geometry:**

$$\begin{aligned}\delta_1 &= \int_0^{10} (\varepsilon_{zz})_1\, dz = (\varepsilon_{zz})_1 L_1\\ &= (.212 \times 10^{-3})(10) = .212 \times 10^{-2}\text{ ft}\\ \delta_2 &= \int_{10}^{20} (\varepsilon_{zz})_2\, dz = (\varepsilon_{zz})_2 L_2\\ &= (.424 \times 10^{-3})(10) = .424 \times 10^{-2}\text{ ft}\end{aligned}$$

The deflections δ_1 and δ_2 are completely **compatible** with each other for all values of δ_1 and δ_2. The total deflection δ_T of point A as the result of the force F is then

$$\delta_T = \delta_1 + \delta_2 = .636 \times 10^{-2}\text{ ft}$$

$$\delta_T = 0.00636\text{ ft}$$

(b) Suppose next that the yield points of the materials are $(\tau_Y)_1 = 45{,}000$ psi and $(\tau_Y)_2 = 60{,}000$ psi, respectively. What is the maximum force possible before yielding takes place in either material?

Considering the materials separately, we have

$$\begin{aligned}(\tau_Y)_1 &= 45{,}000 = \frac{F_1}{\pi(2)^2/4}\\ \therefore F_1 &= 141{,}400\text{ lb}\\ (\tau_Y)_2 &= 60{,}000 = \frac{F_2}{\pi(1)^2/4}\\ \therefore F_2 &= 47{,}100\text{ lb}\end{aligned}$$

Clearly, the maximum load permitted for no yielding would be just smaller than 47,100 lb.

$$F_{\max} = 47{,}100\text{ lb}$$

(c) Next, suppose that a force F is such as to cause a deflection of end A equal to 8×10^{-2} ft. The load is then released. What permanent deflection is thus developed for end A assuming elastic, perfectly plastic behavior of the material?

Example 5.1 (Continued)

The maximum possible deflection of end A with elastic action everywhere occurs when F is just under 47,100 lb, as computed above. We can thus say using **Hooke's law, compatibility,** and **equilibrium** that

$$(\delta_{\max})_{\mathrm{El}} = \frac{(\tau_{zz})_1}{E_1}(L_1) + \frac{(\tau_{zz})_2}{E_2}(L_2)$$

$$= \frac{\dfrac{47{,}100}{\pi(2)^2/4}}{15 \times 10^6}(10) + \frac{\dfrac{47{,}100}{\pi(1)^2/4}}{30 \times 10^6}(10)$$

$$= 1 \times 10^{-2} + 2 \times 10^{-2} = 3 \times 10^{-2} \text{ ft}$$

It is clear that for a deflection of 8×10^{-2} ft there must be yielding for the system. The lower member reaches the yield condition first when the deflection of A is 3×10^{-2} ft and then, with constant stress, "flows" to generate the additional deflection of 5×10^{-2} ft while the upper member undergoes no further change whatever from the elastic extension of 1×10^{-2} ft. This may seem strange, but you must realize that it is a consequence of the "perfectly plastic" idealization we have made. Thus, the force F needed to cause a deflection at end A equal to 8×10^{-2} ft is just over 47,100 lb, causing an elastic elongation in member ① of 1×10^{-2} ft and an elastic elongation in member ② of 2×10^{-2} ft plus a *plastic elongation* in member ② of 5×10^{-2} ft. When the load is released, member ① will completely recover elastically, but member ② will retain a permanent elongation of 5×10^{-2} ft. The force-displacement curve and corresponding stress-strain curve for member ② are shown in Fig. 5.3 for this action.

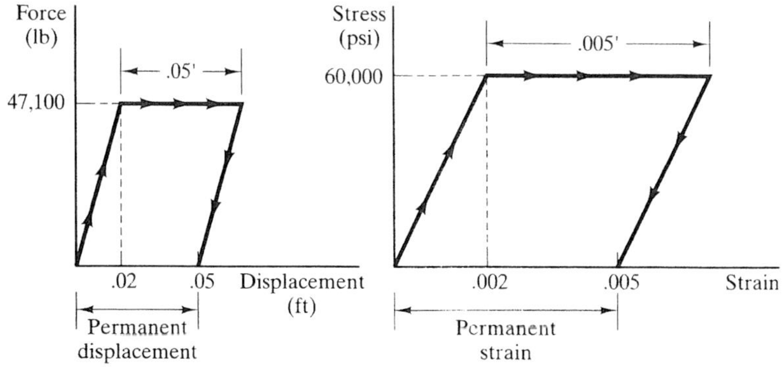

Figure 5.3. Force-displacement and corresponding stress-strain history for member ②.

In Example 5.1 we dealt with two axially loaded uniform prismatic members welded together, each having its own constant stress τ_{zz}. The longation δ of each member was given by the following simple formulation:

$$\delta = \frac{\tau_{zz}}{E} L$$

If we replace τ_{zz} by F/A, we obtain a formula that is easily used for compression or tension in prismatic members and that also is easily remembered. That is,

$$\delta = \frac{FL}{AE} \tag{5.1}$$

In our study of torsion, we will develop a formula for the twist of a circular shaft that is completely analogous to formula (5.1).

Finally, a moment's thought will make it clear that formula (5.1) has the following restrictions for its use:

1. Linear elastic behavior,
2. Constant cross-sectional area, A,
3. Constant force transmitted, F,
4. One-dimensional stress, τ_{zz},
5. Homogeneous material.

For conditions other than the above we must consider an infinitesimal length dL and use calculus. An example problem that does not satisfy the condition above is given next.

Example 5.2

A rigid drum B (see Fig. 5.4) rotates at a speed ω of 500 rpm. Two rods G of length $L = .45$ m are attached to the rigid cylinder B. If the cross-sectional area of each rod is 650 mm^2 and the modulus of elas-

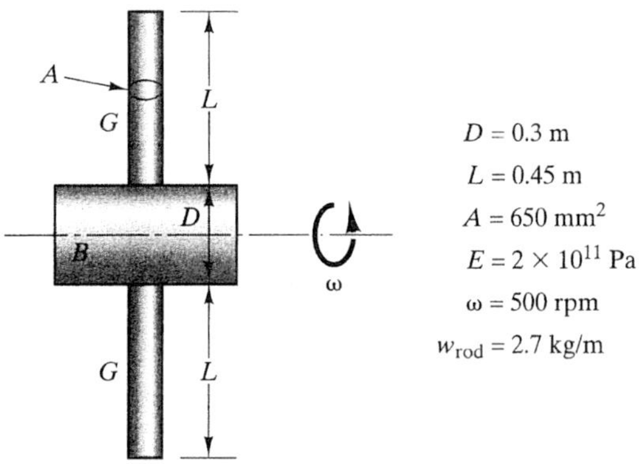

Figure 5.4. Elastic rods G on a rigid drum B.

Example 5.2 (Continued)

ticity is 2×10^{11} Pa, what is the change in length of the rods as a result of the rotation? The rods have a mass per unit length of 2.7 kg/m. The diameter D of drum B is .3 m. Neglect gravitational forces.

We first must find the stress at any section of the rod. To do this, we expose a section by using a free-body diagram as shown in Fig. 5.5, where the section at a distance r from the drum centerline has been exposed. To find the stress at this section, we consider a slice of the rod at some position η from the drum centerline. The centrifugal force associated with this slice is given as follows:

$$df = dm\,\eta\omega^2 = [(2.7)(d\eta)]\eta\left[\frac{500(2\pi)}{60}\right]^2$$

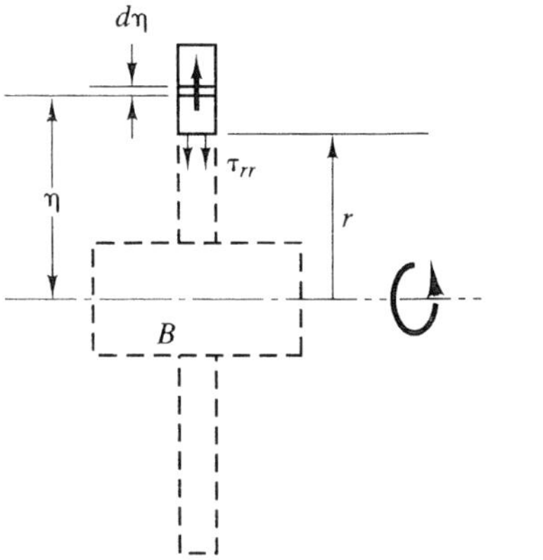

Figure 5.5. Free-body diagram (full lines) exposing a section of a rod at a distance r from centerline.

Newton's law requires that the total force at the section at r be equal and opposite to the integral of the centrifugal forces for all the elemental slices between r and the end of the rod. Thus we have[1]

$$\tau_{rr}A = \int_r^{.3/2+.45} (2.7)\left[\frac{500(2\pi)}{60}\right]^2 \eta\, d\eta$$

Since $A = 650 \times 10^{-6}$ m^2, we get for τ_{rr},

$$\tau_{rr} = \frac{1}{650 \times 10^{-6}}(2.7)\left[\frac{500(2\pi)}{60}\right]^2 \frac{\eta^2}{2}\Bigg|_r^{.6} = 5.69 \times 10^6(.36 - r^2)\text{ Pa}$$

Thus, we see that the normal stress at a section varies with square of the distance from the drum centerline.

Now from **geometry** and a **constitutive law** (Hooke's law) we can calculate the elongation of any slice dr of the entire rod (see Fig. 5.6) as

$$\varepsilon_{rr}\,dr = \frac{\tau_{rr}}{E}dr = \frac{5.69 \times 10^6}{2 \times 10^{11}}(.36 - r^2)dr$$

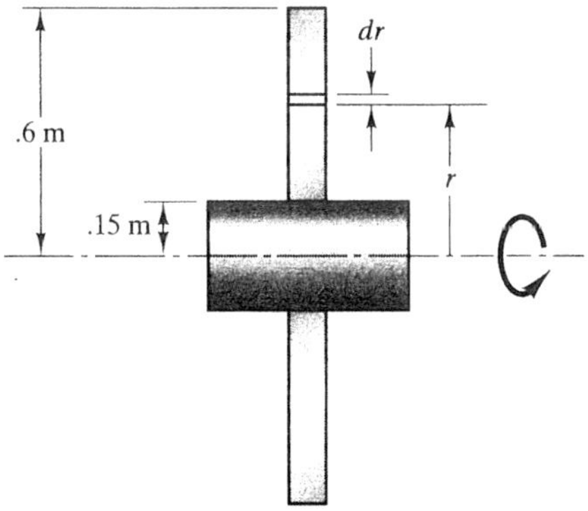

Figure 5.6. Slice dr has elongation $(\tau_{rr}/E)\,dr$.

Integrating to get the total elongation of the rod Δ, we get (**compatibility**)

$$\Delta = \int_{.3/2}^{.6} \varepsilon_{rr}\,dr = \frac{5.69 \times 10^6}{2 \times 10^{11}}\int_{.15}^{.6}(.36 - r^2)dr$$

$$= \frac{5.69 \times 10^6}{2 \times 10^{11}}\left[.36(.6 - .15) - \frac{1}{3}(.6^3 - .15^3)\right]$$

$$= 2.593 \times 10^{-6}\text{ m} = 2.593 \times 10^{-3}\text{ mm}$$

$$\Delta = 0.002593 \text{ mm}$$

[1]Using the D'Alembert force concept, we can think of this as a "statics" problem.

Example 5.3

An elastic cylinder is mounted on a platform in outer space as shown in Fig. 5.7. The platform is decelerating downward at a rate of 25 m/sec^2. A rigid block having a mass of 1000 kg is attached to the top of the cylinder. The mass per unit length of the cylinder is a function of the square of z between the top and the bottom with a value $\overline{m}_T$ of 500 kg/m at the top and a value $\overline{m}_B$ of 200 kg/m at the bottom. The cylinder has a diameter of 200 mm. Determine the stress distribution and the change in length of the cylinder due to the deceleration.

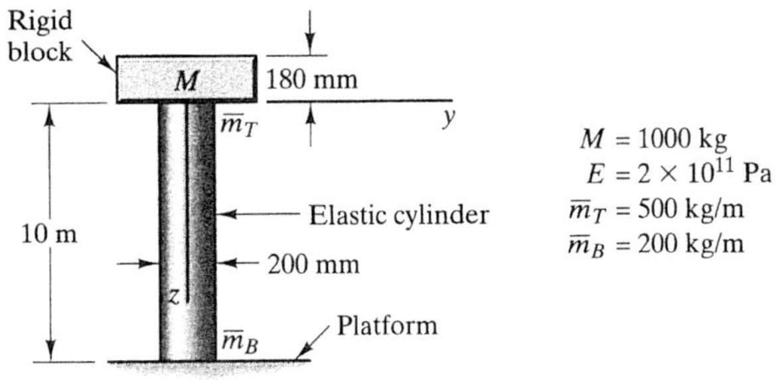

Figure 5.7. System decelerating downward in outer space.

We first draw a free-body diagram exposing the stress τ_{zz} (see Fig. 5.8).

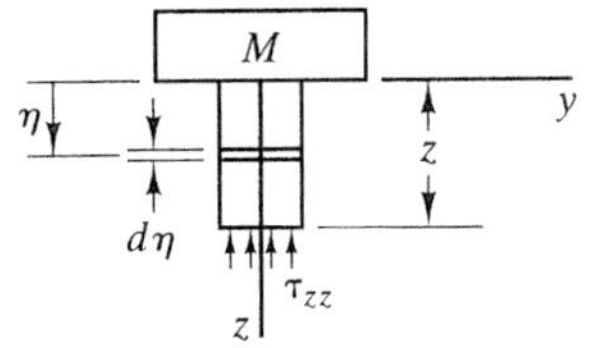

Figure 5.8. Free-body diagram exposing stress in the cylinder.

Next, we compute the mass per unit length, $\overline{m}$, of the cylinder as a function of z. Thus,

$$\overline{m} = az^2 + b$$

$$\text{at } z = 0, \ \overline{m} = 500 \text{ kg/m} \qquad \text{at } z = 10, \ \overline{m} = 200 \text{ kg/m}$$

$$\therefore \overline{m} = 500 - 3z^2 \text{ kg/m}$$

We may now employ **Newton's law** for the free body. Note that weight is not considered since the system is in outer space. Using the dummy variable η, we have

$$-\tau_{zz}\left(\frac{\pi}{4}\right)(.200)^2 = -\int_0^z (500 - 3\eta^2)(d\eta)(25) - (1000)(25)$$

$$\tau_{zz}(.0314) = \left[(500)(25)\eta - (75)\frac{\eta^3}{3}\right]_0^z + 25{,}000$$

$$\tau_{zz} = 3.979 \times 10^5 z - 795.8z^3 + 7.958 \times 10^5$$

Example 5.3 (Continued)

We next use Hooke's law as the **constitutive law.**

$$\varepsilon_{zz} = -\frac{1}{2 \times 10^{11}}\left[3.979 \times 10^5 z - 795.8z^3 + 7.958 \times 10^5\right]$$

Finally, using **geometry** and **compatibility,** we have for the deflection Δ,

$$\Delta = \int_0^{10} \varepsilon_{zz}\,dz = -\frac{1}{2 \times 10^{11}}\left[(3.979 \times 10^5)\left(\frac{100}{2}\right) - 795.8\left(\frac{10^4}{4}\right) + (7.958 \times 10^5)(10)\right]$$

$$\Delta = -1.293 \times 10^{-4}\text{ m} = -.1293\text{ mm}$$

$$\Delta = -.1293\text{ mm}$$

The following example presents a thought-provoking problem that we urge you to examine critically.

Example 5.4

In Fig. 5.9 we have shown a pile (used for foundation preparation in the construction of tall buildings) being driven into earth which is best described as mud. The driving force P_2 is varied in such a way as to maintain a downward movement of the pile at a *constant speed* of 5 ft/sec. A constant resisting force P_1 of 20,000 lb acts at the bottom, while a friction force distribution f having a constant intensity of 2000 lb/in. acts on the surface of the pile in contact with the mud. If the pile enters the mud at time $t = 0$, what is the change in overall length of the pile from that of its undeformed geometry, at the time $t = 2$ sec and at the time $t = 6$ sec? *Hint:* Consider one domain of length h (see Fig. 5.9) where the stress is varying only with position at any time t, and a second domain of length $(50 - h)$ where the stress is uniformly distributed at any time t. Take $h = 0$ ft at time $t = 0$. Take $E = 10 \times 10^6$ psi.

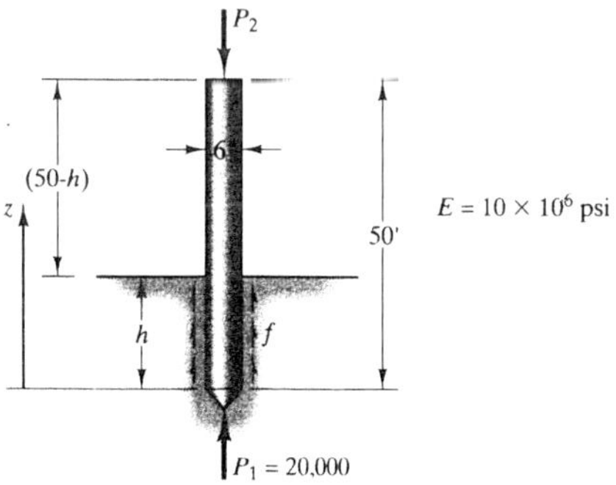

Figure 5.9. Pile is being driven into mud at a constant speed.

Comment: It should be clear that the force P_2 will vary with time and that the *total* friction force will also vary with time. Nevertheless, because of the constant speed of the pile, the equations of equilibrium apply to any and all parts of the pile.

A free-body diagram is shown in Fig. 5.10 wherein the portion of the cylinder subjected to friction on the periphery is shown cross-hatched.

Example 5.4 (Continued)

Part 1: For $z_1 \leq h$ (see Fig. 5.10), we may use **equilibrium** to obtain the force F_1 acting at position z_1 (see Fig. 5.11).

$$F_1 = 20{,}000 + (2000)(12)(z_1) \qquad \text{(keeping } z_1 \text{ in ft)} \qquad \text{(a)}$$

Using Hooke's law as the **constitutive law** and taking an increment of the infinitesimal slice we get (see Fig. 5.10)

$$\delta(dz_1) = \frac{20{,}000 + 24{,}000z_1}{\left[\left(\frac{\pi}{4}\right)(6)^2\right](10 \times 10^6)} dz_1$$

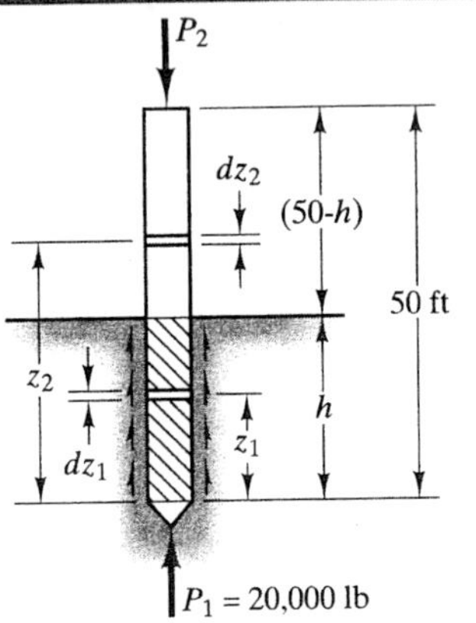

Figure 5.10. F.B.D. of the pile showing slice dz_1 at position z_1 and slice dz_2 at position z_2.

Now using **geometry,** we get for the total shortening Δ_1 of the segment of pile of length h

$$\Delta_1 = \int_0^h \frac{20{,}000 + 24{,}000z_1}{\left[\left(\frac{\pi}{4}\right)(6)^2\right](10 \times 10^6)} dz_1$$

$$\Delta_1 = \frac{1}{\left(\frac{\pi}{4}\right)(36)(10 \times 10^6)}\left[20{,}000h + 24{,}000\frac{h^2}{2}\right]$$

$$= 3.54 \times 10^{-9}\left[20{,}000h + 24{,}000\frac{h^2}{2}\right]$$

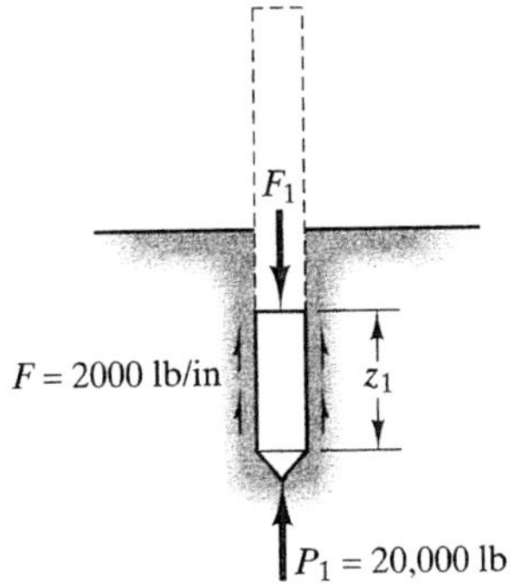

Figure 5.11. F.B.D. of a portion of the pile for $z_1 < h$.

Part 2: For $z_2 \geq h$ (Fig. 5.12), we may use Eq. (5.1) directly, since the stress is uniform for $z_2 \geq h$.

$$\therefore\ \Delta_2 = \frac{P_2(50 - h)}{\left[\left(\frac{\pi}{4}\right)(6)^2\right](10 \times 10^6)} = 3.54 \times 10^{-9}[(20{,}000 + 24{,}000h)(50 - h)]$$

where we used Eq. (a), with z_1 replaced by h, to determine $P_2 = 20{,}000 + 24{,}000h$ (see Fig. 5.10 and Fig. 5.11). From **compatibility,** we have

$$\Delta_{\text{Total}} = \Delta_1 + \Delta_2$$

Since the pile is driven at a constant speed of 5 ft/sec, then at $t = 2$ sec the length h is 10 ft and we get

Example 5.4 (Continued)

$$(\Delta_{\text{Total}})_{t=2} = 3.54 \times 10^{-9}\left[(20{,}000)(10) + 24{,}000\,\frac{100}{2}\right] + 3.54 \times 10^{-9}[20{,}000 + (24{,}000)(10)](50 - 10)$$

$$(\Delta_{\text{Total}})_{t=2} = 4.177 \times 10^{-2}\text{ ft}$$

At $t = 6$ sec, $h = 30$ ft and we get

$$(\Delta_{\text{Total}})_{t=6} = 3.54 \times 10^{-9}\left[(20{,}000)(30) + 24{,}000\,\frac{900}{2}\right] + 3.54 \times 10^{-9}[20{,}000 + (24{,}000)(30)](50 - 30)$$

$$(\Delta_{\text{Total}})_{t=6} = 9.275 \times 10^{-2}\text{ ft}$$

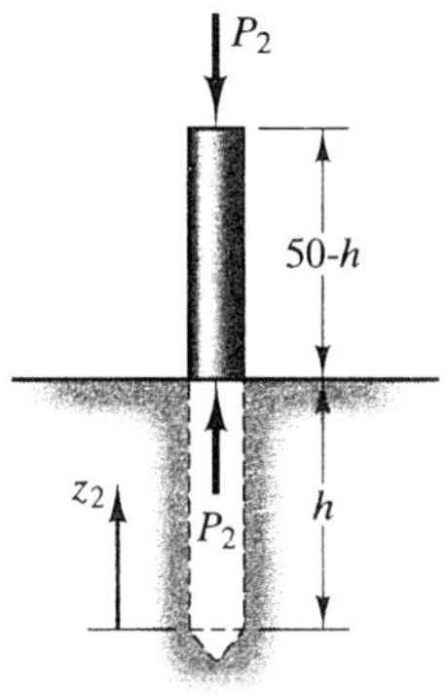

Figure 5.12. F.B.D. of a portion of the pile for $z_2 \geq h$.

5.4 Statically Indeterminate Problems

You will note in Examples 5.1 through 5.4 that the forces transmitted by the members were solvable by "statics" of rigid bodies alone. These problems were thus *statically determinate* problems. In the following four examples the forces cannot be solved by statics of rigid bodies alone. Rather, you will find that such computations must be made simultaneously with *deformation* considerations of the bodies. Thus, we shall for the first time undertake to solve *statically indeterminate* problems.

Example 5.5

A simple structure in Fig. 5.13 is shown wherein three members support load F. We wish to compute the forces in the members and the deflection of joint B. Take the cross-section of each member to be A in.^{2} and the modulus of elasticity to be E lb/in.^{2} for all members.

For **equilibrium** requirements we consider pin B as a free body in Fig. 5.14(a). It is clear that $F_1 = F_3$ and that

$$F_2 + 2F_1 \cos 45° = 10{,}000$$

$$\therefore F_2 + 1.414F_1 = 10{,}000 \qquad \text{(a)}$$

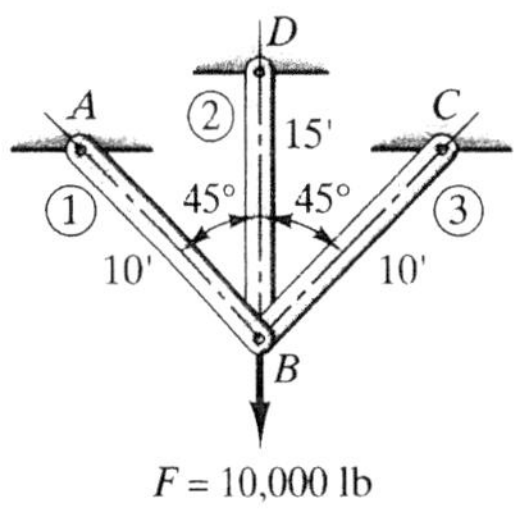

Figure 5.13. Statically indeterminate problem.

Example 5.5 (Continued)

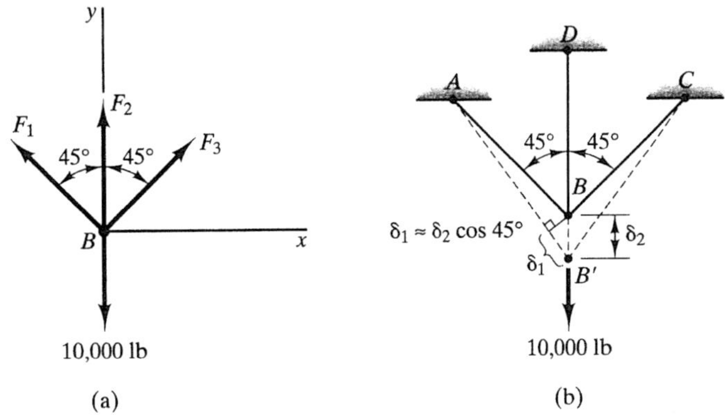

Figure 5.14. Statics and kinematics at joint *B*.

This is as far as rigid-body mechanics takes us. We next employ a **constitutive law** to express the extension of each member in terms of the forces. We have using Eq. (5.1)

$$\delta_1 = \delta_3 = \frac{F_1(10)}{AE} \tag{b}$$

$$\delta_2 = \frac{F_2(15)}{AE} \tag{c}$$

Not all values of δ_1 and δ_2 are **compatible** with each other. In order for the displacement of pin *B* to be *single-valued,* it is necessary that the displacements δ_1 and δ_2 be so related as to give the *same vertical deflection* of pin *B*.[2] Considering Fig. 5.14(b), we may conclude that assuming that $\measuredangle\ AB'D = 45°$ as a result of small deformation

$$\delta_1 = \delta_2 \cos 45° = \delta_2(.707) \tag{d}$$

Now solving for the forces in Eqs. (b) and (c) and substituting into Eq. (a), we have a second equation in δ_1 and δ_2. That is,

[2]Notice in Fig. 5.14(b) that member *AB* must rotate and elongate in a manner that will move the pin *B* when it associated with member *AB* to the very same location that pin *B* reaches when it associated with member *DB*. The elongation of *AB* clearly is found by dropping a perpendicular from *B* to the deformed member *AB′* since this perpendicular slices off a length of *AB′* equal (up to first order of magnitude) to the undeformed length of *AB*.

Example 5.5 (Continued)

$$\frac{AE}{15}\delta_2 + 1.414\frac{AE}{10}\delta_1 = 10{,}000$$

$$\therefore \delta_2 + 2.12\delta_1 = \frac{150{,}000}{AE} \qquad \text{(e)}$$

Accordingly, solving Eqs. (d) and (e) simultaneously, we have for δ_2

$$\delta_2 = \frac{60{,}000}{AE}$$

This, then, is the desired deflection of pin B. The forces F_1, F_2, and F_3 can now be determined from Eqs. (b) and (c).

Although the solution of this problem was straightforward and simple, the determination of the deflections for more complex structures becomes unwieldy by the methods used here. In Chapters 6 and 18 we shall present some very powerful *energy methods* that can be used effectively for such calculations.

Example 5.6

A boat is moored to a dock as shown in Fig. 5.15 by five lines which are all just taut. The lines have a modulus of elasticity of 2×10^8 Pa including the effects of stretching and unwinding. To make adjustments on the outboard motor, it is run at half-throttle, while the boat is tethered as described above, and develops a thrust of 450 N. What are the forces in the lines if the diameter under load is 12.5 mm? What is the distance Δ that the boat advances as a result of the thrust? Take the lines as oriented in a horizontal plane.

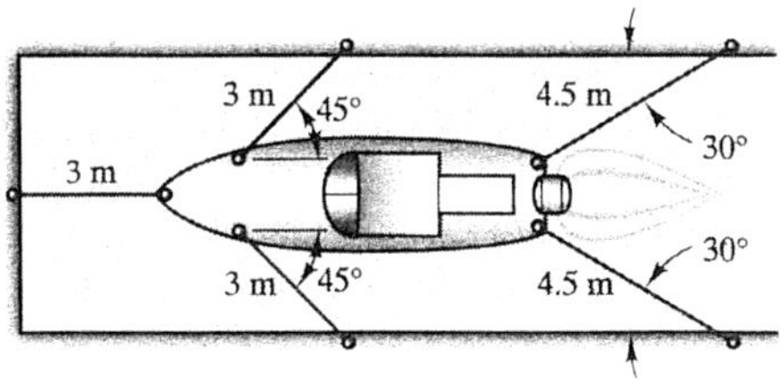

Figure 15.15 A tethered boat under power from an outboard motor.

In Fig. 5.16 we have a shown a free-body diagram of the boat exposing the forces in the lines. Also, we have listed pertinent data for the problem. From **equilibrium,** we can say that

Example 5.6 (Continued)

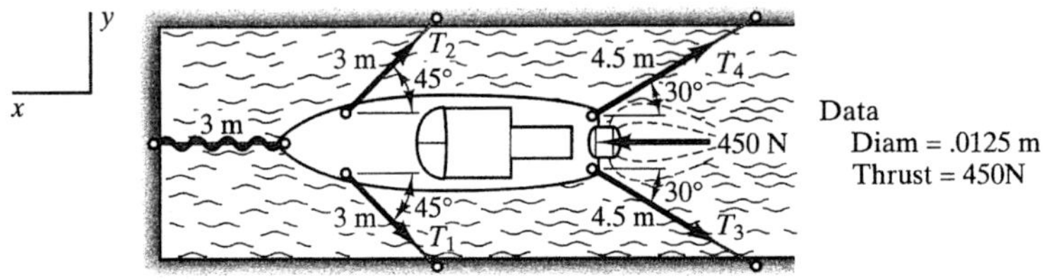

Figure 5.16. Free-body diagram of the boat under load.

$$T_1 = T_2 \qquad T_3 = T_4$$

$$450 - 2T_1 \cos 45° - 2T_3 \cos 30° = 0 \qquad \text{(a)}$$

For **compatibility,** examine Fig. 5.17 showing the rotation and the elongation needed by the lines having tensions T_1 and T_3 in order to accommodate in a compatible manner the movement Δ of the boat when under thrust.

$$\Delta_1 = \Delta_2 = (.707)\Delta \qquad \Delta_3 = \Delta_4 = (.866)\Delta$$

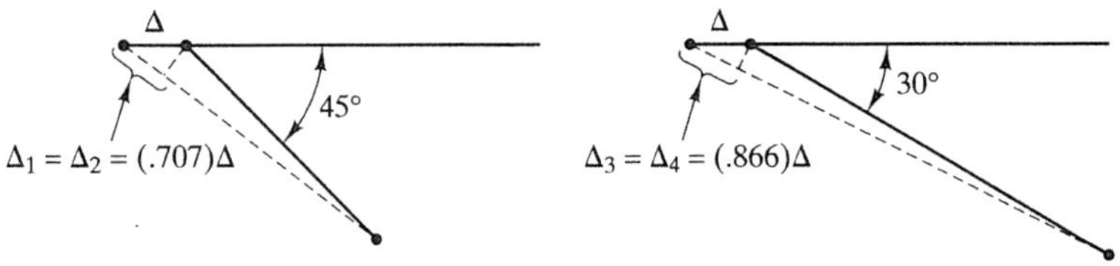

Figure 5.17. Compatible elongations and rotations of the lines.

Next we use a linear elastic **constitutive law.** Thus for T_1

$$T_1 = \tau_1 A = \varepsilon_1 EA = \frac{\Delta_1}{L_1} EA = \frac{(\Delta)(.707)(2 \times 10^8)(\pi)(.0125)^2/4}{3} = 5784\Delta$$

Similarly, we can say

$$T_3 = 4723\Delta$$

Substituting for T_1 and T_3 from above into Eq. (a), we may directly solve for Δ. Thus

$$\Delta = .0275 \text{ m}$$

Example 5.6 (Continued)

and clearly we have

$$T_1 = T_2 = (5784)(.0275) = 159.1 \text{ N}$$
$$T_2 = T_3 = (4723)(.0275) = 129.9 \text{ N}$$

Example 5.7

An aluminum shaft A and a steel sleeve B are fixed (see Fig. 5.18) at the immovable wall C at one end and are welded to a stiff plate D at the other end. A force F_2 of 45 kN is applied to plate D and a force F_1 of 20 kN is applied to the steel sleeve (only) as shown in the diagram. The elastic moduli for the members are given in the diagram. What is the elongation of the system as the result of the loads F_1 and F_2?

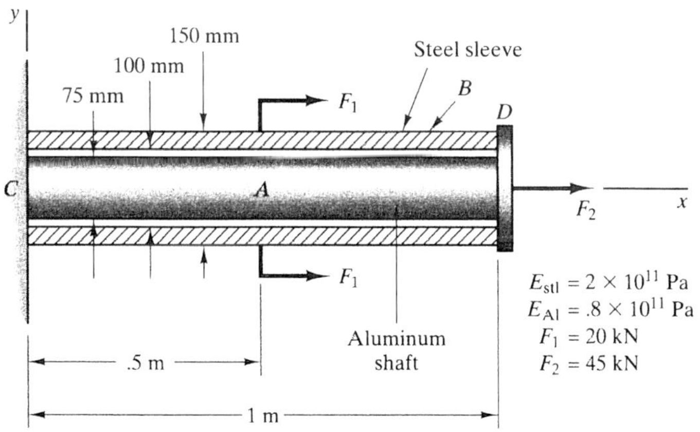

Figure 5.18. Statically indeterminate one-dimensional problem.

Our first step is to consider plate D as a free body (see Fig. 5.19) to expose the force in member A (i.e., F_A) and the force in member B to the right of load F_1 (Fig. 5.18), which we shall denote as F_B. Employing **equilibrium** for this free body, we get

$$45{,}000 - F_A - F_B = 0 \tag{a}$$

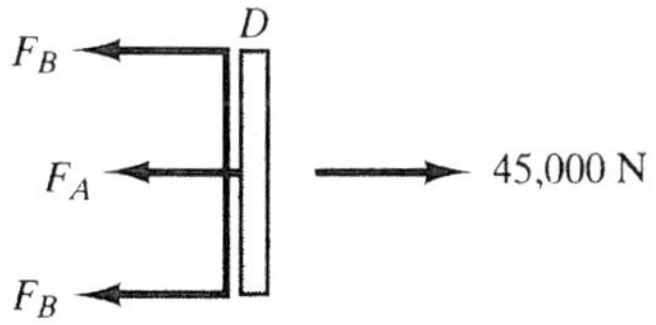

Figure 5.19. Free body of plate D.

Example 5.7 (Continued)

This is the only independent equation of equilibrium, so we see that the problem is statically indeterminate. We now use Eq. (5.1) and thus the **constitutive law** to get the elongation of the aluminum rod and the steel sleeve. In the case of the rod there is a constant tensile force F_A so that we can say

$$\Delta_A = \frac{F_A L_A}{A_A E_A} = \frac{F_A(1)}{[\pi(.075)^2/4](.8 \times 10^{11})} \tag{b}$$

As for the sleeve, it is clear from Fig. 5.19 that for x > .5 m there is force F_B, but for $x \leq .5$ we see from the free body of the sleeve cut to expose the force in this region (see Fig. 5.20) that the force there is $(F_B + 20{,}000)$ N. Hence, for Δ_B we have

$$\Delta_B = \frac{(F_B + 20{,}000)(.5)}{(\pi/4)(.150^2 - .100^2)(2 \times 10^{11})} + \frac{F_B(.5)}{(\pi/4)(.150^2 - .100^2)(2 \times 10^{11})} \tag{c}$$

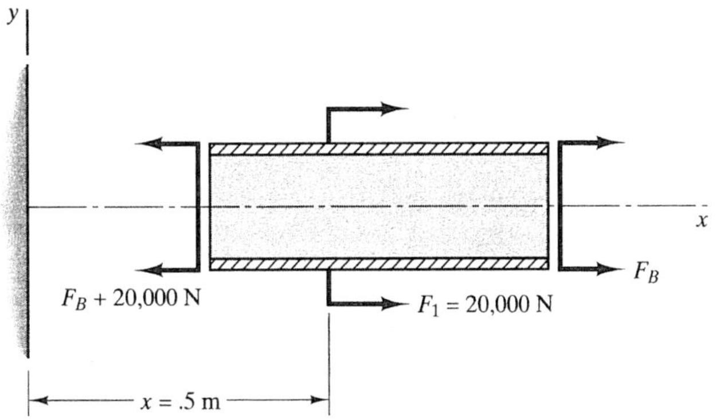

Figure 5.20. Free-body diagram of the sleeve, exposing sections for $0 < x < .5$ m.

From **compatibility** we can now say that

$$\Delta_A = \Delta_B$$

$$\therefore \frac{F_A(1)}{(\pi/4)(.075^2)(.8 \times 10^{11})} = \frac{(F_B + 20{,}000)(.5)}{(\pi/4)(.150^2 - .100^2)(2 \times 10^{11})} + \frac{F_B(.5)}{(\pi/4)(.150^2 - .100^2)(2 \times 10^{11})} \tag{d}$$

We then get from the above,

$$222.2F_A = 40F_B + 4 \times 10^5 \tag{e}$$

Solving Eqs. (e) and (a) simultaneously, we get

Example 5.7 (Continued)

$$F_A = 8390 \text{ N}$$
$$F_B = 36{,}610 \text{ N}$$

From Eq. (b) we can directly get the desired elongation Δ.

$$\Delta = \frac{(8390)(1)}{[\pi(.075^2)/4](.8 \times 10^{11})} = 2.374 \times 10^{-5}\text{m} = .02374 \text{ mm}$$

$$\mathbf{\Delta = 0.02374 \text{ mm}}$$

Example 5.8

A light rigid member AB (see Fig. 5.21) is pinned at A and is further supported by light elastic rods CD and GF. Member AB supports a 5000 lb load at B. The elastic modulus for the two elastic rods CD and GF is given as $E = 30 \times 10^6$ psi and the cross-sectional area for each of these rods is 1 in.². What are the forces in the two supporting elastic rods as a result of the loading?

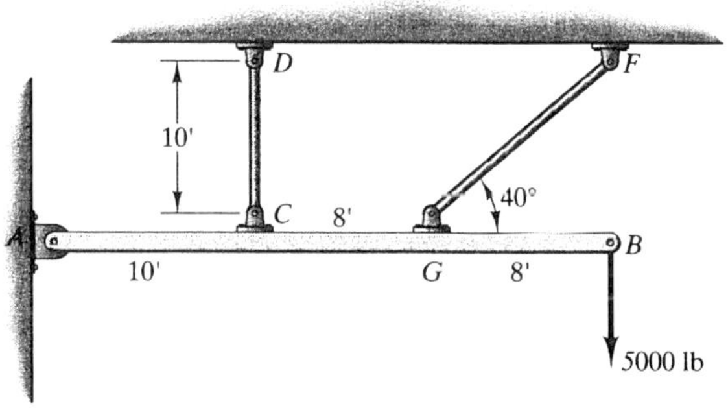

Figure 5.21. Suspended rigid bar with load.

We first show a free-body diagram of member AB in Fig. 5.22. Setting moments about point A equal to zero in accordance with **equilibrium,** we have

$$\sum M_A = 0$$

$$(F_{CD})(10) + (F_{GF})(\sin 40°)(18) - (5000)(26) = 0$$

$$\therefore F_{CD} + 1.157F_{GF} = 13{,}000 \qquad \text{(a)}$$

Example 5.8 (Continued)

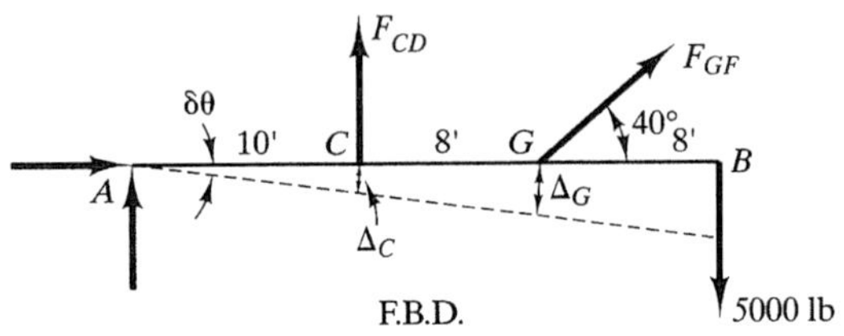

Figure 5.22. Free-body diagram of bar *AB*.

Now we make use of a **constitutive law** for linear elastic behavior of the supporting rods. Using the familiar elongation formula, we obtain the axial elongations of the rods:

$$\Delta_{CD} = \frac{(F_{CD})(10)}{(1)(30 \times 10^6)} \qquad \text{(b)}$$

$$\Delta_{GF} = \frac{(F_{GF})(10)/\sin 40°}{(1)(30 \times 10^6)} \qquad \text{(c)}$$

Compatibility considerations for the rigidity of rod *AB* dictate that the first order *vertical* displacements at points *C* and *G* are given by (see Fig. 5.22)

$$\Delta_C = 10(\delta\theta) \quad \text{and} \quad \Delta_G = 18(\delta\theta)$$

Hence, we can say

$$\Delta_C = \frac{10}{18}(\Delta_G)$$

Now using **compatibility** considerations at pin *G* (see Fig. 5.23) and realizing that $\Delta_C = \Delta_{CD}$, we can say

$$\Delta_C \approx \Delta_{CD} = \frac{10}{18}\Delta_G = \frac{10}{18}(\Delta_{GF}/\cos 50°)$$

Now going to Eqs. (b) and (c) and inserting the displacements into the above equation, we get

$$\frac{(F_{CD})(10)}{(1)(30 \times 10^6)} = \frac{10}{18}\,\frac{(F_{GF})(10)/\sin 40°}{(1)(30 \times 10^6)\cos 50°}$$

$$\therefore\ F_{CD} = 1.345F_{GF}$$

Going to Eq. (a) and replacing F_{CD} we get

$$1.345F_{GF} + 1.157F_{GF} = 13{,}000$$

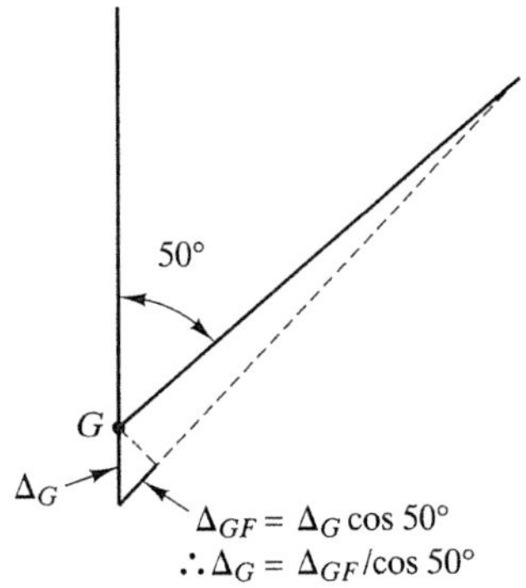

Figure 5.23. Compatibility consideration at pin *G*.

Example 5.8 (Continued)

Solving, we get the desired results

$$F_{GF} = 5197 \text{ lb}$$
$$F_{CD} = 6990 \text{ lb}$$

5.5 Residual Stress Problems

Consider a statically indeterminate structure which is stressed on application of external loads to a point where yielding occurs in all or part of the structure. It is then highly probable that on unloading the structure there will be left in the structure a stress distribution which can be appreciable in value. This is called a *residual stress* distribution. In the following examples we compute such a distribution.

Example 5.9

Shown in Fig. 5.24 are an aluminum alloy rod A ($E_{\text{Al}} = 15 \times 10^6$ psi) and a steel sleeve B ($E_{\text{Stl}} = 30 \times 10^6$ psi) of equal length welded to stiff end plates C and D. An axial tensile force F is applied to the system. If we wish to cause an elongation equal to .025 ft for the system, what value F is needed to accomplish this? Also, compute the stresses in the members after this external load is released. Assume that the materials behave in an elastic, perfectly plastic manner with $Y_{\text{Al}} = 60{,}000$ psi and $Y_{\text{Stl}} = 200{,}000$ psi (**constitutive law**).

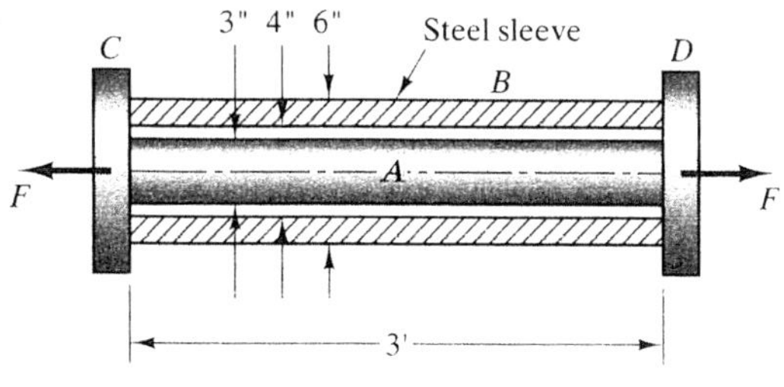

Figure 5.24. Residual stress problem.

We must ascertain whether we shall have plastic deformation in either or both of the members. Consider the alloy member first. The maximum elongation for elastic behavior $(\delta_{\text{Al}})_{\text{El}}$ is

Example 5.9 (Continued)

$$(\delta_{\mathrm{Al}})_{\mathrm{El}} = \frac{60{,}000}{15 \times 10^6}(3) = .012 \text{ ft}$$

Clearly, plastic deformation must exist in the alloy member for the desired elongation of .025 ft. Now consider the steel member similarly. We have

$$(\delta_{\mathrm{Stl}})_{\mathrm{El}} = \frac{200{,}000}{30 \times 10^6}(3) = .020 \text{ ft}$$

Again we have plastic deformation. We can now give the force F required as follows:

$$F = (200{,}000)\frac{\pi}{4}(6^2 - 4^2) + 60{,}000\frac{\pi(3)^2}{4} = 3.57 \times 10^6 \text{ lb}$$

$$F = 3.57 \times 10^6 \text{ lb}$$

Now suppose that the system is released from this load. In the unloading process from the plastic range the materials behave *elastically* with elastic moduli that are *different* from each other. As a result of the difference in elastic moduli, there will be mutual interference between the members in the unloading process. Also, because of the plastic deformation, the final unloaded length of the system will not be the original 3 ft. To see this more clearly, examine Fig. 5.25 where we have shown the stress-strain diagrams for both the steel and the alloy members. Points A and B represent the final loading state of the members corresponding to a total deflection of .025 ft, giving each member a strain of .025/3 = .00833. Now on unloading from A and B, each specimen follows a curve that is parallel to the elastic portion of the loading curve, as has been shown by dashed lines in the diagram. As the unloading occurs, the strain in both members decreases but at any given load must be the *same* for both members (**compatibility**). Also for slow unloading, **equilibrium** must be maintained at all times. There exists a point ε_R on the abscissa which corresponds to the condition of *zero load.* This point occurs where the stress in the steel, $(\tau_R)_{\mathrm{Stl}}$ at point C, times the area of the steel sleeve in accordance with the dictates of equilibrium, gives a force equal and opposite to the force in the alloy coming from $(\tau_R)_{\mathrm{Al}}$ at point D, times the area of the alloy rod. The aforementioned stresses thus remain when the load is removed and are the *residual stresses.* The quantity ε_R times the length is then the permanent elongation of the system. Notice that the alloy ends up in compression, while the steel ends up in tension.

Example 5.9 (Continued)

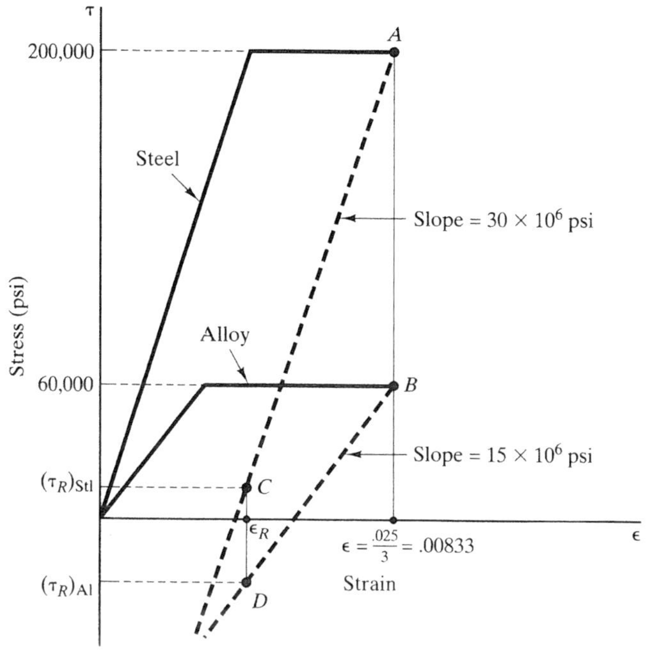

Figure 5.25. Stress-strain histories of sleeve and shaft.

To evaluate the residual stresses we note from **equilibrium** that for the final configuration,

$$[(\tau_{zz})_R]_{\text{Stl}} \frac{\pi}{4}(6^2 - 4^2) + [(\tau_{zz})_R]_{\text{Al}} \frac{\pi}{4}(3^2) = 0$$

$$\therefore\ 5[(\tau_{zz})_R]_{\text{Stl}} + 2.25[(\tau_{zz})_R]_{\text{Al}} = 0 \tag{a}$$

Compatibility next demands that the *decrease* in length of the members during unloading must be the same. This means that the strains "released" on unloading must be the same for the steel and aluminum members. Hence

$$\frac{200{,}000 - (\tau_R)_{\text{Stl}}}{30 \times 10^6} = \frac{60{,}000 - (\tau_R)_{\text{Al}}}{15 \times 10^6}$$

$$\therefore\ (\tau_R)_{\text{Stl}} - 2(\tau_R)_{\text{Al}} = 80{,}000 \tag{b}$$

Solving Eqs. (a) and (b) simultaneously, we have

Example 5.9 (Continued)

$$(\tau_R)_{\text{Stl}} = 14{,}693 \text{ psi}$$
$$(\tau_R)_{\text{Al}} = -32{,}653 \text{ psi}$$

These are the desired residual stresses.

Example 5.10

Rigid bar B.D.B. in Fig. 5.26 weighs 35,000 lb.

(a) What is the maximum applied force $(F_{\text{Max}})_{\text{Elastic}}$ for elastic behavior everywhere?

(b) What is the maximum force F_{Max} that can be applied to the system?

(c) When the maximum force from Part (b) is released directly after it is first reached, what are the residual stresses?

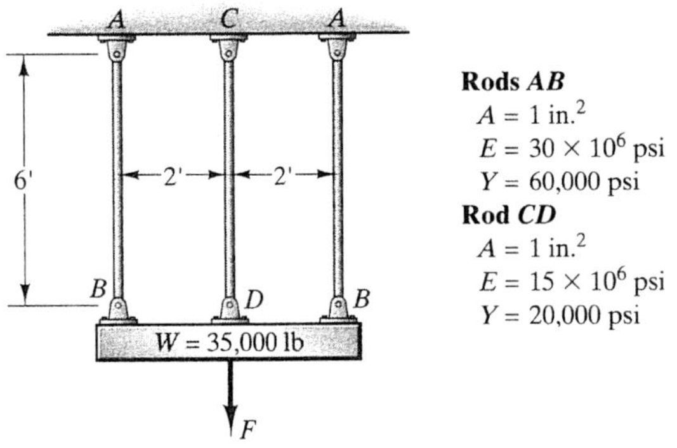

Figure 5.26. A rigid bar supported by three members.

(a) In order to determine which member yields first, we now calculate the yield strains. Thus

$$(\varepsilon_{AB})_Y = \frac{60{,}000}{30 \times 10^6} = 2 \times 10^{-3}$$

$$(\varepsilon_{CD})_Y = \frac{20{,}000}{15 \times 10^6} = 1\frac{1}{3} \times 10^{-3}$$

Since the strains in the members must be the same at all times (**compatibility**), we may conclude that member CD yields first (see Fig. 5.27). The stress in members AB when member CD just begins to yield is obtained from the **constitutive law** as (also see Fig. 5.27).

$$\tau_{AB} = (30 \times 10^6)\left(1\frac{1}{3} \times 10^{-3}\right) = 40{,}000 \text{ psi}$$

Example 5.10 (Continued)

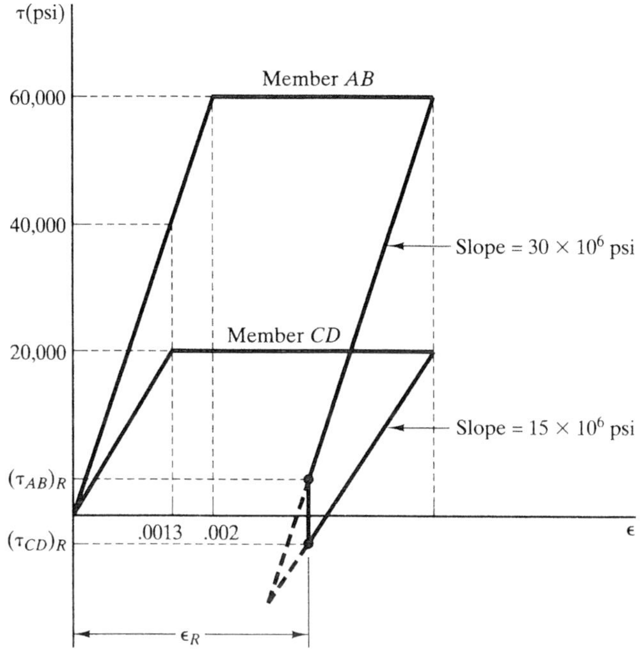

Figure 5.27. Stress-strain histories of members AB and CD.

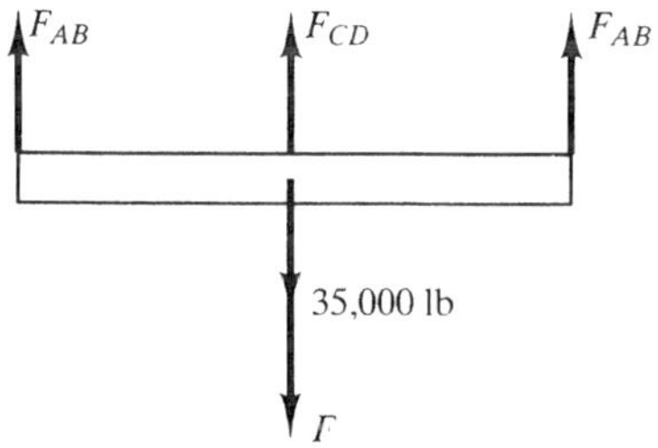

Figure 5.28. Free-body diagram of the rigid bar.

From the free body of the rigid bar shown in Fig. 5.28, the equation of **equilibrium** is written as

$$2F_{AB} + F_{CD} = 35{,}000 + F \tag{1}$$

Using the stress obtained above for members AB and the yield stress for member CD, we get

$$2(40{,}000)(1) + (20{,}000)(1) = 35{,}000 + (F_{\text{Max}})_{\text{Elastic}} \tag{2}$$

$$\therefore\ (F_{\text{Max}})_{\text{Elastic}} = 65{,}000 \text{ lb}$$

(b) The maximum force that can be applied to the system is obtained when all the members are at yield. Thus from the equation of **equilibrium** we obtain

$$2(60{,}000)(1) + (20{,}000)(1) = 35{,}000 + F_{\text{Max}}$$

$$\therefore\ F_{\text{Max}} = 105{,}000 \text{ lb} \tag{3}$$

Example 5.10 (Continued)

Note that this force is first achieved at a strain corresponding to the yield strain of members AB [i.e., $\varepsilon = 0.002$ as calculated in Part (a) and shown in Fig. 5.27].

(c) For the residual stresses we can say from **compatibility** that the strain released during unloading to zero load F must be the same for all members.

$$\frac{60{,}000 - (\tau_{AB})_R}{30 \times 10^6} = \frac{20{,}000 - (\tau_{CD})_R}{15 \times 10^6}$$

Hence,

$$20{,}000 = (\tau_{AB})_R - 2(\tau_{CD})_R \qquad (4)$$

Considering **equilibrium** with $F = 0$ for the unloaded case we have from Eq. (1)

$$2(\tau_{AB})_R(1) + (\tau_{CD})_R(1) = 35{,}000 \qquad (5)$$

Substituting for $(\tau_{AB})_R$ into Eq. (5) from Eq. (4) we then get

$$2[20{,}000 + 2(\tau_{CD})_R] + (\tau_{CD})_R = 35{,}000$$

The residual stress is

$$(\tau_{CD})_R = -1000 \text{ psi}$$

And from Eq. (4), we get the other residual stress

$$(\tau_{AB})_R = 18{,}000 \text{ psi}$$

5.6 Design Problem

The problems undertaken in the series of examples thus far exemplify various analyses. Note that the geometry is entirely established at the outset and we found such quantities as forces and deflections. In the next example we shall consider a simple design problem where under certain operating conditions we are required to decide on the diameter of members for safe economical design.

Example 5.11

A rigid horizontal block W weighing 500,000 N is to be maintained in position by two pin-ended rods AB and CD, as shown in Fig. 5.29. The rods are to have the same diameter and are to be made of the same material having a Young's modulus of 1.2×10^{11}

Example 5.11 (Continued)

Pa. If after connection of the rods and release of the block from the horizontal position, the maximum allowable rotation of the block is to be .05° and if a safety factor for stress of 2.5 is employed, what must be the minimum diameter of the rods for a yield stress of 4.8×10^8 Pa?

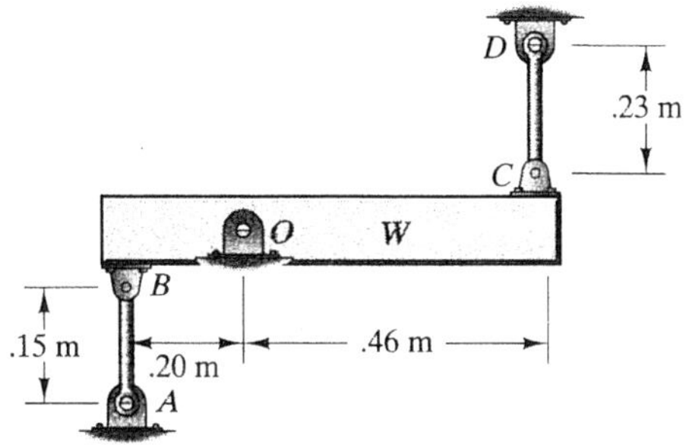

Figure 5.29. Simple design problem.

By permitting the block to rotate after loading the full amount of .05°, we shall be ensuring a *minimum* diameter of the rods. However, in such a procedure we must check to see if the stresses in the rods exceed the allowable stress for the design. We shall proceed in the aforestated manner.

From **equilibrium** in setting moments about point O equal to zero (see Fig. 5.30) we get

$$F_{AB}(.20) + F_{CD}(.46) = (500{,}000)(.13) \qquad \text{(a)}$$

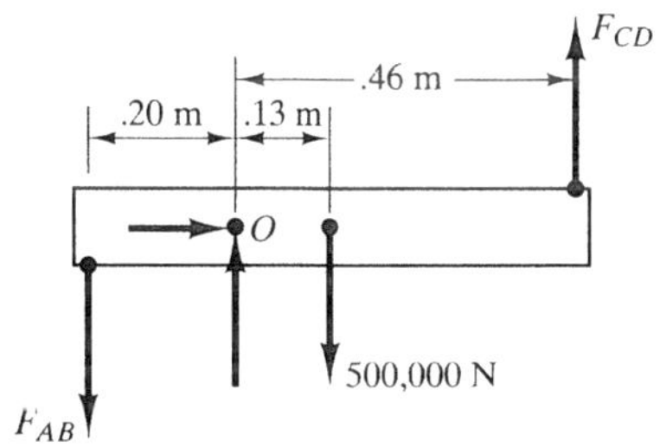

Figure 5.30. Free-body diagram of the block.

Using the **constitutive law** for elastic behavior, we have

$$\delta_{AB} = \frac{F_{AB}}{AE}L_{AB} = \frac{F_{AB}}{A}\left(\frac{.15}{1.2 \times 10^{11}}\right) \text{m} \qquad \text{(b)}$$

$$\delta_{CD} = \frac{F_{CD}}{AE}L_{CD} = \frac{F_{CD}}{A}\left(\frac{.23}{1.2 \times 10^{11}}\right) \text{m} \qquad \text{(c)}$$

(We suggest that you check units to show that the units of A must be in meters.) For maximum allowable rotation of the block **compatibility** requires that

$$\delta_{AB} = (.20)\left(\frac{.05}{360}\right)(2\pi) = .0001745 \text{ m} \qquad \text{(d)}$$

$$\delta_{CD} = (.46)\left(\frac{.05}{360}\right)(2\pi) = .0004014 \text{ m} \qquad \text{(e)}$$

Example 5.11 (Continued)

Combining Eqs. (b) and (d) and Eqs. (c) and (e), we get

$$\frac{F_{AB}}{A} = 1.396 \times 10^8 \text{ Pa} \qquad \text{(f)}$$

$$\frac{F_{CD}}{A} = 2.094 \times 10^8 \text{ Pa} \qquad \text{(g)}$$

Equations (a), (f), and (g) comprise three equations for the three unknowns F_{AB}, F_{CD}, and A. Solving for F_{AB} and F_{CD} in terms of A in Eqs. (f) and (g) and substituting these results into Eq. (a), we get

$$(1.396 \times 10^8)(A)(.20) + (2.094 \times 10^8)(A)(.46) = (500{,}000)(.13)$$

$$\therefore A = 5.232 \times 10^{-4}\text{m}^2$$

The corresponding diameter is then

$$D = .02581 \text{ m}$$

Now compute the forces F_{AB} and F_{CD}. From Eqs. (f) and (g) we get

$$F_{AB} = (1.396 \times 10^8)(5.232 \times 10^{-4}) = 7.304 \times 10^4 \text{ N}$$

$$F_{CD} = (2.094 \times 10^8)(5.232 \times 10^{-4}) = 1.096 \times 10^5 \text{ N}$$

The maximum stress occurs in rod CD and is

$$\tau_{\max} = \frac{1.096 \times 10^5}{5.232 \times 10^{-4}} = 2.094 \times 10^8 \text{ Pa}$$

This is greater than the allowable stress of

$$\frac{4.80 \times 10^8}{2.5} = 1.920 \times 10^8 \text{ Pa}$$

so we conclude that *we cannot permit the deflection of .05°*. The *stress level* rather than the *deflection* becomes the controlling factor.

We then assign the stress in rod CD to be 1.920×10^6 Pa.[3] Equations (a), (b), and (c) with $F_{CD}/A = 1.920 \times 10^6$ Pa still apply, but in Eqs. (d) and (e) we must replace the angle .05° by an unknown quantity β. These equations are now restated as follows:

[3]By assigning the maximum permissible stress to rod CD, we shall be minimizing the diameter of the rods under the constraint that the allowable stress not be exceeded.

Example 5.11 (Continued)

$$.2F_{AB} + .46F_{CD} = (500{,}000)(.13) \tag{h}$$

$$\delta_{AB} = \frac{F_{AB}}{AE}L_{AB} = \frac{F_{AB}}{A}\left(\frac{.15}{1.2 \times 10^{11}}\right)\text{m} \tag{i}$$

$$\delta_{CD} = \frac{F_{CD}}{AE}L_{CD} = \frac{(1.920 \times 10^{8})(.23)}{1.2 \times 10^{11}} = 3.68 \times 10^{-4}\text{ m} \tag{j}$$

$$\delta_{AB} = (.2)\left(\frac{\beta}{360}\right)(2\pi) = 3.49 \times 10^{-3}\beta\text{ m} \tag{k}$$

$$\delta_{CD} = (.46)\left(\frac{\beta}{360}\right)(2\pi) = 8.029 \times 10^{-3}\beta\text{ m} \tag{l}$$

We have here as unknowns F_{AB}, F_{CD}, δ_{AB}, δ_{CD}, and β–five unknowns for which we have five equations. From Eqs. (k) and (l) we see on equating values of β that

$$\delta_{CD} = \frac{8.029 \times 10^{-3}}{3.49 \times 10^{-3}}\delta_{AB} = 2.30\delta_{AB}$$

Using the result of Eq. (j) for δ_{CD}, we get for δ_{AB} from above,

$$\delta_{AB} = 1.600 \times 10^{-4}\text{ m}$$

From Eq. (i) we get

$$1.600 \times 10^{-4} = \frac{F_{AB}}{A}\left(\frac{.15}{1.2 \times 10^{11}}\right)$$

$$\therefore \frac{F_{AB}}{A} = 1.280 \times 10^{8} \tag{m}$$

In Eq. (h) we now divide through by A. We thus have

$$(.2)\frac{F_{AB}}{A} + (.46)\frac{F_{CD}}{A} = \frac{(500{,}000)(.13)}{A}$$

Replacing F_{AB}/A by 1.280×10^8 as per Eq. (m) and F_{CD}/A by 1.920×10^8 Pa (the maximum stress allowed) in the equation above, we get

$$(.2)(1.280 \times 10^{8}) + (.46)(1.920 \times 10^{8}) = \frac{(500{,}000)(.13)}{A}$$

$$\therefore A = 5.706 \times 10^{-4}\text{ m}^2$$

Hence, the minimum diameter is

$$D_{\min} = .02695\text{ m} = 26.95\text{ mm}$$

In practice we use the closest stock diameter larger than the diameter above.

5.7 Thermoelastic Problems

Up to this time we have been considering isothermal, one-dimensional problems. We shall now turn our attention to so-called thermoelastic problems. In Chapter 4 we formulated a one-dimensional constitutive law for Hookean materials in the presence of a temperature field $T(x,y,z)$. What happens to the other laws for this kind of thermoelastic problem? Newton's law is unchanged, since it is based on pure mechanical considerations. The strain-displacement relations are unchanged since they are based on purely geometrical considerations. Accordingly, the procedures for thermoelastic problems described in Chapter 4 are solved in the same way as the preceding isothermal problems. We shall now consider two one-dimensional thermoelastic problems.

Example 5.12

A steel pipe is held by two fixed supports as shown in Fig. 5.31. When mounted, the temperature of the pipe was 15°C. In use, however, cold fluid moves through the pipe, causing it to cool considerably. If we assume that the pipe ends up with a uniform temperature of −18°C and if we take the coefficient of linear expansion to be 11.7 × 10^{-6}/°C for the temperature range involved, determine the force on the wall as a result of this cooling. We shall neglect gravitational forces.

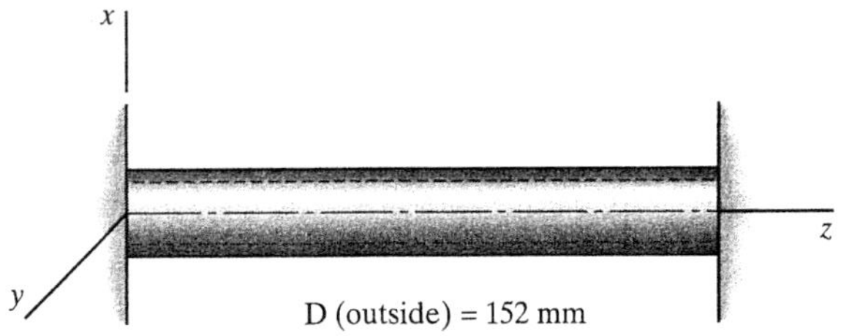

Figure 5.31. Cold water is sent through pipe.

Away from the end supports, we can assume that the stress resulting from the temperature change is a one-dimensional stress distribution in the z direction. Because of the constraint in the z direction imposed by the walls and because of the uniformity of temperature and geometry in the z direction, **compatibility** requires that $\varepsilon_{zz} = 0$. Hence, using the one-dimensional temperature-dependent Hooke's law [Eq. (4.17)] as the **constitutive law,** for τ_{zz} we get

$$\tau_{zz} = -E\alpha\Delta T$$

Since the temperature drops, ΔT is negative and τ_{zz} is positive (tensile). This makes sense since as the pipe is cooled it tries to shrink

Example 5.12 (Continued)

but the fixed supports constrain this action and accordingly induce tensile stress in the pipe.

The force P acting on the supports is now available from **equilibrium** considerations. Thus,

$$P = \tau_{zz} A = -E\alpha\Delta TA$$

We may evaluate τ_{zz} and P numerically. Taking $E = 1.8 \times 10^{11}$ Pa for steel, we have

$$\tau_{zz} = -(1.8 \times 10^{11})(11.7 \times 10^{-6})(-33) = 6.95 \times 10^7 \text{ Pa}$$

$$\therefore\ P = 6.95 \times 10^7 \frac{\pi(.152^2 - .127^2)}{4} = 3.807 \times 10^5 \text{ N}$$

$$P = 3.807 \times 10^5 \text{ N}$$

We see that considerable forces can be developed by thermal effects.

Example 5.13

In Fig. 5.32 is shown a steel bolt and nut and an aluminum sleeve initially touching with no stress. The bolt has 16 threads/in., and when the material is at 60°F, the nut is tightened one-eighth turn. The temperature is then raised from 60°F to 100°F. Determine the stresses in both the bolt and the sleeve. Neglect gravitational forces and take the coefficient of linear expansion for steel and aluminum to be 6.5×10^{-6}/°F and 12×10^{-6}/°F, respectively. The modulus E for aluminum is 10×10^6 psi and that for steel is 30×10^6 psi.

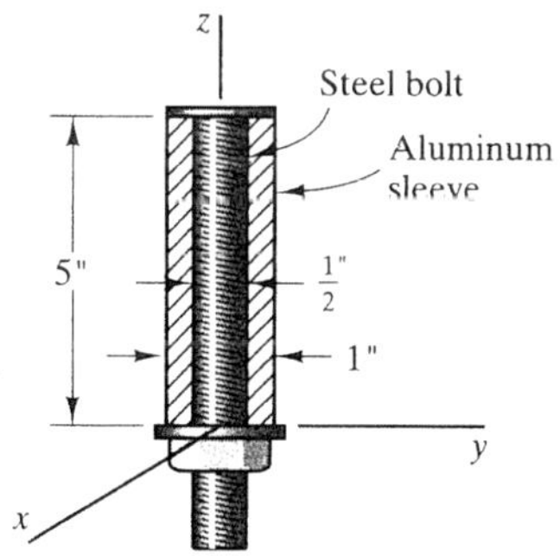

Figure 5.32. Nut–bolt–sleeve assembly.

We shall do this problem in two phases. We shall first compute stresses caused by turning the nut only. Such stresses will be denoted by subscript 1. Then we shall compute the stresses caused by thermal action only. We shall denote the thermal stresses with subscript 2.

Accordingly, we note that turning the nut results in a tensile stress $[(\tau_{zz})_1]_B$ in the bolt. Using Hooke's law as the **constitutive law,** the strain in the bolt is then

$$[(\varepsilon_{zz})_1]_B = \frac{[(\tau_{zz})_1]_B}{E_B} \tag{a}$$

As for the sleeve, let us employ **compatibility** considerations at this time. The change in length of the sleeve must equal the distance moved by the nut. Were there no strain in the steel bolt, the nut

Example 5.13 (Continued)

would move a distance equal to the number of turns of the nut times the advance of the nut per turn (i.e., the so-called lead). However, the bolt will extend by an amount

$$[(\tau_{zz})_1]_B L / E_B$$

and thus decrease the movement of the nut by an equal amount. Thus the net movement of the nut upward is

$$\{(turns)(lead) - [(\tau_{zz})_1]_B L / E_B\}$$

Turning the nut compresses the sleeve and thus the net change in length of the sleeve must accordingly be

$$\Delta L = -\left\{(turns)(lead) - \frac{[(\tau_{zz})_1]_B}{E_B} L\right\} = -\left(\frac{1}{8}\right)\left(\frac{1}{16}\right) + \frac{[(\tau_{zz})_1]_B}{30 \times 10^6} L \quad \text{(b)}$$

The strain in the sleeve is then found by dividing Eq. (b) by L. Thus

$$[(\varepsilon_{zz})_1]_S = \frac{\Delta L}{L} = -.001563 + \frac{[(\tau_{zz})_1]_B}{30 \times 10^6} \quad \text{(c)}$$

The stress in the sleeve is now easily expressed using *Hooke*'s law as

$$[(\tau_{zz})_1]_S = \left\{-.001563 + \frac{[(\tau_{zz})_1]_B}{30 \times 10^6}\right\} 10 \times 10^6 = -15{,}630 + \frac{1}{3}[(\tau_{zz})_1]_B \quad \text{(d)}$$

After the nut has been tightened one-eighth of a turn, it is in **equilibrium** under the force from the sleeve, F_S, and the force from the bolt, F_B (see Fig. 5.33), and we can say

$$-F_S + F_B = 0$$

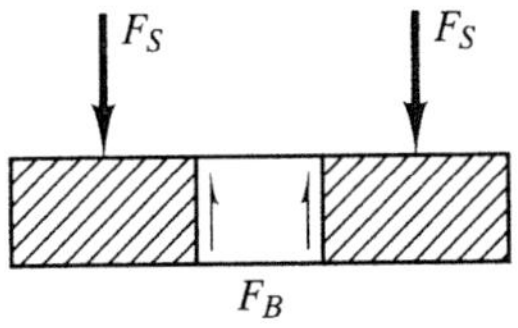

Figure 5.33. Free-body diagram of the nut.

Note that the stress in the sleeve given by Eq. (d) already accounts for the direction of the stress. Thus, we may write

$$[(\tau_{zz})_1]_S \frac{\pi}{4}\left[1^2 - \left(\frac{1}{2}\right)^2\right] + [(\tau_{zz})_1]_B \frac{\pi}{4}\left(\frac{1}{2}\right)^2 = 0 \quad \text{(e)}$$

Substituting from Eq. (d) for $[(\tau_{zz})_1]_S$, on canceling terms we get

$$3\left\{-15{,}630 + \frac{1}{3}[(\tau_{zz})_1]_B\right\} + [(\tau_{zz})_1]_B = 0$$

Example 5.13 (Continued)

Solving for $[(\tau_{zz})_1]_B$ we get

$$[(\tau_{zz})_1]_B = 23{,}445 \text{ psi} \qquad \text{(f)}$$

From Eq. (d) we get

$$[(\tau_{zz})_1]_S = -15{,}630 + 7815 = -7815 \text{ psi}$$

As expected, the stress in the bolt is positive (tensile) and the stress in the sleeve is negative (compression).

Next, consider the stress owing to temperature change. We have the following equations using the **constitutive law** given by Eq. (4.17), with $(\tau_{zz})_2$ representing here only the thermally induced stress:

For bolt:

$$\begin{aligned}[(\varepsilon_{zz})_2]_B &= \frac{1}{E_B}[(\tau_{zz})_2]_B + \alpha_B \Delta T \\ &= \frac{[(\tau_{zz})_2]_B}{30 \times 10^6} + 2.6 \times 10^{-4}\end{aligned} \qquad \text{(g)}$$

For sleeve:

$$[(\varepsilon_{zz})_2]_S = \frac{[(\tau_{zz})_2]_S}{10 \times 10^6} + 4.8 \times 10^{-4} \qquad \text{(h)}$$

Compatibility requires that

$$[(\varepsilon_{zz})_2]_B = [(\varepsilon_{zz})_2]_S \qquad \text{(i)}$$

Hence, from Eqs. (g) and (h), we can say that

$$\frac{[(\tau_{zz})_2]_B}{30 \times 10^6} + 2.60 \times 10^{-4} = \frac{[(\tau_{zz})_2]_S}{10 \times 10^6} + 4.80 \times 10^{-4}$$

This leads to the following equation:

$$[(\tau_{zz})_2]_B = 3[(\tau_{zz})_2]_S + 6600 \qquad \text{(j)}$$

Furthermore, **equilibrium** requires for the nut (Fig. 5.33) that

$$[(\tau_{zz})_2]_S \frac{\pi}{4}\left[1^2 - \left(\frac{1}{2}\right)^2\right] + [(\tau_{zz})_2]_B \frac{\pi}{4}\left(\frac{1}{2}\right)^2 = 0$$

Example 5.13 (Continued)

which can be reduced to

$$3[(\tau_{zz})_2]_S = -[(\tau_{zz})_2]_B \tag{k}$$

Now, substituting from Eq. (k) into Eq. (j), we have

$$[(\tau_{zz})_2]_B = -[(\tau_{zz})_2]_B + 6600$$

Hence,

$$[(\tau_{zz})_2]_B = 3300 \text{ psi} \tag{l}$$

Also from Eq. (k), we have

$$[(\tau_{zz})_2]_S = -1100 \text{ psi} \tag{m}$$

Thus, the increase in temperature results in tensile stress in the bolt and compressive stress in the sleeve.

Note that the effect of temperature is not as readily apparent as in the previous example. In that example, the supports were fixed and thus the thermally induced strains were zero. Thus, a drop in temperature indicates at the outset that the pipe is in tension and vice versa. By contrast, for the problem at hand, the result of a change in temperature is that the thermally induced strains are not zero. It is only through careful analysis involving the use of equilibrium, compatibility, and the constitutive law that we can correctly assess the effect of a temperature change in such problems.

The *total* stress for the bolt and sleeve can now be given. We get

$$[(\tau_{zz})_B]_{\text{total}} = 23{,}445 + 3300 = 26{,}745 \text{ psi}$$
$$[(\tau_{zz})_S]_{\text{total}} = -7815 + 1100 = -8915 \text{ psi}$$

5.8 Closure

The study of one-dimensional stress problems in this chapter has served a number of purposes. First, we have examined a class of problems that has considerable practical importance. Second, we have been able to use the basic laws of solid mechanics in a concerted, coordinated manner on problems still simple enough to permit a reasonable overview of the computation process. Finally, we were able to examine a number of effects side by side, namely linear-elastic behavior, plastic action, and thermal stress phenomena.

In subsequent chapters we shall proceed to multidimensional stress problems. The overall approach in such undertakings will be essentially what has been presented in this chapter. However, because of rapidly escalating complexity encountered in such undertakings, we shall perforce have to carefully choose topics to be studied. For instance, we shall have to limit severely our work in thermal stress.

As a first step we next consider stress-strain relations for a multi-dimensional state of stress.

*5.9 A Look Ahead: Basic Laws of Continua

In your dynamics course, three alternate approaches were explored. They were, broadly speaking:

1. Direct applications of Newton's laws.
2. Energy methods.
3. Linear momentum methods and moment-of-momentum methods.

These all emanate from a *common* source (i.e., Newton's law) and so we can use any one for particle and rigid-body problems. In deformable solid mechanics, we have thus far used Newton's law of the above trio. In the next chapter, we will begin to study energy methods as an alternative way of attacking problems. Energy methods will be an important part of this text and will appear in a number of chapters.

Later, when you study more complex continua such as a flowing fluid with heat transfer and compression, you will have to satisfy **four basic laws.** These basic laws are:

1. Conservation of mass.
2. Linear and moment of momentum (these are, at this point in your studies, Newton's law).
3. First law of thermodynamics.
4. Second law of thermodynamics.

You will study these laws in both your upcoming Thermodynamics course and your Fluid Mechanics course. For more general continua that you will contend with, the above four basic laws are generally **independent** of each other in contrast to our Solids course where they provide generally alternate approaches. As indicated in another Look Ahead section, we pointed out that these laws could be applied to a system (a free body) or more likely to a control volume.

For the **linear momentum equation,** we take the vector sum of external forces acting on the control surface plus the vector sum of forces acting on anything inside the control volume, and we equate this to the rate of flow of linear momentum through the control surface plus the rate of change of linear momentum inside the control volume.

Using a similar approach, for the **moment of momentum,** we sum the moments about a fixed point in inertial space of any external forces acting on the control surface and also of any external forces acting on anything inside the control volume, and we equate this total moment with the rate of flow of angular momentum through the control surface plus the rate change of angular momentum inside the control volume.

For the **first law of thermodynamics,** we add the rate of heat flow through the control surface, plus the rate of work from fluids passing through the control surface, plus the equivalent rate of work from electric currents in wires passing through the control surface, and, finally, plus the rate of work transmitted by shafts or any other devices passing through the control surface, all at time t. This is then equated to the rate of flow of mechanical and internal energy through the control surface plus the rate of change of the mechanical and internal change inside the control volume at time t.

For **conservation of mass,** we equate the net efflux rate of mass through the control surface with the rate of decrease of mass inside the control volume.

Although we can make a similar approach for the **second law of thermodynamics,** it is not as useful in this form. You will spend much time in your Thermodynamics course studying this vital and sophisticated law.

Highlights (5)

In solving problems in solid mechanics, we may use the **St. Venant principle,** which states that if the resultant of a force distribution acting on the boundary of a body is used, the results of the calculations will be valid away from the boundary. The formulations that we use for solving solid mechanics are then:

Equilibrium **Constitutive law** **Compatibility**

For certain problems, there may also be the need to consider **thermal effects.**

One-dimensional problems in solid mechanics have only one nonzero stress in a free body. The problem may be statically determinate or statically indeterminate. For the former, the stresses can be solved by using only equilibrium. In the latter, the deformation must simultaneously be considered. Carefully chosen free-body diagrams are essential in attacking problems even for the simpler one-dimensional problems.

In problems involving bodies of different materials where stress beyond the yield stress has occurred in one or more members of the system, there is the possibility that, when the loads have been removed, stresses remain in the members. This stress is called **residual stress.** It results from the fact that during the unloading process the members are moving along different stress-strain curves, which causes the members to interfere with each other's decreasing deformation somewhat like two people holding on to each other but each trying to move at a different speed. When the movement of the two people stops, a force system is still left between them. In solid mechanics, this force system causes residual stresses.

In design, there may be a number of conflicting requirements that must be addressed while attempting to reach an optimal solution. In Example 5.11, we required the smallest diameter members that still satisfied a given safety factor for stress and had a maximum allowable deformation. First, the maximum deformation was considered since that would lead to the minimum diameter of the rods. The first analysis was then carried out to determine the stresses in the members. We then realized that we exceeded the allowable stress. A second analysis was carried out, this time allowing for the maximum allowable stress and thus restricting deformation. We used this design as it satisfied all the requirements of the problem.

Finally, for thermal stress problems there may be a thermal stress contribution to that stress which is due to mechanical constraint. We must use a constitutive law encompassing the thermal effects. If the mechanical properties are not seriously affected by the temperature and if the deformation does not itself create temperature effects (the usual situation), then the analysis may proceed as in an isothermal problem.

PROBLEMS

5.1. [5.3] In Fig. P.5.1 are three solid cylindrical bars welded together. Each bar is assumed elastic, perfectly plastic in its behavior having E as 2×10^{11} Pa and a yield point of 4.2×10^{8} Pa. What tensile force is needed to move end B relative to A a distance of .3 mm? What tensile force is needed to move end B a distance of 1.8 mm relative to end A?

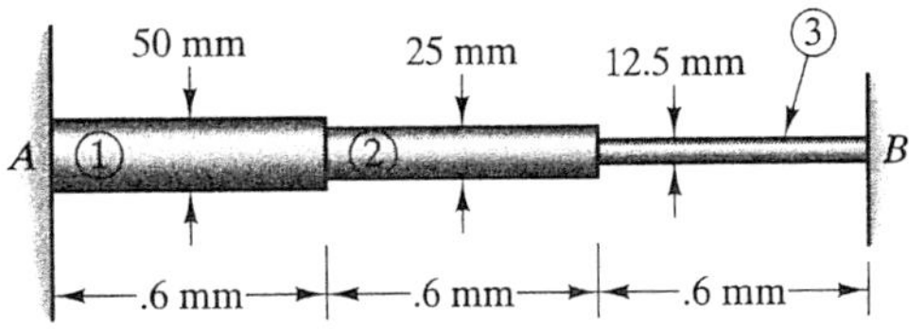

Figure P.5.1.

5.2. [5.3] A 45,000 N force (Fig. P.5.2) acts on a rod A having a modulus of elasticity E of 7×10^{10} Pa. The diameter of A is 50 mm and the length is .3 m. A sleeve B restrains A through a rigid plate C. The mean radius of B is .1 m and its thickness is 2.5 mm. The length of B is .2 m. If B has a modulus of elasticity in compression of 1.4×10^{11} Pa, what is the movement of end D? What is the movement of the plate C?

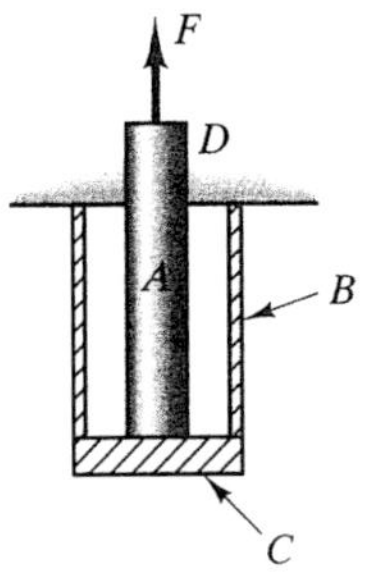

Figure P.5.2.

5.3. [5.3] If $Y_A = 160{,}000$ psi and $Y_B = 200{,}000$ psi in Problem 5.2, what are the stresses in the members if end D is moved upward .3 in.? If D is then released, what is the permanent set? Assume that the bodies act like they are elastic, perfectly plastic.

5.4. [5.3] A cylinder having nonlinear, elastic behavior is shown in Fig. P.5.4(a) hanging by its own weight. If the stress-strain diagram for the material is as shown in Fig. P.5.4(b), what is the deflection of end A of the cylinder? Take γ as a constant equal to 5.76×10^{4} N / m³. Take $D = .15$ m and $L = 1.5$ m.

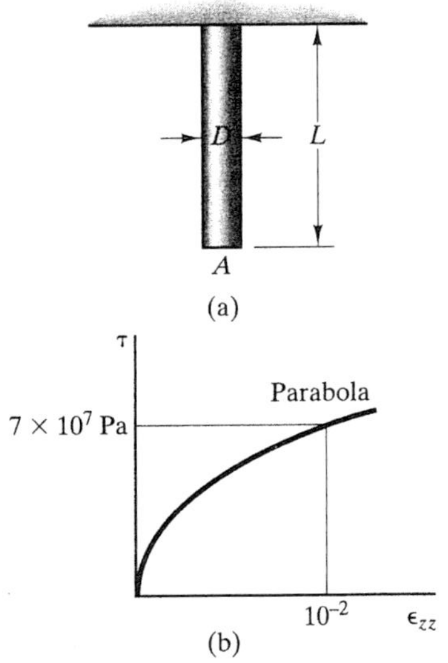

Figure P.5.4.

5.5. [5.3] Do Problem 5.4 for the case where γ varies linearly from 2.88×10^4 N/m^3 at the top to 5.7×10^4 N/m^3 at the bottom. Use the other data as given.

5.6. [5.3] Derive an expression for the elongation of a tapered rod under an axial load, as shown in Fig. P.5.6. Set up the integral for Δ. If computer software is available show that

$$\Delta = \frac{4PL}{\pi d_1 d_2 E}$$

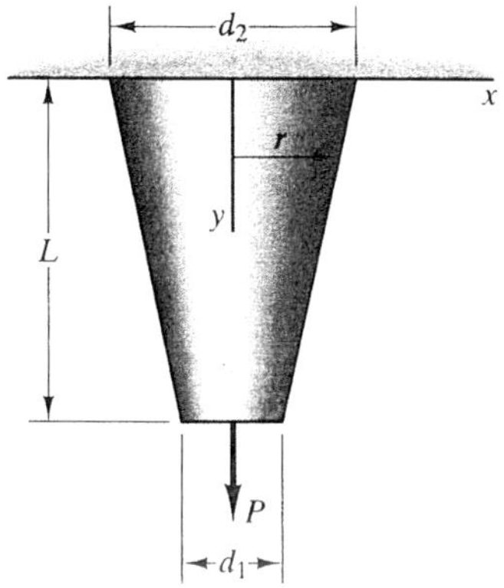

Figure P.5.6.

5.7. [5.3] Using the result from the previous problem, compare, in Fig. P.5.7, the end displacement of a tapered circular rod with that of a uniform rod. The tapered rod has a base diameter of 2 in. and a tip diameter of 1 in. The uniform rod has a diameter of 1.5 in., which is the average diameter of the tapered rod. Take $E = 10 \times 10^6$ psi. From Problem 5.6,

$$\Delta = \frac{4PL}{\pi d_1 d_2 E}$$

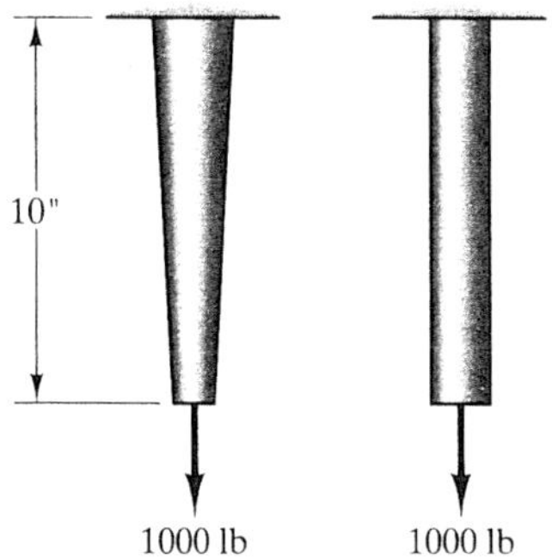

Figure P.5.7.

P.5.8. [5.3] A standard climbing rope has a diameter of 10.5 mm and is 50 m long. If it stretches 5.5% over 1 m when a mass of 80 kg is suspended by it, what is its apparent elastic modulus? Neglect the change in diameter from stretching.

5.9. [5.3] For a femoral bone (upper leg) of 35 cm in length, how much will it stretch before fracture if the ultimate stress in tension is assumed to be 130 MPa and E is 14 GPa?

5.10 [5.3] Consider a composite bar in Fig. P.5.10 made up of isotropic fibers (E_f, v_f) in an isotropic material called a matrix (E_m, v_m). If in the x direction the cross-sectional areas are A_f and A_m for the fibers and the matrix, respectively, determine the effective elastic modulus of the composite in the x direction.

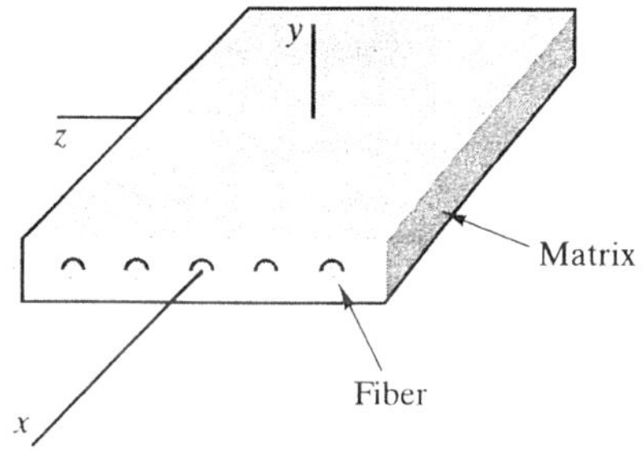

Figure P.5.10.

5.11. [5.3] What is the highest pressure above atmosphere before there is yielding in the cylinder wall A or in the piston rod B (see Fig. P.5.11)? A rigid end plate and reinforcing rings absorb radial effects of the pressure.

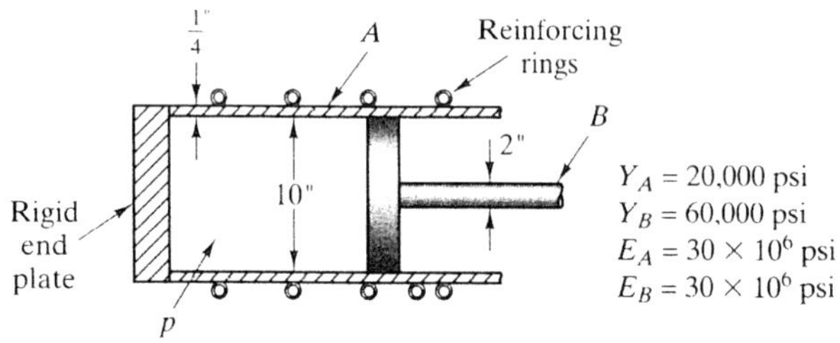

Figure P.5.11.

5.12. [5.3] A one-dimensional specimen is made of a material which, like copper, has a stress-strain relation that can be approximated analytically by the so-called *sinh law* given as follows:

$$\varepsilon = \varepsilon_0 \sinh \frac{\tau}{\tau_0}$$

where ε_0 and τ_0 are constants dependent on the particular material. If a solid cylinder of such a material (see Fig. P.5.12) is subject to a force F of 1000 lb, what is the new length of the cylinder in terms of ε_0 and τ_0? Take $D = 2$ in. and $L = 5$ ft.

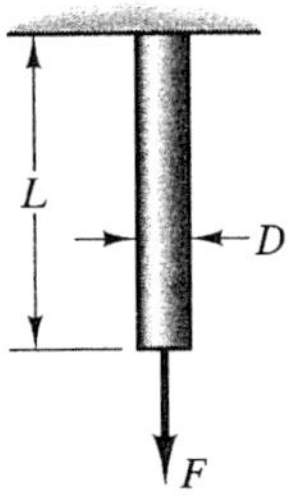

Figure P.5.12.

5.13. [5.3] A plate of thickness t is shown hanging by its own weight in Fig. P.5.13. An xz plane is shown in which the profile of the body is that of a parabola wherein at A the side of the parabola is vertical. Take γ as 200 lb/ft^3 and E as 8×10^6 psi. What is the deflection of end A owing to the weight? First set up the quadrature. Then, if computer software is available, carry out the calculation.

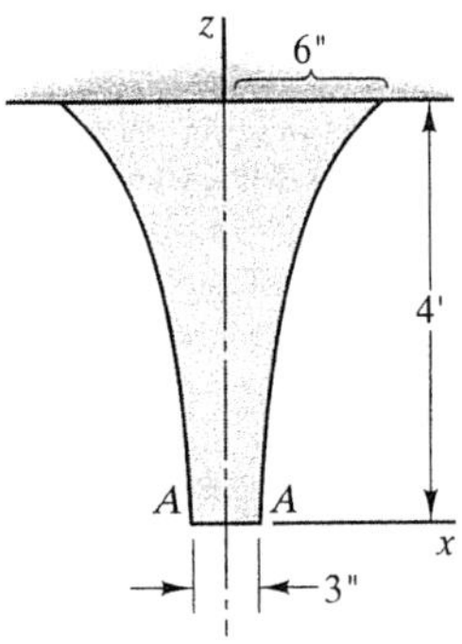

Figure P.5.13.

***5.14. [5.3]** Do Problem 5.13 for the case where it is a body of revolution. From previous problem, $x = .001953z^2 + 1.5$.

5.15. [5.3] A system of two cylinders A and B and a rigid block C in Fig. P.5.15 are to be given an acceleration a upward by support D. What is the maximum acceleration in g's that can be given to the system in outer space without causing yielding anywhere in the cylinders?

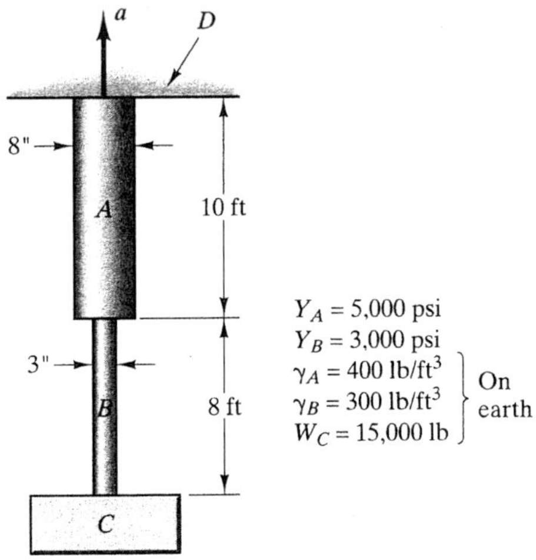

Figure P.5.15.

5.16. [5.3] A rod having a diameter of 6 in. and a Young's modulus of 20×10^6 psi is shown in Fig. P.5.16 loaded at sections A and B by loads $F_1 = 10{,}000$ lb and $F_2 = 5000$ lb. What is the displacement at the end of the rod as a result of the loads?

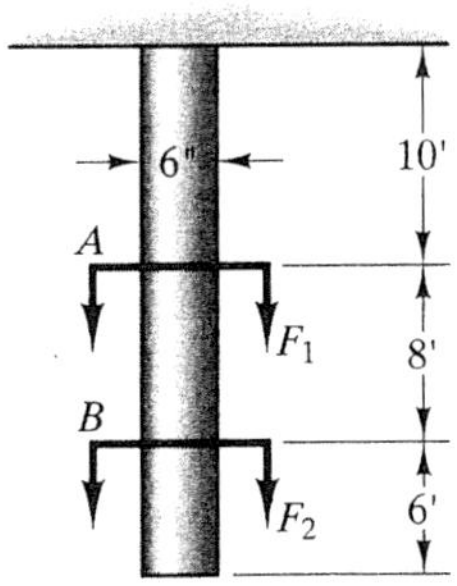

Figure P.5.16.

5.17. [5.3] In Problem 5.16, if we require that $F_2 = .673\ F_1$, what is the maximum deflection δ_{El} possible for which the body behaves completely elastically? Take $Y = 100{,}000$ psi.

5.18. [5.3] Shown in Fig. P.5.18(a) is a hoop composed of an aluminum inner part and a steel outer part wrapped around a rigid cylinder having a smooth surface. A slice of width 7.5 mm has been cut from the hoop as shown. The two portions of the hoop are free to slide relative to each other. If the gap is forced closed by attaching lugs [see Fig. P.5.18(b)] and tightening the nut-bolt arrangement, what are the stresses in the aluminum and steel members? Take $E_{Al} = 7 \times 10^{10}$ Pa and $E_{Stl} = 1.9 \times 10^{11}$ Pa.

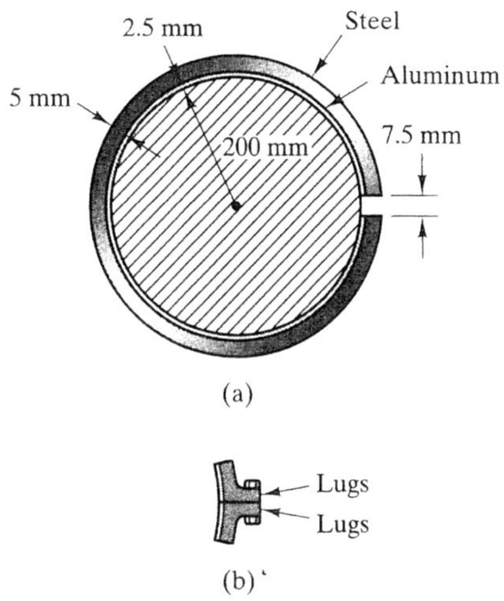

Figure P.5.18.

5.19. [5.3] In Problem 5.18, find the pressure exerted on the rigid cylinder as a result of tightening the hoop.

5.20. [5.3] A material has a yield point of 1×10^9 Pa and is assumed to behave as an elastic, plastic material with linear strain hardening. Take E in the elastic range as 1×10^{11} Pa. If the stress of 1.3×10^9 Pa causes a strain of .045, what stress is needed to reach a strain of .05? If the body is released from this stress, what is the permanent strain induced? For a rod of diameter 50 mm and a length of .6 m, what force is required to cause an elongation of 48 mm if the material behaves as described above? When the rod is released, what is the new length?

5.21. [5.3] A wooden pile is being driven into a hole in the ground as shown in Fig. P.5.21 by a force F = 100,000 lb. A friction force of f lb/unit length opposes this force. The force intensity f varies as the square of the distance z from the top surface, being zero at the top. If the modulus E of the pile is 2×10^6 psi, L = 30 ft, and D = 12 in., how much has the pile shortened at the given loading conditions?

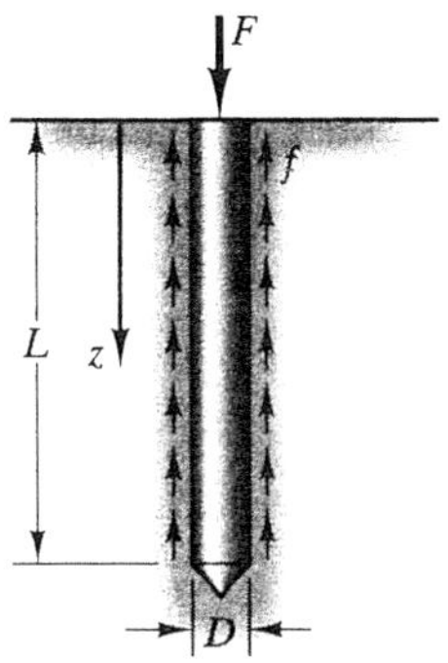

Figure P.5.21.

5.22. [5.3] A rigid cylinder B (see Fig. P.5.22) rotates at a speed ω of 500 rpm. Two rods G of length L = .45 m are attached to the rigid cylinder B. If the cross-sectional area of each rod is 6.25×10^{-4} m^2 and the modulus of elasticity is 1.4×10^{11} Pa, what is the change in length of the rods as a result of deformation? The rods have a mass per unit length of 3 kg/m. Take D = .3 m. If the yield stress is 6×10^8 Pa, what is the maximum speed ω for the system for no yielding anywhere?

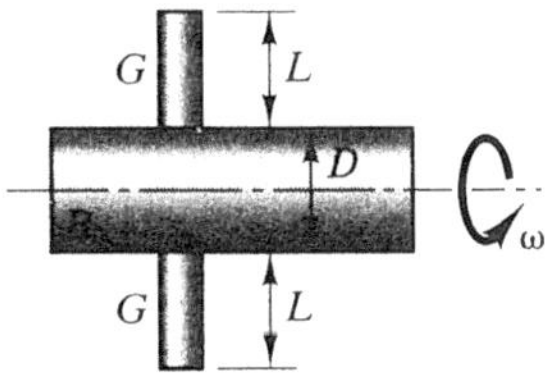

Figure P.5.22.

5.23. [5.3] In Problem 5.22, what is the maximum allowable speed if the yield stress is 1×10^9 Pa and a safety factor of 1.8 is used?

5.24. [5.3] Find Δ_{GA} and Δ_{CD} as a result of the uniform movement of the system up the incline in Fig. P.5.24.

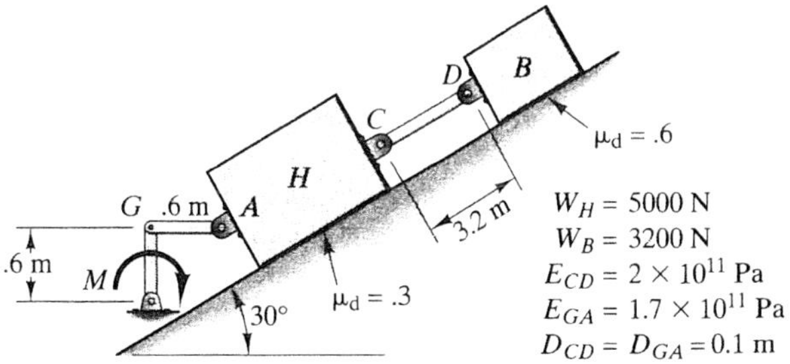

Figure P.5.24.

5.25. [5.3] A cylindrical vertical column in Fig. P.5.25 has a specific weight $\gamma_0 = 400$ lb/ft^3 at the top and a specific weight $\gamma_A = 600$ lb/ft^3 at the bottom. Furthermore, γ varies as the *square* of z. If $E = 20 \times 10^6$ psi, what is the deflection due to gravity of the top? (*Hint:* Watch units; suggest that you work everything in feet.)

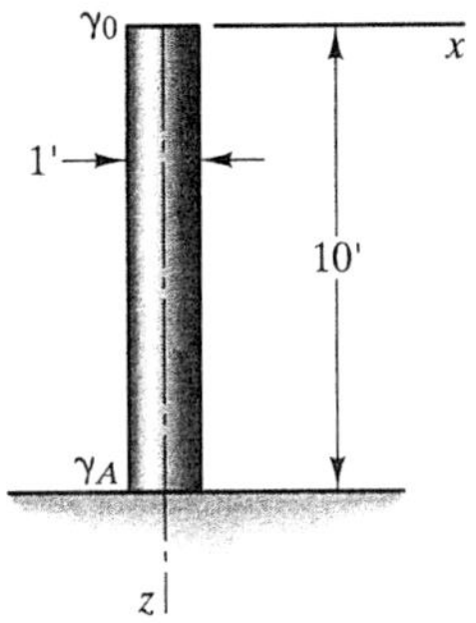

Figure P.5.25.

5.26. [5.3] Three mountain climbers (Fig. P.5.26) are trying to get one of their company as far down a glacier as possible. The climbers weight 150, 175, and 220 lb, respectively. They have three lines of nylon for their use; one line is 30 ft unstretched, while the other two are 20 ft unstretched each. The nylon cord has an effective modulus of elasticity of 4×10^4 psi and has a diameter of about 1/2 in. unloaded. What arrangement of climbers and lines should be used, and what is the maximum distance down that the group can reach if an arrangement like the one shown in the diagram is utilized? What is the key assumption that weakens your solution? Explain in what direction the solution is wrong as a result of your assumption.

Figure P.5.26.

5.27. [5.3] A block A weighs 90,000 N (Fig. P.5.27) and is to be supported by three steel members DC, CE, and EF. What is the approximate horizontal and vertical deflection of pins C and F as a result of gravity? Take the block A as rigid and take $E = 1.93 \times 10^{11}$ Pa for each of the members. (*Hint:* What does the state of strain in member CE imply about the vertical and horizontal movements of pin C? Also neglect vertical movements of C and F resulting from rotation of DC and EF.)

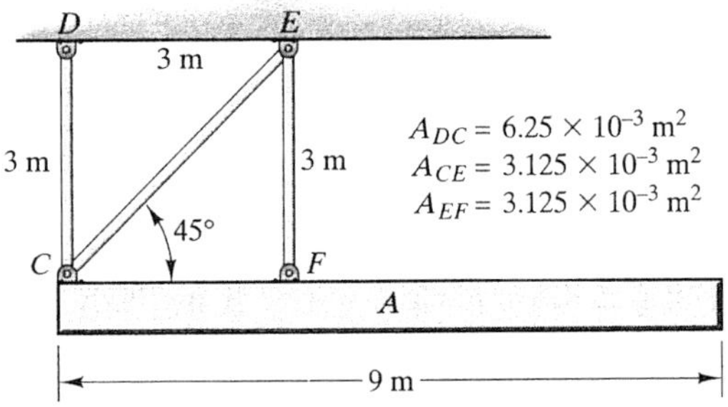

Figure P.5.27.

5.28. [5.3] An elastic rod is welded to a rigid drum in Fig. P.5.28 which rotates at a speed $\omega = 100$ rpm. Two rigid discs slide along the rod and are then welded to the rod at locations shown. What is the radial extension of end A of the rod as a result only of the rotation? Consider that where the rigid discs enclose the rod, there is no elongation of the rod making these parts of the rod rigid. Hence these parts of the rod contribute rigid mass to that of each disc each of whose mass by itself is 4.5 kg.

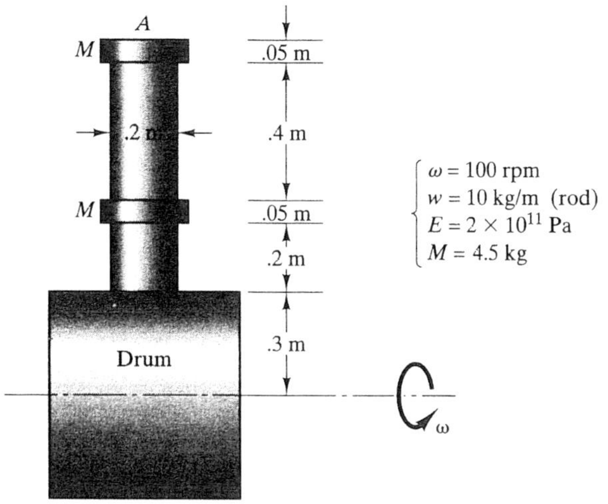

Figure P.5.28.

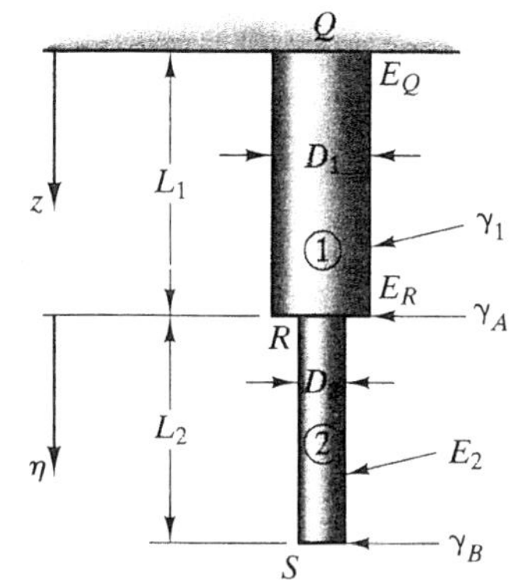

Figure P.5.29.

5.29. [5.3] A composite member in Fig. P.5.29 is made up of two parts welded together as shown in the diagram. The following are the data:

Member (1). E varies as the *square* of the distance from Q going from a value of E_Q at position Q to a value of E_R at position R. $\gamma = \gamma_1$ a constant.

Member (2). E is constant and equal to E_2. γ varies as the *cube* of the distance from R going from γ_R to γ_S.

Show that the deflection Δ_R of the section at R from gravity is

$$\Delta_R = \int_0^{L_1} \frac{\gamma_1(L_1 - z) + D_2^2 L_2(3\gamma_R + \gamma_S)/4D^2}{\left[E_Q + \left(\dfrac{z}{L_1}\right)^2(E_R - E_Q)\right]}\, dz$$

Suggestion: Make use of the two references shown.

5.30. [5.3] A piston assembly, made from a rod and a hollow sleeve of different materials, is fixed to rigid plates at both ends. The piston is installed in a long cylinder as shown in Fig. P.5.30.

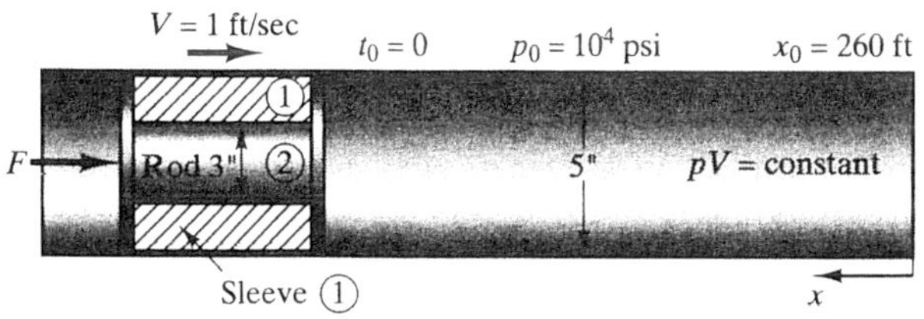

Figure P.5.30.

An external force F is applied to the piston to keep it compressing the air at a constant velocity of V = 1 ft/sec. For the air in the cylinder we have pV = constant, and at t_0, $p_0 = 10^4$ psi and x_0 = 260 ft.

Up to what time will both the rod and the sleeve experience elastic behavior? Take $E_1 = 15 \times 10^6$ psi and $E_2 = 30 \times 10^6$ psi and Y_1 = 49,000 psi and Y_2 = 60,000 psi.

Note that V is used for volume and for velocity. The context should make it clear as which meaning is to be used.

5.31. [5.4] A typical belay anchor in ice climbing consists of three ice screws placed as shown in Fig. P.5.31. A nylon rope called a cordellete is strung through the screws as shown. For an applied force P, what are the forces on the screws?

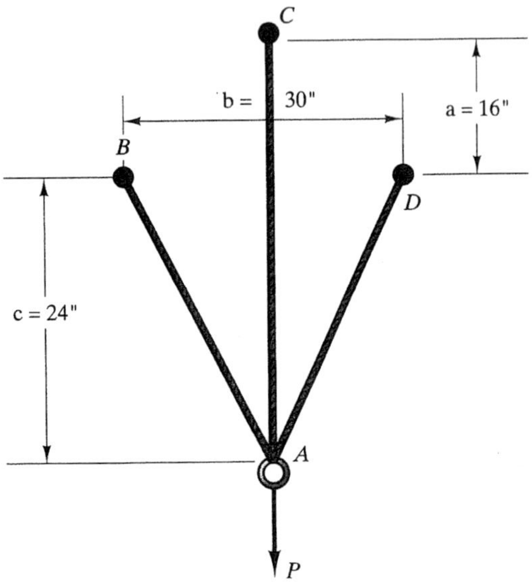

Figure P.5.31.

5.32. [5.4] In the previous problem, where should the anchor screw C be located so as to give equal force to each screw (see Fig. P.5.32)? Report the distance a as measured from point D as shown.

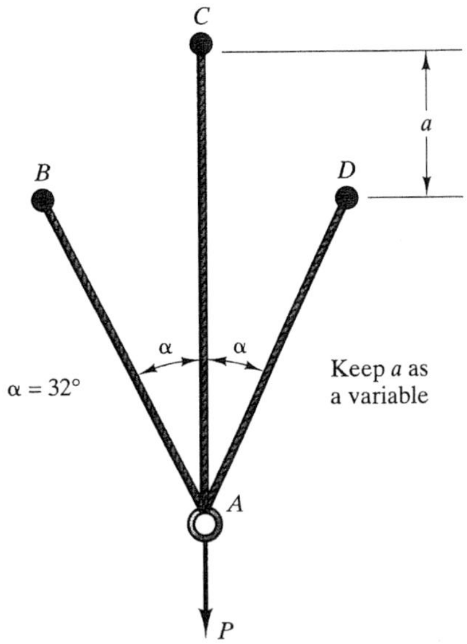

Figure P.5.32.

5.33. [5.4] A rigid beam AB is supported by a steel rod and an aluminum rod as shown in Fig. P.5.33. A load P of 675,000 N is applied to A, causing AB to deflect downward. Now the screw at B is adjusted so that B is depressed to a position where the rod is once again horizontal. How much must the screw at B descend from its initial orientation to achieve this?

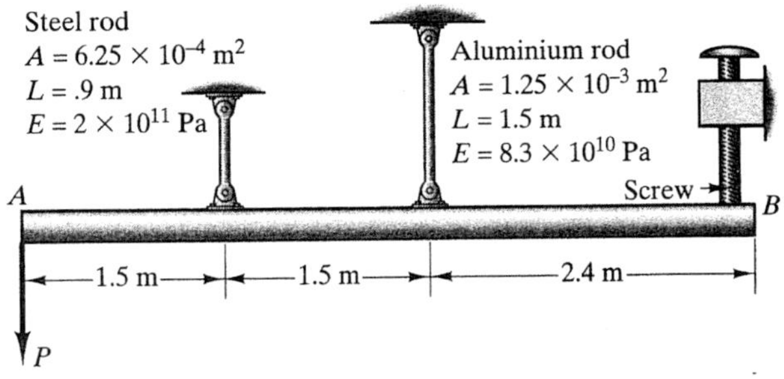

Figure P.5.33.

5.34. [5.4] A rod B of diameter 1 ft is suspended from a support G (see Fig. P.5.34) and is connected to a sleeve A by a rigid connection, CD. If $\gamma_B = 460$ lb/ft^3, $E_B = 30 \times 10^6$ psi, $\gamma_A = 256$ lb/ft^3, and $E_A = 10 \times 10^6$ psi, what is the deflection of DC resulting from the action of gravity on sleeve and cylinder? (*Hint:* Solve for $\int_0^{10} \tau_A dz$ using basic procedures and then get Δ.)

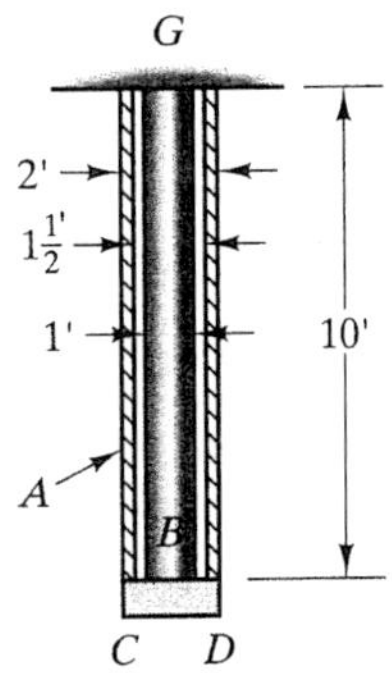

Figure P.5.34.

5.35. [5.4] In Problem 5.34 take $Y_B = 100{,}000$ psi and $Y_A = 60{,}000$ psi. Using a safety factor of 4, how many g's of acceleration in the vertical direction at G can be imposed on the device? Neglect mass of plate CD. Include gravity.

5.36. [5.4] A steel sleeve (3) and an aluminum shaft (1) and (2) are held between rigid end plates A and B (Fig. P.5.36). A force F_2 is applied to end plate B. A second force F_1 is applied to the sleeve (only). What is the elongation of the system as a result of these loads? Use the following data. Neglect the weight.

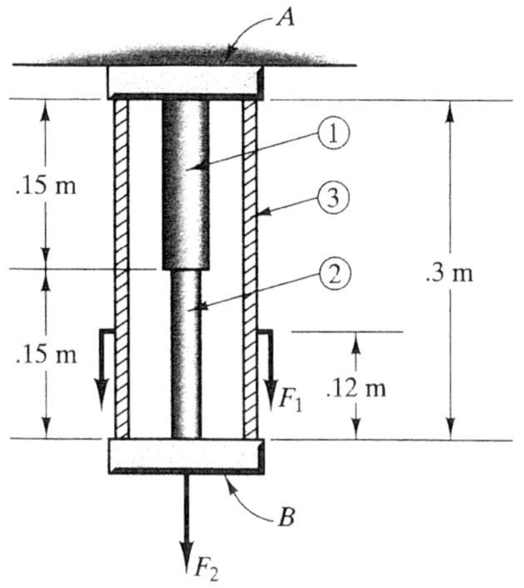

Figure P.5.36.

Sleeve (steel)	Shaft (aluminum)	
$E_{St} = 2 \times 10^{11}$ Pa	$E_{Al} = 1 \times 10^{11}$ Pa	$F_1 = 1000$ N
$D_{outside} = .12$ m	$D_1 = .07$ m	$F_2 = 5000$ N
Thickness = 10 mm	$D_2 = .04$ m	

A word to the wise: Make separate free-body diagrams of *plate A*, *plate B*, and the *sleeve*.

5.37. [5.4] A horizontal *stiff* bar AC in Fig. P.5.37 is pivoted at B and connects to a thin cylindrical rod DG and to a spring at C. A force of 5000 lb is applied at A. What is the rotation of bar AC assuming that it is rigid? $K = 18{,}000$ lb/in. and E for DG is 30×10^6 psi. Neglect the weights of members.

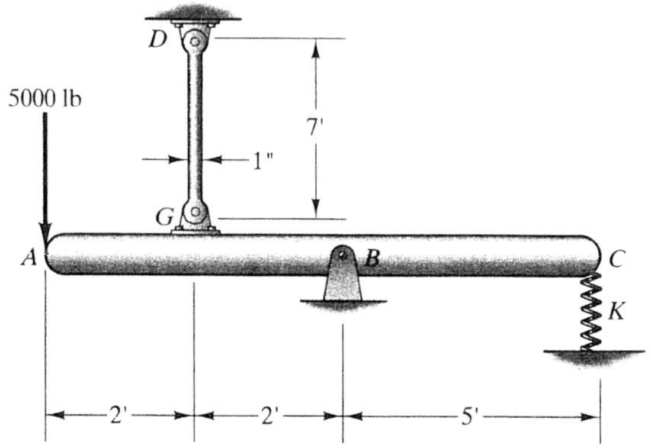

Figure P.5.37.

5.38. [5.4] Rigid bar AB supports a load $P = 10{,}000$ N and is supported by two elastic rods (Fig. P.5.38). The rigid bar weighs 4000 N. Find the value of a for the rigid bar to be horizontal after the 10,000-N load is applied.

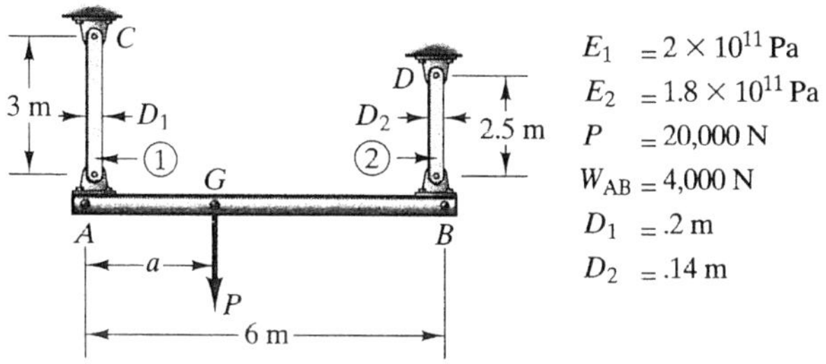

Figure P.5.38.

5.39. [5.4] In Problem 5.38, what is the maximum force P for elastic action in the supporting rods? What is the downward movement of the center of gravity and the rotation of AB with this load? Take $Y_1 = 3 \times 10^8$ Pa and $Y_2 = 2 \times 10^8$ Pa. Also take $a = 2$ m and neglect the weight.

5.40. [5.4] A rigid block weighing 8000 N is supported by rods (see Fig. P.5.40) having each a diameter of 100 mm and a Young's modulus of 2×10^{11} Pa. What are the forces in the members resulting from the weight of the block and the 5000-N force? Compute the downward movement of the block.

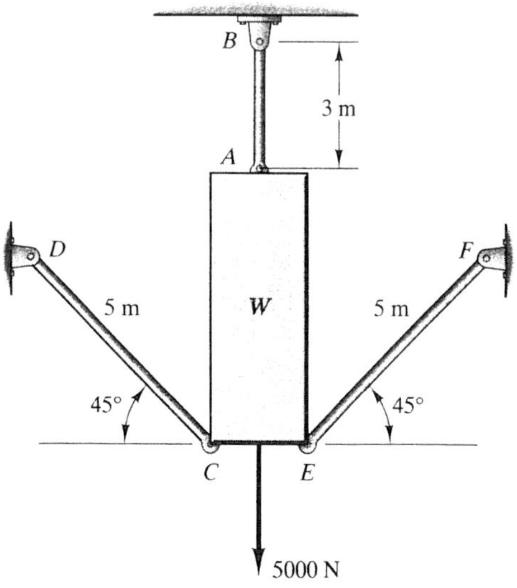

Figure P.5.40.

5.41. [5.4] A rigid block A in Fig. P.5.41 is attached to two identical linear elastic rods inside a space vehicle which is accelerating upward in outer space at a rate of 9 g's relative to inertial space. The block weighs on earth 5000 N and the rods have a weight per unit length on earth of 100 N/m. The modulus of elasticity of the rods is 2×10^{11} Pa. There is no gravitational pull on the system. What is the change in length of the rods as a result of the acceleration?

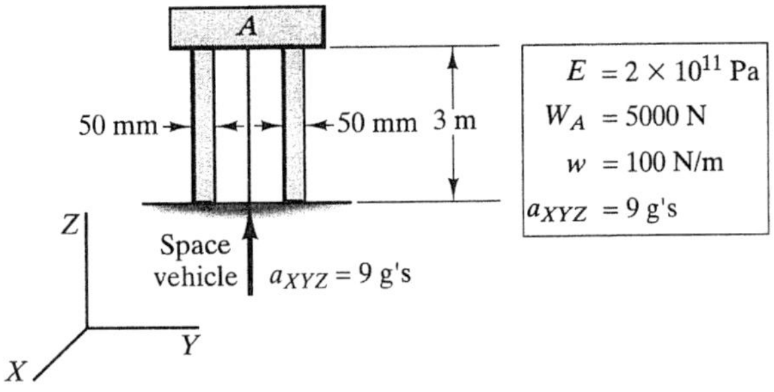

Figure P.5.41.

5.42. [5.4] Evaluate the vertical deflection δ of rigid rod AB in Fig. P.5.42 resulting from an external shear force distribution $f(z)$ given as $f = 100z^3$ kN/m. Do not consider weight. The supporting rods are linear elastic with the same cross-sectional area A and the same elastic modulus E.

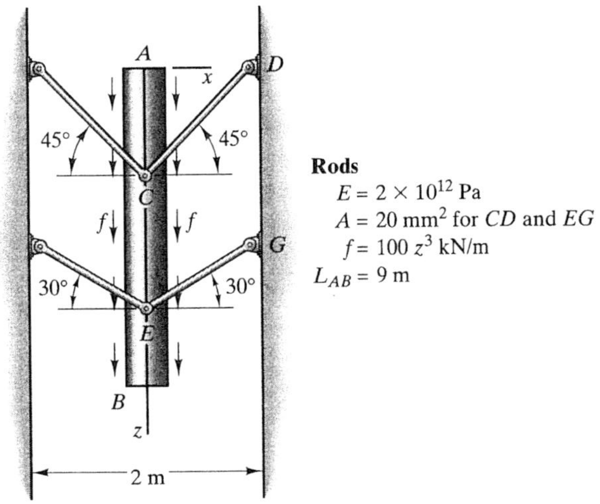

Figure P.5.42.

5.43. [5.4] A rigid bar AB in Fig. P.5.43 is assembled in a vertical plane with an unstretched spring, a force F of 5000 N, and a vertical elastic rod. If unloaded it is at an angle of 30°, what is the amount of rotation when the force of gravity and F are applied? The spring constant is 9×10^8 N/m. The weight of the rod is 3000 N.

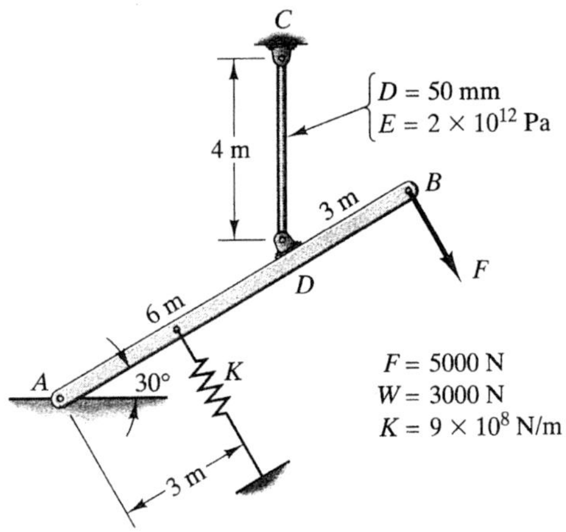

Figure P.5.43.

5.44. [5.4] A weight W (Fig. P.5.44) is supported by two rods, AB and CD, having cross-sectional areas of 1/2 in.^{2} each. A column G with a cross-sectional area of 1 in.^{2} is now placed under the weight so as just to touch without developing any stress in the column. What additional load can be added to W if we have a yield stress in all the members of 180,000 psi and a Young's modulus of 24×10^6 psi? Use a safety factor of 3. What is the deflection of W as a result of this additional load? Take $W = 10{,}000$ lb.

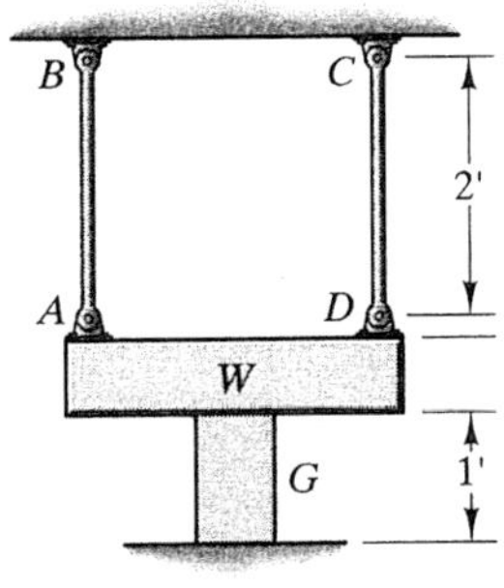

Figure P.5.44.

5.45. [5.4] A 450,000-N force is applied to member AG which is pinned at A (Fig. P.5.45) and which may be considered a rigid body. Find the forces in members DB and FC assuming linear, elastic behavior of these members. Member DB has a modulus of elasticity of 7×10^{10} Pa, and member

FC has a modulus of elasticity of 1.75×10^{11} Pa. The cross-sectional area of each vertical member is 6.25×10^{-5} m^2. Neglect the weight of the members.

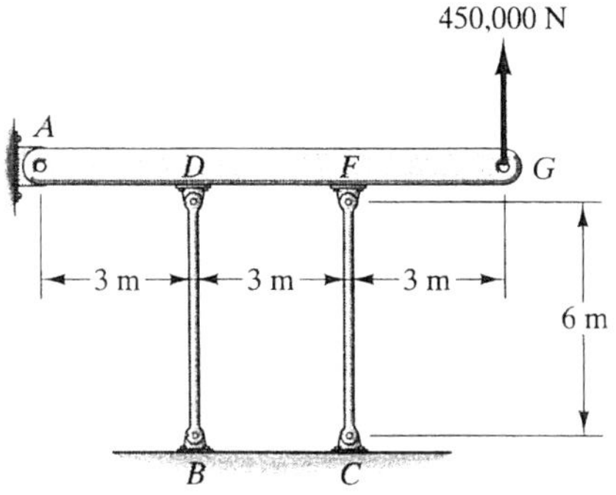

Figure P.5.45.

5.46. [5.4] In Problem 5.45, take the yield stress for member BD as 4.2×10^8 Pa and that of member CF as 7×10^8 Pa. What is the maximum vertical movement of point G for linear elastic behavior of the vertical members? What force is required to cause a vertical movement of point G of 1-1/2 times the value above? Assume elastic, perfectly plastic behavior.

5.47. [5.4] A rigid body AB weighing 100 lb is acted on by force F as shown in Fig. P.5.47. If the springs are unstretched when AB is horizontal, what rotation is experienced by AB when F is applied and AB is released from a horizontal orientation? Rods GH and JM are linear elastic.

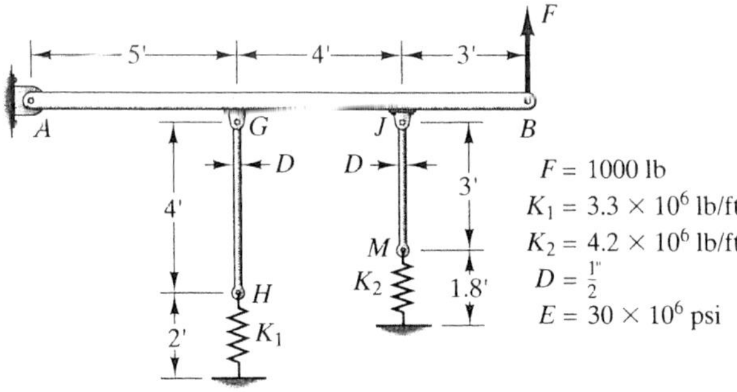

Figure P.5.47.

5.48. [5.4] A 1000-lb block A is to be supported vertically by two thin rods and a linear spring with spring constant K (see Fig. P.5.48). The following data apply:

length of rods = 10 ft each
diameter of rods = 3 in. each
spring constant = 10,000 lb/in.
$E = 30 \times 10^6$ psi

What is the distance d that the block will descend after it is assembled? Initially the spring is undeformed.

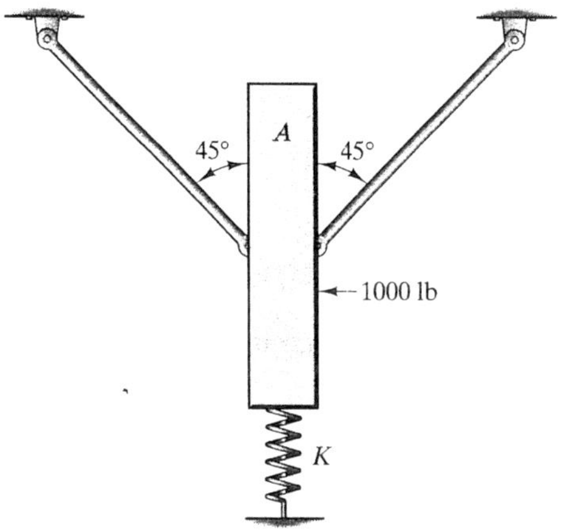

Figure P.5.48.

5.49. [5.4] What is the deflection of pin D? Members have the same cross-sectional area A and the same modulus of elasticity E. See Fig. P.5.49.

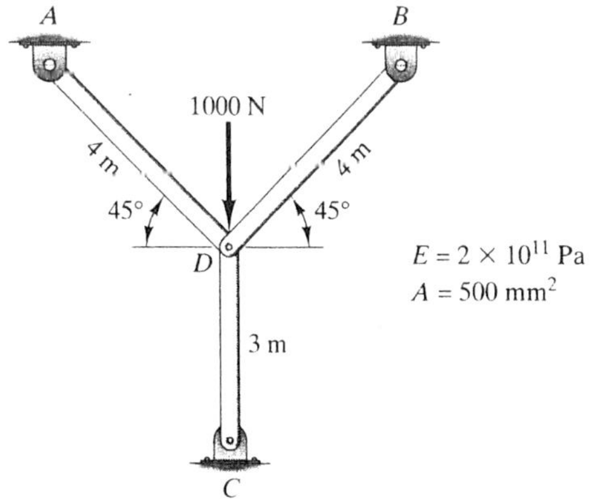

Figure P.5.49.

5.50. [5.4] A 10,000-N load acts on a rigid bar EF in Fig. P.5.50. If a maximum rotation of .08° for rod AC is allowed, what is the minimum diameter of rods AB and CD? The diameters are to be equal and E is 2×10^{11} Pa. Neglect weights.

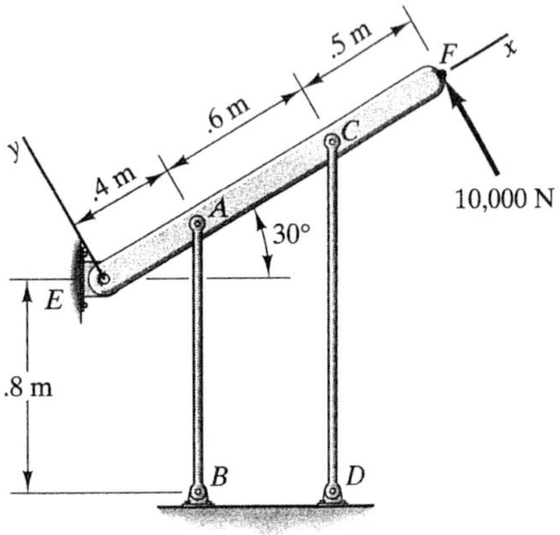

Figure P.5.50.

5.51. [5.4] Find the elongation of rods attached to a rigid drum A (Fig. P.5.51). The modulus of elasticity of the rods varies linearly from 2×10^{11} Pa to 1.2×10^{11} Pa as one goes from the base of the rods to the end tips of the rods. The following data apply:

$$\omega = 10{,}000 \text{ rpm}$$
$$D_{\text{rods}} = 0.08 \text{ m}$$
$$w_{\text{rods}} = 15 \text{ kg/m}$$

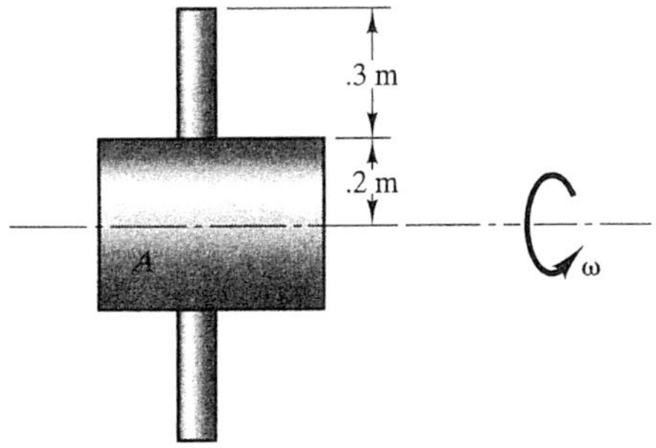

Figure P.5.51.

Set up an integral that should have the following format:

$$\Delta = \int_{.2}^{.5} \frac{(\quad - \quad)(z^2 - .25)}{z - (\quad - \quad)}\,dz$$

5.52. [5.4] Three bars support a rigid 450,000-N block (Fig. P.5.52). If the bars have a modulus of elasticity of 2×10^{11} Pa and a yield stress of 1.4×10^9 Pa, determine the minimum cross-sectional area of the rods (take these to be equal to each other) for a safety factor of 3. What is the deflection of the block for that situation?

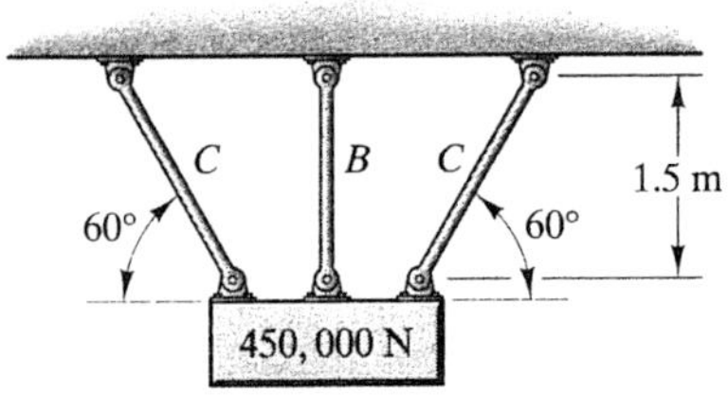

Figure P.5.52.

5.53. [5.5] A force F is applied at a section on a 3-in.-diameter rod held between two immovable walls as shown in Fig. P.5.53. Assume that E for compression is 30×10^6 psi and for tension $E = 20 \times 10^6$ psi and that the yield stress is 150,000 psi for compression and 100,000 psi for tension. If the applied force is such that it moves .100 ft, what is the residual stress when the force is released? The material is elastic, perfectly plastic.

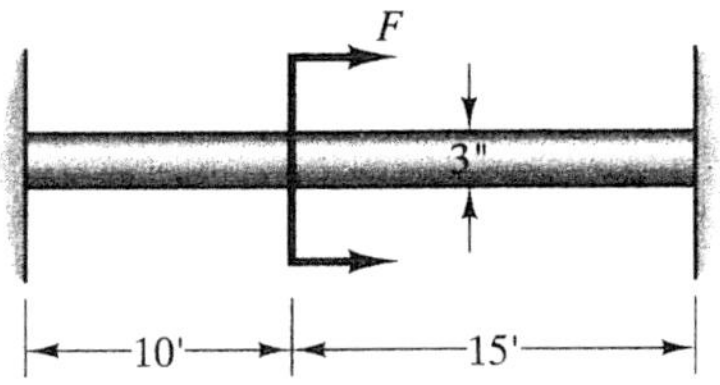

Figure P.5.53.

5.54. [5.5] Using the diagram and data of Problem 5.52, what is the maximum weight that can be supported by the rods for elastic, perfectly plastic behavior wherein the outer rods have just reached yield stress? What is the deflection downward of the block for this condition? If the load is released but the block is maintained in contact with the rods, what are the residual stresses in the rods? Take A as 1.25×10^{-3} m^2 for the rods and assume that the rods are restrained from buckling. From Problem 5.52, we know that $C < B$.

5.55. [5.6] Two rods AB and CD support a 225,000-N rigid block as shown in Fig. P.5.55. If the block is not to move more than .08° as a result of deformation of the rods, what minimal diameter would you use for rod AB when using steel for the material having a modulus of elasticity of 2×10^{11} Pa and working stress of 4.2×10^8 Pa? Assume that the diameter of AB is .75 the diameter of DC.

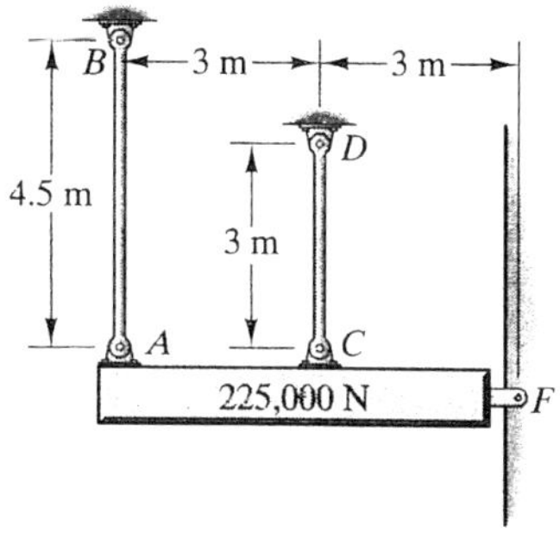

Figure P.5.55.

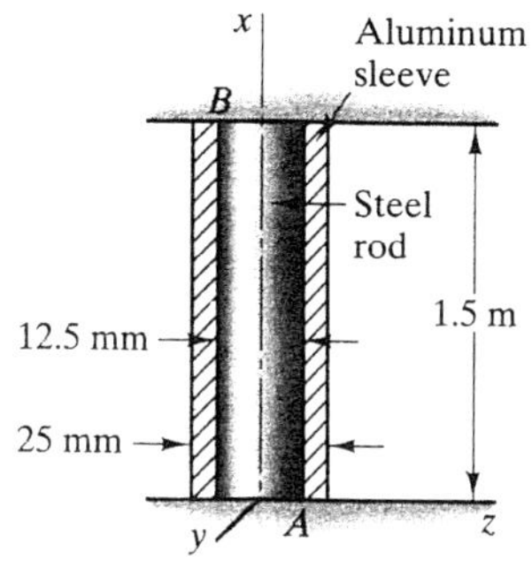

Figure P.5.58.

5.56. [5.6] Using the data of Problem 5.55, design rods so that they have the smallest diameters while maintaining the same stress at all times in each bar. Adjust the position of DC for this design. Take $D_{BA} = D_{CD}$.

5.57. [5.6] A load F is applied to a rigid vertical post as shown in Fig. P.5.57. This member is restrained by rods AB and CD both having a Young's modulus of 30×10^6 psi and a yield stress of 200,000 psi. If the diameter of rod CD is 1 in., what should the length and the minimum diameter of the upper rod be to allow an .008° rotation of the post upon application of the force F? The force F is 10,000 lb. What are the stresses in the rods? Use a safety factor of 2.

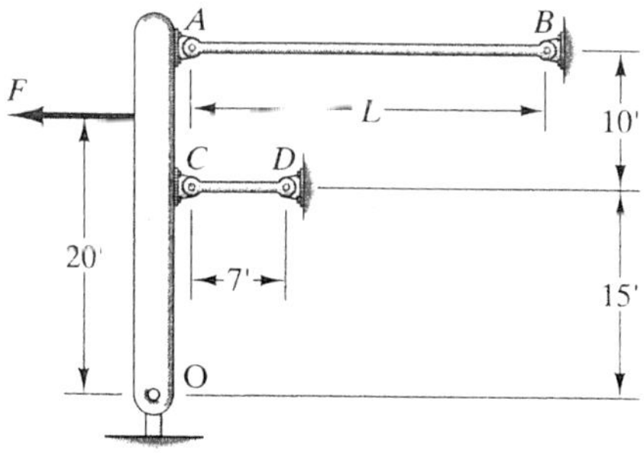

Figure P.5.57.

5.58. [5.7] Fig. P.5.58 shows a steel rod and an aluminum sleeve held between two immovable supports A and B. If the temperature is raised from 15°C to 38°C, what are the thermal stresses in the materials, and what is the force developed on the supports? Take α to be 1.17×10^{-5}/°C for the steel rod and 2.16×10^{-5}/°C for the aluminum sleeve. E for the rod is 2×10^{11} Pa and for the sleeve is 7×10^{10} Pa.

5.59. [5.7] In Fig. P.5.58 assume that the temperature varies linearly from 38°C at A to 15°C at B. What are the thermal stresses in the members? Take α and E as given in Problem 5.58. The rod and sleeve are originally at 15°C.

5.60. [5.7] In Fig. P.5.58 suppose that support B is elastic, having a spring constant K of 1.8×10^7 N/m. What is the elongation of the members if the temperature is changed from 15°C to 38°C? Take α and E as given in Problem 5.58.

5.61. [5.7] In Fig. P.5.58 suppose that support B is elastic, having a spring constant K of 1.8×10^7 N/m. What is the thermal stress in each member if the temperature varies according to the formula $T = ax^2 + b$, with $T = 38$°C at A and 15°C at B? (Take x in the formula in meters.) Take α and E as given in Problem 5.58. The temperature originally is 15°C.

5.62. [5.7] Shown in Fig. P.5.62 is a composite square bar securely fastened at the ends to immovable walls. If the temperature drops from 100°F to 30°F, what is the thermal stress? Take α and E for aluminum to be 12×10^{-6}/°F and 10×10^6 psi, and for steel 6×10^{-6}/°F and 30×10^6 psi, respectively.

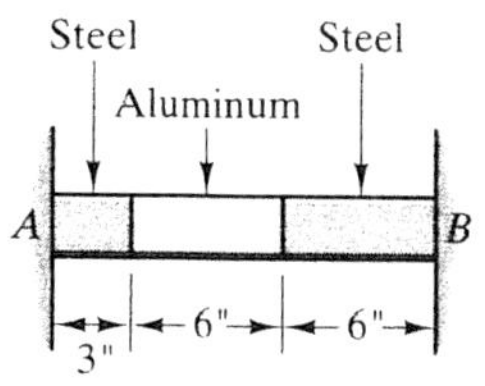

Figure P.5.62.

5.63. A thin-walled tank (see Fig. P.5.63) containing compressed air at a gage pressure of 20 psi is suspended from a ceiling. The wall thickness of the tank is 1/4 in. throughout. The outside diameter is 4 ft and the overall length is 10 ft. The bottom plate weighs 80 lb and the circular cylindrical wall weighs 200 lb per ft. The material is linear elastic with $E = 30 \times 10^6$ psi.

(a) Compute the normal stress in *psi* on horizontal cross sections of the cylinder as a function of z with z in *ft*.

(b) What is the overall elongation of the tank as a result of weight and air pressure?

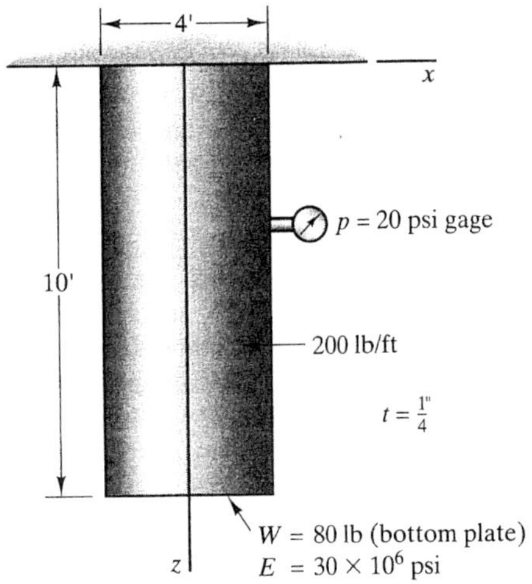

Figure P.5.63.

5.64. Two elastic rods are welded together and are fixed to a rigid cylinder rotating at a steady speed of 1000 rpm as shown in Fig. P.5.64. If $E = 2 \times 10^{11}$ Pa, and if the mass density is 5600 kg/m^3

(a). What is the stress τ_{rr} as a function of r for the upper thin member?

(b). What is the stress τ_{rr} as a function of r for the lower larger-diameter member?

(c). Give the elongation of each member.

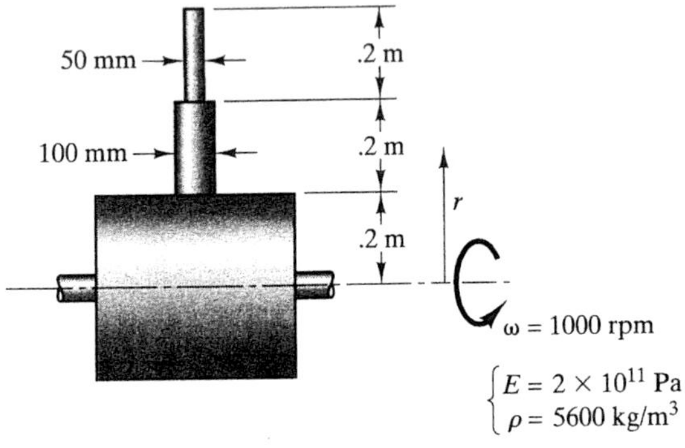

Figure P.5.64.

5.65. An antenna is rigged by four guy wires as shown in Fig. P.5.65. Each guy wire is made of steel having a modulus of elasticity of 30×10^6 psi. Each is tightened from a just-snug connection by turning a turnbuckle 30 times. The turnbuckle has 30 threads/in. The pitch is then 1/30 in. The guy wire has a diameter of 1/2 in. What is the compressive force on the antenna if it is assumed rigid? Note that a turnbuckle is a nut with screw threads at *each end.* Hence, one rotation causes a change in length of the wire two times the pitch of the threads. Neglect the effects of gravity in this problem.

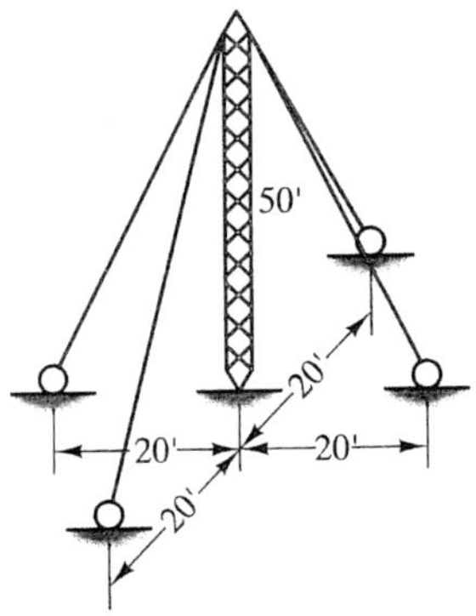

Figure P.5.65.

5.66. Do Problem 5.65 for the case where the antenna shortens .1 in. per 50,000 lb of axial compression.

5.67. A load $F = 45{,}000$ N is applied at a section on a 75-mm-diameter rod held between two immovable walls as shown in Fig. P.5.67. If $E = 1.4 \times 10^{11}$ Pa for the material, what are the restraining forces at the wall? For complete elastic behavior, what is the maximum force F that can be applied if the yield stress for the material is 1.4×10^9 Pa for tension and compression?

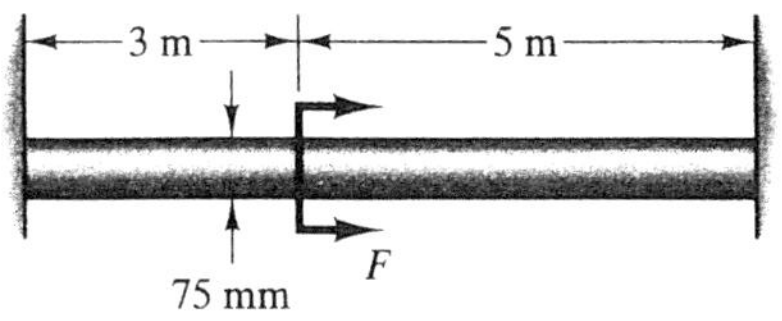

Figure P.5.67.

5.68. Using the geometric data of Problem 5.67, compute the maximum distance δ that F can move for perfectly elastic behavior. Assume that for compression $E = 2 \times 10^{11}$ Pa and for tension $E = 1.3 \times 10^{11}$ Pa and that the yield stress is 1×10^9 Pa for compression and 7×10^8 Pa for tension. What are the supporting forces for this situation?

5.69. Determine the total change in length of the cylinders in Fig. P.5.69 as a result of the applied forces and the weight. The two cylinders are welded together at the 12,000-N load.

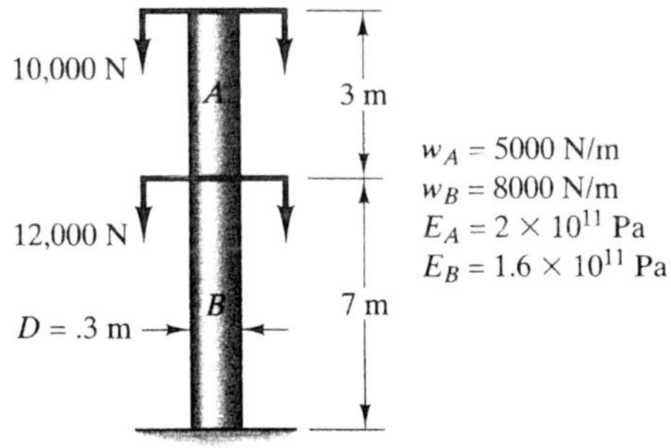

Figure P.5.69.

5.70. Determine the vertical movement δ of the 800-N block after it is carefully put in place (see Fig. P.5.70). This movement is due to the weight of the block. Both members have the same E and A. Neglect friction.

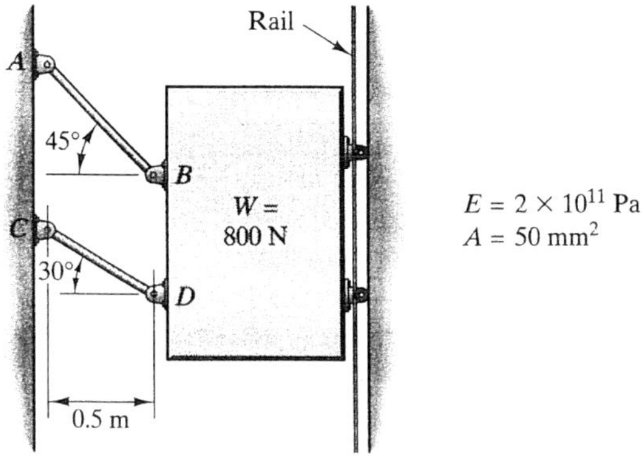

Figure P.5.70.

5.71. Find the elongation of rod AB in Fig. P.5.71 due to centrifugal force for $\omega = 10{,}000$ rpm. The material is linear elastic with $E = 20 \times 10^6$ psi. The specific weight $\gamma = 500$ lb/ft^3. The rods are cylindrical. Neglect the diameter of the horizontal shaft GD.

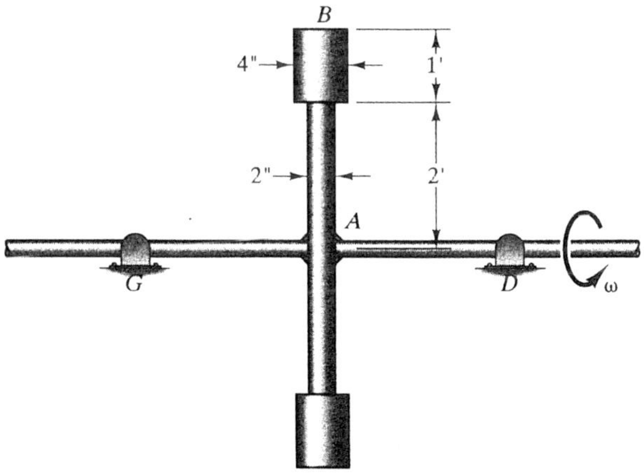

Figure P.5.71.

5.72. In the design of electronic components, an important factor is the relative displacement between a component and the circuit board. Determine this relative expansion for the system shown in Fig. P.5.72 subject to a temperature change of ΔT. Neglect the effect of the solder joints in resisting the deformation.

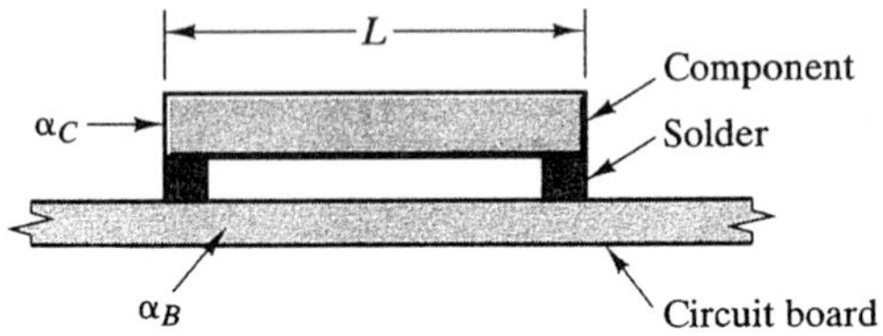

Figure P.5.72.

5.73. In the previous problem, determine the average shear stress in the solder for the following data:

$$\alpha_C = 6 \times 10^{-6}/°\text{F}$$
$$\alpha_B = 30 \times 10^{-6}/°\text{F}$$
$$\Delta T = 200°\text{F}$$
$$L = 1/2 \text{ in.}$$
$$E_{\text{solder}} = 1 \times 10^6 \text{ psi}$$
$$\nu = 0.4$$

The solder joint is a cube 0.1 in. on a side. Note that

$$\tau_{shear} = \frac{1}{2}\frac{E}{1+\nu}\gamma_{shear}$$

5.74. In Fig. P.5.74 is shown a composite bar having a diameter of 2 in. Material a is to be considered a linear, elastic material with a Young's modulus of 20×10^6 psi and a linear coefficient of expansion $6 \times 10^{-6}/°\text{F}$. Material b, however, is to be considered to behave in an elastic, perfectly plastic manner with $E = 30 \times 10^6$, with a yield stress of 50,000 psi and with a linear coefficient of expansion equal to $12 \times 10^{-6}/°\text{F}$. If the temperature is changed from 60°F to 100°F, what is the thermal stress? What is the maximum stress possible for this system?

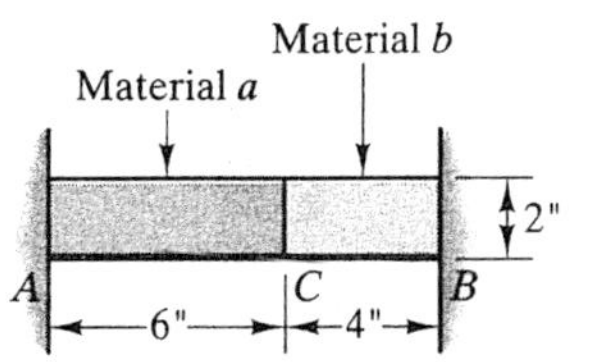

Figure P.5.74.

5.75. Member GH in Fig. P.5.75 is to be taken as a perfectly rigid member. Members AB and CD are linear-elastic members having moduli of elasticity of 1.4×10^{11} Pa and 2×10^{11} Pa, respectively, with equal cross-sectional areas of 6.25×10^{-4} m². What are the supporting forces at H? Neglect the weights of the members. Take $F = 9 \times 10^5$ N.

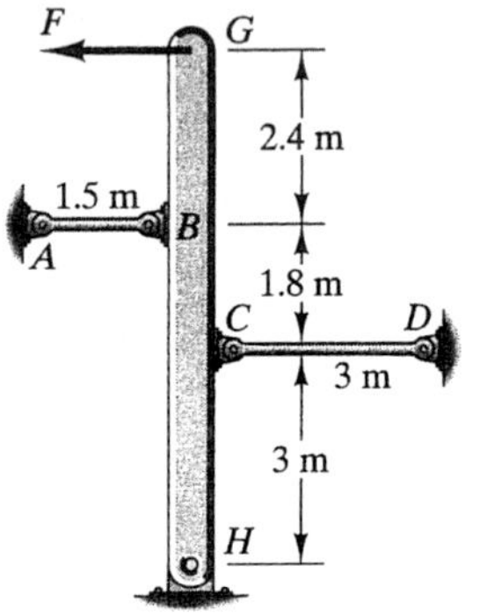

Figure P.5.75.

5.76. What deflection on the rigid block in Fig P.5.76 is caused by the weight of the block? The modulus of elasticity E for each rod is 30×10^{10} Pa, and the cross-sectional area for each rod is 5×10^{-4} m².

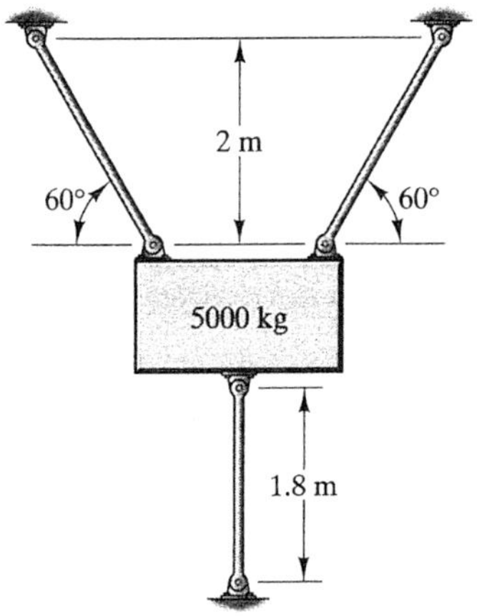

Figure P.5.76.

5.77. In Problem 5.74, what is the temperature at which yielding takes place? What is then the displacement of section C in terms of further temperature increases? Use the result

$$\tau_{xx} = -(\alpha_A L_A + \alpha_B L_B)\Delta T/(L_A/E_A + L_B/E_B)$$

5.78. Fig. P.5.78 shows a simple pin-connected truss. The members are steel and have the same cross section. If member AB is raised to a temperature of 100°C from a temperature of 60°C and if member AC remains at 60°C, what are the stresses in the members resulting from this temperature change? Assume that the members remain straight. Take $\alpha = 10 \times 10^{-6}/°\text{C}$. Take $E = 2 \times 10^{11}$ Pa.

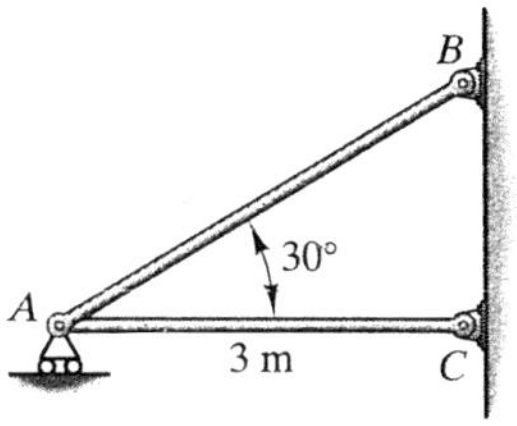

Figure P.5.78.

5.79. A rigid cylinder (Fig. P.5.79) of diameter $D = 2$ ft rotates with angular speed ω of 400 rpm. Two rods A of length $L = 2$ ft are attached to the cylinder. These rods have an outer sleeve B also attached to the cylinder. The sleeve has an elastic modulus of 15×10^6 psi, a mass per unit length of 10 lbm/ft, and a cross-sectional area of 1/2 in.^{2}. Each inner rod A is attached rigidly to sleeve B at H at the ends and has an elastic modulus of 30×10^6 psi, a mass per unit length of 15 lbm/ft, and a cross-sectional area of 1/2 in.^{2}. What is the change in length of the composite members as a result of deformation?

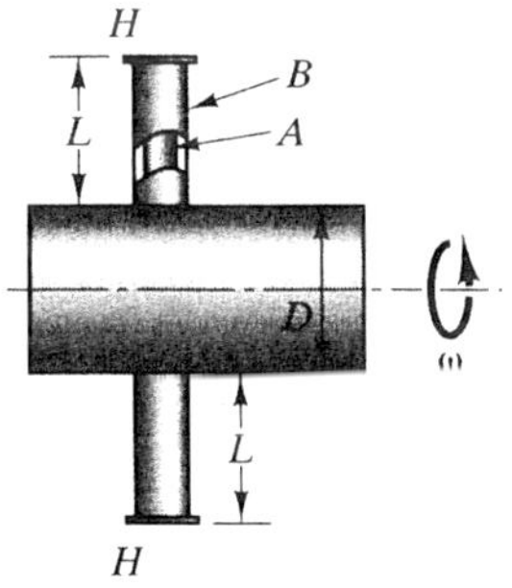

Figure P.5.79.

5.80. In Problem 5.79, determine the minimum speed ω which will cause yielding in both members of the composite rod. Take $Y_{\text{sleeve}} = 150{,}000$ psi and $Y_{\text{rod}} = 200{,}000$ psi. Materials are elastic, perfectly plastic.

5.81. Three rods are pinned together as shown in Fig. P.5.81. The members maintain a 22,500-N force. What is the force developed in each member? The members are perfectly elastic with a modulus of elasticity E for all members. Take the cross-sectional area as constant and equal to A.

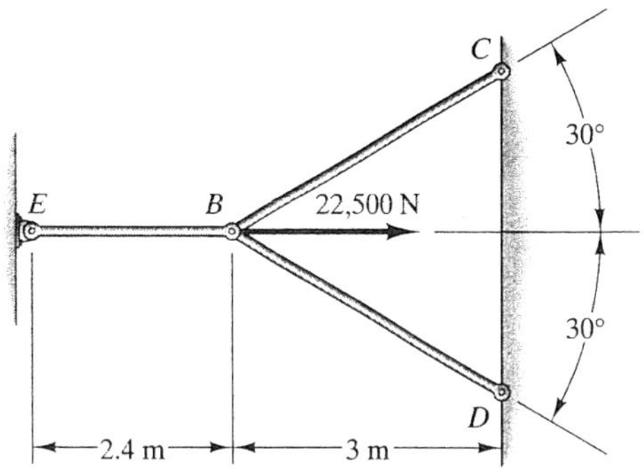

Figure P.5.81.

5.82. In Fig. P.5.82 is shown a system of pin-connected elastic bodies supporting a force of 10,000 lb. If each member has a cross-sectional area of .2 in.^{2} and a modulus of elasticity of 30×10^6 psi, what is the deflection of pin A? If the yield stress is 100,000 psi, what is the largest load the bars can support for a safety factor $n = 2$?

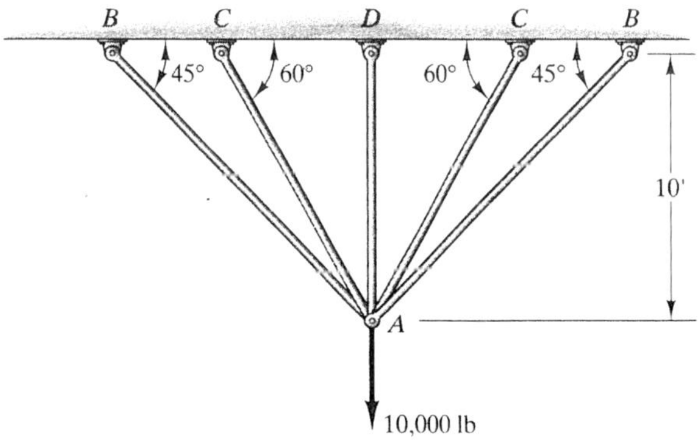

Figure P.5.82.

5.83. A truss supports a 15,000-lb load (see Fig. P.5.83). Each member has a cross-sectional area of 1 in.^{2}. Members AB and DB have a Young's modulus of 30×10^6 psi and a coefficient of expansion $\alpha = 6 \times 10^{-6}$/°F. What are the forces in the members from the load F and a change in temperature from $T = 30$°F for all members to $T = 110$°F for all members?

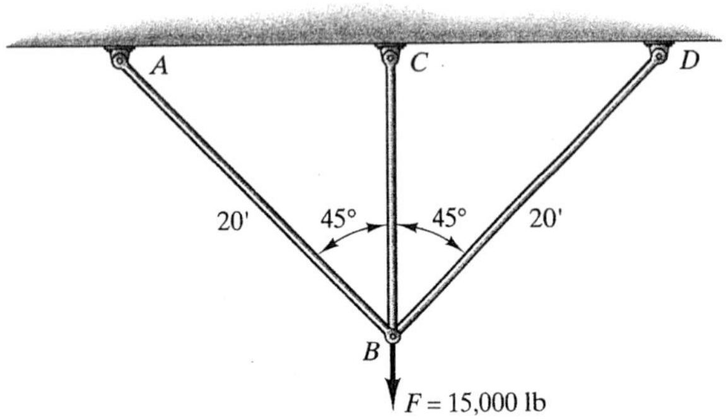

Figure P.5.83.

5.84. In Fig. P.5.84 is shown a set of thin rings that just fit together when the temperature is 100°F. The outer ring is aluminum and the inner ring is steel. Approximate the stress in the rings when the temperature drops from 100°F to 20°F. Take α and E for steel to be 6.5×10^{-6}/°F and 30×10^6 psi, respectively, and for aluminum 10×10^{-6}/°F and 15×10^6 psi, respectively.

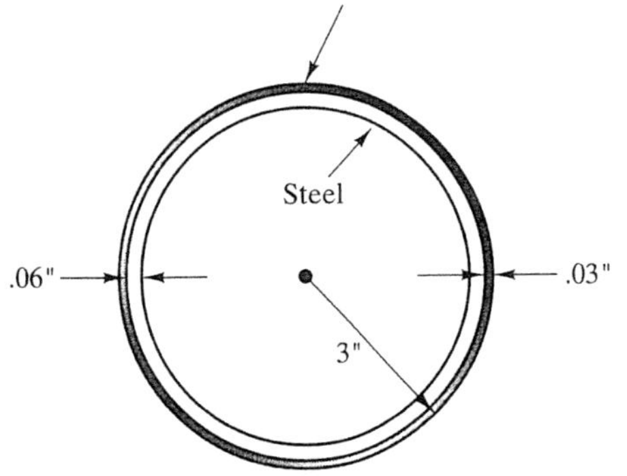

Figure P.5.84.

5.85. A steel hoop is to be *shrunk-fit* onto a steel shaft. That is, the hoop will be heated so as to increase the internal diameter so that the hoop fits onto the shaft. Once on the shaft the hoop is cooled. If the diameter of the shaft is 50.000 mm and the initial internal diameter of the hoop before heating is 49.950 mm, how high a temperature is required to heat the hoop in order to have a clearance of .025 mm when assembling the system? Take the coefficient of linear expansion α to be a constant equal to 10×10^{-6}/°C.

5.86. Concrete is a material that can withstand little tensile stress in contrast to its ability to withstand compressive stress. For this reason steel reinforcing rods are inserted in concrete. Sometimes these rods are deliberately stressed so as to develop compressive stress in regions of the concrete where, on loading, there is expected tensile stress in the concrete. This is called *prestressed concrete.* We show a concrete tank in Fig. P.5.86, where reinforcing rods and reinforcing hoops are shown. The hoops can be tightened to prestress the concrete into compression. If R_i = 1 ft and R_o = 1.2 ft, what should the spacing a be for the hoops to just avoid tension in the concrete if each hoop is stressed to 100,000-psi tension and the inside tank pressure is 200 psi gage? The diameter d of the hoop is 1/2 in.

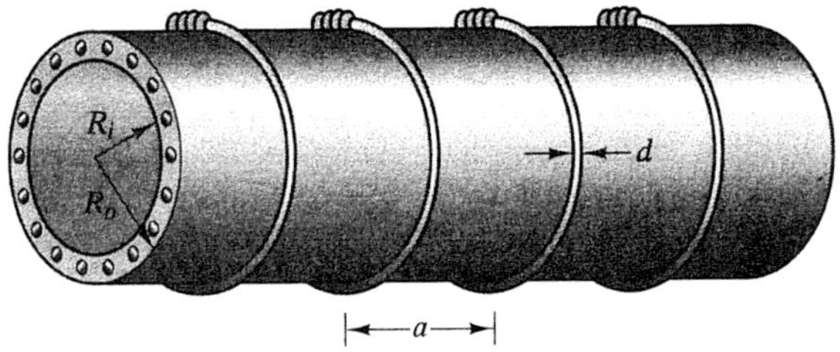

Figure P.5.86.

5.87. A simple vise is shown in Fig. P.5.87 in the process of pressuring rigid body C. In fixed jaw B there is a thrust bearing e, while fixed in the moving jaw A is a nut d. A compressive force of 1000 lb is desired for body C. What rotation of the arm D is needed to accomplish this after contact has just been made by the jaws on C? The screw has a pitch of .2 in./revolution and has a modulus E of 30×10^6 psi.

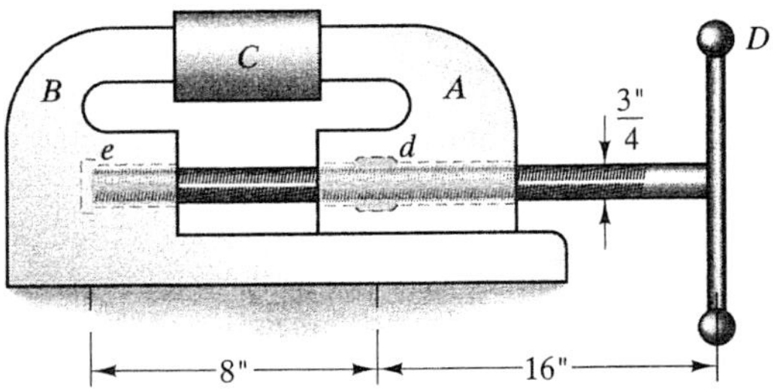

Figure P.5.87.

5.88. In Problem 5.87 body C is replaced by a linear coil spring having a spring constant of 1 million lb/in. How many turns are needed for developing a force of 1000 lb on the spring?

5.89. Using the data of Problem 5.55, design rods so that they have the smallest diameters while maintaining the same stress at all times in each bar. Adjust the position of DC for this design. Take $D_{BA} = D_{CD}$.

5.90. A load F is applied to a rigid vertical post as shown in Fig P.5.90. This member is restrained by rods AB and CD having a Young's modulus of 30×10^6 psi and a yield stress of 200,000 psi. If the diameter of rod CD is 1 in., what should the length and the minimum diameter of the upper rod be to allow a .008° rotation of the post upon application of the force F? The force F is 10,000 lb. What are the stresses in the rods? Use a safety factor of 2.

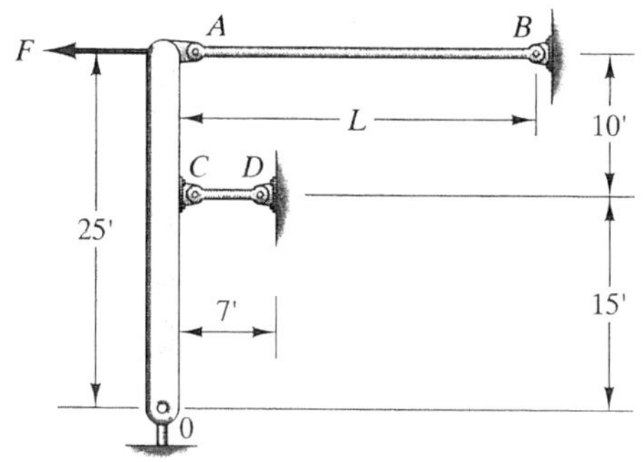

Figure P.5.90.

5.91. Compute the supporting forces for the 3-rod system shown in Fig. P.5.91. All members have the same AE.

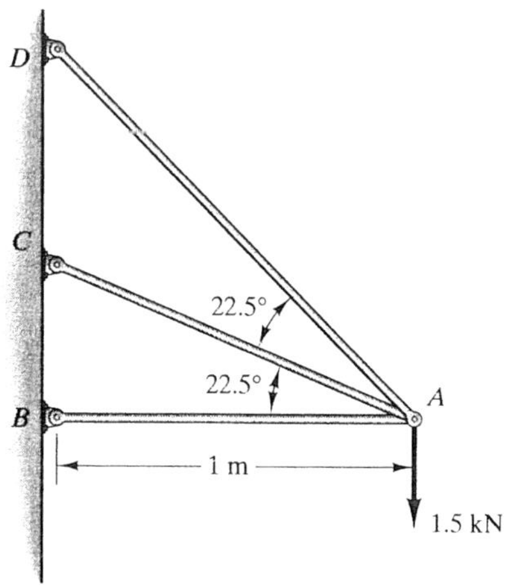

Figure P.5.91.

5.92. Two bars of square cross section are constrained between immovable walls (see Fig. P.5.92). The cross sections are 2 in. on an edge. Body A has a modulus of elasticity of 15×10^6 psi, while body B has a modulus of elasticity of 30×10^6 psi. Forces of 5000 and 10,000 lb are shown applied at sections of each body. What are the supporting forces at the wall? What is the movement of the interface between the bodies as a result of the loads?

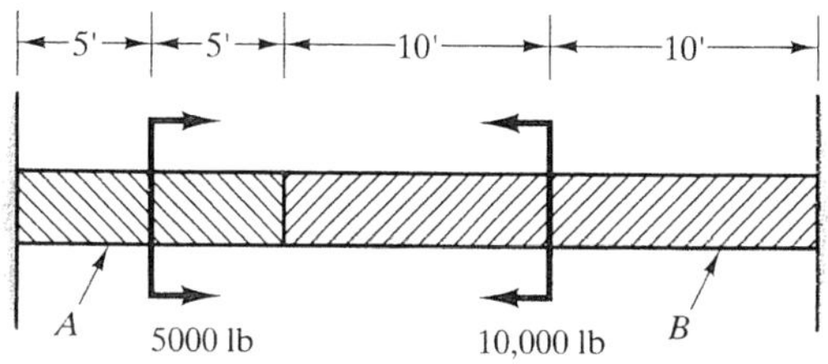

Figure P.5.92.

5.93. A rigid plate A has three legs, as shown in Fig. P.5.93. One leg is .025 mm short. A force of 180,000 N is placed on plate A. If it is placed so that A remains horizontal, what is the deflection of A? The cross-sectional area of the legs is 625 mm^2, and the modulus of elasticity is 2×10^{11} Pa.

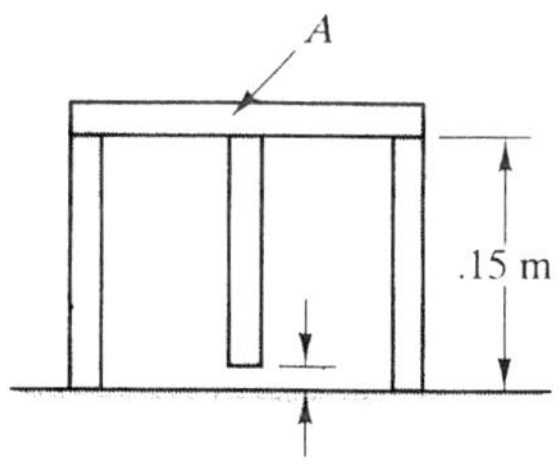

Figure P.5.93.

5.94. In Problem 5.93, give the minimum cross-sectional area of the legs (assuming that one leg is short, as has been shown) to support a 2.25×10^6 N load with a deflection of A not to exceed .075 mm. Take the yield stress of the legs as 7×10^8 Pa. Use a safety factor of 2.5. Take the cross-section of the legs as square.

5.95. Determine the displacement of point B when the 10,000-N force is applied. Rod AB is rigid and is horizontal initially (see Fig. P.5.95). Spring K is undeformed initially. Neglect weight.

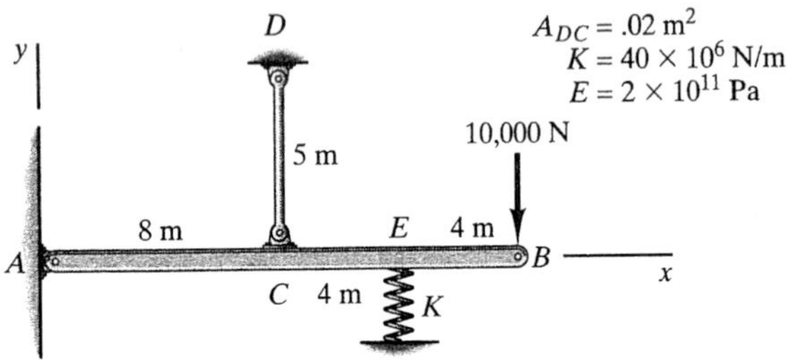

Figure P.5.95.

5.96. A flat smooth plate is rotated about the z axis. A block is held in place (see Fig. P.5.96) by a stop and an elastic cord, as shown. What prestretch of the cord is required to just hold the block in contact with the stop when $\omega = 1000$ rpm? The cord is 0.25 m long and has a diameter of 10 mm and a Young's modulus of 1 GPa. The block has a mass of 10 kg.

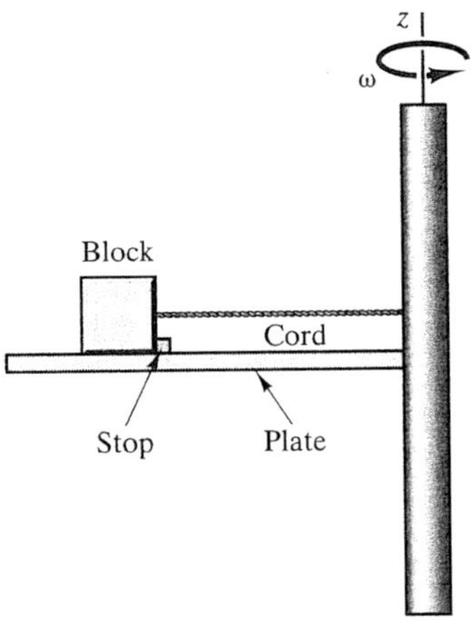

Figure P.5.96.

5.97. Consider the flexible rod with a rigid flange shown in Fig. P.5.97. A ring-shaped slider is held at rest above the flange. Determine the stress in the rod when the slider impacts the flange and elongates the rod to its maximum displacement Δ (*Hint:* Use conservation of energy and let the spring stiffness of the rod $K = AE/L$.)

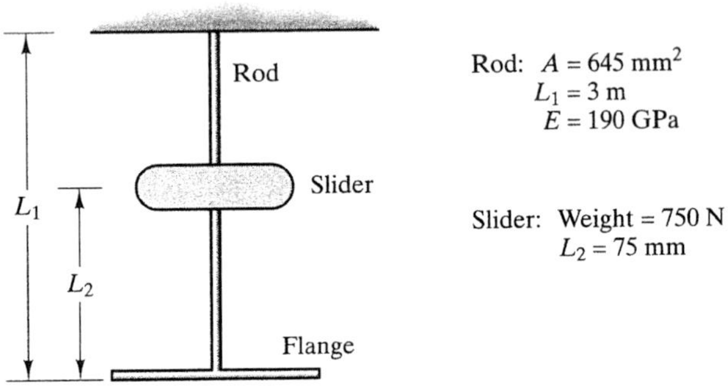

Figure P.5.97.

5.98. A cylinder in Fig. P.5.98 is mounted on a platform A in outer space. This platform is decelerating downward at the rate of 25 m/sec^2. A rigid block is attached to the end of the cylinder. This rigid block has a mass of 1000 kg. The rod has a mass of 500 kg per meter at the top and 200 kg per meter at the bottom. This mass distribution is proportional to the square of z. The cylinder has a diameter of 200 mm.

(a). Determine the stress in the elastic member.
(b). Determine the change in length of the shaft due to the deceleration.
(c). If $Y = 3.5 \times 10^7$ Pa, what is M for plastic flow anywhere in the cylinder?

There is no gravitational affect for the body.

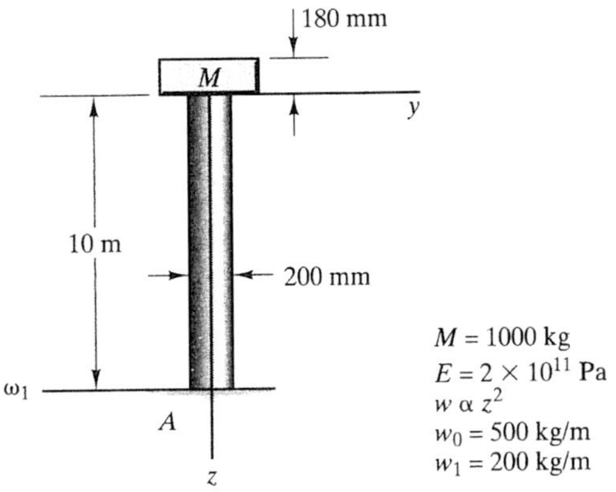

Figure P.5.98.

5.99. For the two-bar structure shown in Fig. P.5.99, determine the configuration for the structure with least weight. First, show that if the members are stressed to their allowable working value τ_{allow}, the weight of the members is given by:

$$W = \frac{PL\rho g}{\tau_{\text{allow}}}\left[\frac{1 + \cos^2\theta}{\sin\theta\cos\theta}\right]$$

Next, plot this function and graphically determine θ for $0 \le \theta \le 90°$ to give the least-weight design. (*Hint:* Ignore the weight of the members compared to the load P.)

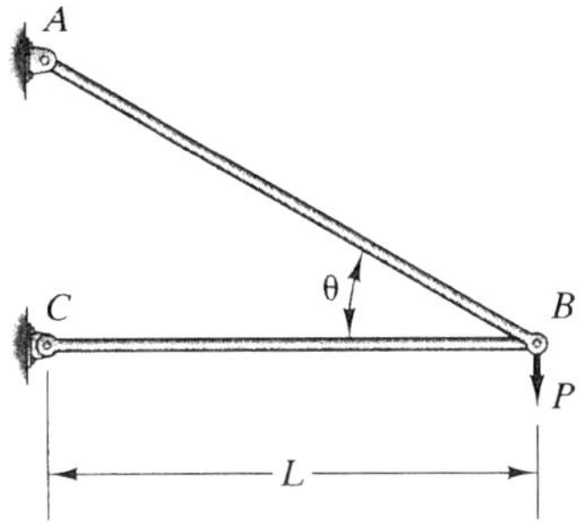

Figure P.5.99.

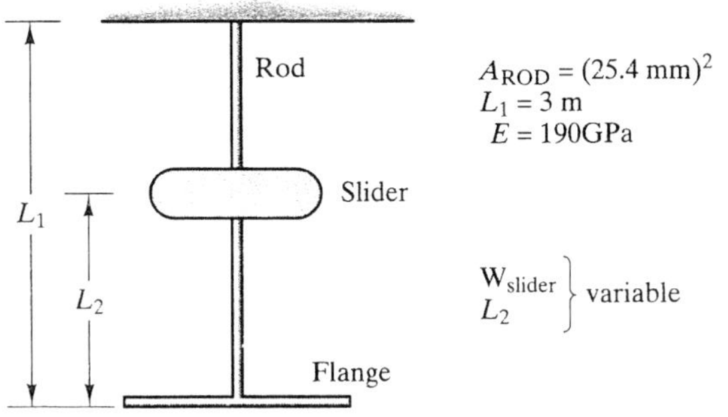

Figure P.5.100.

5.100. For the rod impact problem considered in Problem 5.97 (see Fig. P.5.100), plot the normalized slider drop height L_2/L_1 versus the maximum elongation, Δ, of the very thin rod. Consider slider weights of 325 N, 750 N, and 1500 N. Also show on the plot the limiting drop height ratio for elastic behavior for the following three steel alloys:

High Strength Steel	$Y = 1000$ MPa
Tool Steel	$Y = 500$ MPa
Structural Steel	$Y = 340$ MPa

Neglect all masses except for that of the slider. Consider the rod as a spring. From your plot, comment on the displacement versus drop height. Is the response linear?

5.101. For the 3-bar support structure shown in Fig. P.5.101, determine the forces in the members for $0 \le \theta \le 90°$ and member length ratio L_1/L_2. Plot your results (normalized load F_1/W and F_2/W versus θ). Consider length ratios of $L_1/L_2 = 5$, 2, 1, 1/2, and 1/5. Take AE constant.

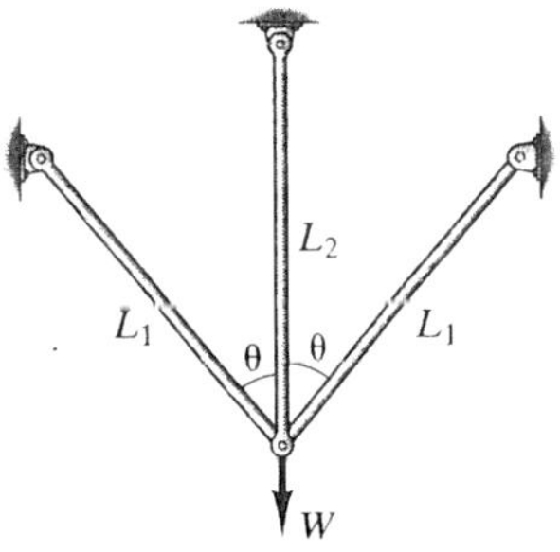

Figure P.5.101.

5.102. A load is to be supported as shown in Fig. P.5.102. What angle θ will result in the least-weight design if both members are stressed to yield Y? Plot the normalized weight

$$\frac{WY}{PL\rho g}$$

of the structure versus θ, for $0° \le \theta \le 90°$.

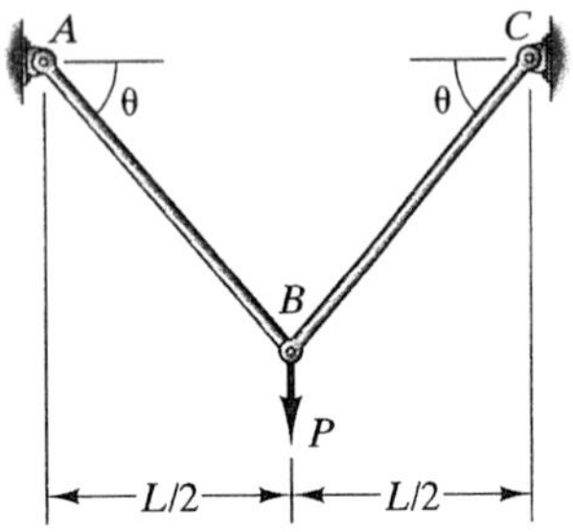

Figure P.5.102.

Data:

For all rods $E = 207$ GPa Diam $B = 1.5$* Diam of A
$Y = 450$ MPa Diam $C = 2.0$* Diam of A
Diam $D = 2.5$* Diam of A

Safety factors:

1, 1.5, 2, and 2.5

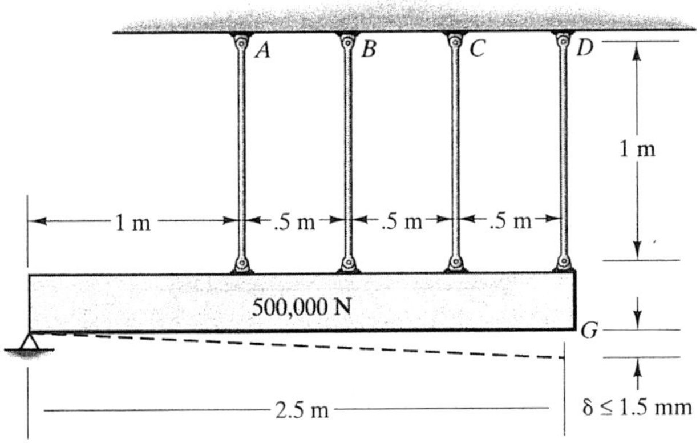

Figure P.5.103.

Project 5 Simple Design Problem

Find the minimum diameters for the supports such that the deflection at the tip G of the 500,000-N rigid block in Fig. P.5.103 does not exceed 1.5 mm when loaded, and/or such that the working stress in the four rods for various factors of safety n is not exceeded.

CHAPTER 6

Generalized Hooke's Law and Introduction to Energy Methods

6.1 Introduction

In Chapter 4, we considered various constitutive laws applicable to a one-dimensional state of stress. In **Part A** of this chapter, we extend one of them, namely Hooke's law, to encompass three-dimensional behavior for isotropic[1] materials. In doing this, we present three key constants that are used to characterize the mechanical behavior of such solids—namely, the Young's modulus, the shear modulus, and Poisson's ratio. Following this, there is a starred section considering briefly stress-strain relations for nonisotropic materials to serve the purpose, for the interested student, of being an introduction to this topic. The resulting key formulation is often called the *generalized Hooke's law*. We shall not consider such materials in our subsequent studies of solid mechanics in this book. However, when studying fluid mechanics, the student will most likely come across Stokes' viscosity law. This constitutive law is very similar to the generalized Hooke's law and the student may wish to come back for another look at this starred section. We close Part A in yet another short, starred section in which we use the equations of equilibrium from Chapter 2; the strain-displacement relations of Chapter 3; and the three-dimensional Hooke's law of this chapter, where we set forth the key equations of the theory of linear-elastic, isothermal elasticity. By having this theory available to us, we will be able to check the assumptions made in setting forth the simpler formulations to be fea-

[1]An isotropic material has properties which are independent of direction. Also note that we shall not be able to extend other constitutive laws presented earlier to three dimensions because of the rapidly escalating difficulty in such undertakings.

tured in this text called strength of materials. This will be done in certain starred problems with the aid of detailed, clear directions.

In **Part B** of this chapter, we will present an introductory but very useful treatment of *energy methods.* This is done by first presenting the interesting Maxwell-Betti reciprocal relation and directly from this the *second Castigliano theorem.* We feel certain that the reader will be amazed at the simplicity of solving displacements for statically determinate truss problems and for forces in statically indeterminate truss problems all in this chapter. This will constitute a leap forward in the student's knowledge of structural analysis. In later chapters, we shall continue to apply the second Castigliano theorem for solving beam and then torsion problems. Finally, for those readers wishing a more complete and thorough treatment of energy methods, we present in starred Chapter 18 the three *displacement methods* and the three *force methods* forming the basis of the general energy methodology. Incidentally, the second Castigliano theorem of this chapter is the third formulation in the system of force methods. Also, these energy methods form the basis of the finite element procedures now used routinely in much engineering design and analysis.

Part A. Simple Constitutive Relations

6.2 Three-Dimensional Hooke's Law for Isotropic Materials

In the discussion of this section, we assume that the properties of the materials are *not dependent on the direction.* That is, we use the same constant E when extending one-dimensional test data in the x, y, and z directions.

Materials having no directional variation in their properties at a point are called *isotropic* materials. Individual crystals of structural materials are actually not isotropic (i.e., they are *anisotropic*), but because of the usual random orientation of the crystals and their large number, the macroscopic behavior of many structural materials is usually considered isotropic. However, as a result of working a metal, such as in a rolling operation, the crystals do attain at times a certain preferential alignment, and we then must take into account anisotropic effects.

To formulate the three-dimensional Hooke's law, we shall consider a vanishingly small rectangular parallelepiped. We shall apply to the element successively three normal stresses and three shear stresses. We shall evaluate the strain for each stress, and then for isotropic material we can superpose the results to relate stress and strain for the general case at the point.

Thus, in Fig. 6.1(a) we show an element subject only to a normal stress τ_{xx}. Using the results from our study of one-dimensional stress, we can say that

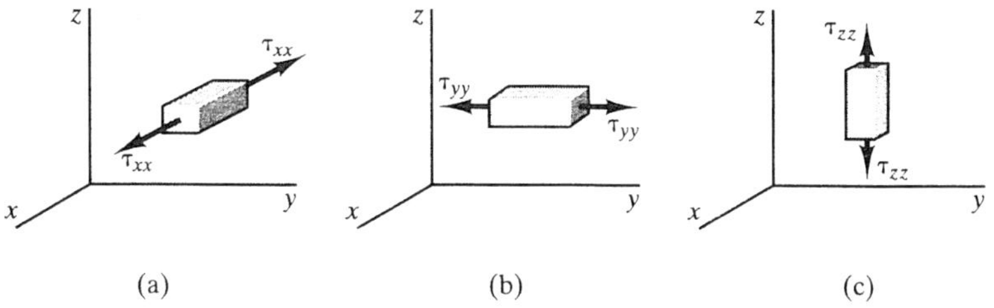

Figure 6.1. Tensile elements at a point.

$$\varepsilon'_{xx} = \frac{\tau_{xx}}{E}$$

where we use primes to denote different strain contributions. We also learned from the one-dimensional test that while the element extends in the x direction it will contract in the directions perpendicular to the x direction.[2] This is the so-called *Poisson effect.* The *Poisson ratio* ν is a measure of this lateral effect and can be determined from the one-dimensional test data. Thus, we can say that

$$\varepsilon'_{yy} = -\nu\varepsilon'_{xx} = -\nu\frac{\tau_{xx}}{E}$$

$$\varepsilon'_{zz} = -\nu\varepsilon'_{xx} = -\nu\frac{\tau_{xx}}{E}$$

At this point we indicate that for isotropic materials, such as the Hookean material we are working with, normal stresses cannot develop shear strains (see Appendix I). If next we apply a normal stress to the element in the y direction [see Fig. 6.1(b)], the following new strains are developed:

$$\varepsilon''_{yy} = \frac{\tau_{yy}}{E}$$

$$\varepsilon''_{xx} = -\nu\varepsilon''_{yy} = -\nu\frac{\tau_{yy}}{E}$$

$$\varepsilon''_{zz} = -\nu\varepsilon''_{yy} = -\nu\frac{\tau_{yy}}{E}$$

Finally, a stress τ_{zz} yields [see Fig. 6.1(c)]

[2]Conversely, if the element contracts in the x direction because of compressive stress τ_{xx}, it will extend in the direction perpendicular to the x direction.

$$\varepsilon_{zz}''' = \frac{\tau_{zz}}{E}$$

$$\varepsilon_{xx}''' = -\nu\varepsilon_{zz}''' = -\nu\frac{\tau_{zz}}{E}$$

$$\varepsilon_{yy}''' = -\nu\varepsilon_{zz}''' = -\nu\frac{\tau_{zz}}{E}$$

We note (see Appendix I) that for isotropic materials shearing stresses will not affect the normal strains. Hence, superposing the effects of the normal stresses we have the following relation between normal stress and normal strain for the general state of stress at a point:

$$\begin{aligned}\varepsilon_{xx} &= \frac{1}{E}[\tau_{xx} - \nu(\tau_{yy} + \tau_{zz})]\\ \varepsilon_{yy} &= \frac{1}{E}[\tau_{yy} - \nu(\tau_{xx} + \tau_{zz})]\\ \varepsilon_{zz} &= \frac{1}{E}[\tau_{zz} - \nu(\tau_{xx} + \tau_{yy})]\end{aligned} \tag{6.1}$$

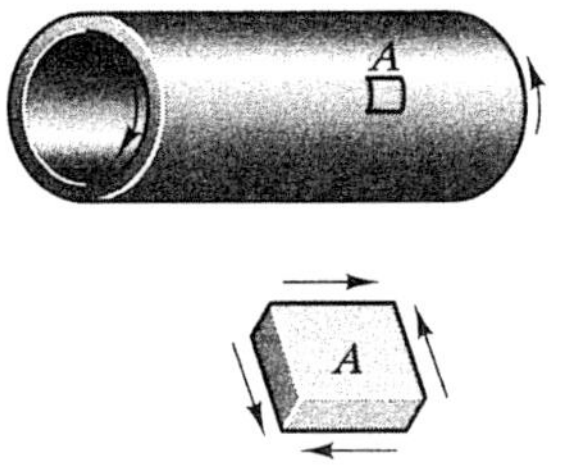

Figure 6.2. Element from a torsion test.

We have based the previous conclusions on results of the tensile test where we were able to relate directly normal stress and normal strain. To experimentally investigate the relation between shear stress and strain, we could use the simple torsion test. Here a cylinder (see Fig. 6.2) is twisted at the ends. We shall be able to show in our study of torsion (Chapter 14) that an element shown at A is subject only to pure shear stress τ. Experiments indicate for Hookean materials that the shear stress τ is proportional to the engineering shear strain γ up to a shear yield stress. (We shall relate this shear yield stress in Chapter 9 for a given material to the yield stress from the tensile test experiment.) Thus, we can say for stresses less than the shear yield stress

$$\gamma = \frac{\tau}{G} \tag{6.2}$$

where the proportionality constant G is called the *shear modulus.*

We have shown in Appendix I that shear stress can only generate the corresponding shear strain for isotropic material. Hence in the three-dimensional case, we may extend the above relation [Eq. (6.2)] separately to each of the shear stresses since they act independently of each other. Thus,

$$\begin{aligned}\gamma_{xy} &= \frac{\tau_{xy}}{G}\\ \gamma_{xz} &= \frac{\tau_{xz}}{G}\\ \gamma_{yz} &= \frac{\tau_{yz}}{G}\end{aligned} \tag{6.3}$$

The three-dimensional Hooke's law can now be given as follows:

$$\begin{aligned}
\varepsilon_{xx} &= \frac{1}{E}[\tau_{xx} - \nu(\tau_{yy} + \tau_{zz})] \\
\varepsilon_{yy} &= \frac{1}{E}[\tau_{yy} - \nu(\tau_{xx} + \tau_{zz})] \\
\varepsilon_{zz} &= \frac{1}{E}[\tau_{zz} - \nu(\tau_{xx} + \tau_{yy})] \\
\gamma_{xy} &= \frac{1}{G}\tau_{xy} \\
\gamma_{xz} &= \frac{1}{G}\tau_{xz} \\
\gamma_{yz} &= \frac{1}{G}\tau_{yz}
\end{aligned} \tag{6.4}$$

6.3 Relation Between the Three Material Constants

There are three constants given in the relations above—E, G, and ν. Actually, only two of the constants are independent; any two may be taken as the independent pair. We shall next show the interdependence of these constants.

Consider a state of plane stress for the square plate of thickness t shown in Fig. 6.3 wherein a uniform tensile stress loading S is applied in the y direction, while a uniform compressive stress loading S is applied in the x direction. For each element in the plate we can say that

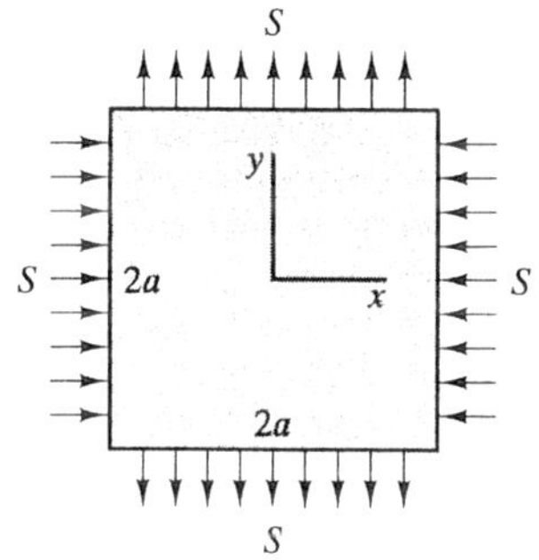

Figure 6.3. Square plate under stress.

$$\begin{aligned}
\tau_{xx} &= -S \\
\tau_{yy} &= S \\
\tau_{xy} &= 0
\end{aligned}$$

In Fig. 6.4 we consider next a portion of the square plate formed by cutting along a diagonal. We may imagine that a line along the cut diagonal is an x' axis, while a line normal to the cut diagonal is a y' axis. This has been shown in the diagram. Consequently, the normal stress and shear stress along the cut section are, respectively, $\tau_{y'y'}$ and $\tau_{y'x'}$. Summing forces in the x' direction, from **equilibrium** we have noting that $\cos 45° = 1/\sqrt{2}$

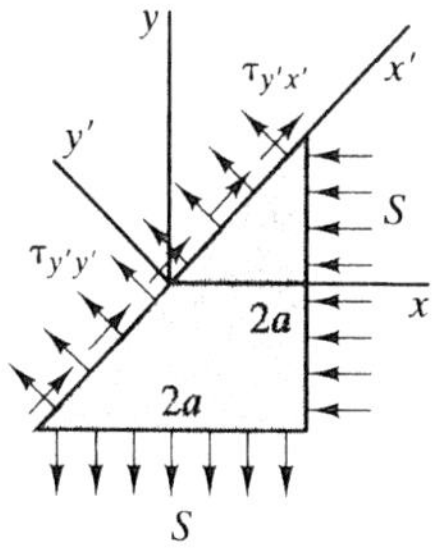

Figure 6.4. Half of square plate.

$$(\tau_{y'x'})(2a)(\sqrt{2})t - 2\left[(S)(2a)\left(\frac{1}{\sqrt{2}}\right)t\right] = 0 \tag{6.5}$$

$$\therefore\ \tau_{y'x'} = \tau_{x'y'} = S$$

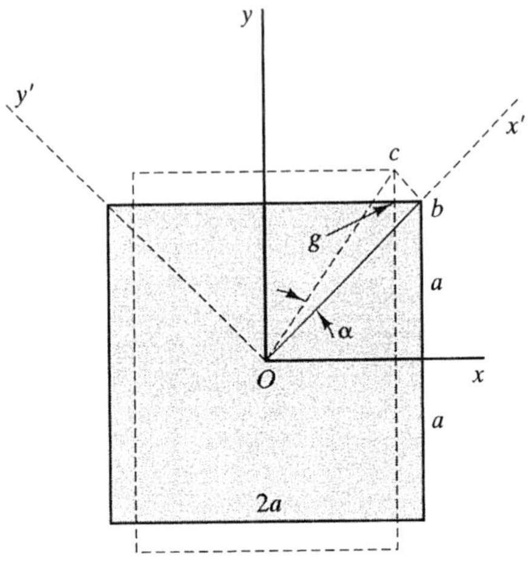

Figure 6.5. Deformation of square plate.

Now consider the deformation of the square plate, which is shown by the dashed line enlarged in Fig. 6.5. Again we have shown reference xy as well as reference $x'y'$, the latter along the diagonals of the undeformed plate. We desire the shear strain $\varepsilon_{x'y'}$. Observing Fig. 6.5 we can conclude that $\varepsilon_{x'y'}$ equals the angle α between the diagonal in the undeformed geometry and the diagonal in the deformed geometry. Again observing Fig. 6.5, we can say that

$$\varepsilon_{x'y'} = \frac{\overline{cb}}{\overline{Ob}} \tag{6.6}$$

Now $\overline{cb}$ is the hypotenuse of a 45°, 45°, 90° triangle gbc, where leg $\overline{gb}$ is half the change in length of the plate in the x direction, while leg $\overline{gc}$ is half the change in length in the y direction. The change in length $\overline{gb}$ is a result of two effects. First there is that due to the compressive stress S in the x direction giving a contribution $(S/E)(a)$. Added to this is a change in length due to the Poisson effect from the strain coming from the tensile stress S in the y direction. This contribution is then $\nu[(S/E)a]$. Accordingly, for $\overline{gb}$ we can say that

$$\overline{gb} = \frac{S}{E}a + \nu\frac{S}{E}a = \frac{Sa}{E}(1+\nu)$$

We can then conclude for triangle gbc on using the result above that

$$\overline{cb} = \frac{\overline{gb}}{\cos 45°} = \sqrt{2}\,\overline{gb} = \sqrt{2}\frac{Sa}{E}(1+\nu) \tag{6.7}$$

Finally, noting from Fig. 6.5 that

$$\overline{Ob} = \frac{a}{\cos 45°} = \sqrt{2}\,a$$

we have from Eqs. (6.6) and (6.7) and the above result that

$$\varepsilon_{x'y'} = \frac{\sqrt{2}(Sa/E)(1+\nu)}{\sqrt{2}\,a} = \frac{S}{E}(1+\nu) \tag{6.8}$$

Hooke's law states next that

$$\tau_{x'y'} = G(2\varepsilon_{x'y'}) \tag{6.9}$$

Substituting for $\tau_{x'y'}$ from Eq. (6.5) and for $\varepsilon_{x'y'}$ from Eq. (6.8), for the equation above, we get

$$S = G\left[2\frac{S}{E}(1+\nu)\right]$$

$$\therefore\ G = \frac{E}{2(1+\nu)} \tag{6.10}$$

We thus arrive at a relation between the three constants G, E, and ν. This relation was reached by looking at a very special state of stress for reference xy and then evaluating the stress $\tau_{x'y'}$ and strain $\varepsilon_{x'y'}$ for a set of primed axes rotated 45° from reference xy. Even though the state of stress was very special, the resulting equation [Eq. (6.10)], relating as it does *only constants*, must then be valid for any state of stress. Thus, for any linear-elastic, isotropic material, knowing any two of the three constants we can determine the third constant from Eq. (6.10).

Example 6.1

The following state of stress exists at a point in a body:

$$\tau_{ij} = \begin{pmatrix} 3000 & 0 & 0 \\ 0 & 2000 & -500 \\ 0 & -500 & 0 \end{pmatrix} \text{psi}$$

What are the strain components at this point? A one-dimensional tensile test gives $E = 30 \times 10^6$ psi and $\nu = .3$ for the body.

We first ascertain the strains for reference xyz of the problem. Thus, using Eq. (6.4), for the normal strains we have

$$\varepsilon_{xx} = \frac{1}{30 \times 10^6}[3000 - (.3)(2000)] = 80 \times 10^{-6}$$

$$\varepsilon_{yy} = \frac{1}{30 \times 10^6}[2000 - (.3)(3000)] = 36.7 \times 10^{-6}$$

$$\varepsilon_{zz} = \frac{1}{30 \times 10^6}[-(.3)(3000 + 2000)] = -50 \times 10^{-6}$$

To get the shear strains we must next determine G. Thus, using Eq. (6.10), we have

$$G = \frac{30 \times 10^6}{2(1 + .3)} = 11.55 \times 10^6 \text{ psi}$$

For the shear strains we then have

$$\varepsilon_{xy} = 0$$

$$\varepsilon_{xz} = 0$$

$$\varepsilon_{yz} = \frac{1}{2}\left(\frac{-500}{11.55 \times 10^6}\right) = -21.6 \times 10^{-6}$$

The strain tensor is then

$$\varepsilon_{ij} = \begin{pmatrix} 80 & 0 & 0 \\ 0 & 36.7 & -21.6 \\ 0 & -21.6 & -50 \end{pmatrix} \times 10^{-6}$$

The three-dimensional Hooke's law developed here for isotropic materials was reached by extending one-dimensional stress-strain relations in an intuitive manner. To proceed to set forth a constitutive law such as this in a more rigorous manner would require us to get involved with atomic structure—that is, with solid-state physics. Although this field of inquiry has developed rapidly in recent years, it is still not feasible to proceed along this avenue in presenting constitutive laws for solid behavior of materials. One must make hypotheses regarding stress-strain relations with the hope that the resulting relation will portray reasonably accurately the behavior of classes of materials of interest to the engineer.

The equations that we have presented actually do represent reasonably well the behavior of many structural materials such as steel, aluminum, and glass. They will accordingly be for us the key constitutive equations to be used in this text provided the proportional limit has not been reached.

6.4 Nonisothermal Hooke's Law

Our remarks in this section have thus far been restricted to *isothermal* linear elastic behavior of materials. When we have thermal effects present as a result of a nonuniform temperature field and/or external constraint to thermal expansion of the material, we must express the stress-strain relations in a more general manner. This is simple to do by merely extending the formulations presented in Section 4.8 to apply to the three-dimensional state of stress. Thus, we can say, noting that ε'_{xx}, ε'_{yy}, and ε'_{zz} are the free expansion terms, that

$$\begin{aligned}
\varepsilon_{xx} &= \varepsilon'_{xx} + \frac{1}{E}[\tau_{xx} - \nu(\tau_{yy} + \tau_{zz})] \qquad \gamma_{xy} = \frac{\tau_{xy}}{G} \\
\varepsilon_{yy} &= \varepsilon'_{yy} + \frac{1}{E}[\tau_{yy} - \nu(\tau_{xx} + \tau_{zz})] \qquad \gamma_{xz} = \frac{\tau_{xz}}{G} \\
\varepsilon_{zz} &= \varepsilon'_{zz} + \frac{1}{E}[\tau_{zz} - \nu(\tau_{xx} + \tau_{yy})] \qquad \gamma_{yz} = \frac{\tau_{yz}}{G}
\end{aligned}$$

Noting that $\varepsilon'_{xx} = \varepsilon'_{yy} = \varepsilon'_{zz} = \alpha\Delta T$ we can then give the stress-strain relations above as follows:

$$
\begin{aligned}
\varepsilon_{xx} &= \frac{1}{E}[\tau_{xx} - \nu(\tau_{yy} + \tau_{zz})] + \alpha\Delta T \\
\varepsilon_{yy} &= \frac{1}{E}[\tau_{yy} - \nu(\tau_{xx} + \tau_{zz})] + \alpha\Delta T \\
\varepsilon_{zz} &= \frac{1}{E}[\tau_{zz} - \nu(\tau_{xx} + \tau_{yy})] + \alpha\Delta T \\
\gamma_{xy} &= \frac{1}{G}\tau_{xy} \\
\gamma_{xz} &= \frac{1}{G}\tau_{xz} \\
\gamma_{yz} &= \frac{1}{G}\tau_{yz}
\end{aligned}
\qquad (6.11)
$$

Example 6.2

A square steel plate sits in a square opening of a rigid flat body as is shown in Fig. 6.6. What uniform temperature increase from a uniform initial temperature is needed to make the steel plate just fill the gap on all sides? The gap around the sides is 0.005 in. Next consider a change in temperature ΔT this time of 800°F above the initial temperature. What is then the stress state of the plate? Take $E = 30 \times 10^6$ psi, $\nu = 0.3$, and $\alpha = 6.5 \times 10^{-6}/°F$.

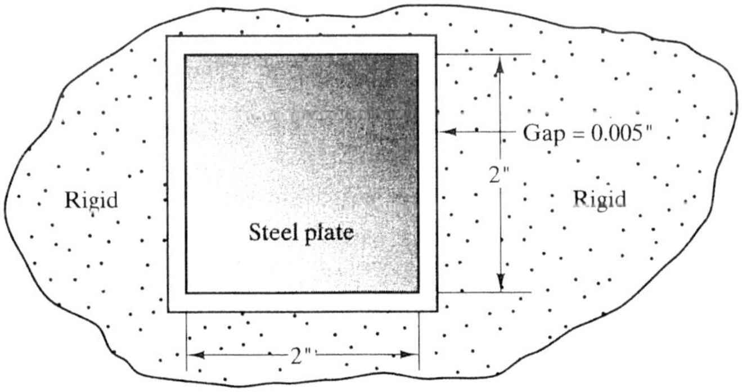

Figure 6.6. Steel plate in a flat, rigid enclosure.

We have here a case of uniform normal strain for ΔT. It is clear that we can get the normal strains from **geometry** immediately

$$
\varepsilon_{xx} = \frac{(2)(0.005)}{2} = 0.005 \qquad \varepsilon_{yy} = \frac{(2)(0.005)}{2} = 0.005
$$

Since the plate just touches the rigid walls there is no stress. And so from the **constitutive law** we can say for the steel isothermal plate,

Example 6.2 (Continued)

$$\varepsilon_{xx} = 0 + \alpha\Delta T = 0 + (6.5 \times 10^{-6})\Delta T$$
$$\varepsilon_{yy} = 0 + \alpha\Delta T = 0 + (6.5 \times 10^{-6})\Delta T$$

We can now compute the desired temperature increase.

$$\Delta T = 769° \text{ F}$$

Now we consider a ΔT of 800°F above the initial temperature. The strains remain $\varepsilon_{xx} = \varepsilon_{yy} = 0.005$ **(geometry)**. We may use the linear-elastic, nonisothermal **constitutive law.** Thus,

$$\varepsilon_{xx} = 0.005 = \frac{1}{E}(\tau_{xx} - \nu\tau_{yy}) + \alpha\Delta T$$

$$\varepsilon_{yy} = 0.005 = \frac{1}{E}(\tau_{yy} - \nu\tau_{xx}) + \alpha\Delta T$$

Substituting $\Delta T = 800°\text{F}$ and $\alpha = 6.5 \times 10^{-6}/°\text{F}$, we can solve for the stresses.

$$\tau_{xx} = \tau_{yy} = -8570 \text{ psi} \quad \tau_{xy} = 0$$

*6.5 Nonisotropic, Linear, Elastic Behavior: Generalized Hooke's Law

Thus far we have heuristically set forth the three-dimensional Hooke's law for isotropic materials. We found that such behavior could be expressed in terms of two independent constants. In the case of *non-isotropic linear-elastic* behavior such as may occur in crystals, we may postulate the following stress-strain relations, which we call the *generalized Hooke's law:*

$$\tau_{xx} = C_{11}\varepsilon_{xx} + C_{12}\varepsilon_{yy} + C_{13}\varepsilon_{zz} + C_{14}\gamma_{xy} + C_{15}\gamma_{yz} + C_{16}\gamma_{xz} \quad (6.12a)$$
$$\tau_{yy} = C_{21}\varepsilon_{xx} + C_{22}\varepsilon_{yy} + C_{23}\varepsilon_{zz} + C_{24}\gamma_{xy} + C_{25}\gamma_{yz} + C_{26}\gamma_{xz} \quad (6.12b)$$
$$\tau_{zz} = C_{31}\varepsilon_{xx} + C_{32}\varepsilon_{yy} + C_{33}\varepsilon_{zz} + C_{34}\gamma_{xy} + C_{35}\gamma_{yz} + C_{36}\gamma_{xz} \quad (6.12c)$$
$$\tau_{xy} = C_{41}\varepsilon_{xx} + C_{42}\varepsilon_{yy} + C_{43}\varepsilon_{zz} + C_{44}\gamma_{xy} + C_{45}\gamma_{yz} + C_{46}\gamma_{xz} \quad (6.12d)$$
$$\tau_{yz} = C_{51}\varepsilon_{xx} + C_{52}\varepsilon_{yy} + C_{53}\varepsilon_{zz} + C_{54}\gamma_{xy} + C_{55}\gamma_{yz} + C_{56}\gamma_{xz} \quad (6.12e)$$
$$\tau_{xz} = C_{61}\varepsilon_{xx} + C_{62}\varepsilon_{yy} + C_{63}\varepsilon_{zz} + C_{64}\gamma_{xy} + C_{65}\gamma_{yz} + C_{66}\gamma_{xz} \quad (6.12f)$$

where the terms C_{ij} form a 6×6 matrix of elastic moduli whose values depend on the material. We are saying in this formulation that each stress at a point is *linearly related* to *all* the strains at the point. We can show that Eqs. (6.12) will degenerate to Eqs. (6.4) when we consider *isotropic,* linear-elastic materials. The guiding principle in doing this is that for isotropy the generalized Hooke's law for some new reference $x'y'z'$ rotated relative to xyz must retain the same constants C_{ij} as when the law is expressed for the xyz reference, as it is done in Eqs. (6.12). Thus, the equation for $\tau_{x'x'}$ would be given as

$$\tau_{x'x'} = C_{11}\varepsilon_{x'x'} + C_{12}\varepsilon_{y'y'} + C_{13}\varepsilon_{z'z'} + C_{14}\gamma_{x'y'} + C_{15}\gamma_{y'z'} + C_{16}\gamma_{x'z'} \tag{6.13}$$

where the terms C_{ij} have the *same values* as in Eqs. (6.12). We can impose a sequence of rotations of the axes xyz requiring each time that the stress-strain relations for the new orientation have the same form as for the original set of axes. Consider, for example, a 180° rotation about the z axis as shown in Fig. 6.7. The direction cosines between the various axes are given in the following tabular form:

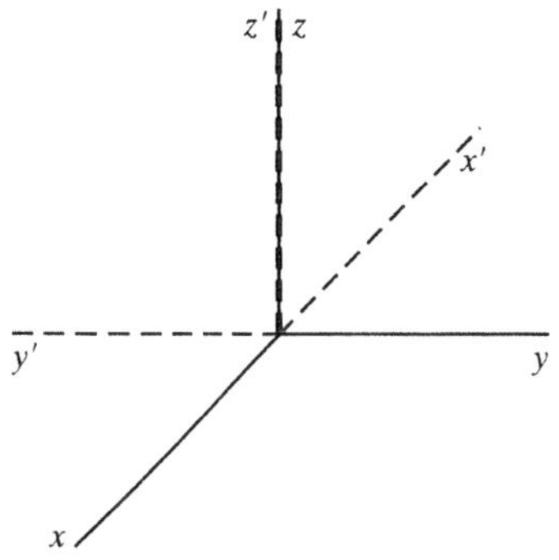

Figure 6.7. 180° rotation about z axis.

	x	y	z
x'	−1	0	0
y'	0	−1	0
z'	0	0	1

We have pointed out several times earlier that once the stress tensor is known for three orthogonal interfaces at a point, that is, for a reference xyz, then one can determine the stresses on any interface at the point. Accordingly, knowing τ_{ij} for reference xyz at a point, we can determine the stress tensor having components for any reference $x'y'z'$ by considering faces of an infinitesimal rectangular parallelepiped having edges parallel to axes $x'y'z'$ at the point. We shall later (Chapter 15) be able to show that the stresses for axes $x'y'z'$ of Fig. 6.7 are given in terms of stresses for xyz as follows:

$$\begin{aligned} \tau_{x'x'} &= \tau_{xx} & \tau_{x'y'} &= \tau_{xy} \\ \tau_{y'y'} &= \tau_{yy} & \tau_{y'z'} &= -\tau_{yz} \\ \tau_{z'z'} &= \tau_{zz} & \tau_{x'z'} &= -\tau_{xz} \end{aligned} \tag{6.14}$$

Similarly, knowing the strain tensor for reference xyz at a point we shall be able later (Chapter 16) to determine strains for any reference $x'y'z'$ rotated arbitrarily relative to xyz at the point. In particular, we shall be able to show for the pair of references shown in Fig. 6.7 that

$$\begin{aligned} \varepsilon_{x'x'} &= \varepsilon_{xx} & \gamma_{x'y'} &= \gamma_{xy} \\ \varepsilon_{y'y'} &= \varepsilon_{yy} & \gamma_{y'z'} &= -\gamma_{yz} \\ \varepsilon_{z'z'} &= \varepsilon_{zz} & \gamma_{x'z'} &= -\gamma_{xz} \end{aligned} \tag{6.15}$$

Next replace the stresses and strains in Eq. (6.13) using the results given by Eqs. (6.14) and (6.15). We get

$$\tau_{xx} = C_{11}\varepsilon_{xx} + C_{12}\varepsilon_{yy} + C_{13}\varepsilon_{zz} + C_{14}\gamma_{xy} - C_{15}\gamma_{yz} - C_{16}\gamma_{xz} \tag{6.16}$$

Comparing Eq. (6.16) with Eq. (6.12a) we see that a necessary requirement for isotropy, that is, for these equations to be identical, is that

$$C_{15} = C_{16} = 0 \tag{6.17}$$

Similarly, examining the other stresses in this manner we conclude additionally that for isotropy the following conditions must hold:

$$\begin{array}{lllll} C_{25} = 0 & C_{36} = 0 & C_{51} = 0 & C_{54} = 0 & C_{63} = 0 \\ C_{26} = 0 & C_{45} = 0 & C_{52} = 0 & C_{61} = 0 & C_{64} = 0 \\ C_{35} = 0 & C_{46} = 0 & C_{53} = 0 & C_{62} = 0 & \end{array} \tag{6.18}$$

The matrix of elastic moduli then simplifies to the form

$$\begin{array}{cccccc} C_{11} & C_{12} & C_{13} & C_{14} & 0 & 0 \\ C_{21} & C_{22} & C_{23} & C_{24} & 0 & 0 \\ C_{31} & C_{32} & C_{33} & C_{34} & 0 & 0 \\ C_{41} & C_{42} & C_{43} & C_{44} & 0 & 0 \\ 0 & 0 & 0 & 0 & C_{55} & C_{56} \\ 0 & 0 & 0 & 0 & C_{65} & C_{66} \end{array} \tag{6.19}$$

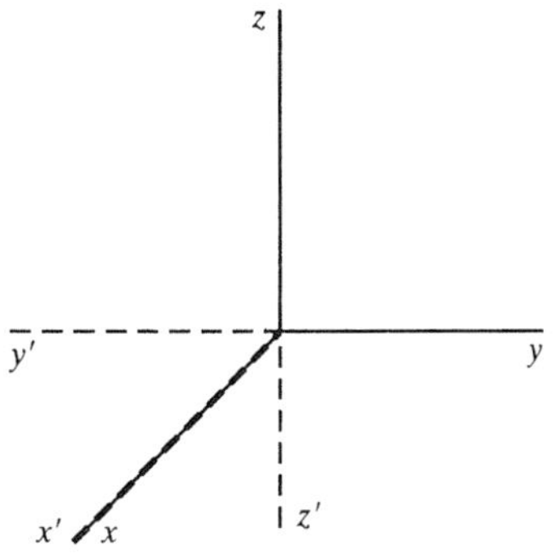

Figure 6.8. 180° rotation about x axis.

Observing Fig. 6.7, it should be clear that, for the foregoing matrix, mechanical behavior in the x direction is the same as corresponding mechanical behavior in the x' direction and that this is also true for the y and y' directions. Thus, we say that there is *elastic symmetry about the planes yz and xz.* To achieve symmetry about plane xy as well, we impose the condition of isotropy for a 180° rotation of axes about the x axis, as shown in Fig. 6.8. We can show that the matrix is reduced to the following form:

$$\begin{array}{cccccc} C_{11} & C_{12} & C_{13} & 0 & 0 & 0 \\ C_{21} & C_{22} & C_{23} & 0 & 0 & 0 \\ C_{31} & C_{32} & C_{33} & 0 & 0 & 0 \\ 0 & 0 & 0 & C_{44} & 0 & 0 \\ 0 & 0 & 0 & 0 & C_{55} & 0 \\ 0 & 0 & 0 & 0 & 0 & C_{66} \end{array} \tag{6.20}$$

This matrix of moduli then represents the case of elastic symmetry about *three orthogonal planes.* This means that there is no change in mechanical behavior when the x,y,z directions are reversed. A material behaving in this way is called *orthotropic.* Examples of orthotropic materials are wood and many crystals. Notice that the number of elastic moduli has been reduced from 36 to 12. By other rotations of axes we can show that for isotropic materials there are no more and no less than two elastic moduli and that Eqs. (6.12) can be reduced to Eqs. (6.4).[3]

[3]See I. H. Shames, *Mechanics of Deformable Solids* (R. E. Krieger, Melbourne, FL, 1964), for a more detailed discussion of this process.

*6.6 A Look Ahead: Fluid Mechanics II

It may come as a surprise to the student that solid mechanics and fluid mechanics have a number of similarities which become more distinct when one looks into these subjects with some depth. We have already discussed strains and strain rates for these continua. Now we will say a word about certain general forms of the constitutive laws.

In starred section 6.5, we presented the generalized Hooke's law with 36 elastic moduli for nonisotropic, linear-elastic solids. By continually rotating axes and imposing isotropy each time, we ended up with two independent constants characterizing the mechanical behavior of the material. Used most were the Poisson ratio ν and the elastic modulus E. The corresponding equations for Eq. (6.12) for an anisotropic (not isotropic) fluid is called Stokes's viscosity law. This constitutive law relates linearly each stress to all six independent strain rate terms through a set of 36 constants called viscosity coefficients. The following is a representation of these equations:

$$\begin{aligned}
\tau_{xx} &= -p + C_{11}\dot{\varepsilon}_{xx} + C_{12}\dot{\varepsilon}_{yy} + C_{13}\dot{\varepsilon}_{zz} + C_{14}\dot{\varepsilon}_{xy} + C_{15}\dot{\varepsilon}_{yz} + C_{16}\dot{\varepsilon}_{xz} \\
\tau_{yy} &= -p + C_{21}\dot{\varepsilon}_{xx} + C_{22}\dot{\varepsilon}_{yy} + C_{23}\dot{\varepsilon}_{zz} + C_{24}\dot{\varepsilon}_{xy} + C_{25}\dot{\varepsilon}_{yz} + C_{26}\dot{\varepsilon}_{xz} \\
\tau_{zz} &= -p + C_{31}\dot{\varepsilon}_{xx} + C_{32}\dot{\varepsilon}_{yy} + C_{33}\dot{\varepsilon}_{zz} + C_{34}\dot{\varepsilon}_{xy} + C_{35}\dot{\varepsilon}_{yz} + C_{36}\dot{\varepsilon}_{xz} \\
\tau_{xy} &= C_{41}\dot{\varepsilon}_{xx} + C_{42}\dot{\varepsilon}_{yy} + C_{43}\dot{\varepsilon}_{zz} + C_{44}\dot{\varepsilon}_{xy} + C_{45}\dot{\varepsilon}_{yz} + C_{46}\dot{\varepsilon}_{xz} \\
\tau_{yz} &= C_{51}\dot{\varepsilon}_{xx} + C_{52}\dot{\varepsilon}_{yy} + C_{53}\dot{\varepsilon}_{zz} + C_{54}\dot{\varepsilon}_{xy} + C_{55}\dot{\varepsilon}_{yz} + C_{56}\dot{\varepsilon}_{xz} \\
\tau_{xz} &= C_{61}\dot{\varepsilon}_{xx} + C_{62}\dot{\varepsilon}_{yy} + C_{63}\dot{\varepsilon}_{zz} + C_{64}\dot{\varepsilon}_{xy} + C_{65}\dot{\varepsilon}_{yz} + C_{66}\dot{\varepsilon}_{xz}
\end{aligned}$$

where p is the pressure. Fluids satisfying the above constitutive law are called *Newtonian fluids*. As in the case of solids we can impose isotropy by considering rotations of axes. Again we end up with two constants which are called the first and second coefficients of viscosity. Using further considerations beyond isotropy, we can generally eliminate the second viscosity coefficient in terms of the familiar first viscosity coefficient.

Part B. Introduction to Energy Methods

6.7 Strain Energy

Consider a solid maintained in equilibrium in space by a system of supporting forces N_i (Fig. 6.9). A system of external surface tractions and body forces are applied in a quasi-static manner.[4] The *first law of thermodynamics* requires that

$$Q - W_L \text{ [illegible] } E_1$$

[4]Applied so slowly that dynami[illegible] vibrations, can be neglected.

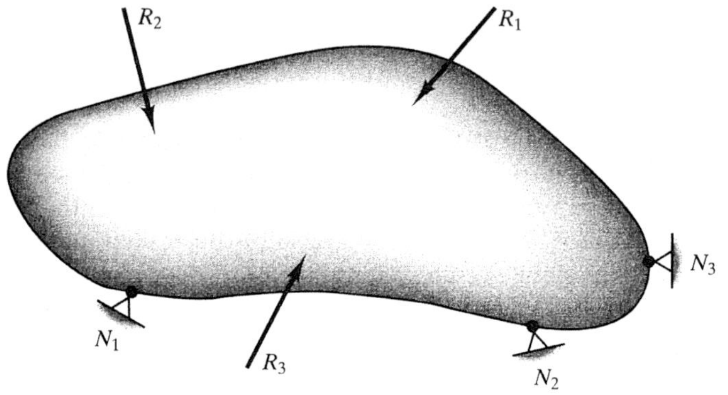

Figure 6.9. Body in equilibrium.

where Q is the heat transfer from the surroundings to the body during the loading process, $-W_k$ is the work done by the external forces during the loading process, and $E_2 - E_1$ is the change of stored energy of the body from heating and loading. It should be clear that the constraining forces do no work since they do not move.

If the process is close to being adiabatic (no heat transfer), then we can drop the term Q. Also, if the work of gravity is negligible during the process (i.e., the center of gravity does not appreciably change position during the loading process), then the change in stored energy, $E_2 - E_1$, is merely the change in internal energy of the material resulting from the work of the surface tractions and body forces excluding gravity. Now if the material is *elastic* (not necessarily linear-elastic), there is no hysteresis, and the body will perform an equal but opposite amount of work on the surroundings, during the unloading process, as was done by the surroundings on it during the loading process. Thus, we can consider for such bodies that energy has been *stored* in the body (hence the name used) as a result of the deformation, and this stored energy equals the external work done on the body. This stored energy, which is available as work on releasing the applied loads in a quasi-static manner, in elastic solids is called *strain energy of deformation*. We shall employ U to represent the strain energy for a given body and we shall employ $\mathcal{U}$ to represent the strain energy per unit volume (the strain-energy density).

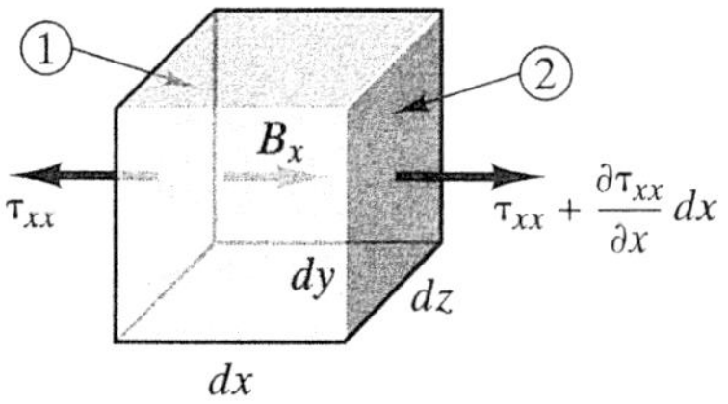

Figure 6.10. Element under pure tension.

At this time we wish to formulate means of computing the strain energy directly from the stress-strain distribution in the body and not from the work done by the external forces as was previously the case. Consider first an infinitesimal rectangular parallelepiped of material under the action of simple tension as shown in Fig. 6.10. The displacement of face 1 in the x direction we denote as u_x and the displacement of face 2 in the x direction can then be given as $u_x + (\partial u_x / \partial x)dx$. The increment of work done by the stresses τ_{xx} on face 1 and by $\tau_{zz} + (\partial \tau_{xx} / \partial x)dx$ on face 2 during infinitesimal displacements du_x at 1 and $d[u_x + (\partial u_x / \partial x)dx]$ at 2 is given as follows on using body force component B_x,

$$
\begin{aligned}
&- (\tau_{xx}) du_x\, dy\, dz + \left(\tau_{xx} + \frac{\partial \tau_{xx}}{\partial x} dx\right) d\left(u_x + \frac{\partial u_x}{\partial x} dx\right) \\
&\quad \times dy\, dz + B_x(dx\, dy\, dz)\, d\left(u_x + \alpha \frac{\partial u_x}{\partial x} dx\right)
\end{aligned} \tag{6.21}
$$

where α is some fraction. Collecting terms, for the foregoing expressions we have

$$
\begin{aligned}
&\tau_{xx} d\left(\frac{\partial u_x}{\partial x}\right) dx\, dy\, dz + \frac{\partial \tau_{xx}}{\partial x} du_x dx\, dy\, dz + \frac{\partial \tau_{xx}}{\partial x} d\left(\frac{\partial u_x}{\partial x}\right) dx^2 dy\, dz \\
&\quad + B_x du_x dx\, dy\, dz + \alpha B_x d\left(\frac{\partial u_x}{\partial x}\right) dx^2 dy\, dz
\end{aligned} \tag{6.22}
$$

We may drop the third term and the last term, since they are of higher order than the other terms in the expression.[5] Thus, we may rewrite the preceding expression in the following form:

$$
\left\{\tau_{xx} d\left(\frac{\partial u_x}{\partial x}\right) + \left(\frac{\partial \tau_{xx}}{\partial x} + B_x\right) du_x\right\} dx\, dy\, dz \tag{6.23}
$$

But according to equilibrium requirements, the expression $\partial \tau_{xx} / \partial x + B_x$ must be zero [see Eq. (2.5)]. Replacing $\partial u_x / \partial x$ by ε_{xx} (thereby satisfying **compatibility** for strain), we can say that the work increment done on the element in simple tension is

$$
\tau_{xx} d\varepsilon_{xx} dv \tag{6.24}
$$

By considering normal stresses and strains in the y and z directions, we may form similar expressions for the work increments done on the element for simple tension in those directions. For a state of stress on the element, wherein τ_{xx}, τ_{yy}, and τ_{zz} act simultaneously, we may compute the work increment done on the element by superposing these results. For an elastic material, this represents the strain-energy increment of the element resulting from normal strains, and we have

$$
(\tau_{xx} d\varepsilon_{xx} + \tau_{yy} d\varepsilon_{yy} + \tau_{zz} d\varepsilon_{zz}) dv \tag{6.25}
$$

As a next step, let us consider strain energy associated with shear strain. We consider for this purpose an infinitesimal rectangular parallelepiped of material subject this time to a shear stress τ_{xy} as shown in Fig. 6.11. The dashed configuration shows the deformed geometry, it being assumed for simplicity that the bottom plane does not move. Clearly, the work done can be attributed to the shear stress on the top surface, and so for the work increment, we use for the strain differentials $d\gamma_{xy}$ at A and $d[\gamma_{xy} + (\partial \gamma_{xy}/\partial x) dx]$ at B a value somewhere between A and B. That is,

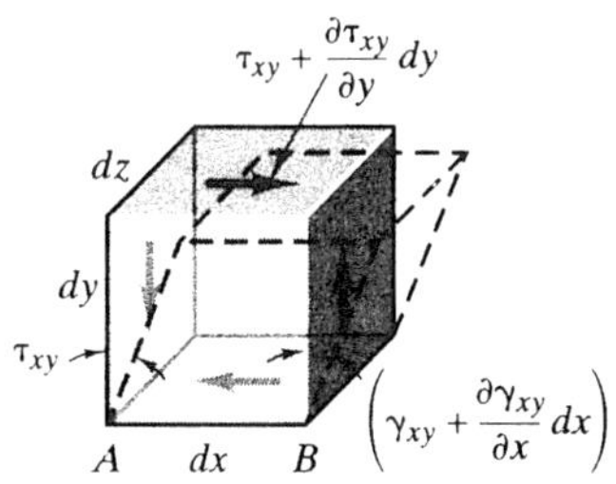

Figure 6.11. Element under pure shear.

$$
\left[\left(\tau_{xy} + \frac{\partial \tau_{xy}}{\partial y} dy\right) dz\, dx\right] \left\{ d\left[\gamma_{xy} + \left(\frac{\partial \gamma_{xy}}{\partial x}\right) \beta\, dx\right] dy \right\} \tag{6.26}
$$

[5]"Higher order" means that the term has more differentials and therefore is much smaller in value.

where β is some fraction. Expanding the expressions, we get

$$\left[\tau_{xy}d\gamma_{xy} + \tau_{xy}d\left(\frac{\partial \gamma_{xy}}{\partial x}\right)\beta\, dx + \frac{\partial \tau_{xy}}{\partial y}\, d\gamma_{xy}\, dy + \frac{\partial \tau_{xy}}{\partial y}\, d\left(\frac{\partial \gamma_{xy}}{\partial x}\right)\beta\, dx\, dy\right] dx\, dy\, dz \tag{6.27}$$

We can drop the last three expressions in the brackets, since they are of higher order when compared with the first expression. We may then express the increment of the work done by shear stress on the element as

$$(\tau_{xy}d\gamma_{xy})\, dx\, dy\, dz \tag{6.28}$$

By similarly considering the increment of the work done by shear stress, τ_{xz} and τ_{yz}, and noting that we can superpose the results[6] for an isotropic elastic material, we can then say that the total strain-energy increment resulting from shear deformation for the element is then

$$(\tau_{xy}d\gamma_{xy} + \tau_{xz}d\gamma_{xz} + \tau_{yz}d\gamma_{yz})\, dv \tag{6.29}$$

Since normal stresses do no work as a result of shear strains, and since shear stresses do no work as a result of normal strains, we can superpose Eqs. (6.25) and (6.29) to get, for an elastic isotropic material, the total strain-energy increment for an element under a general state of stress. Thus, we have

$$(\tau_{xx}d\varepsilon_{xx} + \tau_{yy}d\varepsilon_{yy} + \tau_{zz}d\varepsilon_{zz} + \tau_{xy}d\gamma_{xy} + \tau_{xz}d\gamma_{xz} + \tau_{yz}d\gamma_{yz})dv$$

Dividing by dv and calling the result $d\mathcal{U}$ the differential of the strain energy per unit volume at a point or simply the differential of the strain energy density, we then have

$$d\mathcal{U} = \tau_{xx}d\varepsilon_{xx} + \tau_{yy}d\varepsilon_{yy} + \tau_{zz}d\varepsilon_{zz} + \tau_{xy}d\gamma_{xy} + \tau_{xz}d\gamma_{xz} + \tau_{yz}d\gamma_{yz} = \sum_i \sum_j \tau_{ij}\, d\varepsilon_{ij}$$

Hence, we have, using primes in $d\varepsilon_{ij}$ for better mathematical form,

$$\mathcal{U} = \int_0^{\varepsilon_{ij}} \sum_i \sum_j \tau_{ij}\, d\varepsilon'_{ij} \tag{6.30a}$$

and

$$U = \iiint_V \int_0^{\varepsilon_{ij}} \sum_i \sum_j \tau_{ij}\, d\varepsilon'_{ij}\, dv \tag{6.30b}$$

Up to this point the formulations are valid for isotropic materials having *arbitrary elastic* behavior. We shall now consider the special case of isotropic, linear-elastic behavior. We may compute $\mathcal{U}$ for this case by substituting for the strain in Eq. (6.30(a)) using Eqs. (6.4) and then integrating. Thus, as a first step we have

[6]This means that the work done by τ_{xy} accompanying a deformation γ_{xz} or γ_{yz} is zero. That is, for isotropic materials a shear stress τ_{ij} only does work during the deformation γ_{ij} (see Appendix I). Note also that the *body force work* for Eqs. (6.26) and (6.28) is of higher order (as you may yourself verify) and has not been included.

$$d\mathcal{U} = \frac{\tau_{xx}}{E}[d\tau_{xx} - \nu(d\tau_{yy} + d\tau_{zz})] + \frac{\tau_{yy}}{E}[d\tau_{yy} - \nu(d\tau_{xx} + d\tau_{zz})] + \frac{\tau_{zz}}{E}[d\tau_{zz} - \nu(d\tau_{xx} + d\tau_{yy})] + \frac{\tau_{xy}}{G}d\tau_{xy} + \frac{\tau_{xz}}{G}d\tau_{xz} + \frac{\tau_{yz}}{G}d\tau_{yz} \tag{6.31}$$

Realizing that the stresses are independent of each other we can carry out the integration as follows:

$$\mathcal{U} = \frac{1}{2E}(\tau_{xx}^2 + \tau_{yy}^2 + \tau_{zz}^2) - \frac{\nu}{E}(\tau_{xx}\tau_{yy} + \tau_{xx}\tau_{zz} + \tau_{yy}\tau_{zz}) + \frac{1}{2G}(\tau_{xy}^2 + \tau_{xz}^2 + \tau_{yz}^2) \tag{6.32}$$

The fact that the constant of integration above is set equal to zero indicates that we have zero strain energy present at a point when all the stresses are zero there.

The preceding equations give in terms of the stress distribution the strain energy per unit volume at a point. As an exercise, show that $\mathcal{U}$ in terms of the strains becomes

$$\mathcal{U} = \frac{1}{2}\frac{E\nu}{(1+\nu)(1-2\nu)}(\varepsilon_{xx} + \varepsilon_{yy} + \varepsilon_{zz})^2 + G(\varepsilon_{xx}^2 + \varepsilon_{yy}^2 + \varepsilon_{zz}^2) + \frac{1}{2}G(\gamma_{xy}^2 + \gamma_{yz}^2 + \gamma_{xz}^2) \tag{6.33}$$

The total strain energy U for a given body of volume V can then be given in the following ways:

$$U = \iiint_V \left[\frac{1}{2E}(\tau_{xx}^2 + \tau_{yy}^2 + \tau_{zz}^2) - \frac{\nu}{E}(\tau_{xx}\tau_{yy} + \tau_{yy}\tau_{zz} + \tau_{xx}\tau_{zz}) + \frac{1}{2G}(\tau_{xy}^2 + \tau_{xz}^2 + \tau_{yz}^2)\right]dv \tag{6.34}$$

$$U = \iiint_V \left[\frac{E\nu}{2(1+\nu)(1-2\nu)}(\varepsilon_{xx} + \varepsilon_{yy} + \varepsilon_{zz})^2 + G(\varepsilon_{xx}^2 + \varepsilon_{yy}^2 + \varepsilon_{zz}^2) + \frac{1}{2}G(\gamma_{xy}^2 + \gamma_{yz}^2 + \gamma_{xz}^2)\right]dv \tag{6.35}$$

Example 6.3

Shown in Fig. 6.12 is a long prismatic bar loaded at the ends by axial forces along the centroidal axis of the bar. The stress distribution as measured from the xyz reference may be given by $\tau_{xx} = F/A$ with all other stresses having zero value. Using Eq. (6.34) the strain energy in terms of the force F then may be given as

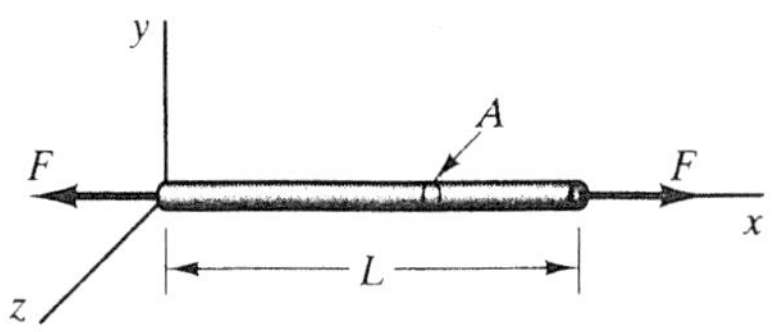

Figure 6.12. Long prismatic rod.

Example 6.3 (Continued)

$$U = \iiint_V \frac{1}{2E}(\tau_{xx}^2)dv = \frac{1}{2E}\int_0^L \left(\frac{F}{A}\right)^2 A\,dx = \frac{F^2L}{2AE} \tag{6.36}$$

wherein we have taken E as constant for the entire bar.

If an elongation δ is imposed on the rod, we can compute the strain energy using Eq. (6.35). For this case we have

$$\begin{aligned} \varepsilon_{xx} &= \frac{\delta}{L} & \gamma_{xy} &= 0 \\ \varepsilon_{yy} &= -\nu\frac{\delta}{L} & \gamma_{yz} &= 0 \\ \varepsilon_{zz} &= -\nu\frac{\delta}{L} & \gamma_{xz} &= 0 \end{aligned}$$

Substituting these strains into Eq. (6.35), we have

$$U = \iiint_V \left[\frac{E\nu}{2(1+\nu)(1-2\nu)}\left(\frac{\delta}{L}\right)^2(1-2\nu)^2 + G\left(\frac{\delta}{L}\right)^2(1+2\nu^2)\right]dv$$

This becomes

$$U = \left[\frac{E\nu(1-2\nu)}{2(1+\nu)} + G(1+2\nu^2)\right]\left(\frac{\delta}{L}\right)^2 LA$$

Replacing G by $E/2(1+\nu)$ and collecting terms, we may get

$$U = \frac{EA\delta^2}{2L} \tag{6.37}$$

Let us consider the strain-energy density function $\mathcal{U}$ for a one-dimensional state of stress along the x axis. Integrating the first term on the right side of Eq. (6.30(a)) we have for this case

$$\mathcal{U} = \int_0^{\varepsilon_{xx}} \tau_{xx}\,d\varepsilon'_{xx} \tag{6.38}$$

A stress-strain curve for a one-dimensional tensile test of an elastic specimen (not necessarily linear) is shown in Fig. 6.13. It is clear from Eq. (6.38) that the *strain energy-density function* $\mathcal{U}$ represents the "area" between the curve and the ε_{xx} axis (shown with a light shading). We now define the *complementary strain-energy density function* $\mathcal{U}^*$ as the area between the curve and the τ_{xx} axis (shown with a darker shading). Thus,

$$\mathcal{U}^* = \int_0^{\tau_{xx}} \varepsilon_{xx}\, d\tau_{xx'} \tag{6.39}$$

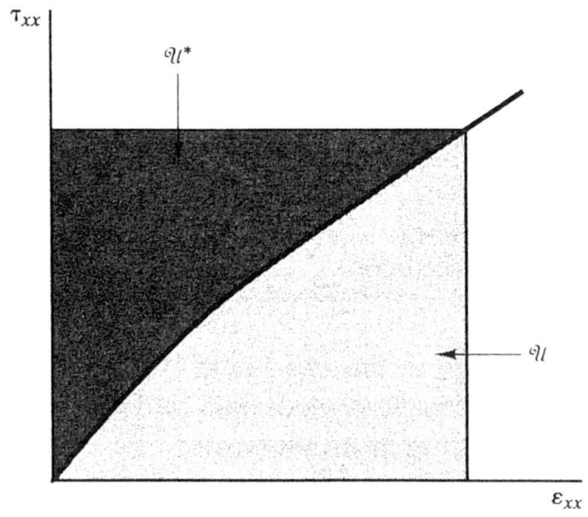

Figure 6.13. Stress-strain diagram.

Just as $\mathcal{U}$ for the general state of stress may be extrapolated from Eq. (6.38) by summing all the stress and differential strain products (see Eq. 6.30a), so will we say for $\mathcal{U}^*$ in the general case,

$$\mathcal{U}^* = \int_0^{\tau_{ij}} \sum_i \sum_j \varepsilon_{ij}\, d\tau'_{ij} \tag{6.40}$$

where we sum over i and j in the integrand. The complementary strain energy U^* is then given as follows:

$$U^* = \iiint_V \mathcal{U}^* dv = \iiint_V \left(\int_0^{\tau_{ij}} \sum_i \sum_j \varepsilon_{ij}\, d\tau'_{ij} \right) dv \tag{6.41}$$

For linear-elastic behavior the curve in Fig. 6.13 becomes a straight line and the "area" on both sides of the curve becomes equal. Hence, we can conclude on extrapolating from the one-dimensional case that for *linear elastic behavior* we arrive at the important result

$$\mathcal{U} = \mathcal{U}^* \tag{6.42}$$

6.8 Castigliano's Second Theorem (Energy Methods 1)

At this time we shall present the *second Castigliano theorem* as a very useful means of solving an array of structures problems. We shall do this by first presenting a simple theorem called the *Maxwell Betti recip-*

rocal theorem. We shall use this theorem only to present the second Castigliano theorem although the latter has its own uses in structures. This will be a simple introduction to energy methods for those students who do not have time to consider starred Chapter 18, which is a rather thorough rigorous treatment including the following pair of analogous sets of methods:

Displacement methods	***Force methods***
Virtual work	Complementary virtual work
Total potential energy	Total complementary energy
First Castigliano theorem	Second Castigliano theorem

Students are urged to go to this chapter if at all possible, since it is the gateway to advanced structural mechanics and also to much used computer formulations (such as finite elements).

Meanwhile, we shall use the second Castigliano theorem in this chapter to solve truss deflection problems in statically determinate trusses and forces in statically indeterminate trusses. These topics are beyond what you did in statics where you only solved for forces in statically determinate trusses. Also, we shall use this theorem when we study beams and torsion in later chapters.

We shall now develop the Maxwell-Betti reciprocal theorem. For this purpose, consider a *linear-elastic* body constrained in space and acted on by point forces as shown in Fig. 6.14. The total number of point forces both external and reactive will be taken as n.

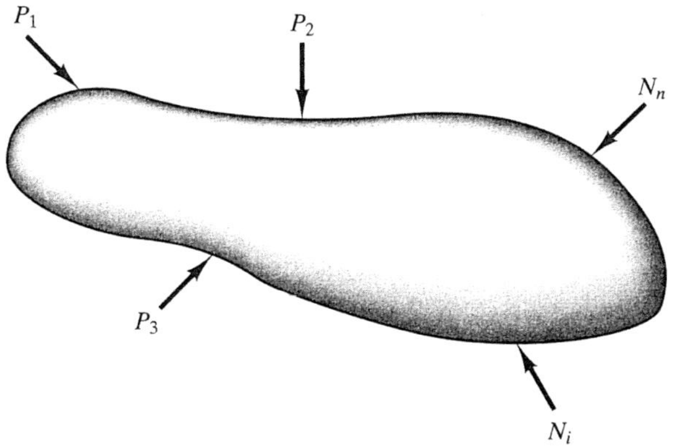

Figure 6.14. A constrained linear-elastic body.

Now let δ_i be the displacement of the point of application of load P_i, remembering that this displacement is due to the total action of all the n forces and not just to P_i. We have shown in Appendix III that the strain energy from all the loads, which we denote as U_I, is given by the following simple formula

$$U_I = \sum_{i=1}^{n} \tfrac{1}{2}\boldsymbol{P}_i \cdot \boldsymbol{\delta}_i \tag{6.43}$$

We might point out that the terms corresponding to rigid supports need not be included in the foregoing expression, since by virtue of the constraints, the displacements are zero at these points.

As a next step, consider *another* loading process for the body wherein a new set of point forces, which we denote with the subscript II, is applied. Including the supporting force there are m forces for this loading process. We may readily compute U_{II} to be

$$U_{II} = \sum_{j=1}^{m} \tfrac{1}{2}\boldsymbol{P}_j \cdot \boldsymbol{\delta}_j \tag{6.44}$$

Now investigate the strain energy of the body under the combined action of *both* force systems I and II. The principle of superposition is assumed to prevail here so that the order of application of the loading systems should not alter the final result. We therefore apply loading system I first, producing strain energy U_I given by

$$U_I = \tfrac{1}{2}\sum_{i=1}^{n} (\boldsymbol{P}_i)_I \cdot (\boldsymbol{\delta}_i)_I \tag{6.45}$$

Now apply load II to the system. In ascertaining the next increase in strain energy, we must remember to include the work done by force system I as a result of deformations induced by force system II. We shall refer to this contribution as $U_{I,II}$. Since force system I remains constant during the application of force system II, we compute $U_{I,II}$ as follows (note 1/2 is omitted):

$$U_{I,II} = \sum_{i=1}^{n} (\boldsymbol{P}_i)_I \cdot (\boldsymbol{\delta}_i)_{II} \tag{6.46}$$

where $(\boldsymbol{\delta}_i)_{II}$ represents the displacement resulting from force system II at the position of the point of application of force $(\boldsymbol{P}_i)_I$ of system I. The remaining portion of the increase in strain energy U_{II} is simply

$$U_{II} = \tfrac{1}{2}\sum_{j=1}^{m} (\boldsymbol{P}_j)_{II} \cdot (\boldsymbol{\delta}_j)_{II} \tag{6.47}$$

where $(\boldsymbol{\delta})_{II}$ is the deflection at the position of application of $(\boldsymbol{P}_j)_{II}$ due to $(\boldsymbol{P}_j)_{II}$. We may give the total strain energy as

$$U = U_I + U_{I,II} + U_{II} \tag{6.48}$$

Now let us reload the body, this time in the *reverse* order. The strain energy U' becomes, in accordance with our previous development,

$$U' = U_{II} + U_{II,I} + U_I \tag{6.49}$$

wherein the quantity $U_{\text{II,I}}$ is given as[7]

$$U_{\text{II,I}} = \sum_{j=1}^{m} (\boldsymbol{P}_j)_{\text{II}} \cdot (\boldsymbol{\delta}_j)_{\text{I}} \tag{6.50}$$

However, according to the principle of superposition the final results are the same and so U and U' must have the same value. It then becomes apparent on comparing the right-hand sides of Eqs. (6.49) and (6.50) that

$$U_{\text{I,II}} = U_{\text{II,I}} \tag{6.51}$$

More precisely, we may then say that (**compatibility**)

$$\sum_{i=1}^{n} (\boldsymbol{P}_i)_{\text{I}} \cdot (\boldsymbol{\delta}_i)_{\text{II}} = \sum_{j=1}^{m} (\boldsymbol{P}_j)_{\text{II}} \cdot (\boldsymbol{\delta}_j)_{\text{I}} \tag{6.52}$$

This is the Maxwell-Betti reciprocal theorem. In words, it says that *the work done by one system of loads owing to displacement caused by a second system of loads equals the work done by the second system of loads owing to displacements caused by the first system of loads.*

We may now use the above theorem to develop the second Castigliano theorem. Consider a *linear-elastic* body loaded by n concentrated external independent forces $\boldsymbol{P}_i$ as shown in Fig. 6.15. We shall

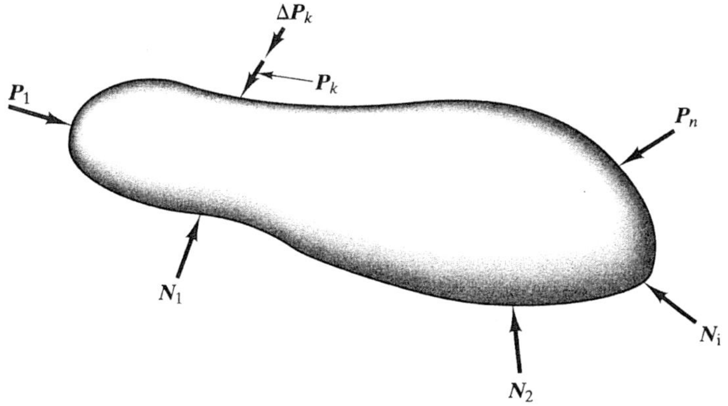

Figure 6.15. Linear-elastic body with load increment $\Delta \boldsymbol{P}_k$.

refer to the supporting forces as $\boldsymbol{N}_i$, and we shall denote the system of forces $\boldsymbol{P}_i$ and $\boldsymbol{N}_i$ as loading system I. Now increase $\boldsymbol{P}_k$ by a small amount $\Delta\boldsymbol{P}_k$ in the direction of $\boldsymbol{P}_k$ (Fig. 6.15).[8] To maintain equilibrium, some or all of the supporting forces will change slightly as a result of $\Delta\boldsymbol{P}_k$. We call the load increment $\Delta\boldsymbol{P}_k$ and the associated force increments at the supports loading system II.

Employing the Maxwell-Betti theorem, for the two load systems we get

[7]Note that $(\delta_j)_{\text{I}}$ are deflections due to load I at points of application of load II.
[8]Since the $\boldsymbol{P}_i$ are independent of each other, the forces other than $\boldsymbol{P}_k$ remain constant.

$$\sum_{i=1}^{n} (\boldsymbol{P}_i)_{\mathrm{I}} \cdot (\Delta\boldsymbol{\delta}_i)_{\mathrm{II}} = (\Delta\boldsymbol{P}_k)_{\mathrm{II}} \cdot (\boldsymbol{\delta}_k)_{\mathrm{I}} \tag{6.53}$$

where each force $(\boldsymbol{P}_i)_{\mathrm{I}}$ of system I acts on its point of application i and moves through a displacement component of i in the direction of force $(\boldsymbol{P}_i)_{\mathrm{I}}$. This displacement component of point i is entirely due to the second force system which is the *single* force $(\Delta\boldsymbol{P}_k)_{\mathrm{II}}$ acting at point k. On the right side of the equation the single force $(\Delta\boldsymbol{P}_k)_{\mathrm{II}}$ acts on its point of application k and moves through a displacement component of point k in the direction of the force $(\Delta\boldsymbol{P}_k)_{\mathrm{II}}$. This displacement component of k is due entirely to *all* the forces in the force system I.

Next formulate the *increase* in strain energy ΔU resulting from the application of the load increment $\Delta\boldsymbol{P}_k$. Recalling our work in the previous section, we have

$$\Delta U = \underbrace{\tfrac{1}{2}(\Delta\boldsymbol{P}_k)_{\mathrm{II}} \cdot (\Delta\boldsymbol{\delta}_k)_{\mathrm{II}}}_{\text{work done by loading II}} + \underbrace{\sum_{i=1}^{n} (\boldsymbol{P}_i)_{\mathrm{I}} \cdot (\Delta\boldsymbol{\delta}_i)_{\mathrm{II}}}_{\text{work done by loading I}} \tag{6.54}$$

We may replace the last term in the preceding equation using Eq. (6.53). Thus,

$$\Delta U = \tfrac{1}{2}(\Delta\boldsymbol{P}_k)_{\mathrm{II}} \cdot (\Delta\boldsymbol{\delta}_k)_{\mathrm{II}} + (\Delta\boldsymbol{P}_k)_{\mathrm{II}} \cdot (\boldsymbol{\delta}_k)_{\mathrm{I}} \tag{6.55}$$

In this equation, we no longer need the subscripts I and II since the presence of Δ is now sufficient for identification with system II. Dividing through by (ΔP_k) we then have

$$\frac{\Delta U}{\Delta P_k} = \frac{1}{2}\frac{\Delta\boldsymbol{P}_k}{\Delta P_k} \cdot \Delta\boldsymbol{\delta}_k + \frac{\Delta\boldsymbol{P}_k}{\Delta P_k} \cdot \boldsymbol{\delta}_k \tag{6.56}$$

We now go to the limit as $\Delta P_k \to 0$. The term on the left-hand side becomes simply $\partial U / \partial P_k$, whereas the first term on the right-hand side goes to zero since $\Delta\boldsymbol{\delta}_k$ vanishes in the limit. We may thus say that

$$\frac{\partial U}{\partial P_k} = \lim_{\Delta P_k \to 0}\left(\frac{\Delta\boldsymbol{P}_k}{\Delta P_k} \cdot \boldsymbol{\delta}_k\right) \tag{6.57}$$

But $\Delta\boldsymbol{P}_k / \Delta P_k$ is the unit vector in the direction of $\boldsymbol{P}_k$. The dot product of $\boldsymbol{\delta}_k$ with this unit vector gives the *component* of displacement of the *point of application* of $\boldsymbol{P}_k$ in the *direction* of the force $\boldsymbol{P}_k$. Denoting this component simply as Δ_k we arrive at the following equation:

$$\frac{\partial U}{\partial P_k} = \Delta_k \tag{6.58}$$

Thus, *the rate of change of the strain energy for a body with respect to any statically independent force P_k gives the deflection component of the point of application of this force in the direction of the force.*

By using a similar argument we can also show that

$$\frac{\partial U}{\partial M} = \theta \tag{6.59}$$

which says that the rate of change of strain energy with respect to a statically independent point couple M gives the amount of rotation at the point of application of the point couple about an axis collinear with the couple moment.

Equations (6.58) and (6.59) constitute what is commonly referred to as *Castigliano's second theorem.*

We shall now illustrate the use of the second Castigliano theorem for statically determinate truss problems. Primarily we shall solve for the movement of specified joints of a truss.

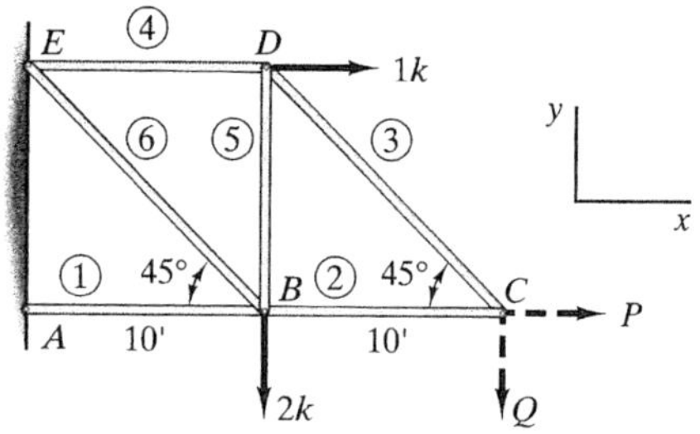

Figure 6.16. Simple plane truss.

Case A: Statically Determinate Truss

Example 6.4. Shown in Fig. 6.16 is a statically determinate simple truss, loaded by concentrated loads at pins D and B. What is the total deflection of pin C as a result of these loads?

We shall assume a horizontal force component P and a vertical force component Q at pin C so as to permit us to use Castigliano's theorem for the horizontal and vertical displacement components there. Our first step is then to determine the strain energy of the system from the 1-kip load[9] at D, the 2-kip load at B, and the loads P and Q at C. By the method of joints, for the forces in the members of the truss we have (**equilibrium**)

$$\begin{aligned} AB &= P - 2(1 + Q) \text{ tension} \qquad & DE &= 1 + Q \text{ tension} \\ BC &= P - Q \text{ tension} & DB &= Q \text{ compression} \\ CD &= \frac{Q}{.707} \text{ tension} & EB &= \frac{2 + Q}{.707} \text{ tension} \end{aligned} \tag{a}$$

We now determine U in the following way [see Eq. (6.36)] (**constitutive law**):

$$U = \sum_i \frac{F_i^2 L_i}{2A_i E_i} \tag{b}$$

Taking A_i and E_i as having the same value for each member, we get

[9]A kip is 1000 lb.

$$U = \frac{1}{2AE}\left\{[P - 2(1 + Q)]^2(10) + (P - Q)^2(10) + \left(\frac{Q}{.707}\right)^2\left(\frac{10}{.707}\right) + (1 + Q)^2(10) + Q^2(10) + \left(\frac{2 + Q}{.707}\right)^2\left(\frac{10}{.707}\right)\right\} \quad \text{(c)}$$

We may now compute the horizontal and vertical components of pin C by first taking partial derivatives of U with respect to P and with respect to Q, respectively, and then letting P and Q equal zero. Thus, for the horizontal component Δ_H we have **(compatibility)**

$$\Delta_H = \left(\frac{\partial U}{\partial P}\right)_{P=Q=0} = \frac{1}{2AE}[2(P - 2 - 2Q)(10) + 2(P - Q)10]_{P=Q=0}$$

$$= \frac{1}{2AE}[(2)(-2)(10)] = -\frac{20}{AE}\text{ ft} \quad \text{(d)}$$

Clearly, the horizontal deflection component of the pin is opposite in sense to the direction of the force P shown in Fig. 6.16. Now we get Δ_V. Thus,

$$\Delta_V = \left(\frac{\partial U}{\partial Q}\right)_{P=Q=0} = \frac{1}{2AE}\Big[2(P - 2 - 2Q)(10)(-2) + 2(P - Q)(10)(-1) + \frac{2Q}{(.707)^2}\left(\frac{10}{.707}\right) + 2(1 + Q)(10) + 20Q + \frac{2(2 + Q)}{(.707)^2}\frac{10}{.707}\Big]_{P=Q=0}$$

$$\therefore\ \Delta_V = \frac{1}{2AE}(80 + 20 + 113.2) = \frac{106.6}{AE}\text{ ft} \quad \text{(e)}$$

The deflection at pin C can then be given as

$$\Delta_C = \frac{1}{AE}(-20\mathbf{i} - 106.6\mathbf{j})\text{ ft} \quad \text{(f)}$$

For a deflection in a truss problem, we would have a very difficult time were we to proceed without availing ourselves of the second Castigliano theorem. To appreciate the power of the energy approach here we suggest you try to solve this problem using the elongations, shortenings, and rotations of the members from strength of materials calculations plus using geometry.

Case B: Statically Indeterminate Trusses

From your study of statics, you will recall that a "just-rigid" truss is a statically determinate truss where the removal of any one member will cause the truss to collapse (sometimes referred to as "forming a kinematic mechanism"). If m represents the number of members and j the number of pins (or joints), the requirements for the just-rigid condition are

Plane truss: $m = 2j - 3$ (6.60a)

Space truss: $m = 3j - 6$ (6.60b)

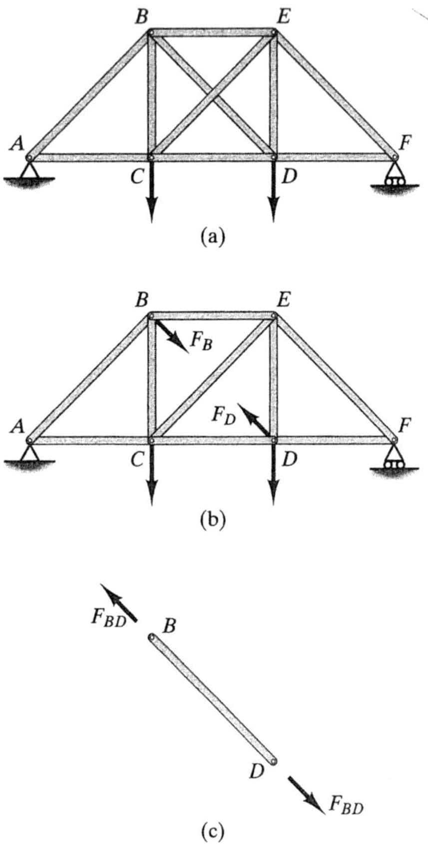

Figure 6.17. (a) Redundant truss structure, (b) Base structure, (c) Free-body diagram for member BD.

If m exceeds the value for the just-rigid truss specified by Eqs. (6.60) then the truss is internally statically indeterminate. Each additional member above that required to keep the structure just-rigid is a *redundant member.*

We first determine a "base structure" that is just-rigid. We do this by eliminating redundant members and replacing them with corresponding forces at the pins. Thus, in Fig. 6.17(a) we have a plane truss with a single redundant member. We have *arbitrarily* chosen member BD as the redundant and we have shown in Fig. 6.17(b) the base structure with BD removed and corresponding forces F_B and F_D applied at the pins. From the free-body-diagram of member BD, shown in Fig. 6.17(c), it is clear that F_B and F_D are of equal magnitude and opposite direction and henceforth will generally be called F_{BD}. The next step is to *solve via statics for all the other member forces as a function of the redundant force* F_{BD} (note that the external loads are included in the base truss) (**equilibrium**). This ensures that the stress field imposed by the redundant internal force(s) will be statically admissible. We now develop U^* for the base truss. Assuming linear-elastic material response, we know from Eq. (6.36) that the energy in each member can be expressed as (**constitutive law**).

$$U = U^* = \int_0^L \frac{F^2}{2AE}\,dx$$

Since F, A, and E are not functions of x for our truss members, we can express the above relation as

$$U = \frac{F^2 L}{2AE} \tag{6.61}$$

Note, each member has a force proportional to F_{BD} plus a force F_i as determined from *equilibrium* of the base truss. Hence, recognizing that A, E, and L are constants, we can write the strain energy for each member i simply as $U_i = (\alpha_i F_i + \beta_i F_{BD})^2$ where α_i and β_i are separate constants for each member. The strain energy for the entire structure is then of the form

$$U = \sum_i (\alpha_i F_i + \beta_i F_{BD})^2$$

We now vary the force F_{BD} while keeping all other forces fixed as per our development of the second Castigliano theorem. Noting that F_{BD} has inputs at pins B and D, we can then say

$$\Delta_{BD} = \frac{\partial U}{\partial F_{BD}}$$

The relative movement between pins B and D we may take as Δ_{BD}. But this elongation is precisely the change in length of the redundant member BD (**compatibility**). Noting that the force on the member is the reaction to the force on pins B and D as per Newton's third law, we can then say

$$\Delta_{BD} = \frac{\partial U}{\partial F_{BD}} = -\frac{F_{BD}L_{BD}}{(AE)_{BD}} \tag{6.62}$$

In a truss having n redundant members, we can make n such statements as Eq. (6.62). Thus, if members 1 through n are redundant, we have

$$\begin{aligned} \frac{\partial U}{\partial F_1} &= -\frac{F_1 L_1}{A_1 E_1} \\ &\vdots \\ \frac{\partial U}{\partial F_n} &= -\frac{F_n L_m}{A_n E_m} \end{aligned} \tag{6.63}$$

where U is computed as earlier in terms of the external loads and the redundant forces $F_1 \ldots F_n$. Each equation results in the solution for one of the redundants.

Although we have illustrated the method by using a plane truss, it should be clearly understood that the method is equally valid and useful for three-dimensional truss structures. We now consider the solution of internally statically indeterminate plane truss structures.

Example 6.5

Shown in Fig. 6.18 is a linear-elastic roof truss acted on by concentrated forces. Note that the 40-kip load is perpendicular to member BC (i.e., it is at 36.9° from vertical, as you may readily verify). Compute the forces in the truss members. Take AE constant.

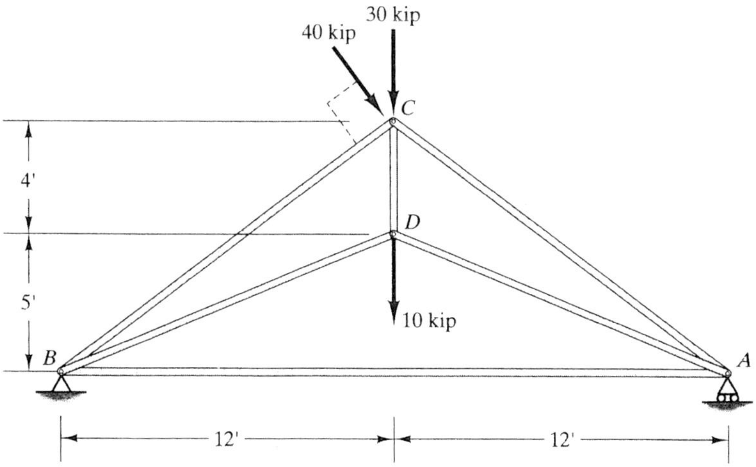

Figure 6.18. Statically indeterminate truss.

Example 6.5 (Continued)

Our first step is to determine the degree of redundancy of the truss (if any). In this case, we have four joints and six members. From Eq. (6.60a) we see that we have one redundant member. We could accordingly remove a member, such as AB, without causing the truss to collapse due to lack of rigidity. By taking AB as the redundant, we have determined the just-rigid truss configuration (or base structure) that will form the basis of our analysis. Fig. 6.19(a) shows the base structure and reaction forces F_{AB} while Fig. 6.19(b) shows a free-body diagram for member AB. Note that the redundant member force is included in both free-body diagrams. We can think of this structure as a determinate truss acted on by external loads plus unknown forces F_{AB}.

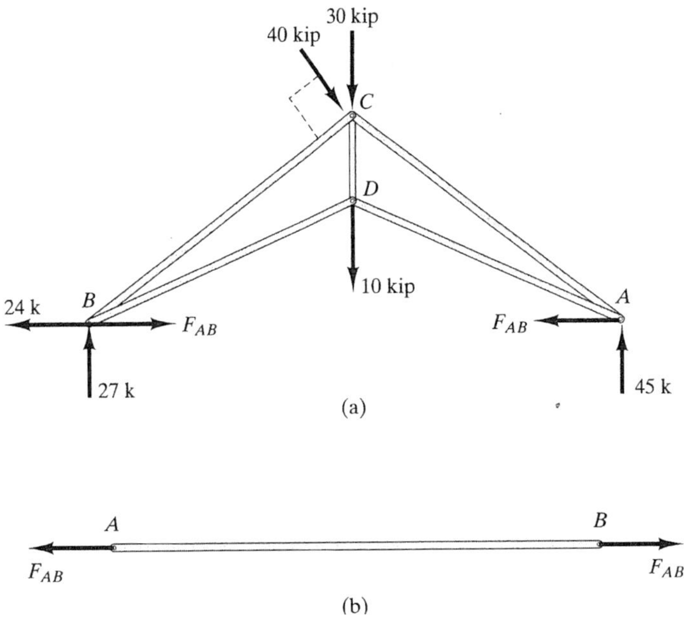

Figure 6.19. Free-body diagrams of base truss.

The next key step is to compute the member forces in the base structure. This can be done by any convenient method. Keep in mind that the force in member AB must be included in the calculations—this insures us that any choice of F_{AB} will result in an equilibrium stress field in the truss. The results are (**equilibrium**)

$$
\begin{aligned}
F_{AD} &= -146.2 + 2.437\,F_{AB} \\
F_{AC} &= 168.8 - 1.562\,F_{AB} \\
F_{BC} &= 138.8 - 1.562\,F_{AB} \\
F_{DC} &= -122.5 + 1.874\,F_{AB} \\
F_{BD} &= -146.2 + 2.437\,F_{AB}
\end{aligned}
\tag{a}
$$

Example 6.5 (Continued)

The strain energy in the base structure is written as [see Eq. (6.61)] (**constitutive law**).

$$U^* = U = \frac{1}{2AE}\Big[(F_{AD})^2(13) + (F_{AC})^2(15) + (F_{DC})^2(4) + (F_{BD})^2(13) + (F_{BC})^2(15)\Big] \tag{b}$$

Substituting in the member forces from Eq. (a) we get

$$U = \frac{1}{2AE}\Big[(-146.2 + 2.437F_{AB})^2(13) + (168.8 - 1.562F_{AB})^2(15) + (-122.5 + 1.874\, F_{AB})^2(4) + (138.8 - 1.562\, F_{AB})^2(15) + (-146.2 + 2.437\, F_{AB})^2(13)\Big] \tag{c}$$

Using Eq. (6.62), we have (**compatibility**)

$$\frac{\partial U}{\partial F_{AB}} = -\frac{F_{AB}L_{AB}}{(AE)_{AB}} \tag{d}$$

Upon substitution of Eq. (c) into Eq. (d), we can compute the force in AB, thus

$$[(-146.2 + 2.437\, F_{AB})(2.437)(13) + (168.8 - 1.562\, F_{AB})(-1.562)(15) + (-122.5 + 1.874\, F_{AB})(1.874)(4) + (138.8 - 1.562\, F_{AB})(-1.562)(15) + (-146.2 + 2.437\, F_{AB})(2.437)(13)] = -(F_{AB})(24)$$

$$\therefore\quad F_{AB} = 65.6 \text{ kips}$$

The forces in the other members are now readily available through Eq. (a).

*6.9 Basic Equations of Elasticity

Any continuum must satisfy the following four basic laws:

1. Newton's law
2. First law of thermodynamics
3. Second law of thermodynamics
4. Conservation of mass

In addition to the basic laws certain constitutive laws must be satisfied; the nature of the constitutive law depends on the material. (Thus, for linear-elastic material we have already presented the three-dimensional Hooke's law.)

We shall consider only bodies in **equilibrium.** Accordingly, we have as one of the basic sets of equations:

$$\begin{aligned}
\frac{\partial \tau_{xx}}{\partial x} + \frac{\partial \tau_{xy}}{\partial y} + \frac{\partial \tau_{xz}}{\partial z} + B_x = 0 \\
\frac{\partial \tau_{yx}}{\partial x} + \frac{\partial \tau_{yy}}{\partial y} + \frac{\partial \tau_{yz}}{\partial z} + B_y = 0 \\
\frac{\partial \tau_{zx}}{\partial x} + \frac{\partial \tau_{zy}}{\partial y} + \frac{\partial \tau_{zz}}{\partial z} + B_z = 0
\end{aligned} \tag{6.64}$$

These equations should, strictly speaking, be applied to the deformed geometry. However, by restricting ourselves to small deformation we may use the *undeformed* geometry in applying these equations.

In considering the other three basic laws for the isothermal, linear-elastic medium which we shall consider in our basic theory, we note that there will be neither heat transfer nor internal friction present. You will recall from earlier courses in mechanics that under such circumstances the energy approach and Newton's law yield the same results for problems of particle and rigid-body mechanics. Similarly here, Newton's law (equilibrium equations) and the first law of thermodynamics are equivalent to each other. That is, once one is satisfied, the other is automatically satisifed. We shall generally use the equation of equilibrium except when we present the approach for beams in Chapter 12 and shafts in Chapter 14 where we also use energy. And we will use it in a more general way in Chapter 18 where we shall employ the elegant energy methods. Next, because of the lack of internal friction and heat transfer in the model and because we shall consider the loadings to be reversible,[10] the second law of thermodynamics places no restriction for the calculations of such models. Therefore, we may proceed knowing that the second law is intrinsically satisfied as a result of the simplifications associated with the model. Finally, by keeping the mass of the body constant at all times we satisfy the conservation of mass requirement. Hence, of the four basic laws only the equations of equilibrium need be used for isothermal simple linear elasticity.[11]

[10]To be applied slowly so that the system is at all times infinitesimally close to equilibrium.

[11]When you study fluid mechanics and heat transfer you will see that the other basic laws must be employed as independent equations along with Newton's law. Also other constitutive laws will be needed. If electric and magnetic effects are important, the continuum must satisfy additional basic laws (called Maxwell's equations), namely Gauss' law, Faraday's law, the generalized Ampère's law, and the law requiring that the divergence of the magnetic induction $\boldsymbol{B}$ be zero.

The **constitutive** law that will play a key role in the theory is, as already noted, the three-dimensional Hooke's law for isotropic bodies, which we present again as

$$
\begin{aligned}
\varepsilon_{xx} &= \frac{1}{E}\left[\tau_{xx} - \nu(\tau_{yy} + \tau_{zz})\right] \\
\varepsilon_{yy} &= \frac{1}{E}\left[\tau_{yy} - \nu(\tau_{xx} + \tau_{zz})\right] \\
\varepsilon_{zz} &= \frac{1}{E}\left[\tau_{zz} - \nu(\tau_{xx} + \tau_{yy})\right] \\
\gamma_{xz} &= \frac{1}{G}\tau_{xz} \qquad \gamma_{xy} = \frac{1}{G}\tau_{xy} \qquad \gamma_{yz} = \frac{1}{G}\tau_{yz}
\end{aligned}
\tag{6.65}
$$

We thus have two of the three formulations presented in simple form in Chapter 5 for dealing with one-dimensional problems. The third, you will recall, is the geometric **compatibility** condition which requires the strains to stem from a single-valued, continuous displacement field. This can be guaranteed if the strains are formed from continuous, single-valued functions $\boldsymbol{u}(x,y,z,t)$ by the following relations, first presented in Section 3.4. Thus, for small strain,[12]

$$
\begin{aligned}
\varepsilon_{xx} &= \frac{\partial u_x}{\partial x} \qquad & \gamma_{xy} &= \frac{\partial u_x}{\partial y} + \frac{\partial u_y}{\partial x} \\
\varepsilon_{yy} &= \frac{\partial u_y}{\partial y} \qquad & \gamma_{xz} &= \frac{\partial u_x}{\partial z} + \frac{\partial u_z}{\partial x} \\
\varepsilon_{zz} &= \frac{\partial u_z}{\partial z} \qquad & \gamma_{yz} &= \frac{\partial u_y}{\partial z} + \frac{\partial u_z}{\partial y}
\end{aligned}
\tag{6.66}
$$

Note now that Eqs. (6.64), (6.65), and (6.66) comprise 15 independent equations for 15 unknowns—6 stresses τ_{ij}, 6 strains ε_{ij}, and 3 displacement field components u_i. In addition to having the dependent variables satisfy the above partial differential equations, it is also necessary that they satisfy the *boundary conditions* of the problem. If the tractions are specified, then the stresses must correspond to these tractions as we get to the surface of the body. (This is called a boundary-

[12]If the strains are not related directly to $\boldsymbol{u}$ as per Eqs. (6.66) and the displacements field $\boldsymbol{u}$ does not explicitly appear in the formulations, then the strains must satisfy the so-called compatibility equations as given by Eqs. (3.13).

value problem of the first kind.) If the displacements on the boundary are specified, we must again ensure that $\boldsymbol{u}$ correspond to these displacements as we get out to the boundary. (This is called a boundary-value problem of the second kind.) Finally, we may prescribe tractions over part of the boundary and displacements over the remaining part of the boundary. (This is called a mixed boundary-value problem.)

The system of 15 equations and unknowns comprises the basic structure of the theory of linear isothermal elasticity. The number of closed-form solutions to this theory is very small. In later chapters we shall present further simplifications that will result in the body of knowledge we call strength of materials. Here not all equations and conditions presented in this section are satisfied. Nevertheless, the formulations of strength of materials are so useful that they comprise the basis of much of present-day engineering analysis and design of structures.

6.10 Closure

At the outset of Part A of this chapter we extrapolated Hooke's law from the one-dimensional stress case to the general state of stress. We did this heuristically for isotropic materials by superposing hypothetical results of one-dimensional tension tests and simple torsion tests for the *xyz* directions at a point. Next we started with an anisotropic material and outlined how by more formal procedure we could reduce the equations to that of the isotropic case. This was done by requiring no changes of the constitutive law on undergoing a series of rotations of axes. We thus made direct use of the concept of isotropy, which demands no directional changes in property at a point.

We then set forth in Part B the concept of strain energy, starting with elementary thermodynamics as learned in freshman physics. We carefully developed the very useful second Castigliano theorem and solved for the deflections of the pins of a statically determinate truss for linear elastic materials. Then we proceeded to solve for forces in statically indeterminate truss problems, again for linear-elastic materials. Both of these kinds of calculations are beyond the student's background as learned in statics. Here we see a more mathematical, more obscure development has led to very practical simple applications to important problems. We shall continue to use the second Castigliano theorem when we study beams and torsion as we move along in the text. The same simple approach that we have seen here for trusses will again be in evidence in these areas.

As a next step we present the chapter on plane stress. Here we will see in a relatively simple but useful environment how stress varies when you rotate the axes at a point. This will lead you to a very simple introduction to the concept of the tensor. Bon voyage!

*6.11 A Look Ahead: Variational Methods

In your statics course, you may have studied the method of virtual work and derived from this method limited to **rigid** bodies the method of total potential energy. Now both methods are covered in Chapter 18 but we can get a useful overview at this time and get a fleeting introduction to the so-called variational methods thereby getting a glimpse at the importance and use of this kind of mathematics.

You may recall that for establishing necessary and sufficient conditions for equilibrium, we could extremize the potential energy (denoted as V) associated with the active, conservative forces of the system. That is, to ensure equilibrium, we could set

$$\delta V = 0$$

where δ is a hypothetical, infinitesimal change of V behaving much like a differential operator and called the variation operator. We have a similar formulation for the case of an **elastic** (not necessarily linearly elastic) body whereby we can guarantee equilibrium. This more general principle is derivable from a more general virtual work principle to be presented in Chapter 18. As will be shown, we must extremize a more complicated quantity which we call π (no relation to 3.1416 . . .) instead of V. The expression π is what we call a *functional,* wherein for the substitution of a function, such as $y(x)$, into the functional, a numerical value is established. A simple example of a functional I is as follows:

$$I = \int_{x_1}^{x_2} F\left(x,y,\frac{dy}{dx}\right) dx$$

where F is a function of x (the so-called independent variable), y, and dy/dx. Substitution of a function $y(x)$ into F followed by an integration between the fixed limits yields for this function $y(x)$ a number for I. Functionals pervade the field of mechanics and most other analytic fields of knowledge. A vital step is to find the function $y(x)$ that will extremize I. This function then becomes known as the *extremal function.* The calculus for doing this is called the *calculus of variations.* The particular functional for the method of total potential energy is given as

$$\pi = -\iiint_V \boldsymbol{B} \cdot \boldsymbol{u}\, dv - \oint\!\!\oint_S \boldsymbol{T} \cdot \boldsymbol{u}\, dA + U$$

The function to be adjusted to extremize π is $\boldsymbol{u}(x,y,z)$ taking the place of $y(x)$ in the preceding functional, and now the independent variables are x, y, and z in place of just x in the preceding functional. The expression U is the energy of deformation presented earlier as

$$\iiint_V \sum_i \sum_j \tau_{ij}\, \delta\varepsilon_{ij}\, dv$$

The principle of total potential energy for the case of elastic bodies is written in the following deceivingly simple looking formulation:

$$\delta\pi = 0$$

In starred chapter 18 in Part A, we show that, when we extremize π with respect to deformation fields $\boldsymbol{u}$ that have the following characteristics

a. They satisfy compatibility
b. Do not violate the constraints of the problem
c. Are related to an elastic (not necessarily linear) constitutive law

then the particular stress field that is linked to this extremal function $\boldsymbol{u}$ via the aforementioned constitutive law must perforce satisfy equilibrium. All other deformation fields will have associated stress fields that do not satisfy equilibrium. We say that $\delta\pi = 0$, properly understood and properly used, is the most powerful equation in solid mechanics. Why do we say this? The reason is this. We satisfy **compatibility** by using only single valued continuous displacement fields. Each such deformation is linked to stress via a **constitutive law.** And finally, satisfying the equation then assures the satisfaction of **equilibrium.** Behold, our three pillars of solid mechanics are all contained in the proper use of this innocent looking equation!

Hopefully, this little discussion will motivate you to look at Chapter 18 during your present course in solid mechanics or at a more appropriate time later in your studies.

We have touched on a broad area of study, namely variational methods, parts of which you will encounter in many of your studies. For example, just the method of total potential energy has the following major uses:

1. It can be used profitably to derive the *proper equations* and the *boundary conditions* for many areas of vital importance such as plate theory, elastic stability theory, dynamics of plates and beams, torsion theory, and so on.
2. From it we can develop a number of vital approximation methods. The most prominent of these methods is the method of *finite elements.*

Highlights (6)

The importance of constitutive laws should now be very clear to the reader. At the outset of the chapter, we were apprised of the possible complexity of constitutive laws. First, there was the three-dimensional linear elastic case. Then some readers may have taken a peek at the non-isothermal linear elastic case where we assumed that the elastic moduli remained constant for modest temperature variations. Furthermore, we assumed that changes in the stress state did not have a linkage to the temperature field (unlike a fluid). For very high stress fields which are time-dependent, this assumption would have to be carefully reconsidered. Next, we took a quick look at nonisotropic linear-elastic materials where there may be many elastic moduli. The case of composite materials now becoming more prevalent in structural design is an important case involving a deliberate lack of isotropy for achieving more efficient design. Finally, for this short list we mention constitutive laws involving plastic deformation. Here, the constitutive law will usually be history dependent. That is, the properties may depend on the stress history preceding the loading being considered. In short, there is much to study for the interested student.

Our next main consideration was the energy methods. We pointed out that the second Castigliano theorem is just one of six energy theorems—actually the last one. We present these in starred Chapter 18 where we present three so-called **displacement theorems,** each one derivable from the preceding theorem. The last and simplest of this trio is the first Castigliano theorem. There is then an analogous trio of theorems each derived from a predecessor and called the **force theorems.** And the simplest, as you might already anticipate, is our second Castigliano theorem. This chapter (18) is not normally covered in a first solids course. Yet, on a few occasions, we have ventured forth into much of this chapter. We can tell you emphatically that it is not beyond sophomore capability. Most important, this is the chapter that will give you the background for more advanced work in structures and most particularly will open the door for you to undertake the method of finite elements. As a bonus, some of the concepts are transferable to other fields of study.

PROBLEMS

6.1. [6.3] The stress tensor at a given point in a body is given as follows

$$\begin{pmatrix} 1000 & 2000 & 0 \\ 2000 & 5000 & -2000 \\ 0 & -2000 & 0 \end{pmatrix} \times 10^4 \text{ Pa}$$

What are the strains at the point for a Poisson ratio of .2 and a Young's modulus of 2×10^{11} Pa?

6.2. [6.3] For a shear modulus of 10×10^6 psi and an elastic modulus of 25×10^6 psi, compute the strain tensor for the following state of stress:

$$\tau_{ij} = \begin{pmatrix} 1000 & -5000 & 0 \\ -5000 & 500 & 500 \\ 0 & 500 & -2000 \end{pmatrix} \text{ psi}$$

6.3. [6.3] For a Poisson's ratio of .30 and a Young's modulus of 30×10^6 psi, determine the strain tensor for the following state of stress:

$$\tau_{ij} = \begin{pmatrix} 0 & 1000 & -2000 \\ 1000 & 500 & -3000 \\ -2000 & -3000 & 1000 \end{pmatrix} \text{ psi}$$

6.4. [6.3] A plate (Fig. P.6.4) with an elliptical hole is being examined to find the stress concentration factor K (for Section 4.7). It is in a state of plane stress (see Section 2.4). By a method called photoelasticity, we can measure stress τ_{yy} at A at the edge of the ellipse to be 1.4×10^7 Pa. What is the strain tensor at A? Take $E = 3.5 \times 10^{10}$ Pa and $\nu = .2$.

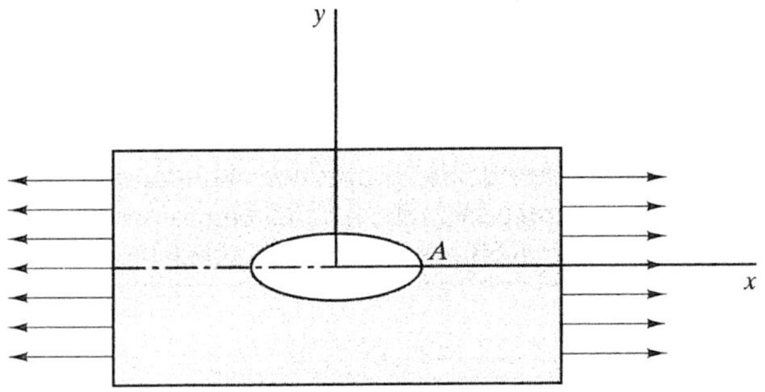

Figure P.6.4.

6.5. [6.3] A thin-walled shaft (Fig. P.6.5) is subject to a torque of 500 ft lb and an axial force of 1000 lb. Find the strains ε_{xx}, ε_{yy}, and ε_{xy} at point P on the surface for a material having $\nu = .3$ and $E = 30 \times 10^6$ psi. Neglect air pressure. Plane xy is just tangent to the pipe surface.

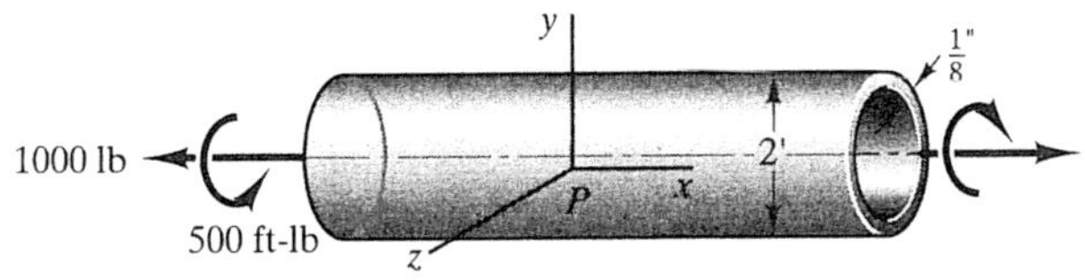

Figure P.6.5.

6.6. [6.3] In the preceding problem the shaft is in a barotropic chamber with a pressure of 5 atmospheres. What is the strain tensor at P?

6.7. [6.3] It is necessary to put a small hole into a large, thin epoxy plate that is subjected to a uniform pressure as shown in Fig. P.6.7. Determine the thickness of a steel ring liner that can be used so that the stress in the epoxy is unperturbed by the hole. Take: $E_{epoxy} = 1 \times 10^6$ psi, $\nu_{epoxy} = 0.4$, $E_{steel} = 30 \times 10^6$ psi, $\nu_{steel} = 0.3$, $r = 4$ in.

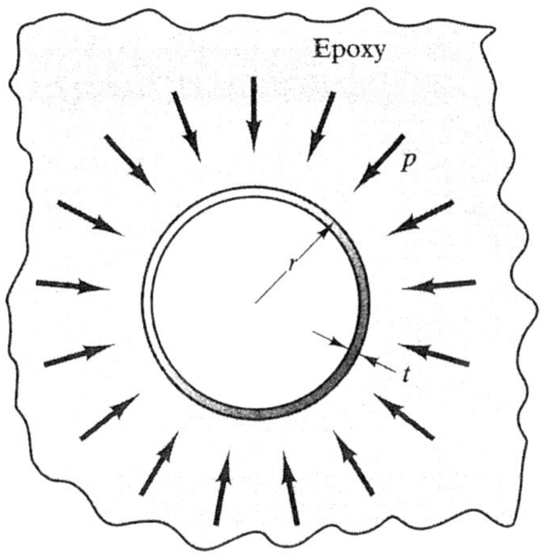

Figure P.6.7.

6.8. [6.3] A thin-walled pipe with a right-angle bend is shown in Fig. P.6.8. The outside diameter D of the pipe is 50 mm and the thickness is 1.5 mm. A load of 450 N is applied at the tip of the system. A strain gauge, which measures normal strains on a surface (as will be explained in more detail later), is oriented in the x direction at A on top of the pipe surface and gives a strain $\varepsilon_{xx} = .00346$. A second strain gage at A oriented in the z direction gives a strain $\varepsilon_{zz} = .00137$. If we neglect air pressure, what are stress components τ_{xx}, τ_{zz}, and τ_{xz} at point A? Poisson's ratio is .25 and $E = 1.4 \times 10^{11}$ Pa.

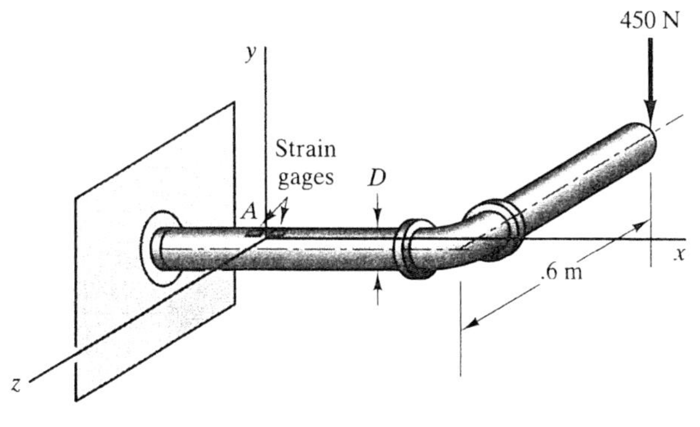

Figure P.6.8.

6.9. [6.3] A 100-N force acts on a rod welded to thin-walled cylinder which is acting as a cantilever beam (Fig. P.6.9). The force is parallel to the xy plane. Decompose the force into rectangular components and move each force to go through point B along the x axis while maintaining rigid-body equivalence for the cylinder. Compute the following stresses on an element at A (se Fig. P.6.9).

1. $(\tau_{xx})_1$ due to simple tension.
2. $(\tau_{xx})_2$ due to bending about the z axis. This you will learn is given as

$$(\tau_{xx})_2 = -\frac{M_z(.05)}{I_{zz}}$$

where M_z is the moment about the z axis at the base and I_{zz} is the second moment of area of the cross section about the z axis.

3. τ_{xz} is the shear stress due to torque about the x axis.

All other stresses are zero. For $E = 2 \times 10^{11}$ Pa and $\nu = .3$, what are the nine strains at point A for reference xyz?

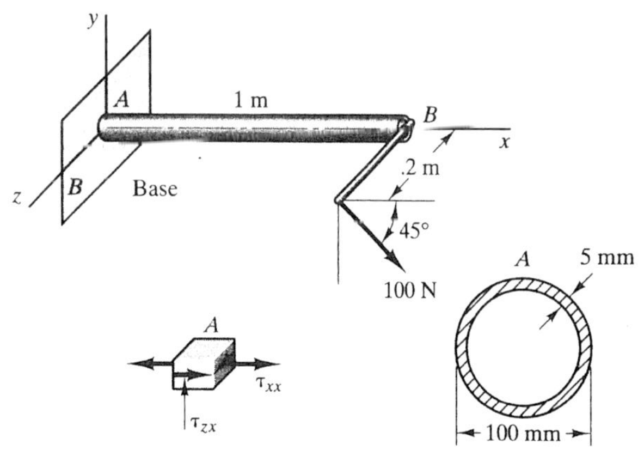

Figure P.6.9.

6.10. [6.3] The volume change per unit volume of a material is given as $e = \varepsilon_{xx} + \varepsilon_{yy} + \varepsilon_{zz}$. Show that for the case of hydrostatic stress ($\tau_{xx} = \tau_{yy} = \tau_{zz} = \bar{\tau}$), the unit volume change can be expressed as $e = \bar{\tau} / K$ where K (called the bulk modulus) is given by $E / [3(1-2\nu)]$ for an isotropic material.

6.11. [6.3] A plate is placed between two rigid, frictionless surfaces (see Fig. P.6.11). For the given load, determine the stress state.

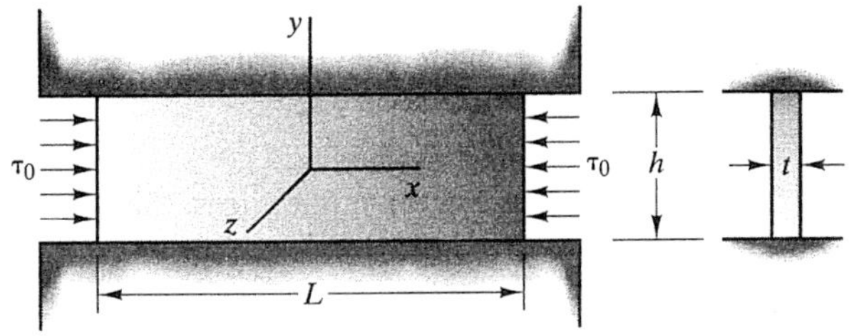

Figure P.6.11.

6.12. [6.3] A sample of an elastic material is subjected to a uniform pressure p in a rigid die as shown in Fig. P.6.12. The die is frictionless. (a) What is the stress state in the elastic material? (b) What is the strain state in the elastic material?

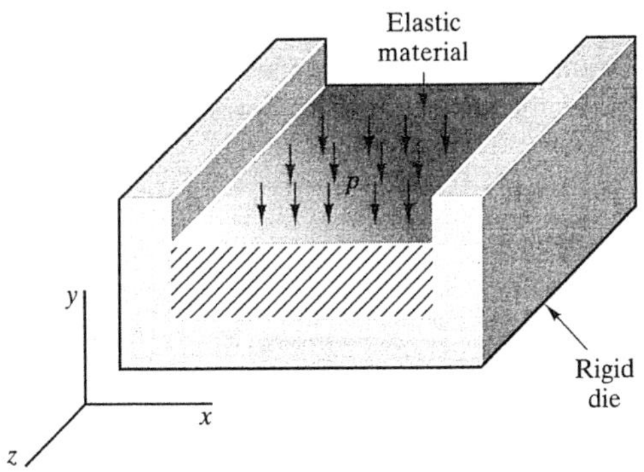

Figure P.6.12.

6.13. [6.3] A force $\boldsymbol{F} = 1000\mathbf{i} + 2000\mathbf{j} + 4000\mathbf{k}$ N acts on a rod welded to a solid circular shaft (see Fig. P.6.13). Move the force $\boldsymbol{F}$ to the center of the cross section at point B, maintaining rigid-body equivalence for the rod. Compute the following stresses on an element of the rod at A.

1. $(\tau_{xx})_1$ due to simple tension.
2. $(\tau_{xx})_2$ due to bending about the z axis given as $-M_z(.15/2) / I_{zz}$, where M_z is the moment about an axis parallel to the z axis taken at the center of the section at A and I_{zz} is the second moment or area about the z axis of the cross section.
3. τ_{xz} due to torque about the x axis given as $M_x(.15/2) /J$, where J is the polar moment of area of the cross section taken at the center of the cross section.

Take G = 8×10^{10} Pa and E = 2×10^{11} Pa. If we take all other stresses for xyz as zero, find the strain tensor at A.

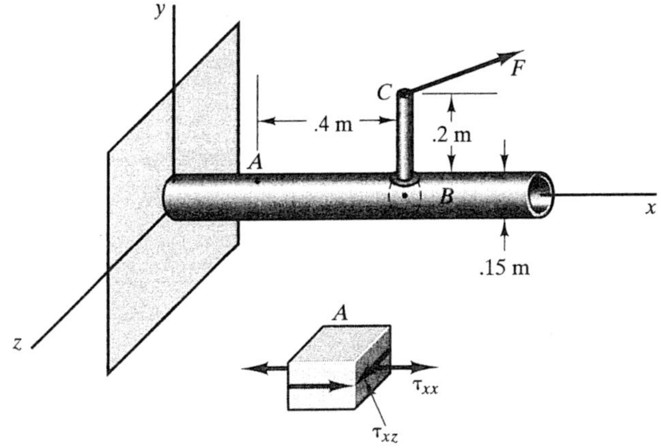

Figure P.6.13.

6.14. [6.3] A tube weighing 100 N/m (Fig. P.6.14) contains air at a pressure of 300,000 Pa above atmosphere. If $E = 2 \times 10^{11}$ Pa and $\nu = .3$, determine the strain tensor at A on the outside surface. Note that the 1000-N force is in the $-z$ direction. The thickness t of the tube is 20 mm. Compute first the following nonzero stresses at A.

1. (τ_{yy}) due to gravity and air pressure. Note that the force from a uniform pressure equals the pressure times the projected area on which it acts, where the projection is taken normal to the direction of the desired force.
2. (τ_{yy}) due to bending about the x axis given as $+M_x(.4/2)/I_{xx}$. (Use the proper sign for the moment M_x about the x axis.)
3. (τ_{yx}) due to torque about the y axis.

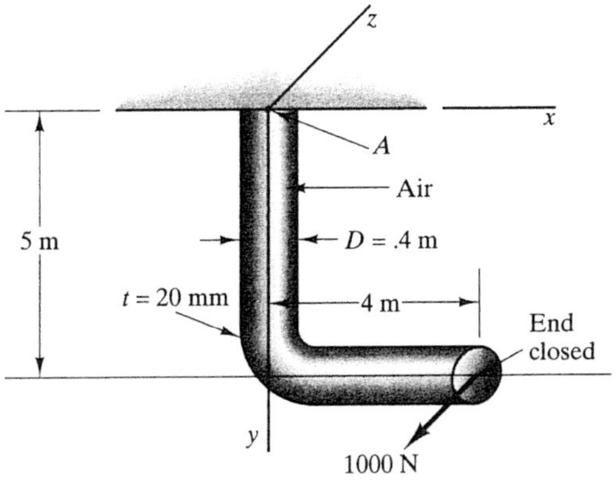

Figure P.6.14.

6.15. [6.3] Determine the equation for the lateral displacement Δ of the rubber shear mount shown in Fig. P.6.15. Then determine the height dimension h for a shear mount with G = 0.75 MPa and a limit stress of 0.50 MPa. The allowable displacement of the mount is 5 mm.

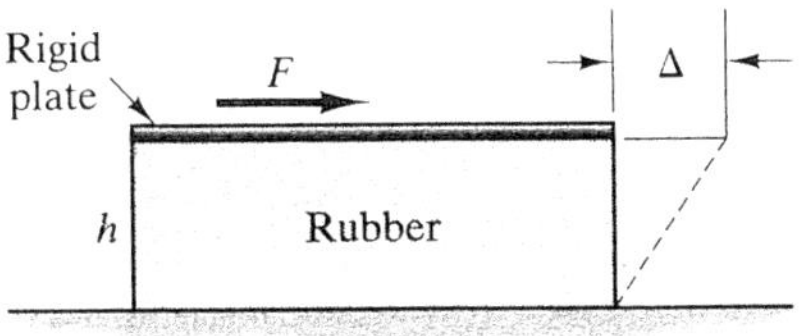

Figure P.6.15.

6.16. [6.3] How should the stresses for cylindrical coordinates τ_{rr}, $\tau_{\theta\theta}$, τ_{zz}, $\tau_{r\theta}$, $\tau_{z\theta}$, and τ_{zr} be related to strains ε_{rr}, $\varepsilon_{\theta\theta}$, ε_{zz}, $\varepsilon_{r\theta}$, $\varepsilon_{z\theta}$, and ε_{zr} for a linear isotropic Hookean material?

6.17. [6.3] From the theory of elasticity the radial and transverse stresses in a thick-walled cylinder (see Fig. P.6.17) subject to an internal pressure of p_i gage are given as follows:

$$\tau_{rr} = \frac{a^2 p_i}{b^2 - a^2}\left(1 - \frac{b^2}{r^2}\right)$$

$$\tau_{\theta\theta} = \frac{a^2 p_i}{b^2 - a^2}\left(1 + \frac{b^2}{r^2}\right)$$

What are the strains ε_{rr}, $\varepsilon_{\theta\theta}$, and $\varepsilon_{r\theta}$ at position r = 2 ft for a cylinder having radii a = 1 ft and b = 3 ft? Take $E = 30 \times 10^6$ psi and ν = .3. The pressure p_i is 500 psi. Assume that the ends are free so that τ_{zz} = 0. Do Problem 6.16 first.

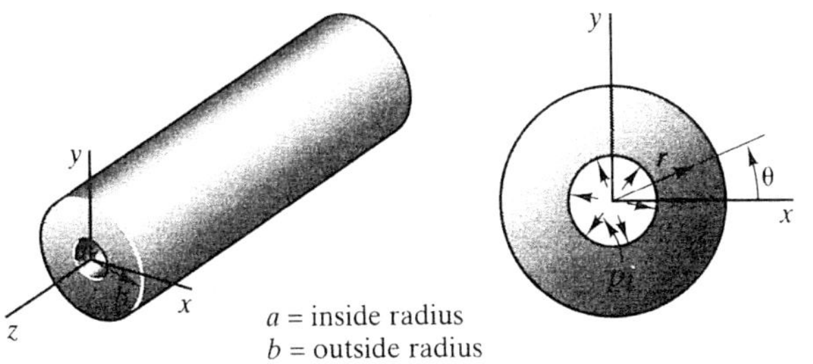

Figure P.6.17.

6.18. [6.3] For the design of hydrophones, a material with a low bulk modulus ($K = E/3(1-\nu)$) is desired since such a material is very sensitive to hydrostatic pressure. Under development are micro-machines and engineered materials that exhibit negative Poisson's ratio. Compare the bulk modulus of a material with ν = 1/3 with that of ν = –1.

6.19. [6.3] A sample of material is placed into a frictionless, rigid cavity with a snug fit (see Fig. P.6.19). A uniform pressure is applied. (a) What is the stress state of the material? (b) What happens as ν approaches 1/2?

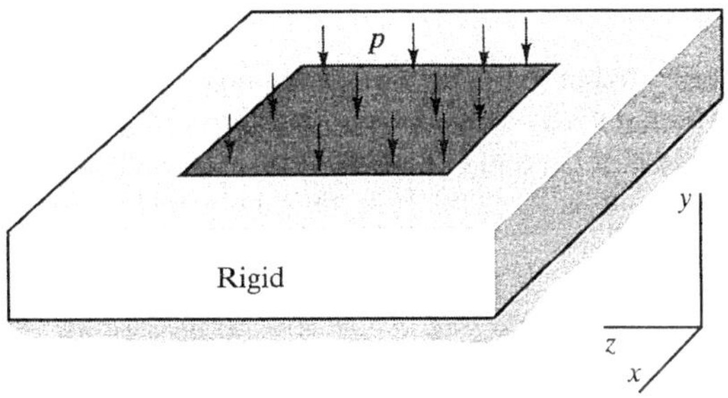

Figure P.6.19.

6.20. [6.4] The aluminum ring, shown in Fig. P.6.20, of thickness 3/4 in. just fits on a 3-in.-diameter rigid cylinder. A gap of 0.0015 in. exists around the outside of the ring. What increase in temperature of the ring will cause the ring to close the gap? For the ring, take $E = 10 \times 10^6$ psi, $\alpha = 10 \times 10^{-6}$/°F, and ν = 0.25. (*Hint:* First show that $\varepsilon_{\theta\theta} = u/r$.)

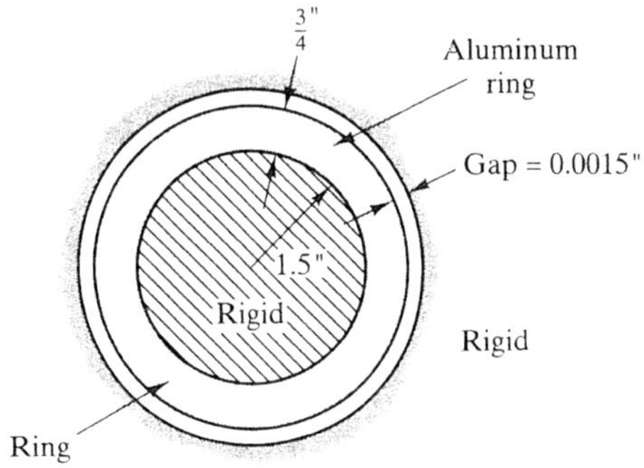

Figure P.6.20.

6.21. [6.4] A 100-in. × 30-in. linear elastic plate ($E = 10 \times 10^5$ psi, ν = 1/3, $\alpha = 10 \times 10^{-6}$/°F) is set in a rigid, frictionless cavity (see Fig. P.6.21). If the temperature is raised by 350°F, determine (a) the strain state in the plate, and (b) the stress state in the plate.

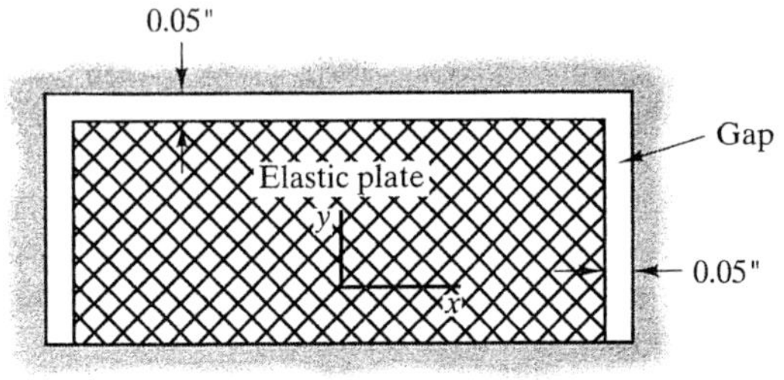

Figure P.6.21.

6.22. [6.4] What is the percentage change in diameter of a quarter in Fig. P.6.22 heated so as to achieve a rise in temperature of 200°F? Take the coefficient of expansion to be constant and equal to 9.5×10^{-6}/°F. First show that

$$\varepsilon_{\theta\theta} = \varepsilon_{xx} = \frac{u}{r}$$

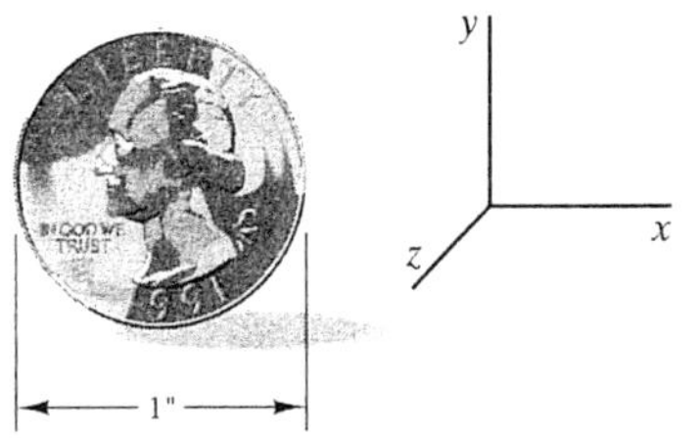

Figure P.6.22.

6.23. [6.5] In Eqs. (6.18), show that $C_{25} = C_{26} = 0$.

6.24. [6.5] Consider a material that has orthotropic symmetry. The elastic moduli are given as

$$C_{ij} = \begin{pmatrix} 3 & 4 & 2 & 0 & 0 & 0 \\ 2 & 1 & 4 & 0 & 0 & 0 \\ 7 & 6 & 3 & 0 & 0 & 0 \\ 0 & 0 & 0 & 4 & 0 & 0 \\ 0 & 0 & 0 & 0 & 2 & 0 \\ 0 & 0 & 0 & 0 & 0 & 3 \end{pmatrix} \times 10^6 \text{ psi}$$

What is the stress tensor corresponding to the following strain tensor:

$$\varepsilon_{ij} = \begin{pmatrix} 6 & 2 & 1 \\ 2 & -3 & 4 \\ 1 & 4 & 2 \end{pmatrix} \times 10^{-3}$$

6.25. [6.7] Suppose that a material is so constituted as to have the following stress-strain relation:

$$\tau_{xx} = C(\varepsilon_{xx})^{3/2}$$

What is the strain energy of a bar of such a material (see Fig. P.6.25) in terms of C and t when subject to a uniform tensile load of 2250 N?

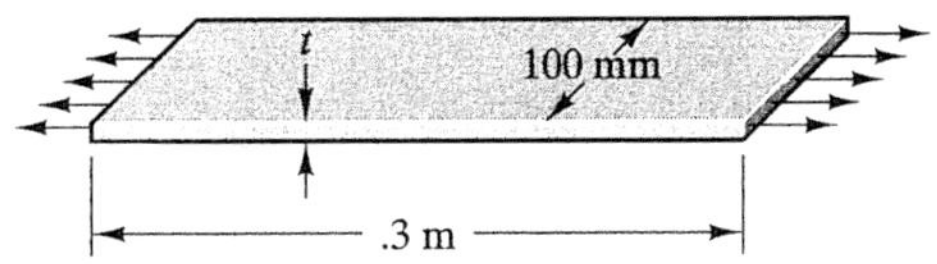

Figure P.6.25.

6.26. [6.7] What is the strain-energy density in each of the two members shown in Fig. P.6.26 welded together and supporting a load of 1000 lb? Take the following data:

$$E_1 = 30 \times 10^6 \text{ psi}$$
$$\nu_1 = .3$$
$$E_2 = 15 \times 10^6 \text{ psi}$$
$$\nu_2 = .2$$

What is the total strain energy of the system? Neglect the weight of the members.

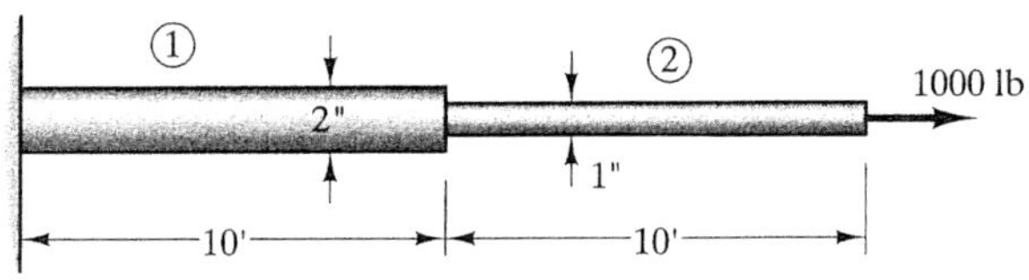

Figure P.6.26.

6.27. [6.7] A force distribution $f(x)$ given as $10x^2$ lb/in., where x is in inches acts on the cylinder shown in Fig. P.6.27. If it creates a one-dimensional state of stress, what is the strain energy for $E = 30 \times 10^6$ psi and $\nu = .3$? Neglect the weight of the member.

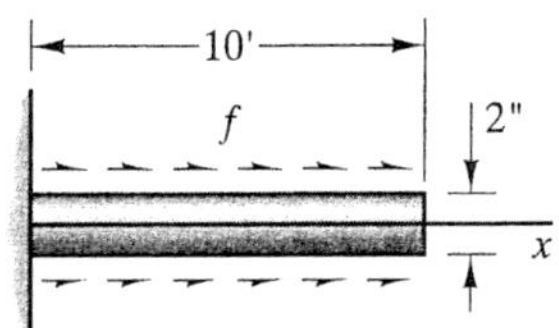

Figure P.6.27.

6.28. [6.7] Compute the strain energy of the simple truss shown in Fig. P.6.28. Take $E = 30 \times 10^6$ psi and $\nu = .3$. Take the cross-sectional area of the member as 20 in.^{2}.

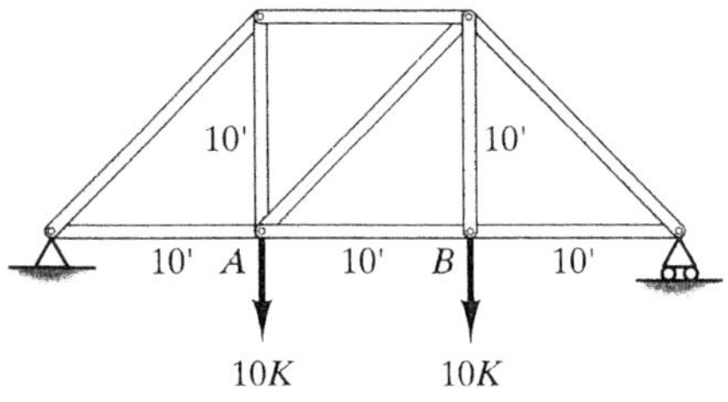

Figure P.6.28.

6.29. [6.7] A thin-walled cylinder (Fig. P.6.29) is subject to a torque T of 7500 N-m. The wall thickness is 6.25 mm and the diameter and length are .6 m and 3 m, respectively. What is the total strain energy for the cylinder? What is the relative angle of twist θ between the ends of the cylinder as a result of the loading? $G = 1 \times 10^{11}$ Pa. (*Hint:* Work of the torque is $1/2T\theta$.)

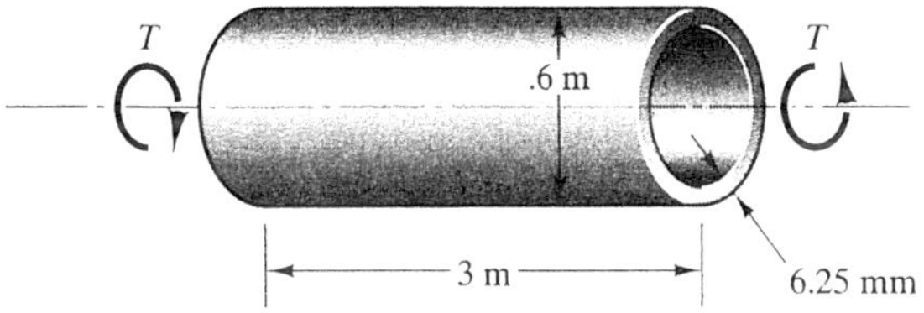

Figure P.6.29.

6.30. [6.7] If the cylinder in Problem 6.29 is twisted 1°, what is the strain energy and what is the torque?

6.31. [6.7] A thin-walled cylinder (see Fig. P.6.31) is subject to a uniform external torque distribution T_0 of 10 ft-lb/ft. The cylinder has an inside diameter of 4 in. and a wall thickness of .1 in. If $G = 15 \times 10^6$ psi, compute the strain energy of the cylinder.

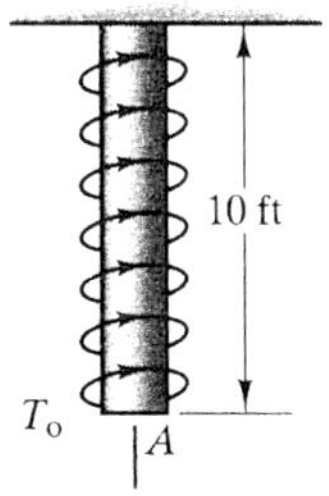

Figure P.6.31.

6.32. [6.7] In Problem 6.31 an axial force is also applied at end A with a value of 1000 lb. Now compute the strain energy from the torque distribution and the axial load. Take $E = 30 \times 10^6$ psi.

6.33. [6.8] What is the vertical deflection of pin A for the truss shown in Fig. P.6.33? All members have the same AE.

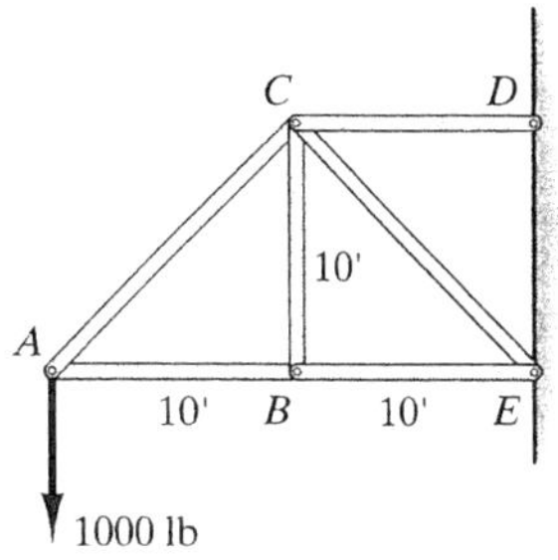

Figure P.6.33.

6.34. [6.8] What is the vertical deflection of pin C for the truss shown in Fig. P.6.34? All members have the same AE. Take $P_1 = 20$ kN, $P_2 = 30$ kN, and $P_3 = 15$ kN.

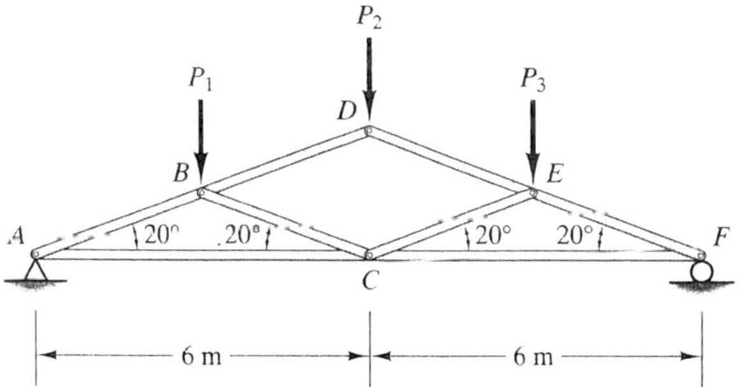

Figure P.6.34.

6.35. [6.8] Find the horizontal deflection of pin D in Fig. P.6.35. All members have the same AE.

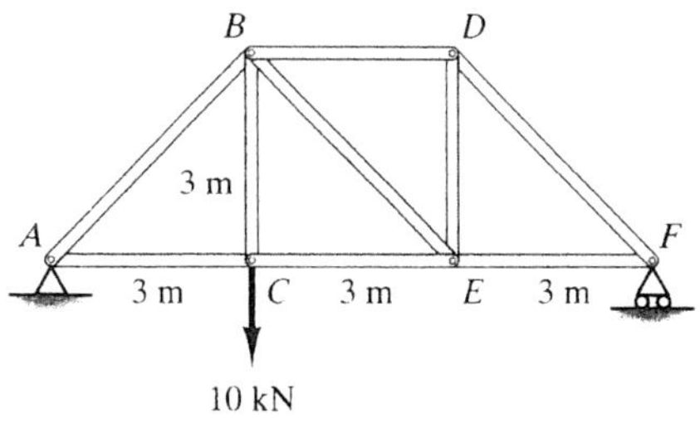

Figure P.6.35.

6.36. [6.8] Find the vertical deflection of pin B in Fig. P.6.36. All members have the same AE.

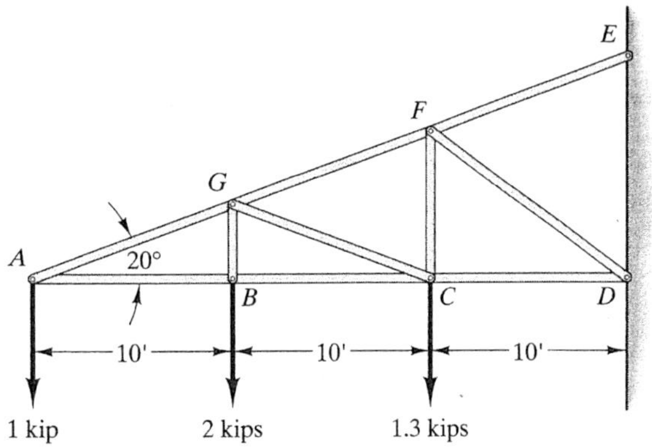

Figure P.6.36.

6.37. [6.8] Find the forces in the members of the truss shown in Fig. P.6.37. Take AE as constant.

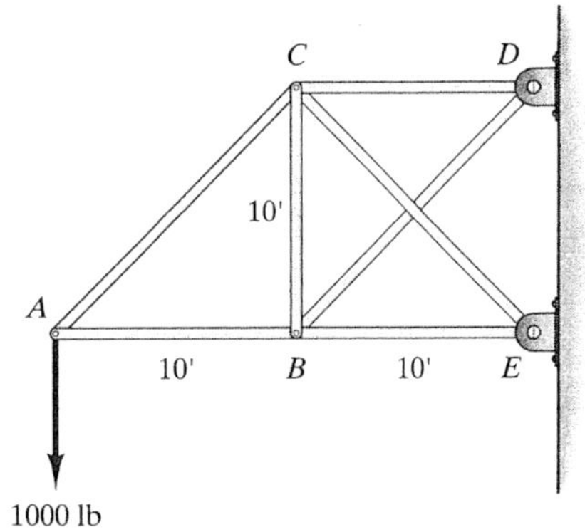

Figure P.6.37.

6.38. [6.8] Find the forces in the members of the truss shown in Fig. P.6.38. Take AE as constant.

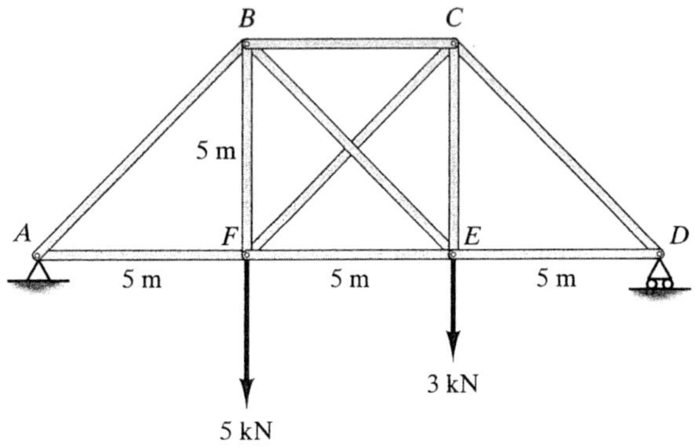

Figure P.6.38.

6.39. Determine the relationship between the strain, measured by a gage at position A as shown in Fig. P.6.39, and the pressure in the thin-walled tank. Use the result that

$$\tau_{xx} = \frac{pr}{t}$$

and

$$\tau_{yy} = \frac{pr}{2t}$$

Take t as the wall thickness.

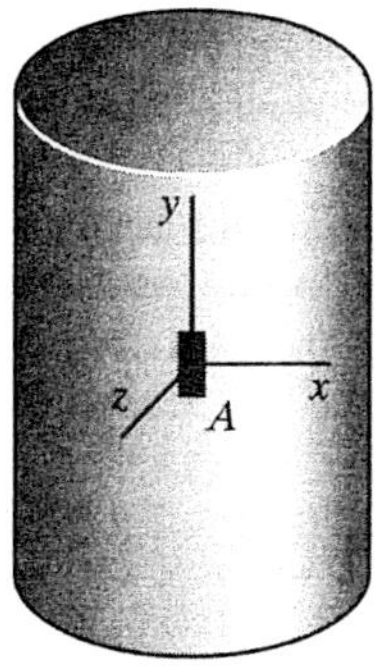

Figure P.6.39.

6.40. What is the stress state in an open-ended, thin-walled tube held between two rigid, frictionless surfaces and pressurized to p gage (see Fig. P.6.40). Recall that the hoop stress for a pressurized, thin-walled cylinder is

$$\tau_{\text{hoop}} = \frac{pr}{t}$$

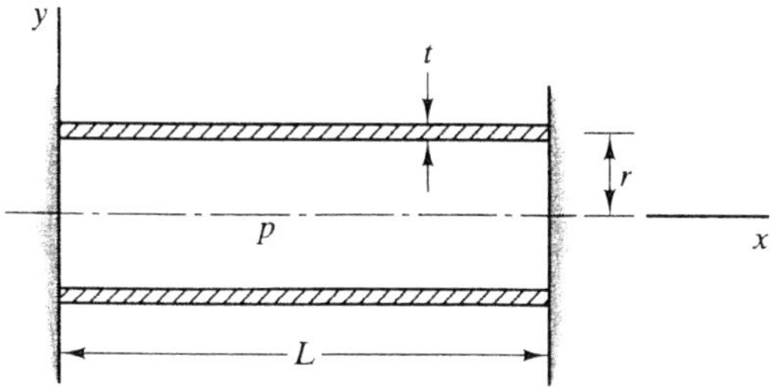

Figure P.6.40.

6.41. Determine the vertical displacement at the load point for the truss shown in Fig. P.6.41 using the second Castigliano theorem. Take for all members $A = 6.45 \times 10^{-4}\text{m}^2$ and $E = 200$ GPa.

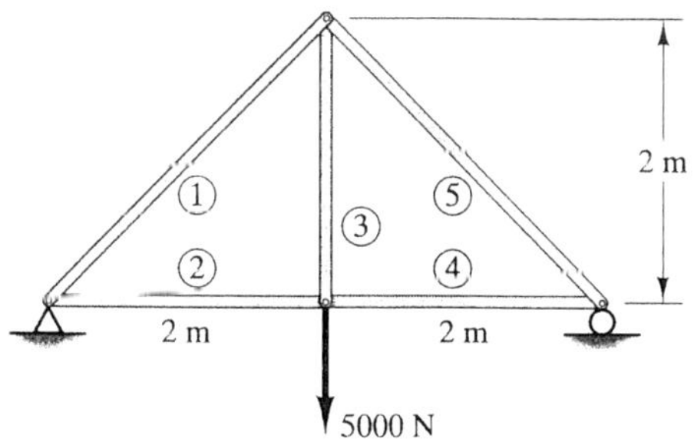

Figure P.6.41.

6.42. A rigid material is formed into a long semicircular channel and lined with a deformable material as shown in Fig. P.6.42. The liner is at first level with the top of the channel. The channel is then filled with a liquid having a specific weight γ lb/in.^{3} so that the liner then extends an amount δ above the channel. Neglecting friction, find δ. *Note:* The pressure in a liquid equals γ times the depth below the free surface of the liquid.

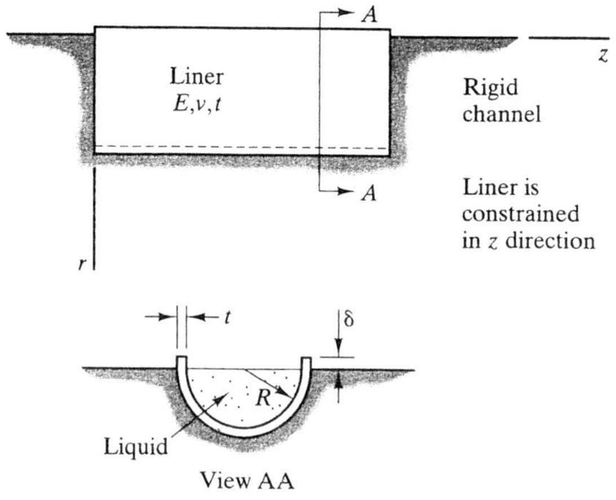

Figure P.6.42.

6.43. Use Castigliano's second theorem to determine the vertical displacement at the point of application of the load as shown in Fig. P.6.43. For all members, take $A = 6$ in.^{2} and $E = 30 \times 10^6$ psi. The *members are numbered as shown.*

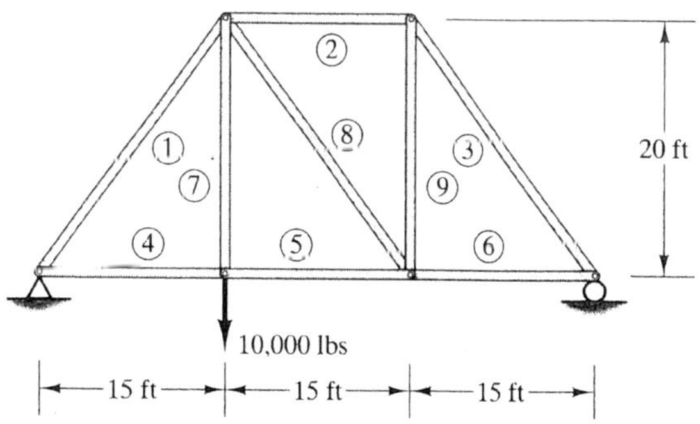

Figure P.6.43.

6.44. What is the strain energy per unit volume at a point for a linear-elastic material having at this point the following state of stress:

$$\begin{pmatrix} 1000 & -500 & 2000 \\ -500 & 2000 & -400 \\ 2000 & -400 & -1000 \end{pmatrix} \text{psi}$$

Take $E = 30 \times 10^6$ psi and $G = 15 \times 10^6$ psi.

6.45. What is the strain energy per unit volume at a point for a linear, elastic material having the following state of strain at this point:

$$\varepsilon_{ij} = \begin{pmatrix} .001 & -.0005 & .003 \\ -.0005 & .002 & -.002 \\ .003 & -.002 & -.001 \end{pmatrix}$$

The shear modulus $G = 1 \times 10^{11}$ Pa and Poisson's ratio is .3.

6.46. A strain gage is to be attached to an aluminum beverage can in Fig. P.6.46. If the wall thickness is 0.10 mm, estimate the change in strain in the hoop direction if 5% of the wall thickness is removed during surface preparation prior to attaching the gage. Recall that for a thin-walled cylinder

$$\tau_{\text{hoop}} = \frac{pr}{t}$$

and that

$$\tau_{\text{long}} = \frac{pr}{2t}$$

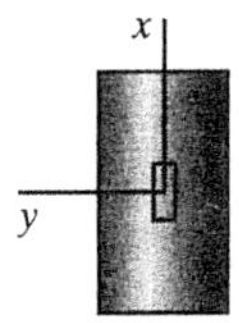

Figure P.6.46.

6.47. What is the change in length of an internally pressurized thin-walled cylinder of length L, radius r, and thickness t (see Fig. P.6.47)? (*Hint:* Use the result that for t much smaller than r, we have:

$$\tau_{xx} = \frac{pr}{2t}$$

and

$$\tau_{yy} = \frac{pr}{t}$$

near A.)

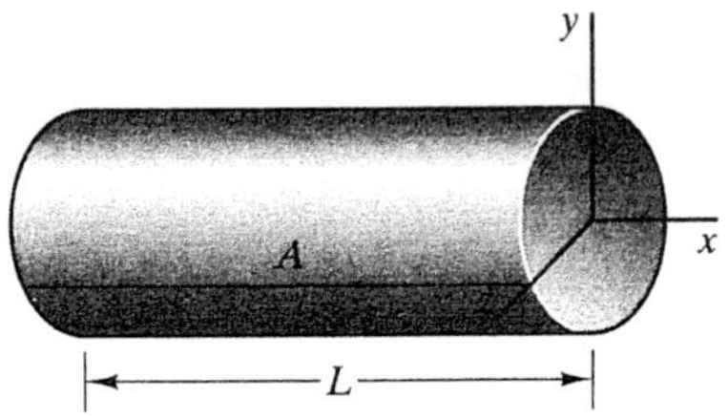

Figure P.6.47.

6.48. A thin-walled air tank is shown in Fig. P.6.48 wherein a pressure of 3.5×10^6 Pa gage is maintained. The thickness of the wall is 6.25 mm. What are the axial and transverse normal strains on the outside surface of the tank away from the ends? What is the change in diameter away from the ends? Take $\nu = .3$ and $E = 2 \times 10^{11}$ Pa.

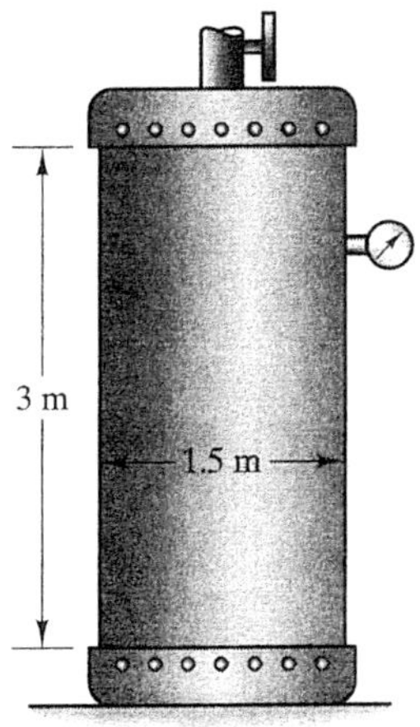

Figure P.6.48.

6.49. In stress-analyzing an elbow (see Fig. P.6.49) we know from strain gages (soon to be discussed in more detail) that at point A on the surface of the elbow

$$\varepsilon_{xx} = .00342$$
$$\varepsilon_{yy} = -.00342$$

The coordinates xy form a tangent plane to the surface at A. What are the normal stresses τ_{xx}, τ_{yy}, and τ_{zz} at A? Take $\nu = .2$ and $E = 1.7 \times 10^{11}$ Pa. What are strains γ_{zy} and γ_{zx}?

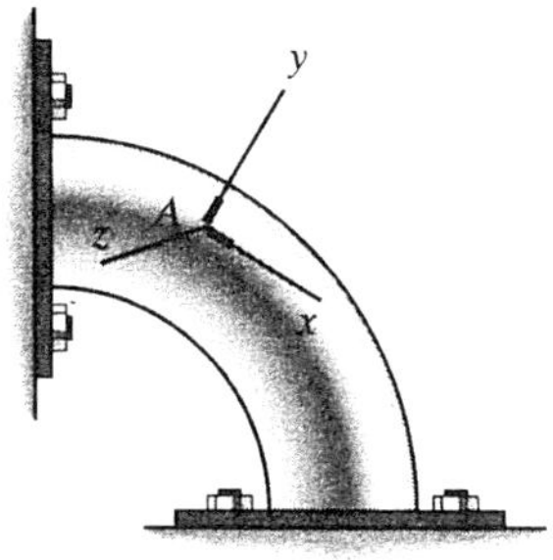

Figure P.6.49.

6.50. A thin-walled cylindrical tank of radius r and wall thickness t is capped with hemispherical ends as shown in Fig. P.6.50. Compute the change in radius at the cylinder-end cap junction due to an internal pressure p. Do this by considering the cylinder and end-cap separately. For a steel tank, what is the displacement ratio between the cylinder and end-cap? Comment on your results.

Figure P.6.50.

6.51. A rectangular 2-in. × 3-in. steel plate sits in a rigid cavity as shown in Fig. P.6.51. (a) For what temperature increase of the plate does it first touch the cavity wall? (b) If a temperature increase of 300°F is imposed, what are the stresses in the plate?

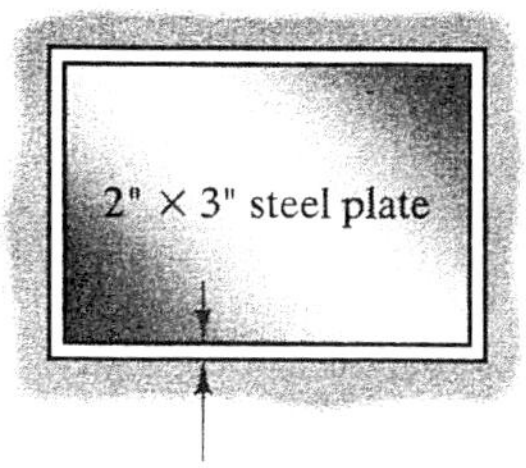

Figure P.6.51.

6.52. In Problem 6.51, determine the temperature increase when the gap is just closed all around the plate.

****6.53.** For the thick-walled cylinder under internal pressure considered in Problem 6.17, plot the stress distribution versus the radius for the data given below. What are the stress boundary conditions at $r = a$ and $r = b$? Are these shown on your plot? Take $a = 1$ ft, $b = 3$ ft, and $p_i = 500$ psi and use the following relationships:

$$\tau_{rr} = \frac{a^2 p_i}{b^2 - a^2}\left(1 - \frac{b^2}{r^2}\right)$$

$$\tau_{\theta\theta} = \frac{a^2 p_i}{b^2 - a^2}\left(1 + \frac{b^2}{r^2}\right)$$

Project 6

Write a computer program to compute the strain tensor given the stress tensor, Young's modulus, and Poisson's ratio. Test your program with the following data:

$$[\tau_{ij}] = \begin{bmatrix} 15000 & 700 & -300 \\ 700 & -2000 & 0 \\ -300 & 0 & -5000 \end{bmatrix} \text{psi}$$

$$E = 30 \times 10^6 \text{ psi}$$

$$\nu = 0.30$$

CHAPTER 7

Plane Stress

7.1 Introduction

In Example 2.6 we introduced a simple state of stress—the so-called one-dimensional state of stress, where for a given reference having axis z along the centerline of the prism, there is only one nonzero stress τ_{zz}. We showed that on interfaces whose normals are at an angle α with the z axis, we get normal and shear stresses that vary with α.

In this chapter we look at a more complex state of stress which we call *plane stress,* introduced in Section 2.4. Recall that here the stresses having one of the coordinates–say, the z axis–as a subscript are all zero. That is, the only possible nonzero stresses are τ_{xx}, τ_{yy}, and τ_{xy}. This state of stress is also termed the *two-dimensional* state of stress, to distinguish it from the one-dimensional state of stress, which we have already examined in considerable detail, and the *general* state of stress, which we shall refer to as the *three-dimensional* state of stress (to be studied later).

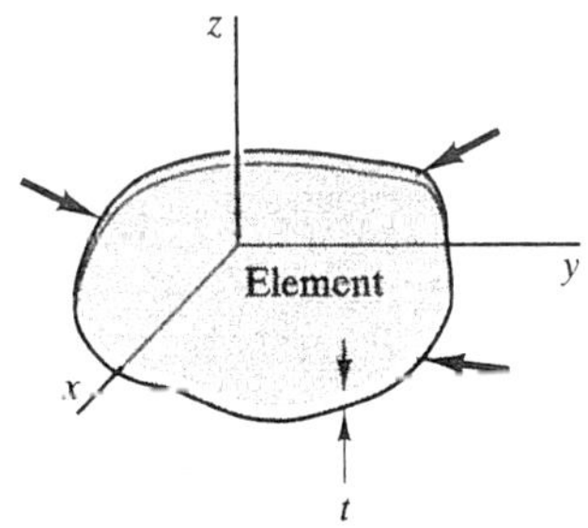

Figure 7.1. Plate under plane stress.

When does plane stress occur? Consider a thin flat plate loaded in the midplane of the plate as is shown in Fig. 7.1. On the upper and lower surface of plate the surface traction is a compressive stress from atmospheric pressure. We shall neglect this as being very small (14.7 psi) compared to stresses in engineering structures which when fully loaded generally exceed 20,000 psi. Because the plate is rather thin, we conclude that since $\tau_{zz} \approx 0$ at the upper and lower surfaces, it is also zero inside. With zero shear traction on the upper and lower bounding surfaces from air pressure, we can similarly conclude that $\tau_{zx} = \tau_{zy} = 0$ throughout the entire plate. We are accordingly left with τ_{xx}, τ_{yy}, and τ_{xy} (which you will note are stresses on interfaces in the plate and which interfaces are parallel to the z axis) as the only stresses that can be nonzero. Furthermore, we shall consider these stresses not to vary over the thickness of the plate.

Now it takes no long-time engineering experience to realize that plane stress problems are of considerable practical importance dealing as they do with many plates loaded in the midplane.[1] Hence, we will be studying material of great utilitarian value. Equally important to us is that we shall at the same time be investigating how stress changes on an interface parallel to the z axis as we change this interface by rotating it about an axis parallel to the z axis. We have already done this, as we have already pointed out, in Example 2.6 for the one-dimensional state of stress. However, that case is too simple to illustrate certain vital concepts relating to stress variations at a point. On the other hand, we shall make good inroads in studying these vital concepts for the case of plane stress. This foundation will enable us to go smoothly to the three-dimensional state of stress for a full development of these vital concepts.

Accordingly, in the next section we start examining stress relations at a point for plane stress. The practical application will be deferred to Chapter 10, when we study beams, and thereafter.

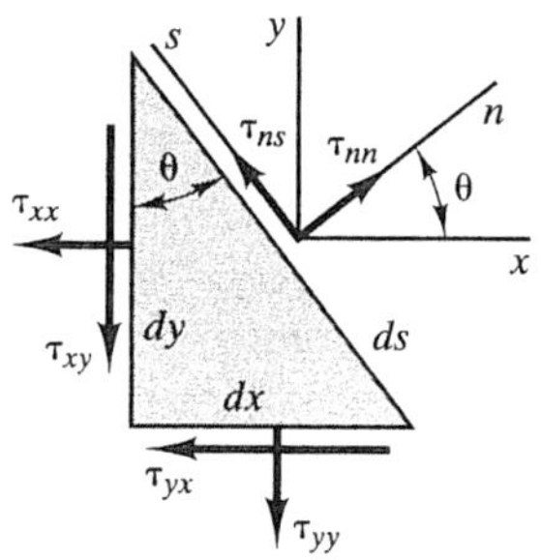

Figure 7.2. Triangular element of the plate.

7.2 Stress Transformations at a Point for Plane Stress

Let us consider that we know τ_{xx}, τ_{yy}, and τ_{xy} at an interface at a point in the plate. What must be the shear stress and the normal stress on an interface at this point where the interface normal is rotated by an angle θ counterclockwise from the x axis? This geometry is most simply pictured by considering an infinitesimal triangular element (see Fig. 7.2) cut from the plate. Notice that the normal axis to the inclined interface has been denoted as n and the tangent axis is denoted as s. Furthermore note that θ is the angle measured *counterclockwise as positive* from the x axis to the n axis. Taking the thickness of the plate as t and summing forces in the direction n, we have, from **equilibrium,**

$$\underline{\Sigma F_n = 0}:$$

$$\tau_{nn} t\, ds - \tau_{xx} t\, dy \cos\theta - \tau_{yy} t\, dx \sin\theta - \tau_{xy} t\, dy \sin\theta - \tau_{yx} t\, dx \cos\theta = 0$$

Dividing through by $t\, ds$ and noting from Fig. 7.2 that $dy/ds = \cos\theta$ and that $dx/ds = \sin\theta$, we get, on solving for τ_{nn},

$$\tau_{nn} = \tau_{xx}\cos^2\theta + \tau_{yy}\sin^2\theta + 2\tau_{xy}\sin\theta\cos\theta \tag{7.1}$$

[1]We would like to point out that plane stress need *not* always be considered in a plate. Indeed, it can exist at a *point* in other more complex geometries. Thus in a thin-walled tank or cylinder we may consider plane stress at a point in the wall wherein z is taken normal to the wall and where τ_{zz} from the atmosphere or from pressure in the tank is neglected. We will ask you to look at several problems of this type in the problems section.

Similarly, summing forces in the s direction, we get

$$\tau_{ns} = -\tau_{xx} \sin\theta \cos\theta + \tau_{yy} \sin\theta \cos\theta + \tau_{xy}(\cos^2\theta - \sin^2\theta) \quad (7.2)$$

It is helpful now to introduce the following trigonometric identities:

$$\cos^2\theta = \tfrac{1}{2}(1 + \cos 2\theta)$$
$$\sin^2\theta = \tfrac{1}{2}(1 - \cos 2\theta) \quad (7.3)$$
$$\sin\theta\cos\theta = \tfrac{1}{2}\sin 2\theta$$

whereupon Eqs. (7.1) and (7.2) become

$$\tau_{nn} = \frac{\tau_{xx} + \tau_{yy}}{2} + \frac{\tau_{xx} - \tau_{yy}}{2}\cos 2\theta + \tau_{xy}\sin 2\theta \quad (7.4)$$

$$\tau_{ns} = \frac{\tau_{yy} - \tau_{xx}}{2}\sin 2\theta + \tau_{xy}\cos 2\theta \quad (7.5)$$

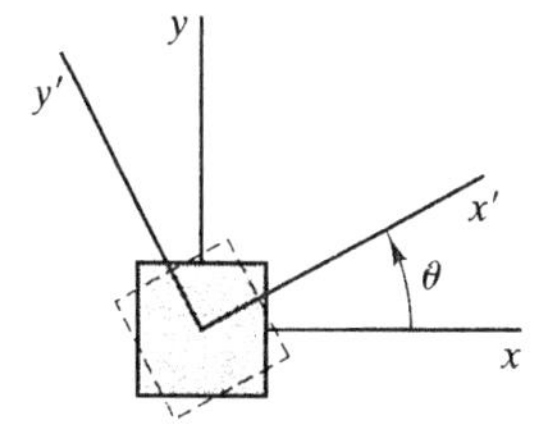

Figure 7.3. Rotation of axes.

We can formulate, from the equations above, stresses for interfaces having as normals x' and y' rotated an angle θ from xy in a counterclockwise manner (see Fig. 7.3) by first letting n be replaced by x'. We thus have

$$\tau_{x'x'} = \frac{\tau_{xx} + \tau_{yy}}{2} + \frac{\tau_{xx} - \tau_{yy}}{2}\cos 2\theta + \tau_{xy}\sin 2\theta \quad (7.6)$$

To get $\tau_{y'y'}$ from Eq. (7.4) we replace n by y' and use the angle $\theta + (\pi/2)$ to orient y' properly relative to the x axis. Thus, we have

$$\tau_{y'y'} = \frac{\tau_{xx} + \tau_{yy}}{2} + \frac{\tau_{xx} - \tau_{yy}}{2}\cos(2\theta + \pi) + \tau_{xy}\sin(2\theta + \pi)$$

This becomes

$$\tau_{y'y'} = \frac{\tau_{xx} + \tau_{yy}}{2} - \frac{\tau_{xx} - \tau_{yy}}{2}\cos 2\theta - \tau_{xy}\sin 2\theta \quad (7.7)$$

Finally, for $\tau_{x'y'}$ we need only replace n by x' and s by y' to get

$$\tau_{x'y'} = \frac{\tau_{yy} - \tau_{xx}}{2}\sin 2\theta + \tau_{xy}\cos 2\theta \quad (7.8)$$

We thus have the *transformation equations* for plane stress in a simple form.

Example 7.1

Beams with rectangular cross sections subjected to loading in a plane of symmetry may be considered as being in a state of plane stress. The results of a given beam analysis indicate that, at a certain point P in the beam (Fig. 7.4) the nonzero stress components are

$$\tau_{xx} = 3000 \text{ psi} \qquad \tau_{yy} = 0 \qquad \tau_{xy} = -400 \text{ psi}$$

What are the nonzero stresses for axes $x'y'$ rotated 30° clockwise from xy at point P?

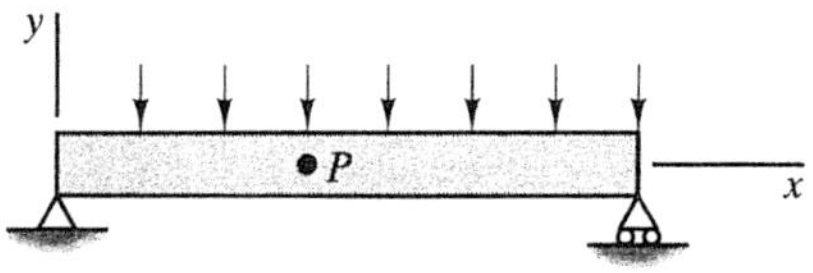

Figure 7.4. Beam with rectangular cross section.

An infinitesimal rectangle depicting interfaces for reference xy in the neighborhood of P is shown in Fig. 7.5(a), while an infinitesimal rectangle is shown in Fig. 7.5(b) depicting interfaces for $x'y'$ in the same neighborhood.

From Eqs. (7.6) and (7.7) we have

$$\tau_{x'x'} = \frac{3000 + 0}{2} + \frac{3000 - 0}{2} \cos(-60°) + (-400) \sin(-60°)$$

$$\tau_{x'x'} = 2596 \text{ psi}$$

$$\tau_{y'y'} = \frac{3000 + 0}{2} - \frac{3000 - 0}{2} \cos(-60°) - (-400) \sin(-60°)$$

$$\tau_{y'y'} = 404 \text{ psi}$$

Finally, from Eq. (7.8) we get

$$\tau_{x'y'} = \frac{0 - 3000}{2} \sin(-60°) + (-400) \cos(-60°)$$

$$\tau_{x'y'} = 1099 \text{ psi}$$

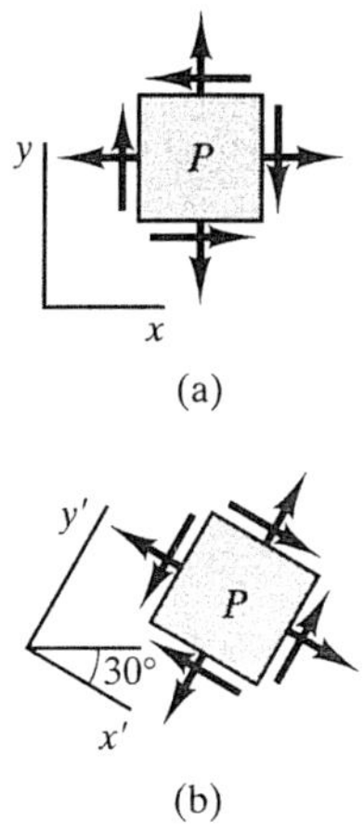

Figure 7.5. Stress at P for different axes.

Notice in the example that $\tau_{xx} + \tau_{yy}$ equals 3000 psi and that $\tau_{x'x'} + \tau_{y'y'}$ also equals the same value. This was no fortuitous accident. It is easy to see by summing Eqs. (7.6) and (7.7) that the sum of orthogonal normal stresses at a point has the same value at the point for all orientations of the axes. That is,

$$\tau_{x'x'} + \tau_{y'y'} = \tau_{xx} + \tau_{yy} \tag{7.9}$$

showing that the sum of orthogonal normal stresses at a point is a function only of the state of stress at a point and not of the orientation of the axes at the point. We call such a quantity an *invariant* at a point.[2] There are also other combinations of stress components at a point that do not depend on the orientation of the reference axes at a point. Such invariants will be pointed out later.

7.3 A Pause and A Comment

In the preceding section we merely used Newton's law for elementary rigid-body mechanics in a very simple procedure. Yet (and this may come as a surprise to you) we have begun the study of a profound subject which will have long-range implications as you move deeper into the subject of solid mechanics in particular and in mechanics of other continua in general. Is there any hint of this in the preceding section? Let us explore this possibility.

For this purpose, note that we started with nine stress terms in the three-dimensional case, which we shall again write as follows:

$$\tau_{ij} = \begin{bmatrix} \tau_{xx} & \tau_{xy} & \tau_{xz} \\ \tau_{yx} & \tau_{yy} & \tau_{yz} \\ \tau_{zx} & \tau_{zy} & \tau_{zz} \end{bmatrix} \tag{7.10}$$

where $\tau_{ij} = \tau_{ji}$. We then looked at a two-dimensional subset of τ_{ij} called plane stress involving as the only possible nonzero stresses τ_{xx}, τ_{yy}, and τ_{xy}. We then derived Eqs. (7.6)–(7.8), which we shall henceforth identify as the *transformation equations*, by which, knowing the plane stress components for an unprimed reference xy, we can easily compute the stresses for a primed reference $x'y'$ rotated relative to xy.

We will stop here now and go back to our study of rigid-body mechanics where we introduced nine inertia terms—six of them products of inertia and three of them moments of inertia. We can list them as follows:

$$I_{ij} = \begin{bmatrix} I_{xx} & -I_{xy} & -I_{xz} \\ -I_{yx} & I_{yy} & -I_{yz} \\ -I_{zx} & -I_{zy} & I_{zz} \end{bmatrix} \tag{7.11}$$

where $I_{ij} = I_{ji}$. You may not have used minus signs in your earlier study, but as you will learn, they should be inserted for products of inertia, for reasons soon to be made clear. You may recall that the *two-dimensional*

[2]Vectors also have their invariants. Thus, the length of a vector is independent of what reference one may use. Hence, for two references xyz and $x'y'z'$ rotated relative to each other at a point, we can say that

$$A_x^2 + A_y^2 + A_z^2 = (A_x')^2 + (A_y')^2 + (A_z')^2$$

More formally, we may say that the sum of the squares of the rectangular components of a vector is invariant with respect to a rotation of axes.

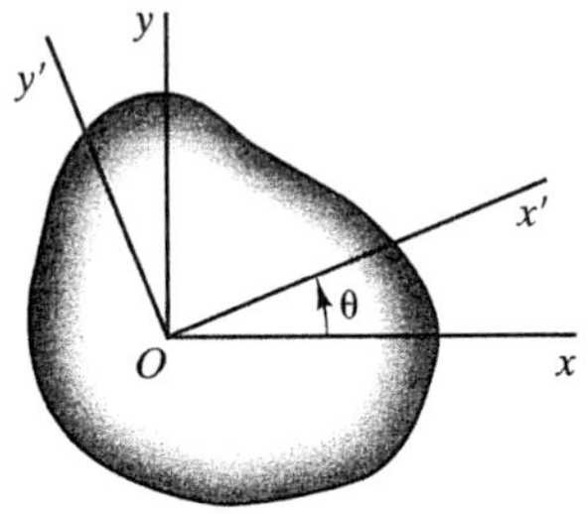

Figure 7.6. Plane area showing two sets of axes rotated relative to each other.

simplification of I_{ij} with the only possible nonzero values corresponding to I_{xx}, I_{yy}, and I_{xy} represented moments and products of a *plane area.* The so-called *transformation equations* to get moments and product of a plane area at a point O for reference $x'y'$ rotated relative to reference xy (see Fig. 7.6) are given as follows:

$$I_{x'x'} = \frac{I_{xx} + I_{yy}}{2} + \frac{I_{xx} - I_{yy}}{2} \cos 2\theta + (-I_{xy}) \sin 2\theta \tag{7.12a}$$

$$I_{y'y'} = \frac{I_{xx} + I_{yy}}{2} - \frac{I_{xx} - I_{yy}}{2} \cos 2\theta - (-I_{xy}) \sin 2\theta \tag{7.12b}$$

$$-I_{x'y'} = \frac{I_{yy} + I_{xx}}{2} \sin 2\theta + (-I_{xy}) \cos 2\theta \tag{7.12c}$$

These transformation equations for the two-dimensional version of I_{ij} are *identical* to the transformation equations for the two-dimensional version of τ_{ij} as presented in Eqs. (7.6)–(7.8). This is so despite *completely different* physical meanings of stress and moments of inertia. This may seem very strange to you at this stage of our study of mechanics. We see here that despite physical differences, stresses, and inertia terms (at least the two-dimensional subsets studied thus far) *respond identically* in the way their components change when you rotate axes at a point.

Why is this of importance to us? A hint of this is present in our modest undertaking in Section 7.2. Note that from the transformation equation for plane stress we could easily demonstrate Eq. (7.9) where we see that however much or little you rotate axes $x'y'$, the sum $(\tau_{x'x'} + \tau_{y'y'})$ does not change. Or in more elegant mathematical language, the sum of the normal stresses in plane stress is *invariant* with respect to a rotation of axes. Now we know that natural phenomena proceed without the assistance of man-made axes. Therefore, if you find a formulation in science that is unaffected by a rotation of axes, there is the strong possibility that the formulation has some physical significance. Indeed, if you look at the three-dimensional sum of the normal stresses, $\tau_{xx} + \tau_{yy} + \tau_{zz}$, you will later see that the invariance with respect to rotation of axes is present here as in the case of plane stress. When you study the mechanics of fluids, you will see that one-third of the sum corresponds to *thermodynamic pressure.*[3] But this invariance of $(\tau_{xx} + \tau_{yy})$ was available from the transformation equations. By the same token the invariant sum $(I_{xx} + I_{yy})$, which you may recall equals the important concept of the *polar moment of area, is deducible from the transformation equations (7.12). Thus knowing how components change when we rotate axes at a point permits us to deduce meaningful invariants and indeed many other important and vital properties and concepts.*

[3]See I. H. Shames, *Mechanics of Fluids,* Third Ed. (McGraw-Hill, New York, 1992), Chap. 2.

We may now give a preliminary definition of a symmetric, second-order Cartesian tensor T_{ij} as follows. If all two-dimensional subsets of T_{ij} transform on a rotation of axes by equations of the form of Eqs. (7.6), (7.7), and (7.8) and/or Eqs. (7.12), the T_{ij} is a second-order Cartesian tensor and has all the many properties of second-order Cartesian tensors. We note that the word "symmetric" in this connection means that $T_{ij} = T_{ji}$. In Chapter 15 a definition for symmetric second-order Cartesian tensors is given for the case of three-dimensional representation of T_{ij}.

In the remaining sections of this chapter we shall set forth some of the aforementioned properties using the case of plane stress as a vehicle. Those readers that go on to Chapter 15 and beyond will continue to expand and generalize certain of these results.

7.4 Principal Stresses and Principal Axes

Now that we have established the transformation equations for plane stress at a point, it will be of interest to evaluate the *extreme* values of stress components at a point in the xy plane. Accordingly, we may find the values of θ for which τ_{nn} is a relative maximum or minimum in the xy plane by setting

$$\frac{\partial \tau_{nn}}{\partial \theta} = 0$$

Using Eq. (7.4), this leads us to (using $\tilde{\theta}$ now since we are limiting the angle to those which satisfy the above equation)

$$\tan 2\tilde{\theta} = \frac{2\tau_{xy}}{\tau_{xx} - \tau_{yy}} \tag{7.13}$$

Now, since the tangent function repeats itself every π rad, there are obviously two values of $2\tilde{\theta}$ in the interval $0 \leq 2\tilde{\theta} \leq 2\pi$ which satisfy the above equation. Hence, the desired values of $\tilde{\theta}$ differ by $\pi/2$. We shall call θ_a the angle corresponding to the normal stress τ_a, and θ_b the angle corresponding to the other stress τ_b. Because $\tau_{xx} + \tau_{yy}$ is invariant at a point, one of these stresses is the algebraically largest and hence the maximum normal stress in the xy plane, while the other must be algebraically the smallest[4] normal stress in the xy plane at a point. To find these maximum and minimum normal stresses we simply substitute the

[4]The algebraic ordering of stress means that the stresses are ordered according to their position on the number scale. Accordingly, the further down the negative number scale, the "smaller" is the stress, algebraically speaking. By such an ordering, a stress of zero value is larger than a stress of −100,000 psi. This ranking is only a mathematical convenience and should not give the reader the absurd idea that zero stress is more dangerous than a stress of −1 million psi because the former is "algebraically larger."

$\tilde{\theta}$'s from Eq. (7.13) into Eq. (7.4) and solve for normal stresses. We do not know a priori which angles from Eq. (7.13) will produce the maximum normal stress in the xy plane and which will produce the minimum normal stress in the xy plane.

We shall be able to easily show in the next section that the extreme stresses τ_a and τ_b for plane stress in the xy plane can be ascertained directly from a set of known stresses τ_{xx}, τ_{yy}, and τ_{xy} at the point. The equations for this purpose are as follows:

$$\tau_a = \frac{\tau_{xx} + \tau_{yy}}{2} + \sqrt{\left(\frac{\tau_{xx} - \tau_{yy}}{2}\right)^2 + \tau_{xy}^2} \tag{7.14a}$$

$$\tau_b = \frac{\tau_{xx} + \tau_{yy}}{2} - \sqrt{\left(\frac{\tau_{xx} - \tau_{yy}}{2}\right)^2 + \tau_{xy}^2} \tag{7.14b}$$

If we form as shown here a right triangle with legs $2\tau_{xy}$ and $\tau_{xx} - \tau_{yy}$ and angle $2\tilde{\theta}$ such that Eq. (7.13) is satisfied, we can say:

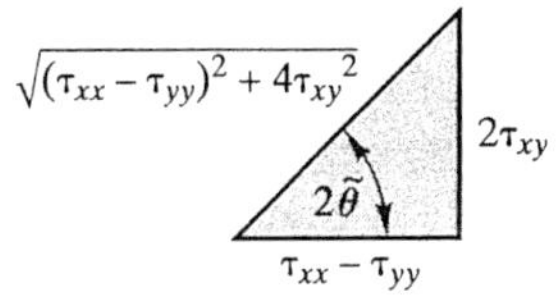

$$\sin 2\tilde{\theta} = \frac{2\tau_{xy}}{\sqrt{(\tau_{xx} - \tau_{yy})^2 + 4\tau_{xy}^2}}$$
$$\cos 2\tilde{\theta} = \frac{\tau_{xx} - \tau_{yy}}{\sqrt{(\tau_{xx} - \tau_{yy})^2 + 4\tau_{xy}^2}}$$

If you then substitute these results into Eq. (7.5), you may easily demonstrate (Problem 7.18) that we get a zero value for the shear stress . This shows that the axes, on whose normal planes we have extreme normal stress, are at the same time *axes on whose normal planes we have zero shear stress.* These axes are called *principal axes,* and the associated normal stresses are called *principal stresses.* Principal axes and their stresses are important in the study of solid mechanics, as you will soon see.

We may now obtain the *extreme* values of shear stress τ_{ns} in the xy plane by setting

$$\frac{\partial \tau_{ns}}{\partial \theta} = 0 \tag{7.15}$$

Using Eq. (7.5), this leads to the result

$$\tan 2\theta_{c,d} = -\frac{\tau_{xx} - \tau_{yy}}{2\tau_{xy}} \tag{7.16}$$

giving us two angles θ_c and θ_d at right angles to each other for this condition. Because the corresponding faces for these angles are orthogonal to each other, we know from the complementary property of shear that the shear stresses on these faces must be equal to each other. Hence, we need use only one of the angles θ_c or θ_d to find the extreme shear stress. We may substitute one of these angles into Eq. (7.5) and solve for the extreme shear stress τ_{ns} for this purpose.[5]

Notice, in comparing Eqs. (7.13) and (7.16), that when we express $\tilde{\theta}$ as θ_a we have

$$\tan 2\theta_a = -\frac{1}{\tan 2\theta_c} \tag{7.17}$$

That is, the tangent of the angles $2\theta_a$ and $2\theta_c$ are negative reciprocals of each other. This means that angles $2\theta_a$ and $2\theta_c$ are separated by 90° and we can conclude that θ_a and θ_c differ by 45°. Thus, the pair of orthogonal axes for the extreme shear stress in the xy plane is rotated 45° from the principal axes. A typical situation is shown in Fig. 7.7. (We could continue each axis around through 360° in the diagram; however, this would correspond to examining the stress components of the "back side" of sections, already included.)

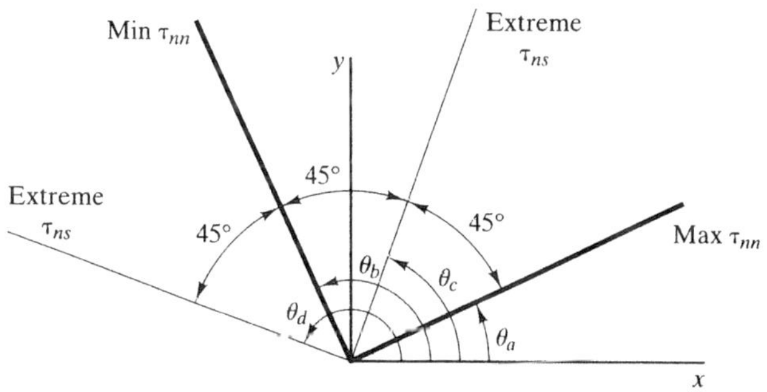

Figure 7.7. Axes of extreme normal stress and extreme shear stress.

It is also to be pointed out that the normal stresses for axes having extreme shear stress are not zero generally. That is, we cannot carry out an analogy between principal stresses and extreme shear stresses in this regard. We shall see this more clearly in Section 7.5 when we study the Mohr circle.

We shall now illustrate the formulations above in the following example.

[5]Unlike normal stress, where the properties of a material may differ markedly for compression and tension, the sign of the shear stress has no relevance as to the manner of behavior of a material other than directional.

Example 7.2

Find the maximum normal and shear stresses in the xy plane for point P of the beam in Example 7.1.

We first find the appropriate directions from Eqs. (7.13). Thus,

$$\tan 2\theta_{a,b} = \frac{2(-400)}{3000 - 0} = -.2667$$

$$\therefore 2\theta_{a,b} = -14.92°, 165.08°$$

Hence,

$$\theta_{a,b} = -7.46°, 82.54°$$

Also,

$$\theta_{c,d} = \theta_{a,b} + 45°$$

$$\therefore \theta_{c,d} = 37.54°, 127.54°$$

We may now go to Eq. (7.4). Hence:

For $\tilde{\theta} = -7.46°$:

$$\tau_{nn} = \frac{3000 + 0}{2} + \frac{3000 - 0}{2} \cos(-14.92°) + (-400) \sin(-14.92°)$$

$$(\tau_{nn})_1 = 3052 \text{ psi}$$

For $\tilde{\theta} = 82.54°$:

$$\tau_{nn} = \frac{3000 + 0}{2} + \frac{3000 - 0}{2} \cos(165.08°) + (-400) \sin(165.08°)$$

$$(\tau_{nn})_2 = -52 \text{ psi}$$

Thus, the maximum normal stress is 3052 psi acting on the section whose normal forms a clockwise angle of 7.46° with the x axis.[6]

We may also use Eq. (7.14) to get the principal stresses directly. Thus, we have

$$\tau_{a,b} = \frac{3000 + 0}{2} \pm \sqrt{\left(\frac{3000 - 0}{2}\right)^2 + (-400)^2}$$

$$\therefore \tau_a = 1500 + 1552 = 3052 \text{ psi}$$

$$\tau_b = 1500 - 1552 = -52 \text{ psi}$$

[6]We wish to point out a short cut when computing a pair of orthogonal normal stresses. After expressing the transformation equation for the first stress, all one need do to get the second stress is to rewrite the first transformation equation changing the signs of the second and third expressions on the right side of the equation.

Example 7.2 (Continued)

Finally, to get the extreme shear stress in the xy plane, substitute either 37.54° or 127.54° into Eq. (7.5). Thus, using $\theta_b = 37.54°$, we have

$$(\tau_{ns})_{\text{extreme}} = \frac{0 - 3000}{2}\sin(75.08°) + (-400)\cos(75.08°)$$

$$(\tau_{ns})_{\text{extreme}} = -1552 \text{ psi}$$

Since cross-shears are equal, for $\theta_c = 127.54°$ we also get

$$\tau_{ns} = -1552 \text{ psi}$$

Hence, the extreme shear stress is -1552 psi acting on the sections whose normals form angles of 37.54° and 127.54° with the x axis.

We have seen that for plane stress there exist two extreme values of normal stress in the xy plane which we have referred to as *principal stresses.* The axes along the direction of these stresses are called principal axes, and they have been shown to be mutually perpendicular. Furthermore, the shear stress components are zero on the sections whose normals are principal axes. In the general, three-dimensional stress state, the following situation exists, as will later be shown. *There are three mutually perpendicular principal axes,* corresponding to normals to sections having *zero shear-stress components.* The normal stresses on these sections are called *principal stresses,* one of which is the algebraic maximum normal stress component at the point, another of which is the algebraic minimum normal stress at the point. In the case of plane stress the third principal axis is always the axis normal to the xy plane. This follows since there is zero shear stress on this plane in accordance with the definition of plane stress. Thus, $\tau_{zz} = 0$ is always the third principal stress for such cases. When dealing with a general state of stress we usually order the principal stresses algebraically as τ_1, τ_2, and τ_3 with τ_1 the algebraically largest stress and τ_3 the algebraically smallest stress.

7.5 Mohr's Circle

A convenient graphical representation called *Mohr's circle* conveys the state of plane stress at a point. To employ Mohr's circle, we must set forth additional sign conventions. Consider the infinitesimal rectangular element shown in Fig. 7.8 where we have shown positive shear and nor-

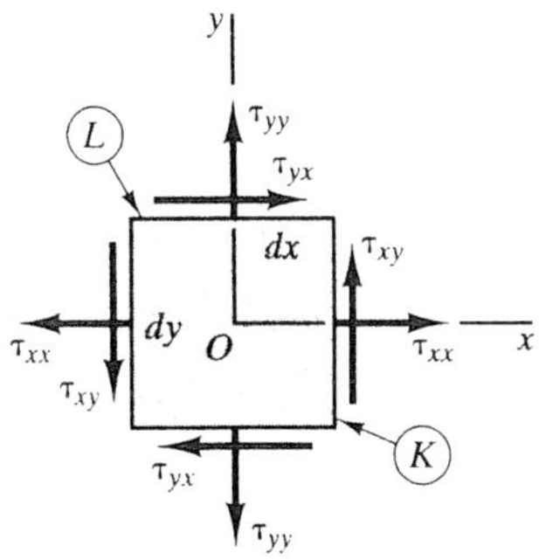

Figure 7.8. Plane stress element; physical plane.

mal stresses. For the Mohr circle construction, we shall employ the following rule for shear stress.[7]

> A shear stress will be taken as positive on a face of an element in regard to Mohr's circle when it yields a clockwise moment about the center point O at the element. A shear stress yielding a counterclockwise moment about point O will then be taken as negative.

Thus, τ_{xy} in Fig. 7.8 will be considered negative and τ_{yx} will be considered positive. There is no change in convention for normal stress when dealing with Mohr's circle.

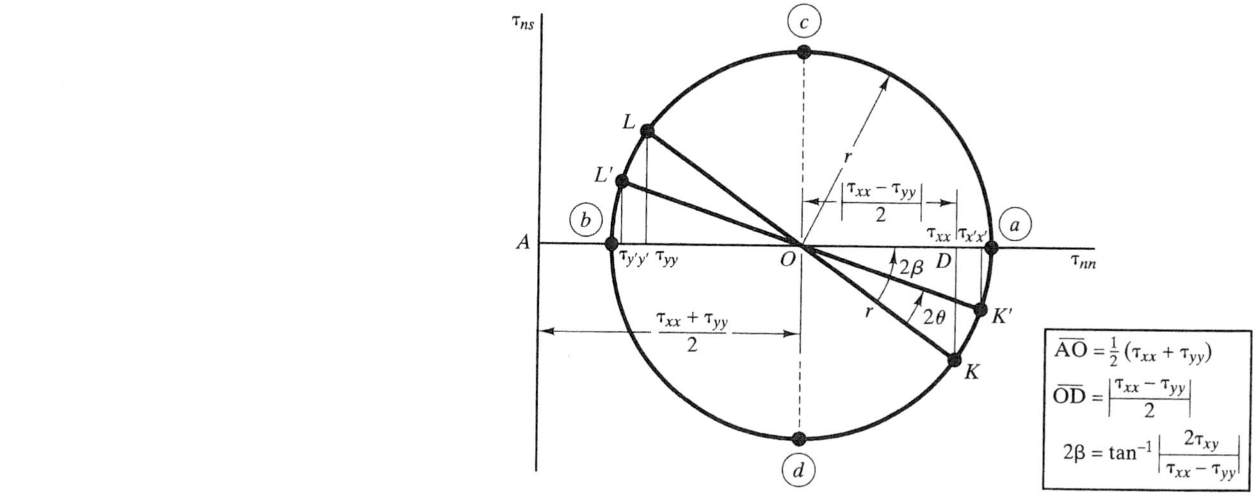

Figure 7.9. Stress plane for Mohr's circle.

Let us now introduce *stress coordinates*, as shown in Fig. 7.9, where τ_{nn} (i.e., normal stress) is the abscissa and τ_{ns} (i.e., shear stress) is the ordinate. We call this the *stress plane;* we call the xy plane the *physical plane.* To draw Mohr's circle on the stress plane, we first plot the stresses for two orthogonal adjacent faces of an element in the physical plane, as, for example, faces K and L in Fig. 7.8, using the sign convention for shear stress developed in the previous paragraph. The mapped points of faces K and L (see Fig. 7.9) are also denoted in the stress diagram as K and L, respectively. Now, connect these points with a straight line so as to intersect the τ_{nn} axis at O. With O as a center, draw a circle through points K and L as shown in the diagram. This is the celebrated Mohr circle.

We shall now show how we can use the Mohr circle and then we shall justify the procedure. Suppose that we wish to know the stress for reference $x'y'$ rotated through an angle θ from xy in the physical plane

[7]The reader should be cautioned that the convention used here for Mohr's circle is one of several in common usage.

as shown in Fig. 7.10.[8] To find point K' in the stress plane corresponding to face K' in the physical plane, we draw a line from point O in the stress plane rotated from the OK axis by an angle 2θ in the *same direction* as in the physical plane. (Remember that rotation in the stress plane is twice that in the physical plane.) The coordinates of the intersection K' of this axis with Mohr's circle then give the shear and normal stress corresponding to face K' in the physical plane. Now, by extending OK' backward, we may also form point L' on Mohr's circle. Since OL' is rotated 180° in the stress plane, point L' corresponds to the state of stress on face L' which is rotated 90° from face K' in the physical plane. We thus have the state of stress for face K' and face L' corresponding to the x' and y' axes, respectively.

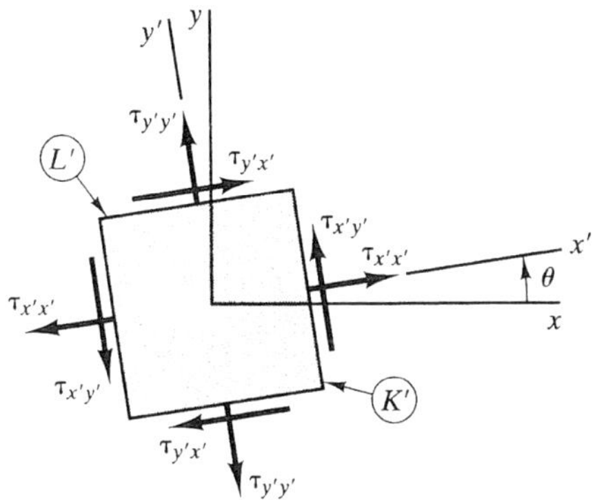

Figure 7.10. Positive rotation of axes.

Thus far we have shown how to construct Mohr's circle and how to employ it. Our next step is proving the validity of the method. To do this, we must show that stresses deduced from Mohr's circle satisfy Eqs. (7.6)–(7.8). Inspection of Mohr's circle (see Fig. 7.9) leads one to the following conclusions:

$$\overline{AO} = \frac{\tau_{xx} + \tau_{yy}}{2} \qquad \overline{OD} = \left|\frac{\tau_{xx} - \tau_{yy}}{2}\right| \qquad 2\beta = \tan^{-1}\left|\frac{\tau_{xy}}{\dfrac{\tau_{xx} - \tau_{yy}}{2}}\right| \tag{7.18}$$

$$r = \sqrt{\overline{OD}^2 + \overline{DK}^2} = \sqrt{\left(\frac{\tau_{xx} - \tau_{yy}}{2}\right)^2 + \tau_{xy}^2} \tag{7.19}$$

These quantities have been shown in the diagram. Using geometrical reasoning for Mohr's circle, we can say for the stress $\tau_{x'x'}$, which corresponds to point K' on the stress diagram, that

$$\tau_{x'x'} = \overline{OA} + r\cos(2\beta - 2\theta) \tag{7.20}$$

Replacing $\overline{OA}$ in accordance with Eq. (7.18) and expanding the cosine term, we get

$$\tau_{x'x'} = \frac{\tau_{xx} + \tau_{yy}}{2} + r(\cos 2\beta \cos 2\theta + \sin 2\beta \sin 2\theta) \tag{7.21}$$

But from the stress diagram it is clear on inspection that

$$\cos 2\beta = \frac{\tau_{xx} - \tau_{yy}}{2r} \tag{7.22}$$

$$\sin 2\beta = \frac{\tau_{xy}}{r} \tag{7.23}$$

Substituting these results into Eq. (7.21), we get

$$\tau_{x'x'} = \frac{\tau_{xx} + \tau_{yy}}{2} + \frac{\tau_{xx} - \tau_{yy}}{2}\cos 2\theta + \tau_{xy}\sin 2\theta \tag{7.24}$$

[8]Be sure to note that positive is from x to y (counterclockwise).

which is the proper relation that we derived earlier [Eq. (7.6)].[9]

In a similar manner, we first show for $\tau_{x'y'}$ that

$$\tau_{x'y'} = -r\sin(2\beta - 2\theta) = -r(\sin 2\beta\cos 2\theta - \cos 2\beta\sin 2\theta) \quad (7.25)$$

Using Eqs. (7.22) and (7.23), we get

$$\tau_{x'y'} = -\tau_{xy}\cos 2\theta + \frac{\tau_{xx} - \tau_{yy}}{2}\sin 2\theta \quad (7.26)$$

This equation checks with Eq. (7.8) when the shear convention of Mohr's circle is taken into account. Thus, we have fully justified the construction and proposed use of the Mohr circle.[10]

We may now readily verify some of the results of the previous section by employing the Mohr circle diagram. Thus, the maximum and minimum normal stresses in the xy plane (i.e., the principal stresses) are at positions a and b, respectively, on the Mohr circle (see Fig. 7.9). Note that the shear stresses are zero at these points, so on the principal planes there is only a normal stress present. Also, it is clear that

$$\tau_a = \frac{\tau_{xx} + \tau_{yy}}{2} + r = \frac{\tau_{xx} + \tau_{yy}}{2} + \sqrt{\left(\frac{\tau_{xx} - \tau_{yy}}{2}\right)^2 + \tau_{xy}^2} \quad (7.27)$$

$$\tau_b = \frac{\tau_{xx} + \tau_{yy}}{2} - r = \frac{\tau_{xx} + \tau_{yy}}{2} - \sqrt{\left(\frac{\tau_{xx} - \tau_{yy}}{2}\right)^2 + \tau_{xy}^2} \quad (7.28)$$

which justify Eqs. (7.14). Furthermore, the extreme values of shear stress occur at positions c and d on the Mohr circle. We can say from the diagram that

$$(\tau_{ns})_{\text{extreme}} = \pm r = \pm\sqrt{\left(\frac{\tau_{xx} - \tau_{yy}}{2}\right)^2 + \tau_{xy}^2} \quad (7.29)$$

Note that position c is rotated 90° from position a on the Mohr circle. Similarly, d is rotated 90° from b. Hence, the planes of extreme shear stress are rotated 45° from the principal planes in the physical diagram. Finally, in the planes of extreme shear stress the normal stress is seen from the Mohr circle to be

$$\tau_{nn} = \frac{\tau_{xx} + \tau_{yy}}{2} \quad (7.30)$$

showing that it is generally nonzero.

We shall not generally use Mohr's circle to get numerical values for particular stresses by strictly graphical means, although one could

[9]Equation (7.7) for $\tau_{y'y'}$ follows directly by increasing θ by π rad in order to get to L'.

[10]We can now state that the Mohr circle is the graphical representation of the two-dimensional symmetric tensor. We could very well now say that a second-order tensor is defined as the two-dimensional subset of a nine-term symmetric array which follows the Mohr circle construction for determining component terms along any coplanar axes.

employ the construction for this purpose. Rather, we shall sketch Mohr's circle approximately to scale so as to get a visual picture of how the stress varies at a point for plane stress. Then, using the circle along with simple trigonometric calculations deduced from the Mohr circle, we can calculate many quantities of interest pertaining to stress at a point. We shall illustrate the latter procedure in the following example.

Example 7.3

Using Mohr's circle for a state of plane stress (see Fig. 7.11) given as

$$\tau_{xx} = 4000 \text{ kPa} \qquad \tau_{yy} = 1000 \text{ kPa} \qquad \tau_{xy} = -500 \text{ kPa}$$

find the principal stresses as well as the stresses for a set of axes $x'y'$ rotated 30° clockwise about xy.

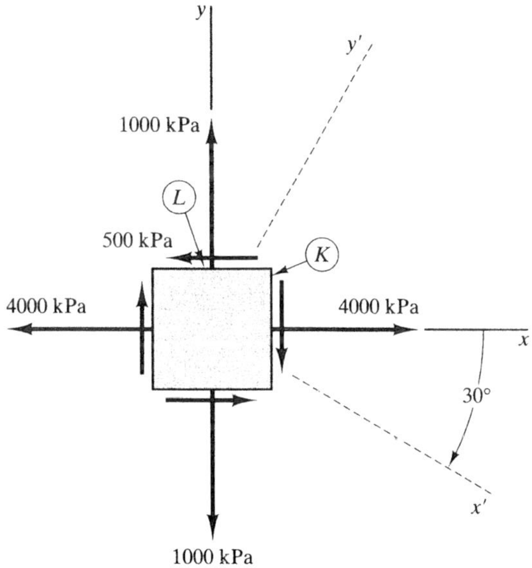

Figure 7.11. Physical diagram.

The point on the stress diagram (Fig. 7.12) corresponding to face K in the physical diagram is found with coordinates having an abscissa of 4000 kPa and an ordinate of +500 kPa. The coordinates on the stress diagram meanwhile for face L are 1000 kPa and −500 kPa. Connecting K and L, we can now locate O, the center of Mohr's circle, and we can then draw the circle. The maximum stress τ_a has a position on Mohr's circle rotated 2β clockwise from OK (see Fig. 7.12). One might determine this angle from the diagram with the

Example 7.3 (Continued)

aid of a protractor and thus find β to be about 9°. However, noting that $GO = OD = |(\tau_{xx} - \tau_{yy})/2| = 1500$ kPa, and using the diagram, we shall compute β more accurately as follows:

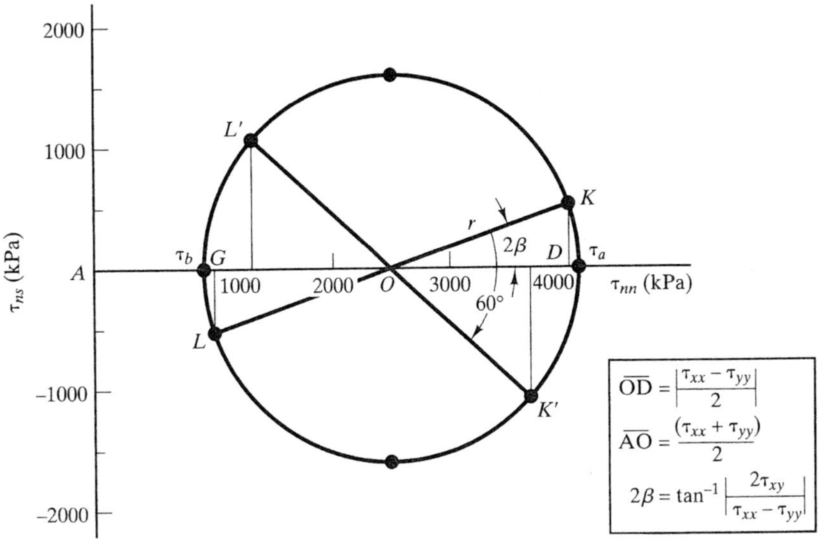

Figure 7.12. Stress diagram.

$$\tan(2\beta) = \frac{500}{1500}$$

$$\therefore \beta = 9.22° \qquad \text{(a)}$$

The values of τ_a and τ_b are seen from the diagram to be, respectively, a little over 4000 kPa and a little less than 1000 kPa. We again use the diagram in giving analytic results here by noting from the diagram that the radius r is given as

$$r = \sqrt{1500^2 + 500^2} = 1581 \text{ kPa} \qquad \text{(b)}$$

Hence,

$$\tau_a = \overline{AO} + r = \frac{\tau_{xx} + \tau_{yy}}{2} + 1581 = 4081 \text{ kPa}$$

$$\tau_b = \overline{AO} - r = \frac{\tau_{xx} + \tau_{yy}}{2} - 1581 = 919 \text{ kPa}$$

Example 7.3 (Continued)

It is also apparent directly from the diagram that the extreme shear stress is a little above 1500 kPa. We also may note that this shear stress must equal the radius of Mohr's circle and from Eq. (b) above we get 1581 kPa for this shear stress as an analytic result.

Now to get stresses $\tau_{x'x'}$, $\tau_{y'y'}$, and $\tau_{x'y'}$ for $\theta = 30°$ we employ a 60° clockwise rotation in Mohr's circle from OK to get point K' in Fig. 7.12. We might then read off a normal stress $\tau_{x'x'}$ of 3700 kPa, a normal stress $\tau_{y'y'}$ of 1300 kPa, and a shear stress $\tau_{x'y'}$ of -1100 kPa. More precisely, we note from the diagram that

$$\begin{aligned}\tau_{x'x'} &= \overline{AO} + r\cos(60° - 2\beta)\\ &= \frac{\tau_{xx} + \tau_{yy}}{2} + r\cos(60° - 2\beta)\\ &= 2500 + 1581\cos(60° - 18.44°) = \quad 3683 \text{ kPa}\end{aligned}$$

$$\tau_{y'y'} = \overline{AO} - r\cos(60° - 2\beta) = \quad 1315 \text{ kPa}$$

$$\tau_{x'y'} = -r\sin(60° - 2\beta) = \quad -1049 \text{ kPa}$$

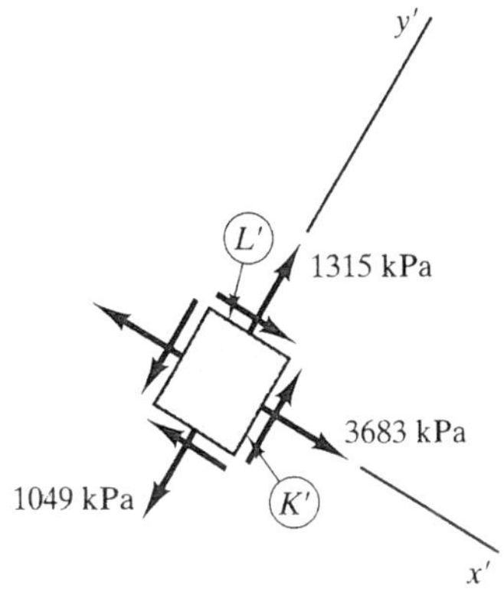

Figure 7.13. Physical plane.

From Fig. 7.13 we see that the negative sign for $\tau_{x'y'}$ from Mohr's circle calls for a *counterclockwise* moment of the shear stress on face K' about the center of the element. Clearly, by the *general convention* outside of Mohr's circle we have a *positive state* of shear stress $\tau_{x'y'}$ present, and this is so indicated in the diagram.

We see from this exercise that we can use the Mohr circle construction to give, by strictly graphical means, all the pertinent information about stress at a point for plane stress. Or it can be used to provide, through simple trigonometric considerations, analytic computations for plane stress characteristics at a point. In the latter case we do not need to use the transformation formulas as well as other derived equations presented earlier. And, importantly, we have a graphical picture of the relative values and positions of the pertinent quantities concerning plane stress at a point for the case at hand.

Since the Mohr circle construction depicts graphically all the values of a special case of a second-order tensor and stems entirely from the basic transformation equations for this special case, we could expect such a construction to similarly depict analogous simplifications of other second-order tensors. Indeed, in Chapter 8 we shall consider Mohr's circle for plane strain. Furthermore, you may have already used Mohr's

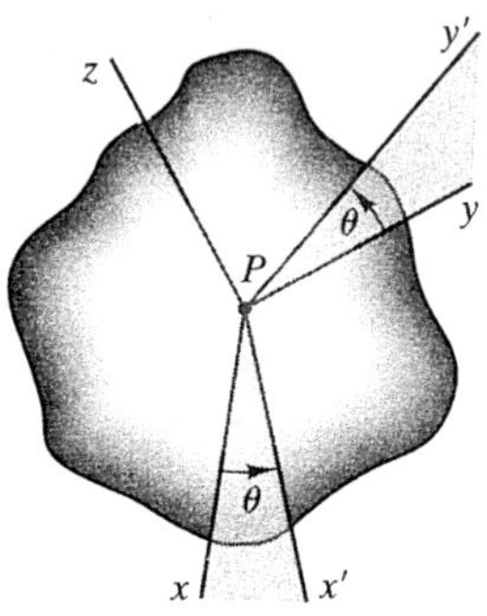

Figure 7.14. Rotation of axes about z at an arbitrary point P.

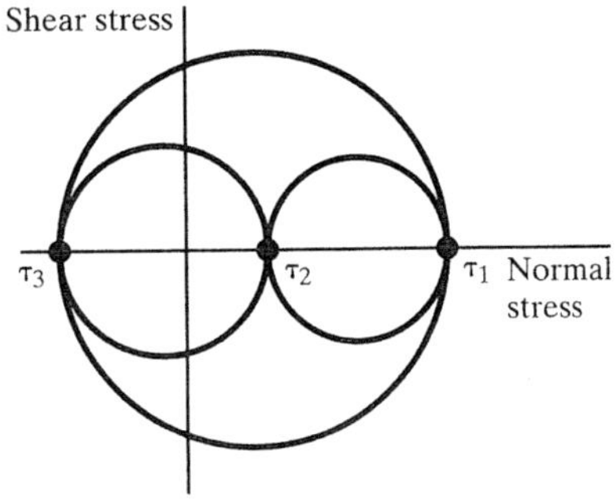

Figure 7.15. Mohr circle representation for general state of stress.

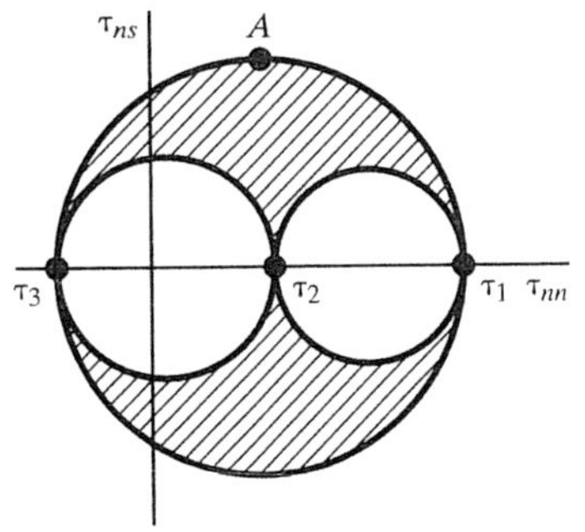

Figure 7.16. Crosshatched region represents possible stress.

circle for second moments and products of area which arise from the aforementioned simplification of the inertia tensor of rigid-body dynamics.

There is also a Mohr circle construction for three-dimensional stress, which we shall now discuss briefly. Consider first a general state of stress at point P (Fig. 7.14). Choose any set of axes xyz at the point and form axes $x'y'$ by rotating about the z axis an angle θ. Clearly, the transformation equations for $\tau_{x'x'}$, $\tau_{y'y'}$, and $\tau_{x'y'}$ have the *same form* as Eqs. (7.6)–(7.8). Accordingly, we can draw a *Mohr's circle* for the states of stress $\tau_{x'x'}$, $\tau_{y'y'}$, and $\tau_{x'y'}$. We have done this thus far for the case of plane stress where the z axis was normal to the plate and consequently was a principal axis. For an *arbitrary* z axis such as at point P, such a Mohr circle reveals the extreme normal stresses $\tau_{x'x'}$ and $\tau_{y'y'}$ at point P for the axes perpendicular to the z axis. These stresses, however, are *not* principal stresses unless the z direction is a principal direction. Let us suppose now that the z axis corresponds to a principal axis. We shall label the three principal stresses as τ_1, τ_2, τ_3, with τ_1 corresponding to the algebraically largest stress and τ_3 the algebraically smallest stress. If the z axis corresponds to τ_3, then Mohr's circle gives τ_1 and τ_2 as the other principal stresses. If, for the same state of stress, we choose the z axis to coincide with the direction of the principal stress τ_2, we get a second Mohr circle with τ_1 and τ_3 as principal stresses. Finally, putting the z axis in the direction of the principal stress τ_1, we get a third Mohr circle yielding τ_2 and τ_3 as principal stresses. These Mohr circles have been shown on a single set of axes in Fig. 7.15. Such a set of circles is the Mohr circle representation for the general state of stress. One can show[11] that for interfaces oriented at some angle inclined toward the principal axes, the point representation of the state of stress must lie in a region *between* all three circles, such as the crosshatched region shown in Fig. 7.16. The largest circle is called the *principal* circle and gives the largest shear stress (points A) for the state of stress at the point. Note that the largest shear stress equals *one-half the difference between the algebraically largest principal stress and the algebraically smallest principal stress.*

7.7 Closure

In this chapter our thrust has been twofold. First we introduced the useful and practical plane stress simplification. Applications will occur in Chapters 10 and 11 when we study the behavior of beams. As a second major consideration we discussed stress relations at a point for plane stress. And with this as a vehicle, we showed how the manner in which components of stress change as we rotate the axes as incorporated in

[11]See I. H. Shames, *Mechanics of Deformable Solids* (R. E. Krieger, Melbourne, FL, 1964), App. I, p. 431.

the transformation equations was sufficient to give rise and meaning to a variety of vital concepts. We can point to invariants, principal stress, principal axes, and Mohr's circle—all derivable from the transformations equations (7.6)–(7.8). We thus eased you into a nodding acquaintance with second-order Cartesian tensors. More complete acquaintance is available in the chapters on three-dimensional stress and strain. But we do not have to wait until then to benefit from the introduction to tensors of this chapter.

In the next chapter we shall introduce the two-dimensional simplification of strain called plane strain. We will show that the transformation equations on rotation of axes are identical to those of the two-dimensional simplifications of τ_{ij} and I_{ij}. Having done this we can present without further formal development the concepts of strain invariants, principal strains, principal axes, and Mohr's circle. All two-dimensional simplifications of second-order, symmetric tensors, however different the physical picture they represent, have the same aforementioned vital properties. Via this approach we hopefully have the happy circumstance of saving time, increasing understanding, and minimizing confusion.

Highlights (7)

A two-dimensional subset of stress distributions, namely τ_{xx}, τ_{yy}, τ_{xy}, and τ_{yx}, is called **plane stress** and represents stresses in a flat plate loaded in its midplane. Transformation equations give the stresses for axes rotated relative to xy at a point. These equations are identical to transformation equations for the second moments and products of area at a point. **Second-order symmetric tensors** may be defined as those sets of nine components, every two-dimensional subset of which obeys these very transformation equations. From these transformation equations, we set forth for stress:

Principal stresses and principal axes
Invariants with respect to rotation of axes
Extreme shear stress
Mohr's circle

Analogous concepts appear for other tensor fields. Thus, merely knowing that a two-dimensional subset satisfies the aforementioned transformation equations makes these vital formulations and concepts available to us immediately. It will later be pointed out that scalars and vectors are special simpler tensors making for greater unity in our mathematics. And in a starred Looking Ahead section in Chapter 15, we will present a powerful notation for the handling of tensors. This notation is called "tensor notation" or "index notation."

PROBLEMS

Plane Stress Transformation	7.1–7.7
Principal Stresses	7.8–7.18
Mohr's Circle	7.19–7.24
Unspecified Section Problems	7.25–7.35
Computer Problems	7.36
Programming Project 7	

7.1. [7.2] Given the following stresses for plane stress

$$\begin{aligned}\tau_{xx} &= 3.5 \times 10^6 \text{ Pa}\\ \tau_{yy} &= -7 \times 10^6 \text{ Pa}\\ \tau_{xy} &= 1.4 \times 10^6 \text{ Pa}\end{aligned}$$

what is the normal stress and the shear stress for an interface having a normal with the following direction:

$$\boldsymbol{\varepsilon}_1 = .707\mathbf{i} + .707\mathbf{j}$$

7.2. [7.2] Given the following stress at a point,

$$\tau_{xx} = 1000 \text{ psi} \qquad \tau_{yy} = 0 \qquad \tau_{xy} = -500 \text{ psi}$$

what are the stresses for interfaces formed by axes $x'y'$ rotated 30° about the z axis in a counterclockwise direction as one looks along the z axis toward the origin?

7.3. [7.2] A plate of thickness 25 mm (Fig. P.7.3) is loaded uniformly by forces F_1 and F_2 having values of 2250 N and 9000 N, respectively. What is the normal stress normal to the diagonals? What are the shear stresses for a pair of axes rotated 30° counterclockwise from the horizontal and vertical directions?

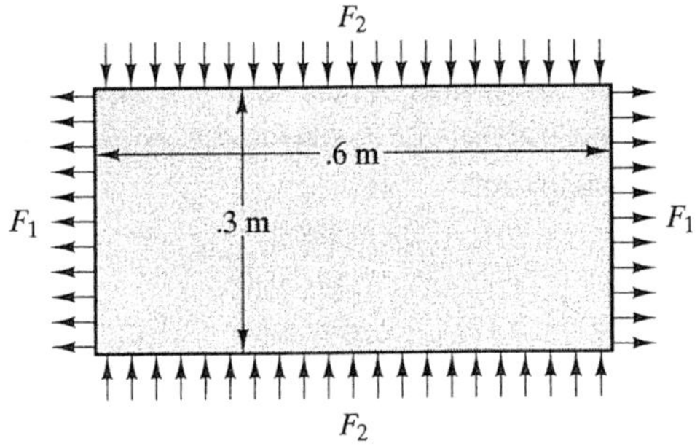

Figure P.7.3.

7.4. [7.2] The stresses at a point in a wooden beam are shown in Fig. P.7.4. What are the stresses normal and parallel to the grain?

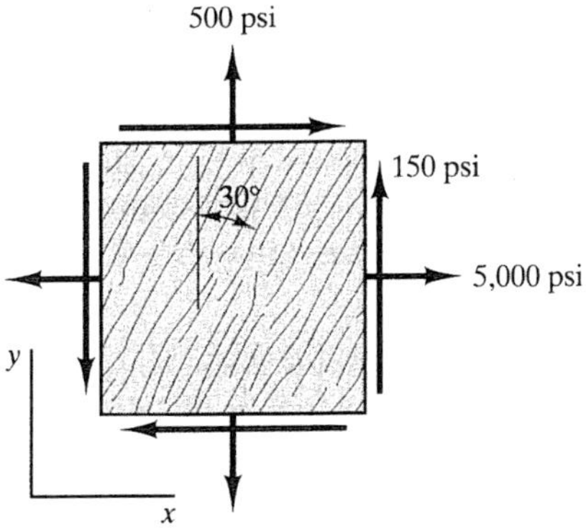

Figure P.7.4.

7.5. [7.2] A thin-walled cylinder (Fig. P.7.5) is under the action of tensile force P of 1000 lb. The outside diameter of the cylinder is 3 in. and the wall thickness is 1/8 in. What is the shear stress and normal stress at point A on a section of the cylinder formed by a cutting plane oriented $\alpha = 40°$ to the cylinder, as has been shown in the diagram?

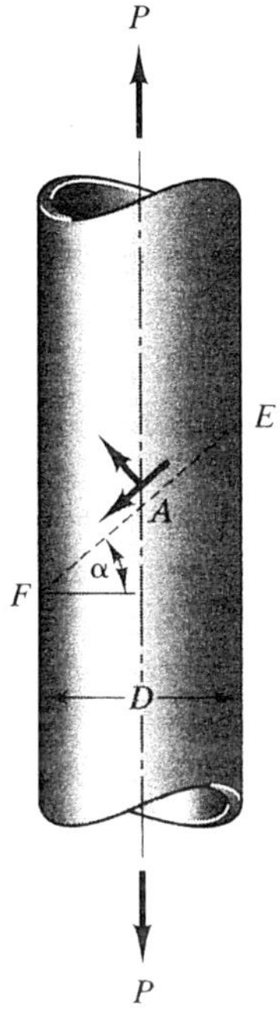

Figure P.7.5.

7.6. [7.2] If the angle α of cut EF in Problem 7.5 can be varied, at what value will the shear stress on an interface of A have its maximum value? What is this value and the value of the normal stress on this interface?

7.7. [7.2] A thin-walled tank is shown in Fig. P.7.7 having an internal pressure above atmosphere of 3.5×10^6 Pa. The outside diameter of the tank is 0.6 m and the thickness of the cylindrical portion of the wall is 12.5 mm. What are the normal stress and the shear stress on interface C formed by a cut AB which is inclined 30° to the tank, as shown in the diagram, and away from the ends? Neglect the compressive stress coming directly from the 3.5×10^6 Pa pressure onto the interface parallel to the cylindrical wall.

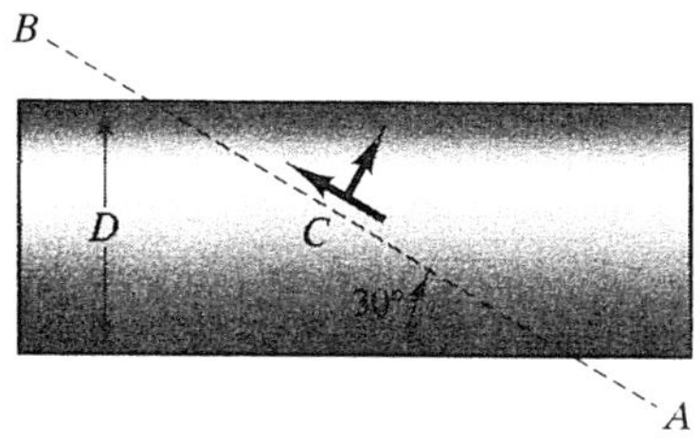

Figure P.7.7.

7.8. [7.4] Given the following stresses,

$$\tau_{xx} = -7 \times 10^6 \text{ Pa}$$

$$\tau_{yy} = 3.5 \times 10^6 \text{ Pa}$$

$$\tau_{xy} = 5.5 \times 10^6 \text{ Pa}$$

what are the principal axes and the principal stresses?

7.9. [7.4] Find the algebraic minimum normal stress at a point where $\tau_{xx} = 500$ psi, $\tau_{yy} = -500$ psi, and $\tau_{xy} = 1000$ psi. What is the direction of the normal to the interface for this stress?

7.10. [7.4] The principal stresses at a point are given as 7×10^6 Pa and 3.5×10^6 Pa. What are the stresses along axes rotated 30° clockwise from the axes?

7.11. [7.4] A thin cylinder in a vacuum is fixed at one end and a torque T of 50 lb-ft is applied at the other (Fig. P.7.11). If $\gamma = 450$ lb/ft^3 for the cylinder and if the stresses do not vary with radial distance from the axis of the cylinder and are symmetric about the axis of the cylinder, what is the algebraic maximum normal stress that is developed in the member? (*Hint:* Consider element A.)

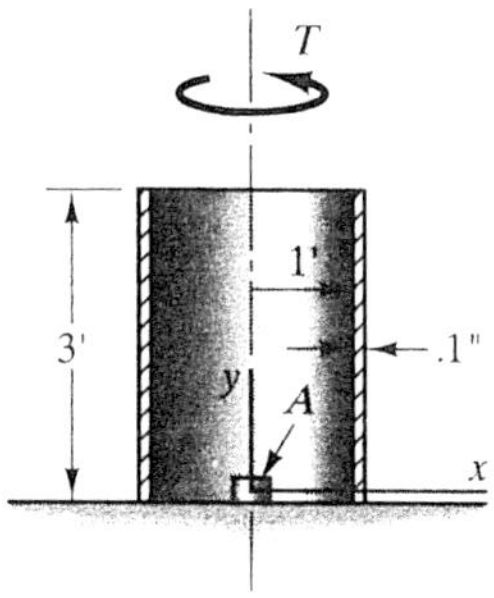

Figure P.7.11.

7.12. [7.4] In Problem 7.8 determine the extreme shear stress and the direction of the axes for this stress.

7.13. [7.4] Find the extreme shear stress in Problem 7.10.

7.14. [7.4] A thin cylindrical tank (Fig. P.7.14) is subject to an internal gage pressure of 10 psi and is at the same time subject to a twisting torque T of 200 lb-ft. If the outside diameter $D = 1$ ft and the thickness of the cylinder is .1 in., what is the maximum normal stress at A away from the ends? What is the maximum shear stress away from the ends? (*Hint:* Since the compression stress coming directly from the internal pressure is so very small compared to the other stresses, we neglect it, thus making the stress at a point on the cylinder wall away from the ends a plane stress problem.)

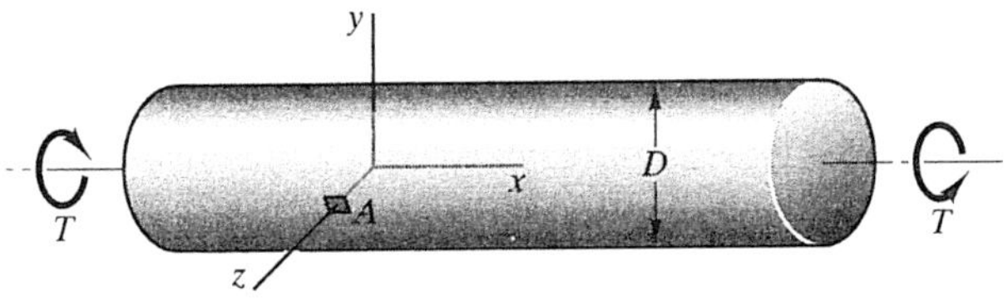

Figure P.7.14.

7.15. [7.4] If the allowable normal stress in Problem 7.14 is 40,000 psi, what is the maximum pressure that you can use in the tank for a torque of 2000 lb-ft?

7.16. [7.4] Shown in Fig. P.7.16 is a curved beam having a thickness t. The beam has an inner radius a and an outer radius b. A pair of couples M is applied at the ends of the beam as shown. The theory of elasticity indicates that the following is a correct stress distribution:

$$\tau_{rr} = -\frac{4M}{kt}\left(\frac{a^2b^2}{r^2}\ln\frac{b}{a} + b^2\ln\frac{r}{b} + a^2\ln\frac{a}{r}\right)$$

$$\tau_{\theta\theta} = \frac{4M}{kt}\left(-\frac{a^2b^2}{r^2}\ln\frac{b}{a} + b^2\ln\frac{r}{b} + a^2\ln\frac{a}{r} + b^2 - a^2\right)$$

$$\tau_{r\theta} = 0$$

where $k = (b^2 - a^2)^2 - 4a^2b^2\,[\ln(b/a)]^2$. Examining the diagram, what should the stress τ_{rr} be on the circular portions of the boundary? Check to see if the solution satisfies these conditions. What is the normal stress in the direction $\theta = 30°$ at $r = 3$ m? Take $a = 2$ m, $b = 4$ m, $M = 5 \times 10^7$ N-m, and $t = .2$ m. What is the maximum normal and the extreme shear stress here?

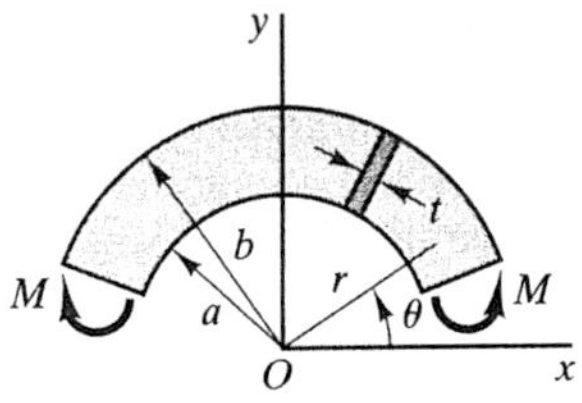

Figure P.7.16.

7.17. [7.4] A vertical line load acts on an infinitely large plate (see Fig. P.7.17). The total load from this distribution is P. The plate is under a state of plane stress which, using cylindrical coordinates, is given from the theory of elasticity as

$$\tau_{rr} = -\frac{2P}{\pi}\frac{\cos\theta}{r}$$

$$\tau_{\theta\theta} = 0$$

$$\tau_{r\theta} = \tau_{\theta r} = 0$$

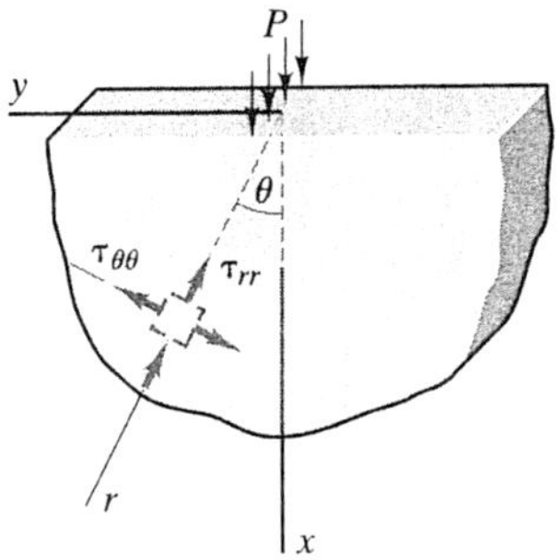

Figure P.7.17.

Why must we exclude the point of application from the solution? What are the stresses τ_{xx}, τ_{yy}, and τ_{xy}, at any position r, θ?

7.18. [7.4] Show that for axes found using Eq. (7.13) (i.e., the principal axes), the shear stress must always be zero.

7.19. [7.5] Draw Mohr's circle for the case where $\tau_{xx} = 0$, $\tau_{yy} = 0$, and $\tau_{xy} = 500$ kPa. What are the principal stresses, and what are the axes for these stresses?

7.20. [7.5] Given the following state of stress,

$$\tau_{xx} = 500 \text{ psi} \qquad \tau_{yy} = -800 \text{ psi} \qquad \tau_{xy} = -300 \text{ psi}$$

show on Mohr's circle the stresses on an interface rotated 30° clockwise from the x axis.

7.21. [7.5] For stresses

$$\tau_{xx} = 1.4 \times 10^7 \text{ Pa}$$
$$\tau_{yy} = -7 \times 10^6 \text{ Pa}$$
$$\tau_{xy} = -3.5 \times 10^6 \text{ Pa}$$

find the stresses on axes rotated 45° clockwise from the xy axes. Find the principal stresses. Use Mohr's circle as an aid for these calculations.

7.22. [7.5] A thin-walled tank in Fig. P.7.22 has air inside at a pressure of 100 psi gage. A torque of 50,000 ft-lb is applied at each end as shown. Using *Mohr's circle,* what is the maximum stress at point A? The wall thickness is .01 ft.

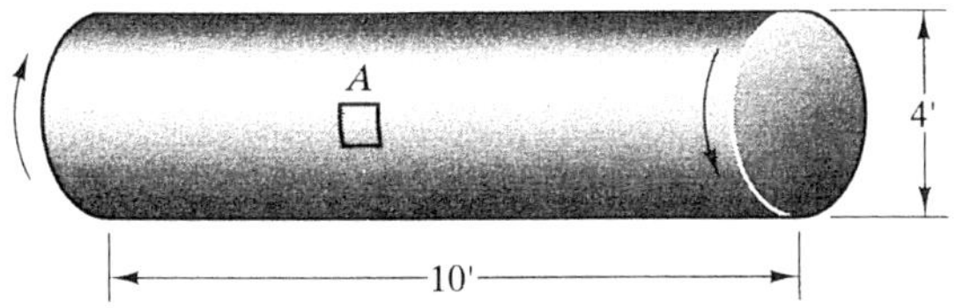

Figure P.7.22.

7.23. [7.5] Given the following stresses at a point,

$$\tau_{xx} = -3000 \text{ psi} \qquad \tau_{yy} = 4000 \text{ psi} \qquad \tau_{xy} = 1000 \text{ psi}$$

what angle must we rotate from the x axis to get an axis with a normal stress of − 2000 psi and a positive shear stress? Use *Mohr's circle.*

7.24. [7.5] If we have for principal stresses at a point $\tau_1 =$ 5000 kPa, $\tau_2 =$ 2000 kPa, and $\tau_3 =$ 1000 kPa, draw *Mohr's circle* for the state of stress at the point. What is the maximum shear stress at the point?

> The following series of four problems will introduce the reader, without much difficulty, to an approach to plane stress via the theory of elasticity.

7.25. Express the equations of *equilibrium* for no body forces in the case of plane stress. Now show that if the stresses are defined in terms of the function $\Phi(x,y)$ as follows:

$$\tau_{xx} = \frac{\partial^2 \Phi}{\partial y^2} \qquad \tau_{yy} = \frac{\partial^2 \Phi}{\partial x^2} \qquad \tau_{xy} = \frac{\partial^2 \Phi}{\partial x \partial y}$$

these equations of equilibrium are identically satisfied. Φ is called the *Airy function.*

7.26. Consider an isotropic material with E, G, and ν constant throughout. Start with the following equation of *compatibility* given from Eq. (3.13) for plane stress:

$$\frac{\partial^2 \varepsilon_{xx}}{\partial y^2} + \frac{\partial^2 \varepsilon_{yy}}{\partial x^2} = \frac{\partial^2 \gamma_{xy}}{\partial x \partial y} \tag{1}$$

Replace the strains using *Hooke's law.* Replace G in terms of E and ν. Now go to the equations of *equilibrium* for plane stress for no body force. Get $\partial^2\tau_{xy}/\partial x \partial y$ in terms of a second derivative of τ_{xx} and also get $\partial^2\tau_{xy}/\partial x\partial y$ in terms of a second derivative of τ_{yy}. Show that

$$\frac{\partial^2 \tau_{xy}}{\partial x \partial y} = -\frac{1}{2}\left(\frac{\partial^2 \tau_{xx}}{\partial x^2} + \frac{\partial^2 \tau_{yy}}{\partial y^2}\right)$$

Substitute for $\partial^2\tau_{xy}/\partial x\partial y$ from Eq. (1) and Hooke's law and show that

$$\left(\frac{\partial^2}{\partial x^2} + \frac{\partial^2}{\partial y^2}\right)(\tau_{xx} + \tau_{yy}) = 0$$

This becomes, using the *Laplacian operator* ∇^2 defined for two dimensions as $(\partial^2/\partial x^2 + \partial^2/\partial y^2)$,

$$\nabla^2(\tau_{xx} + \tau_{yy}) = 0 \tag{2}$$

This is the well-known *Laplace equation* for the function $(\tau_{xx} + \tau_{yy})$, which is then termed a *harmonic function.*

7.27. Given the following state of plane stress:

$$\tau_{xx} = 20{,}000 \text{ psi}$$
$$\tau_{yy} = 40{,}000 \text{ psi}$$
$$\tau_{xy} = -10{,}000 \text{ psi}$$

(a). What are the principal stresses?
(b). What are the stresses for axes $x'y'$ rotated 10° clockwise from xy?
(c). If $E = 30 \times 10^6$ psi and $\nu = .3$, what are the principal strains in the xy plane?

Use *Mohr's circle.*

7.28. Given (see Fig. P.7.28).

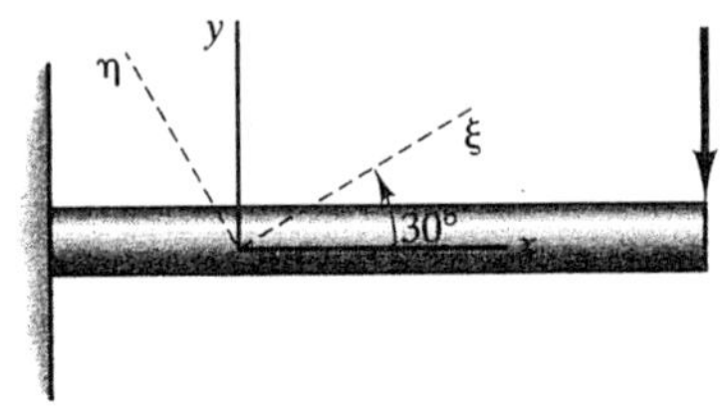

Figure P.7.28.

$$\tau_{\xi\xi} = 1.2 \times 10^8 \text{ Pa}$$
$$\tau_{\eta\eta} = -2 \times 10^8 \text{ Pa}$$
$$\tau_{\xi\eta} = -2 \times 10^8 \text{ Pa}$$

find τ_{xx}, τ_{yy}, *and* τ_{xy} using *Mohr's circle* as an aid.

7.29. Given the following nonzero stresses at a point,

$$\tau_{xx} = -3000 \text{ psi}$$
$$\tau_{yy} = 4000 \text{ psi}$$
$$\tau_{xy} = 2000 \text{ psi}$$

find the normal stress for an axis rotated 22.5° clockwise from the x axis. What is the maximum shear stress at the point? Use *Mohr's circle* as an aid.

7.30. A force F is applied uniformly to the sides of a square plate (see Fig. P.7.30) such that on two opposite faces there is tension and on the other pair of opposite faces there is compression. Along what faces in the plate (seen as lines in the diagram) do we have no normal stress? What is the shear stress for such faces?

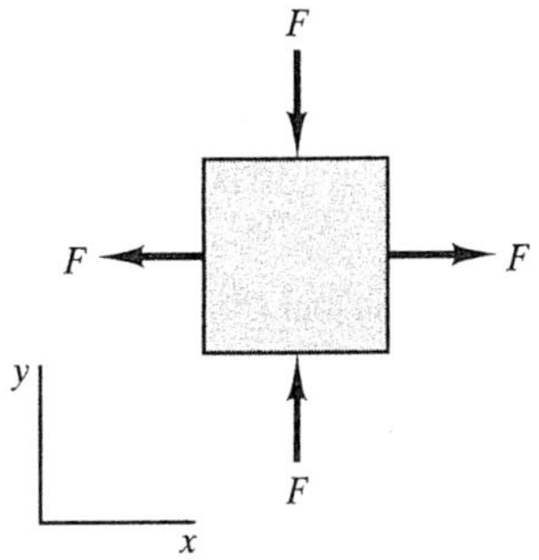

Figure P.7.30.

7.31. A thin cylinder is twisted along its axis by torque T as shown in Fig. P.7.31. What is the stress at interface H on cross section E of the cylinder away from the ends if we assume that the stresses do not vary over thickness t? What are the stresses on interface G on section F away from the ends inclined 45° to the xz plane? Take T = 65 N-m, t = 5 mm, and mean D = 75 mm.

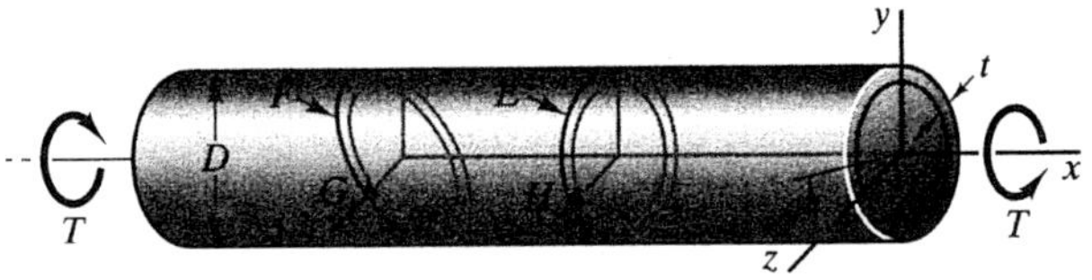

Figure P.7.31.

7.32. Do Problem 7.31 in the case where a tensile axial load of 450 N is also applied at the ends.

7.33. Given stresses

$$\tau_{xx} = 1000 \text{ psi}$$
$$\tau_{yy} = 3000 \text{ psi}$$
$$\tau_{xy} = -1500 \text{ psi}$$

draw *Mohr's circle*. Get the principal stresses. What are the stresses for axes $x'y'$ rotated 20° clockwise from xy using *Mohr's circle* as an aid?

7.34. Given (see Fig. P.7.34)

$$\tau_{xx} = -1.2 \times 10^8 \text{ Pa}$$
$$\tau_{yy} = 2 \times 10^8 \text{ Pa}$$
$$\tau_{xy} = -1 \times 10^8 \text{ Pa}$$

find, using *Mohr's circle* as an aid,

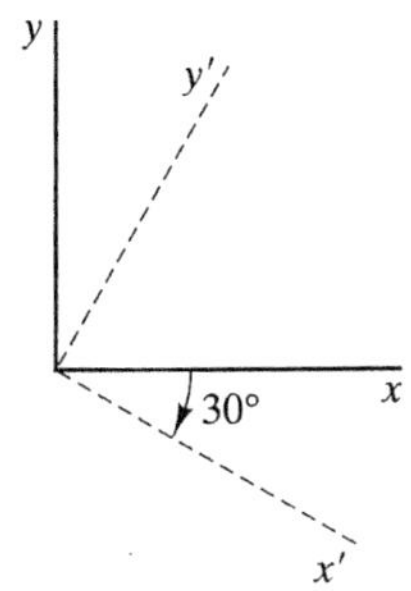

Figure P.7.34.

(a). $\tau_{x'x'}, \tau_{y'y'}, \tau_{x'y'}$.
(b). Principal axes.
(c). Principal stresses.
(d). Extreme shear.

7.35. A thin plate in plane stress is loaded uniformly around its edges as shown in Fig. P.7.35. Determine the principal stresses.

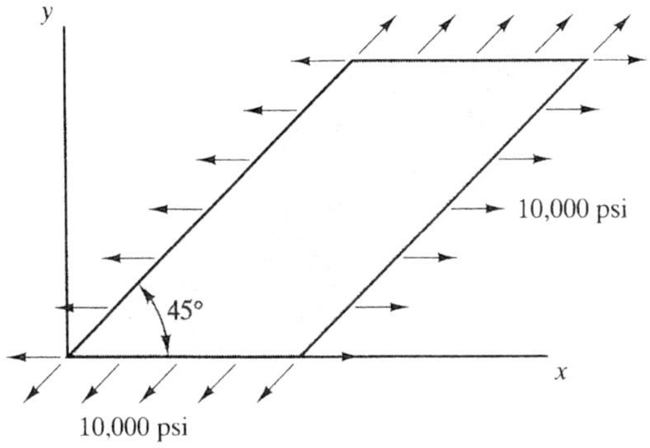

Figure P.7.35.

****7.36.** For a thin-walled pressure vessel with radius r and thickness t (shown in Fig. P.7.36), plot the variation of $\tau_{x'x'}$, $\tau_{y'y'}$, and $\tau_{x'y'}$ for $0 \le \theta \le 90°$ at position A. From your plot, determine the angle that gives the principle values for the stress. Use the simplified formulae for the stress in a thin-walled vessel.

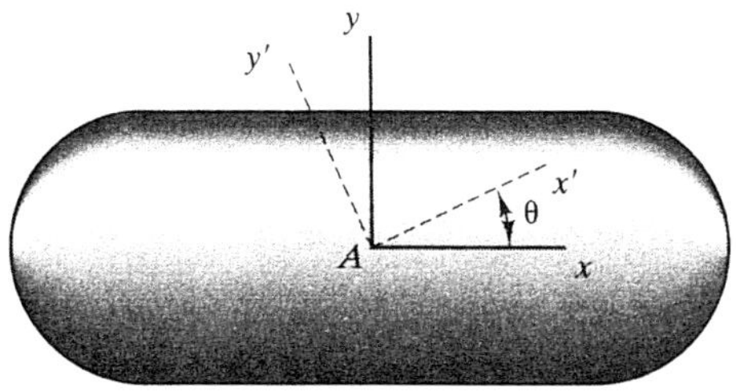

Figure P.7.36.

Project 7 Two-Dimensional Principal Stress Problem

For the following stresses

$$\tau_{xx} = (.05x^{3/2}/y^{1/2}) \times 10^3 \text{ psi} \qquad x \text{ in ft}$$

$$\tau_{yy} = (.02e^{2x} + \ln \sqrt{5y}) \times 10^3 \text{ psi} \qquad y \text{ in ft}$$

$$\tau_{xy} = \sqrt{3x^{1/2} + y^{3/2}} \times 10^3 \text{ psi}$$

and assuming that we have the proper body force distribution for equilibrium, what are the principal stresses in a region

$$0 < x < 11 \text{ ft}$$

$$0 < y < 11 \text{ ft}$$

evaluated in a 10×10 grid of points equally speed in the domain above. Print out as a matrix A_{ij}, where at each point i, j give τ_1, τ_2 for principal stresses in the xy plane.

CHAPTER 8

Plane Strain

8.1 Introduction

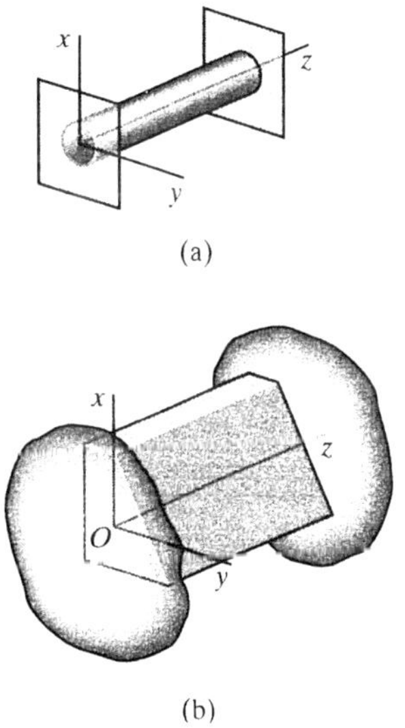

Figure 8.1. Examples of plane strain.

In Chapter 3 we introduced the nine components of the strain tensor ε_{ij} for small deformation. Just as we have done for the stress and the inertia tensor, we now examine the two-dimensional case of strain where the only permissible nonzero strains are ε_{xx}, ε_{yy}, and ε_{xy}. We call this state of strain *plane strain*. For what kind of geometry and loading is plane strain useful? Plane strain portrays the case of a prismatic body constrained not to bend, stretch, or shorten. Furthermore, loads must be applied always in a direction normal to the centerline with no variation allowed in the direction of the axis of the prism taken as the z direction. Examples are shown in Fig. 8.1, where we have shown in (a) a weightless pipe fixed between immovable walls and subject to an external pressure while in (b) we have shown a simple dam with uniform cross section and constrained between immovable barriers. Under these circumstances, it should be clear that the strains *will not depend on the coordinate* z. It is convenient, accordingly, to consider a unit slice of the body, taken normal to the z axis, when discussing strains (Fig. 8.2).

Suppose that we know ε_{xx}, ε_{yy}, and ε_{xy} at some point P in the slice and wish to evaluate strains for a set of axes $x'y'$ rotated about the z axis by an angle θ (positive measure from x to y) as shown in the diagram. In short, what we want are the *transformation equations* for plane strain giving us the components of strain $\varepsilon_{x'x'}$, $\varepsilon_{y'y'}$, and $\varepsilon_{x'y'}$ for a reference $x'y'$ rotated relative to reference xy, for which we have ε_{xx}, ε_{yy}, and ε_{xy}. We shortly will show that these transformation equations are identical to those for plane stress and also for those of moments and product of area. When this has been accomplished, we can consider that plane strain is the two-dimensional subset of a second-order symmetric Cartesian tensor ε_{ij}. What is more, all the properties developed in

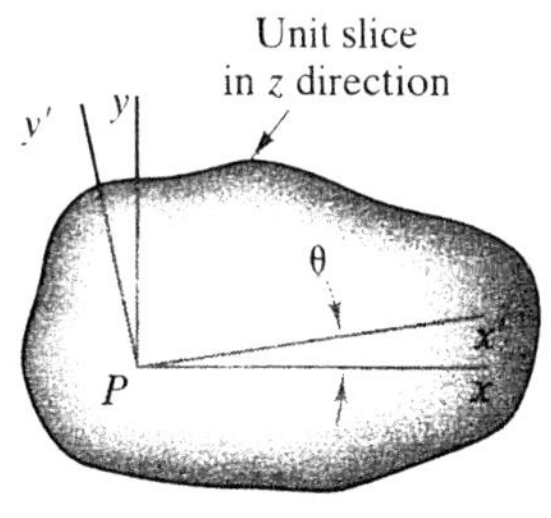

Figure 8.2. Rotation of axes for plane strain.

Chapter 7 for plane stress become immediately applicable to plane strain without further work. We will do this in a manner that will make it easy for you later to develop the full three-dimensional transformation equation for strain.

8.2 A Look Back: Taylor Series and Directional Derivatives

We shall be using *Taylor series* and *directional derivatives* in this chapter and in Chapter 16. For this reason we offer a simple review that will suffice in this text.

Suppose that we wish to determine $f(b)$ in the accompanying plot of $f(x)$ versus x in terms of $f(a)$, at a distance Δx away. We can say first that

$$f(b) = f(a) + eh$$

But eh can be given as

$$eh = eg + gh = (\tan\alpha)\Delta x + gh$$
$$= \left(\frac{df}{dx}\right)_a \Delta x + gh$$

Hence,

$$f(b) = f(a) + \left(\frac{df}{dx}\right)_a \Delta x + gh$$

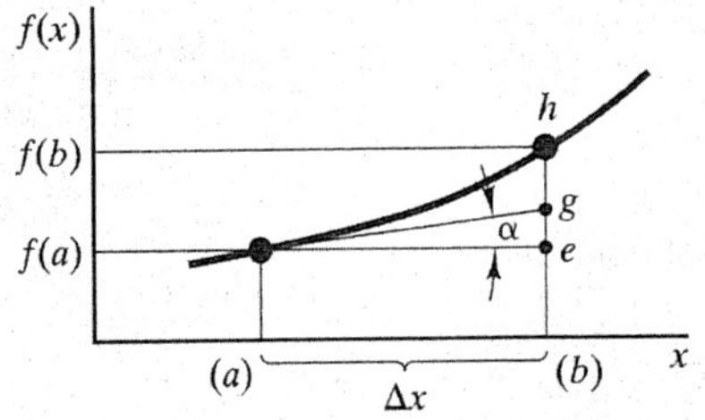

We can further express gh in terms of higher-order derivatives of f with respect to x at a times Δx to higher powers. You may recall that

$$f(b) = f(a) + \left(\frac{df}{dx}\right)_a \Delta x + \frac{1}{2!}\left(\frac{d^2 f}{dx}\right)_a \Delta x^2 + \frac{1}{3!}\left(\frac{d^3 f}{dx^3}\right)_a \Delta x^3 + \cdots$$

This is the so-called Taylor series in *one variable.* We will use this series for variables x and y. The extension of the above for this case is then

$$[f(x,y)]_b = [f(x,y)]_a + \left(\frac{\partial f}{\partial x}\right)_a \Delta x + \left(\frac{\partial f}{\partial y}\right)_a \Delta y + \cdots$$

where we have expressed the series up to first partial derivatives.

As for the *directional derivative,* we are simply using the *chain rule* for differentiation here. Thus, if you have a function $\phi(x,y,z)$ and you wish the partial derivative with respect to a new spatial variable η, you first take $\partial\phi/\partial x$ and multiply by the rate of change of x with respect to η, dx/dn, and then go to $\partial\phi/\partial y$, etc. The result,

$$\frac{\partial\phi}{\partial\eta} = \frac{\partial\phi}{\partial x}\frac{dx}{d\eta} + \frac{\partial\phi}{\partial y}\frac{dy}{d\eta} + \frac{\partial\phi}{\partial z}\frac{dz}{d\eta}$$

is the so-called *directional derivative.*

If $\eta = s$ along a boundary as shown in the accompanying diagram, we have for $\partial\phi/\partial s$

$$\frac{\partial\phi}{\partial s} = \frac{\partial\phi}{\partial x}\frac{dx}{ds} + \frac{\partial\phi}{\partial y}\frac{dy}{ds}$$

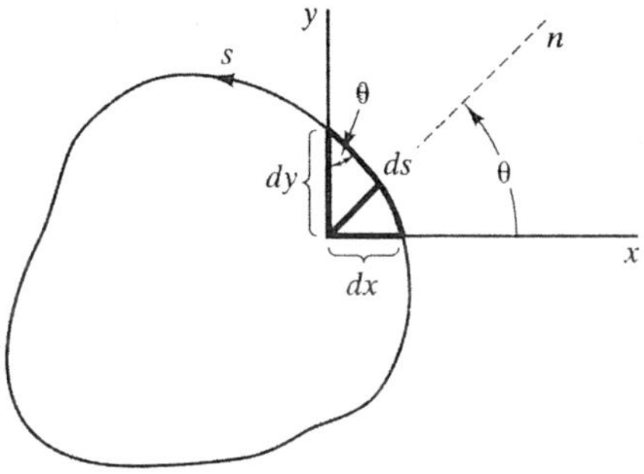

Note that

$$\frac{dy}{ds} = \cos\theta$$

$$\frac{dx}{ds} = -\sin\theta$$

The negative sign above obtains because in moving in the positive s direction in the diagram we are perforce moving in the negative x direction so as to develop $-dx$ to accompany $+ds$.

8.3 Transformation Equations for Plane Strain

Consider a unit slice of a prismatic body in the undeformed geometry in Fig. 8.3 wherein a reference xy is shown at some arbitrary point P in the body. A line segment Δn at an angle θ with x axis is shown connecting P to an arbitrary point Q. Now consider that a loading system has been applied so as to deform the prismatic body in a plane strain manner.

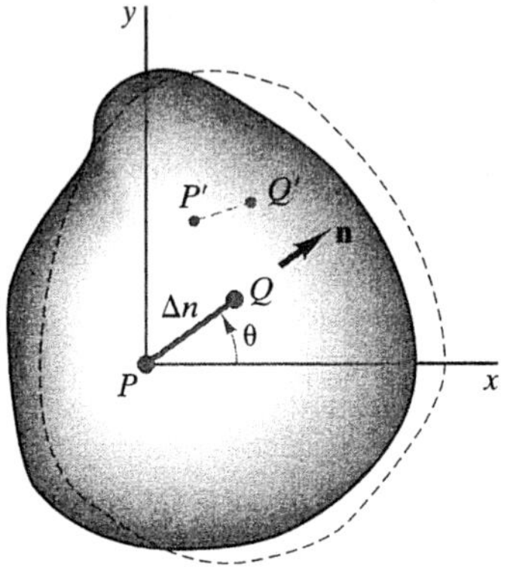

Figure 8.3. Displacement of points P and Q for a plane strain deformation.

The dashed line shows the deformed boundary of the slice and the points P and Q which have moved to P' and Q', respectively.

Let us start by expressing the displacement field component $u_n(x,y)$everywhere in terms of $u_x(x,y)$ and $u_y(x,y)$. Thus,

$$u_n(x,y) = \mathbf{u}(x,y) \cdot \mathbf{n} = u_x(\mathbf{i \cdot n}) + u_y(\mathbf{j \cdot n})$$
$$\therefore u_n(x,y) = u_x \cos \theta + u_y \sin \theta \tag{8.1}$$

We are here just projecting u_x and u_y onto the direction $\mathbf{n}$ to get u_n. Since u_n is a function of x and y, we can express u_n at Q [i.e., $(u_n)_Q$] in terms of u_n and its derivatives at P by using Taylor series expansion. That is,

$$(u_n)_Q = (u_n)_P + \left(\frac{\partial u_n}{\partial x}\right)_P \Delta x + \left(\frac{\partial u_n}{\partial y}\right)_P \Delta y + \cdots \tag{8.2}$$

We next bring $(u_n)_P$ to the left side of the equation and divide by Δn:

$$\frac{(u_n)_Q - (u_n)_P}{\Delta n} = \left(\frac{\partial u_n}{\partial x}\right)_P \frac{\Delta x}{\Delta n} + \left(\frac{\partial u_n}{\partial y}\right)_P \frac{\Delta y}{\Delta n} + \cdots \tag{8.3}$$

Next from Eq. (8.1) substitute for u_n on the right side of Eq. (8.3) to get

$$\frac{(u_n)_Q - (u_n)_P}{\Delta n} = \left[\frac{\partial}{\partial x}(u_x \cos \theta + u_y \sin \theta)\right]_P \frac{\Delta x}{\Delta n}$$
$$+ \left[\frac{\partial}{\partial y}(u_x \cos \theta + u_y \sin \theta)\right]_Q \frac{\Delta y}{\Delta n} + \cdots$$

If we take the limit in the equation above as $\Delta n \to 0$ we get, on remembering that Δx and Δy go to zero as Δn goes to zero,

$$\lim_{\Delta n \to 0} \frac{(u_n)_Q - (u_n)_P}{\Delta n} = \left(\frac{\partial u_x}{\partial x} \cos \theta + \frac{\partial u_y}{\partial x} \sin \theta\right) \frac{dx}{dn} +$$
$$\left(\frac{\partial u_x}{\partial y} \cos \theta + \frac{\partial u_y}{\partial y} \sin \theta\right) \frac{dy}{dn}$$

The left-hand side of the equation should be recognizable as ε_{nn}, the normal strain in the direction $\mathbf{n}$. On the right-hand side of the equation the higher-order terms disappear since each such expression has an excess of Δ's in the numerator over that of the denominator and hence goes to zero in the limit. Finally, the subscript P becomes superfluous as points O and P coalesce in the limit and the ratio $\Delta x/\Delta n$ becomes dx/dn, and so on. From trigonometry we can replace dx/dn by $\cos \theta$ and dy/dn by $\sin \theta$.[1] We thus get

$$\varepsilon_{nn} = \frac{\partial u_x}{\partial x} \cos^2 \theta + \frac{\partial u_y}{\partial x} \sin \theta \cos \theta + \frac{\partial u_x}{\partial y} \cos \theta \sin \theta + \frac{\partial u_y}{\partial y} \sin^2 \theta \tag{8.4}$$

From our strain-displacement relations of Chapter 3 [see Eq. (3.12)] we can write Eq. (8.4) as

[1]See the last figure in Section 8.2 to justify these results.

$$\varepsilon_{nn} = \varepsilon_{xx} \cos^2\theta + \varepsilon_{yy} \sin^2\theta + 2\varepsilon_{xy} \sin\theta \cos\theta \tag{8.5}$$

Looking back at Eq. (7.1) we see that this equation and Eq. (8.5) are identical in form, so using trigonometric identities we can say that

$$\varepsilon_{nn} = \frac{\varepsilon_{xx} + \varepsilon_{yy}}{2} + \frac{\varepsilon_{xx} - \varepsilon_{yy}}{2} \cos 2\theta + \varepsilon_{xy} \sin 2\theta \tag{8.6}$$

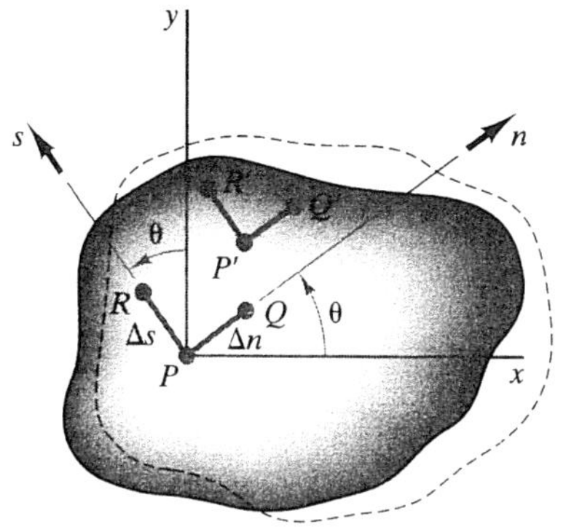

Figure 8.4. Displacement of line segments PR and PQ originally orthogonal for plane strain.

Now consider two arbitrary orthogonal directions at point P (Fig. 8.4) denoted as directions n and s. Line segments PQ and PR of lengths Δn and Δs, respectively, are shown along these directions in the undeformed state. When the body deforms, the endpoints of these segments move to R', Q', P', and there is most likely a change in right angles, giving rise to engineering shear strain γ_{ns}. It will be convenient for us to consider n and s as coordinate axes at this time. Accordingly, from our work in Section 3.4 on strain displacement relations, we can give ε_{ns} as follows:

$$\varepsilon_{ns} = \frac{1}{2}\left(\frac{\partial u_n}{\partial s} + \frac{\partial u_s}{\partial n}\right) \tag{8.7}$$

We can express u_n and u_s in terms of displacement components in the coordinate directions as we did earlier by expanding $\mathbf{u}\cdot\mathbf{n}$ and $\mathbf{u}\cdot\mathbf{s}$ respectively, thus

$$u_n = u_x \cos\theta + u_y \sin\theta \tag{8.8a}$$

$$u_s = -u_x \sin\theta + u_y \cos\theta \tag{8.8b}$$

Substituting the results above into Eq. (8.7), we have

$$\varepsilon_{ns} = \frac{1}{2}\left[\frac{\partial}{\partial s}(u_x \cos\theta + u_y \sin\theta) + \frac{\partial}{\partial n}(-u_x \sin\theta + u_y \cos\theta)\right] \tag{8.9}$$

From Section 8.2, we see that $\partial/\partial s$ and $\partial/\partial n$ are *directional derivatives* defined as follows in two dimensions:

$$\frac{\partial}{\partial s} = \frac{\partial}{\partial x}\left(\frac{dx}{ds}\right) + \frac{\partial}{\partial y}\left(\frac{dy}{ds}\right) = -\sin\theta \frac{\partial}{\partial x} + \cos\theta \frac{\partial}{\partial y} \tag{8.10a}$$

$$\frac{\partial}{\partial n} = \frac{\partial}{\partial x}\left(\frac{dx}{dn}\right) + \frac{\partial}{\partial y}\left(\frac{dy}{dn}\right) = \cos\theta \frac{\partial}{\partial x} + \sin\theta \frac{\partial}{\partial y} \tag{8.10b}$$

Substituting the directional derivative formulations in Eqs. (8.10) into Eq. (8.9), we get the following:

$$\varepsilon_{ns} = \frac{1}{2}\left(-\frac{\partial u_x}{\partial x} \sin\theta \cos\theta + \frac{\partial u_x}{\partial y} \cos^2\theta - \frac{\partial u_y}{\partial x} \sin^2\theta + \frac{\partial u_y}{\partial y} \cos\theta \sin\theta\right.$$

$$\left.-\frac{\partial u_x}{\partial x} \cos\theta \sin\theta - \frac{\partial u_x}{\partial y} \sin^2\theta + \frac{\partial u_y}{\partial x} \cos^2\theta + \frac{\partial u_y}{\partial y} \sin\theta \cos\theta\right)$$

$$\therefore\ \varepsilon_{ns} = -\frac{\partial u_x}{\partial x}\sin\theta\cos\theta + \frac{1}{2}\left(\frac{\partial u_y}{\partial x}\cos^2\theta - \frac{\partial u_y}{\partial x}\sin^2\theta + \frac{\partial u_x}{\partial y}\cos^2\theta - \frac{\partial u_x}{\partial y}\sin^2\theta\right) + \frac{\partial u_y}{\partial y}\sin\theta\cos\theta$$

This becomes, on inserting proper strains on the right side of the equation and rearranging terms,

$$\varepsilon_{ns} = -\varepsilon_{xx}\sin\theta\cos\theta + \varepsilon_{yy}\sin\theta\cos\theta + \varepsilon_{xy}(\cos^2\theta - \sin^2\theta) \quad (8.11)$$

Now go back to Eq. (7.2) and compare this equation with the equation above. *Voilà*—they are the same! Hence, using trigonometric identities, we get for ε_{ns}

$$\varepsilon_{ns} = \frac{\varepsilon_{yy} - \varepsilon_{xx}}{2}\sin 2\theta + \varepsilon_{xy}\cos 2\theta \quad (8.12)$$

Now letting n and s represent x' and y' axes, we get the familiar transformation equations of Chapter 7 that characterize the two-dimensional subset of symmetric, second-order Cartesian tensors.

$$\varepsilon_{x'x'} = \frac{\varepsilon_{xx} + \varepsilon_{yy}}{2} + \frac{\varepsilon_{xx} - \varepsilon_{yy}}{2}\cos 2\theta + \varepsilon_{xy}\sin 2\theta \quad (8.13a)$$

$$\varepsilon_{y'y'} = \frac{\varepsilon_{xx} + \varepsilon_{yy}}{2} - \frac{\varepsilon_{xx} - \varepsilon_{yy}}{2}\cos 2\theta - \varepsilon_{xy}\sin 2\theta \quad (8.13b)$$

$$\varepsilon_{x'y'} = \frac{\varepsilon_{yy} - \varepsilon_{xx}}{2}\sin 2\theta + \varepsilon_{xy}\cos 2\theta \quad (8.13c)$$

8.4 Properties of Plane Strain

We can now simply extend the results of Chapter 7 on plane stress to plane strain without further ado. Thus, to locate principal axes in the xy plane, we may employ the equation

$$\tan 2\tilde{\theta} = \frac{2\varepsilon_{xy}}{\varepsilon_{xx} - \varepsilon_{yy}} \quad (8.14)$$

which is the analog of Eq. (7.13). The principal strains are then found by substituting the appropriate values of $\tilde{\theta}$ into Eqs. (8.13) or they may be found from the analog of Eqs. (7.14), which we state below:

$$\varepsilon_{a,b} = \frac{\varepsilon_{xx} + \varepsilon_{yy}}{2} \pm \sqrt{\left(\frac{\varepsilon_{xx} - \varepsilon_{yy}}{2}\right)^2 + \varepsilon_{xy}^2} \quad (8.15)$$

Example 8.1

Given the following state of plane strain at a point,

$$\begin{aligned} \varepsilon_{xx} &= .002 \\ \varepsilon_{yy} &= -.001 \\ \varepsilon_{xy} &= .003 \end{aligned}$$

what are the principal strains and their directions?

We may employ Eq. (8.14) as follows:

$$\tan 2\tilde{\theta} = \frac{2\varepsilon_{xy}}{\varepsilon_{xx} - \varepsilon_{yy}} = \frac{(2)(.003)}{(.002) - (-.001)} = 2$$

Hence,

$$2\tilde{\theta} = 63.4°, 243.4°$$

$$\therefore \tilde{\theta} = 31.7°, 121.7° \qquad \text{(a)}$$

Substituting $\theta = 31.7°$ into Eqs. (8.13a) and (8.13b) will give us the principal strains:

$$\varepsilon_a = \frac{(.002) + (-.001)}{2} + \frac{(.002) - (-.001)}{2} \cos 63.4° + .003 \sin 63.4°$$

$$\varepsilon_a = .00385$$

$$\varepsilon_b = \frac{(.002) + (-.001)}{2} - \frac{(.002) - (-.001)}{2} \cos 63.4° - .003 \sin 63.4°$$

$$\text{(b)}$$

$$\varepsilon_b = -.00285$$

As a check, notice that $\varepsilon_a + \varepsilon_b = .001 = \varepsilon_{xx} + \varepsilon_{yy}$.

We can also compute the principal strains directly from Eq. (8.15). Thus,

$$\varepsilon_{a,b} = \frac{(.002) + (-.001)}{2} \pm \sqrt{\left[\frac{(.002) - (-.001)}{2}\right]^2 + .003^2}$$

$$= .0005 \pm .00335 \qquad \text{(c)}$$

$$\varepsilon_{a,b} = .00385, -.00285$$

It should be clear, since $\varepsilon_{xz} = \varepsilon_{yz} = 0$, that the z axis is the third principal axis with the third principal strain as zero.

Since the transformation equations are the same, we may employ *Mohr's circle* for plane strain exactly as was done for plane stress. Thus, we form a strain plane (as we did earlier for a stress plane) with normal strain as the abscissa and shear strain as the ordinate. As in the case of plane stress, we must formulate a proper sign convention for plotting shear strains. Consider an element having a shear angle γ_{xy}. If γ_{xy} is positive, according to the general convention presented earlier there is a decrease in the angle between segment dx and dy at a point, as shown in Fig. 8.5(a). A negative shear angle γ_{xy} indicates an increase in the angle between dx and dy, as shown in Fig. 8.5(b). For Mohr's circle we imagine the element positioned so that the angles α and β are equal. Then we plot ε_{xy} as positive if line segment dx rotates away from y (clockwise here) in forming α and negative if line segment dx rotates toward y (counterclockwise here). Also, we plot ε_{yx} as positive if line segment dy rotates toward x (clockwise here) to form β and negative if line segment dy rotates away from x (counterclockwise here). This is shown diagrammatically in Fig. 8.6.

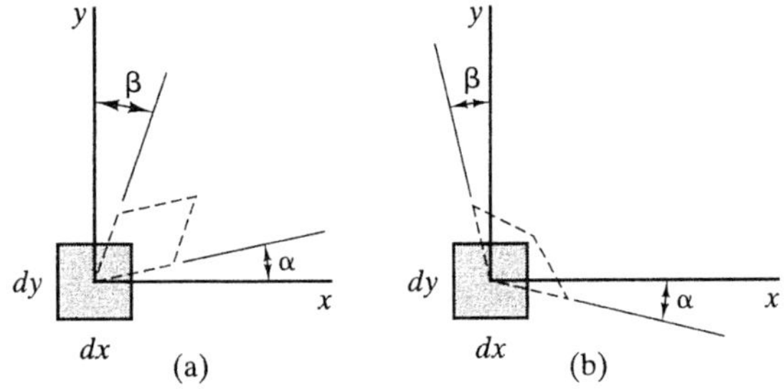

Figure 8.5. Positive and negative shear strain.

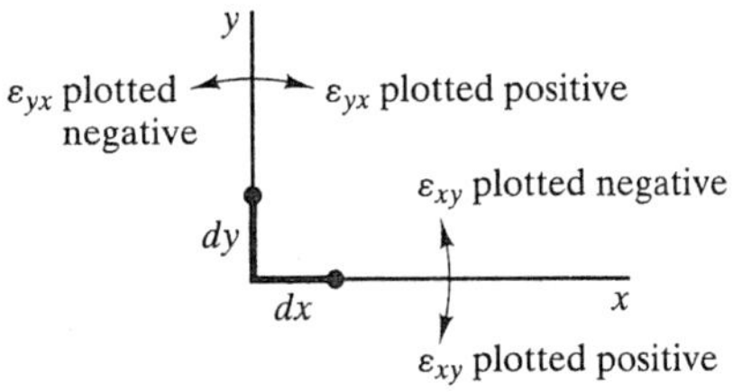

Figure 8.6. Sign convention for Mohr's circle.

The following example illustrates the construction and use of Mohr's circle for strain.

Example 8.2

Draw Mohr's circle for a state of plane strain given as follows:

$$\varepsilon_{xx} = 500 \times 10^{-6} \qquad \varepsilon_{yy} = 1000 \times 10^{-6} \qquad \gamma_{xy} = 600 \times 10^{-6}$$

Find strains for axes rotated 30° counterclockwise from xy.

We first examine segment dx at the point. We set off on the normal strain axis the value 500×10^{-6}, while for the shear strain we consider the quantity $\varepsilon_{xy} = 300 \times 10^{-6}$. The sign is established by consulting Fig. 8.7 and noting, for a decreasing angle as a result of the basic sign convention for ε_{xy}, that we have counterclockwise rotation of segment dx toward y. Thus, we plot -300×10^{-6} on the shear strain axis. We thus arrive at a point which we denote as x (see Fig. 8.8). For the line segment dy we plot $\varepsilon_{yy} = 1000 \times 10^{-6}$ on the

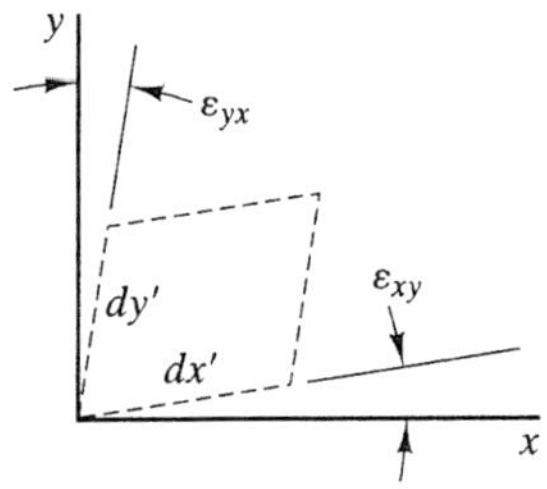

Figure 8.7. Physical plane.

Example 8.2 (Continued)

normal strain axis and, consulting Fig. 8.7 again, since the line segment rotates clockwise, we plot ε_{yx} as $+300 \times 10^{-6}$ on the shear strain axis, establishing point y. Connecting these points, we find the center O for Mohr's circle. We can then draw *Mohr's circle.*

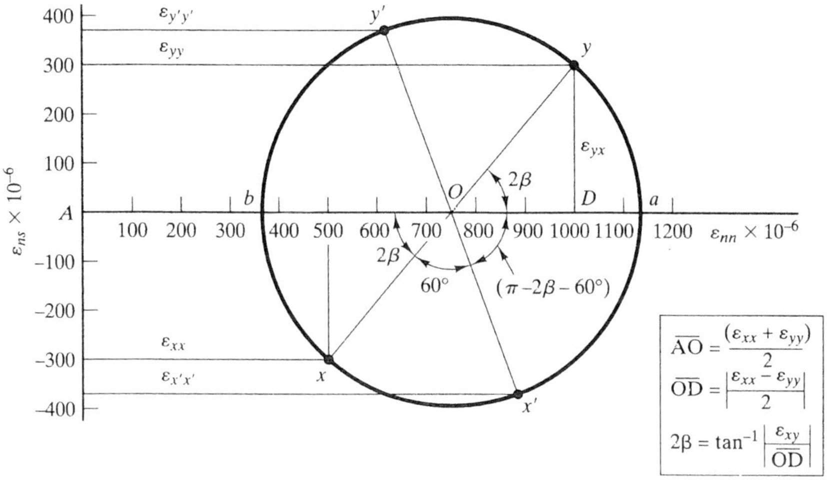

Figure 8.8. Mohr's circle.

To find the normal and shear strains $\varepsilon_{x'x'}$ and $\varepsilon_{x'y'}$ for segments along axes rotated by an angle of 30° counterclockwise from xy (see Fig. 8.9), we rotate in the same direction an angle of 60° from $\overline{Ox}$ in the strain plane. We thus establish a point which we denote as x'. The strain $\varepsilon_{x'x'}$ is seen to be about 880×10^{-6}, while the shear strain $\varepsilon_{x'y'}$ is read off as about -380×10^{-6}. The result for shear deformation has been shown in Fig. 8.9, taking into account the shear strain sign convention for Mohr's circle. From Fig. 8.9 we can now conclude that (approximately) $\gamma_{x'y'} = +760 \times 10^{-6}$, in accordance with our general basic sign convention for shear strain.

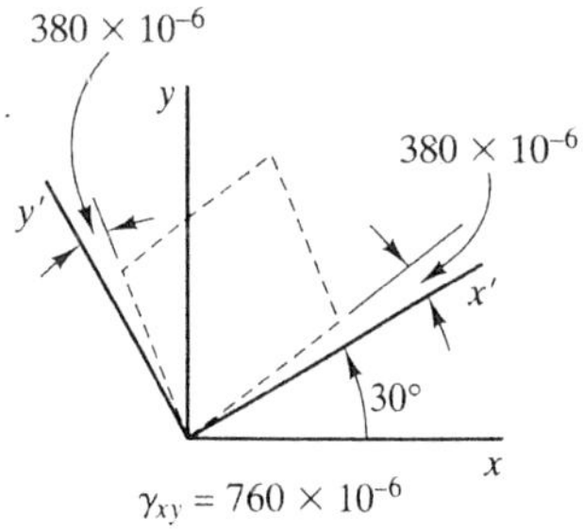

Figure 8.9. Deformation for $x'y'$ plane.

As pointed out in our discussions of Mohr's circle in connection with plane stress, we generally use simple trigonometric considerations taken in connection with Mohr's circle to find $\varepsilon_{x'x'}$ and $\varepsilon_{x'y'}$ rather than depend on reading off these values from the plot. [Note the formulations of $\overline{AO}$, $\overline{OD}$, and 2β in Fig. (8.8).] Thus, the radius of the circle is given as (see Fig. 8.8)

$$r = [\overline{OD}^2 + \varepsilon_{xy}^2]^{1/2} = \left[\left(\frac{1000 - 500}{2}\right)^2 + (300)^2\right]^{1/2} \times 10^{-6}$$

$$= 390 \times 10^{-6}$$

Example 8.2 (Continued)

Also, (noting formulations for $\overline{AO}$, $\overline{OD}$, and 2β in Fig. (8.8))

$$\tan 2\beta = \frac{\varepsilon_{xy}}{\overline{OD}} = \frac{300}{(\varepsilon_{yy} - \varepsilon_{xx})/2} = \frac{300}{250} = 1.200$$

$$\therefore 2\beta = 50.2^\circ$$

Hence,

$$\varepsilon_{x'x'} = \overline{AO} + r\cos(\pi - 2\beta - 60^\circ)$$

$$= \left[\frac{1000 + 500}{2} + 390\cos(69.8^\circ)\right] \times 10^{-6}$$

$$\varepsilon_{x'x'} = 884 \times 10^{-6}$$

$$\varepsilon_{y'y'} = \overline{AO} - r\cos(\pi - 2\beta - 60^\circ)$$

$$\varepsilon_{y'y'} = 615 \times 10^{-6}$$

Also,

$$\varepsilon_{x'y'} = -r\sin(\pi - 2\beta - 60^\circ)$$

$$= -(390)\sin 69.8^\circ \times 10^{-6}$$

$$\varepsilon_{x'y'} = -366 \times 10^{-6}$$

As pointed out earlier, the proper general convention sign for $\varepsilon_{x'y'}$ is positive (see Fig. 8.9, showing the approximate value of $\varepsilon_{x'y'}$ as read off Mohr's circle diagram).

The principal strains are seen from the diagram to be about 1150×10^{-6} and 360×10^{-6}. The third principal strain is $\varepsilon_{zz} = 0$. More accurately, we can say that

$$\varepsilon_a = \overline{AO} + r = \left(\frac{1000 + 500}{2} + 390\right) \times 10^{-6}$$

$$\varepsilon_a = 1140 \times 10^{-6}$$

$$\varepsilon_b = \overline{AO} - r = \left(\frac{1000 + 500}{2} - 390\right) \times 10^{-6}$$

$$\varepsilon_b = 360 \times 10^{-6}$$

To get to the principal axis for ε_a we rotate counterclockwise on Mohr's circle an angle of $(\pi - 2\beta) = 129.8^\circ$ from $\overline{Ox}$. Hence, in the

Example 8.2 (Continued)

physical diagram we rotate counterclockwise from the x axis an angle of 64.9°. The other principal axis is 90° from the aforementioned axis.

8.5 A Pertinent Comment

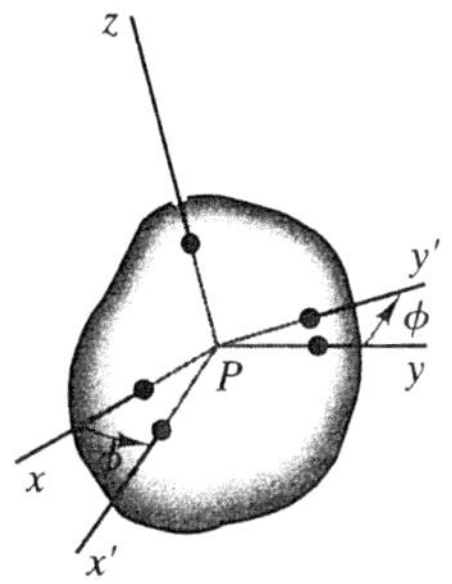

Figure 8.10. Rotation of axes at P.

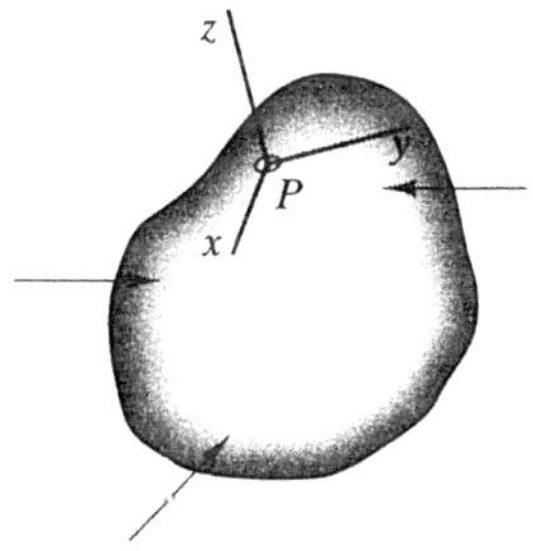

Figure 8.11. Axes tangent and normal to the body surface.

Consider for a moment a general state of strain at a point P inside a body (Fig. 8.10). Choose any plane xy and rotate a set of axes $x'y'$ about the z axis an angle ϕ. Clearly, the transformation equations for normal and shear strains $\varepsilon_{x'x'}$, $\varepsilon_{y'y'}$, and $\varepsilon_{x'y'}$ have the *same form* as Eqs. (8.13). Furthermore, the equation for finding the axes having algebraically the *maximum* and *minimum* normal strains in the xy plane is precisely Eq. (8.14). However, the axes found by such a computation are *not* necessarily principal axes at the point—they are merely axes of extreme normal strain compared to all other normal strains in the xy plane.

We shall now consider the state of strain at a point P on the *surface* of a body. We set up a reference xyz at the point (see Fig. 8.11) such that the z axis is normal to the tangent surface at point P. We can use Eqs. (8.13) and (8.14) to find the maximum and minimum normal strains in the xy plane as discussed in the previous paragraph. Now for a Hookean material, shear stress τ_{zx} is proportional to shear strain γ_{zx}. This is similarly true for the other shear stresses and corresponding shear strains. Consequently, if there is no external loading tangent to the surface at point P, we can conclude that the shear strains γ_{zx} and γ_{zy} must be zero at this point. This means that a line segment dz at point P undergoes no rotation toward either the x axis or the y axis. Actually, it can only elongate or shorten. Hence, the z axis must be a *principal axis* at point P The other principal axes for strain at point P must then lie in the xy plane (i.e., the plane of the tangent surface at point P). Thus Eqs. (8.13) and (8.14) *now* give the other two *principal strains.*

However, you should *not* conclude that we have plane strain at point P; the strain ε_{zz} *need not be zero* even though $\tau_{zz} = 0$. To see this, merely consider Hooke's law for ε_{zz}[2].

$$\varepsilon_{zz} = \frac{1}{E}[-\nu(\tau_{xx} + \tau_{yy})]$$

Clearly, we have a case of plane stress here. Using Eqs. (8.13) and (8.14) we then can get more than just the extreme strains in the xy plane—we actually get the values of the *two principal strains in the xy*

[2]Note that the stresses τ_{xx} and τ_{yy}, as we approach the surface of the body from the inside, will generally not be zero and so the strain ε_{zz} will generally not be zero thus precluding the state of plane strain.

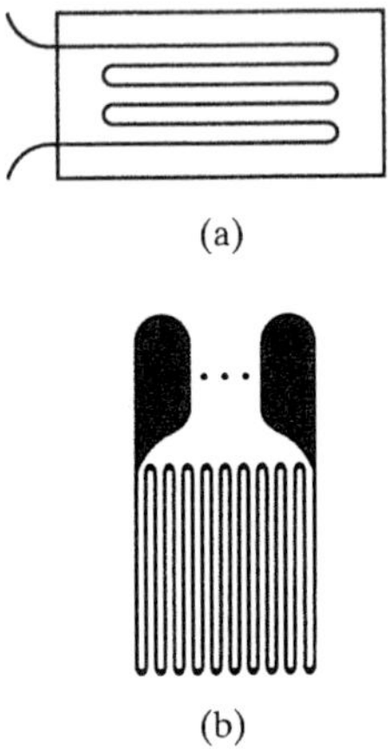

Figure 8.12. Strain gages of two types, wire and foil.

plane and their directions. The third principal strain, as pointed out, is in the z direction.

We shall see how we make good use of these results in the following section when we discuss strain gages.

8.6 Strain Gages

The *strain gage* is the most common device for measuring strain. It may be constructed from a thin wire filament (about .001 in. in diameter) wound so as to orient the bulk of the length of the wire in a single direction [see Fig. 8.12(a)]. Such gages are called *wire* gages to distinguish them from another class of gages called *foil* gages having a similar grid but constructed by photoetching methods from a metal foil having a thickness of .0002 in. [see Fig. 8.12(b)]. The wire or foil is usually bonded to thin paper or plastic. This paper or plastic is then firmly glued to the surface of the body where we are interested in strain. The gage deforms with the body when loads are applied, and if there is strain in the direction of the gage (i.e., along the primary directon of the wires), there results a slight change of electrical resistance of the gage from the change in length of the wire or foil. Using a Wheatstone bridge this slight change in resistance can be measured, and from this the normal strain in the direction of the gage ascertained. With proper protection, gages can be used underwater or in regions of elevated temperature (up to 2000°F for special gages).

Since a single gage will yield only a normal strain in the direction of the gage, we must employ clusters of gages or strain *rosettes* to give pertinent information about the state of strain (such as principal strains) at a point. An arrangement such as shown in Fig. 8.13(a) is called a *rec-*

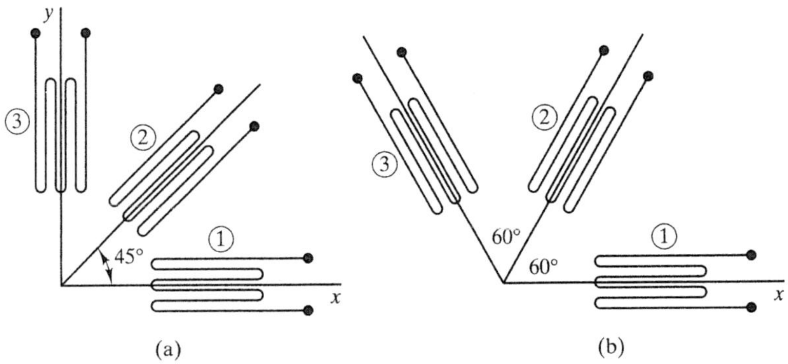

Figure 8.13. (a) Rectangular rosette; (b) equiangular rosette.

tangular rosette, while the arrangement shown in Fig. 8.13(b) is called an *equiangular rosette.* It should be understood that the rosette must be small enough relative to the body in order that the measurements correspond to "strain at a point" on the surface.

Example 8.3

Consider that for a rectangular rosette [see Fig. 8.13(a)] we get the following measurements for strain on the surface of a body:

$$\text{gage } 1 = .002$$
$$\text{gage } 2 = .001$$
$$\text{gage } 3 = -.004$$

What are the principal strains at the point?

As noted earlier we can use the transformation equations (8.13) for strain in the tangent plane of the rosette. Accordingly, we can say for a rotation θ about xy axes in the tangent plane that

$$\varepsilon_{x'x'} = \frac{\varepsilon_{xx} + \varepsilon_{yy}}{2} + \frac{\varepsilon_{xx} - \varepsilon_{yy}}{2} \cos 2\theta + \varepsilon_{xy} \sin 2\theta \tag{a}$$

We let x correspond to the direction of gage 1 and we let y correspond to the direction of gage 3 [see Fig. 8.13(a)]. We need ε_{xy} in order to get any particular strain in the plane of the rosette. We accordingly use the result for gage 2 with $\theta = 45°$ in Eq. (a) as follows for this purpose:

$$.001 = \frac{.002 + (-.004)}{2} + \frac{.002 - (-.004)}{2} \cos 90° + \varepsilon_{xy} \sin 90°$$

$$\therefore \varepsilon_{xy} = .002 \tag{b}$$

To get the principal strains in the plane of the rosette we may first use Eq. (8.14) to find the direction of the principal axes in the plane of the rosette as per our earlier discussion. We may then employ Eqs. (8.13) to get the principal strains. Thus, we have

$$\tan 2\tilde{\theta} = \frac{(2)(.002)}{.002 + .004} = .667$$

$$\therefore \tilde{\theta}_{a,b} = 16.8°, 106.8° \tag{c}$$

where we are using the subscripts a and b to identify the principal axes in the xy plane. The principal strains in the xy plane are these:

Example 8.3 (Continued)

$$\varepsilon_{a,b} = \frac{\varepsilon_{xx} + \varepsilon_{yy}}{2} + \frac{\varepsilon_{xx} - \varepsilon_{yy}}{2} \cos 2\begin{Bmatrix} \tilde{\theta}_a \\ \tilde{\theta}_b \end{Bmatrix} + \varepsilon_{xy} \sin 2\begin{Bmatrix} \tilde{\theta}_a \\ \tilde{\theta}_b \end{Bmatrix}$$

$$= \frac{.002 - .004}{2} + \frac{.002 + .004}{2} \cos \begin{Bmatrix} 33.6° \\ 213.6° \end{Bmatrix} + .002 \sin \begin{Bmatrix} 33.6° \\ 213.8° \end{Bmatrix} \quad \text{(d)}$$

$$\varepsilon_{a,b} = .00261, \; - .00461$$

To get the *third principal strain,* which must be in the z direction, we employ Hooke's law for normal stresses and strains at the point. Noting that principal axes for stress and strain *coincide* for a Hookean material,[3] we may say, employing subscripts $a,b,$ and c to denote the principal stresses and strains, that

$$\begin{aligned} \varepsilon_a &= \frac{1}{E}[\tau_a - \nu(\tau_b + \tau_c)] \\ \varepsilon_b &= \frac{1}{E}[\tau_b - \nu(\tau_a + \tau_c)] \\ \varepsilon_c &= \frac{1}{E}[\tau_c - \nu(\tau_a + \tau_b)] \end{aligned} \quad \text{(e)}$$

We know ε_a and ε_b from the previous calculations, and with no external loading on the surface at the point of interest we can say that $\tau_{zz} = \tau_c = 0$. We have the three equations with three unknowns. If we take $E = 30 \times 10^6$ psi and $\nu = .3$, we can then determine ε_c. We get

$$\varepsilon_c = .000857 \quad \text{(f)}$$

The magnitude of the strain normal to the surface is very small, as is usually the case, and generally need not be considered. Going to our usual notation for principal strains we can then say that

$$\begin{aligned} \varepsilon_1 &= .00261 \\ \varepsilon_2 &= .000857 \\ \varepsilon_3 &= -.00461 \end{aligned} \quad \text{(g)}$$

[3]For Hookean materials $\tau_{xy} = G\gamma_{xy}$, $\tau_{xz} = G\gamma_{xz}$, and $\tau_{yz} = G\tau_{yz}$. Hence, since principal axes correspond to zero off-diagonal terms of the tensor, principal axes of stress and strain coincide for such materials. This is not true, however, for other materials.

As for the *equiangular rosette* in Fig. 8.13(b), we must use via our transformation equations the known values of ε_2 and ε_3 in order to evaluate ε_{yy} and ε_{xy} at the rosette location so that with these data along with $\varepsilon_{xx} = \varepsilon_1$ we can proceed as in Example 8.3 to get principal strains and other vital information.

8.7 Closure

In Chapters 7 and 8 we have presented the very useful and practical cases of plane stress and plane strain. We shall have ample opportunity to make use of these models as we move on.

When one of your authors as a student first studied stress and strain in his first course in solid mechanics (or strength of materials), he was greatly puzzled by the fact that stress, strain, and second moments and products of area had identical characteristics, yet measured completely different effects. In a wish to spare the reader of this confusion, we have made great efforts in this chapter to explain the mathematical linkage between stress, strain, and moments and products of area at least for the much simpler two-dimensional versions of these quantities. Let us reiterate the key idea in this regard. We found that the *manner* in which the two-dimensional components *change* when axes xy are rotated was sufficient in itself to lead to the concepts of principal stresses and strains; to physically meaningful term groupings that were invariant with respect to a rotation of axes; and of course, to the very useful Mohr circle construction. Admittedly, observing the manner in which components of a set of terms T_{ij} change when you rotate axes to learn vital things about the nature of T_{ij} as it pertains to a science is perhaps quite a novel approach for you. But it can eliminate the consternation alluded to earlier for a thinking student. Furthermore, engineers are making much use of the computer in solid mechanics whereby complex problems of structural mechanics are being solved by the method of *finite elements*, which demands a deeper understanding of fundamentals.[4] The gentle introduction to tensors presented in this and the preceding chapters is a move in this direction and thus has the added benefit of preparing you for more advanced undertakings where the nature of transformation equations is of vital importance.

We shall come back later to stress and strain transformation in three-dimensional states for those wishing a deeper understanding of this vital subject.

We have thus far studied one-dimensional tension and/or compression problems in Chapter 4 and 5, where yielding coupled with a factor of safety were the criteria for design. In the next chapter we shall look more carefully at failure criteria for three-dimensional stress.

[4]See the author's text with coauthor C. L. Dym, entitled *Energy and Finite Element Methods in Structural Mechanics* (Taylor & Francis Inc.) 1985.

Highlights (8)

Plane strain is the two-dimensional subset of the nine strain components. It describes deformation of prismatic bodies having uniform loading oriented normal to the axis such as to cause no bending of the body. Hence, the stresses do not vary along the axis. Accordingly, we may use a unit slice of a plane strain member for consideration. When rotating the axes about the longitudinal axis, the transformation equations are exactly those of plane stress. The formulations for principal strains and axes, the invariants, and Mohr's circle follow exactly as presented for plane stress. When employing Mohr's circle, remember that when dx under deformation rotates away from the y axis, then ε_{xy} is positive on the strain plane and when dy rotates under deformation toward the x axis, ε_{yx} is positive on the strain plane. In short, clockwise rotation of dx and/or dy under deformation yields positive shear strain when plotted on the strain plane, while counterclockwise rotation of dx and/or dy under deformation yields negative shear strain when plotted on the strain plane.

Strain gages give strain only along the axis of the strain gage and thus to get the strains in any direction along a surface of a body, one must use a combination of at least three gages, set at different angles and called a rosette. With the three strain readings, it is then necessary to use the transformation equations to arrive at those strains needed for a full set of values of ε_{xx}, ε_{yy}, and ε_{xy} to be able to get such information as the principal strains, and so on. For Hookean materials, the principal axes for strain coincide with the principal axes of stress. Therefore, for a body free of surface tractions at the location of the strain gage, the normal direction to the surface must be a principal direction of stress and consequently of strain. This normal stress is zero but this normal strain will generally **not** be zero. To get this normal strain, we use Hooke's law with the one known principal stress and with two principal strains found from calculations emerging from the rosette formulations. Thus, at the surface of the body, plane stress conditions exist while plane strain conditions generally do not exist.

PROBLEMS

8.1. [8.3] Expand the function $F(x,y)$, given below, as a Taylor series up to terms having squares and second-order products of Δ terms. Get the function at B in terms of the function and derivative at A where $x = 1$, $y = 3$ rad.

$$F(x,y) = x^3 + x^2 \sin y + 20x$$

8.2. [8.3] Given a function $\phi(x,y,z)$ such that

$$\phi(x,y,z) = 2x^2y + 3yx + (z^2 + y^2)$$

find the directional derivative at (2, 1, 3) in a direction $\mathbf{n}$ given as

$$\mathbf{n} = .3\mathbf{i} + .4\mathbf{j} + .866\mathbf{k}$$

8.3. [8.3] For the dam shown in Fig. 8.1(b), we have the following strains at a point away from the ends:

$$\varepsilon_{xx} = -.0003$$
$$\varepsilon_{yy} = -.0005$$
$$\varepsilon_{xy} = .0001$$

What are the strains for axes $x'y'$ rotated 30° about the z axis in a counterclockwise direction as you look along the z axis toward O?

8.4. [8.4] In Problem 8.3, determine the principal axes and the principal strains in the xy plane.

8.5. [8.4] Consider the thick-walled cylinder of Fig. P.8.4 constrained by immovable walls at the ends and subject to an internal pressure so as to undergo plane strain. It is best here to use cylindrical coordinates. The strains are considered for an element shown on a slice in Fig. P.8.4 as follows:

$\varepsilon_{rr} \equiv$ normal strain in radial direction
$\varepsilon_{\theta\theta} \equiv$ normal strain in transverse direction
$\varepsilon_{r\theta} \equiv$ shear strain between line segments dr and $r\,d\theta$ of the element shown

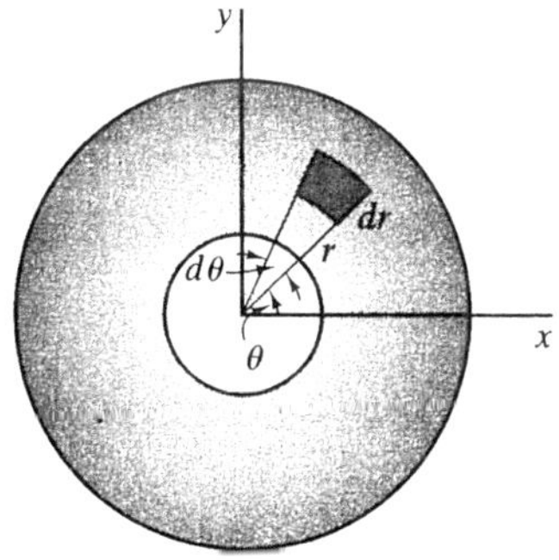

Figure P.8.4.

If the following data are known for a point in the pipe at $\theta = 30°$,

$$\varepsilon_{rr} = -.002$$
$$\varepsilon_{\theta\theta} = .003$$
$$\varepsilon_{r\theta} = .001$$

what are the strains ε_{xx} and ε_{yy} at the point?

8.6. [8.4] In Problem 8.5, what are the principal strains in the xy plane at the point of interest?

8.7. [8.4] Given the following nonzero strains at a point,

$$\varepsilon_{xx} = .0045$$
$$\varepsilon_{yy} = -.0037$$
$$\varepsilon_{xy} = .0046$$

(a). Draw *Mohr's circle.*
(b). Determine the strains at axes rotated 30° clockwise from axes xy.
(c). Show a check for your answer.

8.8. [8.4] A thin-walled cylinder in outer space (see Fig. P.8.8) is rotating at a high speed ω of 500 rpm. It is in a hyperbaric chamber where the pressure is 10 atmospheres (1 atm is 101,325 Pa).

(a). What are the strains ε_{xx} and ε_{yy} at point A on the inside surface?
(b). What are strains ε_{nn} along the direction at 30° to the x axis rotated in a plane at A parallel to the xy plane toward the y axis?

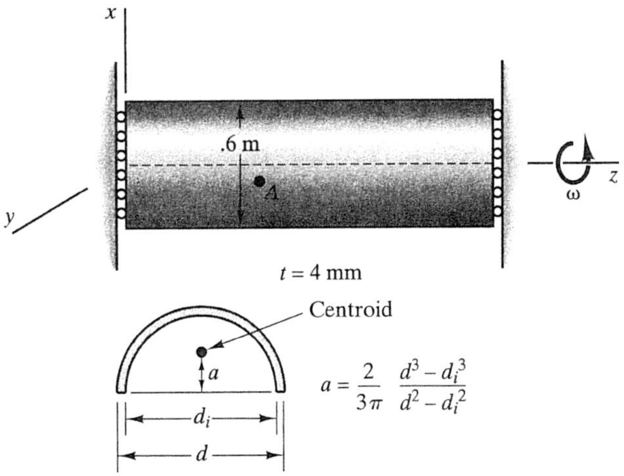

Figure P.8.8.

The cylinder is open at the ends and is fully constrained by the walls shown in the diagram. The density of the cylinder is 5650 kg/m³. Also, $E = 2 \times 10^{11}$ Pa and $\nu = .3$. The thickness of the cylinder is 4 mm. (*Hint:* Look at an infinitesimal rectangular parallelepiped at A and consider stress-strain relations using Hooke's law.)

8.9. [8.4] A slice of a body under plane strain has the following strains:

$$\varepsilon_{xx} = .005 \qquad \varepsilon_{yy} = -.002 \qquad \varepsilon_{xy} = .001$$

What are the principal axes and the principal strains? Use *Mohr's circle* as an aid.

8.10. [8.4] Given

$$\varepsilon_{xx} = -.0003$$
$$\varepsilon_{yy} = .0002$$
$$\varepsilon_{xy} = -.0001$$

Using *Mohr's circle* as an aid, find the principal strains and the strains for axes $x'y'$ rotated 30° counterclockwise from axes xy.

8.11. [8.6] A rectangular strain gage rosette (Fig. P.8.11) gives the following results on the surface of a body:

gage 1 = .002
gage 2 = −.001
gage 3 = −.001

What is the algebraically maximum normal strain in a direction tangent to the surface of the body? What is its direction?

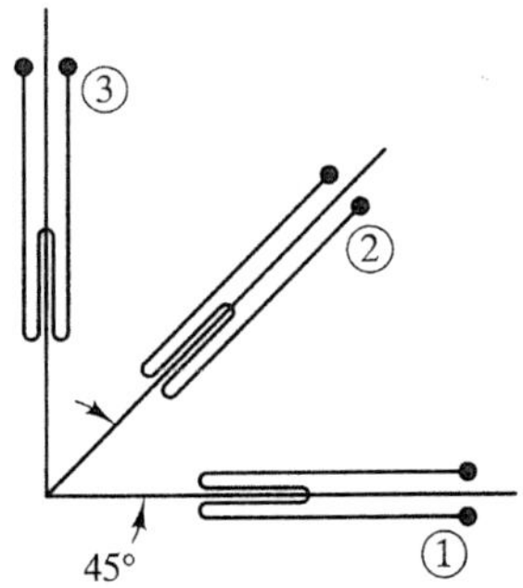

Figure P.8.11.

8.12. [8.6] A rosette gives the following strain readings in the xy plane (see Fig. P.8.12):

gage 1 = .002
gage 2 = .003
gage 3 = −.001

What is ε_{xy}? Get principal strains using formulae and *Mohr's circle.*

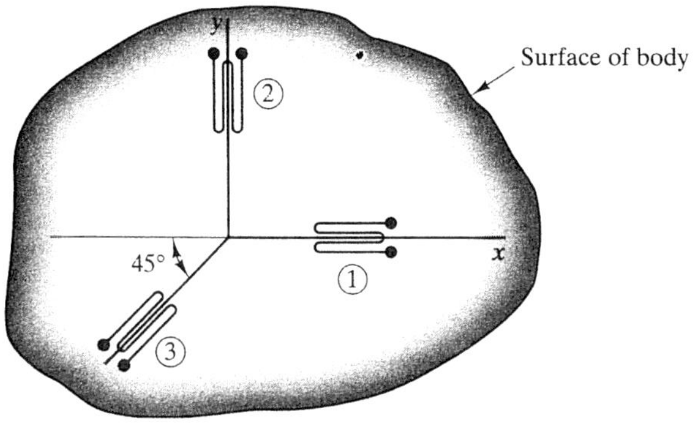

Figure P.8.12.

8.13. [8.6] The following three strains in Fig. P.8.13 are recorded by the strain gages of the rectangular rosette:

$$\varepsilon_1 = .003$$
$$\varepsilon_2 = -.002$$
$$\varepsilon_3 = .004$$

(a). Make necessary calculations to enable you to draw *Mohr's circle.*
(b). Draw *Mohr's circle* and using trigonometry, find principal strains in the *xy* plane and their axes.
(c). For $E = 2 \times 10^{11}$ Pa and $\nu = .3$, what are the principal stresses if the rosette is on a surface exposed just to atmosphere?

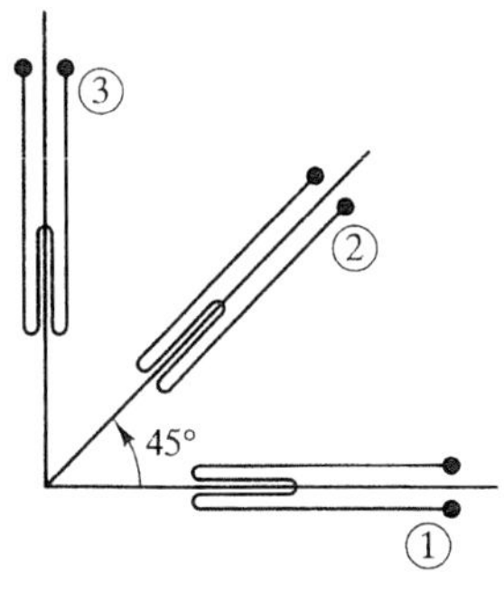

Figure P.8.13.

8.14. [8.6] An equiangular strain gage rosette mounted on the surface of a body (see Fig. P.8.14) gives the following readings:

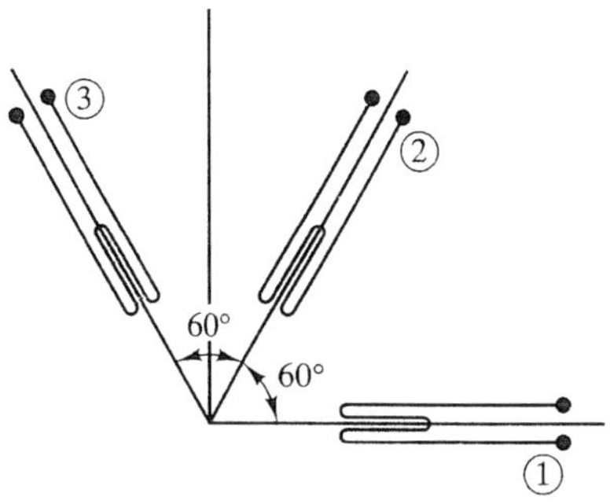

Figure P.8.14.

gage 1 = .001
gage 2 = .003
gage 3 = −.002

What is the algebraically maximum normal strain in a direction tangent to the surface and its direction?

8.15. [8.5] Determine the general relation for the two principal strains in the plane of a rectangular strain gage rosette. State your answer in terms of the measured strains ε_1, ε_2, and ε_3. Show that

$$\varepsilon_{a,b} = \frac{\varepsilon_1 + \varepsilon_2}{2} \pm \sqrt{\frac{1}{2}[(\varepsilon_1 - \varepsilon_2)^2 + (\varepsilon_2 - \varepsilon_3)^2]}$$

gives the correct result. Assume that there is no out-of-plane loading at the location of the rosette and that the material is isotropic.

8.16. [8.6] For the triangular rosette in Fig. P.8.16, the following information is known:

$$\varepsilon_1 = .0050$$
$$\varepsilon_2 = .0035$$
$$\varepsilon_3 = -.0015$$

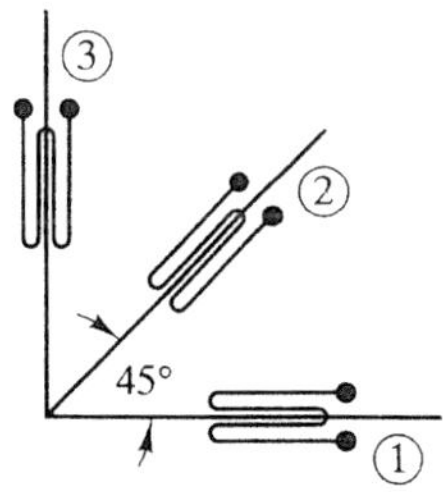

Figure P.8.16.

(a). Using the transformation equations find the principal strains.
(b). Do we have here plane stress or plane strain? Explain.
(c). Set up three equations for determining the principal stresses and the third principal strain.

8.17. A pavement sub-base is modeled as two layers of different materials surrounded by rigid walls as shown in Fig. P.8.17. The top layer fits snugly between the rigid walls. The bottom layer has a 1/2-in. gap on each side. Find the change in height of the two layers when a uniform pressure is applied such that the gap just closes. Assume plane strain and ignore friction. Take $E_1 = 10{,}000$ psi, $v_1 = 1/2$, $E_2 = 6{,}000$ psi, and $v_2 = 1/4$. [*Hint:* First show from Hooke's law for material ① that $\tau_{xx} = \tau_{zz} = -p$ and that $\boldsymbol{\varepsilon}_{yy} = 0$ (i.e., all deformation is in material ②). Taking $\tau_{yy} \approx -p$, show that $p = E_2/199\boldsymbol{v}_2\,(1 + \boldsymbol{v}_2)$ from the constitutive law for $\boldsymbol{\varepsilon}_{xx}$ and $\boldsymbol{\varepsilon}_{zz}$. Now compute $\boldsymbol{\varepsilon}_{yy}$ from the remaining constitutive equations and get the desired result Δ.]

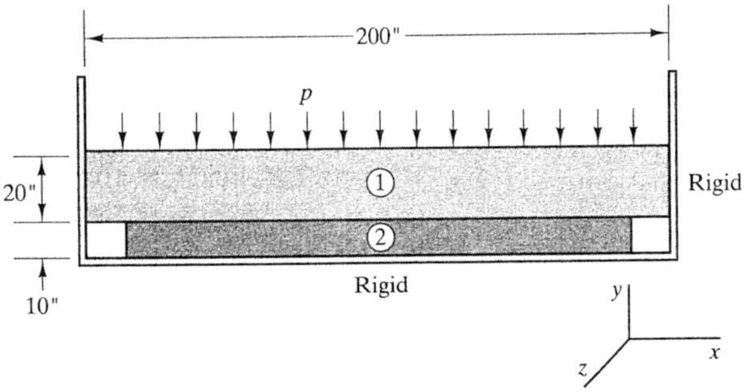

Figure P.8.17.

8.18. Draw *Mohr's circle* and determine the shear strain for axes rotated 30° **clockwise** from the x axis. The following data are given:

$$\varepsilon_{xx} = .002$$
$$\varepsilon_{yy} = -.001$$
$$\varepsilon_{xy} = -.003$$

8.19. A body under plane strain has the following strains:

$$\varepsilon_{xx} = .002 \qquad \varepsilon_{yy} = 0 \qquad \varepsilon_{xy} = -.003$$

What are the strains for axes rotated 30° clockwise from xy? Verify your results on *Mohr's circle*.

8.20. Given the following state of strain,

$$\varepsilon_{ij} = \begin{bmatrix} .005 & .003 & 0 \\ .003 & .001 & 0 \\ 0 & 0 & 0 \end{bmatrix}$$

using *Mohr's circle* as an aid, find the principal strains in the xy plane. Give the directions of the principal strains (all three).

8.21. The strains measured at point A for a beam (Fig. P.8.21) are

$$\varepsilon_{xx} = .003$$
$$\varepsilon_{yy} = -.001$$
$$\varepsilon_{xy} = -.004$$

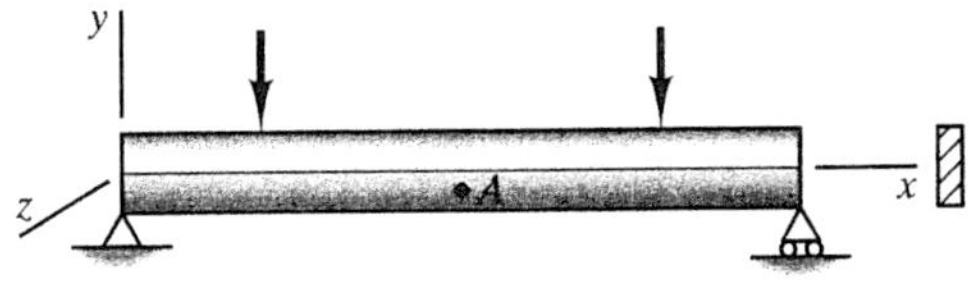

Figure P.8.21.

(a). Using *Mohr's circle* as an aid, compute
1. Principal strains and orientation of principal axes.
2. Strains for axes $x'y'$ at A rotated 30° clockwise.
3. Extreme shear strain at A.

(b). If $E = 2 \times 10^{11}$ Pa and $v = .3$, compute the principal stresses at A.

8.22. In Fig. P.8.22 find the principal strains and principal stresses. Use the transformation equations.

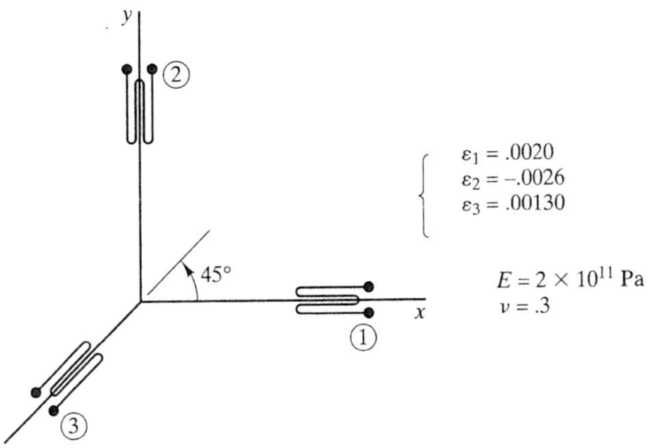

Figure P.8.22.

8.23. Strain gages are installed at various locations and orientations along a railroad rail. As the day warms, thermal strains are induced. Would you expect the thermal strains from the gages to be the same or not? Explain.

8.24. A thin-walled cylinder (see Fig. P.8.24) is rotating at a speed ω of 200 rpm. A rectangular strain rosette mounted on the shaft at A transmits through slip rings the following strain measurements:

$$\varepsilon_1 = .0002$$
$$\varepsilon_2 = .0004$$
$$\varepsilon_3 = -.0001$$

What torque and what power are being transmitted by the cylinder? Take $G = 1 \times 10^{11}$ Pa.

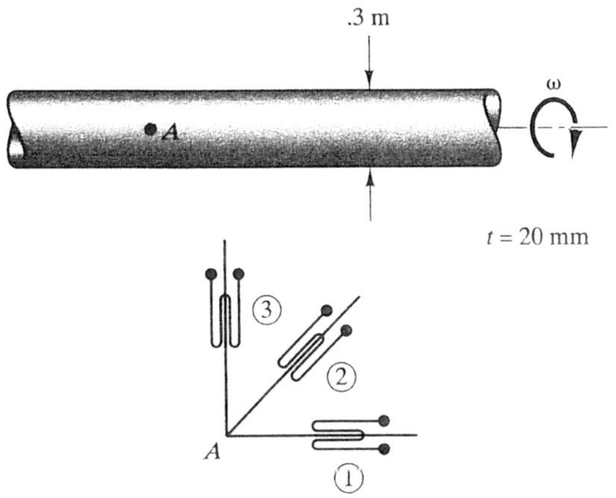

Figure P.8.24.

8.25. Given the following state of strain at a point,

$$\varepsilon_{ij} = \begin{pmatrix} .002 & .003 & .001 \\ .003 & -.002 & .005 \\ .001 & .005 & -.004 \end{pmatrix}$$

what are the maximum and minimum normal strains in the xy plane at the point?

8.26. In Problem 8.25, what are the maximum and minimum normal strains in the yz plane?

8.27. For the strain rosette on a thick-walled pressure vessel under a pressure of 30 psi, we get the indicated readings for the strain gages shown in Fig. P.8.27. What would the maximum longitudinal negative strain be if the pressure in the tank is tripled to 90 psi?

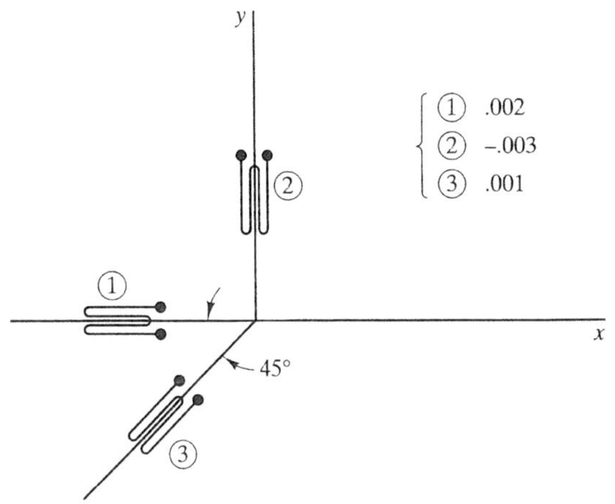

Figure P.8.27.

8.28. Determine ε_{yy} and ε_{xy} from the data taken for a nonconventional strain rosette shown in Fig. P.8.28.

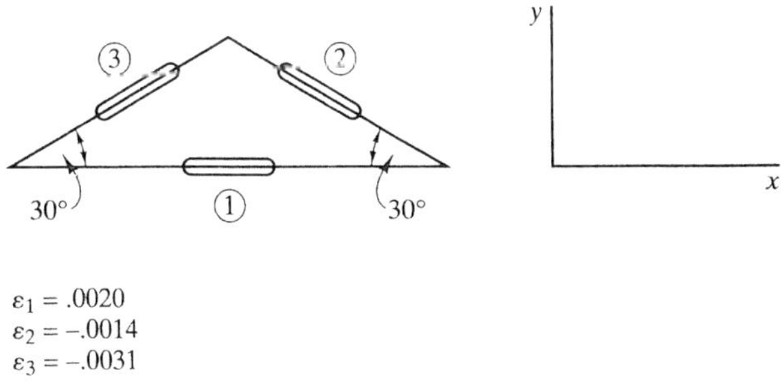

Figure P.8.28.

8.29. Calculate **two principal strains** in Fig. P.8.29 and set up equations (but do not solve) for the third principal strain and the **three principal stresses.** There is negligible pressure on the surface.

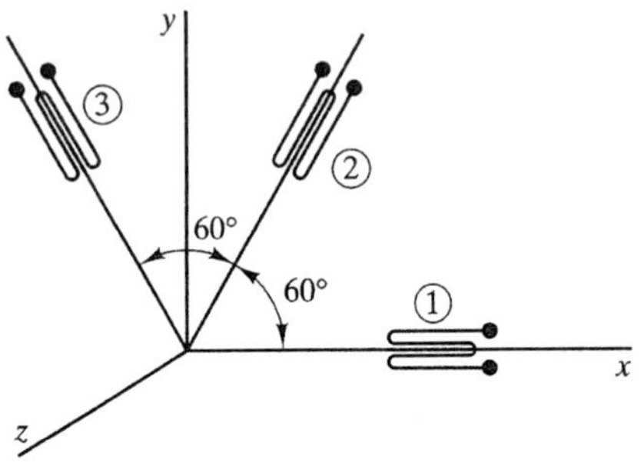

$$\begin{cases} \varepsilon_1 = .002 \\ \varepsilon_2 = .001 \\ \varepsilon_3 = -.004 \\ E = 11 \times 10^{11} \text{ Pa} \\ \nu = .3 \end{cases}$$

Figure P.8.29.

****8.30.** For the strain state given by:

$$\varepsilon_{xx} = -.0003$$
$$\varepsilon_{yy} = -.0005$$
$$\varepsilon_{xy} = 0001$$

plot $\varepsilon_{x'x'}$, $\varepsilon_{y'y'}$, and $\varepsilon_{x'y'}$ for $0° \leq \theta \leq 90°$ where θ is the rotation from x' to x. From your plot, find the angle that $\varepsilon_{x'y'} = 0$ corresponds to. What are the normal strains at this angle? What is the maximum magnitude for ε_{xy}? What are $\varepsilon_{x'x'}$ and $\varepsilon_{y'y'}$ at this rotation?

Project 8: Strain Rosette Problem

In Fig. P.8.30 find the *principal strain values, principal stress values,* and *principal axes* for plane stress at a point for a linear elastic material. The user supplies the following data:

ε_1 α (degrees) $10° < \alpha < 340°$
ε_2 β (degrees) $20° < \beta < 320°$
ε_3 E, modulus of elasticity (Pa)
ν, Poisson's ratio

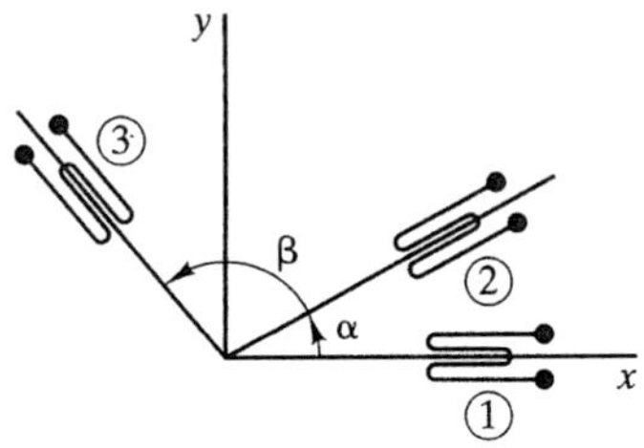

Figure P.8.30.

Print out.

PROBLEM DATA: $\varepsilon_1 =$ ______, $\varepsilon_2 =$ ______, $\varepsilon_3 =$ ______
$\alpha =$ ______, $\beta =$ ______
$E =$ ______, $\nu =$ ______

PRINCIPAL STRAINS: $\varepsilon_1 =$ ______, $\varepsilon_2 =$ ______, $\varepsilon_3 =$ ______

PRINCIPAL STRESSES: $\tau_1 =$ ______ Pa, $\tau_2 =$ ______ Pa,
$\tau_3 =$ ______ Pa

PRINCIPAL AXES: $\pm$ ______ ° from x axis
$\pm$ ______ ° from x axis

Run the program for the following data:

$\varepsilon_1 = .003$ $\quad \nu = .3$
$\varepsilon_2 = -.001$ $\quad \alpha = 45°$
$\varepsilon_3 = .002$ $\quad \beta = 90°$
$E = 2 \times 10^{11}$ Pa

CHAPTER 9

Failure Criteria

9.1 Introduction

In Chapter 4 on the one-dimensional state of stress we discussed *yielding* for ductile materials with the eventual possibility of fracture at the ultimate stress. For brittle materials such as glass we had only the action of *fracture*, which is called brittle fracture. These conditions are generally considered *failure conditions* because of the excess deformation present or because of the actual breakage of parts. For the one-dimensional state, simply a comparison of some measure of the member with that of a one-dimensional test specimen of the same material at yielding or at fracture is sufficient to inform us as to whether for a particular state of stress there is likely to be a failure present. For instance, what we can do is to compare the one-dimensional stress member with the test specimen under yield conditions considering any of the following quantities:

1. Maximum normal stress
2. Maximum shear stress
3. Maximum normal strain
4. Maximum shear strain
5. Maximum distortion strain-energy density[1]

If any of these quantities for the one-dimensional stress member equals the corresponding quantity in the test specimen *at yielding* and away from the necking zone, we can expect yielding in the member. This is similarly true also for predicting brittle fracture but with considerably less accuracy.

[1]This topic was introduced in Part B of Chapter 6 and will be covered in more detail in Chapter 18.

In the case of *multidimensional stress* at a point we have a more complicated situation present. For instance, one may have, for a given general state of stress at a point, a value of one of the five listed quantities equal to that found on a one-dimensional test specimen *at yielding*. However, the other four listed quantities for that state of stress will have values that may *not* correspond to those at yield in the one-dimensional test specimen. Thus, *while one criterion may predict yielding at a point, the others may not.* We have here for the general case at least five possible *different* tests that we can use for predicting yielding or possible brittle fracture. (Actually, there are a number of additional other tests one can employ. We shall not discuss these.[2]) The tests that are most significant to us and that we shall discuss for ductile materials are the maximum shear-stress theory and maximum distortional energy theory (also called the octahedral shear theory). This will be done in the following section. For brittle fracture in brittle materials we shall present in Section 9.4 the maximum normal stress theory.

It is apparent that we are following rather rough empirical approaches here by comparing a general state of stress with the much simpler one-dimensional state of stress. The results accordingly are approximate. Generally one theory is more apropos for certain materials than for others. We wish to point out that research is under way at the present time among metallurgists, material scientists, and engineers to better understand yielding phenomena and brittle fracture so that better predictors of these events can be formulated.

An important concept that we need to discuss before getting to the details of the aforementioned failure theories is the concept of the *hydrostatic* state of stress. Suppose that the three principal stresses are along axes xyz at a point and are mutually equal to each other, each having the value τ. Then going back to Fig. 7.15 for the three-dimensional Mohr circle we see that the circles degenerate to a point. We can then make these conclusions for any pair of orthogonal directions at the point:

1. All normal stress $\tau_{nn} = \tau$
2. All shear stresses $\tau_{ns} = 0$

We thus have a *constant normal stress on all interfaces at a point with zero shear stress on all such interfaces at the point.* For Hookean materials there then is no shear strain at such a point, and only a dilatation or shrinkage can occur. Such a state of stress is the aforementioned *hydrostatic* stress. The name obtains because water under static equilibrium conditions has such a state of stress at all points.[3] We are now ready to examine the key failure theories under present use.

[2]See J. Marin, *Mechanical Behavior of Engineering Materials* (Prentice Hall, Englewood Cliffs, N.J., 1962), Chap. 3.

[3]This is *Pascal's law*, as studied in earlier physics courses.

9.2 Yield Criteria for Ductile Isotropic Materials

Unfortunately, there is no theoretical formulation that gives the state of stress for the condition of yielding in the general case. Experiments do indicate, however, that the *yield condition at a point tends to be independent of hydrostatic stress at the point.* We shall present two empirical criteria for the prediction of yielding that satisfy the required insensitivity to hydrostatic stress. One criterion, called the *Tresca condition,* was first presented in 1864. This condition states that yielding will begin to occur if one-half the largest difference between the principal stresses[4] reaches a certain critical value, τ_{Tr}, which depends on the nature of the material and not on the state of stress at yielding. We say that τ_{Tr} for a material acting in strict accordance with the Tresca criterion is stress-state independent. We may then state the criterion as follows:

$$\text{For onset on yielding} \qquad \frac{1}{2}(\tau_{\max} - \tau_{\min}) = \tau_{\text{Tr}} \tag{9.1}$$

How do we get τ_{Tr}? One way is to examine a state of pure shear stress at yielding. This can be accomplished most simply as you will later see in a simple torsion test. The value of τ_{Tr} is then simply the yield stress in pure shear. This value is often given as k in the literature. Because of the stress-state independence of τ_{Tr} for behavior in strict accordance with the Tresca criterion, we can also use the simple tensile test to give us the value of τ_{Tr}. Accordingly, at yielding,

$$\frac{1}{2}(\tau_Y - 0) = \tau_{\text{Tr}}$$

$$\therefore \ \tau_{\text{Tr}} = \frac{\tau_Y}{2}$$

To ascertain whether we have yielding at a point for a given state of stress we compare the quantity $1/2\ (\tau_{\max} - \tau_{\min})$ with $\tau_Y/2$, resulting from a one-dimensional test of the same material at yielding. If

$$\frac{1}{2}(\tau_{\max} - \tau_{\min}) \leq \frac{\tau_Y}{2}$$

we have no yielding, and if

$$\frac{1}{2}(\tau_{\max} - \tau_{\min}) \geq \frac{\tau_Y}{2}$$

[4]This is the maximum shear stress at a point. Note that this quantity does not change value when the state of stress is changed hydrostatically. Also note that maximum and minimum values in this formula signify *algebraic* maximum and minimum values.

we can expect yielding according to the Tresca criterion.[5] For the *onset* of yielding we accordingly have

$$\frac{1}{2}(\tau_{\max} - \tau_{\min}) = \frac{\tau_Y}{2} \tag{9.2}$$

The reader should be cautioned that we have been discussing *ideal* Tresca behavior wherein τ_{Tr} is stress-state independent. In real materials, it may be that τ_{Tr} from the torsion test at yielding may be at some variance with τ_{Tr} from the tension test at yield. We shall generally use the value of τ_{Tr} from the tensile test at yielding.

We turn next to the *maximum distortion strain-energy theory* (case 5, Section 9.1), which is concerned with the *difference* between the strain-energy density of deformation at a point $\mathcal{U}_{\text{total}}$ and that from a hydrostatic state of stress having the value τ equal to the average of the principal stresses at the point ($\tau = 1/3\ [\tau_1 + \tau_2 + \tau_3]$). Accordingly, from Eq. (6.32) the corresponding strain-energy densities are given as

$$\mathcal{U}_{\text{total}} = \frac{1}{2E}[(\tau_1^2 + \tau_2^2 + \tau_3^2) - 2\nu(\tau_1\tau_2 + \tau_1\tau_3 + \tau_2\tau_3)] \tag{9.3}$$

$$\mathcal{U}_{\text{hydro}} = \frac{3}{2E}\tau^2(1 - 2\nu) = \frac{1}{6E}(\tau_1 + \tau_2 + \tau_3)^2(1 - 2\nu) \tag{9.4}$$

Hence, for the distortion strain energy we have

$$\begin{aligned}\mathcal{U}_{\text{distort}} &= \mathcal{U}_{\text{total}} - \mathcal{U}_{\text{hydro}} \\ &= \frac{1}{6E}(1 + \nu)[(\tau_1 - \tau_2)^2 + (\tau_1 - \tau_3)^2 + (\tau_2 - \tau_3)^2]\end{aligned} \tag{9.5}$$

Note that $\mathcal{U}_{\text{distort}}$ is unchanged when the state of stress is altered hydrostatically.[6] The theory asserts that when $\mathcal{U}_{\text{distort}}$ equals or exceeds a value dependent on the particular material, yielding will occur at the point. Again, we point out that $\mathcal{U}_{\text{distort}}$ ideally is stress-state independent. We can determine its value from a simple torsion test at yielding or a simple tension test at yielding. We shall here use the tensile test where for yielding $\tau_1 = \tau_Y$, $\tau_2 = 0$, $\tau_3 = 0$. Thus, from Eq. (9.5),

$$(\mathcal{U}_{\text{distort}})_{\text{tens}} = \frac{1}{3E}(1 + \nu)\tau_Y^2 \tag{9.6}$$

[5]Note that if $1/2(\tau_{\max} - \tau_{\min}) = 1/2(\tau_Y)$ and there is *unloading* taking place, clearly we do not have yielding.

[6]Recall that a hydrostatic state of stress involves no shear stress at a point and hence for isotropic Hookean materials there is no shear strain. Accordingly, for hydrostatic stress there takes place only expansion or shrinkage with *no change in shape.* It can then be concluded that $\mathcal{U}_{\text{distort}}$ is the strain-energy density resulting from change of shape at a point, as the name *distortion* implies. The theory argues that it is this energy density that is significant for determining whether there will be yielding.

The procedure is to compare $\mathcal{U}_{\text{distort}}$ as given by Eq. (9.5) for a given state of stress at a point with $(\mathcal{U}_{\text{distort}})_{\text{tens}}$ as given above for the same material. If $\mathcal{U}_{\text{distort}}$ is less than $(\mathcal{U}_{\text{distort}})_{\text{tens}}$ the theory predicts no yielding at the point. If, on the other hand, $\mathcal{U}_{\text{distort}}$ equals or is greater than $(\mathcal{U}_{\text{distort}})_{\text{tens}}$ the theory predicts yielding at the point. At the *onset* of yielding we have, accordingly,

$$\frac{1}{6E}(1+\nu)[(\tau_1-\tau_2)^2+(\tau_1-\tau_3)^2+(\tau_2-\tau_3)^2]=\frac{1}{3E}(1+\nu)\tau_Y^2$$

$$\therefore \quad \frac{\sqrt{2}}{2}[(\tau_1-\tau_2)^2+(\tau_1-\tau_3)^2+(\tau_2-\tau_3)^2]^{1/2}=\tau_Y \tag{9.7}$$

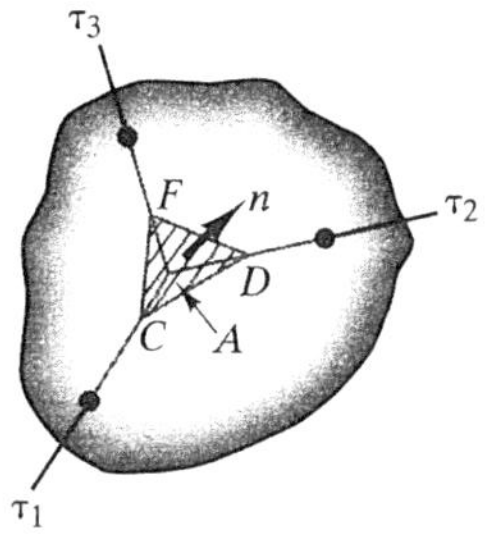

Figure 9.1. Principal stress axes.

We can present the *same criterion* in yet a different manner; in this next form it is called the *Mises criterion.* In this regard consider a point A inside a body (See Fig. 9.1). The principal axes have been shown for the point, and in addition an interface CDF has been drawn infinitesimally close to A with an outward normal oriented so that it is equally inclined to the principal axes. We call this plane at point A the *octahedral* plane. Of interest to us will be the maximum shear stress on this plane and we shall call this shear stress the *octahedral shear stress,* τ_{oct}. In Chapter 15 you will be able to show that this stress can be given in terms of the principal stresses by the following formula:

$$\tau_{\text{oct}}=\tfrac{1}{3}\sqrt{(\tau_1-\tau_2)^2+(\tau_1-\tau_3)^2+(\tau_2-\tau_3)^2} \tag{9.8}$$

Notice, as in the case of $\mathcal{U}_{\text{distort}}$, that the octahedral shear satisfies the basic characteristic of yielding in that it is not affected by changes of stress which are hydrostatic. The Mises criterion, just as the other criteria, assumes ideally stress-state independence. It now asserts that when the octahedral shear stress at a point reaches a particular value dependent on the material only, we have the onset of yielding at this point. The value of the octahedral shear for the onset of yielding we denote as $(\tau_{\text{oct}})_Y$, and we can ascertain the appropriate value through the familiar tensile test. Hence, from the yielding condition for this test we can say that

$$(\tau_{\text{oct}})_Y=\frac{1}{3}\sqrt{(\tau_Y-0)^2+(\tau_Y-0)^2+(0-0)^2}$$

$$\therefore\ (\tau_{\text{oct}})_Y=\frac{\sqrt{2}}{3}\tau_Y \tag{9.9}$$

Now if τ_{oct} for a given state of stress at a point as given by Eq. (9.8) is less than $(\sqrt{2}/3)\tau_Y$ for the same material, the theory indicates that we do not have yielding at the point. On the other hand, if $\tau_{\text{oct}} \geq (\tau_{\text{oct}})_Y$, then the theory predicts yielding. For the *onset* of yielding we have

$$\frac{1}{3}\sqrt{(\tau_1 - \tau_2)^2 + (\tau_1 - \tau_3)^2 + (\tau_2 - \tau_3)^2} = \frac{\sqrt{2}}{3}\tau_Y$$

$$\therefore \frac{\sqrt{2}}{2}\sqrt{(\tau_1 - \tau_2)^2 + (\tau_1 - \tau_3)^2 + (\tau_2 - \tau_3)^2} = \tau_Y \qquad (9.10)$$

Comparing Eqs. (9.7) and (9.10) we see that we arrive at the same yielding criterion via the octahedral stress concept or by the maximum distortion energy theory.

Before going to examples, let us relate the Tresca and Mises yield criteria. We can choose the τ_{Tr} and the octahedral stress for yielding by considering yielding for the *tensile test.* Obviously, both criteria will predict the same stress then for a one-dimensional states of stress. But because the criteria are different from each other, the Tresca yield condition in pure shear for ideal Tresca yield behavior will be *different* from the yield condition in pure shear for ideal Mises yield behavior. Indeed, the two criteria will *agree only for the case of pure tension.*

Alternatively, we can determine τ_{Tr} and the octahedral shear for yielding by considering the case of *pure shear* (the simple torsion test). Now both criteria will predict the same yield state for pure shear. But the criteria will predict *different* tensile stresses for yielding. Again there will be no agreement for yielding in any stress state except pure shear. We shall use the tensile test to establish τ_{Tr} and τ_{oct} for yielding.

We shall now illustrate the use of these criteria in the following examples.

Example 9.1

Consider the following state of plane stress:

$$\begin{aligned} \tau_{xx} &= 10{,}000 \text{ psi} \\ \tau_{yy} &= -20{,}000 \text{ psi} \\ \tau_{xy} &= 5000 \text{ psi} \end{aligned} \qquad (a)$$

If the yield stress τ_Y from a one-dimensional tensile test is 30,000 psi for the materials used, will there be yielding according to the Tresca criterion and according to the maximum distortion strain-energy criterion?

To find the principal stresses in the xy plane we employ Eq. (7.13). Thus,

$$\tan 2\tilde{\theta} = \frac{2\tau_{xy}}{\tau_{xx} - \tau_{yy}} = \frac{10{,}000}{30{,}000} = \frac{1}{3}$$

$$\therefore \tilde{\theta}_{a,b} = 9.22°, 99.22° \qquad (b)$$

Example 9.1 (Continued)

The principal stresses are then

$$\tau_{a,b} = \frac{\tau_{xx} + \tau_{yy}}{2} + \frac{\tau_{xx} - \tau_{yy}}{2} \cos \begin{Bmatrix} 18.43° \\ 198.43° \end{Bmatrix} + \tau_{xy} \sin \begin{Bmatrix} 18.43° \\ 198.43° \end{Bmatrix}$$

$$= \frac{10{,}000 - 20{,}000}{2} + \frac{10{,}000 + 20{,}000}{2} \begin{Bmatrix} .948 \\ -.948 \end{Bmatrix} + (5000) \begin{Bmatrix} .316 \\ -.316 \end{Bmatrix} \quad \text{(c)}$$

$$= 10{,}790, \ -20{,}790 \text{ psi}$$

The other principal stress is clearly 0. Hence, for this case we have

$$\begin{aligned} \tau_1 &= 10{,}790 \text{ psi} \\ \tau_2 &= 0 \qquad \text{(d)} \\ \tau_3 &= -20{,}790 \text{ psi} \end{aligned}$$

For the *Tresca* test we now compare

$$\frac{1}{2}(\tau_{max} - \tau_{min}) = \frac{1}{2}[10{,}790 - (-20{,}790)] = 15{,}790 \text{ psi} \quad \text{(e)}$$

with τ_{Tr} which is $\tau_Y/2 = 15{,}000$ psi. Clearly, the Tresca condition predicts yielding.

As for the *distortional strain-energy* test we compare

$$\begin{aligned} \mathcal{U}_{distort} &= \frac{1}{6E}(1 + \nu)[(\tau_1 - \tau_2)^2 + (\tau_1 - \tau_3)^2 + (\tau_2 - \tau_3)^2] \\ &= \frac{1}{6E}(1 + \nu)[(10{,}790 - 0)^2 + (10{,}790 + 20{,}790)^2 + (0 + 20{,}790)^2] \\ &= \frac{1}{6E}(1 + \nu)(116 \times 10^6 + 998 \times 10^6 + 430 \times 10^6) \qquad \text{(f)} \\ &= \frac{1}{6E}(1 + \nu)(1546) \times 10^6 \end{aligned}$$

with $(\mathcal{U}_{distort})_{tens}$, which from the tensile test is

$$(U_{distort})_{tens} = \frac{1}{3E}(1 + \nu)\,\tau_Y^2 = \frac{1}{6E}(1 + \nu)\,1800 \times 10^6 \quad \text{(g)}$$

Example 9.1 (Continued)

The distortional strain-energy test predicts no yielding. We see that the Tresca criterion using the tensile test data τ_Y is the more conservative measure. However, the distortional strain-energy criterion does tell us that we are close to yielding, and even using this test we should observe caution here.

Example 9.2

A thin-walled cylindrical tank of radius r and wall thickness t is subject to an internal pressure p above atmosphere. At what value of p will yielding first occur according to each of our two yield criteria? Assume that the ends and joints are well reinforced.

You most likely have seen in a footnote of Chapter 2 that the principal stresses τ_1 and τ_2 for an element of the cylinder away from the ends (see Fig. 9.2) are given approximately as[7]

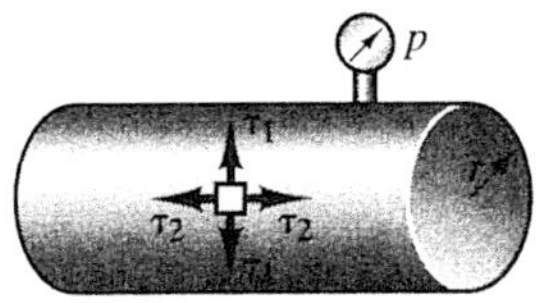

Figure 9.2. Pressurized cylinder.

$$\tau_1 = \frac{pr}{t} \quad \text{(a)}$$

$$\tau_2 = \frac{1}{2}\frac{pr}{t} \quad \text{(b)}$$

$$\tau_3 \approx 0$$

Since $\tau_3 \approx 0$, we may use as the Mises criterion for the *onset* of yielding the following simplified equation formulated from Eq. (9.7):

$$\tau_1^2 - \tau_1\tau_2 + \tau_2^2 = \tau_Y^2 \quad \text{(c)}$$

Eliminating p from Eqs. (a) and (b), we get

$$\tau_1 = 2\tau_2 \quad \text{(d)}$$

Solving Eqs. (c) and (d) simultaneously, we have

[7]Recall from Chapter 2 that the stress τ_3 *normal* to the cylindrical surface and coming directly from the pressure p is negligibly small compared to the stresses τ_2 and τ_1 induced by pressure p in the cylindrical surface in the longitudinal and transverse directions.

Example 9.2 (Continued)

$$\begin{aligned} \tau_2 &= \frac{\tau_Y}{\sqrt{3}} \\ \tau_1 &= \frac{2\tau_Y}{\sqrt{3}} \end{aligned} \tag{e}$$

Going back to Eq. (c) we can conclude that the pressure at which yielding begins to occur according to the *Mises criterion* is

$$p_M = \frac{2\tau_Y t}{\sqrt{3}r} \tag{f}$$

As for the *Tresca* condition, for the onset of yielding we have

$$\begin{aligned} \frac{1}{2}(\tau_{max} - \tau_{min}) &= \frac{\tau_Y}{2} \\ \therefore (\tau_1 - 0) &= \tau_Y \end{aligned} \tag{g}$$

where $\tau_1(= 2\tau_2)$ clearly is the largest normal stress and 0 is the smallest normal stress. Going back to Eq. (a) and using Eq. (g) we may then solve for the pressure corresponding to the onset of yielding. Thus,

$$p_{Tr} = \frac{\tau_Y t}{r} \tag{h}$$

The Tresca criterion predicts a smaller pressure for the onset of yielding. That is,

$$\frac{p_{Tr}}{p_M} = \frac{\sqrt{3}}{2} = .866 \tag{i}$$

Example 9.3

A spherical tank with an outside diameter of 1.2 m is to contain air at a maximum pressure of 3.5×10^6 Pa above the atmosphere. If the yield stress for the material to be used for the tank is 7.8×10^8 Pa.

Example 9.3 (Continued)

design the thickness t for a safety factor of 3. Use both the Tresca and Mises criteria. Take $p_{atm} = 1.013 \times 10^5$ Pa.

We consider an element of the tank as shown in Fig. 9.3(a). Clearly, there are no shear stresses on the faces of the element, and the normal stresses shown are equal. To compute stress τ we take as a free body half the sphere as shown in Fig. 9.3(b). Owing to symmetry, the stress τ is uniform over the cut section. The force F is the resultant force from the inside pressure of 3.5×10^6 Pa. You will recall from earlier work that this force is computed by multiplying the pressure by the projected area in the direction of the desired force. Accordingly, from equilibrium we have

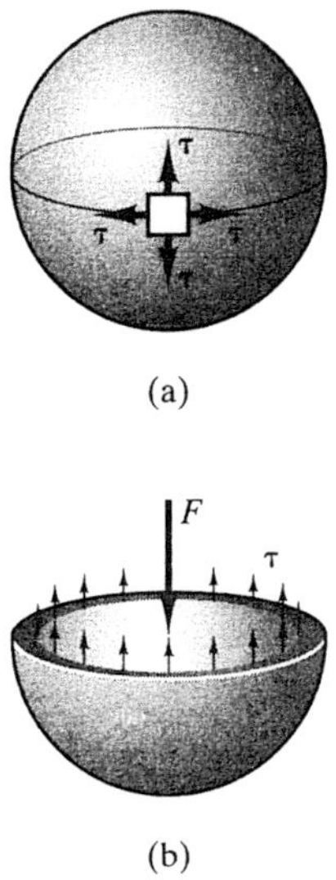

Figure 9.3. Thin-walled spherical tank.

$$\tau(2\pi)\left(r - \frac{t}{2}\right)t = p\pi(r - t)^2$$

$$\therefore \tau = \frac{p(r-t)^2}{2(r - t/2)t} = \frac{3.5 \times 10^6 (.6 - t)^2}{(1.2 - t)t} \qquad \text{(a)}$$

We thus have two principal stresses for the element shown in Fig. 9.3(a). The third principal stress would be some value between 3.6×10^6 Pa on the inside surface and 1.013×10^5 Pa on the outside surface, and accordingly, being so small, we shall consider it to be zero, as in the case of the cylinder.

One guide that we may follow is to find the smallest thickness t that will just do the job. This will be the lightest design and most likely the most economical design. We could, in this regard, have the tank be just below yielding conditions when we have our maximum pressure of 3.5×10^6 Pa. But this would assume complete accuracy of calculations and fabrication—something that is never reached. So instead of using the yield stress τ_Y of 7.8×10^8 Pa we employ, for the given safety factor of 3, a stress of $7.8 \times 10^8 / 3 = 2.6 \times 10^8$ Pa as a hypothetical "yield stress" while using the Tresca and Mises yield criteria.

Considering the *Tresca* criterion first, we shall consider Eq. (9.2). For our case we have

$$\frac{1}{2}(\tau_{max} - 0) = \frac{2.6 \times 10^8}{2}$$

$$\frac{3.5 \times 10^6 (.6 - t)^2}{(1.2 - t)(t)} = 2.6 \times 10^8 \qquad \text{(b)}$$

Example 9.3 (Continued)

Solving for t using the quadratic formula, we get

$$t = 4.00 \text{ mm}$$

This is the thinnest design for a safety factor of 3 according to the Tresca condition.

We next consider the *Mises* condition for this design. Accordingly, we turn to Eq. (9.7) employing for τ_Y the value of 2.6×10^8 Pa. Thus,

$$\frac{\sqrt{2}}{2}\sqrt{2\tau^2} = \tau = \frac{3.5 \times 10^6\,(.6 - t)^2}{(1.2 - t)(t)} = 2.6 \times 10^8$$

$$\therefore \quad t = 4.00 \text{ mm} \tag{c}$$

Hence, in this comparatively simple situation both design criteria give the same result.

We point out that finding the appropriate thickness as we have done is but one part of the design. Further design considerations must be made which depend on the method of construction of this sphere. Conceivably, such considerations may require a thickness other than calculated here.

We have now set forth two criteria which establish those combinations of principal stresses for which a structure may yield at some points in the structure. When such a condition is reached it does not necessarily mean that the structure will fail. Indeed, other parts of the structure may act as constraints to prevent excessive deformation and so the structure is not necessarily in jeopardy, as will be pointed out in our discussion of limit design (Chapter 11).

Finally, we remark again that for yielding action (in ductile materials) we need not be concerned with stress concentrations occurring at small notches, fillets, keyways, and so on. There takes place in these materials local yielding encompassing these regions of stress concentration whereby such local yielding actions usually do no serious harm. At the same time we remind you that this is *not true* for *fatigue* considerations. Here stress concentrations are always important.

*9.3 Yield Surfaces

In this section we shall consider yield surface concepts for both the Tresca and the maximum distortion strain-energy (Mises) criteria. We shall see that the Tresca condition as presented thus far is the more conservative test, as we found in certain specific cases in the previous examples. Let us then consider a stress space having axes corresponding to principal axes τ_1, τ_2, and τ_3. This is shown in Fig. 9.4. Since these princi-

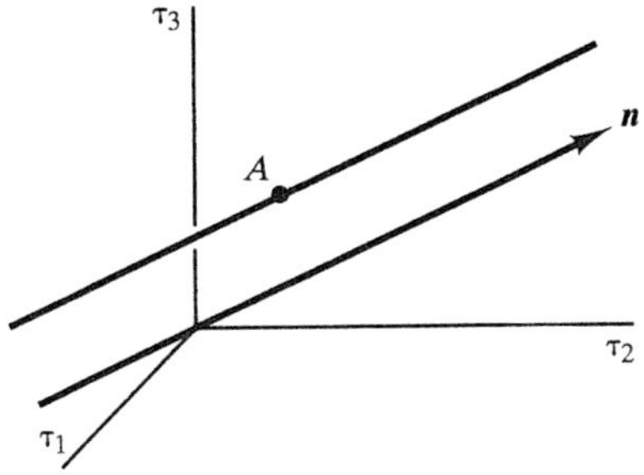

Figure 9.4. Hydrostatic variations in principal stress space.

pal stresses fully establish the state of stress at a point in a body, clearly any point A in the principal stress space represents some unique state of stress. Now a hydrostatic change in stress at point A causes point A to shift along a direction parallel to $\boldsymbol{n}$, which forms *equal angles* with the coordinate axes. This must follow since a hydrostatic stress changes the normal stress in all directions (including the principal directions) by an equal amount. Consider now a plane surface in the principal stress space having a domain (see Fig. 9.5) inside of which the states of stress

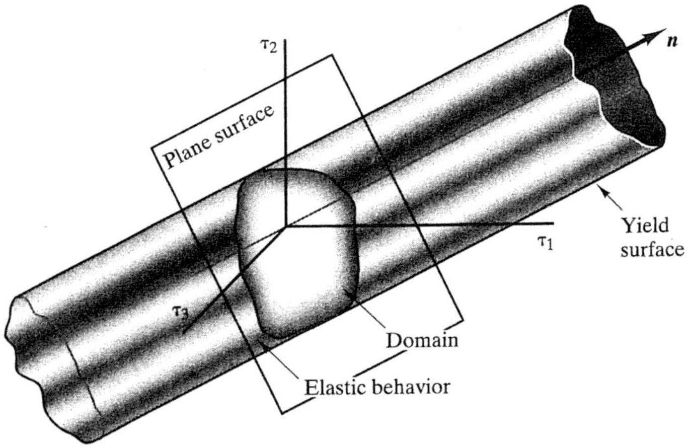

Figure 9.5. Yield surface intersecting a plane normal to $\boldsymbol{n}$.

correspond to elastic behavior of a material and outside of which the states of stress correspond to plastic behavior of this material. The boundary of this domain forms the locus of states in the plane for which there is *onset* of yielding. Taking the yield condition to be independent of hydrostatic stress, this domain can be extended along the direction $\boldsymbol{n}$ so as to form an infinitely long prismatic surface which is the boundary between elastic and inelastic behavior in the principal stress diagram for the particular material. Such a surface is called the *yield surface* and is shown in Fig. 9.5.

Our task now is to establish domains on some plane surface for the Tresca and maximum distortion strain-energy criteria which can then be used to generate yield surfaces for a material. In this way, we can see a geometric representation of the criteria and can then compare them. Consider first the Tresca criterion. For the onset of yielding we have

$$\frac{1}{2}(\tau_{max} - \tau_{min}) = \frac{\tau_Y}{2}$$

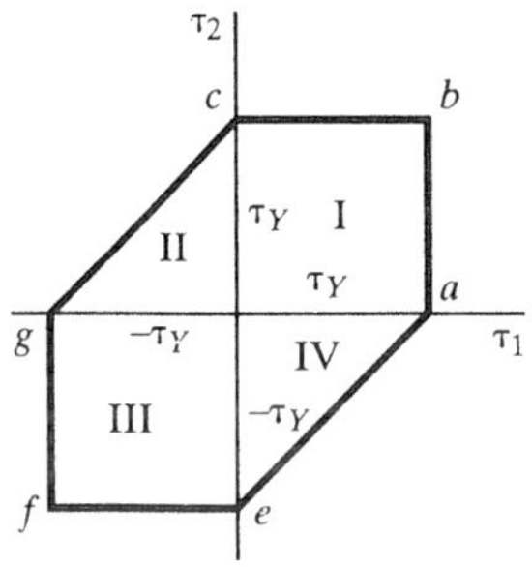

Figure 9.6. Tresca yield locus.

This is actually the mathematical equation of the yield surface that we want to show graphically in the principal stress space. To generate this yield surface most simply we first consider those states of stress for which $\tau_3 = 0$. The resulting curve on the $\tau_1\tau_2$ plane, representing the locus of states for which we can expect the onset of yielding, may be thought of as the intersections of the $\tau_1\tau_2$ plane with the yield surface. Consider Fig. 9.6 for this purpose wherein we have denoted the four quadrants of the reference with Roman numerals. In region I the minimum stress possible is zero value (corresponding to $\tau_3 = 0$). Hence, for onset of yielding, the Tresca condition prescribes the following relation:

$$\tfrac{1}{2}(\tau_{max} - 0) = \tfrac{1}{2}\tau_Y$$

$$\therefore \tau_{max} = \tau_Y$$

Consider first those states of stress where $\tau_1 > \tau_2$. Clearly, τ_1 must be less than τ_Y to prevent yielding. If, on the other hand, $\tau_2 > \tau_1$, we can conclude likewise that τ_2 must be less than τ_Y to prevent yielding. The states of stress for elastic behavior must accordingly be inside the square outer bound shown for region I. The line *abc* then represents a locus of states for onset of yielding in the first quadrant. Now in the third quadrant the maximum principal stress is zero since τ_1 and τ_2 will be negative (except at the origin). Then for the Tresca condition for onset of yielding we have

$$\tfrac{1}{2}(0 - \tau_{min}) = \tfrac{1}{2}\tau_Y$$

$$\therefore \tau_{min} = -\tau_Y$$

Reasoning as we did in quadrant I we can then conclude that *efg* must represent the locus of states for onset of yielding in region III. Consider next region II. The algebraic maximum stress here must be τ_2 (since it is positive) and the algebraic minimum stress must be τ_1 (since it is negative). Hence, the Tresca condition specifies for onset of yielding that

$$\frac{1}{2}(\tau_2 - \tau_1) = \frac{\tau_Y}{2}$$

$$\therefore \tau_2 - \tau_1 = \tau_Y$$

The locus of states for this condition must be a straight line on the $\tau_1\tau_2$ plane with intercepts at points g and c, as you may readily check. Thus, gc is the boundary for elastic behavior in region II, and it should be apparent that by the same reasoning as above ea must be the corresponding boundary in region IV.

Now one may show that the prismatic surface in the direction $\boldsymbol{n}$ that forms the trace shown in Fig. 9.6 when cut by the $\tau_1\tau_2$ plane has as its cross section a regular hexagon. This has been shown in Fig. 9.7.

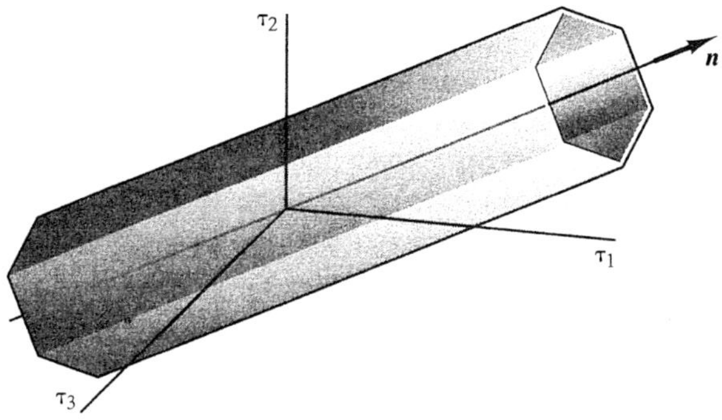

Figure 9.7. Tresca yield surface.

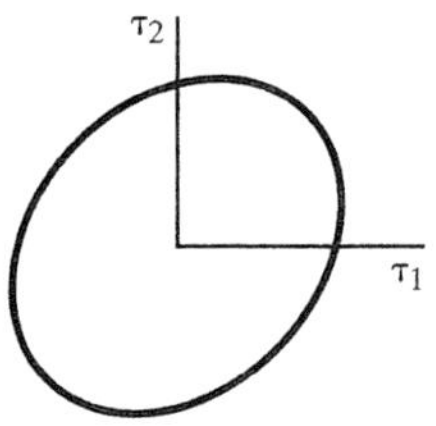

Figure 9.8. Mises yield locus for plane stress.

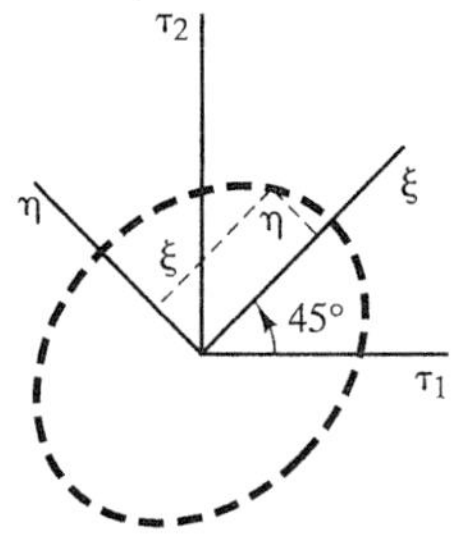

Figure 9.9. To show that Mises yield locus is an ellipse for plane stress.

We next turn to the maximum distortion strain-energy criterion, or, as we shall often say, the Mises criterion. Again as with the Tresca case we consider first those states of stress where $\tau_3 = 0$. If we are interested in the locus of states in the $\tau_1\tau_2$ plane for the onset of yielding, we employ Eq. (9.7) for $\tau_3 = 0$. Thus, we get the following equation:

$$\tau_1^2 - \tau_1\tau_2 + \tau_2^2 = \tau_Y^2 \tag{9.11}$$

The above equation, realizing that τ_Y is constant for a given material, is a second-order curve on the $\tau_1\tau_2$ plane (see Fig. 9.8). Actually it is an *ellipse* with major and minor diameters oriented in directions 45° from the axes τ_1 and τ_2. We can readily show this by expressing the curve for axes ξ and η rotated 45° from τ_1 and τ_2 (see Fig. 9.9). The transformation equations are thus

$$\tau_1 = \zeta \cos 45^\circ - \eta \sin 45^\circ = \frac{\zeta - \eta}{\sqrt{2}}$$
$$\tau_2 = \zeta \sin 45^\circ + \eta \cos 45^\circ = \frac{\zeta + \eta}{\sqrt{2}}$$

Substituting into Eq. (9.11), we get

$$\frac{1}{2}(\zeta^2 + 3\eta^2) = \tau_Y^2$$
$$\therefore \frac{\zeta^2}{2\tau_Y^2} + \frac{\eta^2}{2\tau_Y^2/3} = 1$$

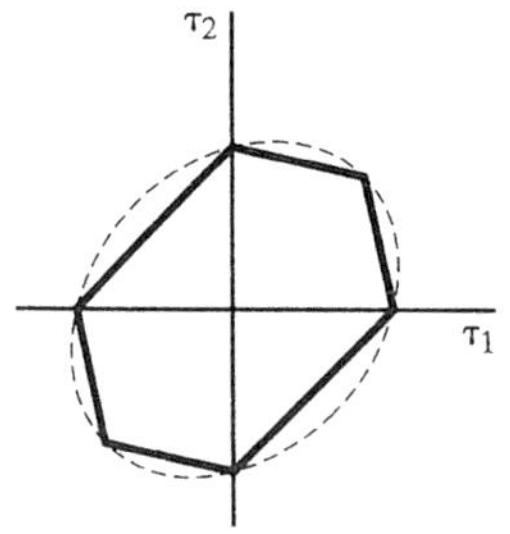

Figure 9.10. Tresca and Mises yield loci for plane stress.

The curve is now clearly established as an ellipse about the axes $\xi\eta$ which are rotated 45° about axes $\tau_1\tau_2$.

For the same material (i.e., a given value of τ_Y), the $\tau_1\tau_2$ plots for the Tresca and the Mises criteria are as shown in Fig. 9.10. Notice that the Tresca condition as presented thus far predicts yielding at a generally lower stress level than does the Mises criterion, and it is thus more conservative. This conclusion stems from the fact that we used the *tensile test* to establish the yield condition for *both* criteria. As pointed out earlier for other states of stress the criteria, being basically different, will predict different yield conditions. That is, they will only give identical results for the one-dimensional state of stress. Hence, in Fig. 9.10 we see that the ellipse of the Mises criterion and the projected hexagon of Tresca coincide at only six states of stress. Four are easily deduced to be the one-dimensional states of τ_1, $-\tau_1$, τ_2, and $-\tau_2$. The other two correspond to one-dimensional states of τ_3 and $-\tau_3$. Thus only for these one-dimensional states do the criteria agree and this is clearly because we used the one-dimensional state of stress at yield to establish both τ_{Tr} and $(\tau_{oct})_Y$. The Mises ellipse is *outside* the Tresca projected hexagon making the latter more conservative in predicting yielding for all states of stress except for the one-dimensional state. As pointed out earlier we could just as well established τ_{Tr} and $(\tau_{oct})_Y$ using the *pure shear* state of stress at yield. One can show in this case that the Tresca ellipse and projected Mises hexagon again touch at only the six states of stress corresponding not surprisingly to the six states of pure shear. In this case the Tresca projected hexagon lies *outside* the Mises ellipse being tangent at the six points corresponding to pure shear. The *Mises* criterion is now the *more conservative* of the two for states of stress other than pure shear. We shall in this text use the one-dimensional state of stress for establishing τ_{Tr} and $(\tau_{oct})_Y$.

The ellipse that we have shown as the Mises criterion is actually the trace between a right circular cylinder, with its axis in the direction $\boldsymbol{n}$, equally inclined to the principal axes, and the $\tau_1\tau_2$ plane, as shown in Fig. 9.11.

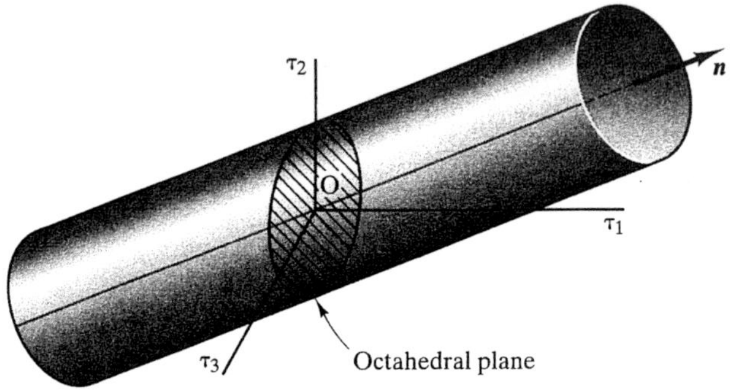

Figure 9.11. Mises yield surface.

If one looks along the axis of the prism for the Tresca and Mises criteria, he/she would see a circle (see Fig. 9.12) enclosing a hexagon.

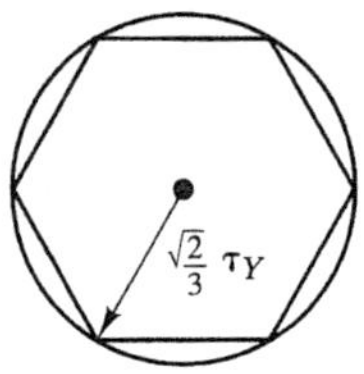

Figure 9.12. Section plane for Tresca and Mises criteria.

9.4 Maximum Normal Stress Theory for Brittle Fracture

To predict brittle fracture we usually employ the *maximum normal stress theory.* This stipulates that fracture in a brittle material such as glass and cast iron will occur whenever the maximum principal stress at a point, if it is tension, equals the ultimate stress from a tensile test, or whenever the algebraic minimum principal stress at a point, if it is compression, equals the ultimate stress from a simple one-dimensional compression test.

In the $\tau_1\,\tau_2$ plane the brittle fracture condition can be shown in Fig. 9.13. Note in the diagram that the material has greater strength in compression than in tension. This is generally the case. Finally, we point out that stress concentrations are *important* for brittle fracture considerations.

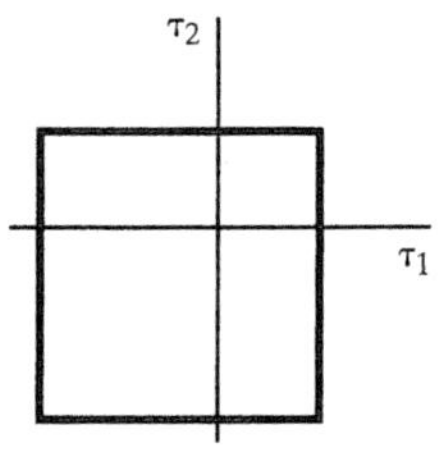

Figure 9.13. Maximum normal stress.

9.5 Comparison of the Theories

In our introduction we presented five possible failure criteria. For ductile materials we considered in some detail the Mises and the Tresca criteria, while for brittle materials we employed the maximum normal stress criterion. How do the theories compare?

In Fig. 9.14 we have shown the five theories plotted for $\tau_3 = 0$ along with data collected from various sources for important structural materials. Notice that the Mises gives the best correlation for steel, copper, and aluminum (ductile materials), while the maximum normal stress theory gives good correlation for cast iron (brittle material).

There are other criteria beyond the five discussed thus far. In closing this discussion we shall briefly discuss one such criterion that is finding use in soil mechanics. In this method we plot the principal Mohr

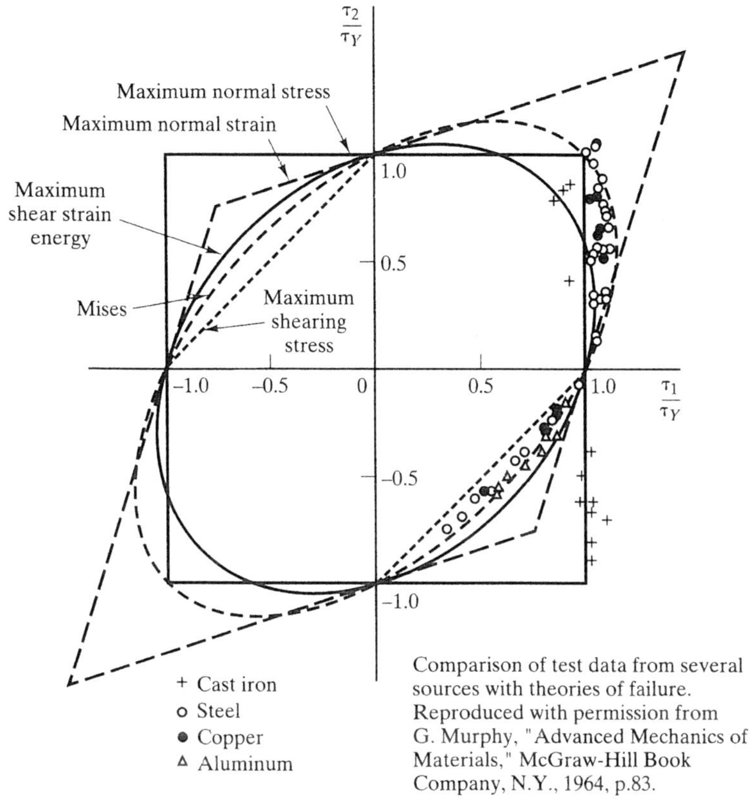

Figure 9.14. Comparison of test data from several sources with theories of failure. [Reproduced with permission from G. Murphy, *Advanced Mechanics of Materials* (McGraw-Hill Book Company, New York, 1964), p. 83.]

circle for a number of states of stress where the material is on the verge of yielding. Usually three states of stress are utilized, namely pure tension, pure compression, and pure shear. An envelope is drawn around the Mohr circles as shown in Fig. 9.15 as dashed curves. We call this the *failure envelope.* Now if for any state of stress the principal Mohr circle cuts through or is tangent to the failure envelope, we can expect failure according to this test.

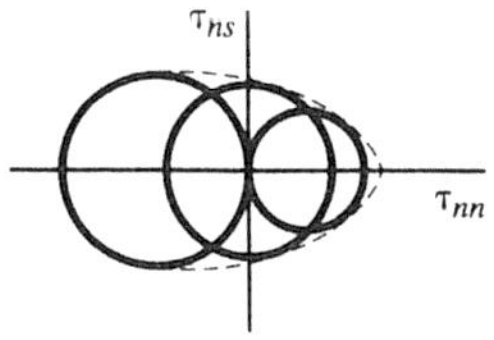

Figure 9.15. Failure envelope.

9.6 Closure

In this chapter we have presented in some detail the Tresca and Mises yield criteria for ductile materials. These are the most widely used by far. At this time we wish to point out that the behavior considered was that of *elastic, perfectly plastic* behavior. There was no *strain hardening* involved and as a result there were fixed yield surfaces. If there is strain hardening, the yield surfaces described would then be *initial yield* surfaces with subsequent yield surfaces that would be obtained once yielding commenced. The shape, movement, and size of subsequent yield surfaces is an important area of study in plasticity theory. Also the relative values of the plastic strain increments are related to the shape of the yield surfaces in the form of *flow laws.* Again you would consider such vital factors when you study plasticity.

Finally, we wish to point out that a behavior pattern that is growing in importance is called *viscoplasticity.* Plastic behavior as we have indirectly pointed out on a number of occasions (e.g., in Section 4.3 on subsequent loadings and unloadings beyond the elastic limit) is *history dependent.* Viscoelastic materials meanwhile have deformations which even for constant loads are *rate dependent.* Viscoplastic materials involve *both* of the aforementioned complexities.[8]

We will have occasion to use the Tresca and Mises yield criteria from time to time as we proceed to study the behavior of various structural elements in subsequent chapters.

Highlights (9)

Hydrostatic stress at a point signifies a uniform normal stress in all directions and zero shear stress on all interfaces at the point. The key idea for failure formulations is that yielding tends to be insensitive to changes in hydrostatic stress. For a state of general stress, there are two widely used empirical formulations that satisfy the above condition. The **Tresca** failure criterion examines the maximum shear stress at a point for any state of stress and predicts the onset of yielding when this maximum shear stress equals a certain value depending only on the material. That value is determined by examining yielding for simple states of stress such as those that occur during the tensile test or a simple torsion test. Unfortunately, this critical value depends somewhat on the test used to establish it. Normally, the tensile test result is used. The other failure criterion is the **maximum distortion criterion** (also called the **Mises** criterion). This criterion examines the difference between the strain energy for the state of stress present and the strain energy for hydrostatic stress using the average of the normal stresses at a point. Since hydrostatic stress results in uniform dilatation, the aforementioned strain-energy difference is associated with the distortion deformation at a point from that of uniform dilatation. The onset of yielding occurs when the distortion strain energy attains a certain value depending on the material. This value is determined using a tensile test or a simple torsion test. The values thus found also will not generally be exactly the same.

Yield surfaces may be shown in principal stress space for each yield theory. Because of the independence of yielding on hydrostatic stress, the yield surfaces form prismatic shapes extending from minus infinity to plus infinity. Each point in principal stress space represents a state of stress. Inside the yield surface, no yielding is predicted for such states of stress and on or outside the yield surface, yielding is predicted. It is important to note that the Tresca and Mises yield criteria can give somewhat different predictions. The more favorable test to use depends on the material being tested.

[8]For further information concerning these complications in behavior, see I. H. Shames and F. A. Cozzarelli, *Elastic and Inelastic Stress Analysis* (Taylor Francis Inc., Washington, D.C.), Chap. 8.

9.7 A Look Back: Equivalent Force Systems

Two force systems, you may recall, are equivalent to each other if they produce the same "push" and the same "turning" action on a free body. The resultant force system on a body is a simpler equivalent force system acting on the body. The simplest resultant force system in the case of a general force system (if we do not consider the wrench) is a single force and a single couple moment. The force vector is the same no matter where you choose to establish this simplest resultant. However, the couple moment vector will depend on the location of the chosen point. For the case of a coplanar force system the simplest resultant will always be either a single force vector with a specific line of action in the plane of the system, or a couple moment normal to the plane of the forces, or, finally, it may be a null vector signifying equilibrium

As we proceed in the next chapter with a consideration of forces and moments on the cross section of a beam, we will generally be dealing with a coplanar force system. Thus, for a two-dimensional cantilever beam, there acts on the embedded portion of the beam at the supporting wall a complex coplanar force system. For the free body of the beam, we can replace this complex force system by the simplest coplanar resultant at the base of the beam just outside of the embedded region. If this resultant is a couple moment, we can put it anywhere on the beam if the entire beam is our free body. If it is a single force, we require a specific line of action for the force. Usually, this line of action is not a convenient one for our study of beams and so we move it to a more convenient parallel line of action. Almost always, we deliberately have the line of action go through the centroid of the cross section. But by doing this, we must include the proper couple moment to insure that the new system is still equivalent to the single force with the specific line of action. Also, whenever we form a free body of a beam and "cut" a section in forming the free body, we will either have a couple moment at the cut section or we will have a force going through the centroid of the section and a couple moment which we call the **bending moment** in beam theory. It is vital to remember that the resultant force system is always associated with a particular free body. In the case of a beam, the resultant force system for the loading of the entire beam is not generally the same as the resultant force system for a portion of the beam.

PROBLEMS

Yield Criteria	9.1–9.11
Yield Surfaces	9.12–9.16
Unspecified Section Problems	9.17–9.18
Computer Problem	9.19
Programming Project 9	

9.1. [9.2] Consider a case of plane stress where

$$\tau_{xx} = 1.4 \times 10^8 \text{ Pa}$$
$$\tau_{xy} = 0.7 \times 10^8 \text{ Pa}$$
$$\tau_{yy} = -1 \times 10^8 \text{ Pa}$$

Is there yielding if $Y = 1.7 \times 10^8$ Pa according to the Mises and the Tresca conditions?

9.2. [9.2] The stress tensor at a point is given as

$$\tau_{ij} = \begin{pmatrix} \tau_{xx} & 0 & 0 \\ 0 & 3.5 & 0 \\ 0 & 0 & 5.5 \end{pmatrix} \times 10^8 \text{ Pa}$$

If $Y = 7 \times 10^8$ Pa, what minimum value should τ_{xx} be for yielding according to the Mises and the Tresca conditions?

9.3. [9.2] Consider the following case of plane stress:

$$\tau_{xx} = 7 \times 10^7 \text{ Pa}$$
$$\tau_{yy} = 14 \times 10^7 \text{ Pa}$$
$$\tau_{xy} = -3.5 \times 10^7 \text{ Pa}$$

If the yield is 1.7×10^8 Pa according to a unidimensional test, determine whether there is yielding according to the Tresca condition and according to the Mises condition.

9.4. [9.2] A thin-walled tank has a pressure of 4.8×10^6 Pa above atmosphere. It is 3 m long, has an outside diameter of .6 m, and has a wall thickness of 3 mm. If Y from a one-dimensional test is 3.5×10^8 Pa, will there be yielding according to the Tresca or the Mises criteria?

9.5. [9.2] Using Tresca, design the thickness of a thin-walled cylinder of Problem 9.4 to withstand the pressure of 4.8×10^6 Pa with a safety factor of 2. Take $Y = 7 \times 10^8$ Pa.

9.6. [9.2] In Problem 6.13 find the maximum vertical distance d from C to the x axis for the force $\boldsymbol{F}$. The yield stress is $Y = 5.727 \times 10^7$ Pa. Note that we have plane stress at A. Use Tresca and a safety factor of 4. Include shear stress from shear force.

9.7. [9.2] Design the thickness of a thin-walled spherical tank to withstand a pressure of 3.5×10^6 Pa above atmosphere, with a safety factor of 2 for a material having $Y = 5.5 \times 10^8$ Pa. The outside diameter of the tank is .6 m.

9.8. [9.2] If $Y = 60{,}000$ psi and a safety factor of 2 is used, what is the maximum pressure p_i in the thick-walled cylinder of Problem 6.17 in accordance with the Mises and the Tresca conditions? Take the following data:

$$a = 1 \text{ in.}$$
$$b = 3 \text{ in.}$$

9.9. [9.2] Test data from a torsion test on a steel specimen indicates that it yielded at 170 MPa. What stress would you expect yielding for the case of pure tension using the Tresca and Mises criteria for yielding?

9.10. [9.2] A tight-fitting air chamber, shown in Fig. P.9.10, is formed between two rigid circular end plates and connected by an aluminum rod. The temperature is raised by 150°F and the air pressure is increased by 20,000 psi. (a) Find the change in length of the rod. (b) Compare the factor of safety by the Tresca and the Mises criteria. Neglect friction at the circular plate boundaries. Take $E = 10 \times 10^6$ psi, $\nu = 0.25$, $\alpha = 20 \times 10^{-6}$/°F, $Y = 100{,}000$ psi.

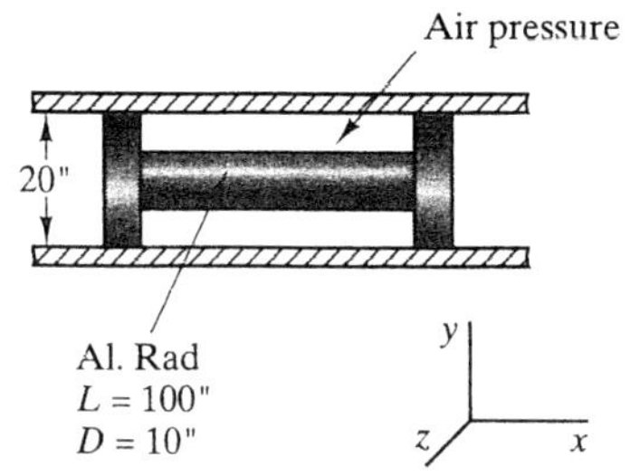

Figure P.9.10.

9.11. [9.2] A uniform gap of 0.002 in. surrounds a steel plate set in a frictionless rigid cavity as shown in Fig. P.9.11. According to the Tresca criterion, at what change in temperature will the plate yield? The yield stress Y is 36,000 psi. What are the principal stresses? Take $E = 30 \times 10^6$ psi, $\nu = 0.3$, $\alpha = 6.5 \times 10^{-6}/°\text{F}$. The plate is not constrained in the out-of-plane direction. (*Hint:* First demonstrate that $\tau_{xx} = 52{,}747 - 278.6\Delta T$ and $\tau_{yy} = 75{,}824 - 278.6\Delta T$. Satisfy Hooke's law for the plate.)

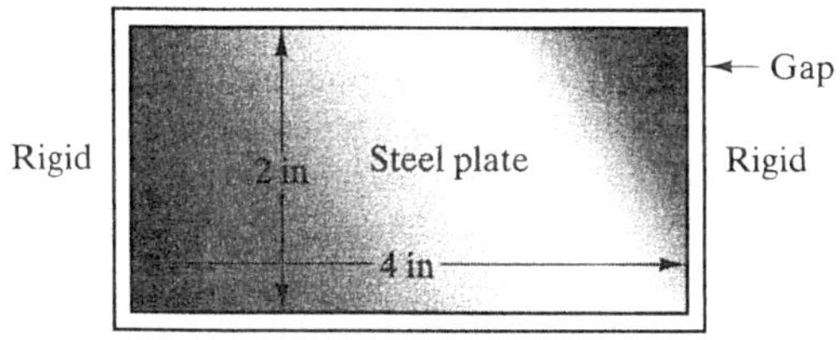

Figure P.9.11.

***9.12. [9.4]** The stress at a point for a brittle material is given as

$$\tau_{ij} = \begin{pmatrix} 90{,}000 & 0 & 0 \\ 0 & 50{,}000 & 20{,}000 \\ 0 & 20{,}000 & 80{,}000 \end{pmatrix} \text{psi}$$

For an ultimate stress of 100,000 psi, will there be failure at the point?

9.13. [9.4] In Problem 8.8 find the maximum value of ω that can be tolerated for a safety factor of 2 and a yield stress of 1.4×10^8 Pa. Use Tresca and Mises criteria.

9.14. [9.4] In Problem 6.9 find the force P to replace the 100-N force so as to cause yielding according to the Tresca criterion. Take $Y = 2 \times 10^8$ Pa. Note from the diagram that we have plane stress.

9.15. [9.4] In Problem 6.14 find the maximum gage pressure p for a yield stress of 5.12×10^8 Pa and a safety factor of 2. Use Tresca. Instead of the 1000-N load, put on a 4000-N-m torque about the y axis. (*Hint:* The smallest normal stress is compressive.)

9.16. [9.4] Glass is a brittle material under normal conditions. However, it has a very high ultimate stress in compression (130,000 psi), making it a valuable material for deep-sea work. (For tension, on the other hand, the ultimate stress is only 10,000 psi.) Design a glass sphere to contain 1 ft^3 of air at approximately atmospheric pressure at a depth of 100 ft below the surface of the ocean. Recall from physics that the pressure in water is given as γd, where γ is the specific weight and d is the depth below the free surface. Take $\gamma = 62.4$ lb/ft^3. Use a safety factor of 4. Do not consider buckling.

9.17. Using thin-walled pressure vessel theory, compute the allowable pressure for a pipe with end caps. Manufacturer's data for the material indicates that it has a maximum allowable stress of 6400 psi. The pipe has an outside diameter of 0.540 in. and an inside diameter of 0.354 in. Do you think that thin-walled theory is applicable for such a pipe? Use the Tresca failure criterion.

9.18. A copper plate 100 mm × 100 mm is placed in a rigid cavity with a 0.10-mm gap around all edges as shown in Fig. P.9.18. According to the Mises failure criterion, at what change in temperature ΔT will yielding occur? Take $E = 1.1 \times 10^{11}$ Pa, $Y = 330$ MPa, $\alpha = 17 \times 10^{-6}/°\text{C}$, and $\nu = 0.35$.

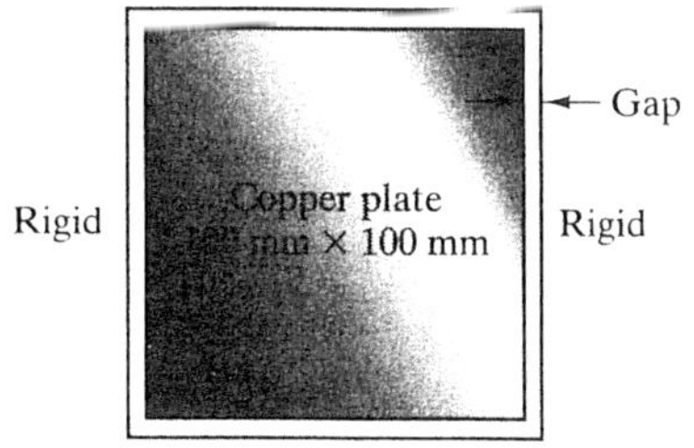

Figure P.9.18.

Other problems will be presented in later chapters involving yield criteria.

****9.19.** When studying the "failure envelope" in plane stress for materials that exhibit yielding, it is common to plot the normalized shear stress $\frac{\tau_{xy}}{Y}$ versus the normalized normal stress $\frac{\tau_{xx}}{Y}$. For both the Tresca and Mises criteria, plot the onset of yielding envelope functions. Develop these by considering the yielding criteria in the form $f\left(\frac{\tau_{xx}}{Y}, \frac{\tau_{xy}}{Y}\right) = 1$. Do they show the same envelope? What happens when you have pure normal stress? Pure shear? (*Hint:* Take τ_{xx} and τ_{xy} as the only non-zero stresses present and Y as the yield stress from a 1D tension test.)

Project 9: Tank Problem II

Find the minimum wall thickness away from the ends for a thin-walled tank subjected in Fig. P.9.19 to an external torque T of 500 ft-lb and an internal pressure p of 500 psi gage. Take the yield stress $Y = 100{,}000$ psi and the safety factor $n = 2$.

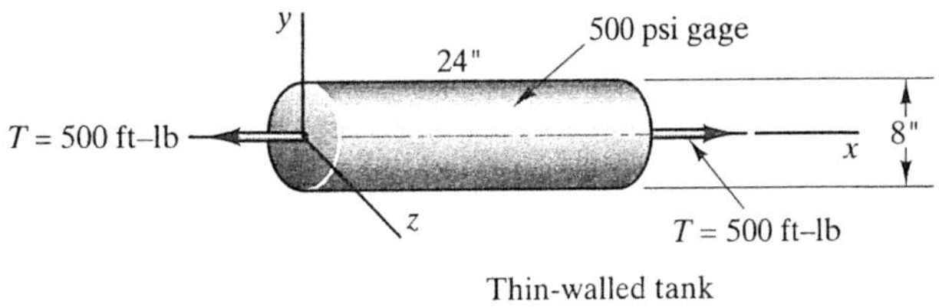

Figure P.9.19.

Use the Mises failure criterion for design. Report the thickness t to the nearest thousandth (0.001) of an inch.

Recall from Chapter 2:

$$\tau_{xx} = \frac{pr}{2t}$$

$$\tau_{yy} = \frac{pr}{t}$$

$$\tau_{xy} = \frac{M}{2\pi r^2 t}$$

Programming notes: Use a *binary search algorithm* to find the wall thickness t. The scheme is to assume upper and lower bounds on the thickness and to take the halfway point between the bounds as a starting thickness. When all calculations (stresses) have been made, we can determine if the true thickness lies above or below the starting point. If the Mises criterion indicates yielding, then using the previous upper limit, the starting point thickness becomes the new lower limit of thickness and a new starting thickness point is found. If there is no yielding, the thickness t lies below the initial starting thickness. The initial starting thickness then becomes the upper limit used to compute a new starting thickness using the previous lower limit. Continue until the difference between successive bound pairs is less than some specified value. The required thickness is then printed out.

Note: Be sure to include some error message if the algorithm does not converge.

CHAPTER 10

Section Forces in Beams

10.1 Introduction

In Chapter 5, we considered a number of problems involving structural members loaded *axially* along their centroidal axes. The resultant force from the applied loads at any cross section of the member was easily established as a force normal to the cross section. Such members are referred to as rods, spars, or axial members. We now consider thin prismatic members loaded *transversely*—that is, perpendicularly to the centerline of the member. Generally, when structural members are loaded transversely we call them *beams*. Of considerable importance will be the components of the resultant force system from the applied loads acting on cross sections of the beam. Knowledge of these resultants will be used in later chapters dealing with stresses and displacements in beams. In this chapter we set our focus on developing methods for computing these force resultants.

10.2 Shear Force, Axial Force, and Bending Moment

Consider first a prismatic beam having a cross-section with a vertical axis of symmetry as shown at the upper right-hand corner of Fig. 10.1(a) (where the y axis is this axis of symmetry). A distributed loading $w(x)$ acts in the vertical plane of symmetry. In addition to this *transverse* loading, we also have an axial force P along the *longitudinal* axis of the beam [see Fig. 10.1(a)]. It will be assumed that the supporting

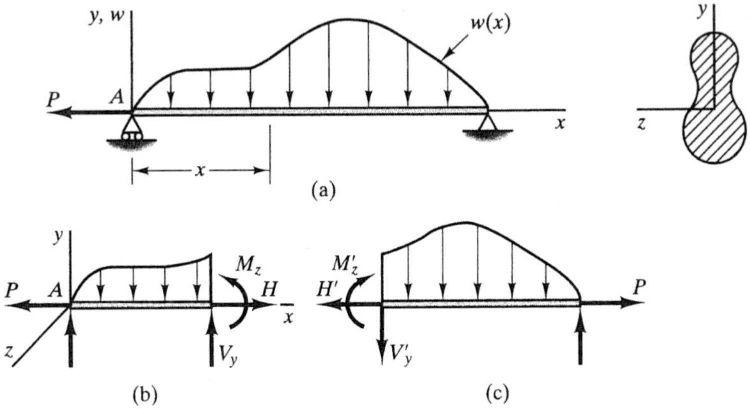

Figure 10.1. Left- and right-hand free-body diagrams exposing a section of the beam.

forces have been determined through statics. The focus of this chapter is to determine the force transmitted across any cross-sectional interface of the beam due to the applied transverse and longitudinal loads. How do we do this? We use our powerful tools of *equilibrium* and *free-body diagrams* to analyze the force resultant at any position along the beam.

We illustrate this in Fig. 10.1(b) where at a position x we make a "cut" in the beam and take a portion of the beam as a free-body so as to "expose" the cross section of the beam at x. Since the external loads acting on the beam are coplanar, it is clear that the load distribution acting on the cross section of the beam at position x must also be coplanar. We know from rigid-body mechanics that we can replace the coplanar load distribution acting on the cross-section of the beam at point x with its simplest resultant—either a single force or a single couple moment in the plane of the load distribution. If the simplest resultant is a single couple moment, it can be positioned arbitrarily. If the simplest resultant is a single force, it must have a particular line of action. However, the position of this line of action is of little interest in beam theory. *Consequently, we deliberately take the position of the resultant force to be at the centroid of the cross-section at all times.* Since we are constraining the resultant force to act through the section's centroid, we must therefore include the proper couple moment M_z to accompany the resultant force as required by static equilibrium. In addition, we decompose the resultant force into orthogonal components on the cross-section: in this case, a vertical force V_y and a horizontal force H. These quantities are shown in Fig. 10.1(b). Since these quantities are used to such a great extent in beam analysis, we have associated names with them. They are:

V_y = *shear-force* component
H = *axial-force* component
M_z = *bending-moment* component

If we had a three-dimensional load, there would have been one additional shear-force component V_z, one additional bending-moment component M_y, and a couple moment M_x along the axis of the beam. We shall call M_x the *twisting-moment* (see Fig. 10.2). We will deal with twisting of beams and shafts in Chapter 14.

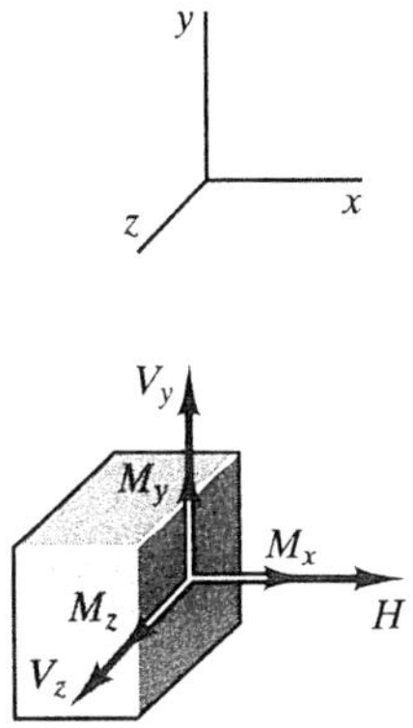

Figure 10.2. Section forces for three-dimensional loading.

Notice in Figure 10.1(c) that a second free-body diagram has been drawn that exposes the "other side" of the cross-section at position x. The shear force, axial force, and bending moment for this section have been primed in the diagram. We know from *Newton's third law* that they should be equal and opposite to the corresponding unprimed quantities in Fig. 10.1(b). We can thus choose for our computations either a left-hand or right-hand free-body diagram. But this poses somewhat of a problem for us when we come to reporting the signs of the transmitted forces and couple moment at a section. We cannot simply use the direction of a force or couple moment at the section since its sense at a section would depend on whether a left-hand or a right-hand free-body diagram was used. Therefore, to associate an unambiguous sign for shear-force, axial-force, and bending-moment at a section, a *sign convention* must be adopted. This brings us to an interesting point in our discussion. Currently, there are two different, but equally effective, sign conventions in popular use for beams. So that the student gets complete coverage of this important issue, we will describe both of them. In doing so, we will use the nomenclature of **positive** and **negative faces.** By a *positive face*, we mean a section of a body having its outward normal from the body in a positive coordinate direction. A *negative face* refers to a section of area having its outward normal from the body pointing in a negative coordinate direction.

For the first sign convention, called the **stress sign convention,** we adopt the following:

> The axial force, shear force, or bending moment acting on a beam cross-section is positive if it acts on a positive face and is directed in a positive coordinate direction.

We note that these components are also positive if they act on a negative face and are directed in a negative coordinate direction. Otherwise, the components are considered to be negative. The above assumes that a right-handed xyz coordinate system is used (as was shown in Fig. 10.1). Note that this is the same stress tensor sign convention that we introduced in Chapter 2. It is identical to that commonly used in the theory of elasticity for the stress at a point and this is why we refer to this convention as the *stress sign convention.*

Consider Fig. 10.1. For the left-hand free-body diagram, the area vector for a beam section at a distance x from the origin points in the positive x coordinate direction. Note also that H, V_y and M_z, also point in positive directions of the xyz axes. Hence, according to the stress convention we have shown a *positive* shear force, a *positive* axial force and a *positive* bending moment at the section at x. For the right-hand free-body diagram the cross-sectional area vector points in the negative x coordinate direction. Since H', V_y' and M_z' point in negative directions of the xyz axes, these components are again positive for the section at x according to our convention. Clearly, only one sign is needed at a section for any of the section force resultants thus avoiding the aforementioned possibility of ambiguity.

For the second sign convention, called the **structural sign convention,** we adopt the following:

> An axial force or bending moment acting on a beam cross-section is positive if it acts on a positive face and is directed in a positive coordinate direction. The shear force is positive if it act in the *negative* coordinate direction on a *positive* face.

We see that the only difference between the two conventions is the sign associated with the shear force. Therefore, the axial-force and bending-moment components are also positive if they act on a negative face and are directed in a negative coordinate direction. For shear force, it is considered *negative* if it acts in the positive coordinate direction on a positive face. Again, we have assumed that a right-handed xyz coordinate system is used. Since this convention is often used in structural (civil) engineering, we refer to it as the *structural sign convention.*

Again consider Fig. 10.1 For the left-hand free-body diagram, the outward area vector for a beam section located a distance x from the origin points in the positive x direction. According to the structural sign convention we have shown a *positive* axial force, a *positive* bending moment, and a *negative* shear force at the section at x. The same is true for the right-hand free-body diagram.

As pointed out, we can solve for V_y, H, and M_z at section x using rigid-body mechanics for either a left-hand or a right-hand free-body diagram provided that we know all the external forces. The quantities V_y, H, and M_z will depend on x and so we can form equations for these quantities in terms of x. Also, it is the practice to sketch shear-force and bending-moment diagrams to convey this information for the entire beam. We shall now consider the computation of the shear force and bending moment as functions of x in the following problems. You are urged to examine them carefully. Later, we shall consider sketching these results in diagrams.

We shall use the **structural convention** in this text.

Example 10.1

We shall express the shear-force and bending-moment equations for the simply-supported beam in Fig. 10.3(a), whose weight we shall neglect. The supporting forces are seen by inspection to be 500 N each. Since there are no horizontal loads, there is no horizontal force at any cut in the beam.

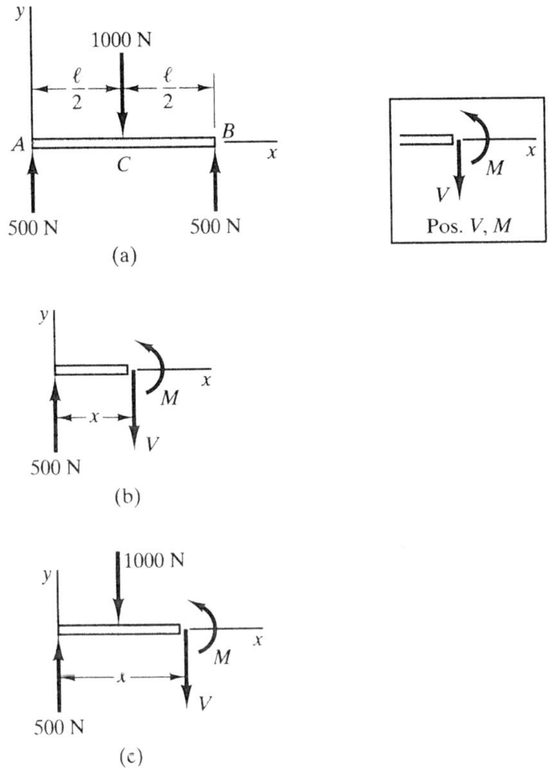

Figure 10.3. Simple beam problem requiring two domains.

To get the shear force and bending moment at a section x, we isolate the left or right side of the beam at x and employ the equations of equilibrium on the resulting free body. If x lies between A and C of the beam, then the only noninternal force present for a left-hand free body is the left supporting force [see Fig.10.3(b)]. Notice we have used directions for V and M (there is no need for subscripts in this simple problem) corresponding to the *positive* states from the point of view of the structural convention. Clearly, the *algebraic* sign we get for these quantities from equilibrium calculations where *we use the right-hand rule* will then correspond to the *convention* sign. If x is between C and B, for such a free body two external forces

Example 10.1 (Continued)

appear [see Fig. 10.3(c)]. Therefore, if the shear force and bending moment are to be expressed as a function of x, clearly *separate equations* covering the *two ranges* or *domains* are necessary.

Let us first consider the domain between A and C. It should be clear that at positions $x = 0$ and $x = l/2$ the shear force is indeterminate since at these positions there is a *point force*.[1] Therefore, in developing the shear-force equations we must *exclude* these points and thus the AC domain will be $0 < x < l/2$. Hence we can say for the AC domain on observing Fig. 10.3(b):

$0 < x < l/2$:

$$500 - V = 0$$
$$\therefore V = 500 \text{ N}$$

We thus have positive shear force according to our structural convention for shear force. Next, consider the bending moment in the AC domain. It will be our practice in considering equilibrium to sum moments about the centroid of the cross-section at the cut section (located at a position x) in the domain of interest. In this way, the shear-force V contributes zero moment in our considerations at all times. Now examine Fig. 10.3(b). We see that the magnitude of the bending-moment M equals $(500)(x)$ N-m and that there is no problem of an indeterminacy of M at $x = 0$. That is, the value of M at $x = 0$ in this problem must be zero and nothing else. The value of M at $x = l/2$ also suffers no indeterminacy. Therefore, the domain for which M can be computed here includes $x = 0$ and $x = l/2$. Thus, from equilibrium, where we use *the right-hand rule,* we can say:

$0 \leq x \leq l/2$:

$$-500x + M = 0$$
$$\therefore M = 500x \text{ N-m}$$

If we had a couple-moment C at $x = 0$, the bending moment would be zero on the left face cross section at A (i.e., at $x = 0^-$) and would be equal in magnitude to the magnitude of C on the opposite face of this cross-section (i.e., at $x = 0^+$), thereby introducing a discontinuity of M at $x = 0$, with the result that M would be indeterminate at $x = 0$.[2] Clearly, the domain for M would *exclude* $x = 0$ and we would have $0 < x \leq l/2$.

[1] This will be seen more clearly when we draw shear-force diagrams.

[2] This also will be more clear when we draw the bending-moment diagrams.

Example 10.1 (Continued)

For simplicity when specifying V and M in a domain, we shall specify the limits of the domain to exclude discontinuities in *either* V and/or M. Thus, we can say for the AC domain:

<u>$0 < x < l/2$:</u>

$$V = 500 \text{ N} \tag{a}$$
$$M = 500x \text{ N-m} \tag{b}$$

Now we are ready for the second domain of our problem. Thus, we can say on observing Fig. 10.3(c):

<u>$l/2 < x < l$:</u>

$$500 - 1000 - V = 0$$
$$\therefore V = -500 \text{ N} \tag{c}$$
$$-500x + 1000(x - l/2) + M = 0$$
$$\therefore M = 500(l - x) \text{ N-m} \tag{d}$$

Notice from Eqs. (a) and (c) how as x passes point C, where we have a point force (1000 N), the shear force *jumps* by -1000 N, showing why we excluded the position $x = l/2$ from the domain for shear force. Notice, however, that from Eqs. (b) and (d) no such jump occurs for M at $x = l/2$. Thus we have

<u>$0 < x < l/2$</u>

$$V = 500 \text{ N}$$
$$M = 500\text{x N-m}$$

<u>$l/2 < x < l$:</u>

$$V = -500 \text{ N}$$
$$M = 500(l - x) \text{ N-m}$$

Example 10.2

Determine the shear-force and bending-moment equations for the simply-supported beam in Fig. 10.4. Neglect the weight of the beam.

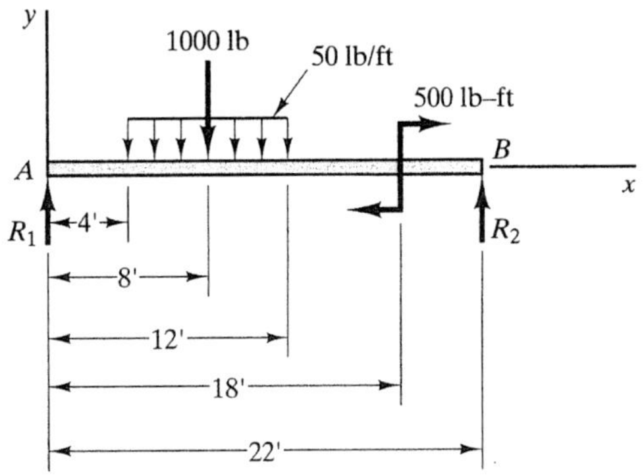

Figure 10.4. Simply supported beam.

By summing to zero moments about each end of the beam, we readily can find from statics

$$R_1 = 868 \text{ lb} \qquad R_2 = 532 \text{ lb}$$

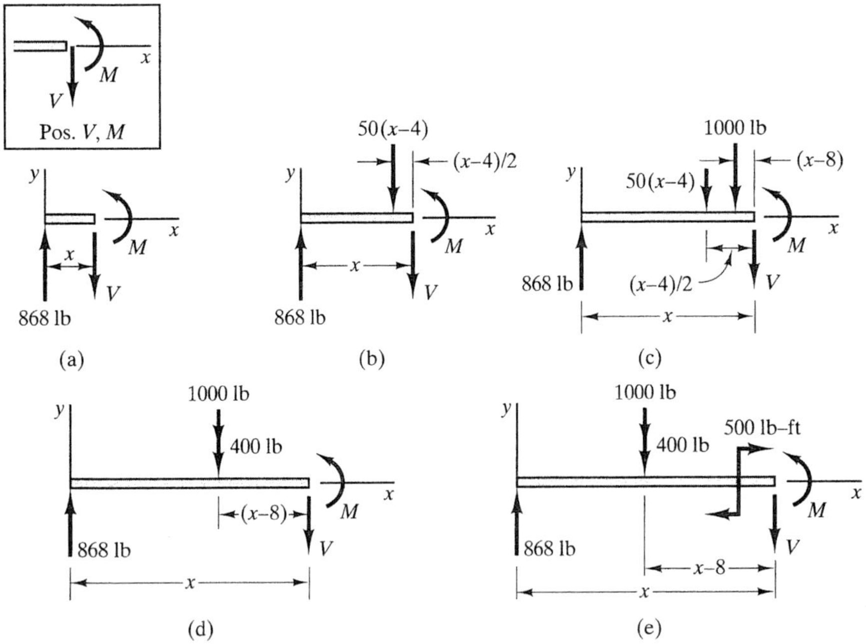

Figure 10.5. Free-body diagram for different domains.

Example 10.2 (Continued)

In Fig. 10.5(a) we have shown a free-body diagram exposing sections between the left support and the uniform load. Summing forces and taking moments about a point in the section, where we have drawn V and M as positive according to the structural convention, we get:

$\underline{0 < x \leq 4:}$

$$868 - V = 0$$
$$\therefore V = 868 \text{ lb}$$
$$-868x + M = 0$$
$$\therefore M = 868x \text{ ft-lb}$$

The next interval is between the beginning of the uniform load and the point force. Thus, observing Fig. 10.5(b), we get using the resultant of the uniform loading to the left of x:

$\underline{4 \leq x < 8:}$

$$868 - 50(x - 4) - V = 0$$
$$\therefore V = -50x + 1068 \text{ lb}$$
$$-868x + \frac{50(x-4)^2}{2} + M = 0$$
$$\therefore M = -25x^2 + 1068x - 400 \text{ ft-lb}$$

We now consider the interval between the 1000-lb point force and the end of the uniform load. Thus, observing Fig. 10.5(c), we get:

$\underline{8 < x \leq 12:}$

$$868 - 50(x - 4) - 1000 - V = 0$$
$$\therefore V = -50x + 68 \text{ lb}$$
$$-868x + \frac{50(x-4)^2}{2} + 1000(x - 8) + M = 0$$
$$\therefore M = -25x^2 + 68x + 7600 \text{ ft-lb}$$

The next interval is between the end of uniform loading and the point couple. We can now replace the uniform loading by its total resultant of 400 lb, as shown in Fig. 10.5(d). Thus:

$\underline{12 \leq x < 18:}$

$$868 - 400 - 1000 - V = 0$$
$$\therefore V = -532 \text{ lb}$$
$$-868x + 1400(x - 8) + M = 0$$
$$\therefore M = -532x + 11{,}200 \text{ ft-lb}$$

The last interval goes from the point couple to the right support. It is to be pointed out that the point couple does not contribute directly

Example 10.2 (Continued)

to the shear force, so we could have used the formulation for V above for the interval $18 < x < 22$. However, the couple does contribute directly to the bending moment and thus the additional interval is required. Accordingly, using Fig. 10.5(e), we get:

$\underline{18 < x < 22:}$

$$V = -532 \text{ lb (as in previous interval)}$$

$$-868x + 1400(x - 8) - 500 + M = 0$$

$$\therefore M = -532x + 11{,}700 \text{ ft-lb}$$

In this example, we have carried out all the algebraic operations, as would be your natural inclination. Actually, you will be glad to know that it is more useful *not* to do this. For instance, for the range $8 < x \leq 12$, we suggest that you write the results as follows:

$\underline{8 < x \leq 12:}$

$$V = -50(x - 4) - 132 \text{ lb}$$

$$M = 868x - 25(x - 4)^2 - 1000(x - 8) \text{ ft-lb}$$

We shall henceforth follow this practice.

Before we proceed further, it must be emphatically pointed out that the replacement of a distributed load, such as the uniform loading in Example 10.2, by a single resultant force is meaningful only for the *particular free* body on which the force distribution acts. For example, to compute the reactions for an *entire* beam taken as a free body, we can, if we include the weight, replace the *weight distribution* by the *total* weight at position $l/2$. For the bending moment at x, the resultant of the weight for the free body becomes wx and is *midway* at position $x/2$. In other words, in making shear-force and bending-moment equations and diagrams, we cannot replace loading distributions over the entire beam by a resultant and then proceed; there is inherent in these equations an *infinite number* of free bodies, each smaller than the beam itself, which makes the above-mentioned replacements invalid for shear-force and bending-moment considerations.

Finally, we wish to point out that for triangular loading distributions you should make good use of the well-known geometrical characteristics of the right triangle. Thus the resultant force is the area, which is 1/2 the base a times the height b. Furthermore, the resultant has a line of action which is at a distance $a/3$ from the height b, as has been indicated in Appendix IV-E

In the following example, we shall consider a cantilever beam carrying a triangular load.

Example 10.3

Determine the shear and bending-moment equations for the cantilever beam shown in Fig. 10.6 carrying a triangular load.

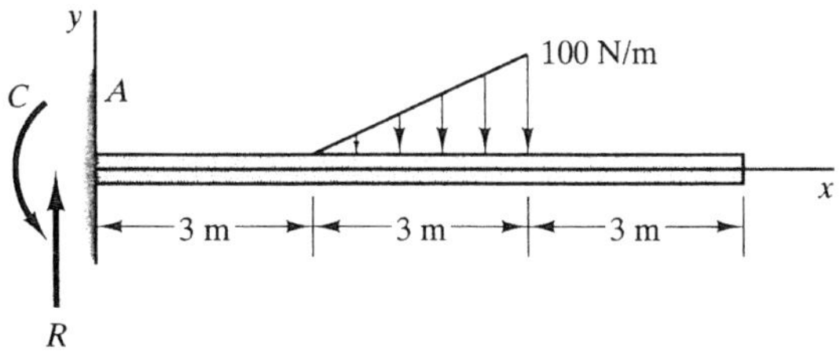

Figure 10.6. Cantilever beam with a triangular load.

The supporting force system shown by C and R is first established using as the resultant force from the triangular loading the area of the triangle with the resultant going through the centroid of the area. Thus,

$\underline{\sum F_y = 0}$:

$$R - \tfrac{1}{2}(3)(100) = 0$$

$$R = 150 \text{ N}$$

$\underline{\sum M_A = 0}$: (using the right-hand rule of statics)

$$C - [\tfrac{1}{2}(3)(100)][3 + \tfrac{2}{3}(3)] = 0$$

$$C = 750 \text{ N-m}$$

We now consider the first domain [see Fig. 10.7(a)].

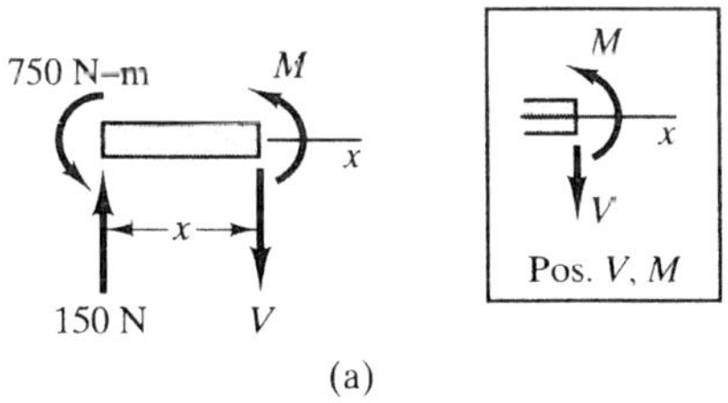

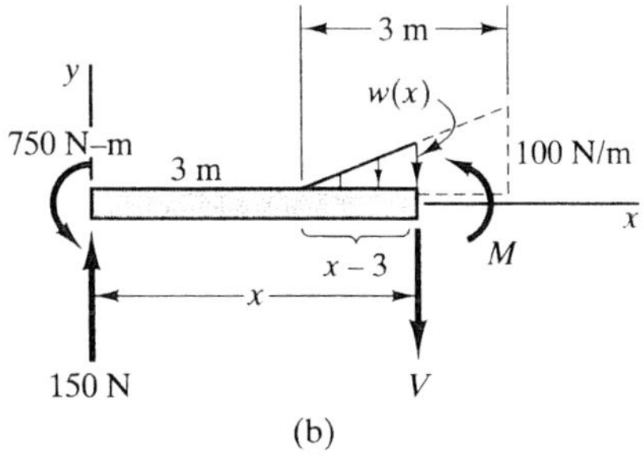

Figure 10.7. Free bodies for two domains of the beam.

Example 10.3 (Continued)

$0 < x \leq 3$:

$$150 - V = 0$$

$$V = 150 \text{ N}$$

$$750 - (150)(x) + M = 0$$

$$M = 150x - 750 \text{ N-m}$$

For the second domain [see Fig. 10.7(b)], we need $w(x)$. From similar triangles, we can say that

$$\frac{w(x)}{100} = \frac{x-3}{3}$$

$$\therefore\ w(x) = \frac{100}{3}(x-3) \text{ N/m}$$

Hence:

$3 \leq x \leq 6$:

$$150 - \frac{1}{2}(x-3)\left[\frac{100}{3}(x-3)\right] - V = 0$$

$$V = 150 - \frac{50}{3}(x-3)^2 \text{ N}$$

$$750 - (150)(x) + \frac{1}{2}(x-3)\left[\frac{100}{3}(x-3)\right]\left[\frac{1}{3}(x-3)\right] + M = 0$$

$$M = -750 + 150x + \frac{50}{9}(x-3)^3 \text{ N-m}$$

Finally, it should be clear by inspection (consider right-hand free-body diagrams for $x > 6$) that the last domain is free of shear and bending moment. Thus,

$6 \leq x \leq 9$:

$$V = M = 0$$

10.3 Direct Formulations of Shear-Force and Bending-Moment Equations

As an alternative approach to that presented in Section 10.2 one can write the shear-force and bending-moment equations in a more direct manner without drawing the simple but bothersome free-body diagrams. The idea is to take one load at a time and mentally satisfy equilibrium while adhering to the sign convention for shear and bending moment. Although this procedure takes a little time to master, it can result in significant overall savings of time. The procedure is as follows. Consider an upward force of magnitude P [see Fig. 10.8(a)]. To maintain this force in equilibrium on sections to the right of it requires a

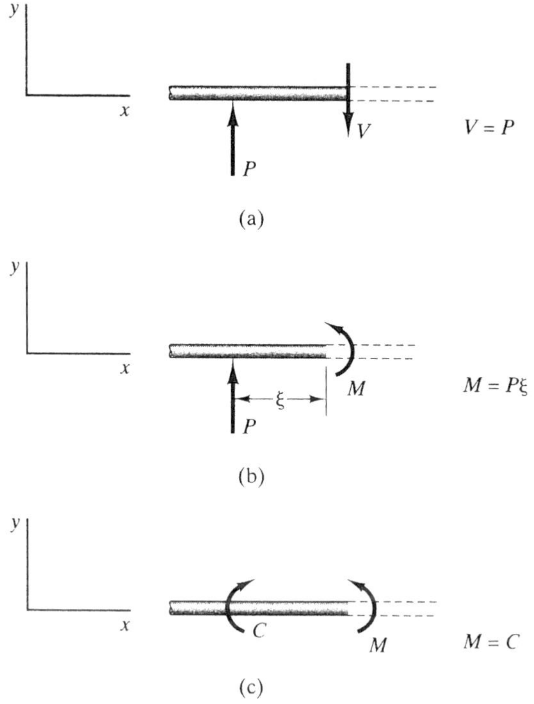

Figure 10.8. Shear forces and bending moments induced by an external force and an external couple moment.

downward shear force of the same magnitude. For our convention, this will be a positive shear force [see Fig. 10.8(a)]. Similarly, a downward force P will induce to the right of it a negative shear force according to our convention. Also, to equilibrate the upward external force at any section a distance ξ to the right of the load, a counterclockwise couple moment of magnitude $P\xi$ is needed as shown in [Fig. 10.8(b)]. The sign for this bending-moment M is clearly positive. For a downward force P, the sign would be negative. Next, a clockwise external couple moment

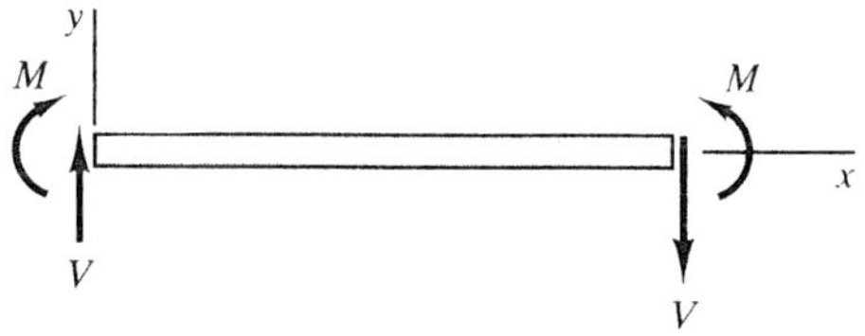

Figure 10.9. Positive shear forces and bending moments.

of magnitude C will require an opposite couple moment to the right of it [see Fig. 10.8(c)]. The induced bending moment will be positive according to our convention. A counterclockwise couple moment clearly will induce a negative bending moment.

We urge the student to use this more efficient procedure when possible for writing shear-force and bending-moment equations. Fig. 10.9 shows positive shear forces and positive bending moments at each end of a beam.

Example 10.4

Evaluate the shear-force and bending-moment equations for the beam shown in Fig. 10.10(a) *without* the aid of free-body diagrams in forming these equations.

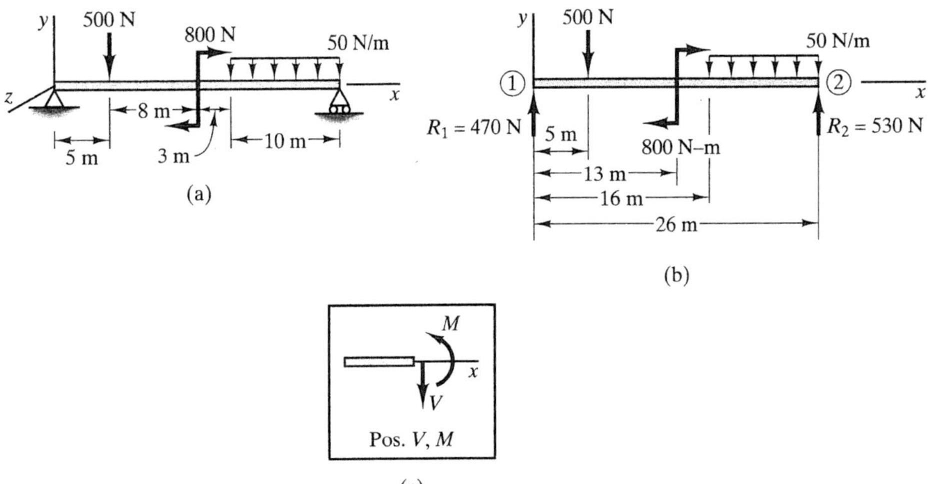

Figure 10.10. Simply supported beam problem showing positive convention for V and M.

We will however use the free-body diagram of the entire beam [Fig. 10.10(b)] to evaluate the supporting forces R_1 and R_2. Equating moments to zero at the ends of the beam, we get, using the right-hand rule of statics,

$$R_1 = 470 \text{ N} \qquad R_2 = 530 \text{ N}$$

We may now directly give the shear-force V and bending-moment M while viewing Fig. 10.10(b). Thus:

$0 < x < 5$:

$$V = R_1 = 470 \text{ N}$$
$$M = 470x \text{ N-m}$$

Example 10.4 (Continued)

$5 < x < 13$:

$$V = 470 - 500 = -30 \text{ N}$$
$$M = 470x - 500(x - 5) \text{ N-m}$$

$13 < x \leq 16$:

$$V = -30 \text{ N (same as previous interval)}$$
$$M = 470x - 500(x - 5) + 800 \text{ N-m}$$

$16 \leq x < 26$:

$$V = 470 - 500 - 50(x - 16) = -30 - 50(x - 16) \text{ N}$$
$$M = 470x - 500(x - 5) + 800 - \frac{50(x - 16)^2}{2} \text{ N-m}$$

Note we have not carried out the arithmetic for M. As indicated earlier, it is more useful to leave the bionomial expressions intact.

Example 10.5

Write the shear and bending-moment equations for the simply supported overhanging beam shown in Fig. 10.11.

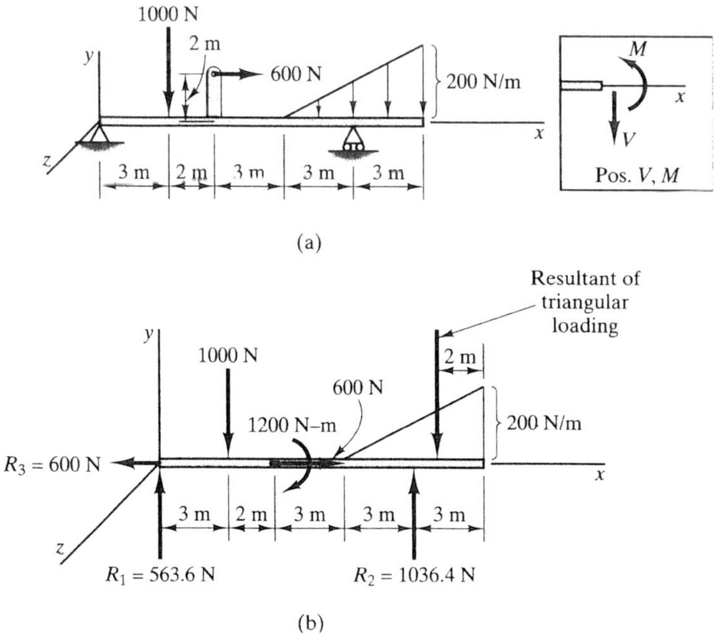

Figure 10.11. Simply supported beam with overhang.

Example 10.5 (Continued)

We shall first find the supporting force system using the free-body diagram shown in Fig. 10.11(b) and using the right-hand rule of statics.

$\Sigma M_{R_2} = 0$:

$$-R_1(11) + (1000)(8) - 1200 - \tfrac{1}{2}(6)(200)(1) = 0$$
$$R_1 = 563.6 \text{ N}$$

$\Sigma M_{R_1} = 0$:

$$R_2(11) - (1000)(3) - 1200 - \tfrac{1}{2}(6)(200)(12) = 0$$
$$R_2 = 1036.4 \text{ N}$$

$\Sigma F_x = 0$:

$$-R_3 + 600 = 0$$
$$R_3 = 600 \text{ N}$$

We now proceed with the shear-force and bending-moment equations for the various domains.

$0 < x < 3$:

$$V = 563.6 \text{ N}$$
$$M = 563.6x \text{ N-m}$$

$3 < x < 5$:

$$V = 563.6 - 1000 = -436.4 \text{ N}$$
$$M = 563.6x - 1000(x - 3) \text{ N-m}$$

$5 < x \leq 8$:

$$V = -436.4 \text{ N}$$
$$M = 563.6x - 1000(x - 3) + 1200 \text{ N-m}$$

For the next two domains we will need $w(x)$ of the triangular loading, for which we refer you to Fig. 10.12. From similar triangles we can see that

$$\frac{w(x)}{200} = \frac{x - 8}{6}$$
$$w(x) = \frac{200}{6}(x - 8)$$

Example 10.5 (Continued)

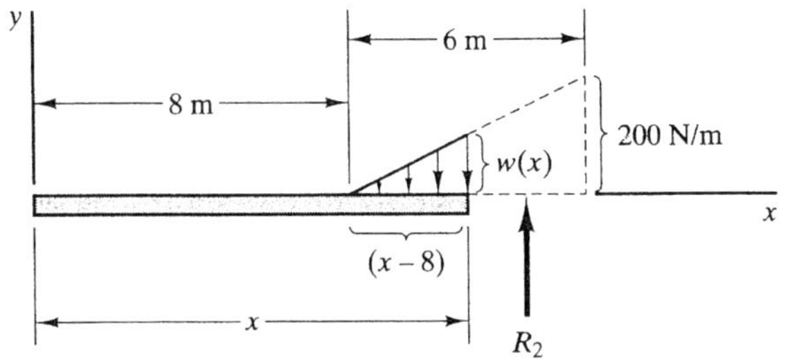

Figure 10.12. Diagram showing triangular loading contributions for domain $8 \le x < 11$.

Now continuing:

$8 \le x < 11$:

$$V = -436.4 - \frac{1}{2}(x-8)\left[\frac{200}{6}(x-8)\right]$$

$$= -436.4 - \frac{50}{3}(x-8)^2 \text{ N}$$

$$M = 563.6x - 1000(x-3) + 1200 - \frac{1}{2}(x-8)\left[\frac{200}{6}(x-8)\right]\left[\frac{1}{3}(x-8)\right]$$

$$= 563.6x - 1000(x-3) + 1200 - \frac{50}{9}(x-8)^3 \text{ N-m}$$

$11 < x \le 14$:

$$V = -436.4 - \frac{50}{3}(x-8)^2 + 1036.4$$

$$= 600 - \frac{50}{3}(x-8)^2 \text{ N}$$

$$M = 563.6x - 1000\,(x-3) + 1200 - \frac{50}{9}(x-8)^3 + 1036.4\,(x-11) \text{ N-m}$$

10.4 Differential Relations for Bending Moment, Shear Force, and Load

In the design of structures, we must often be able to quickly determine the magnitude and location of peak stresses. As we will see in the next chapter, the stresses in beams are a function of (among other variables) the bending moment and shear force at a given cross section.

Consequently, the rapid determination of the maximum values of M_z and V_y in a beam is essential to the design of these structures. From the methods presented in the previous section, we have at our disposal the equations describing the shear force and bending moment at any position along the beam. These equations may be plotted or examined mathematically to determine the location and magnitude of the maximum M_z and V_y in the beam. However, this can be a time consuming exercise. In what follows, we present *differential* relationships between the applied load, shear force, and bending moment. These differential relationships and their integrals may then be used to rapidly *sketch* shear-force and bending-moment diagrams for simple but common loads, allowing for the quick assessment of critical values of M_z and V_y along the beam.

We start by examining an infinitesimal slice of the beam Δx, shown in Fig. 10.13. We adopt the standard structural convention that a *downward* loading w on the beam is *positive.* Furthermore, we shall use

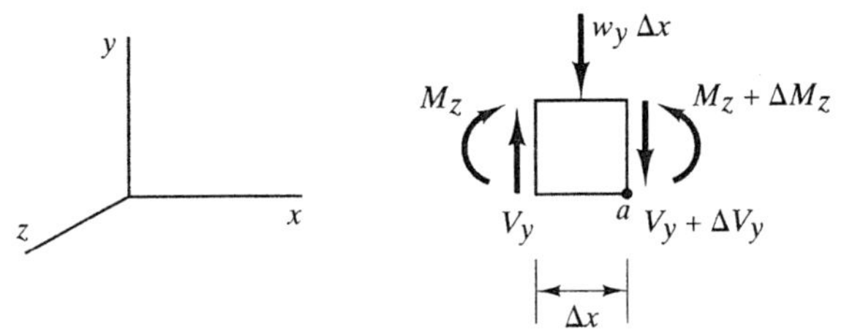

Figure 10.13. Beam element; structural convention.

a subscript for indicating the direction of the loading. Thus, we employ positive w_y for the loading intensity directed along the negative y axis. We have approximated the resultant of the loading w_y over the slice by the quantity $w_y \Delta x$ where w_y is a value of the loading intensity somewhere in the interval Δx. As Δx becomes smaller, this evaluation becomes more accurate and in the limit, as $\Delta x \to 0$, it becomes exact. We shall assume here that the weight of the beam has been included in the intensity of loading so that all force distributions acting on the element have been shown in the diagram. Again, we have employed the structural sign convention in the free-body diagram so that all forces and couple moments are shown in their positive senses in accord with the structural convention.

We now apply the equations of **equilibrium.** Thus, summing forces in the vertical direction (see Fig. 10.13), we get:

$\underline{\Sigma F_y = 0:}$

$$V_y - (V_y + \Delta V_y) - w_y \Delta x = 0$$

Taking moments about corner a of the element, we get using the right-hand rule of statics

$\Sigma M_a = 0$:

$$\therefore -M_z - V_y\Delta x + (w_y\Delta x)(\beta\Delta x) + (M_z + \Delta M_z) = 0$$

where β is some fraction that, when multiplied by Δx, gives the proper moment arm of the force $w_y\Delta x$ about corner a. These equations can be written in the following manner after we cancel terms and divide through by Δx:

$$\frac{\Delta V_y}{\Delta x} = -w_y$$

$$\frac{\Delta M_z}{\Delta x} = V_y - w_y\beta\Delta x$$

In the limit as $\Delta x \to 0$ we get the following differential equations relating loading, shear-force, and bending moment.[3]

$$\frac{dV_y}{dx} = -w_y \tag{10.1a}$$

$$\frac{dM_z}{dx} = V_y \tag{10.1b}$$

Differentiating Eq. (10.1b) with respect to x and then replacing the right-hand side of the equation using Eq. (10.1a), we get

$$\frac{d^2M_z}{dx^2} = -w_y(x) \tag{10.2}$$

The above are differential equations that relate the distributed load w_y to the shear-force V_y and the shear force to the bending-moment M_z, respectively. We can solve these differential equations by integrating Eqs. (10.1) between two positions on the beam. Thus, we have

[3]We note that Eq. (10.1a) is not valid at the location of a point load since the rate of change of the shear with respect to x is indeterminate at such points. Furthermore, a point load induces a jump in the magnitude of the shear force. For a similar reason, Eq. (10.1b) is not valid at the location of a point couple.

$$(V_y)_2 - (V_y)_1 = -\int_1^2 w_y\, dx$$

$$\therefore \quad (V_y)_2 = (V_y)_1 - \int_1^2 w_y\, dx \tag{10.3}$$

$$(M_z)_2 - (M_z)_1 = \int_1^2 V_y\, dx$$

$$\therefore \quad (M_z)_2 = (M_z)_1 + \int_1^2 V_y\, dx \tag{10.4}$$

Equation (10.3) means that the change in the shear force between two points on a beam equals *minus* the area (which may be positive or negative) under the loading curve between these points, provided there is no point force present in the interval. Similarly, Eq. (10.4) indicates that the change in bending moment between two points on a beam equals the area (which may be positive or negative) of the shear-force curve between these points, provided there are no point couples applied in the interval. In sketching the diagrams for the shear-force and bending-moment diagrams, we shall make good use of Eqs. (10.1), (10.3), and (10.4).

10.5 Sketching Shear-Force and Bending-Moment Diagrams

As we pointed out earlier, we do not wish to exactly plot the shear-force and bending-moment diagrams for the beam. We are mainly interested in the locations of the peak shear force and bending moment and in the general shape of the diagrams. Therefore, we would like to simply *sketch* the diagrams showing the general shape of, and key points for the shear force and bending moment. Here is how we propose to use Eqs. (10.1) through (10.4) to sketch the diagrams. Consider a generic portion of a loading diagram that does not contain any point forces as shown in [Fig. 10.14(a)]. Directly below, we are plotting the shear-force diagram wherein at the outset only the shear at point ① (i.e., V_1) is known. The first step in sketching the shear-force curve from position ① to position ②, is to employ Eq. (10.3). The value of V_2 according to this equation equals V_1 *minus* the area under the loading curve between positions ① and ② (the area A_w). This area is **positive** so that $V_2 = V_1 - (A_w) = V_1 - A_w$. Thus there is a decrease in the value of V as we go from position ① to position ② as has been shown in the diagram. Next, we want to sketch the shape of the shear-force diagram between the points ① and ②. This curve will take on one of the following three

$$\frac{dV}{dx} = -w \qquad V_2 = V_1 - \int_1^2 w\,dx$$

$$\frac{dM}{dx} = +V \qquad M_2 = M_1 + \int_1^2 V\,dx$$

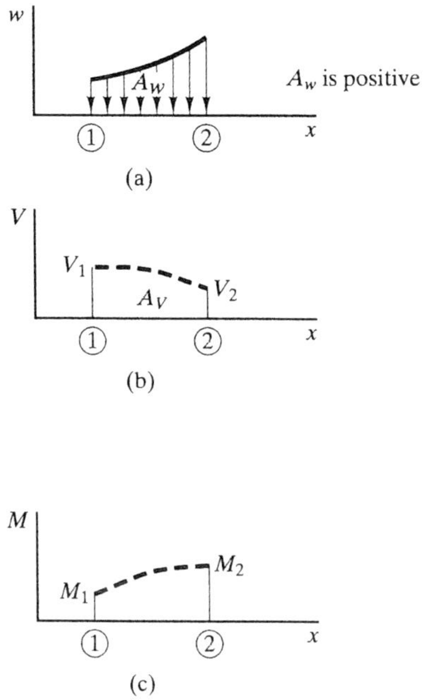

Figure 10.14. Sample construction of shear-force and bending-moment curves.

forms: a **straight line**, a curve that is **steepening**, or a curve that is **flattening**. See Fig. 10.15 for the nomenclature that we use to describe the manner in which the slope of the curve is changing as we move along the beam between the points ① and ②. The nomenclature should be clear if you imagine you are skiing or climbing along these curves. Now, we need to determine which of the three cases we have. From Eq. (10.1), we can examine the magnitude of the slope of the shear curve. That is, we examine the equation

$$\left|\frac{dV_y}{dx}\right| = |-w_y|$$

So, as we move along the beam, we can get the magnitude of the slope of the shear-force diagram from the magnitude of the loading diagram directly above. Thus, we can say that if between points ① and ②:

(1). the magnitude of loading w_y is *constant,* then the magnitude of the slope of the shear-force curve is *constant;* that is, the curve is a *straight line.*

(2). the magnitude of loading w_y is *increasing,* then the magnitude of the slope of the shear-force curve is *increasing;* that is, the curve is *steepening.*

(3). the magnitude of the loading is *decreasing,* then the magnitude of the slope of the shear-force curve is *decreasing;* that is, the curve is *flattening.*

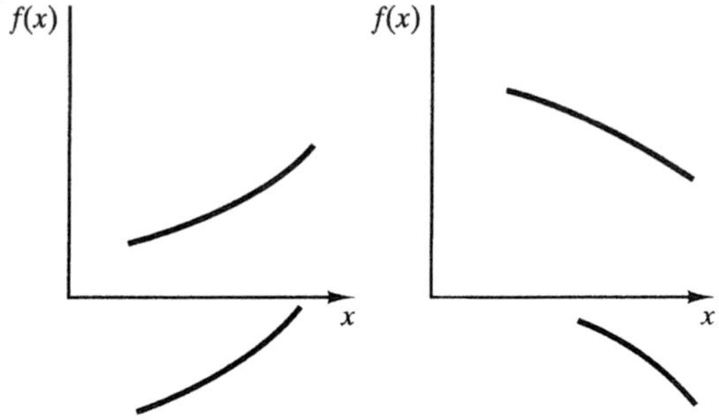

(a). $\left|\frac{df}{dx}\right|$ is *increasing*; slope is *steepening*

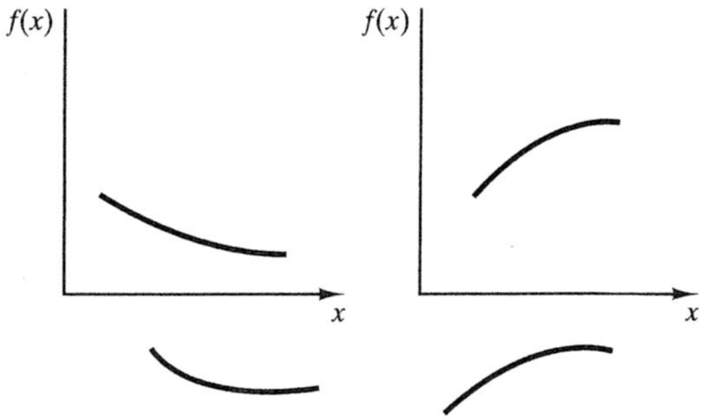

(b). $\left|\frac{df}{dx}\right|$ is *decreasing*; slope is *flattening*

Figure 10.15. Our nomenclature for the way a slope is changing as we increase x.

We see that in Fig. 10.14(a) $|w|$ is increasing. Hence, we connect point ① and ② with a *steepening* curve.

We now move to the bending-moment diagram and we will use a similar approach for sketching it as we did for the shear-force diagram. We start at point ① where M_1 is known and we assume that there are no point couples in the region from ① to ②. From the shear-force diagram in Fig. (10.14), we compute the area under the shear-force diagram between points ① and ②. We have called this area A_V. To sketch this portion of the bending-moment diagram, we use Eq. (10.4) to get the bending moment at position ②, (i.e., M_2). We have a positive area in the shear-force diagram between position ① and ② and so we must *add* this area to M_1 to get M_2. Thus, M_2 is larger than M_1 as shown in Fig. 10.14(c). Next, we must get the general shape of the bending-moment curve between points ① and ②. For this purpose, we use [Eq. 10.1(b)]. Using magnitudes, this equation can be given as follows:

$$\left|\frac{dM_z}{dx}\right| = |V_y|$$

So, as we move along the beam, we get the magnitude of the slope of the bending-moment curve from the magnitude of the shear-force curve directly above. Using the nomenclature of Fig. 10.15, we can say that if between points ① and ②:

1. the magnitude of the shear-force V_y is constant, then the magnitude of the slope of the bending-moment curve is *constant;* that is, the curve is a *straight line.*
2. the magnitude of the shear-force V_y is *increasing,* then the magnitude of the slope of bending-moment curve is *increasing;* that is, the curve is *steepening.*
3. the magnitude of the shear-force V_y is *decreasing,* then the magnitude of the slope of the bending-moment curve is *decreasing;* that is, the curve is *flattening.*

When the above procedure is used, we obtain in Fig. (10.14) a flattening curve for the bending moment between points ① and ②.

We now examine a beam problem using the approach set forth for sketching the shear-force and bending-moment diagrams for beams. In particular, note for the loading diagrams we shall insert point forces and point couples with their given or statically determined directions. It is the **loading** $\boldsymbol{w}$ that must be shown with the proper sign in accordance with the convention.

Example 10.6

Determine the shear-force and bending-moment equations for the simply supported beam shown in Fig. 10.16 and sketch the corresponding diagrams.

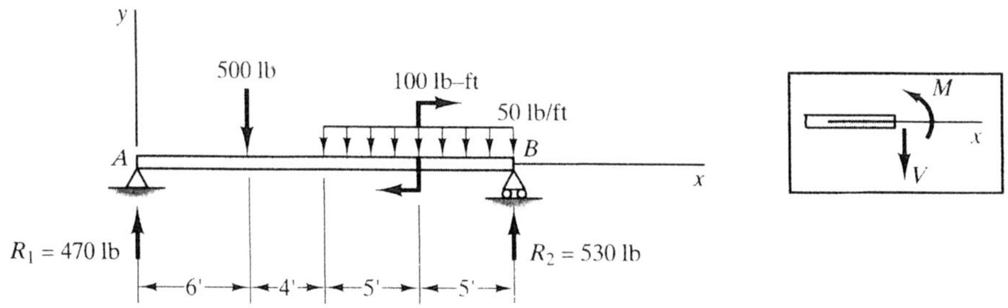

Figure 10.16. Simply supported beam.

We can directly state the shear-force and bending-moment equations as follows on observing Fig. 10.16:

Example 10.6 (Continued)

$0 < x < 6$:

$$V = R_1 = 470 \text{ lb}$$
$$M = R_1 x = 470x \text{ ft-lb}$$

$6 < x \leq 10$:

$$V = 470 - 500 = -30 \text{ lb}$$
$$M = 470x - 500(x - 6) \text{ ft-lb}$$

$10 \leq x < 15$:

$$V = 470 - 500 - 50(x - 10) \text{ lb}$$
$$M = 470x - 500(x - 6) - 50(x - 10)^2 / 2 \text{ ft-lb}$$

$15 < x < 20$:

$$V = 470 - 500 - 50(x - 10) \text{ lb}$$
$$M = 470x - 500(x - 6) - 50(x - 10)^2 / 2 + 100 \text{ ft-lb}$$

In Fig. 10.17, we show the loading diagram for the beam. Note that the *downward* 50 lb/ft loading is drawn as *positive* in the diagram in accordance with the structural convention.

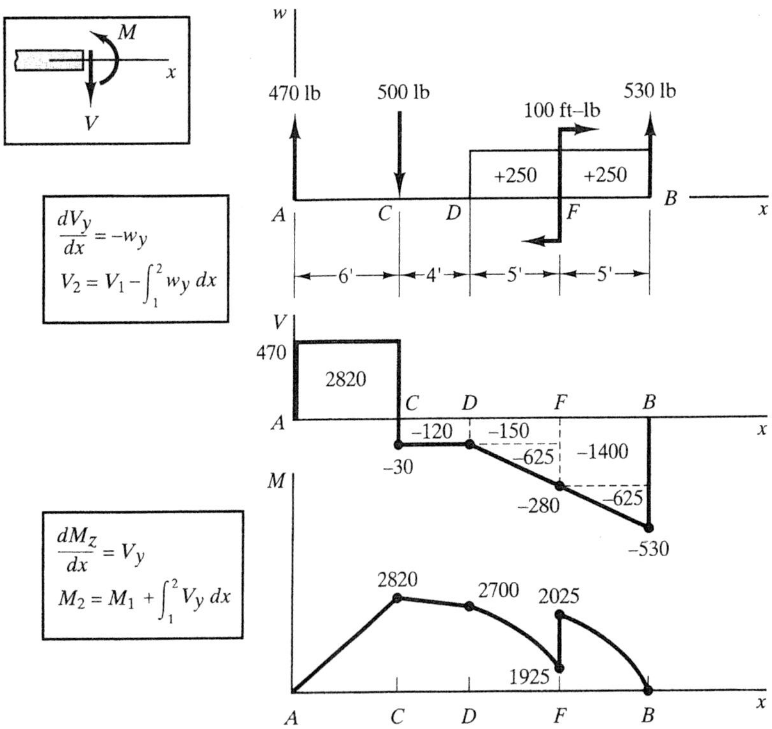

Figure 10.17. Shear and bending moment diagrams.

Example 10.6 (Continued)

Now going to the shear-force diagram, note that the upward force at the left end induces a downward shear force to the right of it. This downward shear force must be positive according to our sign convention and so the shear force jumps upward. Thus, *the jumps in shear force for point forces are in the same direction as the point forces.*[4] Since there is zero loading between A and C, the shear force is constant until we reach point C where there is a sudden downward drop equal to the 500 lb. load. From C to D, there is zero loading and thus the shear force in this region remains constant. Thus, we have horizontal lines for V from A to C and from C to D. Going to B, we have a decrease in the shear force equal to the positive area of 500 in the loading diagram between points D and B. Since w_y is positive and constant in the interval, then clearly the slope of the shear-force curve will be negative and constant in the interval (see formulas in first box). Finally, we undergo an upward jump of 530 lb. to reach a zero value at the end of the beam as must be the case if we have made no errors. Notice we have computed areas in the shear-force diagram for use later in drawing the bending-moment diagram.

Now, we proceed to the bending-moment diagram. Going from point A to point C, we increase M by 2820 ft-lb. The shear force is positive and constant in the range and hence the bending-moment curve will have a constant positive slope (see formulas in second box). From point C to point D we lose 120 ft-lb and we fill in with a straight line with a negative slope since V is negative and constant in this interval. From point D to point F, we lose some more bending moment in the amount of 775 ft-lb. Between points D and F, the shear force is negative and is *increasing in magnitude*. This means that the bending-moment curve will have a negative slope and will be *steepening*. At F, there is a clockwise couple moment and, as a result, the bending-moment curve jumps upward by 100 ft-lb. Finally, between F and B, we subtract 2025 ft-lb. We end up, as expected, with a zero bending moment at the end. Since V is negative with an increasing magnitude, the bending-moment diagram will have a negative slope and will be steepening.

10.6 Problems Requiring Equations and Diagrams

In Example 10.7 we got the equations and the diagrams of shear force and bending moment independently of each other. With simple

[4]It is important to note that while the shear force goes up or down in the direction corresponding to the direction of point loads when you are moving from left to right (i.e., in the direction of the positive x coordinate), the **opposite** occurs for the shear force when you are moving from right to left (i.e., in the **negative** x coordinate direction). Also, we suggest you use formulas in the boxes next to the curves.

loadings such as point forces, point couples, and uniform distributions, this can readily be done. Indeed, this covers many problems that occur in practice. Usually all that is needed is the labeled diagrams of the kind that we set forth in the previous example. In problems with more complex loadings we usually set forth the equations in the customary manner and then sketch the curves using the *equations* to give key values of V and M [the areas for the various curves are no longer the simple familiar ones, thus precluding advantageous use of Eqs. (10.3) and (10.4)]; the key points are then connected by curves sketched by making use of the slope relations, as we have done in Example 10.6.

Example 10.7

Shown in Fig. 10.18(a) is a cantilever beam, with a parabolic loading as well as a point force and couple moment. Shear-force and bending-moment equations and diagrams are to be formulated for the exposed portion of the beam.

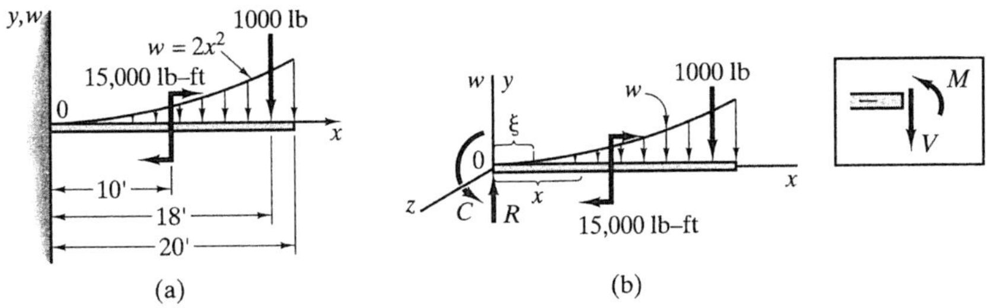

Figure 10.18. Cantilever beam and its loading diagram.

We shall solve for the force and couple moment on the section of the beam adjacent to the wall using the right-hand rules of statics and we shall consider these as external supports for the beam [see Fig. 10.18(b)]. Because these quantities are to be handled as external loads, we do not apply the sign convention of shear and bending moments to them. Thus,

$\underline{\Sigma F_y = 0:}$

$$R - 1000 - \int_0^{20} 2x^2 dx = 0$$

$$\therefore R = 6333 \text{ lb}$$

$\underline{\Sigma M_0 = 0:}$

$$C - 15{,}000 - (1000)(18) - \int_0^{20} 2x^3 dx = 0$$

$$\therefore C = 113{,}000 \text{ ft-lb}$$

Example 10.7 (Continued)

We may next give the shear-force and bending-moment equations using ξ as a dummy variable [see Fig. 10.18(b)] to locate any position along the beam up to position x. We then may say:

$0 < x < 10$:

$$V = 6333 - \int_0^x 2\xi^2 \, d\xi$$

$$= -\frac{2x^3}{3} + 6333 \text{ lb} \qquad \text{(a)}$$

$$M = -113{,}000 + 6333x - \int_0^x 2\xi^2(x - \xi)\, d\xi$$

Keeping x constant during integration, we have

$$M = -\frac{x^4}{6} + 6333x - 113{,}000 \text{ ft-lb} \qquad \text{(b)}$$

$10 < x < 18$:

$$V = 6333 - \int_0^x 2\xi^2 \, d\xi$$

$$= -\frac{2x^3}{3} + 6333 \text{ lb} \qquad \text{(c)}$$

$$M = -113{,}000 + 6333x - \int_0^x 2\xi^2(x - \xi)\, d\xi + 15{,}000$$

$$= -\frac{x^4}{6} + 6333x - 98{,}000 \text{ ft-lb} \qquad \text{(d)}$$

$18 < x \leq 20$:

$$V = 6333 - \int_0^x 2\xi^2 \, d\xi + 1000$$

$$= -\frac{2x^3}{3} + 5333 \text{ lb} \qquad \text{(e)}$$

$$M = -113{,}000 + 6333x - \int_0^x 2\xi^2(x - \xi)\, d\xi + 15{,}000 - 1000(x - 18)$$

$$= -\frac{x^4}{6} + 5333x - 80{,}000 \text{ ft-lb} \qquad \text{(f)}$$

To sketch the diagrams and thus ascertain the positions of the key points, we use the above equations and the slope relations. The

Example 10.7 (Continued)

loading diagram has been redrawn (Fig. 10.19a). The shear force is 6330 lb to the right of the support at A and from Eq. (c) we see that, just to the left of where the 1000-lb load is applied at C (i.e., at $x = 18$), the shear force is 2445 lb (see Fig. 10.19b). Between A and C the loading w is positive and increasing in magnitude. Accordingly, the shear-force curve has a negative slope which is steepening. At C the 1000-lb load contributes a sudden -1000 shear-force increment. According to Eq. (e), at D (i.e., at $x = 20$), the shear force must be zero. Between C and D the loading is positive and increasing in magnitude. The slope of the shear-force diagram then is negative and is steepening. Clearly, the greatest shear force is 6330 lb at the base.

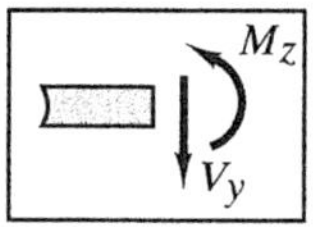

$$\frac{dV_y}{dx} = -w_y$$

$$V_2 = V_1 - \int_1^2 w_y \, dx$$

$$\frac{dM_z}{dx} = V_y$$

$$M_2 = M_1 + \int_1^2 V_y \, dx$$

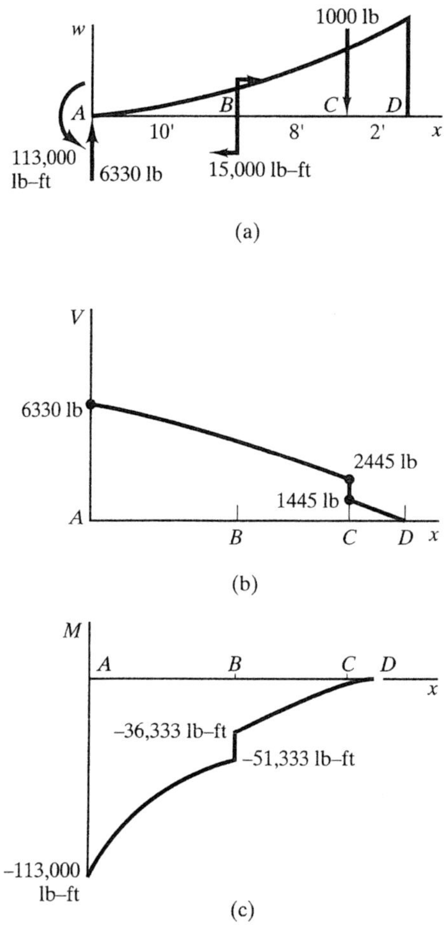

Figure 10.19. (a) Loading diagram; (b) shear-force diagram; (c) bending-moment diagram.

Example 10.7 (Continued)

As for the bending-moment diagram we have at A the instantaneous contribution of $-113{,}000$ ft-lb from the couple at the support. At B to the left of the point couple (at $x = 10$) we have from Eq. (b) the value of $-51{,}333$ ft-lb [see Fig. 10.19(c)]. Between A and B the shear force is positive and decreasing in magnitude. Accordingly, the slope of the bending moment is positive and flattening. As we cross B there is a sudden $+15{,}000$-ft-lb. increment of bending moment resulting from the applied point couple. At D (i.e., at $x = 20$) the bending moment is clearly zero. Between B and D the shear force is positive and decreasing in magnitude with a discontinuous jump at C. Accordingly, between B and D the slope of the bending-moment diagram is positive and flattening with a discontinuity in slope (a cusp) at C. The greatest bending moment, like the shear force, is at the base.

10.7 Additional Considerations

You will note that in the previous examples the key points of the shear-force and bending-moment diagrams were evaluated and labeled. From these diagrams, the maximum positive and negative shear force and bending moment in the beam were identified. We wish to point out that Eq. (10.1a) relates the *derivative* of the shear force to the load. Clearly, if we want to find the extreme values of a function (i.e., relative maximum and minimum), a first step would be to set the first derivative equal to zero. Referring back to Eq. (10.1a), this implies that locations where the load diagram passes through zero should be checked as possible locations of extreme shear force. Likewise, Eq. (10.1b) relates the derivative of the bending moment to the shear force. Again, locations where the shear force is zero (i.e., where the derivative of the bending-moment curve is zero) should be checked as possible locations of extreme bending-moment. Therefore, any position where the loading diagram or shear-force diagram passes through zero should be marked and evaluated (see Figs. 10.20 and 10.21). In the following example, we illustrate this procedure.

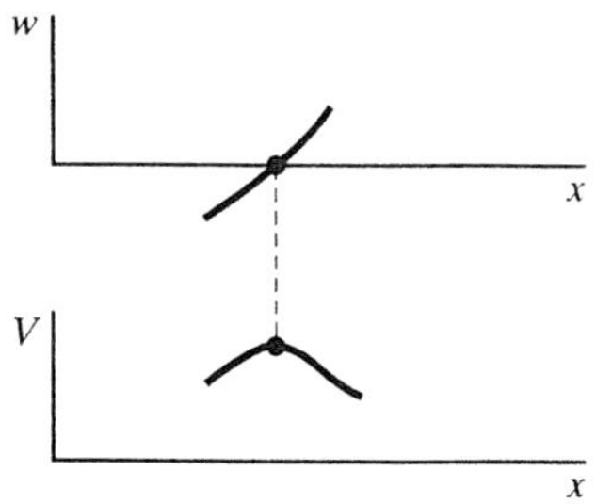

Figure 10.20. Possible maximum shear.

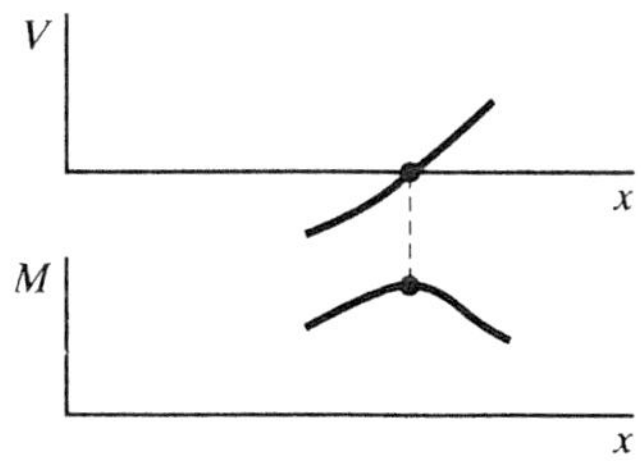

Figure 10.21. Possible maximum bending moment.

Example 10.8

Sketch the shear-force and bending-moment diagrams, being sure to locate and evaluate the bending moment having the largest value for the simply-supported beam in Fig. 10.22.

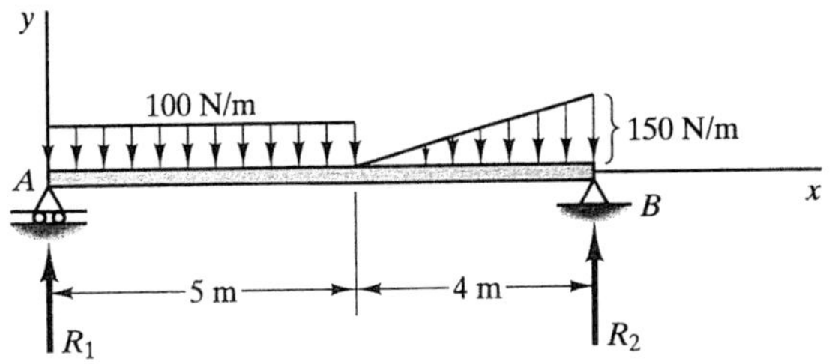

Figure 10.22. Simply-supported beam.

We shall first find the supporting forces, making use of the fact that the resultant force from the triangular loading equals the area under the loading with a line of action passing through the centroid of the triangle.

From **equilibrium** considerations we have

$\Sigma M_A = 0$:

$$-(100)(5)(2.5) - \tfrac{1}{2}(4)(150)[5 + \tfrac{2}{3}(4)] + 9(R_2) = 0$$

$$R_2 = 394.4 \text{ N}$$

$\Sigma M_B = 0$:

$$(100)(5)(6.5) + \tfrac{1}{2}(4)(150)(\tfrac{4}{3}) - 9R_1 = 0$$

$$R_1 = 405.6 \text{ N}$$

We can now sketch the loading, shear-force, and bending-moment diagrams. We do this in the usual way (see Fig. 10.23) but notice that the shear-force curve crosses the x axis at some point E at a distance d to the right of A. We must find the value of d because, as pointed out earlier, we may have a maximum bending moment at this point. Using the property of similar triangles, we can say:

$$\frac{d}{405.6} = \frac{5 - d}{94.4}$$

$$\therefore d = 4.056 \text{ m}$$

Example 10.8 (Continued)

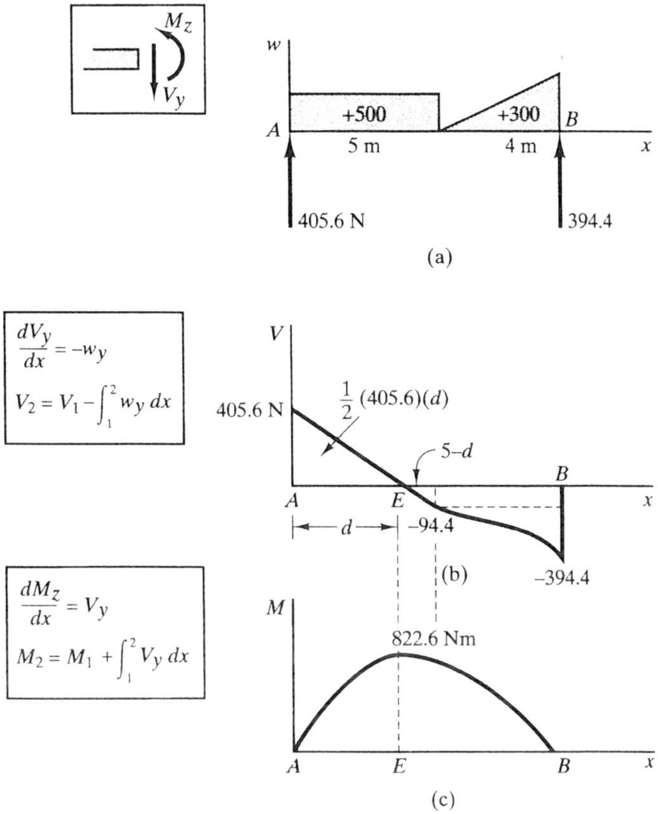

Figure 10.23. Shear-force and bending-moment diagrams showing maximum bending moment at location of zero shear force.

Now going to the bending-moment diagram it is clear that the maximum bending moment is indeed at point E and has the value

$$M_{\max} = (\tfrac{1}{2})(405.6)(4.056) = 822.6 \text{ N-m}$$

Before closing this section, we would like to point out a simplification that you may wish to use in conjunction with triangular loads of the orientation shown in Fig. 10.24. If we add a triangular loading of identical shape but increasing in value with x, we can form the uniform loading of value f_o shown in Fig. 10.25. Now to maintain equivalence we insert in

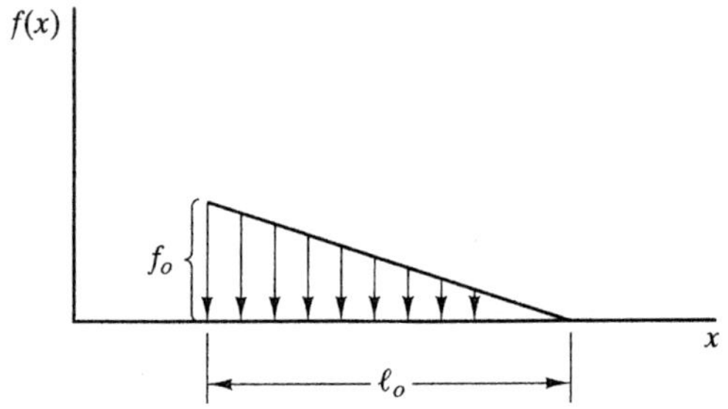

Figure 10.24. Triangular loading.

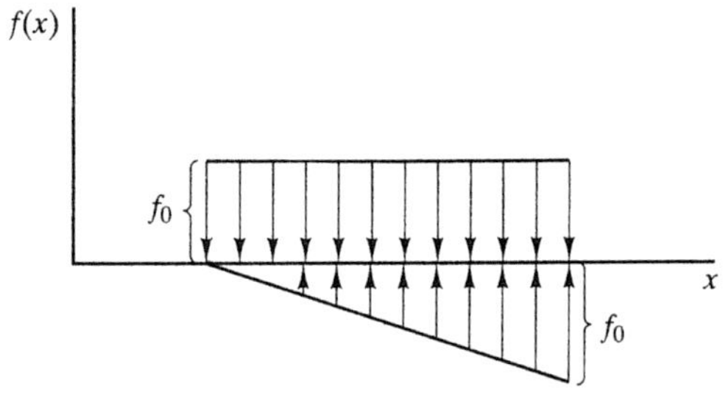

Figure 10.25. Replacement loading.

Fig. 10.25 the same triangular loading that was added in Fig. 10.25, but instead of going downward it should now be going upward, as has been shown in Fig. 10.25. We now have a uniform loading and a triangular loading, the latter in a much simpler orientation. Of course, you need not resort to this procedure; it is offered simply as a suggestion.

10.8 Closure

We have set forth in this chapter the methodology of writing equations and sketching diagrams for the shear force and bending moment at sections along a beam once the supporting forces are known. You will note that only the formulations of rigid-body statics were used in this chapter. In addition, a commonly used sign convention for these tasks was introduced. We hope that you have developed skill in writing the shear-force and bending-moment equations and also with sketching the corresponding diagrams with both speed and accuracy. We will be needing these capabilities as we move forward to the next chapters to study stresses and deformations of beams.

We finish by presenting the key results for the two sign conventions. The authors allow students to use this display on tests.

Stress Convention	**Structural Convention**
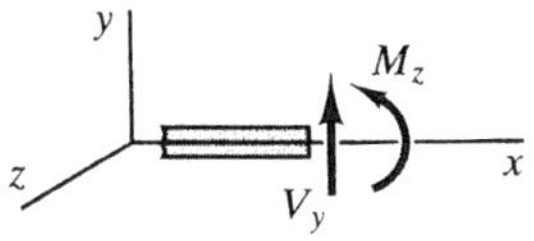	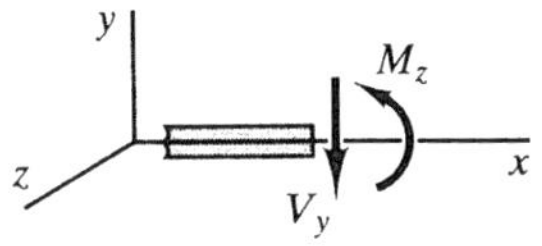
w is positive up	w is positive down
$\dfrac{dV_y}{dx} = -w_y$	$\dfrac{dV_y}{dx} = -w_y$
$V_2 = V_1 - \int_1^2 w_y dx$	$V_2 = V_1 - \int_1^2 w_y dx$
$\dfrac{dM_z}{dx} = -V_y$	$\dfrac{dM_z}{dx} = V_y$
$M_2 = M_1 - \int_1^2 V_y dx$	$M_2 = M_1 + \int_1^2 V_y dx$

10.9 A Look Back

In your statics course, you were introduced to a set of measures that defined the distribution of a plane area relative to a reference. These quantities were called first moments of area and second moments and products of area. In particular, the **first moment**, M_x, of an area A about the x axis is defined as

$$M_x = \iint_A y dA'$$

The centroid is the position where one could concentrate the entire area so that the first moment about an axis would be given as Ad where d is the perpendicular distance from the centroid to the axis (see Fig. 10.26). The position of the centroid is a property of the area. Finally, it follows that the first moment about a centroidal axis is always zero (since $d = 0$).

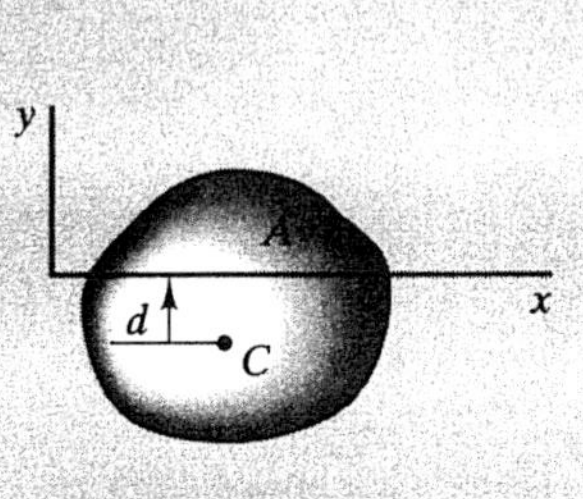

Figure 10.26. The first moment of area A about the x-axis is Ad. C is the centroid of area A.

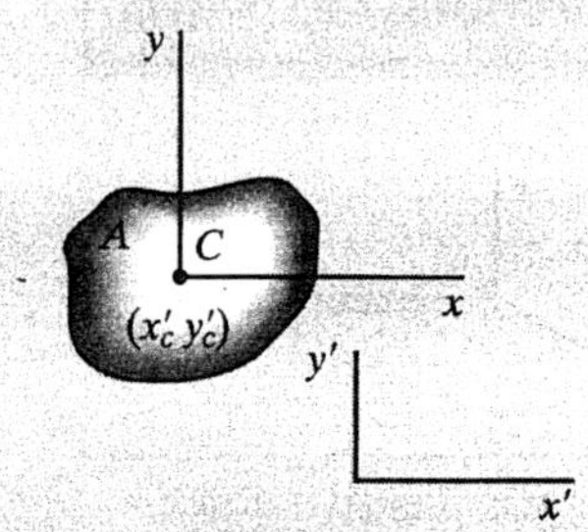

Figure 10.27.

$$I_{x'x'} = I_{xx} + A(y_c')^2$$
$$I_{y'y'} = I_{yy} + A(x_c')^2$$
$$I_{x'y'} = I_{xy} + A(x_c')(y_c')$$

The second moments of area are defined for axes x and y as

$$I_{xx} = \iint_A y^2 dA' \qquad I_{yy} = \iint_A x^2 dA'$$

while the product of area is defined as

$$I_{xy} = \iint_A xy dA'$$

These quantities can be obtained using the mass inertia tensor. This may be done by starting with a thin plate and then letting the thickness t go to zero while at the same time letting the density ρ go to infinity such that the product ρt becomes unity in the limit.[6]

In using the second moments and products of area, we can make good use of the so-called **parallel axis theorems** whereby the second moment of area about some axis x' equals the second moment of area about a parallel axis going through the centroid plus the product of the area times the perpendicular distance squared from the centroid to axis x'. The product of area for a pair of axes $x'y'$ (Fig. 10.27) equals the product of area about the parallel set of axes at the centroid plus the area multiplied by the coordinates of the centroid relative to the axes $x'y'$.

We shall have use of these formulations in the next chapter.

[6]See Shames, I.H., *Engineering Mechanics*, Fourth Edition, 1997, Prentice Hall Inc., Section 9.3.

Highlights (10)

We introduced conventions for the loading w_y, the shear force V_y, and the bending moment M_z at a section in order to have a representation of these quantities at the section that was independent of whether we used a right-hand or a left-hand free-body diagram to expose the section of interest. We have used in this text the structural convention. To determine the aforementioned quantities, we use the equations of statics either formally or mentally and employ **positive** section forces and moments according to our structural convention in this process. By this procedure, the algebraic sign emerging for the section forces and moments by using equations of static equilibrium would then have the proper convention sign. Remember that a positive couple moment for the right-hand rule will correspond to the case where a standard screw will advance in a positive coordinate direction when turned with the same sense as the couple moment.

We write the equations for V_y and M_z for the domains of the beam being sure to exclude positions where there is a point force or a point couple because of the indeterminacy of either V_y and/or M_z at such positions. Additionally, it is important that the student develop skill and speed in sketching shear and bending moment diagrams. The use of the *steepening* and the *flattening* terminology will be very helpful in this procedure. Remember, an increase in the magnitude of the loading function w, means a steepening slope of the corresponding shear curve and a decrease in the magnitude of the loading function w, means a flattening of the slope of the corresponding shear diagram. The very same thinking is followed when using the shear diagram to draw the bending moment diagram.

PROBLEMS

Shear Force and Bending Moment Eqs. Using Free-Body Diagrams	10.1–10.16
Direct Formulation of Shear-Force and Bending-Moment Eqs.	10.17–10.23
Sketching Shear-Force and Bending-Moment Diagrams	10.24–10.43
Unspecified Section Problems	10.44–10.65
Computer Problems	10.66–10.67
Programming Project 10	

In Problems 10.1 to 10.16, use free-body diagrams as an aid.

10.1. [10.2] For the beam shown in Fig. P.10.1, what is the shear force and bending moment at the following positions?

(a). 5 ft from the left end
(b). 12 ft from the left end
(c). 5 ft from the right end

Write the shear force and bending moment as a function of x for the beam.

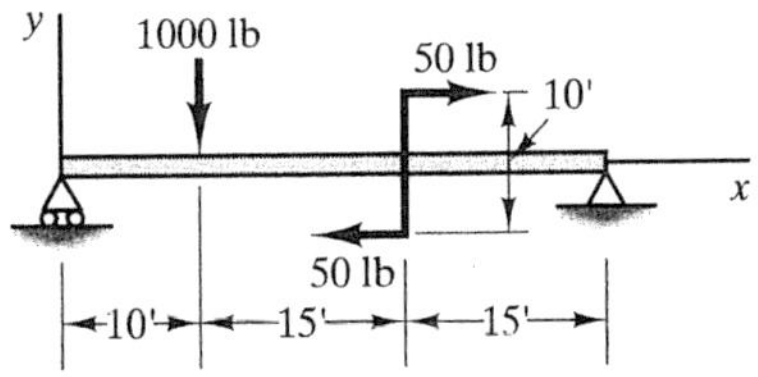

Figure P.10.1.

10.2. [10.2] A hoist can move along a beam (see Fig. P.10.2) while supporting a 10,000-N load. If the hoist starts at the left and moves from $\bar{x} = 3$ m to $\bar{x} = 12$ m, determine the shear force and bending moment at A in terms of $\bar{x}$. At what position $\bar{x}$ do we get the maximum shear force at A and the maximum bending moment at A. What are their values?

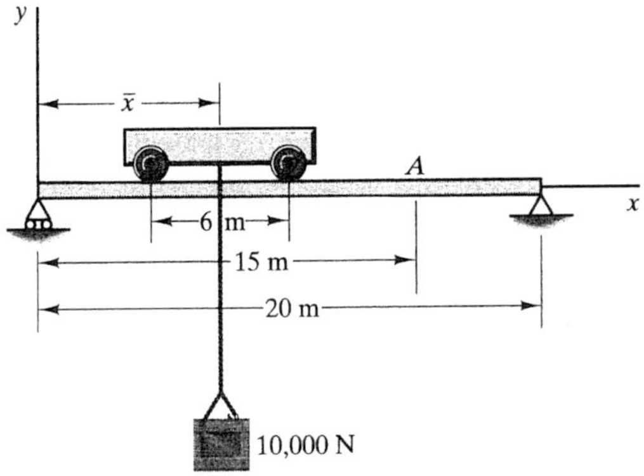

Figure P.10.2.

10.3. [10.2] Formulate the shear-force and bending-moment equations for the simply supported beam shown in Fig. P.10.3. Do not include the weight of the beam.

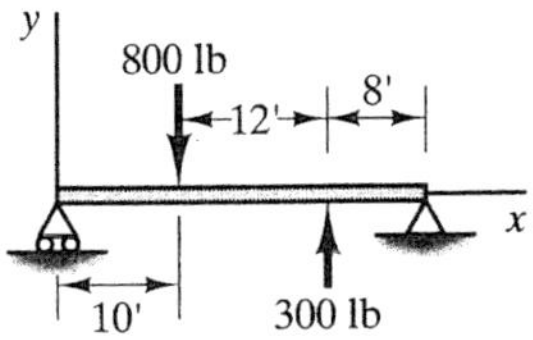

Figure P.10.3.

10.4. [10.2] Formulate the shear-force and bending-moment equations for the cantilever beam shown in Fig. P.10.4. Do not include the weight of the beam.

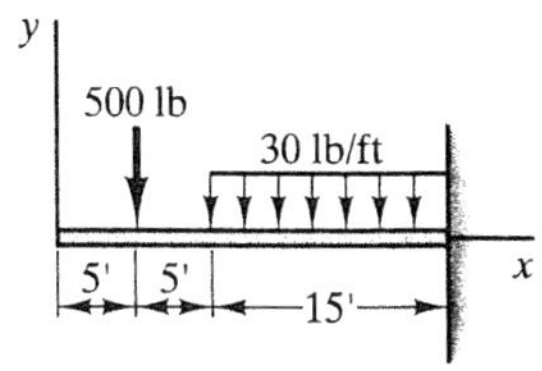

Figure P.10.4.

10.5. [10.2] Determine the shear-force and bending-moment equations for the beam in Fig. P.10.5.

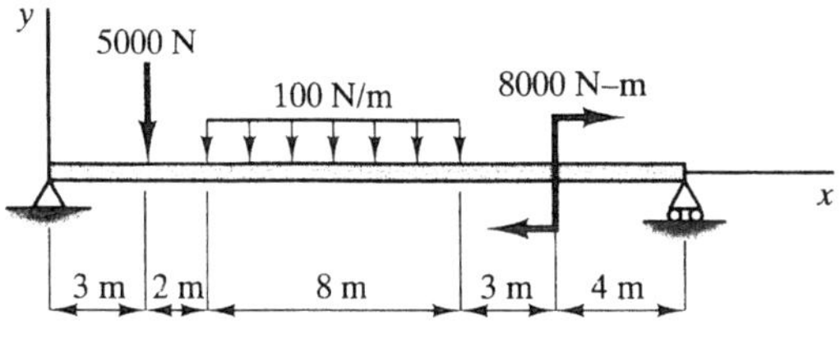

Figure P.10.5.

10.6. [10.2] *BC* Is cantilevered at *C* and pinned to *AB* at *B* (see Fig. P.10.6). Formulate the shear-force and bending-moment equations.

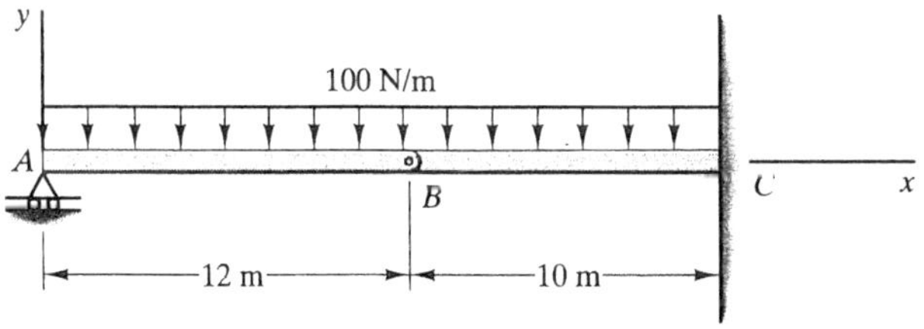

Figure P.10.6.

10.7. [10.2] Formulate shear-force and bending-moment equations for the simply supported overhanging beam (see Fig. P.10.7).

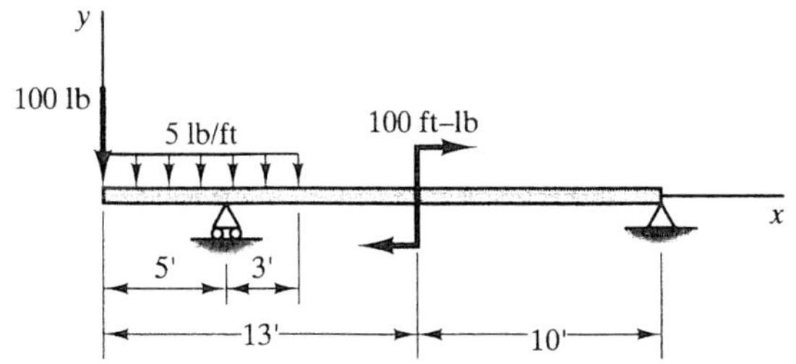

Figure P.10.7.

10.8. [10.2] A cantilever beam *AB* is pinned at point *A* to member *CA* as shown in Fig. P.10.8. Compute the shear-force and the bending-moment equations.

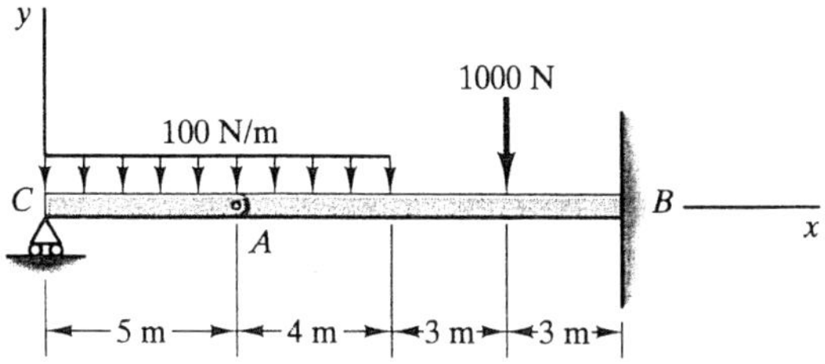

Figure P.10.8.

10.9. [10.2] Formulate the shear-force and bending-moment equations for the cantilever beam *AD* (Fig. P.10.9). *BC* is welded to beam *AD*.

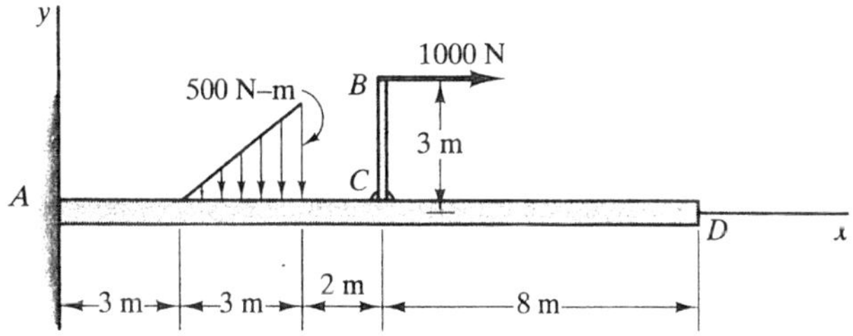

Figure P.10.9.

***10.10. [10.2]** Compute shear force, axial force, and bending moments from *A* to *C* for the bent beam shown in Fig. P.10.10 as functions of *s* along the centerline of the beam.

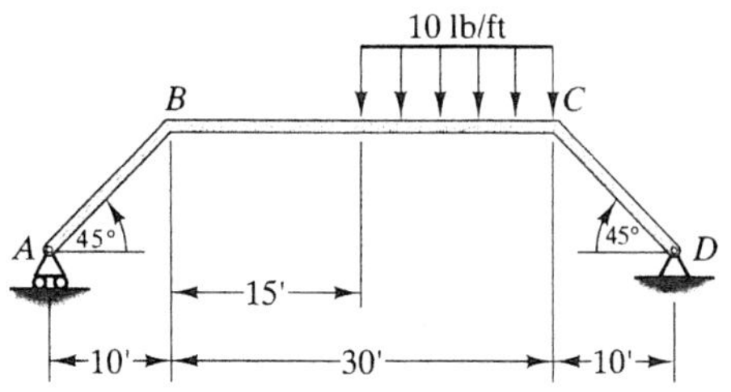

Figure P.10.10.

10.11. [10.2] Determine the shear force, bending moment, and axial force as functions of θ for the circular beam shown in Fig. P.10.11.

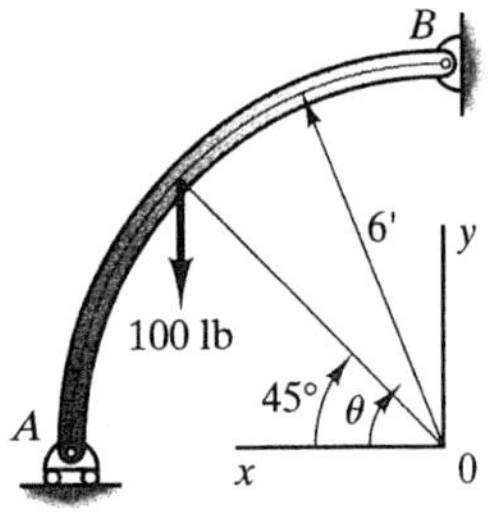

Figure P.10.11.

> Problems 10.12 to 10.16 involve three-dimensional loading for computations of shear forces, bending moments, and twisting moments over and above the V and M for the two-dimensional cases of previous problems.

*__10.12. [10.2]__ Shown in Fig. P.10.12 is a simply supported beam loaded in two planes. This means there will be shear-force components V_y and V_z and bending-moment components M_z and M_y. Compare these as functions of x. The beam is 40 m in length.

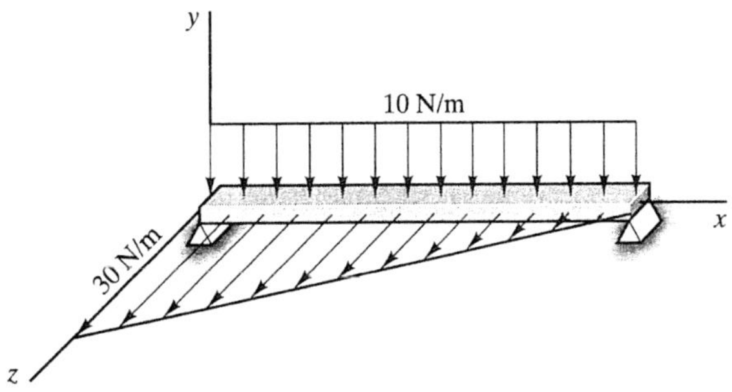

Figure P.10.12.

*__10.13. [10.2]__ Oil flows from a tank through a pipe AB (Fig. P.10.13). The oil weighs 40 lb/ft^3 and, in flowing, develops a drag on the pipe of 1 lb/ft. The pipe has an inside diameter of 3 in. and a length of 20 ft. Flow conditions are assumed the same along the entire length of the pipe. What are the shear force, bending moment, and axial force along the pipe from the oil? z is parallel to the ground and the xy plane is perpendicular to the ground.

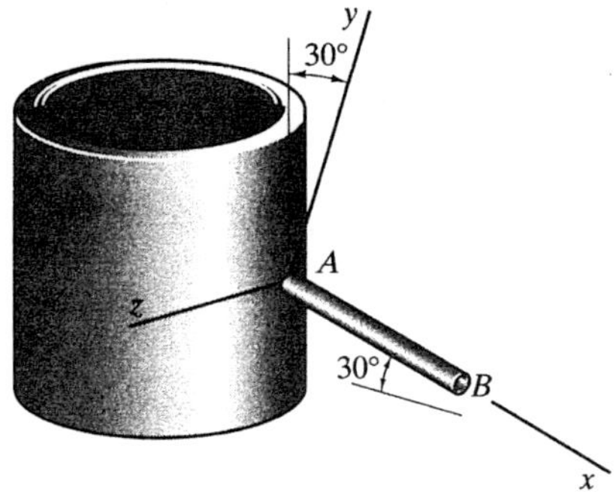

Figure P.10.13.

*__10.14. [10.2]__ Friction forces applied to a cantilever beam (see Fig. P.10.14) result in a uniform torque per unit length from A to B having the value 10 N-m/m while from C to D the friction forces result in an axial load distribution given by f starting from zero at C and increasing linearly until this loading intensity is 20 N/m at D. What are the various moment and force components transmitted through the beam?

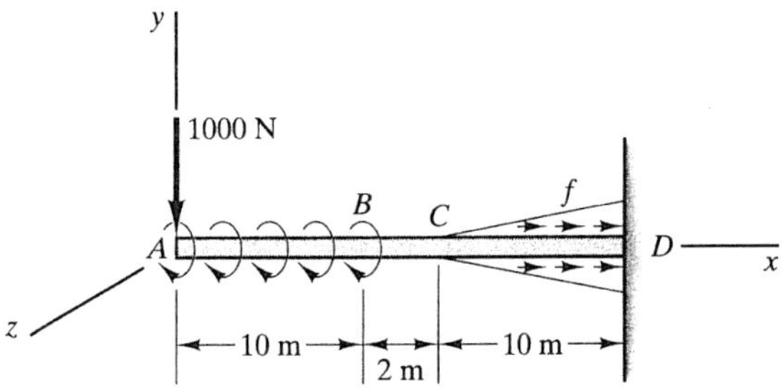

Figure P.10.14.

*__10.15. [10.2]__ A pipe is shown in Fig. P.10.15 closed at D. It weighs 10 lb/ft and has an inside diameter of 2 in. If it is full of water and the pressure of the water is that of the atmosphere at the entrance A, compute the shear force, axial force, and bending moment of the pipe from A to B.

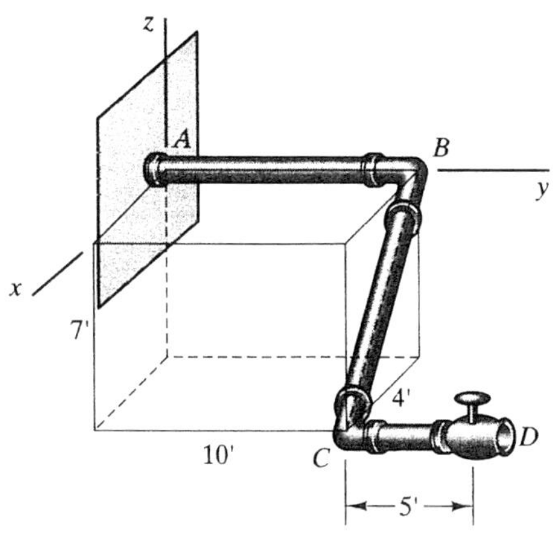

Figure P.10.15.

*10.16. [10.2] Consider a one-cylinder compressor as shown in Fig. P.10.16. The pressure at the piston face is 200 psia. The bore is 3 in. and the stroke is 4 in., while the length of the connecting rod is 6 in. For the position shown, plot V_y and V_z for member AB. Also plot M_y and M_z for member AB. EB is horizontal at instant of interest.

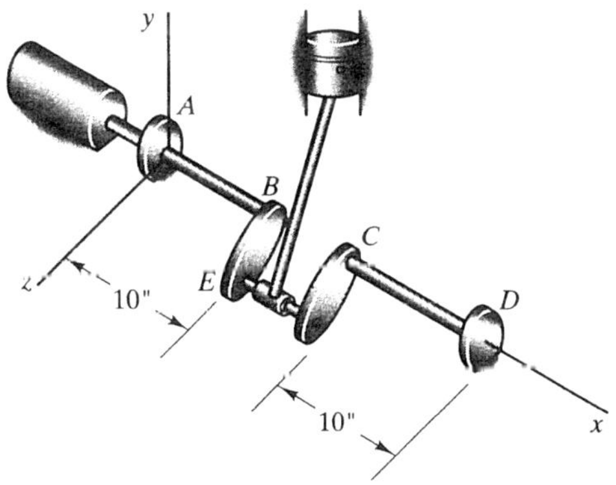

Figure P.10.16.

10.17. [10.3] Do Problem 10.3 without free-body diagrams for computing V and M.

10.18. [10.3] Do Problem 10.4 without free-body diagrams for computing V and M.

10.19. [10.3] Do Problem 10.5 without the use of free-body diagrams for computing V and M.

10.20. [10.3] Do Problem 10.6 without using free bodies for computing V and M.

10.21. [10.3] Do Problem 10.7 without using free bodies for finding V and M equations.

10.22. [10.3] Do Problem 10.8 not using free-body diagrams for getting V and M equations.

10.23. [10.3] Do Problem 10.9 without the aid of free-body diagrams for finding V and M equations.

In Problems 10.24 to 10.31 write shear-force and bending-moment equations without the aid of free-body diagrams. Then sketch shear and bending-moment diagrams labeling key points.

10.24. [10.5]

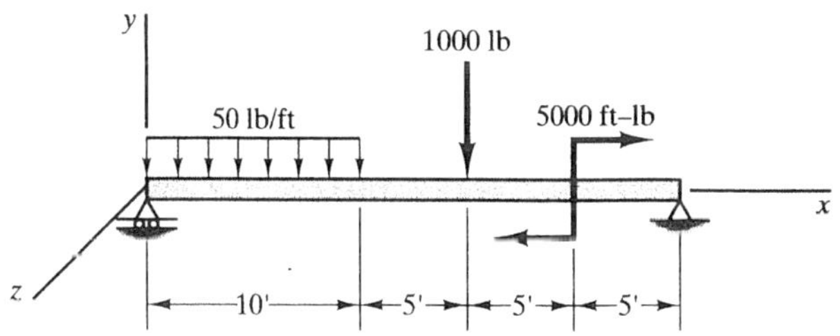

Figure P.10.24.

10.25. [10.5]

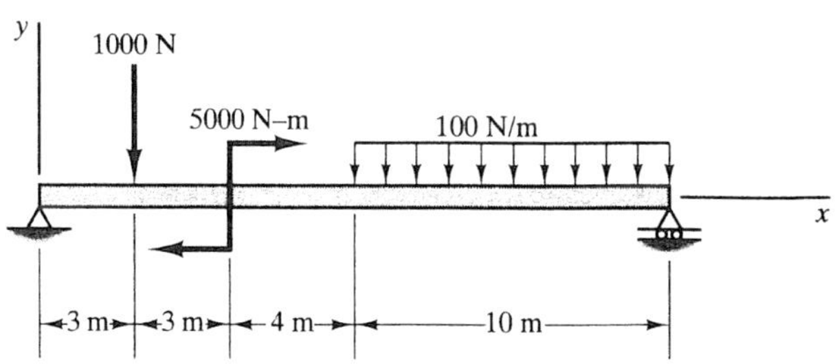

Figure P.10.25.

10.26. [10.5]

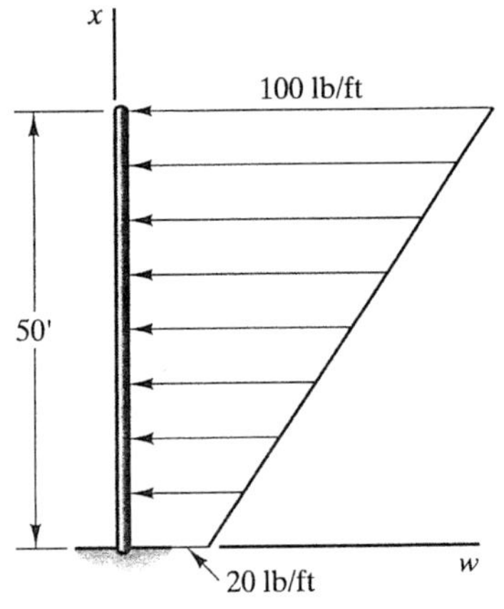

Figure P.10.26.

10.27. [10.5]

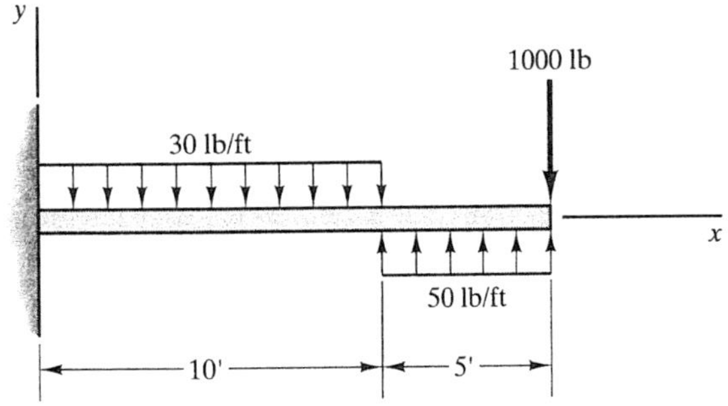

Figure P.10.27.

10.28. [10.5]

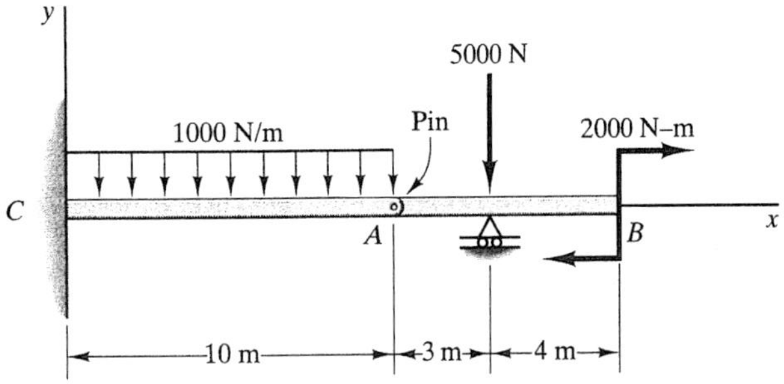

Figure P.10.28.

10.29. [10.5]

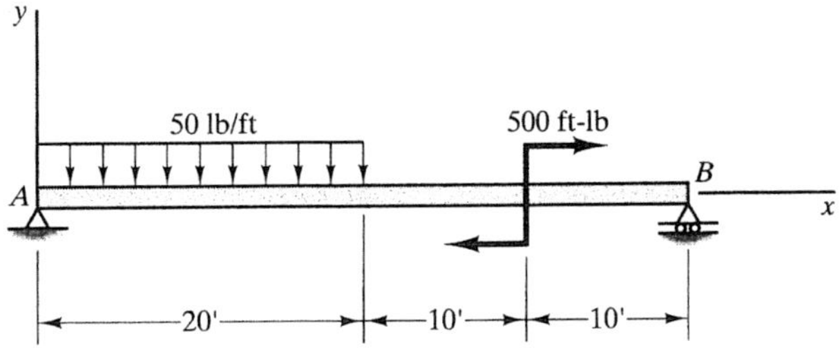

Figure P.10.29.

10.30. [10.5]

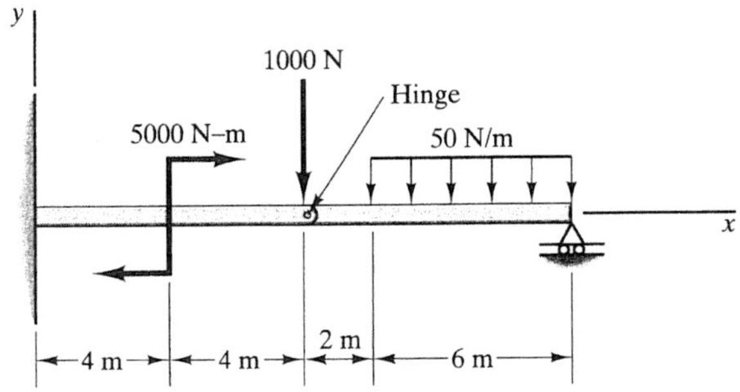

Figure P.10.30.

10.31. [10.5] Note there is a pin at B in Fig. P.10.31. The 1000-N load and 500-N-m couple-moment are applied to a rigid attachment welded to the beam $\overline{AB}$.

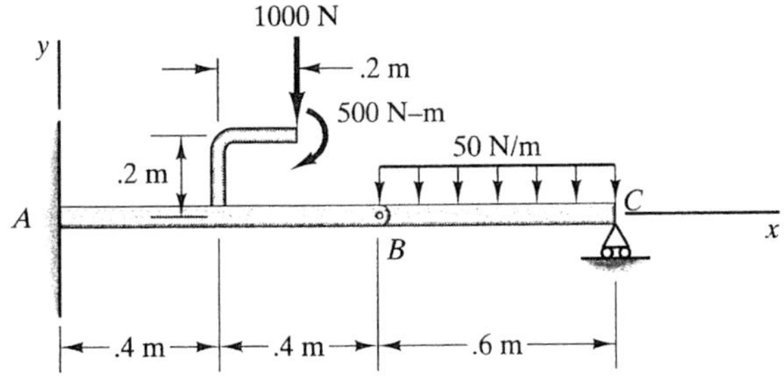

Figure P.10.31.

10.32. [5.5] For the testing of materials in bending, the "Four-Point Bend Test" is sometimes used, as shown in Fig. P.10.32. Compute the shear-force and bending-moment diagram for this configuration. List some reasons as to why this setup is of particular use. (*Hint:* What is of interest in the bending-moment distribution?)

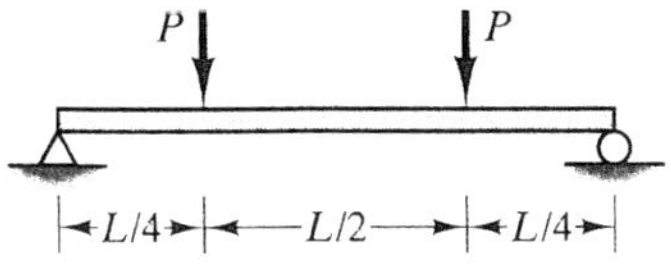

Figure P.10.32.

10.33. [10.5] A temporary elevator, shown in Fig. P.10.33, is erected at a construction site. The fully loaded elevator weighs 4500 lb and it accelerates upward uniformly from rest to a steady velocity of 2000 ft/min in 10 seconds. Draw the bending-moment diagram of the beam for the maximum load case. Take the weight of winch and motor as 500 lb and positioned as shown.

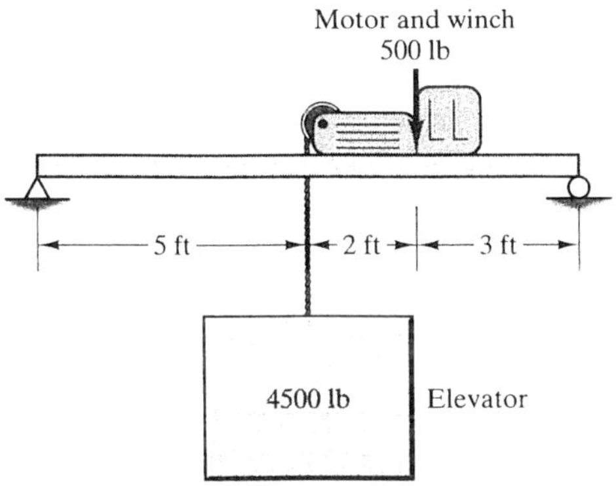

Figure P.10.33.

10.34. [10.5] Find the supporting forces for the simply supported beam in Fig. P.10.34. Then sketch the shear-force and bending-moment diagrams, labeling key points. 1K = 1000 lb.

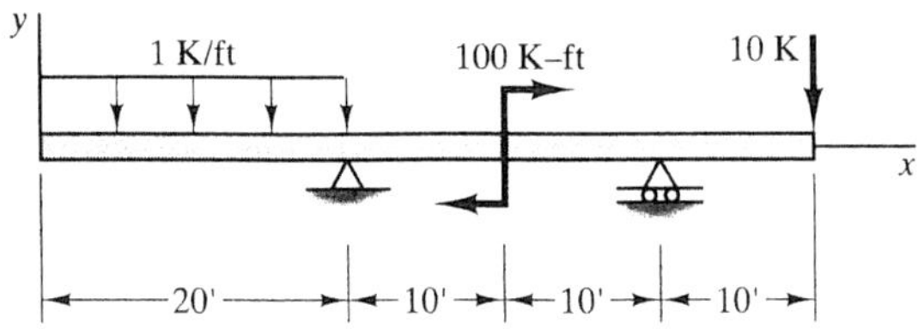

Figure P.10.34.

10.35. [10.5] A steam roller moving across the 30-ft span exerts forces on it as shown in Fig. P.10.35. Find magnitude and location of the bending moment, and $\bar{x}$ for which this moment is an absolute maximum of all possible positions of the vehicle. (*Hint:* Sketch the bending-moment diagram for any position $\bar{x}$ and decide where M_{max} might be.)

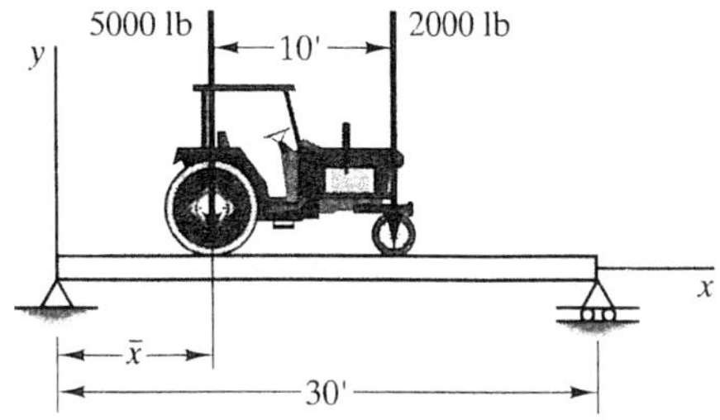

Figure P.10.35.

10.36. [10.5] A simply supported I beam is shown in Fig. P.10.36. A hole must be cut through the web to allow passage of a pipe which runs horizontally at right angles to the beam.

(a). Where, within the marked 24-ft section, would the hole least affect the moment-carrying capacity of the beam?
(b). In the same 24-ft section, where should the hole go to least affect the shear-carrying capacity of the beam?

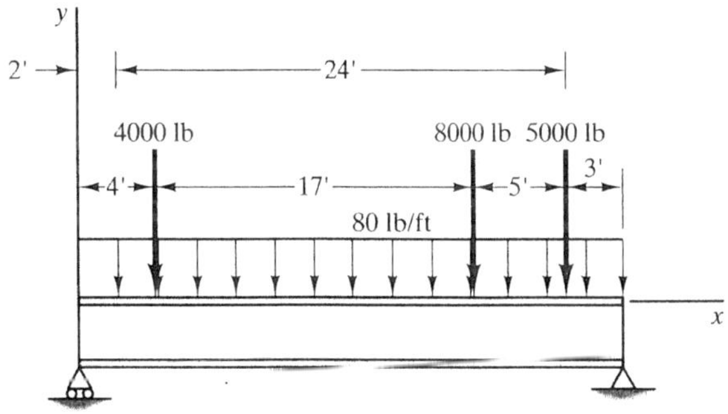

Figure P.10.36.

10.37. [10.5] In Fig. P.10.37, sketch the shear-force and bending-moment diagrams and label the key points.

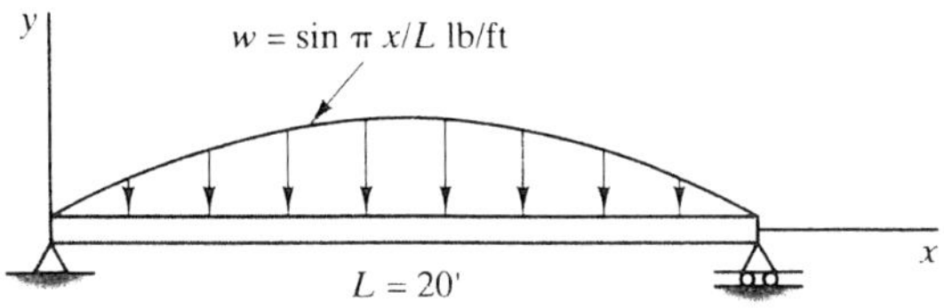

Figure P.10.37.

10.38. [10.5] After finding the supporting forces of the cantilever beam (see Fig. P.10.38) sketch the shear-force and bending-moment diagrams labeling key points.

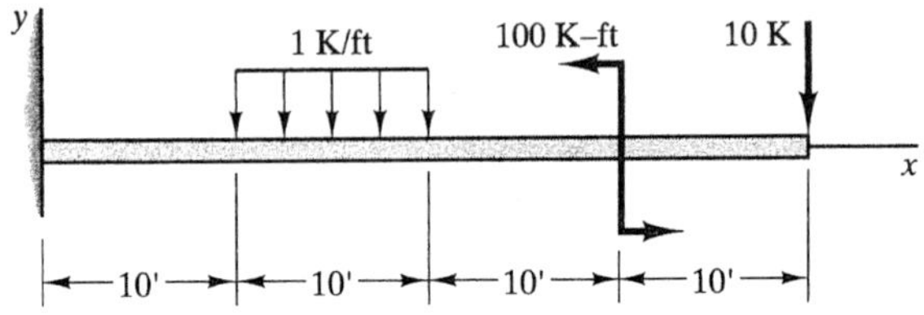

Figure P.10.38.

In Problems 10.39 to 10.42, merely sketch shear-force and bending-moment diagrams, labeling key points. Use whatever numerical calculations are necessary.

10.39. [10.8]

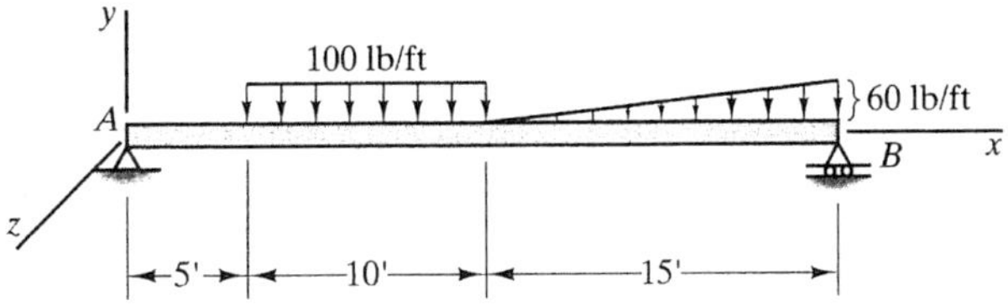

Figure P.10.39.

10.40. [10.8]

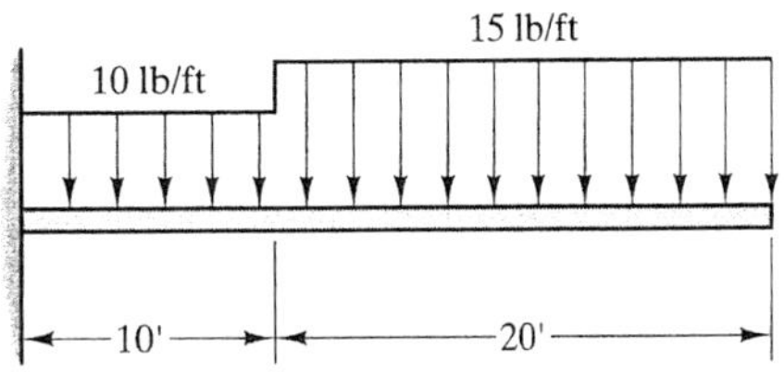

Figure P.10.40.

10.41. [10.8]

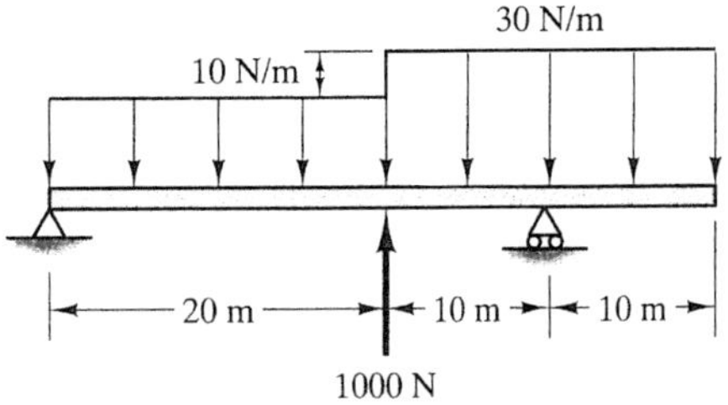

Figure P.10.41.

10.42. [10.8]

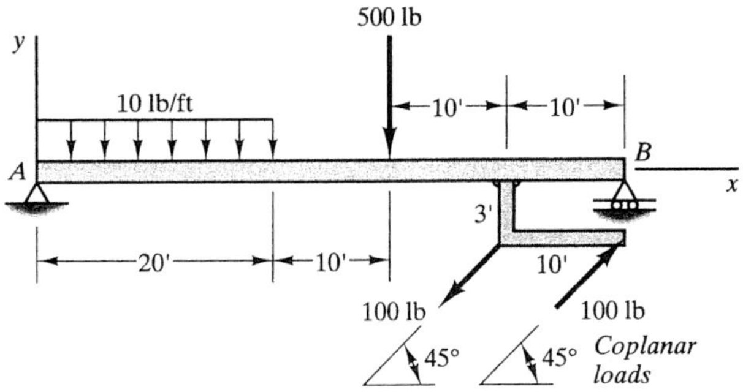

Figure P.10.42.

10.43. [10.8] A simply supported beam AB is shown in Fig. P.10.43. A bar CD is welded to the beam. After determining the supporting forces, sketch the shear-force and bending-moment diagram and determine the maximum bending moment.

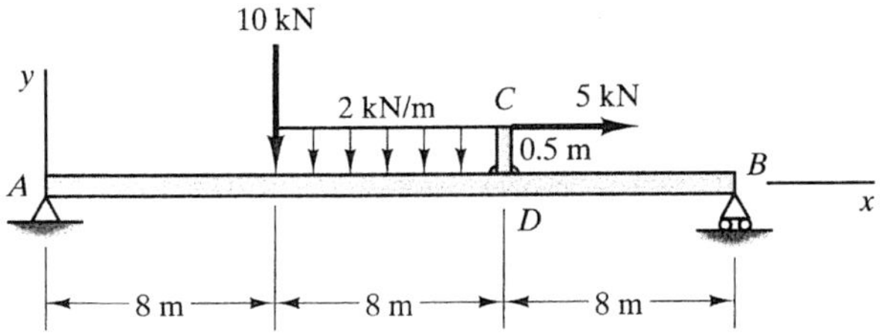

Figure P.10.43

Problems 10.44, 10.45, and 10.46 are challenging problems for the interested student looking for something to sink his or her teeth into. If available, use software for plots of variables.

*__10.44.__ A heavy, 50-ft long pole is to be lifted as shown in Fig. P.10.44. At what distance b should the crane attachment be located in order to impose the minimum flexural stress (caused directly by the bending moment) on its circular cross section if the weight of the pole is uniformly distributed along its length? (*Hint:* Consider the maximum bending moments in each of the two spans in terms of b. Roughly plot the magnitudes of these moments versus b. At what b will the maximum possible moment be the least value?)

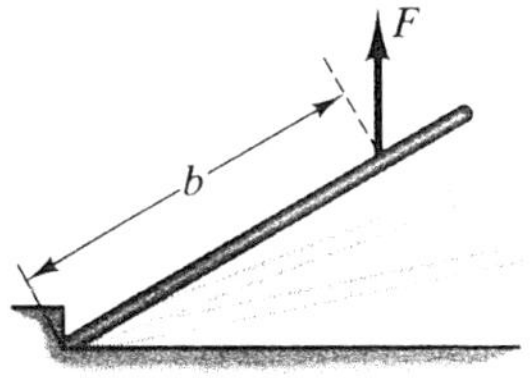

Figure P.10.44.

*$*$***10.45.** A 15-in.-o.d. (outside diameter) steel pipe is to be lifted so as to remain horizontal. For this purpose two rigidly attached yokes separated a distance a are attached to the pipe (see Fig. P.10.45). Two guy wires of length 10 ft each then connect the yokes to a hook. The pipe weighs 80 lb/ft and has a length of 20 ft. Find the proper distance a so that the maximum bending moment in the pipe is minimized. (*Hint:* Compute the maximum bending moment in each of the spans in terms of a. Roughly plot these as functions of a. At what a will the maximum possible moment be the least value?)

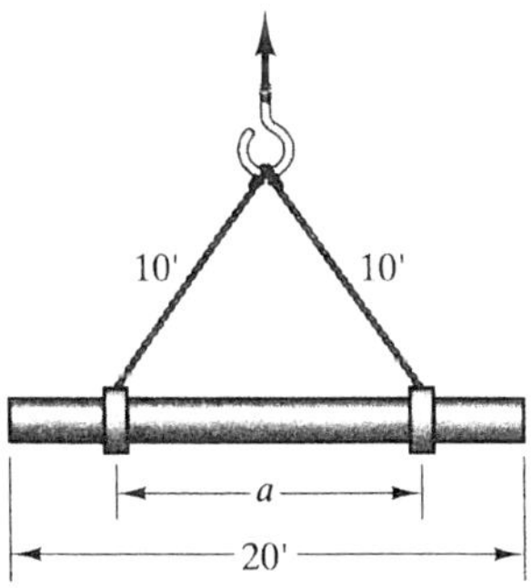

Figure P.10.45.

$*$**10.46.** A NO PARKING sign is made of thin rigid plastic and is supported as shown in Fig. P.10.46. If a wind blows against the sign, a uniform pressure p_0 above that of the stationary atmosphere will develop on the plate. To minimize the maximum bending moment from this pressure, at what position $\bar{x}$ should the lower support be placed? Give result as a fraction of h. (*Hint:* Compute the maximum bending moment in each of two spans in terms of $\bar{x}$ and h. Roughly plot $M_{\max}$ versus $\bar{x}$ for a given h. At what $\bar{x}$ will the maximum possible moment be the least value?)

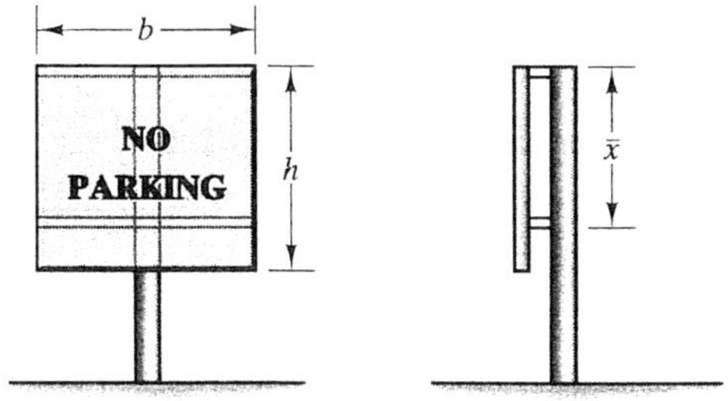

Figure P.10.46.

10.47. In Fig. P.10.47, what is the maximum negative bending moment in the region between the supports for the simply supported beam?

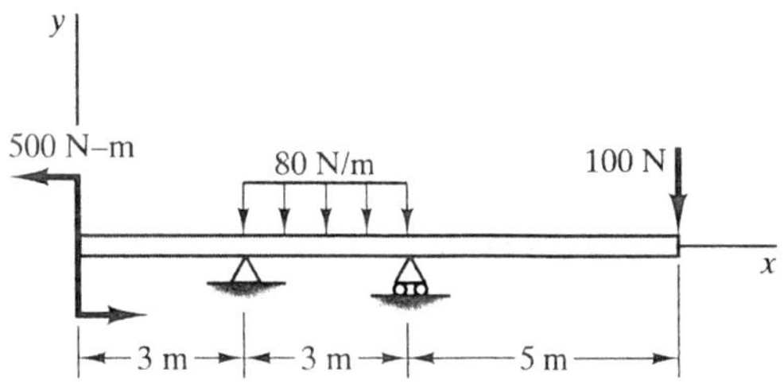

Figure P.10.47.

10.48. A 20-hp motor running at 1750 rpm drives two 1-ft diameter pulleys as shown in Fig. P.10.48. The belt tensions are given as $T_1 = 200$ lb, $T_2 = 100$ lb, and $T_3 = 3T_4$. The coupling C transmits only torque.

(a). Sketch the twisting torque diagram for M_x along the shaft AB.
(b). Sketch the shear-force diagrams for V_y and V_z.
(c). Sketch the bending-moment diagrams for M_y and M_z.

(*Hint:* As you may recall from physics the power transmitted by a shaft is torque times angular velocity in radians per unit time. Also, 1 hp = 33,000 ft-lb/min.)

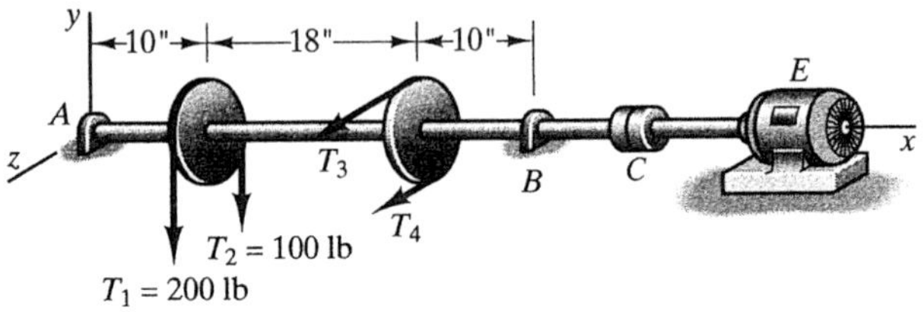

Figure P.10.48.

10.49. Shown in Fig. P.10.49 is suspension of one of the front wheels of an automobile. The pavement exerts a force of 1000 lb on the tire. The tire, brakes, and so on, weigh 100 lb; the center of gravity is taken at the center plane of the tire. Determine the force in the spring and the compression force in *CD*. Now sketch the axial force, shear force, and bending moment for member *AC*.

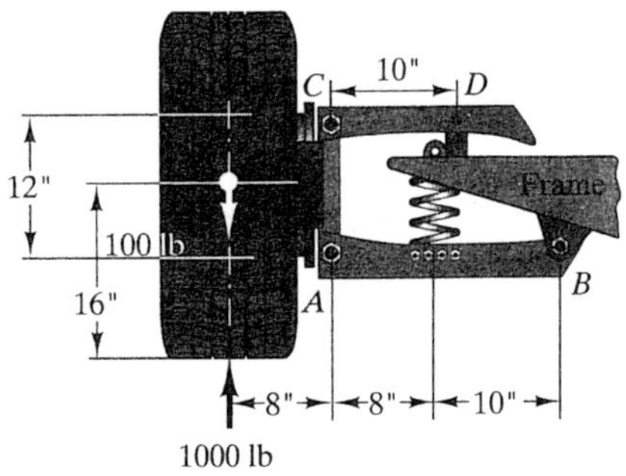

Figure P.10.49.

10.50. Sketch shear-force and bending-moment diagrams for Problem 10.5 after finding supporting forces.

10.51. Sketch shear-force and bending-moment diagrams for Problem 10.7 after finding supporting forces.

10.52. Sketch shear-force and bending-moment diagrams in Problem 10.8 after finding supporting forces and torques.

10.53. For the beam shown in Fig. P.10.53,

(a). Write the shear-force and bending-moment equations.
(b). Sketch the shear force and bending-moment diagrams labeling key points.

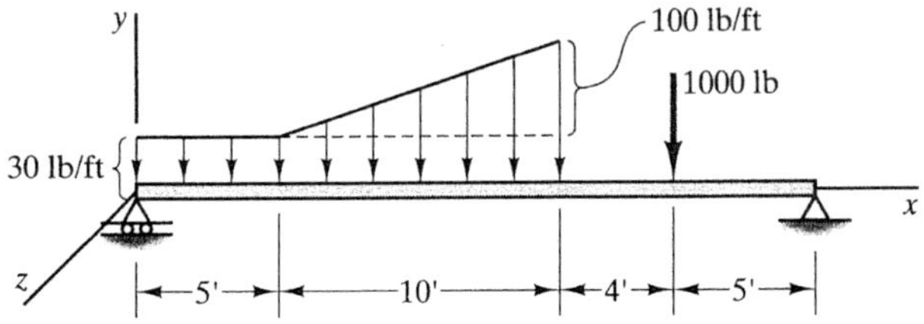

Figure P.10.53.

***10.54.** For the problem shown in Fig P.10.54:

(a). Find the supporting force system.
(b). Write the shear-force and bending-moment equations.
(c). Sketch the shear-force and bending-moment diagrams.

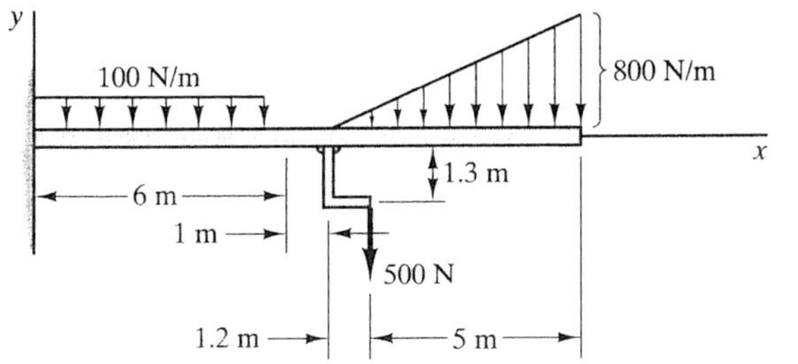

Figure P.10.54.

10.55. Write the equations and sketch the shear-force and bending-moment diagrams for the beam in Fig. P.10.55.

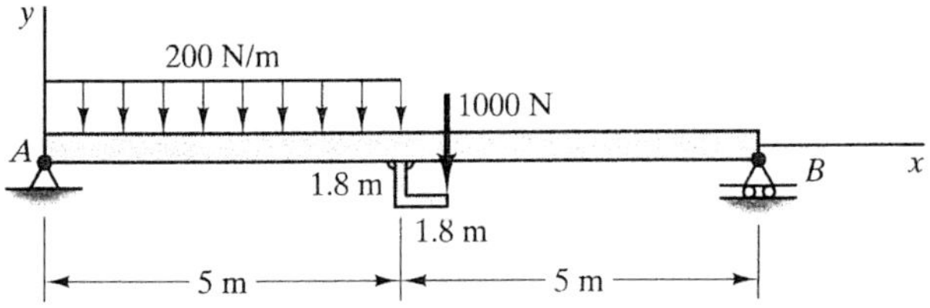

Figure P.10.55.

10.56. Given the shear-force diagram in Fig. P.10.56, draw the corresponding loading diagram. There is a clockwise couple moment of 5000 N-m at $x = 4$ m, and at $x = 0$ there is a counterclockwise couple moment at the base (it is a cantilever beam).

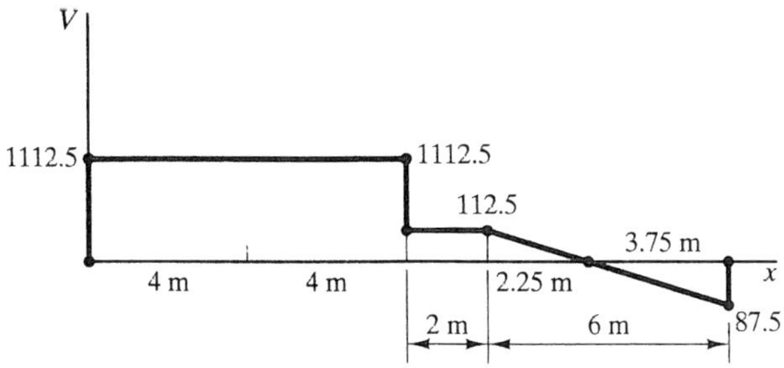

Figure P.10.56.

10.57. For the beam in Fig. P.10.57,

(a). What is the axial force on AB? Use three domains for this.
(b). Sketch the shear-force and bending-moment diagrams.

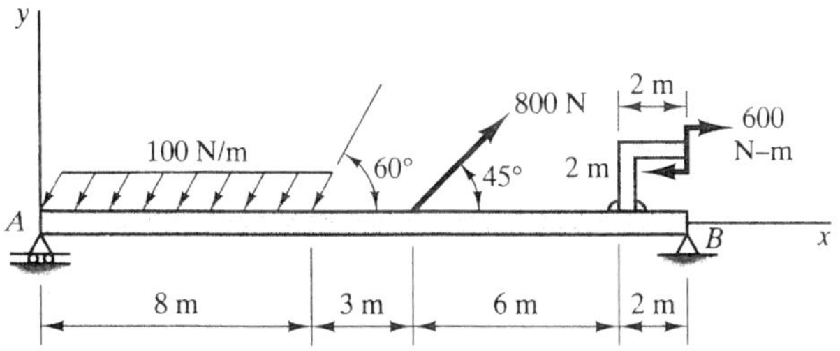

Figure P.10.57.

10.58. Determine the shear-force and bending-moment equations for the beam in Fig. P.10.58. Then sketch the diagrams using the aforementioned equations if necessary to ascertain key points in the diagrams, such as the position between the supports where $V = 0$. What is the bending moment there?

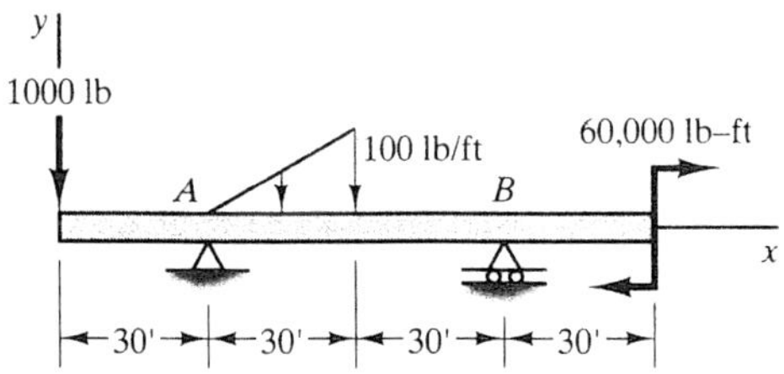

Figure P.10.58.

10.59. A cantilever beam (Fig. P.10.59) supports a parabolic and a triangular load. What are the shear-force and bending-moment equations? Sketch the shear-force and bending-moment diagrams. See the suggestion at the end of Section 10.7 regarding the triangular load.

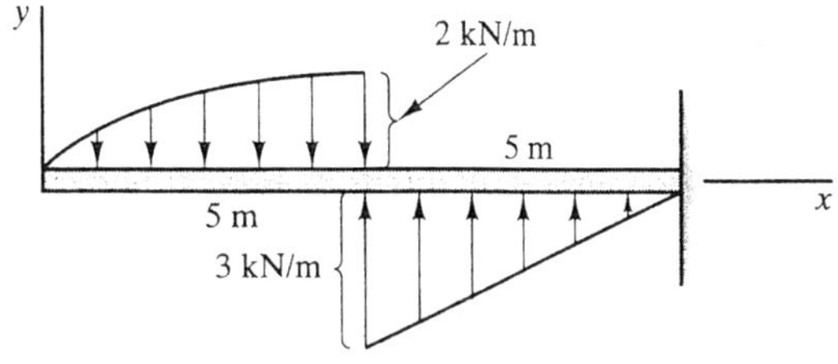

Figure P.10.59.

10.60. Formulate the shear-force and bending-moment equations and sketch the shear-force and bending-moment diagrams labeling the key points.

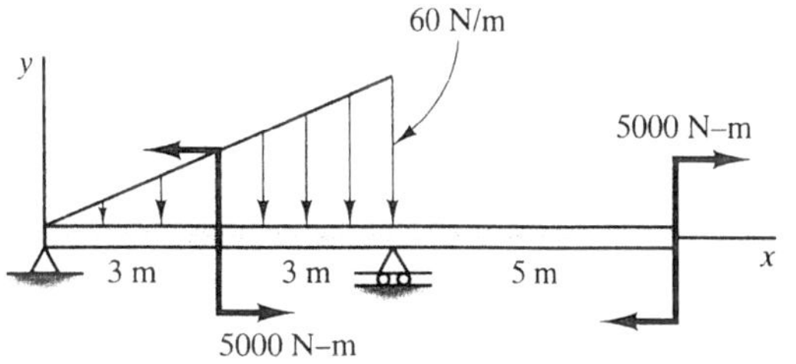

Figure P.10.60.

10.61. Formulate the shear-force and bending-moment equations for the beam in Fig. P.10.61. Sketch the shear-force and bending-moment diagrams.

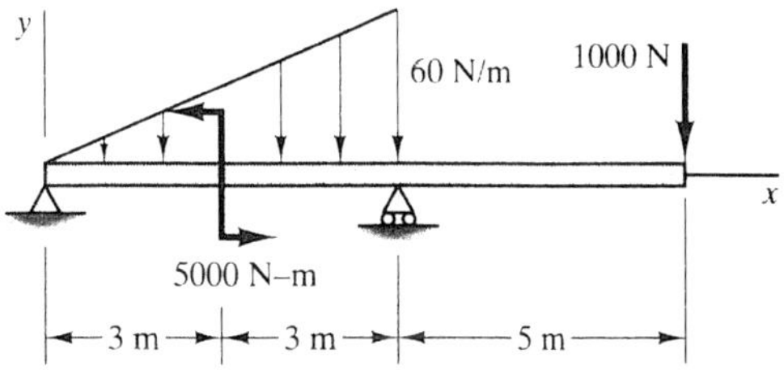

Figure P.10.61.

10.62. Sketch and label shear-force and bending-moment diagrams for the overhanging beam in Fig. P.10.62.

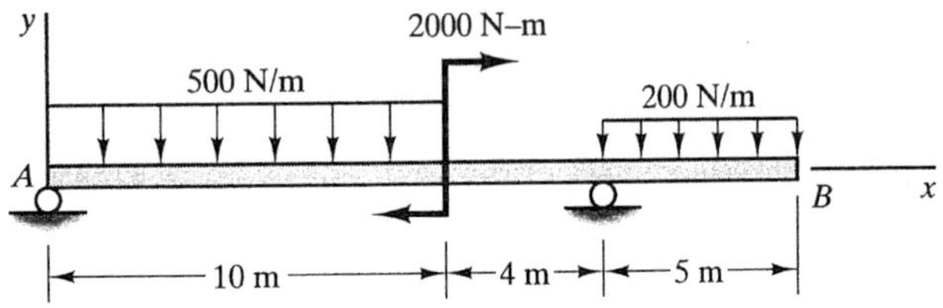

Figure P.10.62.

10.63. Given the bending-moment diagram as shown in Fig. P.10.63 sketch the corresponding shear-force diagram.

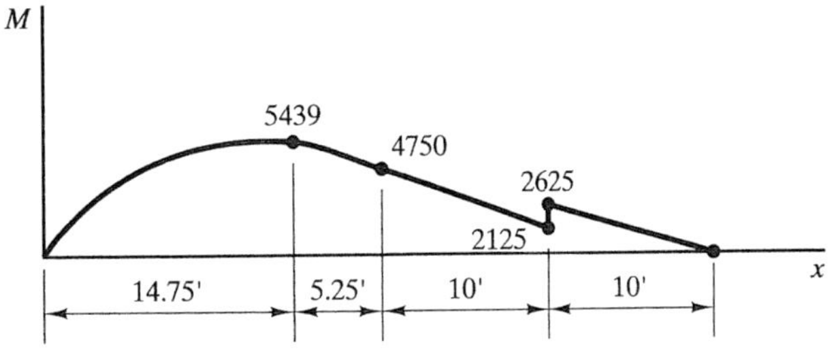

Figure P.10.63.

10.64. The structure shown in Fig. P.10.64 is proposed as a calibration fixture for an experiment. The cantilevered beam DE has an inverted "**T-shaped**" *rigid* bracket attached to its tip, as shown, allowing the placement of the point load P at any location along ABC. Sketch the resulting bending-moment diagrams for the beam DE for the load placed at A, B, and C.

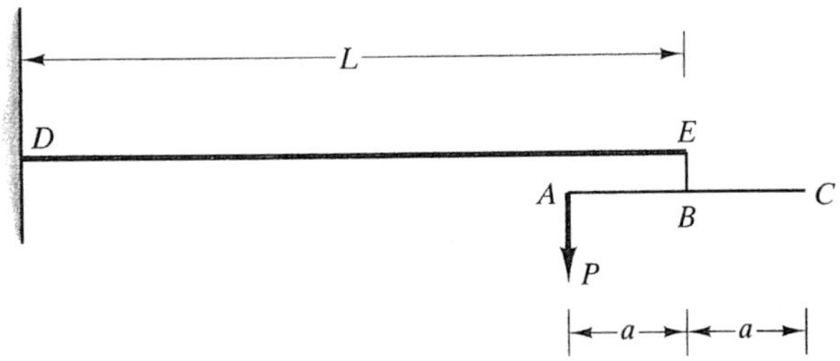

Figure P.10.64.

10.65. A design for a truss bridge is based on the top and bottom members formed into a parabolic shape as shown in Fig. P.10.65. The roadway is suspended from the truss by hangers. Why is the truss shaped this way? (*Hint:* Consider the bending moment from a uniform load.)

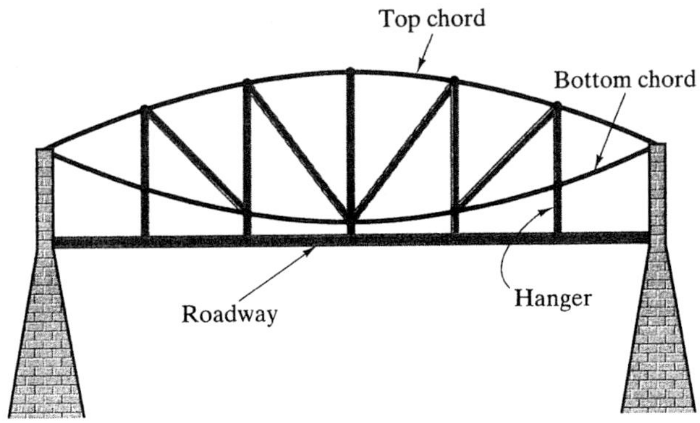

Figure P.10.65.

****10.66.** Consider the simply-supported beam shown in Fig. P.10.66. Plot the location, (x/L), of the maximum bending moment in the beam versus the loading parameter α for $0 \leq \alpha \leq 10$. What should you get when $\alpha = 0$? For the range considered, what is the error associated with approximating the maximum moment location as $L/2$ for $\alpha = 2$ and $\alpha = 10$?

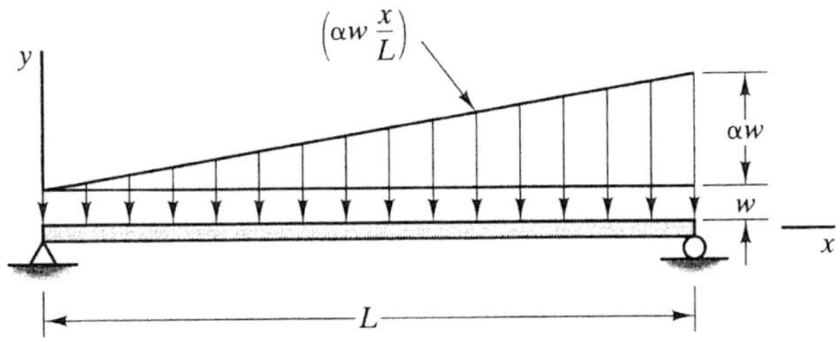

Figure P.10.66.

****10.67.** Consider the uniformly loaded beam shown in Fig. P.10.67 with an overhang. Plot the location, (x/L), of the maximum positive moment versus α. Check your solution for $\alpha = 0$. What happens when $\alpha = 1$? Also, plot $M_{max}/(wL^2)$ for positive and negative bending moments versus α. At what value of α are the two magnitudes equal?

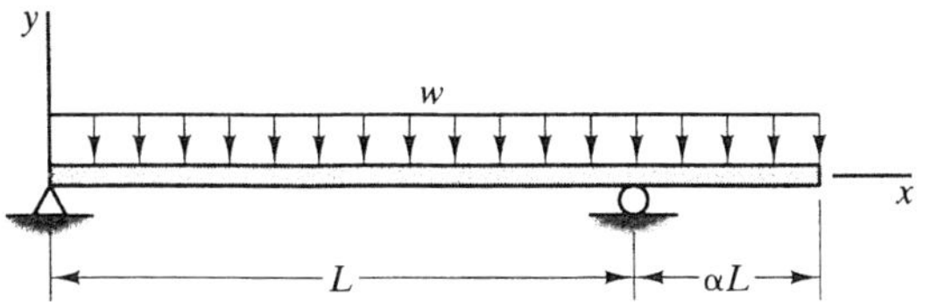

Figure P.10.67.

Project 10: Shear-Force and Bending-Moment Computation

This program is for *shear-force and bending-moment* computation for any loading, for a simply supported beam shown in Fig. P.10.68. The user defines the particular loading function for his or her program. The program should be able to handle *any loading that the user may want to insert.* Using Simpson's rule, *compute V and M at any point x.* Accept values for *L, h, b,* and *x* interactively. Also, *find the maximum shear stress at x.*

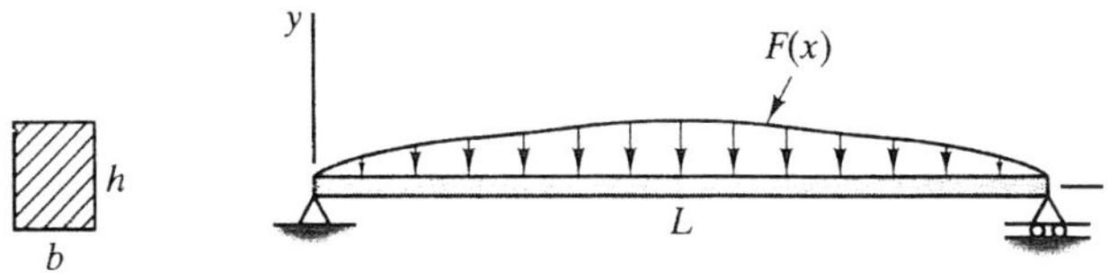

Figure P.10.68.

Work out for the following data in Fig. P.10.69:

$L = 10$ m $F(x) = 50$ N/m
$h = 0.1$ m $x = 4$ m
$b = 0.1$ m

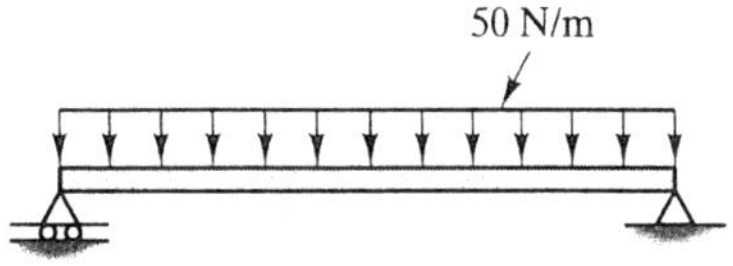

Figure P.10.69.

CHAPTER 11

Stresses in Beams

11.1 Introduction

In Chapter 5 we considered a number of *one-dimensional* stress problems both statically determinate and statically indeterminate. In that study, we made an assumption as to the nature of the stress distribution and then proceeded to employ equations of **equilibrium**, a **constitutive law**, and considerations of **compatibility** of **deformation**, to compute stresses and deflections of the system. The stresses thus computed had to match the external tractions on most of the boundary surface of the body. At certain places the stresses at the surface matched only the rigid-body *statical equivalent* of the external tractions there and we resorted to the St.-Venant principle to assure us that the results were valid away from these areas.

We shall follow a similar procedure in this chapter to find formulations for stress distributions in long prismatic members loaded always at right angles to the axes of the members. Such members are used extensively in engineering practice and as noted in Chapter 10 are called *beams.* More precisely, the procedure will be to make certain reasonable assumptions as to the manner of deformation as well as to values of some of the stresses for the beam. Then using *equilibrium* considerations and a *constitutive law* we shall set forth stress distributions that reasonably match, at the boundary, the external tractions or their statical equivalents. Formulations developed along such lines are called *strength of materials* and are usually approximations of corresponding formulations developed from the more exact approaches of the *theory of elasticity.* Primarily, the ad hoc assumptions made as to deformation and stress, reasonable as they appear, may not result in *compatible* displacement relations or may not result in a good enough match with the external tractions. In the more exact approaches of the theory of elasticity, fewer assumptions as to deformation patterns are made and the equations guaranteeing *compatibility* of deformation and the boundary conditions (at least up to statical equivalency) are satisfied as an integral part of the development. However, such solutions are extremely difficult to carry out in any but very simple problems. Yet such solutions, few as they are, are most valuable to us to check the validity of simpler, handy formulations from strength of materials.

In this chapter we shall concentrate on the linear-elastic behavior of beams. We shall point out when the strength of materials approach gives results identical to those stemming from the theory of elasticity and when the strength of materials approach gives results somewhat different from that of the theory of elasticity. We shall also consider, although briefly, linear-elastic, perfectly-plastic behavior of beams.

Part A. Basic Considerations

11.2 Pure Bending of Symmetric Beams

Let us consider a weightless beam loaded by couples of magnitude M_z at the ends as shown in Fig. 11.1, where the undeformed geometry of

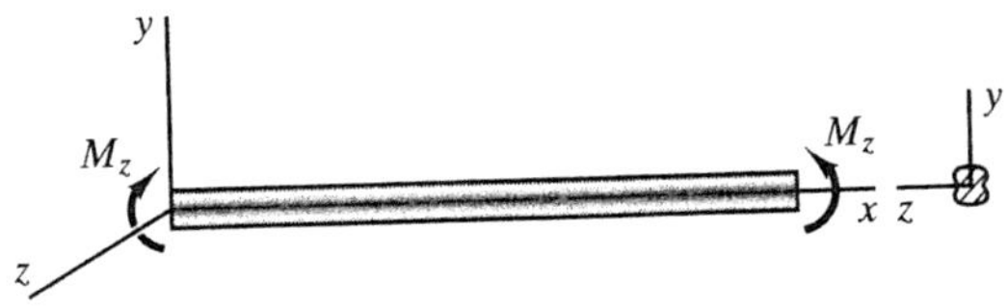

Figure 11.1. Pure bending.

the beam is drawn. As in earlier analyses, we shall usually use unprimed coordinates to refer to undeformed geometry of the beam and primed coordinates for the deformed geometry. Note that the cross section of the beam is *symmetric* about an axis which we have chosen as the y axis. The x axis has been shown running parallel to the centerline of the beam. Later we shall indicate precisely what position the x axis is to occupy. The z axis forms a right-handed triad with our x and y axes.

From **equilibrium** considerations we note that the shear force for V_y is zero for the beam and that the bending moment is a constant equal to M_z for the entire length of the beam.[1] This information, coupled with

[1]Note that the M_z being brought into the discussion here occurs at section inside the beam and so we shall use the convention presented in Chapter 10 (in this case the structural convention) for M_z as it progresses throughout the derivation. The final formulations then require the use of the convention for the bending moment as you will be reminded.

the fact that the cross section of the beam is uniform, permits us to form a simple picture of the deformation geometry of the beam. In Fig. 11.2,

Figure 11.2. Elements of beam having length Δx before deformation.

we show the beam in an exaggerated deformed condition wherein we have separated several portions of the beam. These sections were of length Δx in the undeformed geometry. Clearly, each section is subject to the same bending moment, and away from the ends of the beam, it seems reasonable to assume that all such elements deform exactly the same way. Furthermore, there should be symmetry of the deformed geometry of each element about a transverse center plane for each element. Thus, in Fig. 11.3(a), *A-A* and *B-B* are edges of such center planes for the two elements shown in an enlarged fashion.

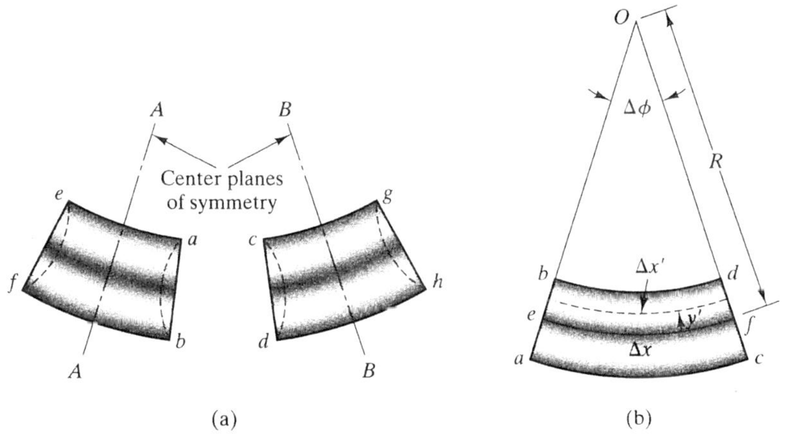

Figure 11.3. (a) Plane sections remain plane; (b) center of curvature.

By accepting the assumption that the aforementioned elements of the beam have the same deformed geometry and also have transverse planes of symmetry for each element, we must accept the fact that the end surfaces of the elements are *plane surfaces*. Thus, if *ab* in Fig. 11.3(a) were imagined not to be plane and instead to be a warped surface, such as the dashed curve, then *ef* must have a mirror image shape because of the plane of symmetry. Now because the elements must deform in exactly the same way, surfaces *cd* and *gh* must warp inward as shown in the diagram. But here we reach an impasse in that the two elements can no longer fit together without creating a void. This would violate compatibility. The same dilemma occurs when we consider that the cross section warps in an outward fashion. Thus, **compatibility** forces the conclusion that sections of the elements must be plane. Since the elements of the beam were of arbitrary length, we can form the generalization that *cross sections of the beam remain plane upon deformation of the beam by the action of pure end couples.*

What further conclusions can we draw as to stress and strain? Consider a single element of the beam as shown in Fig. 11.3(b). As before, we say that the element originally had a length Δx. It should be clear that above a certain surface shown as line *ef*, the *longitudinal fibers* of the beam are compressed, whereas below this surface, the fibers are elongated. Thus, the curved line *ef* undergoes zero extension and must then have the length Δx. This line is the edge of the so-called *neutral surface* of the beam along which no change in length occurs during the deformation of the beam.

Next we extend lines *ab* and *cd* in Fig. 11.3(b) to intersect at O. Since all elements of the beam have the same geometry for a given length Δx, and since they are contiguous to each other, it is clear that all such lines should intersect at a common point O when extended in the manner described. Furthermore, a change in the value of Δx would not change the position of point O in this discussion. We can then conclude that due to the aforementioned transverse plane of symmetry the lines bO and dO are of equal length, as are all such lines for any value of Δx. Consequently, all points on the top or bottom surface of the beam must be equidistant from O. This is similarly true for all surfaces which in the undeformed geometry are parallel to the upper surface of the beam. Thus, *these surfaces, including the undeformed neutral surface, become, when deformed, surfaces whose edges observed in the z direction are concentric circular arcs.* We define the *radius of curvature* for the beam to be the distance R from point O to the neutral surface [see Fig. 11.3(b)]. Also, we now position the x coordinate in Fig. 11.1 so that the neutral surface in the undeformed geometry is in the xz plane. (This means that the x axis coincides with the intersection of the undeformed neutral surface and the longitudinal plane of symmetry of the beam.)

We now introduce y' as the distance from the neutral surface toward point O in the deformed geometry [see Fig. 11.3(b)] to an arc of the beam whose length we denote as $\Delta x'$. Observing this diagram, we can then conclude that

$$\Delta\phi = \frac{\Delta x}{R} = \frac{\Delta x'}{R - y'} \tag{11.1}$$

Rearranging Eq. (11.1), we now form the relation

$$\Delta x' = \frac{R - y'}{R}\Delta x \tag{11.2}$$

Subtracting Δx from both sides of the equation, we then get

$$\Delta x' - \Delta x = \left(\frac{R - y'}{R} - 1\right)\Delta x = -\frac{y'}{R}\Delta x \tag{11.3}$$

Dividing through by Δx, taking the limit as $\Delta x \to 0$, and using the geometrical interpretation of normal strain, we may reach the following result:

$$\lim_{\Delta x \to 0} \frac{\Delta x' - \Delta x}{\Delta x} \equiv \varepsilon_{xx} = -\frac{y'}{R} \tag{11.4}$$

That is, since Δx is the original length of all fibers of the element, ε_{xx} is the normal strain of the fibers at position y' in the deformed geometry. But for *small deformations,* the distance y' from the neutral surface in the deformed geometry to a beam fiber should differ little from the distance y in the *undeformed geometry*[2] to the same beam fiber. We can then give ε_{xx} in terms of the undeformed geometry as follows:

$$\varepsilon_{xx} = -\frac{y}{R} \tag{11.5}$$

Note that the preceding formula properly gives compressive strain on the side of the neutral surface toward the center of curvature and tensile strain on the side of the neutral surface away from the center of curvature. Up to this time, we have considered only deformation. Next we shall consider stress.

It should be understood that our conclusions up to this point are valid for beams composed of *any type* of homogeneous solid material.[3] We shall now employ a **constitutive law** and stipulate that we have a linear-elastic material. Using Hooke's law, for ε_{xx} we have

$$\varepsilon_{xx} = \frac{1}{E}\left[\tau_{xx} - \nu(\tau_{yy} + \tau_{zz})\right]$$

With no physical constraints in the y and z directions, we assume that the stresses $\tau_{yy} = \tau_{zz} = 0$. Solving for τ_{xx} in the preceding equation and using Eq. (11.5) to substitute for ε_{xx}, we then have for the normal stresses

$$\tau_{xx} = -E\frac{y}{R} \tag{11.6a}$$

$$\tau_{yy} = 0 \tag{11.6b}$$

$$\tau_{zz} = 0 \tag{11.6c}$$

[2] Thus, we shall use the undeformed geometry henceforth.

[3] This makes Eq. (11.5) particularly valuable since it is applicable to other than Hookean materials for small deformation.

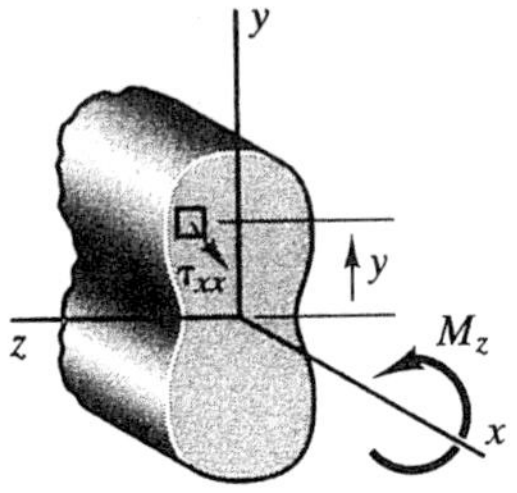

Figure 11.4. Cross section of beam showing stress.

Finally, in the absence of shear force and twisting moment, we shall assume that the shear stresses for our reference system are zero—that is,

$$\tau_{xy} = \tau_{yz} = \tau_{xz} = 0 \tag{11.7}$$

We have thus established the stress distributions at any section of the beam in terms of R, the radius of curvature. To replace R by more desirable terms, we note that the resultant of the stress distribution at any section must be a couple moment of value M_z. In this regard, note in Fig. 11.4 that a positive normal stress τ_{xx} at a position corresponding to a positive value of y forms a *negative* bending moment according to our sign convention for bending moments (see Chapter 10). Hence, we can say for a section that

$$-M_z = \iint_A \tau_{xx} y\, dA$$

Replacing τ_{xx} using Eq. (11.6a), we get

$$M_z = \iint_A \frac{Ey^2}{R}\, dA = \frac{E}{R} \iint_A y^2 dA = \frac{EI_{zz}}{R} \tag{11.8}$$

where I_{zz} is the second moment of area of the cross-sectional area about the *neutral axis*.[4] Solving for R, we get

$$R = \frac{EI_{zz}}{M_z} \tag{11.9}$$

Thus, the radius of curvature is now available in terms of the applied load, the geometry of the beam, and Young's modulus. Substituting for R in Eq. (11.6a) using the foregoing formula, we can now give the stress τ_{xx} in terms of the loading and the beam geometry. Thus,

$$\tau_{xx} = -\frac{M_z y}{I_{zz}} \tag{11.10}$$

This is the well-known *flexure formula,* which forms one of the key formulas of strength of materials. From this formula we see that stress τ_{xx} varies linearly along the y direction, being zero at the neutral axis and

[4]The *neutral axis* is the intersection of the neutral surface and the cross section in the undeformed geometry.

attaining extreme values at the outer fibers. The sign of M_z must be in accordance with the sign convention for M_z.

We have yet to locate the neutral axis at cross sections of the beam. We shall show that the neutral axis is a horizontal centroidal axis at cross sections of the beam. Since we have no axial force along the beam, we require in accordance with **equilibrium** that

$$\iint_A \tau_{xx} dA = 0 \tag{11.11}$$

Substituting for τ_{xx} from Eq. (11.6a), we get

$$-\frac{E}{R}\iint_A y\,dA = 0 \tag{11.12}$$

The preceding equation tells us that the first moment of the cross-sectional area about the neutral axis is zero. This requires that the neutral axis be a *centroidal axis*. We have thus finally positioned the x axis; it must be the intersection of the vertical plane of symmetry and the horizontal plane passing through the horizontal centroidal axes of the sections.

We now point your attention to the fact that as a result of the Poisson effect,

$$\varepsilon_{zz} = -\nu\varepsilon_{xx}$$

$$\varepsilon_{yy} = -\nu\varepsilon_{xx}$$

This means that in the compressed region, the beam will expand laterally, while, in the tensile region, the beam will contract laterally. (You may readily see this by bending a rubber eraser.) As may be shown from the theory of elasticity, this action causes lines parallel to the z direction to become arcs of circles having a large radius of curvature equal to R/ν. We show this action in Fig. 11.5. In particular, note that the neutral surface actually has a double curvature—one in the xy plane and one in the zy plane. The latter curvature is called *anticlastic* curvature. It is to be pointed out that the details discussed in this paragraph, although interesting, have little significance in most practical problems.

We will now illustrate the use of the formulas developed in this section.

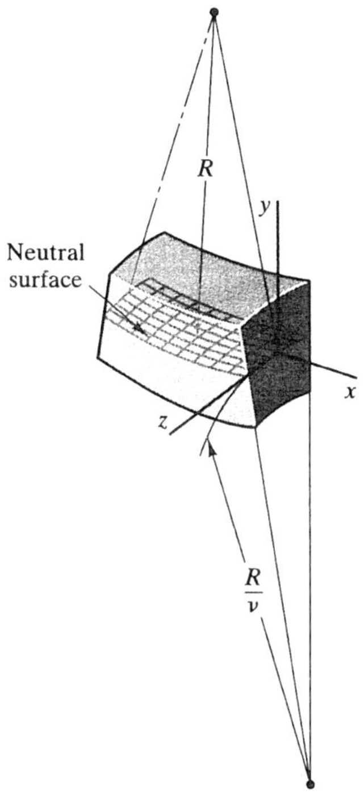

Figure 11.5. Anticlastic curvature.

Example 11.1

Compute the maximum stress and the radius of curvature of the beam shown in Fig. 11.6(a) due only to the couples at the ends (i.e., do not consider the weight of the beam). The modulus of elasticity E for the material is 30×10^6 psi.

Example 11.1 (Continued)

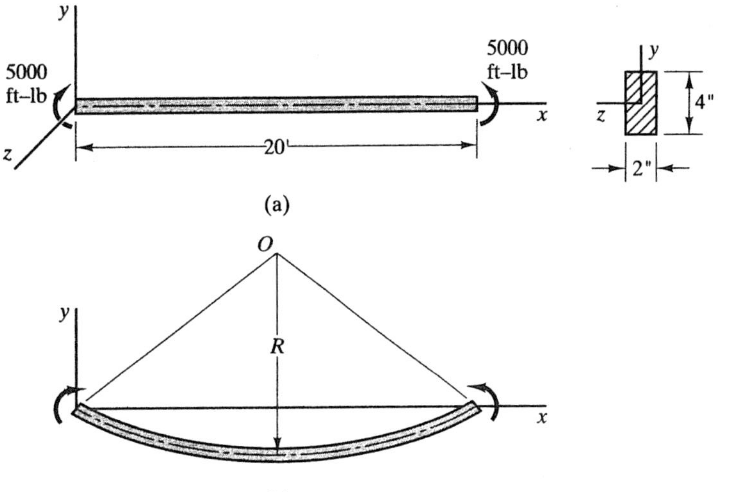

Figure 11.6. Pure bending problem.

To get the stress, we need only use the flexure formula, Eq. (11.10). For the maximum compressive and tensile stress, we set $y = \pm 2$ in this formula, which means we are computing the normal stress on the extreme upper and lower fibers of the beam. Thus we have, using inches

$$(\tau_{xx})_{\max} = -\frac{M_z y_{\max}}{I_{zz}} = -\frac{(5000)(12)(\pm 2)}{(\frac{1}{12})(2)(4)^3} = \pm\ 11{,}250 \text{ psi}$$

$$(\tau_{xx})_{\max} = \pm\ 11{,}250 \text{ psi}$$

Thus we have 11,250 psi compression on the top fibers and 11,250 psi tension on the lower fibers.

Next we compute the radius of curvature for the beam. The center of curvature of the deformed beam has been shown in the diagram at O. The radius of curvature R can be easily computed using Eq. (11.9). Thus again using inches,

$$R = \frac{EI_{zz}}{M_z} = \frac{(30 \times 10^6)(\frac{1}{12})(2)(4^3)}{(5000)(12)} = 5333 \text{ in.}$$

$$R = 5333 \text{ in.}$$

Example 11.2

Figure 11.7(a) shows a cantilever beam of length L loaded by a point couple. We shall neglect the weight of the beam and shall compute the maximum stress at sections along the beam.

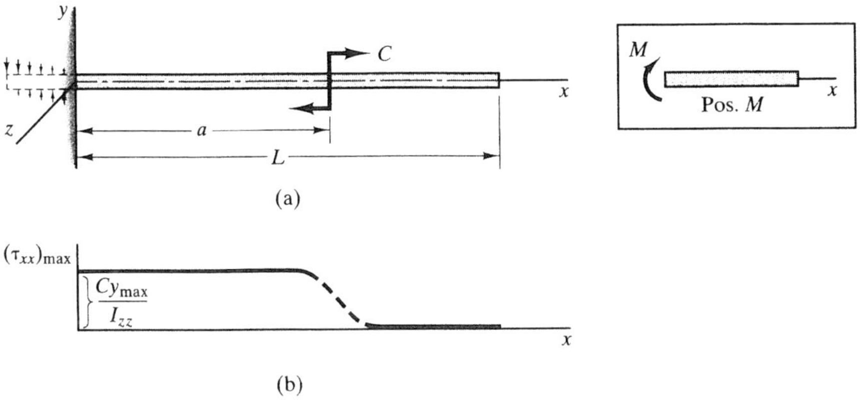

Figure 11.7. Cantilever beam loaded with a couple.

For sections to the right of the point couple C there is a zero resultant force system and so considering (mentally) right-hand free bodies, it is clear that these sections are stress-free. To the left of the couple, we have the case of pure bending and so the maximum stress can be given as

$$(\tau_{xx})_{\max} = -\frac{(-C)y_{\max}}{I_{zz}} = \frac{Cy_{\max}}{I_{zz}} \qquad \text{(a)}$$

It should be clear that we cannot make accurate statements as to the stress distribution near the point couple since we know that the point couple is only an approximation of some applied force distribution spread over a comparatively small region.

Now we go to the support. There we have the resultant couple C of a force distribution spread over the portion of the beam embedded in the wall. Our theory gives for the normal stress τ_{yy} the value of zero up to the wall. However, the supporting couple moment C will arise mainly from compressive loads on parts of the upper surface and the lower surface of the beam [see Fig. 11.7(a)] so that we shall certainly get nonzero stresses τ_{yy} for the portion of the beam inside the wall and immediately outside. This is similarly true for the shear stresses τ_{xy}. These normal and shear stresses will not, however, appreciably affect the stresses τ_{xx} predicted by the theory at sections *right up to the wall.* That is, for normal stress τ_{xx}, we can use the flexure formula right up to the wall face, but for sections at the wall face

Example 11.2 (Continued)

there may be nonzero stresses for τ_{yy} and τ_{xy} not accounted for by the theory.

We have shown a plot of the maximum stress at a section, $(\tau_{xx})_{max}$ versus x, in Fig. 11.7(b). The dashed line portion of the curve indicates lack of knowledge of the value of the stress $(\tau_{xx})_{max}$ near the couple and is put in to represent some kind of continuous variation that must exist in this region in the actual physical problem.

Example 11.3

Find the minimum height h of the cross section of the cantilever beam shown in Fig. 11.8 for a material with a yield stress in both tension and compression of 3.5×10^8 Pa. Use a safety factor n of 2.

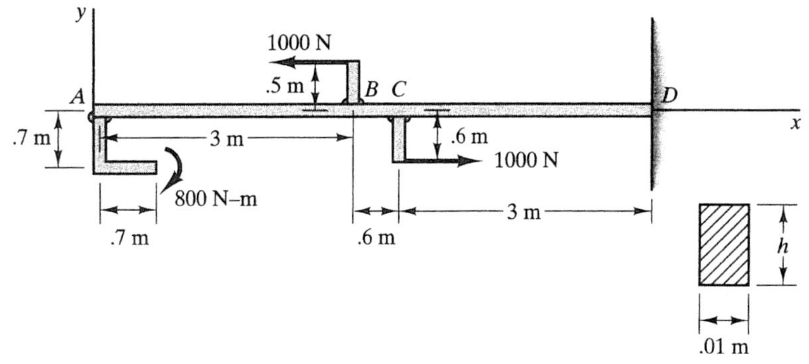

Figure 11.8. Cantilever beam: design problem.

We start by drawing a free-body diagram with the forces acting on the extension rods welded to the beam replaced by equivalent force systems acting directly on the beam (see Fig. 11.9).

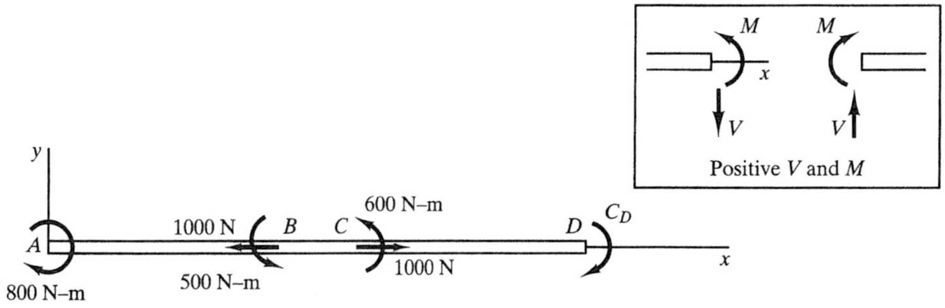

Figure 11.9. Free-body diagram with equivalent systems acting directly on the beam.

Example 11.3 (Continued)

We shall consider the **flexure formula** in three distinct regions of the beam.

Region AB

Note that in accordance with the convention for bending moment, $M_z = +800$ N-m. To then get the desired maximum tensile or compressive stress, clearly, on using the safety factor n and the yield stress Y

$$\frac{Y}{n} = \left|\frac{-M_z y}{I_{zz}}\right|$$

$$\therefore \quad \frac{3.5 \times 10^8}{2} = \left| -\frac{(800)(\pm h/2)}{\left(\frac{1}{12}\right)(.01)(h^3)} \right| \qquad h = .05237 \text{ m}$$

Region BC

We consider for this region the maximum *tensile* stress

$$\frac{3.5 \times 10^8}{2} = -\frac{(300)(-h/2)}{\left(\frac{1}{12}\right)(.01)(h^3)} + \frac{1000}{(.01)(h)}$$

$$(1.75 \times 10^8)h^2 - 1.8 \times 10^5 - 10^5 h = 0$$

$$(1.75 \times 10^3)h^2 - h - 1.8 = 0$$

Using the quadratic formula, we get

$$h = \frac{1 + \sqrt{1^2 + (4)(1.75 \times 10^3)(1.8)}}{(2)(1.75 \times 10^3)} = .0324 \text{ m}$$

Region CD

From equilibrium, $M_{CD} = -300$ N-m. Hence,

$$\frac{3.5 \times 10^8}{2} = \left| -\frac{(-300)(\pm h/2)}{\left(\frac{1}{12}\right)(.01)(h^3)} \right|$$

$$h = .0321 \text{ m}$$

Example 11.3 (Continued)

Hence the minimum h is

$$h = .05237 \text{ m} = 52.37 \text{ mm}$$

It can readily be shown that the stress distribution for pure bending is "exact" (i.e., it is the same solution as one gets from the full theory of elasticity) so the many assumptions made are valid.[5] In the next section we shall consider the case where we have transverse loads along the beam *in addition* to the possibility of end couples with the result that the shear force will not generally be zero along the beam. The results that we shall reach will generally then be *approximate* in that they *do not check exactly* with the corresponding results from the theory of elasticity—we are strictly in the domain of strength of materials.

11.3 Bending of Symmetric Beams with Shear: Normal Stress

We shall now consider symmetric beams under the action of *arbitrary loadings* which are in the plane of symmetry and oriented normal to the center line of the beam. Also, we shall include couples in the plane of symmetry.

We shall *assume* for such cases that the formulas

$$\tau_{xx} = -\frac{M_z y}{I_{zz}} \tag{11.13a}$$

$$R = \frac{EI_{zz}}{M_z} \tag{11.13b}$$

still hold wherein *local* values of M_z and R are utilized. Thus, we take the exact solution of the pure bending case as the starting point for the study of more general beam problems. This is sometimes called the *Euler-Bernoulli theory* or the *technical theory* of bending. In general we can say again that M_z is not constant and the distribution of τ_{xx} will be a function both of y and x. Furthermore, R will be a *local* radius of curvature (i.e., a function of x). Thus, in evaluating $(\tau_{xx})_{max}$, we must choose the section having the *greatest bending moment* and the points in the section *farthest* from the neutral axis. Perhaps our work on bending-moment diagrams in Chapter 10 will now be more fully appreciated. We can show by resorting to the general theory of elasticity that Eqs.

[5]We have asked you to show this as an exercise in Problem 11.12.

(11.13) give results which are highly accurate for *long, slender beams undergoing small deformation.*[6]

We shall now illustrate the use of the simple formulas in the extended range of applicability.

Example 11.4

What is the maximum normal tensile stress τ_{xx} and the minimum radius of curvature for the cantilever beam shown in Fig. 11.10?

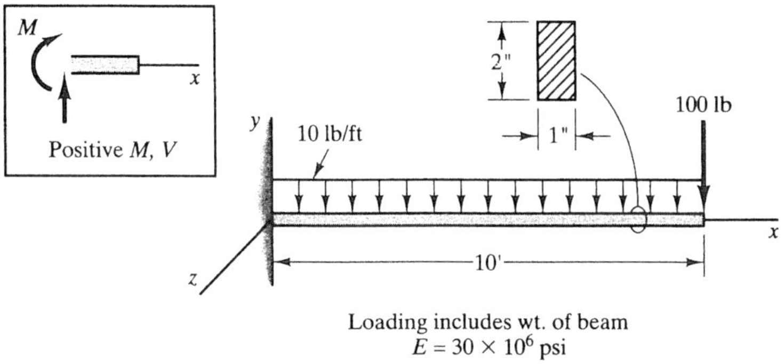

Figure 11.10. Cantilever beam.

It is easily seen that the section having the largest moment (and hence the section having the largest value of τ_{xx}) is at the wall. Imagining a right-handed free body of the beam cut at the wall and using the proper sign convention for bending moments, we have for the bending moment at the wall,[7]

$$M_{\max} = -(10)(100) - \int_0^{10} x10\,dx$$
$$= -1500 \text{ lb-ft} = -18{,}000 \text{ in.-lb}$$

The maximum tensile stress occurs at the uppermost fibers of the beam. We then get

$$(\tau_{xx})_{\max} = -\frac{(-18{,}000)(1)}{(\frac{1}{12})(1)(2^3)} = 27{,}000 \text{ psi}$$

$$(\tau_{xx})_{\max} = 27{,}000 \text{ psi}$$

The minimum radius of curvature occurs where the bending moment is largest (at the wall) and is computed as

[6]Fortunately for strength of materials, these are the types of beam problems generally encountered in practice, for obvious reasons.

[7]In simple problems such as this, one need not worry about the proper sign of M since one can see by inspection whether the desired stress is a tension or a compression stress. The appropriate sign for τ_{xx} may be established in this less formal manner.

Example 11.4 (Continued)

$$R = \frac{(30 \times 10^6)(\frac{1}{12})(1)(2^3)}{(18{,}000)} = 1111 \text{ in.}$$

$$R = 1111 \text{ in.}$$

Example 11.5

We have shown a simply supported beam in Fig. 11.11. This is the same beam problem as that shown in Fig. 10.16. We are to find the maximum tensile stress in the beam and the minimum radius of curvature. Take $E = 30 \times 10^6$ psi.

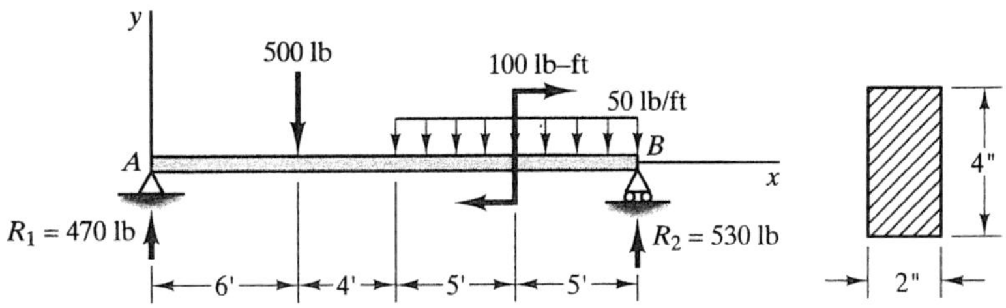

Figure 11.11. Simply supported beam.

The stress we seek must occur at the section having the maximum bending moment M_{max}. Our first step then would be to sketch the shear-force and bending-moment diagrams to determine the position and value of M_{max}. We have already done this for this problem (see Fig. 10.17). We see that M_{max} is 2820 ft-lb at a position 6 ft to the right of the left support. Hence, we can conclude that using inches

$$(\tau_{xx})_{max} = -\frac{[(2820)(12)](-2)}{\left(\frac{1}{12}\right)(2)(4^3)} = 6{,}345 \text{ psi}$$

$$(\tau_{xx})_{max} = 6{,}345 \text{ psi}$$

The minimum radius of curvature is

$$R_{min} = \frac{(30 \times 10^6)\left(\frac{1}{12}\right)(2)(4^3)}{(2{,}820)(12)} = 9{,}456 \text{ in.}$$

$$R_{min} = 9{,}456 \text{ in.}$$

Example 11.6

A cantilever beam is shown in Fig. 11.12 loaded on two of its faces. What is the maximum compressive normal stress $(\tau_{xx})_{max}$?

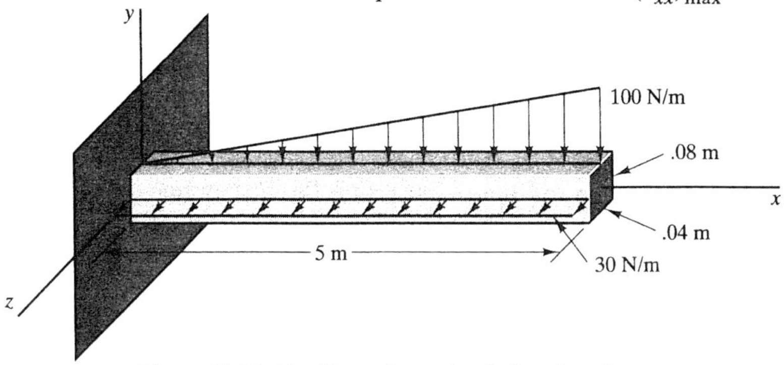

Figure 11.12. Cantilever beam loaded on two faces.

It should be clear that the maximum compressive stress τ_{xx} will occur somewhere on the cross section of the beam at its base at the wall. We will consider each loading separately, and then, using our physical understanding of the problem, we will find where the superposition yields the largest compressive stress on the cross section at the wall.

First consider the triangular load. It has a bending moment at the wall which equals (5)(100)(1/2)(2/3)(5) = 833.3 N-m in magnitude. Hence, the maximum compressive stress will be at the bottom line AB (see Fig. 11.13) of the cross section, given in magnitude as

$$|(\tau_{xx})_{max}|_1 = \frac{(833.3)(.04)}{(\frac{1}{12})(.04)(.08)^3} = 1.953 \times 10^7 \text{ Pa}$$

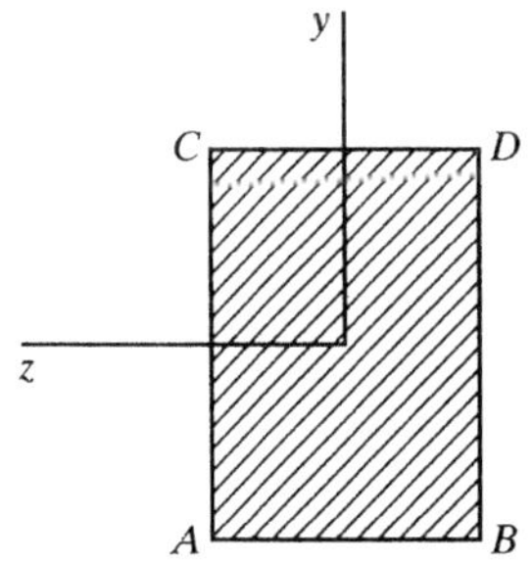

Figure 11.13. Cross section of cantilever at the base.

As for the uniform loading, it has a bending moment at the base about the y axis, which is (5)(30)(2.5)=375 N-m in magnitude. A moment's thought should reveal that the maximum compressive stress will be along line AC of the base cross section and have a magnitude given as

$$|(\tau_{xx})_{max}|_2 = \frac{(375)(.02)}{(\frac{1}{12})(.08)(.04)^3} = 1.758 \times 10^7 \text{ Pa}$$

The maximum compressive stress from the combined loading must occur at point A in Fig. 11.13 where we have the coincidence of maxima of compressive stresses from the two loads.

$$\therefore (\tau_{xx})_{max} = -1.953 \times 10^7 - 1.758 \times 10^7 = \mathbf{-3.711 \times 10^7 \text{ Pa}}$$

Notice in this problem that we used physical feel and common sense freely rather than more formal approaches. You are urged to do likewise when it is possible, as was the case in this problem.

11.4 Bending of Symmetric Beams with Shear: Shear Stress

In the case of the more general loading, we may have a *nonzero shear force* at sections of the beam and we would then expect to have a *shear-stress distribution* τ_{xy} over a section *in addition* to the normal stress distribution τ_{xx}. We shall now present formulations for determining certain *average values* of this shear-stress distribution using the *Euler-Bernoulli* theory of bending.

Accordingly, we have shown in solid lines in Fig. 11.14 a portion of beam of length dx. Note that positive normal stresses τ_{xx} and shear

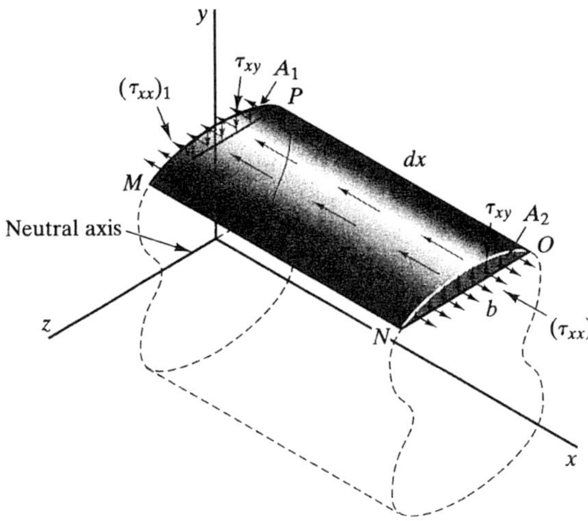

Figure 11.14. Element of a beam showing shear stress.

stress τ_{xy} have been shown on faces A_1 and A_2, whereas on the lower face, taken parallel to the xz plane, we have shown shear stress τ_{yx}. We would expect such a shear stress in order to equilibrate the forces from the normal stresses at the ends of the element, since these forces will differ because of the variation of the bending moment along the beam. The width of the bottom surface has been denoted as b. Since we shall consider equilibrium only in the x direction here, we have not shown external distributed loading to avoid cluttering the diagram. From **equilibrium** considerations in the x direction we thus have

$$-\bar{\tau}_{yx}b\,dx - \iint_{A_1}(\tau_{xx})_1 dA + \iint_{A_2}(\tau_{xx})_2 dA = 0 \qquad (11.14)$$

where $\bar{\tau}_{yx}$ is the *average shear stress* on the lower face of the element over the finite width b. We may express the normal stress at section 1 as

$(\tau_{xx})_1 = (-M_z y/I_{zz})_1$ and, using a two-term Taylor expansion, for section 2, which is an infinitesimal distance dx from section 1, we have[8]

$$(\tau_{xx})_2 = \left(\frac{-M_z y}{I_{zz}}\right)_1 + \left(\frac{d(-M_z y/I_{zz})}{dx}\right)_1 dx \tag{11.15}$$

Accordingly, replacing the stresses in Eq. (11.14) using the *flexure formula* as indicated above and noting that areas A_1 and A_2 are identical, we then obtain on cancelling two terms

$$-\bar{\tau}_{yx} b\, dx - \iint_{A_1} \left(\frac{d(M_z y/I_{zz})}{dx}\right)_1 dx\, dA = 0 \tag{11.16}$$

where we remind you that I_{zz} is the second moment of area about the neutral axis of the *entire* undeformed cross section stemming from the use of the flexure formula.

We have shown in Chapter 10 that $dM_z / dx = V_y$ and noting that y does not in any way depend on x here, we can then say for the quantity in the parenthesis in Eq. (11.16) that

$$\frac{d(M_z y/I_{zz})}{dx} = \frac{dM_z}{dx}\frac{y}{I_{zz}} = \frac{V_y y}{I_{zz}} \tag{11.17}$$

On canceling dx, Eq. (11.16) can now be written as

$$-\bar{\tau}_{yx} b - \frac{V_y}{I_{zz}} \iint_{A_1} y\, dA = 0 \tag{11.18}$$

The integral in Eq. (11.18) is the *first moment of area* A_1 above the cut along b [see Fig. (11.15b)] about the neutral axis. Calling this quantity Q_z and using magnitudes for now we get the following formula called Jourawski's formula:

$$|\bar{\tau}_{yx}| = \left|\frac{V_y Q_z}{I_{zz} b}\right|$$

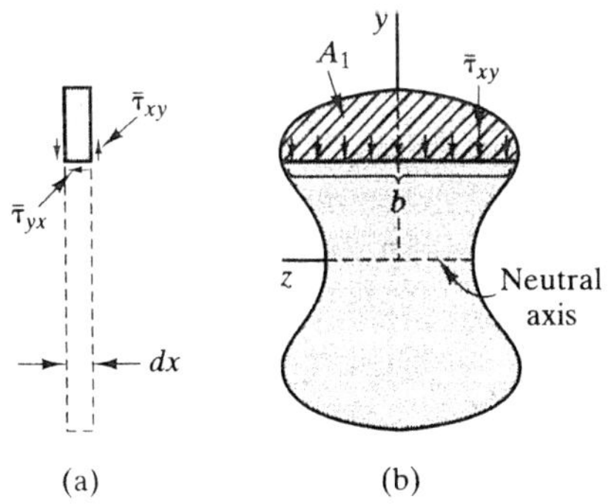

Figure 11.15. End view of beam element.

We shall consider the sign of the shear stress as a separate topic directly after this section. At that time, we will present a simple procedure for determining the proper sign. Meanwhile, we shall continue to discuss salient aspects of the magnitude of the shear stress.

Once again, we point out that the shear stress we have computed is the *average shear stress* over width b on infinitesimal face $MNOP$ of the beam parallel to the xz plane (see Fig. 11.14). We have shown this stress in Fig. 11.15(a) in a *side view* of the beam element of Fig. 11.14, this time drawn more to scale with respect to the length dx of the ele-

[8]We typically let M_z increase in the positive coordinate direction as is usual in derivations.

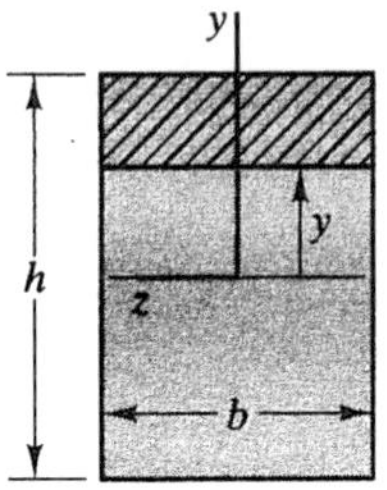

(a)

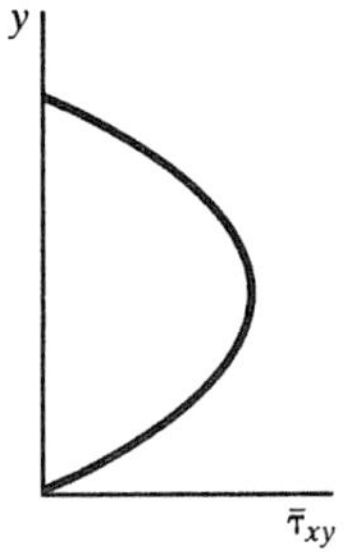

(b)

Figure 11.16. Rectangular cross section showing shear stress distribution.

ment. The average shear stress $\bar{\tau}_{yx}$ that we have evaluated is shown in this diagram. Now using the *complementary property of shear,* we see that we *also* have available from this formulation the average shear stress $\bar{\tau}_{xy}$ along the width b on the *cross section* of the beam as shown in Fig. 11.15(b). Thus,

$$|\bar{\tau}_{xy}| = \left|\frac{V_y Q_z}{I_{zz} b}\right| \tag{11.19}$$

We emphasize that Q_z is the first moment about the neutral axis of the area [shown crosshatched in Fig. 11.15(b)] *above* the line b at which we are computing $\bar{\tau}_{xy}$. Thus, we see that $\bar{\tau}_{xy}$ will start out with a zero value at the top fibers of the beam since Q_z will be zero. For a rectangular cross section [Fig. 11.16(a)] (i.e., for fixed b), $\bar{\tau}_{xy}$ will then increase in value as one proceeds downward. Once the neutral axis is passed, there will be negative contributions to Q_z from the area that lies below the neutral axis. Thus, we see that *the maximum average shear stress, $\bar{\tau}_{xy}$, for a rectangular cross section will occur at the neutral axis.* When we reach the bottom of the cross section, we are computing Q_z for the entire section of the beam. Since the neutral axis is a *centroidal axis,* we see that $Q_z = 0$ when we get to the bottom fibers and so we again have zero shear stress $\bar{\tau}_{xy}$ at this location for any symmetric cross section.

Example 11.7

What is the magnitude of the average shear-stress distribution $|\bar{\tau}_{xy}|$ as a function of y for the rectangular cross section shown in Fig. 11.16(a) in terms of V_y and the dimensions of the cross section?

At any position y above the neutral axis, for Q_z we have

$$Q_z = b\left(\frac{h}{2} - y\right)\left(y + \frac{h/2 - y}{2}\right) = \frac{b}{2}\left[\left(\frac{h}{2}\right)^2 - y^2\right] \tag{a}$$

The magnitude of the shear stress $|\bar{\tau}_{xy}|$ at any position y then becomes

$$|\bar{\tau}_{xy}| = \frac{|V_y|}{(\frac{1}{12})(bh^3)b}\left(\frac{b}{2}\right)\left[\left(\frac{h}{2}\right)^2 - y^2\right]$$

$$|\bar{\tau}_{xy}| = \frac{6|V_y|}{bh^3}\left[\left(\frac{h}{2}\right)^2 - y^2\right] \tag{b}$$

Example 11.7 (Continued)

We see that the shear stress $\bar{\tau}_{xy}$ has a parabolic distribution as shown in Fig. 11.16(b).[9]

Example 11.8

In Example 11.4 determine the magnitude of the shear stress $|\bar{\tau}_{xy}|$ at the section 3 ft from the wall at a position 1/2 in. from the top surface.

The shear force V_y at the section of interest is 170 lb (see Fig. 11.10). Hence, we get

$$|\bar{\tau}_{xy}| = \frac{(170)[(\frac{1}{2})(1)(\frac{3}{4})]}{[(\frac{1}{12})(1)(2^3)](1)} = 95.6 \text{ psi}$$

$$|\bar{\tau}_{xy}| = 95.6 \text{ psi}$$

Example 11.9

Shown in Fig. 11.17 is the cross-sectional area of a wide-flange 14 WF202 I beam.[10] What is $|\bar{\tau}_{xy}|$ as a function of y?

For the upper *flange* region we have [see Fig. 11.17(c)]

$$|\bar{\tau}_{xy}| = \frac{|V_y|}{I_{zz}(15.75)} \underbrace{\left[(15.75)\left(\frac{15.63}{2} - y\right)\right]}_{\text{area above } y} \underbrace{\left(y + \frac{15.63/2 - y}{2}\right)}_{\substack{\text{moment arm from} \\ \text{centroid of } A \text{ to} \\ \text{neutral axis}}}$$

$$|\bar{\tau}_{xy}| = \frac{|V_y|}{5080}(7.815^2 - y^2) \qquad \text{(a)}$$

where we have used $I_{zz} = 2540$ in.^{4} from the handbooks. Next, for the *web* region, we get [see Fig. 11.17(d)], considering the flange and web portions separately,

[9]Thus the maximum shear stress $\bar{\tau}_{xy}$ occurs where the normal stress τ_{xx} is zero, and vice versa.

[10]The geometries of structural shapes can be found in handbooks, such as those available from the American Institute of Steel Construction, Chicago, IL. Also, tables are presented in Appendix IV-A.

Example 11.9 (Continued)

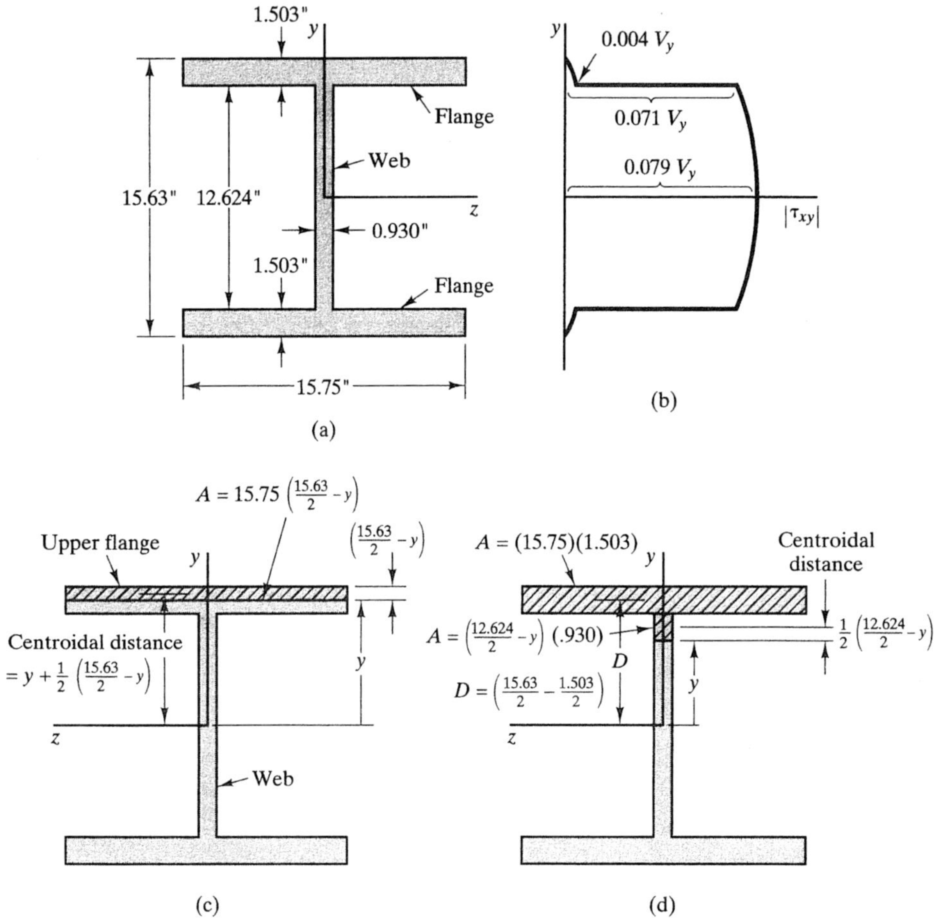

Figure 11.17. Cross section of wide-flanged I beam.

$$|\overline{\tau}_{xy}| = \frac{|V_y|}{I_{zz}(.930)}\left[\underbrace{(15.75)(1.503)}_{\text{area of flange}}\underbrace{\left(\frac{15.63}{2} - \frac{1.503}{2}\right)}_{\substack{\text{moment arm } D \text{ of} \\ \text{flange centroid} \\ \text{about neutral axis}}}\right.$$

$$\left. + \underbrace{\left(\frac{12.624}{2} - y\right)(.930)}_{\text{area of web above } y}\underbrace{\left(y + \frac{12.624/2 - y}{2}\right)}_{\substack{\text{moment arm of} \\ \text{web area above } y \\ \text{about neutral axis}}}\right]$$

Example 11.9 (Continued)

$$|\bar{\tau}_{xy}| = \frac{|V_y|}{2540}\left[180 + \frac{1}{2}(6.312^2 - y^2)\right] \tag{b}$$

A plot of $|\bar{\tau}_{xy}|$ for the section is shown in Fig. 11.17(b), which represents the case where fillets between flanges and web are assumed to have zero radius, thus giving a curve with discontinuities of slope.

Note from this curve that the shear stress magnitude $|\bar{\tau}_{xy}|$ in the flange is very *small* compared to the shear stress $|\bar{\tau}_{xy}|$ in the web. Furthermore, the shear stress magnitude in the web is approximately uniform in value over the entire web. For such problems, it is often assumed for simplicity that the *web carries the shear load with a uniform shear stress* given by

$$|\bar{\tau}_{xy}| = \frac{|V_y|}{A} = \frac{|V_y|}{(.930)(12.64)} = .085|V_y|$$

This result is about 7.6% in error above the maximum in Fig. 11.17(b) making this formulation acceptable in many applications.

We should perhaps emphasize the limitations of the shear stress formula before moving on to other examples. Hence, consider a cross section such as is shown in Fig. 11.18. The total shear stress must have a direction *parallel* to the edge of the section at locations along the edge of the section. Not to have such an alignment would give components of shear stress *normal* to the boundary of the section. And owing to the complementary property of shear, this would require external tangential loads in the x direction on the *lateral surface* of the beam. Since our external loads have been assumed along the y direction, this condition would violate boundary conditions. Hence the shear stress on the cross section must be oriented tangent to the boundary.

Examine Fig. 11.18 further. Because the total shear stress must undergo considerable change in direction as we go from one side of the section to the other, we can expect considerable variation of the vertical component of shear stress. Also, there will be an appreciable horizontal component of shear τ_{xz} near the sides of the section but little near the center. It should be clear that the quantity $\bar{\tau}_{xy}$ has very little meaning for such cross sections. We see here that the formulations of strength of materials give results of very limited usefulness for the computation of shear stress for such sections, in contrast to the computation of normal stress τ_{xx}.[11] For more meaningful results, one must resort to the theory of elasticity.

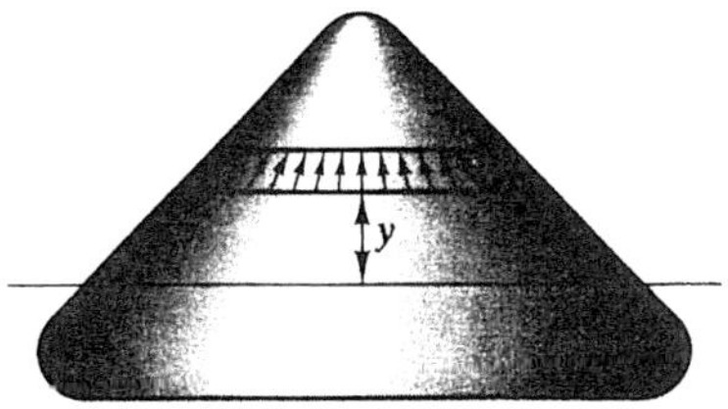

Figure 11.18. $\bar{\tau}_{xy}$ not useful at y.

[11]We wish to point out that where the horizontal line, along which one is measuring $\bar{\tau}_{xy}$, meets boundary lines which are normal to it at its extremities, such as those shown in Fig. 11.19, the computed values of $\bar{\tau}_{xy}$ should be reasonably meaningful for such lines.

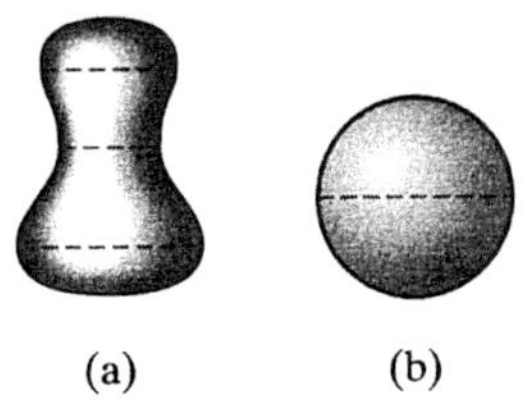

Figure 11.19. $\bar{\tau}_{xy}$ useful along dashed lines.

We can conclude that using the shear formula for a section with straight sides in the y and z directions, such as the rectangular section of Fig. 11.16 or the I-beam section of Fig. 11.17, yields reasonably accurate and meaningful results for the shear-stress distributions. We shall refer to such cross sections as *rectangular-like* cross sections.

In a later section we shall consider the total shear stress in sections corresponding to open *thin-walled* members of arbitrary shape whereby using the methods of strength of materials we will be able to get meaningful results.

Example 11.10

Consider a beam composed of three members of identical material screwed together firmly as shown in Fig. 11.20(b). The beam is simply supported, carrying a total uniform load of 1500 N/m and is 6 m in length [see Fig. 11.20(a)]. If the allowable shear force f per screw to cut the screw in two at the plane of contact between the members is 1000 N, what is the largest safe spacing of the screws?

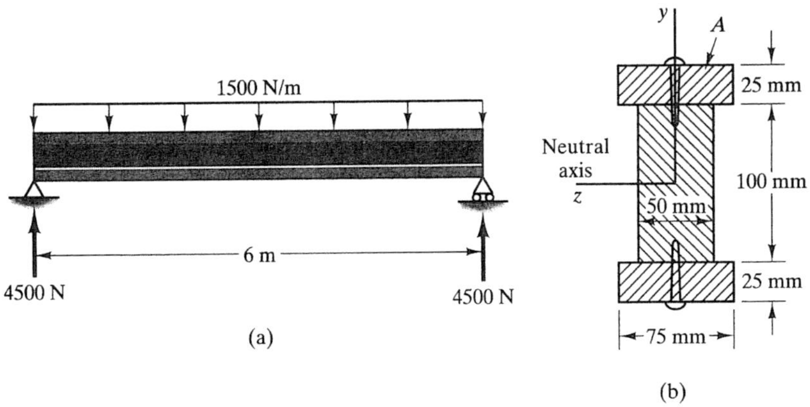

Figure 11.20. Built-up beam.

Consider a free body of a portion of the upper member as shown in Fig. 11.21(a). We have shown only stresses τ_{xx} and force f from the screws in the x direction. Note in contrast to the solid single-member beam of Fig. 11.17 that it is the forces from the screws rather than shear stress over the bottom face that resists the force in the x direction stemming from nonuniform normal stress τ_{xx} in that direction. Were the entire beam *one solid member* we would develop a shear force dF_s on the bottom, cross-hatched face, in Fig. 11.21(b) of

Example 11.10 (Continued)

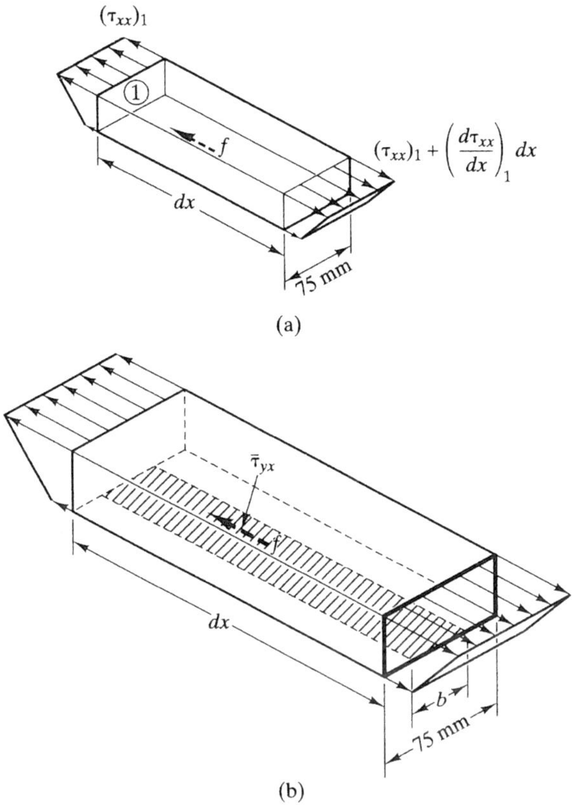

Figure 11.21. Free bodies of upper element of beam. (a) Top element restrained by screws generating force f. (b) Enlargement of top element, imagining that it has been cut from a single solid beam. Shear stress $\bar{\tau}_{yx}$ now restrains this element.

$$dF_s = |\bar{\tau}_{yz} b\ dx|$$

where b is the width of the *contact surface* (50 mm). Based *per unit length* along the beam, we would develop a shear force shear intensity $\mathfrak{F}_s$ on the bottom cross-hatched face in Fig. 11.21(b) of

$$\mathfrak{F}_s = \frac{dF_s}{dx} = |\bar{\tau}_{yx} b\ | \quad \text{(a)}$$

Using Eq. (11.19), this force intensity magnitude can be given as

$$|\mathfrak{F}_s| = \left|\frac{V_y Q_z}{I_{zz}}\right| \quad \text{(b)}$$

Example 11.10 (Continued)

where Q_z is the first moment of cross section of the top member about the neutral axis of the entire beam. In this problem, the force intensity $\mathcal{F}_s$ is developed by the shear forces from the screws. If f is the resisting force from one screw and there are n screws per unit length of beam at the region of interest, we can then imagine a force intensity (based on per unit length) arising from the screws and given as nf. To satisfy **equilibrium,** we equate the resisting force intensity from the screws with the developed force intensity as given by Eq. (b) to obtain

$$nf = \left| \frac{V_y Q_z}{I_{zz}} \right| \tag{c}$$

Using the largest value of V_y over the beam and the allowable value of f, we can then solve for n, the minimum number of screws per unit length of beam. For the problem at hand we have

$$|(V_y)_{\max}| = 4500 \text{ N}$$

$$Q_z = \frac{(25)(75)(62.5)}{10^9} = 1.172 \times 10^{-4} \text{ m}^3$$

$$I_{zz} = \left\{ \frac{1}{12}(50)(100)^3 + 2\left[\frac{1}{12}(75)(25)^3 + (75)(25)(62.5)^2 \right] \right\} 10^{-12} \text{ m}^4$$

$$= 1.901 \times 10^{-5} \text{ m}^4$$

Hence,

$$n = \frac{1}{1000} \frac{(4500)(1.172 \times 10^{-4})}{1.901 \times 10^{-5}} = 27.7/\text{m}$$

We thus require 27.7 screws/m or, in other words, the screws should be spaced $1/n = 36.1$ mm apart at sections where the shear force is greatest. This is also true for the bottom member.

Before leaving this example, it might be instructive to ponder the question as to the advantage of screwing the three members together as opposed to simply letting each one slide freely over the other without the constraints of the screws. Omitting the screws and neglecting friction, each member then acts as a *separate beam* with its *own neutral axis* at its *own centroid.* The total value of I_{zz} for the unattached system is as a result the sum of the second moments of area each taken about the horizontal centroidal axis of the separate members. This total will be much less than that of the connected members since in the attached case all second moments are taken

Example 11.10 (Continued)

about the horizontal centroidal axis of the *entire* cross section. The increase in I for the connected beams then would stem from the *transfer terms* for the outer member cross sections to the centroidal axis of the entire system [the transfer distance would be 1/2(.025) + 1/2(.100) m]. Thus, connecting the beams stiffens the system by increasing the effective second moment of area. In doing problems of unattached beams, note that at a position x we can take the *radius of curvature* as the same for *all members* in the system.

11.5 Determination of the Sign of the Shear Stress

Three cases will be discussed in this section for determining the proper sign of the shear stress. Let us consider a rectangular cross section of a beam subject to a vertical shear force as is shown in Fig. 11.22(a). We first examine the case where a horizontal cut is taken above the neutral axis and the area formed from the horizontal cut is also above the neutral axis. We can then say the following:

$$V_y \text{ is negative}$$

$$\frac{dM_z}{dx} = V_y$$

$$\therefore \quad \frac{dM_z}{dx} < 0$$

$$\text{Hence, } dM_z < 0 \qquad \text{(i.e., it is negative)}$$

$$\text{But} \qquad \tau_{xx} = -\frac{M_z y}{I_{zz}} \text{ and } y \text{ is positive}$$

$$d\tau_{xx} = -\frac{dM_z y}{I_{zz}}$$

$$\text{Thus } d\tau_{xx} > 0 \qquad \text{(i.e., it is positive)}$$

In Fig. 11.22(b) we show a net positive stress increment $d\tau_{xx}$ on the area above the horizontal cut and an average shear stress increment $d\bar{\tau}_{yx}$ on the bottom face of the infinitesimal element. This increment of shear stress is needed to equilibrate the net normal stress increment. Now in Fig. 19.22(c), we show the complementary shear stress increment at the horizontal cut. Clearly, the shear stress must be positive in accordance with our basic stress sign convention.

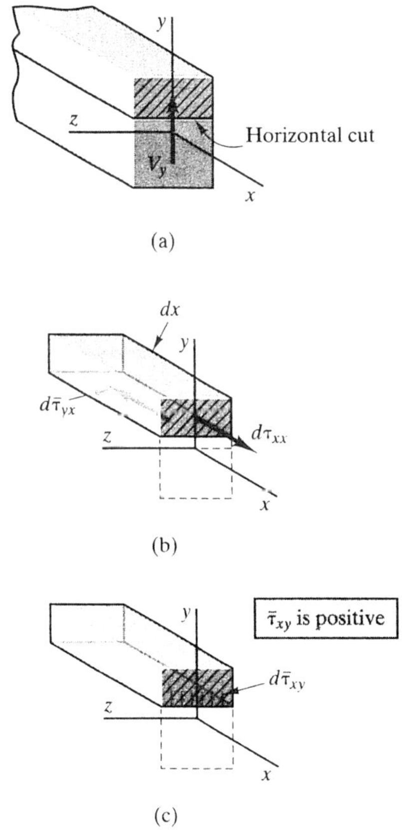

Figure 11.22. (a) Rectangular section with a vertical shear force. (b) Positive increment of normal stress yields a positive increment of shear stress. (c) Positive vertical shear stress on the cross section at the horizontal cut.

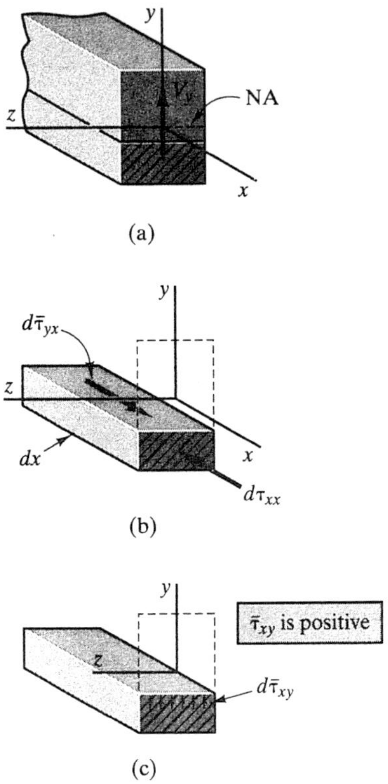

Figure 11.23. (a) An upward shear force on a rectangular cross section. (b) A negative normal stress increment results in a positive shear stress on the upper surface of the element. (c) A positive shear stress on the cross section at the cut.

Next, we look at the rectangular cross section again in Fig. 11.23(a) where we consider the area below a horizontal cut for the determination of the proper shear-stress sign. The horizontal cut in this case is below the neutral axis. We can then say:

$$V_y \text{ is negative}$$
$$\frac{dM_z}{dx} = V_y$$
$$\therefore \frac{dM_z}{dx} < 0$$
$$\text{Hence, } dM_z < 0 \quad \text{(i.e., it is negative)}$$
$$\text{But} \quad \tau_{xx} = -\frac{M_z y}{I_{zz}} \text{ and } y \text{ is negative}$$
$$\text{Thus, } d\tau_{xx} < 0 \quad \text{(i.e., it is negative)}$$

In Fig. 11.23(b), we have the increment of normal stress oriented in the negative direction as well as the increment of shear stress on the top face needed to equilibrate this stress increment. Finally, in Fig. 11.23(c), we show a positive shear stress resulting from the complementary property of shear at the cut on the cross section of the beam. Clearly the vertical shear stress is positive as it was for the case using an area above the cut.

We shall look at one more case and that is the case of the structural cross section shown in Fig. 11.24(a) where we have a horizontal cut on the right wing of the cross section. Accordingly, we have

$$V_y \text{ is positive}$$
$$\frac{dM_z}{dx} = V_y$$
$$\therefore \frac{dM_z}{dx} > 0$$
$$\text{Hence } dM_z > 0 \quad \text{(i.e., it is positive)}$$
$$\text{But} \quad \tau_{xx} = -\frac{M_z y}{I_{zz}} \text{ and } y \text{ is positive}$$
$$\text{Thus } d\tau_{xx} < 0 \quad \text{(i.e., it is negative)}$$

On the top surface of the element in Fig. 11.24(b) we have a positive shear increment $d\bar{\tau}_{yx}$ to equilibrate the negative normal stress increment $d\tau_{xx}$. Finally, with the complementary property of shear stress we see in Fig. 11.24(c) that we arrive at a positive shear stress on the cross section at the cut.

In the preceding examples, the sign to be assigned y was obvious for the determining the sign of $\bar{\tau}_{xy}$. When the area isolated by a cut for shear calculations has regions above and below the neutral axis, proceed by determining the sign of the average vertical distance from the neutral axis to the extremities of the areas. Then use the sign for this value of y.

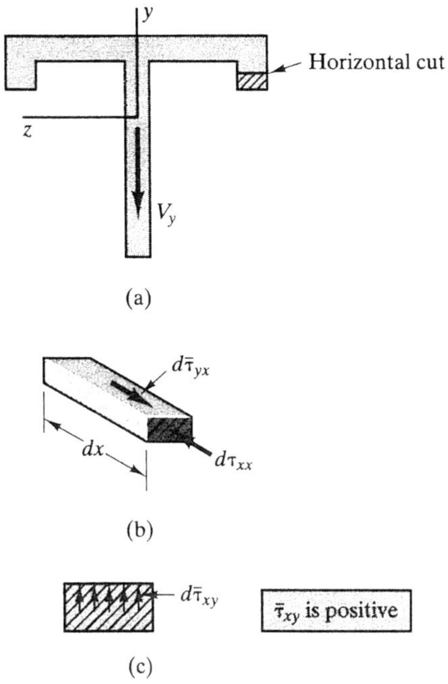

Figure 11.24. (a) A structural cross section with a downward shear force. (b) A negative normal stress increment results in a positive shear stress on the upper surface. (c) A positive shear stress on the cross section at the cut.

Example 11.11

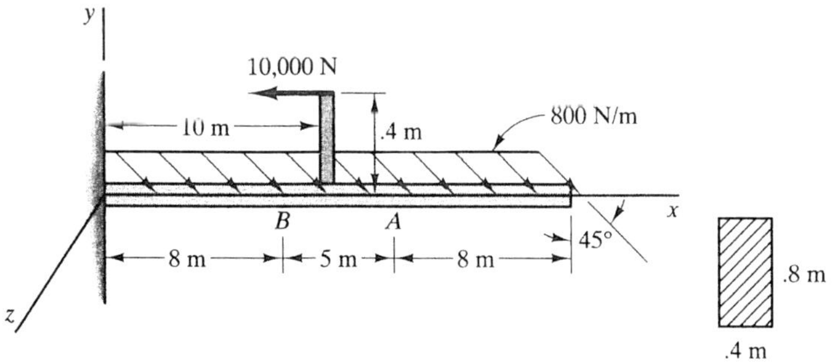

Figure 11.25. A cantilever beam with various loads.

A cantilever beam is shown in Fig. 11.25 with an inclined uniform load acting on the x axis of the beam and a vertical rod on which a 10,000-N force acts. For sections A and B of the beam, do the following:

Example 11.11 (Continued)

(a). Determine the maximum tensile stress.
(b). Determine the magnitude of the maximum shear stress.
(c). Determine the correct sign for the shear stress.

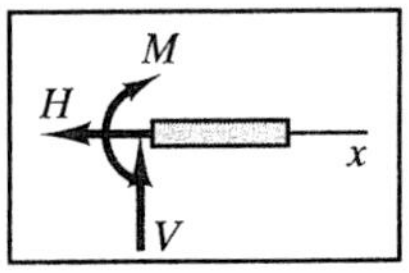

At section *A*:

Part (a)

$$M = -[(800)(\sin 45°)(8)](4) = -1.810 \times 10^4 \text{ N-m}$$

$$V = (800)(\sin 45°)(8) = 4525 \text{ N}$$

$$H = (800)(\cos 45°)\ (8) = 4525 \text{ N}$$

$$(\tau_{xx})_{\max} = -\frac{M y_{\max}}{I} + \frac{H}{A} = -\frac{(-1.810 \times 10^4)(.4)}{\left(\frac{1}{12}\right)(.4)(.8)^3} + \frac{(4525)}{(.4)(.8)} = 4.384 \times 10^5 \text{ Pa}$$

$$(\tau_{xx})_{\max} = 4.384 \times 10^5 \text{ Pa}$$

Part (b)

$$|\bar{\tau}_{xy}|_{\max} = \left|\frac{VQ}{Ib}\right| = \frac{(4525)(.4)(.4)(.2)}{\left(\frac{1}{12}\right)(.4)(.8)^3(.4)} = 2.12 \times 10^4 \text{ Pa}$$

$$|\tau_{xy}|_{\max} = 2.12 \times 10^4 \text{ Pa}$$

Part (c)

For the sign of $\bar{\tau}_{xy}$ we can say

$$V \text{ is positive}$$

$$\therefore \frac{dM}{dx} > 0$$

$$\therefore dM > 0$$

Example 11.11 (Continued)

$$d\tau_{xx} = -\frac{dMy}{I} \qquad y \text{ is positive (for area above the neutral axis)}$$

$d\tau_{xx}$ is negative

From Fig. 11.26, it is clear that the shear stress $\bar{\tau}_{xy}$ is negative.

$\bar{\tau}_{xy}$ is negative

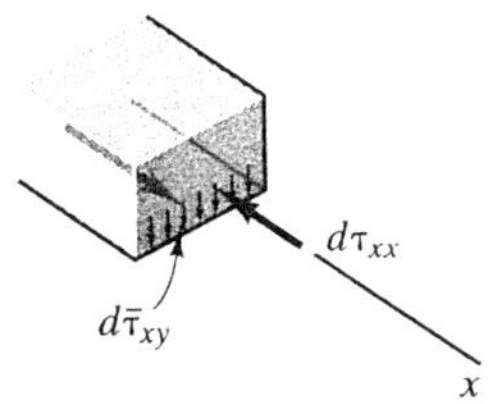

Figure 11.26. Determining the proper sign for the shear stress $\bar{\tau}_{xy}$.

At section *B*:

Part (a)

$$M = -(800)(\sin 45°)(13)\left(\frac{13}{2}\right) + (10{,}000)(.4) = -4.379 \times 10^4 \text{ N-m}$$

$$V = (800)(\sin 45°)(13) = 7{,}353.8 \text{ N}$$

$$H = (800)(\cos 45°)(13) - 10{,}000 = -2647 \text{ N}$$

$$\tau_{xx} = -\frac{My_{\max}}{I} + \frac{H}{A} = -\frac{(-4.379 \times 10^4)(.4)}{\left(\frac{1}{12}\right)(.4)(.8)^3} + \frac{(-2{,}647)}{(.4)(.8)} = 1.018 \times 10^6 \text{ Pa}$$

$$\tau_{xx} = 1.018 \times 10^6 \text{ Pa}$$

Part (b)

$$|\bar{\tau}_{xy}|_{\max} = \left|\frac{VQ}{Ib}\right| = \frac{(7353.8)(.4)(.4)(.2)}{\left(\frac{1}{12}\right)(.4)(.8)^3(.4)} = 3.447 \times 10^4 \text{ Pa}$$

$$|\bar{\tau}_{xy}|_{\max} = 3.447 \times 10^4 \text{ Pa}$$

Part (c)

The formulation of the sign of the shear stress is the same as for section *A* (i.e., $\bar{\tau}_{xy}$ is negative).

$\bar{\tau}_{xy}$ is negative

11.6 Consideration of General Cuts

We have thus far considered only horizontal cuts for the computation of the average vertical shear stresses over the cut. In this section, we shall consider vertical cuts in order to compute average horizontal shear stresses for symmetrical beams loaded in the plane of symmetry.

Consider again the wide-flanged I-beam of Example 11.9 shown again in Fig. 11.27(a) acting as a simply supported beam. In Fig. 11.27(b),

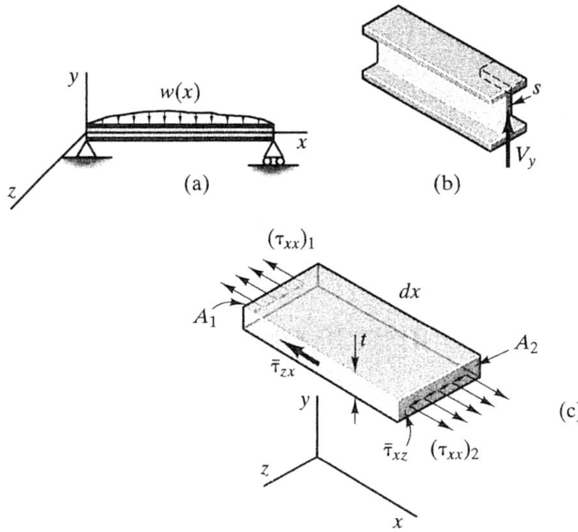

Figure 11.27. Wide-flanged I-beam.

we show a portion of the beam with an element (shown dashed) of infinitesimal length to be cut out for examination. Note that the element presents a *vertical cut* for consideration of shear force in the I-beam flange. Also, note that we use the measure s from the edge of the flange to locate this cut. We have shown this element as a free body in Fig. 11.27(c)[12] where we show a positive stress τ_{xx} which is increasing in the positive coordinate direction.[13] Because of the transverse loading, the bending moment along the beam varies, causing different normal stresses over the cross sections of the I-beam. It is this difference in stress that gives rise, as a result of equilibrium, to the average shear stress $\bar{\tau}_{zx}$ on the cross-hatched face of the element in Fig. 11.27(c). (Note that $-\bar{\tau}_{zx}$ is actually shown in Fig. 11.27(c)). We now consider **equilibrium** of this element in the x direction. Thus

[12]Note carefully for subsequent application of the formulations to be developed that we must be able to form a *meaningful free body* when using a vertical cut.

[13]As we have done earlier, we have M_z and thus τ_{xx} increase in the positive coordinate direction as is typical in derivations.

$$- \bar{\tau}_{zx} t\, dx - \iint_{A_1} (\tau_{xx})_1 dA + \iint_{A_2} (\tau_{xx})_2 dA = 0 \qquad (11.20)$$

where $\bar{\tau}_{zx}$ is the average stress over the thickness t of the cut. We express $(\tau_{xx})_2$ as a Taylor series expansion of $(\tau_{xx})_1$ at A_1 in the following manner:

$$(\tau_{xx})_2 = (\tau_{xx})_1 + \left(\frac{\partial \tau_{xx}}{\partial x}\right)_1 dx \qquad (11.21)$$

where we have used a two-term expansion. Substituting into Eq. (11.20) and canceling terms, we have

$$-\bar{\tau}_{zx} t\, dx + \iint_A \left(\frac{\partial \tau_{xx}}{\partial x}\right) dx\, dA = 0 \qquad (11.22)$$

where we have deleted the no-longer-needed subscripts 1and 2. Now we can replace τ_{xx} by $-M_z y/I_{zz}$, where I_{zz} is the second moment of area of the *entire* cross section (before the cut was made) about the neutral axis. We thus have, on canceling dx,

$$-\bar{\tau}_{zx} t - \iint_A \frac{\partial (M_z y/I_{zz})}{\partial x} dA = 0 \qquad (11.23)$$

Note that only M_z is a function of x. Furthermore,

$$dM_z / dx = V_y$$

Hence, we have

$$-\bar{\tau}_{zx} - \frac{V_y}{I_{zz} t} \iint_A y dA = 0 \qquad (11.24)$$

The integral $\iint_A y dA$ is the first moment (as was the case of horizontal cuts) of the area of the cross section formed *after* cutting the body and isolating the element [Fig. 11.27(c)], where this moment is again taken about the neutral axis. As before, we denote the first moment as Q_z. We will use magnitude signs for the preceding equation at this time and we shall illustrate the determination of the sign later in this section using the same approach that was presented earlier for vertical shear stress.

$$|\bar{\tau}_{zx}| = \left| \frac{V_y Q_z}{I_{zz} t} \right| \qquad (11.25)$$

Employing the complementary property of shear, we see from Fig. 11.27(c) that in this development we have found the average *horizontal* shear-stress magnitude $|\bar{\tau}_{xz}|$ over the thickness t of the flange cross-section at position s of the flange. This is shown in Fig. 11.28(a). Q_z is the

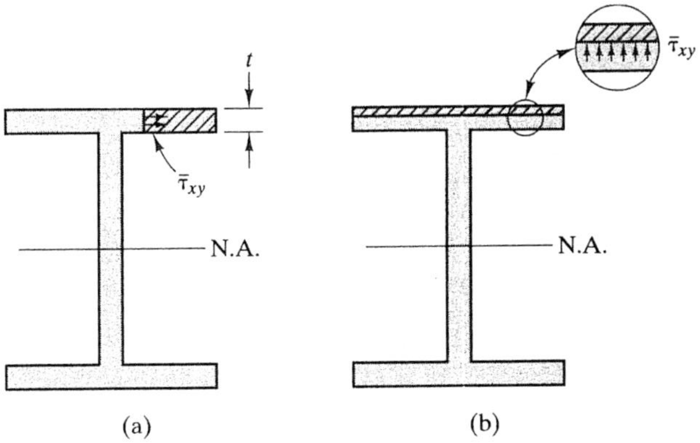

Figure 11.28. Wide-flanged I-beam showing vertical and horizontal cuts.

first moment, taken about the neutral axis N.A., of the cross-hatched area to the right of the cut. Thus a vertical cut yields the average horizontal shear stress along the cut. In Fig. 11.28(b),we have shown for comparison the horizontal cut used earlier in Example 11.7. There, a horizontal cut in the flange generates the average vertical shear stress over the length of the cut. The area for Q_z is that *above* the cut, but the first moment is still about the neutral axis.[14]

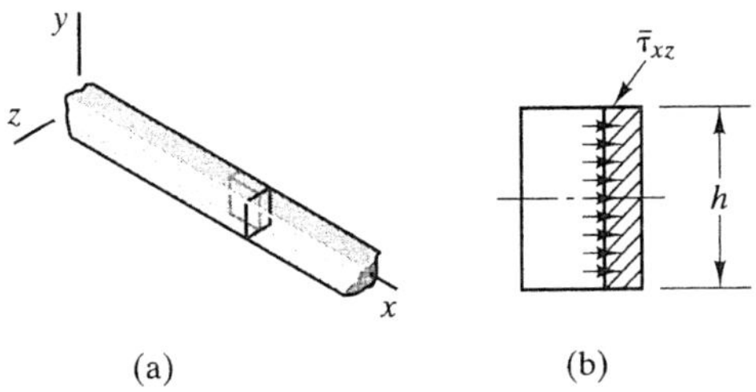

Figure 11.29. Rectangular cross section with vertical cut.

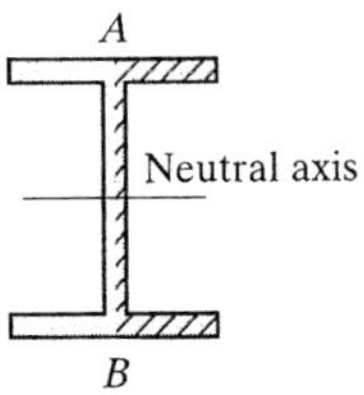

Figure 11.30. Vertical cut through web of I-beam.

In Fig. 11.29 we have shown a rectangular cross section with a vertical cut exposing an average horizontal shear stress. Here, it is clear Q_z is zero, so we conclude that $|\bar{\tau}_{xz}| = 0$, as our instincts may already tell us. In Fig. 11.30 we have shown a vertical cut going through the flanges and web of an I-beam. Clearly, $Q_z = 0$, so the average horizontal shear-stress magnitude is zero. In the rectangular cross-section we could conclude that $|\bar{\tau}_{xz}|$ was zero everywhere along the vertical cut. For the I-beam, because of the complex geometry, the fact that the average horizontal shear-stress magnitude is zero is of limited use to us. Actually, there will be large horizontal shear stress near the junctures A and

[14]Recall that the vertical shear stress was negligibly small in the flanges and was neglected (see Example 11.9).

B of the I-beam cross section, but they will be of opposite senses and so will cancel each other in the averaging process. In the web we can say that the *horizontal shear* is then *zero*.

We shall now examine the I-beam in greater detail in the following example.

Example 11.12

In Example 11.9 we found, using horizontal cuts, that the *vertical* shear stress τ_{xy} in the flanges of the I-beam was very small—so small [see Fig. 11.17(b)] that we could neglect such stress. We wish now to compute the *horizontal* shear stress τ_{xz} in the flanges—that is, the shear stress at right angles to the aforementioned negligibly small shear stress τ_{xy}.

We may do this by employing Eq. (11.25) in conjunction with Fig. 11.31, where we show the cross section of the I-beam in detail. Notice that we use s to measure the location of the desired shear stress in the flange. We get, noting that t_1 is the thickness of the flange and that D is the distance from centroid of the exposed area to the neutral axis

$$|\bar{\tau}_{xz}| = \left|\frac{V_y}{I_{zz}t_1}(st_1D)\right| = \left|V_ys\left(\frac{7.06}{2540}\right)\right| = \left|\frac{V_ys}{360}\right| \qquad \text{(a)}$$

where $I_{xx} = 2540$ in.^{4} from Appendix IV-1.

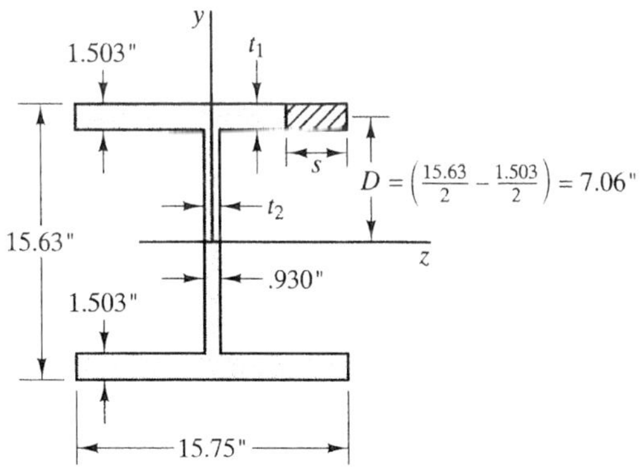

Figure 11.31. Vertical cut in the flange of an I-beam.

The maximum value of $|\bar{\tau}_{xz}|$ occurs when $s = 15.75/2$, (i.e. at the web juncture)[15] for which we get

[15]We have pointed out earlier that we could not learn much about horizontal shear stresses by taking a vertical cut at the web juncture (Fig. 11.30). We should cut just to the right of this juncture. However, using s equal to one half the width of the flange gives us a somewhat larger value of shear stress which is a more conservative result. Also, as you will see, a very useful result will be reached regarding shear flow.

Example 11.12 (Continued)

$$|(\bar{\tau}_{xz})_{\max}| = |(\bar{\tau}_{xz})^{\text{flange at}}_{\text{juncture}}| = |.0219V_y| \tag{b}$$

To get the shear stress on the left side of the flange we would measure from the left to right. Clearly, we would get the same distribution as on the right side except for direction (see Fig. 11.32).

The vertical shear stress $\bar{\tau}_{xy}$ was worked out for the *web* in Example 11.9 (the horizontal shear stress $\bar{\tau}_{xz}$ we know is zero here). The vertical shear stress magnitude in the web at the juncture with the flange is [see Fig. 11.17(b)]

$$|(\bar{\tau}_{xy})^{\text{web at}}_{\text{juncture}}| = |.071V_y| \tag{c}$$

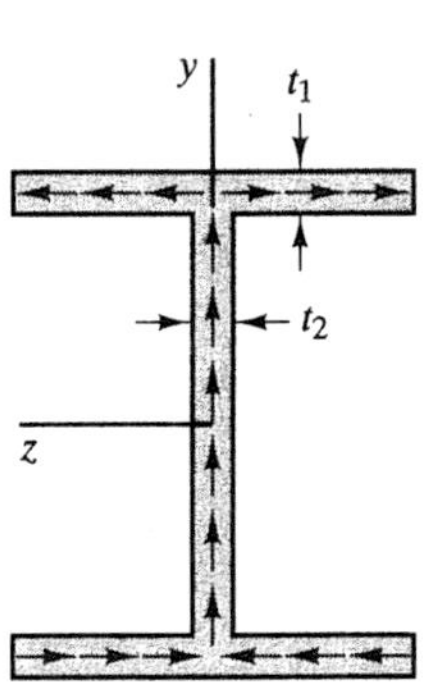

Figure 11.32. Shear "flowing" in I-beam section.

If one considers results from Eqs. (c) and (b) which deal with conditions at the juncture of web and flange, one will find that

$$|(\bar{\tau}_{xy})^{\text{web at}}_{\text{juncture}}|(t_2) = |(\bar{\tau}_{xz})^{\text{flange}}_{\text{at juncture}}|(2t_1) \tag{d}$$

where t_2 is the thickness of the web and t_1 is the thickness of the flange. Now the product ($\bar{\tau}_{\text{shear}}t$) is called *shear flow*, and we see that the shear flow into the juncture from the web equals the shear flow from the juncture into the flange portions (see Fig. 11.32). We can conclude that at the juncture shear flow behaves like the flow of an incompressible fluid within the same confines. In Chapter 14 we shall find that shear flow plays a critical role in the torsion of thin-walled closed cylinders.

Let us next consider the evaluation of the sign for the horizontal shear stress in the flange of the wide-flanged I-beam. Essentially we proceed exactly as we did for the vertical stresses. Thus considering Fig. 11.27(b) we can say the following:

$$V_y \text{ is negative}$$

$$dM_z / dx = V_y$$

$$\text{Hence} \quad dM_z < 0 \quad \text{(i.e., it is negative)}$$

$$\text{But} \quad \tau_{xx} = -\frac{M_z y}{I_{zz}} \quad y \text{ is positive}$$

$$\text{Thus} \quad d\tau_{xx} > 0 \quad \text{(i.e., it is positive)}$$

With a positive increment of normal stress on the area A_1, equilibrium then requires a negative shear increment on the cross-hatched side surface as is shown in Fig. 11.27(c). Consequently using the complementary property of shear at the edge of the vertical cut, we conclude that we must have a negative shear stress on the area A_2.

Example 11.12 (Continued)

This is the proper stress that has been shown in the diagram. The above procedure works for sections exposed to the left or to the right, and for sections exposed above or below the neutral axis. There will be a number of problems in this chapter that will require the determination of the proper sign for shear stress.

Example 11.13

Consider a cantilever beam supporting a load of 1000 lb, as shown in Fig. 11.33. It is constructed of various members of the same material screwed together. Determine the minimum number of screws n needed per unit length to fasten members B and D to the horizontal member C assuming that each screw can withstand a shear force f of 150 lb.

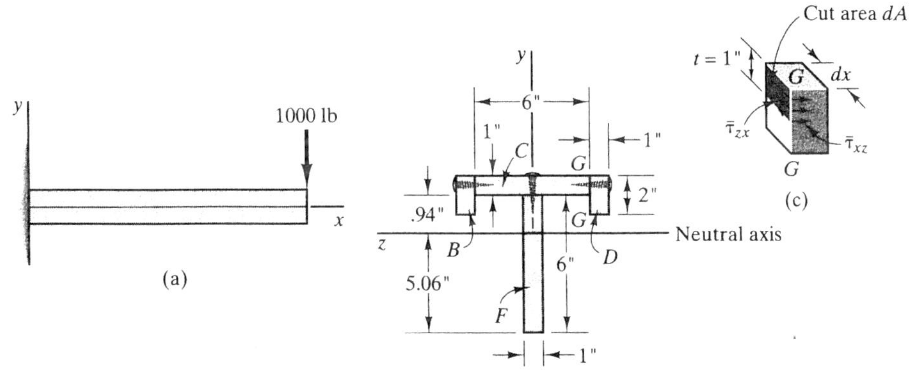

Figure 11.33. Built-up cantilever beam.

Assume that the beam is whole without screws and compute the average shear-stress magnitude $|\bar{\tau}_{xz}|$ at a *vertical* cut along G-G separating off section D [see Fig. 11.32(b)]. Now multiply this stress by the area dA generated by the *cut* over a length dx of the beam to obtain the shear-force increment, dF_s [see Fig. 11.33(c)]. Then, disregarding signs

$$dF_s = \bar{\tau}_{zx} t\, dx$$

and the shear force intensity (force/length), $\mathscr{F}_s$, is given as

$$\mathscr{F} = \frac{dF_s}{dx} = \bar{\tau}_{zx} t = nf \qquad \text{(a)}$$

Example 11.13 (Continued)

The neutral axis for the cross section is first found to be a distance 5.06 in. from the bottom as shown in the diagram, and the second moment of area of the entire cross section about this axis is then easily found to be 61.2 in.^{4} The magnitude of the average shear stress along cut G-G can then be given as follows.[16]

$$|\bar{\tau}_{zx}| = \left|\frac{V_y Q_z}{I_{zz} t}\right| = \left|\frac{(1000)[(2)(1)(.94)]}{(61.2)(1)}\right| = 30.7 \text{ psi} \qquad \text{(b)}$$

where (.94) is the distance between the centroid of D and the neutral axis. Now going to Eq. (a), per inch of beam we have for n on using the above result

$$n = \frac{1}{150}(30.7)(1) = .205$$

$$n = .205 \text{ screws/in.}$$

Thus, we have .205 screws/in. or, using $1/n$, we should space the screws 4.89 in. apart.

Example 11.14

In Fig. 11.34 we have shown a cross section of a beam made up of three members welded together. A vertical downward shear force is present on this section having the magnitude of 50 kN. What approximately is the total shear stress at point A?

As a first step, we shall compute I_{zz} at the neutral axis. Thus,

$$I_{zz} = \frac{1}{12}(.06)(.120)^3 + 2\left[\frac{1}{12}(.100)(.04)^3 + (.04)(.100)(.08)^2\right]$$

$$= 6.091 \times 10^{-5} \text{ m}^4$$

To get $|\bar{\tau}_{xy}|$ at A, we take a horizontal cut through A as shown in Fig. 11.35(a) thus generating a meaningful free body. We then have for $|\bar{\tau}_{xy}|$

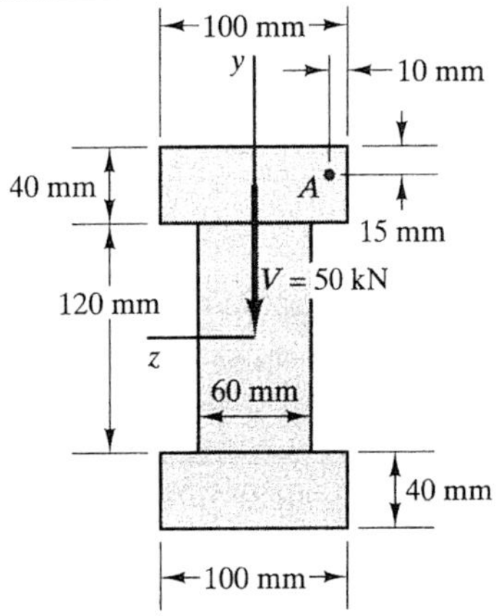

Figure 11.34. Cross section with vertical shear load.

[16]If there were a vertical loading acting directly on member D, equilibrium in the vertical direction would require an additional vertical shear stress on the screws along section G-G.

Example 11.14 (Continued)

$$|\bar{\tau}_{xy}| = \left|\frac{(50)[(.015)(.100)(.085 + .015/2)]}{(6.091 \times 10^{-5})(.100)}\right|$$

$$= 1139 \text{ kPa}$$

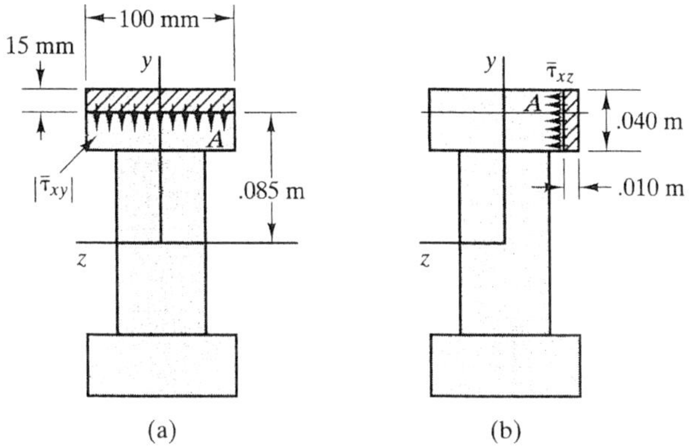

Figure 11.35. Horizontal and vertical cuts at A.

Next, to get the horizontal shear-stress magnitude at A, namely $\bar{\tau}_{xz}$, we take a vertical cut through A [see Fig. 11.35(b)] again generating a meaningful free body. We can now say that

$$|\bar{\tau}_{xz}| = \left|\frac{(50)(.010)(.040)(.080)}{(6.091 \times 10^{-5})(.040)}\right|$$

$$= 657 \text{ kPa}$$

The total shear stress is then[17]

$$\bar{\tau}_{\text{total}} = \sqrt{\bar{\tau}_{xy}^2 + \bar{\tau}_{xz}^2} = (1139^2 + 657^2)^{1/2} = 1315 \text{ kPa}$$

$$\bar{\tau}_{\text{total}} = 1315 \text{ kPa}$$

Suppose now that you had in addition to the vertical shear force a horizontal shear force also going through the centroid of the cross section. In that case, you could find additional shear stresses at A by again taking horizontal and vertical cuts as we did in this example. But now you would take *Q and I about the y axis rather than the z axis* as we did earlier. We suggest that you work this additional step.

[17]We note that what we get is an approximation to the magnitude of the shear stress at A.

Example 11.15

A tip-loaded cantilever beam is shown in Fig. 11.36(a). The cross section of the beam is shown in Fig. 11.36(b). To aid in the ensuing calculations, we have inserted some dashed lines to divide the cross section into rectangles. The following information is to be found.

(a). The maximum tensile stress, $(\tau_{xx})_{\max}$.
(b). The magnitude and then the sign of $\bar{\tau}_{xz}$ at point A.
(c). The magnitude and then the sign of $\bar{\tau}_{xy}$ at point B.

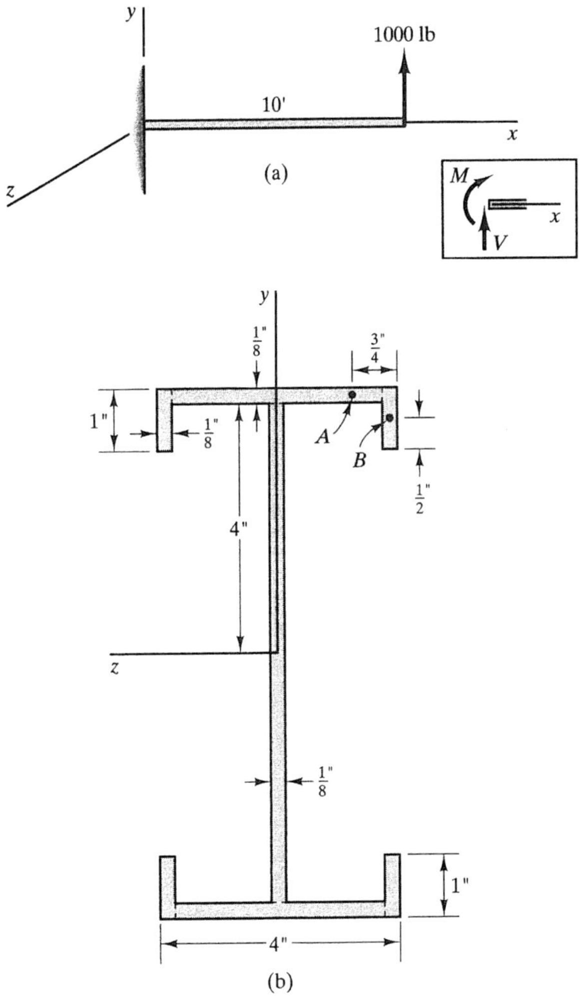

Figure 11.36. (a). Tip-loaded cantilever beam. (b). Cross-section of a winged I beam.

Example 11.15 (Continued)

Part (a)

We will first calculate I_{zz} about the neutral axis. Thus, using the rectangles formed in the cross-section we get

$$I_{zz} = 4\left[\left(\frac{1}{12}\right)\left(\frac{1}{8}\right)(1)^3 + \left(\frac{1}{8}\right)(1)\left(4 + \frac{1}{8} - \frac{1}{2}\right)^2\right] + \left(\frac{1}{12}\right)\left(\frac{1}{8}\right)(8)^3 + 2\left[\left(\frac{1}{12}\right)\left(4 - \frac{1}{4}\right)\left(\frac{1}{8}\right)^3 + \left(4 - \frac{1}{4}\right)\left(\frac{1}{8}\right)\left(4 + \frac{1}{16}\right)^2\right] = 27.42 \text{ in}^4$$

Now we can determine the maximum tensile stress which occurs at the base of the cantilever beam. Thus:

$$(\tau_{xx})_{\max} = -\left(\frac{M_z y}{I_{zz}}\right)_{\max} = -\frac{(10{,}000)(12)(-4\frac{1}{8})}{27.42} = 18{,}050 \text{ psi}$$

$$(\tau_{xx})_{\max} = 18{,}050 \text{ psi}$$

Part (b)

Next, we consider the magnitude of the shear stress $\bar{\tau}_{xz}$ at point A. Using a vertical cut as shown in Fig. 11.37(a), we get

$$|\bar{\tau}_{xz}| = \left|\frac{VQ}{It}\right| = \left|\frac{(-1000)\left[(1)\left(\frac{1}{8}\right)\left(4 + \frac{1}{8} - \frac{1}{2}\right) + \left(\frac{1}{8}\right)\left(\frac{3}{4} - \frac{1}{8}\right)\left(4 + \frac{1}{16}\right)\right]}{(27.42)\left(\frac{1}{8}\right)}\right|$$

$$|\bar{\tau}_{xz}| = 225 \text{ psi}$$

Now we determine the correct sign for this shear stress. We proceed as follows:

$$V_y \text{ is negative}$$

$$dM_z / dx = V_y$$

$$dM_z < 0$$

$$\tau_{xx} = -\frac{M_z y}{I_{zz}} \quad (y \text{ is positive})$$

$$\therefore\ d\tau_{xx} > 0$$

Example 11.15 (Continued)

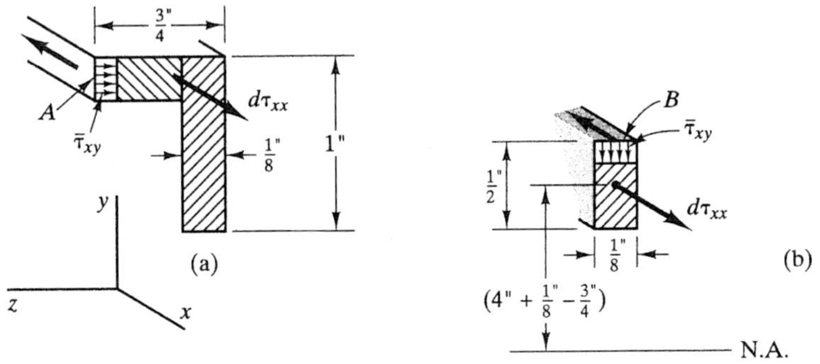

Figure 11.37. Sketches for the determination of the proper sign for shear stresses.

Going to Fig. 11.37(a), we see from the complementary property of shear stress that the shear stress $\bar{\tau}_{xz}$ is negative in accordance with our basic stress sign convention. Hence,

$$\bar{\tau}_{xz} = -225 \text{ psi}$$

Part (c)

Finally, consider point B. For the magnitude of $\bar{\tau}_{xy}$, we have on using a horizontal cut [see Fig. 11.37(b)]

$$|\bar{\tau}_{xy}| = \left|\frac{VQ}{Ib}\right| = \left|\frac{(-1000)\left(\frac{1}{8}\right)\left(\frac{1}{2}\right)\left(4 + \frac{1}{8} - \frac{3}{4}\right)}{(27.42)\left(\frac{1}{8}\right)}\right| = 61.54 \text{ psi}$$

We next find the proper sign for this shear stress.

$$V_y \text{ is negative}$$

$$dM_z / dx = V_y$$

$$dM_z < 0$$

$$\tau_{xx} = -\frac{M_z y}{I_{zz}} \quad (y \text{ is positive})$$

$$\therefore\ d\tau_{xx} > 0$$

Example 11.15 (Continued)

In Fig. 11.37(b), we can see from the above results that $\bar{\tau}_{xy}$ is negative. And so we have

$$\bar{\tau}_{xy} = -61.54 \text{ psi}$$

Part B. Special Topics

*11.7 Composite Beams

Up to this time we have been considering the elastic behavior of beams of *uniform composition.* There are occasions when *composite beams* are used. Such beams are fabricated by riveting, gluing, or clamping together along their length members of *different* materials. This occurs, for example, in concrete beams where steel rods are embedded in the concrete or in wooden beams where metal straps are attached along the beam. We shall now analyze the *elastic* behavior of composite beams.

For this purpose we have shown in Fig. 11.38(a) a composite beam composed of materials 1 and 2 and undergoing pure bending. The cross section of the composite beam is shown in Fig. 11.38(b). Since the beam is under the action of pure bending, the physical arguments used earlier to conclude that *plane sections remain plane* can again be used here since this was not dependent on the constitutive law. As a result, we have a *neutral axis AA* [Fig. 11.38(b)]. However, the neutral axis is *no longer* along the centroidal axis of the cross section, as was the case for beams of uniform composition. In Fig. 11.38(c) we have shown the linear strain variation of ε_{xx} required for such a deformation. The angle β indicated is related to strain as follows:

$$\varepsilon_{xx} = y \tan \beta \tag{11.26}$$

The stress variation is shown in Fig. 11.38(d). Notice that there is a *discontinuous* change in stress as you go from the material with modulus E_1 to the material with modulus E_2 (assumed here to be higher than E_1). The resultant forces F_1 and F_2 from the compressive and tensile stress distributions, respectively, are also shown.

We cannot use the flexure formula here because the neutral axis is not the centroidal axis and because the stress is not a linear function of y. What we shall do is to formulate *another equivalent* problem for

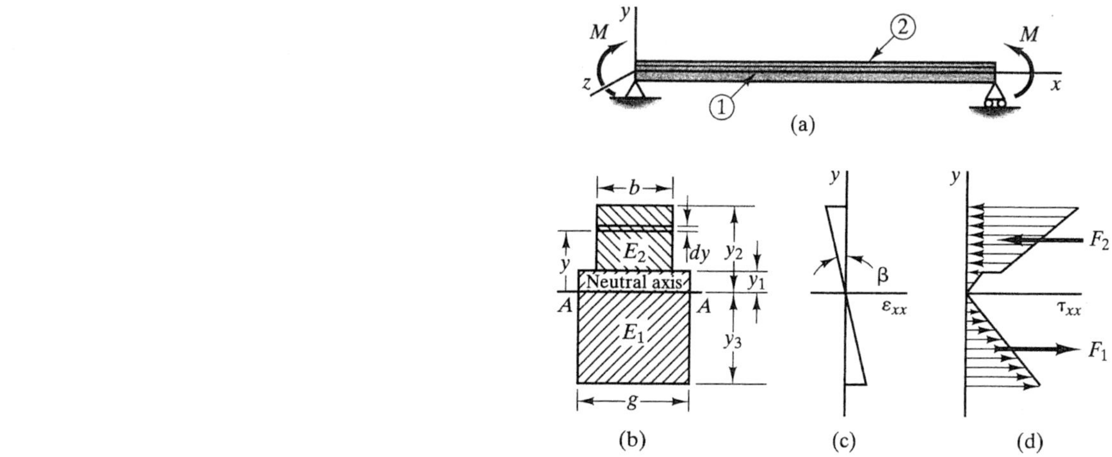

Figure 11.38. Composite beam.

which the flexure formula *is* valid and which at the same time is *simply related* to the problem at hand. We then use the flexure formula on the *new* problem and deduce the stresses on the *actual* problem through the simple relations between the problems. This method is called the method of *equivalent sections.*

In the method of equivalent sections we replace the actual section by a *mechanically equivalent* one such that

1. All materials are assumed to have some common *modulus of elasticity*—say E_1, in this case.
2. The *same strain* ε_{xx} is maintained in the equivalent section as that in the actual section for each value of y—clearly, we then have the same β and the *same neutral axis* for the two sections. However, we do not know yet where the neutral axis is; we repeat it is not at the centroid of the actual section.
3. We next change the *width* of strips dy in the material that has undergone a change of elastic modulus such that for the above deformation there remains the *same force* df between the actual strip and the equivalent strip. Because of this, the equivalent section has the same forces F_1 and F_2 as the actual section and the same moments. The sections are thus, as we say, mechanically equivalent to each other.

We now show that to accomplish condition 3 we must change the width such that the new width equals the *ratio* of the old modulus of the material to the new modulus of the material times the old width as shown in Fig. 11.39(a) for the upper member. Thus, for the corresponding strips having the same y shown in Figs. 11.38(b) and 11.39(a) we use $(\tau_{xx})_{\text{Act}}$ as the stress in the *given composite beam* and $(\tau_{xx})_{\text{Eq}}$ as the stress in the *equivalent beam*. The following equations for df on each strip then apply:

$$(df)_{\text{Act}} = (\tau_{xx})_{\text{Act}}(b\,dy) = (\varepsilon_{xx}E_2)(b\,dy) = (y\tan\beta)(E_2)\,(b\,dy) \quad (11.27a)$$

$$(df)_{\text{Eq}} = (\tau_{xx})_{\text{Eq}}\left(\frac{E_2}{E_1}b\right)dy = (\varepsilon_{xx}E_1)\left(\frac{E_2}{E_1}b\right)dy \quad (11.27b)$$

$$= (y\tan\beta)(E_2)\,(b\,dy)$$

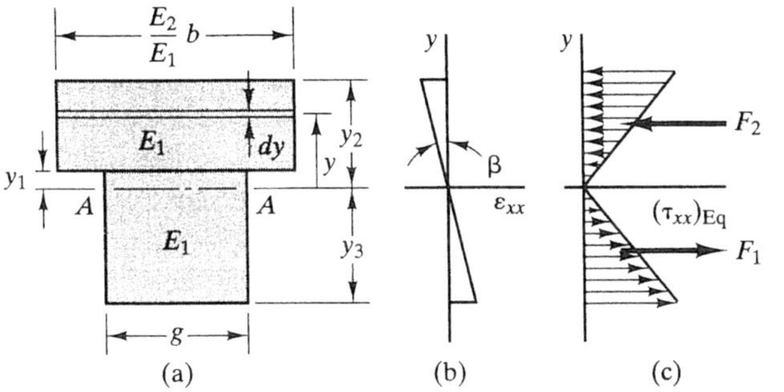

Figure 11.39. Mechanically equivalent section.

We see examining the right-hand sides of the equations above that the proposed change in width results in the same force increments $(df)_{\text{Act}}$ and $(df)_{\text{Eq}}$ as required by condition 3. It should be clear, on looking at the first expressions on the right sides of the above equations, that, where there is a change in modulus, the stress in the equivalent section is *different* from the stress at the corresponding elevations y in the actual section.

What have we then gained by introducing the equivalent section? To answer this query we first compute the forces F_1 and F_2 for the equivalent section in Fig. 11.39(c). Clearly, these forces are identical to those in Fig. 11.38(d) because of mechanical equivalence between sections. Thus noting that the actual width of the lower section is g, on observing Fig. 11.39(a) we have

$$F_2 = \int_0^{y_1}(\tau_{xx})_{\text{Eq}}\,g\,dy + \int_{y_1}^{y_2}(\tau_{xx})_{\text{Eq}}\left(\frac{E_2}{E_1}b\right)dy$$

$$= \int_0^{y_1}(y\tan\beta)(E_1)\,g\,dy + \int_{y_1}^{y_2}(y\tan\beta)(E_1)\left(\frac{E_2}{E_1}b\right)dy$$

$$F_1 = \int_{-y_3}^{0}(\tau_{xx})_{\text{Eq}}\,g\,dy = -\int_{-y_3}^{0}(y\tan\beta)(E_1)g\,dy$$

We require that $F_1 = F_2$ for pure bending, so we may state that

$$\int_0^{y_1} (y \tan \beta)(E_1) g\, dy + \int_{y_1}^{y_2} (y \tan \beta)(E_1) \frac{E_2}{E_1} b\, dy +$$

$$\int_{-y_3}^{0} (y \tan \beta)(E_1) g\, dy = 0$$

Canceling $\tan \beta E_1$ and rearranging, we get

$$\int_{-y_3}^{0} yg\, dy + \int_0^{y_1} yg\, dy + \int_{y_1}^{y_2} y\left(\frac{E_2}{E_1} b\right) dy = 0$$

But this tells us that the *first moment* of the area of the equivalent cross section is zero about the axis A-A, the *neutral axis.* Thus, we see that the *neutral axis coincides with the centroidal axis of the equivalent section.* Accordingly, we can readily find the neutral axis by locating the centroidal axis of the equivalent section.

We now have for the equivalent section a linear strain distribution as measured from the centroidal axis and a uniform composition of material which is Hookean and on which we have a linearly varying stress. These are precisely the conditions that led earlier to the flexure formula for stress and so we can say here for the equivalent section that

$$(\tau_{xx})_{\text{Eq}} = -\frac{My}{I_{\text{Eq}}} \tag{11.28}$$

where M is the same bending moment as that occurring on the actual section (owing to the equality of force elements df between the sections), I_{Eq} is the second moment of area of the *equivalent section about its centroidal axis,* and finally, y is measured from the centroidal axis of the equivalent section.

To get the stress on the actual beam we now make use of the mechanical equivalence built into the equivalent section in relation to the actual section. Where there has been no change in modulus of elasticity the stress for the actual section is the same as the stress for the equivalent section. Where there has been a change in modulus of elasticity, we see on equating the first expressions on the right side of Eq. (11.27a) and Eq. (11.27b) that in order to maintain mechanical equivalence

$$(\tau_{xx})_{\text{Act}} = \frac{E_2}{E_1} (\tau_{xx})_{\text{Eq}} \tag{11.29}$$

Thus, the actual normal stresses for the composite beam are readily computed once the straightforward calculations have been made for determining the equivalent normal stresses.

Example 11.16

Shown in Fig. 11.40 is a composite beam composed of a middle member of wood having a modulus of elasticity of 2×10^6 psi with metal alloy outer straps both having the same modulus of elasticity of 14×10^6 psi. The three members are firmly bolted together to form a unit. If a positive bending moment of 8×10^5 in.-lb is applied, what is the maximum tensile bending stress in each material?

Our first task is to make an *equivalent cross section.* We shall use the modulus of the wood for that purpose. (As an exercise you may want to do this problem using the modulus of the metal alloy.) The equivalent cross section is shown in Fig. 11.41.

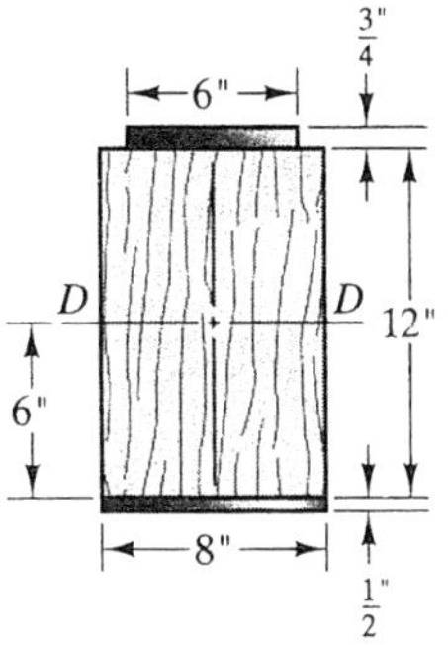

Figure 11.40. Wood-metal alloy composite beam.

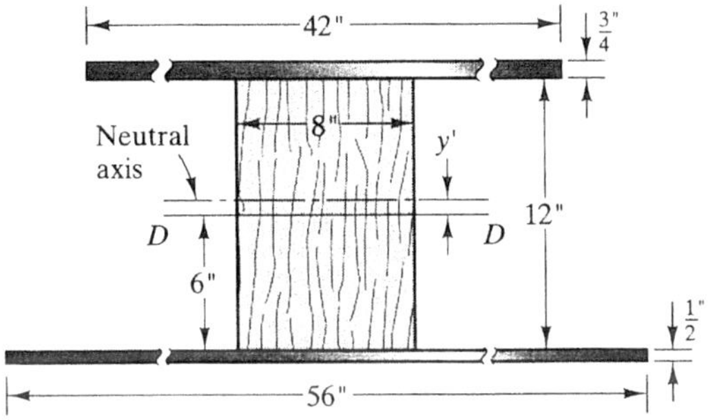

Figure 11.41. Equivalent section.

The position of the neutral axis is first computed. We do this by equating the first moment of the total area A of the equivalent section about a convenient axis, with the sum of the first moments of the three areas representing the two alloy straps and the wooden beam about this axis. We choose as the axis about which to take moments the center axis D-D of the wooden section (Fig. 11.40). Thus, we have

$$Ay' = \left[(56)\left(\frac{1}{2}\right) + (42)\left(\frac{3}{4}\right) + (12)(8)\right]y' = \sum_i A_i y_i$$

$$= -(56)\left(\frac{1}{2}\right)(6.25) + (42)\left(\frac{3}{4}\right)(6.375) + 0$$

$$\therefore y' = .1660 \text{ in.}$$

Thus, the *neutral axis* is at a distance .1660 in. above the axis D-D. Next we shall need I_{Eq} about the neutral axis. Accordingly,

Example 11.16 (Continued)

$$I_{Eq} = \frac{1}{12}(42)\left(\frac{3}{4}\right)^3 + (42)\left(\frac{3}{4}\right)(6 + .375 - .1660)^2$$

$$+ \frac{1}{12}(8)(12)^3 + (8)(12)(.1660)^2$$

$$+ \frac{1}{12}(56)\left(\frac{1}{2}\right)^3 + (56)\left(\frac{1}{2}\right)(6 + .250 + .1660)^2$$

$$= 3524 \text{ in.}^4$$

It is now a simple matter to compute the normal stresses in the equivalent section. For the wood we have for the maximum tensile normal stress

$$[(\tau_{xx})_{Eq}]_{max} = -\frac{M(-6.166)}{3524} = \frac{(8 \times 10^5)(6.166)}{3524} = 1400 \text{ psi}$$

This is the same stress as that at the corresponding value of y in the actual section. Thus, for the wooden portion of the composite member we have

$$[(\tau_{xx})_{wood}]_{max} = 1400 \text{ psi} \qquad \text{(a)}$$

As for the alloy strips, it is easily seen that the bottom member will be in tension and that the maximum stress occurs at the lowest fibers. Accordingly, for the maximum equivalent tensile stress we have

$$[(\tau_{xx})_{Eq}]_{max} = -\frac{M(-6.67)}{3524} = \frac{(8 \times 10^5)(6.67)}{3524} = 1513 \text{ psi}$$

The corresponding stress for the actual section is then computed as

$$[(\tau_{xx})_{Al}]_{max} = \frac{E_{Al}}{E_{wood}}(1513) = (7)(1513) = 10{,}591 \text{ psi} \qquad \text{(b)}$$

$$[(\tau_{xx})_{Al}]_{max} = 10{,}591 \text{ psi}$$

Example 11.17

Concrete is a material that is not able to withstand appreciable tensile stress in contrast to its ability to withstand high compressive stress. For this reason concrete slabs are reinforced with steel rods in regions where there is expected to be tensile stress. A cross section of a reinforced concrete member is shown in Fig. 11.42. The loading must be applied so that we have tension on the lower fibers. There are three steel reinforcing rods shown, each having a cross-sectional area of 625 mm^2. We take $E_{\text{Stl}} = 2.1 \times 10^{11}$ Pa and E of concrete in compression to be 2.4×10^{10} Pa. What is the maximum stress in the concrete and in the steel for a bending moment of 1.2×10^5 N-m?

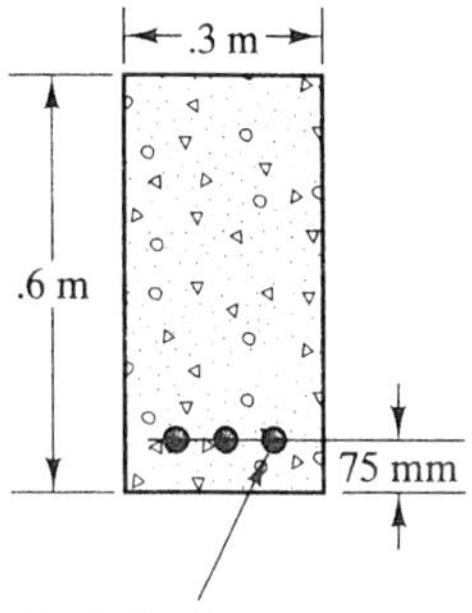

Figure 11.42. Reinforced concrete beam.

Since the concrete cannot be depended on to offer much resistance in tension, we shall assume that in the tensile zone the concrete merely holds the reinforcing rods in place. Accordingly, we shall consider that only the part of the concrete which is *above* the neutral axis (where we have compression) is effective in resisting bending moment. Also, we shall simplify the cross section of the reinforcing rods into that of a narrow rectangle having an area of 1875 mm^2. This effective cross section is shown in Fig. 11.43(a) where we consider that only the cross-hatched areas resist bending moment. For the *equivalent section* [see Fig. 11.43(b)], we shall change the thin rectangular section of area 1875 mm^2 to one of 16,406 mm^2, thereby assigning a modulus E of 2.4×10^{10} Pa (that of concrete) for the equivalent section. Note that the neutral axis is at the bottom of the compression zone at some yet unknown distance d below the top surface. We can determine the location of the neutral axis, and thus determine d, by finding the centroid of the equivalent section. Thus, taking moments about the neutral axis we have

$$[(.3)(d)]\frac{d}{2} - (16{,}406 \times 10^{-6})(.525 - d) = 0$$

$$\therefore \quad d = 0.191 \text{ m}$$

The second moment of area of the equivalent section about the neutral axis using only the transfer term for the steel rods is given as follows where we regard the term corresponding to the second moment of area of the steel about its own axis as negligibly small.

$$I_{\text{Eq}} = \frac{1}{12}(.3)(.191)^3 + (.3)(.191)\left(\frac{.191}{2}\right)^2 + (16{,}406 \times 10^{-6})(.525 - .191)^2$$
$$= 2.53 \times 10^{-3} \text{ m}^4$$

The equivalent maximum stress in the concrete and steel may be given as follows:

Example 11.17 (Continued)

$$[(\tau_{xx})_{\text{Eq}}]_{\text{Conc}} = -\frac{(1.2 \times 10^5)(.191)}{2.53 \times 10^{-3}} = -9.059 \times 10^6 \text{ Pa}$$

$$[(\tau_{xx})_{\text{Eq}}]_{\text{Stl}} = -\frac{(1.2 \times 10^5)[-(.525 - .191)]}{2.53 \times 10^{-3}} = 15.84 \times 10^6 \text{ Pa}$$

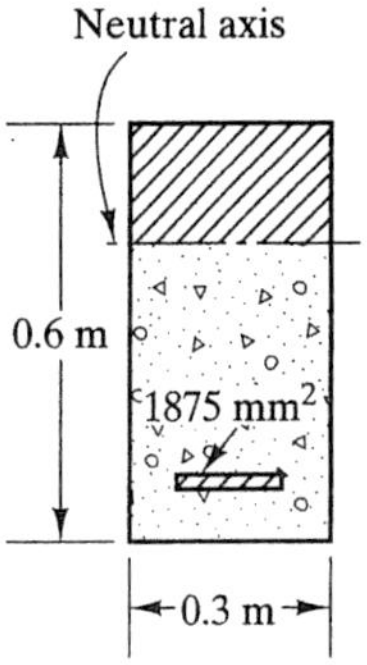

(a)

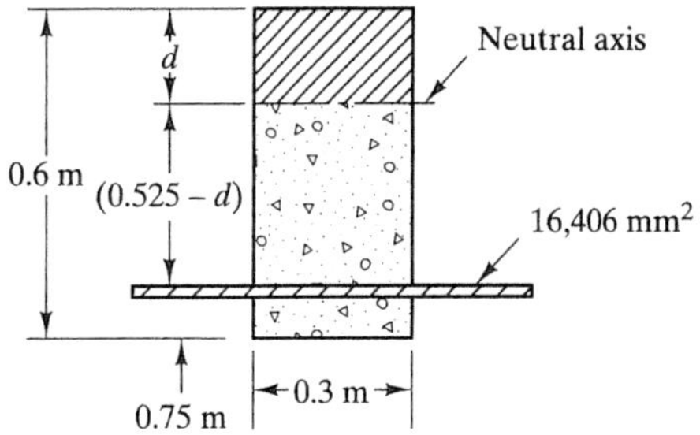

(b)

Figure 11.43. (a) Effective section. (b) Equivalent section.

The corresponding actual stresses are then

$$[(\tau_{xx})_{\text{Act.}}]_{\text{Conc}} = -9.059 \times 10^6 \text{ Pa}$$

$$[(\tau_{xx})_{\text{Act.}}]_{\text{Stl}} = \frac{2.1 \times 10^{11}}{2.4 \times 10^{10}}(15.84 \times 10^6) = 138.6 \times 10^6 \text{ Pa}$$

Notice that we have been somewhat approximate in the calculations in the way we handled the geometrical aspects of the reinforcing rods. Since in actual construction these rods are positioned in the concrete in a manner only approximating the specifications, the simplifications we have made are reasonable.

Just as we used the results from pure bending as the starting point for more general cases in homogeneous beams, so can we extend the theory, thus far presented, to the case of composite beams with shear. We must then simply use the local bending moment in the formulation for stress τ_{xx}. Also, there will now be shear stress present. If the members are *continuously and firmly attached* to each other along their entire contact surface (for instance, the members are glued to each other), then we can formulate, as we did earlier, a means of computing the average shear stress, $\bar{\tau}_{xy}$, that would be quite meaningful for rectangular-like cross sections. However, since such stresses are generally quite small, we shall leave this development and its applications to the problems at the end of the chapter.

*11.8 Case of Unsymmetric Beams

In the previous sections we have been considering the bending of *symmetric beams* wherein the transverse external loads were in the plane of symmetry of the beam and oriented normal to the centerline of the beam. We shall now consider the *pure bending* of a beam having an *arbitrary cross section.*

Accordingly, we consider a cross section of *arbitrary shape* wherein the bending moment M_z is first taken parallel to a *principal axis z* at the centroid C (see Fig. 11.44). All physical arguments of Section 11.2 leading to the conclusion that plane sections remain plane for small deformations can be presented again. This means that there will exist a plane in the beam, namely the neutral surface, oriented originally parallel to the geometric longitudinal axis of the beam, and on this plane we have no stretching during deformation of the beam. The intersection of the neutral surface and the cross section of the beam is then a *neutral axis* and is shown as *H-H* in Fig. 11.44. Clearly, at distance η normal to this axis the stress will be given by the following equation, which is the counterpart of Eq. (11.6a):

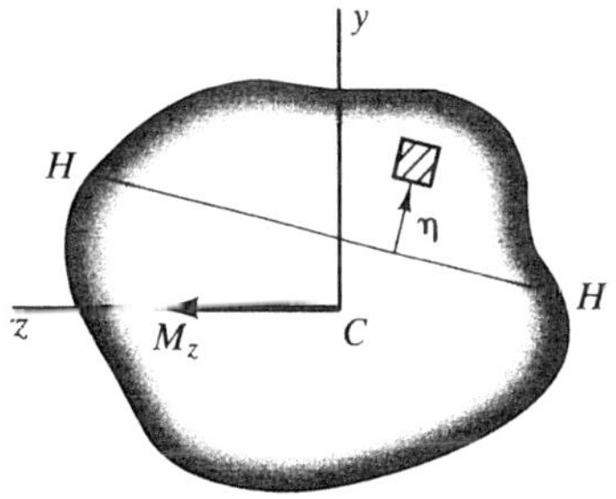

Note: y and z are principal axes

Figure 11.44. Unsymmetric section.

$$\tau_{xx} = -\frac{E\eta}{R} \tag{11.30}$$

where R is again the radius of curvature of the beam. We next integrate the preceding normal stress distribution over the cross-sectional area A. Setting this equal to zero in accordance with the requirements of **equilibrium**, we have

$$-\frac{E}{R}\iint \eta\, dA = 0 \tag{11.31}$$

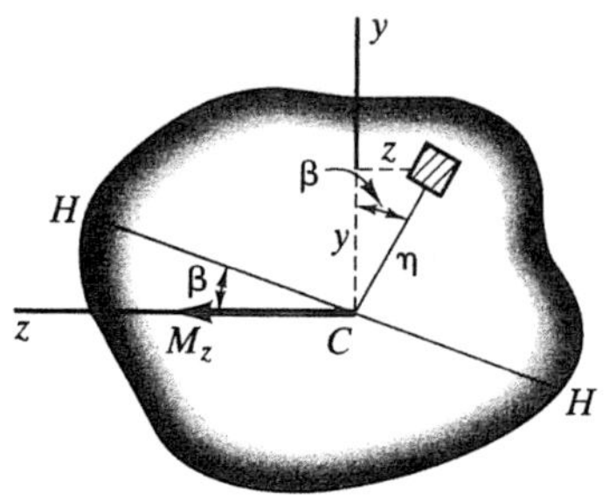

Figure 11.45. Neutral axis goes through centroid.

This indicates that axis *H-H* must be a *centroidal axis,* such as is shown in Fig. 11.45, where *H-H* is drawn inclined at some angle β to the z axis. Next we take moments of the normal stress distribution about the y and z axes. In accordance with the requirement that M_z be the resultant of the distribution, we get, using Eq. (11.30),

$$
\begin{aligned}
-M_z &= \iint \tau_{xx}\, y\, dA = -\iint \frac{\eta E}{R} y\, dA \\
0 &= \iint \tau_{xx}\, z\, dA = -\iint \frac{\eta E}{R} z\, dA
\end{aligned}
\tag{11.32}
$$

We next express η in terms of the y and z coordinates in the following manner (see Fig. 11.45):

$$\eta = y \cos\beta - z \sin\beta \tag{11.33}$$

Substituting Eq. (11.33) into Eqs. (11.32), we get

$$M_z = \frac{E}{R}\cos\beta \iint y^2\, dA - \frac{E}{R}\sin\beta \iint yz\, dA \tag{11.34a}$$

$$0 = -\frac{E}{R}\cos\beta \iint yz\, dA + \frac{E}{R}\sin\beta \iint z^2\, dA \tag{11.34b}$$

Since the yz axes are principal axes, these equations simplify to the following pair:

$$M_z = \frac{E}{R} I_{zz} \cos\beta \tag{11.35a}$$

$$0 = \frac{E}{R} I_{yy} \sin\beta \tag{11.35b}$$

We can satisfy the second of the foregoing equations only by setting $\beta = 0$. This means that the neutral axis must coincide with the principal axis along which we have applied the bending moment. Setting $\beta = 0$ in Eq. (11.35a), we now get, on solving for R,

$$R = \frac{EI_{zz}}{M_z} \tag{11.36}$$

This formula is identical to Eq. (11.9) for the symmetric case. And, by substituting R from the preceding equation into Eq. (11.30) we get, on noting that η is now y, the familiar flexure formula $\tau_{xx} = My/I_{zz}$.

In summary, we can conclude that *if the bending moment M_x is parallel to a principal axis at the centroid of a cross section, the neutral axis coincides with this principal axis and the flexure formula is valid* *with I_{zz} measured about this principal axis and y measured normal to this principal axis.*

If we have pure bending stemming from applied couple moments oriented at an *inclination* to the principal directions at the centroid, such as shown in Fig. 11.46 where y and z are the principal axes, we can

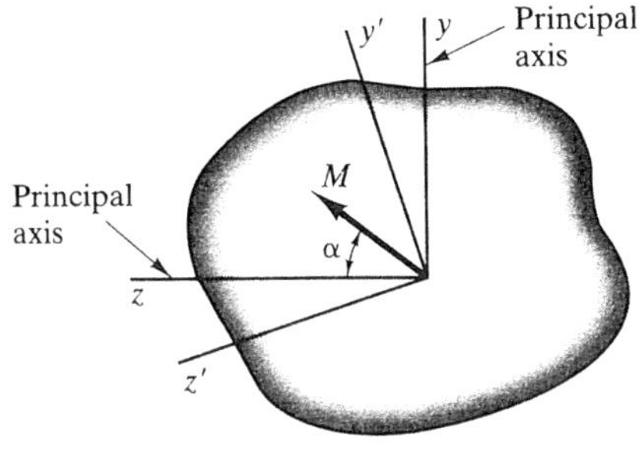

Figure 11.46. M is inclined to principal axes.

proceed by decomposing the couple moment M into components along the principal directions. Using the principle of superposition, we can employ the flexure formula for each component of M in a manner described in the preceding paragraph. Thus, with M oriented at an angle α with the z axis as shown in the diagram, we can compute the normal stress distribution τ_{xx} in the following way:[18]

$$\tau_{xx} = -\frac{(M\cos\alpha)y}{I_{zz}} + \frac{(M\sin\alpha)z}{I_{yy}} \tag{11.37a}$$

There may be occasions when one may prefer *not* to work with the principal axes in an analysis. Let us then consider that we shall prefer

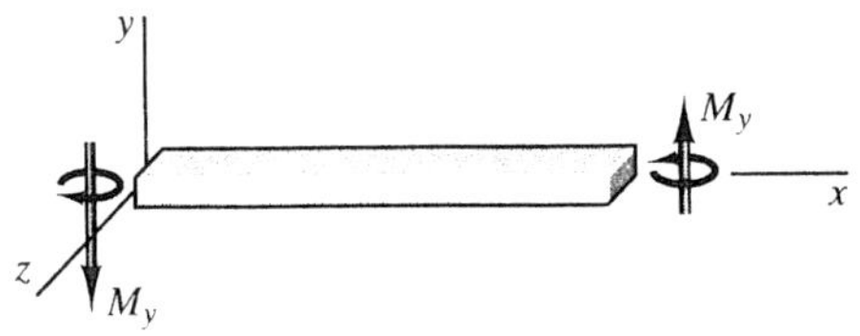

Figure 11.47. Pure bending in xz plane.

[18]Following our established sign convention for bending moments, we end up for pure bending in the xz plane (see Fig. 11.47) with a flexure formula:

$$\tau_{xx} = +\frac{M_y z}{I_{yy}}$$

Note that we now have a plus sign on the right-hand side of the equation. Justify this in your own mind by examining the diagram and visualizing the stresses.

axes $y'z'$ in Fig. 11.46 rather than principal axes yz. Since the stress distribution τ_{xx} is a linear function of the coordinates y and z as given by Eq. (11.37a) we may be sure that for the new coordinate system $y'z'$ we must have $\tau_{x'x'}$ as a linear function of the coordinates y' and z'—since the $y'z'$ axes are merely rotated relative to the yz axes. We may thus say that

$$\tau_{x'x'} = a + by' + cz' \tag{11.37b}$$

The terms a, b, and c are constants to be found by the following equations required between the resultant and the corresponding distribution of stress at a section:[19]

$$-M_{z'} = \iint_A y'\tau_{x'x'}\,d\sigma \tag{11.38a}$$

$$M_{y'} = \iint_A z'\tau_{x'x'}\,d\sigma \tag{11.38b}$$

$$F_R = 0 = \iint_A \tau_{x'x'}\,d\sigma \tag{11.38c}$$

(Can you justify the choice of signs for the first two of the foregoing equations?) Next substitute into Eqs. (11.38) the formulation for $\tau_{x'x'}$ given by Eq. (11.37b). Thus, we get

$$-M_{z'} = a\iint_A y'd\sigma + b\iint_A (y')^2 d\sigma + c\iint_A y'z'd\sigma \tag{11.39a}$$

$$-M_{y'} = a\iint_A z'd\sigma + b\iint_A y'z'd\sigma + c\iint_A (z')^2 d\sigma \tag{11.39b}$$

$$0 = a\iint_A d\sigma + b\iint_A y'd\sigma + c\iint_A z'd\sigma \tag{11.39c}$$

Noting that y' and z' are centroidal axes, we see in Eq. (11.39c) that the last two integrals are zero leaving us to conclude $a = 0$. Employing the usual notation for second moments and products of area, the remaining equations become

$$-M_{z'} = bI_{z'z'} + cI_{y'z'} \tag{11.40a}$$

$$M_{y'} = bI_{y'z'} + cI_{y'y'} \tag{11.40b}$$

Solving for b and c from the foregoing equations, we get

[19]We will use σ to replace A in dA because of the integration limit A in order to have good mathematical form. Also we will later use A' in dA.

$$b = -\frac{M_{z'}I_{y'y'} + M_{y'}I_{z'y'}}{I_{z'z'}I_{y'y'} - I_{z'y'}^2}$$

$$c = \frac{M_{y'}I_{z'z'} + M_{z'}I_{z'y'}}{I_{z'z'}I_{y'y'} - I_{z'y'}^2}$$

Returning to Eq. (11.37b) we can now formulate the bending-stress distribution. We may drop the primes from the notation at this time since our result is valid for any set of axes yz at the centroid. Thus,

$$\tau_{xx} = \frac{(M_yI_{zz} + M_zI_{zy})z - (M_zI_{yy} + M_yI_{zy})y}{I_{zz}I_{yy} - I_{zy}^2} \tag{11.41}$$

The preceding equation is called the *generalized flexure formula.* When zy are principal axes, Eq. (11.41) degenerates to Eq. (11.37a). We have arrived at this formulation for pure bending by physical argument and the use of Newton's law and Hooke's law. However, we can show that for small deformation, this is an *exact solution* according to the theory of elasticity. For the case of more general bending we can use the generalized flexure formula *locally* for long slender beams to get very good approximate results. We shall now illustrate the use of the generalized flexure formula.

Example 11.18

What is the maximum tensile stress τ_{xx} in the cantilever beam shown in Fig. 11.48? The line of action of the 100-lb force goes through the centroid of the cross section.

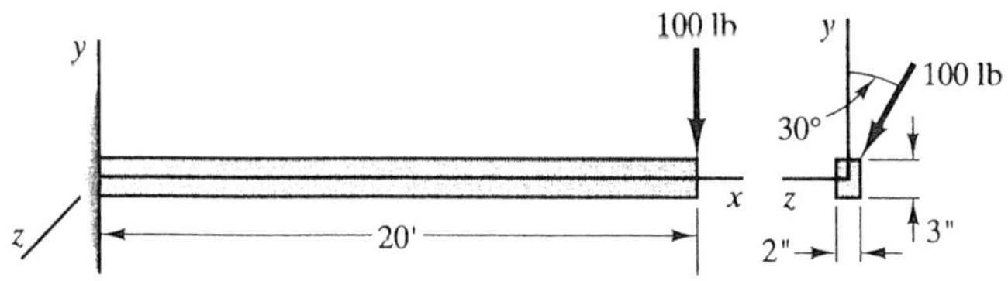

Figure 11.48. Cantilever with unsymmetric loading.

Clearly, the maximum bending moment occurs at the base of the cantilever beam, and we show a free-body diagram in Fig. 11.49 exposing this section. Note that *positive* bending moments *conventionwise* for M_x and M_y have been shown in the diagram. Clearly, the algebraic sign from equilibrium equations using the right-hand rule

Example 11.18 (Continued)

will then correspond to the convention sign for the bending moments. Hence, from **equilibrium** we have

$\Sigma M_y = 0$:

$$-M_y - (100)(\sin 30°)(20) = 0$$

$$\therefore M_y = -1000 \text{ ft-lb}$$

$\Sigma M_z = 0$:

$$-M_z - (100)(\cos 30°)(20) = 0$$

$$\therefore M_z = -1732 \text{ ft-lb}$$

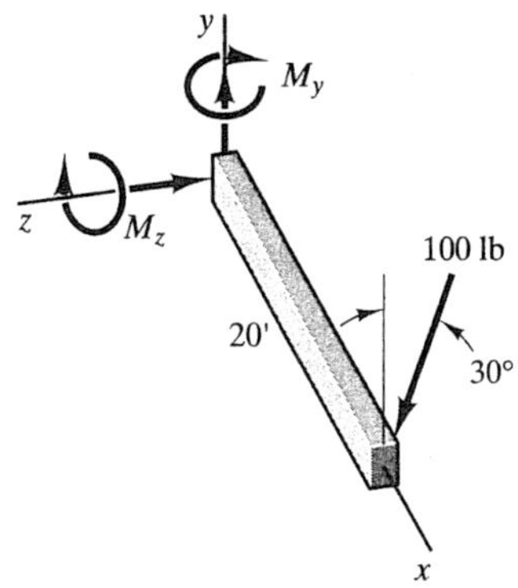

Figure 11.49. Free-body diagram of cantilever.

We can employ Eq. (11.41) here as follows:

$$\tau_{xx} = \frac{[(-1000)(12)(\frac{1}{12})(2)(3^3) + (-1732)(12)(0)]z - [(-1732)(12)(\frac{1}{12})(3)(2^3) + (-1000)(12)(0)]y}{(\frac{1}{12})(2)(3^3)(\frac{1}{12})(3)(2^3) - (0)^2} \quad \text{(a)}$$

[Since the axes yz are principal axes, we could also have used Eq. (11.37a) directly.] Carrying out the numerical work, we get

$$\tau_{xx} = -(6000z - 4619y) \text{ psi} \quad \text{(b)}$$

The maximum tensile stress τ_{xx} occurs when $z = -1$ and $y = 1.5$, giving

$$(\tau_{xx})_{\max} = 6000 + 6928 = 12{,}928 \text{ psi} \quad \text{(c)}$$

$$(\tau_{xx})_{\max} = 12{,}928 \text{ psi}$$

We have solved this problem in a very formal manner being very careful with the signs for the bending moments. In complex problems you will have to proceed in this careful deliberate manner and that is why we solved this problem this way. However, when the problem is as easy as this, we encourage you to proceed in a simple direct manner using common sense and "physical feel." Thus for the vertical component of the force, namely (100) cos 30°, you should be able to visualize that we will get a maximum tensile stress at the *top* fibers AB at the base cross section (see Fig. 11.50). Calling this stress $(\tau_{xx})_1$, we can say

$$|(\tau_{xx})_1| = \frac{[(100)(\cos 30°)(20)(12)](1.5)}{(\frac{1}{12})(2)(3^3)} = 6928 \text{ psi}$$

As for the horizontal component of the 100-lb force, namely (100) (sin 30°), we will get a maximum tensile stress along side BC (see Fig.

Example 11.18 (Continued)

11.50), as you should again be able to visualize. Calling this stress $(\tau_{xx})_2$, we have

$$|(\tau_{xx})_2| = \frac{[(100)(\sin 30°)(20)(12)](1)}{(\frac{1}{12})(3)(2^3)} = 6000 \text{ psi}$$

The total maximum tensile stress is then 12,928 psi at corner *B*.

$$(\tau_{xx})_{\max} = 12{,}928 \text{ psi}$$

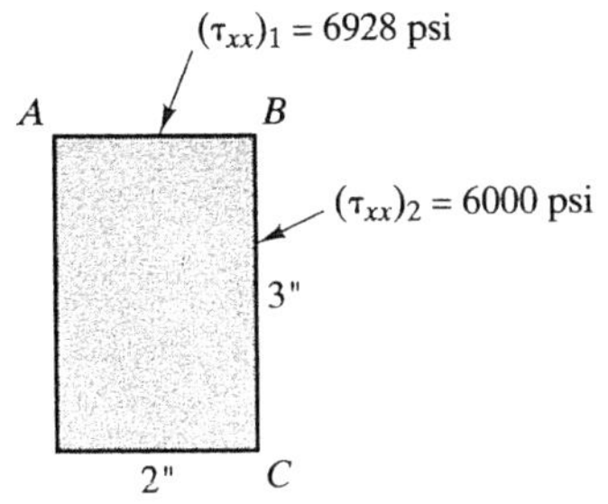

Figure 11.50. Maximum tensile stress τ_{xx} is at *B*.

Now that we have formulated the generalized flexure formula for computing the normal stress τ_{xx}, the next logical step would be to present a similar generalization for the computation of shear stress. Unfortunately, we shall be unable to do this. However, for *rectangular-like cross sections having two orthogonal axes of symmetry,* we can employ previously developed formulas *when the plane of loading passes through the centroid at each section.* The procedure is simply to decompose the shear force into components along the axes of symmetry of the cross section and then employ Eq. (11.19) and Eq. (11.25) for each respective load. This will yield shear-stress distribution $\bar{\tau}_{xy}$ and $\bar{\tau}_{xz}$ for the cross section. The total shear at any point in the cross section may then be reasonably approximated as

$$\sqrt{\bar{\tau}_{xy}^2 + \bar{\tau}_{xz}^2}$$

What happens when the plane of loading does *not* go through the centroid of a rectangular-like cross section having two orthogonal axes of symmetry? To answer this, first consider again the case of loading in a plane of symmetry, which we shall take as the *xy* plane. For such a loading, the cross sections may rotate about the *z* axis as planes and in addition may have superposed a nonuniform movement in the *x* direction giving rise to a *warping* of the cross sections, the latter resulting from nonzero shear stress. What will *not* take place is a rotation of the cross sections about the *x* axis (i.e., there will be *no twisting about the centerline of the beam*). Surely we can exclude this possibility because of the fact that the loading is in the plane of symmetry. Also, for beams with cross sections having *two* orthogonal axes of symmetry and *loaded so that the plane of loading goes through the centerline* there will again be no twisting about the centerline. To see this, merely decompose the loading into components in the planes of symmetry of the beam and employ the previous argument for each component of the loading. Consider next a coplanar loading distribution that does *not* go through the centerline of a beam having a rectangular-like cross section with two axes of symmetry. If this loading is shifted to go through the centerline of the beam, we must add a torque distribution M_x to the loading in accordance with the dictates of rigid-body mechanics. We can compute shear stresses in this situation from the loading at the centerline by the methods described earlier, but this will not give the entire story on shear stresses, for we shall also have to account for shear stresses from the *twisting action of* M_x. The shear stress from the loading at the centerline results from bending deflection only (i.e., there is no rotation of the cross section about the centerline of the beam), whereas the other shear stress results from the aforementioned twisting action and is called *torsional shear stress.* We have considered only shear stress from bending deformation in this chapter. In Chapter 14 we shall examine the case of pure torsion.

If the cross section is of *nonrectangular* shape and has two axes of symmetry, we first have the difficulty that the shear stresses resulting from bending, as formulated from strength of materials, may have little meaning for reasons pointed out in Section 11.4. On lacking two axes of symmetry, it is difficult to ascertain whether for a given loading there will be appreciable twisting or not. (Requiring the loading to go through the centroid will exclude twisting *only* for the special case where there is an axis of symmetry present and the loading is in this axis of symmetry.) In general cases, to find the particular line of action of a loading to ensure bending *without* twisting, we must resort to the theory of elasticity.

We shall see, however, in the next section that, for *open thin-walled cross sections,* we can formulate shear-stress distributions resulting from bending without twisting and can establish the criterion for this kind of deformation.

*11.9 Shear Stress in Beams of Narrow Open Cross Section

In Section 11.8, we formulated means of computing the *normal stress* distribution τ_{xx} in *any* long, slender uniform beam under the action of loads along the periphery which are normal to the centerline of the beam. We *cannot* proceed with comparable accuracy to compute meaningful *shear stresses* in such general cases by extending the methods now available. However, if the beam has a cross section which is an *open thin-walled section,* such as the one shown in Fig. 11.51, we can determine shear stresses, under certain conditions, with reasonably good accuracy.

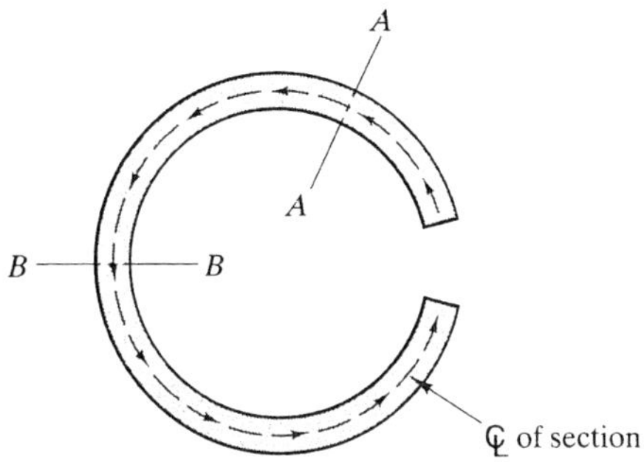

Figure 11.51. Open thin-walled section.

We shall now present the simplifying condition that permits this calculation for such sections. With no shear stresses acting on the lateral boundary in the form of surface tractions, we know from the complementary property of shear stress that the *shear stresses at the boundaries in a cross section of the beam must be tangent to the boundaries.* And since the section is thin, we can assume with reasonably good accuracy, first, that the shear stress is uniform along lines normal to the boundaries, such as AA and BB in Fig. 11.51, and second, that the direction of the shear stress is that of the centerline of the section shown dashed in the diagram.

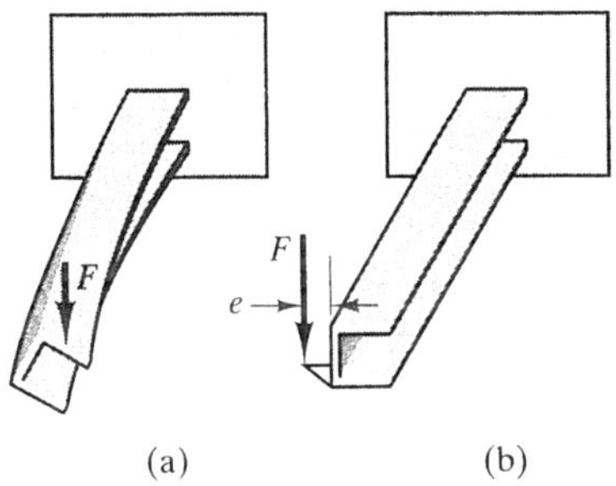

Figure 11.52. Tip-loaded cantilever beam: (a) combined bending and twisting; (b) bending only.

Now that we have established the direction of the shear-stress distribution and simplified its variation transverse to the centerline of the cross section, we must next determine how the stress varies *along* the centerline of the cross section. In this regard, we must carefully point out that the shear stresses to be investigated are those *stemming solely from the bending of beams and not those that may be attributed to the torsion of beams.* It was perfectly clear in our discussion of shear stress in symmetric beams that loading systems in the plane of symmetry could not induce torsional strain. For nonsymmetric shapes and thus for the arbitrary thin-walled cross sections that we are now to consider, it is not obvious whether there will be appreciable twisting or not for a given load. Thus, if a channel acts as a cantilever beam to support a force F going through the centroid of the channel section as shown in Fig. 11.52(a) there will be the twisting action indicated in the diagram. We shall show later that the force F must be applied as shown in Fig. 11.52(b) at an appropriate distance e in order to prevent twisting. Clearly, this is not something that is physically obvious.

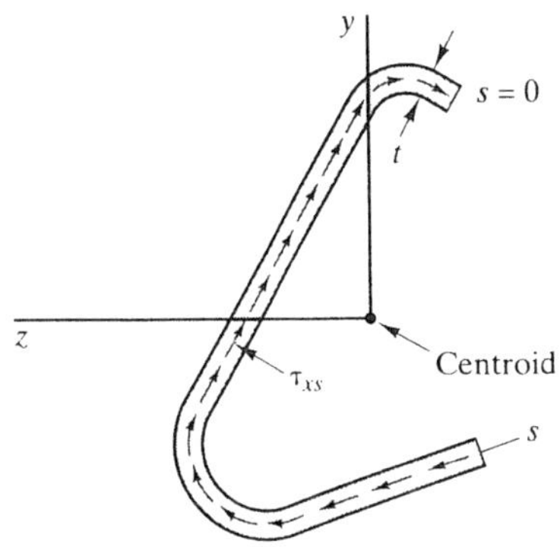

Figure 11.53. Arbitrary, open thin-walled section showing shear stress.

We shall now formulate the shear-stress distribution for *bending without twisting* for an arbitrary, thin-walled, open cross section, such as shown in Fig. 11.53. The thickness at any section is noted as t; s gives the position along the centerline of the section. Note that t may possibly be a function of s. We shall assume that there is no torsion in the beam and that *some coplanar loading distribution parallel to the xy plane induces only bending deformation.* The shear-stress distribution has been shown as τ_{xs}. It is a function of s and x and is oriented along the centerline in the manner described earlier. We now draw the free-body diagram of an infinitesimal portion of the beam shown as full lines in Fig. 11.54.

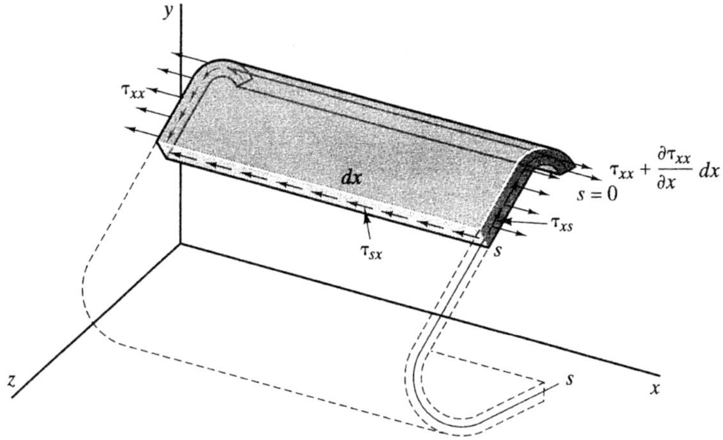

Figure 11.54. Element of thin-walled beam.

Summing forces in the x direction, as a result of **Newton's law,** we have

$$-\tau_{sx} t_c \, dx - \int_0^s \tau_{xx} t \, ds + \int_0^s \left(\tau_{xx} + \frac{\partial \tau_{xx}}{\partial x} dx \right) t \, ds = 0 \qquad (11.42)$$

where t_c is the thickness at the cut at s. Canceling terms and replacing s as an integration variable by σ for better mathematical form, we get on solving for τ_{sx}

$$\tau_{sx} = \frac{1}{t_c} \int_0^s \frac{\partial \tau_{xx}}{\partial x} t \, d\sigma \qquad (11.43)$$

Now we replace τ_{xx} using Eq. (11.41) from the Euler-Bernoulli theory of bending wherein we set $M_y = 0$ for the particular loading system that we have applied. Thus, we get

$$\tau_{sx} = \frac{1}{t_c} \int_0^s \frac{\partial}{\partial x} \left\{ \frac{z M_z I_{zy} - y M_z I_{yy}}{I_{zz} I_{yy} - I_{zy}^2} \right\} t \, d\sigma \qquad (11.44)$$

wherein we have second moments of area and products of area of the cross section about the y and z axes at the centroid. Note from our earlier work that $\partial M_z / \partial x = V_y$. Hence, using magnitudes we have

$$|\tau_{sx}| = \left| \frac{V_y}{t_c(I_{zz} I_{yy} - I_{zy}^2)} \left(I_{zy} \int_0^s zt \, d\sigma - I_{yy} \int_0^s yt \, d\sigma \right) \right| \qquad (11.45)$$

The integrals in the preceding formulation represent *first moments* of that portion of the cross section between $s = 0$ and s, where these moments are taken about the z and y axes, respectively. Hence, we can express Eq. (11.45) in the following manner:

$$|\tau_{sx}| = \left| \frac{V_y}{t_c(I_{zz}I_{yy} - I_{zy}^2)} [I_{zy}Q_y - I_{yy}Q_z] \right| \tag{11.46}$$

Note that t_c is the thickness at the position where the stress τ_{sx} is measured and that, owing to the complementary property of shear at a point, the foregoing shear stress equals τ_{xs}, the shear stress at a position s on the cross section.

Next consider for the same beam a hypothetical coplanar load parallel to the xz plane and in the z direction. This would induce a shear force V_z and a bending moment M_y. By the same procedure used in the previous development, except to replace $|\partial M_y/\partial x|$ by $|V_z|$ we can formulate the following shear-stress distribution for the case where there is only bending and no torsion:

$$|\tau_{sx}| = \left| \frac{V_z}{t_c(I_{zz}I_{yy} - I_{yz}^2)} \left(I_{yz} \int_0^s yt\, d\sigma - I_{zz} \int_0^s zt\, d\sigma \right) \right| \tag{11.47}$$

This equation may also be written as

$$|\tau_{sx}| = \left| \frac{V_z}{t_c(I_{yy}I_{zz} - I_{yz}^2)} (I_{yz}Q_z - I_{zz}Q_y) \right| \tag{11.48}$$

where Q_y and Q_z are first moments of the area to the right of s about the y and z axes, respectively.

We shall now illustrate the use of the theory above for the case of the channel in the following example.

Example 11.19

We are to determine the shear stresses in the channel cantilever beam shown in Fig. 11.55 loaded by a single force F for the condition of *no twisting* in the beam.

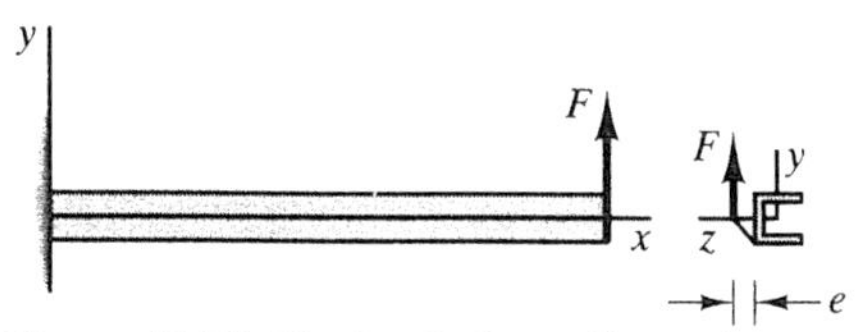

Figure 11.55. Tip-loaded cantilever beam.

Example 11.19 (Continued)

In Fig. 11.56 we have shown an enlargement of the cross section having its normal in the plus x direction of the channel. The zy axes are centroidal axes drawn parallel to the sides of the channel. These axes also happen to be principal axes of the section. The shear stress τ_{xy} is shown corresponding to an upward shear force V_y. It must be pointed out that the thicknesses t_1 and t_2 are small compared to h, the height of the channel, making the thin-walled beam theory applicable.

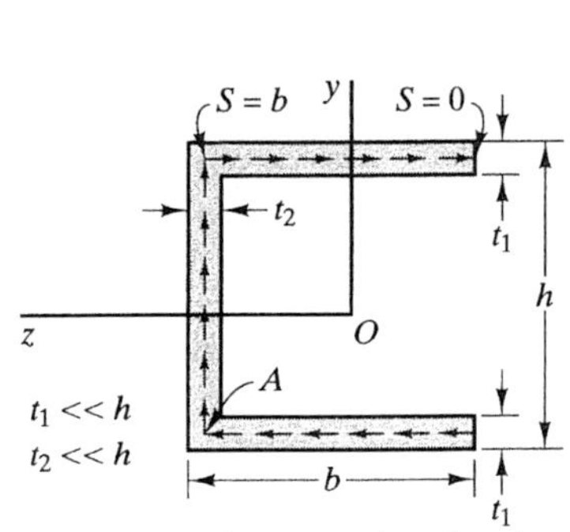

Figure 11.56. Cross section showing shear stress.

Accordingly, Eq. (11.45) becomes, on remembering that y and z are principal axes,

$$|\tau_{xs}| = \left|\frac{V_y}{t_c I_{zz} I_{yy}}\left(-I_{yy}\int_0^s yt\,d\sigma\right)\right| = \left|\frac{V_y Q_z}{I_{zz} t_c}\right| \tag{a}$$

Considering the *upper flange* to include the left top corner, as has been shown in Fig. 11.57(a), we compute τ_{xs} at position s in the upper flange in the following manner:

$$|(\tau_{xs})_{\text{flange}}| = \frac{|V_y|[(h - t_1)/2]s}{I_{zz}} \tag{b}$$

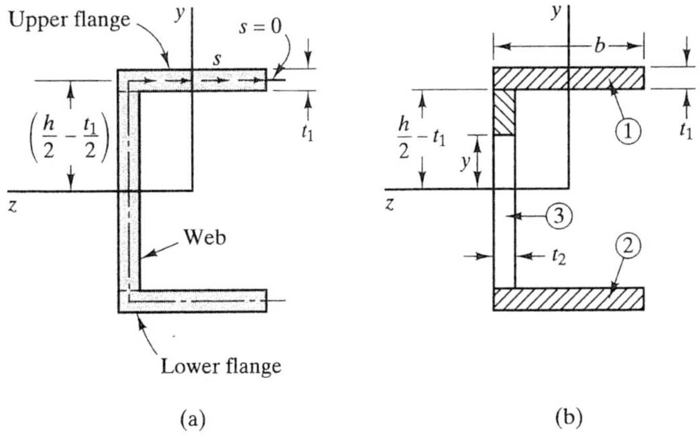

Figure 11.57. Channel section.

The term I_{zz} is most easily computed by decomposing the cross section into two flanges, shown as 1 and 2 in Fig. 11.57(b), with the web shown as 3. Thus,

Example 11.19 (Continued)

$$I_{zz} = \frac{1}{12}t_2(h - 2t_1)^3 + 2\left[\frac{1}{12}bt_1^3 + bt_1\left(\frac{h}{2} - \frac{t_1}{2}\right)^2\right] \quad \text{(c)}$$

The shear stress in the *web* is taken so as not to include the inside fillets of the channel. Observing Fig. 11.57(b), the term Q_z is computed considering the upper flange and the web portion separately as follows:

$$Q_z = bt_1\left(\frac{h - t_1}{2}\right) + t_2\left(\frac{h}{2} - t_1 - y\right)\left[y + \frac{(h/2 - t_1) - y}{2}\right] \quad \text{(d)}$$

The shear stress can then be given as

$$|(\tau_{xs})_{\text{web}}| = \left|\frac{V_y}{I_{zz}t_2}\left\{\frac{bt_1}{2}(h - t_1) + \frac{t_2}{2}\left[\left(\frac{h}{2} - t_1\right)^2 - y^2\right]\right\}\right| \quad \text{(e)}$$

where I_{zz} is given by Eq. (c) and $V_y = F_1$.

It must be pointed out that at the corners, where the direction of the cross section centerline is rapidly changing from horizontal to vertical, we have not really evaluated the shear stress. It should be apparent that our simple theory cannot give a good account of shear stress at this region where the geometry is comparatively complex. Unfortunately, large stress concentrations arise at such corners. Such regions must be examined by numerical techniques or by methods of experimental stress analysis. As explained in Chapter 4, such studies often result in a dimensionless stress concentration factor which is used to multiply the stress formulated from simplified theory so as to yield a more realistic result. Thus, the simplified theory is still useful in such situations. Also, despite the inadequacy of the theory at the corners, we shall still be able to formulate a reasonably accurate value of e (see Fig. 11.55) for no twisting.

As a check let us now compute the shear force V_y from the shear-stress distribution. Since the forces in the flanges are equal and opposite, we need consider only the web for computing V_y. Thus,

$$V_y = \int_{-[(h/2)-t_1]}^{[(h/2)-t_1]} \frac{V_y}{I_{zz}t_2}\left\{\frac{bt_1}{2}(h - t_1) + \frac{t_2}{2}\left[\left(\frac{h}{2} - t_1\right)^2 - y^2\right]\right\}t_2\,dy \quad \text{(f)}$$

Integrating and putting in limits, we get

Example 11.19 (Continued)

$$V_y = \frac{V_y}{I_{zz}}\left[(bt_1)(h - t_1)\left(\frac{h}{2} - t_1\right) + \frac{1}{12}t_2(h - 2t_1)^3\right] \qquad \text{(g)}$$

The bracketed expression on the right-hand side of Eq. (g) has a value close but not quite equal to I_{zz}, as can be readily seen when we express I_{zz}—see Eq. (c)—in the following way:

$$I_{zz} = bt_1(h - t_1)\left(\frac{h}{2} - \frac{t_1}{2}\right) + \frac{1}{12}t_2(h - t_1)^3 + \frac{1}{6}bt_1^3 \qquad \text{(h)}$$

If the right-hand side of Eq. (h) were the same as the bracketed quantity on the right-hand side of Eq. (g), we would end up with the identity $V_y = V_y$, thus justifying the steps we have taken. The discrepancy arises in the uncertainty in the upper and lower corners of the channel. In this analysis, they have arbitrarily been included as part of the flanges, but they could just as well have been made part of the web, which would have resulted in a different formulation of shear stress there. If t_1 and t_2 are much smaller than h and b, we can drop t_1 and t_2 when added or subtracted to h or $h/2$ and we can delete the expression

$$\tfrac{1}{6}bt_1^3$$

as small compared to the remaining terms in Eqs. (g) and (h). We see that the proper requirement of $V_y = V_y$ in Eq. (g) is then reached.

*11.10 A Note on the Shear Center for Thin-Walled Open Members

In the preceding section we set forth shear-stress distributions for bending *without twisting.* We shall now comment on the criterion for this state of deformation. For this purpose we shall employ the concept of the *applied shear force,* which is the simplest resultant force of all forces coming onto a free body exposing a section of the beam, except for the shear force itself at the section. (Thus, in Fig. 11.58 the applied shear force for a free body exposing a section at x is simply $w_0(L-x)$ using for simplicity free body II.) The shear-stress distribution computed on the basis of bending without twisting must relate properly to the applied

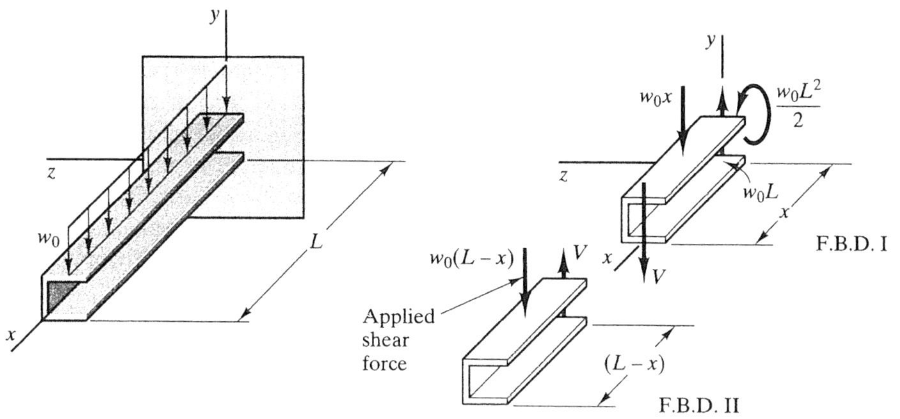

Figure 11.58. Illustration of applied shear force.

shear force from the standpoint of equilibrium. In particular, the following must occur to eliminate the possibility of twisting at a section:

> The applied shear force must have a line of action such that it develops an equal but *opposite* twisting moment about any axis parallel to the centerline of the beam as that stemming from the shear stress distribution at the exposed section computed on the basis of bending action only.

We may work backward to find the desired line of action of the load. We first ascertain the shear-stress distribution from the assumption of bending without twisting. With this information we then go back to find the appropriate line of action of the applied shear force for the section to avoid twisting. Thus, in Fig. 11.59(a) we show a stress distribution in a thin-walled section stemming from an applied shear force in the minus y direction.[20] Using Eq. (11.45) to compute τ_{xs} and relating twisting moments about O as described above between the applied shear force and the stress distribution, we can determine $\bar{z}$ for the applied shear force to ensure a condition of no twisting. Similarly, for an ap-plied shear force in the minus z direction for a section [see Fig. 11.59(b)] we can find a shear-stress distribution τ_{xs} using Eq. (11.47) and thus find $\bar{y}$ for no twisting at a section. The distances $\bar{z}$ and $\bar{y}$ will be dependent only on the cross section geometry and not the particular value of the load. Clearly, the value of stress at a point in a section is

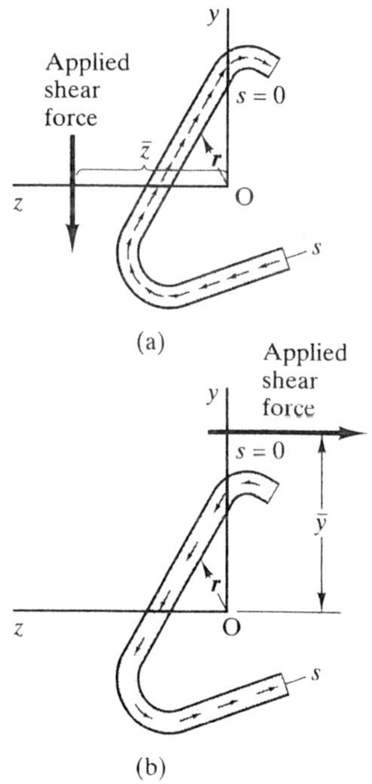

Figure 11.59. Vertical and horizontal applied shear forces.

[20]Note that the shear force itself, namely V_y, as a result of equilibrium will be directed upwards for the section shown which has a normal in the plus x direction.

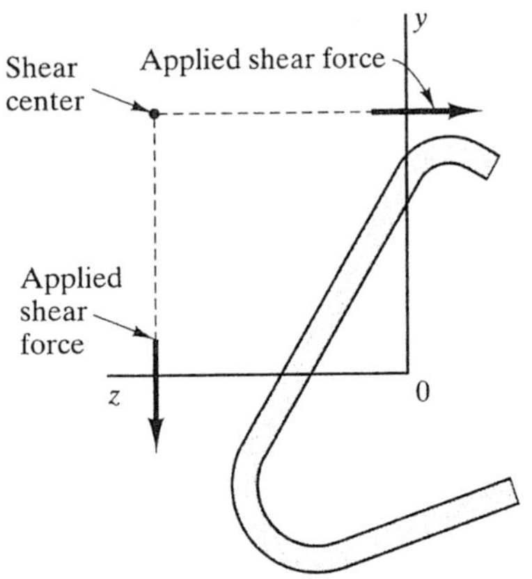

Figure 11.60. Shear center of a section.

directly proportional to the applied shear force for that section. Hence, once a line of action is found for no twisting this will be the same line of action for no twisting at that section no matter what the particular value of load is. Now the intersection of the lines of action of the applied shear forces discussed above is shown in Fig. 11.60 and is called the *shear center.* An applied shear force going through the shear center for a section induces bending without twisting at the section. This must be true since we can decompose this force into components having the appropriate lines of action which are at distances $\bar{y}$ and $\bar{z}$ from 0, determined above, to develop only bending at the section.

We shall now illustrate the procedure described by going back to Example 11.19 to find the proper line of action of the applied shear force F to avoid twisting in the channel.

Example 11.20

Find the proper value of e in Example 11.19 to have bending without twisting of the channel section everywhere along the channel. Here the load F is taken downward.

We have redrawn the figure for this example in Fig. 11.61(a). Furthermore, in Fig. 11.61(b) we have shown a free body exposing a section at x having a normal in the minus x coordinate direction. Finally, in Fig. 11.61(c) this section is shown as viewed from $x = 0$ with a shear-stress distribution corresponding to an upward shear force V_y. The applied shear force for this section is F (downward).

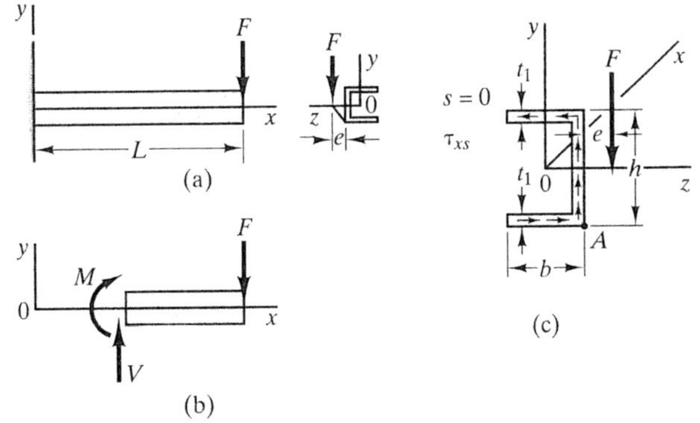

Figure 11.61. Shear force in cantilever beam.

To determine e, we set the twisting moment of the applied shear force F about an axis parallel to the x axis and going through point A [see Fig. 11.61(c)] equal and opposite to the moment about this axis of the shear-stress distribution of the section at position x. For such an axis we clearly need be concerned only with the stresses

Example 11.20 (Continued)

in the upper flange. Accordingly, we can say, using the proper directional signs for the moments, that

$$Fe - \int_0^b [(h - t_1)\tau_{xs}]t_1\, ds = 0 \qquad \text{(a)}$$

Employing Eq. (b) of Example 11.19 for τ_{xs} we get

$$Fe = \left| \int_0^b t_1 \frac{V_y(h - t_1)^2}{2I_{zz}} s\, ds \right|$$

$$= \left| \frac{t_1 V_y (h - t_1)^2}{2I_{zz}} \left[\frac{s^2}{2} \right]_0^b \right| = \frac{t_1 |V_y| (h - t_1)^2 b^2}{4I_{zz}} \qquad \text{(b)}$$

Now examining Fig. 11.61(b), we can readily conclude that $|V_y| = |F|$. We can now cancel F in the preceding equation and solve for e. Thus,

$$e = \frac{t_1(h - t_1)^2 b^2}{4I_{zz}} \qquad \text{(c)}$$

Since t_1 and t_2 are small compared to b, we shall approximate I_{zz} employing Eq. (h) of Example 11.19 for this purpose. We may thus say for $t \ll h$ that

$$I_{zz} \approx \frac{bt_1 h^2}{2} + \frac{1}{12} t_2 h^3 \qquad \text{(d)}$$

Now using the foregoing expression for I_{zz} in Eq. (c) and dropping t_1 in the squared bracket of that equation, we get on canceling h^2

$$e \approx \frac{t_1 b^2}{2bt_1 + t_2 h/3} \qquad \text{(e)}$$

*11.11 Inelastic Behavior of Beams: The Elastic, Perfectly-Plastic Case

Our discussion up to this time has been restricted to linear-elastic materials. We shall now turn to the case of elastic, perfectly-plastic behavior. (Later we shall consider the more general case where the material may have any arbitrary stress-strain law.)

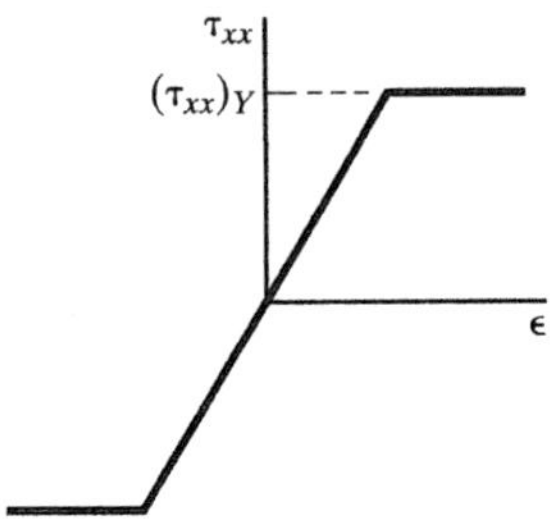

Figure 11.62. Elastic, perfectly-plastic behavior.

We shall assume that the material behaves in compression in the same way that it does in tension (see Fig. 11.62). This idealization is meaningful for certain ductile structural materials such as structural steel.

The physical arguments that we employed for pure bending to show that plane sections remain plane were presented without reference to the nature of the stress-strain law for the particular material. Thus, we can conclude that the normal strain ε_{xs} at a position y from the neutral surface in the undeformed geometry can be expressed as

$$\varepsilon_{xx} = -\frac{y}{R} \tag{11.49}$$

provided that we stay within small deformation limitations. You may recall that it was only at this stage of development of the theory that we used Hooke's law in Section 11.2.

For simplicity, we shall consider a beam having a rectangular cross section with the applied couples in a plane of symmetry, as shown in

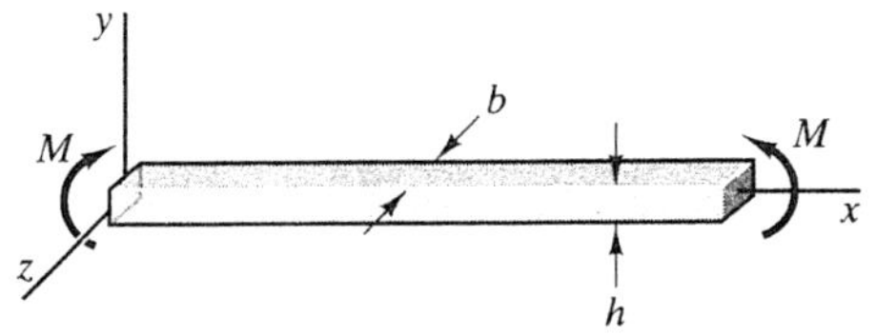

Figure 11.63. Pure bending.

Fig. 11.63. The maximum moment M_E that can be applied without plastic deformation may be computed using the flexure formula as follows:

$$\tau_Y = -\frac{M_E(-h/2)}{I}$$

$$\therefore\ M_E = \frac{I\tau_Y}{h/2} \tag{11.50}$$

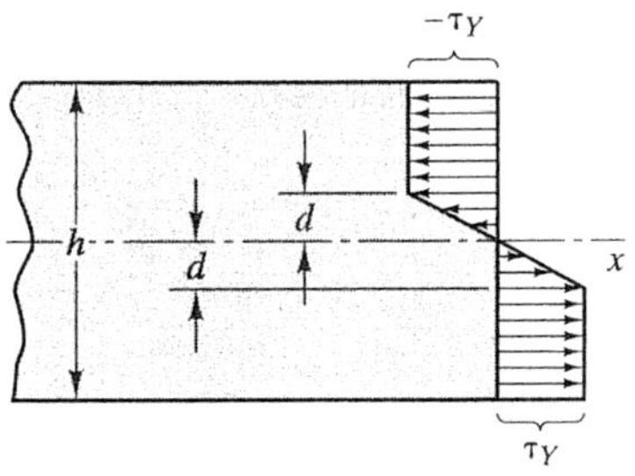

Figure 11.64. Stresses at a section.

where τ_Y is the *yield stress* of the material. If the applied couple is increased *above* this value, plastic deformation occurs first at the outer fibers of the beam. These fibers cannot develop greater stress magnitude than that of τ_Y and so give no increased resistance to further increases in load. Thus, as the load is increased, adjacent fibers closer to the neutral axis reach a stress with magnitude τ_Y and remain at this stress with the result that regions of constant stress magnitude τ_Y, penetrate from the outer fibers toward the neutral surface in a manner suggested by Fig. 11.64. We denote as d the distance from the neutral surface to the position where the yield stress begins. Clearly, in this case d must be the same above and below the neutral surface, as indicated in the diagram.

We may use Hooke's law between $-d \leq y \leq d$, so that the strain at $y = d$, which we denote as $(\varepsilon_{xx})_Y$, can be given as

$$(\varepsilon_{xx})_Y = -\frac{1}{E}\tau_Y \tag{11.51}$$

Substituting this result into Eq. (11.49) for the position $y = d$, we then get for $1/R$,

$$\frac{1}{R} = \frac{\tau_Y}{Ed} \tag{11.52}$$

We thus have the radius of curvature in terms of τ_Y and E, which are properties of the material, and d, which depends on the applied load of M. The stress distribution is also known in terms of these quantities. Thus, for $d \leq y \leq h/2$ (i.e., the plastic region on top) we have $\tau_{xx} = -\tau_Y$, and for $-d \geq y \geq -h/2$ (i.e., the plastic region at the bottom) we have $\tau_{xx} = \tau_Y$. Furthermore, for the elastic region between $-d \leq y \leq d$, we have

$$\tau_{xx} = E\varepsilon_{xx} = -E\frac{y}{R} \tag{11.53}$$

wherein we have used Hooke's law and Eq. (11.49). Now replacing $1/R$ in accordance with Eq. (11.52), for τ_{xx} we have

$$\tau_{xx} = -\frac{\tau_Y y}{d} \tag{11.54}$$

Equation (11.54) merely shows the linear variation of stress in the elastic range and could have been determined by merely inspecting Fig. 11.64.

Our task is now to determine d in terms of the load M. First we note that the stress distribution must have a zero resultant force for this problem. This requires that

$$\int_{-h/2}^{-d} \tau_Y \, b \, dy + \int_{-d}^{+d} \tau_{xx} b \, dy + \int_{d}^{h/2} [-\tau_Y] b \, dy = 0 \tag{11.55}$$

Clearly, the first and last integrals cancel each other. Substituting for τ_{xx} using Eq. (11.54), we then have

$$-\frac{\tau_Y}{d}\int_{-d}^{+d} yb \, dy = 0 \tag{11.56}$$

This shows that the neutral axis is the centroidal axis for this case. *For less simple stress-strain laws the coincidence of the neutral axis and the centroidal axis need not occur.*

Next we take moments of the distribution about the neutral axis. Setting this equal to $-M$ for reasons explained in Section 11.2, we have

$$\int_{-h/2}^{-d} \tau_Y \, yb \, dy + \int_{-d}^{d} \tau_{xx} \, yb \, dy + \int_{d}^{h/2} [-\tau_Y] yb \, dy = -M \qquad (11.57)$$

Substituting for τ_{xx} using Eq. (11.54) and integrating, we get

$$b\,\tau_Y \frac{y^2}{2}\bigg]_{-h/2}^{-d} - \frac{b\tau_Y}{d}\frac{y^3}{3}\bigg]_{-d}^{d} - b\,\tau_Y \frac{y^2}{2}\bigg]_{d}^{h/2} = -M \qquad (11.58)$$

Putting in the limits and solving for M, we get

$$M = b\,\tau_Y\left(\frac{h^2}{4} - \frac{d^2}{3}\right) \qquad (11.59)$$

We can solve for d for Eq. (11.59) for any given applied load M so that we now have a complete statement as to the stress distribution for this problem.

The maximum moment that the beam can withstand occurs when $d = 0$, that is, when the elastic core has shrunk to zero thickness and plastic action pervades the entire section. This bending moment is called the *plastic-hinge moment,* M_H. Setting $d = 0$ in Eq. (11.59), we see that M_H must have the following value for the rectangular cross section under consideration:

$$M_H = \frac{b\,\tau_Y h^2}{4} \qquad (11.60)$$

Until now, we have considered the case of pure bending. As in our earlier work, we can extend the results of these formulations to loadings which include shear force, provided that we use *local* values of M. Furthermore, the radius of curvature R computed from Eq. (11.52) would then be a *local radius of curvature.*

In the following section we shall see how we can make use of the plastic-hinge concept.

*11.12 A Note on the Failure of a Structure: Limit Design

For certain purposes one may consider that a structure fails when, for a linear, elastic material, the yield stress is reached somewhere so that permanent deformation is developed there. It may well be that the structure, despite this permanent deformation, can still be capable of withstanding even greater loads. Hence, if the presence of a reasonable amount of permanent deformation is acceptable in a structure,[21] then one can design on the basis of larger loads than those required by exclusion of plastic deformation. We shall now introduce the concept of *limit design* as a means of ascertaining acceptable maximum loads for a structure in which reasonable plastic deformation is permitted.

To do this let us consider the simply supported beam, shown in Fig. 11.65(a) loaded by a uniform loading w whose value we shall

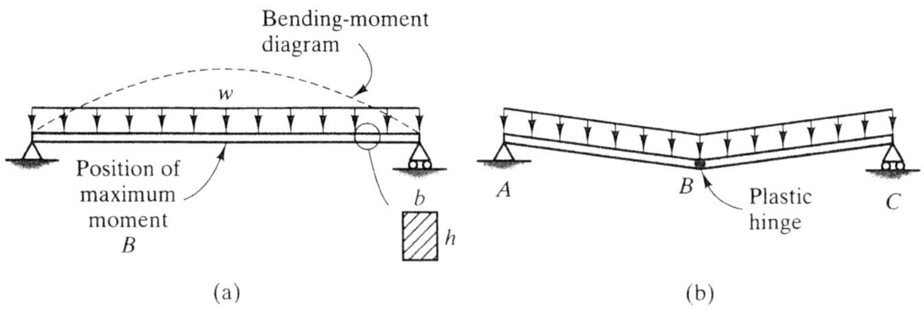

Figure 11.65. Development of a plastic hinge.

assume can be varied. The bending-moment diagram has been shown as a dashed curve in this diagram. The maximum bending moment occurs at the center of the beam. Clearly, if w is increased sufficiently, the maximum possible moment M_H will be reached at point B at the center of the beam. For further increases in external load, point B acts as a "hinge," offering no increased resistance to bending. In our simply supported beam there would then be a collapse of the beam under such an increase of load. We would say that the beam becomes at this time a *mechanism* and is no longer fulfilling its mission of withstanding the external loading. This has been shown in an exaggerated manner in Fig. 11.65(b). Note that the load has been increased to a degree where the beam acts like two members pinned at points A, B, and C. With C free to move horizontally, the beam then acts like a mechanism having certain constraints.

In more complex systems the achievement of the moment M_H at a particular point may not mean that the structure has been reduced to a mechanism. The structure may still continue to withstand increased loads with acceptably small deflections. In that case the moment at the aforementioned position remains at the value M_H, but moments *elsewhere* increase in value as the load is increased further. Clearly, another position in the structure will then eventually reach the value M_H. If a *mechanism* is formed at this stage, the structure is considered to have failed. If not, one continues to increase the load *until enough plastic hinges have been formed so as to render the structure a mechanism.* The

[21]Ordinary buildings are in this category.

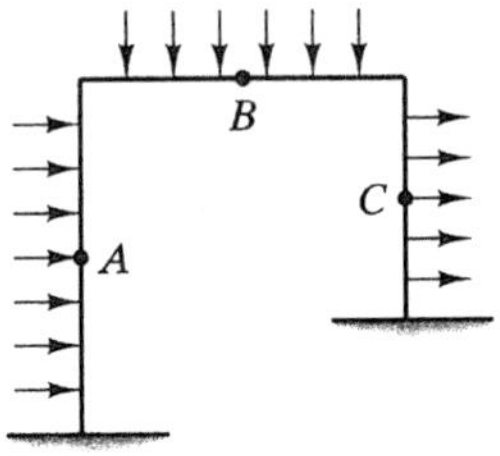

Figure 11.66. Frame with three hinges.

structure "fails" at this particular load.[22] Thus, consider the frame shown in Fig. 11.66. On increasing the loading we first form a plastic hinge at A. However, the frame can continue to maintain increased loads. A second hinge at B eventually appears—again without robbing the structure of its load-carrying capacity. Finally, it is clear on further increased loading that a third hinge at C forms, thus creating a mechanism. We then have failure of the frame to support the loading.

The following example illustrates the difference in maximum loads possible according to the two failure criteria presented here.

Example 11.21

Compare the maximum intensity of loading w for the simply supported beam shown in Fig. 11.65 when the limit-design criterion is used and when the elastic-limit criterion is used.

From rigid-body mechanics the bending moment M at the center of the beam is

$$M = \frac{wl^2}{8} \tag{a}$$

Hence, for elastic action, we have for the maximum stress at the center section of the beam

$$\tau_{xx} = \frac{(wl^2/8)(h/2)}{I_{zz}} = \frac{wl^2h}{16I_{zz}} \tag{b}$$

The maximum load possible with the presence of only elastic deformation is then

$$(w_{\max})_1 = \frac{16\tau_Y I_{zz}}{hl^2} = \frac{4}{3}\frac{b\tau_Y h^2}{l^2} \tag{c}$$

Now let us compute the maximum loading for the limit-design criterion. From Eq. (11.60), for M_H we have

$$M_H = \frac{b\,\tau_Y\,h^2}{4} \tag{d}$$

Substituting the preceding result for M_H into Eq. (a) and solving for $(w_{\max})_2$, we have

[22]See F. G. Hodge, *Plastic Analysis of Structure* (McGraw-Hill, New York, 1959), for a detailed study of the *limit-design* approach.

Example 11.21 (Continued)

$$(w_{max})_2 = \frac{2b\,\tau_Y\,h^2}{l^2} \qquad \text{(e)}$$

The ratio of the maximum loads according to the two criteria is then

$$\frac{(w_{max})_2}{(w_{max})_1} = \frac{[2b\,\tau_Y\,h^2]l^2}{\frac{4}{3}[b\,\tau_Y\,h^2]l^2} = \frac{3}{2}$$

Thus, the limit-design criterion allows for considerably more load. It is up to the designer to decide on what design criterion he or she must base his or her work.

*11.13 Inelastic Behavior of Beams: Generalized Stress-Strain Relation

In Section 11.11 we considered materials whose stress-strain diagram could be idealized for certain problems as the elastic, perfectly-plastic case. It was pointed out that structural steel can often be treated in this manner. Because such an important material can often be handled this way, there is considerable use for the limit-design approach.

There are, however, structural materials which have stress-strain diagrams for which our idealized stress-strain diagrams have little meaning. Accordingly, let us consider some general stress-strain diagrams, such as in Fig. 11.67. As explained in Section 11.11, we may employ the familiar relation for small deformation

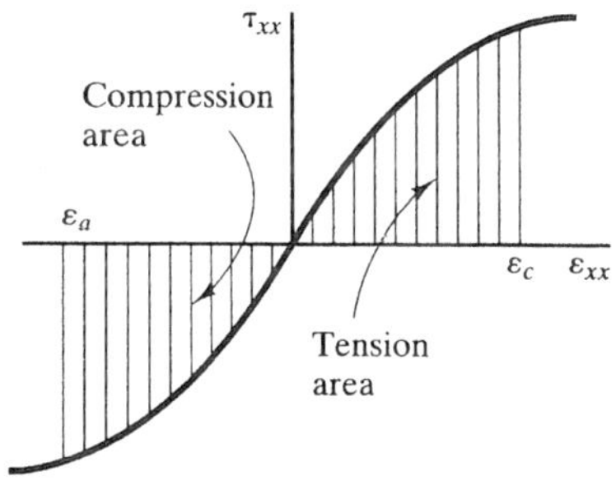

Figure 11.67. General constitutive law.

$$\varepsilon_{xx} = -\frac{y}{R} \qquad (11.61)$$

If a and c represent the distances from the neutral surface to the outermost fibers, as shown in Fig. 11.68, we can accordingly say for the extreme strains ε_a and ε_c that

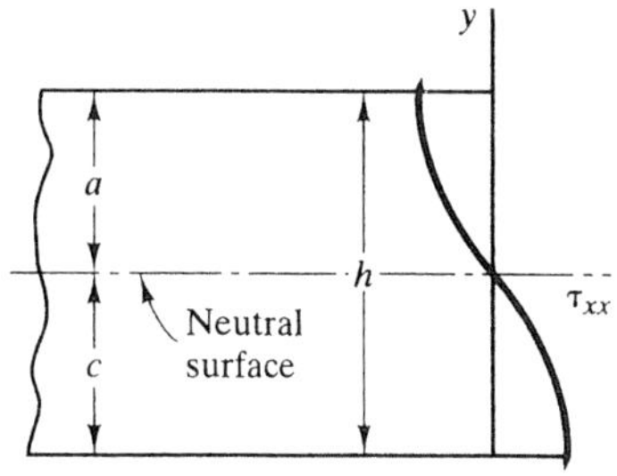

Figure 11.68. Stress at a section.

$$\varepsilon_a = -\frac{a}{R} \qquad (11.62)$$

$$\varepsilon_c = \frac{c}{R} \qquad (11.63)$$

Keep in mind, however, that we do not know the position of the neutral surface (it does not necessarily pass through a centroidal axis of the cross section). As for the stress distribution, we require that τ_{xx} satisfy the following conditions for pure bending:

$$\iint_A \tau_{xx}\, dA' = 0 \tag{11.64a}$$

$$\iint_A y\tau_{xx}\, dA' = -M \tag{11.64b}$$

We have now presented the basic formulations for the determination of the stress τ_{xx} anywhere in the section.

As a simple case, we consider a beam having a rectangular cross section of width b and height h. Noting that $dA' = bdy$ for this case and using the differential of Eq. (11.61) to replace dy by $-Rd\varepsilon_{xx}$, we now rewrite Eq. (11.64a) as follows:

$$\iint_A \tau_{xx} dA = \int_{-c}^{a} \tau_{xx} b\, dy = -\int_{\varepsilon_c}^{\varepsilon_a} \tau_{xx} bR\, d\varepsilon_{xx}$$

$$= -bR\int_{\varepsilon_c}^{\varepsilon_a} \tau_{xx} d\varepsilon_{xx} = 0 \tag{11.65}$$

The last of the preceding equations may also be expressed as

$$\int_{\varepsilon_c}^{\varepsilon_a} \tau_{xx} d\varepsilon_{xx} = \int_{\varepsilon_c}^{0} \tau_{xx} d\varepsilon_{xx} + \int_{0}^{\varepsilon_a} \tau_{xx} d\varepsilon_{xx} = 0 \tag{11.66}$$

This means that the magnitude of the compression area in the stress-strain diagram must equal the magnitude of the tension area in the stress-strain diagram. These regions have been crosshatched in Fig. 11.67. Thus, for any ε_a we can evaluate ε_c by graphical or numerical procedures. Now in Eq. (11.62) we can eliminate R by dividing Eq. (11.62) into Eq. (11.63). We get

$$\frac{\varepsilon_a}{\varepsilon_c} = -\frac{a}{c} \tag{11.67}$$

Furthermore, using the relation

$$a + c = h \tag{11.68}$$

we can now determine a and c for the chosen ε_a.

Next consider Eq. (11.64b). Using Eq. (11.61) for y and its differential form for dy we may write Eq. (11.64b) as

$$\iint_A y\tau_{xx} dA' = \int_{-c}^{a} y\tau_{xx} b\, dy = \int_{\varepsilon_c}^{\varepsilon_a} R^2 \varepsilon_{xx}\tau_{xx} b\, d\varepsilon_{xx}$$
$$= bR^2 \int_{\varepsilon_c}^{\varepsilon_a} \varepsilon_{xx}\tau_{xx}\, d\varepsilon_{xx} = -M \tag{11.69}$$

The last integral represents the first moment of the shaded area in the stress-strain diagram about the τ_{xx} axis. By numerical or graphical methods we may then find from this equation the value of M that corresponds to the chosen value ε_a.

By the preceding formulations, we can thus set forth a *trial-and-error procedure,* choosing values of ε_a until we get the proper bending moment M in accordance with Eq. (11.69). When this has been attained, we have available ε_c as described above, and the position of the neutral surface in terms of the value of a or c as determined from Eqs. (11.67) and (11.68). The radius of curvature R is now available from Eqs. (11.62) and (11.63). The strain at any elevation y from the neutral axis is computed from Eq. (11.61), and the corresponding stress may be read off the stress-strain diagram. Thus, we have for a value of M at a section complete information as to stress and strain at that section.

11.14 Stress Concentrations for Bending

In Chapter 4 we discussed stress concentrations resulting from small cuts in a member. You will recall that for brittle materials or where fatigue is a problem such considerations are vital for proper design. In beams having sudden changes in geometry or having notches or other cuts, we again have stress concentrations present. We have available concentration factors K for various cuts, and we use them in conjunction with the simple strength of materials formulas that would be valid were there no cuts or sudden geometry changes. These factors are available in handbooks.[23] We illustrate a particular case in Fig. 11.69 where we show a beam under pure bending with a sudden change in width from H to h through a fillet having radius r. Note that K is plotted as a function of r/h for various values of H/h. The maximum normal stress for this case is then

$$\tau_{\max} = K\frac{M(h/2)}{I} \tag{11.70}$$

where the flexure formula is used for the smaller section.

[23]See footnote on page 99.

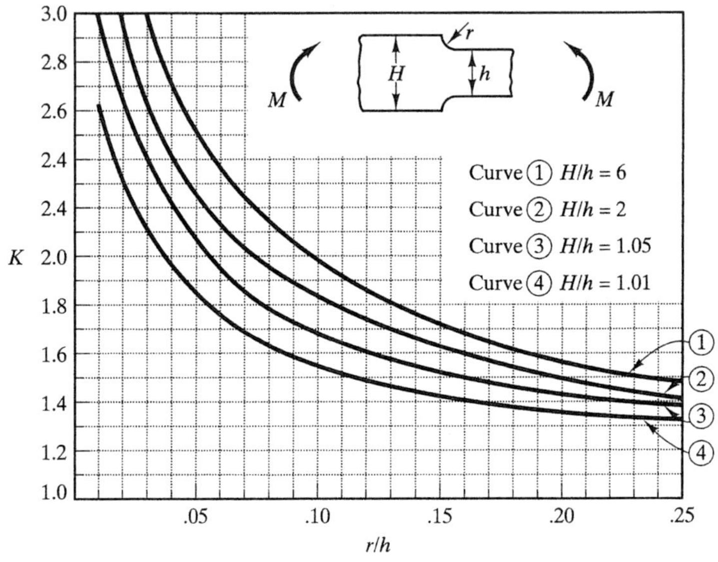

Figure 11.69. Stress concentration factor curves.

*11.15 Bending of Curved Beams

Let us now consider a weightless beam loaded by equal end moments M_z. The beam is initially curved and the cross section is symmetric about the loading plane. For this beam, we will use a cylindrical coordinate system centered at the center of curvature of the undeformed beam as shown in Fig. 11.70.

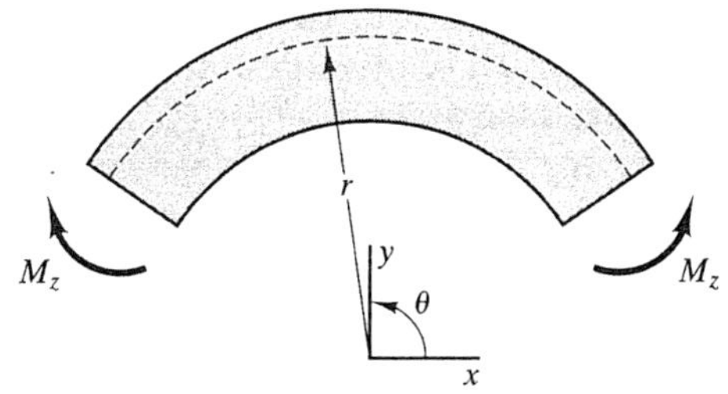

Figure 11.70. Curved beam with end moments.

As in the case of a straight beam in pure bending, we can again assume that cross sections of the beam prior to loading remain plane after loading. This assumption assures that each section will deform in a **compatible** way with its neighboring sections. In addition, we will assume that all the strains are small.

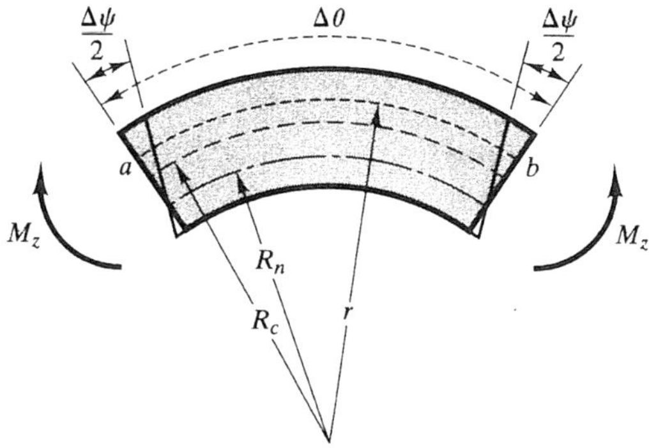

Figure 11.71. Section of curved beam under load.

We have shown in Fig. 11.71 a portion of the beam defined by the angle $\Delta\theta$. Note that the location of the neutral axis R_n is different from the location of the centroidal axis R_c of the section. As in the case of straight beams, we will not assume a location for the neutral axis but rather we will use the equations of equilibrium to locate its position. For now, we allow that the centroidal and neutral axes may not coincide. Also shown in Fig. 11.71 is the rotation of the section due to the applied moments. We can see from the figure that the total angular change of our curved beam segment is $\Delta\psi$.

The length of a beam fiber ab located at a distance r from the center of curvature of the undeformed beam is given by

$$ab = r\Delta\theta \tag{11.71}$$

The change in length of this fiber (note that the fiber has shortened) is given as

$$\delta = -(r - R_n)\Delta\psi \tag{11.72}$$

where $\Delta\psi/2$ is the rotation of the end of the cross sections of the beam element about the neutral axis end points as a result of deformation. Using Eq. (11.71), we can now write the strain $\varepsilon_{\theta\theta}$ in the circumferential direction as

$$\begin{aligned}\varepsilon_{\theta\theta} &= \lim_{\Delta\to 0}\frac{\delta}{ab}\\ &= \lim_{\Delta\to 0}\frac{-(r - R_n)\Delta\psi}{r\Delta\theta}\\ &= \left(\frac{R_n}{r} - 1\right)\frac{d\psi}{d\theta}\end{aligned} \tag{11.73}$$

Our conclusions up to this point are valid for beams undergoing small strains and composed of any type of homogeneous material. Now we introduce a **constitutive law.** For a linear-elastic material, Hooke's law gives for the normal strain $\varepsilon_{\theta\theta}$

$$\varepsilon_{\theta\theta} = \frac{1}{E}[\tau_{\theta\theta} - \nu(\tau_{rr} + \tau_{zz})] \tag{11.74}$$

In the absence of physical constraints in the r and z directions, we can assume that stresses τ_{rr} and τ_{zz} are zero.[24] Therefore, the circumferential stress field is given by

$$\tau_{\theta\theta} = E\varepsilon_{\theta\theta} = E\frac{d\psi}{d\theta}\left(\frac{R_n}{r} - 1\right) \tag{11.75}$$

Invoking **equilibrium,** we see that we have two equations for the stress distribution on the cross section: summation of forces in the θ direction (i.e., normal to the cross section) and summation of moments about the z axis. From these two equations we can establish the position of the neutral axis and the relationship between the applied moment and normal stress in the beam. First, let's consider the summation of forces.

$\underline{\sum F_\theta = 0}$:

$$\therefore \iint_A \tau_{\theta\theta}\, dA' = E\frac{d\psi}{d\theta}\iint_A \left(\frac{R_n}{r} - 1\right) dA' = 0 \tag{11.76}$$

Since E and $d\psi/d\theta$ are not zero, we concluded from Eq. (11.76) that

$$\iint_A \left(\frac{R_n}{r} - 1\right) dA' = 0 \tag{11.77}$$

By rearranging Eq. (11.77), we arrive at

$$R_n = \frac{A}{\displaystyle\iint_A \frac{dA'}{r}} \tag{11.78}$$

We see in Eq. (11.78) that the position of the neutral axis is a function of the cross-sectional shape only. Its location can be tabulated for various cross sections. For cross sections of complex shape, the neutral axis can be found by dividing the cross section into simpler regions and the summation Eq. (11.78) gives the location. This is expressed as

[24]In fact, due to the curvature of the beam, a radial stress component τ_{rr} will be generated. For solid sections, this component can often be neglected. See A. C. Ugural and S. K. Fenster, *Advanced Strength and Applied Elasticity,* 3rd ed., Prentice-Hall, 1995.

$$R_n = \frac{\sum A}{\sum \iint_A \frac{dA'}{r}}$$

Our second **equilibrium** equation requires that the resultant of the stress distribution at any section be a couple moment of value M_z. Thus we have

$$\iint_A \tau_{\theta\theta}(R_n - r)dA' = M_z \tag{11.79}$$

In the above we see that for an assumed positive normal stress, a value of r larger than R_n (i.e., above the neutral axis) results in a negative external moment while values of r smaller than R_n result in a positive external moment. This is analogous to the case for straight beams as shown in Fig. 11.4. Introducing the normal stress from Eq. (11.75) into Eq. (11.79) we get

$$\begin{aligned} M_z &= E\frac{d\psi}{d\theta}\iint_A \frac{(R_n - r)^2}{r}dA' \\ &= E\frac{d\psi}{d\theta}\iint_A \left(\frac{R_n^2 + r^2 - 2R_n r}{r}\right)dA' \\ &= E\frac{d\psi}{d\theta}\iint_A \left(\frac{R_n^2}{r} + r - 2R_n\right)dA' \end{aligned} \tag{11.80}$$

With a substitution from Eq. (11.78), the first term in the above integral can be expressed as

$$\iint_A \frac{R_n^2}{r}dA' = R_n^2\iint_A \frac{dA'}{r} = R_n A \tag{11.81}$$

The second term in the integral of Eq. (11.80) is equivalent to the first moment of the area about the center of curvature. This is expressed as $R_c A$ where R_c is seen to be the distance from the center of curvature of the undeformed beam to the centroid of the section. The last of the three integrals is simply $2R_n A$. We can now rewrite Eq. (11.80) as

$$M_z = AE\frac{d\psi}{d\theta}[R_c - R_n] \tag{11.82}$$

Solving the above for $d\psi/d\theta$ gives

$$\frac{d\psi}{d\theta} = \frac{M_z}{AE(R_c - R_n)} \tag{11.83}$$

Equation (11.83) yields significant information regarding the location of the neutral axis. Note that for positive values of $d\psi$, M_z is also positive. Therefore, the term $(R_c - R_n)$ in the denominator must be positive. Consequently, R_n must always be smaller than R_c. In other words, we have the important result that *the neutral axis of a curved beam will always lie closer to the center of curvature than the centroid.*

Substitution of Eq. (11.83) into Eq. (11.75) gives

$$\tau_{\theta\theta} = \frac{M_z(R_n - r)}{Ar(R_c - R_n)} \tag{11.84}$$

Let $R_c - R_n = e$ where e is called the *eccentricity*. Note that the distance between the centroid and the neutral axis is always positive, thus giving us a positive eccentricity. We then get for the stress[25]

$$\tau_{\theta\theta} = \frac{M_z(R_n - r)}{Are} \tag{11.85}$$

Eq. (11.85) represents the circumferential stress over the cross section of a curved beam. As a check, we envision a positive moment applied to a curved beam such as the one shown in Fig. 11.70. Such a load would produce compressive stress in beam fibers above the neutral axis and tensile stress in fibers below the neutral axis. Referring to Eq. (11.85), and noting that A, r, R_n, and e are always positive, we see that if we investigate the region above the neutral axis, then $r > R_n$ and the stress is compressive. Likewise, if $r < R_n$ then the stress below the neutral axis is tensile, as was expected.

We can now outline a procedure for investigating the stresses in a curved beam of known cross section. First, we compute the area A and the distance to the centroidal axis R_c for the beam. These are readily computed from the geometry. Next, the distance to the neutral axis R_n is computed for the cross section from Eq. (11.78). The eccentricity, e, is then determined. From the moment and the geometrical parameters just computed, Eq. (11.85) provides the stress. Note that, as in the case of straight beams, we assume that the formulas developed for pure bending still apply when we have a general loading on the beam. In such a case, we may also develop significant normal stresses in the beam due to axial loads. Such stress may be superimposed on the bending stress determined from the preceding analysis.

[25]Equation 11.85 was developed by E. Winkler in 1858.

Example 11.22

A semicircular curved beam is loaded by equal end moments as shown below in Fig. 11.72. The cross section of the beam is trapezoidal and the subscripts i and o refer to the inside and outside surfaces of the beam, respectively. Determine the maximum tensile and compressive stresses in the beam.

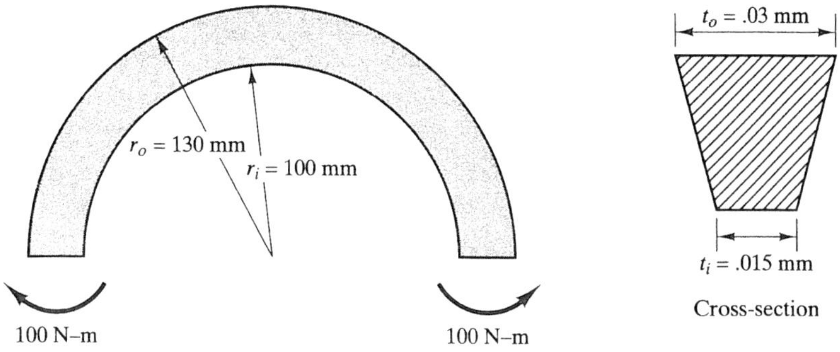

Figure 11.72. Curved beam with equal end moments.

First, we compute the area:

$$A = \frac{1}{2}(t_o + t_i)(r_o - r_i)$$
$$= \frac{1}{2}(0.030 + 0.015)(0.130 - 0.100) = 6.75 \times 10^{-4}\, m^2$$

Next, we resolve the location of the centroid of the cross section relative to the center of curvature. This is determined using the first moment of the area about the center of curvature. For the trapezoid, it is given by

$$R_c = r_i + \frac{1}{3}(r_o - r_i)\left(\frac{2t_o + t_i}{t_o + t_i}\right) \tag{a}$$

In the above, the first term on the right-hand side is simply the distance from the center of curvature to the base of the trapezoid. The second term locates the centroid of the trapezoid relative to the base. Substituting the known parameters from the problem geometry gives

$$R_c = 0.100 + \frac{1}{3}(0.130 - 0.100)\left(\frac{2(0.030) + 0.015}{0.030 + 0.015}\right) = 0.116667 \text{ m}$$

We next compute the distance to the neutral axis R_n. In Fig. 11.73, we show a differential slice of the area of the section located a distance r from the center of curvature. The strip has a differential area dA given by $t\,dr$.

Example 11.22 (Continued)

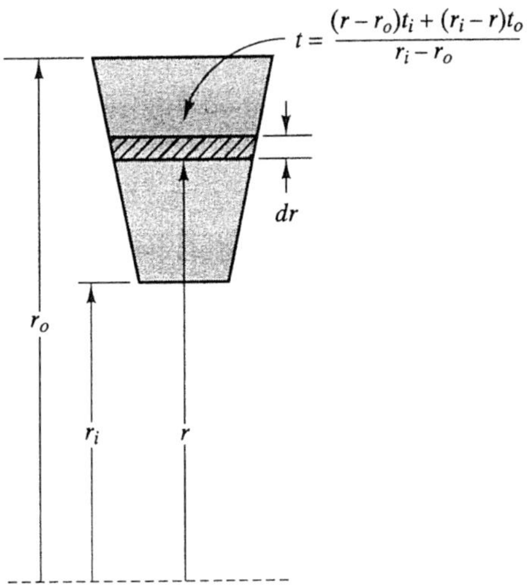

Figure 11.73.

Therefore, we can derive the following:

$$\iint_A \frac{dA'}{r} = \int_{r_i}^{r_o} \frac{(r - r_o)t_i + (r_i - r)t_o}{(r_i - r_o)r}\, dr$$

$$= \frac{1}{r_o - r_i}\left[(r_o - r_i)(t_o - t_i) + (r_o t_i - r_i t_o)\ln\left(\frac{r_o}{r_i}\right)\right] \qquad \text{(b)}$$

For the problem at hand, this is computed to be

$$\iint_A \frac{dA'}{r} = \frac{1}{(0.130 - 0.100)}[(0.130 - 0.100)(0.030 - 0.015) + (0.003 - 0.00195)\ln(1.3)]$$

$$= 5.81725 \times 10^{-4}\, m$$

From Eq. (11.78), we can now compute the distance to the neutral axis as

$$R_n = \frac{A}{\iint_A \frac{dA'}{r}} = \frac{6.75 \times 10^{-4}}{5.817251 \times 10^{-4}} = 0.116034\, m$$

The distance between the centroid and the neutral axis (i.e., the eccentricity) is computed as

$$e = 0.116667 - 0.116034 = 6.33 \times 10^{-4}\, m$$

Example 11.22 (Continued)

Notice that sufficient digits must be carried in the computation of R_c and R_n since the difference of these two near-equal numbers is required. Computing the normal stresses from Eq. (11.85), we get for the *inner surface*

$$\tau_{\theta\theta} = \frac{M_z(R_n - r)}{Are}$$

$$= \frac{100(0.116 - 0.100)}{(6.75 \times 10^{-4})(0.100)(6.33 \times 10^{-4})} = 37.4\ MPa$$

and for the *outer surface*

$$\tau_{\theta\theta} = \frac{100(0.116 - 0.130)}{(6.75 \times 10^{-4})(0.130)(6.33 \times 10^{-4})} = -25.2\ MPa$$

If we compare these stresses to those computed ignoring the curvature, we get

$$(\tau_{\theta\theta})_i = -\frac{My}{I} = -\frac{(100)(0.0167)}{4.875 \times 10^{-8}} = 34.3 MPa$$

$$(\tau_{\theta\theta})_o = -\frac{(100)(-0.0133)}{4.875 \times 10^{-8}} - 27.3 MPa$$

For this case, the results are in error by less than 10%. If the radius of curvature of the beam decreases with respect to the depth of the beam, this error will increase. Note that the general form of the results shown in Eqs. (a) and (b) can be applied to a triangular cross section (by setting either t_i or t_o to zero) or to a rectangle (by setting t_i equal to t_o).

11.16 Closure

In our study of one-dimensional stress problems in an earlier chapter we employed a **constitutive law**, equations of **equilibrium**, and **compatibility of deformation** to solve each particular problem. In the present chapter we used these ingredients completely or partially to develop *theories* that gave us stresses for certain classes of beam problems. The pure bending of elastic beams involved the complete use of the basic laws and resulted in exact results for normal stress. The Euler-Bernoulli or technical theory for beams in which shear stresses are present invoked approximations from the exact theory but nevertheless, when used appropriately, gives good results, particularly for normal stress. The totality of these very useful approximate theories is part of what we call strength of materials in contrast to the theory of elasticity where a more exact approach is maintained.

Now that the stress analysis of a number of classes of beams has been examined, we shall continue our study of strength of materials next by considering the deflection of straight beams in the following chapter.

Highlights for Part A (11)

The two key formulas for the **Euler-Bernoulli** theory for the bending of linear-elastic, symmetric beams are the flexure formula and the radius of curvature formula,

$$\tau_{xx} = -\frac{M_z y}{I_{zz}} \quad \text{and} \quad R = \frac{EI_{zz}}{M_z}$$

This theory gives good results for long slender beams undergoing small deformation. The y coordinate in the flexure formula is measured from the **neutral axis** along whose surface there is no axial extension or compression. The neutral axis coincides with the horizontal centroidal axis of the cross section for symmetric, linear-elastic beams. You must remember to use the proper convention sign for M_z. In complex problems, you will need to sketch the shear and bending moment diagrams to establish the extreme values of M_z and V_y from which the extreme values of normal and shear stresses can be determined.

Another important but less precise formula is the shear formula which for the magnitude of the average values of shear stresses over horizontal and vertical cuts, respectively, are given by:

$$|\overline{\tau}_{xy}| = \left|\frac{V_y Q_z}{I_{zz} b}\right| \quad \text{and} \quad |\overline{\tau}_{xz}| = \left|\frac{V_y Q_z}{I_{zz} t}\right|$$

where Q_z is the first moment about the neutral axis of the area in the section exposed by the cut. Horizontal cuts are used to obtain vertical shear stresses while vertical cuts give you horizontal shear stresses. Remember that these formuli are useful only when the cut intersects the boundaries with an orientation normal to the boundaries. To get the correct sign for the shear stress, we make the following statements:

sign of V
with $dM_z / dx = +V_y$
get sign of dM_z
and from flexure formula get sign of $d\,\tau_{xx}$

Finally, with a simple sketch using the complementary property of shear, we get the proper sign of the shear stress.

A concept that you will again see in your later studies is that of **shear flow,** which is the product of the shear stress in a narrow section times the width of the section.

PROBLEMS

Pure Bending	11.1–11.9
Bending with Shear*	11.10–11.30
Shear from Vertical Cuts	11.31–11.41
Composite Beams	11.42–11.47
Unsymmetric Bending	11.48–11.52
Shear Stress in Beams Having Narrow Open Sections	11.53–11.57
Shear Center	11.58–11.61
Inelastic Behavior	11.62–11.67
Curved Beams	11.68–11.69
Unspecified Section Problems	11.70–11.106
Computer Problems	11.107–11.111
Programming Project 11	

11.1. [11.2] What is the maximum tensile normal stress τ_{xx} in the simply supported beam shown in Fig. P.11.1 resulting only from the end couples? What is the normal stress τ_{xx} resulting from these couples 25 mm from the top of the section? Finally, determine the radius of curvature of the neutral surface. (The material is steel having a modulus of elasticity of 2×10^{11} Pa.)

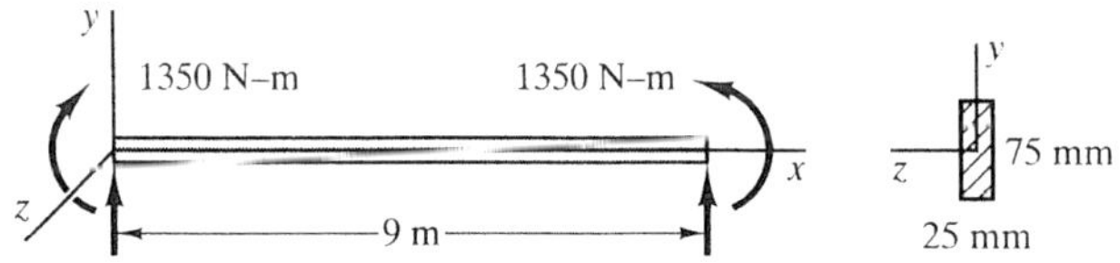

Figure P.11.1.

11.2. [11.2] In Fig. P.11.2 is shown a cantilever beam loaded by equal and opposite 450 N forces. On what portion of the beam is the flexure formula, as developed in Section 11.2, valid? Compute the maximum tensile normal stress τ_{xx} in this region developed by the forces. What is the state of strain at this position? (Take $E = 2 \times 10^{11}$ Pa and $\nu = .30$.)

*Unless requested otherwise, compute only *shear stress magnitude* in this chapter.

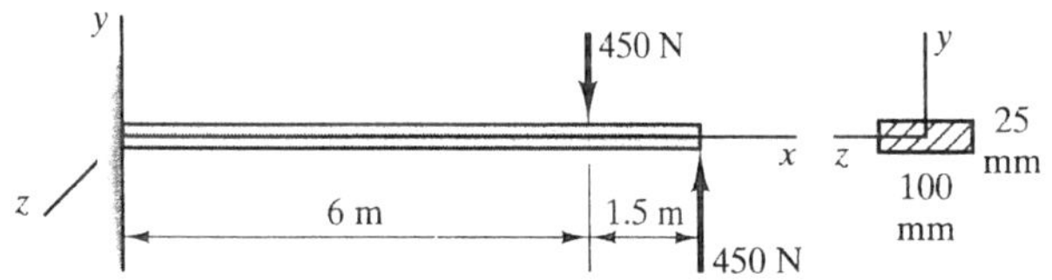

Figure P.11.2.

11.3. [11.2] Consider the cross section shown in Fig. P.11.3 where the upper boundary in the first of four identical quadrants is given as $y^3 = ax + c^3$. What is the flexure formula for pure bending at this section about the x axis?

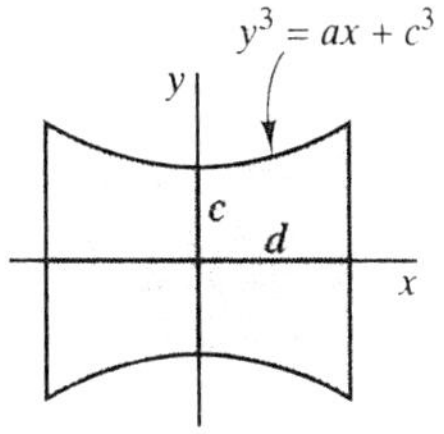

Figure P.11.3.

11.4. [11.2] Compute the maximum tensile normal stress τ_{xx} resulting from the couple acting on the cantilever beam shown in Fig. P.11.4.

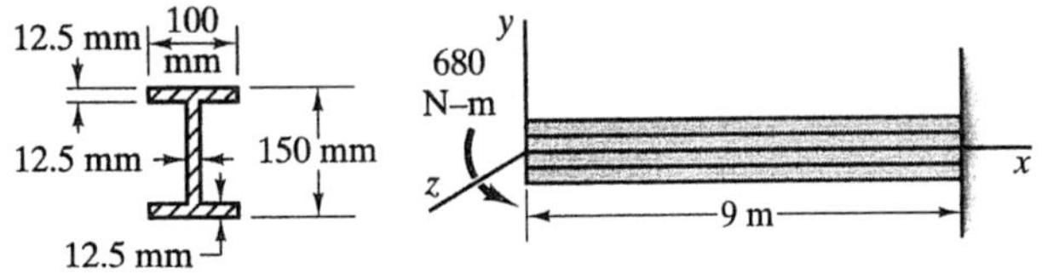

Figure P.11.4.

11.5. [11.2] In structural design, the second moment of area for an **I**-section (see Fig. P.11.5) is often approximated by neglecting the contribution of the web and considering only the two flanges. Is this a conservative approximation? For the typical **I**-shape shown, compare I_{zz} computed exactly and by the approximation.

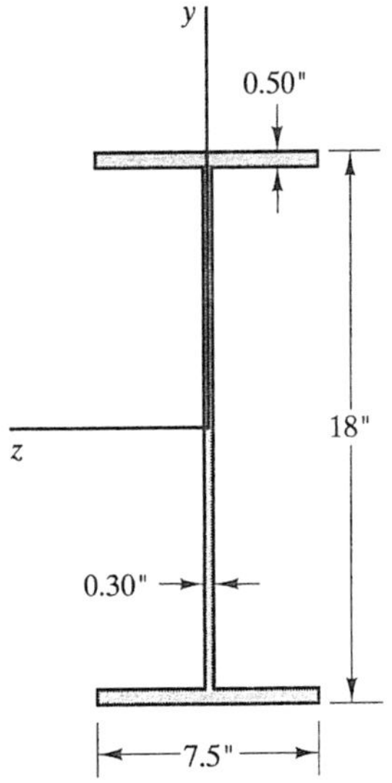

Figure P.11.5.

11.6. [11.2] A cantilever beam (see Fig. P.11.6) is loaded at the tip by couples M_y and M_z having the values 56 N-m and 40 N-m, respectively. What is the maximum tensile normal stress τ_{xx}? Take $E = 2 \times 10^{11}$ Pa. Use physical reasoning to locate maximum stress.

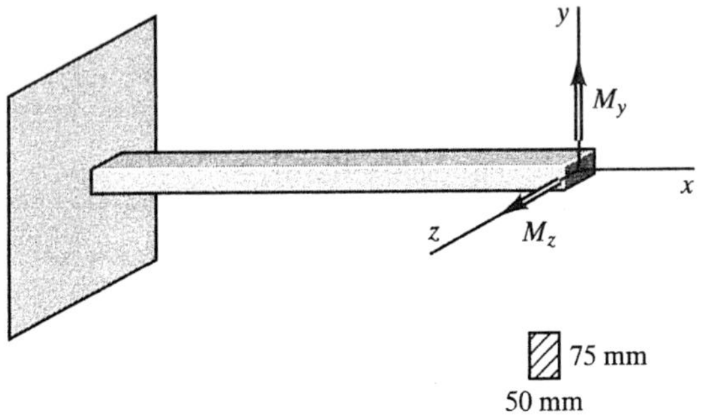

Figure P.11.6.

11.7. [11.2] Shown in Fig. P.11.7 is a cantilever beam loaded by a couple of 135 N-m at the end. Determine the vertical deflection of point A of the beam if the modulus of elasticity is 1.5×10^{11} Pa. [*Hint:* It will simplify computations if you approximate $\cos\theta$ by a two-term power series expansion, $(1 - \theta^2/2)$.]

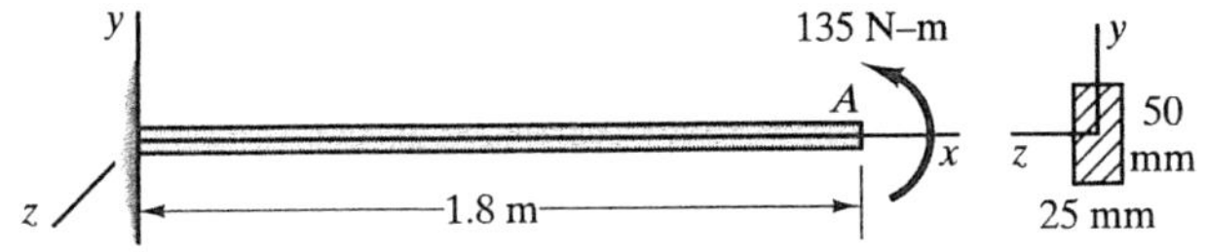

Figure P.11.7.

11.8. [11.2] A 2 in. × 2 in. × 13 ft cantilever beam having an elastic modulus of 30×10^6 psi is shown in Fig. P.11.8. It is loaded by two couples. What is the maximum normal stress 2 ft from the left end? 8 ft from the left end?

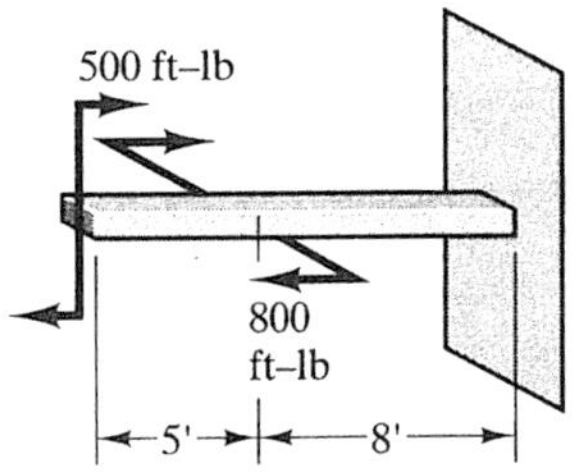

Figure P.11.8.

11.9. [[1.2] Shown in Fig. P.11.9 is a simply supported beam loaded by end couples of 1400 N-m. The beam is made of two separate beams one laid over the other as shown. Assuming they can slide one over the other freely, determine the radius of curvature of the beams. What is the maximum stress resulting from the applied loads?

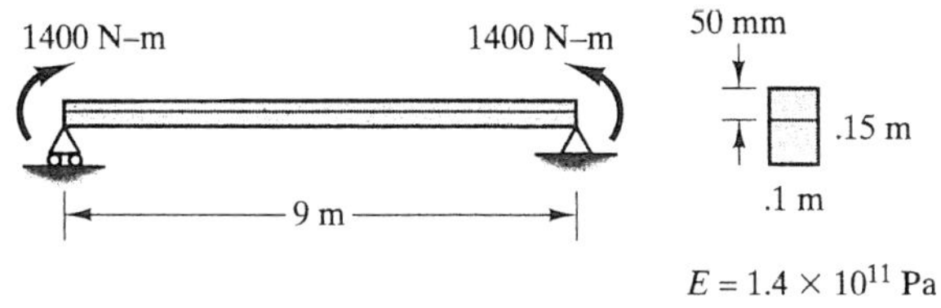

Figure P.11.9.

Unless requested otherwise, only compute shear stress magnitude.

11.10. [11.4] Compute the maximum normal stress τ_{xx} and the maximum shear stress τ_{xy} for the cantilever beam shown in Fig. P.11.7. (Include the weight of the beam 150 N/m.)

11.11. [11.4] Determine the maximum normal stress τ_{xx} and the maximum shear stress τ_{xy} at a section 5 ft from the right support for the simply supported beam shown in Fig. P.11.11.

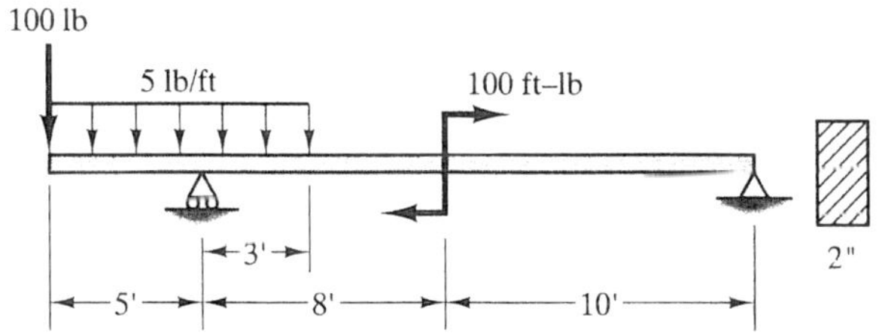

Figure P.11.11.

***11.12. [11.4]** Show that the *pure bending* solution is an *exact* solution in accordance with the theory of elasticity. Thus, first, show that, for no body forces, the equations of equilibrium [Eqs. (6.64)] are satisfied; that Hooke's law [Eqs. (6.65)] is satisfied; that the compatibility equations [Eqs. (3.13)] are satisfied; and, finally, that the boundary conditions within the limits of St.-Venant's principle are satisfied.

11.13. [11.4] Compute the maximum normal tensile stress τ_{xx} and the maximum shear stress τ_{xy} at a section 3 m from the left support for the simply supported beam shown in Fig. P.11.13.

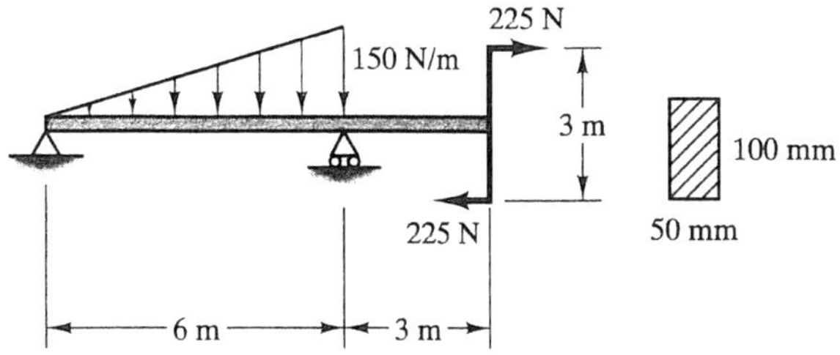

Figure P.11.13.

11.14. [11.4] For a shear force V_y of 45,000 N for the section shown in Fig. P.11.14, what is the maximum shear stress τ_{xy} away from areas of stress concentration (at corners)?

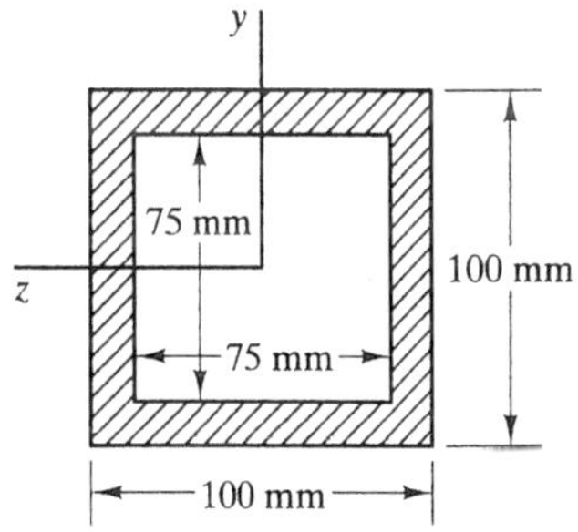

Figure P.11.14.

11.15. [11.4] Compute the maximum tensile normal stress τ_{xx} and the maximum shear stress τ_{xy} for the beam shown in Fig. P.11.2 at $x = 3.75$ m. (Include the weight of the beam, 150 N/m.)

11.16. [11.4] To help in the prevention of cracks in load-bearing masonry walls, no tensile stress should be permitted. Show that if a compressive load P acts along the center third (i.e., $|e| = h/6$) of the principal axes of a rectangular cross section (see Fig. P.11.16), then no tensile stresses will exist.

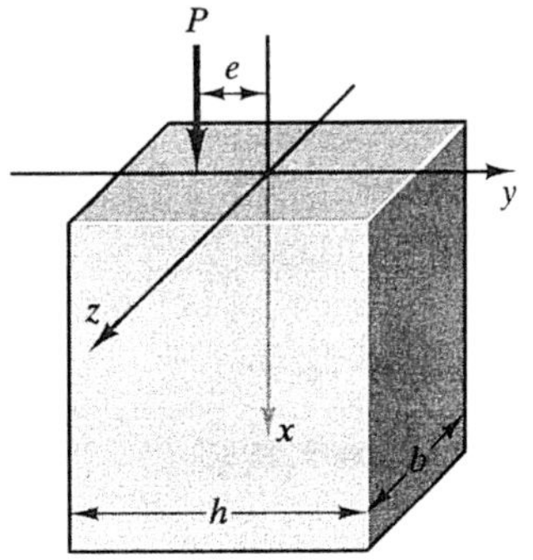

Figure P.11.16.

11.17. [11.4] A gymnast rotates about a horizontal bar at a uniform rate taking 1.5 seconds per revolution. If the center of mass of the gymnast is 1.1 m from the centerline of the bar, and if she has a mass of 65 kg, what is the maximum bending stress in the bar? The length of the bar is 2.5 m. Assume that the gymnast is at the center of the length of the bar and that the bar has a diameter of 40 mm.

11.18. [11.4] What force, positioned at the midspan, will cause the femur bone shown in Fig. P.11.18 to fracture? Assume the bone is simply-supported and that the fracture stress is 250 MPa. Delete the effects of bone marrow and consider that the bone and the marrow cavity are circular.

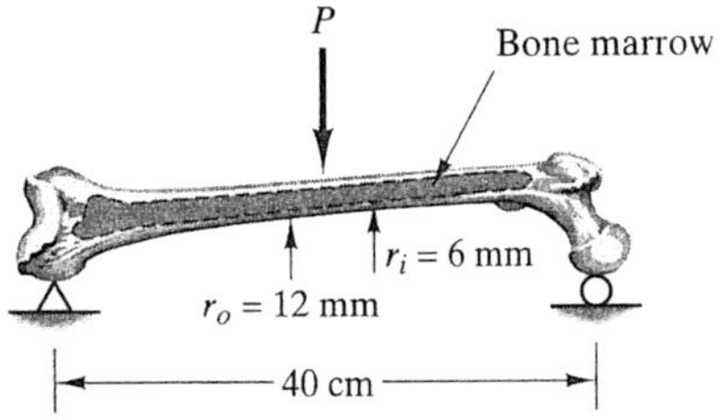

Figure P.11.18.

11.19. [11.4] In Problem 11.11, compute the maximum shear stress τ_{xy} for the entire beam, as well as the maximum normal stress τ_{xx} for the entire beam.

11.20. [11.4] In Problem 11.13, compute the maximum shear stress τ_{xy} for the entire beam as well as the maximum tensile normal stress τ_{xx} for the beam.

11.21. [11.4] Shown in Fig. P.11.21 is a cantilever beam. Compute the maximum normal compressive stress τ_{xx} for the beam and the maximum shear stress τ_{xy} for the beam.

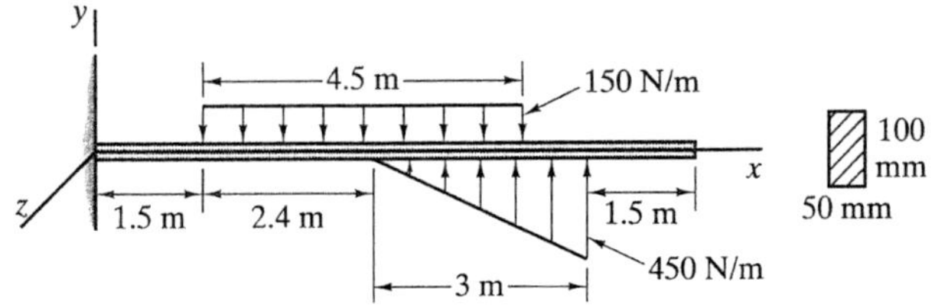

Figure P.11.21.

11.22. [11.4] For the cantilever beam shown in Fig. P.11.22, compute the stresses τ_{xx} and τ_{xy} at a section 1.5 m from the left support and 25 mm from the top edge of the cross section.

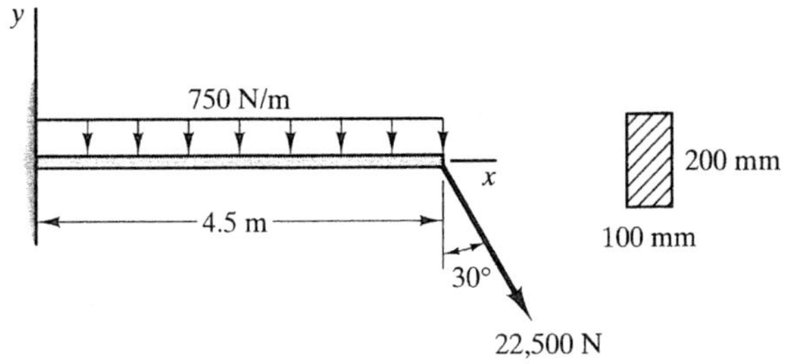

Figure P.11.22.

11.23. [11.4] A beam BD is welded to a simply supported beam AB (see Fig. P.11.23). The yield in tension in each beam is 3.5×10^8 Pa. Find the heights h_1 and h_2 for a safety factor of 1.8 for each beam, disregarding stress concentrations. Beam BED is bent at E.

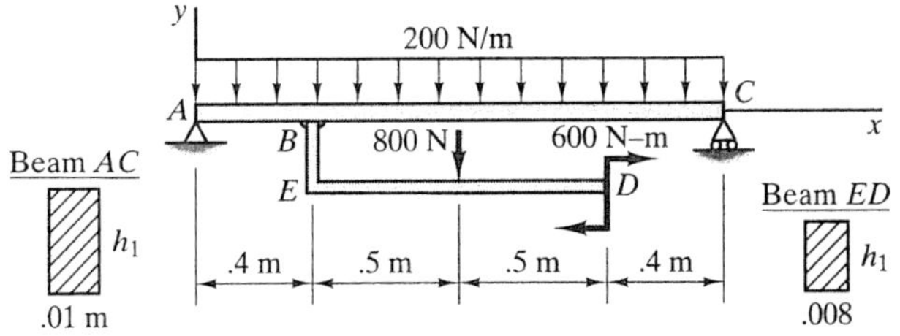

Figure P.11.23.

11.24. [11.4] A material has a tensile working stress of 12,000 psi and a compressive working stress of 7,500 psi. For a trapezoidal cross section (see Fig. P.11.24), what should be the ratio a/b so that the tensile and compressive faces simultaneously reach their respective working stresses? Note that for the neutral axis:

$$\bar{y} = \frac{h}{3}\frac{2a + b}{a + b}$$

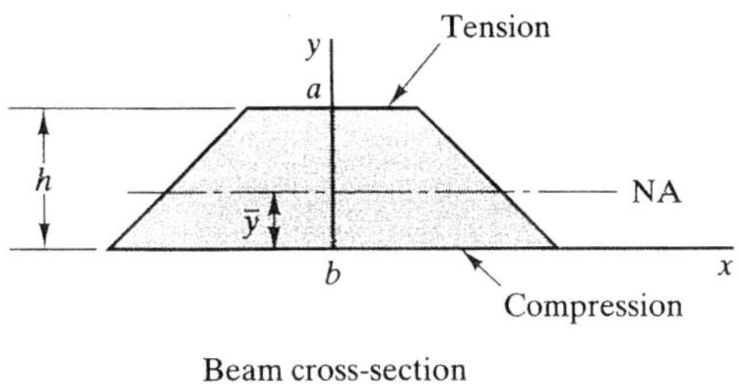

Figure P.11.24.

11.25. [11.4] How could a simple arrangement of strain gages be used to "cancel out" unwanted bending strains in a tensile test of a material?

11.26. [11.4] Two bars are free to slide over each other and carry a uniform load w_0 (see Fig. P.11.26). If $E_1 = 20 \times 10^6$ psi and $E_2 = 30 \times 10^6$ psi, what loading w_0 will cause a maximum tensile bending stress τ_{xx} of 20,000 psi? Where does it occur? (*Hint:* How are the radii of curvature of the beams related?)

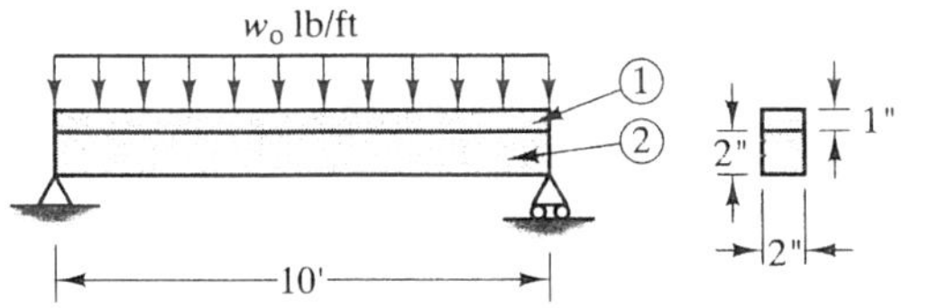

Figure P.11.26.

11.27. [11.4] Three 50 mm × 100 mm wooden planks are glued together (see Fig. P.11.27) and used as a simply supported beam of 3 m to support a uniform load of 750 N/m. What maximum shear stress do you estimate the glue will be subject to for these conditions?

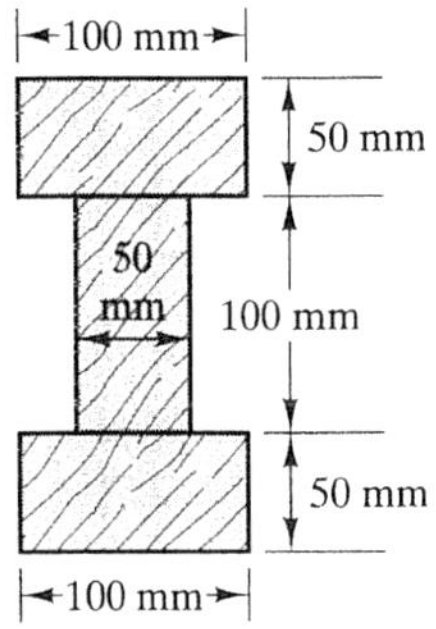

Figure P.11.27.

11.28. [11.4] An American standard I beam 18I70 (I_{zz} = 917.5 in.^{4} and d = 18″) is shown in Fig. P.11.28 suspended from a support. When a load P is applied, a strain gauge oriented vertically on the surface of the I-beam at A registers a strain of 300×10^{-6}. What is the load P? Take $E = 29 \times 10^6$ psi.

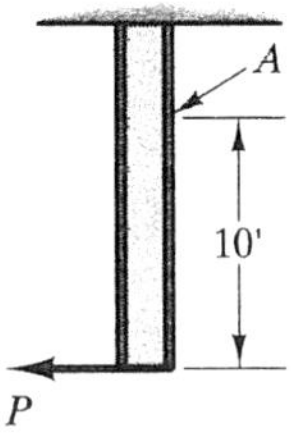

Figure P.11.28.

11.29. [11.4] Two wooden planks are nailed together as shown in Fig. P.11.29(a). If the bending moment for the beam has the diagram shown in Fig. P.11.29(b) and if the nails can handle 900 N each in shear, what should the maximum spacing be for the nails?

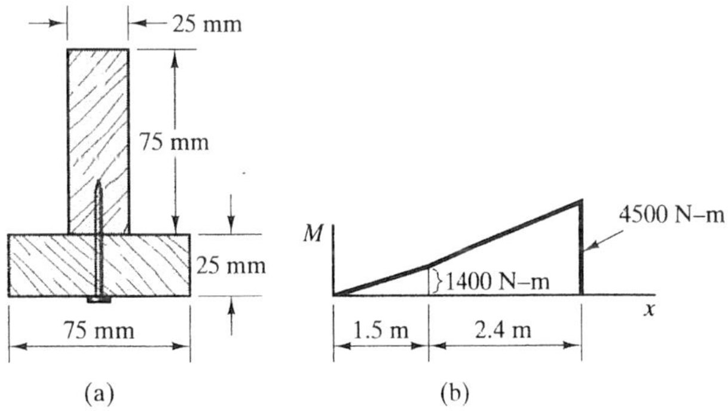

Figure P.11.29.

11.30. [11.4] In Fig. P.11.30,

a. Determine the supporting forces.
b. Determine the maximum tensile stresses at sections A and B.
c. Determine the maximum shear stress magnitudes at sections A and B.

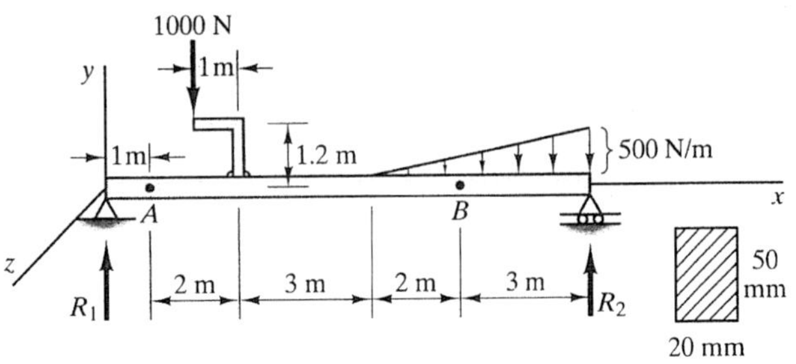

Figure P.11.30.

11.31. [11.6] The cross section of a channel member is shown in Fig. P.11.31. Determine the magnitudes of the horizontal shear stress at A and the vertical shear stress at B for a shear-force magnitude of 10,000 lb in the positive y direction. There is no torsion in this member—only bending.

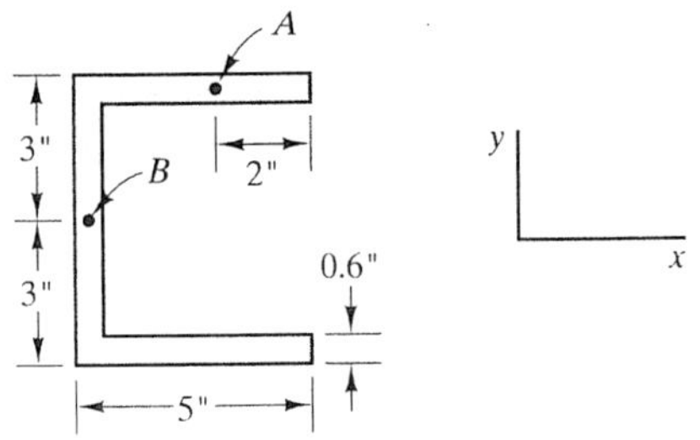

Figure P.11.31.

11.32. [11.6] Shear force $V = 1000$ N goes through centroid C and is oriented along a diagonal of the cross section (see Fig. P.11.32). Determine the approximate total shear stress at A.

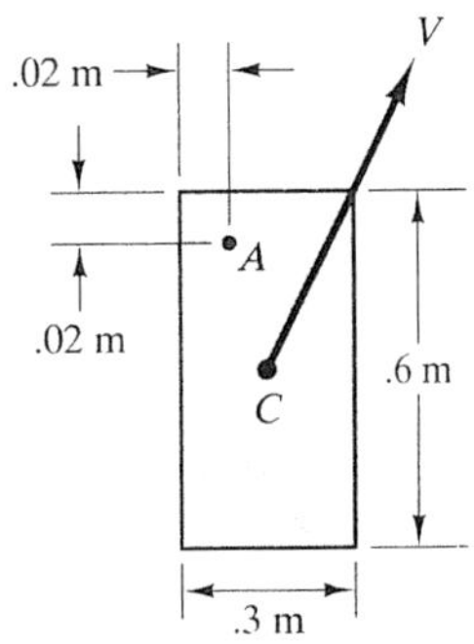

Figure P.11.32.

11.33. [11.6] Find the stress τ_{xx} and the magnitude of the total shear stress at point A in a section shown as B-B in Fig. P.11.33.

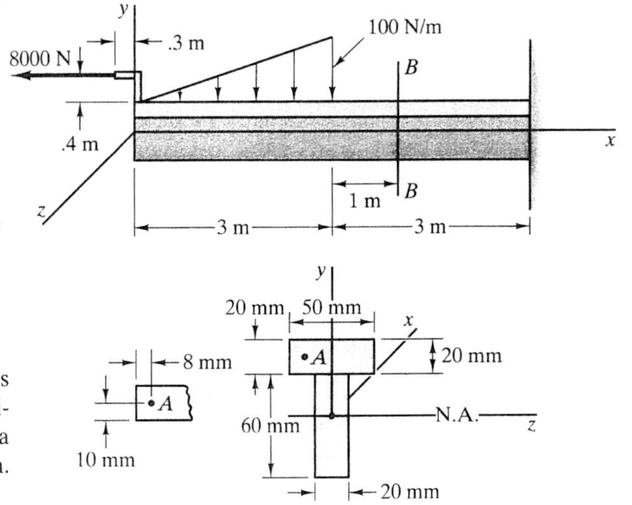

Figure P.11.33.

11.34. [11.6] Find the maximum vertical shear stress in Fig. P.11.34.

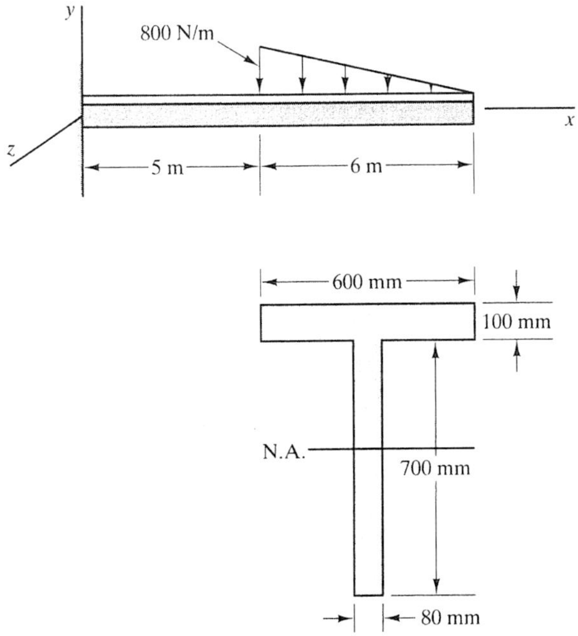

Figure P.11.34.

11.35. [11.6] At point A at the base of the cantilever beam in Fig. P.11.35,

a. Determine the stress τ_{xx}.
b. Compute the total shear-stress magnitude at A in the cross-section of the beam.

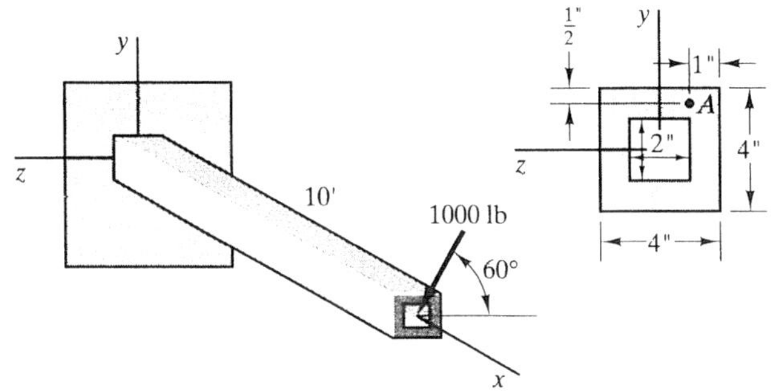

Figure P.11.35.

11.36. [11.6] Shown in Fig. P.11.36 is the positive face of a structural section.

a. Find the magnitude of the total shear at A from $V = 10{,}000$ N.
b. Find τ_{xx} at A for $M = -8000$ N-m.

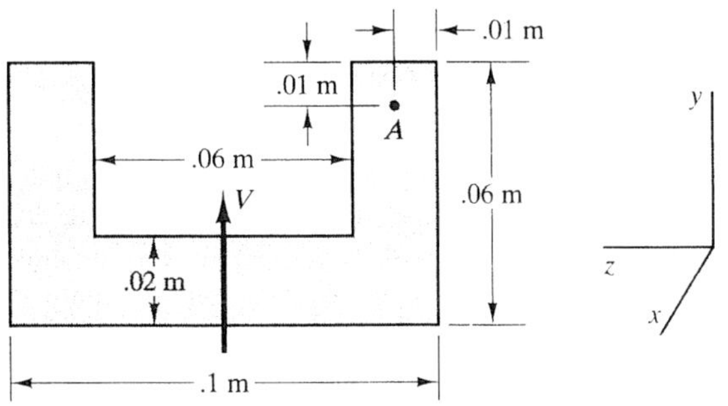

Figure P.11.36.

11.37. [11.6] At $x = 7$ ft for point A in Fig. P.11.37:

a. Find the magnitude of the vertical shear stress.
b. Determine the sign of the vertical shear stress. Use the circled diagram as an aid in establishing this direction.
c. Determine the normal stress at A.

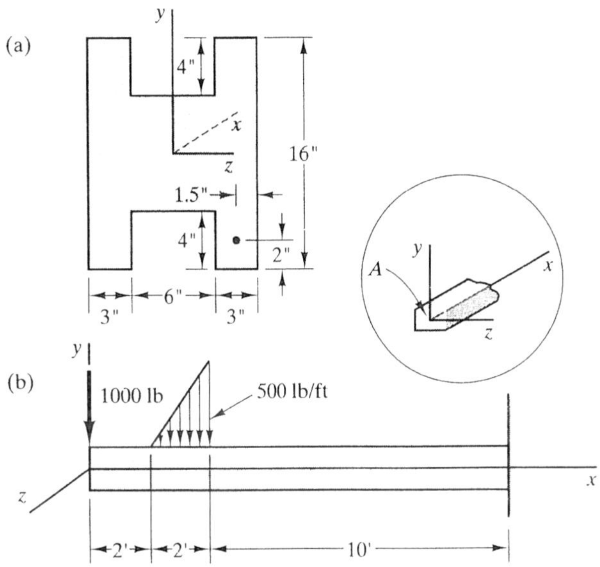

Figure P.11.37.

11.38. [11.6] In Fig. P.11.38:

a. Find maximum tensile normal stress τ_{xx} from the triangular load.
b. Find the maximum $|\bar{\tau}_{xy}|$ (vertical shear) from the triangular load.
c. Find the maximum $|\bar{\tau}_{xz}|$ (horizontal shear) in the upper flange. Disregard stress concentrations at corners and the fact that some corners are rounded.

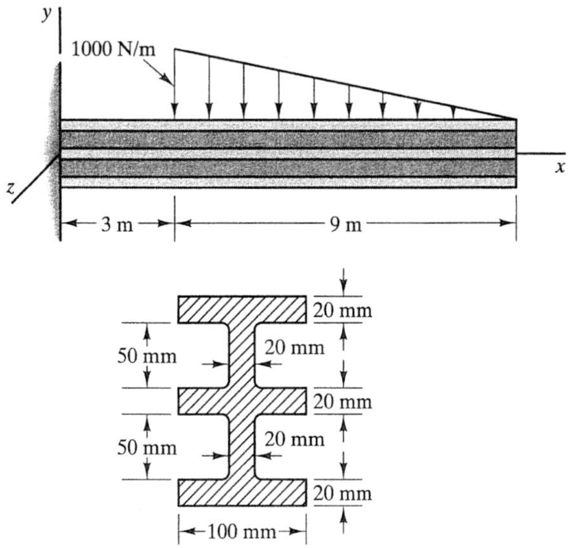

Figure P.11.38.

11.39. [11.6] Estimate the maximum shear stress in the flange of a 16WF78 wide-flange beam supporting a force of 50,000 lb.

11.40. [11.6] A beam is to be made up from two standard channels (see Fig. P.11.40) to support a load of 500 lb/ft each. The yield stress is 102,000 psi and the safety factor is 1. Pick the smallest standard channel for the job. Include the weight of the channels. Use Appendix IV-B.

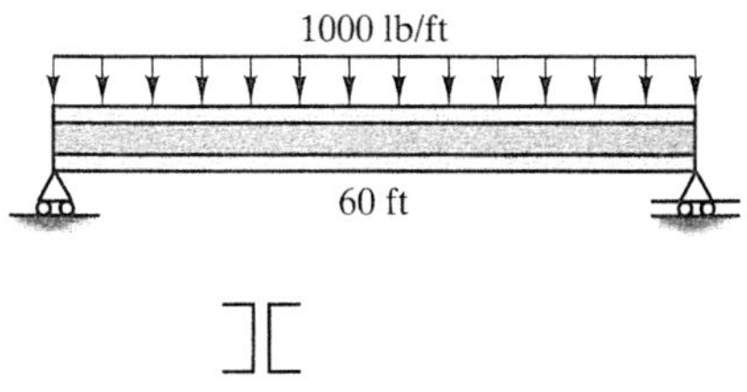

Figure P.11.40.

11.41. [11.6] In Problem 11.40 find the smallest and lightest wide-flange I beam for a yield stress of 80,000 psi and a safety factor of 1.5. The shear stress in the web must not exceed 12,000 psi.

11.42. [11.7] In the cantilever shown in Fig. P.11.42, determine the load P needed to develop a maximum tensile bending stress of 5000 psi in the system. The system is composed of two materials fastened firmly together to form a composite beam. Member A has $E = 8 \times 10^6$ psi and member B has $E = 16 \times 10^6$ psi.

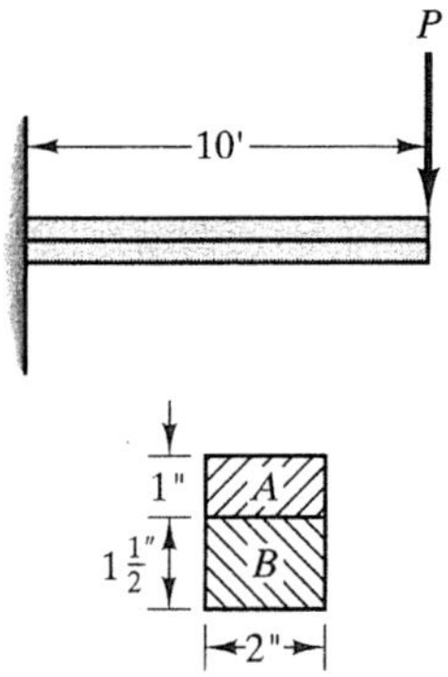

Figure P.11.42.

11.43. [11.7] Shown in Fig. P.11.43 is a composite beam where part B extends only 10 ft along the beam but is firmly fastened to A. If $E_A = 3.5 \times 10^{10}$ Pa and $E_B = 7 \times 10^{10}$ Pa, what is the maximum tensile bending stress in the cross section along the beam? Indicate where this stress exists.

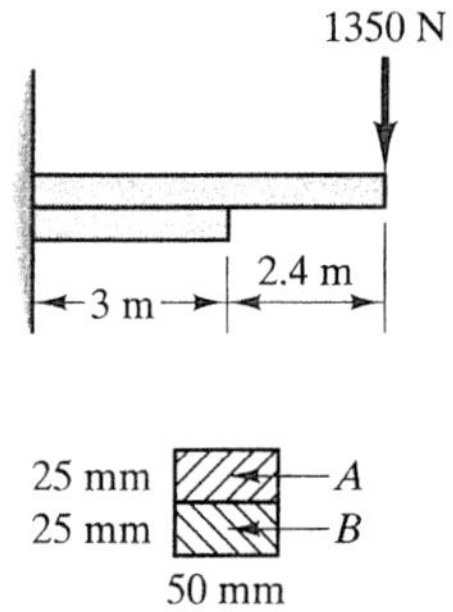

Figure P.11.43.

11.44. [11.7] In Fig. P.11.44 is shown a simply supported steel reinforced concrete beam loaded by a uniform load of 5000 lb/ft. There are four steel rods each of diameter 1 in., as shown in the enlarged cross section in the diagram. Compute the maximum compressive stress in the beam, taking $E = 4 \times 10^6$ psi for concrete in compression and disregarding any resistance from the concrete in tension. (Take $E = 30 \times 10^6$ psi for the steel rods.)

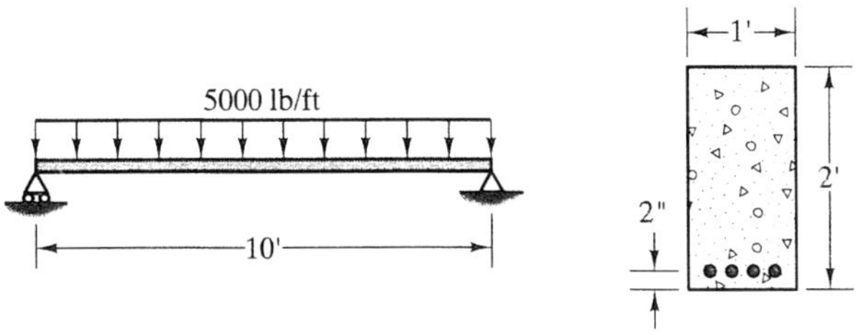

Figure P.11.44.

11.45. [11.7] A composite beam is shown in Fig. P.11.45. Member A has a modulus of elasticity of 1.4×10^{10} Pa and member B has a modulus of elasticity of 2.8×10^{10} Pa. What are the maximum bending stresses in the section both in tension and compression?

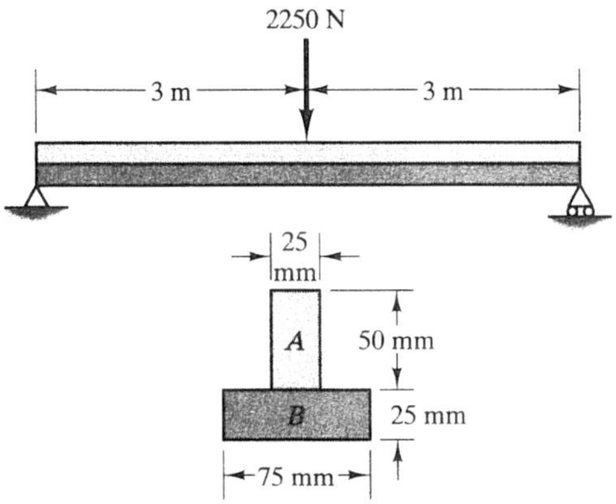

Figure P.11.45.

***11.46. [11.7]** Consider a composite rod which consists of three intimately connected rods and which is suspended vertically as shown in Fig. P.11.46.

(a). As was done in the text for bending of composite beams, start with the fact that normal strains ε_{xx} have to be equal at the interface between the different materials. Hence, ε_{xx} is uniform over the entire section, while stress can be discontinuous. Develop an *equivalent section* made of *one* of the materials such that the same force is transmitted across the equivalent section as across the actual section. Thus, for an equivalent section using material 1, show that

$$(A_{\text{Eq}})_2 = A_2\frac{E_2}{E_1} \tag{1}$$

and the equivalent stress is then

$$\tau_{\text{Eq}} = \frac{F}{A_{\text{Eq}}}$$

where A_{Eq} is the total equivalent cross-sectional area. Also, the actual stresses are

$$(\tau_1)_{\text{Act}} = (\tau_3)_{\text{Act}} = \tau_{\text{Eq}} \qquad (\tau_2)_{\text{Act}} = \frac{E_2}{E_1}\tau_{\text{Eq}} \tag{2}$$

(b). Neglecting the weight of the member, compute the tensile stresses developed by the load $F = 100{,}000$ N at section A-A. Use $E_1 = E_3 = 7 \times 10^{10}$ Pa and $E_2 = 1 \times 10^{11}$ Pa.

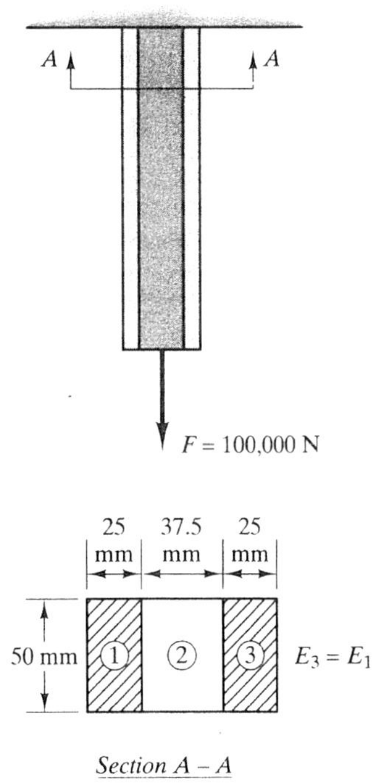

Figure P.11.46.

11.47. [11.7] The composite rod shown in Fig. P.11.47 is made of two metals. Compute the normal stresses in the cross section. See Problem 11.46 before doing this problem.

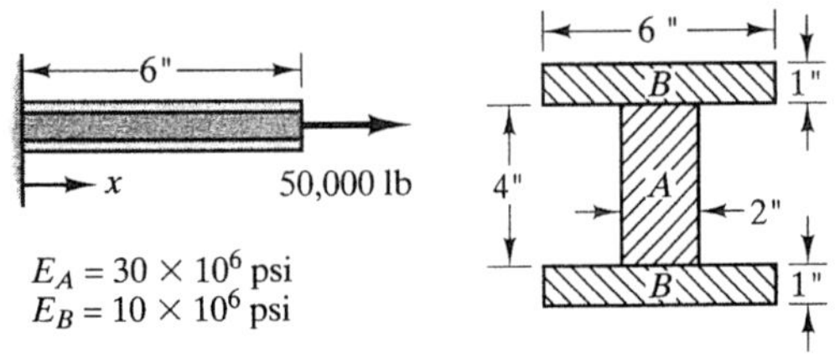

Figure P.11.47.

11.48. [11.8] In Fig. P.11.48 is shown a cantilever beam acted on by a force given as

$$P = 100\mathbf{i} - 50\mathbf{j} + 50\mathbf{k} \text{ lb}$$

What is the stress τ_{xx} at position

$$\mathbf{r} = 1.5\,\mathbf{j} + \tfrac{1}{2}\mathbf{k} \text{ in.}$$

What is the shear stress τ at this position? Give the direction of this shear stress. Get τ_{xx} by a formal procedure and by a "commonsense" approach.

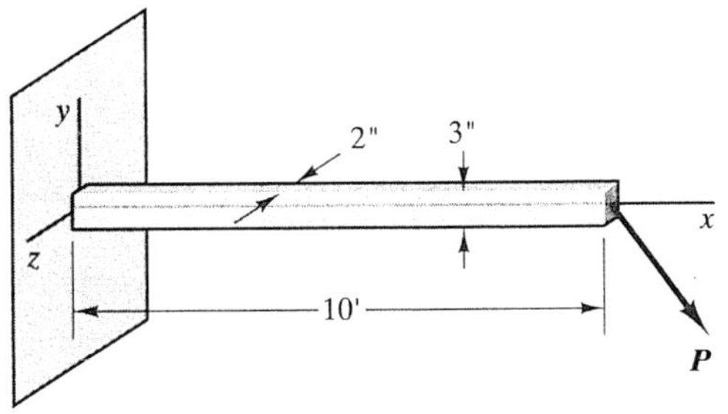

Figure P.11.48.

11.49. [11.8] In Fig. P.11.49,

a. Find the total normal stress τ_{xx} at the base section at A. Do formally and then via common sense.
b. Find the total shear stress in the base cross section at A.

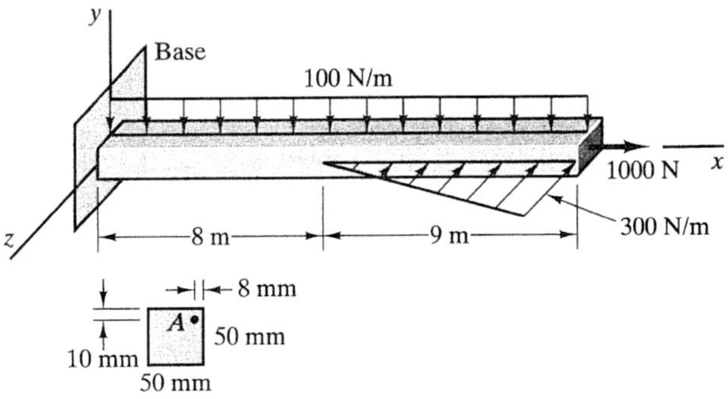

Figure P.11.49.

11.50. [11.8] At the base section of the cantilever beam in Fig. P.11.50:

a. Find τ_{xx} at A. Do formally and then via common sense.
b. Find the total shear stress at A in the yz plane.

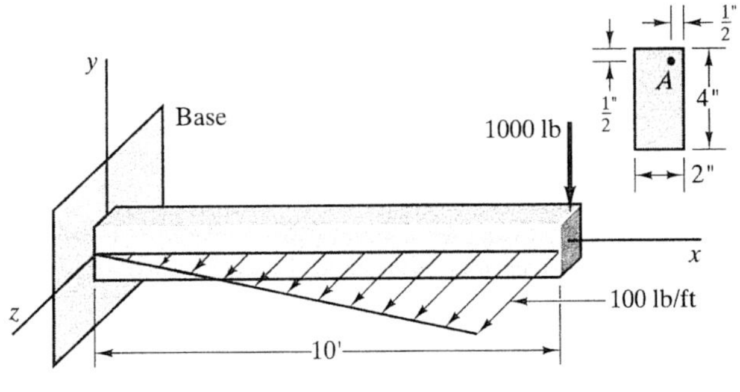

Figure P.11.50.

11.51. [11.8] In Fig. P.11.51:

a. Find maximum tensile stress τ_{xx}. Do formally and then via common sense.
b. Find maximum stress $\bar{\tau}_{xy}$.

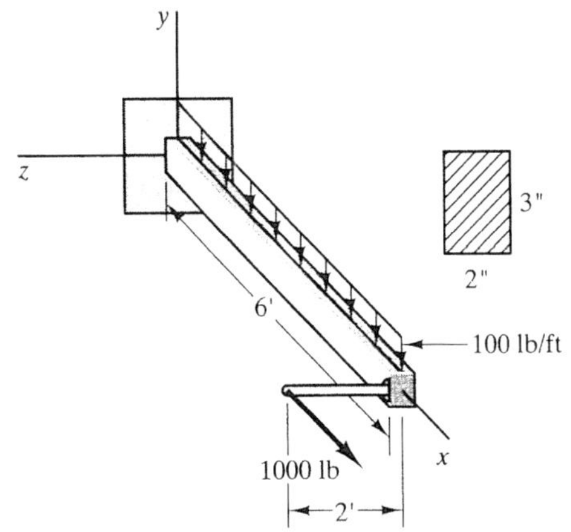

Figure P.11.51.

11.52. [11.8] Find the maximum tensile stress τ_{xx} and the extreme vertical and horizontal shear stresses on the cross section at the midspan of the beam shown in Fig. P.11.52 supported at the end with ball joints.

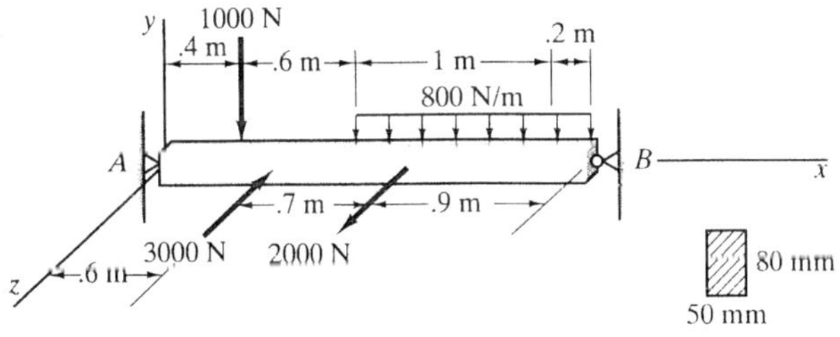

Figure P.11.52.

11.53. [11.9] For a shear force $V_y = 10{,}000$ lb, compute the shear stress in the top flange (see Fig. P.11.53) and at a position A in the center of the vertical web. Assume that the loading is such as to cause no twisting. The following additional data are given:

$$I_{zz} = 2.74 \text{ in.}^4$$
$$I_{yy} = 2.89 \text{ in.}^4$$
$$I_{zy} = 1.25 \text{ in.}^4$$

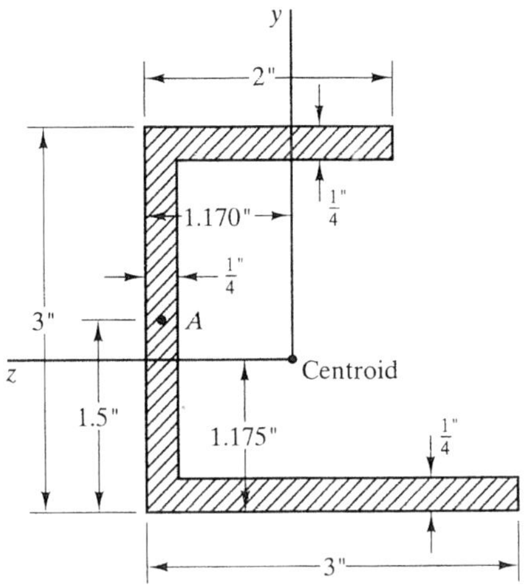

Figure P.11.53.

11.54. [11.9] Find the shear stress at position A and at position B as shown in Fig. P.11.54 for a shear force $V_y = 10{,}000$ lb. The following data are given:

$$I_{zz} = 2.509 \text{ in.}^4$$
$$I_{yy} = 5.16 \text{ in.}^4$$
$$I_{zy} = -.566 \text{ in.}^4$$

The shear loading is at a position such that there is no twisting.

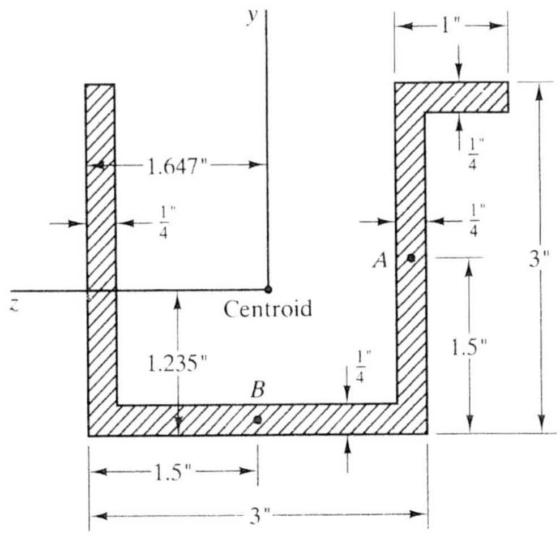

Figure P.11.54.

11.55. [11.9] What is the bending shear-stress distribution in the upper flange for the z-shaped section shown in Fig. P.11.55 for a shear force V_y? Take $a = 100$ mm, $b = 125$ mm, and $t = .005$ m.

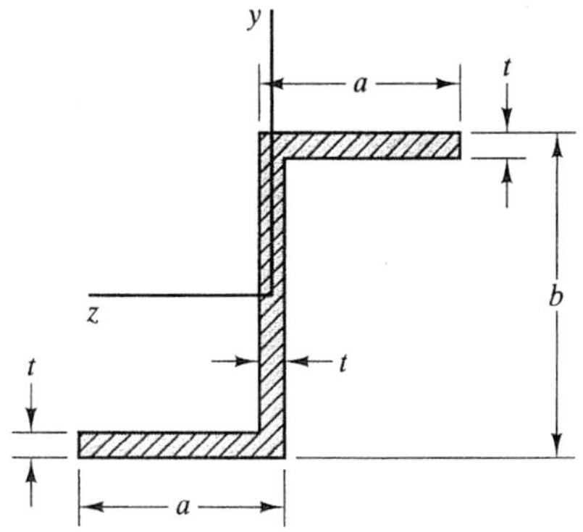

Figure P.11.55.

11.56. [11.9] Solve for the bending shear-stress distribution in the upper flange of the z-shaped member of Problem 11.55 for a shear force V_z.

11.57. [11.9] What is the bending shear-stress distribution in the section shown in Fig. P.11.57 for V_y? Use θ as the independent variable. Take $t \ll R$.

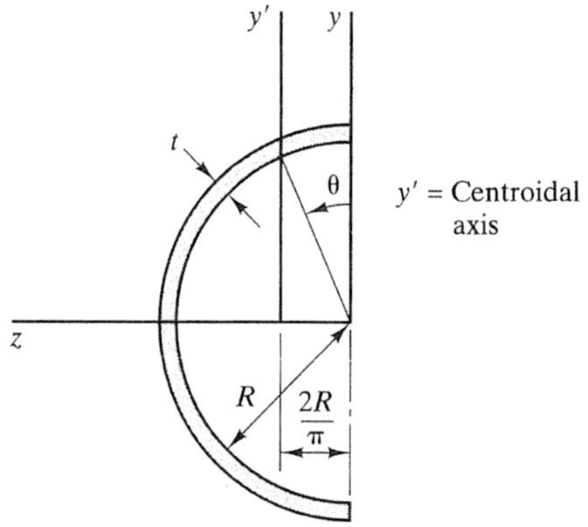

Figure P.11.57.

11.58. [11.10] By inspection, indicate where the shear center should be for the cross section shown in Fig. P.11.58. Explain.

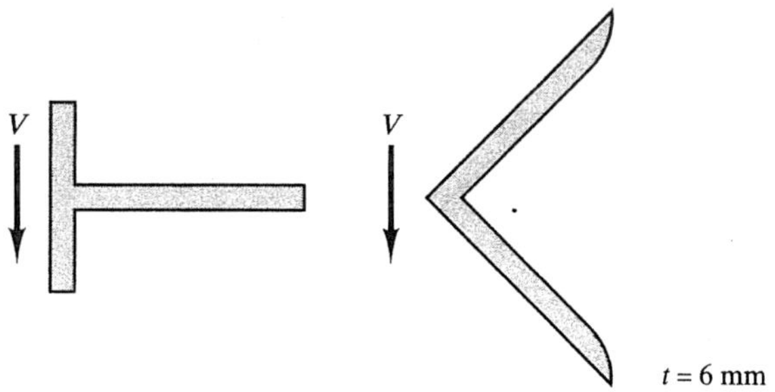

Figure P.11.58.

11.59. [11.10] In Problem 11.53, at what position from the left side should the applied external shear load proceed for no twisting if it is in the y direction?

11.60. [11.10] Find the position of the shear center for the section shown in Fig. P.11.60. Note $I_{yy} = 1.902 \times 10^{-6}$ m^4, $I_{zz} = 1.526 \times 10^{-5}$ m^4.

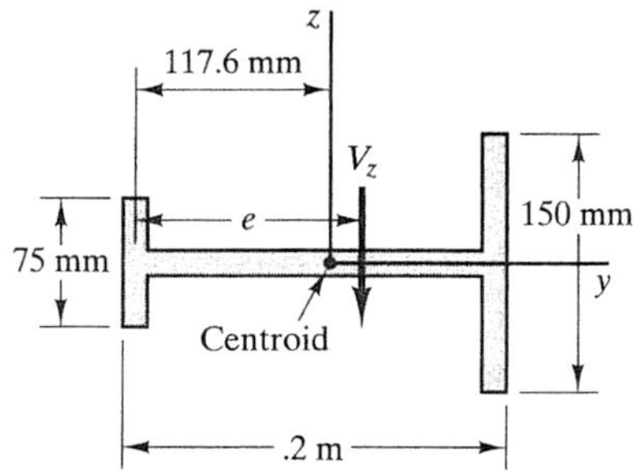

Figure P.11.60.

11.61. [11.10] Show that the shear center for the cross section in Fig. P.11.61 has a value e given as $4R/\pi$.

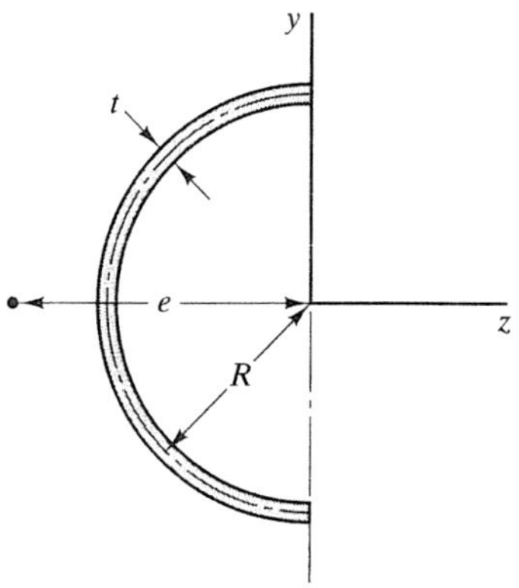

Figure P.11.61.

11.62. [11.11] What is the penetration of plastic behavior in an elastic, perfectly plastic beam if the bending moment is three-fourths of the plastic-hinge moment for the beam? The beam is rectangular having a width $b = 50$ mm and $h = 100$ mm. Take $E = 2 \times 10^{11}$ Pa and $Y = 4 \times 10^8$ Pa. What is the radius of curvature of the beam at this point?

11.63. [11.11] What is the maximum loading w_0 (see Fig. P.11.63) for elastic behavior everywhere in the beam? If this loading is increased 25%, what is the maximum radius of curvature in the beam? Take $E = 20 \times 10^6$ psi and $Y = 60{,}000$ psi.

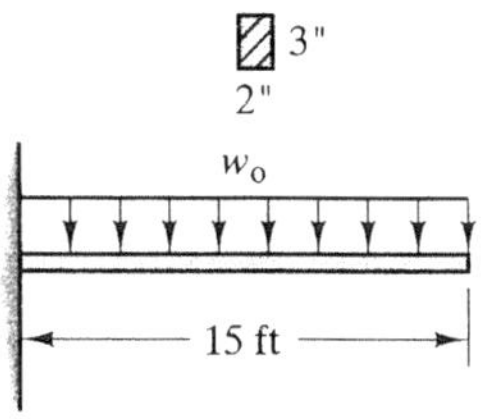

Figure P.11.63.

11.64. [11.11] If the yield stress is 4×10^8 Pa for the cantilever beam shown in Fig. P.11.64, what is the load P required to cause the formation of a plastic-hinge moment?

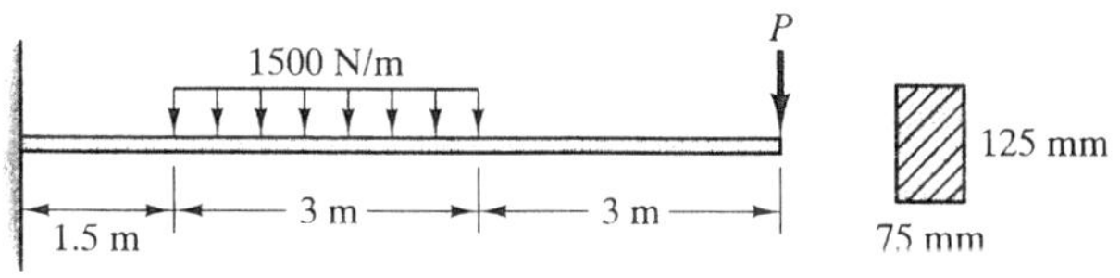

Figure P.11.64.

11.65. [11.12] What is the maximum load P that can be put on the simply-supported beam shown in Fig. P.11.65 if the yield stress is 3.5×10^8 Pa?

a. From a limit-design criterion.
b. From a perfectly elastic behavior criterion.

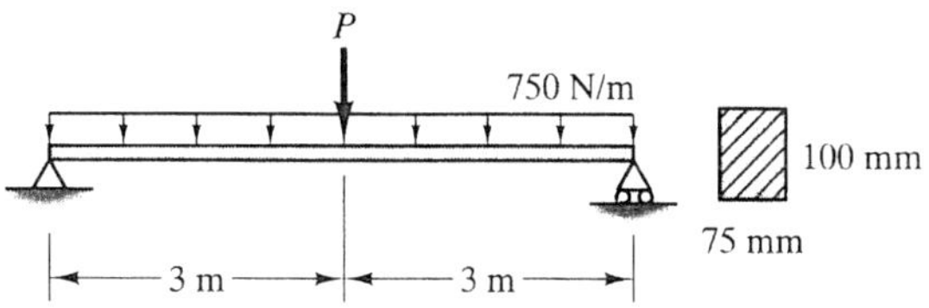

Figure P.11.65.

11.66. [11.12] A bent beam having a 25 mm × 25 mm section and a modulus of elasticity of 2×10^{11} Pa is shown loaded in Fig. P.11.66. What is the maximum value of w_0 for a yield stress of 6×10^8 Pa from a limit-design criterion and from a perfectly elastic behavior criterion?

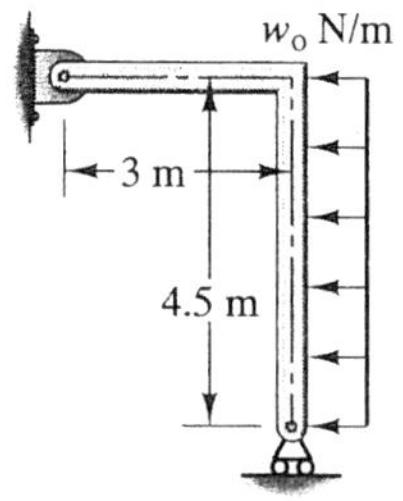

Figure P.11.66.

11.67. [11.12] Shown in Fig. P.11.67 is a cantilever beam with a triangular loading distribution. For a yield stress of 4×10^8 Pa and a modulus of elasticity of 2×10^{11} Pa, what is the maximum loading w_0 for an elastic behavior criterion and for a limit-design criterion? Neglect the effects of the weight of the beam.

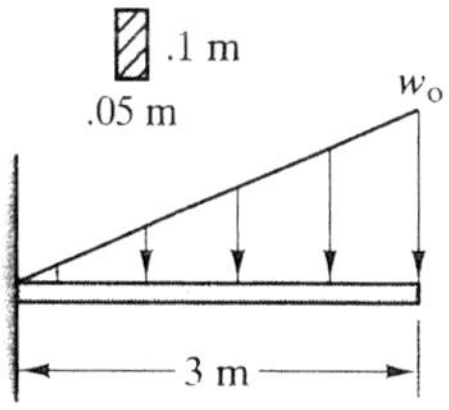

Figure P.11.67.

11.68. [11.15] What is the maximum stress in the section A-A from the load $P = 1000$ lb (see Fig. P.11.68)? Use curved beam theory and compare results with that from straight beam theory.

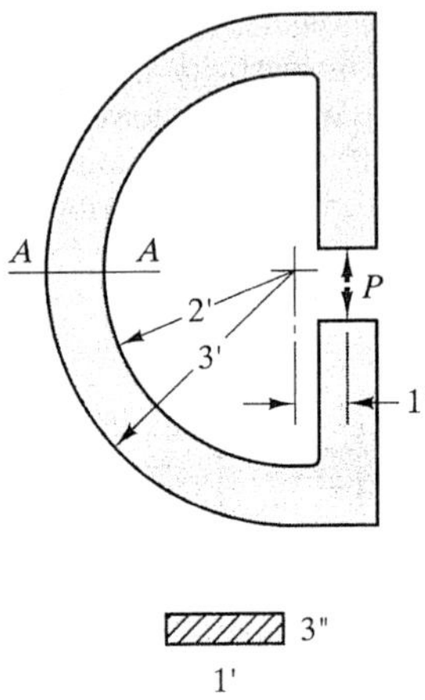

Figure P.11.68.

11.69. [11.15] What is the maximum tensile stress in the hook section *A-A* shown in Fig. P.11.69? Take $P = 2250$ N. Compare the result with that taken using straight beam theory.

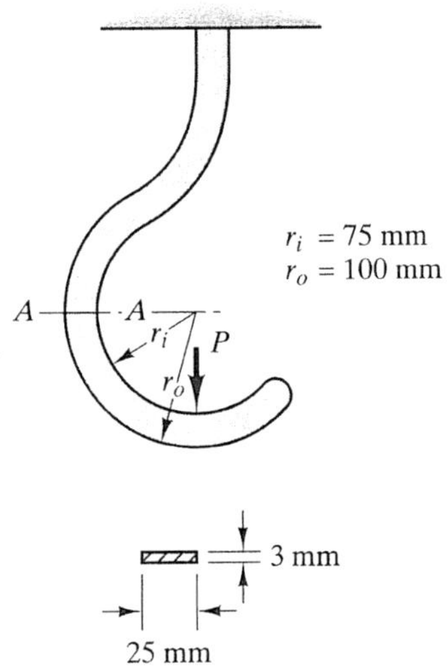

Figure P.11.69.

11.70. For the beam shown in Fig. P.11.70, determine the orientation of the neutral axis. (*Hint:* First show using Eq. (11.37a) that

$$\frac{y}{z} = M_y I_{zz} \,/\, M_z I_{yy}$$

then solve for the angle that the neutral axis makes with the z-axis.)

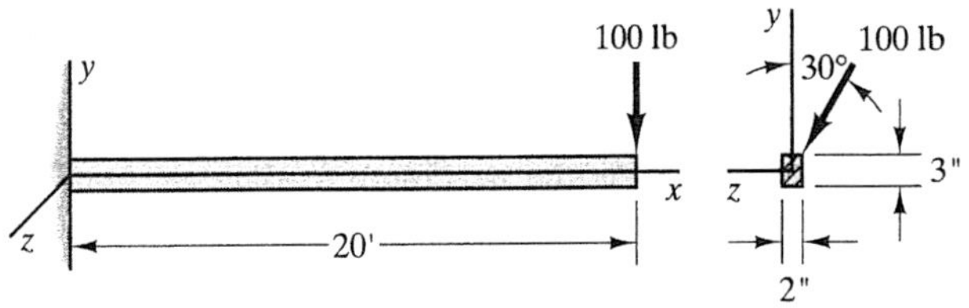

Figure P.11.70.

11.71. In Fig. P.11.71,

a. Give the normal stress at point A at a section of the beam at $x = 4m$.
b. Give the magnitude of the vertical shear stress at point A.
c. Give the proper sign of the shear stress at point A. Do **not** do this intuitively. Start with the shear force and proceed from there.

Note that the cross-section shown is as seen looking from left to right along the x axis.

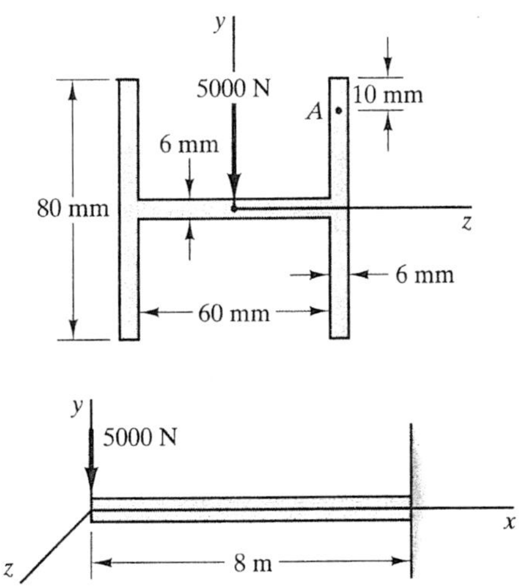

Figure P.11.71.

11.72. [11.4] A structure is shown in Fig P.11.72 involving a 3-in. standard steel pipe (see Appendix IV-D). Member CB is pinned to AB at B and is pinned to member CD, which is, in turn, pinned to a welded attachment to member AB. When a horizontal force P is applied at C a strain gauge at position G reads 200×10^{-6} positive strain. What is the force P if E for all members is 30×10^6 psi? Neglect weights.

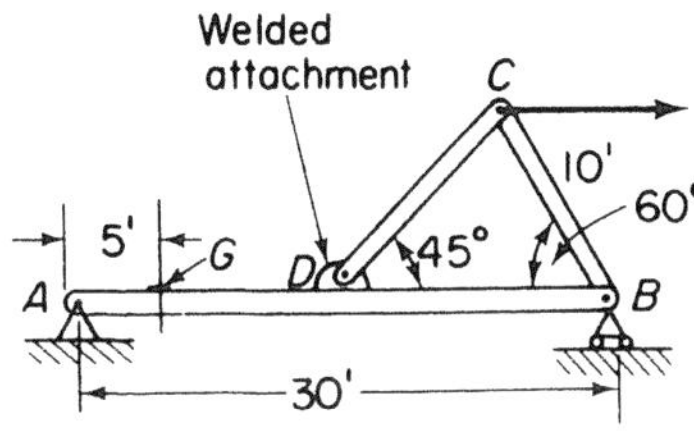

Figure P.11.72.

11.73. A torque arm for an automobile is shown in Fig. P.11.73. At section AA, compute the maximum bending stress. If the yield stress of the material is 500 MPa, does it fail in bending?

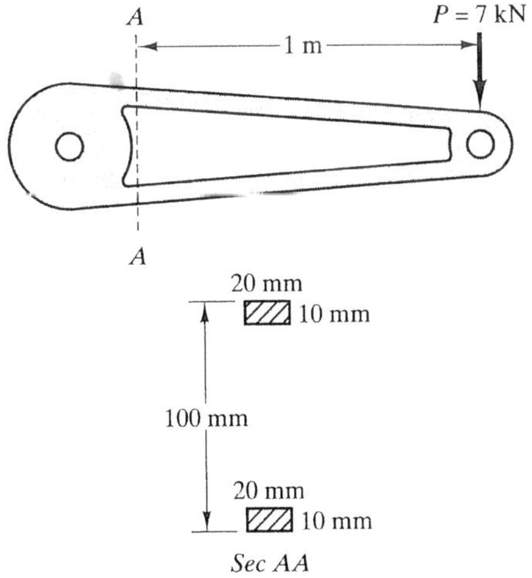

Figure P.11.73.

11.74. Shown in Fig. P.11.74 is a wide-flanged I-beam cantilevered at C.

(a). Find the maximum stress τ_{xx} in the cross section. Do formally and then via common sense.

(b). Find the shear stress at point A 3 in. from the top.

(c). What is the extreme shear stress at A? Use Mohr's circle. A is at the cross section at the base.

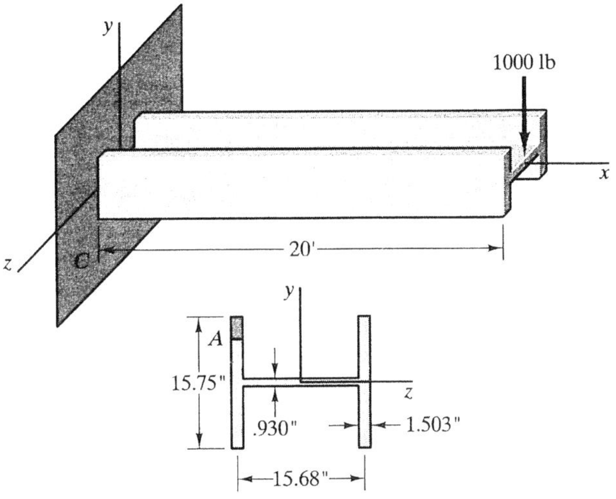

Figure P.11.74.

11.75. Find the maximum stress at A in Fig. P.11.75.

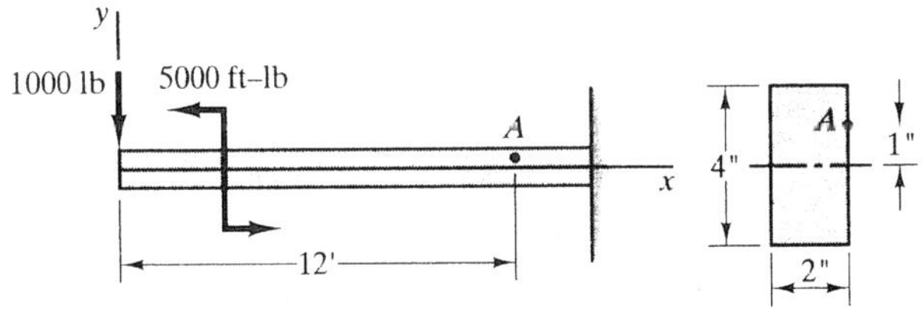

Figure P.11.75.

11.76. [11.4] An 8-in. WF40 wide-flange beam has steel plates of $\frac{1}{2}$ in. in thickness, as shown in Fig. P.11.76. The rivets are $\frac{3}{4}$ in. in diameter and can stand a shearing stress of 8000 psi. If the member is used as a cantilever beam to support a load of 100,000 lb at the tip and has a length of 10 ft, how many equally spaced rivets should be used?

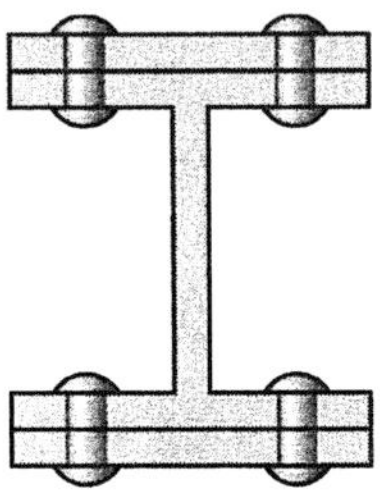

Figure P.11.76.

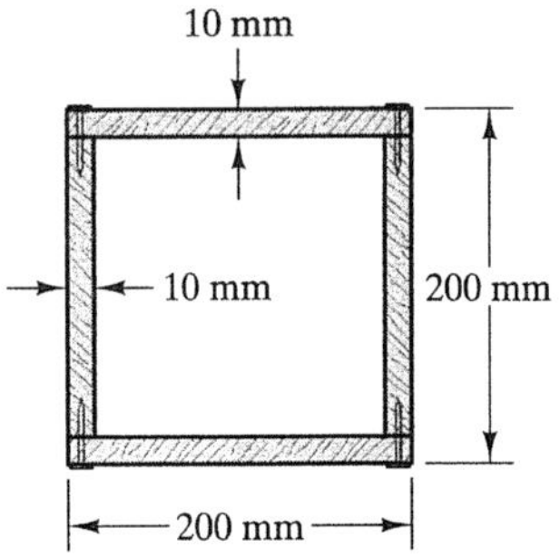

Figure P.11.77.

11.77. [11.4] For a shear force of 2250 N, what should the spacing be for nails shown in the wooden built-up beam shown in Fig. P.11.77? Each nail carries a shear force of 225 N.

11.78.

(a). Compute the stress τ_{xx} at A at the base of the cantilever in Fig. P.11.78 from bending about y and z axes. Do formally and then via common sense.

(b). Compute the total shear stress τ in the cross section at point A. Consider V_y and V_z.

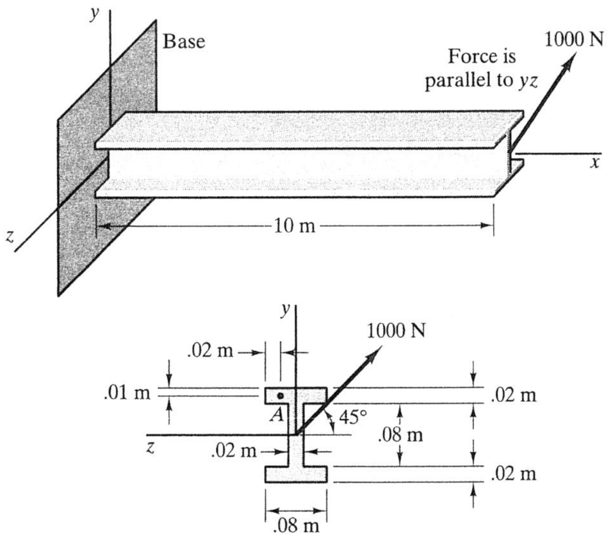

Figure P.11.78.

11.79. A square beam is held vertically by ball joints (see Fig. P.11.79). It weighs 2000 N/m. What is the maximum compressive stress and the extreme shear stress on a section .3 m from A?

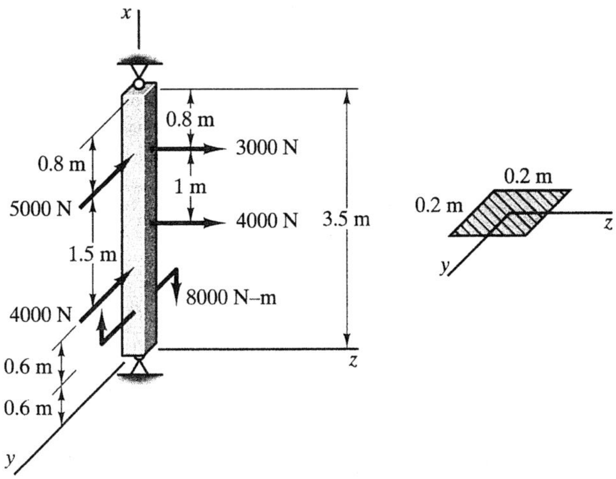

Figure P.11.79.

11.80. Split-rings are commonly used as retainers on shafts. Using symmetry, plot the moment diagram for the split-ring and load shown in Fig. P.11.80. Is the stress constant for a prismatic split-ring cross section? How could the thickness t be varied as a function of θ to give a uniform bending stress throughout the ring? The width w of the split-ring is constant.

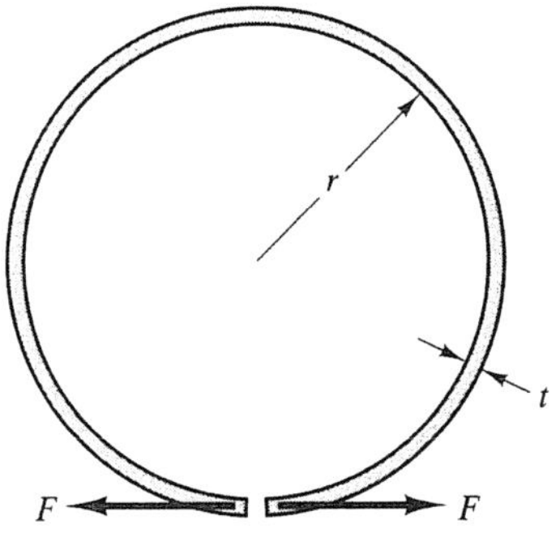

Figure P.11.80.

11.81. Shown in Fig. P.11.81 is a composite beam. Determine the neutral axis for this case. Determine the maximum normal stress.

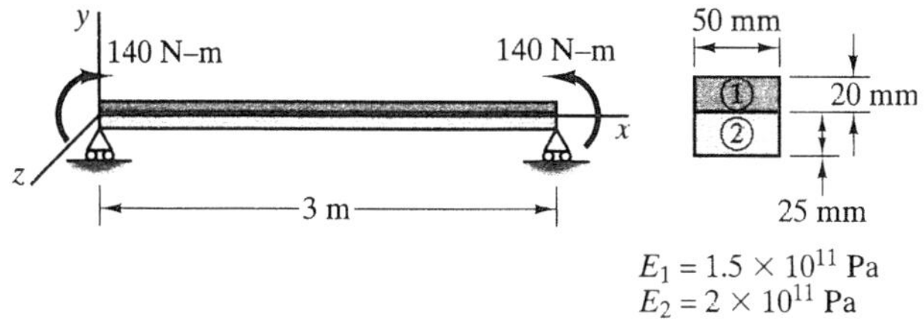

Figure P.11.81.

11.82. Shown in Fig. P.11.82 is a simply-supported beam under the action of pure couples. The beam is made of two materials that are entirely joined together. What is the maximum stress in each material of the beam?

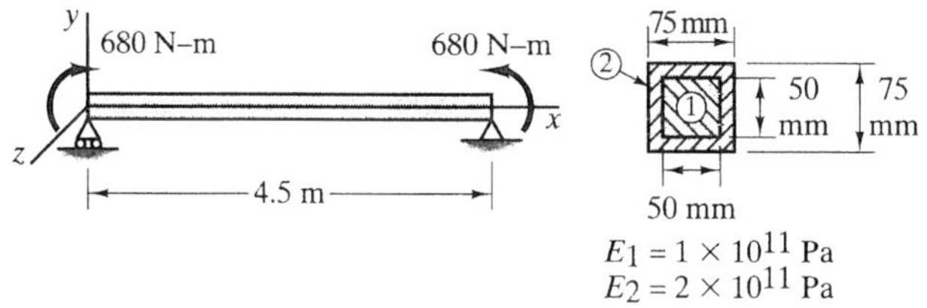

Figure P.11.82.

11.83. Determine the required force in an internal prestressing steel cable placed eccentrically as shown in Fig. P.11.83 so as to eliminate tensile stresses on the bottom face of the beam. The eccentricity e is 5 in. The beam is 24 in. deep by 12 in. wide.

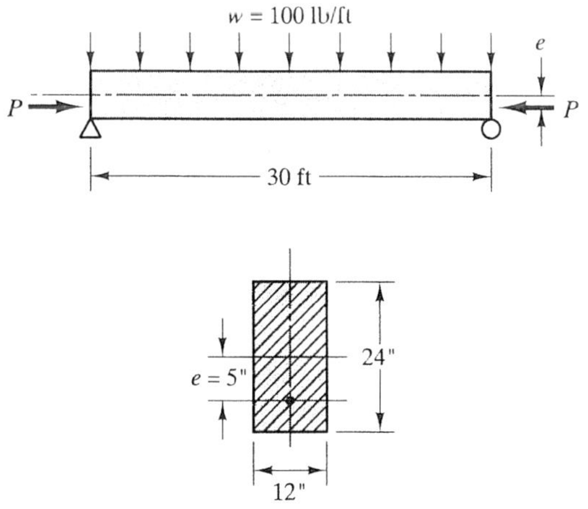

Figure P.11.83.

11.84. Consider a structural cross section in Fig. P.11.84:

(a.) For a vertical shear force on a positive face of 10,000 N, what is the magnitude of the total shear stress at A?

(b). For a positive bending moment of 80,000 N-m, what is the normal stress at A?

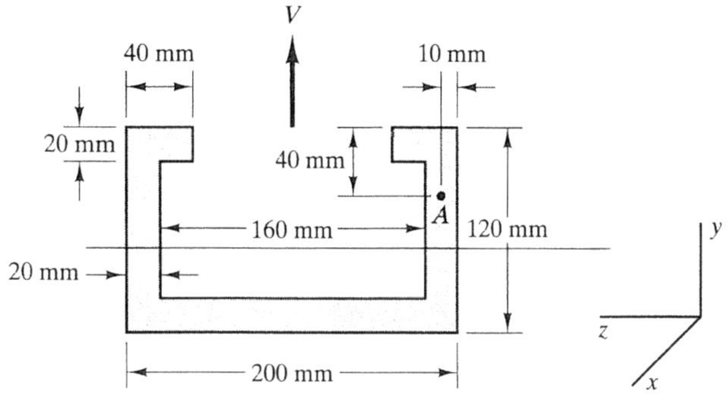

Figure P.11.84.

11.85. Choose a wide-flanged I-beam from Appendix IV–A for beam AB in Fig. P.11.85. The yield stress is 40,000 psi. Use a safety factor of 2. Disregard stress concentrations.

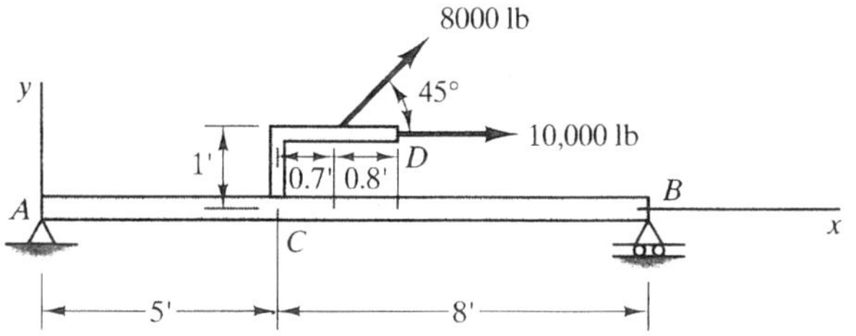

Figure P.11.85.

11.86. Shown in Fig. P.11.86 is a C clamp. If a force P of 200 lb is developed by the clamp, what is the maximum stress at section M-M?

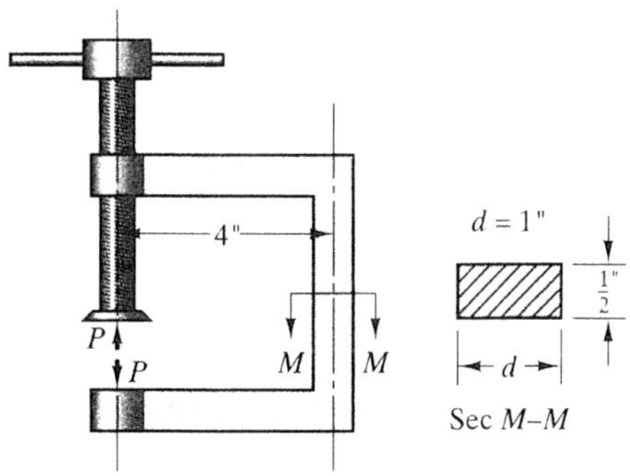

Figure P.11.86.

11.87. If the maximum force P to be expected on a C clamp of the type shown in Fig. P.11.86 is 400 lb and the allowable normal stress at section M-M is not to exceed 20,000 psi, determine the minimum length d required for this purpose.

11.88. Two bars are connected to a solid base but otherwise are free to slide over each other with negligible resistance (see Fig. P.11.88). If $E_A = 2 \times 10^{11}$ Pa and $E_B = 7 \times 10^{10}$ Pa, what is the maximum tensile stress τ_{xx} in each bar? (*Hint:* How are the radii of curvature of the beams related?)

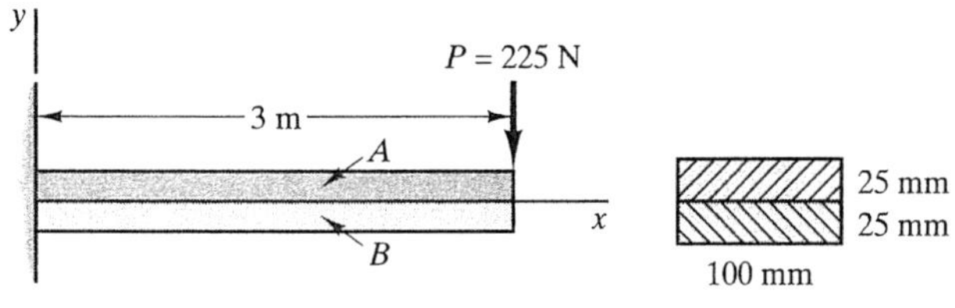

Figure P.11.88.

11.89. In Example 11.13, determine the number of screws needed to connect members C and F using the same data of this example.

11.90. An I-beam is formed by welding three bars together as shown in Fig. P.11.90. For a shear force V of 45,000 N, what is the maximum shear stress τ_{xy}? Also determine the total shear stress at point A.

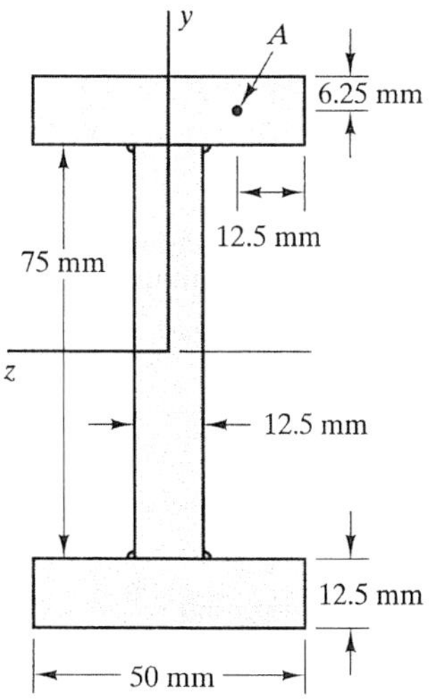

Figure P.11.90.

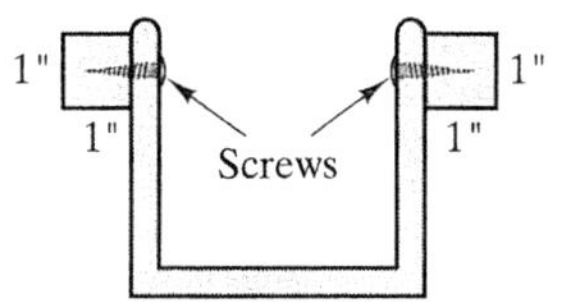

Figure P.11.91.

11.91. In Fig. P.11.91 is shown a 12-in. 30-lb/ft standard channel (see Appendix IV–B). Steel bars 1 by 1 in. are screwed to the sides of the channel. For a shear force of 500 lb, what should the maximum spacing of the screws be if each screw can take a 200-lb shear force?

11.92. In Fig. P.11.92,

(a). Find maximum tensile stress τ_{xx} and maximum shear stress τ_{xy}.

(b). What is the maximum load that the beam can hold for a yield stress of 50,000 psi tension and a safety factor of 2?

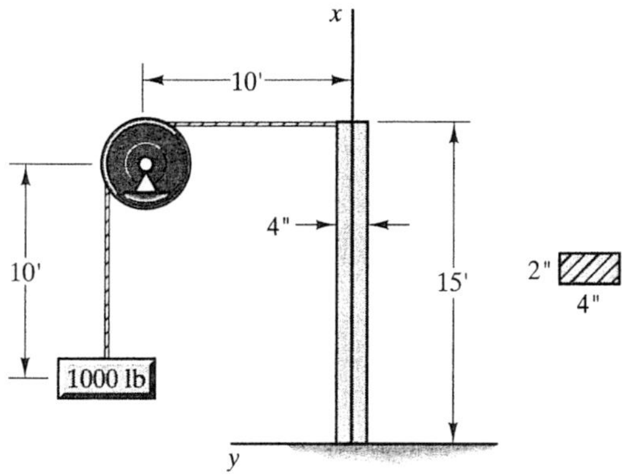

Figure P.11.92.

11.93. Shown in Fig. P.11.93 is a simply supported beam with a triangular loading distribution. What is the maximum tensile normal stress τ_{xx} and maximum shear stress τ_{xy}, resulting from this loading, at the midpoint of the beam?

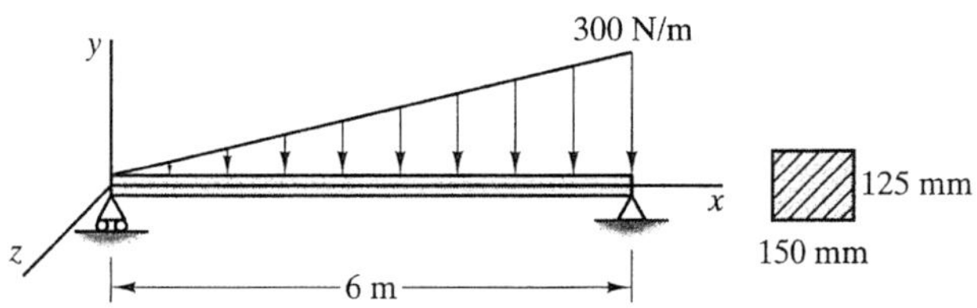

Figure P.11.93.

11.94. In Problem 11.93, determine for the entire beam the *maximum tensile normal stress* τ_{xx} and the *maximum shear* τ_{xy} *stress* resulting from the loading. Be sure to state location for these stresses.

11.95. Determine the maximum normal stress τ_{xx} and the maximum shear stress τ_{xy} resulting from the loadings for the cantilever beam shown in Fig. P.11.95.

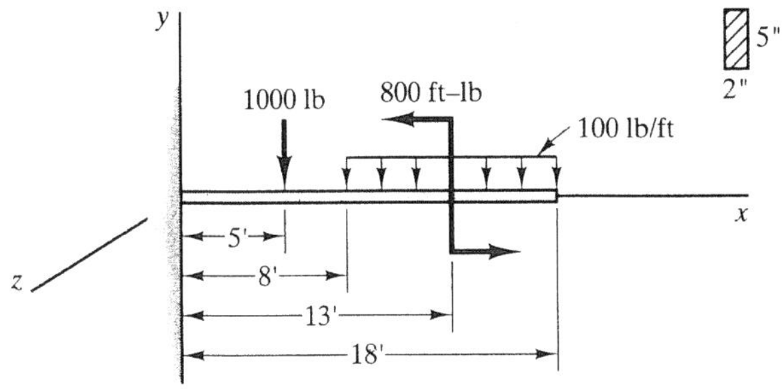

Figure P.11.95.

11.96. An extrusion process is typically used to manufacture plastic PVC pipe. This can lead to seams in the pipe where the material flowed and then rejoined around the die. To test the strength of these seams, the following test is proposed (see Fig. P.11.96). What is the stress state at point A?

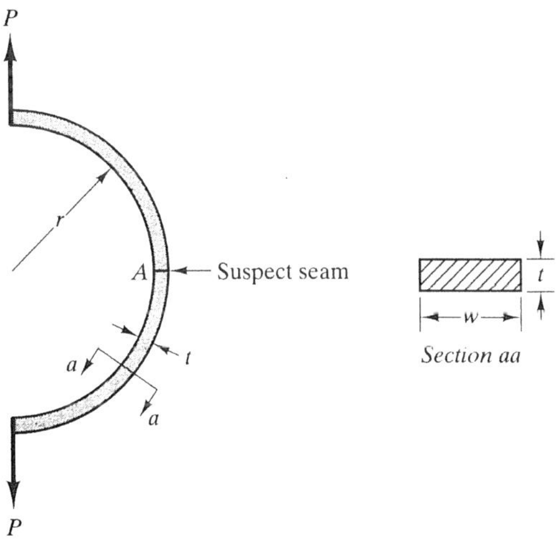

Figure P.11.96.

11.97. At Section B-B in Fig. P.11.97,

(a). Compute the bending moment and the magnitude of the shear force.
(b). Compute τ_{xx} at A.
(c). Compute $|\bar{\tau}_{xy}|$ at A.
(d). Compute $|\bar{\tau}_{xz}|$ at A.

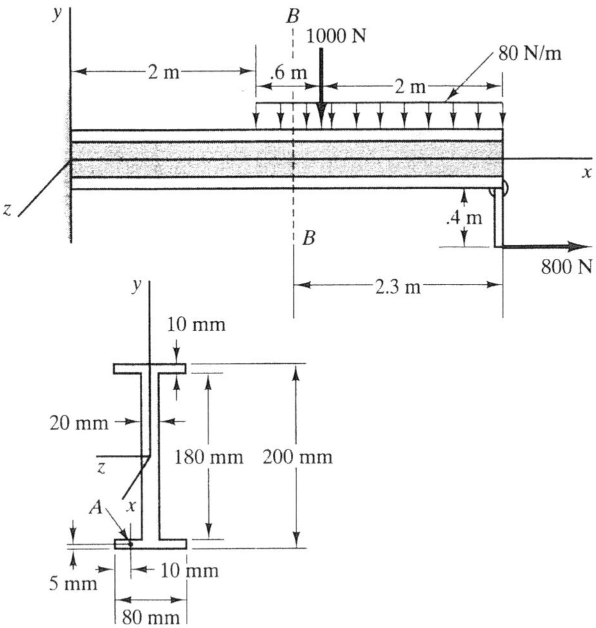

Figure P.11.97.

11.98. Determine in Fig. P.11.98:

(a). The second moment of area I_{zz}.
(b). The stress τ_{xx} at B.
(c). The magnitude of τ_{xz} at A.

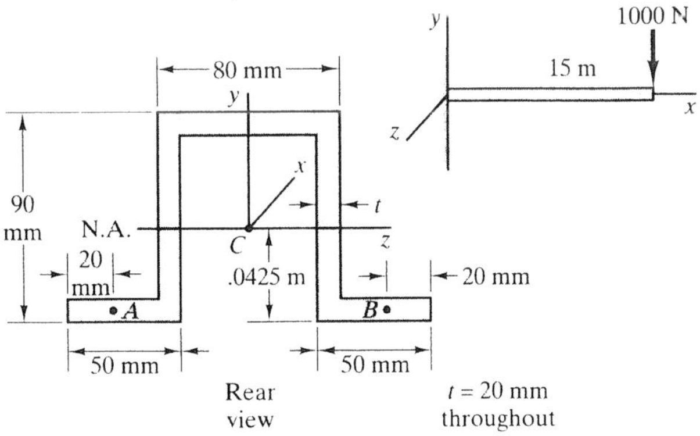

Figure P.11.98.

11.99. Which cross section shown in Fig. P.11.99 will have greater shear strength? Each cross section is composed of plates of equal thickness that are glued together as shown. The thickness of each part is 25 mm and the load is applied in the yz plane.

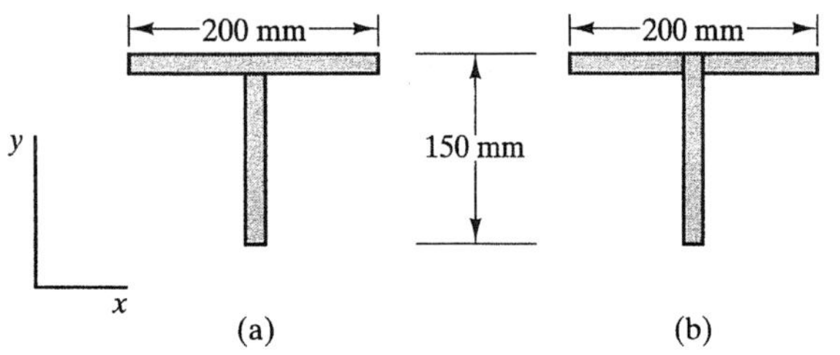

Figure P.11.99.

11.100. Consider a flexible circuit card composed of two thin symmetrically spaced copper conducting layers separated by a dielectric material, as shown in Fig. P.11.100. Determine the percent elongation in the bottom copper layer if the circuit card is bent around a mandrel of radius R_m. By definition, the percent elongation is

$$\left|\frac{y}{R}\right| (100)$$

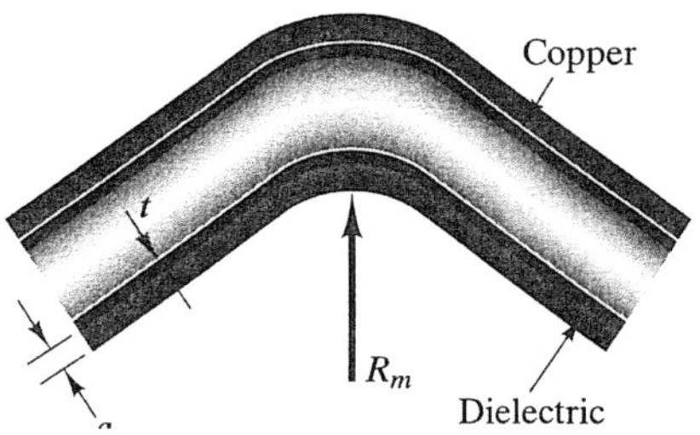

Figure P.11.100.

11.101. Aluminum beverage containers (i.e., soda cans) are typically manufactured by extruding them in the shape of a thin cylindrical shell. However, sometimes they are formed with segmented flat sides making a container with a polygonal cross section. Discuss how this new shape would change the analysis of these common pressure vessels (see Fig. P.11.101).

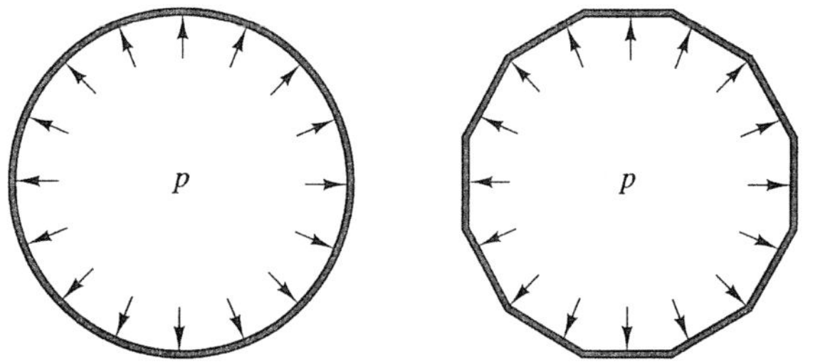

Figure P.11.101.

11.102. Electrical cables for elevators are often hung beneath the car as shown in Fig. P.11.102. For a 1/32-in. diameter copper wire ($E = 16 \times 10^6$ psi), what is the maximum bending stress when the cable is curled as the car moves? Take the radius $r = 1.5$ ft.

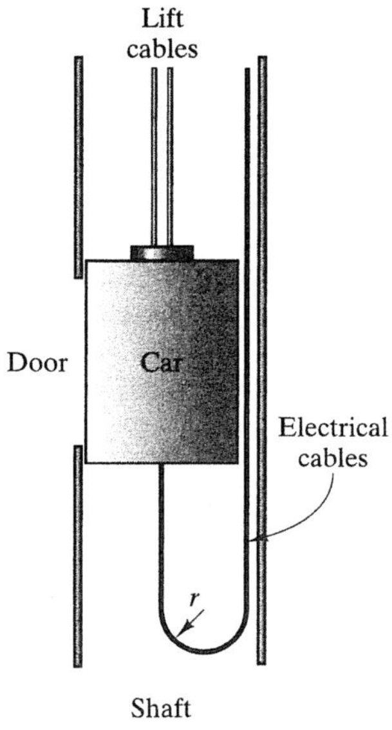

Figure P.11.102.

11.103. For a given prismatic, beam with the y-axis as the axis of symmetry, the orientation of the neutral axis passing through the centroid can be found at any position by setting the normal stress equal to zero in the flexure formula and solving for y/z of the axis of zero stress. Does this axis orientation change along the beam length? [*Hint:* Can you use Eq. 11.37(a)?]

11.104. A gate of length $L = 8$ ft and thickness $d = 2$ in. has width (into the page) of 6 ft (see Fig. P.11.104). This gate restrains water in a static condition at a depth $h = 3$ ft. If the pressure from the water is equal to γ which is the specific weight times the distance below the free surface, what is the maximum tensile stress in the gate at a position g which is 1 ft from A along the gate? Neglect the weight of the gate.

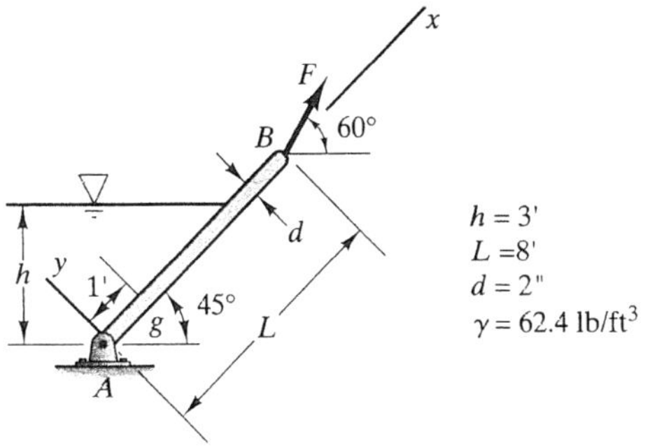

Figure P.11.104.

$*$**11.105.** Compute the distance to the neutral axis for a curved beam with the cross-sectional geometry shown in Fig. P.11.105. [*Hint:* See Example 11.22, Eq. (b).]

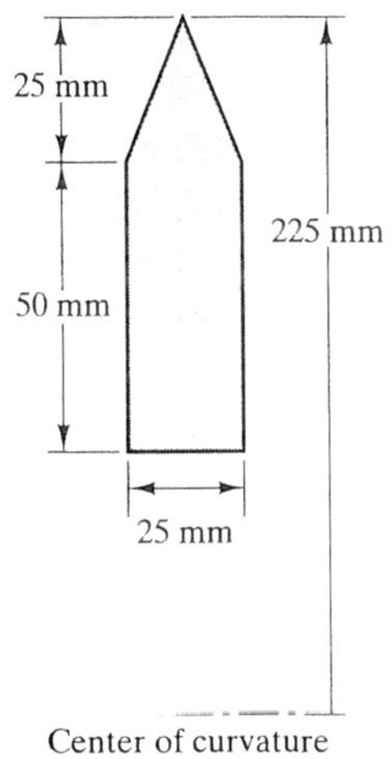

Figure P.11.105.

$**$**11.106.** As an *approximation* to the bending stresses in a shear pin, a four-point load distribution as shown in Fig. P.11.106 is sometimes assumed. If P/LD gives the direct *bearing stress* on the pin, show that the ratio of the maximum bending stress to the direct bearing stress on the pin is given as:

$$\frac{\tau_{\text{bending}}}{\tau_{\text{direct}}} = \frac{4[2 + (L/t)]}{\pi(L/t)}(L/D)^2$$

Develop a design aide by plotting this stress ratio for $1/2 \le L/t \le 5$ and $L/D = 0.5, 1.0, 1.5,$ and 2.0.

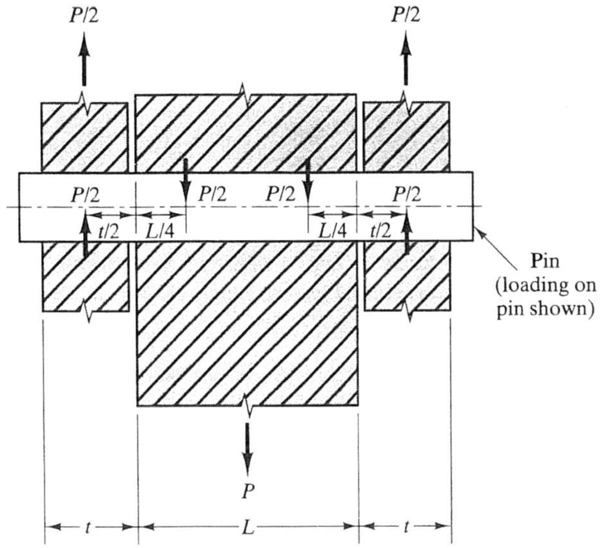

Figure P.11.106.

$**$**11.107.** From the theory of elasticity, it can be shown that the solution for the bending normal stress at midspan of the beam shown in Fig. P.11.107 is given as:

$$\tau_{xx} = \frac{8w}{bh}\left(\frac{L}{h}\right)^2\left[-0.75y + y\left(0.075\left(\frac{h}{L}\right)^2 - 0.5\left(\frac{y}{L}\right)^2\right)\right]$$

For $1 \le L/h \le 10$, compare the maximum τ_{xx} given above with that obtained via the flexure formula. Plot the percent error between the elasticity solution and the flexure formula versus L/h. At what ratio of beam span length $(2L)$ to beam depth (h) do we get 5% error in the bending normal stress if we use the flexure formula? Take the beam width as b.

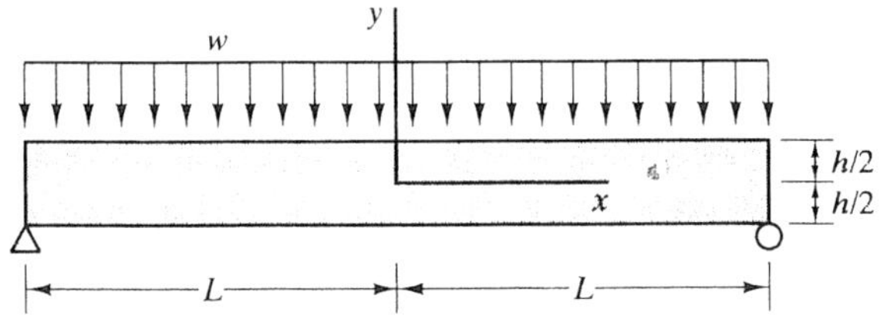

Figure P.11.107.

$**$**11.108.** Consider a simply-supported beam of rectangular cross section with height h and width b under self-weight only. Compute the beam depth h versus span L. Compare the results for $0.10 \text{ m} \le L \le 10 \text{ m}$ and the following materials:

Material	*Density (kg/m³)*	*Yield Stress (MPa)*
Oak	650	40
Structural Steel	7850	340
Alum. 7075-T6	2800	480
Hard Copper	8900	330
Magnesium Alloy	1800	250

Rank the materials according to required beam depth. Comment on the depth required versus length for any material.

****11.109.** For a symmetric cross section loaded at an angle β through the centroid (see Fig. P.11.109), plot the angle of the neutral axis α for $0 \le \beta \le 45°$. What does this suggest regarding large variations in the ratio of I_{zz}/I_{yy}? Show this by considering I_{zz}/I_{yy} ratios of 50, 25, 10, 5, 2, 1, 1/2, 1/5, 1/10, 1/25, and 1/50. (*Hint:* Consider the two-coordinate general flexure formula.)

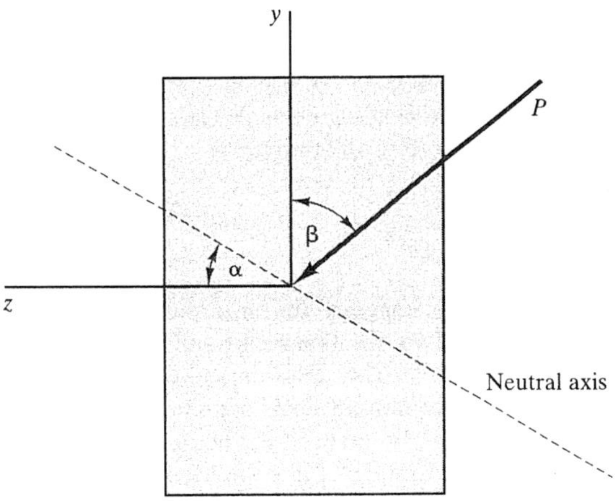

Figure P.11.109.

****11.110.** In the design of wooden bows used in archery, a trapezoidal cross section (see Fig. P.11.110) is often used since wood has different working stresses in tension and compression. This cross section allows the neutral axis to be shifted so that the working stress in tension and in compression may be reached simultaneously. What is the maximum of tensile/compressive stress ratio so that the limits are simultaneous? Plot the ratio y_{bot} to y_{top} versus b/a.

$$\text{Hint: } y_{bot} = \frac{h}{3}\,\frac{2a + b}{a + b}$$

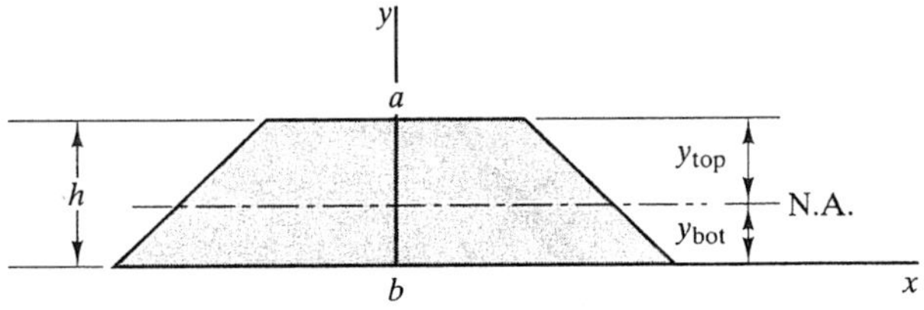

Figure P.11.110.

Project 11 Bending Stress Problem

Find the value of the maximum tensile stress in the cantilever beam shown in Fig. P.11.111 for the following user-supplied data:

F (N)	
θ (degrees)	0–360° in 10° intervals
a (mm)	
b (mm)	
L (m)	

Run the program for the following data:

F = 1000 N	L = *10 m*
a = 20 mm	θ = 0–360° in 10° intervals
b = 40 mm	

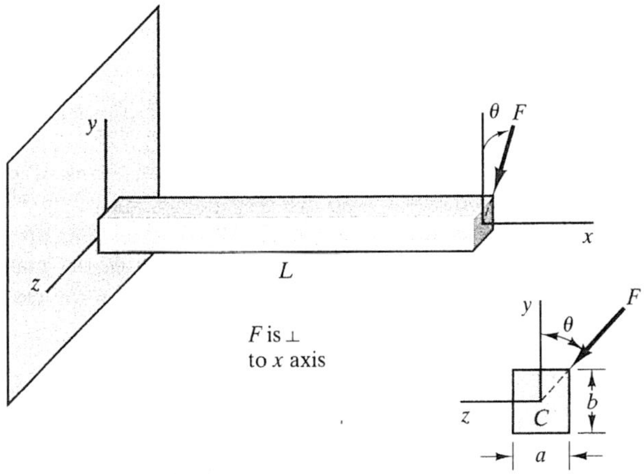

Figure P.11.111

CHAPTER 12

Deflection of Beams

12.1 Introduction

In Chapter 11 we focused our attention on the stresses in beams after proposing a particular mode of deformation for pure bending. We shall now examine the deflection of certain classes of beam problems using key results of Chapter 11.

Before embarking on this study it is to be pointed out that the deflection of a beam is an important design consideration. In addition, deflection considerations are essential for the handling of beams supported in a statically indeterminate manner. We shall consider statically indeterminate beams in detail later in the chapter.

12.2 Differential Equations for Deflection of Symmetric Beams

We shall consider here the deflection of symmetric beams loaded vertically and also with couples all within the plane of symmetry. You will recall from our analysis of the case of pure bending that for such beams the neutral surface deforms into a circular strip having a radius of curvature given by the formula

$$R = \frac{EI_{zz}}{M_z} \tag{12.1}$$

If we consider more general loadings with arbitrary force distributions which are normal to the centerline of the beam and that are in the plane of symmetry, we may still employ the preceding relation, realizing that it gives the *local* radius of curvature of the neutral surface. Also, it is important to note that the deflection to be computed by using Eq. (12.1) is a result of the variation of *normal stress* over a section and is called the *bending deformation;* here *normal strains* ε_{xx} having opposite signs above and below the neutral surface result in deflection of the neutral surface. There is also a *shear deformation* owing to shear stresses at a section. Here shear strains ε_{xy} result in the aforementioned shear deformation (see directly below), which is independent of the lon-

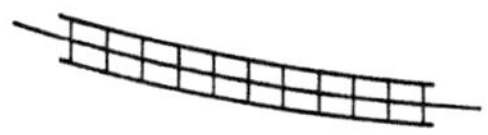

gitudinal strains ε_{xx} of the beam fibers. We shall see in Section 12.6 that deflections due to shear can usually be neglected.

It was pointed out earlier that, since the applied couples for the case of pure bending cannot generally be expected to have force distributions required by the theory, we can expect, in accordance with St.-Venant's principle, that the formulation of Eq. (12.1) becomes less correct the shorter the beam is in relation to its cross-sectional dimensions. The same conclusion follows for more general loadings. We shall thus consider in this section long slender beams for which Eq. (12.1) can be used locally with reasonable accuracy for small deformations. The bending moment M_z in this equation is a function of position *x,* and for beams supported in a statically determinate manner, it is computed by methods set forth in Chapter 10.

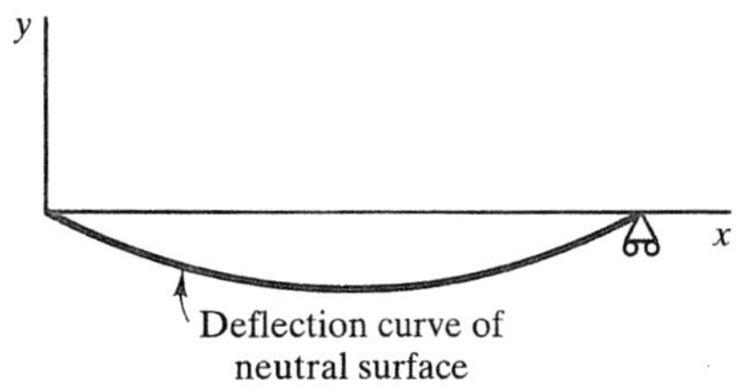

Figure 12.1. Deflection curve of neutral surface.

In our discussion of the deflection of a beam we shall refer to the deflection of the neutral surface and we shall use *v taken as a positive pointing upward* to give this deflection at each position along the *x* axis.[1] This has been shown in Fig. 12.1, where we have a grossly exaggerated deflection curve for a simply supported beam. We can express the radius of curvature *R* in terms of the deflection curve by the following well-known formulation from analytic geometry:

$$\frac{1}{R} = \frac{d^2v/dx^2}{[1 + (dv/dx)^2]^{3/2}} \tag{12.2}$$

Since we are restricted to small deformations, the slope dv/dx will be small, so that the term $(dv/dx)^2$ can be neglected when compared to unity. The foregoing relation then becomes

$$\frac{1}{R} = \frac{d^2v}{dx^2} \tag{12.3}$$

[1]The value *y* used in Chapter 11 gives the distance of an element of the beam from the neutral axis in the undeformed geometry. Recall that this distance was assumed unchanged during deformation.

Substituting for R in Eq. (12.1) using the preceding relation gives us

$$\frac{d^2v}{dx^2} = \frac{M_z}{EI_{zz}} \tag{12.4}$$

This is the basic differential equation for the deflection curve in terms of the bending moment M_z. We shall next replace M_z in Eq. (12.4) by employing Eq. (10.2), which we now restate as

$$\frac{d^2M_z}{dx^2} = -w_y \tag{12.5}$$

Thus, multiply Eq. (12.4) by EI_{zz} and then take the second derivative of both sides with respect to x. We then get, on employing Eq. (12.5), the differential equation of the deflection curve in terms of the loading function w_y. Thus,

$$\frac{d^2}{dx^2}\left(EI_{zz}\frac{d^2v}{dx^2}\right) = \frac{d^2M_z}{dx^2} = -w_y \tag{12.6}$$

If I_{zz} and E are constants, Eq. (12.6) may be written as follows:

$$\frac{d^4v}{dx^4} = -\frac{w_y}{EI_{zz}} \tag{12.7}$$

We thus have a differential equation now in terms of the loading function in addition to the one in terms of the bending moment. Which does one use? Generally it is Eq. (12.4) that is used, and we shall illustrate such usage in the following pair of examples. However, the use of the singularity function (Chapter 13) favors the equation involving the loading function. We shall illustrate this approach in Chapter 13.

Example 12.1

A cantilever beam is uniformly loaded over its span as shown in Fig. 12.2. Find the deflection curve $v(x)$.

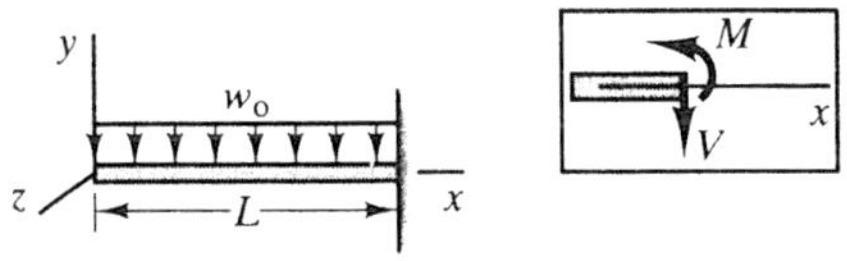

Figure 12.2. Cantilever beam.

Example 12.1 (Continued)

The bending moment $M(x)$ is clearly seen by inspection to be $-w_0x^2/2$. Hence, for the region $0 \le x < L$ we can say,[2] using Eq. (12.4),

$$\frac{d^2v}{dx^2} = \frac{1}{EI}\left(-\frac{w_0x^2}{2}\right) \quad \text{(a)}$$

Integrating twice, we get

$$\frac{dv}{dx} = \frac{1}{EI}\left(-\frac{w_0x^3}{6} + C_1\right) \quad \text{(b)}$$

$$v = \frac{1}{EI}\left(-\frac{w_0x^4}{24} + C_1x + C_2\right) \quad \text{(c)}$$

where C_1 and C_2 are constants of integration. When $x = L$ we have $v = dv/dx = 0$. Accordingly, we may determine the constants of integration as follows:

$$C_1 = \frac{w_0L^3}{6}$$

$$C_2 = \frac{w_0L^4}{24} - \frac{w_0L^4}{6} = -\frac{1}{8}w_0L^4$$

We thus have as a final result

$$v = \frac{1}{EI}\left(-\frac{w_0x^4}{24} + \frac{w_0L^3}{6}x - \frac{1}{8}w_0L^4\right) \quad \text{(d)}$$

The next problem is more complex and will require us to consider *domains* of the beam so that we can properly express bending moments and can thus express the differential equations for the deflection curve. To simplify computations, we now consider the integration of the expression $P(x-a)$. The first impulse to do this would yield

$$\int P(x - a)\,dx = \frac{Px^2}{2} - Pax + C_0 \quad \text{(12.8a)}$$

[2]We must exclude positions such as $x = L$, where we have a point couple rendering M indeterminate at such points. If M is continuous at a point at the extremity of a domain we include such a point for Eq. (12.4) even if the shear force is discontinued there.

where C_0 is an arbitrary constant of integration. Now we *formally* do the integration, treating $(x-a)$ as the running variable, to get the expression

$$\int P(x-a)\,dx = \frac{P(x-a)^2}{2} + C_1 \tag{12.8b}$$

Expanding Eq. (12.8b), we get the preceding result [Eq. (12.8a)], with C_0 replaced by $(Pa^2/2 + C_1)$. Since *both* C_0 and $(Pa^2/2 + C_1)$ are *arbitrary,* as is required by the integration, we may then properly use Eq. (12.8b), where C_1 is taken as the constant of integration. Similarly, we may directly demonstrate that

$$\int P(x-a)^2\,dx = \frac{P(x-a)^3}{3} + C_2$$

In general,

$$\int P(x-a)^n\,dx = \frac{P(x-a)^{n+1}}{n+1} + C_n \tag{12.9}$$

The use of the integration form in Eq. (12.9) will make it much simpler to find the equations to be solved for determining the constants of integration. We illustrate this in the next example.

Example 12.2

Shown in Fig. 12.3 is a simply supported beam with a variety of loadings. We will determine the deflection curve.

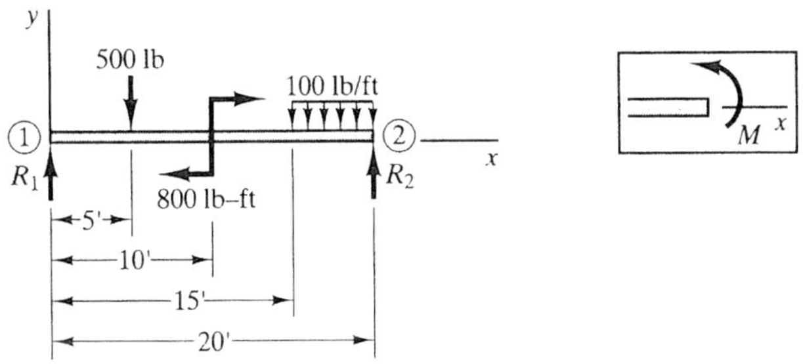

Figure 12.3. Simply supported beam.

Example 12.2 (Continued)

As a first step, we compute the supporting forces by rigid-body mechanics.

$\underline{\sum M_2 = 0}$:

$$-R_1(20) + (500)(15) - 800 + (500)(2.5) = 0$$

$$\therefore R_1 = 397.5 \text{ lb}$$

$\underline{\sum M_1 = 0}$:

$$20R_2 - (500)(5) - 800 - (500)(17.5) = 0$$

$$\therefore R_2 = 602.5 \text{ lb}$$

As a check, we can sum forces in the vertical direction. Thus

$$\sum F_y = 0 = 397.5 + 602.5 - 500 - (5)(100)$$

$$0 = 0$$

We now consider a series of domains.

$\underline{0 \le x \le 5}$:

$$M = 397.5x$$

$$\therefore \frac{d^2v}{dx^2} = \frac{1}{EI}(397.5x)$$

$$\frac{dv}{dx} = \frac{1}{EI}\left(397.5\frac{x^2}{2} + C_1\right) \qquad \text{(a)}$$

$$v = \frac{1}{EI}\left(397.5\frac{x^3}{6} + C_1x + C_2\right) \qquad \text{(b)}$$

$\underline{5 \le x < 10}$:

$$M = 397.5x - 500(x - 5)$$

$$\therefore \frac{d^2v}{dx^2} = \frac{1}{EI}[397.5x - 500(x - 5)]$$

Integrating and using Eq. (12.9), we get

$$\frac{dv}{dx} = \frac{1}{EI}\left[397.5\frac{x^2}{2} - 500\frac{(x-5)^2}{2} + C_3\right] \qquad \text{(c)}$$

$$v = \frac{1}{EI}\left[397.5\frac{x^3}{6} - 500\frac{(x-5)^3}{6} + C_3x + C_4\right] \qquad \text{(d)}$$

Example 12.2 (Continued)

$10 < x \leq 15$:

$$M = 397.5x - 500(x - 5) + 800$$

$$\therefore \frac{d^2v}{dx^2} = \frac{1}{EI}[397.5x - 500(x - 5) + 800]$$

$$\frac{dv}{dx} = \frac{1}{EI}\left[397.5\frac{x^2}{2} - 500\frac{(x-5)^2}{2} + 800x + C_5\right] \quad \text{(e)}$$

$$v = \frac{1}{EI}\left[397.5\frac{x^3}{6} - 500\frac{(x-5)^3}{6} + 800\frac{x^2}{2} + C_5x + C_6\right] \quad \text{(f)}$$

$15 \leq x \leq 20$:

$$M = 397.5x - 500(x - 5) + 800 - 100\frac{(x-15)^2}{2}$$

$$\therefore \frac{d^2v}{dx^2} = \frac{1}{EI}\left[397.5x - 500(x - 5) + 800 - 100\frac{(x-15)^2}{2}\right]$$

$$\frac{dv}{dx} = \frac{1}{EI}\left[397.5\frac{x^2}{2} - 500\frac{(x-5)^2}{2} + 800x - 100\frac{(x-15)^3}{6} + C_7\right] \quad \text{(g)}$$

$$v = \frac{1}{EI}\left[397.5\frac{x^3}{6} - 500\frac{(x-5)^3}{6} + 800\frac{x^2}{2} - 100\frac{(x-15)^4}{24} + C_7x + C_8\right] \quad \text{(h)}$$

The *boundary conditions* for the problem will be considered next.

1. When $x = 0$, $v = 0$.

From Eq. (b), we can conclude that $C_2 = 0$. Also:

2. When $x = 20$, $v = 0$.

From Eq. (h), we get

$$0 = \frac{(397.5)(20^3)}{6} - \frac{(500)(15^3)}{6} + \frac{(800)(20^2)}{2} - \frac{(100)(5^4)}{24} + 20C_7 + C_8$$

$$\therefore 20C_7 + C_8 = -406{,}145 \quad \text{(i)}$$

We must next properly *patch* the equations between the domains. That is, the slope and the deflection at the end of one

Example 12.2 (Continued)

domain must, respectively, equal the slope and deflection at the beginning of the next domain. Thus between the first and second domains we require **(compatibility)** that

$$\left[\frac{dv(5)}{dx}\right]_{\text{Eq.(a)}} = \left[\frac{dv(5)}{dx}\right]_{\text{Eq.(c)}}$$

$$\therefore \frac{1}{EI}\left(397.5\frac{5^2}{2} + C_1\right) = \frac{1}{EI}\left(397.5\frac{5^2}{2} + 0 + C_3\right)$$

[It is the appearance of the zero on the right-hand side that is the simplification resulting from the use of Eq. (12.9).] Hence,

$$C_1 = C_3 \tag{j}$$

Also,

$$[v(5)]_{\text{Eq.(b)}} = [v(5)]_{\text{Eq.(d)}}$$

$$\frac{1}{EI}\left(397.5\frac{5^3}{6} + 5C_1\right) = \frac{1}{EI}\left(397.5\frac{5^3}{6} - 0 + 5C_3 + C_4\right)$$

$$\therefore 5C_1 - 5C_3 - C_4 = 0 \tag{k}$$

For the next two domains, we have just to the left of $x = 10$ (i.e., $x = 10^-$) and just to the right of $x = 10$ (i.e., $x = 10^+$).

$$\left[\frac{dv(10^-)}{dx}\right]_{\text{Eq.(c)}} = \left[\frac{dv(10^+)}{dx}\right]_{\text{Eq.(e)}}$$

$$\frac{1}{EI}\left[397.5\frac{(10^2)}{2} - \frac{(500)(5^2)}{2} + C_3\right] =$$

$$\frac{1}{EI}\left[397.5\frac{(10^2)}{2} - \frac{(500)(5^2)}{2} + (800)(10) + C_5\right]$$

$$\therefore C_3 - C_5 = 8000 \tag{l}$$

$$[v(10^-)]_{\text{Eq.(d)}} = [v(10^+)]_{\text{Eq.(f)}}$$

$$\frac{1}{EI}\left[397.5\frac{(10^3)}{6} - 500\frac{(5^3)}{6} + 10C_3 + C_4\right] =$$

$$\frac{1}{EI}\left[397.5\frac{(10^3)}{6} - 500\frac{(5^3)}{6} + 800\frac{(10^2)}{2} + 10C_5 + C_6\right]$$

$$\therefore 10C_3 + C_4 - 10C_5 - C_6 = 40{,}000 \tag{m}$$

Example 12.2 (Continued)

Finally, we have

$$\left[\frac{dv(15)}{dx}\right]_{\text{Eq.(e)}} = \left[\frac{dv(15)}{dx}\right]_{\text{Eq.(g)}}$$

$$\frac{1}{EI}\left[397.5\frac{(15^2)}{2} - 500\frac{(10^2)}{2} + (800)(15) + C_5\right] =$$

$$\frac{1}{EI}\left[397.5\frac{(15^2)}{2} - 500\frac{(10^2)}{2} + (800)(15) + 0 + C_7\right]$$

$$\therefore C_5 = C_7 \qquad \text{(n)}$$

$$[v(15)]_{\text{Eq.(f)}} = [v(15)]_{\text{Eq.(h)}}$$

$$\frac{1}{EI}\left[397.5\frac{(15^3)}{6} - 500\frac{(10^3)}{6} + 800\frac{(15^2)}{2} + 15C_5 + C_6\right] =$$

$$\frac{1}{EI}\left[397.5\frac{(15^3)}{6} - 500\frac{(10^3)}{6} + 800\frac{(15^2)}{2} + 0 + 15C_7 + C_8\right]$$

$$\therefore 15C_5 + C_6 - 15C_7 - C_8 = 0 \qquad \text{(o)}$$

In forming the patching equations you need not write everything down as we have done. By inspection you can arrive directly at the proper equations such as Eq. (o), since many of the terms are either zero or cancel, as can be easily observed from the domain equations. We now rewrite the equations for the constants.

$$C_2 = 0$$

$$20C_7 + C_8 = -406{,}145 \qquad \text{(i)}$$

$$C_1 = C_3 \qquad \text{(j)}$$

$$5C_1 - 5C_3 - C_4 = 0 \qquad \text{(k)}$$

$$C_3 - C_5 = 8000 \qquad \text{(l)}$$

$$10C_3 + C_4 - 10C_5 - C_6 = 40{,}000 \qquad \text{(m)}$$

$$C_5 = C_7 \qquad \text{(n)}$$

$$15C_5 + C_6 - 15C_7 - C_8 = 0 \qquad \text{(o)}$$

The equations are rather simple to handle. By substituting for C_3 in Eq. (k), using Eq. (j), we obtain

Example 12.2 (Continued)

$$5C_1 - 5C_1 - C_4 = 0$$
$$\therefore C_4 = 0 \tag{p}$$

Next going to Eq. (m), we get

$$10(C_3 - C_5) - C_6 = 40{,}000 \tag{q}$$

But $(C_3 - C_5) = 8000$, from Eq. (l). Hence, we may solve for C_6.

$$C_6 = 40{,}000 \tag{r}$$

Now we go to Eq. (o). Replacing C_5, using Eq. (n), we get

$$15C_7 + C_6 - 15C_7 - C_8 = 0$$
$$\therefore C_8 = C_6 = 40{,}000 \tag{s}$$

Going to Eq. (i), we may determine C_7:

$$20C_7 + 40{,}000 = -406{,}145$$
$$\therefore C_7 = -22{,}307 \tag{t}$$

From Eq. (n),

$$C_5 = -22{,}307 \tag{u}$$

From Eq. (l),

$$C_3 = 8000 - 22{,}307 = -14{,}307$$

From Eq. (j),

$$C_1 = -14{,}307 \tag{v}$$

The example is thus complete.

You will note that even with the suggested simplifying steps, Example 12.2 was rather long and laborious. Those readers who study singularity functions in Chapter 13 will find a method that will greatly reduce the amount of labor necessary for these deflection problems. Also, they will become familiar with a very useful system of functions that find their way into many areas of engineering science and applied mathematics. Finally, one may make good use of mathematical software here.

12.3 Additional Problems

We will now look at some additional problems of interest. First we will look for the maximum deflection for a beam. We shall consider a simple problem for this purpose.

Example 12.3

Find the maximum deflection of the simply supported beam shown in Fig. 12.4. Take $E = 30 \times 10^6$ psi. The beam is a W8 × 20 wide-flange I-beam (see Appendix IV-A). Include the weight of the beam.

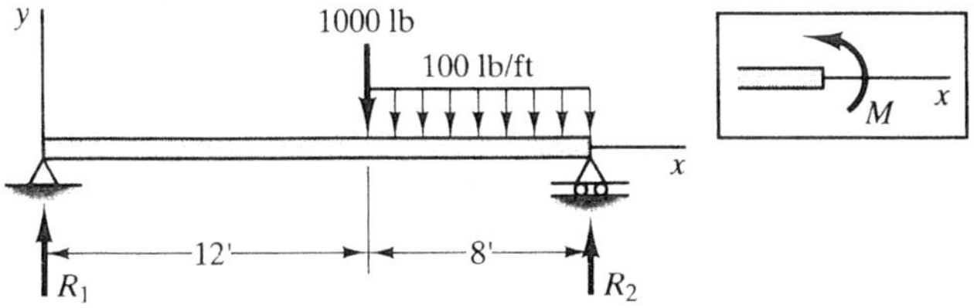

Figure 12.4. Simply supported beam.

We first find R_1.

$\Sigma M_{R_2} = 0$:

$$-R_1(20) + (20)(20)(10) + (1000)(8) + (800)(4) = 0$$

$$\therefore R_1 = 760 \text{ lb}$$

Now find the *deflection* curve.

$0 \leq x \leq 12$:

$$\frac{d^2v}{dx^2} = \frac{1}{EI}\left(760x - 20\frac{x^2}{2}\right)$$

$$\frac{dv}{dx} = \frac{1}{EI}\left(760\frac{x^2}{2} - 20\frac{x^3}{6} + C_1\right) \tag{a}$$

$$v = \frac{1}{EI}\left(760\frac{x^3}{6} - 20\frac{x^4}{24} + C_1x + C_2\right) \tag{b}$$

$12 \leq x \leq 20$:

$$\frac{d^2v}{dx^2} = \frac{1}{EI}\left[760x - 20\frac{x^2}{2} - 1000(x - 12) - 100\frac{(x-12)^2}{2}\right]$$

Example 12.3 (Continued)

$$\frac{dv}{dx} = \frac{1}{EI}\left[760\frac{x^2}{2} - 20\frac{x^3}{6} - 1000\frac{(x-12)^2}{2} - \frac{100(x-12)^3}{6} + C_3\right] \tag{c}$$

$$v = \frac{1}{EI}\left[760\frac{x^3}{6} - 20\frac{x^4}{24} - 1000\frac{(x-12)^3}{6} - \frac{100(x-12)^4}{24} + C_3x + C_4\right] \tag{d}$$

Boundary conditions are as follows:

1. When $x = 0$, $v = 0$.

$$\therefore C_2 = 0$$

2. When $x = 20$, $v = 0$.

$$\therefore \frac{(760)(20)^3}{6} - \frac{(20)(20^4)}{24} - \frac{(1000)(8^3)}{6} - \frac{(100)(8^4)}{24} + 20C_3 + C_4 = 0$$

We get

$$20C_3 + C_4 = -7.78 \times 10^5 \tag{e}$$

Patch conditions **(compatibility)** are considered next.

$$[v(12)]_{\text{Eq.(b)}} = [v(12)]_{\text{Eq.(d)}}$$

$$\therefore 12C_1 + C_2 = 12C_3 + C_4 \tag{f}$$

$$\left[\frac{dv(12)}{dx}\right]_{\text{Eq.(a)}} = \left[\frac{dv(12)}{dx}\right]_{\text{Eq.(c)}}$$

$$\therefore C_1 = C_3 \tag{g}$$

Noting $C_2 = 0$ and using Eq. (g), we see that $C_4 = 0$, from Eq. (f). Going to Eq. (e), we see that

$$C_3 = -3.89 \times 10^4$$

We shall now examine the extremal condition $dv/dx = 0$ in each domain. Thus, for $0 \le x \le 12$ we have

$$0 = \frac{dv}{dx} = \frac{1}{EI}\left(760\frac{x^2}{2} - 20\frac{x^3}{6} - 3.89 \times 10^4\right) \tag{h}$$

Solving on a programmable calculator or by trial and error, we get the following real root for x satisfying Eq. (h).

$$x = 10.62 \text{ ft}$$

Example 12.3 (Continued)

Since this position is *inside* the prescribed range of Eq. (h), we conclude that we do have zero slope of the deflection curve in this domain and so in this problem we need not look at the second domain. If we got as a real root a value less than zero or greater than 12, we would then examine the second domain for a meaningful root. That is, we would examine the condition

$$0 = \frac{dv}{dx} = \frac{1}{EI}\left[760\frac{x^2}{2} - 20\frac{x^3}{6} - 1000\frac{(x-12)^2}{2} - 100\frac{(x-12)^3}{6} - 3.89 \times 10^4\right]$$

We get for a real root

$$x = 10.73 \text{ ft}$$

Clearly, this is outside the prescribed range of the second domain and may be disregarded.

Hence we get for v_{max} using Eq. (b):

$$v_{max} = \frac{1}{(30 \times 10^6)(69.2)(\frac{1}{144})}\left[760\frac{(10.62)^3}{6} - 20\frac{(10.62)^4}{24} - (3.89 \times 10^4)(10.62)\right]$$

$$\therefore v_{max} = -.01887 \text{ ft} = -.226 \text{ in.}$$

We next consider two beams interconnected by a pin. It is vital to remember for such problems that the *slope* of the deflection curve is undetermined at the pin (the deflection curve itself is continuous at the pin, however), so we shall exclude the position of the pin from our deflection curve. We do this by forming domains to the left and to the right of the pin. The *patch conditions* at the pin include now *only* the continuity of the deflection v (not the slope dv/dx).

Also, as you will see in the following example, we should check the deflection of the pin in ascertaining the maximum deflection. Just looking for zero slopes is not sufficient in such problems in finding the position of maximum deflection.

Example 12.4

Find the maximum deflection of the pin-connected beams shown in Fig. 12.5. The weight of the beam has been included in the 180 N/m uniform loading. Take $E = 2 \times 10^{11}$ Pa.

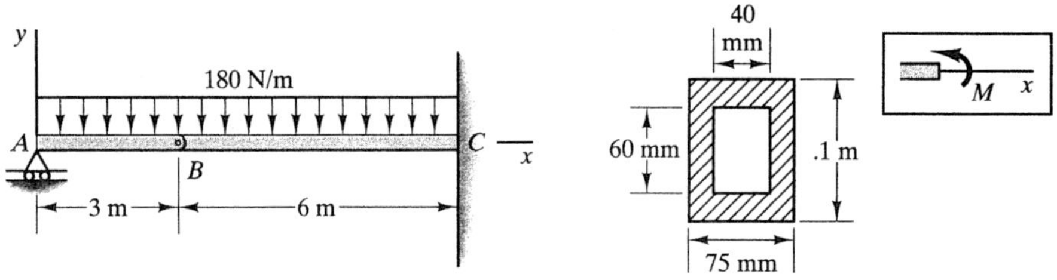

Figure 12.5. Pin-connected beams.

It would at first seem that we have a statically indeterminate support system here, but this is not the case. We can take $\overline{AB}$ as free body with the bending moment zero at the pin at B and solve for the supporting force at A. Thus observing Fig. 12.6, we can say:

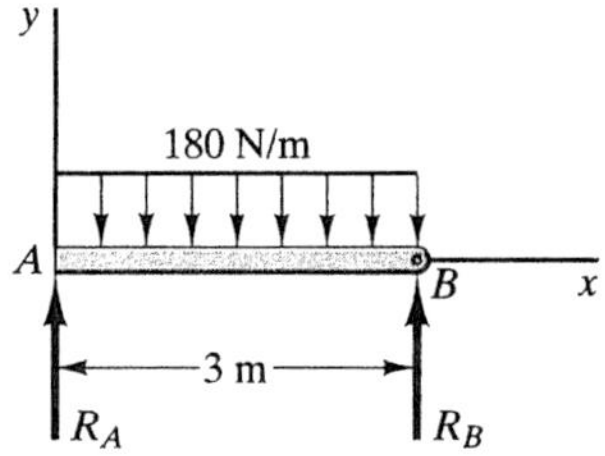

Figure 12.6. Free body of AB.

$\Sigma M_B = 0$:

$$-R_A(3) + (180)(3)\left(\frac{3}{2}\right) = 0$$

$$\therefore R_A = 270 \text{ N}$$

We can now proceed with the deflection curve analysis.

$0 \le x < 3$:

$$\frac{d^2v}{dx^2} = \frac{1}{EI}\left[270(x) - 180\left(\frac{x^2}{2}\right)\right]$$

$$\frac{dv}{dx} = \frac{1}{EI}\left[270\left(\frac{x^2}{2}\right) - 180\left(\frac{x^3}{6}\right) + C_1\right] \quad \text{(a)}$$

$$v = \frac{1}{EI}\left[270\left(\frac{x^3}{6}\right) - 180\left(\frac{x^4}{24}\right) + C_1x + C_2\right] \quad \text{(b)}$$

$3 < x < 9$:

$$\frac{d^2v}{dx^2} = \frac{1}{EI}\left[270(x) - 180\left(\frac{x^2}{2}\right)\right]$$

$$\frac{dv}{dx} = \frac{1}{EI}\left[270\left(\frac{x^2}{2}\right) - 180\left(\frac{x^3}{6}\right) + C_3\right] \quad \text{(c)}$$

Example 12.4 (Continued)

$$v = \frac{1}{EI}\left[270\left(\frac{x^3}{6}\right) - 180\left(\frac{x^4}{24}\right) + C_3 x + C_4\right] \qquad \text{(d)}$$

You will note that except for the constants of integration the deflection equations are identical for this simple problem for both domains.

Boundary conditions:

1. When $x = 0$, $v = 0$.

$$\therefore C_2 = 0$$

2. When $x = 9$, $dv/dx = 0$.

$$270\left(\frac{9^2}{2}\right) - 180\left(\frac{9^3}{6}\right) + C_3 = 0$$

$$\therefore C_3 = 1.094 \times 10^4$$

3. When $x = 9$, $v = 0$.

$$270\left(\frac{9^3}{6}\right) - 180\left(\frac{9^4}{24}\right) + (1.094 \times 10^4)(9) + C_4 = 0$$

$$\therefore C_4 = -8.206 \times 10^4$$

Patch condition:

$$[v(3)]_{\text{Eq.(b)}} = [v(3)]_{\text{Eq.(d)}}$$

$$\therefore C_1(3) + C_2 = C_3(3) + C_4$$

Noting that $C_2 = 0$, $C_3 = 1.094 \times 10^4$, and $C_4 = -8.206 \times 10^4$, we can solve for the remaining unknown constant C_1. That is,

$$3C_1 + 0 = (3)(1.094 \times 10^4) - 8.206 \times 10^4$$

$$\therefore C_1 = -1.641 \times 10^4$$

We now look for *zero slopes* of v in the two domains. Thus for the left domain we have

$$\frac{dv}{dx} = 0 = \frac{1}{EI}\left(270\,\frac{x^2}{2} - 180\,\frac{x^3}{6} - 1.641 \times 10^4\right) \qquad \text{(e)}$$

We find as a real root for this equation,

$$x = -6.92 \text{ m}$$

Example 12.4 (Continued)

Clearly, we discard this result, coming as it does outside the domain of Eq. (e). Look next at the remaining domain.

$$\frac{dv}{dx} = 0 = \frac{1}{EI}\left[270\left(\frac{x^2}{2}\right) - 180\left(\frac{x^3}{6}\right) + 1.094 \times 10^4\right]$$

We get as the only zero-slope position,

$$x = 9.00 \text{ m}$$

This corresponds to the base of the cantilever and represents the trivial condition of a minimum deflection of zero.

We should check the pin. Thus, from Eq. (b) we have

$$v(3) = \frac{1}{EI}\left[270\left(\frac{3^3}{6}\right) - 180\left(\frac{3^4}{24}\right) - (1.641 \times 10^4)(3)\right]$$
$$= -\frac{4.862 \times 10^4}{EI} \text{ m}$$

It should now be clear that the maximum deflection must occur at the pin.

The value of EI is next computed.

$$EI = (2 \times 10^{11})\left[\left(\frac{1}{12}\right)(.075)(.1)^3 - \left(\frac{1}{12}\right)(.040)(.060)^3\right]$$
$$= 1.106 \times 10^6 \text{ N-m}^2$$

The maximum deflection then is

$$v(3) = -\frac{4.862 \times 10^4}{1.106 \times 10^6} = -.0440 \text{ m}$$

12.4 Statically Indeterminate Beams

In Sections 12.2 and 12.3 we examined beams whose supporting forces can be determined solely by methods of rigid-body mechanics. We shall now consider beams with statically indeterminate supports. For such cases, we must take into account the manner of deformation in order to ascertain the supporting forces. Thus, for statically indeterminate beams, the deflection of the beam is determined *simultaneously* with evaluation of the supporting forces rather than independently as was the case for statically determinate beams.

It will be helpful now to define the degree of *redundancy, n,* as the number of supporting forces or torques in excess of that which can be solved by rigid-body mechanics alone for a particular beam problem. Thus, for the beam shown in Fig. 12.7(a) there is one more supporting

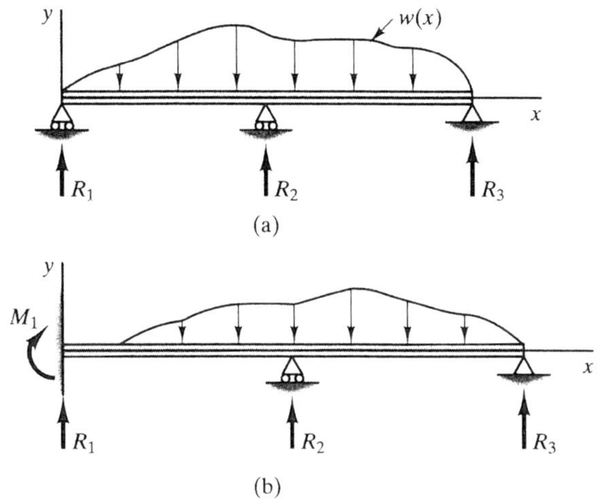

Figure 12.7. Statically indeterminate beams.

force than can be solved by rigid-body mechanics. Thus, the degree of redundancy, or more briefly the redundancy, is unity. For the cantilever beam shown in Fig. 12.7(b), the degree of redundancy is 2. It is clear that there will be additional unknowns for statically indeterminate beams equal in number to the degree of redundancy of the problem. For the problems that we shall consider, there will always be additional boundary conditions equal to the degree of redundancy of the problem. This permits us to solve for the additional unknowns. Thus, in Fig. 12.7(a), if R_3 is considered the redundant constraint, then we have the additional boundary condition that $v = 0$ at the point of contact for this support. Similarly, in Fig. 12.7(b), if R_2 and R_3 are considered the redundant constraints, we have the additional boundary conditions that $v = 0$ at these supports. Or we may consider R_1 and M_1 as the redundant constraints for this case with the additional conditions that $v = dv/dx = 0$ at $x = 0$.

The procedure for solving the supporting forces for statically indeterminate beam problems is to first choose redundant constraints and then formulate the bending-moment distribution in terms of these redundant constraints as well as the given external loads on the beam. To do this we may at times have to first solve for some of the other supporting forces in terms of the redundant constraints via the use of rigid-body mechanics applied to the entire beam. We then integrate Eq. (12.4) for the problem at hand, determining the constants of integration and the unknown redundant constraints by considering the boundary conditions and patch conditions of the problem. Once the redundant constraints are known we can readily determine the other supporting forces and moments by rigid-body statics.

We shall now illustrate this procedure in the following example.

Example 12.5

The cantilever beam shown in Fig. 12.8 supports a uniform loading w_0 of 10 kN/m and a concentrated couple-moment M_0 having the value of 100 kN-m. Find the supporting forces and the deflection curve in terms of EI. The beam is 10 m long.

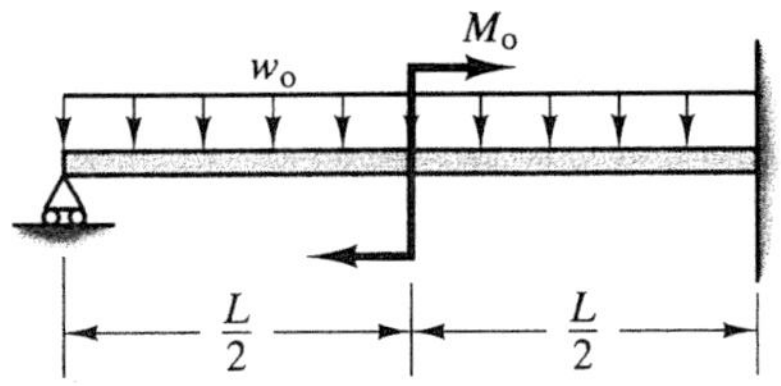

Figure 12.8. Cantilever beam.

The free-body diagram for the entire beam is shown in Fig. 12.9. We shall consider the supporting force R_1 as the redundant constraint in the ensuing computations. We can here compute the bending moment M in terms of R_1 without the necessity of determining other supporting forces or torques in terms of R_1. Accordingly, we shall employ Eq. (12.4) for two spans of the beam as follows:

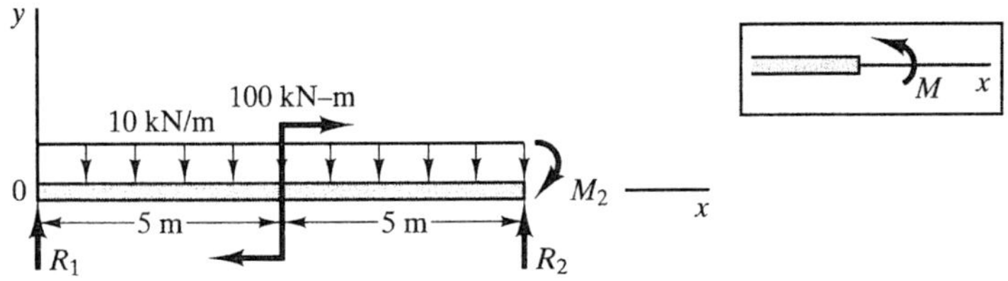

Figure 12.9. Free-body diagram of cantilever beam.

$0 \le x < 5$:

$$\frac{d^2v}{dx^2} = \frac{1}{EI}\left(R_1 x - \frac{10x^2}{2}\right) \quad \text{(a)}$$

$5 < x < 10$:

$$\frac{d^2v}{dx^2} = \frac{1}{EI}\left(R_1 x - 10\frac{x^2}{2} + 100\right) \quad \text{(b)}$$

Integrating twice, for the spans we get

$0 \le x < 5$:

$$\frac{dv}{dx} = \frac{1}{EI}\left(R_1\frac{x^2}{2} - \frac{10x^3}{6} + C_1\right) \quad \text{(c)}$$

$$v = \frac{1}{EI}\left(R_1\frac{x^3}{6} - \frac{10x^4}{24} + C_1 x + C_2\right) \quad \text{(d)}$$

$5 < x < 10$:

$$\frac{dv}{dx} = \frac{1}{EI}\left(R_1\frac{x^2}{2} - \frac{10x^3}{6} + 100x + C_3\right) \quad \text{(e)}$$

Example 12.5 (Continued)

$$v = \frac{1}{EI}\left(R_1\frac{x^3}{6} - \frac{10x^4}{24} + 100\frac{x^2}{2} + C_3x + C_4\right) \qquad \text{(f)}$$

We have four constants of integration plus the unknown R_1 to be determined. We can note that

$$\text{at } x = 0, \qquad v = 0$$

$$\text{at } x = L, \qquad \frac{dv}{dx} = v = 0$$

Applying these conditions, we have

$$C_2 = 0 \qquad \text{(g)}$$

$$C_3 = -\frac{R_1(10)^2}{2} + \frac{(10)(10)^3}{6} - (100)(10) = -50R_1 + 667 \qquad \text{(h)}$$

$$C_4 = -\frac{R_1(10^3)}{6} + \frac{(10)(10^4)}{24} - (100)\frac{(10^2)}{2} - (-50R_1 + 667)(10) \qquad \text{(i)}$$

$$= 333R_1 - 7.51 \times 10^3$$

Next we apply the *patch conditions* (**compatibility**) at $x = 5$. Thus

$$\left[\frac{dv(5^-)}{dx}\right]_{\text{Eq.(c)}} = \left[\frac{dv(5^+)}{dx}\right]_{\text{Eq.(e)}}$$

$$R_1\left(\frac{5^2}{2}\right) - 10\frac{(5^3)}{6} + C_1 = R_1\left(\frac{5^2}{2}\right) - 10\frac{(5^3)}{6} + (100)(5) + C_3$$

$$\therefore C_1 = 500 + C_3 \qquad \text{(j)}$$

Also,

$$[v(5^-)]_{\text{Eq.(d)}} = [v(5^+)]_{\text{Eq.(f)}}$$

$$R_1\left(\frac{5^3}{6}\right) - \frac{(10)(5^4)}{24} + C_1(5) = R_1\left(\frac{5^3}{6}\right) - \frac{(10)(5^4)}{24} + 100\frac{(5^2)}{2} + C_3(5) + C_4$$

$$\therefore 5C_1 = 1250 + 5C_3 + C_4 \qquad \text{(k)}$$

Replacing C_3 and C_4 using Eqs. (h) and (i) in Eqs. (j) and (k), we get the following simultaneous equations for C_1 and R_1:

Example 12.5 (Continued)

$$C_1 = 1167 - 50R_1$$
$$5C_1 = 83R_1 - 2.92 \times 10^3$$

Solving for R_1, we get

$$R_1 = 26.3 \text{ kN}$$

The other supporting forces are now readily available from rigid-body mechanics. Thus,

$\Sigma F_y = 0$:

$$R_1 - (10)(10) + R_2 = 0$$
$$\therefore R_2 = 73.7 \text{ kN}$$

$\Sigma M_0 = 0$:

$$-(100)(5) - 100 + R_2(10) - M_2 = 0$$
$$\therefore M_2 = 137 \text{ kN-m}$$

We have accordingly determined both the deflection equation and the supporting forces simultaneously.

Example 12.6

A uniformly loaded beam with three supports is shown in Fig. 12.10. Find the supporting forces.

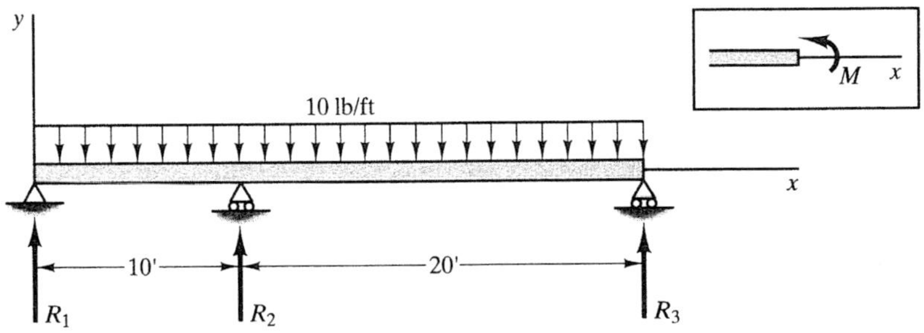

Figure 12.10. Uniformly loaded beam on three supports.

Example 12.6 (Continued)

We have a redundancy of unity and we shall use R_1 as the redundant constraint. As a first step we shall determine force R_2 as a function of R_1 by summing moments about the point at support R_3 to satisfy **equilibrium** for the entire beam taken as a free body. Thus:

$\sum M_{R_3} = 0$:

$$-R_1(30) - R_2(20) + (10)(30)(15) = 0$$

$$\therefore R_2 = -1.5R_1 + 225 \text{ lb} \qquad \text{(a)}$$

Now we can look at the two domains.

$0 \leq x \leq 10$:

$$\frac{d^2v}{dx^2} = \frac{1}{EI}\left(R_1 x - 10\frac{x^2}{2}\right)$$

$$\frac{dv}{dx} = \frac{1}{EI}\left(R_1\frac{x^2}{2} - 10\frac{x^3}{6} + C_1\right) \qquad \text{(b)}$$

$$v = \frac{1}{EI}\left(R_1\frac{x^3}{6} - 10\frac{x^4}{24} + C_1 x + C_2\right) \qquad \text{(c)}$$

$10 \leq x \leq 30$:

Using Eq. (a) for R_2, we have

$$\frac{d^2v}{dx^2} = \frac{1}{EI}\left[R_1 x - \frac{10x^2}{2} + (-1.5R_1 + 225)(x - 10)\right]$$

$$\frac{dv}{dx} = \frac{1}{EI}\left[R_1\frac{x^2}{2} - \frac{10x^3}{6} + (-1.5R_1 + 225)\frac{(x-10)^2}{2} + C_3\right] \qquad \text{(d)}$$

$$v = \frac{1}{EI}\left[R_1\frac{x^3}{6} - \frac{10x^4}{24} + (-1.5R_1 + 225)\frac{(x-10)^3}{6} + C_3 x + C_4\right] \qquad \text{(e)}$$

We now look at *boundary conditions.*

1. When $x = 0$, $v = 0$.

$$\therefore \text{From Eq. (c): } C_2 = 0 \qquad \text{(f)}$$

2. When $x = 10$, $v = 0$.

$$\therefore \text{From Eq. (c): } \frac{R_1(10^3)}{6} - \frac{(10)(10^4)}{24} + C_1(10) = 0 \qquad \text{(g)}$$

Example 12.6 (Continued)

3. When $x = 30$, $v = 0$.

$$\therefore \text{ From Eq. (e): } \frac{R_1(30^3)}{6} - \frac{(10)(30^4)}{24} +$$

$$(-1.5R_1 + 225)\frac{(20^3)}{6} + 30C_3 + C_4 = 0 \quad \text{(h)}$$

Next we consider the *patch conditions* (**compatibility**). At $x = 10$ ft,

$$\left[\frac{dv(10)}{dx}\right]_{\text{Eq.(b)}} = \left[\frac{dv(10)}{dx}\right]_{\text{Eq.(d)}}$$

Only using expressions that do not cancel out, we see from Eqs. (b) and (d) that

$$C_1 = C_3 \quad \text{(i)}$$

Also,

$$[v(10)]_{\text{Eq.(c)}} = [v(10)]_{\text{Eq.(e)}}$$

$$\therefore 10C_1 = 10C_3 + C_4 \quad \text{(j)}$$

Solving Eqs. (f)–(j) simultaneously, we can readily find for R_1 the result

$$R_1 = 12.5 \text{ lb}$$

From Eq. (a) we get $R_2 = 206.25$ lb, and by summing forces we get $R_3 = 81.25$ lb. As a *check,* take moments about R_1.

12.5 Superposition Methods

At times, for certain standard loadings, we may make good use of the *linearity* of the differential equations of deflection. That is, we may *superpose* known solutions for each of the loads on the beam. To aid in this approach, we show in Fig. 12.11 the deflection formulations for several simple types of loadings on simply-supported and cantilevered beams.

It should be clear that we may write the deflection equation immediately for a problem whose loadings are those presented in the abbreviated list of Fig. 12.11. More extensive lists may be found in structural handbooks.

We shall now illustrate the method of superposition as used for a beam supported in a statically determinate manner. Note that a force pointing downward is positive in accord with the structural convention.

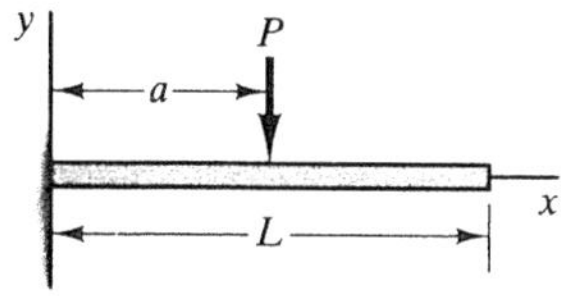

(a)

$0 \leqslant x \leqslant 0$

$$v = \frac{P}{6EI}(x^3 - 3x^2 a)$$

$a \leqslant x \leqslant L$

$$v = \frac{P}{6EI}[-(x-a)^3 + x^3 - 3x^2 a]$$

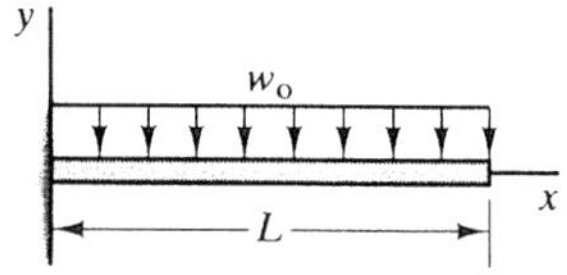

(b)

$$v = \frac{w_o x^2}{24EI}(-x^2 - 6L^2 + 4Lx)$$

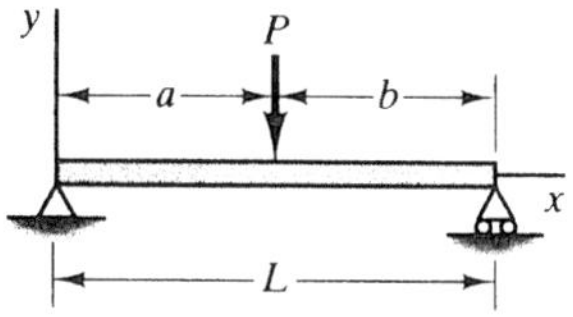

(c)

$0 \leqslant x \leqslant a$

$$v = \frac{Pb}{6LEI}[x^3 - (L^2 - b^2)x]$$

$0 \leqslant x \leqslant L$

$$v = \frac{Pb}{6LEI}\left[x^3 - \frac{L}{b}(x-a)^3 - (L^2 - b^2)x\right]$$

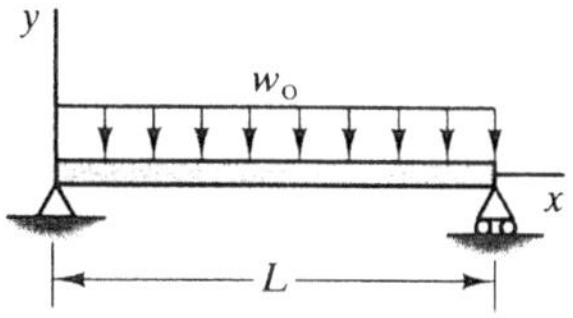

(d)

$$v = \frac{w_o x}{24EI}(-L^3 + 2Lx^2 - x^3)$$

Figure 12.11. Deflection formulas for simple beam loadings.

Example 12.7

Shown in Fig. 12.12 is a simply-supported beam carrying a uniform loading of 50 N/m and a 5000-N concentrated load. What is the deflection at the midpoint of the beam?

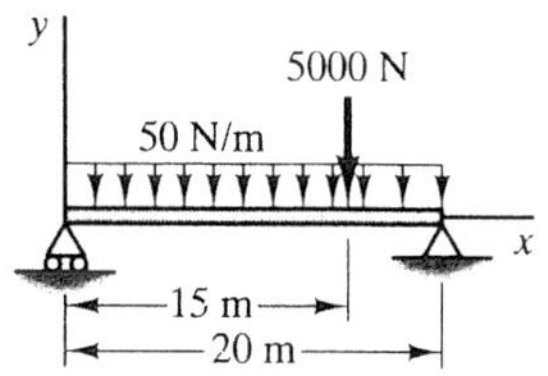

Figure 12.12. Simply-supported beam.

Example 12.7 (Continued)

Noting that the value of x to be used is less than $a = 15$, from Fig. 12.11(c) and 12.11(d) we have, on using the proper domain for the 5000-lb force

$$v = \frac{(5000)(5)}{(6)(20)(EI)}[x^3 - (20^2 - 5^2)x] + \frac{50x}{24EI}[-20^3 + (2)(20)x^2 - x^3]$$

At $x = 10$ we have

$$v(10) = \frac{1}{EI}\left\{\frac{(5000)(5)}{(6)(20)}[1000 - (400 - 25)(10)]\right.$$

$$\left. + \frac{(50)(10)}{24}[-8000 + (2)(20)(100) - 1000]\right\}$$

$$v(10) = -\frac{6.77 \times 10^5}{EI}\text{ m}$$

The superposition method is equally useful for handling statically indeterminate beam problems when the loadings and redundant constraints are those for which solutions are available from a list of formulas. The deflection equation, which is now the superposition of results stemming from individual consideration of the loads and the redundant constraints, is then subject to the additional boundary conditions supplied by the redundant constraints. We can then solve for the redundant constraints from these equations. This also gives us the full evaluation of the deflection curve. We illustrate this in the following examples.

Example 12.8

What is the reaction force at B for the statically indeterminate beam shown in Fig. 12.13?

The supporting force R_B at the right end of the beam is unknown, but we know that the deflection there is zero. Hence, using the formulas from Fig. 12.11 for a cantilever beam, we have

Example 12.8 (Continued)

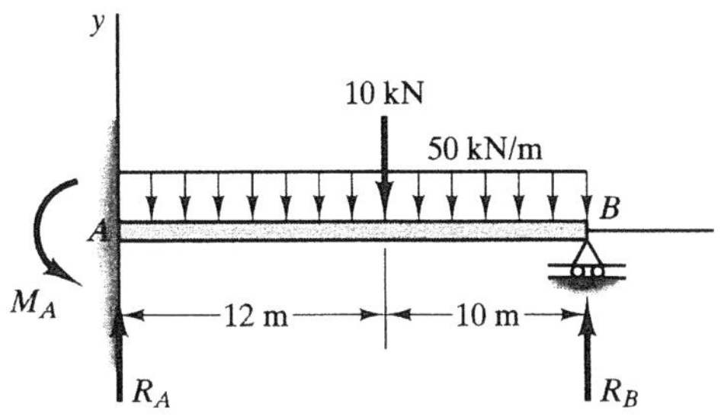

Figure 12.13. Statically indeterminate beam.

$0 < x \leq 12$:

$$v = \left\{ \frac{10}{6EI}[x^3 - (3x^2)(12)] - \frac{(-R_B)}{6EI}[-x^3 - (3x^2)(22)] + \frac{50x^2}{24EI}[-x^2 - (6)(22)^2 + (4)(22)(x)] \right\} 10^3 \text{ m} \tag{a}$$

$12 < x < 22$:

$$v = \left\{ \frac{10}{6EI}[-(x-12)^3 + x^3 - (3x^2)(12)] - \frac{(-R_B)}{6EI}[x^3 - (3x^2)(22)] + \frac{50x^2}{24EI}[-x^2 - (6)(22)^2 + (4)(22)(x)] \right\} 10^3 \text{ m} \tag{b}$$

We can determine R_B from Eq. (b) by setting $v = 0$ when $x = L = 22$ m. Thus,

$$0 = \left\{ \frac{10}{6EI}[-(22-12)^3 + (22)^3 - (3)(22)^2(12)] - \frac{(-R_B)}{6EI}[-(2)(22)^3] + \frac{(50)(22)^2}{24EI}[-22^2 - (6)(22)^2 + (4)(22)(22)] \right\} 10^3$$

Solving for R_B, we get

$$R_B = -416 \text{ kN} \tag{c}$$

Hence the force R_B is directed up according to the structural convention we are using for loads.

Example 12.9

Cantilever beam AB supports a load $P = 50$ lb, as shown in Fig. 12.14. A simply supported beam CD is attached to the end of cantilever beam AB and supports a load of 30 lb/ft. Determine the deflection δ of point B. The beams have the same cross sections and the same modulus of elasticity.

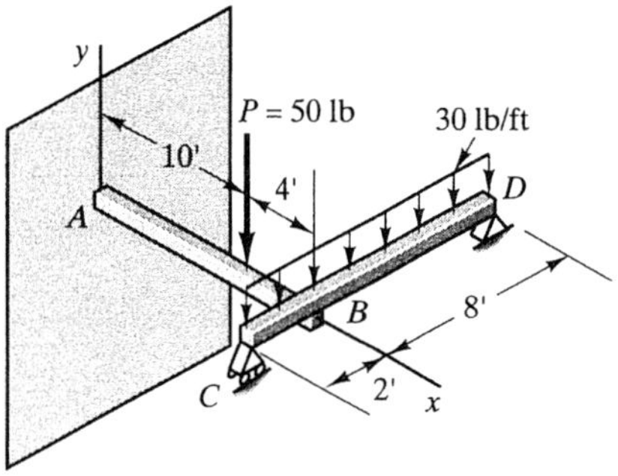

Figure 12.14. System of beams.

A free body of the entire system is shown in Fig. 12.15. This free body involves a parallel force system for which we have three available equations of equilibrium to solve for four unknowns. Clearly, the system is statically indeterminate. To expose the desired force in a free-body diagram, we will have to consider the cantilever beam as a separate free-body diagram. This is shown in Fig. 12.16. Note we have a parallel coplanar system of forces for this free body, so we have only two equations of equilibrium for those unknowns. We shall consider R to be the redundant force, and so using Fig. 12.11, we can give the deflection curve for $10 \leq x \leq 14$.

$$v = \frac{50}{6EI}[-(x-10)^3 + x^3 - (3)(x^2)(10)] + \frac{R}{5EI}[x^3 - (3)(x^2)(14)]$$

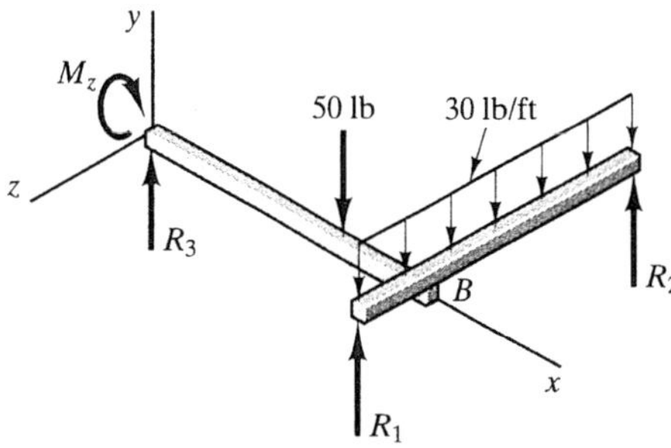

Figure 12.15. System is statically indeterminate.

Example 12.9 (Continued)

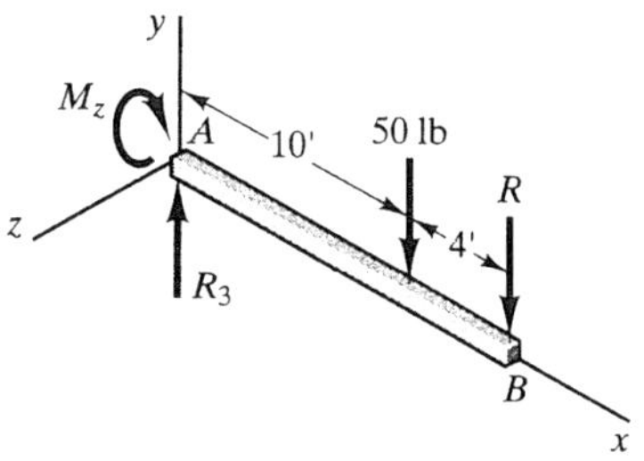

Figure 12.16. Free body of cantilever AB.

We will focus our attention on the redundant constraint. Thus, setting $x = 14$, we have for δ, the deflection at the tip of the cantilever beam,

$$\begin{aligned}\delta &= \frac{50}{6EI}[-(14-10)^3 + 14^3 - (3)(14^2)(10)] + \frac{R}{6EI}[14^3 - 3(14)^3] \\ &= -\frac{1}{EI}[26.7 \times 10^3 + (.915 \times 10^3)R]\end{aligned} \quad \text{(a)}$$

Now **compatibility** requires that the deflection at B for the simply supported beam be exactly the same δ as in Eq. (a). We show this beam in Fig. 12.17 using the reaction to the force R of Fig. 12.16. Again using Fig. 12.11, we can give the deflection at B directly as follows:

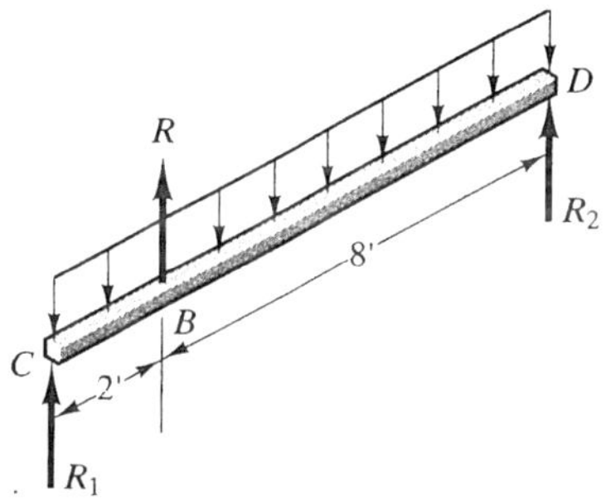

Figure 12.17. Free body of simply supported beam.

$$\begin{aligned}\delta &= \frac{(30)(2)}{24EI}[-10^3 + (2)(10)(4) - 8] \\ &\quad + \frac{(-R)(8)}{(6)(10)(EI)}\left[2^3 - (10^2 - 8^2)(2)\right] \\ &= -\frac{1}{EI}(2.32 \times 10^3 - 8.53R)\end{aligned} \quad \text{(b)}$$

Equating the δ's from Eqs. (a) and (b), we get

$$-26.7 \times 10^3 - .915 \times 10^3 R = -2.32 \times 10^3 + 8.53R$$

$$\therefore R = -26.4 \text{ lb (see footnote 3 below)}$$

We can now get δ from Eq. (a). Thus,

$$\delta = \frac{-1}{EI}[26.7 \times 10^3 + (.915)(-26.4)(10^3)] = \frac{-2.54 \times 10^3}{EI} \text{ ft}$$

[3]The minus sign indicates we have assumed the opposite direction required.

*12.6 Shear Deflection of Beams

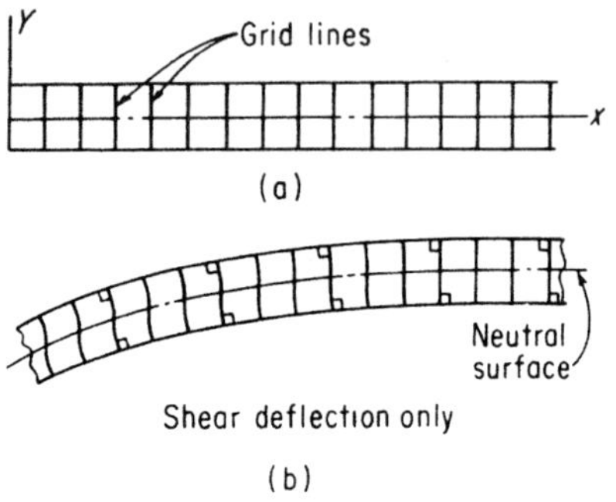

Figure 12.18. Shear deflection.

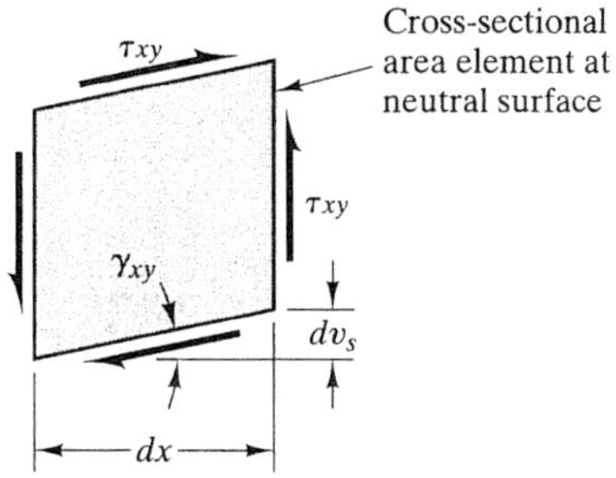

Figure 12.19. Deformed element.

The deflections that we have thus far considered stem from the effects of the bending moment along the beam, as pointed out earlier. There is also the possibility of deflection of a beam owing to the shear forces. We shall soon see that this deflection contribution is usually very small compared to the deflection resulting from the bending moments in the case of long slender beams, and in most instances, this deflection is disregarded.

We shall now consider a symmetric beam loaded in the plane of symmetry. The beam is shown in the undeformed geometry in Fig. 12.18(a) with a system of vertical grid lines. Fig. 12.18(b) shows the shear deformation of the beam and its system of grid lines. Since the shear stress τ_{xy} is zero at the upper and lower fibers, the shear strains γ_{xy} are zero there for a Hookean material and hence the grid lines must remain at right angles to the edges, as has been indicated in the diagram. This means that at some position along each grid line the direction of the grid line must be vertical. We can assume that this point is at the neutral surface so that area elements of the cross section at positions along the neutral surface *remain vertical and merely slide one along another,* as shown in Fig. 12.19. Observing this diagram, we note that the differential equation for the shear deflection v_S of the neutral surface can then be easily formulated as

$$dv_s = (\gamma_{xy})_{\text{N.A.}}\, dx = \frac{(\tau_{xy})_{\text{N.A.}}}{G}\, dx \tag{12.10}$$

where $(\gamma_{xy})_{\text{N.A.}}$ and $(\tau_{xy})_{\text{N.A.}}$ are the engineering shear angle and shear stress, respectively, at the neutral axis. It is customary to express $(\tau_{xy})_{\text{N.A.}}$ in the following form:

$$(\tau_{xy})_{\text{N.A.}} = \frac{\alpha V_y}{A} \tag{12.11}$$

where V_y is the shear force and A is the total cross-sectional area. A computation of $(\tau_{xy})_{\text{N.A.}}$ for the case of a rectangular cross section (see Example 11.7) would indicate to us that $\alpha = 3/2$ for that case. Values of α for other cross sections are available in handbooks. Equation (12.10) can then be written in the following manner:

$$\frac{dv_S}{dx} = \frac{\alpha V_y}{GA} \tag{12.12}$$

We then integrate this differential equation to get v_S.

To get the total deflection, v_T, we superpose the shear deflection, v_S, and the deflection resulting from bending, v_B. In the following example we shall evaluate both shear and bending deflections.

Example 12.10

A cantilever beam having a rectangular cross section and carrying a uniform loading w is shown in Fig. 12.20. Compute the total deflection curve v_T.

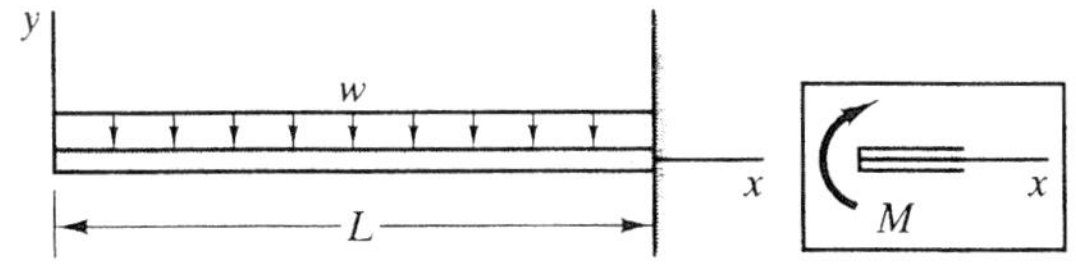

Figure 12.20. Cantilever beam.

We start with *shear deformation* for which in the domain $0 \leq x < L$. We have

$$\frac{dv_S}{dx} = \frac{\alpha V_y}{GA} = \frac{3wx}{2GA} \tag{a}$$

Integrating, we have

$$v_S = \frac{3}{2}\frac{w}{GA}\frac{x^2}{2} + C_1 \tag{b}$$

The boundary condition here is that $v_S = 0$ when $x = L$. Thus,[4]

$$C_1 = -\frac{3}{4}\frac{w}{GA}L^2 \tag{c}$$

Our result is then

$$v_S = \frac{3}{4}\frac{w}{GA}(x^2 - L^2) \tag{d}$$

For the deflection because of bending, we leave it to you to show that

$$v_B = \frac{w}{EI}\left(-\frac{x^4}{24} + \frac{L^3}{6}x - \frac{L^4}{8}\right) \tag{e}$$

The total deflection is then

$$v_T = w\left[\frac{3}{4GA}(x^2 - L^2) + \frac{1}{EI}\left(-\frac{x^4}{24} + \frac{L^3}{6}x - \frac{L^4}{8}\right)\right] \tag{f}$$

[4]Note that $dv_S/dx \neq 0$ at $x = L$. Can you explain why?

Example 12.10 (Continued)

To compare the bending and shear deflections, compute the maximum deflection at $x = 0$ for each contribution. Thus, we have

$$(v_B)_{\max} = -\frac{wL^4}{8EI} \tag{g}$$

$$(v_S)_{\max} = -\frac{3}{4}\frac{wL^2}{GA} \tag{h}$$

Take the cross section of the beam to have a width b and a height h. Then the maxima above can be expressed as

$$(v_B)_{\max} = -\frac{wL^4}{8E(\frac{1}{12})bh^3} = -\frac{3}{2}\frac{wL^4}{Ebh^3} \tag{i}$$

$$(v_S)_{\max} = -\frac{3}{4}\frac{wL^2}{Gbh} \tag{j}$$

Taking the ratio of Eq. (i) over Eq. (j) and noting that E/G is approximately 2.5 for steels, we have

$$\frac{(v_B)_{\max}}{(v_S)_{\max}} = \frac{\frac{3}{2}(wL^4/Ebh^3)}{\frac{3}{4}(wL^2/Gbh)} = .8\left(\frac{L}{h}\right)^2 \tag{k}$$

We see here that for long slender beams the deflection resulting from bending will be *much larger* than that from shear. Thus, if $L = 20h$, the maximum bending deflection is 320 times the shear deflection. The latter is clearly a negligible contribution. For short stubby beams where say $L = 2h$ the maximum bending deflection is only 3.2 times the shear deflection, and we should then not neglect the latter contribution.

In the bulk of problems that we shall deal with, we shall have long slender beams. Indeed, the technical theory of bending is accurate only for such cases, so we shall neglect the deformation resulting from shear.

12.7 Energy Methods for Beams

We now consider the energy stored in a symmetric, prismatic beam of length L due to external loads normal to the centroidal axis. In Chapter 11 we explored the relationships between external loads on a beam and the internal stress resultants. Specifically, we derived the flexure formula that related the bending moment to longitudinal normal stress and Jourawski's formula that related the transverse shear stress to the shear force. Therefore, we can identify two strain energies: one for the bending normal stress, the other for the transverse shear stress.

From our work on symmetric beams with loads normal to centroidal axis, we know using the familiar xyz reference that there is only one nonzero normal stress. If we choose x as the longitudinal axis, the normal stress τ_{xx}, for loading in the xy-plane is given by the flexure formula

$$\tau_{xx} = -\frac{M_z y}{I_{zz}}$$

We can now use Eq. (6.34) to develop an expression for the complementary strain energy in the beam due to *normal stress*, thus

$$U^* = \iiint_V \frac{\tau_{xx}^2}{2E}\,dv = \int_0^L \left[\iint_A \frac{M_z^2 y^2}{2EI_{zz}^2}\,dA'\right] dx = \int_0^L \left[\frac{M_z^2}{2EI_{zz}^2}\iint_A y^2\,dA'\right] dx$$

Noting that $\iint_A y^2\,dA' = I_{zz}$, we then have for the complementary strain energy in a symmetric beam loaded in the xy centroidal plane:

$$U^* = \int_0^L \left[\frac{M_z^2}{2EI_{zz}}\right] dx \tag{12.13}$$

If the loading is in the xz-plane, the complementary strain energy is given as

$$U^* = \int_0^L \left[\frac{M_y^2}{2EI_{yy}}\right] dx \tag{12.14}$$

We now consider the complementary strain energy due to the *shear stress* in the beam resulting from a shear force V_y at a section. For the symmetric beam with a rectangular cross section and, loaded in the xy-plane as discussed above, we have τ_{xy} as the only nonzero shear stress. We again use Eq. (6.34) to develop an expression for the strain energy in the beam due to shear stress, thus

$$U^* = \iiint_V \frac{\tau_{xy}^2}{2G}\,dv$$

Recall that for shear stress arising from bending, the distribution of τ_{xy} at a cross section is not constant; in general, it may vary across the cross section (i.e., a function of both y and z). Determining this distribution for each cross section can be very time-consuming. One way to avoid this is to use tabulated *shape factors*. These shape factors correlate the energy computed using the actual shear stress distribution with that of a cross section with equal area but with a constant distribution of shear stress. The main advantage is that the constant shear stress distribution is readily calculated as that due to *simple shear,* i.e., $\tau_{xy} = V/A$, where A is the cross-sectional area. We can then state

$$U^* = \iiint_V \frac{\tau_{xy}^2}{2G} dv \bigg|_{\tau_{xy} \to \text{actual}} = \alpha_y \iiint_V \frac{\tau_{xy}^2}{2G} dv \bigg|_{\tau_{xy} \to \text{simple}} \tag{12.15}$$

where α_y represents the shape factor for a given cross section. Using the simplified version of the energy equation, we have

$$U^* = \alpha_y \int_0^L \left[\iint_A \frac{V_y^2}{2GA^2} dA' \right] dx = \alpha_y \int_0^L \frac{V_y^2}{2GA} dx$$

So that the *complementary strain energy due to bending induced shear* for loading in the xy-plane is given as

$$U^* = \alpha_y \int_0^L \frac{V_y^2}{2GA} dx \tag{12.16}$$

For loading in the xz-plane we obtain

$$U^* = \alpha_z \int_0^L \frac{V_z^2}{2GA} dx \tag{12.17}$$

The shape factors α_y and α_z are determined based on the shape and orientation of the cross section. The following example illustrates a method for determining these factors.

Example 12.11

For a beam with a rectangular cross section, as shown in Fig. 12.21, compute the shape factor associated with shear strain energy. The beam is prismatic and the loading is in the xy plane.

We start by assuming that the shear stress distribution across the face of the section can be adequately described by Jourowski's formula

$$\tau_{xy} = \frac{V_y Q_z}{I_{zz} b}$$

so that Eq. (12.15) becomes[5]

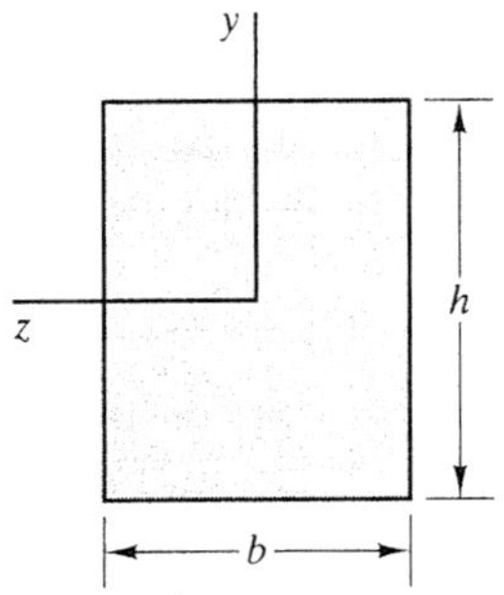

Figure 12.21. Rectangular beam section.

[5] α_y is a *shape factor* equating energy for uniform shear stress to that of the actual shear stress distribution. Note that α_y differs from α (see Eq. 12.11), since the former is based on strain energy and the latter is based on τ_{xy} at the neutral axis.

Example 12.11 (Continued)

$$U^* = \iiint_V \frac{1}{2G}\left[\frac{V_y Q_z}{I_{zz} b}\right]^2 dv = \alpha_y \iiint_V \frac{1}{2G}\left[\frac{V_y}{A}\right]^2 dv \qquad \text{(a)}$$

Canceling terms from both sides of the equation and, recognizing that the beam is prismatic of length L and cross section area A, we integrate with respect to x on both sides and then with respect to A on the right side. Then we get on cancelling L,

$$\frac{1}{I_{zz}^2}\iint_A \left[\frac{Q_z}{b}\right]^2 dA' = \frac{\alpha_y}{A}$$

$$\therefore\ \alpha_y = \frac{A}{I_{zz}^2}\iint_A \left[\frac{Q_z}{b}\right]^2 dA' \qquad \text{(b)}$$

For the rectangle we have

$$A = bh$$
$$I_{zz} = \tfrac{1}{12}bh^3$$
$$Q_z = \frac{b}{2}\left[\left(\frac{h}{2}\right)^2 - y^2\right] \qquad \text{(From Example 11.7)}$$

Substitution of the above parameters into Eq. (b) yields

$$\alpha_y = \frac{bh}{\frac{1}{144}b^2h^6}\int_{-\frac{b}{2}}^{\frac{b}{2}}\int_{-\frac{h}{2}}^{\frac{h}{2}}\left[\frac{\frac{b^2}{4}\left[\left(\frac{h}{2}\right)^2 - y^2\right]^2}{b^2}\right]dy\,dz$$

$$\therefore\ \alpha_y = \frac{6}{5}$$

So we see that the shape factor $\alpha_y = 6/5$ for the rectangular section. Note that we get the same shape factor for loading in the xz plane (i.e., $\alpha_z = 6/5$).[6]

Shape factors for other common cross sections are shown in Fig. 12.22.

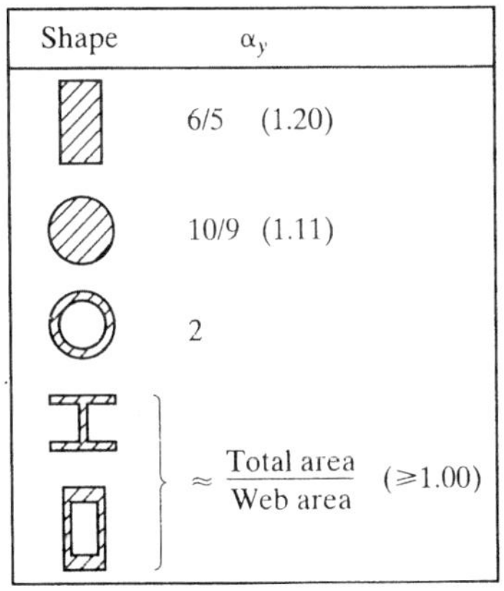

Figure 12.22. Shape factors for common cross sections.

Example 12.12

Determine the displacement at the tip ($x = 0$) of the end-loaded cantilevered beam shown in Fig. 12.23 using Castigliano's

[6] The theory of elasticity predicts a different shear stress distribution for certain ratios of b/h. Therefore, the validity of the shape factor is for those conditions for which Jourawski's formula are valid. See Ugural and Fenster, *Advanced Strength and Applied Elasticity*, Prentice Hall, 1995.

Example 12.12 (Continued)

second theorem. Assume that the beam is prismatic and is made from a linear elastic material.

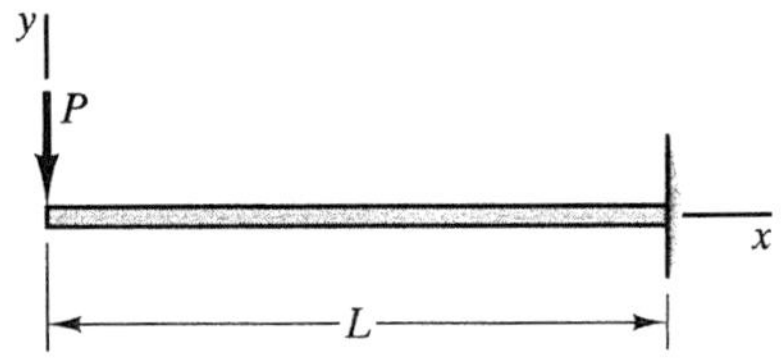

Figure 12.23. Tip loaded cantilevered beam.

We start by assuming that the displacement at the beam tip is due to both bending and shear deformation in the beam. We can write U^* (or U since the material is linear elastic) as the sum of the flexural energy given by Eq. (12.14) and the shear energy given by Eq. (12.17) **(constitutive law)**

$$U^* = \int_0^L \frac{M_z^2}{2EI_{zz}}\,dx + \alpha_y \int_0^L \frac{V_y^2}{2GA}\,dx \tag{a}$$

The bending moment at any position is $M_z(x) = -Px$ and the shear is simply $V_y(x) = -P$ **(equilibrium)**. Substitution into Eq. (a) gives

$$U^* = \int_0^L \frac{(-Px)^2}{2EI_{zz}}\,dx + \alpha_y \int_0^L \frac{(-P)^2}{2GA}\,dx \tag{b}$$

Evaluating the integrals in Eq. (b) we get

$$U^* = \frac{P^2L^3}{6EI_{zz}} + \alpha_y \frac{P^2L}{2GA} \tag{c}$$

Using the second Castigliano theorem given in Eq. (6.58) we can readily determine the displacement Δ at the *location* of and in the *direction* of the load P as **(compatibility)**

$$\frac{\partial U^*}{\partial P} = \Delta = \frac{PL^3}{3EI_{zz}} + \alpha_y \frac{PL}{GA} \tag{d}$$

Since the solution for Δ is positive, the beam deflects in the direction of the load P, as was expected. From our work in this chapter, we recognize the first term on the right-hand side of Eq. (d) as the dis-

Example 12.12 (Continued)

placement at the tip of a point-loaded cantilevered beam. The second term is the additional displacement due to the shear deformation in the beam.

Carrying the calculations further, we can determine the relative influence of the shear deformation on the tip displacement. Assuming a rectangular cross section of height h and width b we have from Example 12.11 [Fig. 12.22] $\alpha_y = 6/5$. The displacement given by Eq. (d) is now expressed as

$$\Delta = \frac{PL}{bh}\left[\frac{4}{E}\left(\frac{L}{h}\right)^2 + \frac{6}{5G}\right] \tag{e}$$

If we assume the beam is made from steel ($\nu = 0.3$) then from Eq. (6.10) we have $E = G/2.6$. Rewriting Eq. (e)

$$\Delta = \frac{PL}{bhE}\left[4\left(\frac{L}{h}\right)^2 + 3.12\right] \tag{f}$$

The first term in the bracket of Eq. (f) is the contribution due to bending, the second term is due to shear. The ratio of beam length to height (L/h) is often called the *aspect ratio* of the beam. We can see from Eq. (f) that for a beam with an aspect ratio of unity (i.e., $L = h$), the bending and shear contributions are of the same order of magnitude; the shear deformation in this case is significant. As the aspect ratio increases, the beam becomes *slender* and the relative contribution to the displacement of the bending and shear terms change. We can see in Eq. (f) that the aspect ratio, which appears in the bending term, is to the second power while the shear term is constant. Clearly, as the length of the beam becomes much greater than its height, the affect of shear on the tip displacement becomes negligible.[7] For example, with an aspect ratio $L/h = 10$, the contribution of shear to the deformation is less than 1%, this drops to less than 0.2% for $L/h = 20$. Similar results are obtained with other beam cross sections. We conclude that for slender beams (say, $L/h \geq 10$), the effect of shear on the displacement may be ignored.

Example 12.13

Consider the simply-supported, linear-elastic beam with a 100-kN force at P its center as shown in Fig. 12.24. We are to determine the

[7]Recall that this was also pointed out in starred section 12.6.

Example 12.13 (Continued)

deflection and rotation at point A as a result of this 100-kN loading. Use Castigliano's second theorem.

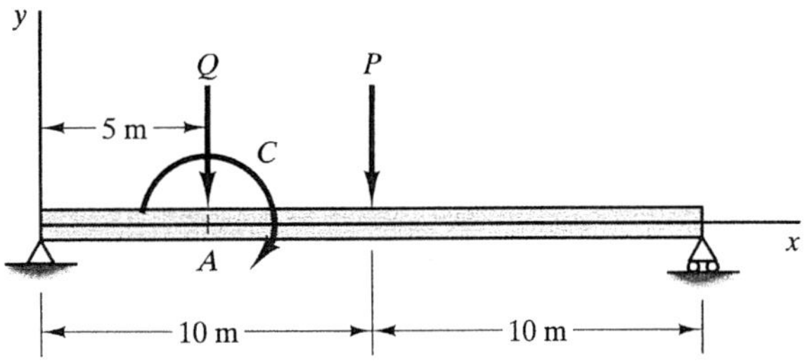

Figure 12.24. Simply Supported Beam.

To apply Castigliano's second theorem, we need to have a generalized point force at the point of interest. In this case, we do not have a point load or a point couple-moment where we want the displacement and rotation. So we proceed as follows: we introduce a *hypothetical* (or "dummy") point force Q and point couple C at A. The effect is that we now have the necessary generalized point forces at A (Q for the displacement and C for the rotation). We keep these dummy forces until we have invoked Castigliano's second theorem; then we can evaluate the resulting expressions with the dummy loads set to zero. Although Q and C are dummy loads, we must treat them as if they really exist. Therefore, we must include them in our equilibrium calculations. We now illustrate this procedure.

If we assume that the beam is slender, we can neglect shear effects on the displacement. Therefore, the complementary strain energy is given in terms of the bending moment only. If we let the applied 100-kN force be represented by P, then the moment is given as **(equilibrium):**

$$0 \le x < 5: \quad M_z(x) = (P/2 + 3Q/4 - C/L)\,x$$
$$5 < x \le 10: \quad M_z(x) = (P/2 + 3Q/4 - C/L)\,x - Q(x-5) + C$$
$$10 \le x \le 20: \quad M_z(x) = (P/2 + 3Q/4 - C/L)\,x - Q(x-5) - P(x-10) + C$$

We now write the energy as **(constitutive law)**

Example 12.13 (Continued)

$$U^* = \int_0^L \frac{M_z^2}{2EI_{zz}}\,dx$$

$$= \int_0^5 \frac{\{[(P/2) + (3Q/4) - (C/L)]x\}^2}{2EI_{zz}}\,dx$$

$$+ \int_5^{10} \frac{\{[(P/2) + (3Q/4) - (C/L)]x - Q(x-5) + C\}^2}{2EI_{zz}}\,dx$$

$$+ \int_{10}^{20} \frac{\{[(P/2) + (3Q/4) - (C/L)]x - Q(x-5) - P(x-10) + C\}^2}{2EI_{zz}}\,dx$$

We do not yet expand the above expressions, nor do we integrate; we *first* apply the second Castigliano theorem to U* to get the displacement Δ_A and rotation θ_A. This will reduce some of the computations. To find Δ_A, we differentiate U* with respect to Q, as **(compatibility)**

$$\Delta_A = \frac{\partial U^*}{\partial Q}$$

$$= \int_0^5 \frac{\{[(P/2) + (3Q/4) - (C/L)]x\}\{3x/4\}}{EI_{zz}}\,dx$$

$$+ \int_5^{10} \frac{\{[(P/2) + (3Q/4) - (C/L)]x - Q(x-5) + C\}\{3x/4 - (x-5)\}}{EI_{zz}}\,dx$$

$$+ \int_{10}^{20} \frac{\{[(P/2) + (3Q/4) - (C/L)]x - Q(x-5) - P(x-10) + C\}\{3x/4 - (x-5)\}}{EI_{zz}}\,dx \qquad \text{(a)}$$

and to find θ_A, we differentiate with respect to C, as

$$\theta_A = \frac{\partial U^*}{\partial Q}$$

$$= \int_0^5 \frac{\{[(P/2) + (3Q/4) - (C/L)]x\}\{-x/L\}}{EI_{zz}}\,dx$$

$$+ \int_5^{10} \frac{\{[(P/2) + (3Q/4) - (C/L)]x - Q(x-5) + C\}\{1 - x/L\}}{EI_{zz}}\,dx \qquad \text{(b)}$$

$$+ \int_{10}^{20} \frac{\{[(P/2) + (3Q/4) - (C/L)]x - Q(x-5) - P(x-10) + C\}\{1 - x/L\}}{EI_{zz}}\,dx$$

Example 12.13 (Continued)

Again, to help reduce the computations, we now set $P = 100$-kN, $Q = 0$, and $C = 0$ in both Eqs. (a) and (b). (Note that the sign of P was included when computing the moment.) We evaluate the expressions at this step having avoided integrating the now unnecessary terms associated with the dummy loads Q and C. Performing the integration gives us the required information:

$$\Delta_A = \frac{34{,}375}{3EI_{zz}} \text{ m}$$

$$\theta_A = \frac{1875}{EI_{zz}} \text{rad}$$

The above values act in the direction of their respective generalized forces, Q and C as shown in Fig. 12.24. We see that both the displacement and rotation are positive, indicating that they act in the same sense as Q and C, respectively. The unit of EI_{zz} is kN-m.^{2}

Example 12.14

Determine the supporting forces for the statically indeterminate, uniformly loaded beam in Fig. 12.25. Use the second Castigliano theorem. Take EI_{zz} as constant for the beam.

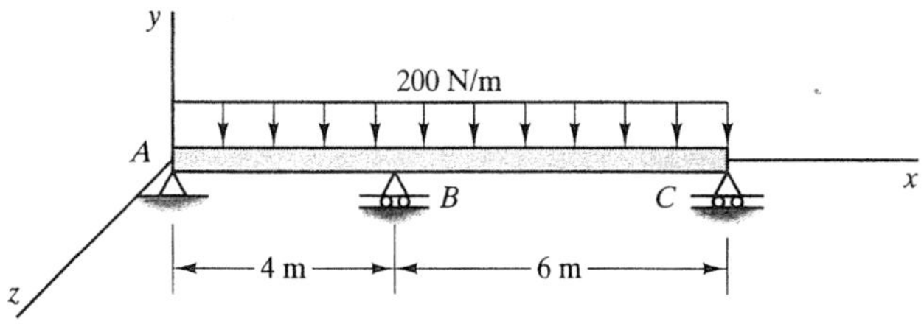

Figure 12.25. Statically indeterminate beam.

We take as the redundant force R_A. Taking moments about support C we have from **equilibrium**

$\underline{\Sigma M_C = 0}$:

$$R_A(10) + (200)(10)(5) - R_B(6) = 0$$

$$\therefore R_B = 1666.7 - 1.6667\, R_A \quad \text{(a)}$$

Example 12.14 (Continued)

We next use the second Castigliano theorem applied to support A. Thus, setting the deflection equal to zero at this point we have (**compatibility** and **constitutive law**)

$$\Delta_A = 0 = \frac{\partial}{\partial R_A}\int_0^4 \frac{\left(R_A x - 200\dfrac{x^2}{2}\right)^2}{2EI_{zz}}\,dx + \frac{\partial}{\partial R_A}\int_4^{10} \frac{[R_A x - 100x^2 + (1666.7 - 1.6667R_A)(x-4)]^2}{2EI_{zz}}\,dx$$

Differentiating under the integral sign, we get on multiplying through by EI_{zz}

$$0 = \int_0^4 (R_A x - 100x^2)(x)dx + \int_4^{10} [R_A x - 100x^2 + (1666.7 - 1.6667R_A)(x-4)][x - (x-4)(1.6667)]dx$$

Carrying out the integration either by long hand or by using computer software we arrive at the following equation:

$$53.53R_A - 1.1862 \times 10^4 = 0$$
$$\therefore R_A = 225 \text{ N}$$

Finally, from equilibrium again, we can determine the other supporting forces to be

$$R_B = 1292 \text{ N} \qquad R_C = 483 \text{ N}$$

12.8 Closure

In this chapter we have presented various means of determining the deflection v of the neutral surface for various beam problems. We used the *domain* approach and we first considered statically determinate beam problems. This then set the stage for the study of statically indeterminate beam problems where the supporting forces were solved *simultaneously* with the deflection curve. We next showed how we could employ the superposition method, encompassing known deflection equations for simple loadings and supports, to expeditiously solve statically determinate and statically indeterminate beam problems. In practice you will probably rely heavily on this method. We then examined the interesting topic of shear deflection as distinct from bending deflections, which, up to this point, we have limited ourselves to in the chapter. We showed that for a long slender beam, we could generally ignore this contribution. Finally, we presented energy methods applied to beams using the second Castigliano theorem.

No doubt you were not too thrilled with the amount of algebra needed to complete a solution for many of the beam problems. There is a means of greatly cutting down on this arithmetic. The method favored in this text is the use of *singularity functions* presented in the next chapter. This is a bona fide area of mathematics that is easily learned. In fact, some of our students, desperate to escape the arithmetic of the present chapter, have on their own read and absorbed this material and, to the delight of the instructors, have used it in homework and even in tests. The use of singularity functions that we present goes far beyond this course appearing in heat transfer, electromagnetic theory, circuit theory, applied mathematics, elasticity, and in the growing area of numerical boundary elements.

*12.9 A Look Ahead: A Closer Look at Beam Deflection and Highlights

When we discuss the deflection of a beam, we normally are referring to the deformed shape of the neutral surface of the beam. There are two essential ways of causing this deformation. By far, the largest contributor to this deformation is due to bending of the beam whereby fibers above the neutral surface either are all shortened or all lengthened. On the other side of the neutral surface the opposite behavior is occurring. To accommodate this action, the neutral surface must change shape giving rise to what we call bending deformation. There is a contribution to deformation of the neutral surface due to shear whereby adjacent infinitesimal vertical lines move vertically relative to each other. The contribution to deflection from shear is usually very small for long slender beams, and thus is often neglected for such beams.

Note that only part of the deformation of the beam has been specified. A good approximation for the deformation of the rest of the beam material is to assume that plane sections of the beam in the undeformed geometry remain plane and rigid and move so as to remain oriented normal to the neutral surface. In the case of plates, we focus on the midplane surface of the plate and give the deflection of this surface. And for the rest of the plate, we usually assume that vertical lines in the undeformed geometry remain rigid and move so as to remain normal to the deformed midplane surface. When you read the "A Look Ahead" section for finite elements in Chapter 17, you will see that the evaluation of the movement of the nodal points using energy methods is analogous to the evaluation of the neutral surface deformation for beams or to the deformation of the midsurface of plates. The relation of the rest of the deformation of the beam material relative to the deformed neutral surface is analogous to choosing interpolation functions for finite elements, and similarly for plates.

More advanced studies of beams will include consideration of the shear deformation along with the bending deformation for beams having more complex cross sections. Beams for which this dual approach is employed are called Timoshenko beams. Other interesting problems which will most likely be examined include continuous elastic foundations and numerical methods of analysis (e.g., the Ritz method). Advanced studies of plates would then follow.

PROBLEMS

12.1. [12.2] In Problem 11.1, determine the deflection curve. What is the maximum deflection? (Take $E = 2 \times 10^{11}$Pa.)

12.2. [12.2] What is the deflection curve for Problem 11.93? Get result in terms of *EI*.

12.3. [12.2] What is the deflection curve for Problem 11.95? Set up equations for constants of integration, but do not solve for them.

12.4. [12.2] What is the deflection curve for Problem 11.13?

12.5. [12.2] In Problem 11.48, compute the displacement vector of the point of application of the force *P*. Take $E = 30 \times 10^6$ psi.

12.6. [12.2] What is the deflection curve for the beam shown in Fig. P.12.6? Set up equations for constants of integration, but do not solve for them.

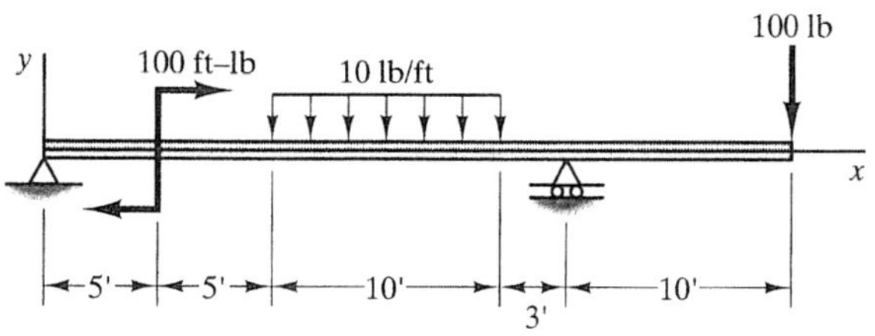

Figure P.12.6.

12.7. [12.2] What is the deflection at $x = 15$ ft in Fig. P.12.7?

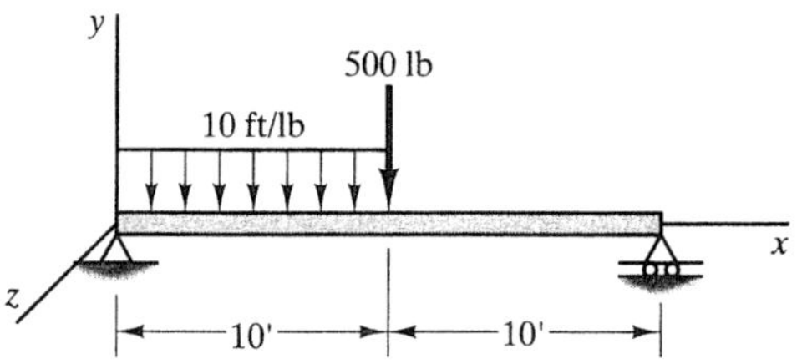

Figure P.12.7.

12.8. [12.2] Find the position where the slope is largest (see Fig. P.12.8). Where is d^3v/dx^3 largest? What is its value? Take $E = 2 \times 10^{11}$ Pa.

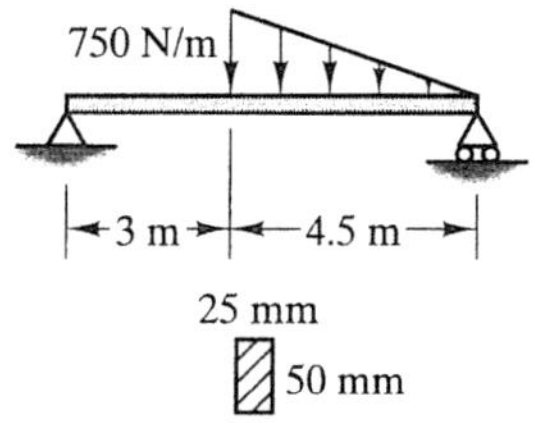

Figure P.12.8.

12.9. [12.2] Determine the deflection curve for the simply supported beam in Fig. P.12.9. Notice that the intensity of loading for the triangular load is given at $x = 11$ m.

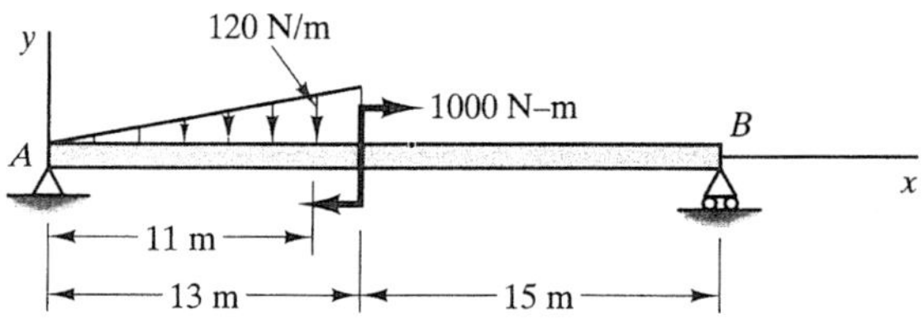

Figure P.12.9.

12.10. [12.2] A steel beam with a cross-sectional height of 100 mm and width of 20 mm is to be replaced by an aluminum beam. If the width remains unchanged, what is the required cross-sectional height of the aluminum beam to keep the stiffness EI, the same as the steel beam? Will the bending stresses in the aluminum beam be higher or lower than in the steel beam? Take $E_{steel} = 210$ GPa and $E_{alum} = 70$ GPa.

12.11. [12.2] For the beam shown in Fig. P.12.11,

(a) Find the supporting forces.
(b) Get the deflection equations for ED and find the deflection of point C.
(c) What is the **total** vertical movement of point A considering CBA to be rigid?

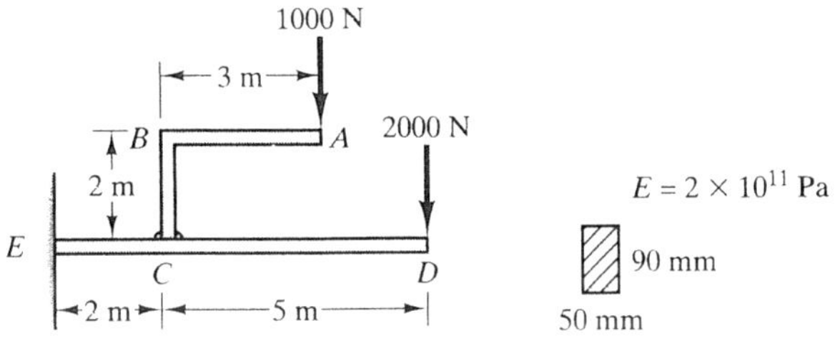

Figure P.12.11.

12.12. [12.2] What load P is needed to cause a deflection of 5 mm at the midpoint of the beam (Fig. P.12.12)? Take $E = 2 \times 10^{11}$ Pa.

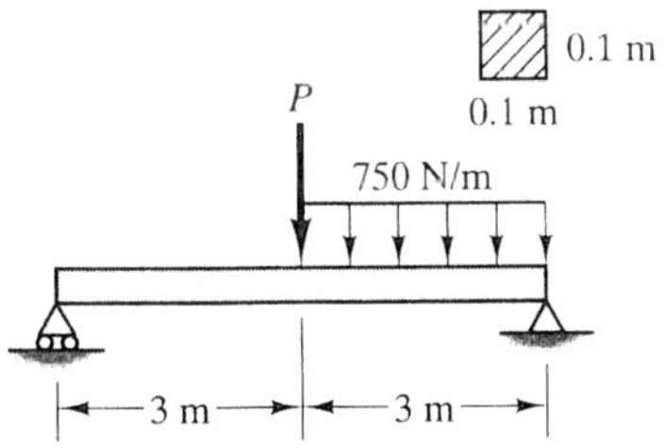

Figure P.12.12.

12.13. [12.2] A beam is shown in Fig. P.12.13 on an *elastic foundation* represented by the series of small arrows at the base. This foundation resists downward deflection in a linear manner measured by the *foundation modulus* k defined as the resisting force per unit length x and per unit deflection v. Write the differential equation for this member to support load $w = \sin(\pi x/L)$. What is essentially different between this equation and the ones that we have been solving up to now?

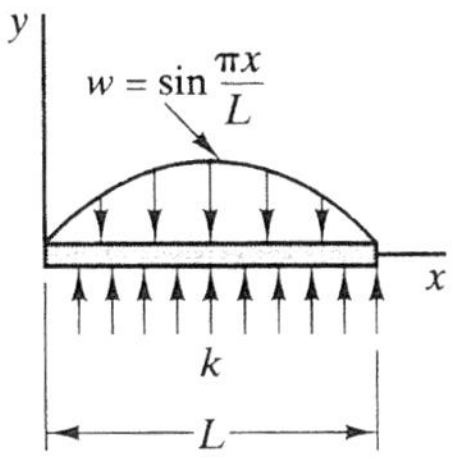

Figure P.12.13.

In Problems 12.14 and 12.15 the beams are not uniform. Assume that the Euler-Bernoulli theory of beams is still valid for such beam problems for approximate results.

12.14. [12.2] A cantilever beam is shown in Fig. P.12.14 wherein each edge forms a curve that is a parabola given as $y^2 = (a^2/L)x$ for the top curve (the bottom curve is a mirror image). What is the deflection curve for a tip load P? The beam is of uniform thickness t throughout and has a constant modulus of elasticity E. Show that the deflection at tip is $-PL^3/(Ea^3t)$.

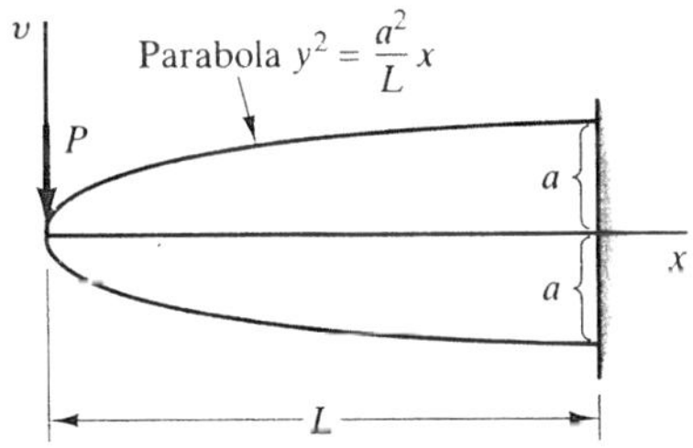

Figure P.12.14.

12.15. [12.2] Shown in Fig. P.12.15 is a cantilever beam having a circular cross section with a linearly varying diameter. Formulate the deflection curve for this beam under its own weight. The material has a specific weight γ and a modulus of elasticity E.

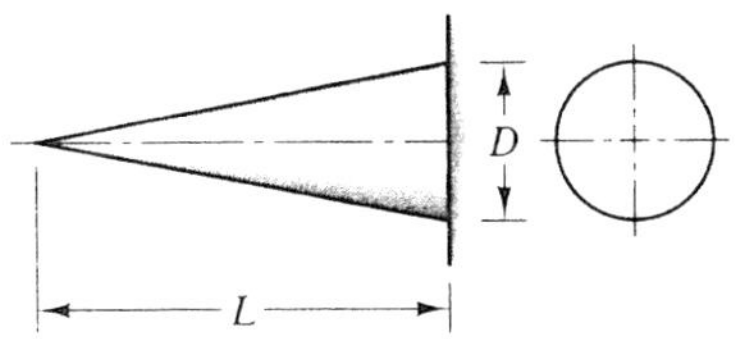

Figure P.12.15.

12.16. [12.2] The elastic modulus E can be determined by a bending test (see Fig. P.12.16), rather than from the familiar one-dimensional tensile or compressive test. (a) Determine an equation for E based on the 3-Point bend test shown. (b) List three reasons why this value may not agree with that obtained from a one-dimensional test. *Hint:* Consider deflection Δ of center and measure Δ.

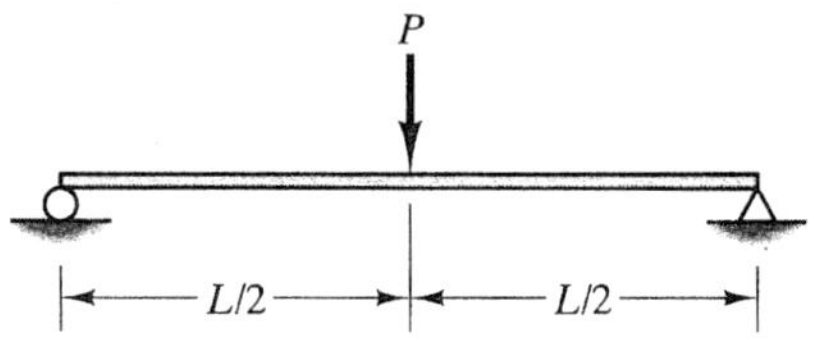

Figure P.12.16.

12.17. [12.3] Find the maximum deflection for the cantilever beam in Fig. P.12.17. Take $E = 30 \times 10^6$ psi.

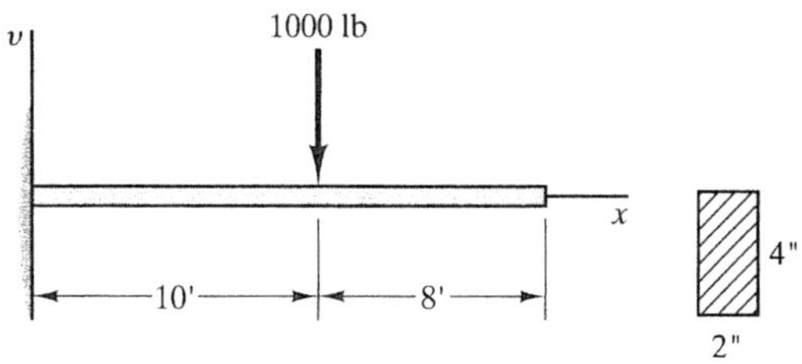

Figure P.12.17.

12.18. [12.3] Find the maximum deflection for the cantilever beam shown in Fig. P.12.18. The beam is an 8WF28 wide-flanged I-beam. Take $E = 2.5 \times 10^{11}$ Pa. Include the weight of the beam. Sketch the deflection curve.

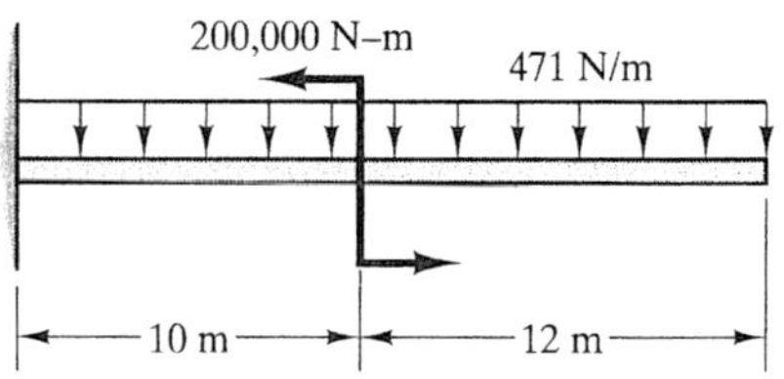

Figure P.12.18.

12.19. [12.3] What is the maximum deflection of the beam shown in Fig. P.12.19? The beam is a wide-flanged 8WF67 I-beam. Include the weight. Take $E = 2 \times 10^{11}$ Pa.

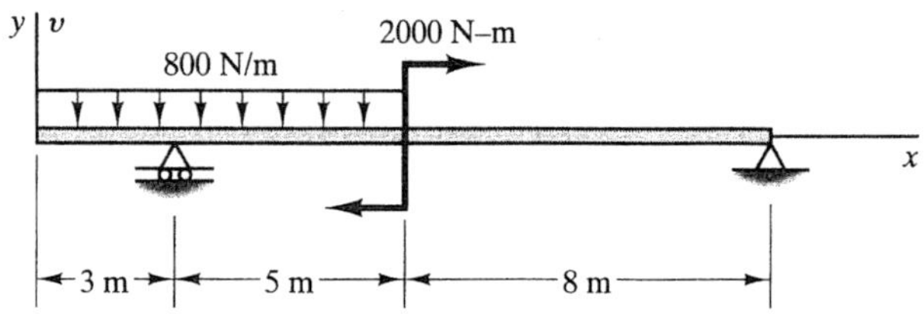

Figure P.12.19.

12.20. [12.4] Compute the supporting forces for the beam shown in Fig. P.12.20.

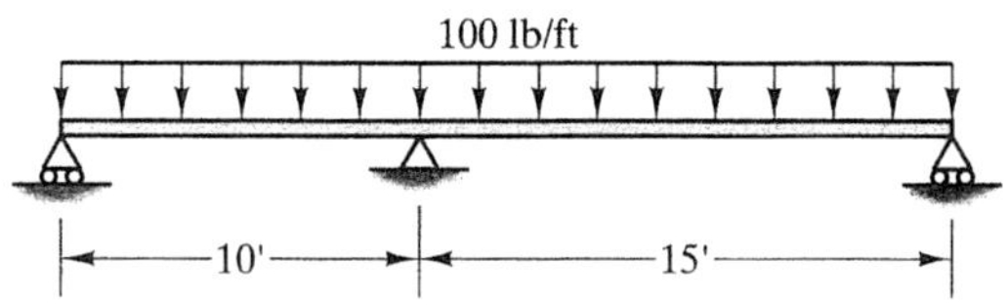

Figure P.12.20.

12.21. [12.4] Find the supporting force system for the beam shown in Fig. P.12.21.

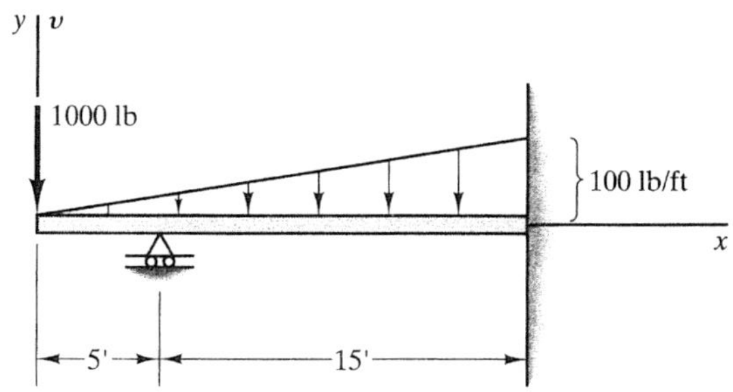

Figure P.12.21.

12.22. [12.4] Find the supporting forces and the deflection at $x = 9$ ft for $E = 30 \times 10^6$ psi (see Fig. P.12.22).

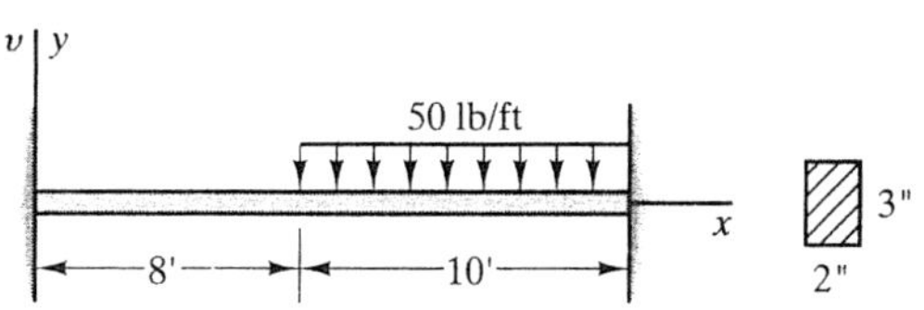

Figure P.12.22.

12.23. [12.4] Determine the supporting forces for the beam shown in Fig. P.12.23.

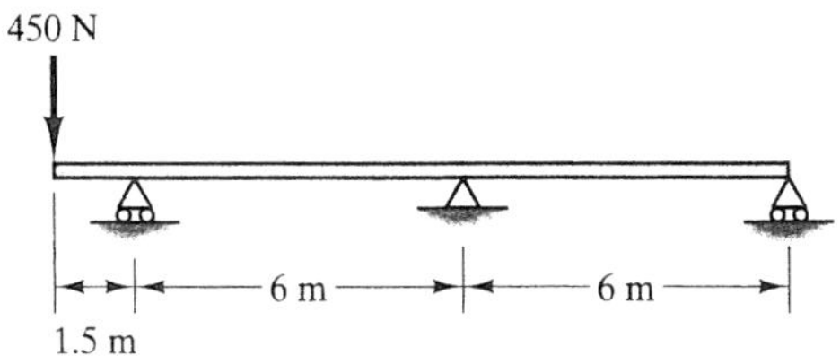

Figure P.12.23.

12.24. [12.4] In Fig. P.12.24 is shown a cantilever beam. Compute the supporting forces.

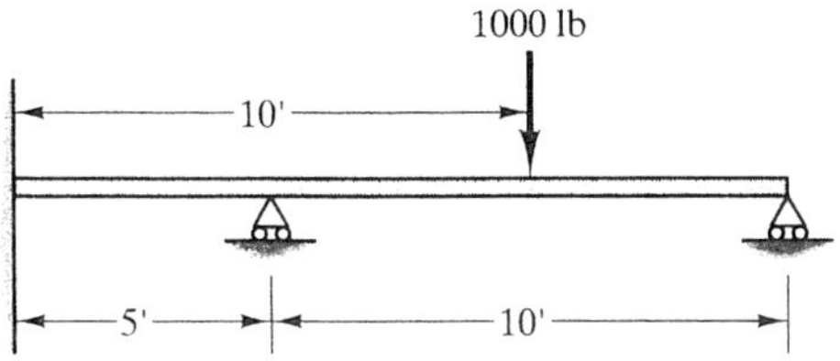

Figure P.12.24.

12.25. [12.4] What is the deflection curve for the beam AB in Fig. P.12.25? Set up all the necessary equations but do not solve for the constants. Give equations in terms of EI.

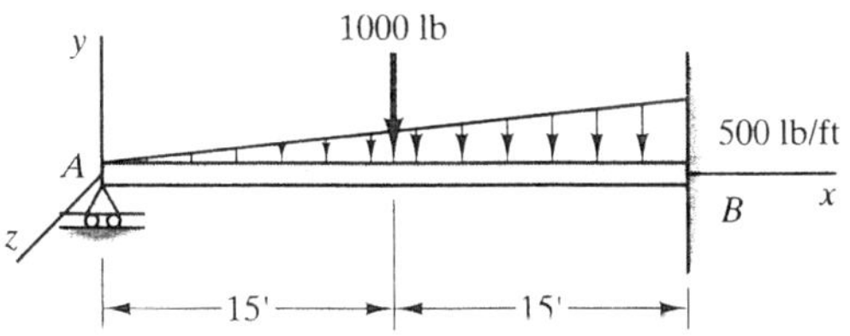

Figure P.12.25.

12.26. [12.4] Find the deflection curve and the supporting forces in terms of EI for the beam in Fig. P.12.26.

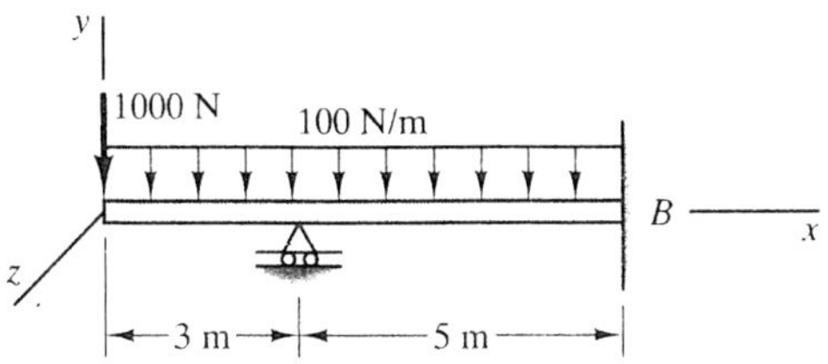

Figure P.12.26.

12.27. [12.4] What is the deflection curve for the beam AB in Fig. P.12.27? Set up all the necessary equations but do not solve for the constants. Give equations in terms of EI.

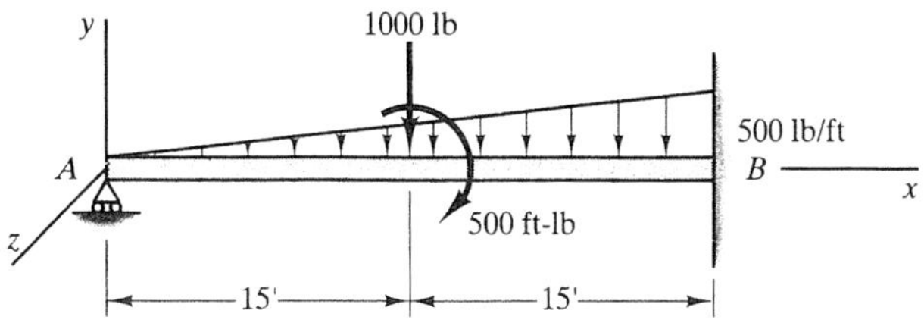

Figure P.12.27.

12.28. [12.4] For the beam AB in Fig. P.12.28,

(a). Find the supporting force at A.
(b). A light pointer CD is attached to the beam. Give the horizontal and vertical movements of end C resulting from the deformation of the beam.

Get your results in terms of EI.

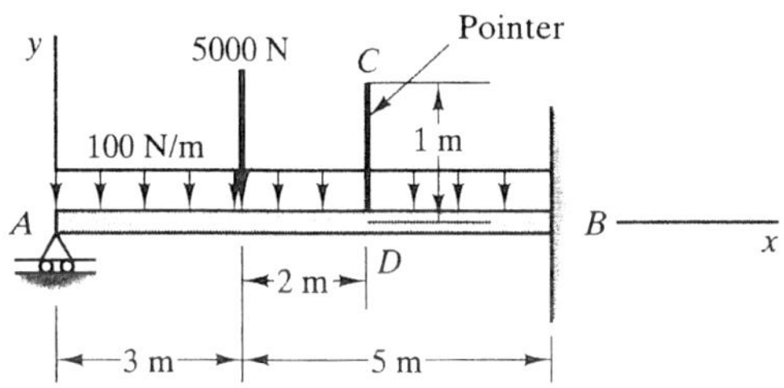

Figure P.12.28.

12.29. [12.4] In Fig. P.12.29, set up and number equations for constants but do not solve for constants.

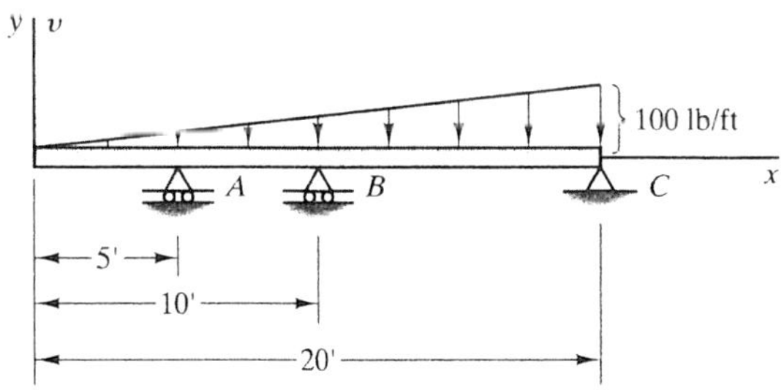

Figure P.12.29.

In Problems 12.30 to 12.34 it is helpful if you use the result (easily derivable) that the end deflection of a tip-loaded cantilever beam is $PL^3/(3EI)$.

12.30. [12.4] A cantilever beam is attached to a wire which at 15°C is just taut (see Fig. P.12.30). What is the maximum stress in the cantilever beam if the temperature in the wire drops to $-18°C$? Take E for the wire to be 2×10^{11} Pa and the coefficient of expansion α to be constant over the temperature range at a value of $1.17 \times 10^{-5}/°C$. The modulus for the beam is 2×10^{11} Pa. The cross-sectional area of the wire is 180 mm^2.

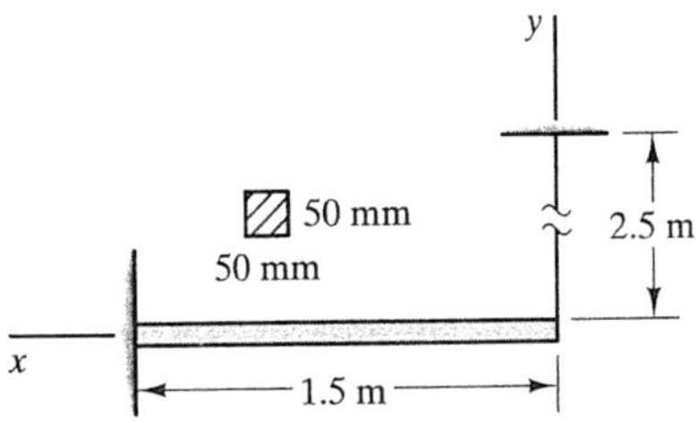

Figure P.12.30.

12.31. [12.4] Two upright cantilever beams are connected at their tips by a wire as shown in Fig. P.12.31. A 450 N load has been applied. What are the deflections of points A and B for the following data? All materials have a modulus of elasticity of 1.7×10^{11} Pa. The cantilever beams have square cross sections 93 mm on edge and the wire has an area at the cross section of 6.25 mm^2.

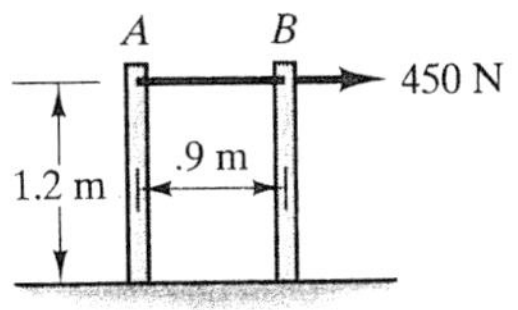

Figure P.12.31.

12.32. [12.4] Do Problem 12.31 for a situation where in addition to the loading there is a temperature drop of 22°C. The coefficient of expansion for the wire is $1.1 \times 10^{-5}/°C$ for the temperature range.

12.33. [12.4] Two upright cantilever beams each having a square cross section of 1 $in.^2$ and a modulus of elasticity of 30×10^6 psi are shown in Fig. P.12.33. A rigid element is welded to one of the cantilever beams, and before the loading there is a gap $\delta = .3$ in. between it and the second cantilever beam. What force P is required to generate a strain $\varepsilon = -.0004$ on a strain gage at the base of the cantilever beam on the right? The length L of each beam is 5 ft.

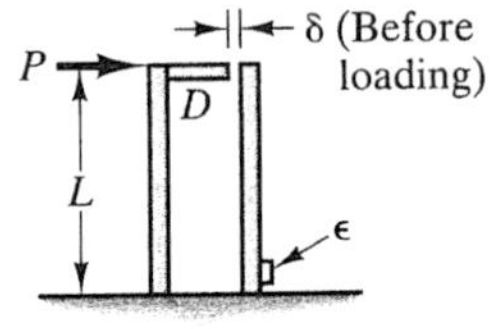

Figure P.12.33.

12.34. [12.4] Disk F (se Fig. P.12.34) rotates uniformly at a speed ω rad/sec. To measure this speed we have placed two cantilever beams into the disk and have connected them with a guy wire which is just taught when $\omega = 0$. When $\omega \neq 0$ a strain will be developed at c where we have a strain gage. If for each cantilever the diameter is 10 mm and $E = 2 \times 10^{11}$ Pa, find ω for a strain of .0002. The guy wire has an area of 60 mm^2 and a modulus of elasticity of 2×10^{11} Pa. Neglect the mass of the guy wire and recall from physics that the centrifugal force of particle of mass m rotating about a point r units away with speed ω is $mr\omega^2$. Finally, the rods each have a weight per meter of 15 N/m.

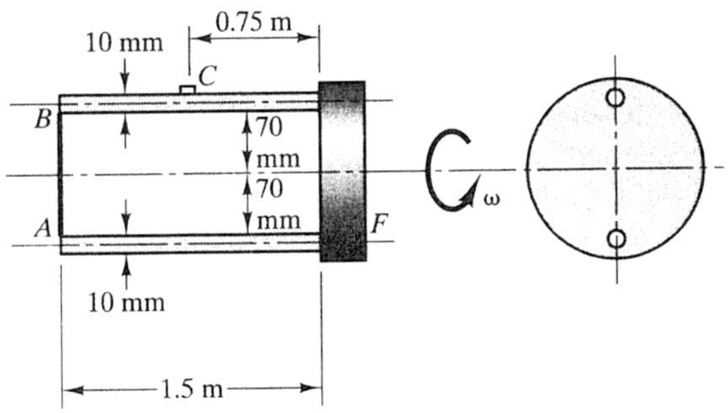

Figure P.12.34.

12.35. [12.5] Two cantilever beams are interconnected by a linear spring having a spring constant K of 9×10^5 N/m (Fig. P.12.35). What is the force in the spring for a load P of 45,000 N? Take I for the beams as 1.4×10^{-4} m^4, $L = 3$ m, and $E = 2 \times 10^{11}$ Pa. Use superposition formulae. For $\Delta_B = -1.25 \times 10^3$ m, what is P?

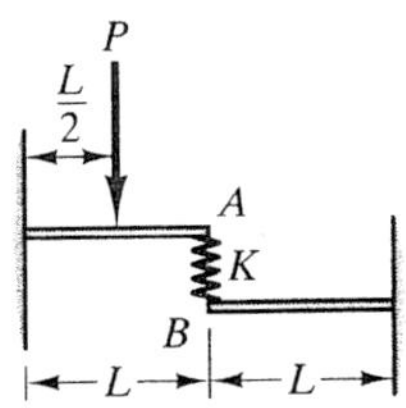

Figure P.12.35.

12.36. [12.5] Using the superposition principle, compute the deflection at a position 1.2 m from the end of the cantilever beam shown in Fig. P.12.36.

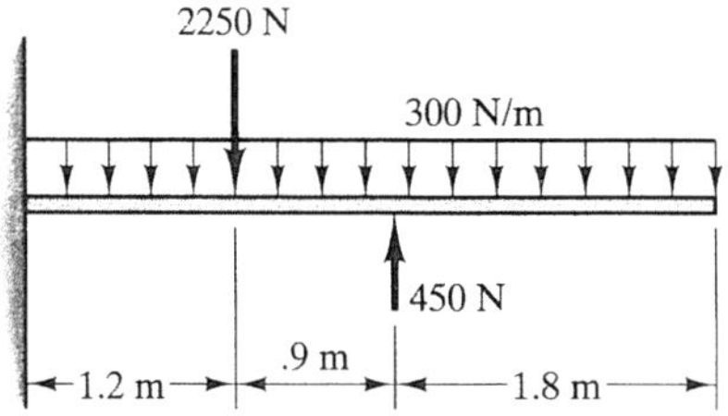

Figure P.12.36.

12.37. [12.5] Using the superposition principle, compute the deflection at the midpoint for the simply supported beam shown in Fig. P.12.37.

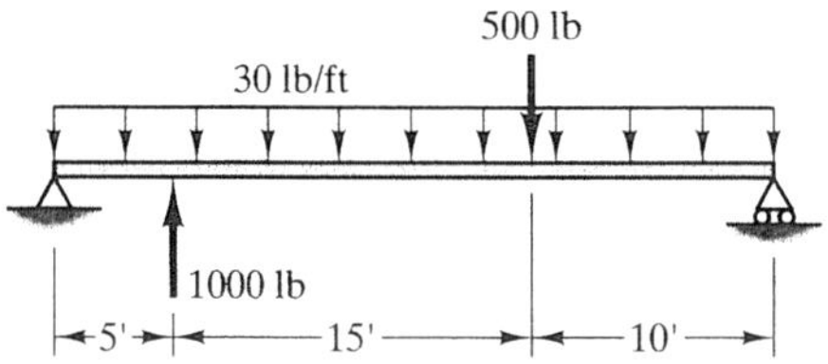

Figure P.12.37.

12.38. [12.5] In Fig. P.12.38 is shown a cantilever beam with a uniform load of 750 N/m. Compute the supporting forces for the beam. (Use the superposition principle.)

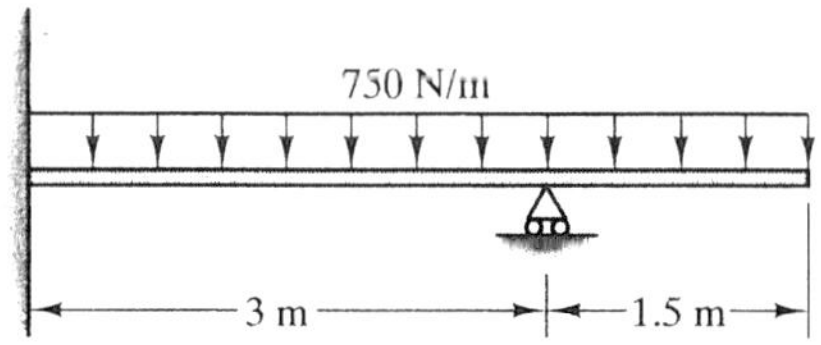

Figure P.12.38.

12.39. [12.5] What is the supporting force system for the beam shown in Fig. P.12.39? (Take $E = 2 \times 10^{11}$ Pa, $I = 1.95 \times 10^{-5}$ m^4, and $K = 3.6 \times 10^5$ N/m. Use the superposition principle.)

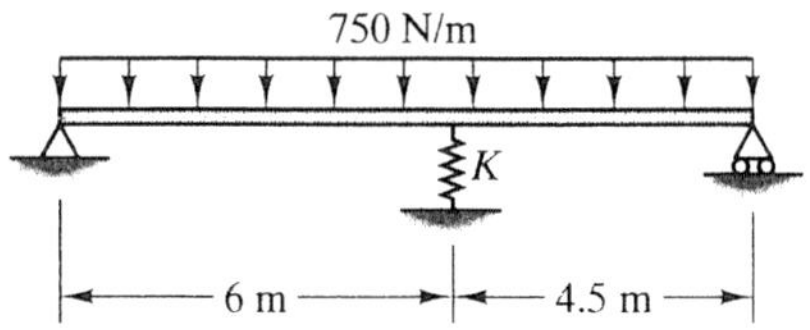

Figure P.12.39.

Other problems involving the superposition principle can be assigned from problems preceding Problem 12.30.

12.40. [12.6] Consider a cantilever beam of length L with a tip load P acting on it. The cross section is rectangular, having a width b and a height h. What is the ratio of maximum deflections of shear and bending for (a) $h/L = 1/300$ and (b) $h/L = 1$? Take $E/G = 2.6$.

12.41. [12.6] In Problem 12.40, find the total deflection at the midlength of the beam from both shear and bending deflections. Use the following data:

$$b = .06 \text{ m} \qquad G = 7 \times 10^{10} \text{ Pa}$$
$$h = .3 \text{ m} \qquad L = .9 \text{ m}$$
$$E = 2 \times 10^{11} \text{ Pa} \qquad P = 45{,}000 \text{ N}$$

12.42. [12.6] Formulate the shear deflection curve for the beam shown in Fig. P.12.36. The section is rectangular.

***12.43. [12.7]** The Maxwell Betti theorem applied to a beam can be written as follows:

$$\int \tau_{xx}^{(1)} \varepsilon_{xx}^{(2)} dx = \int \tau_{xx}^{(2)} \varepsilon_{xx}^{(1)} dx$$

where the integration is carried out over the entire length of the beam. Using Fig. P.12.43, demonstrate the correctness this form of the Maxwell Betti reciprocal theorem for the loads P_1 and P_2 with $E = 2 \times 10^{11}$ Pa.

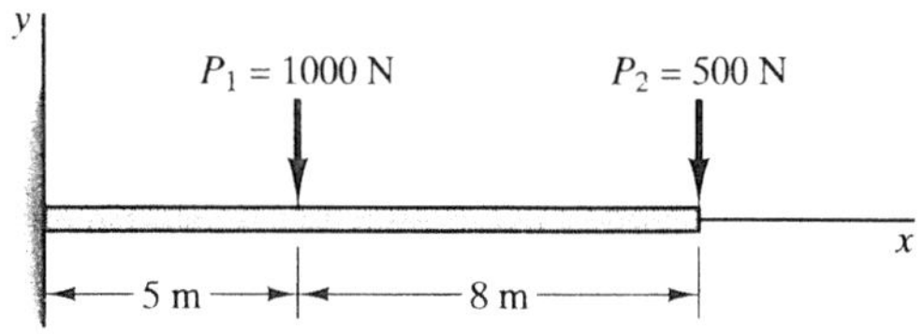

Figure P.12.43.

12.44. [12.7] Assume that the deflection equation for a cantilever beam loaded at the end by force P as shown in Fig. P.12.44(a) is known. Using this result and the Maxwell-Betti reciprocity relation, determine the end deflection of the cantilever beam stemming from a load Q applied a distance a from the tip as shown.

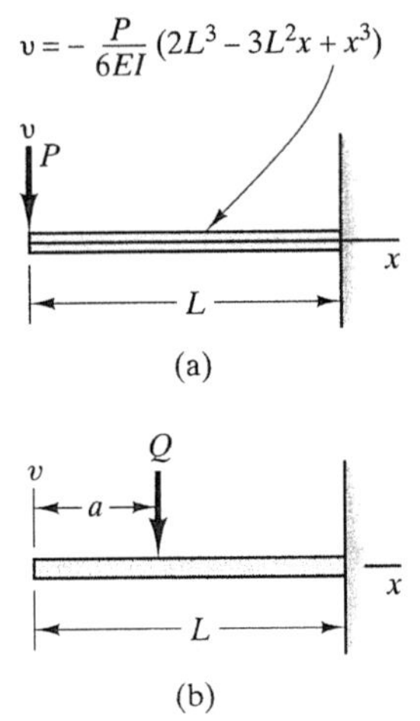

Figure P.12.44.

12.45. [12.7] Determine the displacement at the tip of the uniformly loaded cantilever beam shown in Fig. P.12.45. Use the second Castigliano theorem.

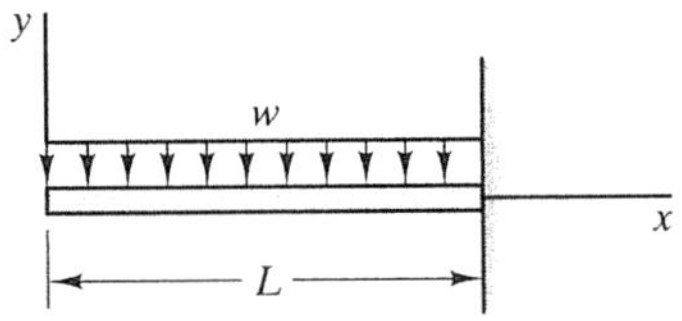

Figure P.12.45.

12.46. [12.7] Using energy methods, compute the vertical deflection of the beam at $x = L/2$ in Problem 12.45.

12.47. [12.7] Find the support system for Problem 12.22 using the second Castigliano theorem.

12.48. [12.8] Determine the supporting forces for the beam shown in Fig. P.12.48 using the second Castigliano theorem.

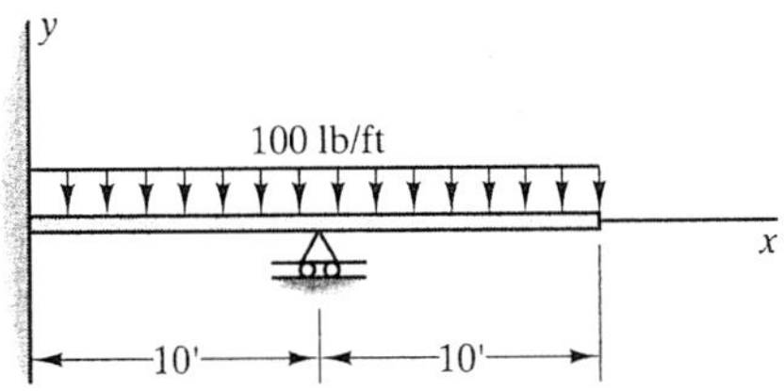

Figure P.12.48.

12.49. [12.7] Determine the reactions for the beam shown in Fig. P.12.49 using energy methods.

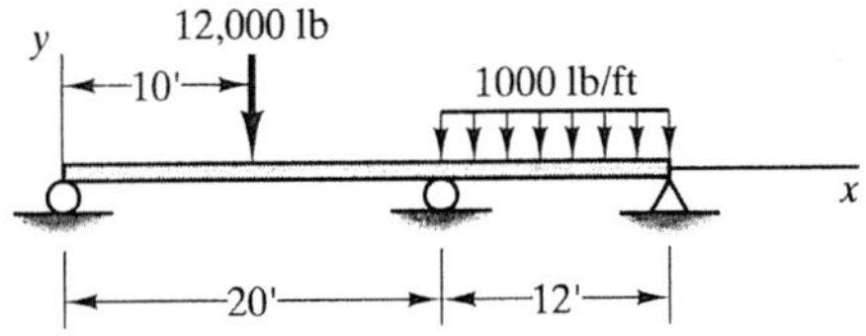

Figure P.12.49.

12.50. [12.7] Redo Problem 12.20. Specifically, compute the supporting forces. Use energy methods.

12.51. [12.7] Redo Problem 12.7, this time using the second Castigliano theorem. Determine the vertical displacement δ when $x = 15$ ft.

12.52. [12.7] Compute the lateral displacement at the tip of the quarter circle arch loaded as shown in Fig. P.12.52. Assume that the arch is thin so that straight beam theory is applicable. Using energy methods, include the effects of shear, axial, and bending energy on the displacement. Compare the relative contributions of each on the total displacement. Take $R = 50$ in., $E = 30 \times 10^6$ psi, and $\nu = 0.30$.

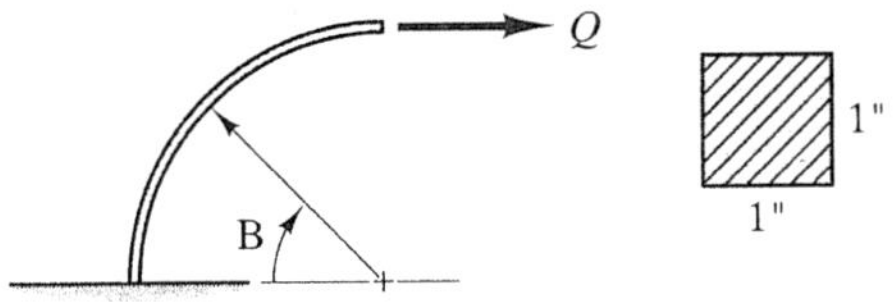

Figure P.12.52.

12.53. [12.7] Determine the horizontal movement of point A on the semicircular hoop shown in Fig. P.12.53. Consider bending energy only.

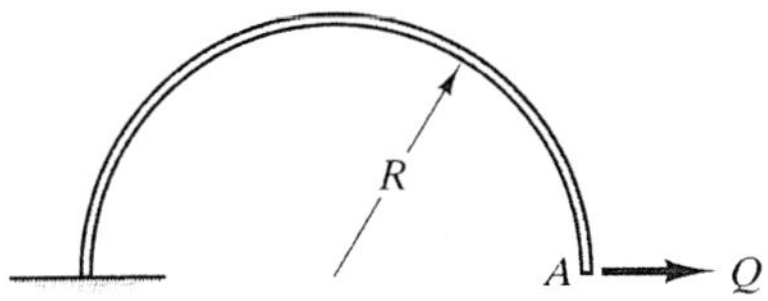

Figure P.12.53.

12.54. [12.7] Using energy methods, solve for the supporting forces of the uniformly loaded beam shown in Fig. P.12.54.

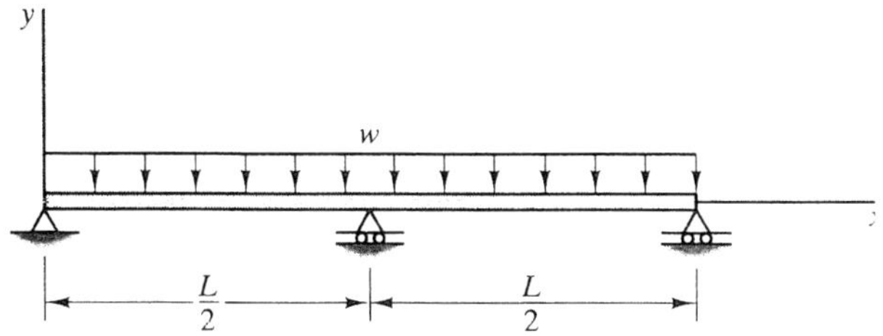

Figure P.12.54.

12.55. [12.7] Determine the deflection at the center of the simply-supported, uniformly-loaded beam shown in Fig. P.12.55. Use the second Castigliano theorem.

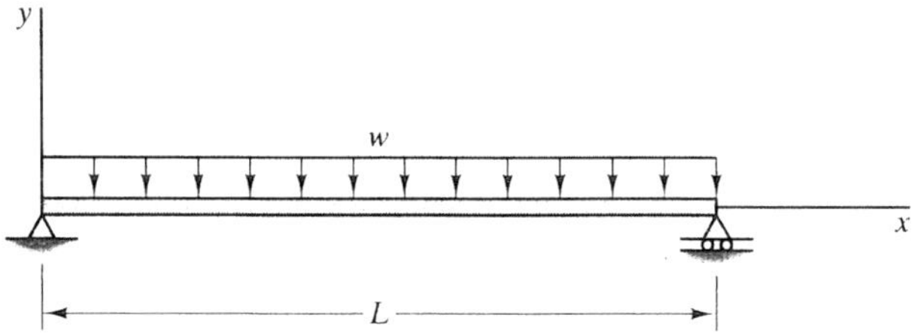

Figure P.12.55.

12.56. In computer hard disk drives, the read/write head is typically attached to a cantilevered, horizontal suspension arm that is linearly tapered, with constant thickness, as shown in Fig. P.12.56. Determine the force-displacement relation for the idealized read/write suspension arm for a point load P at the head acting downward (into the page) on the arm. Use the energy method.

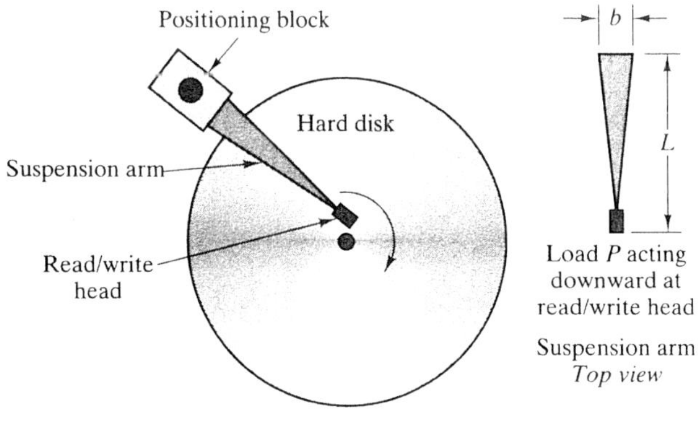

Figure P.12.56.

12.57. A shaft shown in Fig. P.12.57 supports a 4500 N force and is itself simply supported. For $E = 2 \times 10^{11}$ Pa for the entire shaft, what is the deflection curve?

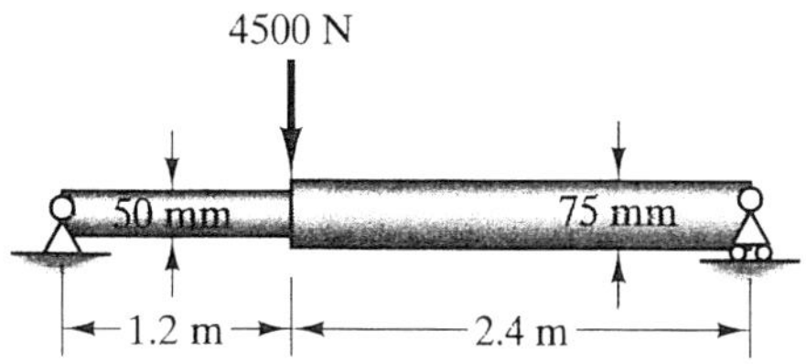

Figure P.12.57.

12.58. Find the vertical movement of point A in Fig. P.12.58. The lower member is welded to the cantilever beam. Take $E = 30 \times 10^6$ psi and I as 20 in.^{4} for the upper beam and the welded lower beam. Consider the connecting member D to be rigid.

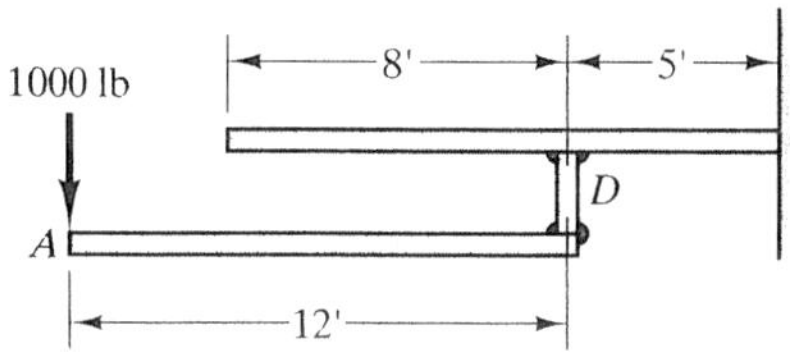

Figure P.12.58.

12.59. Two shafts are welded together as has been shown in Fig. P.12.59. Shaft A has a modulus of elasticity of 2×10^{11} Pa and shaft B has a modulus of elasticity of 1.4×10^{11} Pa. What is the deflection curve and the maximum deflection of the member from the external load?

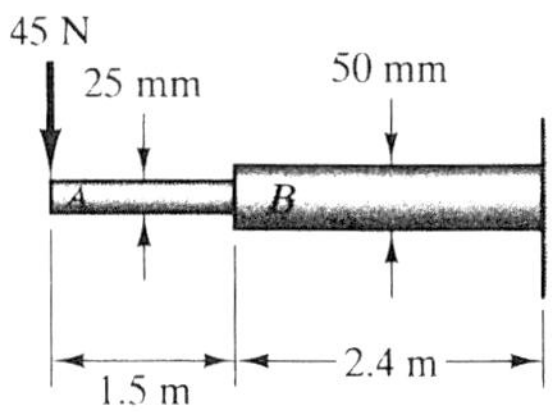

Figure P.12.59.

12.60. Two beams are attached to a firm base (see Fig. P.12.60) but are free to slide over each other with no resistance. What is the deflection at the end of the beams for a uniform loading of 200 lb/ft? Take $E_A = 30 \times 10^6$ psi and $E_B = 15 \times 10^6$ psi.

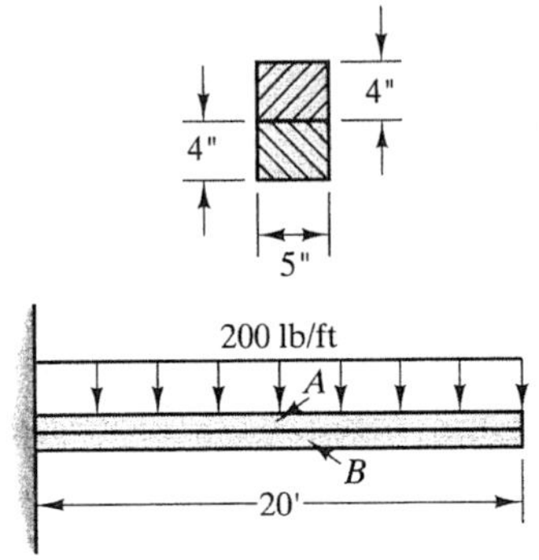

Figure P.12.60.

12.61. What is the tip deflection of the cantilever beam AB (see Fig. P.12.61)? Take $E = 2 \times 10^{11}$ Pa.

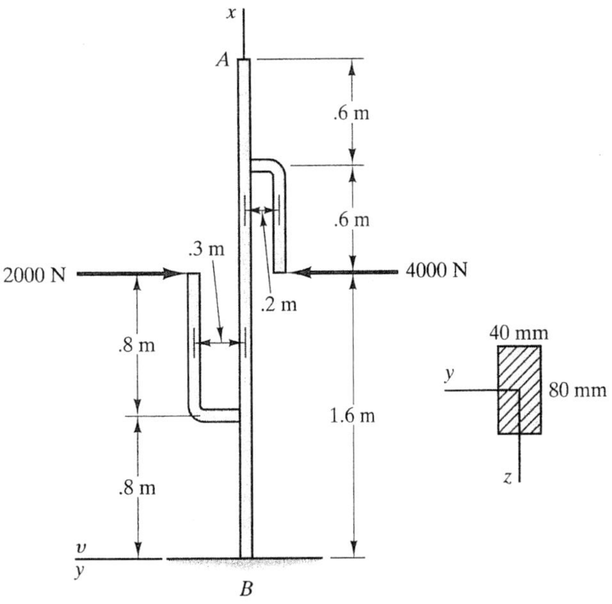

Figure P.12.61.

12.62. The cantilevered beam shown in Fig. P.12.62 supports a 1000 lb load at its tip. A $\frac{1}{4}''$ diameter steel cable is attached to the beam as shown. What is the displacement of the tip of the beam? The elastic modulus of the beam and cable is 30×10^6 psi.

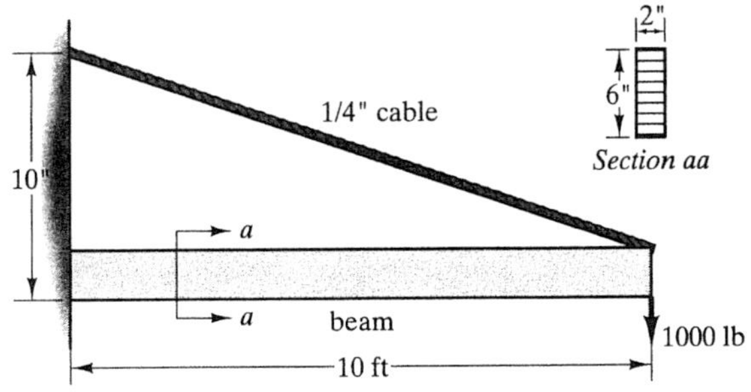

Figure P.12.62.

12.63. Using an energy method approach, compute the supporting force system for the tip-supported cantilever beam shown in Fig. P.12.63.

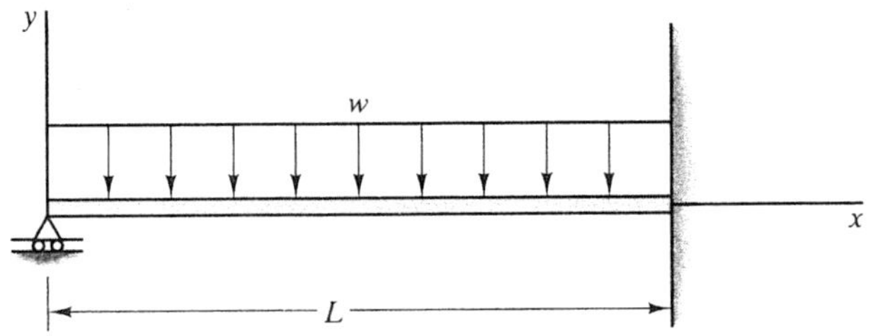

Figure P.12.63.

12.64. Using the second Castigliano theorem compute the deflection at $x = L/2$ in Problem 12.63.

12.65. For beam ABC in Fig. P.12.65 set up five simultaneous equations for the supporting force R_A and four constants. Do not solve for these five unknowns.

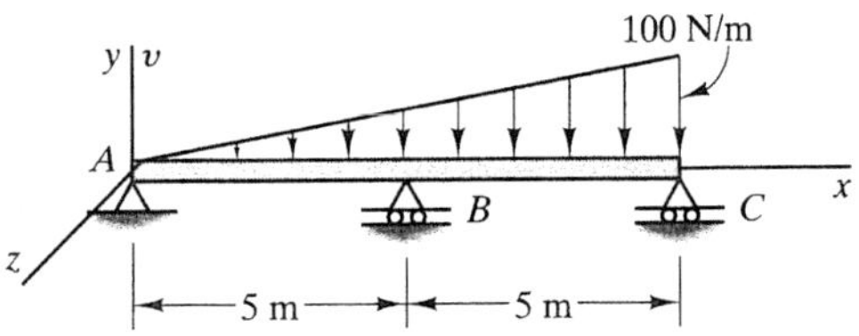

Figure P.12.65.

12.66. What is the force in the spring in Fig. P.12.66? Take $E = 30 \times 10^6$ psi.

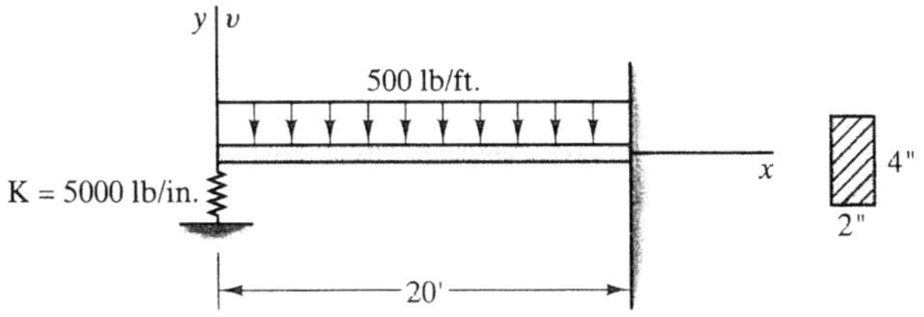

Figure P.12.66.

12.67. Find the deflection v of the cantilever beam in terms of EI for Fig. P.12.67. Determine the complete supporting force system.

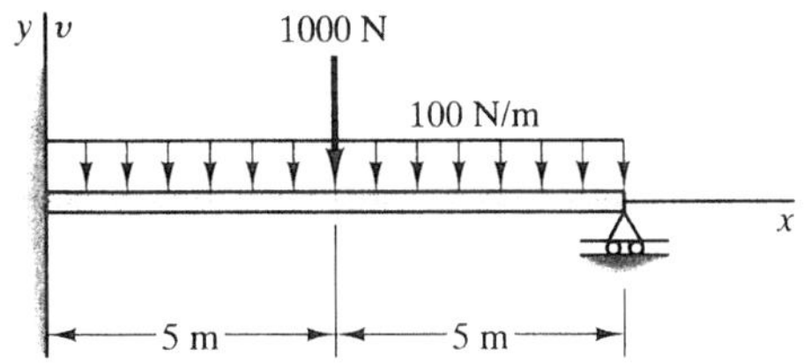

Figure P.12.67.

12.68. Find the supporting force at A for the beam in Fig. P.12.68.

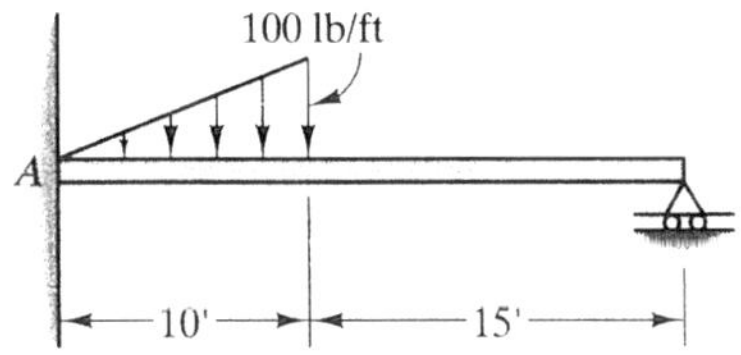

Figure P.12.68.

12.69. Choose the supporting force at A as the redundant constraint in Fig. P.12.69. Set up necessary equations for determining the supporting forces at A, but do not solve for the values of any unknowns.

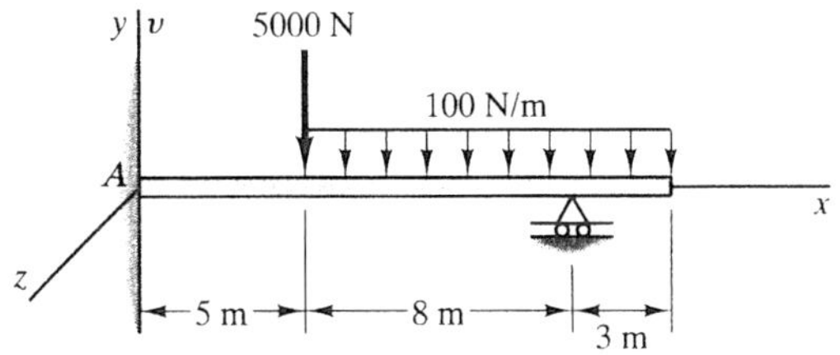

Figure P.12.69.

12.70. Develop the deflection curve of the beam shown in Fig. P.12.70.

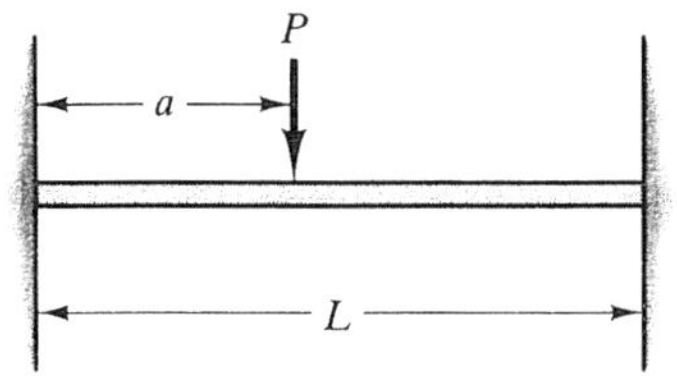

Figure P.12.70.

12.71. Derive an expression for the curvature of a beam (see Fig. P.12.71) subjected to a linearly varying temperature distribution through the depth, but which does not vary along the length of the beam. Show that the governing equation for the displacement is given by

$$d^2v/dx^2 = \frac{\alpha}{h}(T_B - T_T).$$

Take h as the beam height, T_B and T_T as the bottom and top beam surface temperatures, and α as the coefficient of thermal expansion. (*Hint:* determine the relative rotation of 2 sections of a beam dx apart from each other due to a change in temperature as shown in the diagram. What is the relation between $d\Delta/dx$ and the curvature?)

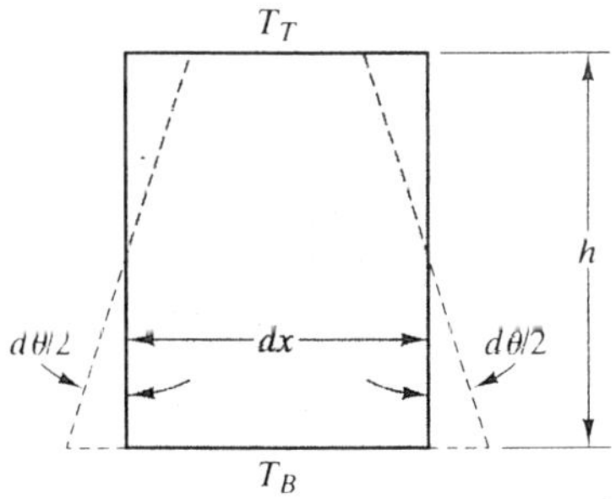

Figure P.12.71.

12.72. In order to erect continuous beam bridges, members may have to be field spliced at locations that are not at the supports. Sometimes the beam sections are limited in length due to transportation, manufacturing, or handling restrictions at the construction site. For the beam shown in Fig. P.12.72, where would you place splices for three beam sections to make a continuous beam? Give two considerations—one from a strength point of view, and one from a construction point of view. *Hint:* Sketch an approximate bending moment diagram and think about construction proceeding from the ends.

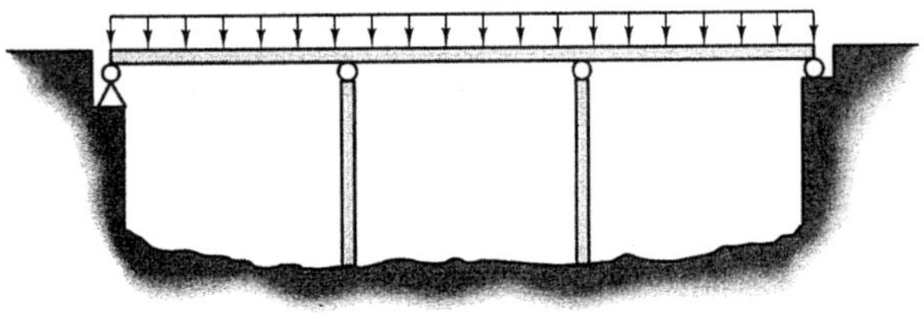

Figure P.12.72.

12.73. A roof beam for an office building is 8″ deep (see Fig. P.12.73). Outside, the temperature is 110°F. Inside, it is 70°F. What is the maximum displacement of the beam due to the temperature difference? Take $E = 30 \times 10^6$ psi and $\alpha = 6.5 \times 10^{-6}$/°F. [*Hint:* From Problem 12.71 the curvature due to the temperature change is

$$v'' = \frac{\alpha}{h}(T_B - T_T).]$$

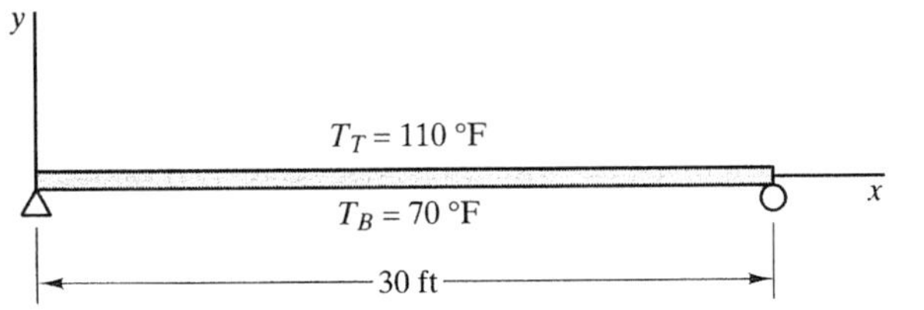

Figure P.12.73.

***12.74.** Shown in Fig. P.12.74 is a sensor that is used for measuring the shear stress in slow-moving fluids, such as air. Such devices are now constructed in silicon using micromachining techniques. Determine the displacement for a given fluid shear stress τ across the sensor. (*Hint:* Treat the floating element as a rigid block that moves due to the shear of the fluid on its surface. Superpose the effects of the four connecting beams called "tethers." Include shear stress acting on tethers. What is the boundary condition at the floating element?) Take E = 170 GPa and the thickness t = 5 μm. = 5×10^{-6} m.

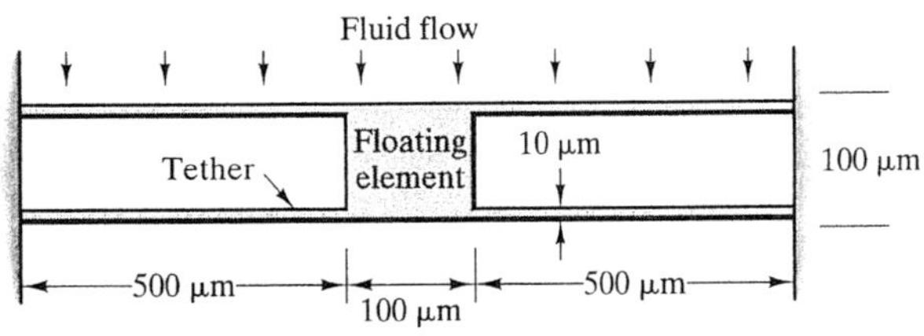

Figure P.12.74.

12.75. What is the deflection curve for the beam shown in Fig. P.12.75? Set up equations for the constants of integration but do not solve. Loadings are all in the xy plane.

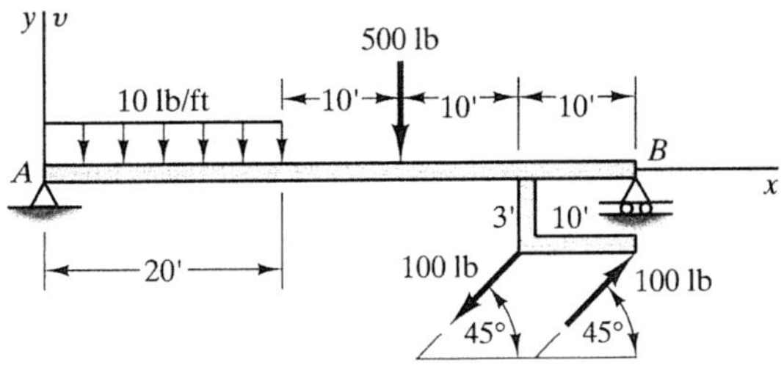

Figure P.12.75.

12.76. What is the deflection curve for the beams connected by a pin in Fig. P.12.76? What is the deflection at x = 8 m? $E = 2 \times 10^{11}$ Pa.

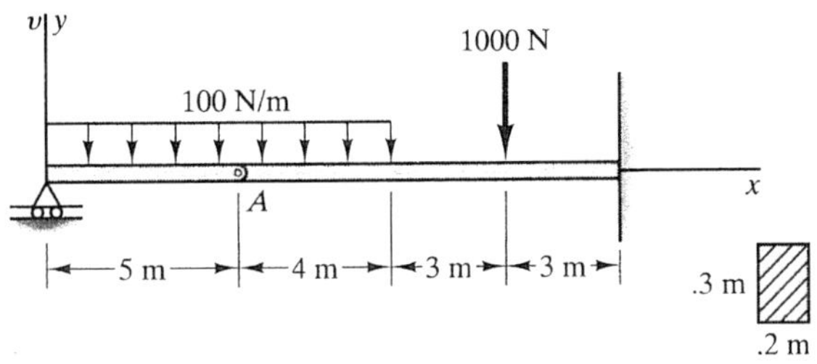

Figure P.12.76.

12.77. What is the deflection of the pin C due only to loads shown (Fig. P.12.77)? Neglect the effect of axial force on deflection. $E = 2.5 \times 10^{11}$ Pa.

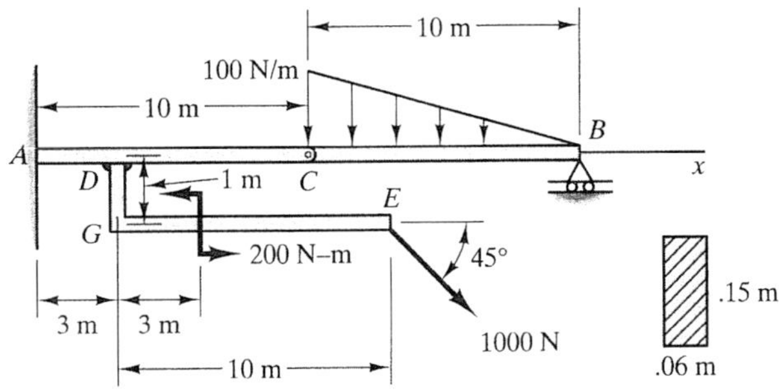

Figure P.12.77.

12.78. Compute the deflection curve for the cantilever beam (Fig. P.12.78) supporting a sinusoidal distribution of loading.

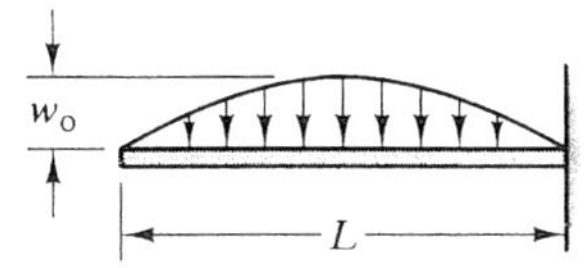

Figure P.12.78.

12.79. Shown is a cantilever beam in Fig. P.12.79 which is pinned at A. Find the supporting forces and moments.

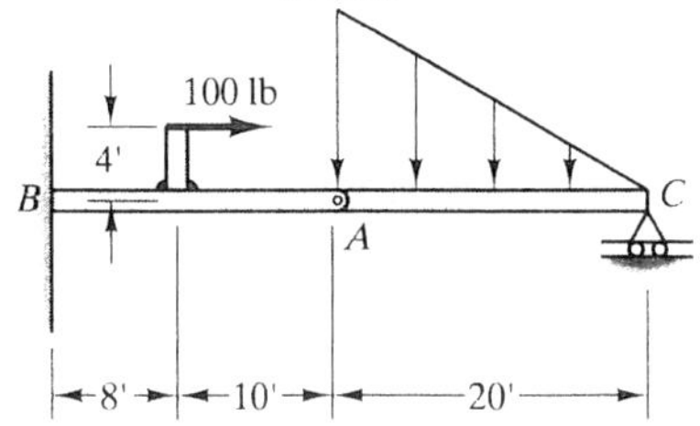

Figure P.12.79.

12.80. Determine the force P required to close the toe nail clippers shown in Fig. P.12.80 where the gap is 0.75 mm. Take $E = 200$ GPa and assume that the boundary condition at "A" can be treated as fixed. Point D is pinned while B just rests on the surface (neglect friction). Use the beam deflection table. The force at the toe nail is 35 N.

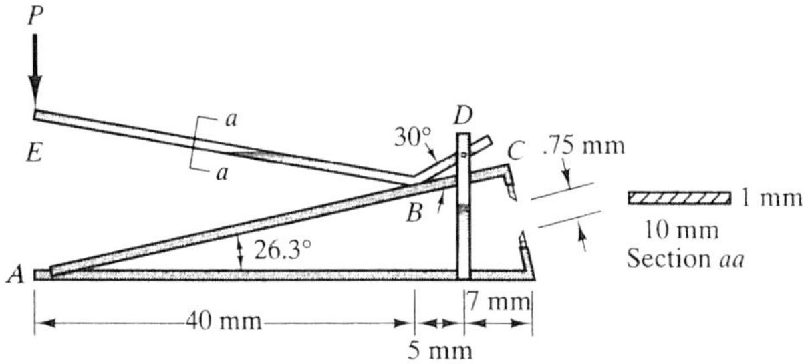

Figure P.12.80.

12.81. Determine a general force-displacement relationship for point A on the U-shaped spring shown in Fig. P.12.81. Use the second Castigliano theorem. Consider only bending strain energy.

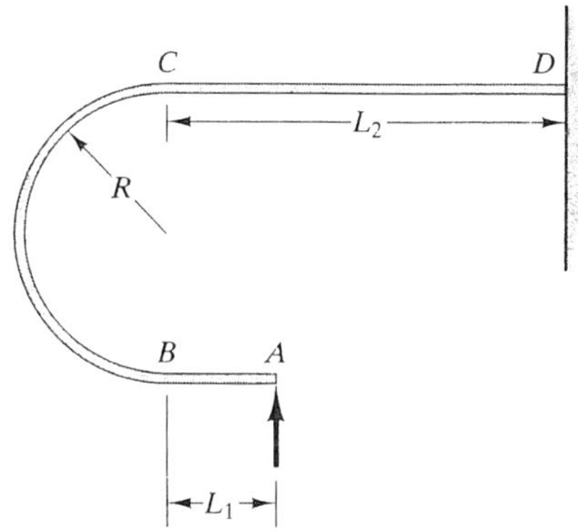

Figure P.12.81.

12.82. A simply-supported beam of length L and uniform load w, has its load replaced by a statically equivalent point load $P = wL$ placed at the center of the span. What error in the maximum stress and displacement is introduced by this approximation?

12.83. To determine the elastic modulus E in bending, a "three-point bend test," as idealized in Fig. P.12.83, is some times used. Show that if the span-to-depth ratio $L/h \geq 14$, the influence of shear deformation on the measured mid-span displacement will be less than 2% of the total displacement. Take $E/G = 2.5$ and assume a rectangular cross section of height h and width b.

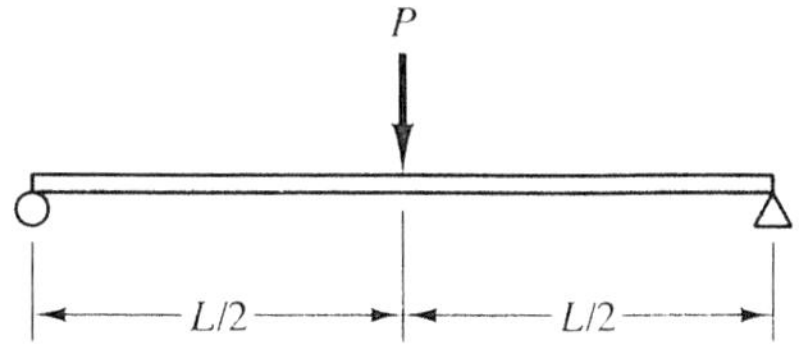

Figure P.12.83.

12.84. Redo Problem 12.12 using the second Castigliano theorem.

12.85. Compute the vertical displacement at the tip of the overhang for the beam shown in Fig. P.12.85. Use energy methods.

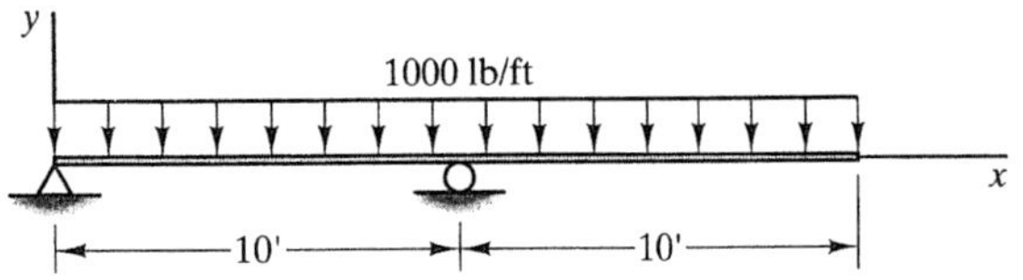

Figure P.12.85.

12.86. Using the result that the beam curvature is given by $d^2v/dx^2 = (\alpha/h)[T_b - T_t]$, where the temperature varies linearly from the bottom surface b to the top surface t but does not vary with x, derive an expression for the tip displacement of a cantilevered beam of length L subjected to different top and bottom surface temperatures as shown in Fig. P.12.86. Take the elastic modulus as E and the thermal expansion coefficient as α. Use energy method.

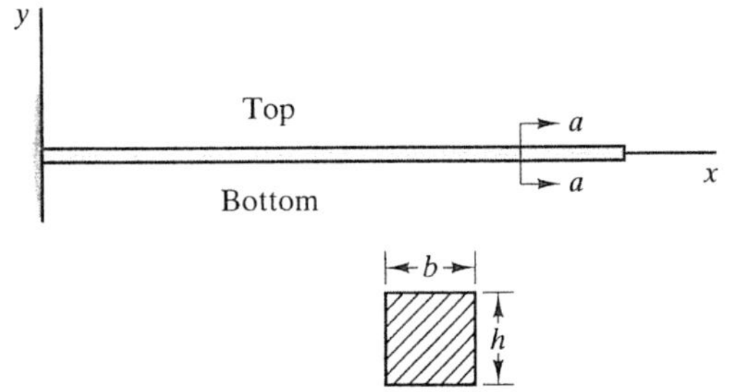

Figure P.12.86.

12.87. Redo Problem 12.24 this time using energy methods.

12.88. What is the displacement of the center of the beam shown in Fig. P.12.88? Use the energy method.

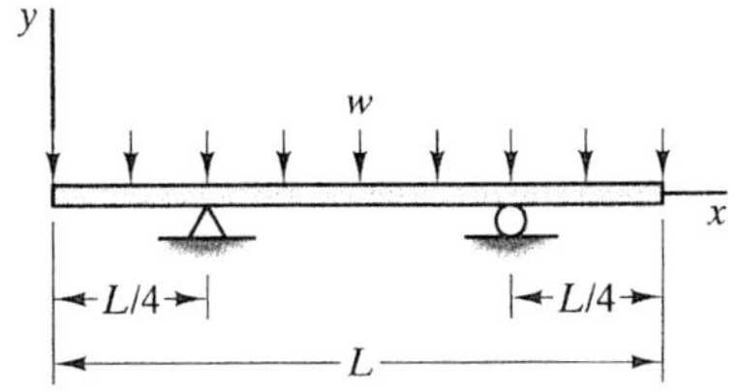

Figure P.12.88.

****12.89.** Develop a general solution for the supporting force system for the beam shown in Fig. P.12.89. Plot the roller reaction versus α. At what point does this reaction equal the total applied load (i.e., when does this reaction $= wL$)?

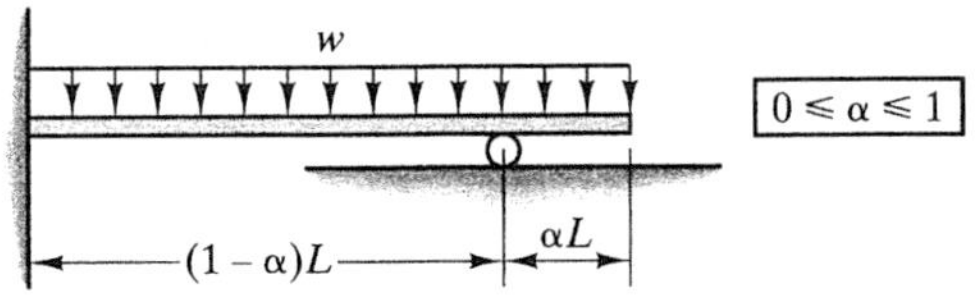

Figure P.12.89.

****12.90.** For Problem (12.62), plot the force in the cable (T) versus cable angle (α) for $0 \le \alpha \le 60$. What angle corresponds to the largest force in the cable? From Problem 12.62, we determined that

$$T = \frac{PL^3 \sin \alpha}{\dfrac{3Il}{A} + L^3 \sin^2 \alpha}$$

where T is the cable tension and l is the length of cable. Take $P = 1000$ lb and $L = 10$ ft.

****12.91.** What is the displacement of the center of the beam shown in Fig. P.12.91 as a function of the positions of the symmetrically spaced supports, given in terms of α? The supports can be placed such that

$$0 \le \alpha \le \tfrac{1}{2}L.$$

Check your results for $\alpha = 0$. Plot the normalized displacement

$$\frac{384EI\Delta}{wL^4}$$

versus α. What is the maximum displacement? Label any key points on your plot.

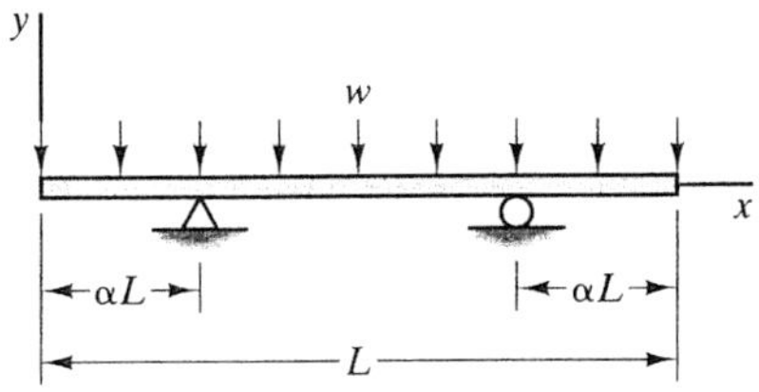

Figure P.12.91.

****12.92.** For the beam shown in Fig. P.12.92, plot the mid-span displacement as a function of the ratio of EI_1/EI_2 for the range $0.10 \leq EI_1 / EI_2 \leq 10$.

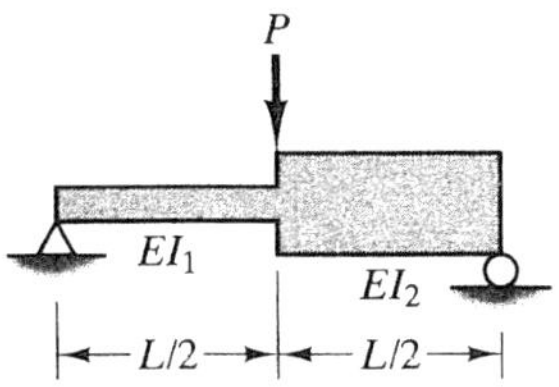

Figure P.12.92.

****12.93.** Formulate the general solution for the three supporting forces of the beam shown in Fig. P.12.93 as a function of α. Next plot R_1 / wL, R_2 / wL, and R_3 / wL versus α. Investigate when the reactions change sign as α is varied.

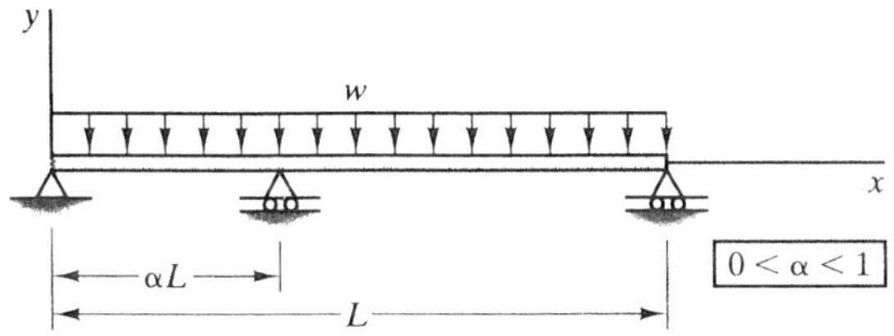

Figure P.12.93.

****12.94.** Build a general solution for tip displacement for the beam shown in Fig. P.12.94. Plot Δ_{TIP} versus α for $0 \leq \alpha \leq 1.0$. Where can the support be moved interior to the beam and result in zero displacement of the tip?

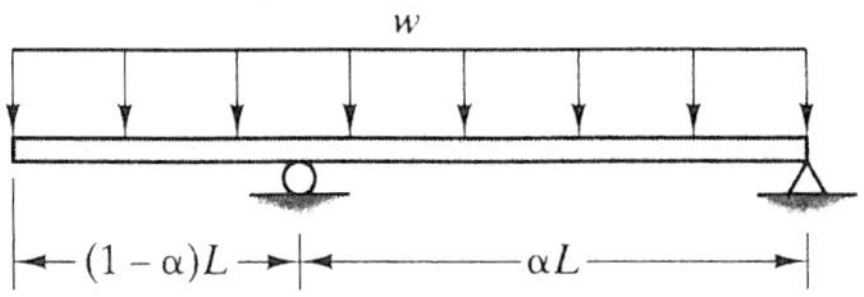

Figure P.12.94.

****12.95.** For the propped cantilever beam shown in Fig. P.12.95, determine the reaction force at the roller for $0 \leq \alpha \leq 3/4$. Plot the normalized reaction force

$$\frac{8R}{wL}$$

versus α for this range. At what value of α will the roller support force be a minimum?

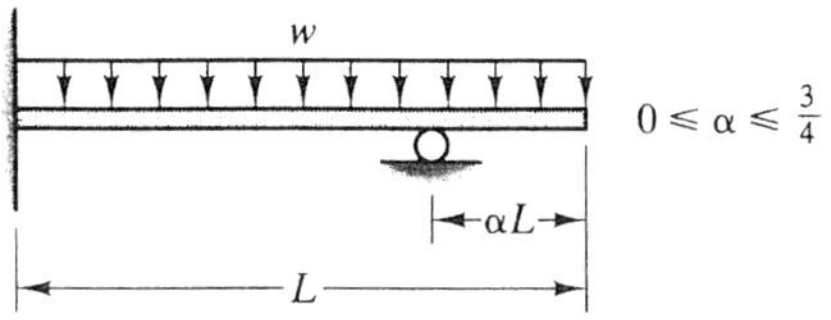

Figure P.12.95.

Project 12 Deflection of a Simply-Supported Beam

Using superposition formulas, for any set of loads P_i(N), any uniform loading w_0 (N/m), and any length L (m), find the deflection of the beam in Fig. P.12.96) at any point x. The program should be able to handle up to 50 point forces.

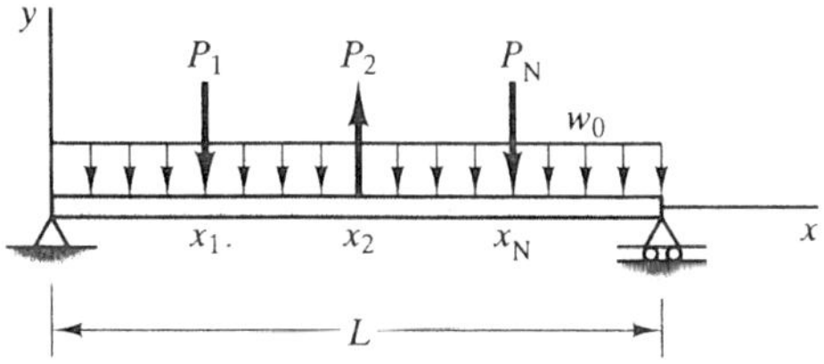

Figure P.12.96.

Run the program for the following data:

$P_1 = -5000$ N	$w_0 = -50$ N/m
$x_1 = 15$ m	$x = 10$ m
$L = 20.0$ m	$EI = 10^7$ N-m^2

*CHAPTER 13

Singularity Functions

13.1 Introduction

We shall introduce the reader to a set of very useful and interesting functions called singularity functions. As you will soon see, these functions do not have the familiar properties of functions, and one sometimes referred to them as "pathological" functions. Yet they can be very useful in simple studies, such as in Chapter 12, in greatly reducing the amount of writing and arithmetic. More than just that, these functions will be found freely used in more advanced studies of solid mechanics and indeed in many other fields.

13.2 Delta Functions and Step Functions

We shall consider three singularity functions, the first of which will be the *Dirac delta* function. This function will permit us to express a point force or any other such concentration in a convenient and useful form. Consider now a function $f(x)$ [see Fig. 13.1(a)] which is zero from $-\infty$ to $(\xi - \Delta)$, then forms an isosceles triangle of height $1/\Delta$ and base 2Δ as shown in the diagram, and finally, is zero again from $(\xi + \Delta)$ to $+\infty$. Note that the area under the curve is unity for *all* values of Δ. A way to define the delta function is to consider this function f in the limit as $\Delta \to 0$. Clearly, the peak of the curve goes to infinity as the base goes to zero, and we end up with an infinite spike at $x = \xi$, enclosing an area of unit value. The latter is true since

$$\lim_{\Delta \to 0}\left(\int_{-\infty}^{+\infty} f\,dx\right) = \lim_{\Delta \to 0}\left[\frac{1}{2}(2\Delta)\left(\frac{1}{\Delta}\right)\right] = 1 \tag{13.1}$$

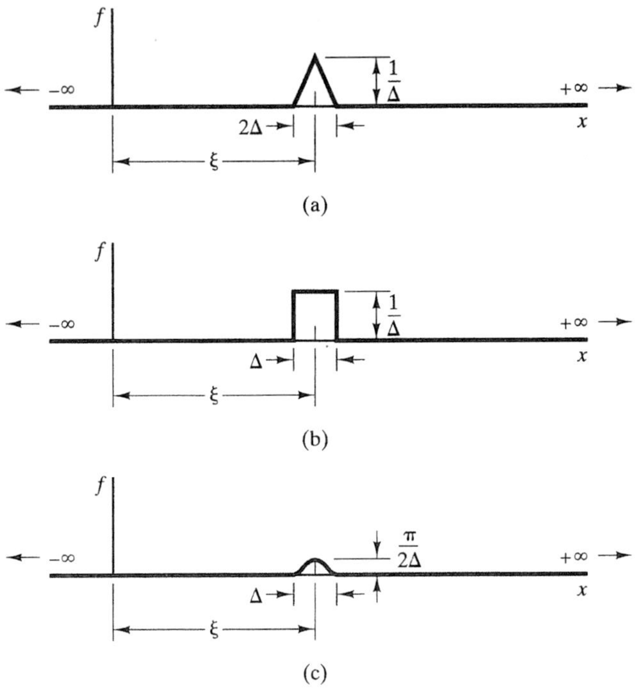

Figure 13.1. When $\Delta \to 0$ we get the delta function in each case.

wherein we have used the formula for the area of a triangle. The function f in the limit is the **Dirac delta** function, and we denote it as $[\delta(x - \xi)]$, where ξ indicates the position of the singularity (i.e., the spike). To summarize, the delta function $[\delta(x - \xi)]$ is zero everywhere except at $x = \xi$, where it is infinite and has an integral across the singularity of unit value.

It should be understood that the delta function could have been presented conveniently in many ways involving functions other than that shown in Fig. 13.1(a). Thus, in Fig. 13.1(b) and (c) we have shown functions f involving, respectively, a rectangular shape centered at ξ and a sinusoidal shape centered at ξ. We could proceed in both cases to arrive at $[\delta(x - \xi)]$ by letting $\Delta \to 0$. In either case we arrive at the singularity at ξ and the unit integral of the function across ξ. The curve f that you use is just a vehicle toward arriving at the delta function and can be discarded once the properties of the delta function are understood.

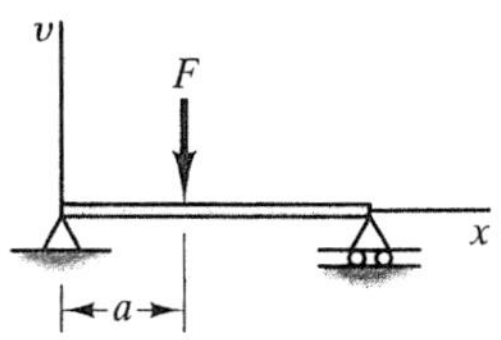

Figure 13.2. Point force at a.

How can we use the delta function? As a simple case, consider that we want to express a point force F lb at some position a (see Fig. 13.2) functionally—that is, we want to have a loading function $w(x)$ lb/ft that fully characterizes the point force.[1] We may do this directly as follows

[1]Note that we are using the familiar structural sign convention for w.

by considering the delta function to have units of $(\text{length})^{-1}$ and employing the value of F in force units:

$$w(x) = F[\delta(x - a)] \tag{13.2}$$

Clearly, $w(x)$ is zero everywhere except at the point of contact of the force F where, as you must expect from a point load, the value of w becomes infinite (remember that w is the intensity of load per unit length). Furthermore, the resultant of w is simply F itself since

$$\int_{-\infty}^{+\infty} w\,dx = F\int_{-\infty}^{+\infty} [\delta(x - a)]dx = (F)(1) = F \tag{13.3}$$

Thus, $F[\delta(x - \xi)]$ is a loading distribution over x that has all the properties of the point force.

Associating other dimensions with the delta function, we can similarly represent other concentrated quantities as distributions. Suppose that we wish to express a point couple, C ft-lb, at position $x = \xi$ [see Fig. 13.3(a)] as a *couple distribution* in terms of the variable x. Then considering $[\delta(x - \xi)]$ to have dimensions of $(\text{foot})^{-1}$ we can give the desired distribution as $C[\delta(x - \xi)]$, where C is in ft-lb. Similarly, consider a one-dimensional heat source Q at position $x = \xi$ [see Fig. 13.3(b)]. By having $[\delta(x - \xi)]$ take the dimensions of $(\text{length})^{-1}$ we can give the appropriate distribution in terms of x as $Q[\delta(x - \xi)]$, where Q is in British thermal units.

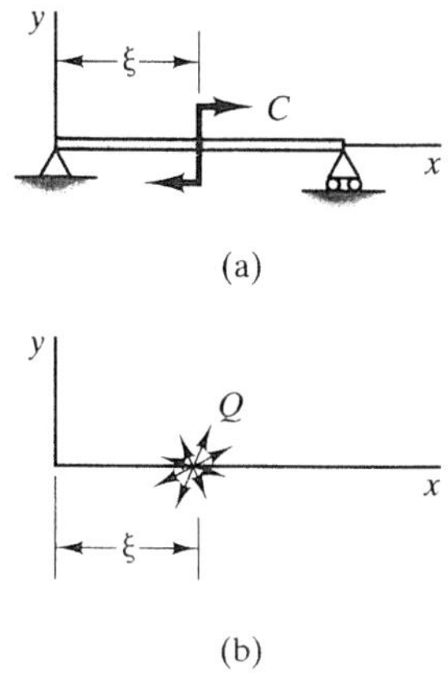

Figure 13.3. Point couple moment and a point heat source.

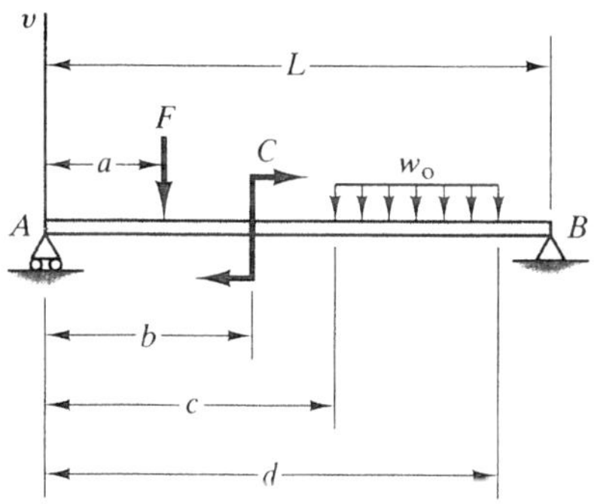

Figure 13.4. Beam with various loadings.

It was pointed out at the beginning of this discussion that we needed also a formulation which would permit the appearance or the abrogation of distributed loads, such as the uniform loading illustrated in Fig. 13.4. For such purposes, we shall now introduce the second of our singularity functions, which we shall call the **unit step function,** $[u(x)]$. This function is defined as *the integral, with a variable upper limit, of a delta function*.[2] That is,

$$[u(x)] = \int_{-\infty}^{x} [\delta(\zeta)]d\zeta \tag{13.4}$$

where ζ is just a running variable that could as well have been given by any convenient symbol. The function $[\delta(\zeta)]$ is the same as $[\delta(\zeta - 0)]$, so we can conclude that the singularity is at the origin in accordance with our previously stated notation. Clearly, then, when the upper limit x is less than zero, the integral is zero. If x exceeds zero, we get the value of unity for any such value[3] of x in accordance with our previous integra-

[2]Those students familiar with probability theory note that the relation between the delta function and the step function is of the same form as the relation between the density function and the distribution function.

[3]Thus, there is a contribution to the integration of unity at the origin but zero contribution at all points beyond the origin.

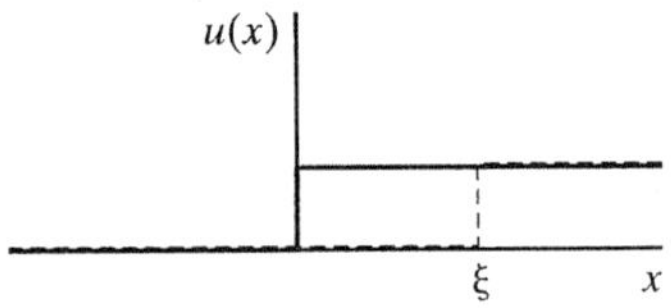

Figure 13.5. Plots of step functions at $x = 0$ and $x = \xi$.

tion of the delta function. Thus, if we plot $[u(x)]$, we get the familiar step function shown in Fig. 13.5 as the full line. It should now be a simple matter to deduce that the unit step function $[u(x - \xi)]$ defined as

$$\int_{-\infty}^{x} [\delta(\zeta - \xi)]d\zeta$$

has its step at position ξ as shown by the dashed line in the diagram.

Using these step functions, it is now a simple procedure to express the loading function for the uniform loading distribution in Fig. 13.4. Thus,[4]

$$w(x) = w_0[u(x - c)] - w_0[u(x - d)] \tag{13.5}$$

where the step functions are taken dimensionless. The first step function introduces the loading distribution at the correct position c, whereas the second step function cancels the contribution of the first expression at the proper position d and to the right. The combination is then the desired uniform loading w_0 from c to d.

Let us next consider the integration of the product of a step function and some function $g(x)$ in the following form:

$$f(x) = \int_{-\infty}^{x} g(\zeta)[u(\zeta - \xi)]d\zeta \tag{13.6}$$

Before proceeding, it should be understood that ξ is a fixed position locating the position of the "step"; ζ is simply an integration variable running from $-\infty$ to x; and x is the variable of the problem and may be less or greater than ξ. Let us assume first that $x < \xi$ or, in other words, that we are integrating from $-\infty$ to a position x *before* the step. In this case $[u(\zeta - \xi)]$ is zero over the range of integration, and $f(x)$ is then correspondingly zero in value. If now x exceeds ξ, we can forget about integration from minus infinity up to the step, since we get zero contribution, and integrate from the step at ξ onward to x. Thus, we can say that

$$f(x) = \int_{\xi}^{x} g(\zeta)[u(\zeta - \xi)]d\zeta \tag{13.7}$$

From position ξ onward, the step function is of constant value unity and may be extracted from the integral in the following manner:

$$f(x) = [u(x - \xi)]\int_{\xi}^{x} g(\zeta)d\zeta \tag{13.8}$$

Thus, we integrate a function $g(\zeta)$ times a step function as follows:

[4]Note w_0 is positive since it is pointing downward (structural convention).

We merely move up to the step, extract the step function from the integral, and integrate the function from the step onward.

It may already be evident that the integration procedures for the singularity functions are quite simple. Thus, a delta function integrates into a step function, whereas a step function is a quantity that is easily integrated by methods presented in the preceding paragraph. This will make for great simplicity in our use of the singularity functions in the determination of the deflection equations.

13.3 Deflection Computations Using Singularity Functions

The approach that we shall now follow for the evaluation of the deflection curve $v(x)$ is to utilize the infinite beam concept. That is, we shall hypothetically extend the beam under study to $-\infty$ on the left end and to $+\infty$ at the right end, as has been shown in Fig. 13.6 for the simply

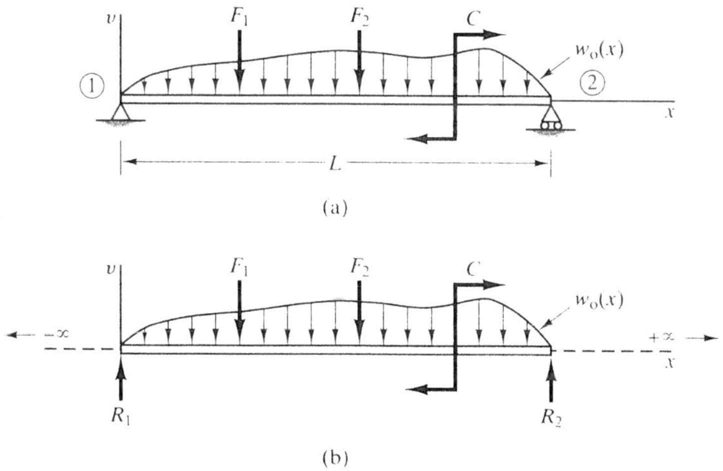

Figure 13.6. Formation of the infinite beam for a simply supported beam.

supported beam. The hypothetical extensions are to be considered weightless. There will be no forces acting on these extensions, so the shear and bending moments will perforce be zero in these hypothetical extensions. As a further example, we have shown (Fig. 13.7) a cantilever extended to infinity both plus and minus in the manner described above. If we establish the origin of the coordinate system at the left end of the actual beam as we have in Figs. 13.6 and 13.7, then just to the left end of the beam, which position we denote as 0^-, we can say that:

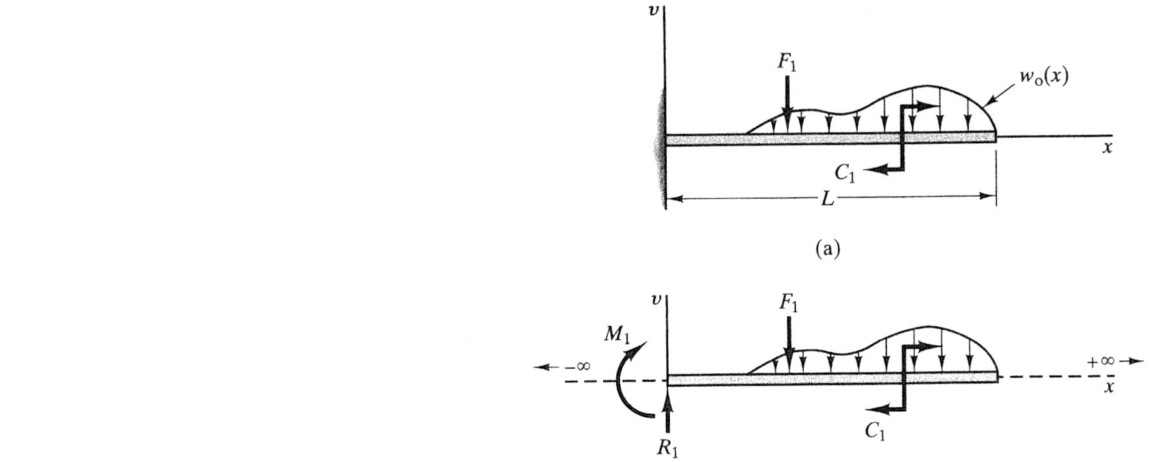

Figure 13.7. Formation of the infinite beam for a cantilever beam.

At $x = 0^-$:

$$V = 0 \tag{13.9a}$$
$$M = 0 \tag{13.9b}$$

and just to the right of the beam whose length we take as L, we can say, using L^+ to indicate the position to which we are referring:

At $x = L^+$:

$$V = 0 \tag{13.10a}$$
$$M = 0 \tag{13.10b}$$

In considering the simply-supported beam with its extensions in Fig. 13.6(b) and the cantilever beam with its extensions in Fig. 13.7(b), we shall consider the supporting forces and couple moment as unknowns to be calculated in the process that we shall detail.

We next formulate a loading function $w_y(x)$ for this entire infinite beam making use of the singularity functions. Using this loading function in Eq. (12.7), we carry out four integrations from $-\infty$ to x to evaluate v in terms of four unknown constants of integration and two unknown forces R_1 and R_2. These six quantities are all evaluated by examining certain boundary conditions associated with the beam. Thus, to the left of the origin, that is, $x = 0^-$, and to the right of the support 2, that is, $x = L^+$, we have zero bending moment and zero shear in our hypothetical beam. From Eq. (12.4), we see that the zero bending moment means that $d^2v/dx^2 = 0$ in these domains. Also, since $dM_z/dx = V_y$, Eq. (12.4) indicates that for $V_y = 0$ it is necessary that d^3v/dx^3 also be zero in these domains. Furthermore, we know that the deflection v is zero at $x = 0$ and $x = L$ as a result of the constraints. Summarizing conditions we must impose on our solution, we have for the simply supported beam

At $x = 0^-$:

$$\frac{d^2v}{dx^2} = 0 \tag{13.11}$$

$$\frac{d^3v}{dx^3} = 0 \tag{13.12}$$

At $x = 0$:

$$v = 0 \tag{13.13}$$

At $x = L^+$:

$$\frac{d^2v}{dx^2} = 0 \tag{13.14}$$

$$\frac{d^3v}{dx^3} = 0 \tag{13.15}$$

At $x = L$:

$$v = 0 \tag{13.16}$$

Thus we have the six conditions that permit us to solve for the six unknowns.

Consider next a cantilever beam as shown in Fig. 13.7. Note that the supporting force system now consists of a force R_1 and a point couple moment M_1 applied at the origin. As before, the loading function w_y is formulated for the entire beam and four integrations are again carried out. The boundary conditions are the same as in the simply supported beam except for Eq. (13.16). Thus, instead of a zero deflection at $x = L$, we have a zero slope of the deflection curve at $x = 0$. The boundary conditions then become:

At $x = 0^-$:

$$\frac{d^2v}{dx^2} = 0 \tag{13.17}$$

$$\frac{d^3v}{dx^3} = 0 \tag{13.18}$$

At $x = 0$:

$$v = 0 \tag{13.19}$$

$$\frac{dv}{dx} = 0 \tag{13.20}$$

At $x = L^+$:

$$\frac{d^2v}{dx^2} = 0 \tag{13.21}$$

$$\frac{d^3v}{dx^3} = 0 \tag{13.22}$$

We thus have again six conditions by which we can compute the four constants of integration as well as the unknown quantities R_1 and M_1.

We shall now illustrate the procedure of finding the deflection curve by considering several simple problems.

Example 13.1

Shown in Fig. 13.8(a) is a simply-supported beam. We are to determine the deflection curve $v(x)$ and the supporting forces. The infinite hypothetical beam for our calculations is shown in Fig. 13.8(b).

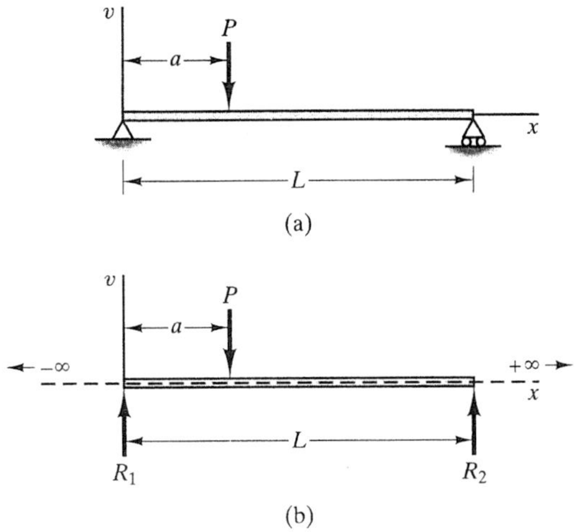

Figure 13.8. Simply-supported beam.

We now write the loading function $w_y(x)$ for the entire infinite beam. Thus,[5]

$$w_y(x) = -R_1[\delta(x)] + P[\delta(x-a)] - R_2[\delta(x-L)] \qquad \text{(a)}$$

Substituting into the governing differential equation (12.7), we have, using the notation I in place of I_{zz},

$$\frac{d^4v}{dx^4} = -\frac{w_y(x)}{EI} = \frac{1}{EI}\{R_1[\delta(x)] - P[\delta(x-a)] + R_2[\delta(x-L)]\} \qquad \text{(b)}$$

We shall now carry out for integrations from $-\infty$ to x. The first two integrations will involve $(d^3v/dx^3)_{-\infty}$ and $(d^2v/dx^2)_{-\infty}$ but since both V and M are zero to the left of $x = 0$, we can conclude that the constants of integrations for the first two integrations will always be zero. The third and fourth integrations involve $(dv/dx)_{-\infty}$ and

[5]Note again that we are using the structural convention of having a downward loading distribution as positive.

Example 13.1 (Continued)

$[v(-\infty) - (C'_3)(\infty)]$ on the left side of the equations. These need not be zero. We shall denote these quantities as C_3 and C_4 (i.e., as unknown constants), and we shall determine them using boundary conditions. We now proceed with the integration as follows:[6]

$$\frac{d^3v}{dx^3} = \frac{1}{EI}\{R_1[u(x)] - P[u(x-a)] + R_2[u(x-L)]\} \tag{c}$$

$$\frac{d^2v}{dx^2} = \frac{1}{EI}\{R_1x[u(x)] - P(x-a)[u(x-a)] + R_2(x-L)[u(x-L)]\} \tag{d}$$

$$\frac{dv}{dx} = \frac{1}{EI}\left\{R_1\frac{x^2}{2}[u(x)] - P\frac{(x-a)^2}{2}[u(x-a)] + R_2\frac{(x-L)^2}{2} \times [u(x-L)] + C_3\right\} \tag{e}$$

$$v = \frac{1}{EI}\left\{R_1\frac{x^3}{6}[u(x)] - P\frac{(x-a)^3}{6}[u(x-a)] + R_2\frac{(x-L)^3}{6}[u(x-L)] + C_3x + C_4\right\} \tag{f}$$

The boundary conditions at $x = L^+$ require that[7]

$$\left(\frac{d^3v}{dx^3}\right)_{L^+} = 0 = \frac{1}{EI}(R_1 - P + R_2) = 0$$

$$\therefore R_1 + R_2 = P \tag{g}$$

$$\left(\frac{d^2v}{dx^2}\right)_{L^+} = 0 = R_1L - P(L-a) = 0$$

$$\therefore R_1 = \frac{P(L-a)}{L} \tag{h}$$

Returning to Eq. (g), we then see that

$$R_2 = P - \frac{P(L-a)}{L} = \frac{Pa}{L} \tag{i}$$

We thus have determined the supporting forces, a job we could have done by inspection for this simple problem.

We shall now proceed to the final boundary conditions requiring v to be zero at $x = 0$ and $x = L$. Note that we get zero at a loca-

[6]We integrate expressions $(x-a)$ and $(x-L)$ in Eq. (d) as was done in Chapter 11. This, recall, is permissible because of the presence of the arbitrary constant C_3.

[7]Note that Eqs. (g) and (h) are the rigid-body equations of equilibrium for the beam corresponding to $\Sigma F_y = 0$ and $\Sigma M_{R_2} = 0$.

Example 13.1 (Continued)

tion to the left of a step. The first condition renders $C_4 = 0$. The second condition becomes

$$0 = \frac{1}{EI}\left[\frac{R_1L^3}{6} - \frac{P(L-a)^3}{6} + C_3L\right] \tag{j}$$

$$\therefore C_3 = \frac{P(L-a)^3}{6L} - \frac{R_1L^2}{6} \tag{k}$$

Now going back to Eq. (f) and putting in the values of R_1, R_2, and the constants of integration, we get

$$\begin{aligned} v = \frac{1}{EI}\Bigg\{ &\frac{P(L-a)}{L}\frac{x^3}{6}[u(x)] - \frac{P(x-a)^3}{6}[u(x-a)] \\ &+ \frac{Pa}{6L}(x-L)^3[u(x-L)] + \left[\frac{P(L-a)^3}{6L} - \frac{PL(L-a)}{6}\right]x\Bigg\} \end{aligned} \tag{l}$$

Example 13.2

What is the deflection of the beam in Fig. 13.9, (a) at $x = 7$ m, and (b) at $x = 13$ m in terms of EI? The loading function is

$$w_y(x) = -R_1[\delta(x)] + 5000[\delta(x-5)] + 500[u(x-9)] - 500[u(x-17)]$$

$$-R_2[\delta(x-22)]$$

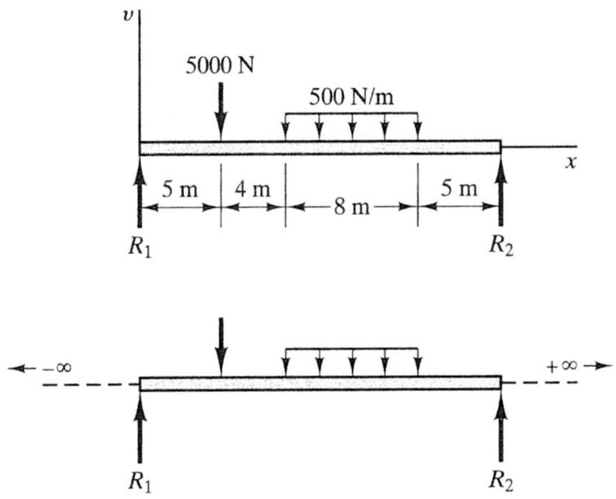

Figure 13.9. Simply-supported beam.

Example 13.2 (Continued)

The differential equation for v and subsequent integrations are as follows:

$$\frac{d^4v}{dx^4} = -\frac{w_y}{EI} = \frac{1}{EI}\{R_1[\delta(x)] - 5000[\delta(x-5)] - 500[u(x-9)] + 500[u(x-17)] + R_2[\delta(x-22)]\}$$

$$\frac{d^3v}{dx^3} = \frac{1}{EI}\{R_1[u(x)] - 5000[u(x-5)] - 500(x-9)[u(x-9)] + 500(x-17)[u(x-17)] + R_2[u(x-22)]\}$$

$$\frac{d^2v}{dx^2} = \frac{1}{EI}\left\{R_1x[u(x)] - 5000(x-5)[u(x-5)] - 500\frac{(x-9)^2}{2}[u(x-9)] + 500\frac{(x-17)^2}{2}[u(x-17)] + R_2(x-22)[u(x-22)]\right\}$$

$$\frac{dv}{dx} = \frac{1}{EI}\left\{R_1\frac{x^2}{2}[u(x)] - 5000\frac{(x-5)^2}{2}[u(x-5)] - 500\frac{(x-9)^3}{6}[u(x-9)] + 500\frac{(x-17)^3}{6}[u(x-17)] + R_2\frac{(x-22)^2}{2}[u(x-22)] + C_3\right\}$$

$$v = \frac{1}{EI}\left\{R_1\frac{x^3}{6}[u(x)] - 5000\frac{(x-5)^3}{6}[u(x-5)] - 500\frac{(x-9)^4}{24}[u(x-9)] + 500\frac{(x-17)^4}{24}[u(x-17)] + R_2\frac{(x-22)^3}{6}[u(x-22)] + C_3x + C_4\right\}$$

Set $d^3v/dx^3 = 0$ when $x = 22^+$. We get

$$R_1 - 5000 - 500(22-9) + 500(22-17) + R_2 = 0$$

$$\therefore R_1 + R_2 = 9000 \qquad \text{(a)}$$

Next set $d^2v/dx^2 = 0$ when $x = 22^+$.

Example 13.2 (Continued)

$$R_1(22) - 5000(22-5) - 500\frac{(22-9)^2}{2} + 500\frac{(22-17)^2}{2} = 0$$

$$\therefore R_1 = 5500 \text{ N}$$

Hence, from Eq. (a), $R_2 = 3500$ N. Now we set $v = 0$ at $x = 0$.

$$\therefore C_4 = 0$$

Finally, we set $v = 0$ at $x = 22$.

$$5500\frac{22^3}{6} - 5000\frac{(22-5)^3}{6} - 500\frac{(22-9)^4}{24} + 500\frac{(22-17)^4}{24} + 22C_3 = 0$$

$$\therefore C_3 = -2.31 \times 10^5$$

The deflection at $x = 7$ m is first calculated.[8]

$$v(7) = \frac{1}{EI}\left\{5500\frac{7^3}{6} - 5000\frac{(7-5)^3}{6} - (2.31\times 10^5)(7)\right\} = -\frac{1.310\times 10^6}{EI}\text{ m}$$

The deflection at $x = 13$ m is next computed.

$$v(13) = \frac{1}{EI}\left\{5500\frac{13^3}{6} - 5000\frac{(13-5)^3}{6} - 500\frac{(13-9)^4}{24} - (2.31\times 10^5)(13)\right\}$$
$$= -\frac{1.422\times 10^6}{EI}\text{ m}$$

13.4 The Doublet Function

We have thus far set forth the delta function to enable us to express a point load as an intensity of loading *w*, and we have set forth the step function to enable us to conveniently express loadings which do not extend over the entire span of the beam as loading functions. Now we shall set forth the third of our singularity functions, the **doublet func-**

[8]We note again that if the position $x = 7$ m is to the *left* of a step, we get *zero* contribution from the expression containing that step, and if $x = 7$ m is to the *right* of a step, we get a contribution from the expression containing that step with the step becoming *unity*.

tion, which will permit us to express the point couple as an intensity of loading w for use in our differential equations.[9]

As in our introduction to the delta function we shall go through a limiting process to arrive at the doublet function. One way to proceed is to consider a loading function of the type shown in Fig. 13.10. Note that

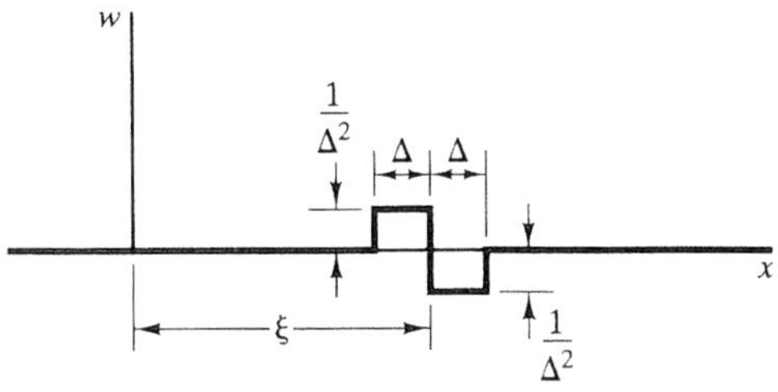

Figure 13.10. Function to be used for the doublet function.

we have a rectangular area of value $1/\Delta$ in the positive region of the plot and an identical area of value $1/\Delta$ in the negative region of the plot. We now replace these areas by directed line segments going through their centroids with senses corresponding to the convention sign of w for the particular area, as shown in Fig. 13.11.[10] The directed line segments

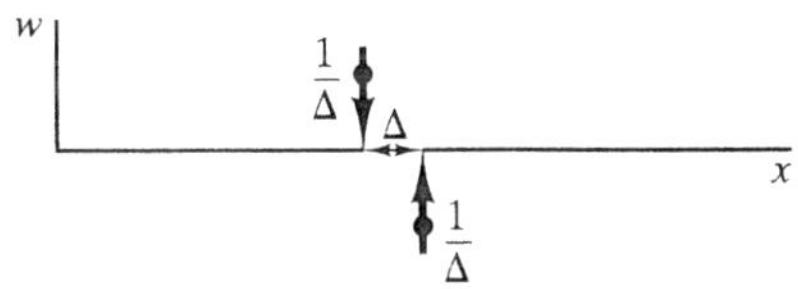

Figure 13.11. Areas replaced by vectors.

clearly must be Δ apart and may be considered to generate a "moment" of value unity with a counterclockwise sense. If we now take the limit as $\Delta \to 0$, the directed line segments elongate without limit while coming ever closer together. However, at all times and in the limit they generate a *unit couple moment.* The limiting function is our doublet function and

[9]You may wonder whether we have not already done this earlier when we discussed the use of the delta function to represent a point couple and a point heat source. At that time, we pointed out that $C[\delta(x-\xi)]$ is the *distribution function* of *applied external moment,* representing a point couple of magnitude C at point ξ on the beam. Also, $Q[\delta(x-\xi)]$ is the *distribution function of heat source,* representing a point heat source a position ξ along the rod. What we are attempting to do here is to set forth the distribution function of *force loading* corresponding to an *applied point couple,* and not the distribution function of *applied external moment* for this couple as was done earlier.

[10]We remind you that a positive w points downward for the structural convention.

it is denoted as $[\eta(x - \xi)]$, with ξ corresponding to the position of the singularity. The doublet function has the following characteristics:

1. It is zero everywhere except at $x = \xi$, the position of the singularity, where it is indeterminate.
2. The distribution generates a unit couple moment at the singularity as described above.

We used the function shown in Fig. 13.10 to generate by a limiting process the above characteristics of $[\eta(x - \xi)]$. Actually many other kinds of functions could have been conveniently used. For instance, in place of the rectangles we could have used appropriate triangles or sinusoids. Thus, the particular function w is a vehicle for arriving at the doublet function; once the goal of the doublet is reached there is no longer a need for the vehicle. Now $[\eta(x - \xi)]$ with units L^{-2} can be used to represent a loading distribution $w(x)$, with all the characteristics of a point couple at position $x = \xi$. Thus if the value of the couple is C (Fig. 13.12), then $C[\eta(x - \xi)]$ gives the force distribution generating the desired point couple. Notice that the unit point couple generated by $[\eta(x - \xi)]$ at $x = \xi$ is a *counterclockwise* couple.

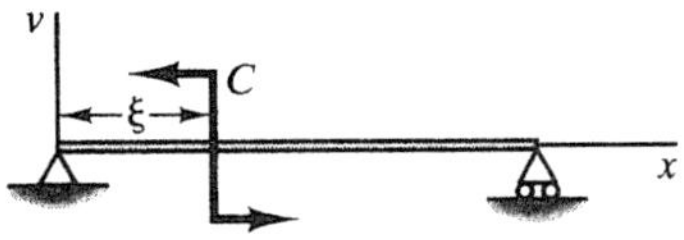

Figure 13.12. Point couple.

Let us next investigate the meaning of the following integration:

$$g(x) = \int_{-\infty}^{x} [\eta(\zeta - \xi)]d\zeta \tag{13.23}$$

For this purpose we can put our "vehicle function" shown in Fig. 13.10 to good use. Using the indicated function w as the integrand we can readily conclude that $g(x)$ turns out to be precisely the plot shown in Fig. 13.1(a). Indeed, if we let $\Delta \to 0$ as we must to reach $[\eta(\zeta - \xi)]$ needed in Eq. (13.23), we see that $g(x)$ becomes simply the *delta function.* Thus, we reach the following happy conclusion:

$$[\delta(x - \xi)] = \int_{-\infty}^{x} [\eta(\zeta - \xi)]d\zeta \tag{13.24}$$

Thus, in our shear and bending-moment considerations the doublet function integrates to the delta function.

We shall now discuss the consistency of the sign adopted for the doublet function and the sign convention for bending moment that we have been using. First consider the loading distribution $C[\eta(x)]$ representing a *counterclockwise* couple at $\xi = 0$ (see Fig. 13.13). The differential equation of equilibrium says that

$$\frac{dV}{dx} = -w(x) = -C[\eta(x)]$$

We can then say that

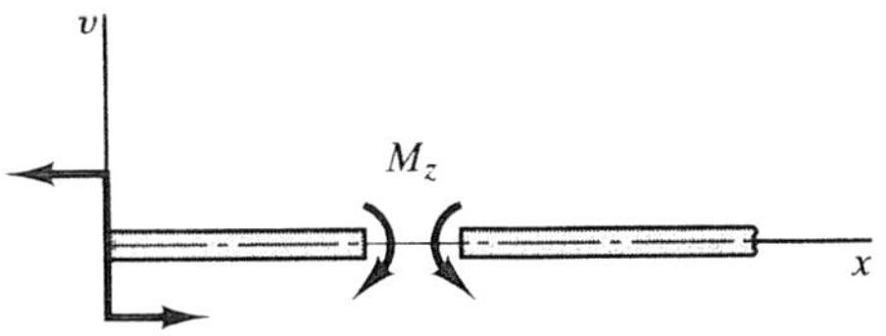

Figure 13.13. Bending moment from counterclockwise couple.

$$V = -\int_{-\infty}^{x} C[\eta(\zeta)]d\zeta = -C[\delta(x)]$$

As for the bending moment we have

$$\frac{dM}{dx} = V = -C[\delta(x)]$$
$$\therefore M = -C[u(x)]$$

Thus, a positive (counterclockwise) doublet function $\eta(x)$ gives a negative bending moment to the right of its position. Now reexamine Fig. 13.13, where we have applied a counterclockwise couple C at $x = 0$ and have shown bending moments M *induced* by C to the right of the origin. Clearly, these bending moments are negative according to our convention. Thus, there is consistency between the convention of using a positive doublet function for counterclockwise couples and the basic convention we have been using for the sign of bending moments.

Example 13.3

Consider a cantilever beam shown in Fig. 13.14 loaded by a uniform load w_0 over a portion of the beam. Formulate the deflection curve.

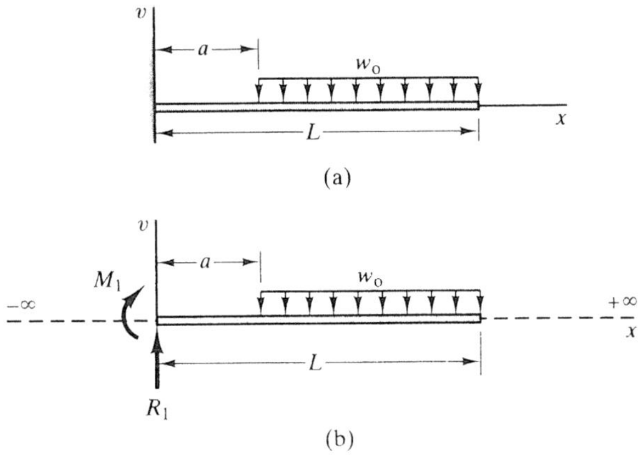

Figure 13.14. Cantilever beam.

Example 13.3 (Continued)

Using an infinite beam, the loading function is

$$w(x) = -M_1[\eta(x)] - R_1[\delta(x)] + w_0[u(x-a)] - w_0[u(x-L)] \tag{a}$$

The basic differential equation becomes

$$\frac{d^4v}{dx^4} = -\frac{w_y}{EI} = \frac{1}{EI}\{M_1[\eta(x)] + R_1[\delta(x)] - w_0[u(x-a)] + w_0[u(x-L)]\} \tag{b}$$

We now perform four integrations from $-\infty$ to x. Thus,

$$\frac{d^3v}{dx^3} = \frac{1}{EI}\{M_1[\delta(x)] + R_1[u(x)] - w_0(x-a)[u(x-a)] + w_0(x-L)[u(x-L)]\} \tag{c}$$

$$\frac{d^2v}{dx^2} = \frac{1}{EI}\left\{M_1[u(x)] + R_1x[u(x)] - w_0\frac{(x-a)^2}{2}[u(x-a)] + w_0\frac{(x-L)^2}{2}[u(x-L)]\right\} \tag{d}$$

$$\frac{dv}{dx} = \frac{1}{EI}\left\{M_1x[u(x)] + R_1\frac{x^2}{2}[u(x)] - \frac{w_0(x-a)^3}{6}[u(x-a)] + \frac{w_0(x-L)^3}{6}[u(x-L)] + C_3\right\} \tag{e}$$

$$v = \frac{1}{EI}\left\{\frac{M_1x^2}{2}[u(x)] + \frac{R_1x^3}{6}[u(x)] - \frac{w_0(x-a)^4}{24}[u(x-a)] + \frac{w_0(x-L)^4}{24}[u(x-L)] + C_3x + C_4\right\} \tag{f}$$

At $x = L^+$, we have $d^3v/dx^3 = d^2v/dx^2 = 0$. Thus,

$$\frac{1}{EI}[R_1 - w_0(L-a)] = 0 \tag{g}$$

and

$$\frac{1}{EI}\left[M_1 + R_1L - \frac{w_0(L-a)^2}{2}\right] = 0 \tag{h}$$

Example 13.3 (Continued)

We may solve for R_1 and M_1 from these equations:

$$R_1 = w_0(L-a) \tag{i}$$

$$M_1 = -w_0L(L-a) + \frac{w_0(L-a)^2}{2} = \frac{w_0}{2}(a^2 - L^2) \tag{j}$$

The boundary conditions at the support requiring $v = dv/dx = 0$ at $x = 0$ clearly render $C_3 = C_4 = 0$. We may then give the deflection curve in the following form:

$$v = \frac{1}{EI}\left\{\frac{w_0(a^2-L^2)}{4}x^2[u(x)] + \frac{w_0(L-a)}{6}x^3[u(x)] - \frac{w_0(x-a)^4}{24}[u(x-a)] + \frac{w_0(x-L)^4}{24}[u(x-L)]\right\} \tag{k}$$

In the following example we consider a statically indeterminate problem.

Example 13.4

Shown in Fig. 13.15(a) is a simply-supported beam with a degree of redundancy of unity. We wish to determine the supporting forces R_1, R_2, and R_3 and the deflection curve, the latter in terms of EI.

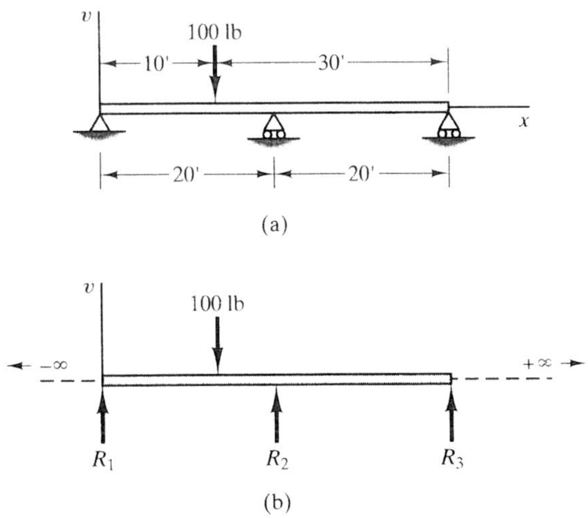

Figure 13.15. Statically indeterminate beam.

Example 13.4 (Continued)

The loading function for the infinite hypothetical replacement beam is given as

$$w_y(x) = -R_1[\delta(x)] + 100[\delta(x-10)] - R_2[\delta(x-20)] - R_3[\delta(x-40)] \quad \text{(a)}$$

The basic differential equation for deflection is then

$$\frac{d^4v}{dx^4} = -\frac{w_y(x)}{EI} = \{R_1[\delta(x)] - 100[\delta(x-10)] + R_2[\delta(x-20)] + R_3[\delta(x-40)]\} \quad \text{(b)}$$

Integrating four times, we get

$$\frac{d^3v}{dx^3} = \frac{1}{EI}\{R_1[u(x)] - 100[u(x-10)] + R_2[u(x-20)] + R_3[u(x-40)]\} \quad \text{(c)}$$

$$\frac{d^2v}{dx^2} = \frac{1}{EI}\{R_1x[u(x)] - 100(x-10)[u(x-10)] + R_2(x-20)[u(x-20)] + R_3(x-40)[u(x-40)]\} \quad \text{(d)}$$

$$\frac{dv}{dx} = \frac{1}{EI}\left\{\frac{R_1x^2}{2}[u(x)] - 100\frac{(x-10)^2}{2}[u(x-10)] + R_2\frac{(x-20)^2}{2}[u(x-20)] + R_3\frac{(x-40)^2}{2}[u(x-40)] + C_3\right\} \quad \text{(e)}$$

$$v = \frac{1}{EI}\left\{\frac{R_1x^3}{6}[u(x)] - 100\frac{(x-10)^3}{6}[u(x-10)] + R_2\frac{(x-20)^3}{6} \times [u(x-20)] + R_3\frac{(x-40)^3}{6}[u(x-40)] + C_3x + C_4\right\} \quad \text{(f)}$$

Since the deflection is zero at the origin, C_4 must also be zero. To the right of the support R_3, we have $d^3v/dx^3 = d^2v/dx^2 = 0$. We then have

$$R_1 - 100 + R_2 + R_3 = 0 \quad \text{(g)}$$

$$40R_1 - 3000 + 20R_2 = 0 \quad \text{(h)}$$

Furthermore, at $x = 20$ and $x = 40$, we have $v = 0$. That gives us the following equations:

Example 13.4 (Continued)

$$\frac{8000}{6}R_1 - \frac{100}{6}(10)^3 + C_3(20) = 0 \qquad \text{(i)}$$

$$\frac{64 \times 10^3}{6}R_1 - \frac{100}{6}(30)^3 + R_2\frac{(20)^3}{6} + C_3(40) = 0 \qquad \text{(j)}$$

Equations (g)–(j) are now rewritten as follows:

$$R_1 + R_2 + R_3 = 100 \qquad \text{(k)}$$

$$2R_1 + R_2 = 150 \qquad \text{(l)}$$

$$1333R_1 + 20C_3 = 16.67 \times 10^3 \qquad \text{(m)}$$

$$10.67 \times 10^3 R_1 + 1333R_2 + 40C_3 = 450 \times 10^3 \qquad \text{(n)}$$

We may substitute for R_2 in Eq. (n), using Eq. (l) and we may then eliminate C_3 from Eqs. (m) and (n). This permits us to solve for R_1. We get

$$R_1 = 40.4 \text{ lb} \qquad \text{(o)}$$

From Eqs. (k) and (l) we get the other supporting forces. Thus,

$$R_2 = 69.3 \text{ lb} \qquad \text{(p)}$$

$$R_3 = -9.64 \text{ lb} \qquad \text{(q)}$$

Finally, from Eq. (m) we can compute C_3 as

$$C_3 = -1859 \qquad \text{(r)}$$

We thus have simultaneously determined the deflection curve and the supporting forces.

13.5 Closure

In this chapter we have presented the very useful singularity functions and showed how with their use we could greatly simplify the integration of differential equations for deflection of beams. We concentrated on loadings involving point forces, point couples, and uniform loadings. These are the most common loadings. For more complex loadings we can still use singularity functions, but with added work and added ingenuity on the part of the engineer.

It is to be pointed out that many authors use Clebsch brackets or "disappearing" brackets, $<x - a>^n$, which are zero when $x < a$ and unity when $x \geq a$. Admittedly, these are easier to learn to use for beam problems than the singularity functions we have presented. However, our singularity functions have *far wider* use in other engineering sciences, applied mathematics, and physics. Thus, you will find our functions used in circuit theory, elasticity, heat transfer, electromagnetic theory, and so on. Your authors feel that it is well worth the reader's small extra effort to become familiar with these widely used functions.

PROBLEMS

Deflection Equations Using Delta and Step Functions	13.1–13.6
The Doublet Function and Its Use	13.7–13.18

13.1. [13.2] Given the following loading function,

$$w(x) = 200[u(x-10)] + 60[\delta(x-15)] + 100[u(x-15)] + 200[\delta(x)] + 100[u(x-20)]$$

what is $w(x)$ at $x = 3$? $x = 12$? $x = 18$?

13.2. [13.3] Determine the deflection curve for the beam shown in Fig. P.13.2. What is the deflection at $x = 15$ ft?

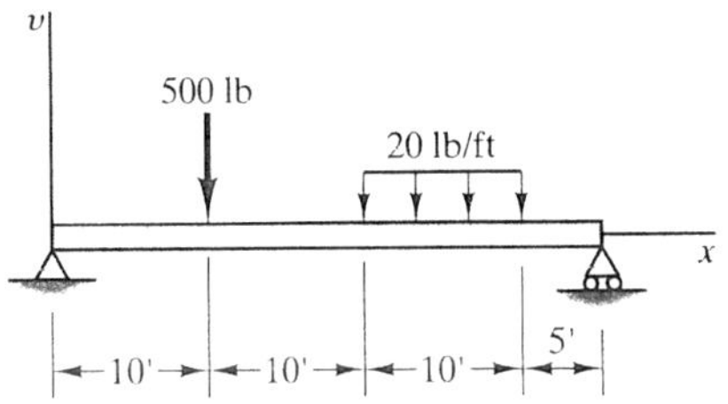

Figure P.13.2.

13.3. [13.3] Do Problem 12.7 using singularity functions. What is the slope of the curve at $x = 20$?

13.4. [13.3] Determine the deflection curve for the beam shown in Fig. P.13.4.

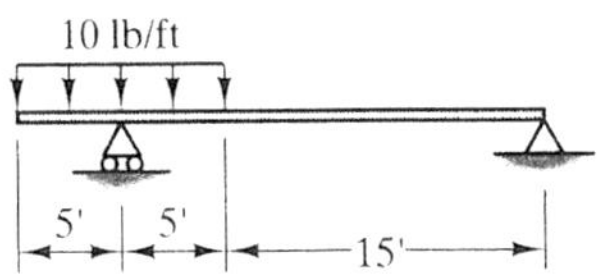

Figure P.13.4.

13.5. [13.3] Find the supporting forces and the deflection at a position 15 ft from the left end of the beam shown in Fig. P.13.5.

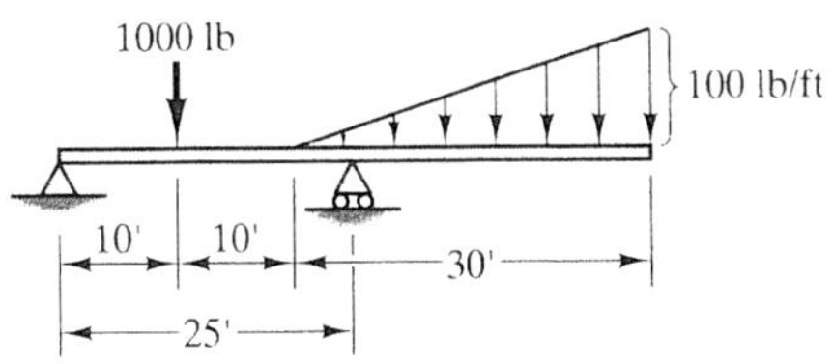

Figure P.13.5.

13.6. [13.3] Do Problem 12.12 using singularity functions.

13.7. [13.4] Do Problem 12.18 using singularity functions.

13.8. [13.4] Solve Problem 12.61 with the use of singularity functions.

13.9. [13.4] Work Problem 12.75 with the aid of singularity functions.

13.10. [13.4] Solve Problem 12.22 using singularity functions.

13.11. [13.4] Solve Problem 12.66. Also determine the supporting force and torque on the right end.

13.12. [13.4] Determine the maximum deflection of the cantilever beam in Fig. P.13.12 using singularity functions. Compute the supporting force system.

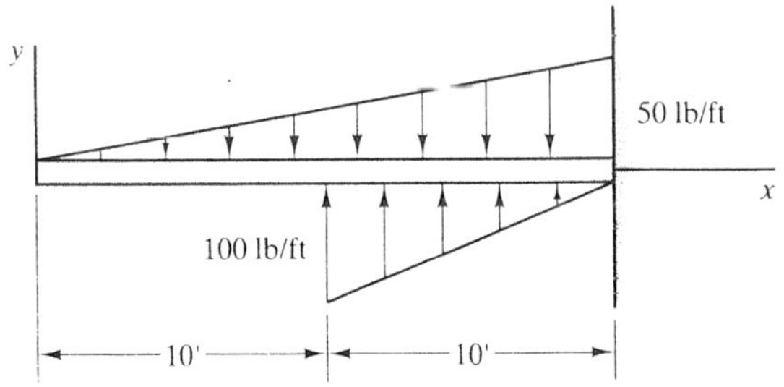

Figure P.13.12.

***13.13. [13.4]** A parabolic load is shown in Fig. P.13.13 acting on the simply-supported beam. Using singularity functions, determine the deflection curve and the supporting forces. (*Hint:* Replace the loading term $(x-10)^2[u(x-20)]$ by

$$\begin{aligned}[(x-10)+10-10]^2[u(x-20)] &= [(x-20)+10]^2[u(x-20)] \\ &= [(x-20)^2[u(x-20)] \\ &\quad + 20(x-20)[u(x-20)] \\ &\quad + 100[u(x-20)]\end{aligned}$$

Why is it desirable to do this?

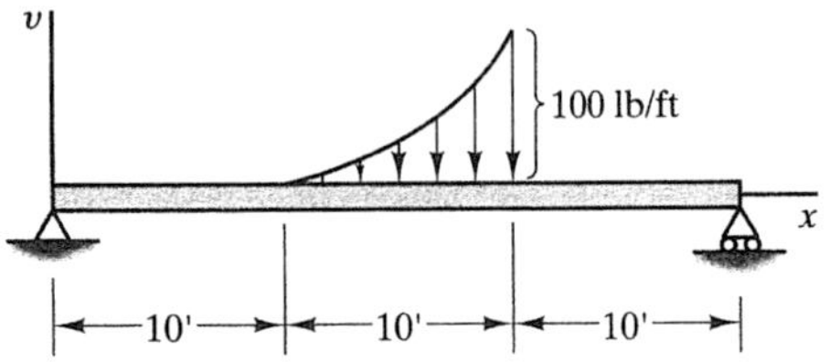

Figure P.13.13.

13.14. [13.4] Using singularity functions, find the deflection curve for the cantilever beam shown in Fig. P.13.14. What is d^3v/dx^3 at $x = 17$ ft?

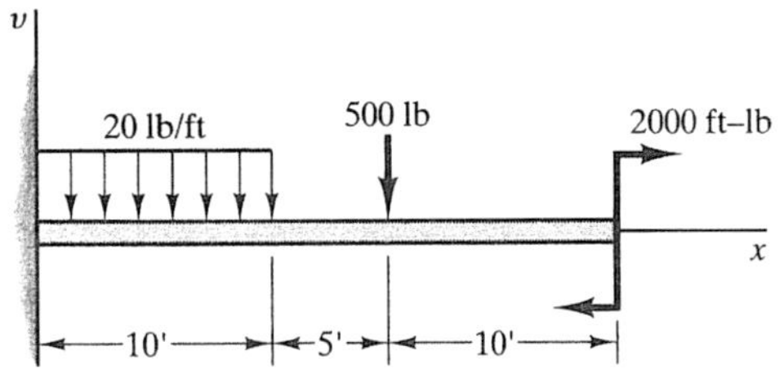

Figure P.13.14.

13.15. [13.4] Compute the supporting forces for the beam shown in Fig. P.13.15.

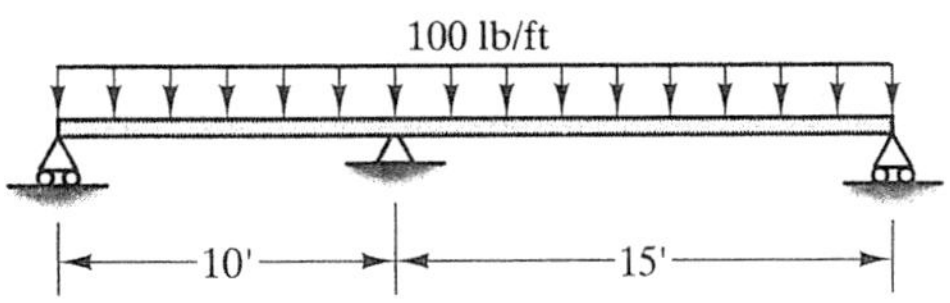

Figure P.13.15.

13.16. [13.4] Determine the supporting forces for the beam shown in Fig. P.13.16.

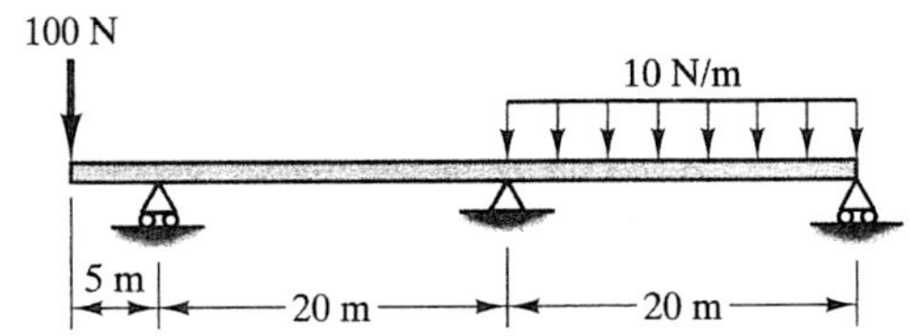

Figure P.13.16.

13.17. [13.4] Develop the deflection curve of the beam shown in Fig. P.13.17.

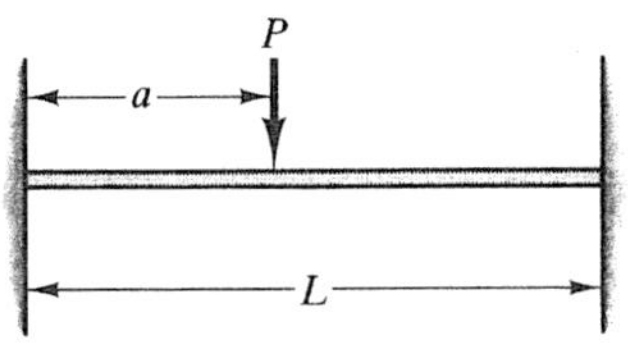

Figure P.13.17.

13.18. [13.4] Shown in Fig. P.13.18 are two beams fastened at A. If both beams are made of the same material, set up equations for determining the supporting force system. Show that $R_1 = .48P$, $R_2 = .640P$, $R_3 = -.120P$, $R_4 = R_5 = .240P$ satisfy your equations.

> Other problems using singularity functions may be assigned from Chapter 12.

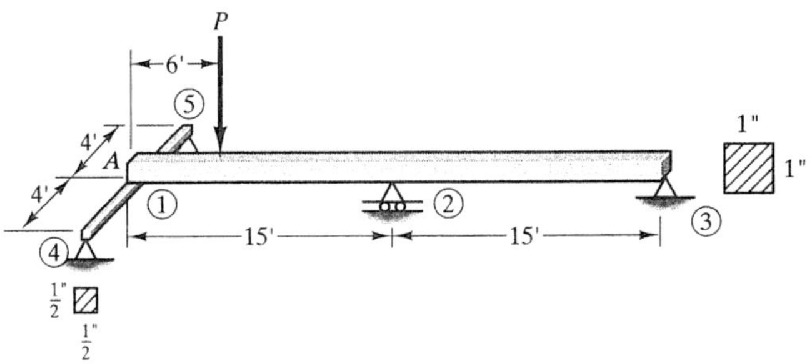

Figure P.13.18.

CHAPTER 14

Torsion

14.1 Introduction

In this chapter we examine the stresses and deformations in certain long, slender prismatic shafts under the action of twisting moments. First we shall consider the case of a *circular* shaft under the action of equal and opposite point couples as shown in Fig. 14.1. As in Chapter 12 on beams, we shall present physical arguments for certain *compatible* deformation characteristics of a circular shaft, and with the use of *Hooke's law* and *Newton's law*, this will lead to stress and strain formulations for the shaft. Unfortunately, we shall not be able to extend our closed-form solutions to noncircular cross sections of solid shafts. One has to resort to the theory of elasticity and/or numerical techniques for such cases. However, with the use of simplifying assumptions, we shall be able to consider approximations for the torsion of *thin-walled, noncircular shafts* later in the chapter. Also, we shall present some empirical formulations for the torsion of shafts with certain classes of composite cross sections. Finally, we shall examine energy methods (the second Castigliano theorem) applied to shafts.

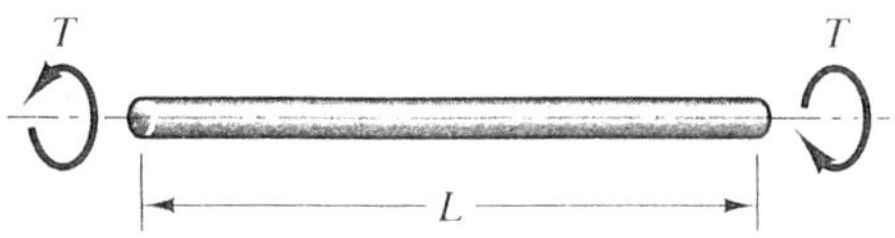

Figure 14.1. Shaft under torsion.

14.2 Circular Shafts

We shall now employ arguments of symmetry in stipulating *compatible* displacement characteristics of a circular shaft loaded as shown in Fig. 14.1. In Fig. 14.2, we show two elements of this shaft each of length Δx.

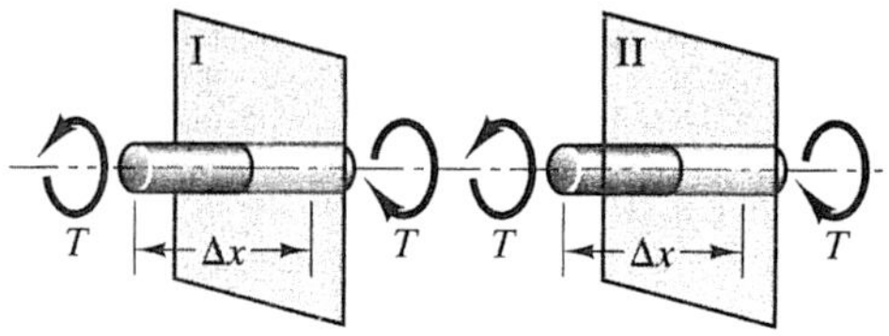

Figure 14.2. Shaft elements with planes of symmetry.

Note that each element is subject to identical sets of equal and opposite couple moments. Clearly, each element must deform in exactly the same manner. Furthermore, for shafts having circular cross sections there is a midplane dividing each shaft element into identical shapes as viewed from the ends of the element. That is, planes I and II in Fig. 14.2 form such mid-planes for the shaft elements. We can justify such an assertion by the following argument: Consider an element of shaft again as shown in Fig. 14.3(a). A z axis has been shown normal to the shaft

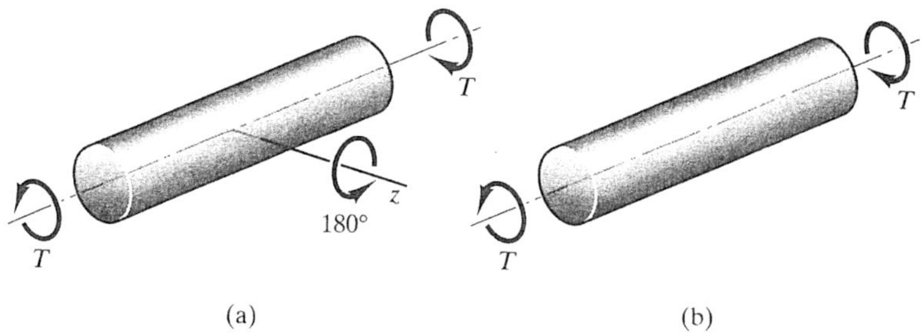

Figure 14.3. Shaft element and torques rotated 180°.

centerline at the midpoint of the element. Now rotate the element and its applied torques 180° about the z axis. The result is shown in Fig. 14.3(b). The element at (b) clearly has identically the *same loading* as at (a). Therefore, the left end of the element at (b) must have the *same deformed* shape as the left end of the element at (a). But the shape of the left end of (b) as viewed from the left has to be *identical* also to the shape of the right end of (a) as viewed from the right since diagram (b) is a simple rotation of diagram (a). Thus, we conclude that the left and right ends of the element have the same shape when viewed from the ends.

Owing to the *axial symmetry* of the original geometry and the applied loads, all radial lines in an undeformed cross section must undergo the same manner of deformation. Thus, an end surface of the elements discussed above must "dish in" or "bulge out" as a surface of revolution, or, possibly, remain a plane surface as a result of the applied torques. Consider that an end bulges out or dishes in. Clearly, this must be true for both ends of the element. And the deformation of one element must be the same as the deformation of the adjacent element. But these elements clearly cannot now fit together in the deformed geometry to form a continuous shaft either having constant diameter and/or not having voids. We can conclude from **compatibility** that the section *must* then remain plane during deformation. Thus, for shafts of *circular cross section, plane sections remain plane during deformation.* Recall that axial *symmetry* about the shaft centerline was a prime requirement for this conclusion and therefore we *cannot* assert that plane sections remain plane for nonaxially symmetric shafts such as those with square sections. We shall consider such shafts briefly later in the chapter.

We can make still further conclusions concerning the deformations *within* the plane of the cross section. For this purpose consider the front end of an element of shaft shown in Fig. 14.4(a) where a family of diameters in the deformed geometry is shown. All such diameters

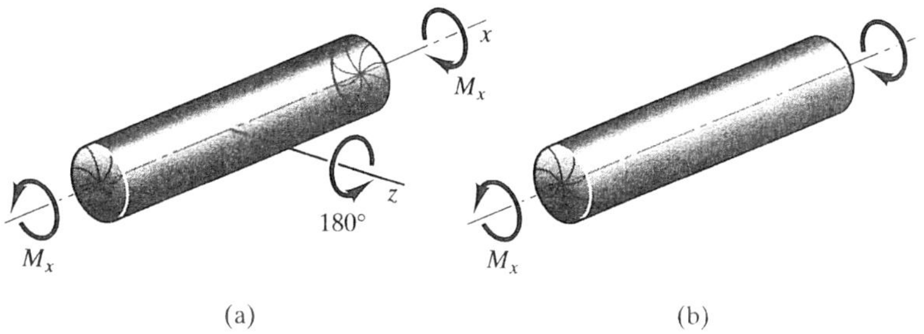

Figure 14.4. Shaft element with possible deformed diameters rotated 180°.

should have the same shape as a result of axial symmetry. Also, at the other end of the element we have shown an oppositely turning family of diameters in the deformed geometry. Now let us rotate the element and its torques by 180° about the indicated z-axis. The rotated geometry is shown in Fig. 14.4(b).[1] Clearly, we have at (b) the same original geometry subjected to the same torques as at (a), so the shape of the deformed geometry should be the same for both diagrams. Thus, the assumed shape of diameters at the right end of (a) gives no contradiction. How-

[1]To verify this picture, fold two sheets of paper as shown below and draw indicated markings on outer faces of the folded sheets. Now rotate one of the sheets 180° about axis AA and then compare the markings on the front faces with each other.

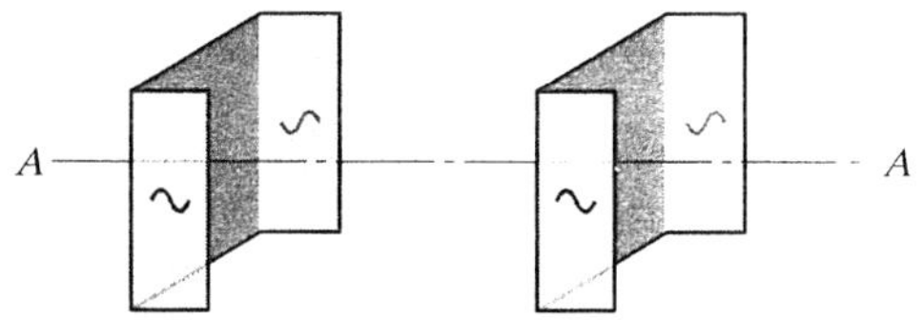

ever, were these elements to be placed together so as to form a continuous shaft, we readily see that the diameter shapes at the right end of (a) are incompatible with the diameter shapes at the left end of (b). To avoid this dilemma, it is clear from **compatibility** that the deformed geometry of the diameters should be straight lines. Thus, we can state the following.

> Radial lines in the cross section of a circular rod merely rotate when there is torsional deformation present. Furthermore, because of axial symmetry each diameter in a section rotates the same amount as the other diameters in the section.

With these conclusions as to the general manner of deformation, we may next proceed to formulate stress and strain distributions for the circular shaft in torsion. We may first say that for small deformations the longitudinal strain ε_{xx} is zero and we shall assume that the radial strain ε_{rr} is zero. Thus, since all diameters in any one cross section rotate as straight lines and with equal angle we can now state with $\varepsilon_{rr} = 0$ that cross sections of the shaft merely rotate about the centerline of the shaft and undergo no other movement. This means that $\varepsilon_{\theta\theta}$, that is, the normal strain in the transverse direction, is zero. Clearly, since cross sections of the shaft merely rotate, infinitesimal line segments in the r and θ directions [see Fig. 14.5(a)] at a point undergo no change in angle

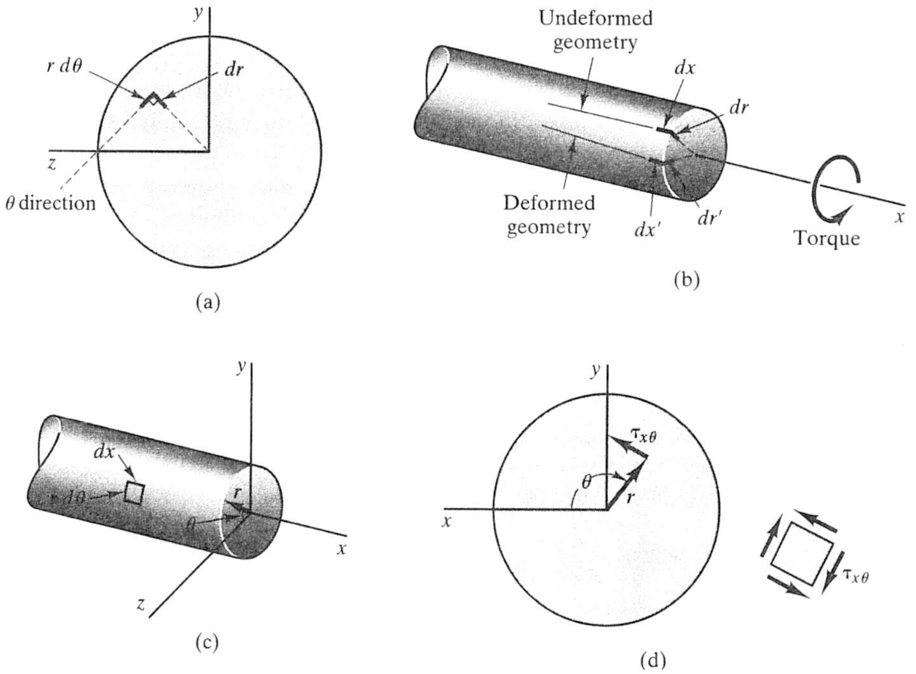

Figure 14.5. Views of shaft section.

between them. Hence, $\gamma_{r\theta}$ is zero. Similarly, infinitesimal line segments in the r and x directions at a point [see Fig. 14.5(b)] remain *essentially orthogonal* during deformation (assumed small). Hence, γ_{rx} is zero. In fact, for a cylindrical coordinate system with x as the cylindrical axis [see Fig. 14.5(c)] *only one strain is nonzero* for the assumed behavior and that is $\gamma_{x\theta}$. This is easily seen by noting [see Fig. 14.5(c)] that line segment dx in the shaft, originally parallel to the axis of the shaft, will rotate as a result of the aforestated deformation to lose perpendicularity with contiguous line segment $rd\theta$. This, then, yields the aforestated shear angle $\gamma_{x\theta}$. This shear angle will be explained further in the next paragraph. This strain is associated with the shear stress $\tau_{x\theta}$, shown in Fig. 14.5(d). This shear stress is in the plane of the section and in the transverse direction.

To compute shear angle $\gamma_{x\theta}$, we have shown an element of the shaft of length Δx and of radius r in Fig. 14.6. According to our state-

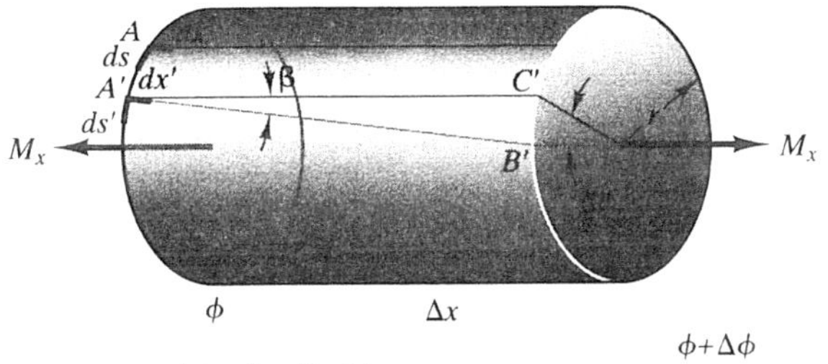

Figure 14.6. Deformed rod element.

ments for the deformation of the shaft, a straight line on the periphery of the shaft parallel to the x axis will deform into a helix. Line AB is a portion of such a straight line, and, when the loads are applied, this line moves and deforms to become $A'B'$ in the diagram. For vanishingly small Δx, the segment $A'B'$ can be considered as a straight-line segment forming the angle β with $A'C'$ taken parallel to AB as shown in Fig. 14.6. We see also that β is the change of right angle between the originally orthogonal segments ds and dx at A so that we can say that

$$\gamma_{x\theta} = \gamma_{\theta x} = \beta \tag{14.1}$$

Furthermore, it is clear from the diagram that

$$\beta = \lim_{\Delta x \to 0}\left(\frac{B'C'}{\Delta x}\right) = \lim_{\Delta x \to 0}\frac{r\Delta\phi}{\Delta x} = r\frac{d\phi}{dx} \tag{14.2}$$

where *$d\phi$ is the angle of twist* between sections dx apart.[2] The derivative $d\phi/dx$ is called the *rate of twist*—that is, the rotation per unit length of shaft along the shaft. It should be clear that the rate of twist for this case is a constant.

Now as a **constitutive law** we employ Hooke's law and from the preceding equations, we have for the stresses

$$\tau_{x\theta} = G\gamma_{x\theta} = Gr\frac{d\phi}{dx} \tag{14.3}$$

Since we could have chosen free bodies inside the shaft of any radius and performed the same calculations leading up to Eq. (14.3), it should be understood that r in this equation is any radius from zero up to one-half the diameter of the shaft. Also, since $\varepsilon_{rr} = \varepsilon_{\theta\theta} = \varepsilon_{xx} = 0$, we see from Hooke's law that $\tau_{rr} = \tau_{\theta\theta} = \tau_{xx} = 0$. We can then conclude, since $\gamma_{r\theta} = \gamma_{rx} = 0$, that the *only nonzero stress* is $\tau_{x\theta}$, given in Eq. (14.3).

From **equilibrium**, considerations the shear-stress distribution must yield the correct resultant twisting moment M_x at a section. That is, using Eq. (14.3),

$$M_x = \iint_A Gr^2\frac{d\phi}{dx}\,dA' = G\frac{d\phi}{dx}\iint_A r^2 dA' = G\frac{d\phi}{dx}J \tag{14.4}$$

where J is the *polar moment of area* of the cross section about the center. This permits us to solve for the rate of twist, $d\phi/dx$, in terms of the applied couple. Thus,

$$\frac{d\phi}{dx} = \frac{M_x}{GJ} \tag{14.5}$$

The total angle of twist $\Delta\phi$ between two sections distance L apart becomes[3]

[2]The reader should not confuse the angles ϕ and θ. The angle ϕ represents *rotation of cross sections* of the shaft, while θ is a cylindrical coordinate locating, with *r, a point in the cross section.* Also note that since no constitutive law was used, Eq. (14.2) is valid for all materials and for plastic deformation.

[3]You will perhaps note that Eq. (14.6) is the analog formula of the elongation δ of a simple, one-dimensional member as given in Chapter 5. That is,

$$\delta = \frac{FL}{AE}$$

Note that

$$\begin{aligned} \Delta\phi &\to \delta \\ F &\to M_x \\ A &\to J \\ E &\to G \end{aligned}$$

$$\Delta\phi = \int_0^L \frac{M_x}{GJ}dx = \frac{M_x L}{GJ} \tag{14.6}$$

Combining Eqs. (14.3) and (14.5) to eliminate $d\phi/dx$, we then have a convenient formula for the shear stress,

$$\tau_{x\theta} = \frac{M_x r}{J} \tag{14.7}$$

which for torsion is the counterpart of the beam flexure formula. Notice that the stress is zero along the centerline and is greatest at the outer fibers.

If we do not have a constant torque at each section of the shaft, then M_x becomes a function of position along the shaft and $d\phi/dx$, given by Eq. (14.5), is no longer a constant. Consequently, for such a case ϕ can be determined (approximately, since the theory is no longer perfectly correct for variable torque) by the integration of Eq. (14.5) with M_x a function of x. Finally, $\tau_{x\theta}$ in Eq. (14.7) will then be a function of *r and x*.

As a measure of *torsional stiffness,* we shall next consider the concept of the torsional spring constant, analogous to the spring constant for linear springs. The *torsional spring constant* is defined as the torque required per unit angle twist of shaft and is thus given as M_x/ϕ. From Eq. (14.6), for the circular shaft we then have

$$\frac{M_x}{\phi} = \frac{GJ}{L} \tag{14.8}$$

The unit of twist is the radian in the usual formulation of the torsional stiffness and the unit length is the foot or meter. This does not mean that a shaft having a torsional stiffness of A ft-lb/rad/ft is expected to undergo for each foot a full radian of twist for A ft-lb of torque. Clearly, this would take us well beyond small deformation limits of the theory and would be most likely to result in the fracture of the test specimen. What this particular measure means is that, if the ratio of torque to twist for a unit length for small deformation (i.e., for small angles of twist) were imagined to be maintained until hypothetically a twist of 1 rad is reached, the value of this theoretical torque would be the measure of the torsional stiffness.

Another useful definition is the *torsional strength* defined as the torque per unit maximum stress. That is,

$$\text{torsional strength} = \frac{M_x}{\tau_{\max}} \tag{14.9}$$

Substituting for $\tau_{\max}$ using Eq. (14.7) with $r = D/2$ and using $J = \pi D^4/32$ for a circular cross section, we then get

$$\text{torsional strength} = \frac{M_x}{(M_x D)/(2J)} = \frac{\pi D^3}{16} \tag{14.10}$$

We can show from the theory of elasticity that the stress distribution proposed by the preceding theory is, for small deformation and constant diameter shafts, an exact theory.

14.3 Torsion Problems Involving Circular Shafts

We shall now consider a series of examples that will illustrate the use of the formulations presented in Section 14.2.

Example 14.1

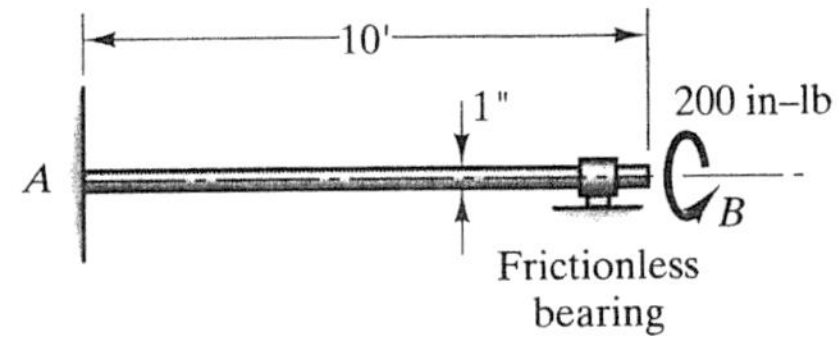

Figure 14.7. Torsion shaft fixed at one end.

Shown in Fig. 14.7 is a circular shaft of diameter 1 in. and of length 10 ft. It is fixed in a wall at A and is subject to 200 in.-lb torque at the other end B, as shown in the diagram. The material is steel having a shear modulus $G = 15 \times 10^6$ psi.

The *maximum shear stress* resulting from torsion at a section (away possibly from the ends) can be computed in the following way using the formula $J = \pi r^4/2$:

$$\tau_{\max} = \frac{M_x(D/2)}{J} = \frac{(200 \text{ in.-lb})(\frac{1}{2} \text{ in.})}{[\pi(\frac{1}{2})^4/2] \text{ in.}^4} \tag{a}$$

$$\tau_{\max} = 1019 \text{ psi}$$

If the diameter of the shaft were halved, the maximum torsional shear stress would be

$$\tau_{\max} = \frac{(200)(\frac{1}{4})}{\pi(\frac{1}{4})^4/2} = 8150 \text{ psi} \tag{b}$$

The ratio of *torsional strength* of the two cases is then

$$\frac{(M_x/\tau_{\max})_a}{(M_x/\tau_{\max})_b} = 8$$

Also, the ratio of the *torsional spring constants* between the two sizes is given as

Example 14.1 (Continued)

$$\frac{(M_x/\phi)_a}{(M_x/\phi)_b} = \frac{(G/L)[\pi(\frac{1}{2})^4/2]}{(G/L)[\pi(\frac{1}{4})^4/2]} = 16 \tag{c}$$

We can thus see that doubling the diameter of a shaft increases the torsional strength 8-fold and the torsional spring constant 16-fold.

For the shaft of diameter 1 in. we have, for the *angle of twist* at cross section B,

$$\phi = \frac{(10)(12)(200)}{(15 \times 10^6)[\pi(\frac{1}{2})^4/2]} = .01630 \text{ rad} \tag{d}$$

Example 14.2

Shown in Fig. 14.8 is a hollow circular shaft having an outer diameter of 5 in. and an inner diameter of 4 in. A torque of 1000 ft-lb is applied at the right end, while the left end is maintained fixed in the wall. What are the maximum and minimum torsional shear stresses in a section of the shaft and what is the angle of twist at the right end? (Take $G = 15 \times 10^6$ psi.)

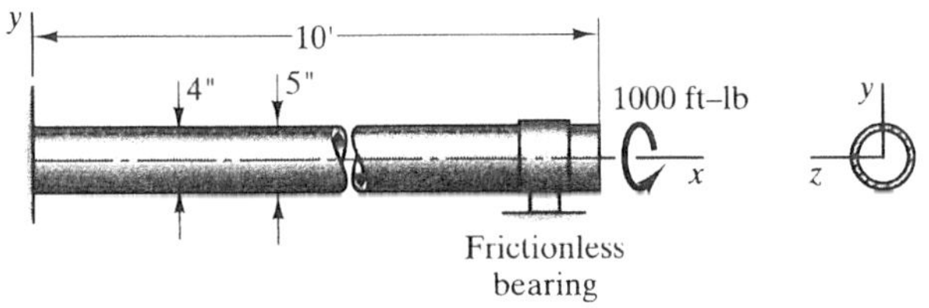

Figure 14.8. Hollow shaft under torsion.

Because of the *symmetry* of this shaft about the x axis we can conclude, upon reconsidering the development in Section 14.2, that the deformation prescribed for the solid circular shaft holds for the hollow circular shaft. Therefore, we can use the resulting equations of that section provided that we insert the proper evaluation of J.

It is then clear that minimum torsional shear stress occurs at the inner radius of the shaft and is given as

$$\tau_{\min} = \frac{M_x r_{\min}}{J} = \frac{(1000)(12)(2)}{(\pi/2)(2.5^4 - 2^4)} = 662 \text{ psi} \tag{a}$$

The maximum shear stress is

Example 14.2 (Continued)

$$\tau_{\max} = \frac{2.5}{2}(662) = 828 \text{ psi} \tag{b}$$

Finally, the angle of twist for the right-end section becomes

$$\phi = \frac{M_x L}{GJ} = \frac{(1000)(12)(10)(12)}{(15 \times 10^6)(\pi/2)(2.5^4 - 2^4)} = .00265 \text{ rad} \tag{c}$$

Example 14.3

Consider a solid shaft having the same cross-sectional area as the hollow shaft in Example 14.2 and subject to the same torque. If the solid shaft is of the same material, compare the torsional strength and the torsional stiffness of the two shafts.

A simple arithmetic calculation will reveal that the diameter of the solid shaft must be 3 in. Both cross sections are shown in Fig. 14.9.

The *torsional strength* for each shaft will first be computed. Thus, employing Eqs. (14.9) and (14.7):

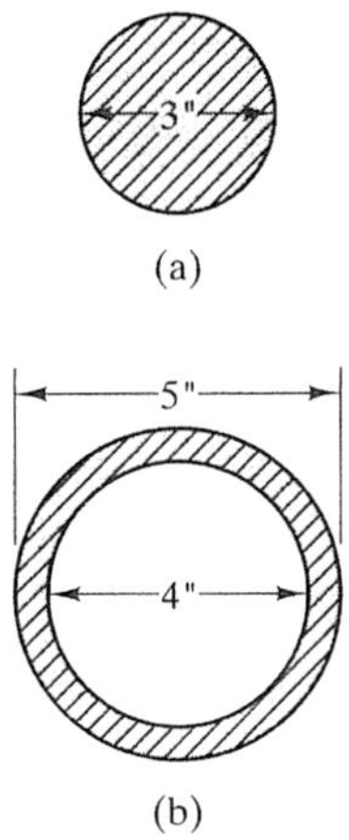

Figure 14.9. Solid and hollow shaft sections with the same area.

$$\left(\frac{M_x}{\tau_{\max}}\right)_a = \left(\frac{J}{r_{\max}}\right)_a = \frac{\pi[(1.5^4)/2]}{1.5} = 5.30 \text{ in.}^3$$

$$\left(\frac{M_x}{\tau_{\max}}\right)_b = \left(\frac{J}{r_{\max}}\right)_b = \frac{\pi[(2.5^4)/2 - 2^4/2]}{2.5} = 14.49 \text{ in.}^3$$

The ratio of the torsional strengths between the two cross sections is 2.73 in favor of the hollow shaft, showing that material has been used more efficiently here.

Next we examine the *torsional stiffness* for the two shafts:

$$\left(\frac{M_x}{\phi}\right)_a = \left(\frac{GJ}{L}\right)_a = \frac{G}{L}\left[\pi\frac{(1.5)^4}{2}\right]$$

$$\left(\frac{M_x}{\phi}\right)_b = \left(\frac{GJ}{L}\right)_b = \frac{G}{L}\left\{\pi\left[\frac{(2.5)^4}{2} - \frac{2^4}{2}\right]\right\}$$

The *ratio* of the two values for any length of material is then

$$\frac{(M_x/\phi)_b}{(M_x/\phi)_a} = \frac{2.5^4 - 2^4}{1.5^4} = 4.56$$

Example 14.3 (Continued)

showing an even greater gain in stiffness for the hollow shaft over the solid shaft than the already significant gain in torsional strength.

It is then quite clear why automobile manufacturers have employed hollow drive shafts over the years.

Example 14.4

In Fig. 14.10 is shown a circular shaft held fixed at the ends and loaded by a torque M_x. We wish to determine the reactive torques $(M_x)_1$ and $(M_x)_2$ developed by the applied torque M_x.

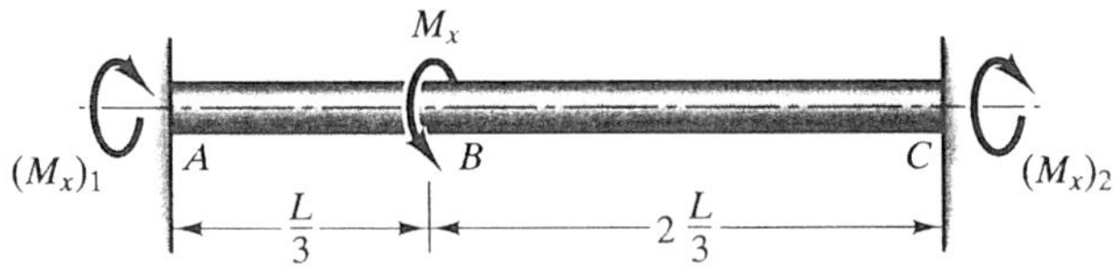

Figure 14.10. Fixed-ended circular shaft.

We have only one nontrivial equation of equilibrium here for the determination of the two unknown torques, so the problem is *statically indeterminate,* requiring the use of strength of materials.

Equilibrium gives us the following relation:

$$(M_x)_1 + (M_x)_2 = M_x \tag{a}$$

We shall next use **compatibility** by adding the twist of the shaft from A to B (see Fig. 14.11), plus that from B to C. The total twist at C for a compatible deformation must equal zero. Thus using a **constitutive law,** we have on mentally forming right-hand free-body diagrams,

$$\frac{[M_x - (M_x)_2](L/3)}{GJ} + \frac{-(M_x)_2[2(L/3)]}{GJ} = (\phi)_c = 0 \tag{b}$$

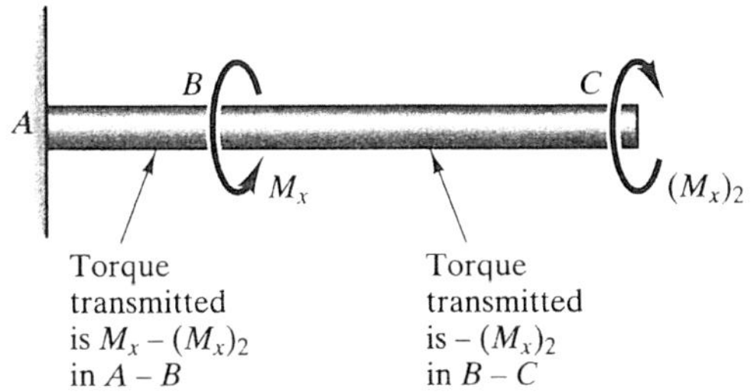

Figure 14.11. Consider twist at C from torques M_x and $(M_x)_2$.

Example 14.4 (Continued)

Canceling GJ, we get

$$(M_x)_2 = \frac{M_x}{3} \tag{c}$$

Hence,

$$(M_x)_1 = \frac{2M_x}{3} \tag{d}$$

The answers we have reached here may be intuitively obvious for this simple problem. For more complicated ones, however, we must rely on the calculations.

The next problem in the sequence is a design problem.

Example 14.5

An oceangoing tanker has twin steam turbines that turn at 1800 rpm while delivering 8200 kW each. The power of each turbine is transmitted by a shaft to a system of reduction gears, which, at an efficiency of 90%, reduces the rotational speed at which the power is transmitted to 107 rpm. A second shaft then transmits this power to a propeller.

(a). If the yield stress in shear for the shafts is 6.9×10^8 Pa, what should the diameters be for the shafts if they are solid? Use a safety factor of 2 and assume that a thrust bearing absorbs the axial force from the propeller before it gets to the shafts in question.

(b). Design the shafts for a safety factor of 2 assuming they are hollow with the inside diameter one half of the outside diameter.

As learned in elementary physics the power transmitted by a torque T is given as

$$\text{power} = T\omega \tag{a}$$

where ω is the angular velocity in radians per unit time. Hence, the torque T_1 in the first shaft (from the turbine) is given as

Example 14.5 (Continued)

$$8.2 \times 10^6 = T_1\left[1800\frac{\text{rev}}{\text{min}}\right]\left(\frac{2\pi \text{ rad}}{\text{rev}}\right)\left(\frac{1 \text{ min}}{60 \text{ s}}\right) \tag{b}$$

$$\therefore T_1 = 43{,}502 \text{ N-m} \tag{c}$$

As for the second shaft (to the propeller) we have a power transmission which is now decreased to 90% of the original power. We then have for the torque T_2 in this shaft

$$[.90][8.2 \times 10^6] = T_2(107)(2\pi)/(60)$$

$$\therefore T_2 = 659{,}000 \text{ N-m} \tag{d}$$

Now the maximum allowable stress is 3.45×10^8 Pa, so for a *solid shaft*, for D_1 and D_2 using the torsional stress formula, we have

$$3.45 \times 10^8 = \left\{\frac{T_1(D_1/2)}{\pi(D_1/2)^4/2}\right\}$$

$$\therefore \quad D_1 = .0863 \text{ m} = 86.3 \text{ mm} \tag{e}$$

$$3.45 \times 10^8 = \left\{\frac{T_2(D_2/2)}{\pi(D_2/2)^4/2}\right\}$$

$$\therefore \quad D_2 = .213 \text{ m} = 213 \text{ mm} \tag{f}$$

As for the *hollow shafts* we proceed as follows:

$$3.45 \times 10^8 = \frac{T_1(D_1/2)}{(\pi/2)[(D_1/2)^4 - (\frac{1}{2}D_1/2)^4]}$$

$$\therefore \quad D_1 = .0881 \text{ m} = 88.1 \text{ mm} \tag{g}$$

$$3.45 \times 10^8 = \frac{T_2(D_2/2)}{(\pi/2)[(D_2/2)^4 - (\frac{1}{2}D_2/2)^4]}$$

$$\therefore \quad D_2 = .218 \text{ m} = 218 \text{ mm} \tag{h}$$

It is also to be pointed out that allowance must be made for stress concentrations, a topic which we shall introduce briefly in Section 14.4.

We now finish this series of examples with one that appears complicated, but succumbs readily to an orderly attack.

Example 14.6

A system of two concentric torsion members is shown in Fig. 14.12 connected at their ends by rigid end disks G. The outside member C is a tube and the inside member consists of shafts A and B of different diameters and of different materials. The shear moduli for A, B, and C are listed in the diagram. At the right end, the rigid disk is held by a rubber grommet that resists rotation acting as a linear torsional spring with a torsional spring constant $(K_{\text{tors}})_D$ equal to 3×10^6 N-m/rad. A torque of 200 N-m is applied to the outside tube and a torque of 500 N-m is applied to the left end disk as shown in the diagram.

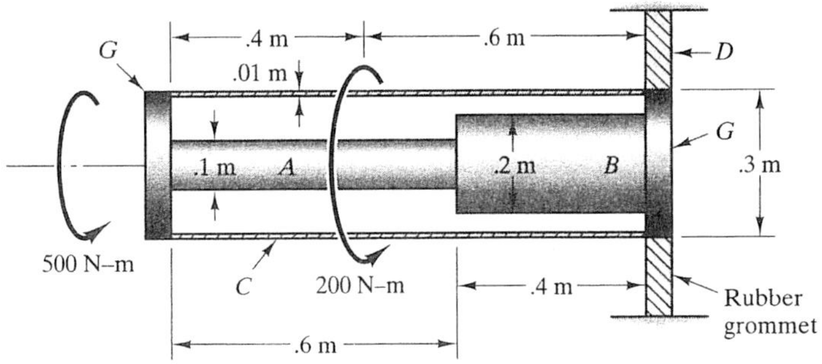

$G_A = 1.8 \times 10^{11}$ Pa
$G_B = 1.4 \times 10^{11}$ Pa
$G_C = 1 \times 10^{11}$ Pa

Figure 14.12. Complex torsion system.

(a). Determine the torques transmitted by each member.
(b). Find the rotation of the left end plate.

We will first compute the *relative rotation* $\Delta\phi$ of the end disks by considering separately the tube and the shaft whose free-body diagrams involving only torques are shown in Fig. 14.13 as F.B.D. II and F.B.D. III. Thus, considering the tube, we can say for $\Delta\phi$ **(constitutive law):**

$$\Delta\phi = \frac{(T_C + 200)(.6)}{(1 \times 10^{11})(\pi/32)(.3^4 - .28^4)} + \frac{T_C(.4)}{(1 \times 10^{11})(\pi/32)(.3^4 - .28^4)} \quad \text{(a)}$$

Now considering the shaft and noting from **compatibility** that we get the same $\Delta\phi$ as above, we have

$$\Delta\phi = \frac{T_A(.4)}{1.4 \times 10^{11}(\pi/32)(.2)^4} + \frac{T_A(.6)}{1.8 \times 10^{11}(\pi/32)(.1)^4} \quad \text{(b)}$$

Example 14.6 (Continued)

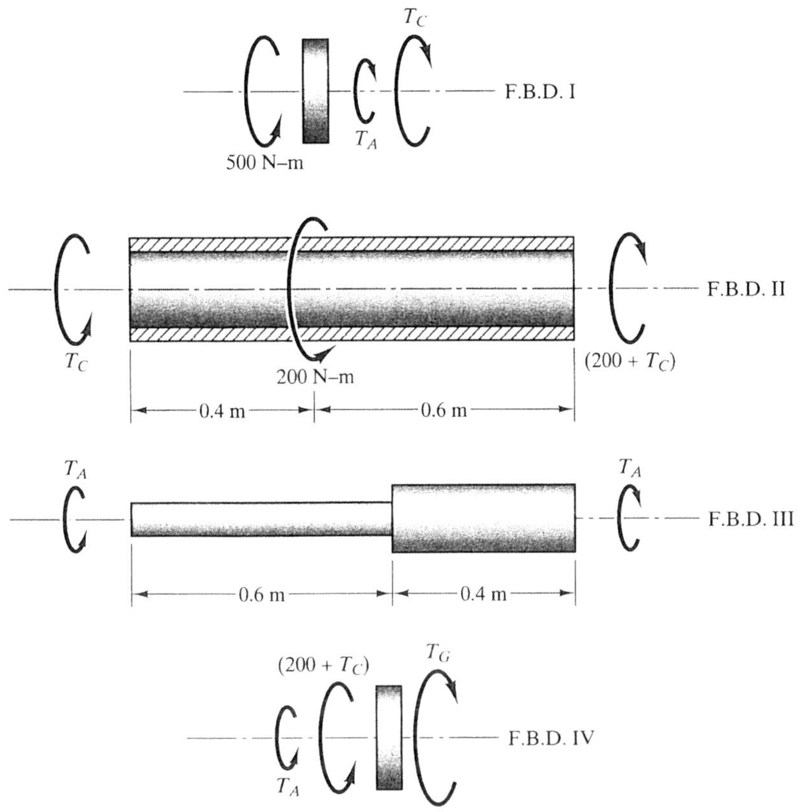

Figure 14.13. Free-body diagrams.

Next considering F.B.D. I showing the external 500-N-m torque and the reactions to the torques T_A and T_C applied at the left end to the tube and shaft, respectively, we have from **equilibrium:**

$$T_A + T_C = 500 \qquad \text{(c)}$$

Solve for T_A above and substitute it into Eq. (b). Now equate the right sides of (a) and (b) and cancel terms common to each expression. We get

$$\frac{(T_C + 200)(.6)}{(1)(.3^4 - .28^4)} + \frac{T_C(.4)}{(1)(.3^4 - .28^4)} = \frac{(500 - T_C)(.4)}{(1.4)(.2^4)} + \frac{(500 - T_C)(.6)}{(1.8)(.1^4)}$$

Solving for T_C we get

$$T_C = 421.12 \text{ N-m}$$

Example 14.6 (Continued)

From Eq. (c) we then have

$$T_A = 500 - T_c = 78.88 \text{ N-m}$$

Finally, we consider F.B.D. IV of the right end disk. We get for the torque T_G coming from the grommet onto the end disk

$$T_A + (200 + T_C) - T_G = 0$$

$$\therefore T_G = (200 + 421.12) + 78.88 = 700 \text{ N-m}$$

To get the total rotation $\delta\phi$ of the left end disk, we add the rotation of the right-hand end disk in its rubber grommet plus the relative rotation $\Delta\phi$ between the end disks. Thus, using Eq. (b) for $\Delta\phi$, we have

$$\delta\phi = \frac{700}{3 \times 10^6} + \left[\frac{(78.88)(.4)}{1.4 \times 10^{11}(\pi/32)(.2)^4} + \frac{(78.88)(.6)}{1.8 \times 10^{11}(\pi/32)(.1)^4}\right]$$

$$\therefore \quad \delta\phi = 2.615 \times 10^{-4} \text{ rad}$$

14.4 Stress Concentrations

We have pointed out that when there is a sudden change in geometry we find a stress concentration. The high local stress in such regions is usually found with the aid of experiment or by use of numerical methods such as finite elements. As pointed out earlier we employ a stress concentration factor K in conjunction with a standard formula from strength of materials to get maximum stress. In the case of torsion of shafts such concentrations could occur at a location of a rapid change in diameter (see Fig. 14.14). Notches, holes, and keyways are other locations on the shaft where there is a concentration of stress to be expected. For ductile materials such as mild steel these stress concentrations are of no great significance since at high loadings *local yielding* at the trouble spots takes place and no great difficulties are encountered. However, for *brittle* materials or where *fatigue* is an important consideration then we must take these stress concentration factors fully into account in design.

In Fig. 14.14 we have shown K plotted against the ratio of the fillet radius r over the smaller diameter d for various ratios of the larger to smaller diameter, D/d. The way to use K for this case is as follows:

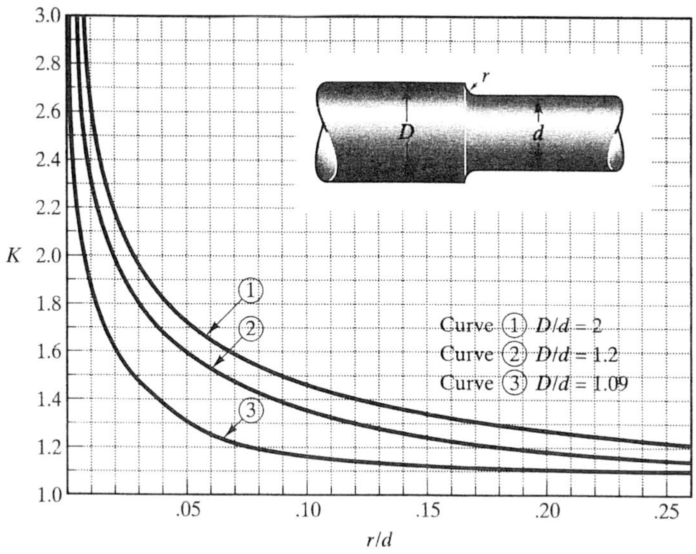

Figure 14.14. Stress concentration factor for reduced shafts.

$$\tau_{\max} = K\frac{M_x(d/2)}{J} \tag{14.11}$$

where J is computed for the smaller diameter d. The same formula holds for the notched shaft shown in Fig. 14.15.

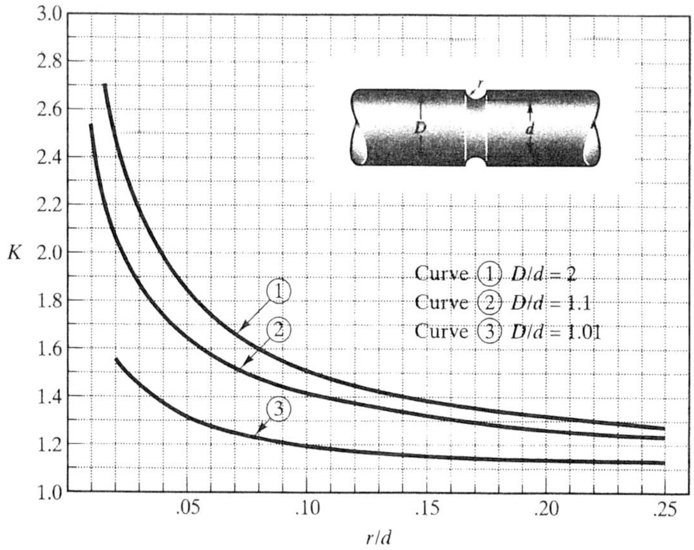

Figure 14.15. Stress concentration factors for notched shafts.

14.5 Torsion of Thin-Walled, Noncircular Closed Shafts

We shall now set forth strength of materials formulations for the analysis of the torsion of thin-walled closed shafts, an example of which is shown in Fig. 14.16. A closed section is one in which the midline of the

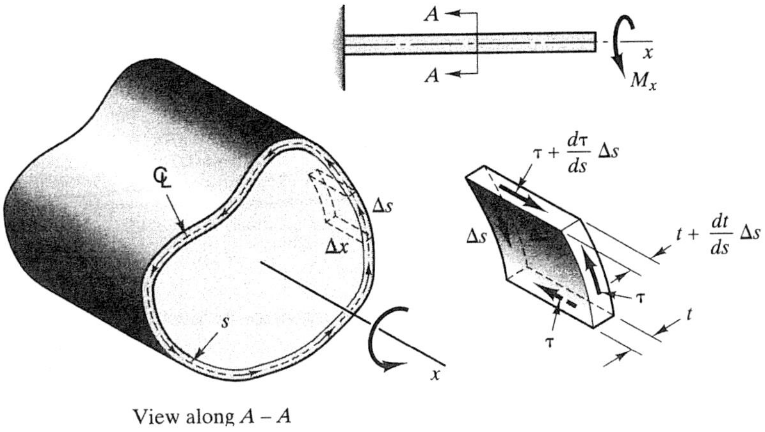

Figure 14.16. Thin-walled noncircular shaft.

cross section forms a closed loop. The analysis will not require the wall to be of uniform thickness, and we may therefore show a wall with a varying thickness. We consider a small element of length Δx and height Δs as a free-body diagram, as indicated in the figure. Owing to the narrow width of the boundary, we shall assume, as we did for bending of thin-walled members, that the shear stress is tangent to the midline between the inner and outer boundaries at a section and has a constant value across the thickness of the wall between the inner and outer boundaries.

Applying the equation of **equilibrium** in the x direction for the element and using the simple notation τ for the shear stress, we get

$$-\tau t\Delta x + \left(\tau + \frac{d\tau}{ds}\Delta s\right)\left(t + \frac{dt}{ds}\Delta s\right)\Delta x = 0$$

where s is measured along the midline between the inner and outer boundaries. Canceling terms, we may reach the following equation:

$$\left(t\frac{dt}{ds} + t\frac{d\tau}{ds}\right)\Delta s + \frac{d\tau}{ds}\frac{dt}{ds}(\Delta s)^2 = 0 \qquad (14.12)$$

Dropping the last expression in Eq. (14.12) as negligible and noting that the first expression in parentheses is the derivative with respect to s of the product τt, we may simplify the result to the following form:

$$\frac{d}{ds}(\tau t) = 0 \tag{14.13}$$

This means that the product τt is a constant for the section. We introduce the notation q for this product and, as in Chapter 11, we call it the *shear flow.* We may thus say that the *shear flow is constant for a thin-walled closed shaft under simple torsion.*

Next consider the requirement that the resultant of the shear-stress distribution at the section equals the torque M_x. That is,

$$M_x\mathbf{i} = \oint \boldsymbol{r} \times (\tau t)\, \boldsymbol{ds} \tag{14.14}$$

where $\boldsymbol{r}$ is measured from some convenient point O in the section to the element vector $\boldsymbol{ds}$. Since we have demonstrated that τt is a constant, we may reformulate the foregoing equation as

$$M_x\mathbf{i} = \tau t \oint \boldsymbol{r} \times \boldsymbol{ds} \tag{14.15}$$

You may easily verify by consulting Fig. 14.17 that $\boldsymbol{r} \times \boldsymbol{ds} = 2d\boldsymbol{A}$, where dA is the area element shown crosshatched in the diagram.[4] Hence, we have

$$\oint \boldsymbol{r} \times \boldsymbol{ds} = 2\boldsymbol{A} = 2A\mathbf{i} \tag{14.16}$$

where $\boldsymbol{A}$ is the total plane area vector of the area enclosed by the mid-line s. Thus, Eq. (14.15) degenerates to the form

$$M_x = 2\tau t A \tag{14.17}$$

Solving for the shear stress at any position s along the section where the thickness is t, we get

$$\tau = \frac{M_x}{2tA} \tag{14.18}$$

In computing A, we often use the area enclosed by the outer boundary of the section for simplicity. Since we are limited by this theory to thin walls, little error should be introduced in this section. However, the re-entrant corners and sharp bends in the cross section can give rise to sig-

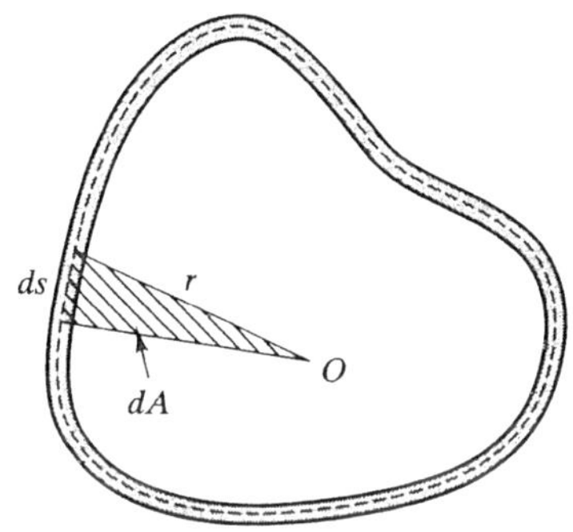

Figure 14.17. Cross section of the thin-walled shaft.

[4]Recall from vector analysis that $|\boldsymbol{C} \times \boldsymbol{B}|$ equals the area of a *parallelogram* having $\boldsymbol{C}$ and $\boldsymbol{B}$ as sides. The crosshatched region in Fig. 14.17 is half the parallelogram formed by $\boldsymbol{ds}$ and $\boldsymbol{r}$. Note that the direction of $d\boldsymbol{A}$ corresponds to $\mathbf{i}$.

nificant stress concentrations. This can be eased somewhat through the use of corner fillets and rounding exterior corners.[5]

We shall now illustrate the use of the preceding formula and check its validity for the case of a thin-walled circular cylinder.

Example 14.7

Compute τ for a thin-walled circular cylinder as shown in Fig. 14.18 and compare the result with the exact theory.

We first employ the *approximate theory.* In this problem, we shall use the mean radius $r_m = (r_i + r_o)/2$ for the computation of A. For τ we then have, using Eq. (14.18),

$$\tau = \frac{M_x}{2At} = \frac{M_x}{2[\pi(r_o + r_i)^2/4](r_o - r_i)} = \frac{2M_x}{\pi(r_o + r_i)^2(r_o - r_i)} \tag{a}$$

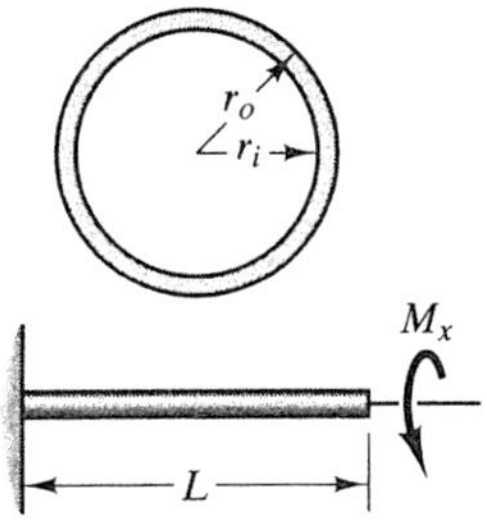

Figure 14.18. Thin-walled circular cylinder.

Now we go to the *exact theory.* The shear stress τ' at the mean radius r_m is

$$\tau' = \frac{M_x r_m}{J} = \frac{M_x(r_i + r_o)/2}{(\pi/2)(r_o^4 - r_i^4)} = \frac{M_x(r_i + r_o)}{\pi(r_o^4 - r_i^4)} \tag{b}$$

To be able to compare Eq. (b) with Eq. (a), we next rewrite Eq. (b) as follows:

$$\tau' = \frac{M_x(r_i + r_o)}{\pi(r_o^2 + r_i^2)(r_o + r_i)(r_o - r_i)} = \frac{M_x}{\pi(r_o^2 + r_i^2)(r_o - r_i)} \tag{c}$$

Taking the ratio of Eq. (a) to Eq. (c), we get

$$\frac{\tau'}{\tau} = \frac{M_x/\pi(r_o^2 + r_i^2)(r_o - r_i)}{2M_x/\pi(r_o + r_i)^2(r_o - r_i)} = \frac{(r_o + r_i)^2}{2(r_o^2 + r_i^2)} \tag{d}$$

We may put the right-hand side of Eq. (d) in a more useful form by dividing numerator and denominator by r_o^2. Thus,

$$\frac{\tau'}{\tau} = \frac{[1 + (r_i/r_o)]^2}{2[1 + (r_i/r_o)^2]} \tag{e}$$

We see from the foregoing formulation that if r_i/r_o approaches unity (i.e., the wall becomes vanishingly thin), then τ'/τ also approaches unity. Thus, the approximate theory becomes increasingly more cor-

[5]See R. D. Cook and W. E. Young, *Advanced Mechanics of Materials,* McMillan, 1985.

Example 14.7 (Continued)

rect as the wall becomes increasingly more thin, as is to be expected from the development of the approximate theory.

If $r_i/r_o = .9$, we then get

$$\frac{\tau'}{\tau} = \frac{(1 + .9)^2}{2(1 + .81)} = .997$$

Clearly, the approximate theory gives very good results for this configuration. We can expect, therefore, at least reasonably good results for noncircular sections.

Since we have linear, elastic behavior here, we may make use of the conservation of the energy law as discussed in Section 6.4 to equate the work done by the external torque M_x to the increase in strain energy of the shaft.[6] That is, for a net rotation ϕ between the ends of the shaft we have, using Eq. (6.34),

$$\frac{1}{2} M_x \phi = \frac{1}{2} \int_0^L \oint \frac{\tau^2}{G} t \, ds \, dx \tag{14.19}$$

where, to cover the entire volume of the shaft, x goes from zero to L and s goes around the entire periphery of the thin-walled section. We replace ϕ by $(d\phi/dx)L$ and integrate the right-hand side with respect to x. On canceling $L/2$, we get

$$M_x\left(\frac{d\phi}{dx}\right) = \frac{1}{G} \oint \tau^2 t \, ds \tag{14.20}$$

Now replace τ^2 using Eq. (14.18). We get

$$M_x\left(\frac{d\phi}{dx}\right) = \frac{1}{G} \oint \frac{M_x^2}{4t^2 A^2} t \, ds$$

$$\therefore \quad \left(\frac{d\phi}{dx}\right) = \frac{M_x}{4A^2 G} \oint \frac{ds}{t} \tag{14.21}$$

If the thickness t is a constant, we get

[6]Readers who have not studied Section 6.4 may delete the remainder of this section with no loss in continuity.

$$\left(\frac{d\phi}{dx}\right) = \frac{M_x S}{4A^2 Gt} \tag{14.22}$$

where S is the *perimeter* of A, or, in other words, the length of the midline between the inner and outer walls.

It is to be noted that Eqs. (14.21) and (14.22) are often referred to as *Bredt's formula.* We shall now illustrate the use of Bredt's formula.

Example 14.8

We shall employ Bredt's formula for the case of the thin-walled circular shaft of Example 14.7. We then compare this result with that of the exact analysis developed earlier.

Using Fig. 14.18, we may give the rate of twist in the following manner in accordance with Bredt's formula

$$\left(\frac{d\phi}{dx}\right)_{\text{Bredt}} = \frac{M_x S}{4A^2 Gt} = \frac{(M_x)[2\pi(r_o + r_i)/2]}{(4)\{\pi[(r_o + r_i)/2]^2\}^2 G(r_o - r_i)} \tag{a}$$

$$\left(\frac{d\phi}{dx}\right)_{\text{Bredt}} = \frac{4M_x}{\pi G(r_o + r_i)^3(r_o - r_i)}$$

Now from the exact analysis, we have

$$\left(\frac{d\phi}{dx}\right)_{\text{exact}} = \frac{M_x}{GJ} = \frac{M_x}{G(\pi/2)(r_o^4 - r_i^4)} \tag{b}$$

To compare the results, let us first take the ratio of the foregoing rates of twist. Thus,

$$\frac{(d\phi/dx)_{\text{Bredt}}}{(d\phi/dx)_{\text{exact}}} = \frac{4M_x/\pi G(r_o + r_i)^3(r_o - r_i)}{2M_x/G\pi(r_o^2 + r_i^2)(r_o + r_i)(r_o - r_i)} = \frac{2(r_o^2 + r_i^2)}{(r_o + r_i)^2} \tag{c}$$

Now divide numerator and denominator by r_o^2. We then have

$$\frac{(d\phi/dx)_{\text{Bredt}}}{(d\phi/dx)_{\text{exact}}} = \frac{2[1 + (r_i/r_o)^2]}{(1 + r_i/r_o)^2} \tag{d}$$

If we let $r_i/r_o \to .9$, we get for the ratio of twist the value 1.003. Thus, for this geometry, the Bredt formula gives highly accurate results.

Example 14.9

As a second example, we consider the case of a thin-walled rectangular shaft shown in Fig. 14.19. We wish the shear-stress distribution and the total twist of the end section at A as a result of the 100-lb-ft torque. The thickness of the wall is uniform and equals .1 in. and the material is steel having a shear modulus G of 15×10^6 psi.

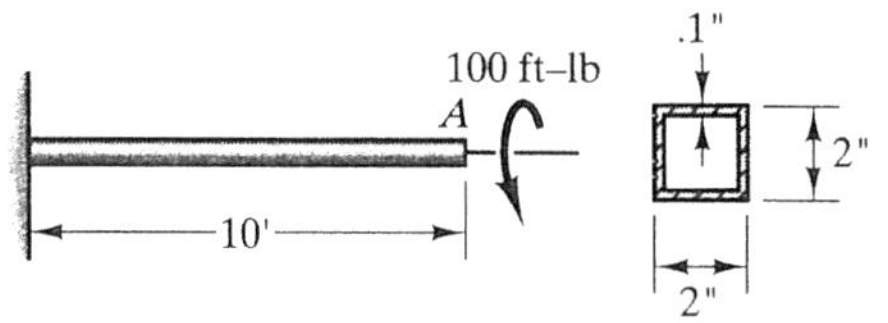

Figure 14.19. Square thin-walled shaft.

Using Eq. (14.18), we have, for the shear stress τ along the midline of the cross section

$$\tau = \frac{M_x}{2tA} = \frac{(100)(12)}{(2)(0.1)(1.9)(1.9)} = 1662 \text{ psi} \qquad \text{(a)}$$

Although the equation above may be quite accurate away from the corners, we must use caution at the corners where there may be appreciable stress concentration depending on the local geometry.

The twist ϕ_A at position A can be given with the aid of Eq. (14.22) as

$$\phi_A = \frac{M_x SL}{4A^2Gt} = \frac{(1200)(4)(1.9)(120)}{(4)(1.9)^4(15 \times 10^6)(0.1)} = 0.01400 \text{ rad} = .802° \qquad \text{(b)}$$

*14.6 Elastic, Perfectly Plastic Torsion

We shall now consider the case of elastic, perfectly plastic torsion of a circular shaft. Such behavior might take place in a shaft made of mild steel.

For the one-dimensional stress problem of Chapter 5 and the bending problems of Chapter 11, we considered yielding to take place when the normal stress evaluated in the problem equaled the yield stress for the material, as determined from a one-dimensional tensile test. (Such a procedure is valid for bending problems since the state of stress at the outermost fibers is one-dimensional—that is, only one stress, a normal stress, is not zero for the reference *xyz*.) You may possibly have considered criteria for yielding for more general states of stress in Chapter 9. In the case of pure torsion there is only one nonzero stress, a shear stress, for reference x, r, θ, where x is along the centerline of the shaft. We can experimentally relate shear stress to shear strain in the way we related normal stress to normal strain in the tensile test by carrying out a simple torsion test on a cylinder. We could in this way find a shear yield stress, τ_s analogous to the tensile yield stress Y (or τ_Y). We have shown that τ_s from the torsion test is close to one-half the value Y from the tensile test for the same material. The idealized plot of shear stress versus shear strain for elastic, perfectly plastic behavior is shown in Fig. 14.20.

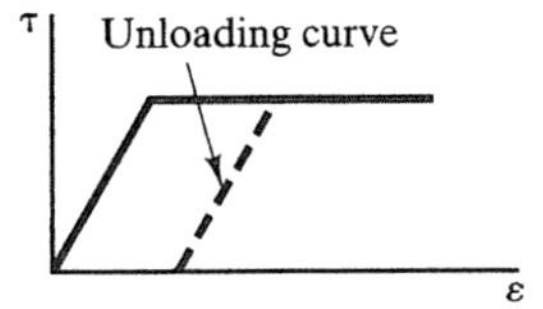

Figure 14.20. Elastic, perfectly-plastic behavior.

As the torque is increased on a shaft, the outer fiber will eventually reach the yield stress τ_s. What happens when the torque is increased further? We answer this by noting that the physical arguments concerning deformation presented in Section 14.2 can be restated for plastic deformation—namely, that plane sections remain plane and sections rotate an angle ϕ, dependent on x, as rigid plates. This means that the shear strain $\gamma_{x\theta}$ will vary linearly with the radius, and, as a consequence, a growing applied torque as already indicated will eventually generate the yield stress, first at the outer boundary followed by a penetration of the plastic zone in toward the centerline. In regions of the section where the yield stress has been reached the material simply flows with no change in stress as the section rotates in response to increasing torque. Thus, in Fig. 14.21(a) we have shown a situation wherein there is a *plastic ring* for $r \geq a$ and an *elastic core* for $r \leq a$. The stress versus radius diagram is shown in Fig. 14.21(b). The torque can theoretically be increased until plastic action pervades the entire shaft section. The shaft can then offer no additional resistance to the applied torque and we reach the conditions of a *torsional hinge* analogous to the *plastic hinge* of bending. It will be of interest to compute the maximum torque for elastic behavior and the torque for the aforestated torsional hinge. Thus, for a shaft of radius R we have[7]

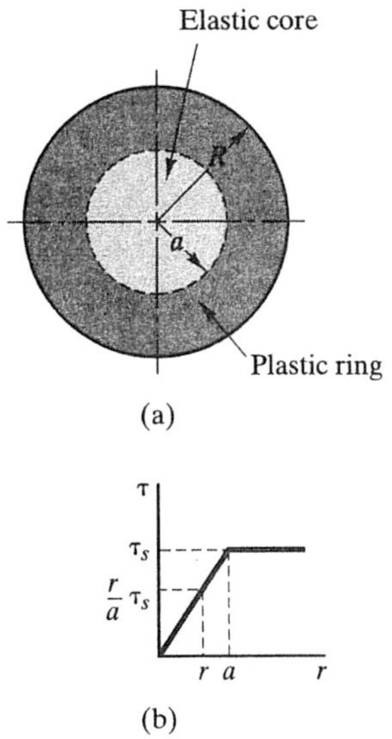

Figure 14.21. Elastic, perfectly-plastic torsion.

$$T_{\text{max El}} = \frac{\tau_s J}{R} = \frac{\tau_s[(\pi/2)R^4]}{R} = \frac{1}{2}\pi R^3 \tau_s \tag{14.23a}$$

$$T_{\text{hinge}} = \int_0^{2\pi}\int_0^R \tau_s(r\,dr\,d\theta)r = 2\pi\tau_s\frac{R^3}{3} \tag{14.23b}$$

The ratio of these quantities is

[7]Note we are simply using the regular torsion formula for the elastic case.

$$\frac{T_{\text{hinge}}}{T_{\text{max El}}} = \frac{\frac{2}{3}\pi R^3 \tau_s}{\frac{1}{2}\pi R^3 \tau_s} = \frac{4}{3} \tag{14.24}$$

showing that once the elastic limit in a shaft has been reached at the outer fibers, there is still a considerable external increase of torque that the shaft can withstand.

Suppose that a torque greater than $T_{\text{max El}}$ and less than T_{hinge} is applied. How do we get the rate of twist? We may use Fig. 14.21 in this regard. The torque at the section is related to the stress distribution as follows:

$$\begin{aligned} T &= \int_0^{2\pi}\int_0^a \left(\frac{r}{a}\tau_s\right) r^2\,dr\,d\theta + \int_0^{2\pi}\int_a^R (\tau_s) r^2\,dr\,d\theta \\ &= \tau_s\left(\frac{2\pi}{a}\right)\frac{a^4}{4} + \tau_s\left(\frac{2}{3}\pi\right)(R^3 - a^3) \\ &= \frac{\pi\tau_s}{3}\left(2R^3 - \frac{1}{2}a^3\right) \end{aligned}$$

The radius a of the elastic core is then found from the equation above to be

$$a = \left(4R^3 - \frac{6T}{\pi\tau_s}\right)^{1/3} \tag{14.25}$$

As for the rate of twist, we may return to Eq. (14.2), which results from the assumed deformation for all constitutive laws as discussed at the outset. Note that from this equation we can say that

$$\begin{aligned} \gamma_{x\theta} &= r\frac{d\phi}{dx} \\ \therefore\ \frac{d\phi}{dx} &= \frac{\gamma_{x\theta}}{r} \end{aligned} \tag{14.26}$$

Considering the elastic core of the shaft, we can employ for $\gamma_{x\theta}$ the result τ_s/G at $r = a$. For the rate of twist $d\phi/dx$ we then have

$$\frac{d\phi}{dx} = \frac{\tau_s}{Ga} \tag{14.27}$$

where a is determined from Eq. (14.25).

Let us next consider what occurs when a shaft, loaded as described, is released. To understand the unloading process we point out first that on a stress-strain diagram an unloading from the plastic region occurs along a straight line parallel to the elastic part of the loading

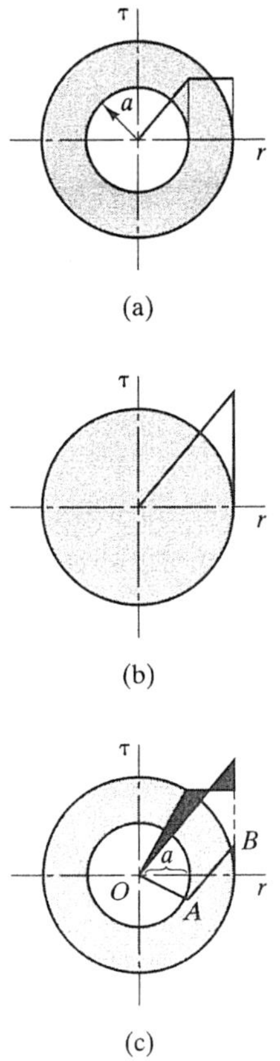

Figure 14.22. Development of residual stress.

curve (see Fig. 14.20). That is, the unloading is *elastic.* (This is exactly the same as for the one-dimensional tensile stress case discussed in Chapter 4.) There is, as a result, the possibility of *residual stress* as well as the possibility of a *residual state of strain* remaining in the shaft upon release of the applied torques. To best understand this, consider Fig. 14.22. In part (a) of this figure we have the stress plotted against r for elastic, perfectly-plastic behavior, as described earlier, for a given torque T. Now make believe that the material behaved purely elastically during the loading by the aforestated torque T. The stress diagram would be as shown in Fig. 14.22(b) with the maximum stress given as $T(D/2)/J$. Now when we unload we are *releasing* the given torque T and, as pointed out, the stress-strain relation must be *elastic* during this action. Hence, when unloading we must *subtract* the stress distribution shown in Fig. 14.22(b) from the one shown in Fig. 14.22(a).[8] The difference between these distributions is shown as the darkened region in Fig. 14.22(c) and is the residual stress. This stress distribution is then shown replotted as OAB along the horizontal radius. Notice that the maximum value occurs at either $r = a$ or at $r = D/2$.

To get the residual rate of twist, we subtract from the rate of twist with the given applied torque T an elastic rate of twist found by the familiar formula T/JG, where T is the given applied torque. Thus,

$$\left(\frac{d\phi}{dx}\right)_{\text{res}} = \left(\frac{d\phi}{dx}\right)_{\text{plastic}} - \frac{T}{JG} \tag{14.28}$$

We shall now illustrate the formulations above in the following example.

Example 14.10

A shaft of radius 25 mm transmits a torque of 20,000 N-m. What is the rate of twist? What is the maximum residual stress on unloading? Finaly, what is the residual rate of twist on unloading? Take $G = 1.035 \times 10^{11}$ Pa and τ_s as 6.895×10^8 Pa.

We must first find out whether we have completely elastic or elastic-plastic behavior or whether we have applied a torque to this shaft in excess of what it can withstand. For this reason we first compute T_{maxEl} and T_{hinge}. Thus, from Eqs. (14.23),

[8]The reader may be perplexed by noting from Fig. 14.22(b) that the shear stress to be subtracted is greater than τ_s at and near the outer fibers. Keep in mind that this is a shear-stress distribution that is being subtracted; it does not of itself exist anywhere. It is the *difference* between the distribution in Fig. 14.22(a) and the distribution in Fig. 14.22(b) that is the *real* distribution [shown as OAB in Fig. 14.22(c)], and it is here that we cannot exceed the value τ_s.

Example 14.10 (Continued)

$$T_{\text{max El}} = \frac{1}{2}\pi R^3 \tau_s = 1.692 \times 10^4 \text{ N-m} \tag{a}$$

$$T_{\text{hinge}} = 2\pi\tau_s\left(\frac{R^3}{3}\right) = 2.256 \times 10^4 \text{ N-m} \tag{b}$$

Clearly, we have both an elastic core and a plastic ring present. The radius of the core a is, using Eq. (14.25),

$$a = \left[(4)(.025)^3 - \frac{6}{\pi}\left(\frac{20{,}000}{6.895 \times 10^8}\right)\right]^{1/3} = .01922 \text{ m} \tag{c}$$

The *rate of twist* from Eq. (14.27) is then

$$\frac{d\phi}{dx} = \frac{\tau_s}{Ga} = \frac{6.895 \times 10^8}{(1.035 \times 10^{11})(.01922)} = .3466 \text{ rad/m} \tag{d}$$

To get the *residual stress distribution* we note first that the following is the stress distribution with the given torque applied:

$0 \leq r \leq .01922$:

$$\tau = \left(\frac{r}{.01922}\right)(6.895 \times 10^8) = 3.587 \times 10^{10} r \text{ Pa} \tag{e}$$

$.01922 \leq r \leq .025$:

$$\tau = 6.895 \times 10^8 \text{ Pa}$$

From this we subtract the stress:

$$\tau = \frac{Tr}{J} = \frac{20{,}000}{\frac{1}{2}(\pi)(.025)^4} r = 3.259 \times 10^{10} r \text{ Pa} \tag{g}$$

Hence, the residual stress is given as follows:

$0 \leq r \leq .01922$:

$$\tau_{\text{res}} = 3.587 \times 10^{10} r - 3.259 \times 10^{10} r = 3.28 \times 10^9 r \text{ Pa} \tag{h}$$

$.01922 \leq r \leq .025$:

$$\tau_{\text{res}} = 6.895 \times 10^8 - 3.259 \times 10^{10} r \text{ Pa} \tag{i}$$

The peak stresses occur at position A [see Fig. 14.22(c)], where $r = .01922$ and at position B, where $r = .025$ m. Thus, using Eqs. (h) and (i),

$$(\tau_{\text{res}})_A = (3.28 \times 10^9)(.01922) = 6.30 \times 10^7 \text{ Pa} \tag{j}$$

$$(\tau_{\text{res}})_B = (6.895 \times 10^8) - (3.259 \times 10^{10})(.025) = -12.53 \times 10^7 \text{ Pa} \tag{k}$$

Example 14.10 (Continued)

Clearly, we have -12.53×10^7 Pa as the maximum residual shear stress.

Lastly, we compute the final *rate of twist* on unloading as follows:

$$\left(\frac{d\phi}{dx}\right)_{\text{res}} = .3466 - \frac{T}{JG}$$

$$= .3466 - \frac{20{,}000}{\frac{1}{2}\pi(.025)^4(1.035 \times 10^{11})} \qquad (1)$$

$$= .3466 - .3149 = \boxed{.0317 \text{ rad/m}}$$

*14.7 Noncircular Cross Sections

We have thus far considered torsion of a circular shaft and torsion of a shaft having a thin wall. You will recall that the simplifying arguments presented for torsion of the circular cross section were based on the *axial symmetry* of geometry. The conclusions reached from this geometry accordingly cannot be extended to an arbitrary cross section where such symmetry is lacking. For instance, we concluded earlier that the maximum shear stress occurs at the outer fibers of the circular cylinder (i.e., at the furthest points from the centerline axis). In the case of a square shaft [see Fig. 14.23(a)] actually the *opposite* is true. Thus, consider

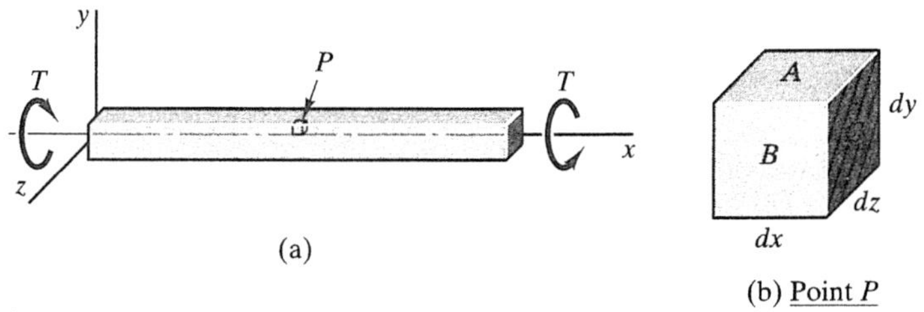

Figure 14.23. Rectangular shaft under torsion.

points P along any of the edges of the shaft. Clearly, these points are farthest from the centroidal axis, x. A vanishingly small element at such a point P is shown in Fig. 14.23(b). Since there is no surface traction on face A, clearly shear stress τ_{yx} must be zero there. From the complementary property of shear this means that τ_{xy} on face C, now in the cross section of the shaft, must also be zero. This is similarly true for shear stress τ_{zx} on face B and shear stress τ_{xz} on face C. This means that the shear stresses on C in the x and y directions are zero; there is then zero shear stress on this interface. Thus, at the edges of the cross section (i.e., at points farthest from the centerline) we have minimum shear stress rather than maximum shear stress for outermost points, as was the case for a circular shaft. Furthermore, the assumption that plane sections remain plane and rotate like rigid platelets must be abandoned. Actually the cross sections *warp* when we do not have the axial symmetry of the circular shaft. This can be seen in Fig. 14.24. If we imagine a fine grid of

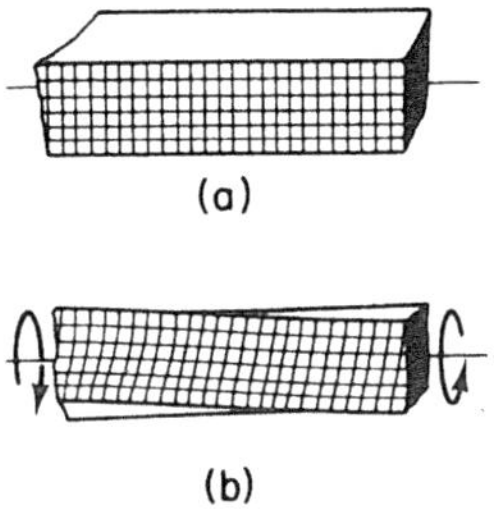

Figure 14.24. Rectangular shaft showing warping during torsion.

orthogonal lines on a rectangular shaft, we can see that as the shaft is twisted, the cross sections distort out of their undeformed plane. This did not occur with a circular section. This out-of-plane warpage is a direct result of the shear stress distribution on the cross section. Since the shear stress must be zero at the corners (as just discussed), it follows that the grid will remain orthogonal near the corners. That is, a small square of the grid at the corner of the untwisted shaft remains a small square after the shaft is twisted. So the unstressed corners of the shaft simply "go along for the ride" as the shaft is twisted. If we look at the grid near the midline, we can see that it distorts from its orthogonal shape due to the presence of shear stress. It is this angular distortion that causes the corners of the shaft to be warped out of the plane of the cross section. As in the case of thermal stress, if we allow the cross section to warp then no *additional* stress is produced. However, if the cross section is prevented from warping, normal stresses are generated.

Using the theory of elasticity, we can obtain closed-form solutions for such noncircular cross sections as ellipses and equilateral triangles as well as approximate solutions for rectangular cross sections.[9] Solutions for other shapes may be obtained using numerical methods.[10] Table 14.1 summarizes key torsional information for some common cross sections.[11] Note that the maximum shear stress (τ_{max}) for noncircular

[9]See S. Timoshenko and J. N. Goodier, *Theory of Elasticity* (McGraw-Hill, 1951).

[10]See I. H. Shames and C. L. Dym, *Energy and Finite Element Methods in Structural Mechanics* (Hemisphere, 1985).

[11]See, for example, the *Handbook of Mathematical, Scientific, and Engineering Formulas* (Research & Education Assoc., 1984).

Table 14.1. Key torsional information for some common cross sections.

	τ_{max}	K_t
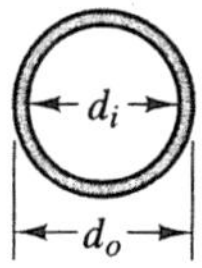	$\frac{16M_x}{\pi d_o^3}$	$\frac{\pi(d_o^4 - d_i^4)}{32}$
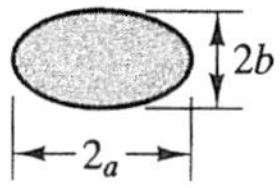	$\frac{2M_x}{\pi ab^2}$ $(a > b)$	$\frac{\pi a^3 b^3}{a^2 + b^2}$
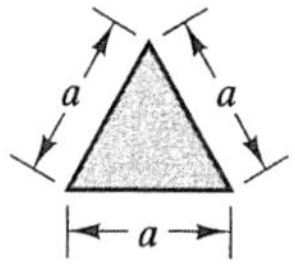	$\frac{20M_x}{a^3}$	$\frac{\sqrt{3}a^4}{80}$
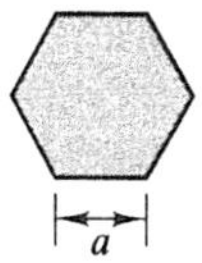	$\frac{1.1M_x}{a^3}$	$1.02\, a^4$
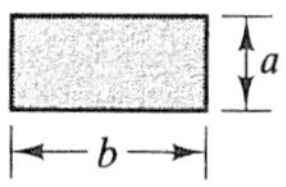	$\frac{M_x}{\kappa ba^2}$	βba^3

b/a	1.0	1.2	1.5	2.0	3.0	5.0	10	50	∞
κ	0.208	0.219	0.231	0.246	0.267	0.292	0.312	0.329	0.333
β	0.141	0.166	0.196	0.229	0.263	0.291	0.312	0.329	0.333

sections occurs in the outer fibers of the cross section nearest the center of twist. The torsional constant K_t is used in the angle-of-twist equation

$$\Delta\phi = \int_0^L \frac{M_x}{GK_t}\, dx \tag{14.29}$$

where we can see the similarity with Eq. (14.6). *Note that only for a circular section is the polar moment of area equal to K_t.* The values of K_t are listed in Table 14.1 for some common shaft cross sections.

Example 14.11

Consider the shaft shown in Fig. 14.25 made from a 10 mm by 25 mm rectangular cross section. The shaft is 1 m long and supports a torque of 50 N-m. The material is a high-strength steel having a

Example 14.11 (Continued)

shear modulus of 75 GPa. Determine the maximum shear stress in the section and the total twist of the free end A.

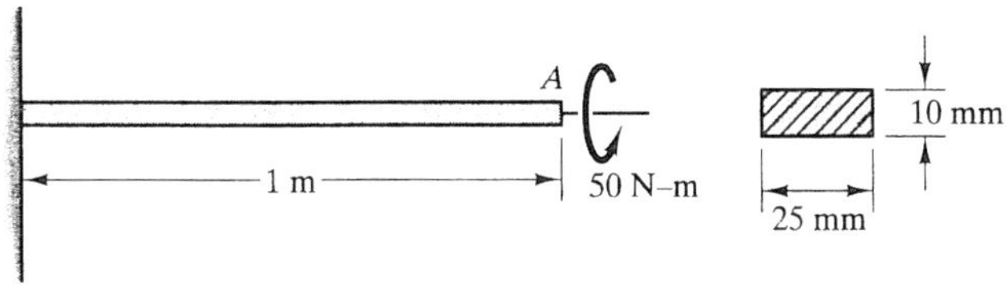

Figure 14.25. Rectangular shaft with an applied torque.

Using the results listed in Table 14.1, we see that we must first determine the aspect ratio (b/a) of the rectangular beam. Note that the cross section's aspect ratio is always greater than or equal to unity. We can readily compute the aspect ratio as 25 mm/10 mm = 2.5. From Table 14.1 we find that $\kappa = 0.257$ (by interpolation) and that the maximum shear stress is given by

$$\tau_{\max} = \frac{M_x}{\kappa b a^2} = \frac{50}{(0.257)(0.025)(0.010)^2} = 77.8 \text{ MPa}$$

This maximum stress occurs along the midline of the longest side of the cross section, that is, at the location on the surface *nearest* the center (not at the furthest distance from the center as is the case with the circular cross section). This location is shown as point B in Fig. 14.26.

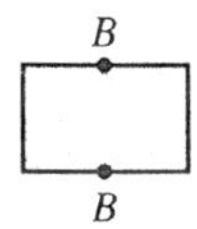

Figure 14.26. Positions of maximum shear stress.

The twist ϕ_A at position A is determined with the aid of Table 14.1. For the aspect ratio of 2.5, we see that $\beta = 0.246$ (by interpolation), and therefore we can compute the torsional constant K_t as

$$K_t = \beta b a^3 = (0.246)(0.025)(0.010)^3 = 6.15 \times 10^{-9} \text{ m}^4$$

Now we can compute the twist angle from Eq. (14.29) as

$$\phi_A = \frac{M_x L}{G K_t} = \frac{(50)(1)}{(75 \times 10^9)(6.15 \times 10^{-9})} = 0.108 \text{ rad} = 6.19°$$

In this next example, we compute the torsional constant K_t for a member that is composed of a number of rectangular sections. Such sections are sometimes referred to as composite sections. The maximum shear stress occurs on the rectangular section that has the largest thickness.

Example 14.12

Consider two identical channel sections, each with an 8″ web, a 2.26″ flange, and weighing 11.5 lb/ft (see Appendix IV-B) that are to be used together for a beam [see Fig. 14.27(a)]. Compare the torsional stiffness for two different means of attaching the channel sections, shown in Fig. 14.27(b) and (c). Recall that the torsional stiffness is the torque required per unit angle of twist [see Eq. (14.8)].

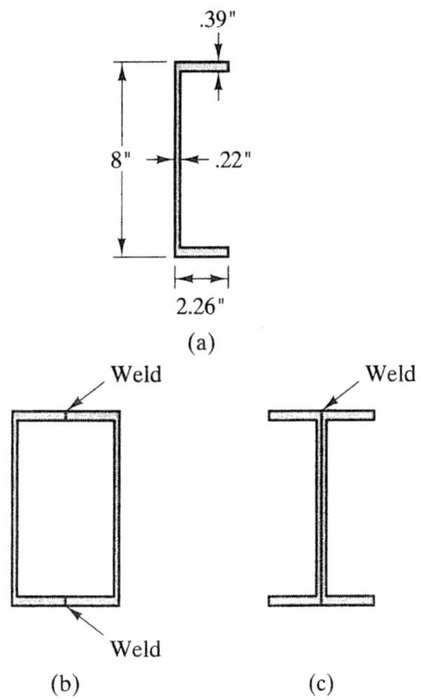

Figure 14.27. Composite sections involving channels.

Case #1:
In this case, the two channels are placed so that their flanges are touching. Welded in this position, they form the *closed* box cross section shown in Fig. 14.27(b). To get the torsional stiffness, we use Bredt's law [Eq. (14.21)], rewritten as

$$\text{torsional stiffness} = \frac{M_x}{\phi} = \frac{4A^2G}{L\oint \frac{ds}{t}} \qquad \text{(a)}$$

We have left $\oint \frac{ds}{t}$ in Eq. (a) because we have a nonuniform wall thickness (i.e., the flange and web have different thicknesses). However, since the thickness is assumed to be constant for the web and flange, respectively, we can replace the integral with $\sum_i \frac{S_i}{t_i}$. For the box cross section, this is given as

$$\sum_i \frac{S_i}{t_i} = \frac{2(8.0 - 0.39)}{0.22} + \frac{2[(2.26)(2) - 0.22]}{0.39} = 91.2$$

The area A enclosed by the midline of the section is computed as $(8.0-0.39)[2.26)(2)-0.22] = 32.7\ \text{in}^2$. From Eq. (a) we get

$$\text{torsional stiffness} = \frac{4(32.7)^2G}{91.2L} = 46.9\frac{G}{L}\ \text{in-lb/radian}$$

Case #2
Now we consider the case wherein we have welded the two channel sections "back-to-back" along their webs as shown in Fig. 14.27(c). By attaching them in this fashion, we have created an *open* cross section. Since we have a shape composed of a series of rectangular sections, we can use the results listed in Table 14.1 to approximate the torsional constant for the section. We do this by summing the tor-

Example 14.12 (Continued)

sional constants for the web and flange of the section, treating each as a rectangle. Eq. (14.29) is rewritten as

$$\text{torsional stiffness} = \frac{M_x}{\phi} = \frac{G}{L}\sum_i (K_t)_i \qquad \text{(b)}$$

We have chosen to treat the web and the four flanges as separate rectangles, as shown in Fig. 14.28(a). It is possible to consider the cross section as two flanges and a web as shown in Fig. 14.28(b). This leads to a slightly stiffer prediction of the torsional stiffness as the reader may verify.

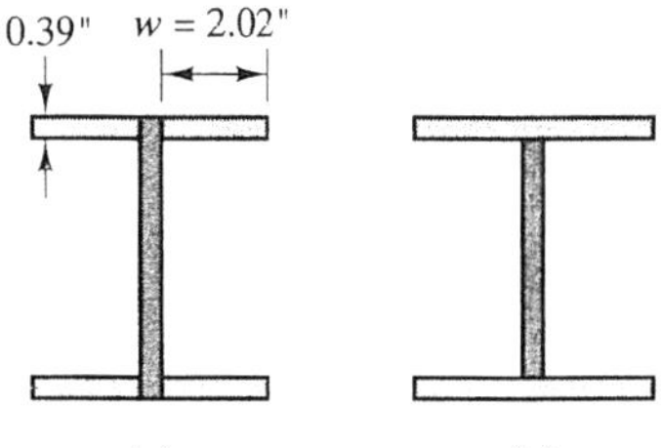

Figure 14.28. Cross sections with web and flange interchanged.

Flange (four):

$$\left.\begin{aligned} t &= 0.39 \text{ in} \\ w &= 2.26 - 0.22 = 2.02 \text{ in.}\end{aligned}\right\} \therefore b/a = 5.18 \quad \beta = 0.292 \text{ (by interpolation)}$$

Web (one):

$$\left.\begin{aligned} t &= 2(0.22) = 0.44 \text{ in.} \\ w &= 8.0 \text{ in.}\end{aligned}\right\} \therefore b/a = 18.2 \quad \beta = 0.315 \text{ (by interpolation)}$$

Eq. (b) then gives us

$$\text{torsional stiffness} = \left[(0.315)(8)(0.44)^3 + (4)(0.292)(2.02)(0.39)^3\right]\frac{G}{L}$$

$$= 0.355\frac{G}{L} \text{ in-lb/rad}$$

By comparison of Cases 1 and 2, we can see that the open section has a significantly lower torsional stiffness than the closed section. This marked difference between the torsional stiffness of open and closed cross sections should be kept in mind when designing systems subjected to twisting actions.

The torsional stiffness of a concentric hollow noncircular shaft can be *approximated* by summing the torsional stiffness of the component shafts.[12] As an illustration, the torsional stiffness of the hollow shaft shown in Fig. 14.29 is estimated by subtracting the torsional stiffness of the hexagonal hole from that of the outer circular contour.

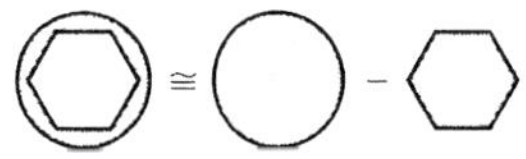

Figure 14.29. Equivalent shaft concept for torsion.

[12]See R. Isakower, *The Shaft Book* (US Army ARRADCOM, MISD User's Manual 80-5, 1980).

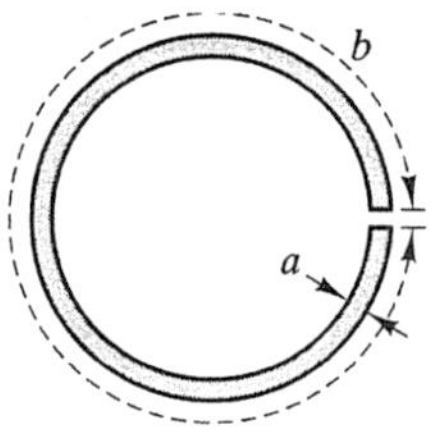

Figure 14.30. Split tube cross section.

For *thin-walled* rectangular sections ($b >> a$), Table 14.1 indicates that we may take β as 1/3. If such a section is then formed into a continuous open section, such as the split tube cross section shown in Fig. 14.30, we can estimate the torsional constant K_t as $ba^3/3$ where a and b are measured as shown in Fig. 14.30.

When n thin-walled rectangular sections are connected together to form an open section, for example as in a T- or I-shape, we can estimate the torsional constant as

$$K_t = \frac{1}{3}\sum_{i=1}^{n} b_i a_i^3 \tag{14.30}$$

As with closed sections, stress concentrations may exist at re-entrant corners and at locations of sudden changes in geometry. These can be reduced by using ample corner fillets when possible.

14.8 Strain-Energy Computations for Twisting

In Chapter 6 we developed the concept of strain energy for elastic bodies. By assuming a linear elastic constitutive law, $U = U^*$, we can write the strain energy in the body as a function of the stress and certain material properties. This was given in Chapter 6 and is repeated here:

$$\begin{aligned} U = U^* = \iiint_V \Bigg[& \frac{1}{2E}(\tau_{xx}^2 + \tau_{yy}^2 + \tau_{zz}^2) \\ & - \frac{\nu}{E}(\tau_{xx}\tau_{yy} + \tau_{xx}\tau_{zz} + \tau_{zz}\tau_{yy}) \\ & + \frac{1}{2G}(\tau_{xy}^2 + \tau_{yz}^2 + \tau_{xz}^2) \Bigg] dv \end{aligned} \tag{6.34}$$

To obtain the strain energy in a shaft due to twisting, we start by considering a prismatic circular shaft of length L and radius r subjected to end torques only. At a given cross section of the shaft, *we have only one nonzero shear stress*[13] $\tau_{x\theta}$. From our study in this chapter, we know that this stress is given by $M_x r/J$. Using Eq. (6.34) we can write the strain energy as

$$U = U^* = \iiint_V \frac{\tau_{x\theta}^2}{2G}\,dv = \int_0^L \left[\iint_A \frac{M_x^2 r^2}{2GJ^2}\,dA \right] dx = \int_0^L \left[\frac{M_x^2}{2GJ^2} \iint_A r^2\,dA \right] dx$$

Noting that $\iint_A r^2 dA = J$, we then have for the strain energy of a linear elastic circular shaft

[13]We are using a cylindrical coordinate system with x as the cylindrical axis along the centerline of the shaft.

$$U = U^* = \int_0^L \frac{M_x^2}{2GJ}\,dx \qquad (14.31)$$

For the case of non-circular cross-sections, the polar moment of area J is replaced by the torsional constant K_t.

As we have shown in Chapter 6 for one-dimensional stress problems and in Chapter 12 for beam deformation, we can again make use of the second Castigliano theorem for computing the relationship between applied loads and deformation, given symbolically by

$$\frac{\partial U}{\partial Q_i} = \Delta_i \qquad (14.32)$$

where Q_i's are the generalized point forces (including torques) and the Δ_i's are the generalized displacements (including rotations). We now illustrate the use of the energy approach with the following examples.

Example 14.13

Re-analyze the shaft studied in Example 14.4, this time by using the second Castigliano theorem (see Fig. 14.31).

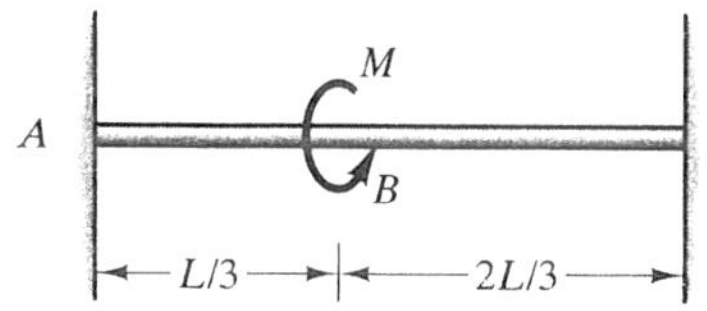

Figure 14.31. Clamped shaft with an applied torque.

Note that this shaft is statically indeterminate. As with other indeterminate structures we have studied using the second Castigliano theorem, we first choose a base structure and redundant support reactions. In this case, we choose the torque at A to be the redundant reaction. The base structure is shown in Fig. 14.32 where the direction of the unknown redundant M_A is assumed to be positive.

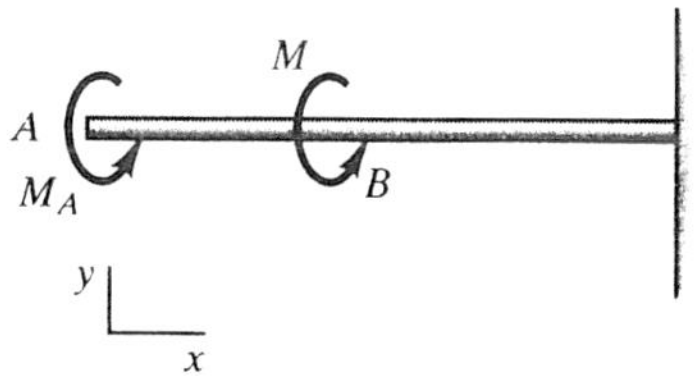

Figure 14.32. Left clamp is released.

The second Castigliano theorem is used to determine the twist at A due to the applied loads M and M_A. For the two point torques, we have two domains, thus (**constitutive laws**)

$0 < x < L/3$:

$$M(x) = -M_A$$

$$\therefore U_1 = \int_0^{L/3} \frac{(-M_A)^2}{2GJ}\,dx$$

$L/3 < x < L$:

$$M(x) = -M_A - M$$

$$\therefore U_2 = \int_{L/3}^{L} \frac{(-M_A - M)^2}{2GJ}\,dx$$

Example 14.13 (Continued)

Applying the second Castigliano theorem (**compatibility**) we can compute the twist at A as

$$\phi_A = \frac{\partial U_1}{\partial M_A} + \frac{\partial U_2}{\partial M_A}$$

$$= \frac{(-M_A)(-1)(L/3)}{GJ} + \frac{(-M_A - M)(-1)(2L/3)}{GJ} = (M_A + \tfrac{2}{3}M)\frac{L}{GJ}$$

Enforcing compatibility, we set $\phi_A = 0$ (since the shaft is prevented from rotating at A). Thus we get

$$0 = (M_A + \tfrac{2}{3}M)\frac{L}{GJ}$$

$$\therefore \quad M_A = -\tfrac{2}{3}M$$

The negative sign tells us that our assumed direction for the reaction M_A is incorrect. From **equilibrium,** we determine that $M_C = -\frac{1}{3}M$, which is the same result obtained previously.

Example 14.14

The "L" shaped frame ABC shown in Fig. 14.33 is fixed at A and carries a vertical load P at point C. The length of BC is L_1, and the length of AB is L_2. Both are prismatic, circular beams of the same diameter and of the same linear elastic material. Determine the vertical displacement at point C.

In this problem, we can see that beam BC will bend due to the applied load P. However, beam AB will both bend and twist since the load P is eccentric to the centerline of AB. Therefore, the vertical displacement at C is composed of the *bending* displacement from AB and BC and the *twisting* of AB (as we have seen in Chapter 12, we can neglect deformation due to shear if the beams are sufficiently slender). To make the solution easier, we will use local coordinate

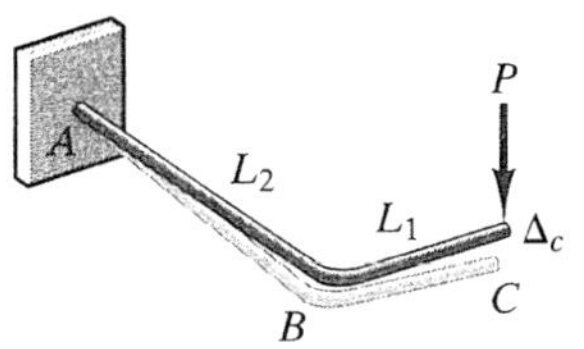

Figure 14.33. L-shaped cantilevered frame.

Example 14.14 (Continued)

systems s_1 and s_2 for the free-body diagrams of BC and AB, shown in Fig. 14.34.

From the free-body diagram exposing a cross section of BC at s_1 [see Fig. 14.34(a)], **equilibrium** tells us that

$$M_{BC} = Ps_1 \tag{a}$$

Similarly, cutting member AB at s_2, we conclude that [see Fig. 14.34(b)]

$$M_{AB} = Ps_2 \tag{b}$$

$$T_{AB} = PL_1$$

We can now build the strain energy expression as the sum of the bending energy in BC and the bending and twisting energy in AB. Thus (**constitutive law**)

$$U^* = \int_0^{L_1} \frac{M_{BC}^2}{2EI} ds_1 + \int_0^{L_2} \frac{M_{AB}^2}{2EI} ds_2 + \int_0^{L_2} \frac{T_{AB}^2}{2GJ} ds_2 \tag{c}$$

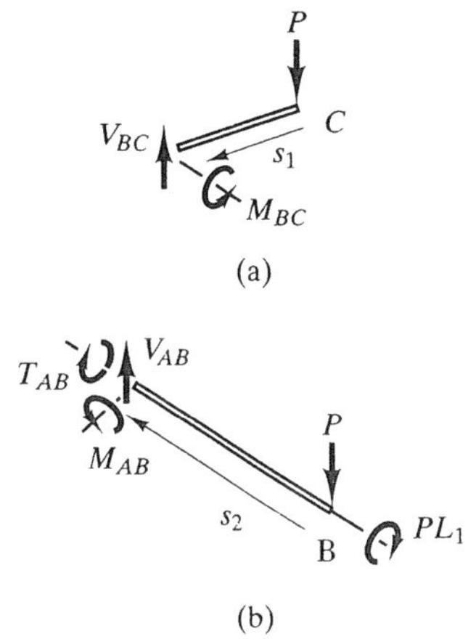

Figure 14.34. Free-body diagrams.

Notice that we have ignored the effect of shear on the displacement. We then substitute Eqs. (a) and (b) into (c) giving

$$U^* = \int_0^{L_1} \frac{(Ps_1)^2}{2EI} ds_1 + \int_0^{L_2} \frac{(Ps_2)^2}{2EI} ds_2 + \int_0^{L_2} \frac{(PL_1)^2}{2GJ} ds_2 \tag{d}$$

Invoking the second Castigliano theorem, we get (**compatibility**)

$$\frac{\partial U^*}{\partial P} = \Delta_C = \int_0^{L_1} \frac{Ps_1^2}{EI} ds_1 + \int_0^{L_2} \frac{Ps_2^2}{EI} ds_2 + \int_0^{L_2} \frac{PL_1^2}{GJ} ds_2 \tag{e}$$

$$\therefore \quad \Delta_C = \frac{PL_1^3}{3EI} + \frac{PL_2^3}{3EI} + \frac{PL_1^2 L_2}{GJ}$$

As a quick check of the solution, we see that if either L_1 or L_2 is zero, the displacement is the same as that of a cantilevered beam with a point load at the tip, as it should be. Also, we notice that the tip displacement is a function of the bending rigidity of both AB and BC as well as the torsional rigidity of AB. To get an idea of the relative contribution of the three terms in Eq. (e), we now consider the case where $L_1 = L_2 = L$ and the beams are a solid circular cross

Example 14.14 (Continued)

section made from steel. From this, we may state that $I = \pi d^4/64$, $J = 2I$, and $G = E/2.6$ so that the Δ_C can be expressed as

$$\Delta_C = \frac{PL^3}{EI}\left[\underbrace{0.333 + 0.333}_{\text{Bending}} + \underbrace{1.30}_{\text{Torsion}}\right]$$

It is somewhat surprising to see that the twisting of beam AB accounts for the majority of the tip displacement (about 66%) while the combined bending of AB and BC accounts for about 34% of the tip displacement. This is valuable information since, from a design perspective, it is often important to know which parts of a structure contribute significantly to the displacement.

Example 14.15

A linear elastic circular shaft supports a 1000 N-m torque [see Fig. 14.35(a)] and is held at the right end by a stiff rubber grommet having a torsional spring constant $K_T = 2 \times 10^{11}$ N-m/rad. Calculate the rotation of the shaft at $x = 2$ m and at $x = 6$ m, the latter at the position corresponding to the section A-A. The shear modulus G for the shaft is 2×10^{11} Pa.

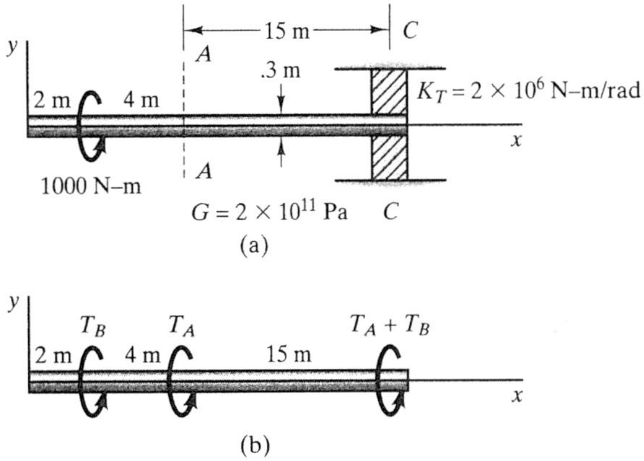

Figure 14.35. Shaft supports a torque and is restrained by a rubber grommet.

We shall insert torque T_A at the section A–A [see Fig. 14.35(b)]. At the appropriate time, we shall set the value of T_A equal to zero.

Example 14.15 (Continued)

Meanwhile, we shall need the polar moment of area J for the cross section of the shaft. Thus

$$J = \frac{\pi}{32} D^4 = \frac{\pi}{32}(.3)^4 = 7.95 \times 10^{-4}\ \text{m}^4$$

We shall use the second Castigliano theorem and we will need U^* for the system. We have for this calculation (**constitutive law** plus **equilibrium**)

$$U^* = U = \int_2^6 \frac{(T_B)^2}{2GJ}\,dx + \int_6^{21} \frac{(T_B + T_A)^2}{2GJ}\,dx + \frac{1}{2}K_T\left[\frac{T_B + T_A}{K_T}\right]^2$$

For θ_B we can then say (**compatibility**)

$$\theta_B = \left(\frac{\partial U^*}{\partial T_B}\right)_{T_A=0} = \left[\int_2^6 \frac{2T_B}{2GJ}\,dx + \int_6^{21} \frac{(T_B + T_A)(2)}{2GJ}\,dx + \frac{1}{2}\frac{(T_A + T_B)(2)}{K_T}\right]_{T_A=0}$$

Inserting proper values for the torques and integrating, we get

$$\theta_B = \frac{1000}{(2 \times 10^{11})(7.95 \times 10^{-4})}(4) + \frac{1000}{(2 \times 10^{11})(7.95 \times 10^{-4})}(15) + \frac{1000}{2 \times 10^6}$$

Hence,

$$\theta_B = .0006195\ \text{rad}$$

Next we go to θ_A.

$$\theta_A = \left(\frac{\partial U^*}{\partial T_A}\right)_{T_A=0} = \left[\int_6^{21} \frac{(T_A + T_B)(2)}{2GJ}\,dx + \frac{1}{2}\frac{(T_A + T_B)(2)}{K_T}\right]_{T_A=0}$$

Inserting numerical values and integrating, we can solve for θ_A.

Example 14.15 (Continued)

$$\theta_A = \frac{1000}{(2 \times 10^{11})(7.95 \times 10^{-4})}(15) + \frac{1000}{2 \times 10^6}$$

$$\theta_A = .0005943 \text{ rad}$$

14.8 Closure

In this chapter we first considered the case of linear, elastic torsion of a circular shaft. Using physical arguments—particularly those stemming from axial symmetry—we were able to arrive at the stress and strain distributions, which, we pointed out, correspond to exact solutions in accordance with the theory of elasticity. We then considered the torsion of closed thin-walled cylinders for which we were able to formulate Bredt's formula as an approximation for the deformation and stress of this class of problems. We then returned to the circular cylinder to discuss the case of the elastic, perfectly-plastic constitutive law. We then gave some practical formulations for certain noncircular cross sections. Finally, we used the second Castigliano theorem to solve torsion problems in our continuing effort to bring the energy method to the attention of the student.

With the conclusion of this chapter, we now have the ability to analyze systems with axial, bending, and torsional loads. Our analyses may lead us to stress components on three orthogonal interfaces at a point. Consequently, we will have to expand our ability to interpret these stress states.

Recall that in Chapter 2 we defined stresses for three orthogonal interfaces at a point. We pointed out in Chapter 2 that once the nine stress components on three orthogonal interfaces were known at a point, we could determine the stresses on *any* interface at that point. And in Chapter 7 we looked at the special case of two-dimensional stress, where only stresses τ_{xx}, τ_{yy}, and $\tau_{xy} = \tau_{yx}$ could be nonzero. Our main goal was to be able to determine the stresses $\tau_{x'x'}$, $\tau_{y'y'}$, and $\tau_{x'y'}$ for axes $x'y'$ rotated relative to xy. The resulting equations were called *transformation equations* and we took great pains to see that the key attributes of plane stress, such as principal axes, Mohr's circle, and invariants, were *directly attributable* to these equations. For those reasons we termed the plane stress components and any other set of quantities A_{xx}, A_{yy}, $A_{xy} = A_{yx}$ having the same aforementioned transformation equations the two-dimensional subsets of *second-order* symmetric *Cartesian tensors*. In the next chapter we shall look at three-dimensional stress at a point and we will accordingly expand on the work on tensors started in Chapter 7.

Highlights (14)

For a uniform circular shaft with opposite torques at the ends, a consideration of **compatibility** leads to the conclusion that, like the simple pure bending of beams, plane sections remain plane for small deformation. Then, considering **equilibrium** and Hooke's law as the **constitutive law,** two key formulas result. First, there is the twist formula

$$\Delta\phi = \frac{M_x L}{GJ}$$

which is the analog of the extension formula for one-dimensional members, namely

$$\Delta L = \frac{FL}{AE}$$

The corresponding terms should be obvious. Next, there is the shear-stress formula

$$\tau = \frac{M_x r}{J}$$

and this one is the analog for the flexure formula for beams, namely

$$\tau_{xx} = \frac{M_z y}{I_{zz}}$$

This theory of torsion depends on the axial symmetry of the shaft and can easily be shown to satisfy the theory of elasticity. That is, it is exact.

For thin-walled tubes having possibly a noncircular shape, one can get an approximation of the shear stress along the midline between the inner and outer boundaries of the cross section by using the shear flow concept along with equilibrium. Next, by using a work strain-energy equation in conjunction with Hooke's law, one can get the rate of twist for the tube using a simple formula called *Bredt's formula* [Eq. (14.22)].

There are approximate infinite series solutions for rectangular cross sections. The closer to a square boundary, the more accurate the solutions can be. There are also empirical approaches possible for more complex cross sections using a composite breakdown of the cross section into simpler constituent areas. Some are presented in this chapter.

For the elastic, perfectly plastic model of material behavior, plastic behavior in a circular cross section first occurs at the outer radius as the applied torque is increased. This plastic ring then gets thicker as the torque is increased until the entire cross section has yielded giving the maximum resisting torque possible for the shaft. The stresses and rate of twist are readily computable during this loading process as has been explained in this chapter. On unloading from a state wherein plastic deformation has been reached, there will result a **residual stress.** The prime reason for this to occur is that the stress-strain curve in the plastic zone on loading is such that the stress remains a **constant stress,** namely the yield stress. But on **unloading,** the stress-strain curve, in what was the plastic zone, becomes that of **elastic** behavior with the same elastic modulus, E, that was present on loading in the elastic range (see Fig. 14.22).

PROBLEMS

Torsion of Circular Shafts	14.1–14.38
Torsion of Thin-Walled Noncircular Closed Shafts	14.39–14.41
Elastic, Perfectly Plastic Torsion	14.42–14.43
Torsion of Shafts With Noncircular Cross Sections	14.44–14.47
Energy Methods for Torsion	14.48–14.49
Unspecified Section Problems	14.50–14.77
Computer Problem	14.78
Programming Project 13	

14.1. [14.3] What is the maximum shear stress for the solid shaft shown in Fig. P.14.1? What is the total twist at A? Determine the torsional spring constant and the torsional strength of the shaft. (Take $G = 1 \times 10^{11}$ Pa.) Length L Is 3 m.

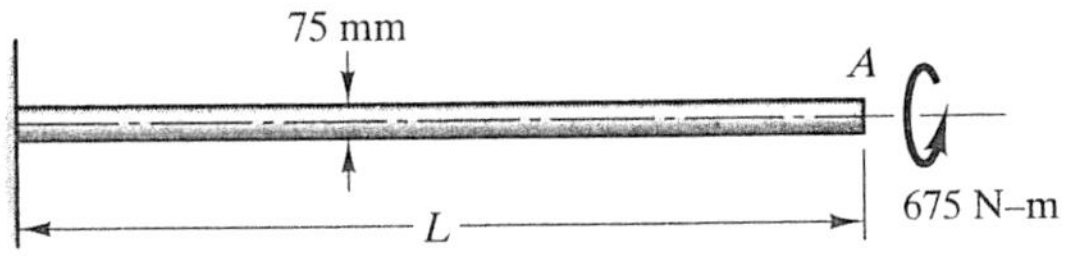

Figure P.14.1.

14.2. [14.3] In Problem 14.1, determine the amount of torque that can be applied to the shaft if the yield stress is 2.8×10^8 Pa and if a safety factor of 2 is employed.

14.3. [14.3] Evaluate the maximum torsional shear stress for the hollow shaft shown in Fig. P.14.3. Determine the angle of twist at a position 3 ft from the left end. (Take $G = 15 \times 10^6$ psi.)

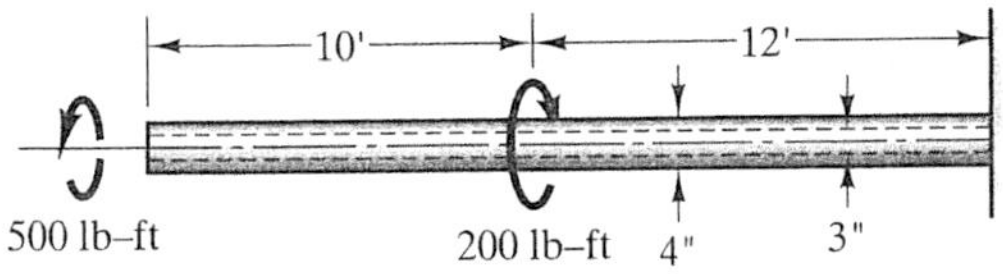

Figure P.14.3.

14.4. [14.3] Shown in Fig. P.14.4 is a solid shaft of two sections welded together at A. The materials for the section have different shear moduli. What is the twist at B at the end of the shaft?

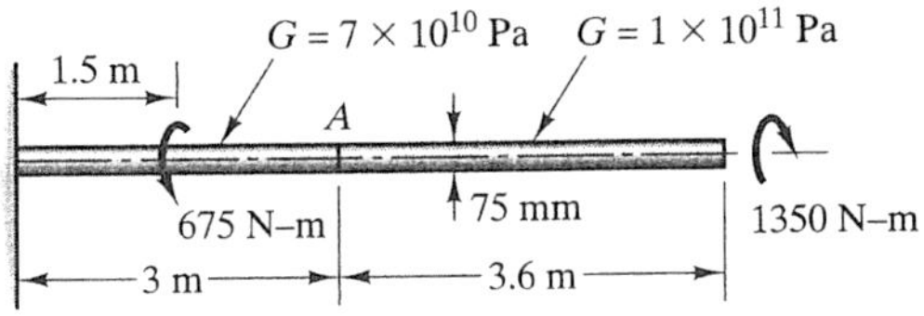

Figure P.14.4.

14.5. [14.3] In Fig. P.14.5 is shown a solid shaft. Find the length of an *equivalent shaft* of the same material having the same torsional stiffness. The diameter of the equivalent shaft is to be 75 mm.

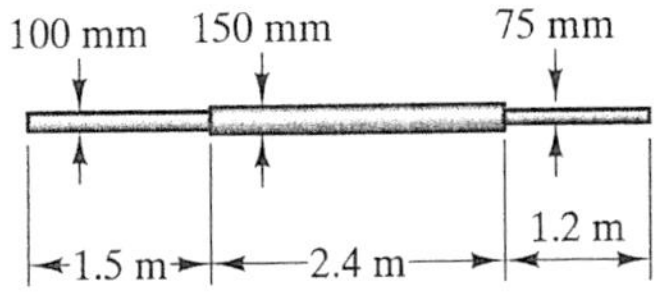

Figure P.14.5.

14.6. [14.3] Plot τ_{max} versus x in the series of shafts welded together as shown in Fig. P.14.6. Disregard stress concentrations at the juncture of the materials. What is the total twist at the left end? Take the following data:

$$G_A = 10 \times 10^6 \text{ psi}$$
$$G_B = 20 \times 10^6 \text{ psi}$$
$$G_C = 15 \times 10^6 \text{ psi}$$

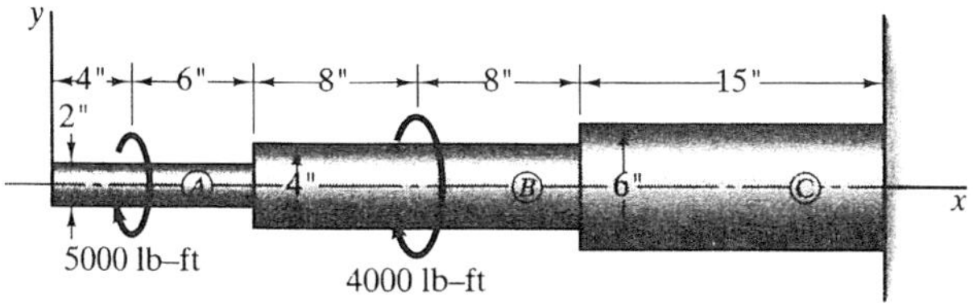

Figure P.14.6.

14.7. [14.3] For the system shown in Fig. P.14.7,

(a). Find the supporting torques at A.
(b). Find the magnitude of the twist $|\Delta\phi|$ at B.

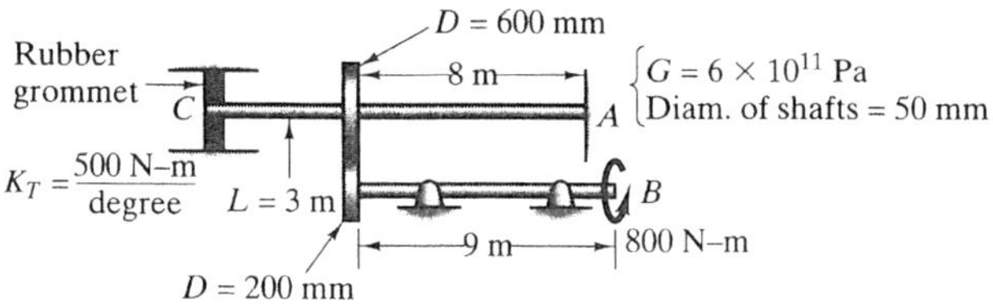

Figure P.14.7.

14.8. [14.3] A turbine is developing 7500 kW at 6000 rpm (see Fig. P.14.8). For an allowable shear stress of 1×10^8 Pa, what should be the diameter D_1 of the shaft transmitting torque from the turbine to the reduction gear system? If the reduction in speed is 100 to 1, what should be the diameter D_2 of the shaft between the reduction gears and the generator?

Figure P.14.8.

14.9. [14.3] What is the limiting speed of a shaft transmitting 7.5×10^5 W if the allowable stress is 1.4×10^8 Pa and the diameter is .15 m? If $G = 1.4 \times 10^{11}$ Pa, what is the minimum length to allow for a rotation of 6° between the ends?

14.10. [14.3] Determine the horsepower transmitted by a 10-in. solid shaft of length 10 ft if it is twisted an angle of 2° over its length. Compute the maximum shear stress. Take $G = 15 \times 10^6$ psi. The shaft rotates at 150 rpm.

14.11. [14.3] A torque T of 150 N-m is transmitted by a shaft A (see Fig. P.14.11) at an angular speed ω_A of 500 rpm. The diameter of shaft A is .15 m. If the same maximum stress is to be maintained in shaft B as in shaft A, what should its diameter be? Take $D_1 = .6$ m and $D_2 = .3$ m.

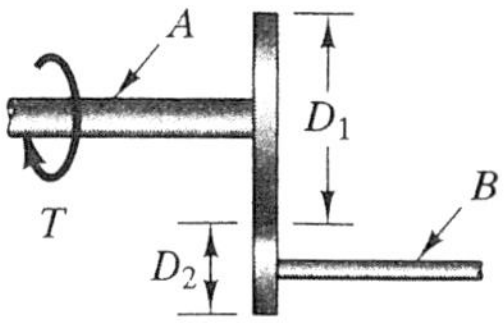

Figure P.14.11.

14.12. [14.3] If the allowable shear stress is 1×10^8 Pa for a hollow shaft having an outside diameter of .6 m and an inside diameter of .3 m, what is the maximum twist over a length of 4.5 m for $G = 1 \times 10^{11}$ Pa? What is the rate of twist?

14.13. [14.3] Torques C_1 and C_2 are applied to two shafts connected to two gears as shown in Fig. P.14.13. End A is connected through a rubber grommet to ground.

(a). Determine the rotation of end B as a result of the torques.
(b). Determine the rotation of the upper shaft.

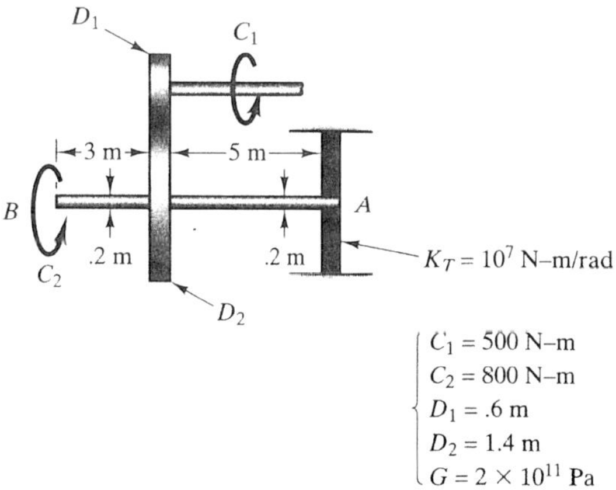

Figure P.14.13.

14.14. [14.3] Compute the supporting torques and the rotation of the left end of the shaft in Fig. P.14.14.

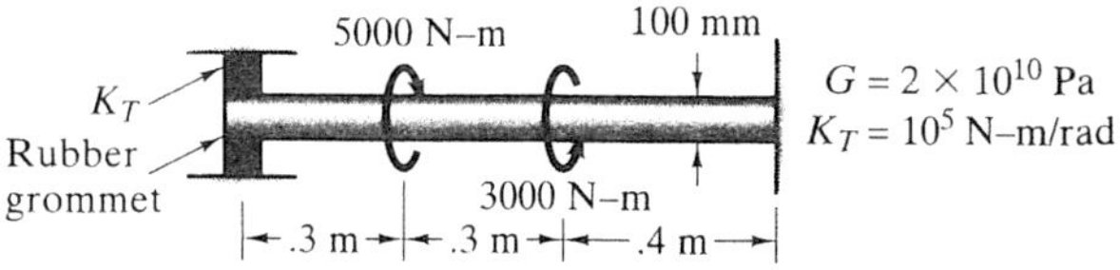

Figure P.14.14.

14.15. [14.3] Determine the horsepower transmitted by a 8-in. solid shaft of length 5 ft if it is twisted a total angle of 1° over its length. Compute the maximum shear stress. Take $G = 10 \times 10^6$ psi. The shaft rotates at 200 rpm.

14.16. [14.3] Find the supporting torque A in Fig. P.14.16. At B the shaft is bonded to a rubber grommet giving a torsional stiffness of 10^6 ft-lb/rad. Shaft C is steel with $G = 10 \times 10^6$ psi. Tube D is welded to shaft C at E through a rigid plate and to the base at A. It is also steel with the same G as C.

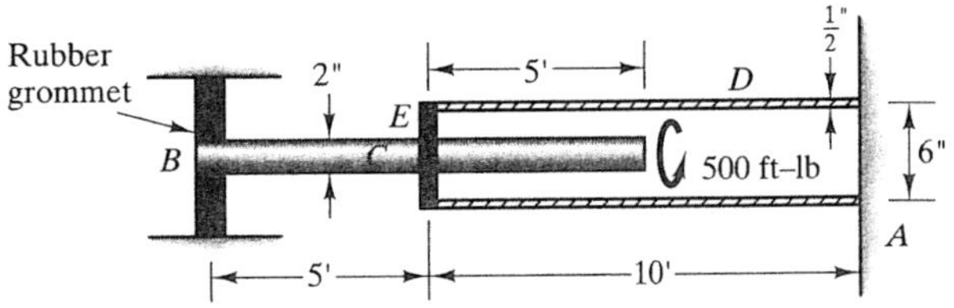

Figure P.14.16.

14.17. [14.3] Find rotation at A from the 500-ft-lb torque (see Fig. P.14.17). Take $G = 15 \times 10^6$ psi for the shafts. Gears connect the two shafts with each other.

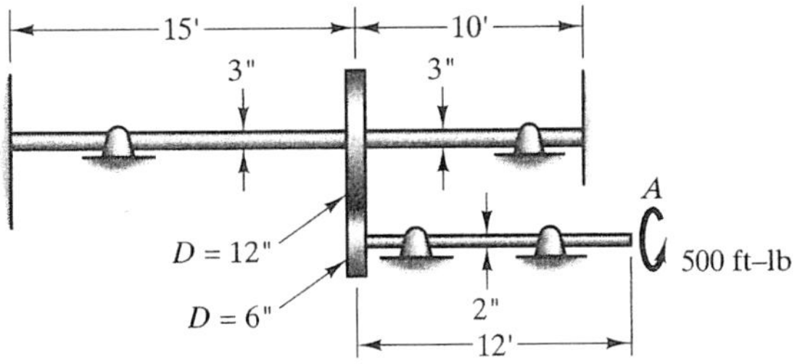

Figure P.14.17.

14.18. [14.3] A sleeve is welded to a shaft at C in Fig. P.14.18. What is the twist angle at D from the 500-N-m torque on sleeve CD? Use $G = 10 \times 10^{11}$ Pa as the shear modulus for both sleeve and shaft. Neglect twisting deformation of that part of D which is perpendicular to AB at C.

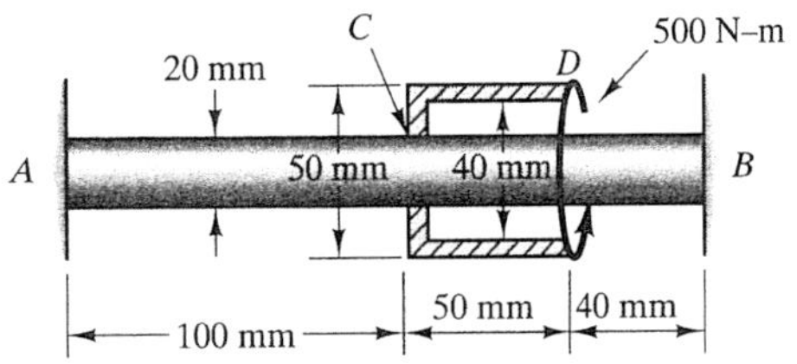

Figure P.14.18.

14.19. [14.3] Find the supporting torques in the shaft shown in Fig. P.14.19. Take $G = 15 \times 10^6$ psi. Make a simple plot of $\phi(x)$ and give maximum value of ϕ. Note the 1" cylindrical hole.

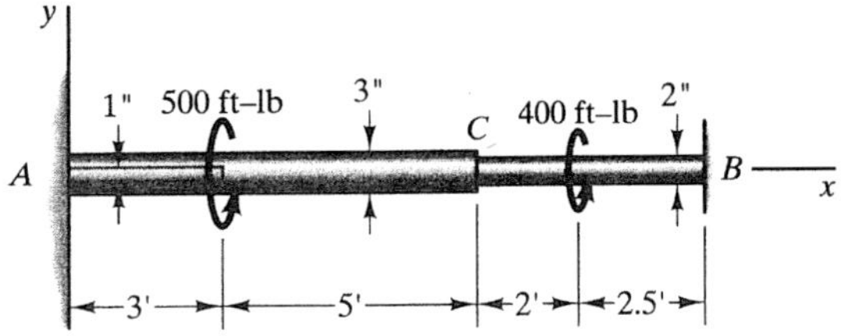

Figure P.14.19.

14.20. [14.3] An aluminum shaft A is fixed to the wall at B and is connected to a steel sleeve E through a rubber elastic grommet at C in Fig. P.14.20. Find the rotation at H due to the torque $T = 500$ N-m. Use the following data:

$$G_{Stl} = 1 \times 10^{11} \text{ Pa}$$
$$G_{Al} = .5 \times 10^{11} \text{ Pa}$$
$$K_T \text{ of grommet} = 5 \times 10^8 \text{ N-m/rad}$$

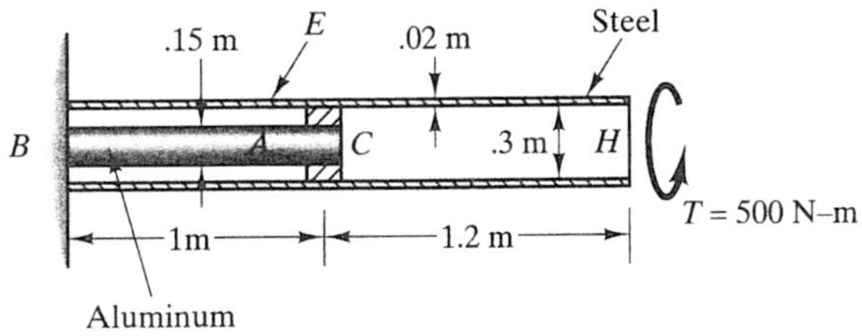

Figure P.14.20.

14.21. [14.3] In Fig. P.14.21 is shown a shaft composed of an inside solid rod having $G = 7 \times 10^{10}$ Pa and an outer sleeve having $G = 1.4 \times 10^{11}$ Pa. What is the maximum stress when shafts are connected at the right end only?

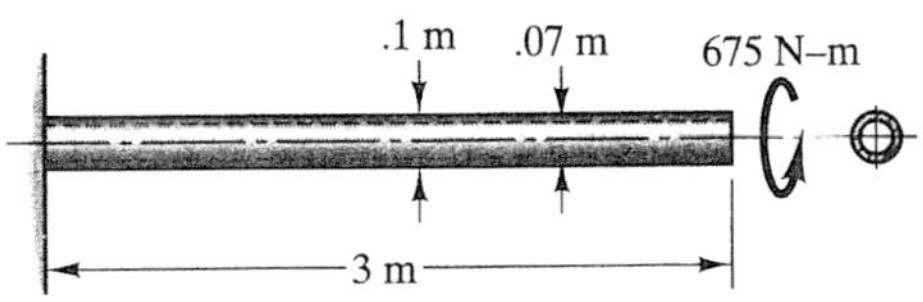

Figure P.14.21.

14.22. [14.3] A steel shaft (Fig. P.14.22) having diameter $D_1 = 6$ in. and an outer sleeve of aluminum having an inner diameter of 12 in. and a thickness t of 2 in. is held by rigid end plates on which a torque T of 5000 ft-lb is applied. What are the largest torsion stresses in each material? Take $L = 10$ ft, $G_{Stl} = 15 \times 10^6$ psi and $G_{Al} = 10 \times 10^6$ psi.

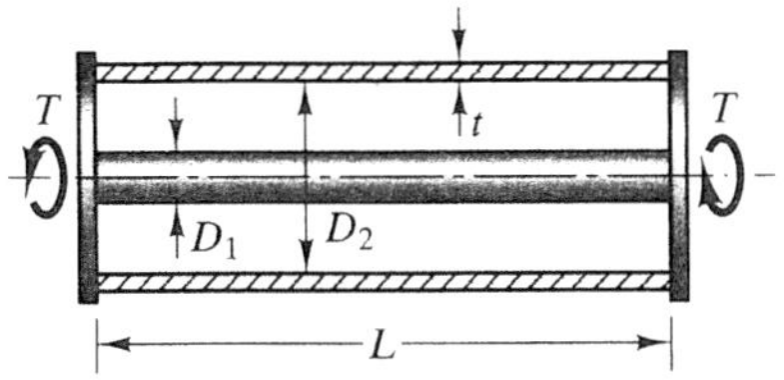

Figure P.14.22.

14.23. [14.3] A steel shaft A and an aluminum sleeve B are connected at the right end (see Fig. P.14.23) by a rigid support and at the left end by a rigid plate C. A torque $T_1 = 500$ ft-lb is applied to the plate and a second torque $T_2 = 800$ ft-lb is applied to the sleeve as shown. What is the maximum torsional shear stress in each member? What is the angle of twist at the rigid plate? Take $G_A = 15 \times 10^6$ psi and $G_B = 10 \times 10^6$ psi. (*Hint:* Consider free-body diagrams of sleeve B, shaft A, and end plate C.)

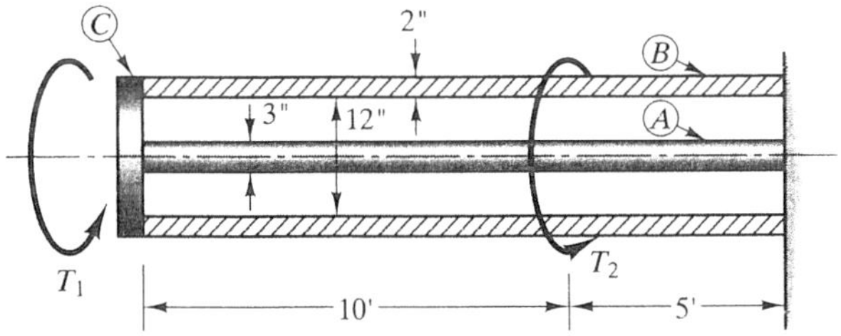

Figure P.14.23.

14.24. [14.3] Consider the close-coiled spring shown in Fig. P.14.24. Show that if the shear *force* in the coil is assumed to give a uniform stress, the maximum shear stress in the coil is given as

$$\tau_{max} = \frac{P}{A} + \frac{PRr}{J}$$

where A is the cross-sectional area of the wire making up the spring and J is its polar moment of area. Next, show that for n complete coils of the spring, the elongation δ of the spring resulting from twisting of the wire is given as

$$\delta = \frac{4PR^3n}{Gr^4}$$

What other deformation contributions give rise to further deflection?

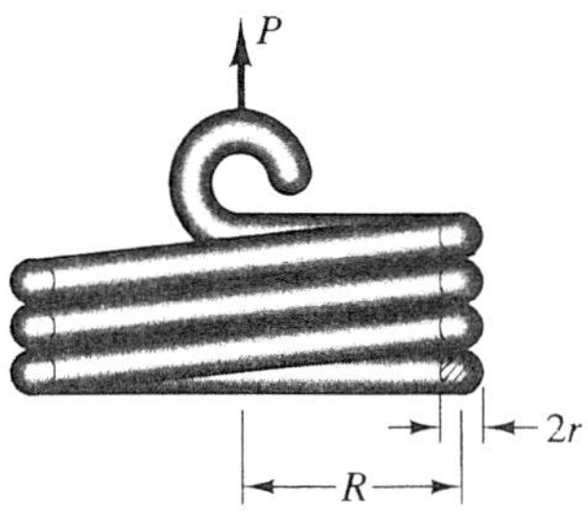

Figure P.14.24.

14.25. [14.3] A steel rod having a shear modulus of 1×10^{11} Pa is used to form a spring having a mean radius of 75 mm. The diameter of the rod is 10 mm. What is the spring constant K if the rod is 1.5 m long? See Problem 14.24 before solving this problem.

14.26. [14.3] A very good formulation for the maximum shear stress in helical springs (see Fig. P.14.24) is the *Wahl formula*, given as follows:

$$\tau_{max} = \frac{2PR}{\pi r^3}\left[\frac{4(R/r) - 1}{4(R/r) - 4} + \frac{.615}{R/r}\right]$$

Determine the shear stress in Problem 14.25 for a 450-N load using the approximate formula of Problem 14.24 and the Wahl formula. What is the percentage error for the approximate formula?

***14.27. [14.3]** Considering torsional stresses only, compute the maximum shear stress in a "slinky" toy shown in Fig. P.14.27 when it is hanging vertically under its own weight. A typical slinky weighs about 1/2 lb, has about 90 coils, and has a diameter D of about 2.7 in. The rectangular cross section of the wire itself is $b = .01$ in. and $a = 0.025$ in. For a rectangular cross-section, we will learn in Section 14.7 that $\tau_{max} = M_x/\kappa ba^2$. Here, $\kappa \approx .280$.

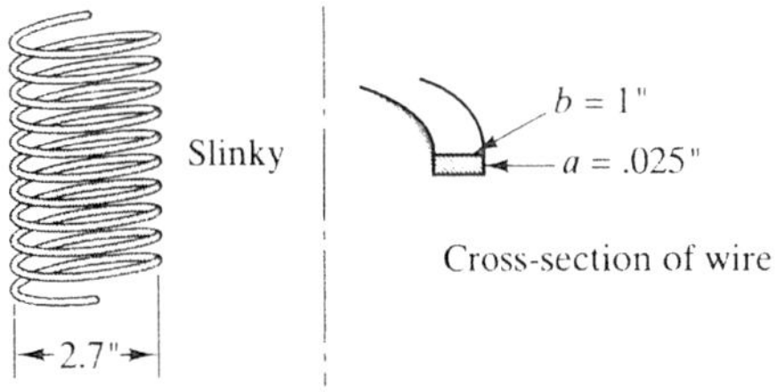

Figure P.14.27.

14.28. [14.3] Shafts 1 and 2 shown in Fig. P.14.28 are circular shafts and have the same diameter. Investigate the effect of doubling the diameter of shaft 1 on the reactions.

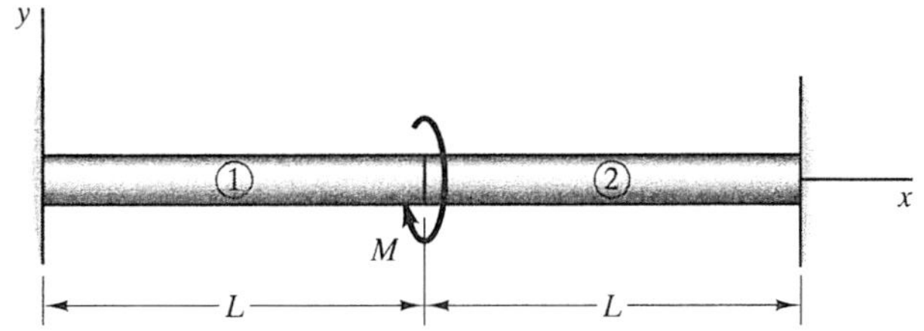

Figure P.14.28.

14.29. [14.3] A shaft B is shrunk-fit over a shaft A (Fig. P.14.29) so as to transmit torque without slippage. Shaft A has a value of $G_A = 15 \times 10^6$ psi, while shaft B has a modulus $G_B = 20 \times 10^6$ psi. What is rate of twist α for a torque $T = 1000$ ft-lb applied at the end as shown? Find the maximum shear stress in each material. See result from Problem 14.62.

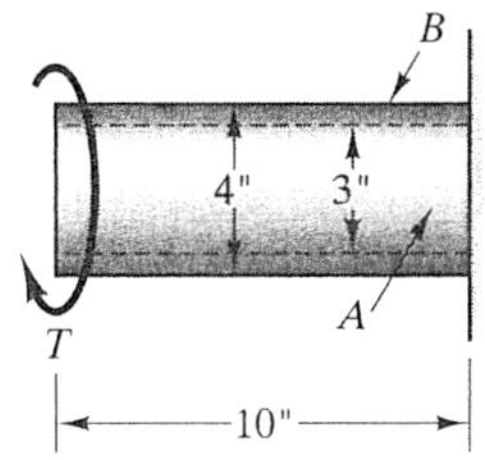

Figure P.14.29.

14.30. [14.3] What is the twist at the end of the shaft shown in Fig. P.14.30 (the sleeve and inside core are made of different materials and are entirely fastened together), when G for the sleeve is 1×10^{11} Pa and G for the core is 7×10^{10} Pa? See result from Problem 14.62.

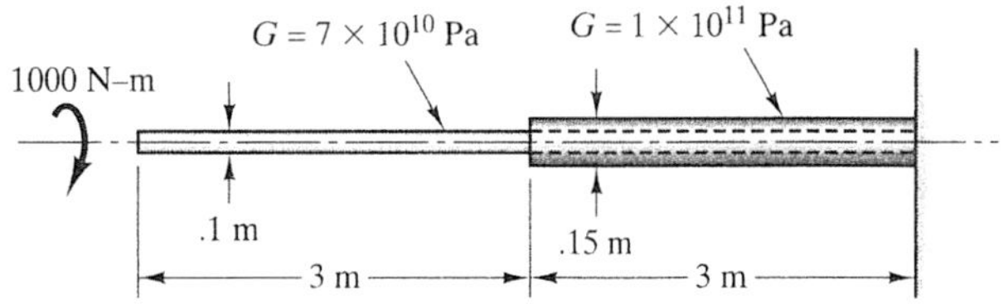

Figure P.14.30.

In Problems 14.31 to 14.37 we have combined torsion and bending. These are sometimes called "combined-stress" problems.

14.31. [14.3] A rod AB is welded onto a solid cylindrical cantilever beam CB in Fig. P.14.31. What is the maximum principal stress at point a in the cantilever beam? Neglect the weights of CB and AB. *Note:* Point a is on top of cantilever. (*Hint:* What kind of stress exists at a?)

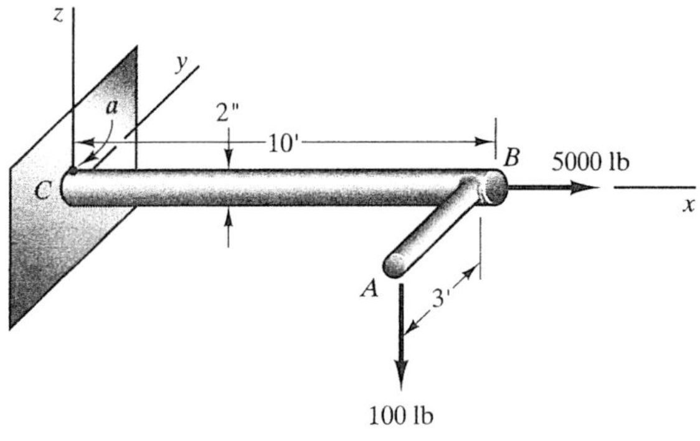

Figure P.14.31.

14.32. [14.3] Standard 10-in. steel pipe (see Appendix IV-F) is shown in Fig. P.14.32. It is full of water. Show Mohr's circle for stresses parallel to the xz plane at point A in the diagram. What is the extremal value of shear stress parallel to the xz plane for this point? Include the weight of the piping in your calculations.

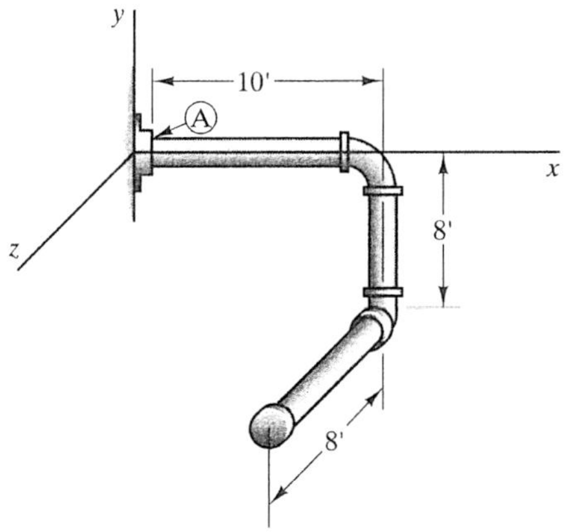

Figure P.14.32.

14.33. [14.3] For Fig. P.14.33, determine the principal stresses at point a, 5 ft from the support.

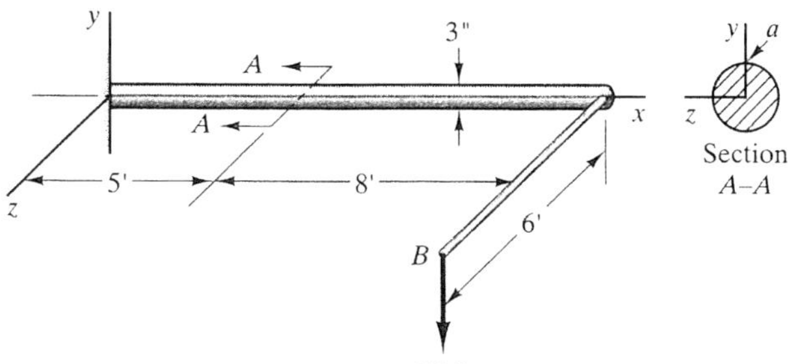

Figure P.14.33.

14.34. [14.3] In Problem 14.33, evaluate the total downward deflection of point B. (The shaft has a Young's modulus of 30×10^6 psi and a Poisson ratio of .3. The rod connecting the load to the shaft has a diameter of 1.5 in. and a Young's modulus of 30×10^6 psi.)

> The following two problems assume that the student has studied Chapter 9.

14.35. [14.3] A thin-walled pipe system is shown in Fig. P.14.35 supporting a force P. If P is 5000 lb, what should the thickness t be for a tensile yield stress of 50,000 psi and a safety factor of 1? Use the octahedral shear criterion. Do not consider weight of the pipe.

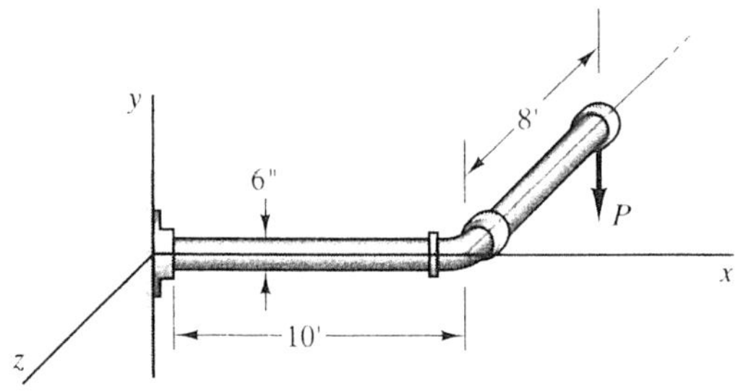

Figure P.14.35.

14.36. [14.3] What standard-size pipe should you employ in Fig. P.14.36 for a tensile yield stress of 80,000 psi and a safety factor of 1? Do not consider the weight of the pipe. Use Tresca yield criterion.

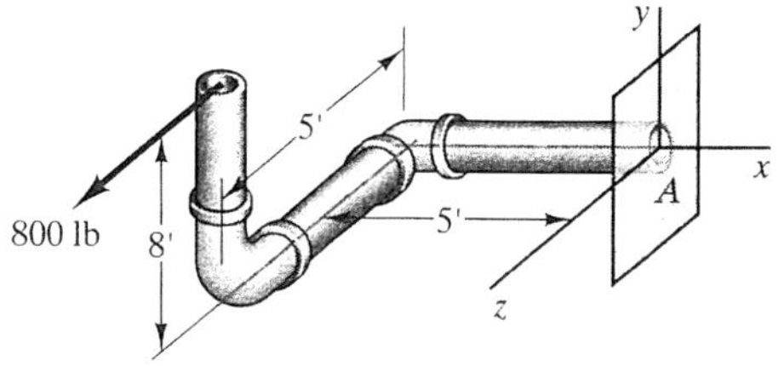

Figure P.14.36.

14.37. [14.3] Shaft AB is supported by four beams all welded together at B as shown in Fig. P.14.37. Each beam is of length L_1 and has a bending rigidity EI. The shaft is of length L_2 and has a torsional rigidity GJ. What is the total twist at A?

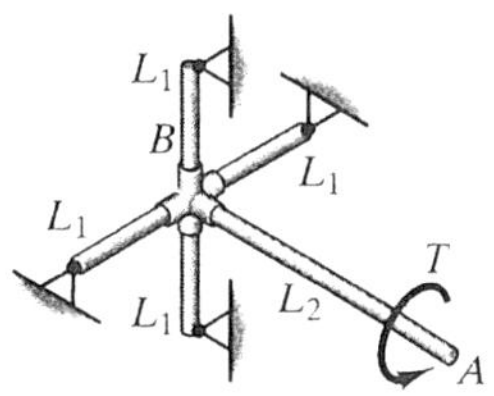

Figure P.14.37.

14.38. [14.3] Redo Problem 14.37, however, with the beams now fixed at the ends as shown in Fig. P.14.38.

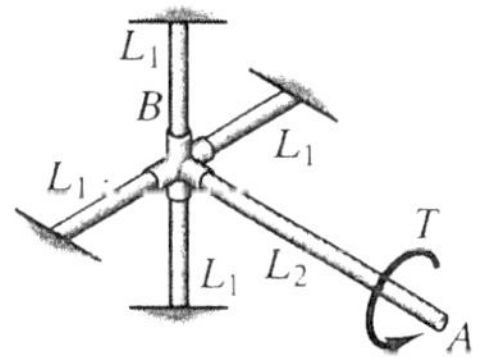

Figure P.14.38.

14.39. [14.5] In Fig. P.14.39 is shown a section of a thin-walled member made from an aluminum alloy. If the thickness is .050 in., what is the stress and rate of twist for a torque of 1000 in.-lb? Take G for this material as 4×10^6 psi. Assume no buckling.

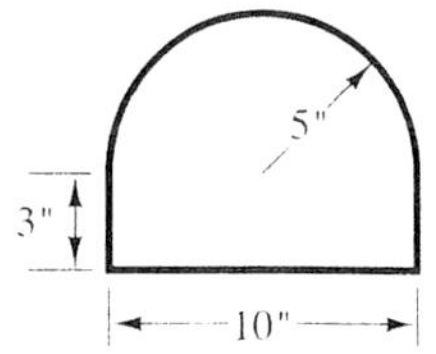

Figure P.14.39.

14.40. [14.5] Shown in Fig. P.14.40 is the cross section of a thin-walled member which is part of an airplane wing. If the allowable shear stress is 5.5×10^7 Pa, find the allowable shear flow for this section and the stresses away from corners. Sides AB and CD have a thickness of 1 mm and sides AC and BD have a thickness of 2 mm. The mean height of the section is 265 mm and G for the material is 2.8×10^{10} Pa. Assume no buckling. Find the allowable torque transmitted and the rate of twist.

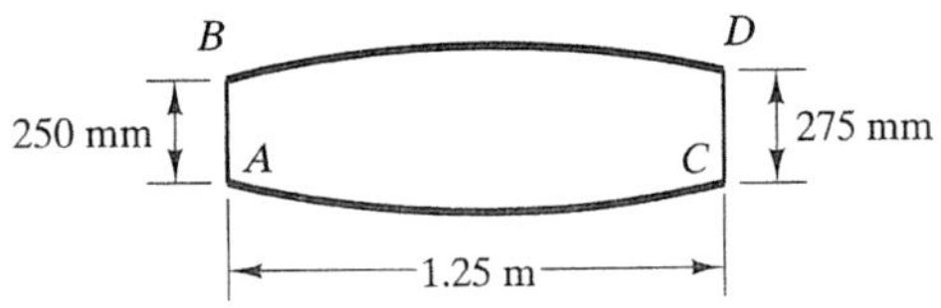

Figure P.14.40.

14.41. [14.5] Shown in Fig. P.14.41 is a thin-walled member having the shape of an ellipse. For $G = 5 \times 10^6$ psi, determine the shear stress and rate of twist for an applied torque of 500 in.-lb. The wall is .050 in. thick.

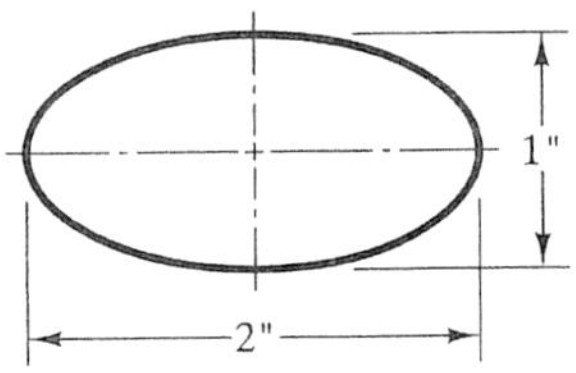

Figure P.14.41.

14.42. [14.6] A torque of 3300 N-m is transmitted through a 50-mm-diameter solid shaft having a yield stress in shear of 1.1×10^8 Pa. Is there plastic deformation for this loading? If so, how far has the plastic action penetrated from the surface assuming elastic, perfectly plastic behavior? What is the maximum possible static moment?

14.43. [14.6] Consider a hollow steel shaft with a 130-mm outside diameter and a 70-mm inside diameter. What is the maximum torque that this shaft can transmit if the shear yield stress is 1.4×10^8 Pa and the behavior is elastic, perfectly plastic?

14.44. [14.7] Compute the torsional stiffness of the hollow shaft shown in Fig. P.14.44.

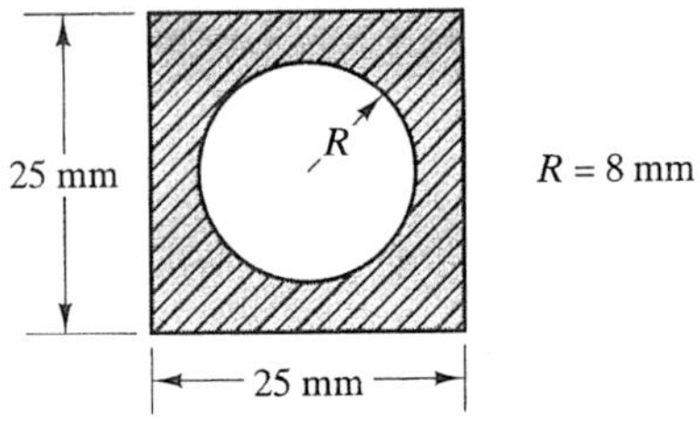

Figure P.14.44.

14.45. [14.7] Compare the torsional stiffness of the shapes shown in Fig. P.14.45. Each has a cross-sectional area of 5 in.^{2}

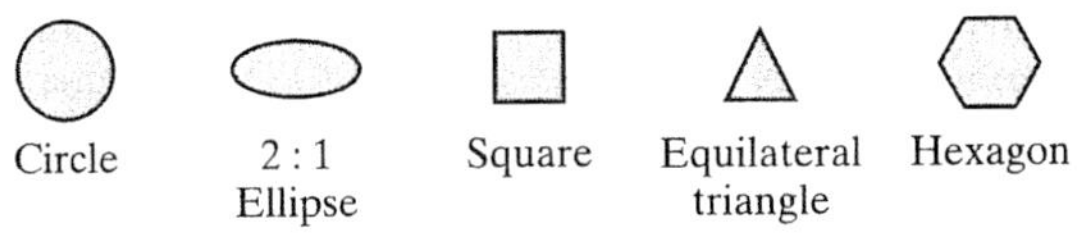

Figure P.14.45.

14.46. [14.7] Compare the torsional stiffness of the two shapes shown in Fig. P.14.46.

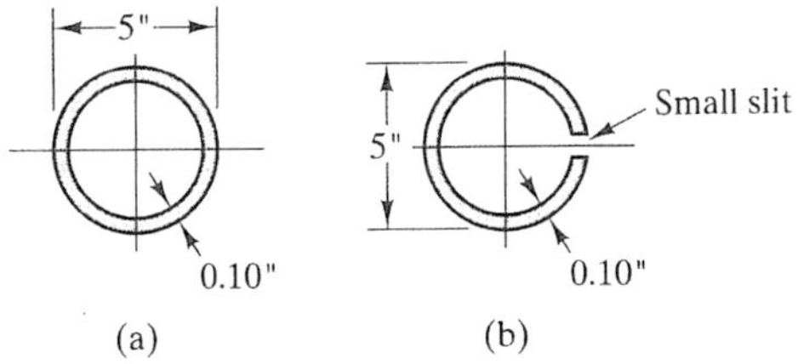

Figure P.14.46.

14.47. [14.7] Compute the torsional stiffness of the beam shown in Fig. 14.27(c) of Example 14.12. Compare the value obtained with that of the example.

14.48. [14.8] Determine the resisting torques at the supports for the shaft shown in Fig. P.14.48 using the second Castigliano theorem.

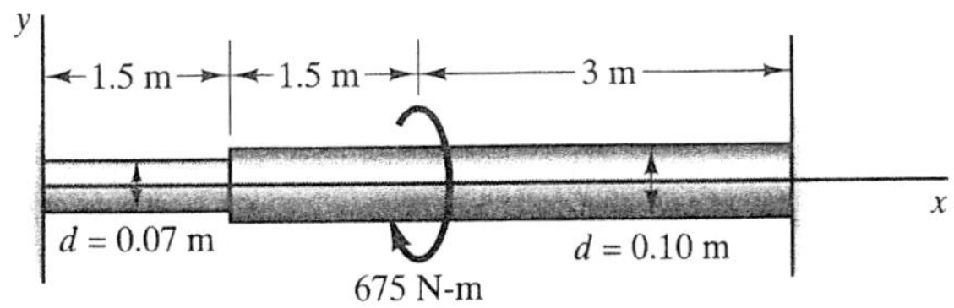

Figure P.14.48.

14.49. [14.8] Determine the relationship between the force F and vertical deflection Δ for the split ring loaded as shown in Fig. P.14.49. Assume that the flexural rigidity is EI and that the torsional rigidity is GK_t. Make use of the second Castigliano theorem.

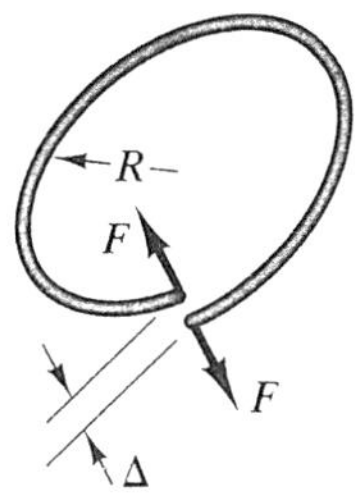

Figure P.14.49.

14.50. Two solid shafts 1 and 2 (see Fig. P.14.50) are rigidly connected to a hollow shaft 3. Shaft 2 is fixed at the right end, while shaft 3 is fixed at the left end. A torque T = 500 ft-lb is applied to the end of the shaft 1. What are the supporting torques? Take the same G for all shafts.

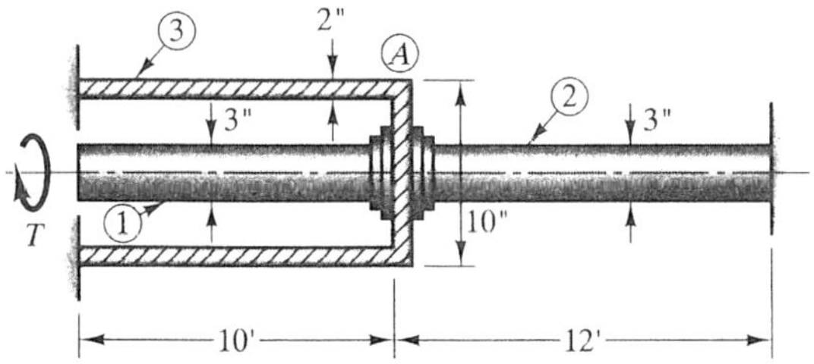

Figure P.14.50.

14.51. Do Problem 14.50 for the case of a single applied torque T_A = 500 ft-lb clockwise at A as you observe along the shaft from left to right. The torque is applied on the connecting plate between shaft 3 and shafts 1 and 2.

14.52. A shaft having a diameter D of 3 in. (see Fig. P.14.52) is acted on by a torque T of 300 ft-lb at one end, while at the other end is connected by a rubber material to a fixed rigid disk. The latter support gives a linear, elastic torsional spring constant K_T to the end having the value of 700 ft-lb/deg. What is the total twist ϕ at end A and the relative twist ϕ_{AB} between ends A and B? The length L is 10 ft and $G = 10 \times 10^6$ psi.

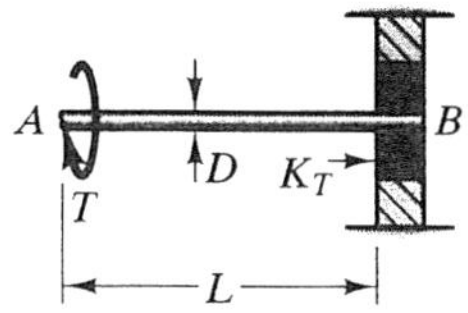

Figure P.14.52.

14.53. A torque T of 675 N-m is applied to a shaft as shown in Fig. P.14.53. The shaft is attached to two rubber grommets, which in turn are attached to fixed disks. Each gives a linear, elastic torsional spring constant K_T = 550 N-m/deg. If $G = 1 \times 10^{11}$ Pa, find the twist of end A and the relative twist between support 1 and support 2.

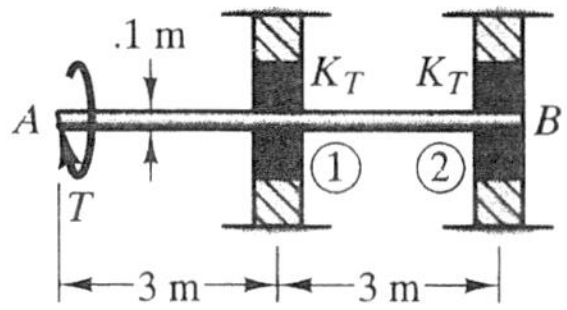

Figure P.14.53.

14.54. Compare the torsional stiffness of the hollow box shaft shown in Fig. P.14.54 by using Bredt's formula and by using the approximate method presented in Section 14.7.

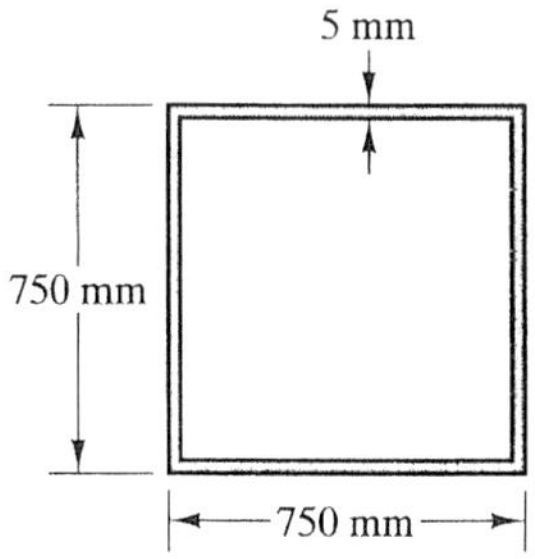

Figure P.14.54.

14.55. Find the supporting torques at A and B (see Fig. P.14.55). What is the rotation of the shaft at A? Take $G = 1.2 \times 10^{11}$ Pa. What is the maximum shear stress in the shaft? Sketch ϕ. vs. position along shaft.

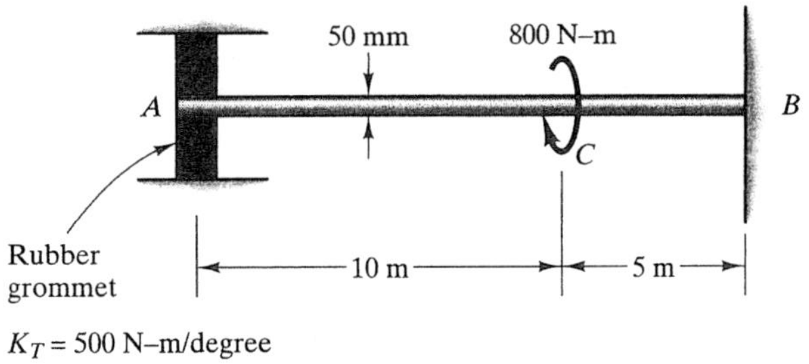

Figure P.14.55.

14.56. Find the supporting torque at A in Fig. P.14.56. At end B, the shaft is held by a rubber bushing giving a resisting torque of 10,000 ft-lb per radian. Take $G = 30 \times 10^6$ psi for the shaft.

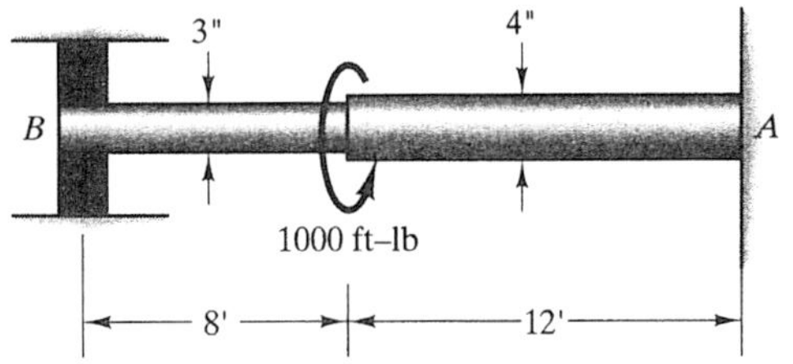

Figure P.14.56.

14.57. What are the supporting torques in terms of d and l at A and B (see Fig. P.14.57)? Determine the length of a drilled hole at a function of diameter to have the torques at A and B equal. Do this for $l < 1.5$ m. Determine l for $d = 60$ mm from your equation for this case.

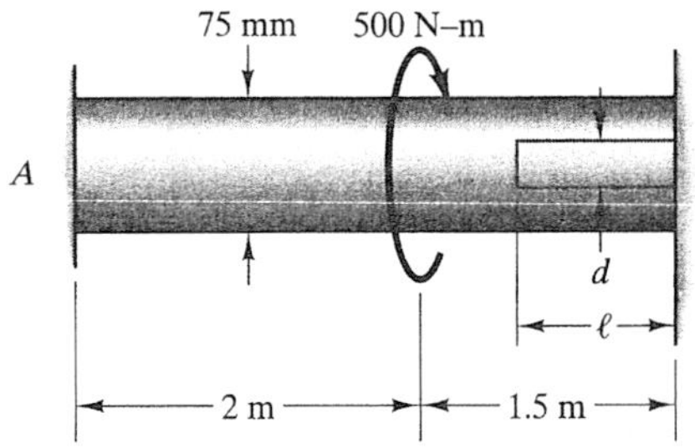

Figure P.14.57.

14.58. The shaft shown in Fig. P.14.58 is a shaft on which is applied a linearly varying torque distribution. What is the twist at end A of the shaft? The diameter of the shaft is .1 m and the modulus of shear is 1×10^{11} Pa. Assume that the theory developed thus far is valid for nonuniform torques.

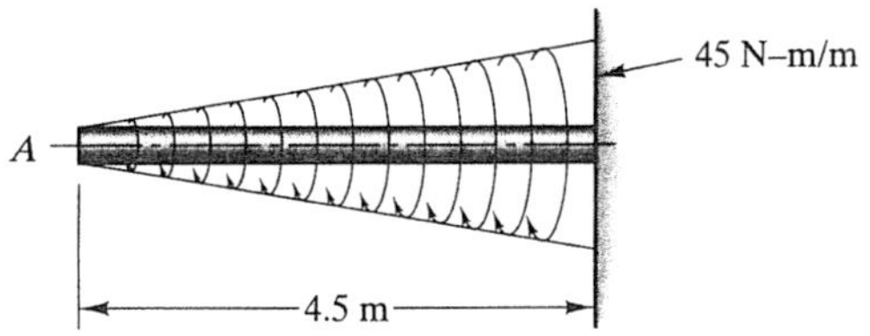

Figure P.14.58.

14.59. If the shaft in Problem 14.58 is loaded as shown in Fig. P.14.59, would you expect the same twist at end A of the shaft? Check to see if your supposition is correct. See comment at end of Problem 14.58.

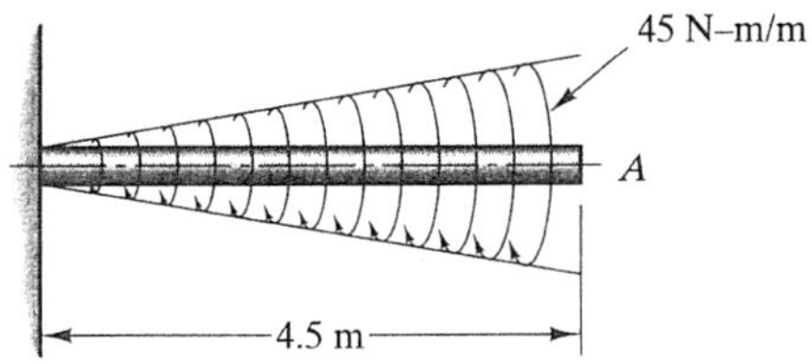

Figure P.14.59.

14.60. Do Problem 14.59 assuming this time that the loading varies parabolically.

14.61. Suppose in Problem 14.59 that the right side of the shaft were fixed in a rigid support. Determine the torques at the supports to resist the linearly varying load.

*__14.62.__ Several shafts are shrink-fitted over a solid shaft (see Fig. P.14.62) so as to form a combined member which can transmit torque with no slippage between the members. Explain why the assumption of cross sections rotating as platelets still applies. If each material is linear and elastic, show using equilibrium, Hooke's law, and the assumption above that

$$T = \alpha\left[\left(\int_0^{R_1} r^2 dA\right)G_1 + \left(\int_{R_1}^{R_2} r^2 dA\right)G_2 + \left(\int_{R_2}^{R_3} r^2 dA\right)G_3\right]$$

$$= \alpha[G_1 J_1 + G_2 J_2 + G_3 J_3]$$

where α is the rate of twist.

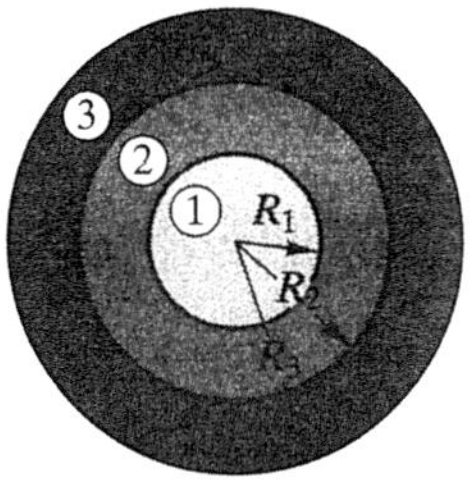

Figure P.14.62.

14.63. What is the maximum torque that can be applied to a $\frac{1}{4}$- in. bolt without yielding? Take the yield stress of the material as 36,000 psi.

14.64. You are to design for an automobile a hollow shaft having an outside diameter of 75 mm to transmit a maximum torque of 1000 N-m. What is the thickness t for a steel shaft with a yield stress of 3.5×10^8 Pa? Employ a safety factor of 5.

14.65. Shown in Fig. P.14.65 are two shafts connected by rigid gears. Determine the maximum shear stress in each shaft and the twist at A. (The lower shaft is supported by two bearings.)

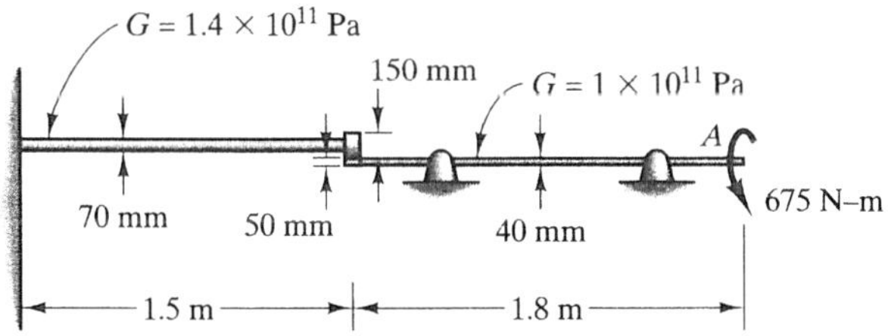

Figure P.14.65.

14.66. What is the twist at A for the shaft shown in Fig. P.14.66? The diameter varies linearly as indicated. Assume that the simple theory for uniform shafts is valid locally here. (Take $G = 20 \times 10^6$ psi.)

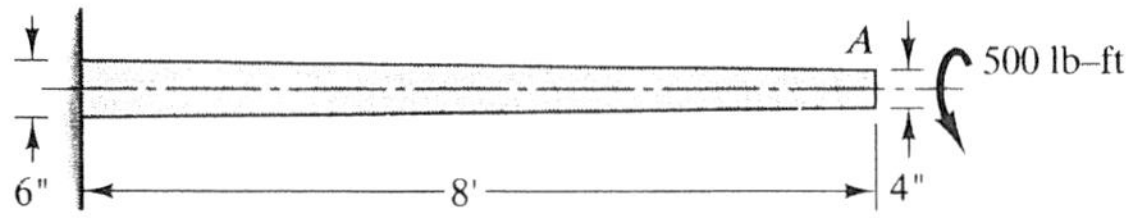

Figure P.14.66.

14.67. In Problem 14.66, assume that a uniform hole 1 in. in diameter has been bored out of the center of the shaft. Compute the maximum stress.

14.68. A shaft is formed by laminating very thin cylinders as shown in Fig. P.14.68. The inside diameter (i.e., of the smallest cylinder) is 1 in. and the outside diameter is 3 in. We shall assume that the shear moduli of the laminae are different and can be expressed as a function of r as follows:

$$G = G_1(1 + .2r)$$

Determine the maximum shear stress and the twist at the end of the shaft for $G_1 = 15 \times 10^6$ psi.

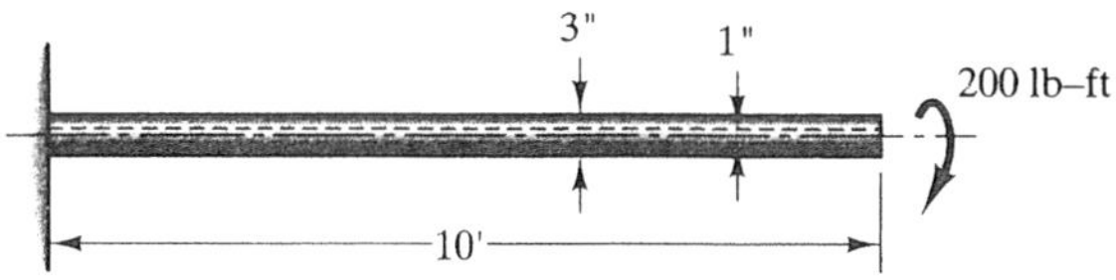

Figure P.14.68.

14.69. Determine the resisting torques at the support for the shaft shown in Fig. P.14.69. (Take $G = 10 \times 10^6$ psi.)

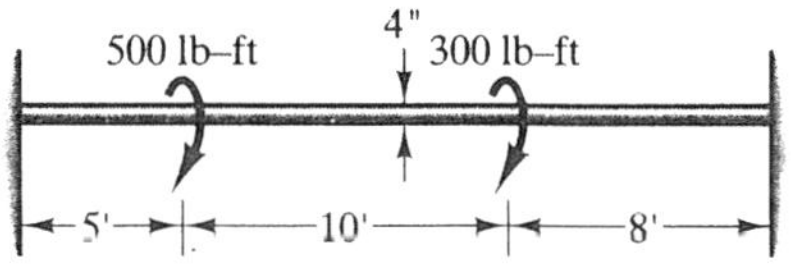

Figure P.14.69.

14.70. Determine the resisting torques at the supports for the shaft shown in Fig. P.14.70.

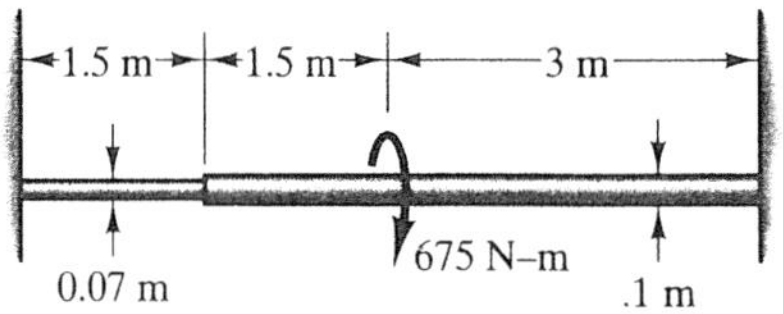

Figure P.14.70.

14.71. In Fig. P.14.71 is shown a shaft made by welding two sections together. The moduli of the materials comprising the sections are given in the diagram. Determine the resisting torques at the supports.

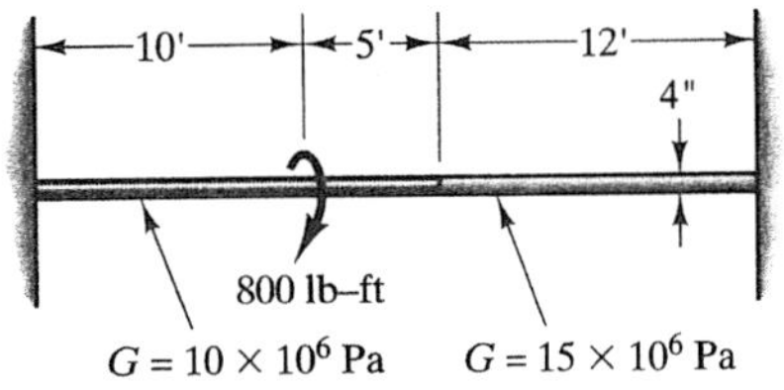

Figure P.14.71.

14.72. Two solid shafts 1 and 2 (see Fig. P.14.72) are rigidly connected to a hollow shaft 3. Shaft 2 is fixed at the right end, while shaft 3 is fixed at the left end. A torque $T = 500$ ft-lb is applied to the end of shaft 1. What are the supporting torques? Take the same G for all shafts. Use energy method and compare with results of Problem 14.50.

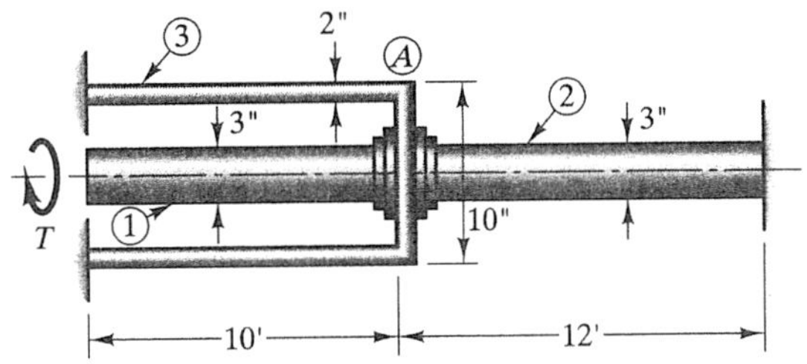

Figure P.14.72.

14.73. What are the principal stresses at the top of section A in Fig. P.14.73 as a result of the force P given as follows:

$$\mathbf{P} = 500\mathbf{i} + 800\mathbf{j} \text{ lb}$$

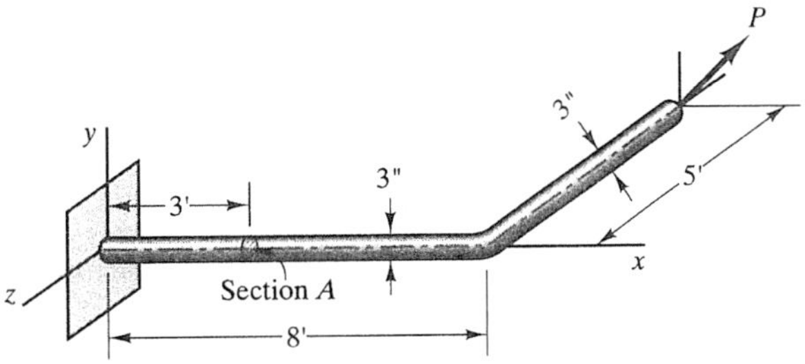

Figure P.14.73.

***14.74.** The "L" shaped frame ABC shown in Fig. P.14.74 is fixed at C and carries a vertical load $P = 100$ lb at point C. The length of BC is 6 in., and the length of AB is 8 in. Both are prismatic, square beams of identical cross section and linear elastic material. Determine the dimension b of the cross section so that vertical displacement at point C is ¼ in. Take $E = 30 \times 10^6$ psi and $\nu = 0.33$. Do not include shear deformation. Use energy method.

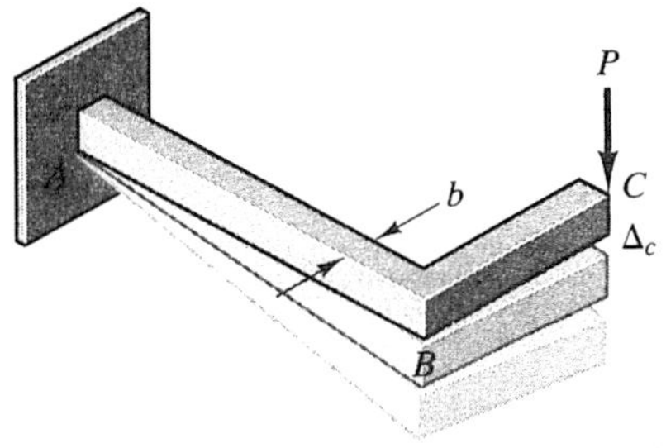

14.75. Compute the torsional constant K_t for the shape shown in Fig. P.14.75 using Bredt's formula and by approximation summing the torsional stiffnesses of outside and inside hexagonal boundaries. *Hint*: First show that

$$K_t = \frac{4A^2 t}{S}$$

using Bredt's formula and Eq. 14.29. Take the cross-sectional area

$$A = \frac{3\sqrt{3}}{2}b$$

Finally, using the midline of the section of length $\bar{b}$ show that for the outside length

$$b = \bar{b} + \frac{a}{\sqrt{3}}$$

and for the inside length

$$b' = \bar{b} - \frac{a}{\sqrt{3}}$$

Using software and Table 14.1 for a hexagon, you can show that

$$K_t = 4.203a\bar{b}^3 + 1.40a^3 b$$

For $\bar{b} \gg a$ explain why

$$K_t = 4.203ab^3$$

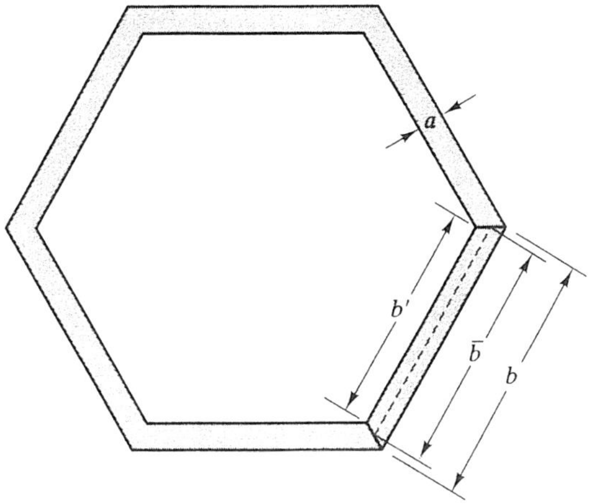

Figure P.14.75.

14.76. Redo Problem 14.34 using the second Castigliano theorem. Consider bending and twisting energy only.

14.77. Redo Problem 14.69 using the second Castigliano theorem.

****14.78.** Plot the reacting torques versus the shaft diameter ratio d_1/d_2 for the system shown in Fig. P.14.78. Consider ratios of the range 1 to 5. At what ratio does shaft 1 take 95% of the load?

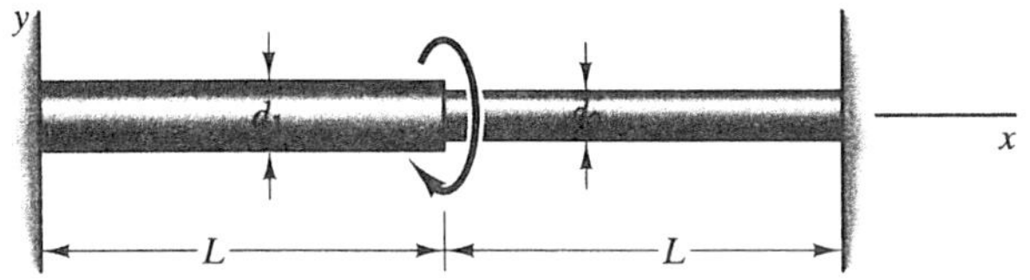

Figure P.14.78.

Project 13 Composite Torsion Problem

For any set of L's, G's, D's, and T, find the resulting angular twist for Fig. P.14.79. *Use S1 units.*

Run the program for the following data:

$L_1 = 1$ m $\quad L_2 = 1$ m $\quad L_3 = 2$ m

$G_1 = 1 \times 10^{11}$ Pa $\quad G_2 = 1.5 \times 10^{11}$ Pa $\quad G_3 = 2 \times 10^{11}$ Pa

$D_1 = 50$ mm $\quad D_2 = 100$ mm $\quad D_3 = 75$ mm

$T = 1000$ N-m

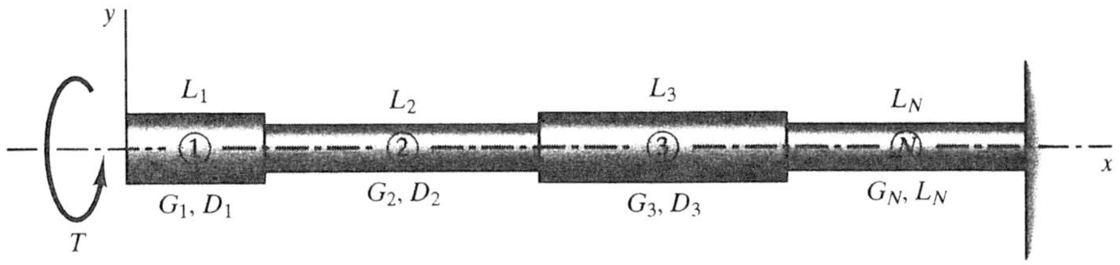

Figure P.14.79.

CHAPTER 15

Three-Dimensional Stress Properties at a Point

15.1 Introduction

In Chapter 2 we set forth three stress components for an infinitesimal interface: namely, a normal stress and two shear stresses. Additionally, we introduced notation that permitted the easy identification of nine stresses for three orthogonal interfaces at a point parallel to the planes of a reference xyz. In Example 2.6, we considered a one-dimensional stress problem and showed how stress varied as we changed the orientation of cuts for a tensile member. We then pointed out that knowing nine stresses for the three orthogonal interfaces at a point mentioned above, we could determine three stresses on interfaces having any arbitrary orientation at the point. In this chapter we present the formulations for carrying this out. In doing so, we shall generalize the concepts presented in Chapter 7, where we considered plane stress. At that time we presented the two-dimensional transformation equations showing how stress varied as we rotated axes at a point thus relating two-dimensional stress components τ_{xx}, τ_{yy}, and τ_{xy} for axes xy with stress components $\tau_{x'x'}$, $\tau_{y'y'}$, and $\tau_{x'y'}$ for rotated reference $x'y'$. At that time we introduced the concept of the Cartesian second-order tensor. A second objective of this chapter is then to present the second-order tensor more fully.

We accordingly take the critical step of finding the three-dimensional stress transformation equation in the next section.

15.2 Three-Dimensional Transformation Formulations for Stress

Consider a point O somewhere inside a solid body (see Fig. 15.1). A reference xyz is shown at this point. Imagine an infinitesimal interface

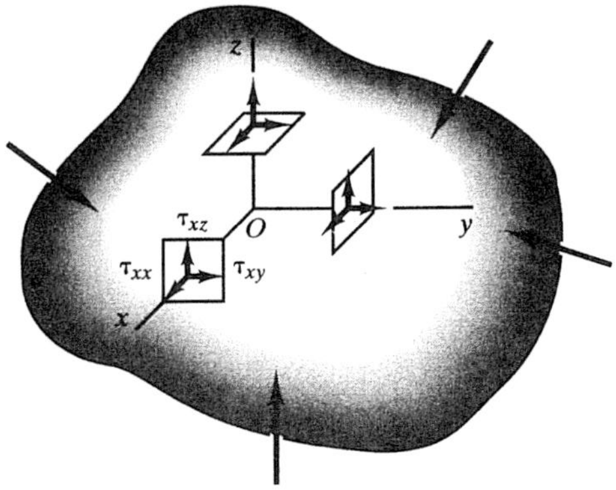

Figure 15.1. Orthogonal interfaces extracted from point O in a body.

at O with a normal in the x direction. Consider stresses τ_{xx}, τ_{yy}, and τ_{xz} on this interface parallel to the axes xyz.

Now move the interface away from point O along the x axis for convenience in viewing, as has been shown in Fig. 15.1. Do the same for interfaces in the y and z directions, as has been shown in the diagram. We will prove that if we know nine such stresses at a point, we can readily determine normal and shear stresses on an interface at point O having any orientation relative to axes xyz (see Fig. 15.2). It is convenient

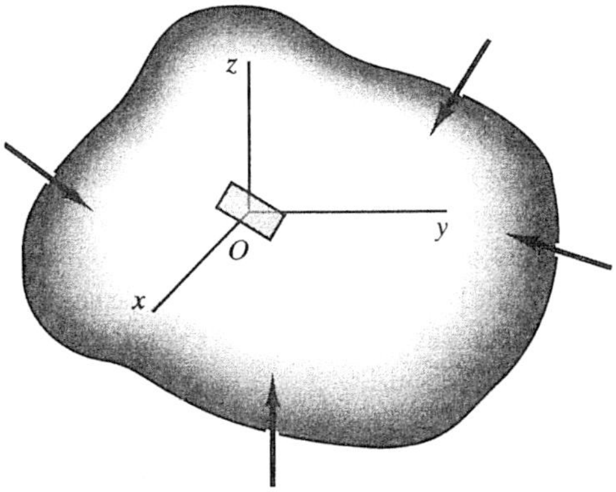

Figure 15.2. Arbitrary interface at O.

to utilize a vanishingly small rectangular parallelepiped of the body at point O to generate the three aforementioned orthogonal interfaces at point O, as has been shown in Fig. 15.3. Additionally, we can use the back faces of an infinitesimal tetrahedron (see Fig. 15.4) for generating

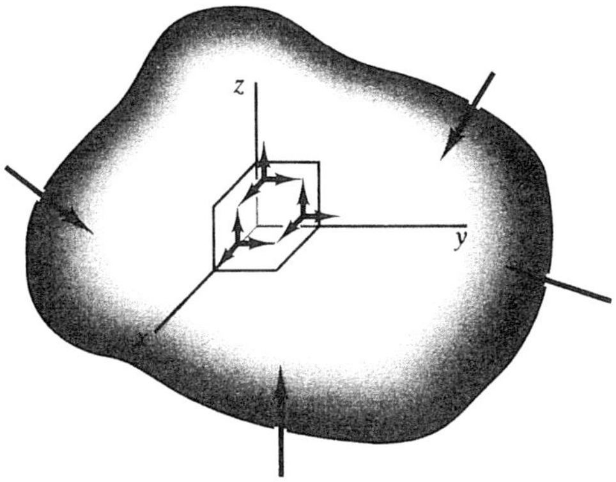

Figure 15.3. Vanishingly small rectangular parallelepiped generating three orthogonal interfaces at O.

the three orthogonal interfaces at point O. We then have the added benefit from this tetrahedron that we can give the front face $\overline{ABC}$ any orientation we please. By using **Newton's law** in the n and s directions, normal and tangential respectively to the interface, we can determine

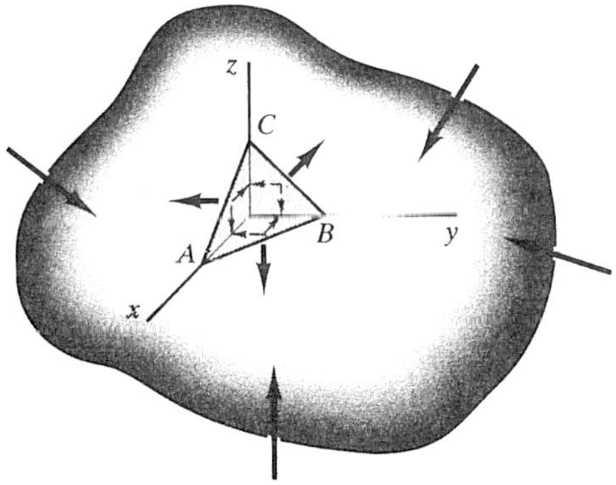

Figure 15.4. Tetrahedron at O generating three orthogonal interfaces at O and face $\overline{ABC}$ with arbitrary orientation.

the stresses τ_{nn} and τ_{ns} on $\overline{ABC}$ in terms of the nine stresses on the back faces of the tetrahedron (see Fig. 15.5). Due to the infinitesimal size of the tetrahedron, stresses τ_{nn} and τ_{ns} may be considered equal up to a differential increment to the corresponding stresses on an interface parallel to $\overline{ABC}$ directly at point O.

In using Newton's law for the tetrahedron, note that we make use of the fact that the angle between two planes (see Fig. 15.6) equals the

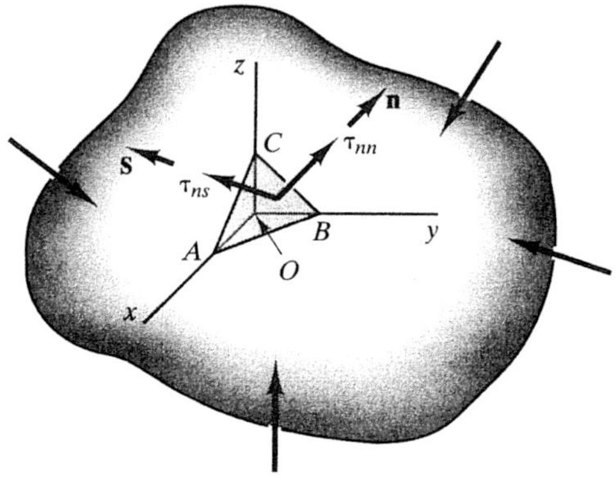

Figure 15.5. Normal and shear stresses on ABC.

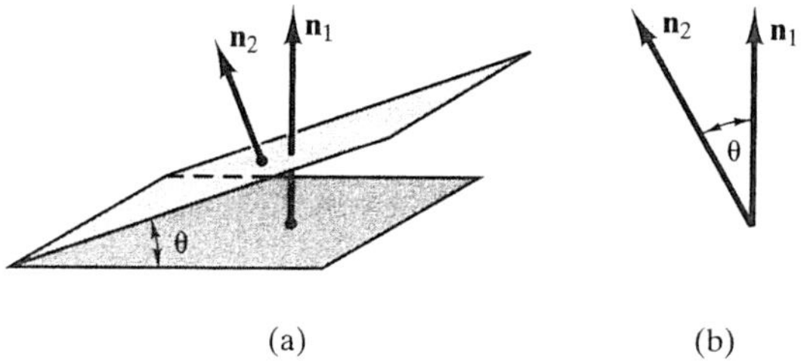

Figure 15.6. Angle θ between two planes.

angle between the normals to the planes. Hence the angle between interface $\overline{ABC}$ and the xz plane (see Fig. 15.5) is the angle between **n** and **j**. The cosine of this angle then is expressible as a_{ny}. The projection of $\overline{ABC}$ onto the xz plane (to form $\overline{AOC}$) is clearly $\overline{ABC}\, a_{ny}$.[1] Similar reasoning is used for the projections of $\overline{ABC}$ onto the other coordinate planes. The area vector components of the *back sides* of the tetrahedron can then be given as follows:

$$\begin{aligned}\overline{AOC} &= -\overline{ABC}\, a_{ny} \\ \overline{BOA} &= -\overline{ABC}\, a_{nz} \\ \overline{COB} &= -\overline{ABC}\, a_{nx}\end{aligned} \tag{15.1}$$

Clearly, a_{nx}, a_{ny}, and a_{nz} are the *direction cosines* of **n** for the x, y, and z axes, respectively.

[1]To see this, project the vector representation of area ABC onto the y axis along which the unit vector normal to the xz plane is **j**.

With this background, we will next consider the small but finite tetrahedron in Fig. 15.7 whose edges are of length Δx, Δy, and Δz,

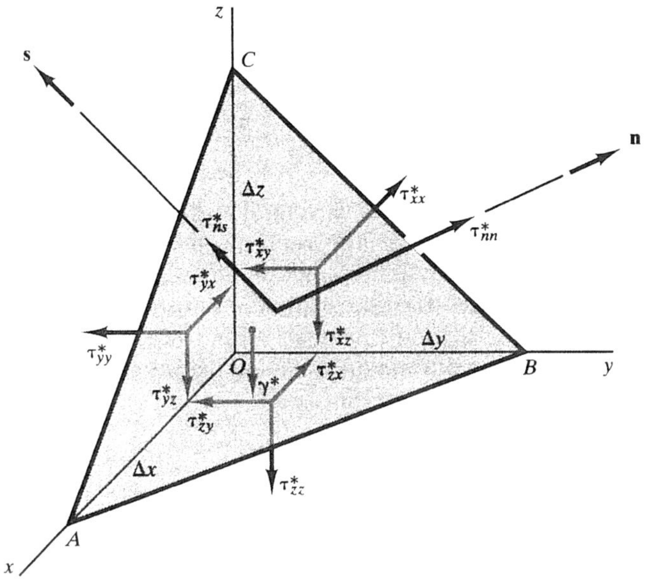

Figure 15.7. Small but finite tetrahedron with average stresses on faces.

respectively. On the back faces, we have shown a set of nine positive average stresses (the asterisk will indicate an average value). As before, the face ABC may have any inclination **n** and has an average normal stress τ^*_{nn} acting on it. An average shear stress τ^*_{ns} in the arbitrary direction **s** in the plane of $\overline{ABC}$ is also shown.

Next we write **Newton's law** in the direction of **n.** Thus, using the *magnitudes* of the area vector components of the back faces of the tetrahedron, we obtain

$$\begin{aligned}\tau^*_{nn}\overline{ABC} &- \tau^*_{xx}|\overline{COB}|a_{nx} - \tau^*_{xy}|\overline{COB}|a_{ny} - \tau^*_{xz}|\overline{COB}|a_{nz} - \\ &\tau^*_{yx}|\overline{AOC}|a_{nx} - \tau^*_{yy}|\overline{AOC}|a_{ny} - \tau^*_{yz}|\overline{AOC}|a_{nz} - \\ &\tau^*_{zx}|\overline{BOA}|a_{nx} - \tau^*_{zy}|\overline{BOA}|a_{ny} - \tau^*_{zz}|\overline{BOA}|a_{nz} - \\ &\gamma^*\frac{\Delta x\Delta y\Delta z}{6}a_{nz} = \rho^*\frac{\Delta x\Delta y\Delta z}{6}\alpha_n\end{aligned} \tag{15.2}$$

where α_n is the acceleration in the direction **n**, γ^* is the average specific weight, and ρ^* is the average density.

We now replace $|\overline{AOC}|$, $|\overline{BOA}|$, and $|\overline{COB}|$, using Eq. (15.1) with minus sign deleted. Next, we divide through by $\overline{ABC}$ to get

$$\begin{aligned}\tau_{nn}^* - \tau_{xx}^* a_{nx}^2 \quad &- \tau_{xy}^* a_{nx} a_{ny} - \tau_{xz}^* a_{nx} a_{nz} - \\ \tau_{yx}^* a_{ny} a_{nx} - \tau_{yy}^* a_{ny}^2 \quad &- \tau_{yz}^* a_{ny} a_{nz} - \\ \tau_{zx}^* a_{nz} a_{nx} - \tau_{zy}^* a_{nz} a_{ny} &- \tau_{zz}^* a_{nz}^2 \quad - \\ \frac{\gamma^*}{6} \frac{\Delta x \Delta y \Delta z}{\overline{ABC}} a_{nz} &= \frac{\rho^*}{6} \frac{\Delta x \Delta y \Delta z}{\overline{ABC}} \alpha_n\end{aligned}$$

Finally, we take the limit of each term as the quantities Δx, Δy, and Δz approach zero. Clearly, the last two terms then vanish, since $\overline{ABC}$ is of the order of magnitude of a product of *two* of the terms Δx, Δy, and Δz leaving in both cases after cancelation a Δ term which subsequently vanishes in the limit. Also, the average values of stress becomes exact values at a point (thus allowing us to delete the asterisks). For τ_{nn} at a point on rearranging terms[2] we thus have

$$\begin{aligned}\tau_{nn} = \tau_{xx} a_{nx}^2 \quad &+ \tau_{xy} a_{nx} a_{ny} + \tau_{xz} a_{nx} a_{nz} + \\ \tau_{yx} a_{ny} a_{nx} + \tau_{yy} a_{ny}^2 \quad &+ \tau_{yz} a_{ny} a_{nz} + \\ \tau_{zx} a_{nz} a_{nx} + \tau_{zy} a_{nz} a_{ny} &+ \tau_{zz} a_{nz}^2\end{aligned} \tag{15.3}$$

We see that the normal stress on *any* interface at a point depends only on the stresses on an orthogonal set of interfaces at the point as well as the direction cosines associated with the desired normal stress. The direction cosines are measured relative to coordinate axes parallel to the aforementioned set of orthogonal interfaces.

Before proceeding further, let us review what we have done. We have employed *Newton's law* on a body so shaped as to relate τ_{nn}^* for any direction **n** with the nine average stresses on orthogonal planes formed from xyz. Now in continuum mechanics Newton's law holds regardless of the size of the body, and so, keeping the orientation of the sides intact, we went to zero size. Equation (15.3) was thus reached, giving us the relation between a stress τ_{nn} on an interface oriented arbitrarily at a point in terms of the nine stresses at the point for three orthogonal interfaces of the xyz reference at the point. The tetrahedron was just a vehicle for accomplishing this via Newton's law.

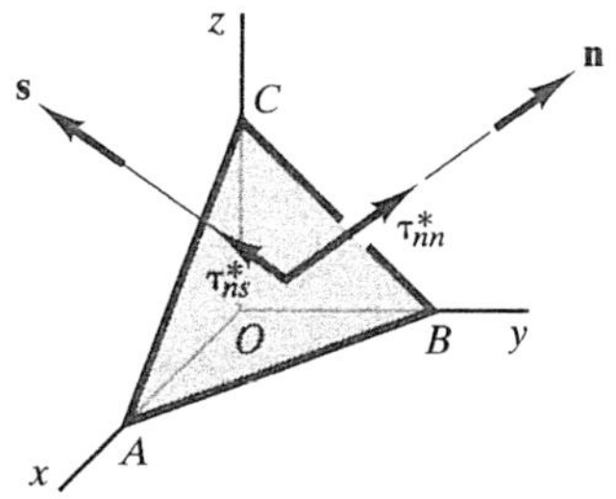

Figure 15.8. Average normal and shear stresses on face ABC.

We now proceed to compute shear stress in some chosen direction **s** on the inclined surface $\overline{ABC}$ of the tetrahedron by a computation similar to that performed for the normal stress. Accordingly, in Fig. 15.8 we have shown the tetrahedron with a shear stress τ_{ns}^* on the inclined face.

[2]Notice that the terms on the right-hand side of the equation have been put in the form of a matrix array of terms for ease in working with the equation.

The direction cosines of direction **s** are given as a_{sx}, a_{sy}, and a_{sz}. Since the normal-stress direction **n** and the shear-stress direction **s** are at right angles to each other, the following equation must be satisfied by the two corresponding sets of direction cosines:

$$a_{sx}a_{nx} + a_{sy}a_{ny} + a_{sz}a_{nz} = 0$$

If we now write Newton's law in the direction of the shear stress, we may proceed in a manner paralleling the development of Eq. (15.3) to form the following equation:

$$\begin{aligned}\tau_{ns} = \tau_{xx}a_{nx}a_{sx} + \tau_{xy}a_{nx}a_{sy} + \tau_{xz}a_{nx}a_{sz} +\\ \tau_{yx}a_{ny}a_{sx} + \tau_{yy}a_{ny}a_{sy} + \tau_{yz}a_{ny}a_{sz} +\\ \tau_{zx}a_{nz}a_{sx} + \tau_{zy}a_{nz}a_{sy} + \tau_{zz}a_{nz}a_{sz}\end{aligned} \tag{15.5}$$

Equations (15.3) and (15.5) thus permit us to compute stresses on *any* given interface at a point provided that we know the nine stresses on an orthogonal set of interfaces at the point. [You should have no difficulty in deriving Eqs. (d) of Example 2.6 as well as Eqs. (7.4) and (7.5) from the equations above.]

Example 15.1

A uniform pressure of 500 psi is developed on faces *EGHF* and *ABCD* of the solid block shown in Fig. 15.9. Also, a uniform tensile force distribution (suction) is maintained on faces *GHCB* and *EFDA* having the value of 100 psi. What are the normal stress and shear stresses on the interface shown on the diagonal surface *GHDA* of the block?

It should be apparent that all interfaces parallel to any one face of the reference *xyz* have constant values of stress. We thus have here a case where the stresses for a given reference are constant throughout the body (i.e., the stress distribution is *uniform*). By inspection we can thus say that at *all points* in the body

$$\begin{aligned}\tau_{xx} &= -500 \text{ psi} & \tau_{xy} &= \tau_{yz} = \tau_{yx} = \tau_{zy}\\ \tau_{yy} &= 100 \text{ psi} & &= \tau_{xz} = \tau_{zx} = 0\\ \tau_{zz} &= 0\end{aligned} \tag{a}$$

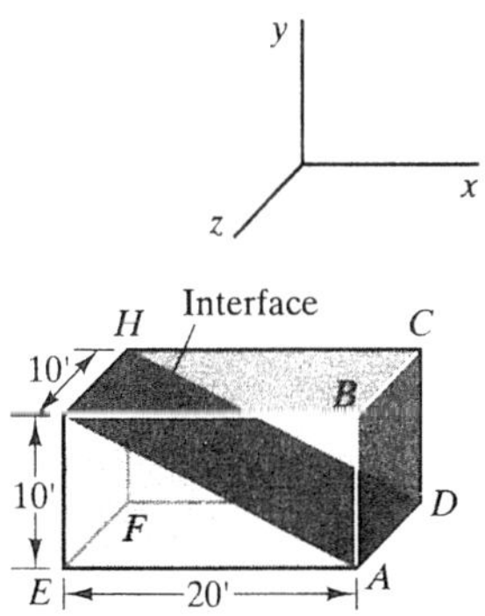

Figure 15.9. Uniform stresses in block.

Hence, at any and all interfaces on surface *GHDA* we have the same normal and shear stresses. Accordingly, we shall examine the entire surface shown on edge in Fig. 15.10. We first ascertain the direction cosines of the normal **n** to this surface. It is clear that

Example 15.1 (Continued)

$$a_{nx} = \cos\beta = \sin\alpha = \frac{10}{22.4} = .446$$
$$a_{ny} = \cos\alpha = \frac{20}{22.4} = .892 \tag{b}$$
$$a_{nz} = \cos 90^\circ = 0$$

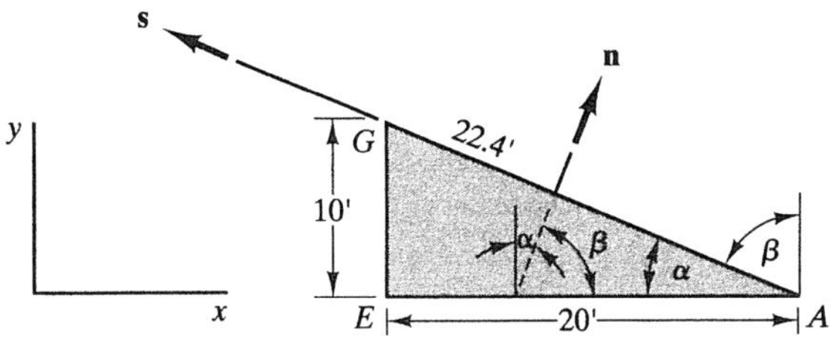

Figure 15.10. View of *GHDA* on edge.

We then have from Eq. (15.3) the following result for τ_{nn} at all points on *GHDA:*

$$\tau_{nn} = \tau_{xx}a_{nx}^2 + \tau_{yy}a_{ny}^2$$
$$= (-500)(.446)^2 + (100)(.892)^2 \tag{c}$$
$$\tau_{nn} = -19.89 \text{ psi}$$

As for the shear stress, we choose for **s** the direction along the inclined surface parallel to side AG (see Fig. 15.9). The direction cosines for **s** are then (see Fig. 15.10)

$$a_{sx} = -\cos\alpha = -.892$$
$$a_{sy} = \cos\beta = \sin\alpha = .446 \tag{d}$$
$$a_{sz} = \cos 90^\circ = 0$$

Hence, using Eq. (15.5) for τ_{ns} at all points on *GHDA,* we get

$$\tau_{ns} = \tau_{xx}a_{nx}a_{sx} + \tau_{yy}a_{ny}a_{sy}$$
$$= (-500)(.446)(-.892) + (100)(.892)(.446) \tag{e}$$
$$\tau_{ns} = 239 \text{ psi}$$

We leave it to you to show that the other shear stress at right angles to τ_{ns} on the interface is zero.

Example 15.1 (Continued)

Because the stresses are uniform, we may easily check the validity of the above results by considering as a free body *EFGHDA* as shown in Fig. 15.11. From **equilibrium** considerations we then have:

$\sum F_s = 0$:

$$-(500)[(144)(10)(10)]\cos\alpha - (100)[(144)(10)(20)]\sin\alpha + \tau_{ns}(144)(10)(22.4) = 0$$

$$\therefore \tau_{ns} = 239 \text{ psi}$$

$\sum F_n = 0$:

$$-(500)[(144)(10)(10)]\sin\alpha - (100)[(144)(10)(20)]\cos\alpha + \tau_{nn}(144)(10)(22.4) = 0$$

$$\therefore \tau_{nn} = -19.89 \text{ psi}$$

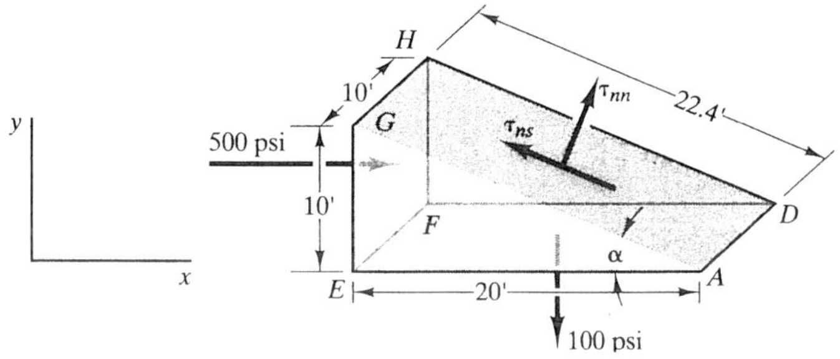

Figure 15.11. Prismatic body showing desired stresses.

Example 15.2

Analysis of a particular body indicates that stresses for orthogonal interfaces associated with reference *xyz* at a given point are, in kPa,

$$\begin{array}{lll} \tau_{xx} = 3000 \text{ kPa} & \tau_{xy} = -1000 \text{ kPa} & \tau_{xz} = 0 \text{ kPa} \\ \tau_{yx} = -1000 \text{ kPa} & \tau_{yy} = 2000 \text{ kPa} & \tau_{yz} = 2000 \text{ kPa} \\ \tau_{zx} = 0 \text{ kPa} & \tau_{zy} = 2000 \text{ kPa} & \tau_{zz} = 0 \text{ kPa} \end{array}$$

Determine the normal stress τ_{nn} on the infinitesimal interface at this point whose unit normal is

$$\mathbf{n} = .60\mathbf{j} + .80\mathbf{k}$$

Example 15.2 (Continued)

Also determine the shear stress τ_{ns} on the same interface in a direction parallel to the x axis.

The direction cosines for **n** and **s** are

$$\begin{aligned} a_{nx} &= 0 & a_{sx} &= 1 \\ a_{ny} &= .60 & a_{sy} &= 0 \\ a_{nz} &= .80 & a_{sz} &= 0 \end{aligned}$$

Using Eq. (15.3), we get

$$\begin{aligned} \tau_{nn} = {} & (3000)(0^2) + (-1000)(0)(.60) + (0)(0)(.80) + \\ & (-1000)(.60)(0) + (2000)(.60)^2 + (2000)(.60)(.80) + \\ & (0)(.80)(0) + (2000)(.80)(.60) + (0)(.80)^2 \end{aligned}$$

$$\therefore \tau_{nn} = 2640 \text{ kPa}$$

Similarly, using Eq. (15.5), we have

$$\begin{aligned} \tau_{ns} = {} & (3000)(0)(1) + (-1000)(0)(0) + (0)(0)(0) + \\ & (-1000)(.60)(1) + (2000)(.60)(0) + (2000)(.60)(0) + \\ & (0)(.80)(1) + (2000)(.80)(0) + (0)(.80)(0) \end{aligned}$$

$$\therefore \tau_{ns} = -600 \text{ kPa}$$

Instead of using n to denote the direction of an interface we shall use x' [see Fig. 15.12(a)]. Then Eq. (15.3) can be used to give $\tau_{x'x'}$ by replacing n by x' as follows:

$$\begin{aligned} \tau_{x'x'} = {} & \tau_{xx} a_{x'x}^2 + \tau_{xy} a_{x'x} a_{x'y} + \tau_{xz} a_{x'x} a_{x'z} + \\ & \tau_{yx} a_{x'y} a_{x'x} + \tau_{yy} a_{x'y}^2 + \tau_{yz} a_{x'y} a_{x'z} + \\ & \tau_{zx} a_{x'z} a_{x'x} + \tau_{zy} a_{x'z} a_{x'y} + \tau_{zz} a_{x'z}^2 \end{aligned} \tag{15.6a}$$

where $a_{x'x}$ is the direction cosine between the x' axis and x axis, and so on. Similarly, for a direction y' normal to x' at the point [see Fig. 15.12(b)], we get from Eq. (15.3)

$$\begin{aligned} \tau_{y'y'} = {} & \tau_{xx} a_{y'x}^2 + \tau_{xy} a_{y'x} a_{y'y} + \tau_{xz} a_{y'x} a_{y'z} + \\ & \tau_{yx} a_{y'y} a_{y'x} + \tau_{yy} a_{y'y}^2 + \tau_{yz} a_{y'y} a_{y'z} + \\ & \tau_{zx} a_{y'z} a_{y'x} + \tau_{zy} a_{y'z} a_{y'y} + \tau_{zz} a_{y'z}^2 \end{aligned} \tag{15.6b}$$

Finally, for a direction z' forming the third axis of $x'y'z'$ at the point [see Fig. 15.12(c)], we get from Eq. (15.3)

$$\begin{aligned}\tau_{z'z'} = \; & \tau_{xx}a_{z'x}^2 + \tau_{xy}a_{z'x}a_{z'y} + \tau_{xz}a_{z'x}a_{z'z} + \\ & \tau_{yx}a_{z'y}a_{z'x} + \tau_{yy}a_{z'y}^2 + \tau_{yz}a_{z'y}a_{z'z} + \\ & \tau_{zx}a_{z'z}a_{z'x} + \tau_{zy}a_{z'z}a_{z'y} + \tau_{zz}a_{z'z}^2\end{aligned} \tag{15.6c}$$

Thus, we have found the formulations above for $\tau_{x'x'}$, $\tau_{y'y'}$, and $\tau_{z'z'}$ directly from Eq. (15.3) by letting n take on successively the identity x', y', and z', and so Eq. (15.3) may be considered the seminal equation for generating normal stress for orthogonal interfaces corresponding to a reference $x'y'z'$ rotated relative to xyz at a point.

In a similar manner, we have shown axes x' and z' in Fig. 15.13(a). The shear stress $\tau_{x'z'}$ can be found from Eq. (15.5) in the manner presented earlier. Thus, we have:

$$\begin{aligned}\tau_{x'z'} = \; & \tau_{xx}a_{x'x}a_{z'x} + \tau_{xy}a_{x'x}a_{z'y} + \tau_{xz}a_{x'x}a_{z'z} + \\ & \tau_{yx}a_{x'y}a_{z'x} + \tau_{yy}a_{x'y}a_{z'y} + \tau_{yz}a_{x'y}a_{z'z} + \\ & \tau_{zx}a_{x'z}a_{z'x} + \tau_{zy}a_{x'z}a_{z'y} + \tau_{zz}a_{x'z}a_{z'z}\end{aligned} \tag{15.7}$$

The shear stress $\tau_{x'z'}$ has been shown in Fig. 15.13(b) by means of a vanishingly small parallelepiped that pictures the interface. By the procedures described above, we can ascertain other shear stresses $\tau_{z'x'}$, $\tau_{x'y'}$, $\tau_{y'x'}$, $\tau_{y'z'}$, and $\tau_{z'y'}$ for interfaces associated with axes $x'y'z'$. All of these shear stresses, then, can be directly expressed by utilizing Eq. (15.5) with n and s replaced by the appropriate primed coordinates. Thus, Eq. (15.5) is the parent equation for generating shear stresses for a reference $x'y'z'$ rotated relative to xyz at a point.[3]

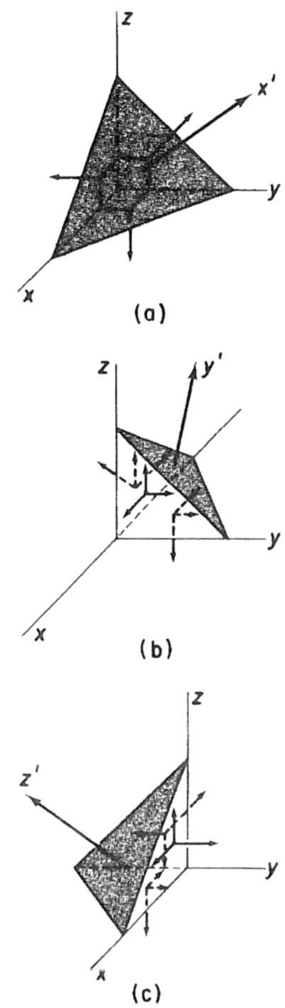

Figure 15.12. Introduction of primed axes.

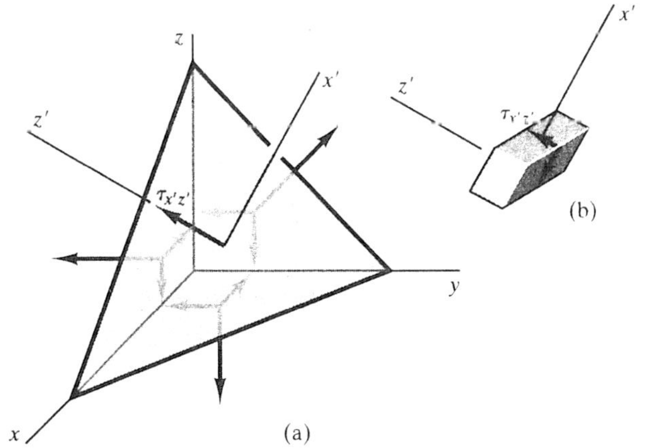

Figure 15.13. Primed axes $x'z'$ with corresponding shear $\tau_{x'z'}$.

[3]Actually, we need use only Eq. (15.5) to generate all nine stresses for the primed axes, since we can let *both* n and s become x' or y' or z' and thereby generate the normal stresses for x', y', z'. We need not further bother with Eq. (15.3) for this purpose.

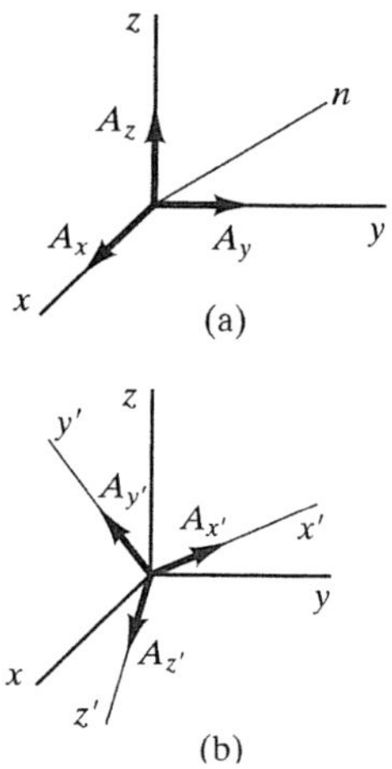

Figure 15.14. Components of A for primed and unprimed axes.

Thus, knowing nine stress components for a given reference xyz, we can compute nine stress components for a reference $x'y'z'$ rotated relative to xyz at a point by making appropriate use of the transformation equations [Eqs. (15.3) and (15.5)].

To better visualize the process we have been considering, we will go through a parallel development with the three components of a vector $\boldsymbol{A}$ (see Fig. 15.14). We can find the component A_n for $\mathbf{A}$, where $\mathbf{n}$ is any given direction, in the following manner:

$$\begin{aligned} A_n &= \mathbf{A} \cdot \mathbf{n} = A_x \mathbf{i} \cdot \mathbf{n} + A_y \mathbf{j} \cdot \mathbf{n} + A_z \mathbf{k} \cdot \mathbf{n} \\ \therefore A_n &= A_x a_{nx} + A_y a_{ny} + A_z a_{nz} \end{aligned} \tag{15.8}$$

If now the n direction corresponds to an x' axis [see Fig. 15.14(b)], then we replace n by x' above to get

$$A_{x'} = A_x a_{x'x} + A_y a_{x'y} + A_z a_{x'z} \tag{15.9a}$$

Similarly, for y' and z' we have the following results:

$$A_{y'} = A_x a_{y'x} + A_y a_{y'y} + A_z a_{y'z} \tag{15.9b}$$

$$A_{z'} = A_x a_{z'x} + A_y a_{z'y} + A_z a_{z'z} \tag{15.9c}$$

Thus, Eqs. (15.9a)–(16.9c) are readily established from Eq. (15.8) by replacing n by x', y', and z', respectively. Thus, Eq. (15.8) is the parent equation for the computation of three new components of a vector for a reference $x'y'z'$ in terms of the three components for reference xyz, where $x'y'z'$ is rotated relative to xyz. Equation (15.2) is accordingly the transformation equation for vectors, and, since it incorporates the parallelogram law and all the properties of vectors, *this equation fully characterizes vectors.* That is, we could define a vector as *one whose components transform in accordance with Eq. (15.8) under a rotation of axes.*

In the case of stress we have nine components that transform in accordance with certain equations that have some resemblance to the simpler transformation equation for vectors. We call the nine quantities which transform in accordance with Eq. (15.5) *second-order tensors.*[4]

In addition to the stress tensor, we shall also find in the following chapter that strain is a second-order tensor. And in the study of rigid-body dynamics one deals with the inertia tensor. Because these quantities transform in a certain way when we rotate coordinate axes at a point, they have certain distinct characteristics which set them apart from other quantities. Moreover, these characteristics are deducible from the transformation equations.

[4]Actually, in a more general context, vectors are first-order tensors and scalars are zero-order tensors.

Example 15.3

The stress tensor at a point in a body is given as follows:

$$\tau_{ij} = \begin{pmatrix} -500 & 0 & -100 \\ 0 & 0 & 0 \\ -100 & 0 & 800 \end{pmatrix} \text{ kPa}$$

for a given reference xyz. A reference $x'y'z'$ is formed by rotating the references about the z axis an angle α, as shown in Fig. 15.5(a). What are the components of stress for the primed reference?

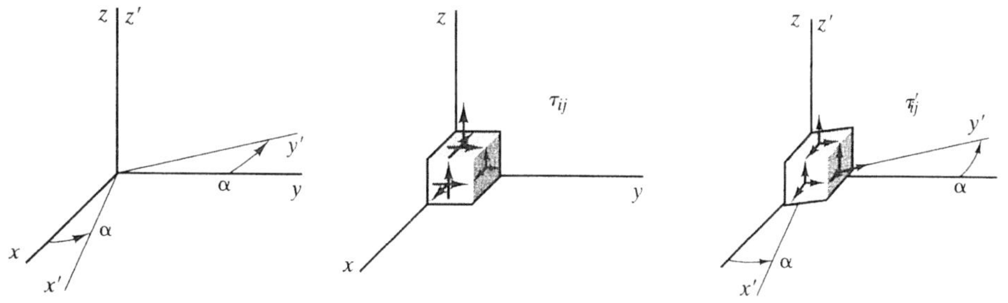

(a)

$$\tau_{ij} = \begin{pmatrix} -500 & 0 & -100 \\ 0 & 0 & 0 \\ -100 & 0 & 800 \end{pmatrix} \text{ KPa} \qquad \tau'_{ij} = \begin{pmatrix} -375 & 216 & -86.6 \\ 216 & -125 & 50 \\ -86.6 & 50 & 800 \end{pmatrix} \text{ KPa}$$

Figure 15.15. Transformation of stress on rotation of axes.

We first need the direction cosines positioning the primed axes relative to the unprimed axes. Thus, we have for $\alpha = 30°$

$$\begin{aligned} a_{x'x} &= \cos 30° = .866 & a_{y'x} &= \cos 120° = -.500 & a_{z'x} &= \cos 90° = 0 \\ a_{x'y} &= \cos 60° = .500 & a_{y'y} &= \cos 30° = .866 & a_{z'y} &= \cos 90° = 0 \\ a_{x'z} &= \cos 90° = 0 & a_{y'z} &= \cos 90° = 0 & a_{z'z} &= \cos 0° = 1 \end{aligned}$$

We can now get the stress components in the primed reference by using the transformation equations. Thus,

$$\begin{aligned} \tau_{x'x'} &= (-500)(.866)^2 + (-100)(0)(.866) + (-100)(.866)(0) + (800)(0)^2 \\ &= -375 \text{ kPa} \\ \tau_{y'y'} &= (-500)(-.500)^2 + (-100)(0)(-.500) + (-100)(.500)(0) + (800)(0)^2 \\ &= -125 \text{ kPa} \\ \tau_{z'z'} &= (-500)(0)^2 + (-100)(1)(0) + (-100)(0)(1) + (800)(1)^2 \\ &= 800 \text{ kPa} \\ \tau_{x'z'} &= (-500)(.866)(0) + (-100)(0)(0) + (-100)(.866)(1) + (800)(0)(1) \\ &= -86.6 \text{ kPa} \end{aligned}$$

Example 15.3 (Continued)

$$\tau_{z'x'} = (-500)(0)(.866) + (-100)(0)(0) + (-100)(1)(.866) + (800)(1)(0)$$
$$= -86.6 \text{ kPa}$$
$$\tau_{y'z'} = (-500)(-.500)(0) + (-100)(0)(0) + (-100)(-.500)(1) + (800)(0)(1)$$
$$= 50.0 \text{ kPa}$$
$$\tau_{z'y'} = (-500)(0)(-.500) + (-100)(0)(0) + (-100)(1)(-.500) + (800)(1)(0)$$
$$= 50.0 \text{ kPa}$$
$$\tau_{x'y'} = (-500)(.866)(-.500) + (-100)(0)(-.500) + (-100)(.866)(0) + (800)(0)(0)$$
$$= 216 \text{ kPa}$$
$$\tau_{y'x'} = (-500)(-.500)(.866) + (-100)(-.500)(0) + (-100)(0)(.866) + (800)(0)(0)$$
$$= 216 \text{ kPa}$$

The stress tensor for the primed reference is then given as follows:

$$\tau'_{ij} = \begin{pmatrix} -375 & 216 & -86.6 \\ 216 & -125 & 50 \\ -86.6 & 50 & 800 \end{pmatrix} \text{kPa}$$

We have shown in Fig. 15.15(b) the nine stresses on three orthogonal interfaces corresponding to reference xyz by using the usual infinitesimal rectangular parallelepiped to generate the desired interfaces at a point. Similarly, the nine stresses on three orthogonal interfaces for reference $x'y'z'$ have also been shown in part (c) of the diagram.

Example 15.4

In Fig. 15.16, we have a case of *uniform* stress wherein the stress tensor is given as

$$\begin{bmatrix} 0 & 3000 & 0 \\ 3000 & -4000 & 0 \\ 0 & 0 & 5000 \end{bmatrix} \text{kPa}$$

(a). What is the shear stress on face ABC along the edge AC?
(b). What is the shear stress on face $DAEC$, also along edge AC?

For both parts of the problem we shall need the displacement vector $\overrightarrow{AC}$. For this purpose note that

$$r_A = 1\mathbf{i} + 3\mathbf{k} + 1\mathbf{j} \text{ mm}$$
$$r_C = 2\mathbf{j} + 3\mathbf{k} \text{ mm}$$

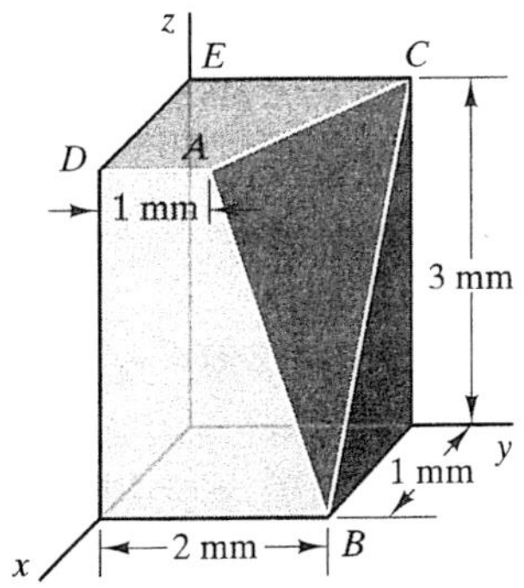

Figure 15.16. A block having a uniform state of stress.

Example 15.4 (Continued)

Hence we can say

$$\overrightarrow{AC} = \mathbf{r}_C - \mathbf{r}_A = (\mathbf{j} - \mathbf{i}) \text{ mm}$$

and that

$$\boldsymbol{\varepsilon}_{AC} = \frac{\mathbf{j} - \mathbf{i}}{(1^2 + 1^2)^{\frac{1}{2}}} = \frac{1}{\sqrt{2}}(\mathbf{j} - \mathbf{i})$$

Therefore, the direction cosines for the above unit vector are

$$a_{sx} = -\frac{1}{\sqrt{2}} = -.707$$

$$a_{sy} = \frac{1}{\sqrt{2}} = .707$$

$$a_{sz} = 0$$

For Part (a) we will need the unit vector **n** normal to face ABC. Thus

$$\mathbf{n} = \frac{\overrightarrow{AB} \times \overrightarrow{AC}}{|\overrightarrow{AB} \times \overrightarrow{AC}|}$$

$$\overrightarrow{AB} \times \overrightarrow{AC} = (1\mathbf{j} - 3\mathbf{k}) \times (\mathbf{j} - \mathbf{i}) = \mathbf{k} + 3\mathbf{i} + 3\mathbf{j}$$

Therefore

$$\mathbf{n} = \frac{\mathbf{k} + 3\mathbf{i} + 3\mathbf{j}}{\sqrt{1 + 9 + 9}} = .2294\mathbf{k} + .6882\mathbf{i} + .6882\mathbf{j}$$

The corresponding direction cosines are

$$a_{nx} = .6882$$

$$a_{ny} = .6882$$

$$a_{nz} = .2294$$

We are now ready to give the shear stress on face ABC along edge AC. Using the basic transformation equation we have

$$\begin{aligned}\tau_{ns} = {} & (0)(a_{nx}a_{sx}) + (3000)(a_{nx}a_{sy}) + (0)(a_{nx}a_{sz}) + \\ & (3000)(a_{ny}a_{sx}) + (-4000)(a_{ny}a_{sy}) + (0)(a_{ny}a_{sz}) + \\ & (0)(a_{nz}a_{sx}) + (0)(a_{nz}a_{sy}) + (5000)(a_{nz}a_{sz})\end{aligned}$$

Inserting values we get

Example 15.4 (Continued)

$$
\begin{aligned}
(\tau_{ns})_1 = \quad & 0 \quad + \quad (3000)(.6892)(.707) \quad + \quad 0 \quad + \\
& (3000)(.6882)(-.707) \quad + (-4000)(.6882)(.707) \quad + \quad 0 \quad + \\
& 0 \quad + \quad 0 \quad + \quad (5000)(.2294)(0)
\end{aligned}
$$

$$(\tau_{ns})_1 = -1.946 \times 10^3 \text{ Pa}$$

Next we go to Part (b) for the shear stress on face $ACDE$ along edge AC. The direction cosines for this shear stress are

$$
\begin{aligned}
a_{nx} &= 0 & a_{sx} &= -.707 \\
a_{ny} &= 0 & a_{sy} &= .707 \\
a_{nz} &= 1 & a_{sz} &= 0
\end{aligned}
$$

Now going again to the transformation equation we get the value 0 as you should verify yourself.[5]

We have shown in Chapter 7 where we considered plane stress, or in other words, the "two-dimensional" stress tensor, that just from the transformation equations on rotating axes we were able to generate such vital concepts as

Principal stresses
Principal axes
Mohr's circle
Certain invariants

We concluded that any set of four quantities A_{xx}, A_{yy}, A_{xy}, A_{yx}, where $A_{yx} = A_{xy}$, having the same aforementioned transformation equations must then possess the above-mentioned characteristics. Because these characteristics are so vital in many areas of engineering science, we have given these quantities a special identification, calling them components of a tensor and study them as an entity. In later sections we shall be able to expand on the properties of second-order tensors for the full three-dimensional case using stress as our vehicle.

[5]Although coming late in the course, this example distinctly shows how along a common line we have two distinct interfaces with different stresses along this line.

*15.3 Principal Stresses for a General State of Stress

We have pointed out in Section 7.5 that there exist three mutually perpendicular axes at a point on whose reference planes there is zero shear stress. On one of these faces the normal stress is the maximum normal stress at the point. These axes are the principal axes. We have shown how we can determine the orientation of principal axes as well as the corresponding stresses for the case of plane stress. We shall now consider the computation of principal stresses and the orientation of principal axes for a general state of stress.

To compute the principal stresses and principal axes, we now refer you to Fig. 15.17, which depicts a tetrahedron bounded by the coordinate planes and an oblique plane $\overline{ABC}$ having a normal in the direction **n.** We shall assign $\overline{ABC}$ the role of being a principal plane in the ensuing calculations by having a normal stress τ as the *only* nonzero stress on this plane. The principal axis corresponding to such an interface then has direction cosines a_{nx}, a_{ny}, and a_{nz}. We shall denote the area of $\overline{ABC}$ as ΔA, and the outside faces of the tetrahedron having normals in the coordinate directions will have areas ΔA_x, ΔA_y, and ΔA_z, respectively. As in previous considerations of the tetrahedron we can relate the aforementioned areas as follows:

$$\begin{aligned}\Delta A_x &= -\,\Delta A\, a_{nx}\\ \Delta A_y &= -\,\Delta A\, a_{ny}\\ \Delta A_z &= -\,\Delta A\, a_{nz}\end{aligned} \tag{15.10}$$

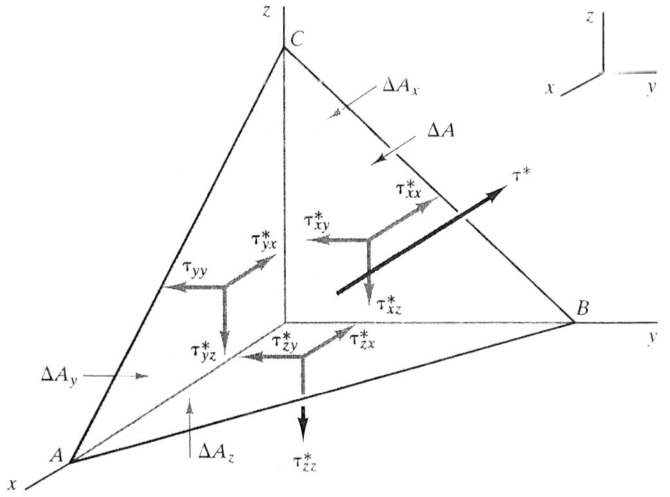

Figure 15.17. Tetrahedron face ABC has only normal stress.

We now consider **Newton's law**, this time in the x, y and z directions, respectively.

Using B_x, B_y, and B_z to be body force components and a_x, a_y, and a_z to be accelerate components we have

$$\tau^*\Delta A\, a_{nx} - \tau^*_{xx}|\Delta A_x| - \tau^*_{yx}|\Delta A_y| - \tau^*_{zx}|\Delta A_z| + B^*_x \frac{\Delta x \Delta y \Delta z}{6} = \rho^* \frac{\Delta x \Delta y \Delta z}{6} a_x$$

$$\tau^*\Delta A\, a_{ny} - \tau^*_{xy}|\Delta A_x| - \tau^*_{yy}|\Delta A_y| - \tau^*_{zy}|\Delta A_z| + B^*_y \frac{\Delta x \Delta y \Delta z}{6} = \rho^* \frac{\Delta x \Delta y \Delta z}{6} a_y \tag{15.11}$$

$$\tau^*\Delta A\, a_{nz} - \tau^*_{xz}|\Delta A_x| - \tau^*_{yz}|\Delta A_y| - \tau^*_{zz}|\Delta A_z| + B^*_z \frac{\Delta x \Delta y \Delta z}{6} = \rho^* \frac{\Delta x \Delta y \Delta z}{6} a_z$$

where the asterisks indicate average values of stresses over the appropriate finite faces. Substituting from Eqs. (15.10) for ΔA_x, ΔA_y, and ΔA_z and then dividing through by ΔA, we may get the following results on multiplying by -1 and rearranging the terms:

$$(\tau^*_{xx} - \tau^*)a_{nx} + \tau^*_{yx}a_{ny} + \tau^*_{zx}a_{nz} - B^*_x \frac{\Delta x \Delta y \Delta z}{6\Delta A} = -\rho^* \frac{\Delta x \Delta y \Delta z}{6\Delta A} a_x$$

$$\tau^*_{xy}a_{nx} + (\tau^*_{yy} - \tau^*)a_{ny} + \tau^*_{zy}a_{nz} - B^*_y \frac{\Delta x \Delta y \Delta z}{6\Delta A} = -\rho^* \frac{\Delta x \Delta y \Delta z}{6\Delta A} a_y \tag{15.12}$$

$$\tau^*_{xz}a_{nx} + \tau^*_{yz}a_{ny} + (\tau^*_{zz} - \tau^*)a_{nz} - B^*_z \frac{\Delta x \Delta y \Delta z}{6\Delta A} = -\rho^* \frac{\Delta x \Delta y \Delta z}{6\Delta A} a_z$$

Now let ΔA approach zero. All the average stresses become exact stresses at a point, and body force and inertia expressions vanish, as explained in Section 15.2. We then have, using the complementary property of shear,

$$(\tau_{xx} - \tau)a_{nx} + \tau_{xy}a_{ny} + \tau_{xz}a_{nz} = 0$$

$$\tau_{yx}a_{nx} + (\tau_{yy} - \tau)a_{ny} + \tau_{yz}a_{nz} = 0 \tag{15.13}$$

$$\tau_{zx}a_{nx} + \tau_{zy}a_{ny} + (\tau_{zz} - \tau)a_{nz} = 0$$

In addition, we have the following *geometric* relationship, which is always valid for a set of direction cosines:

$$a^2_{nx} + a^2_{ny} + a^2_{nz} = 1 \tag{15.14}$$

Having put τ in the role of a principal stress by having zero shear stresses on the corresponding interface, *Newton's law* and *geometry* [Eqs. (15.13) and (15.14), respectively] now *impose* certain conditions on τ and the corresponding direction cosines. In an effort to satisfy these equations we shall first consider Eqs. (15.13) to be a set of linear algebraic equations with a_{nx}, a_{ny}, and a_{nz} as the unknowns. Clearly, if these direction cosines are zero, Eqs. (15.13) are satisfied but then we violate Eq. (15.14). Therefore, this possibility (called a trivial solution in mathematical parlance) must be ruled out. We shall proceed to solve for a_{nx}, a_{ny}, and a_{nz} from Eqs. (15.13) by employing the well-known Cramer's rule as follows:

$$a_{nx} = \frac{\begin{vmatrix} 0 & \tau_{xy} & \tau_{xz} \\ 0 & \tau_{yy} - \tau & \tau_{yz} \\ 0 & \tau_{zy} & \tau_{zz} - \tau \end{vmatrix}}{\begin{vmatrix} \tau_{xx} - \tau & \tau_{xy} & \tau_{xz} \\ \tau_{yx} & \tau_{yy} - \tau & \tau_{yz} \\ \tau_{zx} & \tau_{zy} & \tau_{zz} - \tau \end{vmatrix}}$$

$$a_{ny} = \frac{\begin{vmatrix} \tau_{xx} - \tau & 0 & \tau_{xz} \\ \tau_{yx} & 0 & \tau_{yz} \\ \tau_{zx} & 0 & \tau_{zz} - \tau \end{vmatrix}}{\begin{vmatrix} \tau_{xx} - \tau & \tau_{xy} & \tau_{xz} \\ \tau_{yx} & \tau_{yy} - \tau & \tau_{yz} \\ \tau_{zx} & \tau_{zy} & \tau_{zz} - \tau \end{vmatrix}}$$

$$a_{nz} = \frac{\begin{vmatrix} \tau_{xx} - \tau & \tau_{xy} & 0 \\ \tau_{yx} & \tau_{yy} - \tau & 0 \\ \tau_{zx} & \tau_{zy} & 0 \end{vmatrix}}{\begin{vmatrix} \tau_{xx} - \tau & \tau_{xy} & \tau_{xz} \\ \tau_{yx} & \tau_{yy} - \tau & \tau_{yz} \\ \tau_{zx} & \tau_{zy} & \tau_{zz} - \tau \end{vmatrix}}$$

Upon inspection of the numerators above, we see that they are all zero. Thus, if the denominators (same for all direction cosines) are *other* than zero, we get the forbidden result of $a_{nx} = a_{ny} = a_{nz} = 0$. To avoid this dilemma, we conclude that the denominators must be zero. We conclude that a *necessary condition* for the satisfaction of *Newton's law* and *geometry* in this case is that

$$\begin{vmatrix} \tau_{xx} - \tau & \tau_{xy} & \tau_{xz} \\ \tau_{yx} & \tau_{yy} - \tau & \tau_{yz} \\ \tau_{zx} & \tau_{zy} & \tau_{zz} - \tau \end{vmatrix} = 0 \tag{15.15}$$

We expand the determinant in Eq. (15.15) to obtain, after grouping like powers of τ, the following cubic equation:

$$\tau^3 - A\tau^2 + B\tau - C = 0 \tag{15.16}$$

where

$$\begin{aligned} A &= \tau_{xx} + \tau_{yy} + \tau_{zz} \\ B &= \tau_{xx}\tau_{yy} + \tau_{yy}\tau_{zz} + \tau_{zz}\tau_{xx} - \tau_{xy}^2 - \tau_{yz}^2 - \tau_{zx}^2 \\ C &= \tau_{xx}\tau_{yy}\tau_{zz} + 2\tau_{xy}\tau_{yz}\tau_{xz} - \tau_{xy}^2\tau_{zz} - \tau_{yz}^2\tau_{xx} - \tau_{zx}^2\tau_{yy} \end{aligned} \tag{15.17}$$

When we have solved the cubic equation for the three values of τ we order them from the largest downward algebraically and call these roots τ_1, τ_2, and τ_3. (It can be shown that these stresses must be real.[6]) We see that Newton's law and geometry prescribe that, for the role we have assigned τ, three values are permitted for the given set of stresses τ_{ij} associated with reference xyz. These obviously are the three principal stresses.

To determine the direction of the maximum normal stress, we substitute τ_1 for τ in Eqs. (15.13). We shall find that one of the equations becomes redundant, but the other two equations, together with Eq. (15.14), enable us to solve for a_{nx}, a_{ny}, and a_{nz}. Thus, we have the direction of τ_1. This is one of the principal axes. The other axes are located by substituting τ_2 and τ_3 in a similar manner. The directions so established can be shown to be mutually orthogonal.[7]

Before continuing we shall illustrate in the following examples the computation that we have thus far outlined.

Example 15.5

Given the following state of stress, find the principal stresses and the principal axes:

$$\tau_{xx} = 200 \text{ psi} \quad \tau_{xy} = 100 \text{ psi} \quad \tau_{xz} = 300 \text{ psi} \quad \tau_{yz} = \tau_{yy} = \tau_{zz} = 0$$

Using Eqs. (15.17), we have

$$\begin{aligned} A &= 200 \\ B &= -100{,}000 \\ C &= 0 \end{aligned}$$

[6]See I. H. Shames and C. L. Dym, *Energy and Finite Element Methods in Structural Mechanics* (Taylor & Francis Co., 1997), Sec. 1.6.

[7]See the above footnote.

Example 15.5 (Continued)

Therefore, from Eq. (15.16)

$$\tau^3 - 200\tau^2 - 100{,}000\tau = 0$$

Factoring, we get

$$\tau(\tau^2 - 200\tau - 100{,}000) = 0$$
$$\tau(\tau - 432)(\tau + 232) = 0$$

The obvious roots, arranged in descending order, are

$$\tau_1 = 432 \text{ psi}$$
$$\tau_2 = 0 \text{ psi}$$
$$\tau_3 = -232 \text{ psi}$$

To find the direction of τ_1, substitute 432 for τ in Eqs. (15.13). Thus,

$$(200 - 432)a_{nx} + 100a_{ny} + 300a_{nz} = 0$$
$$100a_{nx} + (0 - 432)a_{ny} + 0 = 0$$
$$300a_{nx} + 0 + (0 - 432)a_{nz} = 0$$

We note that these equations are not all independent since $-.232$ times the second equation plus $-.696$ times the third equation equals the first equation. Hence, we may solve only for ratios. We may use any pair of these equations. Thus, from the second and third equations,

$$a_{ny} = .232a_{nx} \qquad \text{(a)}$$
$$a_{nz} = .695a_{nx} \qquad \text{(b)}$$

As a third relationship we use Eq. (15.14):

$$a_{nx}^2 + a_{ny}^2 + a_{nz}^2 = 1$$

Hence, substituting Eqs. (a) and (b) into Eq. (c) we get

$$a_{nx}^2 + (.232)^2 a_{nx}^2 + (.695)^2 a_{nx}^2 = 1$$
$$\therefore\ a_{nx} = \frac{1}{\sqrt{1 + .232^2 + .695^2}} = .806$$

Hence,

$$a_{ny} = .1870 \qquad a_{nz} = .560$$

Thus, the principal axis corresponding to the *largest principal stress* of 432 psi has a direction **n** given by

Example 15.5 (Continued)

$$\mathbf{n} = .806\mathbf{i} + .1870\,\mathbf{j} + .560\mathbf{k}$$

The directions of τ_2 and τ_3 are found in a similar manner by successively substituting 0 and −232 into Eqs. (15.13), etc.

Example 15.6

In Example 15.5 the cubic equation for principal stresses was easily factored, thus eliminating the task of formally solving the general cubic equation. As an example of another case where we can simplify the procedure, consider the following state of stress:

$$\tau_{ij} = \begin{pmatrix} 200 & 100 & 0 \\ 100 & 0 & 0 \\ 0 & 0 & 300 \end{pmatrix} \text{kPa}$$

Compute the principal stresses.

Here the stresses τ_{zx} and τ_{zy} are zero and so an interface with a normal in the z direction is devoid of shear. This means that the z direction must correspond to a principal axis and that one of the principal stresses is known—it is 300 kPa. To get the other principal stresses divide $(\tau - 300)$ into the left-hand side of the cubic equation for principal stress. We then get as a result a quadratic expression in τ, which may be set equal to zero to generate two more roots. We shall illustrate these steps. The cubic equation is

$$\tau^3 - 500\tau^2 + 50 \times 10^3\tau + 30 \times 10^5 = 0 \qquad \text{(a)}$$

Hence, by polynomial division

$$(\tau - 300)\overline{\big)\tau^3 - 500\tau^2 + 50 \times 10^3\tau + 30 \times 10^5}$$

gives

$$\tau^2 - 200\tau - 10{,}000$$

We can then express Eq. (a) as follows:

$$(\tau - 300)(\tau^2 - 200\tau - 10{,}000) = 0$$

Setting the quadratic expression equal to zero we get

$$\tau^2 - 200\tau - 10{,}000 = 0$$

Example 15.6 (Continued)

Using the quadratic formula we then get for roots two additional stresses τ as follows:

$$\tau = \frac{200 \pm \sqrt{200^2 - (4)(1)(-10{,}000)}}{2}$$
$$= 241.5; \quad -41.5$$

The principal stresses are then

$$\tau_1 = 300 \text{ kPa}$$
$$\tau_2 = 241.5 \text{ kPa}$$
$$\tau_3 = -41.5 \text{ kPa}$$

*15.4 Tensor Invariants

In finding principal stresses and principal axes, we have actually uncovered additional information of significance. We have shown early in the chapter that the normal stress on an interface δA at a point depends entirely on the nine stresses for a set of orthogonal interfaces at a point associated with a reference xyz. Clearly, if we choose to work with orthogonal interfaces at the point associated with some *other* reference $x'y'z'$, we must arrive at the *same* normal stress for the aforementioned interface δA. Accordingly, the principal stresses at a point are not dependent for their values on which particular reference we choose to use to express nine stresses on three orthogonal interfaces at a point. We must get the same results at a point no matter what the orientation of the reference used at the point. This means that all references at a point must yield for this point the very *same* cubic equation for τ. This in turn implies that the coefficients A, B, and C have the same values for all references used at the point. For example, considering a primed and an unprimed set of axes rotated relative to each other we can say that

$$A = A'$$
$$B = B'$$
$$C = C'$$

Another way of stating this conclusion is to say that A, B, and C are *invariant* with respect to a rotation of axes. We call A, B, and C, respectively, the *first, second,* and *third tensor invariants of stress.* Note that the first tensor invariant for stress is simply the generalization in three dimensions of the invariant presented earlier in the discussion of plane stress. Because natural phenomena proceed without regard to man-made coordinate systems and references, we might expect that invariants would have physical significance. This is true here. You will later learn that A is related to the pressure concept used in thermodynamics. We shall later allude to physical significance of B and C as well.[8]

The cubic equation that generated the three tensor invariants stemmed from Newton's law and geometry. But these are the very sources you will recall for the development of the transformation equations for stress that we identified as the defining equations for second-order tensors. Actually we could have arrived at the *very same cubic equations for* τ *by working directly from the transformation equations for stress. Accordingly, all the results of this section and the preceding section are valid for any second-order symmetric tensor* A_{ij}.

*15.5 A Look Ahead: Tensor Notation[9]

You will have noticed that Eqs. (15.3) and (15.5) had very orderly appearances in that the subscripts for the stresses ran from x to y to z downward for the first subscript and horizontally for the second subscript. Also, for the subscripts of the direction cosines, the respective second subscripts of each pair of a's had the same subscripts as the associated stress term. Because of this simplicity and regularity, one can make efficient use of a double index scheme for these subscripts, which is sometimes called Cartesian tensor notation. In so doing, we can better understand the relation between scalars, vectors, and second-order tensors. Also, we shall at the same time be able to readily extrapolate our thinking to include so-called higher-order tensors.

For this purpose we introduce the concept of the *free index*. The free index is any letter which appears only once as a subscript in a group of terms. Thus, i is the free index in the following expressions:

$$V_i; \quad A_{ij}V_j$$

When we have an expression with a free index such as V_i, we can consider that we are presenting *any one* component of the array of terms formed by having i become an x, a y, or a z. This array accordingly is

[8]In Problem 15.16 you are asked to show that B is the sum of minors of the left-to-right diagonal of the stress tensor, while C is the determinant of the stress tensor itself.

[9]See I. H. Shames and C. L. Dym, *Energy and Finite Element Methods in Structural Mechanics*. Taylor and Francis Corp., 1985, Appendix I, pages 709 to 727

$$\begin{pmatrix} V_x \\ V_y \\ V_z \end{pmatrix}$$

Alternatively, V_i can be interpreted to represent the entire *set* of components given above. Clearly, V_i may then be considered a vector since it includes the three rectangular components of the vector. The particular meaning of the free index depends on the *context* of the discussion.[10] Similarly, the expression A_{ij} has two free indices and can represent any one component of the array of terms formed from all possible permutations of the subscripts with i and j taking on the values x, y, and z. This array of terms must then be

$$\begin{pmatrix} A_{xx} & A_{xy} & A_{xz} \\ A_{yx} & A_{yy} & A_{yz} \\ A_{zz} & A_{zy} & A_{zz} \end{pmatrix}$$

Alternatively, the expression A_{ij} can represent the entire set of the above components. Clearly, A_{ij} could possibly in this context represent a second-order tensor.

Next we shall introduce the concept of *dummy indices* by prescribing that when letters i, j, k, l, and m are repeated in an expression we *sum* terms formed by letting the repeated indices take on successively the values x, y, and z. Thus, the expression $A_{ij}V_j$ used earlier may represent a set of three terms each of which is a sum of three expressions. Thus, retaining the free index i we have

$$A_{ij}V_j = A_{ix}V_x + A_{iy}V_y + A_{iz}V_z$$

If we wish to express the set for the free index i, we get

$$A_{ij}V_j \equiv \begin{pmatrix} (A_{xx}V_x + A_{xy}V_y + A_{xz}V_z) \\ (A_{yx}V_x + A_{yy}V_y + A_{yz}V_z) \\ (A_{zx}V_x + A_{zy}V_y + A_{zz}V_z) \end{pmatrix}$$

If we have a double set of repeated indices such as $A_{ij}B_{ij}$, we take on the value x and sum terms with j ranging over x, y, and z; then we add to this three terms found by letting i become y while letting j range again over x, y, and z; and so on. We thus get a *sum* of the nine terms. It should be clear that, since a pair of dummy indices always

[10]Actually, we used the concept of the free index when we used the letter n in Section 15.2 to represent successively x', y', and z' in connection with the development of the transformation equations of stress.

sums out, the particular letters of the set *ijklm* used is immaterial. We can accordingly say that $A_iB_i = A_jB_j = A_mB_m$, and so on.

Now let us go back to Eqs. (15.9a)–(15.9c), which are the transformation equations for a vector. We can give these equations in the following compact manner if we use the notions of free and dummy indices as presented above. Thus,

$$A_{i'} = a_{i'j}A_j \tag{15.18}$$

where i' becomes x', y', and z' in a set.[11] Verify this relation yourself. Once having done this, now go to the transformation equations for the stress components from reference xyz to reference $x'y'z'$. Consider in this regard the following formulation:

$$\tau_{i'j'} = a_{i'k}a_{j'l}\tau_{kl} \tag{15.19}$$

Note that i' and j' are free indices and can represent any and all permutations wherein i' is replaced by x', y', or z' and j' is replaced by x', y', or z'. Consider the case where $i' \equiv x'$ and $j' \equiv x'$. From above, we then get

$$\begin{aligned}\tau_{x'x'} = {} & a_{x'x}a_{x'x}\tau_{xx} + a_{x'x}a_{x'y}\tau_{xy} + a_{x'x}a_{x'z}\tau_{xz} + \\ & a_{x'y}a_{x'x}\tau_{yx} + a_{x'y}a_{x'y}\tau_{yy} + a_{x'y}a_{x'z}\tau_{yz} + \\ & a_{x'z}a_{x'x}\tau_{zx} + a_{x'z}a_{x'y}\tau_{zy} + a_{x'z}a_{x'z}\tau_{zz}\end{aligned}$$

Comparing the above results with Eq.(15.6a) we see that we have identical results. Similarly, if $i' \equiv x'$ and $j' \equiv z'$, from Eq. (15.19) we get

$$\begin{aligned}\tau_{x'z'} = {} & a_{x'x}a_{z'x}\tau_{xx} + a_{x'x}a_{z'y}\tau_{xy} + a_{x'x}a_{z'z}\tau_{xz} + \\ & a_{x'y}a_{z'x}\tau_{yx} + a_{x'y}a_{z'y}\tau_{yy} + a_{x'y}a_{z'z}\tau_{yz} + \\ & a_{x'z}a_{z'x}\tau_{zx} + a_{x'z}a_{z'y}\tau_{zy} + a_{x'z}a_{z'z}\tau_{zz}\end{aligned}$$

The result above is identical to Eq. (15.7). Clearly, using the concepts of the free and dummy indices, Eq. (15.19) represents the transformation of the nine stresses for reference xyz to the nine stresses for reference $x'y'z'$. We can accordingly say that any set of nine quantities A_{ij} that transforms under a rotation of axes as follows

$$A_{i'j'} = a_{i'k}a_{j'l}A_{kl} \tag{15.20}$$

is a second-order tensor. Notice how similar the transformations are for the vectors and second-order tensors: The number of free indices

[11]Note that in an equation the same free indices appear in every expression. This is an important check on the correctness of an equation in index notation.

is increased by 1, and the number of sets of a's in the expression is similarly increased when going from a vector representation to a tensor representation. Accordingly, vectors are often called *first-order* tensors. To get to a scalar we can proceed by extending the above conclusions in the other direction. We should have no free index and no sets of a's. The equation

$$M' = M$$

has these characteristics: scalars are thus *zeroth-order tensors.* Now a third-order tensor will have three free indices and will transform according to an equation having three sets of a's in it. Thus,

$$A_{m'n'p'} = a_{m'i}a_{n'j}a_{p'k}A_{ijk}$$

How many terms are there for a third-order tensor? Can you express the transformation equation for a fourth-order tensor? It must be carefully pointed out that higher-order tensors above the second are not mathematical curiosities. They do exist in mechanics and other fields and are important.

15.6 Closure

In this chapter we showed that knowing the nine stresses on a set of orthogonal interfaces at a point we could find stresses on any interface at the point. Indeed, by knowing nine stresses at each point for interfaces parallel to a reference xyz, that is, knowing the *stress field*,

$$\begin{pmatrix} \tau_{xx} & \tau_{xy} & \tau_{xz} \\ \tau_{yx} & \tau_{yy} & \tau_{yz} \\ \tau_{zx} & \tau_{zy} & \tau_{zz} \end{pmatrix}$$

we effectively have complete information concerning the distribution of force throughout the body. While the importance of stress in mechanics needs no emphasis, the significance of the second-order tensor, formulated while developing the characteristics of stress, perhaps needs additional comment. We developed in this chapter transformation formulas by which, knowing nine stresses for interfaces parallel to reference xyz at a point, we could find nine stresses for interfaces parallel to a reference $x'y'z'$ rotated relative to xyz at the point. There are many other quantities having nine components associated with a reference at a point whereby the components for a new reference rotated relative to the given reference at the point are determined by *exactly the same transformation equations* developed for stress. Because such quantities have many identical mathematical properties owing to the common transformation equation, we consider them as an entity which we call second-order tensors. Some of the aforementioned characteristics that we have investigated in this chapter are principal axes, principal values of the tensor, and tensor invariants. The concepts of the tensor and the techniques for making certain calculations associated with tensors pervade mathematical physics and the engineering sciences. (For instance, electrical engineers will deal with the permittivity tensor in electromagnetics of continua, not to speak of the quadruple tensor in antenna theory and solid-state physics.)

In Chapter 16 we shall show that strain is a second-order tensor quantity. Once the latter is accomplished, all the attributes of second-order tensors established in this chapter are immediately available for strain without further study. It is as if by making the slight extra mathematical effort of considering second-order tensors in Chapter 15 we shall get a "free ride" later in Chapter 16. Those students that go on to study rigid-body dynamics will get another free ride when they study the inertia tensor.

Highlights (15)

An interface is an infinitesimal plane surface on the surface or inside a body. On an interface, there may be a normal force intensity called a normal stress and two orthogonal tangential force intensities called shear stresses. We have shown that, knowing nine stresses on three orthogonal interfaces at a point, we can determine stresses on any interface at the point via **transformation equations.** These equations give the nine stress components for interfaces associated with a reference $x'y'z'$ rotated arbitrarily relative to reference xyz for which nine stresses are known. These transformation equations define what we have been calling a second-order tensor. From these transformation equations come a series of vital results which we now list.

1. There are three orthogonal principal stresses wherein one stress is the algebraically largest stress and another is the algebraically smallest stress.
2. A cubic equation can be written whose roots are the three principal stresses and whose three coefficients are invariants with respect to a rotation of axes. One of these is the sum of the normal stresses for any reference axes at a point. Because natural phenomena are not axes-dependent, these and other invariants often represent physical quantities.
3. The principal axes have zero shear stress on the interfaces normal to these axes.
4. There is the Mohr circle representation for three-dimensional stress presented in Chapter 7.

All symmetric second-order tensors have the above characteristics and so it is very worthwhile to consider such quantities as a separate entity since they occur in all analytic sciences and are a vital area of applied mathematics.

PROBLEMS

3-D Transformation Equations	15.1–15.11
Principal Stresses and Invariants	15.12–15.19
Computer Problem	15.20

15.1. [15.2] The stress components of orthogonal interfaces parallel to xyz at a point are known to be

$$\tau_{xx} = 7 \times 10^6 \text{ Pa} \qquad \tau_{xy} = \tau_{yx} = 1.4 \times 10^6 \text{ Pa}$$
$$\tau_{yy} = -4 \times 10^6 \text{ Pa} \qquad \tau_{xz} = \tau_{zx} = 0$$
$$\tau_{zz} = 0 \qquad \tau_{yz} = \tau_{zy} = -2.8 \times 10^6 \text{ Pa}$$

What is the normal stress in the direction $\boldsymbol{\varepsilon}$ such that

$$\boldsymbol{\varepsilon} = .11\mathbf{i} + .35\mathbf{j} + .93\mathbf{k}$$

15.2. [15.2] Imagine a stress distribution where the stresses τ_{xx}, τ_{yy}, τ_{zz}, τ_{xy}, τ_{xz}, and τ_{yz} are uniform throughout a body. What does this imply for parallel stresses on parallel interfaces at different points? Suppose for such a distribution that we have

$$\tau_{xx} = 1000 \text{ psi} \qquad \tau_{xy} = \tau_{yx} = 0$$
$$\tau_{yy} = -1000 \text{ psi} \qquad \tau_{xz} = \tau_{zx} = 500$$
$$\tau_{zz} = 1000 \text{ psi} \qquad \tau_{yz} = \tau_{zy} = -500 \text{ psi}$$

What is the normal stress on plane $ABCD$ of the parallelepiped in the body shown in Fig. P.15.2?

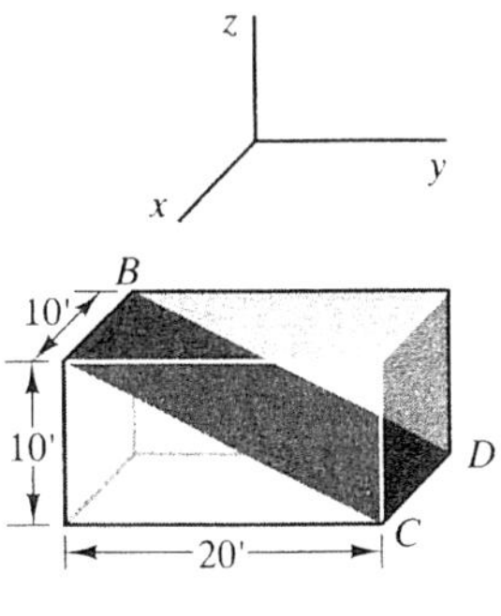

Figure P.15.2.

15.3. [15.2] Derive Eqs. (d) in Example 2.6 using the general transformation equation.

15.4. [15.2] If on three orthogonal interfaces at a point the three normal stresses are equal and the shear stresses are all zero, show that on any interface at that point the normal stress *also* has the same value as the aforementioned normal stress and there is zero shear stress. Such a state of stress is called *hydrostatic* (see Chapter 9) since it exists in a static fluid.

15.5. [15.2] Derive the plane stress transformation equations from the three-dimensional transformation stress equations.

15.6. [15.2] The flow of a fluid is described by the following velocity field:

$$\mathbf{V} = .16x^2\mathbf{i} + .10y^2\mathbf{j} + (.3y - .52z)\mathbf{k} \text{ ft/sec}$$

You will learn later when you study fluid mechanics that stress at a point in certain fluids is related to the velocity field by the following relations (Stokes' viscosity law):

$$\tau_{xx} = \mu\left[2\frac{\partial V_x}{\partial x} - \frac{2}{3}\left(\frac{\partial V_x}{\partial x} + \frac{\partial V_y}{\partial y} + \frac{\partial V_z}{\partial z}\right)\right] - p$$
$$\tau_{yy} = \mu\left[2\frac{\partial V_y}{\partial y} - \frac{2}{3}\left(\frac{\partial V_x}{\partial x} + \frac{\partial V_y}{\partial y} + \frac{\partial V_z}{\partial z}\right)\right] - p$$
$$\tau_{zz} = \mu\left[2\frac{\partial V_z}{\partial z} - \frac{2}{3}\left(\frac{\partial V_x}{\partial x} + \frac{\partial V_y}{\partial y} + \frac{\partial V_z}{\partial z}\right)\right] - p$$
$$\tau_{xy} = \tau_{yx} = \mu\left(\frac{\partial V_x}{\partial y} + \frac{\partial V_y}{\partial x}\right)$$
$$\tau_{xz} = \tau_{zx} = \mu\left(\frac{\partial V_x}{\partial z} + \frac{\partial V_z}{\partial x}\right)$$
$$\tau_{yz} = \tau_{zy} = \mu\left(\frac{\partial V_y}{\partial z} + \frac{\partial V_z}{\partial y}\right)$$

where p is the so-called hydrostatic pressure and μ is the coefficient of viscosity. What is the normal stress on an interface of fluid at position (1, 0, 2) having a normal direction given as

$$\boldsymbol{\varepsilon}_n = .80\mathbf{i} + .60\mathbf{j}$$

Give the result in terms of p and μ.

15.7. [15.2] We have used a Cartesian reference to generate a set of orthogonal interfaces at a point. Actually, using other orthogonal curvilinear coordinates we can similarly generate a set of orthogonal interfaces at a point. For instance, using cylindrical coordinates at a point we may generate the following set of stresses on orthogonal interfaces:

$$\begin{pmatrix} \tau_{rr} & \tau_{r\theta} & \tau_{rz} \\ \tau_{\theta r} & \tau_{\theta\theta} & \tau_{\theta z} \\ \tau_{zr} & \tau_{z\theta} & \tau_{zz} \end{pmatrix} \tag{1}$$

These stresses are shown in Fig. P.15.7. Suppose that you have a set of stresses for (1) given as

$$\begin{pmatrix} 5000 & 0 & -3000 \\ 0 & 2000 & 1000 \\ -3000 & 1000 & 0 \end{pmatrix} \text{psi}$$

representing stress at $r = 6$ in., $\theta = 30°$, $z = 10$ in. in a body. What is the stress τ_{xx} at this point?

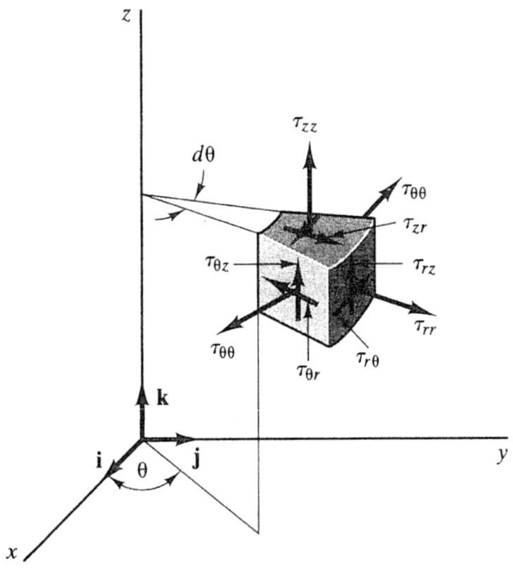

Figure P.15.7.

15.8. [15.2] A thin plate is shown in Fig. P.15.8 loaded at its ends uniformly by a normal force intensity S psi. There is a small hole of radius a at the center of the plate. Without the hole the only nonzero stress for the reference xyz is

$$\tau_{yy} = S$$

With the hole we can show from the theory of elasticity that for cylindrical coordinates the nonzero stresses are given as

$$\tau_{rr} = \frac{S}{2}\left[\left(1 - \frac{a^2}{r^2}\right) + \left(-1 + 4\frac{a^2}{r^2} - 3\frac{a^4}{r^4}\right)\cos 2\theta\right]$$

$$\tau_{\theta\theta} = \frac{S}{2}\left[\left(1 + \frac{a^2}{r^2}\right) + \left(1 + 3\frac{a^4}{r^4}\right)\cos 2\theta\right]$$

$$\tau_{r\theta} = \tau_{\theta r} = \frac{S}{2}\left(1 + 2\frac{a^2}{r^2} - 3\frac{a^4}{r^4}\right)\sin 2\theta$$

The maximum stress occurs at the hole. Show that it is in a direction tangent to the circle with a value of $3S$. Thus, we see that the presence of a small (innocent) hole can cause a threefold increase in stress over what would be the case of no hole present. This is an example of a *stress concentration*—that is, a buildup of stress because of the presence of a hole or possibly, as in other cases, a notch, keyway, and so on.

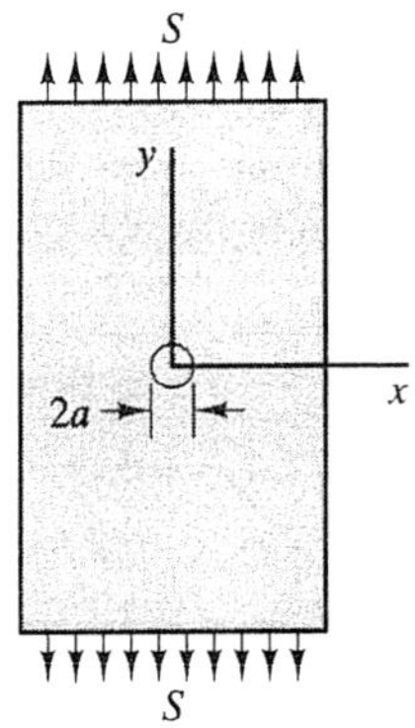

Figure P.15.8.

15.9. [15.2] In Problem 15.8 what is the shear stress τ_s in the xy plane on an interface oriented at 45° to the x and y axes at $r = 2$ ft and $\theta = 30°$ (see Fig. P.15.9). Take $S = 500$ psi and $a = .3$ ft.

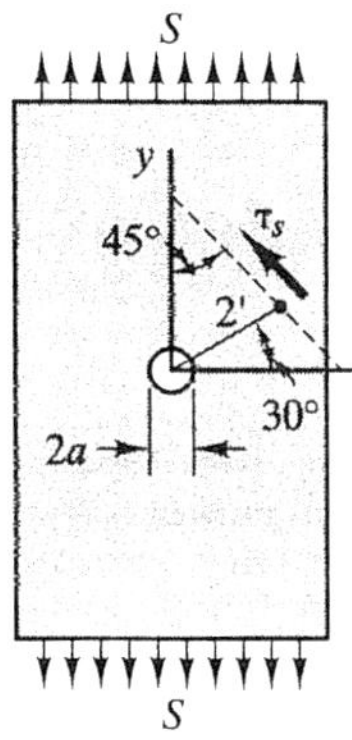

Figure P.15.9.

15.10. [15.2] The stresses at point P for an xyz reference are

$$\tau_{ij} = \begin{pmatrix} 0 & 200 & 300 \\ 200 & 500 & 0 \\ 300 & 0 & 600 \end{pmatrix}$$

What are stresses $\tau_{x'x'}$ and $\tau_{x'y'}$ for a set of axes $x'y'z'$ rotated 30° about the x axis as shown in Fig. P.15.10?

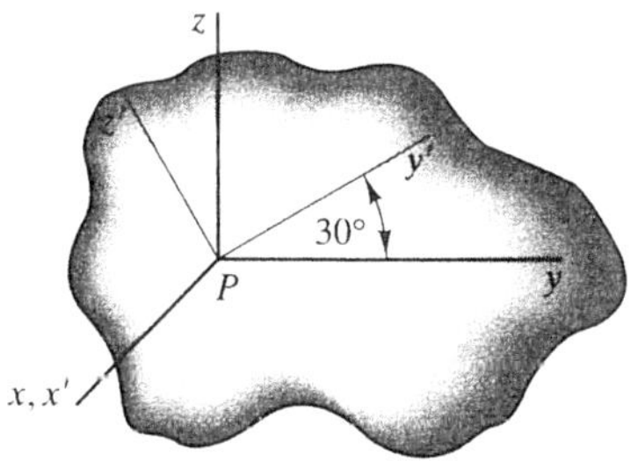

Figure P.15.10.

15.11. [15.2] The stress tensor at a point for a reference xyz is

$$\begin{pmatrix} 0.7 & 0 & 0 \\ 0 & 0 & 0 \\ 0 & 0 & 2 \end{pmatrix} \times 10^6 \text{ Pa}$$

What is the stress tensor for a set of axes at the point rotated 45° about the z axis in a clockwise direction as one looks toward the point?

15.12. [15.3] Given the following state of stress at a point,

$$\tau_{xx} = 0 \text{ psi} \qquad \tau_{xy} = -600 \text{ psi}$$
$$\tau_{yy} = 600 \text{ psi} \qquad \tau_{xz} = 0 \text{ psi}$$
$$\tau_{zz} = 0 \text{ psi} \qquad \tau_{yz} = -300 \text{ psi}$$

what are the principal stresses and the direction of the algebraically minimum stress?

15.13. [15.3] If the following stresses exist at a point,

$$\tau_{xx} = 1000 \text{ psi} \qquad \tau_{yy} = 0 \text{ psi} \qquad \tau_{zz} = 500 \text{ psi}$$
$$\tau_{xz} = 500 \text{ psi} \qquad \tau_{xy} = \tau_{yz} = 0$$

what are the principal stresses at the point and the directions of the principal axes?

15.14. [15.3] Given the stress tensor

$$\tau_{ij} = \begin{pmatrix} 5000 & 0 & -3000 \\ 0 & 2000 & 1000 \\ -3000 & 1000 & 0 \end{pmatrix} \text{ psi}$$

find the principal stresses if one of them is known to be 2188 psi.

15.15. [15.3] One of the principal stresses for the following state of stress is known to be approximately 749 psi:

$$\tau_{ij} = \begin{pmatrix} 0 & 200 & 300 \\ 200 & 500 & 0 \\ 300 & 0 & 600 \end{pmatrix} \text{ psi}$$

What approximately is the largest compressive stress at the point and its direction?

15.16. [15.4] Show that the second tensor invariant for stress is the sum of the minors of the principal diagonal (the diagonal comprising the normal stresses). Show that the third tensor invariant for stress is the determinant of the stress tensor.

15.17. [15.4] A long steel thin-walled tank (Fig. P.15.17) contains compressed air at a pressure of 200 psi gage. Wall thickness is 1 in. In an earthquake the tank is given an acceleration upward of 9 g's, and a torque of 50,000 ft-lb is developed about its longitudinal axis. What is the maximum tensile stress away from the ends? Take the specific weight of steel as 450 lb/ft^3.

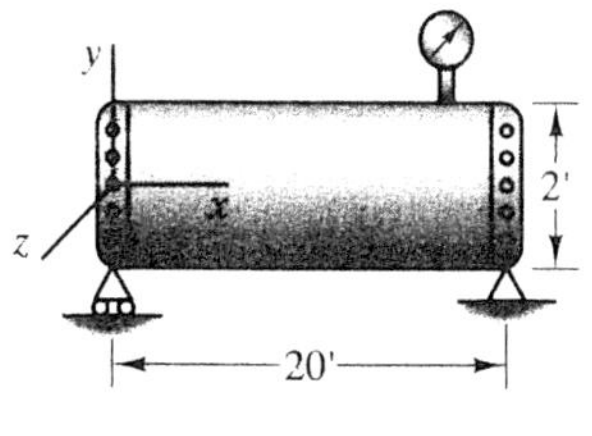

Figure P.15.17.

15.18. [15.4] Using transformation formulations for stress at a point, show that $(\tau_{nn})_{oct}$, the normal stress on the octahedral plane in Fig P.15.18, is given as

$$(\tau_{nn})_{oct} = \frac{1}{3}(\tau_1 + \tau_2 + \tau_3)$$

Now, using Fig. P.15.18, equate the resultant force on the octahedral plane with the resultant force on the coordinate faces of the indicated tetrahedron to show that

$$\tau_{oct}^2 = \frac{1}{9}[(\tau_1 - \tau_2)^2 + (\tau_2 - \tau_3)^2 + (\tau_3 - \tau_1)^2]$$

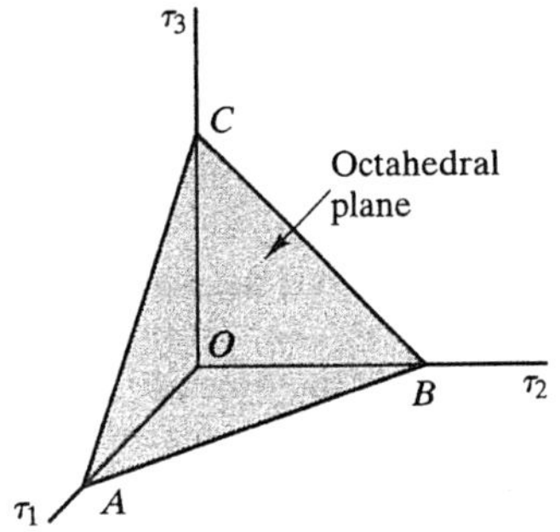

Figure P.15.18.

15.19. [15.4] Show that

$$9\tau_{oct}^2 = 2(A)^2 - 6(B)$$

where A and B are the first and second tensor invariants for stress. (*Hint:* Add and subtract $4\tau_1\tau_2 + 4\tau_1\tau_3 + 4\tau_2\tau_3$.) See Problem 15.18.

****15.20.** For the state of stress given in Problem 15.14, plot the characteristic equation. Determine the principal stresses from the plot.

CHAPTER 16

Three-Dimensional Strain Relations at a Point

16.1 Introduction

In Chapter 3 we introduced the strain components at a point for a reference xyz. Recall that we considered the relative movement of adjacent points along three orthogonal axes at a point. For convenience in this regard we considered the deformation of a rectangular parallelepiped having the aforementioned adjacent points at four of its corners. We pointed out that once the six independent strain components were known for one set of orthogonal axes we could readily determine the strain components for any set of axes rotated relative to the first reference. We show the strains for the primed and unprimed axes in Fig. 16.1. We shall find that the equations giving the primed strains in terms of the six given unprimed strains are identical in form to those for stresses at a point. We shall thus show that strain components transform with a rotation of axes exactly as do the stress components. Accordingly, we shall be able to conclude that strain is a second-order tensor. Hence, many of the conclusions of Chapter 15 will be valid here for strain, as will be seen later in the chapter.

Note that the computations we shall make in this chapter, knowing that strain is a second-order tensor, can be made in the same way for the inertia tensor of rigid body dynamics, the quadruple tensor of antenna theory, the permittivity tensor of solid state physics, and so on.

16.2 Transformation Equations for Strain

Consider now an arbitrary direction $\mathbf{n}$ at point P in a body (Fig. 16.2). A segment PQ of length Δn is shown in the diagram. In the deformed geometry this segment becomes $P'Q'$, as has been indicated. We can

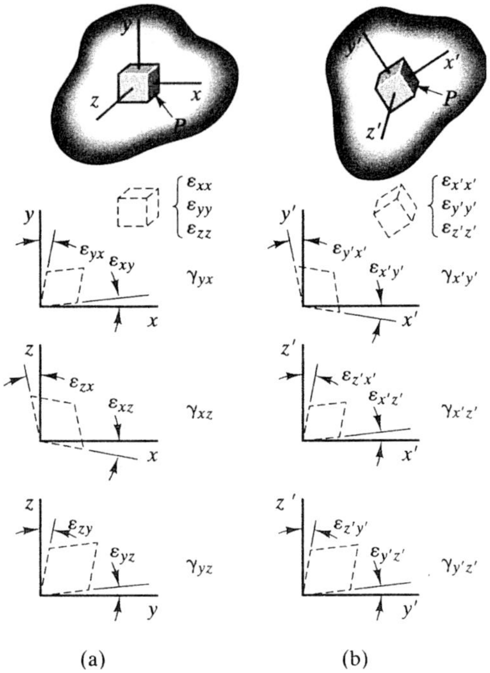

Figure 16.1. Two sets of strains at a point.

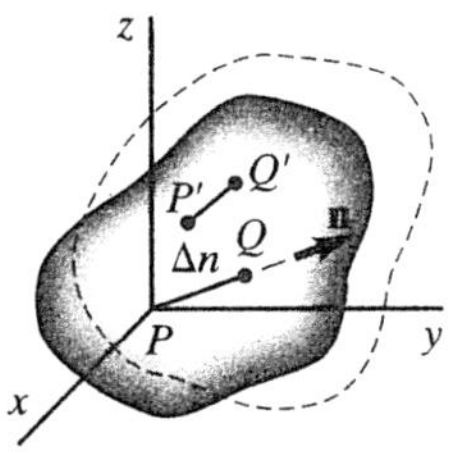

Figure 16.2. Line segment PQ in a deforming body.

give the displacement component u_n in the direction $\mathbf{n}$ at any point in terms of the rectangular components of $\boldsymbol{u}$ at that point as follows:

$$u_n = \boldsymbol{u} \cdot \mathbf{n} = u_x(\mathbf{i} \cdot \mathbf{n}) + u_y(\mathbf{j} \cdot \mathbf{n}) + u_z(\mathbf{k} \cdot \mathbf{n})$$

$$= u_x a_{nx} + u_y a_{ny} + u_z a_{nz} \tag{16.1}$$

where a_{nx}, a_{ny}, and a_{nz} are the direction cosines of $\mathbf{n}$. Since u_n is a function of position we can express u_n at point Q as a Taylor series expansion in terms of u_n at point P as follows:

$$(u_n)_Q = (u_n)_P + \left(\frac{\partial u_n}{\partial x}\right)_P \Delta x + \left(\frac{\partial u_n}{\partial y}\right)_P \Delta y + \left(\frac{\partial u_n}{\partial z}\right)_P \Delta z + \cdots \tag{16.2}$$

where Δx, Δy, and Δz are the rectangular projections of Δn and where higher-order terms in Δx, Δy, and Δz are indicated by the dots. Now substitute above for u_n using Eq. 16.1. We get on subtracting $(u_n)_P$ from both sides of the above equation.

$$\begin{aligned}(u_n)_Q - (u_n)_P = &\left\{\frac{\partial}{\partial x}[u_x a_{nx} + u_y a_{ny} + u_z a_{nz}]\right\}_P \Delta x + \\ &\left\{\frac{\partial}{\partial y}[u_x a_{nx} + u_y a_{ny} + u_z a_{nz}]\right\}_P \Delta y + \\ &\left\{\frac{\partial}{\partial z}[u_x a_{nx} + u_y a_{ny} + u_z a_{nz}]\right\}_P \Delta z + \cdots\end{aligned}$$

If we divide both sides by Δn and take the limit as $\Delta n \to 0$ while noting that Δx, Δy, and Δz also go to zero, we get

$$\begin{aligned}\lim_{\Delta n \to 0} \frac{(u_n)_Q - (u_n)_P}{\Delta n} = &\frac{\partial}{\partial x}[u_x a_{nx} + u_y a_{ny} + u_z a_{nz}]\frac{dx}{dn} + \\ &\frac{\partial}{\partial y}[u_x a_{nx} + u_y a_{ny} + u_z a_{nz}]\frac{dy}{dn} + \\ &\frac{\partial}{\partial z}[u_x a_{nx} + u_y a_{ny} + u_z a_{nz}]\frac{dz}{dn}\end{aligned}$$

The left-hand side of the equation should be recognizable as ε_{nn}, the normal strain in the direction **n**. On the right-hand side of the equation the higher-order terms disappear since each such expression has an excess of Δ's in the numerator over that of the denominator and hence goes to zero in the limit. Finally, the subscript P becomes superfluous as points Q and P coalesce in the limit and the ratio $\Delta x/\Delta n$ becomes dx/dn, and so on. From the Pythagorean theorem we can replace dx/dn, dy/dn, and dz/dn by the direction cosines a_{nx}, a_{ny}, and a_{nz}, respectively. We thus have, carrying out the differential operations and collecting terms in the above equation,

$$\begin{aligned}\varepsilon_{nn} = &\frac{\partial u_x}{\partial x} a_{nx}^2 + \frac{\partial u_y}{\partial y} a_{ny}^2 + \frac{\partial u_z}{\partial z} a_{nz}^2 + \left(\frac{\partial u_x}{\partial y} + \frac{\partial u_y}{\partial x}\right) a_{nx} a_{ny} \\ &+ \left(\frac{\partial u_x}{\partial z} + \frac{\partial u_z}{\partial x}\right) a_{nx} a_{nz} + \left(\frac{\partial u_y}{\partial z} + \frac{\partial u_z}{\partial y}\right) a_{ny} a_{nz}\end{aligned} \tag{16.3}$$

Note that the expression, $[(\partial u_x/\partial y) + (\partial u_y/\partial x)]a_{nx}a_{ny}$, can be given as

$$\frac{1}{2}\left(\frac{\partial u_x}{\partial y} + \frac{\partial u_y}{\partial x}\right) a_{nx} a_{ny} + \frac{1}{2}\left(\frac{\partial u_y}{\partial x} + \frac{\partial u_x}{\partial y}\right) a_{ny} a_{nx}$$

The last two expressions in Eq. (16.3) can be similarly treated. For Eq. (16.3) using these new expressions we then get

$$
\begin{aligned}
\varepsilon_{nn} = {} & \frac{\partial u_x}{\partial x} a_{nx}^2 + \frac{1}{2}\left(\frac{\partial u_x}{\partial y} + \frac{\partial u_y}{\partial x}\right) a_{nx} a_{ny} + \frac{1}{2}\left(\frac{\partial u_x}{\partial z} + \frac{\partial u_z}{\partial x}\right) a_{nx} a_{nz} + \\
& \frac{1}{2}\left(\frac{\partial u_y}{\partial x} + \frac{\partial u_x}{\partial y}\right) a_{ny} a_{nx} + \frac{\partial u_y}{\partial y} a_{ny}^2 + \frac{1}{2}\left(\frac{\partial u_y}{\partial z} + \frac{\partial u_z}{\partial y}\right) a_{ny} a_{nz} + \\
& \frac{1}{2}\left(\frac{\partial u_z}{\partial x} + \frac{\partial u_x}{\partial z}\right) a_{nz} a_{nx} + \frac{1}{2}\left(\frac{\partial u_z}{\partial y} + \frac{\partial u_y}{\partial z}\right) a_{nz} a_{ny} + \frac{\partial u_z}{\partial z} a_{nz}^2
\end{aligned}
$$

We may now introduce the strain terms for reference xyz using the strain displacement relations. We can form the following equation:

$$
\begin{aligned}
\varepsilon_{nn} = {} & \varepsilon_{xx} a_{nx}^2 + \varepsilon_{xy} a_{nx} a_{ny} + \varepsilon_{xz} a_{nx} a_{nz} + \\
& \varepsilon_{yx} a_{ny} a_{nx} + \varepsilon_{yy} a_{ny}^2 + \varepsilon_{yz} a_{ny} a_{nz} + \\
& \varepsilon_{zx} a_{nz} a_{nx} + \varepsilon_{zy} a_{nz} a_{ny} + \varepsilon_{zz} a_{nz}^2
\end{aligned} \tag{16.4}
$$

Note that Eq. (16.4) for ε_{nn} is exactly the same form as Eq. (15.3) for stress τ_{nn}. Since n could be an x', y', or z', we can immediately set forth the normal strain for a set of primed axes rotated relative to xyz in the same manner as was done for normal stress. Thus,

$$
\begin{aligned}
\varepsilon_{x'x'} = {} & \varepsilon_{xx} a_{x'x}^2 + \varepsilon_{xy} a_{x'x} a_{x'y} + \varepsilon_{xz} a_{x'x} a_{x'z} + \\
& \varepsilon_{yx} a_{x'y} a_{x'x} + \varepsilon_{yy} a_{x'y}^2 + \varepsilon_{yz} a_{x'y} a_{x'z} + \\
& \varepsilon_{zx} a_{x'z} a_{x'x} + \varepsilon_{zy} a_{x'z} a_{x'y} + \varepsilon_{zz} a_{x'z}^2 \\[1ex]
\varepsilon_{y'y'} = {} & \varepsilon_{xx} a_{y'x}^2 + \varepsilon_{xy} a_{y'x} a_{y'y} + \varepsilon_{xz} a_{y'x} a_{y'z} + \\
& \varepsilon_{yx} a_{y'y} a_{y'x} + \varepsilon_{yy} a_{y'y}^2 + \varepsilon_{yz} a_{y'y} a_{y'z} + \\
& \varepsilon_{zx} a_{y'z} a_{y'x} + \varepsilon_{zy} a_{y'z} a_{y'y} + \varepsilon_{zz} a_{y'z}^2 \\[1ex]
\varepsilon_{z'z'} = {} & \varepsilon_{xx} a_{z'x}^2 + \varepsilon_{xy} a_{z'x} a_{z'y} + \varepsilon_{xz} a_{z'x} a_{z'z} + \\
& \varepsilon_{yx} a_{z'y} a_{z'x} + \varepsilon_{yy} a_{z'y}^2 + \varepsilon_{yz} a_{z'y} a_{z'z} + \\
& \varepsilon_{zx} a_{z'z} a_{z'x} + \varepsilon_{zy} a_{z'z} a_{z'y} + \varepsilon_{zz} a_{z'z}^2
\end{aligned} \tag{16.5}
$$

We can conclude at this point that the normal strains transform with a rotation of axes *exactly* as do normal stresses.

Now consider two arbitrary orthogonal directions at point P (Fig. 16.3) denoted as directions **n** and **s.** Line segments PQ and PR of lengths Δn and Δs, respectively, are shown along these directions in the undeformed state. When the body deforms, the end points of these segments move to R', Q', P', and there is most likely a change in right angles between the line segments, giving rise to shear strain γ_{ns}. It will be convenient for us to consider n and s as coordinate axes at this time. Accordingly, from our work in Section 3.4 we can give ε_{ns} as follows:

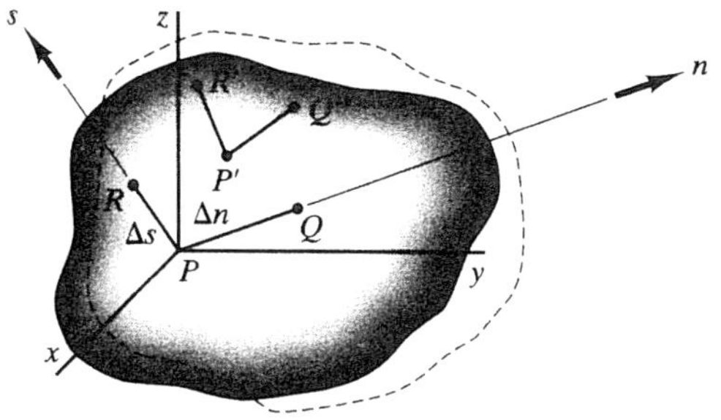

Figure 16.3. Orthogonal line segments PQ and RP in a deforming body.

$$\varepsilon_{ns} = \frac{1}{2}\left(\frac{\partial u_n}{\partial s} + \frac{\partial u_s}{\partial n}\right) \tag{16.6}$$

We can express u_n and u_s in terms of displacement components in the coordinate directions as we did earlier:

$$\begin{aligned} u_n &= \boldsymbol{u} \cdot \mathbf{n} = u_x a_{nx} + u_y a_{ny} + u_z a_{nz} = \sum_i u_i a_{ni} \\ u_s &= \boldsymbol{u} \cdot \mathbf{s} = u_x a_{sx} + u_y a_{sy} + u_z a_{sz} = \sum_i u_i a_{si} \end{aligned} \tag{16.7}$$

where the index i sums over x, y, and z. Substitute the results above into Eq. (16.6) as follows:

$$\begin{aligned} \varepsilon_{ns} &= \frac{1}{2}\left[\frac{\partial}{\partial s}\left(\sum_i u_i a_{ni}\right) + \frac{\partial}{\partial n}\left(\sum_i u_i a_{si}\right)\right] \\ &= \frac{1}{2}\left(\sum_i \frac{\partial u_i}{\partial s} a_{ni} + \sum_i \frac{\partial u_i}{\partial n} a_{si}\right) \end{aligned} \tag{16.8}$$

Now $\partial u_i/\partial s$ and $\partial u_i/\partial n$ are *directional derivatives* and may be given as follows:[1]

$$\begin{aligned} \frac{\partial u_i}{\partial s} &= \frac{\partial u_i}{\partial x}\frac{dx}{ds} + \frac{\partial u_i}{\partial y}\frac{dy}{ds} + \frac{\partial u_i}{\partial z}\frac{dz}{ds} \\ \frac{\partial u_i}{\partial n} &= \frac{\partial u_i}{\partial x}\frac{dx}{dn} + \frac{\partial u_i}{\partial y}\frac{dy}{dn} + \frac{\partial u_i}{\partial z}\frac{dz}{dn} \end{aligned} \tag{16.9}$$

We then have

[1]See Section 8.2.

$$\varepsilon_{ns} = \frac{1}{2}\left[\sum_i\left(\frac{\partial u_i}{\partial x}\frac{dx}{ds} + \frac{\partial u_i}{\partial y}\frac{dy}{ds} + \frac{\partial u_i}{\partial z}\frac{dz}{ds}\right)a_{ni} + \sum_i\left(\frac{\partial u_i}{\partial x}\frac{dx}{dn} + \frac{\partial u_i}{\partial y}\frac{dy}{dn} + \frac{\partial u_i}{\partial z}\frac{dz}{dn}\right)a_{si}\right]$$

Noting again as before that $dx/ds = a_{sx}$, $dx/dn = a_{nx}$, and so on, we get the following equation on carrying out the summations and collecting terms:

$$\begin{aligned}\varepsilon_{ns} = {} & \frac{\partial u_x}{\partial x}a_{nx}a_{sx} + \frac{1}{2}\left(\frac{\partial u_x}{\partial y} + \frac{\partial u_y}{\partial x}\right)a_{nx}a_{sy} + \frac{1}{2}\left(\frac{\partial u_x}{\partial z} + \frac{\partial u_z}{\partial x}\right)a_{nx}a_{sz} + \\ & \frac{1}{2}\left(\frac{\partial u_y}{\partial x} + \frac{\partial u_x}{\partial y}\right)a_{ny}a_{sx} + \frac{\partial u_y}{\partial y}a_{ny}a_{sy} + \frac{1}{2}\left(\frac{\partial u_y}{\partial z} + \frac{\partial u_z}{\partial y}\right)a_{ny}a_{sz} + \\ & \frac{1}{2}\left(\frac{\partial u_z}{\partial x} + \frac{\partial u_x}{\partial z}\right)a_{nz}a_{sx} + \frac{1}{2}\left(\frac{\partial u_z}{\partial y} + \frac{\partial u_y}{\partial z}\right)a_{nz}a_{sy} + \frac{\partial u_z}{\partial z}a_{nz}a_{sz}\end{aligned} \tag{16.10}$$

Employing the strain displacement relations for Eq. (16.10) we then get

$$\begin{aligned}\varepsilon_{ns} = {} & \varepsilon_{xx}a_{nx}a_{sx} + \varepsilon_{xy}a_{nx}a_{sy} + \varepsilon_{xz}a_{nx}a_{sz} + \\ & \varepsilon_{yx}a_{ny}a_{sx} + \varepsilon_{yy}a_{ny}a_{sy} + \varepsilon_{yz}a_{ny}a_{sz} + \\ & \varepsilon_{zx}a_{nz}a_{sx} + \varepsilon_{zy}a_{nz}a_{sy} + \varepsilon_{zz}a_{nz}a_{sz}\end{aligned} \tag{16.11}$$

Note again that this transformation formulation is identical to that of Eq. (15.5) for shear stress. Realizing that the n and s may be considered as a pair of primed coordinate axes of a set rotated relative to xyz, we may express the shear strains for the primed axes quite simply from the above result. Thus, for $\varepsilon_{x'y'}$ we would have

$$\begin{aligned}\varepsilon_{x'y'} = {} & \varepsilon_{xx}a_{x'x}a_{y'x} + \varepsilon_{xy}a_{x'x}a_{y'y} + \varepsilon_{xz}a_{x'x}a_{y'z} + \\ & \varepsilon_{yx}a_{x'y}a_{y'x} + \varepsilon_{yy}a_{x'y}a_{y'y} + \varepsilon_{yz}a_{x'y}a_{y'z} + \\ & \varepsilon_{zx}a_{x'z}a_{y'x} + \varepsilon_{zy}a_{x'z}a_{y'y} + \varepsilon_{zz}a_{x'z}a_{y'z}\end{aligned} \tag{16.12}$$

We can similarly find all other shear strains for the primed axes. We can now conclude that the strain components at a point transform exactly as do the stress components at a point on a rotation of axes. Indeed, we can now say that the *strain terms form a second-order Cartesian tensor.*

Example 16.1

The following displacement field describes the movement of a body under load:

$$\boldsymbol{u} = [(x^2 + y^2)\mathbf{i} + (3 + xz)\mathbf{j} - .6z^2\mathbf{k}] \times 10^{-2}\ \text{ft} \tag{a}$$

Example 16.1 (Continued)

Compute the normal strain at (0, 1, 3) in a direction s given as

$$\mathbf{s} = .60\mathbf{i} + .80\mathbf{j} \tag{a}$$

We shall do this in two ways. The most direct way is to evaluate $\partial u_s/\partial s$ at (0, 1, 3). Thus,

$$u_s = \boldsymbol{u} \cdot \mathbf{s} = [(.60)(x^2 + y^2) + (.80)(3 + xz)] \times 10^{-2} \tag{b}$$

Noting that $\partial u_s/\partial s$ is both a directional derivative and the desired quantity ε_{ss} we have

$$\frac{\partial u_s}{\partial s} = \varepsilon_{ss} = \frac{\partial u_s}{\partial x}\frac{dx}{ds} + \frac{\partial u_s}{\partial y}\frac{dy}{ds} + \frac{\partial u_s}{\partial z}\frac{dz}{ds}$$

Noting that

$$\begin{aligned} \frac{dx}{ds} &= a_{sx} = .60 \\ \frac{dy}{ds} &= a_{sy} = .80 \\ \frac{dz}{ds} &= a_{sz} = 0 \end{aligned} \tag{c}$$

we get for ε_{ss} on taking appropriate partial derivatives of Eq. (b)

$$\begin{aligned} \varepsilon_{ss} &= [(1.20x + .80z)(.60) + (1.20y)(.80) + (.80x)(0)] \times 10^{-2} \\ &= [.72x + .48z + .96y] \times 10^{-2} \end{aligned}$$

At the position of interest (0, 1, 3) we get

$$\begin{aligned} \varepsilon_{ss} &= [(.72)(0) + (.48)(3) + (.96)(1)] \times 10^{-2} \\ \varepsilon_{ss} &= .0240 \end{aligned} \tag{d}$$

We shall now proceed a *second way* by using the transformation equations. The strain tensor field ε_{ij} for the problem at hand is found from Eqs. (3.12) to be

Example 16.1 (Continued)

$$\varepsilon_{ij} = \begin{pmatrix} 2x & y + \frac{z}{2} & 0 \\ y + \frac{z}{2} & 0 & \frac{x}{2} \\ 0 & \frac{x}{2} & -1.2z \end{pmatrix} \times 10^{-2}$$

At the position of interest we have

$$\varepsilon_{ij} = \begin{pmatrix} 0 & 2.5 & 0 \\ 2.5 & 0 & 0 \\ 0 & 0 & -3.6 \end{pmatrix} \times 10^{-2} \qquad \text{(e)}$$

Using the transformation equations in conjunction with the direction cosines (c) we have

$$\varepsilon_{ss} = \{(2.5)(.60)(.80) + (2.5)(.80)(.60) + (-3.6)(0^2)\} \times 10^{-2}$$

$$\varepsilon_{ss} = .0240$$

16.3 Properties of Strain

We shall now examine other conclusions for a *general state of strain* resulting from the fact that strain has been shown to be a second-order tensor. Thus, in accordance with Eqs. (15.17) the following quantities are *invariant* at a point with respect to a rotation of axes:

$$\begin{aligned} \varepsilon_{xx} + \varepsilon_{yy} + \varepsilon_{zz} &= A \\ \varepsilon_{xx}\varepsilon_{yy} + \varepsilon_{xx}\varepsilon_{zz} + \varepsilon_{yy}\varepsilon_{zz} - \varepsilon_{xy}^2 - \varepsilon_{yz}^2 - \varepsilon_{xz}^2 &= B \\ \varepsilon_{xx}\varepsilon_{yy}\varepsilon_{zz} + 2\varepsilon_{xy}\varepsilon_{yz}\varepsilon_{zx} - \varepsilon_{xy}^2\varepsilon_{zz} - \varepsilon_{xz}^2\varepsilon_{yy} - \varepsilon_{yz}^2\varepsilon_{xx} &= C \end{aligned} \qquad (16.13)$$

These are the *strain invariants.* Also knowing strain tensor components at a point for reference xyz we can find the principal strains from the following equation:

$$\begin{vmatrix} \varepsilon_{xx} - \varepsilon & \varepsilon_{xy} & \varepsilon_{xz} \\ \varepsilon_{yx} & \varepsilon_{yy} - \varepsilon & \varepsilon_{yz} \\ \varepsilon_{zx} & \varepsilon_{zy} & \varepsilon_{zz} - \varepsilon \end{vmatrix} = 0 \qquad (16.14)$$

This results in the familiar cubic equation

$$\varepsilon^3 - A\varepsilon^2 + B\varepsilon - C = 0 \tag{16.15}$$

wherein A, B, and C are the invariants for strain as given by Eqs. (16.13).[2] For finding the principal axes we have the following equations:

$$\begin{aligned} (\varepsilon_{xx} - \varepsilon)a_{nx} + \varepsilon_{xy}a_{ny} + \varepsilon_{xz}a_{nz} &= 0 \\ \varepsilon_{yx}a_{nx} + (\varepsilon_{yy} - \varepsilon)a_{ny} + \varepsilon_{yz}a_{nz} &= 0 \\ \varepsilon_{zx}a_{nx} + \varepsilon_{zy}a_{ny} + (\varepsilon_{zz} - \varepsilon)a_{nz} &= 0 \end{aligned} \tag{16.16}$$

$$a_{nx}^2 + a_{ny}^2 + a_{nz}^2 = 1 \tag{16.17}$$

The computation of the principal strains and their directions accordingly are exactly the same as those for principal stresses. We shall illustrate this in the following example.

Example 16.2

Given the following state of strain at a point,

$$\varepsilon_{ij} = \begin{pmatrix} 80 & 0 & 0 \\ 0 & 36.7 & -21.6 \\ 0 & -21.6 & 50 \end{pmatrix} \times 10^{-6}$$

what are the principal strains and the directions of the principal axes?

For this purpose we now go back to Eq. (16.13) to compute the tensor invariants for strain. We thus get

$$A = (80 + 36.7 + 50) \times 10^{-6} = 166.7 \times 10^{-6}$$

$$\begin{aligned} B &= (80)(36.7) \times 10^{-12} + (80)(50) \times 10^{-12} + (36.7)(50) \times 10^{-12} - (-21.6)^2 \times 10^{-12} \\ &= 8305 \times 10^{-12} \end{aligned}$$

$$\begin{aligned} C &= (80)(36.7)(50) \times 10^{-18} - (-21.6)^2(80) \times 10^{-18} \\ &= 109{,}500 \times 10^{-18} \end{aligned}$$

The cubic equation for the principal strain is then

$$\varepsilon^3 - 166.7 \times 10^{-6}\varepsilon^2 + 8305 \times 10^{-12}\varepsilon - 109.5 \times 10^{-15} = 0 \tag{a}$$

Now since ε_{xy} and ε_{xz} are zero, it means that the segment dx of the rectangular parallelepiped (Fig. 16.4) at the point of interest in the

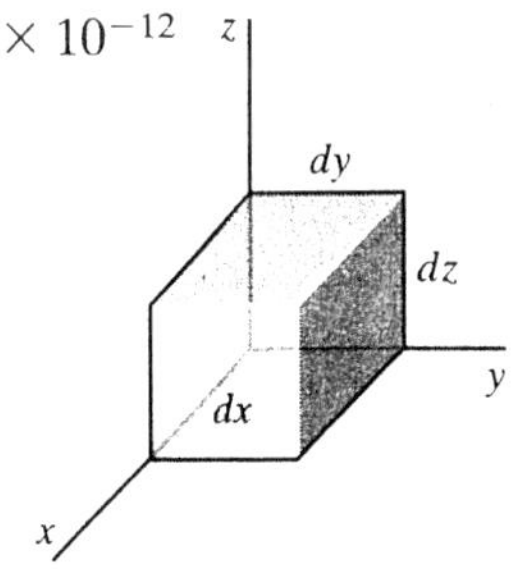

Figure 16.4. Infinitesimal rectangular parallelepiped at the point of interest.

[2]As with stress for the general state we denote the principal strains as ε_1, ε_2, and ε_3 with ε_1 the algebraically largest and with ε_3 the algebraically smallest.

Example 16.2 (Continued)

undeformed geometry does not rotate toward the z-axis or toward the y axis as a result of deformation. Indeed, other than rigid-body movements it can only stretch or shrink. Accordingly, the x direction must be a principal direction and so one of the principal strains at the point is then

$$\varepsilon = 80 \times 10^{-6}$$

We now divide $(\varepsilon - 80 \times 10^{-6})$ into the left-hand side of Eq. (a) to get a quadratic expression $\varepsilon^2 - 86.7 \times 10^{-6}\varepsilon + 1369 \times 10^{-2}$. We can then express Eq. (a) as follows:

$$(\varepsilon - 80 \times 10^{-6})\,(\varepsilon^2 - 86.7 \times 10^{-6}\varepsilon + 1369 \times 10^{-12}) = 0$$

To find the remaining principal strains we set the second parenthesized expression equal to zero. Thus,

$$\varepsilon^2 - 86.7 \times 10^{-6}\varepsilon + 1369 \times 10^{-12} = 0$$

Using the quadratic formula we get

$$\varepsilon = \frac{86.7 \times 10^{-6} \pm \sqrt{86.7^2 \times 10^{-12} - (4)(1369 \times 10^{-12})}}{2}$$

$$\therefore\ \varepsilon = 65.93 \times 10^{-6};\ \ 20.77 \times 10^{-6} \qquad \text{(b)}$$

We list the *principal strains* accordingly as follows:

$$\varepsilon_1 = 80 \times 10^{-6}$$
$$\varepsilon_2 = 65.93 \times 10^{-6}$$
$$\varepsilon_3 = 20.77 \times 10^{-6}$$

To find the principal axis for ε_2 we go to Eqs. (16.16). Remembering that only two of the equations are independent, on substituting $\varepsilon_2 = 65.93$ into the first two equations we get

$$(80 - 65.93) \times 10^{-6}a_{nx} + (0)(a_{ny}) + 0(a_{nz}) = 0$$
$$0(a_{nx}) + (36.7 - 65.93) \times 10^{-6}a_{ny} - 21.6 \times 10^{-6}a_{nz} = 0 \qquad \text{(c)}$$

Example 16.2 (Continued)

Next, noting that

$$a_{nx}^2 + a_{ny}^2 + a_{nz}^2 = 1 \tag{d}$$

we can readily solve Eqs. (c) and (d) simultaneously for the direction cosines. For ε_2 we get

$$a_{nx} = 0$$
$$a_{ny} = -.594$$
$$a_{nz} = .804$$

Similarly, we can show that for ε_3

$$a_{nx} = 0$$
$$a_{ny} = .804$$
$$a_{nz} = .594$$

Note that a rectangular parallelepiped having edges originally along the x, ε_2, and ε_3 directions, respectively, would remain a rectangular parallelepiped after deformation.

16.4 Closure

In this chapter we have shown how, once knowing that a quantity is a symmetric, second-order tensor, we can without further ado set forth a number of important properties of the quantity. Accordingly, when you study the inertia tensor in rigid-body dynamics or the quadruple tensor in electromagnetic theory you should have little difficulty in finding the key properties of these quantities by extending the results of Chapter 15 just as we have done in Chapter 16. The reader will note that in Chapters 15 and 16 we have gone back to extend Chapters 7 and 8.

We now examine a very interesting and important phenomenon called elastic stability. The subject will have ramifications well beyond the application at hand, as you will have opportunity to realize.

Highlights (16)

By having shown that the nine strain terms transform exactly as the defining transformation equations for second-order tensors, we can list all the properties of strain that were presented for stress in the Highlights section at the end of the preceding chapter. It should now be apparent why this study was worthwhile in this course and should be equally valuable as you proceed further in your studies.

PROBLEMS

3-D Transformation Equations	16.1–16.4
Properties of Strain	16.5–16.18

16.1. [16.2] Given the following displacement field,

$$u = [(6x^2)\mathbf{i} + (3 + zx)\mathbf{j} + (-2 + xy)\mathbf{k}] \times 10^{-2}\ \text{m}$$

what is the normal strain at (0, 1, 3) in the direction $\mathbf{s} = .3\mathbf{i} + .4\mathbf{j} + .866\mathbf{k}$? Do this (a) by using the relation $\partial u_s/\partial s = \varepsilon_{ss}$ and (b) by rotation of axes.

16.2. [16.2] Given the following state of strain at a point,

$$\varepsilon_{ij} = \begin{pmatrix} .02 & .01 & 0 \\ .01 & -.02 & .03 \\ 0 & .03 & .04 \end{pmatrix}$$

what is the normal strain along a direction equally inclined to the x, y, and z axes? What is the shear strain for axes having the directions

$$\varepsilon_1 = .6\mathbf{i} + .8\mathbf{j} \qquad \varepsilon_2 = .4\mathbf{i} - .3\mathbf{j} + .866\mathbf{k}$$

16.3. [16.2] Given the following strains at a point for reference xyz,

$$\begin{pmatrix} .02 & .01 & 0 \\ .01 & .03 & -.04 \\ 0 & -.04 & 0 \end{pmatrix}$$

what are the strain components $\varepsilon_{x'x'}$ and $\varepsilon_{x'y'}$ for reference $x'y'z'$ which is reached by rotating about the y-axis an angle of 45° counterclockwise as you look along the y-axis toward the origin?

16.4. [16.2] A plate is shown in Fig. P.16.4 under uniform loadings at the edges. As a result of these loads, faces AB and CD separate by an amount of .1 ft, while BC and AD separate by an amount .2 ft. The thickness of the plate originally is .1 ft. If there is no change in the volume, what are the strains $\varepsilon_{x'x'}$ and $\varepsilon_{z'x'}$ for axes $x'y'z'$ rotated 30° counterclockwise from reference xy shown in the diagram?

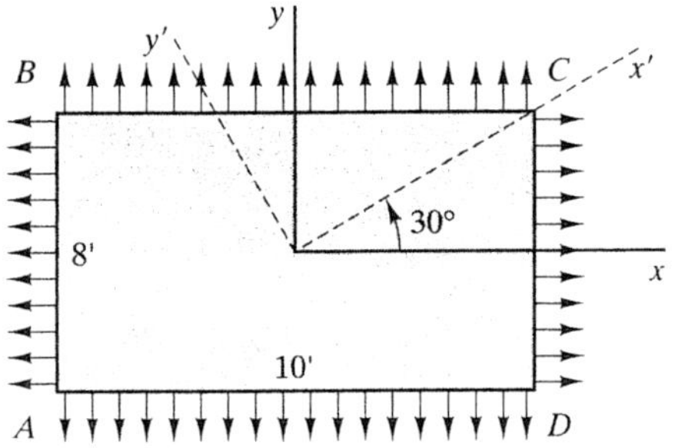

Figure P.16.4.

16.5. [16.3] Show that the principal axes of stress correspond to the principal axes of strain for a Hookean material.

16.6. [16.3] The stress tensor at a point is given as follows:

$$\tau_{ij} = \begin{pmatrix} 3000 & 4000 & 0 \\ 4000 & 5000 & -2000 \\ 0 & -2000 & -7000 \end{pmatrix}$$

If $E = 30 \times 10^6$ and $\nu = .3$, what is the normal strain in the direction

$$\varepsilon = .6\mathbf{i} - .8\mathbf{j}$$

at the point?

16.7. [16.3] Show, using Fig. P.16.7 and the general three-dimensional transformation equations, that for plane strain

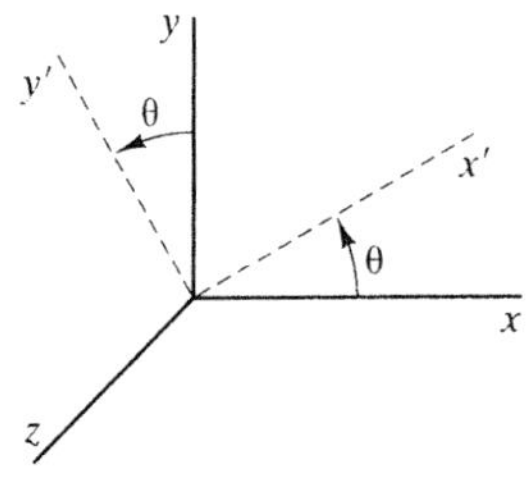

Figure P.16.7.

$$\varepsilon_{x'x'} = \frac{\varepsilon_{xx} + \varepsilon_{yy}}{2} + \frac{\varepsilon_{xx} - \varepsilon_{yy}}{2} \cos 2\theta + \frac{\gamma_{xy}}{2} \sin 2\theta$$

16.8. [16.3] The principal strains at a point are

$$\varepsilon_{xx} = .003$$
$$\varepsilon_{yy} = .001$$
$$\varepsilon_{zz} = 0$$

What is $\gamma_{x'y'}$ for axes rotated 30° about the z-axis as shown in Fig. P.16.8?

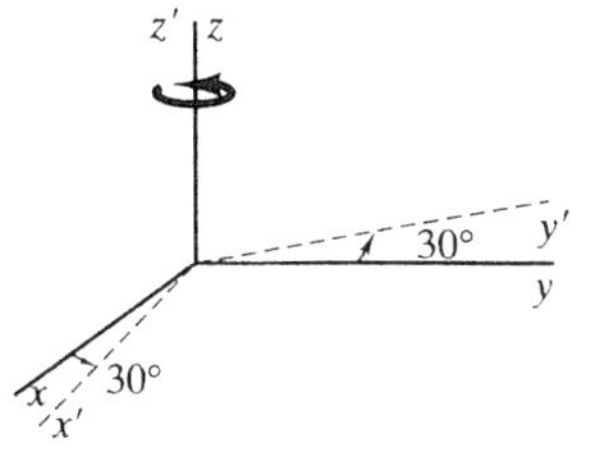

Figure P.16.8.

16.9. [16.3] Given the displacement field

$$\boldsymbol{u} = [(x^2 + 3y)\mathbf{i} + (3y^3 + yz)\mathbf{j} + (3x^2 + z^2)\mathbf{k}] \times 10^{-6} \text{ m}$$

what are the principal strains at position (0, 0, 3) m?

16.10. [16.3] Given the strain tensor

$$\varepsilon_{ij} = \begin{pmatrix} .002 & 0 & 0 \\ 0 & .003 & .002 \\ 0 & .002 & -.001 \end{pmatrix}$$

What are the principal strains?

16.11. [16.3] A submarine is at a depth of 600 m. Strain gauges indicate that along two orthogonal directions the normal strains in directions tangent to the outer surface of the sub are

$$\varepsilon_a = -10 \times 10^{-4}$$
$$\varepsilon_b = -18 \times 10^{-4}$$

Also along a line at 45° to a and b, the normal strain is $\varepsilon = -9 \times 10^{-4}$. What are the principal strains at that point? Take γ of seawater as 10,150 N/m³. Recall that pressure below the free surface is given as γd, where d is the depth below the free surface. Take $p_{\text{atm}} = 101{,}325$ Pa. The sub is stationary relative to the water. Take $E = 2 \times 10^{11}$ Pa and $\nu = .3$.

16.12. [16.3] What are the principal strains at point A close to the base (see Fig. P.16.12)? Take $E = 2 \times 10^{11}$ Pa and $\nu = .3$. A force distribution at A causes a compression stress of $\tau_{zz} = -2 \times 10^7$ Pa.

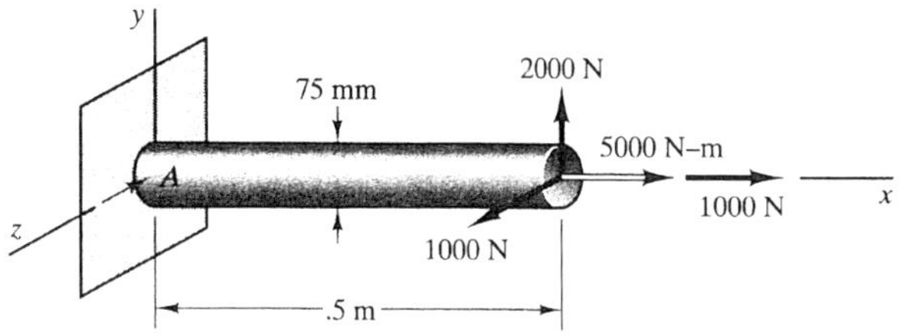

Figure P.16.12.

16.13. [16.3] In Chapter 2 we examined the relative movement of adjacent points (see Fig. P.16.13). We assumed with no loss in generality that A was stationary. Show that the displacements a and b are given as

$$u_a = \frac{\partial u_y}{\partial x} dx + \cdots$$

$$u_b = \frac{\partial u_x}{\partial y} dy + \cdots$$

Hence show that the change of right angles of AC and AB is

$$\varepsilon_{xy} = \frac{1}{2}\left(\frac{\partial u_x}{\partial y} + \frac{\partial u_y}{\partial x}\right)$$

Also note that the average angular velocity of dx and dy about z is

$$\omega_z = \frac{1}{2}\left(\frac{\partial \dot{u}_x}{\partial y} - \frac{\partial \dot{u}_y}{\partial x}\right)$$

Show that

$$\omega_x = \frac{1}{2}\left(\frac{\partial \dot{u}_y}{\partial z} - \frac{\partial \dot{u}_z}{\partial y}\right)$$

$$\omega_y = \frac{1}{2}\left(\frac{\partial \dot{u}_z}{\partial x} - \frac{\partial \dot{u}_x}{\partial z}\right)$$

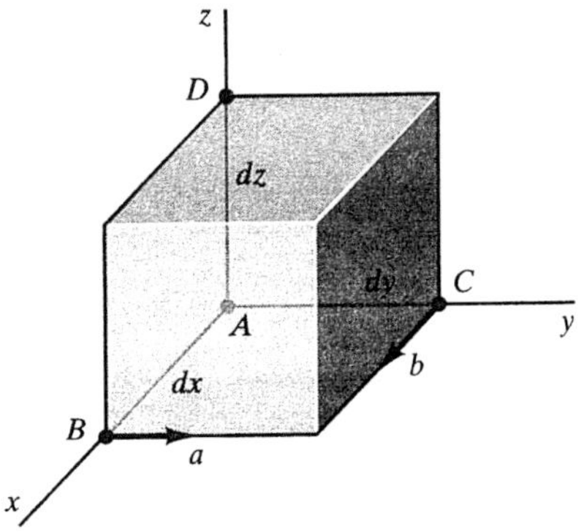

Figure P.16.13.

16.14. [16.3] A plastic is flowing into a mold to form some desired shape. The velocity field at some time t in a small given domain is known to be given as

$$V = [(x^2 + y)\mathbf{i} + (\dot{z}y)\mathbf{j} + (2 + zy)\mathbf{k}] \times 10^{-2}$$

What are the strain rate components $\dot{\varepsilon}_{x'x'}$, $\dot{\varepsilon}_{x'y'}$, for axes rotated counterclockwise 30° about the z-axis as you look toward the origin? Give their values at position (1, 3, 4).

16.15. [16.3] Given the following state of strain at a point,

$$\begin{pmatrix} 0 & .02 & 0 \\ .02 & -.01 & -.03 \\ 0 & -.03 & 0 \end{pmatrix}$$

what are the maximum and minimum values of normal strain there?

16.16. [16.3] Given the following state of strain at a point,

$$\begin{pmatrix} .02 & 0 & -.01 \\ 0 & -.01 & 0 \\ -.01 & 0 & .03 \end{pmatrix}$$

what is the direction of the algebraically minimum normal strain at the point?

For those readers who have examined the *Look Ahead* Section 15.5, the following two exercises are presented involving two of the most important equations of solid mechanics.

16.17. [16.3] Show that the strain-displacement relations of Chapter 3 all can be found from the following equation in index notation.[3]

$$\varepsilon_{ij} = \frac{1}{2}\left(\frac{\partial u_i}{\partial x_j} + \frac{\partial u_j}{\partial x_i}\right) \qquad (16.18)$$

where u_i is single-valued and continuous.

***16.18. [16.3]** Using the strain-displacement relations in index notation from Problem 16.17, form new equations by taking the following second-order partial derivatives:

$$\frac{\partial^2 \varepsilon_{ij}}{\partial x_k \partial x_l} \qquad \frac{\partial^2 \varepsilon_{ij}}{\partial x_k \partial x_i}$$

$$\frac{\partial^2 \varepsilon_{kl}}{\partial x_i \partial x_j} \qquad \frac{\partial^2 \varepsilon_{ki}}{\partial x_l \partial x_j}$$

Show that when we add the left-hand relations and subtract the relations on the right, the right-hand side of the resulting equation is zero. That is,

$$\frac{\partial^2 \varepsilon_{ij}}{\partial x_k \partial x_l} + \frac{\partial^2 \varepsilon_{kl}}{\partial x_i \partial x_j} - \frac{\partial^2 \varepsilon_{ij}}{\partial x_k \partial x_i} - \frac{\partial^2 \varepsilon_{ki}}{\partial x_l \partial x_j} = 0 \qquad (16.19)$$

Because the equations above were developed from Eq. (16.18), where strains were properly related to a single-valued displacement field, they *must be satisfied by compatible strain fields.* In other words, these are the *compatibility equations* in index notation.

[3]Subscripts can take on values x, y, and z.

CHAPTER 17

Introduction to Elastic Stability

17.1 Introduction

The problems that we have been discussing thus far have been those for which there was stable equilibrium. We have learned in mechanics that other types of equilibria are possible, namely, neutral equilibrium and unstable equilibrium. In this chapter we shall consider these cases. Such considerations become important when one deals with structures having stringent weight restrictions, such as aircraft and missile structures.

17.2 Definition of Critical Load

To introduce the concepts of the critical load we shall begin by examining the rod shown in Fig. 17.1(a) pinned at A. Neglecting the weight of the

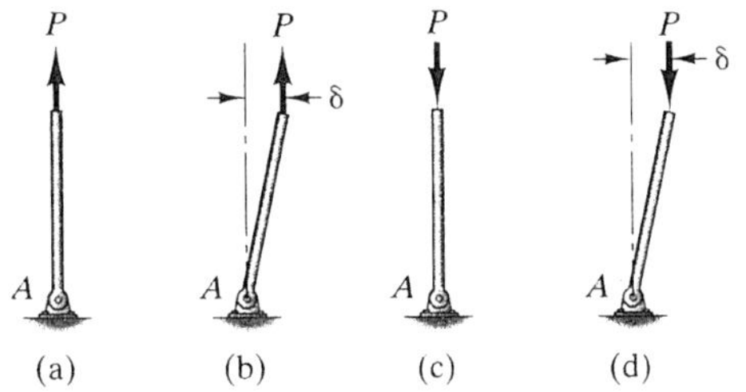

Figure 17.1. Weightless pinned rod for stability consideration.

member it should be clear that the application of a tensile force P will result in a *stable* equilibrium. That is, a small displacement δ to either side [see Fig. 17.1(b)] will induce a *restoring torque* $P\delta$ about A to bring the member back to the vertical orientation. Contrast this to the case where P is reversed, such as shown at Fig. 17.1(c). Here we have *unstable* equilibrium since any deflection δ [see Fig. 17.1(d)] will induce a *nonrestoring moment* about A. There is then set forth a precipitous action with increasing torque and increasing deflection feeding on each other. The case of *neutral* equilibrium is best illustrated by the hard sphere on a flat surface. Here any displacement induces neither restoring nor nonrestoring torques. The preceding stability problems involve no elastic deformations of the rods and are simply the kind of rigid-body stability problems considered in earlier mechanics courses.

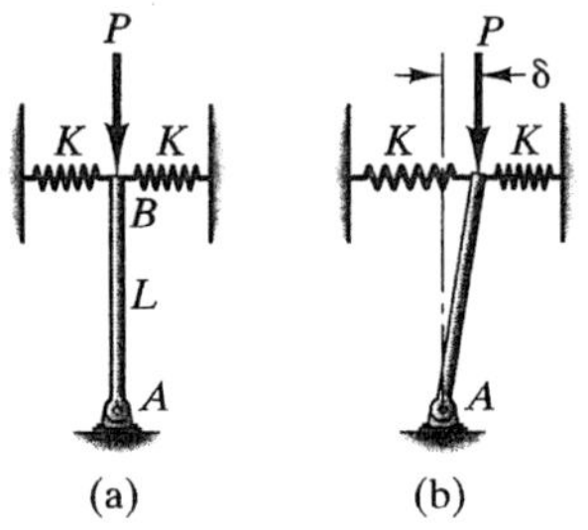

Figure 17.2. Pinned rod with elastic side constraints.

Now let us consider a vertical rod with two horizontal spring forces at B, as shown in Fig. 17.2(a). These springs bring in a certain stabilizing influence on the structure because now, for small enough loads P, there will be introduced for a small deflection δ a net restoring torque about A. There is thus stable equilibrium for a small enough force P. Note that the torque from the springs will be given by $2KL\delta$ in the restoring direction, whereas the load P yields a nonrestoring torque $P\delta$. Hence, if by increasing P we reach a condition where

$$2KL\delta = P\delta \tag{17.1}$$

then clearly we have zero net restoring torque and thus a case of neutral equilibrium. The member can theoretically remain at equilibrium for positions corresponding to any small deflection δ under these conditions.[1] The load which causes this system to attain a condition of neutral equilibrium is termed the *critical* or *buckling load,* and is denoted as P_{cr}. Thus, solving for P_{cr} in the preceding equation, we have

$$P_{cr} = 2KL \tag{17.2}$$

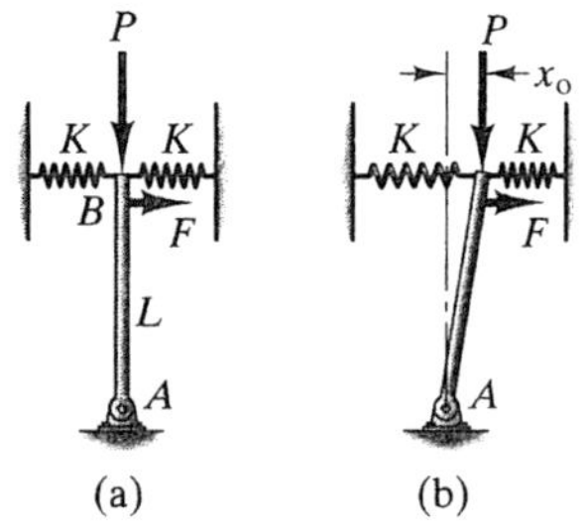

Figure 17.3. Transverse load F added to system.

If now we apply a load P exceeding P_{cr}, we then have unstable equilibrium, since the nonrestoring torque from P is now always larger than the restoring torque from the springs for any deflection δ. Once again we see the possibility of precipitous action described earlier.

We shall now present an important consequence of the above mentioned criticality with the aid of a slightly modified version of the preceding problem. We now apply a transverse force F at B as shown in Fig. 17.3(a). Now clearly the vertical position is no longer a position of equilibrium, but instead some deflected position denoted by x_0 as in Fig. 17.3(b) corresponds to a possible condition of equilibrium. We can ascertain such a position by taking moments about point A of the forces acting on the bar and setting the sum of these moments equal to zero. Thus,

$$Px_0 + FL - 2Kx_0L = 0 \tag{17.3}$$

[1]This should be clear because δ cancels out of Eq. (17.1).

Solving for x_0, we get

$$x_0 = \frac{FL}{2KL - P} = \frac{F}{2K}\frac{1}{1 - P/2KL} \tag{17.4}$$

When P is very small, the deflection x_0 is approximately $F/2K$, a result which does not surprise us. But when P approaches the value $2KL$, then x_0 approaches infinity, since a denominator on the right side of Eq. (17.4) goes to zero. Of course, the "blowing up" of x_0 is only a mathematical phenomenon. Actually the preceding formulation is valid only for very small x_0 and so becomes meaningless when x_0 becomes large. Note, however, that the load P causing the deflection to go mathematically to infinity for any transverse force F, however small, is none other than the critical load formulated for this rod and spring system earlier. We may thus present the important consequence of the *critical load. In the presence of a critical load* P_{cr} *some other transverse force* F, *however small, can cause mathematically infinite deflection.*[2]

A moment's thought will further explain what the mathematics has so vividly portrayed. If we have neutral equilibrium, then clearly there is no way that the system can resist the transverse load since any and all configurations theoretically can be reached without outside effort. Any transverse force, however small, obviously will under these circumstances cause unconstrained deflection.

17.3 A Note on Types of Elastic Instabilities

In the preceding section we described a very simple elastic instability and set forth the concept of the critical load that attended this type of elastic instability. You will note, in retrospect, that a force P generated a nonrestoring moment that (and this is vital) grew with increasing deflection. A restoring torque meanwhile was supplied by springs that were outside the member. In more practical cases we have elastic instabilities wherein the restoring elastic action is supplied by the member itself, not by an outside agent. There are many such types of elastic instabilities of interest to the engineer. They all have the common trait that we examined carefully in Section 17.2 and that is this:

> As a result of certain critical loadings there is a possibility of large dangerous deformations resulting from moments or moment

[2]Actually, what will happen is that as $P \to P_{cr}$, large deflections will be developed as a result of the transverse load F. The approximate theory expresses this by having the deflection blow up. Therefore, critical loads are a signal of possibly dangerous large deflections in actual situations.

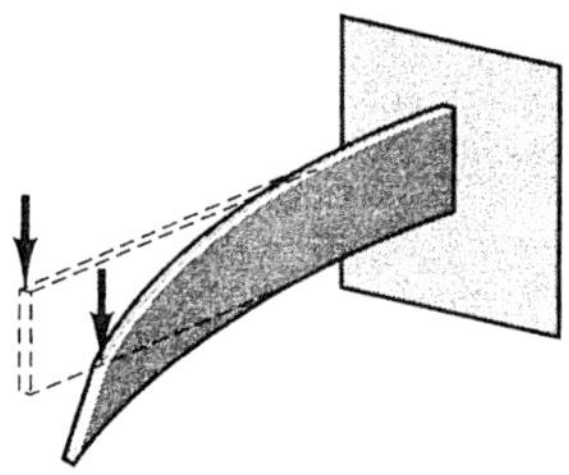

Figure 17.4. Lateral buckling of a cantilever beam.

distributions that grow with the deformations so as to cause the kind of precipitous actions akin to those described earlier in the very simple cases.

We shall list some of the more common types:

1. Compressive buckling of a column
2. Lateral buckling of a deep narrow beam (see Fig. 17.4)
3. Buckling of a cylinder under torsion (see Fig. 17.5)
4. Crumpling of a cylinder under a compressive force (see Fig. 17.6)
5. Twist buckling of a shaft in torsion.

Figure 17.5. Torsion buckling of cylinder.

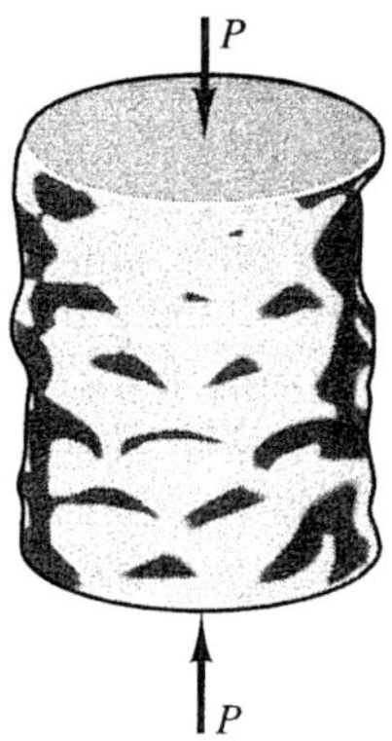

Figure 17.6 Compression buckling of a cylinder.

We shall be able to consider only the first case in this text.

As a preliminary consideration, recall from our study of beams and shafts that a stress exceeding the elastic limit does not necessarily mean failure of the member as a structural element. We have a similar situation pertaining to buckling. We shall compute elastic instabilities on the basis of small deflections as we have already done in the previous section. The theory may indicate as a result of elastic instability that the structure is incapable of carrying the required loads, but we must keep in mind that the thcory is valid only for small deflections. Thus, it may be true that, after the structure has buckled a certain degree in accordance with the prediction of small deformation theory, the new geometry reached may be one which is stable and capable of withstanding further buckling. Hence, to check to see whether a structure will completely fail its mission, we would have to use large deformation theory requiring more advanced techniques beyond the scope of this text.[3]

[3]The further buckling beyond the realm of small deformation is called *postbuckling* in the literature.

Knowing critical loads from small deformation theory, however, is vital for the design engineer since the probability of failure becomes large when actual loads approach computed critical loads.

17.4 Beam-Column Equations

We shall introduce the beam-column as a beam supporting a transverse loading $w(x)$ and at the same time subject to an axial load. We have illustrated this in Fig. 17.7, where the more interesting case of axial

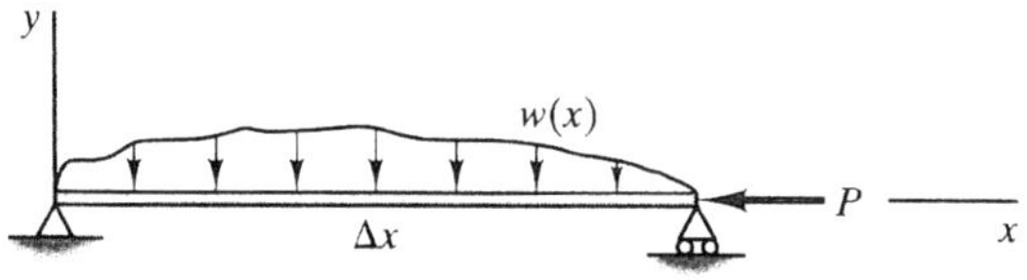

Figure 17.7. Beam-column.

compression rather than axial tension is shown. It should be clear that the moment developed at any section x by the force P will depend on the deflection at that point. For long slender beams, then, we have to employ the *deformed geometry* in writing the equilibrium equations. This will be a "first" for us.

We shall consider the equilibrium of an element Δx of the beam column as shown in Fig. 17.8 in a deformed condition. Note that Δv is the

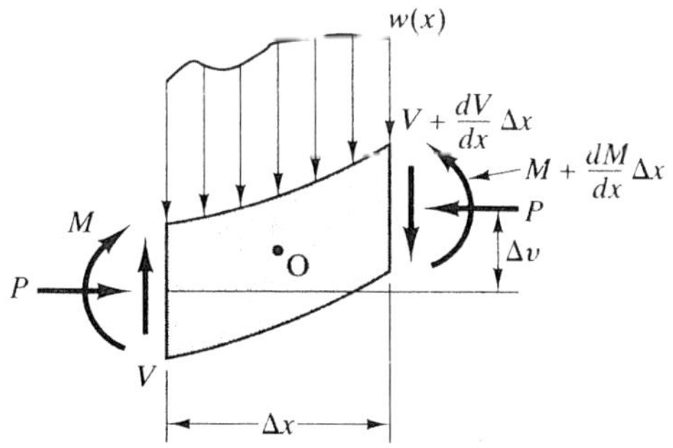

Figure 17.8. Beam-column element.

change in elevation of the neutral surface occurring over the length Δx of the element. Summing forces in the y direction, for *equilibrium* we have

$$V - \left(V + \frac{dV}{dx}\Delta x\right) - [w(x)]_{\text{avg}}\Delta x = 0 \tag{17.5}$$

Simplifying and going to the limit with $\Delta x \to 0$ we get

$$\frac{dV}{dx} = -w \tag{17.6}$$

a result we have formulated in the past. Now taking moments about centerpoint *O*, we get an equation different from what we have had in the past. That is,

$$\left(M + \frac{dM}{dx}\Delta x\right) - M - V\frac{\Delta x}{2} - \left(V + \frac{dV}{dx}\Delta x\right)\frac{\Delta x}{2} + P\Delta v + (w)(\Delta x)(k\Delta x) = 0$$

where k is some fraction. Canceling terms where possible, we next divide by Δx and then go to the limit with $\Delta x \to 0$. We get

$$\frac{dM}{dx} - V + P\frac{dv}{dx} = 0 \tag{17.7}$$

If we assume that the deflection of the beam is due only to bending deformation (i.e., neglect shear deformation), we may express M as [see Eq. (12.4)]

$$M = EI\frac{d^2v}{dx^2} \tag{17.8}$$

Substituting into Eq. (17.7), we get

$$\frac{d}{dx}\left(EI\frac{d^2v}{dx^2}\right) - V + P\frac{dv}{dx} = 0 \tag{17.9}$$

Differentiating once again with respect to x and using Eq. (17.6) we then have

$$\frac{d^2}{dx^2}\left(EI\frac{d^2v}{dx^2}\right) + P\frac{d^2v}{dx^2} = -w(x) \tag{17.10}$$

If EI is constant, we then get for Eq. (17.10)

$$\left(\frac{d^4v}{dx^4}\right) + \frac{P}{EI}\left(\frac{d^2v}{dx^2}\right) = -\frac{w}{EI} \tag{17.11}$$

This is the differential equation of the beam-column. We shall now consider the special case of the column.

17.5 The Column: Buckling Loads

Let us now consider the case where $w(x) = 0$, leaving only the axial load. The beam-column can now be considered as a *column*. This case is shown in Fig. 17.9. As the load grows one can feel intuitively that as it reaches a certain value the column will "give way" suddenly with a large deflection being developed. Everyone who has used as a child a thin stick as a cane has experienced this. The load that permits the aforestated precipitous action is the critical or buckling load, akin to the load described as the critical load for rigid bodies connected to springs described in Section 17.2. What is different here is that the member itself supplies the elastic forces and not external springs. However, we shall see that, as before, the critical load, P_{cr}, will be that load which permits the member to be in neutral equilibrium. This means that for a given load P_{cr} there are an infinite number of small deflection geometries which theoretically can exist in equilibrium. We have shown two such configurations as dashed lines in Fig. 17.9. With this in mind we shall proceed to solve Eq. (17.11) with $w = 0$.

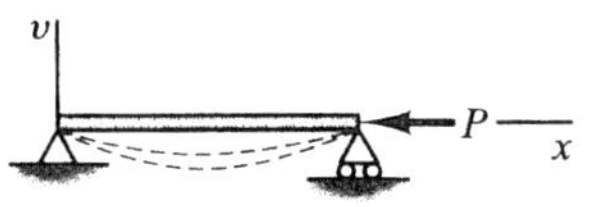

Figure 17.9. Column pinned at ends.

Let us as a first step replace d^2v/dx^2 by m to form the equation

$$\frac{d^2m}{dx^2} + \frac{P}{EI}m = 0 \tag{17.12}$$

This is a familiar equation, the solution of which is

$$m = A \sin\sqrt{\frac{P}{EI}}x + B\cos\sqrt{\frac{P}{EI}}x \tag{17.13}$$

where A and B are constants of integration. Now, replacing m by d^2v/dx^2 and integrating twice, we get

$$v = -A\frac{EI}{P}\sin\sqrt{\frac{P}{EI}}x - B\frac{EI}{P}\cos\sqrt{\frac{P}{EI}}x + C_3x + C_4 \tag{17.14}$$

Replacing $-A(EI/P)$ and $-B(EI/P)$ by C_1 and C_2, respectively, we have as the general solution to the differential equation

$$v = C_1\sin\sqrt{\frac{P}{EI}}x + C_2\cos\sqrt{\frac{P}{EI}}x + C_3x + C_4 \tag{17.15}$$

We now subject this solution to the boundary conditions of the column which is pin-connected at both ends. Thus, we require that[4]

$$\text{When } x = 0 \begin{cases} v = 0 \\ \dfrac{d^2v}{dx^2} = 0 \end{cases} \qquad \text{When } x = L \begin{cases} v = 0 \\ \dfrac{d^2v}{dx^2} = 0 \end{cases} \tag{17.16}$$

Applied to Eq. (17.5) the foregoing conditions lead to the following equations:

$$v(0) = 0 = C_2 + C_4 \tag{17.17a}$$

$$\left(\frac{d^2v}{dx^2}\right)_{x=0} = 0 = -C_2\frac{P}{EI} \tag{17.17b}$$

$$v(L) = 0 = C_1 \sin\sqrt{\frac{P}{EI}}L + C_2\cos\sqrt{\frac{P}{EI}}L + C_3L + C_4 \tag{17.17c}$$

$$\left(\frac{d^2v}{dx^2}\right)_{x=L} = 0 = -C_1\frac{P}{EI}\sin\sqrt{\frac{P}{EI}}L - C_2\frac{P}{EI}\cos\sqrt{\frac{P}{EI}}L \tag{17.17d}$$

From Eq. (17.17b) we see that $C_2 = 0$ and Eq. (17.17a) gives us the result $C_4 = 0$. This leaves us with two equations involving C_1 and C_3. Thus,

$$C_1 \sin\sqrt{\frac{P}{EI}}L + C_3L = 0 \tag{17.18a}$$

$$C_1\frac{P}{EI}\sin\sqrt{\frac{P}{EI}}L = 0 \tag{17.18b}$$

If we set $C_1 = 0$ in Eq. (b), then clearly C_3 must also be zero as per Eq. (a). This gives us the trivial solution that zero deflection is an admissible equilibrium configuration. For a *nontrivial* solution, we require for Eq. (b) that

$$\sin\sqrt{\frac{P}{EI}}L = 0 \tag{17.19}$$

This then requires that $C_3 = 0$ for satisfaction of Eq. (17.18a) and places no restriction on the value of C_1. Hence, we may say that for nontrivial solutions we have the *necessary* requirement that

[4]Note that $M \,\alpha\, d^2v/dx^2$. For pin, $M = 0$, hence $d^2v/dx^2 = 0$.

$$\sqrt{\frac{P}{EI}}L = n\pi, \qquad n = 1,2,\ldots \tag{17.20}$$

so as to render the sine term equal to zero. The value P satisfying this condition is the *critical load,* P_{cr}. We thus have

$$(P_{cr})_n = \frac{n^2\pi^2 EI}{L^2} \tag{17.21}$$

Note again that constant C_1 has been left unscathed in the process of satisfying Eqs. (17.18). We are justified in calling the forces in Eq. (17.21) critical forces since for each $(P_{cr})_n$ there is permitted a deflection curve which has an *arbitrary* but small *amplitude* C_1 and which satisfies *equilibrium* requirements as well as *boundary conditions.* We have accordingly reached neutral equilibrium on arriving at the loads $(P_{cr})_n$. For any n, Eq. (17.15) becomes

$$v_n = C_n \sin\sqrt{\frac{(P_{cr})_n}{EI}}\,x \tag{17.22}$$

where C_n is an arbitrary constant (for which we have just considered C_1), which, of course, must be reasonably small so as not to violate the small deformation limitation of the theory. (We shall shortly explain the physical significance of the critical loads for n exceeding unity.) It is the usual practice to call the lowest critical load $(P_{cr})_1$ the *Euler load,* or the buckling load, and we shall at times denote this load as P_E.

To summarize the results, we shall consider the plot of P/P_E, the actual load over the Euler load, versus the amplitude δ of the deflection curve (see Fig. 17.10). For $P/P_E < 1$ (i.e., for values of P below the Euler load) we have learned that only the trivial solution $v = 0$ is permitted. Hence, P/P_E versus δ coincides with the P/P_E axis, as shown in the diagram. When $P/P_E = 1$ we have the trivial solution $v = 0$ possible as well as, now, an infinity of solutions (half-wave sinusoids) with arbitrary (but small) amplitude. Thus, in Fig. 17.10 at $P/P_E = 1$ there is a horizontal line present. As P/P_E exceeds unity again only the trivial solution $v = 0$ is present.[5] Thus, at $P/P_E = 1$ the plot has two branches—the horizontal branch and the vertical branch along the P/P_E axis. A branch point such as this is called a *bifurcation point.* At the second critical load, with $P/P_E = 4$ [note from Eq.

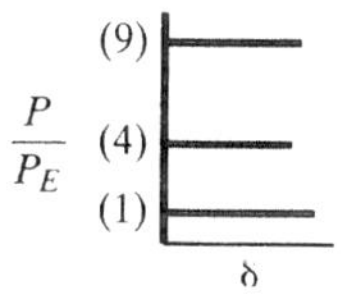

Figure 17.10. Plot of P/P_E versus δ showing bifurcation points.

[5]The reader is cautioned that the column we are dealing with is an *ideal* column. As will be explained more fully later, an actual column will undergo large deflections when P_E is approached and more so when it is exceeded. This is so since a real column, among other things, is never *perfectly straight* as is assumed in our ideal column discussed here.

(17.21) that we have n^2 on the right-hand side], we again have a bifurcation point. This is also the case for the third critical load with $P/P_E = 9$, and so on.

What is the physical significance of the results that we have attained thus far? Let us consider a load P which approaches the lowest critical load $(P_{cr})_1$. It should be clear that we are approaching a neutral equilibrium of the type just described and that any given transverse load will cause increasingly greater deflections as P gets closer to the critical load. Thus, near the critical load there is danger of failure of the member as a load-carrying device. The shape of the family of deflec-

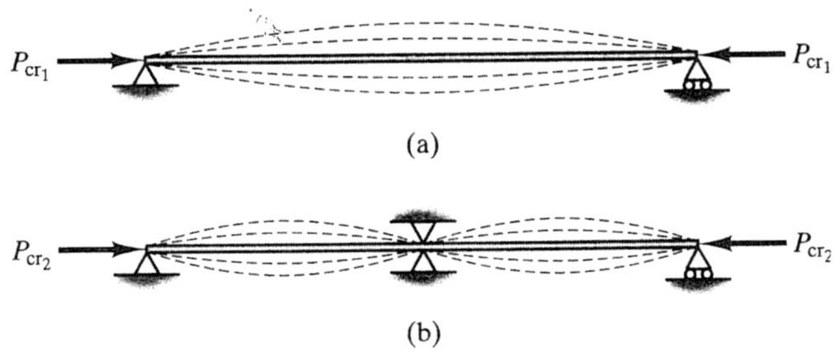

Figure 17.11. First and second buckling modes.

tions of the beam for $n = 1$ is called the *first buckling mode* [see Fig. 17.11(a)] and we say in the case of a failure of the type just described that the member has failed in the first buckling mode.

Let us suppose next that we apply lateral constraints at the midpoint of the beam as shown in Fig. 17.11(b) to prevent any vertical motion of the beam there. Now we may safely increase the load P so as to pass through the first critical value since none of the permitted deformations for neutral equilibrium in the first mode are now possible except $v = 0$. When P equals $(P_{cr})_2$, that is, the second critical load, we see from Eq. (17.21) that

$$v_2 = C_2 \sin \sqrt{\frac{4\pi^2 EI}{L^2 EI}}\, x = C_2 \sin \frac{2\pi}{L} x \tag{17.23}$$

The second buckling modes are a family of full sinusoids with a nodal point in the center, in contrast to the half-waves between the supports shown in Fig. 17.11(a) for the first buckling mode. For this configuration [see Fig. 17.11(b)] there is theoretically no lateral force at the midconstraints and in our setup the midconstraints could theoretically be removed. With the midconstraints in place, it should be clear that with P approaching the second critical load a given transverse load on one of the spans would cause large deflections, and might cause a failure of the system in the second buckling mode.

In a similar way by using additional lateral constraints, we can show the possibility of a neutral equilibrium configuration wherein the beam has two nodal points as shown in Fig. 17.12. Thus, the higher

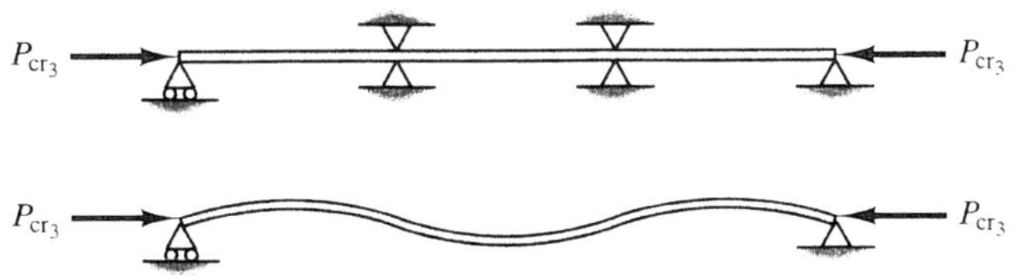

Figure 17.12. Third buckling mode.

buckling loads correspond to more complex buckling modes. Generally, we shall be interested in the lowest buckling mode since it takes considerable care to be able to reach the higher buckling modes.

Before we go any further, it must be emphasized to the reader that the column analyzed was an ideal column not to be found in the real world. We will mention at this time two such degrees of perfection built into the theory:

1. The column is perfectly straight.
2. The axial load is exactly collinear with the centerline of the column.

In mechanics parlance condition 1 requires *no initial curvature* of the column while condition 2 requires *zero eccentricity* of loading. The actual behavior of a real column lacking the perfections above can only approach the behavior we have described for the ideal case as the imperfections are continually decreased. We shall look into these interesting cases later. We shall only point out here that initial curvature and eccentricity of load have the effect of a transverse load. Hence, with the approach of neutral equilibrium as one gets near a critical load one can expect large deformations for real-world columns. Thus, the ideal formulation signals to us when we can expect large deformations for real-world columns. Our calculations, thus far limited as they are to ideal cases, nevertheless have much practical importance to us.

We have thus far considered only the buckling of a pin-ended column. The buckling load, it is to be pointed out, is strongly dependent on the end conditions of the column. By the method followed in this section for computing the buckling load of the pin-ended column we can determine buckling loads for other end conditions. In Fig. 17.13 we

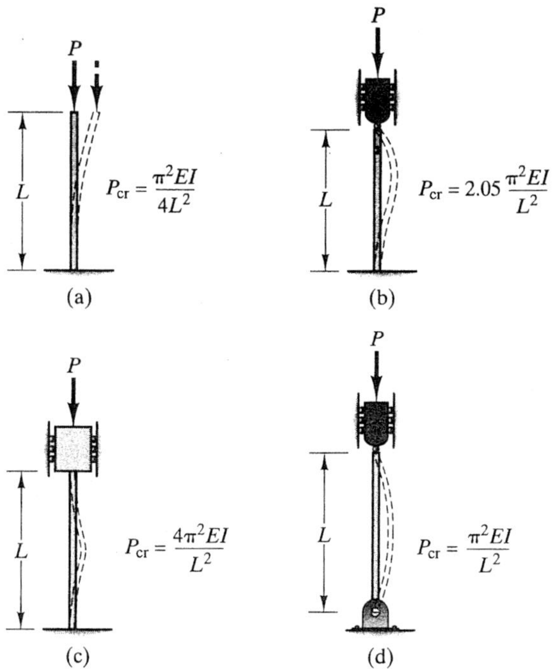

Figure 17.13. Buckling loads for common types of columns: (a) clamped-free column; (b) clamped-hinged column; (c) clamped-clamped column; (d) pin-ended column.

show results for some common end conditions, including the pin-ended case developed in this section.

In the following example we shall illustrate how we may use the ideal results of this section for purposes of design. You will see that we assume ideal behavior but to take into account departures from ideal behavior, such as we have mentioned in this section, plus others we employ a safety factor.

Example 17.1

A steel container D [see Fig. 17.14(a)] is supported in such a way that the system can quickly be disassembled. Specifically, two identical angle irons, AB, pinned at both ends, support one end A of the container, while the other ends B are pinned to the wall. The total weight of the container and contents is 40,000 lb, having a cen-

Example 17.1 (Continued)

ter of gravity at the geometrical center of the container as shown. Choose the proper angle iron size using, as a safety factor, three times the expected load. Take $E = 30 \times 10^6$ psi and $Y = 36{,}000$ psi.

The members AB, being long and slender, must be designed with buckling as the foremost consideration. Also, we want the cheapest angle iron that will do the job safely. To take the safety factor into account, we shall consider that at the total load of 120,000 lb we have incipient buckling in each member. The axial force in each angle iron is seen from the free body of pin A [Fig. 17.14(b)] to be (**equilibrium**)

$$F_{AB}(.707) = 30{,}000$$

$$\therefore F_{AB} = 42{,}433$$

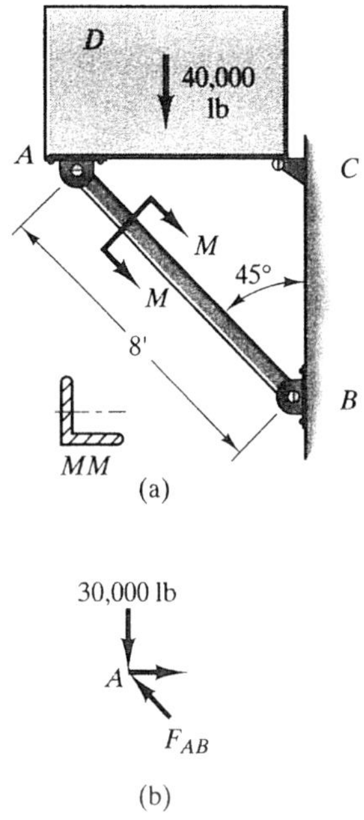

Figure 17.14. Design of angle iron brace.

We shall now suppose that at the load of 42,433 lb we get incipient buckling in the members. Hence, using Eq. (17.21) for the first mode we can say for the required value of I that (**constitutive law**)

$$42{,}434 = \frac{\pi^2(30 \times 10^6)(I)}{(8^2)(144)}$$

$$\therefore \quad I = 1.320 \text{ in.}^4$$

Considering the tables in Appendix IV-C we see that a $3 \times 3 \times \frac{3}{8}$ angle will do the job.

As a final step we check the normal stress in each angle. Using data from Appendix IV-C we can say that

$$\tau = \frac{42{,}433}{2.11} = 20{,}110 \text{ psi}$$

This normal stress is below the yield stress of the material.

Example 17.2

A rigid block G in Fig. 17.15 is guided by a frictionless vertical rod AB and is supported by two vertical rods CD.

Example 17.2 (Continued)

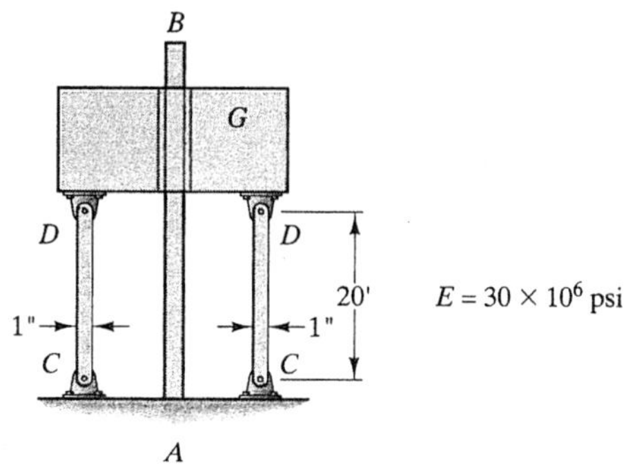

Figure 17.15. Rods CD support the weight G which is guided by column AB.

(a) What is the maximum weight W that the rods can support for a safety factor of 2?
(b) What is the length L of the rods and the weight W of G in order for buckling and crushing to occur simultaneously with no safety factor? Take $Y = 30{,}000$ psi.

Part a.

With a safety factor of 2, the load per member is W. For *buckling* in the first mode we have

$$W = \left[\frac{1^2\pi^2 EI}{L^2}\right] \qquad I = \frac{\pi D^4}{64} = \frac{(\pi)(1^4)}{64} = .0491 \text{ in.}^4$$

Hence, for W we get

$$W = 252 \text{ lb}$$

Part b.

Without the safety factor, we have the normal stress for crushing as

$$\tau = 30{,}000 = \frac{W/2}{\left(\frac{\pi}{4}\right)(1^2)}$$

Hence,

$$W = 47{,}124 \text{ lb}$$

For *buckling* we have

Example 17.2 (Continued)

$$\frac{47{,}124}{2} = \frac{1^2\pi^2 EI}{L^2} = \frac{(\pi^2)(30 \times 10^6)(.0491)}{L^2}$$

with L in inches. The desired length of L in feet is then,

$$L = 2.07 \text{ ft}$$

Example 17.3

An ideal solid rod is installed between two fixed supports as shown in Fig. 17.16 at a uniform temperature T. If the rod material has a constant coefficient of expansion α, what is the increase in temperature at which elastic instability first occurs? Assume that E remains constant.

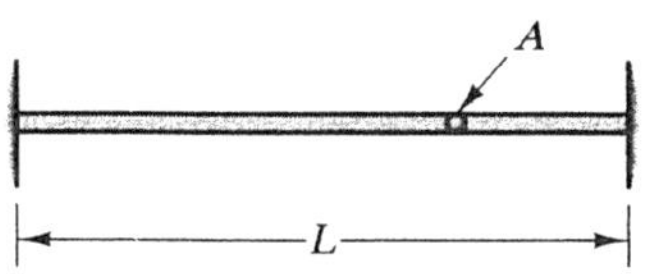

Figure 17.16. Solid rod clamped at both ends to fixed supports.

Since the ends are fixed, **compatibility** requires that $\varepsilon_{xx} = 0$. Hence, using Hooke's law as the **constitutive law** the compressive stress τ in the rod is given as

$$\tau = \alpha(\Delta T)E \tag{a}$$

The critical stress τ_{cr} then is given as

$$\tau_{cr} = \frac{P_{cr}}{A}$$

and so going back to Eq. (a), we have on substituting the above for τ at the critical state

$$\frac{P_{cr}}{A} = \alpha(\Delta T)_{cr}E \qquad \therefore\ (\Delta T)_{cr} = \frac{P_{cr}}{A\alpha E} \tag{b}$$

For a clamped-clamped column we have from Fig. 17.13(c) for the first buckling mode,

$$P_{cr} = 4\frac{\pi^2 EI}{L^2}$$

Going to Eq. (b), we get the desired temperature increase $(\Delta T)_{cr}$ as follows:

$$(\Delta T)_{cr} = \frac{4\pi^2 I}{\alpha A L^2}$$

17.6 Looking Back as Well as Ahead

The general aspects of the formulations that we developed in the preceding section are by no means restricted to the phenomenon of buckling. Actually, we found that certain boundary conditions in our analysis restricted a parameter P in our differential equation to an infinite set of discrete values which we called the critical or buckling loads. In a boundary value problem of this sort, sometimes called a *Sturm–Liouville boundary-value problem,* the set of values assumed by a parameter in the differential equation required for a nontrivial solution is called a set of *eigenvalues;* the corresponding nontrivial solutions of the dependent variable excluding arbitrary constants are called the *eigenfunctions.*

In the study of the vibration of continuous systems, such as strings, membranes, and plates, the eigenvalues turn out to be the natural frequencies of the system, whereas the eigenfunctions are the mode shapes for these vibrations. In the study of the "whirling of shafts" the eigenvalues are the critical angular speeds of the shaft. Thus, you will surely see this kind of mathematical formulation in your other engineering science courses, not to speak of your advanced mathematics courses.

17.7 Solution of Beam-Column Problems

We have pointed out that when P becomes critical for a column then transverse loads cause very large deformations. We can show this and at the same time illustrate the solution of a beam-column problem.

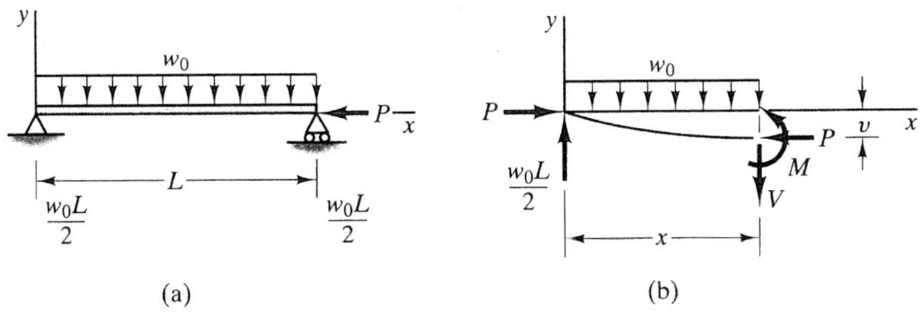

Figure 17.17. Pin-ended beam column problem.

Accordingly, consider the beam column shown in Fig. 17.17(a). We could employ Eq. (17.11) for this case, but because of the simple loading present, we can easily determine the bending moment M as a function of x and having done this go to Eq. (17.8) to work with a second-order rather than a fourth-order differential equation. Hence, the bending moment at any section x [see Fig. 17.17(b)] is seen from statics to be

$$M = \frac{w_0 L}{2} x - \frac{w_0 x^2}{2} - Pv \tag{17.24}$$

Hence, for the deflection equation we have

$$EI \frac{d^2 v}{dx^2} = \frac{w_0 L}{2} x - \frac{w_0 x^2}{2} - Pv$$

$$\therefore \quad \frac{d^2 v}{dx^2} + \frac{P}{EI} v = \frac{w_0}{2EI}(Lx - x^2) \tag{17.25}$$

The general solution to the differential equation may be given as the general *complementary solution* v_c (with the right-hand side set equal to zero) plus a *particular solution* v_p, which is any solution to the full differential equation.[6] Thus,

$$v = v_c + v_p$$

where

$$\frac{d^2 v_c}{dx^2} + \frac{P}{EI} v_c = 0 \tag{17.26}$$

The general solution of the above equation is

$$v_c = C_1 \sin \sqrt{\frac{P}{EI}}\, x + C_2 \cos \sqrt{\frac{P}{EI}}\, x \tag{17.27}$$

where C_1 and C_2 are the required two arbitrary constants of integration. As for the particular solution we shall examine a possibility $v_p = Ax + Bx^2 + C$ and attempt on substitution into Eq. (17.25) to determine A, B, and C to satisfy the full differential equation. Hence, on substituting we get

$$2B + \frac{P}{EI}(Ax + Bx^2 + C) = \frac{w_0}{2EI}(Lx - x^2)$$

Equating coefficients of x and x^2 and the remaining constants respectively, we get

[6]We are using the results of the theory of ordinary differential equations here.

$$\frac{P}{EI}A = \frac{w_0L}{2EI}$$

$$\therefore A = \frac{w_0L}{2P}$$

$$\frac{P}{EI}B = -\frac{w_0}{2EI}$$

$$\therefore B = -\frac{w_0}{2P}$$

$$2B + \frac{P}{EI}C = 0$$

$$\therefore C = -\left(\frac{EI}{P}\right)(2)\left(\frac{-w_0}{2P}\right) = \frac{w_0EI}{P^2}$$

Hence,

$$v_P = \frac{w_0L}{2P}x - \frac{w_0}{2P}x^2 + \frac{w_0}{P^2}EI$$

The general solution then is

$$v = C_1 \sin\sqrt{\frac{P}{EI}}\,x + C_2 \cos\sqrt{\frac{P}{EI}}\,x + \frac{w_0}{2P}(Lx - x^2) + \frac{w_0EI}{P^2}$$

When $x = 0, L,$ we know that $v = 0$. Hence, we have

$$0 = C_2 + \frac{w_0EI}{P^2}$$

$$\therefore C_2 = -\frac{w_0EI}{P^2}$$

$$0 = C_1 \sin\sqrt{\frac{P}{EI}}\,L - \frac{w_0EI}{P^2}\cos\sqrt{\frac{P}{EI}}\,L + \frac{w_0EI}{P^2}$$

$$\therefore C_1 = \frac{1}{\sin\sqrt{P/EI}\,L}\left[\frac{w_0EI}{P^2}\left(\cos\sqrt{\frac{P}{EI}}\,L - 1\right)\right]$$

The problem has now been fully solved as to deflection.

In determining the stress distribution at any section we superpose a uniform compressive stress P/A with a bending stress found using the flexure formula with M given by Eq. (17.24). Note that we use the computed deflection v in evaluating M at any section.

We can now show the effects of lateral loads on a pin-ended column when $P \to P_{cr}$. Look at the formulation for C_1 above. When P approaches the value $\pi^2 EI/L^2$ (i.e., the critical value) the term sin $\sqrt{P/EI}\,L$ goes to zero and the constant C_1 blows up to infinity. The amplitude of the deflection curve becomes infinite. We thus show the effect predicted earlier of a transverse loading in the presence of a buckling load. Of course, the beam cannot actually deflect to infinity—the whole theory that we are using is valid only for small deformation. However, it does signal to us the strong possibility that large deflections—even failure—will be present when P approaches its critical value. We thus see the *first* consequence of a buckling load.

17.8 Initially Bent Member

In Section 17.5, we studied an initially straight weightless member, and by considering the possibility of neutral equilibrium we were able to compute the buckling loads. Now we shall consider a member which is initially slightly bent, having the deflection $v_0(x)$ as shown in Fig. 17.18.

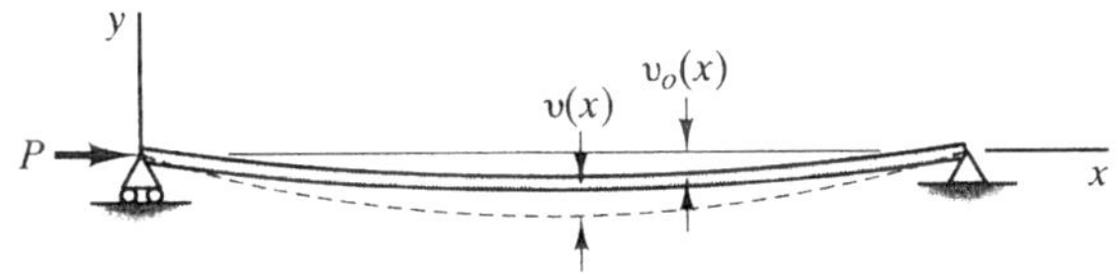

Figure 17.18. Initially bent column.

We shall employ $v(x)$ for deflection induced by the axial load P from the initial bent geometry so that the *total* deflection from the x-axis $v_T(x)$ is

$$v_T(x) = v_0(x) + v(x) \tag{17.28}$$

We may use the differential equation (17.7) developed earlier provided that we use $v_T(x)$. Thus,

$$\frac{dM}{dx} - V + P\frac{dv_T}{dx} = 0 \tag{17.29}$$

Replacing v_T by $v + v_0$ and differentiating with respect to x we have, on rearranging the terms,

$$\frac{d^2M}{dx^2} - \frac{dV}{dx} + P\frac{d^2v}{dx^2} = -P\frac{d^2v_0}{dx^2} \tag{17.30}$$

Now, replacing M using Eq. (17.8) and setting $dV/dx = 0$ in accordance with Eq. (17.6) for zero transverse load, we have for uniform beams

$$\frac{d^4v}{dx^4} + \frac{P}{EI}\frac{d^2v}{dx^2} = -\frac{P}{EI}\frac{d^2v_0}{dx^2} \tag{17.31}$$

For simplicity, we shall now assume that the initial deflection curve v_0 is given as

$$v_0 = \delta_0 \sin \frac{\pi x}{L} \tag{17.32}$$

The differential equation then becomes

$$\frac{d^4v}{dx^4} + \frac{P}{EI}\frac{d^2v}{dx^2} = \frac{P\delta_0}{EI}\frac{\pi^2}{L^2}\sin\frac{\pi x}{L} \tag{17.33}$$

We shall first set forth a *particular* solution to this differential equation. We shall employ the following trial solution:

$$v = C \sin \frac{\pi x}{L} \tag{17.34}$$

Substituting into Eq. (17.33) and canceling $\sin \pi x/L$, we get

$$C\frac{\pi^4}{L^4} - C\frac{P}{EI}\frac{\pi^2}{L^2} = \frac{P\delta_0}{EI}\frac{\pi^2}{L^2} \tag{17.35}$$

Solving for C, we have

$$C = \frac{P\delta_0/EI}{\pi^2/L^2 - P/EI} \tag{17.36}$$

Multiplying the numerator and denominator by EI, we get

$$C = \frac{P\delta_0}{EI\pi^2/L^2 - P} \tag{17.37}$$

Now we see that when $P = EI\pi^2/L^2$ we get an infinite value for C and therefore an infinite deflection v at positions away from the ends. However, this value of P is the first buckling load, as you may recall, for a straight beam with the same supports. We accordingly see a *second* consequence of the buckling load. That is, if there is initial curvature in the column we can expect large deflections as we approach the critical load just as in the case earlier when we had a transverse load. The simplified theory we are using indicates this by having the deflection blow up to infinity when the buckling load is present. Since as already indicated there is no such thing as a "perfectly straight" column we can see why knowing the buckling load is important for design.

Actually, the initial deflection $v_0(x)$ is equivalent to some transverse load on the corresponding straight member. This is clear when one compares Eq. (17.31) with Eq. (17.11) developed for the beam-column. The problem of the column with an initial bent geometry is equivalent to the corresponding initially straight beam with a transverse load given as

$$w = P\frac{d^2v_0}{dx^2} \tag{17.38}$$

Thus, this equivalent loading results in the member attaining an infinite deflection, mathematically speaking, when criticality is reached.

To get the complete solution we must find the complementary solution to the differential equation [Eq. (17.31)]. This solution was set forth in Section 17.5 [see Eq. (17.15)], and we now rewrite it as follows:

$$v_c = C_1 \sin\sqrt{\frac{P}{EI}}\,x + C_2\cos\sqrt{\frac{P}{EI}}\,x + C_3x + C_4 \tag{17.39}$$

The total solution is then

$$v = C_1 \sin\sqrt{\frac{P}{EI}}\,x + C_2\cos\sqrt{\frac{P}{EI}}\,x + C_3x + C_4 + \frac{P\delta_0}{EI\pi^2/L^2 - P}\sin\frac{\pi x}{L} \tag{17.40}$$

To complete the solution we must submit the total solution to the following boundary conditions:

$$\text{When } x = 0,\quad v = \frac{d^2v}{dx^2} = 0 \qquad \text{When } x = L,\quad v = \frac{d^2v}{dx^2} = 0 \tag{17.41}$$

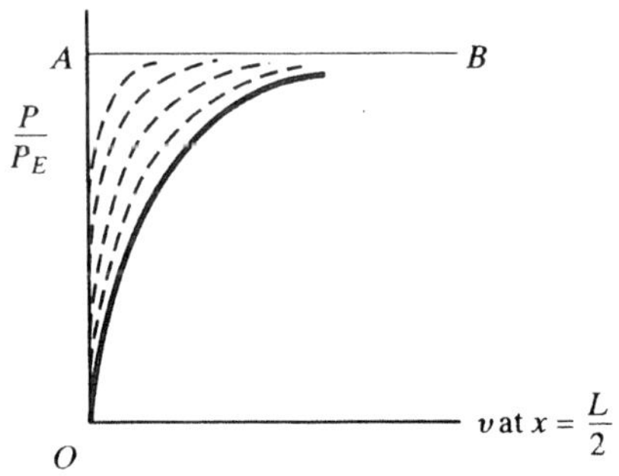

Figure 17.19. Load-deflection curve for various degrees of curvature.

In this way we may determine the four constants of integration. We shall leave it to the reader as an exercise to show that $C_1 = C_2 = C_3 = C_4 = 0$.

In Fig. 17.19, the full line is a plot of the deflection v at position $x = L/2$ as a function of the load P/P_E for a given initial curvature amplitude δ_0. Note that deflections now appear *immediately* on application of the force P. This deflection increases at an increasingly rapid rate as P approaches its critical value. The lines OA and AB represent the mathematical solution for an *initially straight* ideal column arrived at earlier. As the initial bent configuration is lessened (i.e., as δ_0 is decreased), the deflection curves approach OAB, as indicated by the dashed lines in the diagram. Clearly, we cannot reproduce plot OAB in the laboratory because there will always be some initial curvature in the beam. We see,

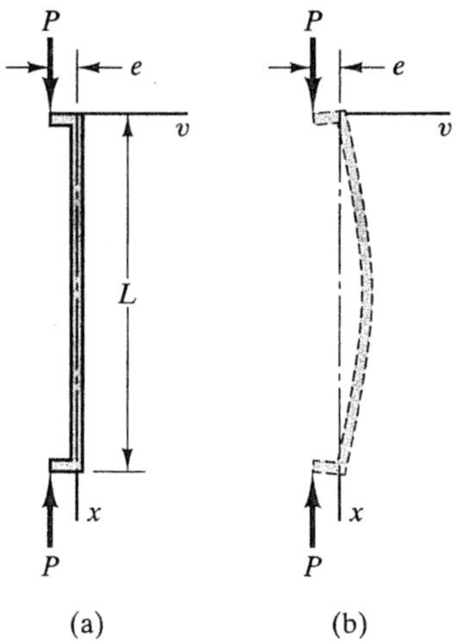

Figure 17.20. Eccentrically loaded column.

however, that the mathematical formulation for a straight beam gives the asymptotes for the various deflection curves for initially bent beams. And it should be evident that for safety axial loads should remain well below the critical load.

*17.9 Eccentrically Loaded Columns

We shall now consider a straight beam once again but this time loaded *eccentrically* as shown in Fig. 17.20(a) in the undeformed geometry and as shown in Fig. 17.20(b) in the deformed geometry. We wish to determine the deformed geometry and the maximum compressive stress for this case.

We shall employ Eq. (17.8) for this case. Thus, we have

$$\frac{d^2v}{dx^2} = \frac{1}{EI}(M) = -\frac{1}{EI}[P(e+v)]$$

$$\therefore \frac{d^2v}{dx^2} + \frac{P}{EI}v = -\frac{Pe}{EI} \qquad (17.42)$$

The complementary plus a particular solution is readily seen to be

$$v = C_1 \sin\sqrt{\frac{P}{EI}}x + C_2 \cos\sqrt{\frac{P}{EI}}x - e \qquad (17.43)$$

To determine C_1 and C_2 note that when

$$x = 0, \qquad v = 0$$

$$x = L, \qquad v = 0$$

Hence, on substituting Eq. (17.43) into the boundary conditions above, we get

$$C_2 = e \qquad (17.44a)$$

and

$$0 = C_1 \sin\sqrt{\frac{P}{EI}}L + C_2 \cos\sqrt{\frac{P}{EI}}L - e$$

$$\therefore C_1 = \frac{e\left(1 - \cos\sqrt{\frac{P}{EI}}L\right)}{\sin\sqrt{\frac{P}{EI}}L} \qquad (17.44b)$$

Using standard trigonometric identities[7] we can give C_1 in a more compact form as follows:

$$C_1 = e\frac{\sin\sqrt{\frac{P}{EI}}\frac{L}{2}}{\cos\sqrt{\frac{P}{EI}}\frac{L}{2}}$$

The general solution then becomes

$$v = e\left(\frac{\sin\sqrt{\frac{P}{EI}}\frac{L}{2}}{\cos\sqrt{\frac{P}{EI}}\frac{L}{2}}\sin\sqrt{\frac{P}{EI}}x + \cos\sqrt{\frac{P}{EI}}x - 1\right)$$

$$\therefore v = e\left(\frac{\cos\sqrt{\frac{P}{EI}}\left(\frac{L}{2} - x\right)}{\cos\sqrt{\frac{P}{EI}}\frac{L}{2}} - 1\right) \quad (17.45)$$

where we have combined the first two terms on the right-hand side of the equation and have used a trigonometric identity. Notice, when P equals the critical load for the pin-connected column

$$\cos\sqrt{\frac{P}{EI}}\frac{L}{2} = 0$$

and v blows up to infinity, as one would expect from earlier discussions. We thus arrive at a *third* consequence of buckling. The maximum value of v occurs at $x = L/2$ and we denote this value of v as δ. Hence, for Eq. (17.45) we get

[7]The following formulas are used:

$$\sin\frac{a}{2} = \left[\frac{1}{2}(1 - \cos a)\right]^{1/2}$$

$$\sin a = 2\sin\frac{a}{2}\cos\frac{a}{2}$$

Later, in Eq. (17.45), we shall use the identity

$$\cos a \cos b + \sin a \sin b = \cos(a - b)$$

$$\delta = e\left(\frac{1}{\cos\sqrt{\dfrac{P}{EI}}\dfrac{L}{2}} - 1\right)$$
$$= e\left(\sec\sqrt{\frac{P}{EI}}\frac{L}{2} - 1\right) \tag{17.46}$$

Hence, the maximum bending moment for the column is

$$M_{\max} = -P(e+\delta) = -P\left[e + e\left(\sec\sqrt{\frac{P}{EI}}\frac{L}{2} - 1\right)\right]$$
$$= -Pe\sec\sqrt{\frac{P}{EI}}\frac{L}{2} \tag{17.47}$$

The maximum compressive stress for the column (possibly away from the ends where there is stress concentration) is then

$$\tau_{\max} = -\frac{P}{A} - \frac{M_{\max}(-c)}{I}$$
$$= -\frac{P}{A} - \frac{Pe\left(\sec\sqrt{\dfrac{P}{EI}}\dfrac{L}{2}\right)c}{A\rho^2}$$

$$\therefore \quad \tau_{\max} = -\frac{P}{A}\left(1 + \frac{ec}{\rho^2}\sec\frac{L}{2\rho}\sqrt{\frac{P}{EA}}\right) \tag{17.48}$$

where c is the distance from the neutral axes to the furthermost fibers and ρ is the radius of gyration of the cross section in the plane of bending. The above formula is known as the *secant column* formula. We shall now illustrate its use.

Example 17.4

A cantilever beam is shown in Fig. 17.21. What is the maximum load P permitted for a working stress in compression of 2.75×10^8 Pa? The cross section of the beam is 50×50 mm^2 with a modulus of elasticity of 2×10^{11} Pa.

The cantilever beam shown in Fig. 17.21 can be considered as one half of the eccentrically loaded column shown in Fig. 17.22 for

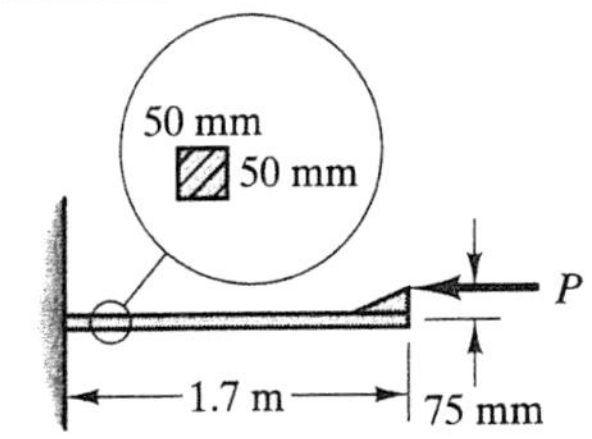

Figure 17.21. Cantilever beam.

Example 17.4 (Continued)

which we can employ the secant formula. We know τ_{max} and we want to find P. Thus, noting that the radius of gyration ρ for this case is $0.025/\sqrt{3}$ we can say for Eq. (17.48) that[8]

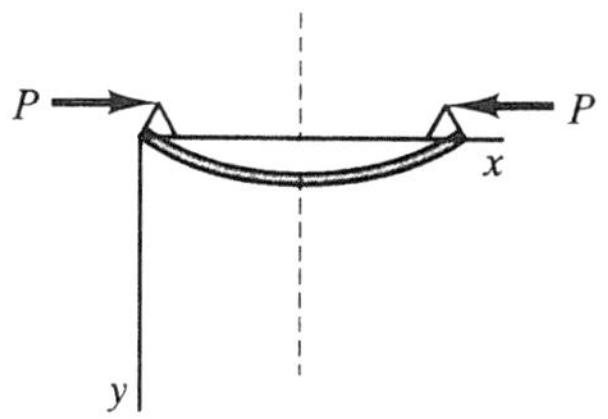

Figure 17.22. Eccentrically loaded column.

$$-2.75 \times 10^8 = -\frac{P}{2.5 \times 10^{-3}}\left\{1 + \frac{(.075)(.025)}{(0.025)^2/3} \times \sec\left[\frac{(1.7)(2)}{2(0.025)/\sqrt{3}}\sqrt{\frac{P}{(2 \times 10^{11})(.05)^2}}\right]\right\} \quad \text{(a)}$$

This becomes

$$6.88 \times 10^5 = P[1 + 9 \sec(5.267 \times 10^{-3}\sqrt{P})] \quad \text{(b)}$$

By trial and error we get the result

$$P = 3.77 \times 10^4 \text{ N} \quad \text{(c)}$$

17.10 General Considerations

In reconsidering the critical loads given in Fig. 17.13, you will note that all such formulas are of the form

$$P_{cr} = C\frac{\pi^2 EI}{L^2} \quad (17.49)$$

where C is a constant dependent on the end conditions. It will be convenient to replace I by $\rho^2 A$, where ρ is the radius of gyration of the cross section, and, furthermore, to replace L^2/C by L_e^2, called the *effective length*. Thus, we have

$$P_{cr} = \frac{\pi^2 EA}{(L_e/\rho)^2} \quad (17.50)$$

[8] $I = \frac{1}{12}bh^3 = (\frac{1}{12})(.05)(.05)^3 = \rho^2 A = \rho^2(.05)^2$

$$\therefore \rho = \frac{.025}{\sqrt{3}}$$

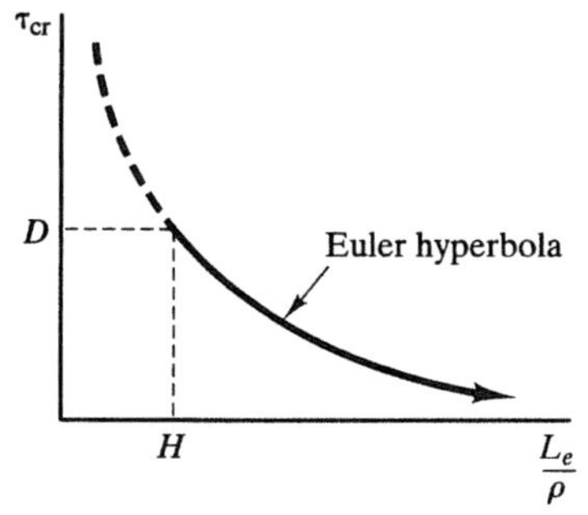

Figure 17.23. Critical stress versus slenderness ratio.

The ratio L_e/ρ is called the *slenderness ratio.* Finally, dividing through by A, we may set forth the *critical stress* τ_{cr} in the following manner:

$$\tau_{cr} = \frac{\pi^2 E}{(L_e/\rho)^2} \tag{17.51}$$

In Fig. 17.23, we have plotted the critical stress τ_{cr} versus the slenderness ratio. Notice that the curve has been shown as a dashed line for stresses τ_{cr} above the value D. This is done because point D represents the *proportional limit,* and consequently the portion of the curve above this point is not meaningful as a buckling criterion since its derivation assumed linear-elastic behavior of the material. This means that, for slenderness ratios greater than point H on the diagram, that is, for long slender columns, we can use the Euler formulation, Eq. (17.51), whereas for slenderness ratios smaller than point H we encounter *inelastic behavior* and therefore require other formulations for finding the buckling load. Of course, if the slenderness ratio is extremely small (a squat compression block), there is the distinct possibility that the material will fail by fracture or by excessive deformation before the buckling load is reached. In the next section, we shall make a brief introductory study of inelastic buckling of columns.

*17.11 Inelastic Column Theory

As pointed out in the previous section, for long slender columns involving materials that are linear-elastic up to a proportional limit stress, we can use the so-called Euler hyperbola (solid line of Fig. 17.23) for stability considerations. To ascertain what to do for slenderness ratios less than the value H, we consider first the idealization of a linear-elastic material with simple strain hardening shown in Fig. 17.24(a) for compression. We may associate two moduli with this behavior. There is, of course, the Young's modulus E_1 for the linear, elastic range up to stress D, and then we can assign E_2 equal to the slope of the curve for the plastic range. In Fig. 17.24(b) we show plots of τ_{cr} versus L_e/e from Eq. (17.51) for both values of E. It should be apparent that for τ_{cr} less than D' the curve for E_1 in Fig. 17.24(b) can be used in stability considerations. This means that for slenderness ratios exceeding H' we use the linear-elastic criterion. Should the stress τ exceed the value D', however, the behavior of the material changes so as to have a new but still constant[9] modulus E_2. Because E_2 is constant, we use the E_2 curve in

[9]This is true at least for loading. On unloading the modulus changes back to E_1, as you may recall from Chapter 4. We shall come back to this point later.

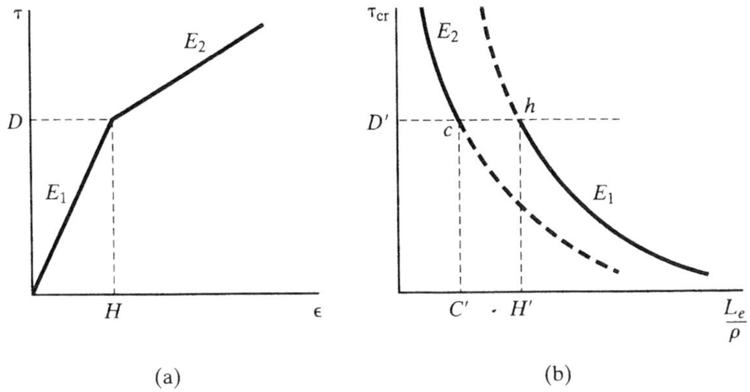

Figure 17.24. Case of simple strain hardening.

Fig. 17.24(b), and so when the slenderness ratio is less than C' we switch to this curve from the E_1 curve. Thus, we can use Eq. (17.51) for two ranges by employing the appropriate E for each range. This leaves a gap between C' and H' for which we have presented no criterion. Actu-

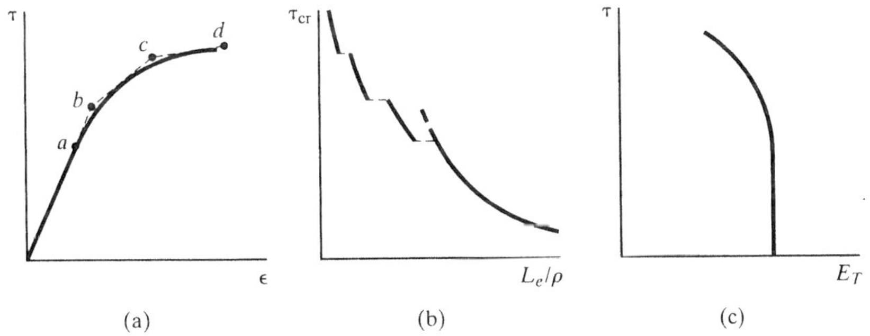

Figure 17.25. Case of general strain hardening.

ally, no real material would exhibit a stress-strain curve with a sharp corner. Instead there will be a curve with a continuously varying slope, such as is shown in Fig. 17.25(a). We, however, can utilize the concepts of the aforementioned idealized case in order to study the stability of more realistic cases whose stress-strain curve is like that in Fig. 17.25(a). For this purpose we approximate this stress-strain curve by a series of tangent lines *ab, bc, cd,* and so on, as shown in the diagram. This means that we arrive at an extension of the stress-strain idealization of Fig. 17.24(a) with many constant values of E. Using the Euler hyperbola for each range of constant E as we did earlier we arrive at a set of curves shown in Fig. 17.25(b). We can accordingly use Eq. (17.51) for each range where we have approximated the stress-strain diagram with a straight line by employing the appropriate E in the formula. By using points ever closer together in Fig. 17.25(a) we may more closely represent the actual stress-strain curve. In the limit we can duplicate the original curve by this process. At each point in the stress-strain curve we define the *tangent modulus*, E_T, as the slope of the curve at that point. In the limit we get a smooth curve for τ_{cr} versus L_e/ρ and can use Eqs. (17.50) and (17.51) over the *entire range* of the stress-strain diagram provided that we use E_T in the equation. This is the *Engesser tangent modulus* theory of buckling.

In using the tangent-modulus theory it is useful to have available a curve of τ versus E_T. This is done most effectively by measuring the slope of the stress-strain curve (and thus E_T) for various values of τ and then making a smooth curve of τ versus E_T as shown in Fig. 17.25(c). To get a critical load now for a given geometry—that is, for a given L_e/ρ, we guess at a value of E_T and from Eq. (17.51) determine τ_{cr}. Next go to the appropriate τ versus E_T curve for your case and read off E_T for the computed τ_{cr}. Now take this value of E_T and go back to Eq. (17.51) to start another sequence over again. When the E_T used in the formula and the E_T read from the τ versus E_T curve are reasonably close you have the proper τ_{cr} and have the proper P_{cr}. By this method we can formulate a value of τ_{cr} (and thus P_{cr}) for any value of slenderness ratio L_e/ρ and hence form the curve τ_{cr} versus L_e/ρ for stress above the proportional limit for a given material. This is shown in Fig. 17.26, where

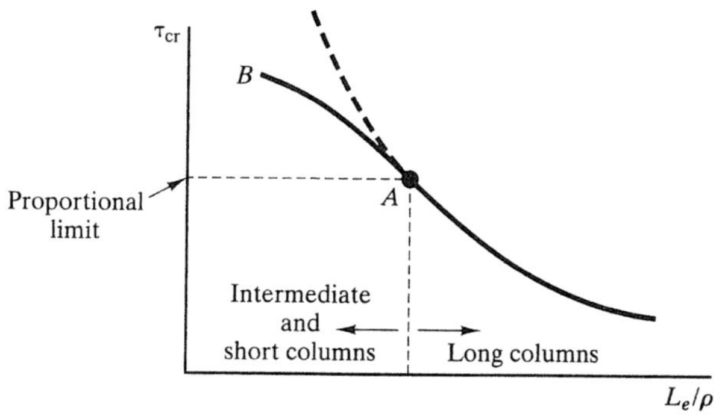

Figure 17.26. τ_{cr} versus L_e/ρ.

AB is the curve found by the aforementioned steps. Experiments check these results for the usual structural materials with good accuracy.

Engesser first presented his theory in 1889.[10] In 1910 von Kármán supported this theory by presenting favorable experimental work on small specimens of rectangular cross section. Despite this, the tangent-

[10] *Zeitschrift für Architektur und Ingenieurwesen*, 1889.

modulus theory was criticized for the following reason: It assumes that for the infinitesimal bent shapes of the column permitted when the critical load has been reached, the tangent modulus is constant throughout the material at a value corresponding to τ_{cr}. Actually, this is not the case. For bent shapes, part of the cross section is subject to an *unloading,* and in this action the material takes on a modulus corresponding to the linear, elastic value E as discussed in Section 4.3. Thus, observing Fig. 17.27 if a corresponds to τ_{cr}, then a bent configuration, however small, would mean that part of the material would unload along ab having a modulus E and part of the material would load farther along ac having modulus E_T. It was argued that the tangent-modulus theory, since it did not take this unloading phenomenon into account, did not properly give the buckling load, and, as a result, the so-called *double-modulus theory* in which an *effective* or *reduced* modulus E_r, dependent both on E and E_T as well as the geometry, was employed.[11] This theory gave buckling loads greater than that predicted by the Engesser theory. Despite the fact that experiments yielded buckling loads between the Engesser load and the double-modulus load but definitely closer to the Engesser load, the double-modulus theory was considered to be the correct theory. In 1946, however, F. R. Shanley successfully explained the relation between the two theories and experimental data.[12] The Engesser theory is now used most since it is somewhat more accurate and is on the conservative side.

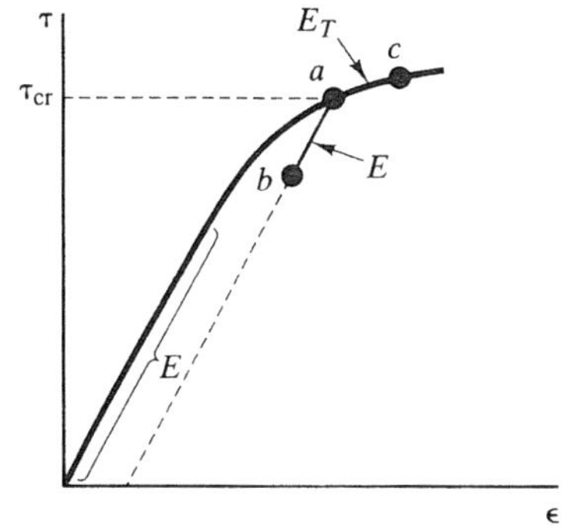

Figure 17.27. Partial linear-elastic unloading at *a*.

*17.12 A Note on Column Formulas

We wish to point out that there are many empirical formulas for relating τ_{cr} with L_e/ρ. These formulas stem from efforts to extend the Euler curve for large slenderness ratios into the region of intermediate and small slenderness ratios. Essentially a straight line or a parabola both generally tangent to the Euler curve (see Fig. 17.28) is used for the range of intermediate to small slenderness ratios. The empirical formulas have the form

$$\tau_{cr} = C_1 + \frac{C_2}{(L_e/\rho)^n}$$

The values of the constants depend on the range of L_e/ρ, the material used, and the particular industry. In design you would use some fraction

[11]See F. Bleich, *Buckling Strength of Metal Structures* (McGraw-Hill, New York, 1952), Sec. 4.

[12]See F. R. Shanley, "The Column Paradox," *Journal of Aeronautical Science,* Vol. 13, No. 5 (Dec. 1946), and "Inelastic Column Theory," *ibid.,* Vol. 14, No. 15 (May 1946).

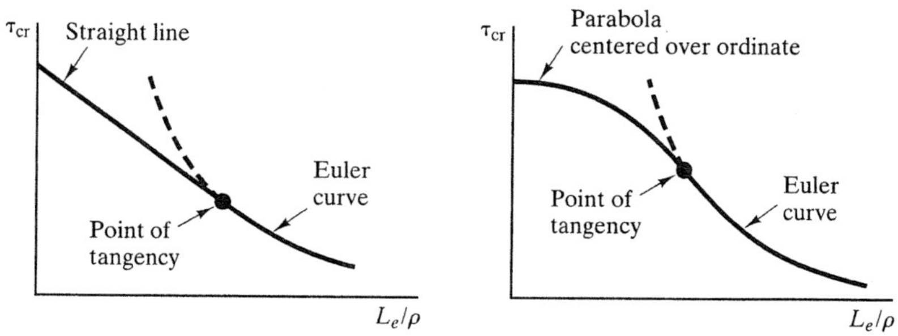

Figure 17.28. Extensions of the Euler curve.

of τ_{cr} for an allowable stress. The safety factor you use depends on the industry you are in. Sometimes the formulas to be used and the safety factor to be employed are determined by government-imposed codes, usually city or state codes. The variations to be found are great, and hence it serves little purpose in being more specific. You are referred to handbooks and building and construction codes for specific information.[13]

17.13 Closure

In this chapter we have introduced the buckling phenomenon. You will note that in this chapter we computed the moment contribution from the axial load by considering the *actual* deflection v of the member. That is, for the only time in this book, we used the *deformed* geometry of the body in our considerations of statics.

We defined the critical load for ideal columns as the value of axial load P at which neutral equilibrium is reached whereby a family of deflection curves (modes) of small amplitude all satisfy the deflection equation. Specifically, we developed our results for a *pin-ended, perfectly straight* column having only an axial load (*no transverse load*) acting, which is *perfectly colinear* with the centerline of the column. We then examined effects of departures from the aforestated idealized conditions. Thus for different end supports we presented critical loads (Fig. 17.13) reached by methods analogous to that used for the pin-ended column. We then showed that when the critical load is theoretically pre-

[13]We list the following references for more specific information:

1. *American Institute of Steel Construction,* Chicago, IL.
2. *ALCOA Structural Handbook,* Aluminum Company of America, Pittsburgh, Pa., 1962.
3. *N.L.M.A. National Design Specification,* National Lumber Manufacturers Association, Washington, D.C., 1962.

sent, other departures from the ideal case, namely initial curvature and eccentricity of loading, result mathematically in infinite deflection in our approximate theory. This is also the case for a transverse load. From an engineering viewpoint, we can thus anticipate large deformation and stress in a column when the axial load is *near* the critical load. This is due to the ever-present initial curvature and eccentricity of loading. Accordingly, we design a column to have a buckling load that is n times the expected axial load, where n may be considered as some safety factor. The value of n to be chosen depends on the nature of the problem and is best learned from experience in the particular industry involved.

In this chapter we encountered the Sturm–Liouville problem where a parameter (in our case P) in the differential equation is restricted to certain discrete values so that nontrivial solutions of the differential equation will exist for the particular boundary conditions of the problem. Similar problems are to be found in other engineering sciences. In particular the study of vibrations of deformable bodies will surely include the Sturm–Liouville problem. In this area, the natural frequency of vibration will play the same role as critical loads for elastic stability.

Thus, this chapter has opened wide areas of study for us. First, there is the very broad and important field of elastic stability; second are the vast areas of applied mathematics that we have briefly introduced in this chapter.

Next, we turn for a more detailed study of the powerful and elegant *energy methods*. These methods most likely will appear strange to you at first reading. Nevertheless, they will be the ones you will turn to more and more as you delve deeper into various vital aspects of mechanics. Also the kind of thinking involved will be found in other engineering sciences and applied mathematics. A word of advice: it is vital to study the derivations very carefully. For those going ahead – bon voyage!

*17.14 A Look Ahead: Finite Elements[14]

With the advent of the computer, a very important numerical procedure for solving all kinds of boundary value problems has come to the fore and is now widely used. That procedure is the method of finite elements. We will look at it from the viewpoint of solid mechanics out of which it first grew.

[14]See I. H. Shames and C. L. Dym, *Energy and Finite Element Methods in Structural Mechanics*, Taylor and Francis Corp., 1985.

In this method, the domain of the problem is decomposed into a grid of small finite areas for plane stress problems or by small finite volumes for three-dimensional problems. For example, the aforementioned areas for plane stress could be triangles, rectangles, or trapezoids. The size of these elements can vary over the domain. One tries to have the grid size near the boundary of the problem small enough to be able to give a good fit with the boundary of the problem. For each area element, certain points are chosen to work with. For a triangular element one might use the corners of the triangle as well as points on or inside the sides of the triangles. These points are called **nodal points.** In this methodology, the displacements of these points are sought. The choice of the area elements and the nodal points depends on the kind of accuracy required and the nature of the problem. Primarily it is up to the discretion of the engineer. Next a function, called an **interpolation function** or **shape function,** is chosen to give the displacement field in the element in terms of the displacements of the nodal points of the element. An important step now requires that the interpolation function used to evaluate the displacement at each of the nodal points give the same displacement as that of the nodal point itself. That is, there must be self-consistency (**compatibility**). Along contiguous boundaries of the elements, certain conditions must be met by the interpolation functions concerning continuity of the displacement field and its normal and tangential derivatives. If these conditions are not met, the process might not tend to converge as the grid is made ever smaller and with ever more elements. This would indicate a lack of success for the analysis.

The finite element method formulations are done with matrix notation. A matrix is chosen to relate strains to the displacement field (**constitutive law**). Satisfying the principle of virtual work (**equilibrium**), a matrix equation is then formed involving the nodal **displacement vector,** a matrix called the **stiffness matrix,** and an external **force vector** for each element. These matrix equations are then combined into global equations for the entire domain. The final procedure is to solve a large system of simultaneous algebraic equations for the nodal displacement quantities. The computer is needed to form the grid system and also to solve the numerous algebraic equations for the nodal displacements. Knowing these, we can get the displacement for the entire domain using the interpolation functions. Thus, energy methods are used to determine the displacements of a set of discrete points (nodal points) after which the displacement of the rest of the domain is known in terms of the nodal point displacements via the interpolation functions.

Highlights (17)

Elastic instability occurs when a force or forces act on the deformed geometry of a member such as to further increase the applied moments and thus to induce yet further deformation, and so on. For the column studied, the force was the axial force P and the deformation caused this force to form a couple moment distribution along the column that further increased the deformation as well as further increased the accompanying couple moment distribution, and so on. That is why we had to use the deformed geometry in this chapter. The buckling condition occurs when we find the load P that causes **neutral equilibrium.** That is the load for which there is an infinity of solutions for the differential equation (which is based on equilibrium) and the boundary conditions. To achieve this condition, we found that we were limited to an infinite set of discrete values of P called **critical loads** or **eigenvalues.** The infinite set of solutions are called the **eigenfunctions.** The reason for the importance of these critical loads is directly related to the neutral equilibrium condition. For, in this state the beam column cannot resist a transverse load nor can it tolerate, without large deformation, an initial curvature of the column nor an eccentricity of loading. For ideal members, the theory gives an infinite deflection in the presence of the buckling load for the above conditions. Realizing that the theory is only valid for small deflections, the prediction of an infinite deflection is really fictitious. However, it does warn us that **real** members that are near buckling load will exhibit large deflections. Instabilities occur in many areas of study of interest to the engineer. The elastic instability described in this chapter is perhaps the most easy to understand and the most graphical.

PROBLEMS

Definition of the Critical Load	17.1–17.4
The Column and the Beam Column; Buckling Loads	17.5–17.27
Initially Bent Columns	17.28–17.30
Eccentrically Loaded Members	17.31–17.33
Unspecified Section Problems	17.34–17.51
Computer Problems	17.52–17.53
Programming Project 15	

17.1. [17.2] Find the critical load for the rigid column of length L shown in Fig. P.17.1. Note that two identical springs are attached to the column from rigid foundations. The column is pinned at the base.

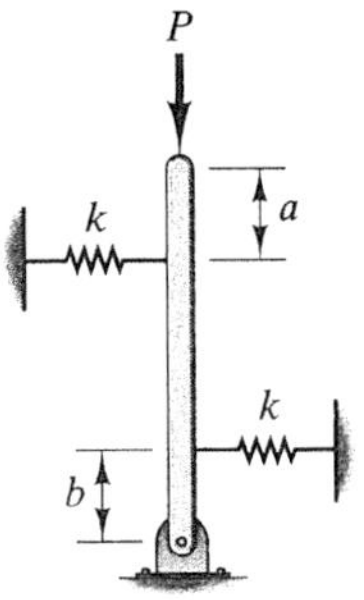

Figure P.17.1.

17.2. [17.2] Shown in Fig. P.17.2 is a system of springs and rigid bodies. What is the critical load P for this system? The bodies are pinned at A.

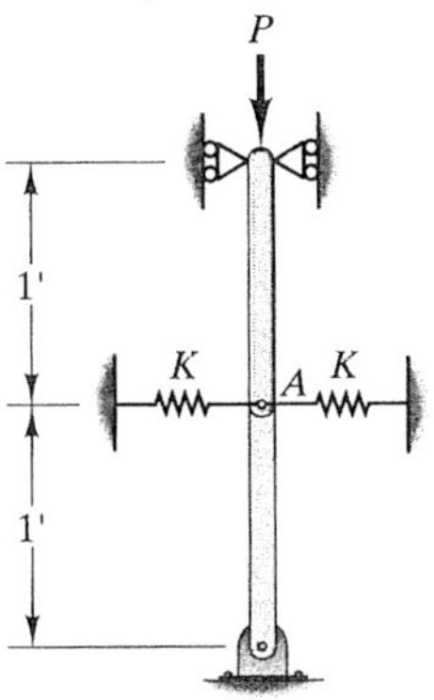

Figure P.17.2.

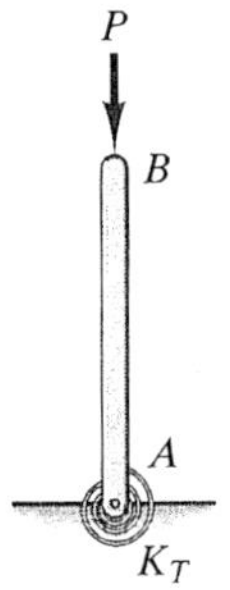

Figure P.17.3.

17.3. [17.2] A rod of length L is pinned at end A (Fig. P.17.3) to a rigid base and is at the same time connected to a torsional spring (resists rotation) having a spring constant of K_T ft-lb/rad. When no load is applied to B, the rod is vertical. What is the critical value of the vertical load P?

17.4. [17.2] Two rods are pinned together at A as shown in Fig. P.17.4. A torsional spring connects the two members at A so as to resist the relative rotation of the two members. It has a spring constant of K_T ft-lb/rad. The bar on the bottom is pinned to the ground at B. When no loads are applied the rods are vertical. Find the critical load for P.

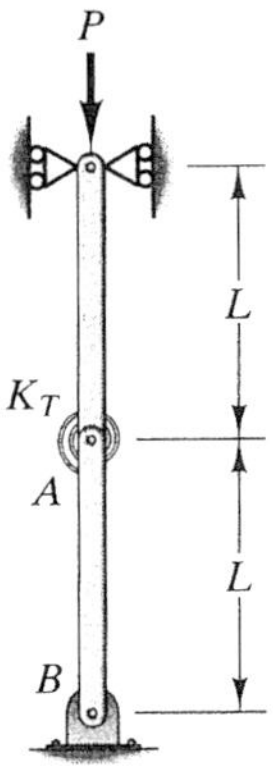

Figure P.17.4.

17.5. [17.5] What is the buckling load for a pin-ended rod, 3 in. in diameter and 20 ft long. (Take $E = 30 \times 10^6$.)

17.6. [17.5] In Fig. P.17.6 is shown a container in which a load of 9×10^4 N is held. Design the angle iron supports. There are four of them, one on each corner, and each has a free length of 3 m, as shown in the diagram. Use a safety factor of 2 and assume that the upper end of the unbraced support is fixed, while the bottom is effectively pin-connected. Take the modulus of elasticity for the steel as 2×10^{11} Pa. Constraints (not shown) prevent top part of device from moving laterally.

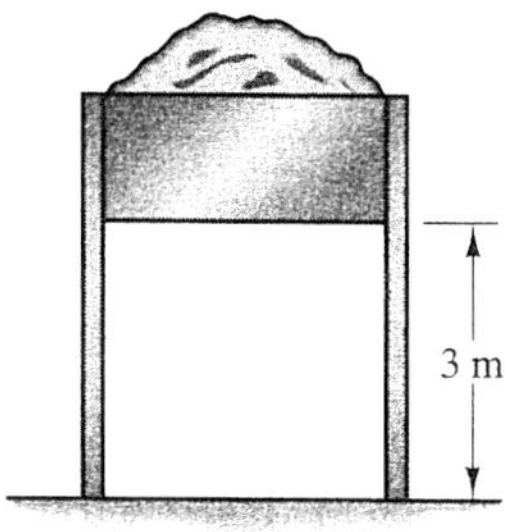

Figure P.17.6.

17.7. [17.5] A rock crusher is shown in Fig. P.17.7. Arms AB and BC are 5 ft in length. A pressure p of 500 psi is developed on the 5-in. piston D. The smallest that α gets is 10°. Determine h of arms $\overline{AB}$ and $\overline{BC}$ noting that the width is 2 in., as shown in the diagram. For safety, design for a 50% overload. Take $E = 30 \times 10^6$ psi for the members.

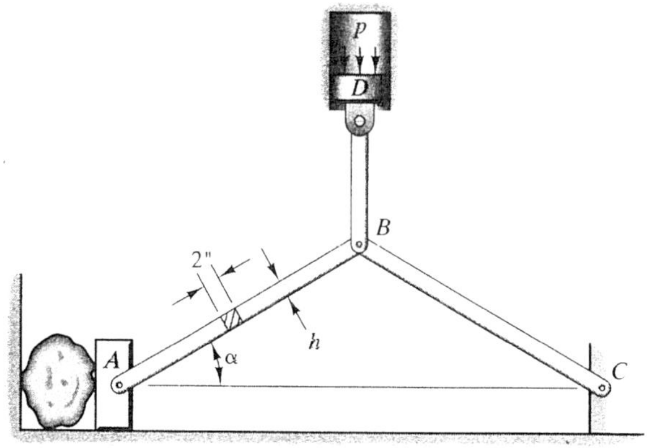

Figure P.17.7.

17.8. [17.5] A steel connecting rod is shown in Fig. P.17.8. It is 2.4 m long and has a cross section of 100 × 50 mm. The piston has a diameter of .635 m. What is the maximum safe pressure p in the cylinder when AB is locked horizontally? AB can be considered pin-ended for bending in the plane of the diagram and fixed ended for bending in the plane normal to the paper. $E = 2 \times 10^{11}$ Pa.

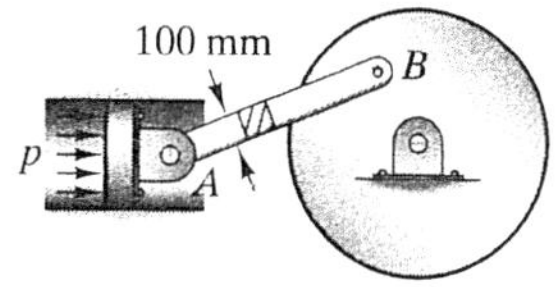

Figure P.17.8.

17.9. [17.5] A rod is installed in fixed supports as shown in Fig. P.17.9 at a temperature T. If the material has a linear coefficient of expansion α, what is the increase in temperature at which elastic instability occurs if we assume that E is constant?

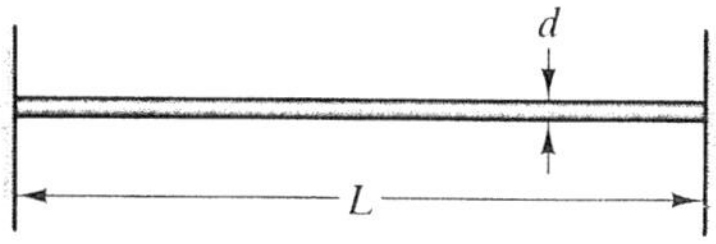

Figure P.17.9.

17.10. [17.5] What weight W and what diameter D will cause simultaneously a crushing failure and a buckling failure of the rod AB in Fig. P.17.10? Neglect all other weights.

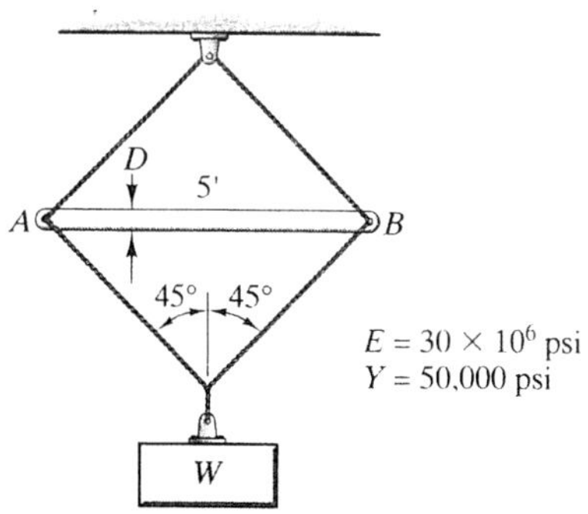

Figure P.17.10.

17.11. [17.4] What force will cause the femur bone shown in Fig. P.17.11 to buckle? Take $E = 14$ GPa. Assume the bone can be modeled as a straight hollow tube and is fixed-free at the ends.

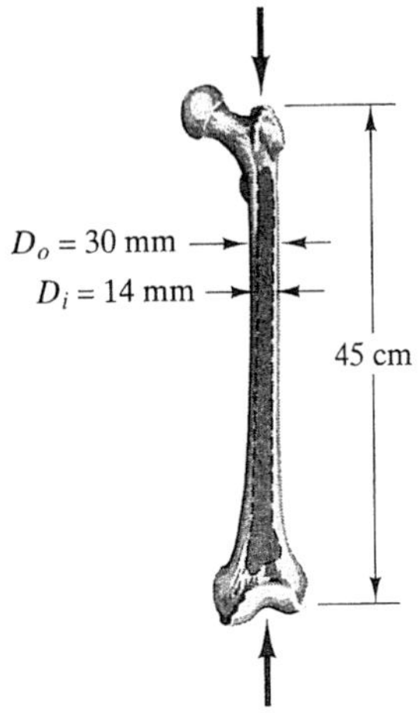

Figure P.17.11.

17.12. [17.5] A space device in Fig. P.17.12 is to accelerate in the Z direction in outer space. How many g's of acceleration can be allowed for the truss supporting a 50-kg mass A? The rods have a diameter of 40 mm and have a modulus $E = 2 \times 10^{12}$ Pa. Also, $Y = 7.5 \times 10^8$ Pa. Use a safety factor of 1.8. Neglect mass of rods.

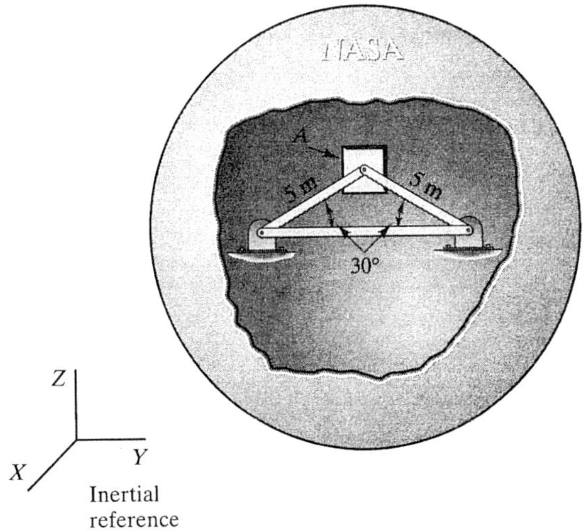

Figure P.17.12.

17.13. [17.5] Given for the flexible column made of aluminum alloy as shown in Fig. P.17.13, a yield stress (compression) = 55,000 psi and $E = 10.5 \times 10^6$ psi, you are asked to check the strength and stability in the xy plane of the column. Discuss possible modes of failure and find the critical load. The column is supported laterally, as shown, by two supports away from the ends.

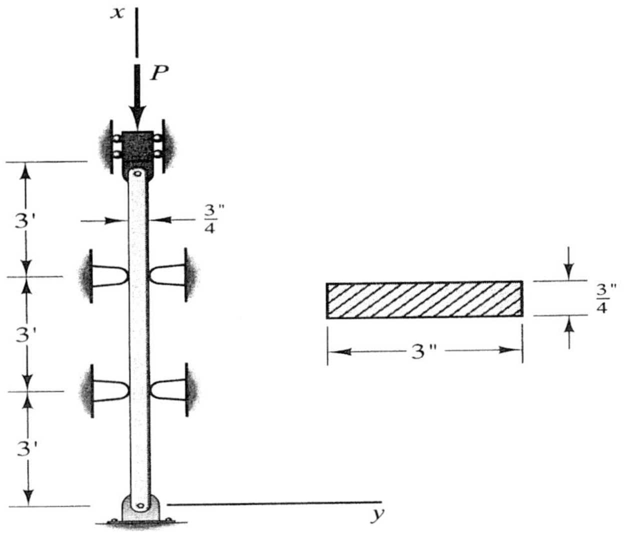

Figure P.17.13.

17.14. [17.5] Shown in Fig. P.17.14 is a simple pin-connected truss. What is the maximum value of P that the system can withstand without buckling occurring in any of its members? (Take $E = 2.1 \times 10^{11}$ Pa. Take the radius of gyration of the cross section of each member for bending in the xy plane as 25 mm and the area of each section as 6.25×10^{-4} m^2.)

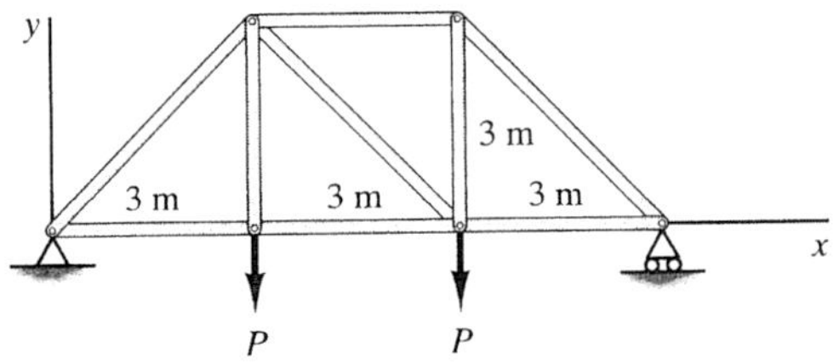

Figure P.17.14.

17.15. [17.5] Consider a column with one end ($x = 0$) fixed (clamped) and the other end ($x = L$) pinned [see Fig. 17.13(b)]. From first principles show that for a nontrivial solution the following equation must be satisfied:

$$\tan\left(\sqrt{\frac{P}{EI}}\,L\right) = \sqrt{\frac{P}{EI}}\,L \tag{1}$$

Demonstrate that

$$P = 2.05\frac{\pi^2 EI}{L^2} \qquad (2)$$

satisfies Eq. (1) and is hence the buckling load.

17.16. [17.5] Consider the device in Fig. P.17.16 undergoing plane motion.

(a) At what angular speed ω will buckling take place in the first mode?
(b) At what angular speed ω will buckling take place in the second mode?

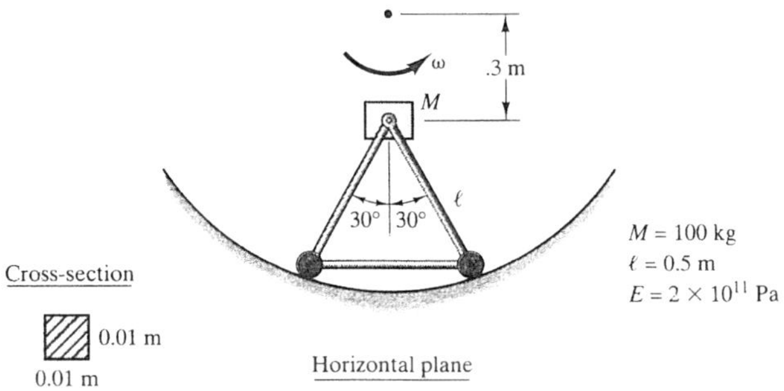

Figure P.17.16.

17.17. [17.5] What is the critical force P_{cr} for the steel column shown in Fig. P.17.17 free at the top and clamped at the bottom? Take $E = 30 \times 10^6$ psi.

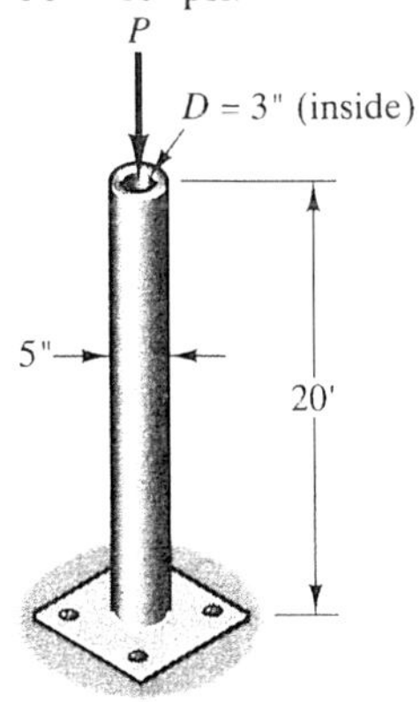

Figure P.17.17.

17.18. [17.5] What is the allowable load P for a safety factor of 2.5 for the steel wide-flanged 8WF20 I-beam in Fig. P.17.18 if we include the weight for possible crushing but disregard it for buckling? Use the following data:

$$E = 30 \times 10^6 \text{ psi} \qquad \text{and} \qquad \tau_Y = 40{,}000 \text{ psi}$$

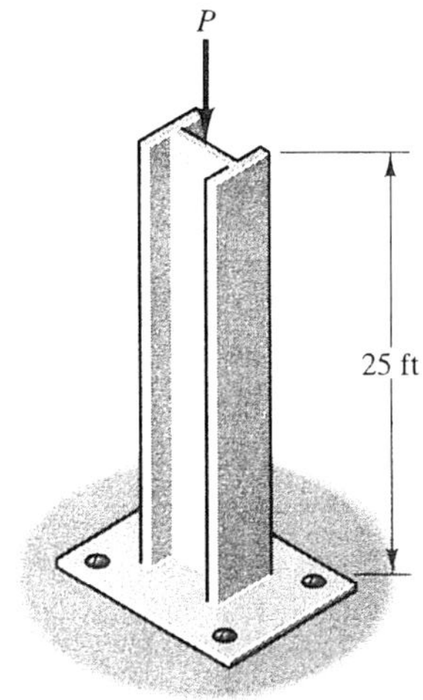

Figure P.17.18.

17.19. [17.5] In Problem 17.18, for a length of 47 ft and for a force P of 10,000 lb, what size wide-flanged I-beam should you use? Take $E = 30 \times 10^6$ psi and use a safety factor of 3.

17.20. [17.5]

(a) A rigid block G is guided by a frictionless vertical rod AB and is supported by two inclined rods CD (see Fig. P.17.20). What is the maximum weight of G that the rods can support using a safety factor of 2?
(b) What should be the length L of the rods and the weight W of G in order to have buckling and crushing occur simultaneously? Take $Y = 30{,}000$ psi. No safety factor here.

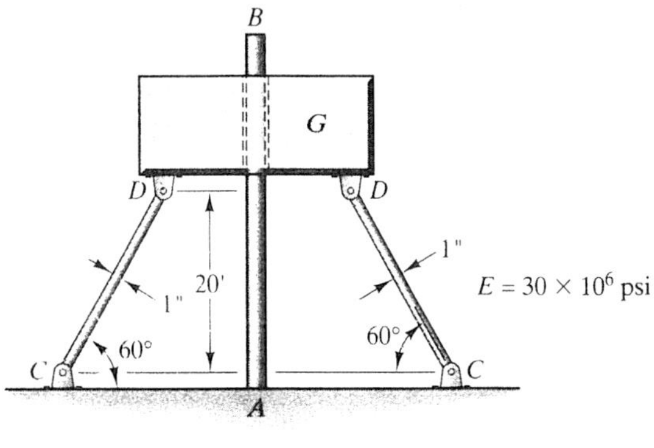

Figure P.17.20.

17.21. [17.5] A rod is shown in an initial configuration dimension-wise for $F = 0$ with the linear spring undeformed (see Fig. P.17.21). The force is then increased steadily until buckling first occurs. What is the value of F when this first occurs? We have neglected the weight of the rod. The variable y is the distance to the top of the rod as F is increased.

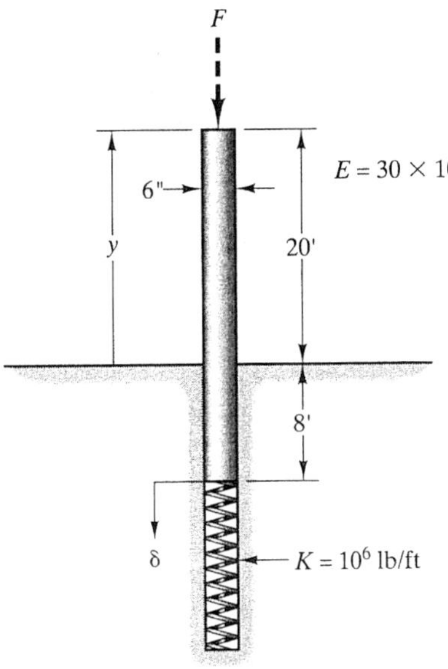

Figure P.17.21.

17.22. [17.5] In the design of columns, the "effective length" nomenclature is often used. This quantity is defined as the length of a pinned-pinned column to take the place of a column with different end constraints so as to have the same buckling load. For each of the four cases shown in Fig. 17.13, compute the effective length, L_e.

17.23. [17.5] An aluminum column, with a cross section as shown in Fig. P.17.23 is held by ball joints at both ends. The column is 8 m long. What is the smallest buckling load? Take $E = 70$ GPa and $\tau_Y = 210$ MPa.

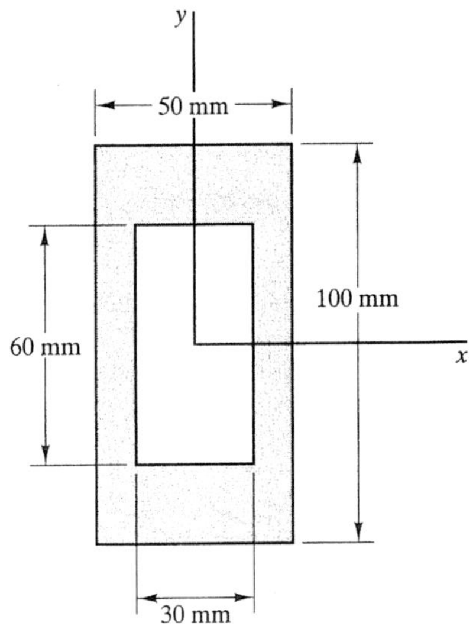

Figure P.17.23.

17.24. [17.5] If in Problem 17.23, the column is restrained in the x direction by pin-connected, rigid rods at mid-height, what is then the smallest buckling load?

17.25. [17.5] A timber with a square cross section supports a platform. It carries a load of 100,000 lb. If $E = 2 \times 10^6$ psi for the timber and its length between supports is 20 ft, design the member to withstand a load five times the expected load for the case where the member is clamped at one end and pinned at the other end.

17.26. [17.5] Two standard channels and two steel slabs are welded together to form a column whose length is 40 ft as shown in Fig. P.17.26. For a safety factor of 3, what is the maximum axial load for a clamped-clamped column made as described? $E = 30 \times 10^6$ psi. Do not consider self weight in computing the buckling load. Consider only buckling along the z axis of the system. The weight of each channel is 18.75 lb/ft and $\tau_Y = 40{,}000$ psi.

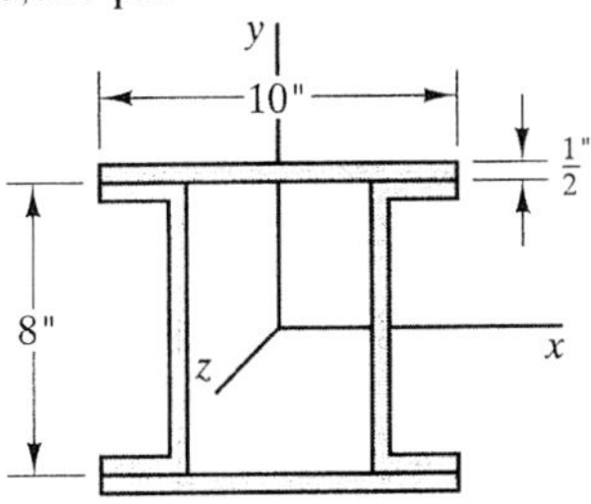

Figure P.17.26.

17.27. [17.5] Get the smallest h for structure ABC to support the weight $W = 2000$ N with a safety factor of 3. Take $E = 2 \times 10^{11}$ Pa. See Fig. P.17.27.

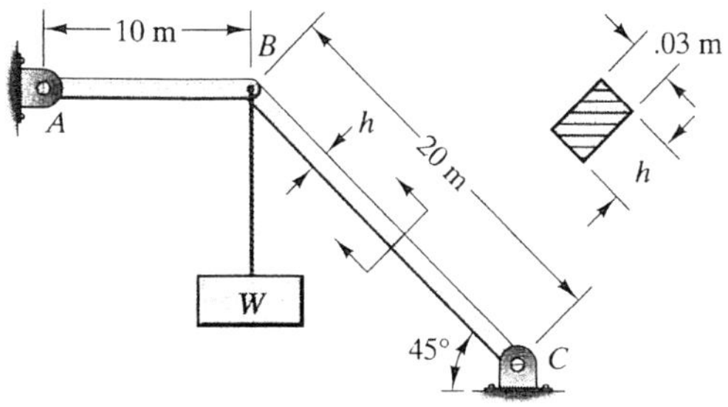

Figure P.17.27.

17.28. [17.7] Consider a column which is pin-connected at the ends supporting an axial load P and end couples M_0 as shown in Fig. P.17.28. What is the deflection curve? Explain why this problem is the same as an eccentrically loaded column considered in Section 17.9.

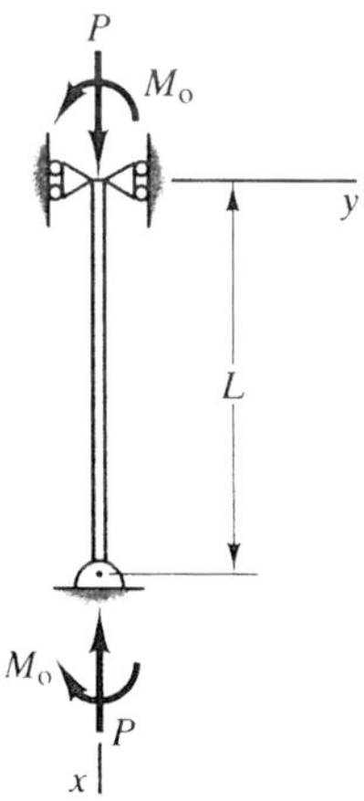

Figure P.17.28.

17.29. [17.8] In the pin-ended initially bent member of Section 17.8, determine the maximum deflection of the member (including the initial bent shape) for a load P which is one fourth of the buckling load. Use the following data:

$L = 10$ ft	$I = 20$ in.^{4}
$E = 30 \times 10^6$ psi	$\delta_0 = 2$ in.

17.30. [17.8] Consider an initially bent column which is fixed at $x = 0$ and free at $x = L$ (see Fig. P.17.30). Assume that the initial shape is that of a sinusoid such that

$$v_{\text{initial}} = \delta_0 \sin \frac{\pi x}{2L}$$

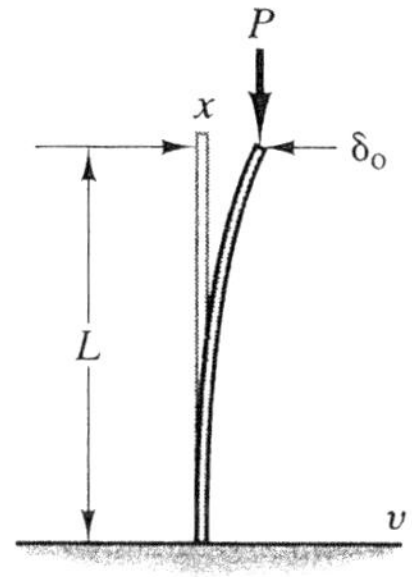

Figure P.17.30.

What is the maximum deflection of the column for a load P which is equal to $.2P_{cr}$? Use the following data:

$L = 1.5$ m	$I = 6 \times 10^{-6}$ m^4
$E = 2 \times 10^{11}$ Pa	$\delta_0 = 25$ mm

(*Hint:* Replace column by a simply-supported column of length $2L$ and work with it using results of Section 17.8.)

17.31. [17.9] Shown in Fig. P.17.31 is an eccentrically loaded column supporting a load P of 500 lb. What is the horizontal deflection of point A? The diameter of the column is 2 in. and the modulus E is 30×10^6 psi. (*Hint:* Make into eccentric column similar to Fig. P.17.18 and use the results of Section 17.9.)

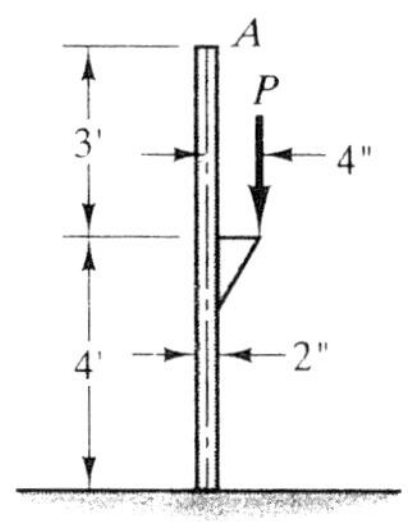

Figure P.17.31.

17.32. [17.9] A column is made of a standard 5-in. steel pipe. The column is 20 ft long. If the pipe is loaded eccentrically with $e = 4$ in., what is the maximum load P for a safety factor of 3? Take the yield stress as 60,000 psi and the modulus of elasticity as 30×10^6 psi. [*Hint:* Set up a relationship of the form $P = f(P)$. Guess at P and insert in $f(P)$. Solve for new P and reinsert, and so on. This is a *recursion* process.]

17.33. [17.9] The secant column formula is the basis for many design code restrictions on compression members. Consider the situation wherein stresses from two different loading cases are formulated separately for a given member. Can these stresses be superposed to determine the "worst case" design load? Explain your answer.

17.34. How far must point A of the linear spring move in Fig. P.17.34 from the unstretched state of the spring before buckling occurs in rod BC? The spring constant is $K = 10^7$ N/m and $E = 2 \times 10^{11}$ Pa for the rod.

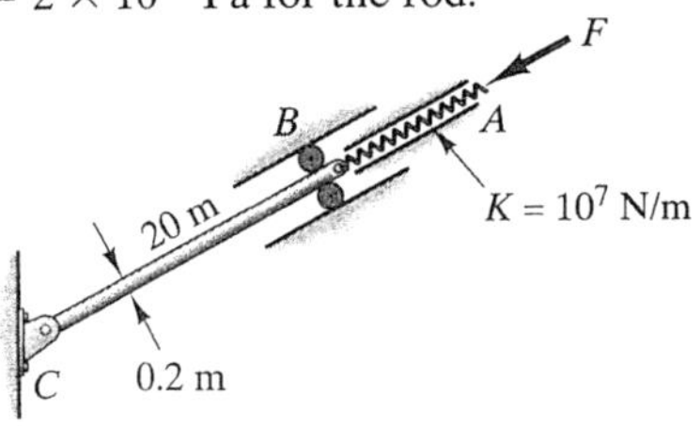

Figure P.17.34.

17.35. [17.5] What is the maximum load F for no buckling for the system shown in Fig. P.17.35? All members are rods of diameter 2 in. and all have $E = 30 \times 10^6$ psi.

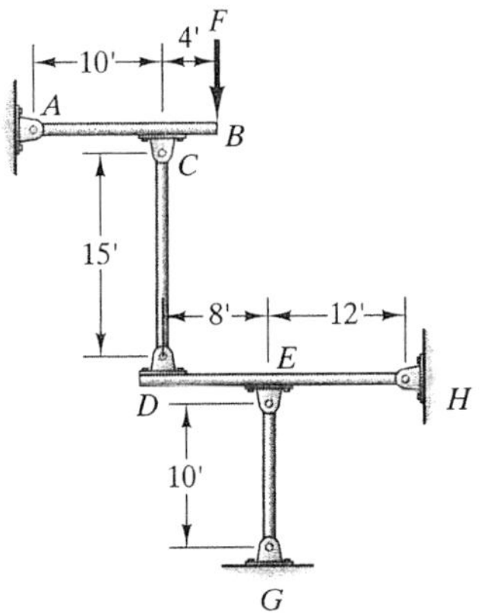

Figure P.17.35.

17.36. What is the maximum intensity of loading w_0 for a safety factor of 2 considering the possibility of buckling for member AB in Fig. P.17.36? Take $E = 2 \times 10^{11}$ Pa. Also, what is the maximum intensity of loading if horizontal lateral constraints are placed at points along one third and along two thirds the length of member AB? Buckling is constrained to the plane of the problem.

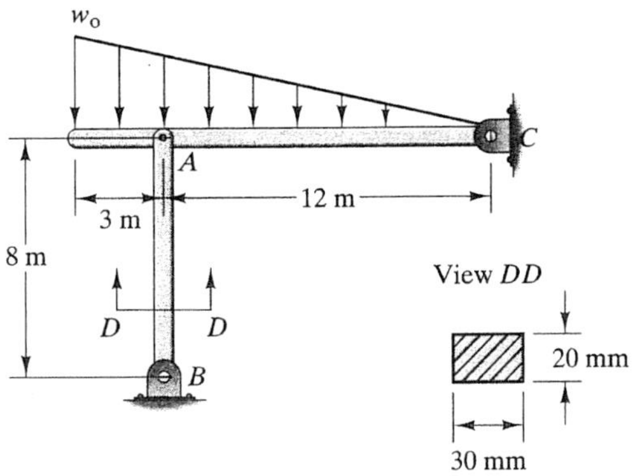

Figure P.17.36.

17.37. For a safety factor n of 3, what is the maximum load P_{max} that can be put on the truss shown in Fig. P.17.37? ρ_z is the radius of gyration of the cross section of each member along an axis parallel to the pins. Neglect weights of members.

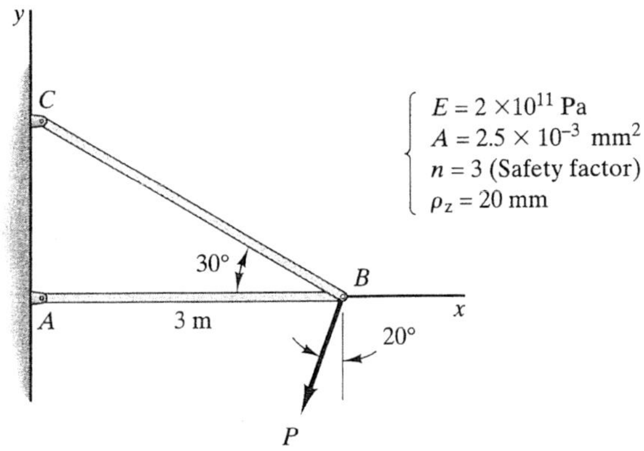

Figure P.17.37.

17.38. Find P_{max} for the data given in Fig. P.17.38.

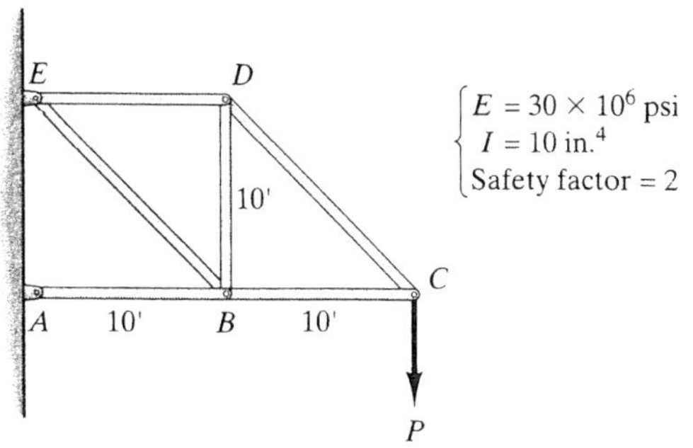

Figure P.17.38.

17.39. Set up general formulation for the permissible load W acceptable to the scissors jack shown in Fig. P.17.39. Use n as a safety factor, I as the second moment of area for the cross section of the members about centroidal axes normal to the plane of the jack. Evaluate W for the case where $L = .3$ m, $\theta = 60°$, $E = 2 \times 10^{11}$ Pa, $n = 3$, and $I = 7.8 \times 10^4$ mm^4. Work out for buckling. Is screw rod AB subject to buckling?

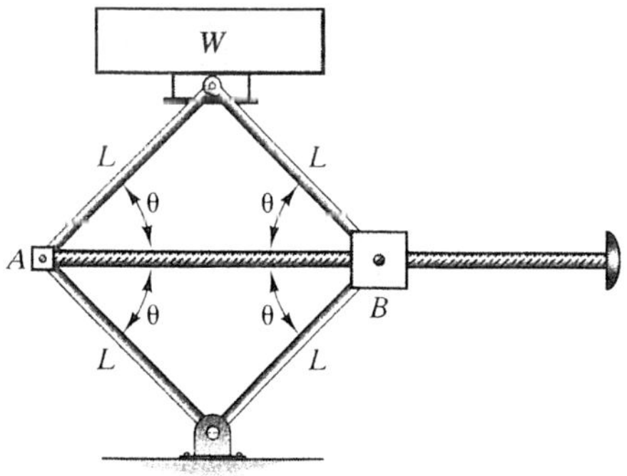

Figure P.17.39.

17.40. What is the relation between W, D, and h and the yield stress Y in Fig. P.17.40 for buckling and crushing loads to be equal? For $D = .1$ m and $Y = 3 \times 10^8$ Pa, what are h and W from your formulations? Take $E = 2 \times 10^{11}$ Pa.

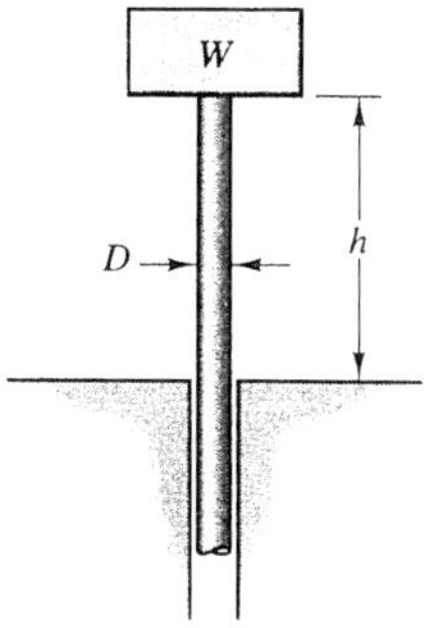

Figure P.17.40.

17.41. In Problem 17.40, what is h for a load of 1000 kN and a safety factor of 2.5, considering buckling as a criterion? Use the data of Problem 17.40.

17.42. What is the maximum loading intensity w_o for the triangular loading to avoid buckling of member AC in Fig. P.17.42? Buckling is in the plane of the problem.

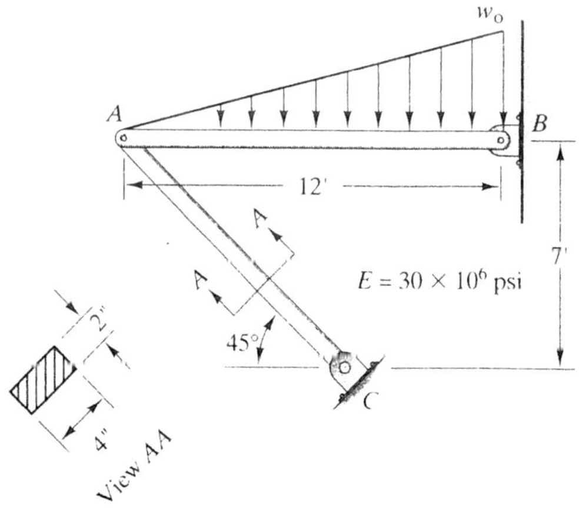

Figure P.17.42.

17.43. Consider a column which is fixed at $x = 0$ and free at $x = L$. What is the formula for the buckling load? Work out from first principles while noting from Eqs. (17.8) and (17.9) that

$$M = EI\frac{d^2v}{dx^2}$$

$$V = -EI\frac{d^3v}{dx^3} - P\frac{dv}{dx}$$

17.44. Four standard 6x6 angle iron members having a thickness of 1 in. are assembled together in Fig. P.17.44 so that they touch along their entire length to form a 30-ft column. If the column is pinned at the ends, what is the buckling load? Now disconnect the members and separate their faces by 1 in. so the sides remain parallel. Then pin each one to the same supports at the ends. What is now the buckling load for the assemblage?

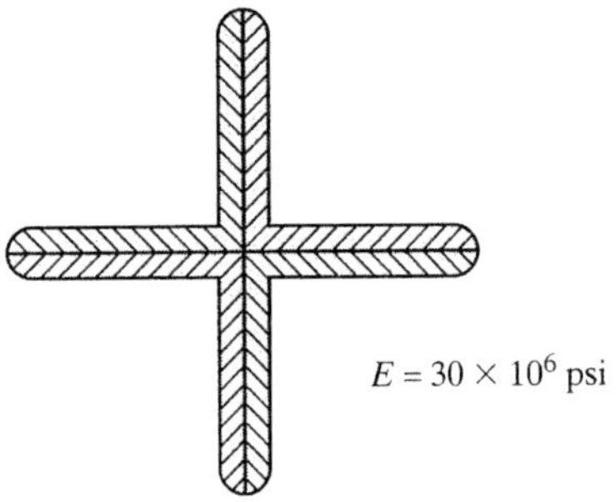

Figure P.17.44.

17.45. Set up the differential equation for the deflection of the axially loaded beam shown in Fig. P.17.45. Determine the deflection curve.

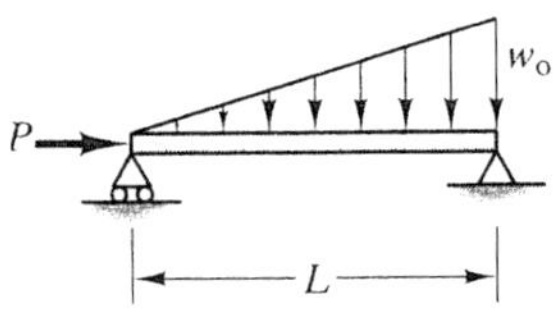

Figure P.17.45.

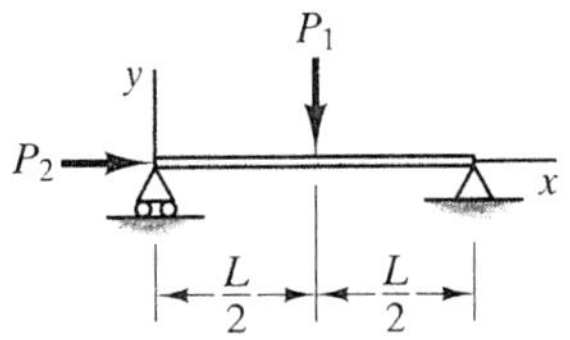

Figure P.17.46.

17.46. Find the maximum deflection for the axially and transversely loaded beam in Fig. P.17.46. (*Hint:* Use half of the beam for consideration.)

17.47. Shown in Fig. P.17.47 is a truss loaded (improperly) with a uniform loading $w_0 = 50$ lb/ft over member AB. What is the maximum vertical deflection of member AB relative to that of pins A and B? All members have the same modulus $E = 30 \times 10^6$ psi and the same second moment of area $I = 50$ in.^{4}.

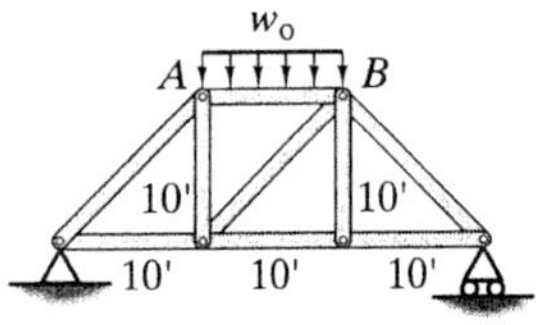

Figure P.17.47.

17.48. If the rod in Problem 17.5 were restrained from lateral motion at two points so placed as to divide the rod into three lengths, what would then be the critical load?

17.49. Shown in Fig. P.17.49 is a vertical rod having a diameter of 2 in. and a modulus of elasticity of 30×10^6 psi. The guy wire has an ultimate stress of 50,000 psi and a diameter of 1/4 in. What is the maximum load P that this system will withstand before failure as a load-carrying system can be expected?

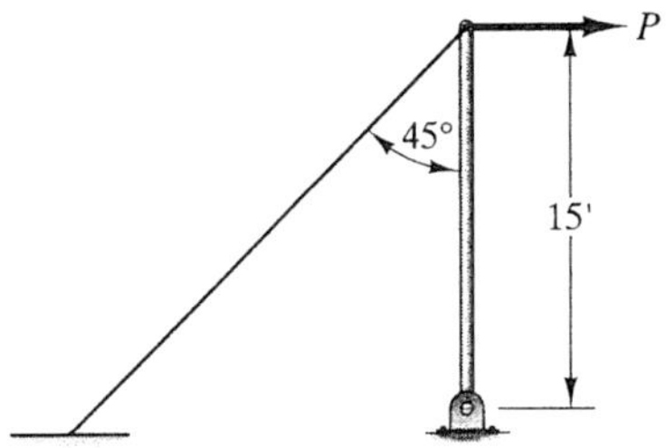

Figure P.17.49.

17.50. What is the buckling load of a column 6 m long with a rectangular cross section 50 mm $\times$ 100 mm? The column is clamped at one end and hinged at the other end so as to have the largest buckling load. Take E as 2.1×10^{11} Pa.

17.51. Shown in Fig. P.17.51 is a column having a rectangular cross section. It is supported at the ends by ball-joint connections. If the buckling is permitted by constraints to take place only in the yz plane, what is the buckling load? If the plane of buckling is restricted to the xz plane, what is the buckling load? Finally, compute the buckling load for a plane at 45° to the xz and yz planes. Take $E = 30 \times 10^6$ psi. The length of the column is 20 ft.

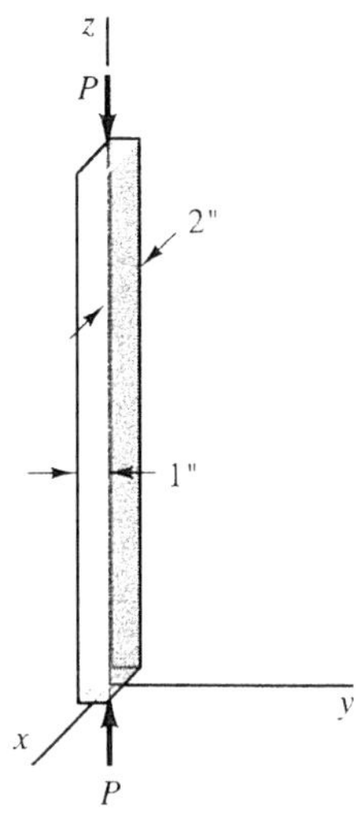

Figure P.17.51.

17.52. Redo Problem 17.39 and plot the normalized load $\frac{wL^2}{EI}$ versus the angle θ for $0 \leq \theta \leq 90°$. What happens when $\theta = 0°$? Why? Let $n = 1$.

17.53. To assist in the design of compression members, it is sometimes useful to consult a "column curve" based on the eccentricity of the load P. Plot such a curve for nominal stress, P/A, versus the slenderness ratio L/ρ. To get the nominal stress go to the secant equation with the left side equal to the yield stress τ_Y. Next, using the right side of the secant equation, find the nominal stress, P/A, which for a given value of ec/ρ^2 and a chosen value of L/ρ give the value of the yield stress (specified on the left side of the secant equation). Note that the secant equation, Eq. (17.48), is nonlinear and requires for this an iterative solution approach. For a structural steel member with $E = 30 \times 10^6$ psi and $\tau_Y = 36{,}000$ psi, plot column curves for $ec/\rho^2 = 0.25, 0.50, 0.75, 1.0$ with L/ρ going from 0 to 200 in increments of 25.

Project 14 Rock Crusher Stability Problem

In Fig. P.17.52, find the gage pressure p that will result in buckling as α goes from 5° to 45° in steps of 1°. Take $h = 3$ in. The following data apply:

Diameter of the piston D, 5 in.

$$E = 30 \times 10^6 \text{ psi}$$

$$AB = BC = 5.0 \text{ ft}$$

Design for a 50% overload.

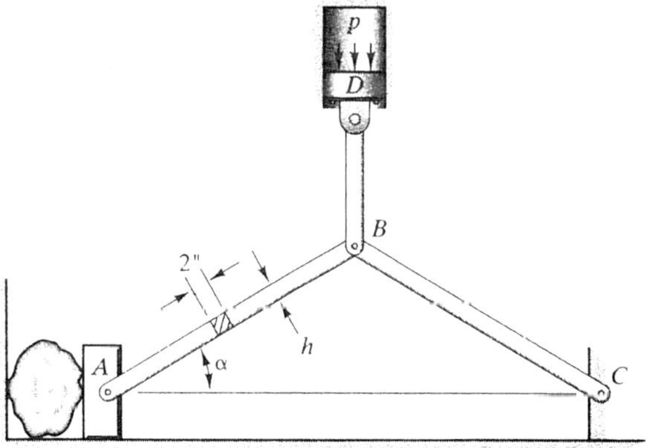

Figure P.17.52.

*CHAPTER 18

Energy Methods

18.1 Introduction

Energy methods, as the term applies in solid mechanics, cover a large family of methods that make use of the energy of the system in order to obtain a solution. As we have seen in Chapter 6, and we will explore further here, the energy in a body can be attributed to external loads and internal stresses and strains. The energy approach is an important and powerful tool for the solution of many problems in solid mechanics; it often provides a direct and simple solution methodology. A descendant of the energy-based analysis methods is the popular *finite element method*. This method is used extensively in engineering to analyze complex problems not only in the field of stress analysis, but also in vibration, heat transfer, fluid flow, and many other areas of vital interest to engineering design. Consequently, a good understanding of energy methods has become even more essential for today's engineers. In addition, many of the techniques presented in this chapter are ideally suited to "hand-calculation" type problems. This makes these techniques very useful for quick design checks, or for smaller, less complicated analyses. In this chapter, we endeavor to present the essential theory underlying certain key aspects of energy methods. Our approach is designed to help the reader understand the background and proper use of these new tools. Furthermore, by presenting a sound theoretical basis, it will make it easier for the reader to understand the more advanced energy methods that will characterize his or her senior-graduate courses.

In Part A of this chapter, we present a number of key energy methods: the so-called *displacement methods*, consisting of virtual work, total potential energy, and the first Castigliano theorem, and in Part B of this chapter an analogous set of principles that are the so-called *force methods* consisting of complementary virtual work, total complementary potential energy, and the second Castigliano theorem.

Before proceeding, it may be wise to orient the reader further as to what he or she is about to get into. Up to now, we have been following a straightforward solution approach in that we used an equation of *equilibrium* in a fairly direct manner. Along with *compatibility* and a *constitutive law,* these three key ingredients led to a solution. In the process, you were aware of these ingredients at all times. We sometimes call this the *vector mechanics* approach.

In what is to follow, it is sometimes more difficult to see exactly where the basic ingredients of *equilibrium, compatibility,* and a *constitutive law* come into play in the solution of a problem. Instead of working directly with an equation of equilibrium or a condition of compatibility, we will find ourselves setting certain conditions on a function or on what we shall define more carefully as a *functional* (for now consider as a function of a function rather than a function of variables). It will be harder for this reason to have a feel, as you may have had up to now, for the physics of the analyses.

The approach that we will now undertake, which includes the so-called energy methods, is called *variational mechanics.* The most intriguing aspect of this new approach is that although it may be less physically obvious and more mathematical in nature, the resulting computational abilities often make for *tremendous simplification* for certain problems and furthermore, lead to a powerful method called *finite elements,* which, we repeat, is now widely used in modern engineering design.

Finally, this change of approach that you will undertake here in solid mechanics has its counterpart in other engineering sciences and they may be linked by a calculus called the *calculus of variations.* Although this is beyond the level of this book, you will be quite ready to undertake its study when you have completed this chapter. Thus, your potential growth at this point stretches far beyond the realm of solid mechanics.

Part A: Displacement Methods

18.2 Principle of Virtual Work

Some students will have studied the method of virtual work for particles and rigid bodies in earlier mechanics courses. The key idea for this method is that of the *virtual displacement.* Such a displacement is considered a *small, arbitrary, hypothetical change in the position of a particle without violating any constraints on the particle and with no attending elapse of real time.* Constraints may arise, for example, from the presence of a specific real surface that the particle must not penetrate during the virtual displacement.

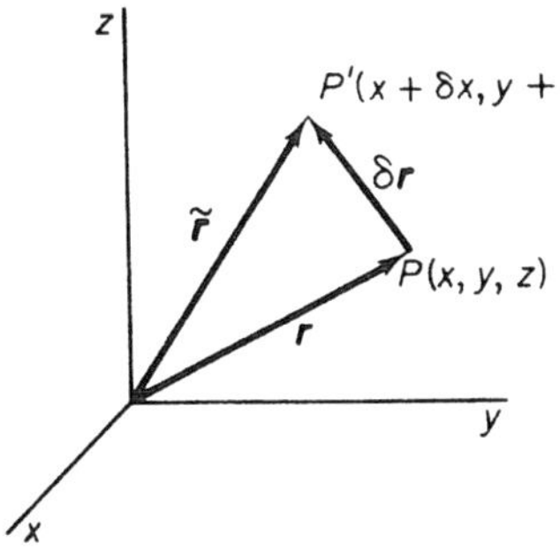

Figure 18.1. Virtual displacement vector.

The change in the spatial coordinates for a virtual displacement we repeat is in no way linked to a change in time. (Such a link would normally exist through Newton's laws when we are working in the field of mechanics.) This has been shown in Fig. 18.1, where we have a virtual displacement $\delta \boldsymbol{r}$ from point $P(x, y, z)$ to point P', which has coordinates $(x + \delta x)$, $(y + \delta y)$, and $(z + \delta z)$. We call the position vector to point P' the *varied position* vector $\tilde{\boldsymbol{r}}$. We can then say that

$$\delta \boldsymbol{r} = \tilde{\boldsymbol{r}} - \boldsymbol{r}$$

We may at this stage consider δ above to be an *operator* acting on $\boldsymbol{r}$ so as to generate the difference vector between the position vector $\boldsymbol{r}$ itself and the varied position vector $\tilde{\boldsymbol{r}}$.

We may also introduce the concept of the *varied function* $\tilde{G}$ such that given a function $G(x, y, z)$ we may form

$$\tilde{G} = G(x + \delta x, y + \delta y, z + \delta z)$$

where δx, δy, and δz are components of $\delta \boldsymbol{r}$.[1] We now define the *variation* of G, denoted as δG, as

$$\delta G = \tilde{G} - G$$

We now extend the concept of a virtual displacement of a *point* to that of a virtual displacement *field.* This field is a single-valued, continuous vector field representing a *hypothetical movement of each point of a deformable body consistent with the constraints present and with no consideration of time.* We shall restrict ourselves here to such *virtual displacement fields,* that result only in infinitesimal deformation.[2] We have shown the gross exaggeration of a virtual displacement field in Fig. 18.2,

[1]The changes in the coordinates x, y, and z are not linked to time through the basic laws of physics as would be the case if we were considering G to represent some physical quantity in some real process.

[2]One need not so restrict oneself. That is, we can work with virtual displacement fields for *finite* deformation and formulate a principle of virtual work. This would take us beyond the scope of this book, however.

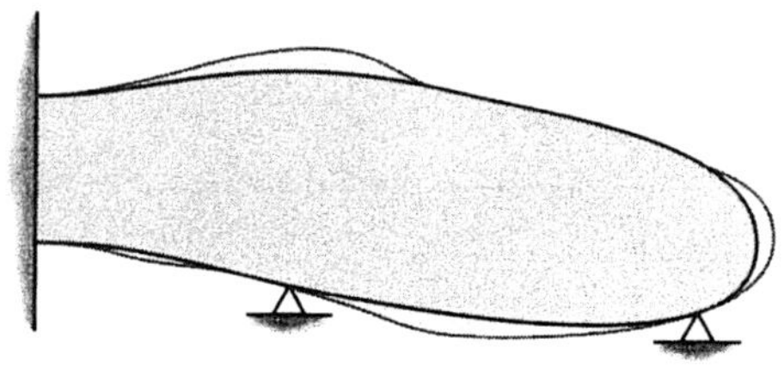

Figure 18.2. Virtual displacement field consistent with constraints.

wherein you will notice that the constraints on the body have not been violated. These constraints are the two point supports (no translation) and the rigid wall (no translation or rotation). We can conveniently set forth a *virtual displacement field* by employing the *variational operator* δ presented earlier. This operator represents an infinitesimal change from an original state, such as the position vector field, to a varied state, namely the varied position vector field. Thus $\delta \boldsymbol{u}$ may be considered as a virtual displacement field giving a hypothetical infinitesimal movement of each point in the body from a given geometry to another infinitesimally close geometry which we can call the varied geometry. The constraints present in the system are taken into account by imposing proper conditions on the variation so that the virtual displacement field does not violate any of the constraints. Mathematically, what we are doing is making an infinitesimal change in the dependent variable $\boldsymbol{u}$ while allowing no change in the independent time variable t, which in a real case would also have to undergo a change. Thus, the movement $\delta \boldsymbol{u}$ takes place conceptually within the engineer's mind. The virtual displacement concept can be used for a rigid body or a deformable body. In the former case, the virtual displacement field must be such as not to violate the rigidity of the body.

We define virtual work as the work performed by the body forces and the traction forces both of which are **maintained constant while the body undergoes the virtual displacement.** With a total body force distribution $\boldsymbol{B}(x, y, z)$ and a total surface force distribution $\boldsymbol{T}(x, y, z)$, we can then write the virtual work expression, denoted as δW_{virt}, as

$$\delta W_{\text{virt}} = \iiint_V \boldsymbol{B} \cdot \delta \boldsymbol{u}\, dv + \oint\!\!\oint_S \boldsymbol{T} \cdot \delta \boldsymbol{u}\, dA \tag{18.1}$$

The two integrals on the right-hand side of Eq. (18.1) are a direct result of the dot product of force and distance. By integrating, over the volume of the body, the dot product of body force (dimensions: $[F/L^3]$) and virtual displacement (dimension: $[L]$) results in work (dimensions: $[FL]$). The same idea holds for the surface traction; however, its dimensions are $[F/L^2]$.

For *rigid bodies,* some may recall the virtual work had to be zero for equilibrium. *For deformable bodies this is no longer true.* We demonstrate this by considering a simple one-dimensional system; the prismatic rod shown in Fig. 18.3. Here, a unidirectional body force dis-

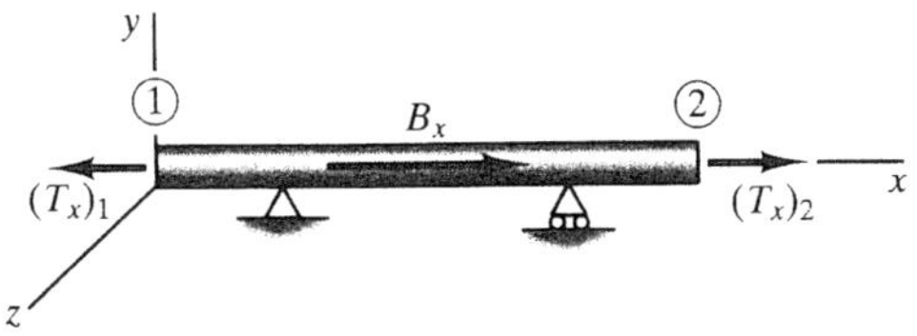

Figure 18.3. One-dimensional system.

tribution $B_x(x)$ acts throughout the rod, and surface traction force distributions $T_x(x)$ act only on the end surfaces of the rod.

We shall find the virtual work, via Eq. (18.1), for the rod and its loading by considering a virtual displacement field $\delta u_x(x)$—that is, a virtual displacement field that is in the x direction and that is a *single-valued, continuous function* only of x. From Eq. (18.1), we see that the second integral on the right-hand side is a product of applied surface forces and the virtual displacement. Since the applied surface traction is zero everywhere except at the ends of the rod, we can rewrite the virtual work expression as

$$\begin{aligned} \delta W_{\text{virt}} &= \iiint_V B_x \delta u_x dv + \left(\iint_A T_x \delta u_x dA\right)_2 - \left(\iint_A T_x \delta u_x dA\right)_1 \\ &= \iiint_V B_x \delta u_x dv + \iint_A [(T_x \delta u_x)_2 - (T_x \delta u_x)_1]\, dA \end{aligned} \tag{18.2}$$

The bracketed expression above (in the surface integral) may be given as follows:

$$[(T_x \delta u_x)_2 - (T_x \delta u_x)_1] = \int_1^2 \frac{d}{dx}(T_x \delta u_x)\, dx \tag{18.3}$$

Note that T_x is the force per unit area normal to the cross section. We recognize this as the stress τ_{xx}. Making this change and substituting Eq. (18.3) into Eq. (18.2), we get

$$\delta W_{\text{virt}} = \iiint_V B_x \delta u_x dv + \iiint_V \frac{d}{dx}(\tau_{xx} \delta u_x)\, dv \tag{18.4}$$

Note that $dxdA$ has been replaced by dv in the last expression. Using the product rule to carry out the differentiation in Eq. (18.4), we may state on collecting terms that

$$\delta W_{\text{virt}} = \iiint_V \left[\left(B_x + \frac{d\tau_{xx}}{dx}\right)\delta u_x + \tau_{xx}\frac{d}{dx}(\delta u_x)\right] dv \tag{18.5}$$

We now apply *two key steps* in the development of the desired virtual work relationship. The *first key step* is the introduction and enforcement of **equilibrium** for our energy expression. Until now, we have not formally stated equilibrium in our development. If we recall the equations of equilibrium in *differential form* [see Eq. (2.5)] we see that for one-dimensional systems, we may delete the first expression in the integral of Eq. (18.5). From here on, the resulting energy formulations contain the *necessary conditions* that they satisfy the *equations of equilibrium.* That concludes the first of the two key steps.

For the *second key step,* we interchange the order of the derivative and the delta operators and arrive at[3]

$$\delta W_{\text{virt}} = \iiint_V \tau_{xx} \delta\left(\frac{du_x}{dx}\right) dv \tag{18.6}$$

We will replace du_x/dx by ε_{xx} and thereby introduce an arbitrary **kinematically compatible** strain field. Why? This field is kinematically compatible since it is formulated directly from a single-valued, continuous displacement field which here, in addition, happens to be arbitrary except for not violating the constraints. Also, by using $du_x/dx = \varepsilon_{xx}$, we are limiting the result to *small deformation*—as you have learned in Chapter 3 on strain. This is the *second* crucial step in the development. The ensuing formulations must now satisfy *equilibrium* as a result of step 1 for any *kinematically compatible strain field* ε_{xx} which has been injected in step 2. We then have the following simple yet crucial expression that

$$\delta W_{\text{virt}} = \iiint_V \tau_{xx} \delta\varepsilon_{xx}\, dv \tag{18.7}$$

We have come to an important result. The virtual work of the external force distribution is clearly *not zero* for equilibrium of a deformable body; it equals what may be considered as the *virtual work of the internal forces.* That is, we can consider the product of stress and virtual strain, integrated over the body, as the internal virtual work. Upon checking the dimensions, they do indeed equal those of work $[FL]$.

The preceding equation represents the principle of virtual work for a simple one-dimensional stress problem. By virtue of the fact that the equilibrium equation was used in its development, we repeat that Eq. (18.7) must then be a *necessary condition for equilibrium.* Note that the virtual displacement field was an arbitrary one-dimensional displacement field having only the requirements of being single-valued,

[3]The δ operator acting on a variable is very much like a differential operator acting on this variable, except that for the delta operator there is a differential change of the variable without the accompanying change in time as required by the differential operator. Because of this similarity, we can interchange the delta operator with a differential or derivative operator.

continuous and of not violating the constraints on the body. We call all such displacement fields for a given problem the *admissible* displacement fields for the problem. Accordingly, consider for the one-dimensional body undertaken in this discussion a force distribution $\boldsymbol{B}$ and a surface traction distribution $\boldsymbol{T}$ that are statically compatible. By that we mean that these forces satisfy rigid body equations of equilibrium as studied in your statics course. We can now say with these forces that the associated one-dimensional stress distribution τ_{xx} *must satisfy Eq. (18.7)* for *any* compatible, one-dimensional deformation in order to ensure the satisfaction of the equilibrium equations.[4]

For a *more general case,* we can extrapolate Eq. (18.7) to include the virtual work done by all normal and shear stresses as

$$\delta W_{\text{virt}} = \iiint_V [\tau_{xx}\delta\varepsilon_{xx} + \tau_{yy}\delta\varepsilon_{yy} + \tau_{zz}\delta\varepsilon_{zz} + \tau_{xy}\delta\gamma_{xy} + \tau_{xz}\delta\gamma_{xz} + \tau_{yz}\delta\gamma_{yz}]dv \tag{18.8}$$

We usually express Eq. (18.8) in the following manner:

$$\iiint_V \boldsymbol{B} \cdot \delta\boldsymbol{u}\, dv + \oiint_S \boldsymbol{T} \cdot \delta\boldsymbol{u}\, dA = \iiint_V \sum_i \sum_j \tau_{ij}\delta\varepsilon_{ij}\, dv \tag{18.9}$$

where we sum over i and j in the last expression. In this general statement of the virtual work principle, we can say that for a given statically compatible body force $\boldsymbol{B}(x, y, z)$ and surface traction distribution $\boldsymbol{T}(x, y, z)$, the correct stress distribution τ_{ij} in the body (i.e., the distribution that satisfies equilibrium requirements) must satisfy the virtual work equation for any admissible deformation field consistent with the constraints. Furthermore, this condition is sufficient for equilibrium. That is, no additional conditions have to be specified. Thus, the virtual work equation can be used *in place of* equilibrium. It should be made clear that *no constitutive law* has been employed in this development; that is, the *virtual work principle is valid for any material.* The only restriction that we have set forth in arriving at Eq. (18.9) is that of infinitesimal (i.e., small) deformation.

The way we use the virtual work principle is as follows. Generally, we know statically compatible loads, generically expressed as $\boldsymbol{B}$ and $\boldsymbol{T}$, for a given problem. We are interested in finding the stress distribution. We then choose some simple kinematically compatible deformation that is consistent with the constraints present and that permits us to

[4]One can also show that this is *sufficient* for equilibrium as in the case of the rigid-body statics.

determine the stresses τ_{ij} from Eq. (18.9). *The stresses so determined must then satisfy equilibrium requirements.* If the displacements are desired, a constitutive law must be introduced to link the actual stresses to the actual displacements. Satisfaction of the principle of virtual work then leads to the actual displacements due to the applied loads because we will have satisfied equilibrium, compatibility, and a constitutive law, the familiar three pillars of solid mechanics used throughout this book. Since we have used a virtual displacement field, this approach is called the *displacement method.* A popular application of the principle of virtual work is called the *dummy displacement method.* It is safe to assume that the "dummy" here refers to the fact that we impose a virtual displacement field.

The following example illustrates the application of the principle of virtual work to a simple two-bar structure. In this example and others to follow, we will be sure to indicate to the reader when equilibrium, compatibility, and an acceptable constitutive law are being used so that the reader does not lose sight of the three pillars of solid mechanics.

Example 18.1

For the two-member pin-jointed truss shown in Fig. 18.4, apply the principle of virtual work to determine the forces in the members. Next, introduce a linear elastic constitutive law and determine the displacement of point A. The cross-sectional area and length are the same for bars AB and AC. Assume small displacements.

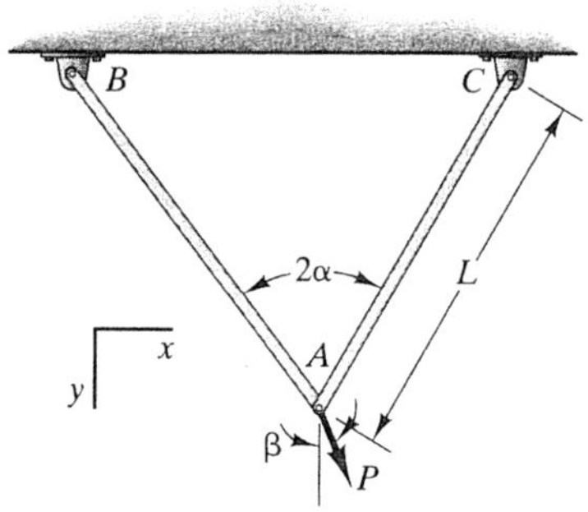

Figure 18.4. Simple truss.

This is a statically determinate problem. The forces in the two members can easily be determined by considering a pin A as a free body and applying equilibrium directly. However, we shall proceed here using the principle of virtual work to illustrate its use under the most simple conditions.

We can generate *compatible* deformations in the system by giving pin A *virtual displacements* in the vertical and horizontal directions. We will start by giving pin A a virtual displacement δv_A as shown in Fig. 18.5.

The accompanying virtual strains in the axial direction for the two members are readily calculated as (**compatibility**)

$$\delta\varepsilon_{AB} = \frac{\delta v_A \cos\alpha}{L}$$
$$\delta\varepsilon_{AC} = \frac{\delta v_A \cos\alpha}{L} \qquad \text{(a)}$$

Example 18.1 (Continued)

We have invoked small displacement theory allowing us to consider $\angle BA'A$ to be essentially the same as α. With no body forces, the principle of virtual work [Eq. (18.9)] for our virtual displacement becomes (**equilibrium**)

$$(P\cos\beta)\delta v_A = \int_0^L \tau_{AB}\delta\varepsilon_{AB} A dl + \int_0^L \tau_{AC}\delta\varepsilon_{AC} A dl \qquad \text{(b)}$$

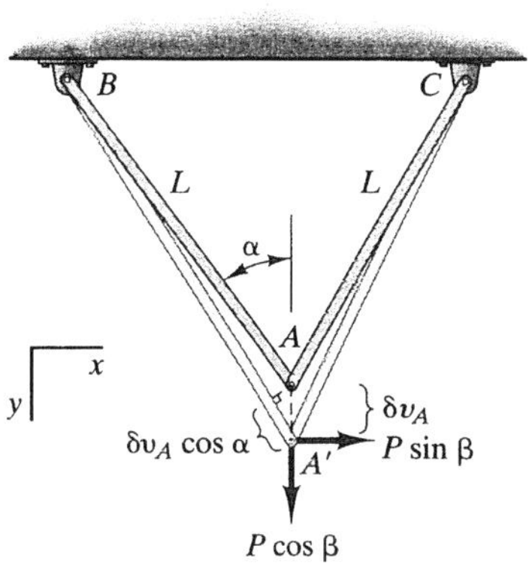

Figure 18.5. Truss showing a vertical virtual displacement.

where A is the cross-sectional area. Substituting the virtual strains [Eq. (a)] into (b) and carrying out the integration gives

$$(P\cos\beta)\delta v_A = \tau_{AB} A\delta v_A \cos\alpha + \tau_{AC} A\delta v_A \cos\alpha \qquad \text{(c)}$$

Canceling out δv_A in Eq. (c); letting the member's stress equal the force divided by area; and, finally, rearranging terms we get

$$P\cos\beta = (F_{AB} + F_{AC})\cos\alpha \qquad \text{(d)}$$

We recognize Eq. (d) as a statement of equilibrium in the vertical direction.

To obtain a second equation, we proceed in a similar fashion, now applying a horizontal virtual displacement δu_A, as shown in Fig. 18.6.

We see from Fig. 18.6 that $A'B$ lengthens by $\delta u_A \sin\alpha$ while similarly $A'C$ shortens by the same amount. The virtual strains are given as (**compatibility**)

$$\delta\varepsilon_{AB} = \frac{\delta u_A \sin\alpha}{L}$$
$$\delta\varepsilon_{AC} = -\frac{\delta u_A \sin\alpha}{L} \qquad \text{(e)}$$

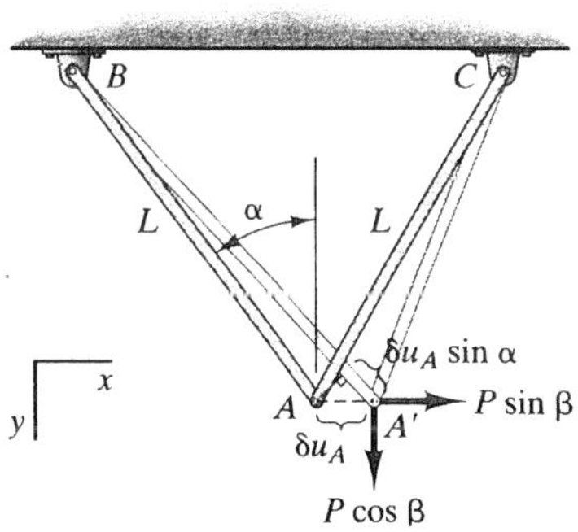

Figure 18.6. Truss showing horizontal virtual displacement.

The principle of virtual work gives (**equilibrium**)

$$(P\sin\beta)\delta u_A = \tau_{AB} A\delta u_A \sin\alpha - \tau_{AC} A\delta u_A \sin\alpha \qquad \text{(f)}$$

Canceling terms in Eq. (f) and letting the member's stress equal the force divided by area, we get

$$P\sin\beta = (F_{AB} - F_{AC})\sin\alpha \qquad \text{(g)}$$

This is our second equilibrium equation. Solving Eqs. (d) and (g) we get

Example 18.1 (Continued)

$$F_{AB} = \tfrac{1}{2}P\left(\frac{\cos\beta}{\cos\alpha} + \frac{\sin\beta}{\sin\alpha}\right)$$

$$F_{AC} = \tfrac{1}{2}P\left(\frac{\cos\beta}{\cos\alpha} - \frac{\sin\beta}{\sin\alpha}\right)$$

By applying the principle of virtual work, we obtained the equations of **equilibrium** that we readily solved to get the member forces.

We now wish to obtain the actual displacement of point A. In order to link force with displacement, we must introduce a constitutive law. Here we assume linear elastic response of the members. To determine the vertical displacement v_A, we return to the principle of virtual work given by Eq. (b). We start with the **constitutive law,** given as

$$\begin{aligned} \tau_{AB} &= E\varepsilon_{AB} = E\frac{v_A \cos\alpha}{L} \\ \tau_{AC} &= E\varepsilon_{AC} = E\frac{v_A \cos\alpha}{L} \end{aligned} \tag{h}$$

where E is the elastic modulus and where we have used Eq. (a) for ε_{AB} with v_A replacing δv_A thereby satisfying **compatibility.** As will soon be evident, v_A will be the actual displacement component of pin A. Substitution of the stresses from Eq. (h) and virtual strains from Eqs. (a) all into the virtual work equation [Eq. (b)] gives

$$\begin{aligned} P\cos\beta\delta v_A = &\int_0^L \left(E\frac{v_A\cos\alpha}{L}\right)\left(\frac{\delta v_A\cos\alpha}{L}\right)A dl \\ &+ \int_0^L \left(E\frac{v_A\cos\alpha}{L}\right)\left(\frac{\delta v_A\cos\alpha}{L}\right)A dl \end{aligned} \tag{i}$$

Canceling δv_A, integrating, and rearranging, we obtain

$$P = \frac{2AE}{L}\frac{\cos^2\alpha}{\cos\beta}v_A$$

$$\therefore \quad v_A = \frac{PL\cos\beta}{2AE\cos^2\alpha}$$

We follow a similar procedure for the horizontal displacement u_A. Thus, for virtual work we have

Example 18.1 (Continued)

$$(P \sin \beta)\delta u_A = \int_0^L \tau_{AB}\delta\varepsilon_{AB}A\,dl + \int_0^L \tau_{AC}\delta\varepsilon_{AC}A\,dl \qquad \text{(j)}$$

We now go to the **constitutive law**, $\tau_{AB} = E\varepsilon_{AB}$ and $\tau_{AC} = E\varepsilon_{AC}$. Use Eqs. (e) with u_A replacing δu_A to get ε_{AB} and ε_{AC} thereby satisfying **compatibility**. We then get

$$\begin{aligned} \tau_{AB} &= E\varepsilon_{AB} = E\frac{u_A \sin \alpha}{L} \\ \tau_{AC} &= E\varepsilon_{AC} = -E\frac{u_A \sin \alpha}{L} \end{aligned} \qquad \text{(k)}$$

Substitute in Eq. (j) for τ_{AB} and τ_{AC} using Eqs. (k) and replace $\delta\varepsilon_{AB}$ and $\delta\varepsilon_{AC}$ using Eq. (e). We then get

$$P \sin \beta\, \delta u_A = \int_0^L \left(E\frac{u_A \sin \alpha}{L}\right)\left(\frac{\delta u_A \sin \alpha}{L}\right) A\,dl + \int_0^L \left(E\frac{u_A \sin \alpha}{L}\right)\left(\frac{\delta u_A \sin \alpha}{L}\right) A\,dl$$

Integrating and rearranging terms we have for u_A

$$u_A = \frac{PL \sin \beta}{2AE \sin^2 \alpha}$$

The displacements v_A and u_A must be the *actual displacements* since they were *admissible* and were related through a *constitutive law* to a stress distribution that satisfied the *virtual work equations* and thus *equilibrium*. So we see that the energy solution does indeed draw upon the corner stones of solid mechanics: equilibrium, compatibility, and constitutive law.

18.3 Method of Total Potential Energy

In the statement of the principle of virtual work [Eq. (18.9)] the applied forces and stresses are in no way related to the virtual displacement and virtual strains. That is, *no constitutive law* is implied in the *development* of the principle of virtual work. We now consider the application of the principle to an *elastic* body (not necessarily linearly elastic) for which the *stresses are now linked to strains through some constitutive law.*

Recall [Eq. (6.30)] that we can form a differential of the strain energy density, $\mathcal{U}$, as follows:

$$d\mathcal{U} = \tau_{xx}d\varepsilon_{xx} + \tau_{yy}d\varepsilon_{yy} + \tau_{zz}d\varepsilon_{zz} + \tau_{xy}d\gamma_{xy} + \tau_{xz}d\gamma_{xz} + \tau_{yz}d\gamma_{yz}$$

In a similar fashion, we can form the variation of the strain-energy density $\mathcal{U}$. Thus,

$$\delta\mathcal{U} = \tau_{xx}\delta\varepsilon_{xx} + \tau_{yy}\delta\varepsilon_{yy} + \tau_{zz}\delta\varepsilon_{zz} + \tau_{xy}\delta\gamma_{xy} + \tau_{xz}\delta\gamma_{xz} + \tau_{yz}\delta\gamma_{yz} \quad (18.10)$$

Taking the integration throughout the volume of the body, we have

$$\begin{aligned}\iiint_V \delta\mathcal{U}\, dv &= \iiint_V [\tau_{xx}\delta\varepsilon_{xx} + \tau_{yy}\delta\varepsilon_{yy} + \tau_{zz}\delta\varepsilon_{zz} \\ &\quad + \tau_{xy}\delta\gamma_{xy} + \tau_{xz}\delta\gamma_{xz} + \tau_{yz}\delta\gamma_{yz}]\, dv \qquad (18.11) \\ &= \iiint_V \sum_i \sum_j \tau_{ij}\delta\varepsilon_{ij}\, dv\end{aligned}$$

Like the differential operator, we can interchange the variation operator with the integral to form the following equation:

$$\iiint_V \delta\mathcal{U}\, dv = \delta\left(\iiint_V \mathcal{U}\, dv\right) = \delta U = \iiint_V \sum \sum \tau_{ij}\delta\varepsilon_{ij}\, dv \quad (18.12)$$

where we used the fact that the strain energy U is simply the integral over the domain of the strain-energy density $\mathcal{U}$. Accordingly, using Eq. (18.12) to replace the right side of Eq. (18.9), we get on rearranging the terms,

$$\delta U - \iiint_V \boldsymbol{B} \cdot \delta\boldsymbol{u}\, dv - \oiint_S \boldsymbol{T} \cdot \delta\boldsymbol{u}\, dA = 0 \qquad (18.13)$$

We shall now *define the potential energy, V,* of the applied loads as a function of the displacement field $\boldsymbol{u}$ and the applied load in the following way:

$$V = -\iiint_V \boldsymbol{B} \cdot \boldsymbol{u}\, dv - \oiint_S \boldsymbol{T} \cdot \boldsymbol{u}\, dA \qquad (18.14)$$

The variation of V is then formed by varying $\boldsymbol{u}$ while holding the loads constant, as is required by the principle of virtual work. Thus we have for δV:

$$\delta V = -\iiint_V \boldsymbol{B} \cdot \delta\boldsymbol{u}\, dv - \oiint_S \boldsymbol{T} \cdot \delta\boldsymbol{u}\, dA \qquad (18.15)$$

Now going back to Eq. (18.13), we have, on using Eq. (18.15),

$$\delta U + \delta V = 0 \tag{18.16}$$

We *define the total potential energy* π as follows:

$$\pi = U + V \tag{18.17}$$

Hence Eq. (18.16) becomes

$$\delta\pi = 0 \tag{18.18}$$

This is the *principle of the total potential energy,* which is a generalization for deformable elastic bodies of the total potential energy of particles and rigid bodies. When we say that $\delta\pi = 0$, we are saying that π does not change when infinitesimal virtual displacements are instituted. $\delta\pi = 0$ implies the *extremizing* of π, which is analogous to setting the first derivative of a function of a single variable equal to zero to find the local minima and maxima.

With this in mind, what else does $\delta\pi = 0$ tell us in light of its derivation? Thinking back to the principle of virtual work from which it is derived, we must choose kinematically compatible deformations that do not violate the constraints to vary U and V and thus π. Now we have a *stress field* associated through some *elastic constitutive law* to each *varied deformation.* The principle of total potential energy tells us here that the one *particular* deformation that *extremizes* π *when compared to values of* π *for all other varied admissible deformations* (and hence completes the process which was started with the virtual work principle), *must then be the deformation whose corresponding stress field satisfies equilibrium.* The latter conclusion thus stems from the parent formulation, virtual work, wherein the stress satisfying the virtual work equation satisfies equilibrium.

The total potential energy principle may be considered to be the most powerful in solid mechanics. Why is this so? By properly extremizing π with respect to admissible strain fields, we are satisfying **equilibrium,** some **elastic constitutive law,** and **kinematic compatibility** in one fell swoop. The three pillars of solid mechanics are embodied in this amazing principle.

The following example illustrates the use of the total potential energy method applied to the same truss studied in Example 18.1.

Example 18.2

Using the method of total potential energy, find the displacement of point A of the truss presented in Example 18.1.

Example 18.2 (Continued)

We need to determine the total potential energy π and then extremize it; that is, take its variation and set it equal to zero, as stated in Eq. (18.18). When we take the variation, we must ensure that the varied deformations do not violate the constraints of the structure. The external potential energy [Eq. (18.14)] of the truss, in the absence of body forces, is given by

$$V = -(P\cos\beta)v_A - (P\sin\beta)u_A \tag{a}$$

where v_A and u_A are the vertical and horizontal displacements of pin A, respectively. We next derive a general expression for the strain energy for an n-member truss, each member assumed to be prismatic, linear elastic, and in a one-dimensional stress state. Thus (**constitutive law**)

$$\begin{aligned} U &= \iiint_V \sum_i \sum_j \tau_{ij}\,d\varepsilon_{ij}\,dv = \sum_{i=1}^{n} A_i L_i \int \tau_i\,d\varepsilon_i \\ &= \sum_{i=1}^{n} A_i L_i \int \frac{\varepsilon_i}{E_i}\,d\varepsilon_i = \tfrac{1}{2}\sum_{i=1}^{n} A_i E_i L_i \varepsilon_i^2 \end{aligned} \tag{b}$$

where A_i is the cross-sectional area, τ_i is the normal stress, and L_i the length of the ith member. For our case, we have two identical members, so the strain energy becomes

$$U = \tfrac{1}{2}AEL(\varepsilon_{AB}^2 + \varepsilon_{AC}^2) \tag{c}$$

We need now to introduce strain-displacement relations that are derived from a kinematically admissible deformation field. This is done by giving pin A displacements in the vertical and horizontal directions. Therefore, we can use the results from Example 18.1, repeated here (**compatibility**)

$$\begin{aligned} \varepsilon_{AB} &= \frac{\cos\alpha}{L}v_A + \frac{\sin\alpha}{L}u_A \\ \varepsilon_{AC} &= \frac{\cos\alpha}{L}v_A - \frac{\sin\alpha}{L}u_A \end{aligned} \tag{d}$$

We then substitute the admissible strain-displacement relations [Eq. (d)] into the strain energy [Eq. (c)] so that we can express the total potential energy [Eq. (18.17)] as

Example 18.2 (Continued)

$$\pi = U + V$$

$$= \tfrac{1}{2}AEL\left[\left(\frac{\cos\alpha}{L}v_A + \frac{\sin\alpha}{L}u_A\right)^2 + \left(\frac{\cos\alpha}{L}v_A - \frac{\sin\alpha}{L}u_A\right)^2\right]$$
$$-(P\cos\beta)v_A - (P\sin\beta)u_A \tag{e}$$

The principle of total potential energy requires that π be an extremum with respect to kinematically admissible deformation fields characterized by v_A and u_A **(equilibrium)**. Hence,

$$\begin{aligned}\delta\pi &= \delta U + \delta V = 0\\ &= \frac{\partial\pi}{\partial v_A}\delta v_A + \frac{\partial\pi}{\partial u_A}\delta u_A = 0\end{aligned} \tag{f}$$

where we have used the chain rule. Since the varied displacements δv_A and δu_A are independent, Eq. (f) requires that

$$\begin{aligned}\frac{\partial\pi}{\partial v_A} &= 0\\ \frac{\partial\pi}{\partial u_A} &= 0\end{aligned} \tag{g}$$

Substituting Eq. (e) into the above and solving for v_A and u_A, we get

$$\begin{aligned}v_A &= \frac{PL\cos\beta}{2AE\cos^2\alpha}\\ u_A &= \frac{PL\sin\beta}{2AE\sin^2\alpha}\end{aligned} \tag{h}$$

This is identical to the result obtained in Example 18.1 by the principle of virtual work. We know that these displacements are the actual displacements because in extremizing π, we find the *one set* of *kinematically compatible displacements* that through an *elastic constitutive law* is related to the *stress field* that in turn must satisfy *equilibrium.*

18.4 A Comment on the Total Potential Energy Method

We now want to point out that we cannot here fully exploit the powerful method of total potential energy, and we want to explain why. In the study of particles and rigid bodies we extremized $V(x, y, z)$, where clearly V is a function. In Example 18.2, involving a simple truss, the total potential energy π turned out also to be a function, but this time of u and v. Now when it comes to more general elastic bodies, such as beams, plates, and so on, we are still involved with extremizing π but it will no longer be a function as described above. Instead, we deal with a quantity that in its *simplest form* may be expressed as

$$\pi = \int_1^2 F\left(x, y(x), \frac{dy(x)}{dx}\right) dx \tag{18.19}$$

where F is some specified function of the independent variable x, and the dependent variables $y(x)$, and $dy(x)/dx$. This expression is called a *funcional.* Our objective with a functional is to find a function, in this case $y(x)$, that when substituted into Eq. (18.19), will *extremize* the functional (i.e., the result will be a number that will represent either the maximum or minimum of the functional). To extremize a functional requires a different kind of calculus, called the *calculus of variations.* To fully understand and appreciate modern methods of structural mechanics, you should study this more advanced topic.[5] We will, however, further illustrate the use of the total potential energy principle when we present the first Castigliano theorem, which is a special, simplified, very useful adaption of the total potential energy principle, in the next section.

18.5 The First Castigliano Theorem

We now consider elastic bodies held by rigid supports (see Fig. 18.7). Acting on this body are only point forces and point couple moments. These are called *generalized forces.* They are independent of each other and are denoted collectively as Q_i. Along each force Q_i and collinear with it is a displacement Δ_i of the point of application. At the point of application of each point torque, Q_i, is a rotation collinear with the couple moment and also designated as Δ_i. We call Δ_i the *generalized displacements.* These displacements stem from all the loads. Thus, we have N generalized displacements for the N generalized forces.

[5]See, for example, I. H. Shames and C. L. Dym, *Energy and Finite Element Methods in Structural Mechanics,* Taylor & Francis, Washington DC, 1997.

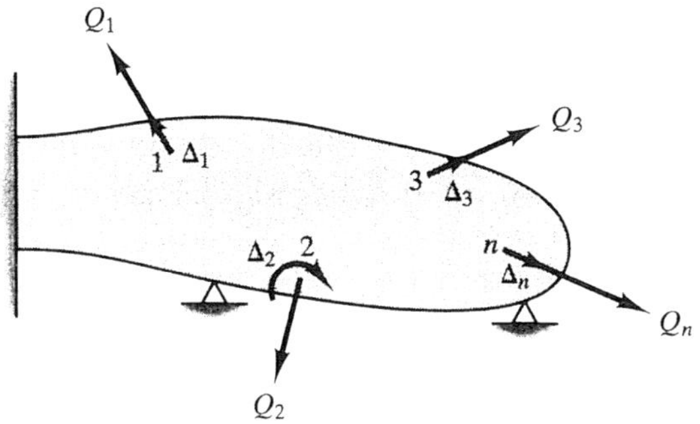

Figure 18.7. Generalized point forces and generalized displacements.

We shall consider this case as a special case of the method of total potential energy. Hence, let us suppose that we can express the potential energy of the internal stresses and strains, (i.e., the strain energy) U, not in terms of strain, as we did in Eq. 6.33, but in terms of the generalized displacements Δ_i. That is,

$$U = \sum_{i=1}^{N} U_i(\Delta_i) \tag{18.20}$$

Now let us express the potential energy of the applied loads, which is given as the sum of the point loads times their respective collinear displacements

$$V = -\sum_{i=1}^{N} Q_i \Delta_i \tag{18.21}$$

We can then build $\pi = U + V$, and by taking its variation and equating it to zero (i.e., $\delta\pi = 0$), as per the method of total potential energy, we get

$$\begin{aligned} \sum_{i=1}^{N} \frac{\partial U}{\partial \Delta_i}\delta\Delta_i - \sum_{i=1}^{N} Q_i \delta\Delta_i &= 0 \\ \therefore \sum_{i=1}^{N} \left(\frac{\partial U}{\partial \Delta_i} - Q_i \right) \delta\Delta_i &= 0 \end{aligned} \tag{18.22}$$

Since we may consider the $\delta\Delta_i$ to be independent of each other, we conclude that

$$\frac{\partial U}{\partial \Delta_i} = Q_i \qquad i = 1,2,\ldots,N \tag{18.23}$$

This is the *first Castigliano theorem*.[6] By computing U in terms of Δ_i we may then determine the required force or torque needed for a particular

[6]So named after A. Castigliano (1847–1884) who published this method in 1873.

generalized displacement Δ_i. Keep in mind that the direction of the point force or couple moment is along the same line as the corresponding generalized displacement. Since the first Castigliano theorem is derived essentially from the principle of virtual work, any solution by the former can be found by the latter with the aid of the proper constitutive law.

Example 18.3

Using the first Castigliano theorem, find the displacement of pin A of the two-bar truss studied in Examples 18.1 and 18.2.

The first Castigliano theorem is expressed mathematically in Eq. (18.23). In order to find the generalized point load related to the generalized displacement, we must take the derivative of the strain energy with respect to the displacement. This displacement must act at the location of interest and the resulting load is along the same line of action as the displacement. From Example 18.2, we have the strain energy as a function of the vertical and horizontal displacements at pin A, repeated below as **(constitutive law** and **compatibility)**

$$U = \tfrac{1}{2}AEL\left[\left(\frac{\cos\alpha}{L}v_A + \frac{\sin\alpha}{L}u_A\right)^2 + \left(\frac{\cos\alpha}{L}v_A - \frac{\sin\alpha}{L}u_A\right)^2\right] \quad \text{(a)}$$

Straightforward application of Castigliano's first theorem will give us the force along the same line of action as the displacement, thus **(equilibrium)**

$$\begin{aligned} \frac{\partial U}{\partial v_A} &= P_v = \left(\frac{2AE}{L}\cos^2\alpha\right)v_A \\ \frac{\partial U}{\partial u_A} &= P_u = \left(\frac{2AE}{L}\sin^2\alpha\right)u_A \end{aligned} \quad \text{(b)}$$

where P_v and P_u are the vertical and horizontal components of the force P applied to pin A. Recognizing from Fig. 18.5 that we can write these components as

$$\begin{aligned} P_v &= P\cos\beta \\ P_u &= P\sin\beta \end{aligned} \quad \text{(c)}$$

Substitution of Eq. (c) into Eq. (b) gives

Example 18.3 (Continued)

$$v_A = \frac{PL\cos\beta}{2AE\cos^2\alpha}$$

$$u_A = \frac{PL\sin\beta}{2AE\sin^2\alpha}$$

This result is identical to that of the previous two examples.

Example 18.4

Determine the relationship between the displacement Δ, of the rigid end cap and the applied load *P*, for the rod (material 1) and sleeve (material 2) shown in Fig. 18.8. Both the rod and sleeve materials are linear elastic. Also, determine the stresses.

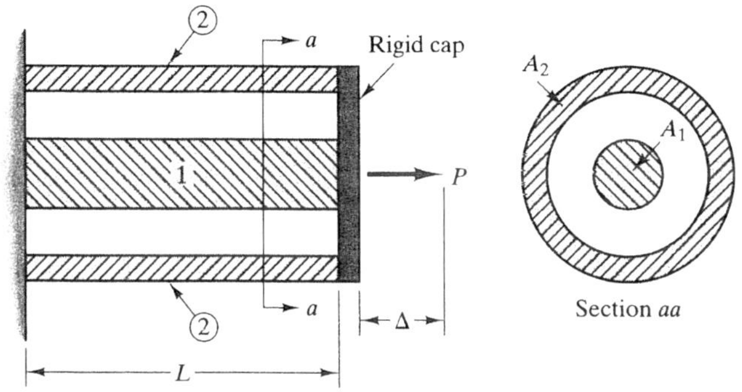

Figure 18.8. Sleeve and cylinder under tensile load.

We will assume a one-dimensional stress state in the rod and sleeve. The displacement of the rigid end cap gives the normal strain in both the rod and sleeve (**compatibility**) that we write as

$$\begin{aligned} \varepsilon_1 &= \Delta/L \\ \varepsilon_2 &= \Delta/L \end{aligned} \tag{a}$$

Where ε_1 and ε_2 are the longitudinal normal strains in material 1 and 2, respectively. From this, we can formulate the strain energy *U* as we did in Example 18.2 as the sum of the energies from the rod and the sleeve, thus using a linear-elastic **constitutive law**. From Eq. 6.37, we have

$$U = \tfrac{1}{2}A_1LE_1\varepsilon_1^2 + \tfrac{1}{2}A_2LE_2\varepsilon_2^2 \tag{b}$$

Example 18.4 (Continued)

Notice that we have subscripted the cross-sectional area A, and elastic modulus E, for the two materials. We substitute the strains from Eq. (a) into the strain energy expression [Eq. (b)]

$$U = \tfrac{1}{2}A_1LE_1\left(\frac{\Delta}{L}\right)^2 + \tfrac{1}{2}A_2LE_2\left(\frac{\Delta}{L}\right)^2 \qquad \text{(c)}$$

and apply the first Castigliano theorem **(equilibrium)**

$$\frac{\partial U}{\partial \Delta} = P = \frac{\Delta}{L}[A_1E_1 + A_2E_2]$$

$$\therefore \quad \Delta = \frac{PL}{A_1E_1 + A_2E_2} \qquad \text{(d)}$$

With the displacement known, the strains in the rod and sleeve are obtained by back substitution of the displacement given by Eq. (d) into the strain-displacement relations, given by Eq. (a). From the **constitutive law,** we then get

$$\tau_1 = E_1\varepsilon_1$$

$$\tau_1 = \frac{PE_1}{A_1E_1 + A_2E_2} = \frac{P}{A_1 + A_2\dfrac{E_2}{E_1}}$$

$$\tau_2 = E_2\varepsilon_2$$

$$\tau_2 = \frac{PE_2}{A_1E_1 + A_2E_2} = \frac{P}{A_1\dfrac{E_1}{E_2} + A_2}$$

Example 18.5

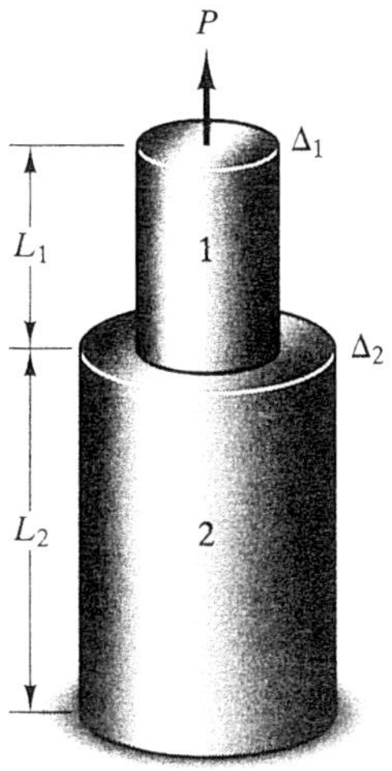

Figure 18.9. Two stacked cylinders under tensile load.

For the two stacked rods shown in Fig. 18.9, determine the total displacement due to the applied load P. Materials 1 and 2 are linear elastic.

Unlike Example 18.4, in this problem we have two independent displacements, Δ_1 and Δ_2. We will use the first Castigliano theorem to determine the displacement in each rod and from this, we can get the total displacement. The strain in each rod is

$$\varepsilon_1 = \frac{\Delta_1}{L_1} \qquad \varepsilon_2 = \frac{\Delta_2}{L_2} \tag{a}$$

The total strain energy for the system is the sum of the energy from each rod. Thus (**constitutive law**)

$$U = \tfrac{1}{2}A_1E_1L_1\left(\frac{\Delta_1}{L_1}\right)^2 + \tfrac{1}{2}A_2E_2L_2\left(\frac{\Delta_2}{L_2}\right)^2 \tag{b}$$

We see that the strain energy is a function of the two displacements Δ_1 and Δ_2. We must apply the first Castigliano theorem to determine Δ_1 and Δ_2 as follows (**equilibrium**):

$$P_1 = \frac{\partial U}{\partial \Delta_1} = \frac{A_1E_1}{L_1}\Delta_1 \qquad P_2 = \frac{\partial U}{\partial \Delta_2} = \frac{A_2E_2}{L_2}\Delta_2 \tag{c}$$

From equilibrium, the force in each rod is P. Therefore, letting $P_1 = P_2 = P$ and enforcing **compatibility** $[\Delta = \Delta_1 + \Delta_2]$ we get from Eq. (c) the desired result for the displacement

$$\Delta = P\left(\frac{L_1}{A_1E_1} + \frac{L_2}{A_2E_2}\right)$$

Part B: Force Methods

18.6 Principle of Complementary Virtual Work

We now present a set of three principles collectively called the *force methods* that are analogous to principles of the displacement methods of Part A. We first present a principle that is the analog of the principle of virtual work, namely the *principle of complementary virtual work.* In this approach, the forces and deformation *swap roles,* in that it is the forces that are varied while the displacements are held constant. Furthermore, concepts analogous to the principle of total potential energy and the first Castigliano theorem will be developed where again, forces are varied rather than displacements. As might be expected, the methods of solution of problems stemming from these principles are called *force methods.*

We begin by defining *complementary virtual work* δW^*_{virt} as follows:

$$\delta W^*_{\text{virt}} = \iiint_V \delta \boldsymbol{B} \cdot \boldsymbol{u}\, dv + \oiint_S \delta \boldsymbol{T} \cdot \boldsymbol{u}\, dA \tag{18.24}$$

where $\delta\boldsymbol{B}$ is a virtual body force distribution and $\delta\boldsymbol{T}$ is a virtual surface traction force distribution. Whereas in our previous work, the admissible virtual displacement field had to result in compatible deformations wherein constraints are not violated, the admissible *virtual force fields* here must satisfy the requirements of *equilibrium* of the body as a whole. Furthermore, the admissible *virtual stresses* $\delta\tau_{ij}$ resulting from the virtual loads must satisfy the *differential equations of equilibrium.*

We now compute δW^*_{virt} for a simple one-dimensional problem where we now employ a unidirectional virtual body force distribution $\delta B_x(x)$ which is a function only of x, and virtual traction forces $[\delta T_x(x)]_1$ and $[\delta T_x(x)]_2$ at the end surfaces as has been shown in Fig. 18.10. Then we can say for this case:

$$\begin{aligned}\delta W^*_{\text{virt}} &= \iiint_V \delta B_x u_x\, dv + \left(\iint_A \delta T_x u_x dA'\right)_2 - \left(\iint_A \delta T_x u_x dA'\right)_1 \\ &= \iiint_V \delta B_x u_x\, dv + \iint_A [(\delta T_x u_x)_2 - (\delta T_x u_x)_1] dA\end{aligned} \tag{18.25}$$

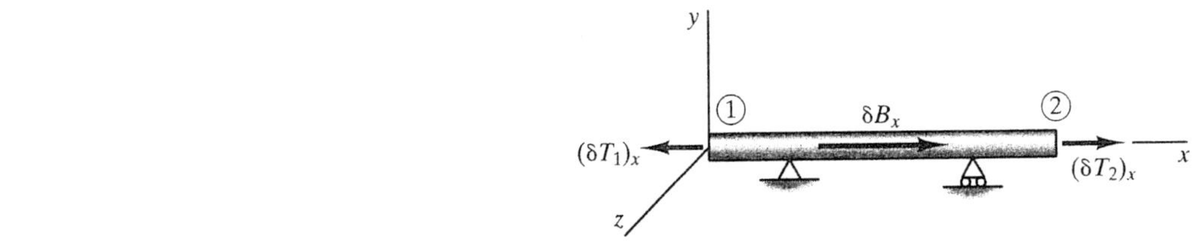

Figure 18.10. One-dimensional force system.

But the expression in brackets in the surface integral can be expressed as:

$$[(\delta T_x u_x)_2 - (\delta T_x u_x)_1] = \int_1^2 \frac{d}{dx}(\delta T_x u_x)dx \tag{18.26}$$

Note that δT_x is the force per unit area normal to the cross section, and we recognize this as the virtual stress $\delta\tau_{xx}$. Making this change in Eq. (18.26) and then substituting it into Eq. (18.25), we get

$$\delta W^*_{\text{virt}} = \iiint_V \delta B_x u_x dv + \iiint_V \frac{d}{dx}(\delta\tau_{xx}u_x)dv \tag{18.27}$$

Note that $dx\, dA$ has been replaced by dv in the last expression. Using the product rule to carry out the differentiation in Eq. (18.27) and then collecting terms, we may state that

$$\delta W^*_{\text{virt}} = \iiint_V \left[\left(\delta B_x + \frac{d\delta\tau_{xx}}{dx}\right)u_x + \delta\tau_{xx}\frac{d}{dx}(u_x)\right] dv \tag{18.28}$$

We now apply *two key steps* in the development of the desired complementary virtual work relationship. The *first key step* is to enforce the requirement that the admissible forces and stresses satisfy the equations of *equilibrium* in our energy expression [Eq. (18.28)]. Recalling the equations of equilibrium in *differential form* [see Eq. (2.5)] we see that for one-dimensional systems, we must delete the first expression in the integral. The resulting formulation becomes

$$\delta W^*_{\text{virt}} = \iiint_V \delta\tau_{xx}\left(\frac{du_x}{dx}\right)dv \tag{18.29}$$

For the *second key step*, we shall replace du_x/dx by ε_{xx} and thereby *guarantee that* ε_{xx} *is a compatible strain field since we are getting it directly from a single-valued, continuous displacement field.* This is the *second* crucial step in the development. We then have the following simple yet critical expression

$$\delta W^*_{\text{virt}} = \iiint_V \delta\tau_{xx}\varepsilon_{xx}\, dv \tag{18.30}$$

The complementary virtual work of the external force distribution equals what may be considered as the *complementary virtual work of the internal forces.* That is, we can consider the product of virtual stress and actual strain, integrated over the body, as the internal complementary virtual work.

Equation (18.30) is the principle of *complementary virtual work* for a one-dimensional body. It is a necessary condition for *compatibility.*

That is, for a one-dimensional strain field to satisfy compatibility, it must satisfy the principle of complementary work for any one-dimensional virtual force and stress distributions that satisfy the requirements of equilibrium.

For a more general case, we can extrapolate Eq. (18.30) to the following form:

$$\delta W^*_{\text{virt}} = \iiint_V [\delta\tau_{xx}\varepsilon_{xx} + \delta\tau_{yy}\varepsilon_{yy} + \delta\tau_{zz}\varepsilon_{zz} + \delta\tau_{xy}\gamma_{xy} + \delta\tau_{xz}\gamma_{xz} + \delta\tau_{yz}\gamma_{yz}]dv \tag{18.31}$$

We usually express the equation above in the following manner:

$$\iiint_V \delta\boldsymbol{B} \cdot \boldsymbol{u}\, dv + \oiint_S \delta\boldsymbol{T} \cdot \boldsymbol{u}\, dA = \iiint_V \sum_i \sum_j \delta\tau_{ij}\varepsilon_{ij} dv \tag{18.32}$$

We can conclude that a necessary condition for having a three-dimensional *kinematically compatible deformation, is the satisfaction of Eq. (18.32) for all admissible virtual force and associated stress distributions.* (One may also show that this condition is sufficient for kinematic compatibility.) Thus, while the *principle of virtual work satisfies equilibrium using virtual (compatible) deformation,* the *principle of complementary virtual work satisfies compatibility using virtual equilibrium forces and stresses.* It should be understood that *no constitutive law has been employed* in the development of this principle. The only restriction that we have set forth in arriving at Eq. (18.32) is that of infinitesimal deformation (stemming from our use of the strain-displacement relation, $\varepsilon_{xx} = du_x/dx$).

One way of using the complementary principle of virtual work is to determine the proper *displacement field* knowing the *actual strains* for the problem. We now explain further. In employing the complementary virtual work principle for a given problem, we first choose a hypothetical system of external loads (the simpler the better) which satisfies equilibrium for the given problem and then compute the resulting internal forces from these loads. The resulting hypothetical stresses for the body are called *virtual stresses.*Now, if we know the proper *actual* strain field from the *actual loads* on the body from other calculations, we can use these strains in the complementary virtual work equation thereby assuring compatibility between the displacement field $\boldsymbol{u}$ in the equation and the proper strain field we are using. This field $\boldsymbol{u}$ must the be the *actual* displacement field, since in this equation $\boldsymbol{u}$ and ε_{ij} are properly related in accordance with compatibility requirements.

We illustrate these remarks in the following example.

Example 18.6

Shown in Fig. 18.11 is a statically determinate simple truss. For linear elastic behavior of the members, all of whom have identical cross sections and elastic moduli, find the vertical deflection Δ of joint C from the given external forces P_1, P_2, and P_3.

In using the principle of virtual complementary work, we must keep in mind that $\delta \boldsymbol{B}$, $\delta \boldsymbol{T}$, and $\delta\tau_{ij}$ represent *any* system of loads and stresses that satisfy the requirements of equilibrium. For our purposes it is best (as you will soon see) to choose a *unit force* at joint C

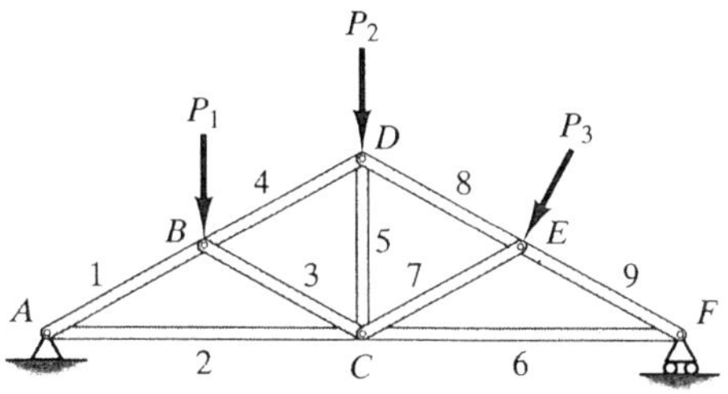

Figure 18.11. Simple truss.

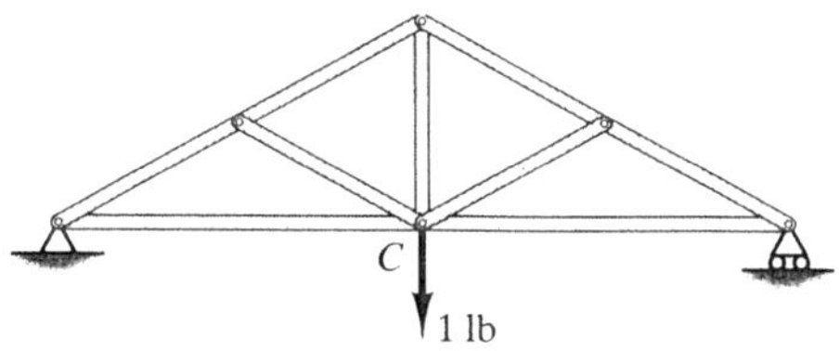

Figure 18.12. Unit dummy load.

(see Fig. 18.12) in the direction of the desired displacement as the only virtual external load. The reactions to this load and the stresses can then be computed by the method of joints. Such stresses since they satisfy **equilibrium** constitute appropriate virtual stresses for the problem. Hence we have for the qth member,

$$\delta\tau_q = \frac{(F_{\text{virt}})_q}{A_q} \tag{a}$$

where $(F_{\text{virt}})_q$ is the force in the qth member from the virtual unit load at C.

As for the strain terms in the principle of total complementary energy, we shall use *actual* strains computed from the *actual* loads

Example 18.6 (Continued)

P_1, P_2, and P_3 by the simple methods of statics. Thus for the qth member (**constitutive law**)

$$\varepsilon_q = \left(\frac{F_{\text{act}}}{AE}\right)_q \tag{b}$$

where $(F_{\text{act}})_q$ is the actual force in the qth member from the actual loads. Employing the principle of complementary virtual work for the actual strains and the aforementioned virtual forces and stresses, we get from Eq. (18.32)

$$(1)(\Delta) = \Delta = \sum_{q=1}^{9} \int_0^{L_q} \left(\frac{F_{\text{virt}}}{A}\right)_q \left(\frac{F_{\text{act}}}{AE}\right)_q dx\, A_q$$

$$\Delta = \sum_{q=1}^{9} \left(\frac{F_{\text{virt}}}{A}\right)_q \left(\frac{F_{\text{act}}}{AE}\right)_q L_q A_q \tag{c}$$

The principle of complementary virtual work demands that Δ as given above be **compatible** with the actual strains in the members: Clearly, it must then be the *actual* displacement of joint C in the direction of the unit load.

The method presented in the example is called the *unit dummy load method.* Recall that this problem could also have been solved easily by using the first Castigliano theorem.

18.7 Complementary Potential Energy Principle

With this background established, let us now reconsider the principle of complementary virtual work. Up to this time the forces and stresses have not been related to deformation field (i.e., no constitutive law has been implied). We now consider the application of this principle to an *elastic* (not necessarily linear elastic) body for which the strains are *linked* to the stresses via *some constitutive law.* Under such circumstances we can say that

$$d\mathcal{U}^* = \varepsilon_{xx}d\tau_{xx} + \varepsilon_{yy}d\tau_{yy} + \varepsilon_{zz}d\tau_{zz} + \gamma_{xy}d\tau_{xy} + \gamma_{yz}d\tau_{yz} + \gamma_{xz}d\tau_{xz} \qquad (18.33)$$
$$= \sum_i \sum_j \varepsilon_{ij}d\tau_{ij}$$

where in Eq. (18.33) we are summing over the indices i and j. Similarly, we can say, since the variation operator is very similar to the differential, that

$$\delta\mathcal{U}^* = \sum_i \sum_j \varepsilon_{ij}\delta\tau_{ij}$$

Accordingly,

$$\delta U^* = \iiint_V \sum_i \sum_j \varepsilon_{ij}\delta\tau_{ij}\,dv \qquad (18.34)$$

We can now replace the expression on the right side of Eq. (18.32) by δU^*, so that we have, on rearranging this equation,

$$\delta U^* - \iiint_V \delta\boldsymbol{B}\cdot\boldsymbol{u}\,dv - \oint\!\!\oint_S \delta\boldsymbol{T}\cdot\boldsymbol{u}\,dA = 0 \qquad (18.35)$$

as the analog to Eq. (18.13). We will make good use of this relationship in the next section.

We now present the remaining steps in the development of the *complementary potential energy* functional. First, we *define* the *complementary potential function, V^**, such that for δV^*

$$\delta V^* = -\iiint_V \boldsymbol{u}\cdot\delta\boldsymbol{B}\,dv - \oint\!\!\oint_S \boldsymbol{u}\cdot\delta\boldsymbol{T}\,dA \qquad (18.36)$$

Recall that when taking the variation of V^*, $\boldsymbol{u}$ must remain constant. Hence, Eq. (18.35) may be written as

$$\delta U^* + \delta V^* = 0 \qquad (18.37)$$

We next define the total *complementary potential energy* π^* as

$$\pi^* = U^* + V^* \qquad (18.38)$$

so that Eq. (18.37) becomes

$$\delta\pi^* = 0 \qquad (18.39)$$

This is the principle of *total complementary energy*. What does it say? Starting with the complementary virtual work from which it is derived, we know that each admissible *force and stress field must satisfy equilibrium*. Now due to an *elastic constitutive law* being present, each such stress field is *linked to a distinct deformation field.* The principle of total complementary energy then says that the *particular stress field that extremizes* π^* (i.e., the total complementary energy) must then be *linked to a deformation field that is kinematically compatible.* This is true since setting the variation equal to zero finishes the process started with complementary work under the additional conditions of linkage between stress and strain via an elastic constitutive law. Any other statically compatible stress field will link to a deformation that is not kinematically compatible.

18.8 Use of the Total Complementary Energy Principle

To illustrate a use of the method of total complementary energy, we consider the *statically indeterminate* truss in Fig. 18.13. We have here linear elastic behavior and we wish to determine the forces in the members.

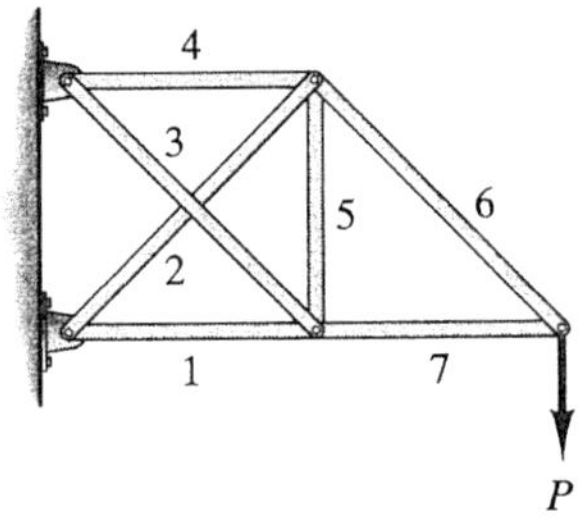

Figure 18.13. Statically indeterminate truss.

To solve this problem, we shall cut either member 2 or member 3, each of which is a redundant member.[7] In Fig. 18.14, we have chosen to cut member 2 and have placed at the surfaces exposed by the cut the forces f_2. Furthermore, we delete all external applied loads such as P. What is then left is a statically determinate truss with equal and opposite forces f_2 applied. We can readily solve for the forces in each of the members by the method of joints (note the supporting forces are zero) and we shall denote these forces as f_i. Clearly, for a linear elastic system, each such force will be directly proportional to f_2. Thus we can say, using proportionality constants C_i, that

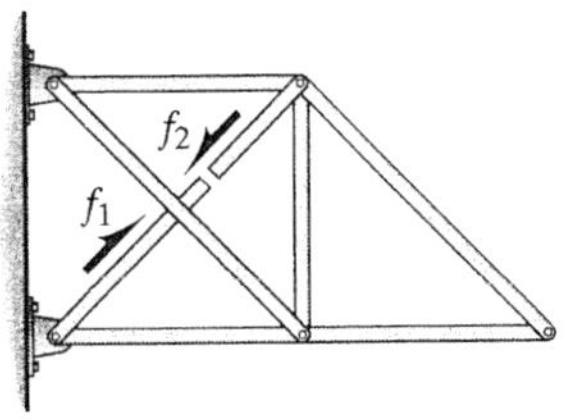

Figure 18.14. Redundant member 2 is cut and external loads removed.

[7]Without a redundant member the truss may continue to function as a supporting system and does not become a "mechanism."

$$f_i = C_i f_2 \qquad i = 1, 2, \ldots, 7 \tag{18.40}$$

whereC_2 = 1. Now delete member 2 (see Fig. 18.15) and solve by the method of joints for the forces in the remaining members stemming from the applied external loads, in this case P **(equilibrium)**. We shall call these forces F_i. Now add the forces F_i to f_i, the latter given by Eq. (18.40). The total force on the pth member that we are seeking is then given as

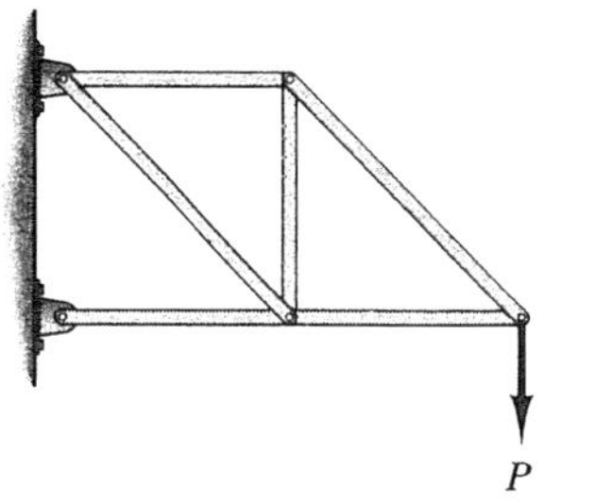

Figure 18.15. Redundant member removed and external load applied.

$$\text{total force in member } p = F_p + C_p f_2 \tag{18.41}$$

Accordingly, the stress τ_p in the pth member is then

$$\tau_p = \frac{F_p + C_p f_2}{A_p} \tag{18.42}$$

The corresponding strain ε_p is **(constitutive law)**

$$\varepsilon_p = \frac{F_p + C_p f_2}{A_p E_p} \tag{18.43}$$

With the steps we have taken, we can now think of our truss as having one redundant member cut so that in addition to the original external traction loads (in this case P), we have two additional equal and opposite traction forces f_2 at the cut surfaces. If we vary f_2, we generate an infinite set of stress fields given for the members by Eq. (18.42), *all of whom satisfy equilibrium.* Thus this family of stress fields is admissible for use of the total complementary energy principle. Before proceeding further, we shall express the complementary strain energy U^* for these admissible stress states. Thus,

$$\begin{aligned} U^* &= \iiint_V \int \sum_i \sum_j \varepsilon_{ij} d\tau_{ij} dv \\ &= \sum_{p=1}^{7} \int \left(\frac{F_p + C_p f_2}{E_p A_p}\right) d\left(\frac{F_p + C_p f_2}{A_p}\right) A_p L_p \end{aligned} \tag{18.44}$$

Extracting E_p from the first expression in parentheses, we can integrate the last integral to get

$$U^* = \sum_{p=1}^{7} \frac{1}{2}\left(\frac{F_p + C_p f_2}{A_p}\right)^2 \frac{1}{E_p} A_p L_p = \frac{1}{2} \sum_{p=1}^{7} \frac{(F_p + C_p f_2)^2}{A_p E_p} L_p \tag{18.45}$$

The total complementary energy principle now becomes

$$\delta U^* - \iint_{S_2} \boldsymbol{u} \cdot \delta \boldsymbol{T} dA = 0 \tag{18.46}$$

The only traction force that is varied for the family of admissible stress states that we have decided to use is f_2 at the surfaces exposed by the cut. However, the surfaces undergo the *same displacement* and since the forces f_2 on the surfaces are *equal and opposite in direction,* the surface integral in Eq. (18.46) must vanish, leaving us with the following equation representing the principle of complementary energy for the truss:

$$\delta U^* = \delta\left[\frac{1}{2}\sum_{p=1}^{7}\frac{(F_p + C_p f_2)^2}{A_p E_p} L_p\right] = 0 \tag{18.47}$$

To extremize the *function* of f_2, that is, the expression in brackets, we simply require that

$$\frac{d}{df_2}\left[\frac{1}{2}\sum_{p=1}^{7}\frac{(F_p + C_p f_2)^2}{A_p E_p} L_p\right] = 0 \tag{18.48}$$

Now note that *any* value of f_2 would have led to an admissible stress field (one that satisfies equilibrium). And each such stress field is linked by the constitutive law to some deformation field of the truss. By having extremized π^*, however, we *single out* the *one value of* f_2 whose associated deformation field *must satisfy compatibility* (the deformation fields for all the other f_2's do not satisfy compatibility). Thus this f_2 must be the correct value because in our procedure we have satisfied

1. Equilibrium
2. The appropriate constitutive law
3. Compatibility

The stresses can then easily be computed from Eq. (18.42).

If there are more than one redundant member, cut the others as we have done here for member 2. Then for n redundant members we would get n equations such as Eq. (18.48) to solve for n unknown f's. We present truss problems, using the above results, as exercises in the Problems section.

In this application of the complementary energy principle the π^* was a *function.* In the study of beams, plates, and so on, π^* will be a *functionl* rather than a function, as has been explained in Section 18.6. Here to extremize π^* we need the methodology of the calculus of variations. You should study this vital area in a more advanced course.[8]

We now look at a special case of the total complementary energy principle and we will see its use for a number of interesting problems.

[8]See I. H. Shames and C. L. Dym, *Energy and Finite Element Methods in Structural Mechanics,* Taylor & Francis Co.

18.9 The Second Castigliano Theorem

To develop the second Castigliano theorem we now consider an elastic body maintained in equilibrium by a system of rigid supports (see Fig.

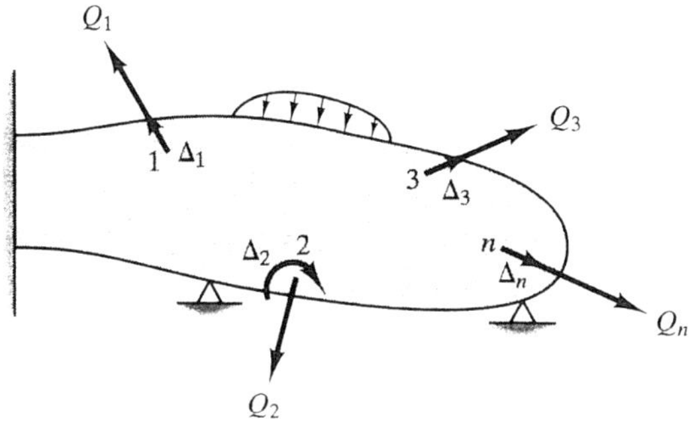

Figure 18.16. Generalized point forces and generalized displacements.

18.16). We consider that this body is acted on by a system of point loads and point couples (referred to as *generalized forces*) that we denote as Q_i, as well as distributed traction and body forces. For each Q_i we have associated a *generalized displacement* Δ_i. *These generalized displacements act along the same line as the corresponding generalized forces.* The Δ_i result from deformation induced by all external forces discrete and continuous.

We assume now that *complementary strain energy* U^* can be given in terms of the applied point forces and torques Q_i, and other applied traction and body force distributions as

$$U^* = \sum_{i=1}^{N} U_i^*(Q_i) + \textit{addition terms} \tag{18.49}$$

The 'additional terms' add to the complementary strain energy but they are not a function of the Q_i's; they are a result of the other loads on the body. Varying *only* the Q_i's, the additional terms (that are not a function of Q_i's) vanish.[9] This is why we separated them in Eq. (18.49). Using the chain rule we get

$$\delta U^* = \sum_{k=1}^{N} \frac{\partial U^*}{\partial Q_k} \delta Q_k \tag{18.50}$$

We now rewrite Eq. (18.35) for this case and we vary only the discrete generalized forces Q_i. We then arrive at the following equation:

[9]We will suppose that we can vary the Q's independent of each other to form δQ_i.

$$\sum_{k=1}^{N} \frac{\partial U^*}{\partial Q_k} \delta Q_k - \sum_{k=1}^{N} \Delta_k \delta Q_k = 0$$
$$\therefore \sum_{k=1}^{N} \left(\frac{\partial U^*}{\partial Q_k} - \Delta_k \right) \delta Q_k = 0 \tag{18.51}$$

Since Q_k are independent, we conclude that

$$\frac{\partial U^*}{\partial Q_k} = \Delta_k \tag{18.52}$$

Thus, Eq. (18.52) gives us the generalized displacement in the direction of the generalized point force. This is called the *second Castigliano theorem.* For linear elastic behavior, $U^* = U$ and we have

$$\frac{\partial U}{\partial Q_k} = \Delta_k \tag{18.53}$$

The second Castigliano theorem is very useful for determining the relation between applied forces and displacements. Examples using the second Castigliano theorem have been presented in Chapters 6, 12, and 14 for trusses, torsion rods, and beams.

18.10 Closure

This chapter has provided us with some remarkable developments. First and foremost, we have seen that the somewhat abstract and mathematical energy approach has often led to a direct and simple computation for a variety of problems. For example, we discovered that extremizing the total potential energy of a system readily produced the equilibrium equations. This was one of two parallel sets of principles that are analogies of each other: the displacement method and the force method. We hope that the beauty and symmetry of these methods is enjoyed by the reader as it is by the authors. What may come as a surprise is that the equations of equilibrium and compatibility play similar but alternate roles in the two approaches.

It was our intent in this chapter to lay a solid foundation in the energy approach to problems. Yet, we did not want the material to be so full of theory that you could not see the application. As a compromise, we have set forth the basis for the understanding of the energy approach and then developed some of the more popular techniques for exploiting this new approach. The method of total potential energy forms the theoretical underpinnings for the popular finite element method. We also presented the Castigliano theorems that are powerful and useful for a wide range of hand-solution problems in solid mechanics.

PROBLEMS

Method of Virtual Work	18.1–18.8
Method of Total Potential Energy	18.9–18.23
First Castigliano Theorem	18.24–18.28
Method of Complementary Work	18.29–18.33
Method of Total Complementary Energy	18.34–18.36
Second Castigliano Theorem	18.37–18.41
Programming Problems	18.42–18.44

18.1. [18.2]

(a). Write two equations for the deflection components of pin D in Fig. P.18.1. All members have the same E and A. Do not solve.

(b). Why will this give a correct solution?

Use the method of virtual work.

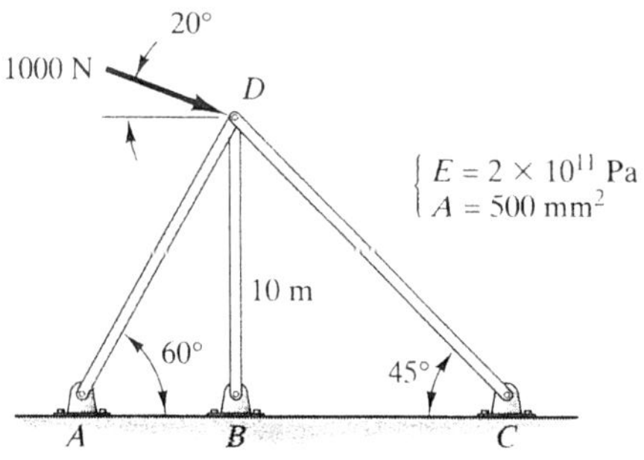

Figure P.18.1.

18.2. [18.2] In Fig. P.18.2:

(a.) Set up two equations for u, v using the method of virtual work, for the actual displacements u_D, v_D. Each member is linear elastic with the same E and the same cross-sectional area.

(b). State why your u_D and v_D must be the actual displacements.

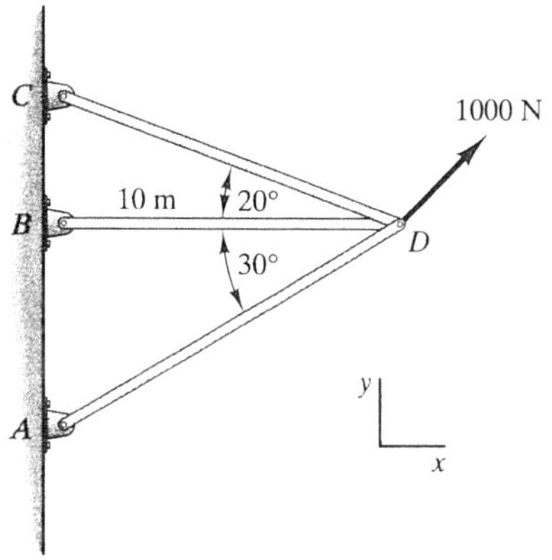

Figure P.18.2.

18.3. [18.2] What is the deflection of joint C due to the 1000-lb load shown in Fig. P.18.3? Both members are linear elastic and have same values of E and A. Do by the method of virtual work.

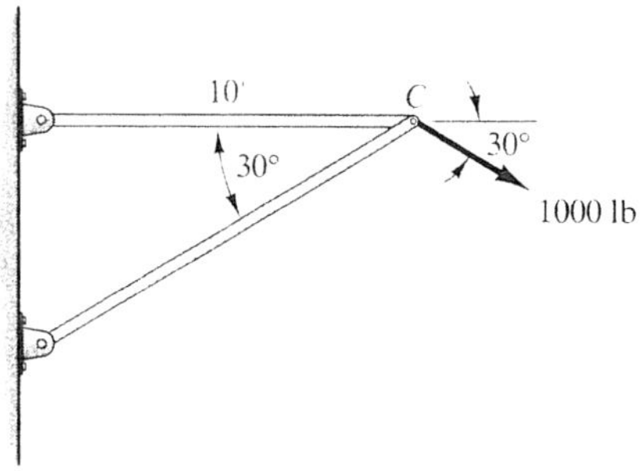

Figure P.18.3.

18.4. [18.2] Using the method of virtual work, find the deflection of joint C of the system of supports shown in Fig. P.18.4. All members are linear elastic having the same properties and cross section.

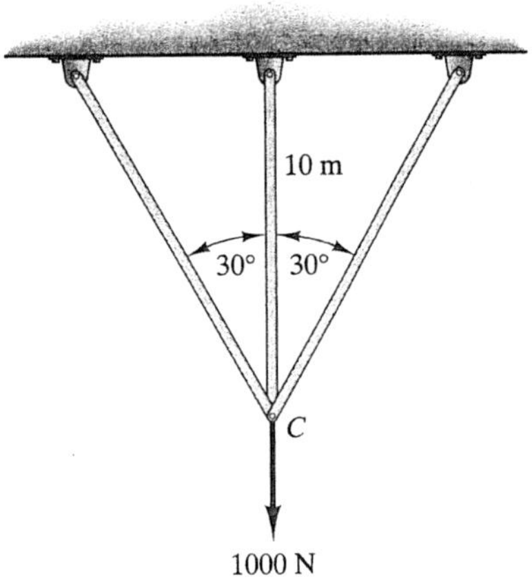

Figure P.18.4.

18.5. [18.2] What is the total deflection of pin C in Fig. P.18.5 from the 1000-lb force? Do by the method of virtual work, assuming that A and E are the same for all members.

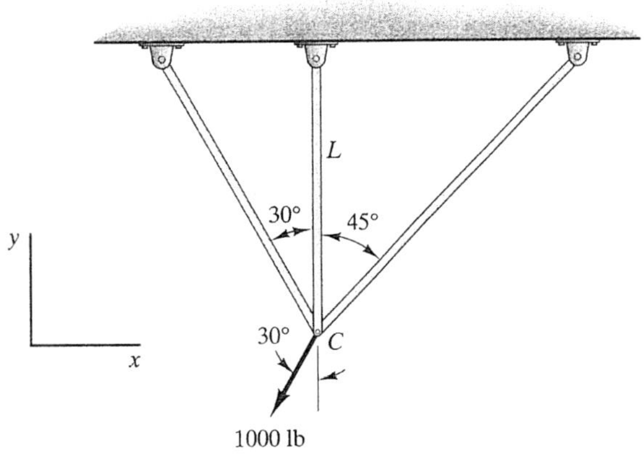

Figure P.18.5.

18.6. [18.2] By the method of virtual work, determine the movement of pins B and C in the truss shown in Fig. P.18.6 as a result of the 1000-N load. Assume that all members have the same E and A and are linear elastic.

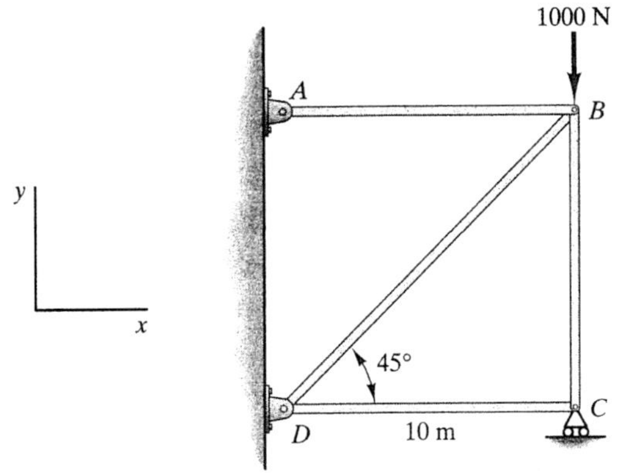

Figure P.18.6.

18.7. [18.2] Find the movement of pins A and B (see Fig. P.18.7) using virtual work. Take all the members to be linear elastic and to have the same values of A and E.

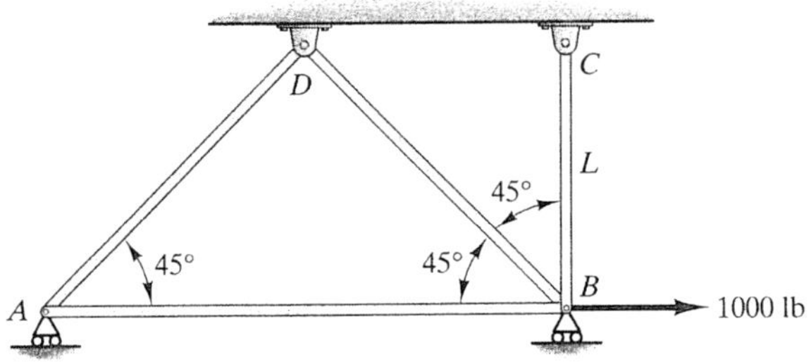

Figure P.18.7.

18.8. [18.2] Do Problem 18.7 for the case where there is a linear elastic member connecting pins A and C in addition to the members shown in Fig. P.18.5. Take all members to have the same values of A and E.

18.9. [18.4] Shown in Fig. P.18.9 is a pin-corrected simple truss. What is the total strain energy of deformation? Each of the members has a cross-sectional area of 15 in.^{2} and a modulus of elasticity of 30×10^6 psi. (See Example 6.3.)

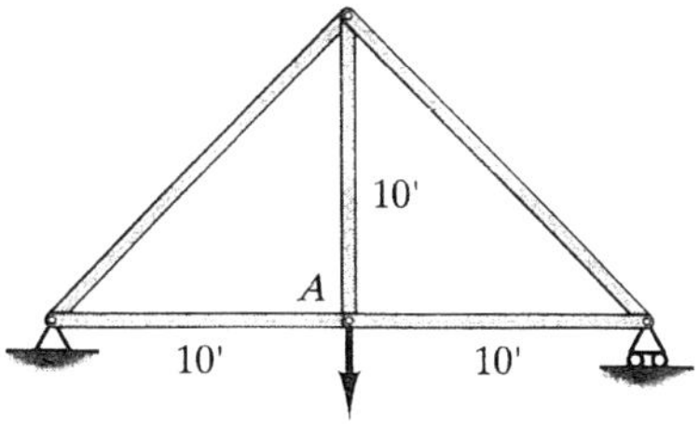

Figure P.18.9.

18.10. [18.4] What is the total strain energy of deformation of the simple truss shown in Fig. P.18.10? Take A of the members as 20 in.^{2} and E as 30×10^6 psi.

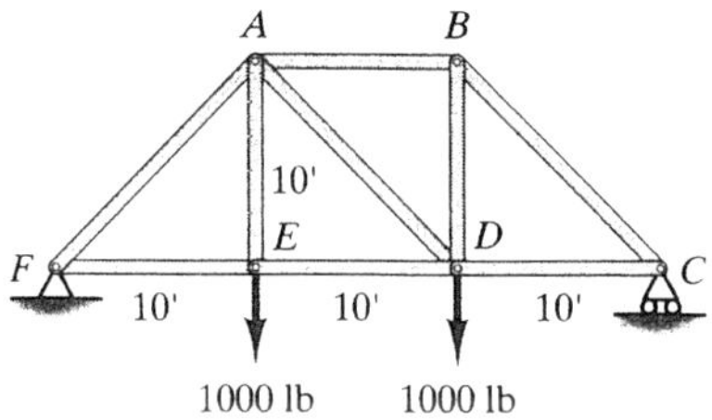

Figure P.18.10.

18.11. [18.4] What is the normal strain energy of deformation for the cantilever beam shown in Fig. P.18.11?

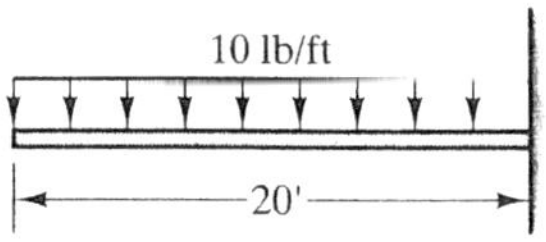

Figure P.18.11.

18.12. [18.4] What is the normal strain energy for the simply supported beam shown in Fig. P.18.12?

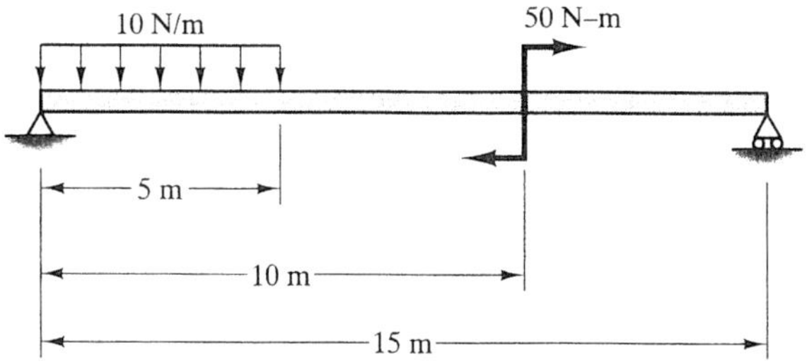

Figure P.18.12.

18.13. [18.4] What is the normal strain energy for the beam shown in Fig. P.18.13? The intensity of loading at the support is 50 lb/ft.

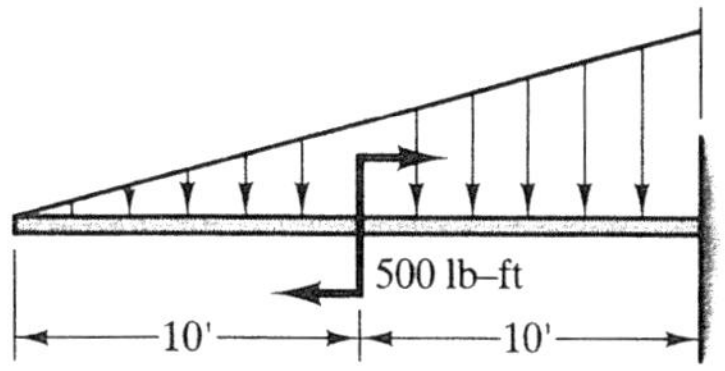

Figure P.18.13.

18.14. [18.4] Compute the shear-deformation energy for the cantilever beam in Problem 18.11. The cross section of the beam is rectangular with a 4-in. base and a 6-in. height.

> In succeeding problems we shall consider shear-strain energy from bending as negligible unless otherwise stated.

18.15. [18.4] What is the strain energy for the simply supported beam shown in Fig. P.18.15?

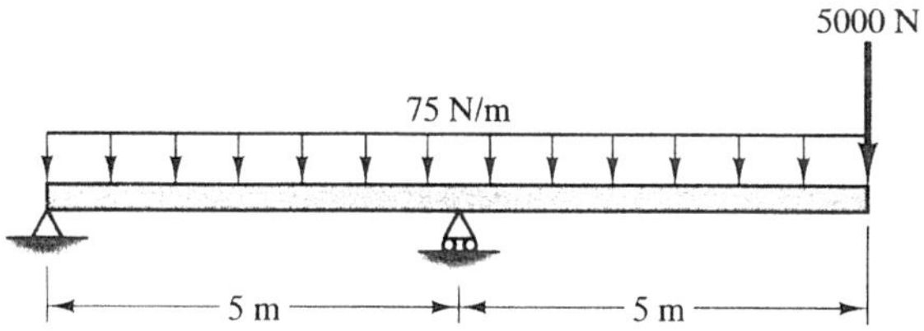

Figure P.18.15.

18.16. [18.4] What is the strain energy for the tube shown in Fig. P.18.16 under the action of a twisting couple and an axial load at the end? Take $E = 30 \times 10^6$ psi and $G = 15 \times 10^6$ psi.

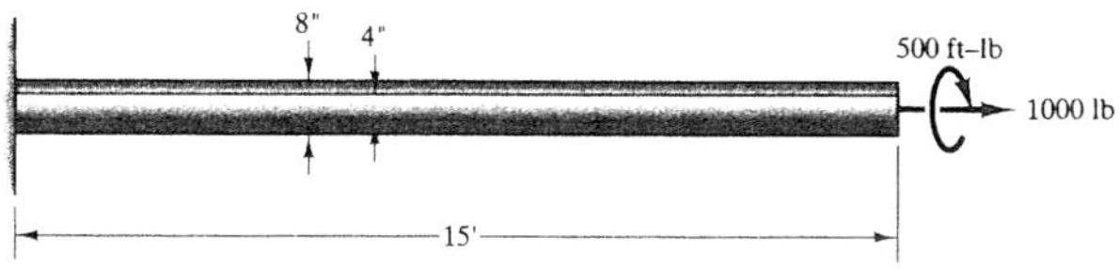

Figure P.18.16.

18.17. [18.4] What is the strain energy of deformation for the shaft shown in Fig. P.18.17?

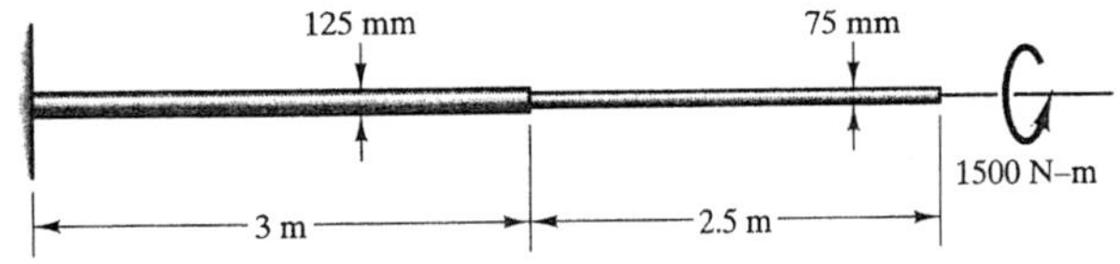

Figure P.18.17.

18.18. [18.4] In Fig. P.18.18:

(a). State the principle of total potential energy.
(b). Find π for the above truss in terms of u, v. (Do not do arithmetic.) Explain how you now get forces. (Do not actually do this.)
(c). Explain why forces found above must be correct.

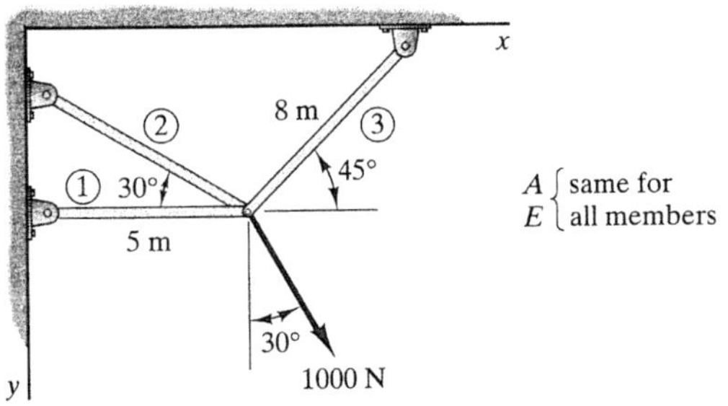

Figure P.18.18.

18.19. [18.4] Do Problem 18.3 by the method of total potential energy.

18.20. [18.4] Do Problem 18.4 by the method of total potential energy.

18.21. [18.4] Do Problem 18.5 by the method of total potential energy.

18.22. [18.4] Do Problem 18.6 by the method of total potential energy.

18.23. [18.4] Do Problem 18.7 by the method of total potential energy.

18.24. [18.5] Do Problem 18.3 using the first Castigliano theorem.

18.25. [18.5] Do Problem 18.4 using the first Castigliano theorem.

18.26. [18.5] Do Problem 18.5 using the first Castigliano theorem.

18.27. [18.5] Do Problem 18.6 using the first Castigliano theorem.

18.28. [18.5] Do Problem 18.7 using the first Castigliano theorem.

18.29. [18.6] What is the vertical deflection of pin A in Fig. P.18.29? All members have the same EA. Use the method of complementary virtual work.

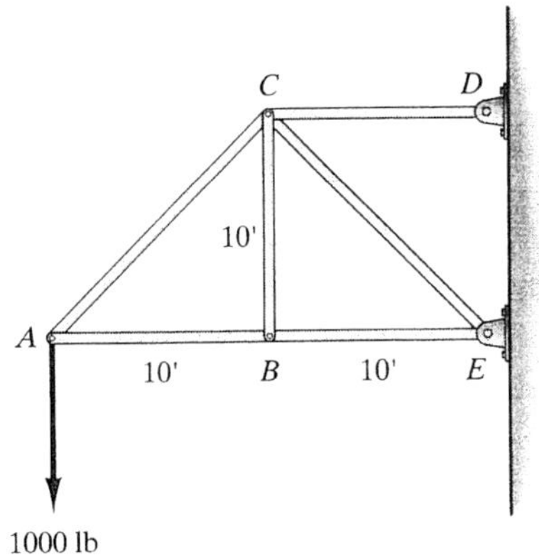

Figure P.18.29.

18.30. [18.6] Find the deflection in the vertical direction of pin C (see Fig. P.18.30). All members have the same EA. The following are the values of the loads:

$$P_1 = 20 \text{ kN}$$
$$P_2 = 30 \text{ kN}$$
$$P_3 = 15 \text{ kN}$$

Use the method of complementary virtual work.

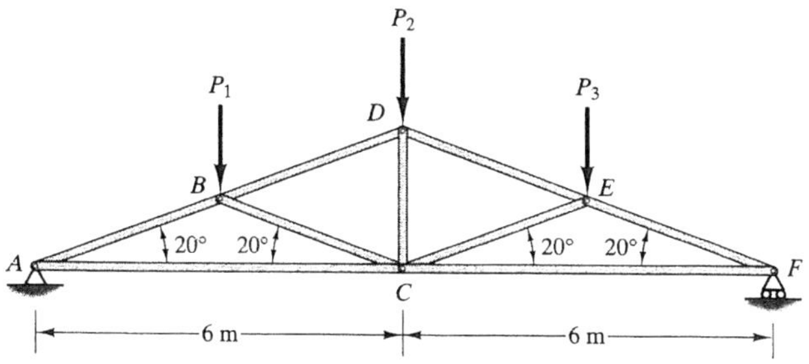

Figure P.18.30.

18.31. [18.6] Find the deflection in the vertical and horizontal directions of pin E in Fig. P.18.31. The members all have the same value of EA. Use the method of complementary virtual work.

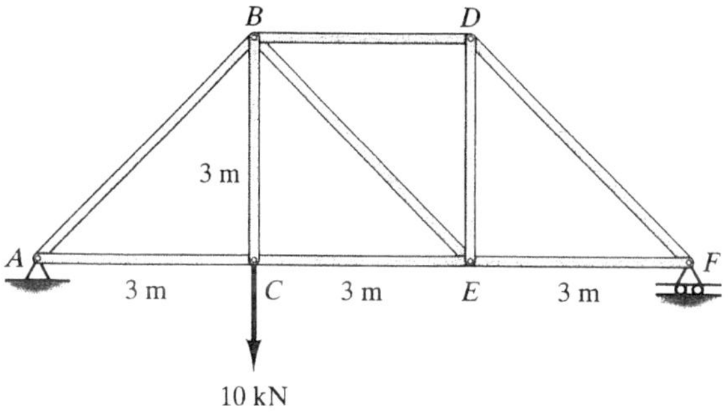

Figure P.18.31.

18.32. [18.6] Find the vertical deflection of pin B in Fig. P.18.32. Taking EA = const. for all members, use the method of complementary virtual work. (1 kip = 1000 lb)

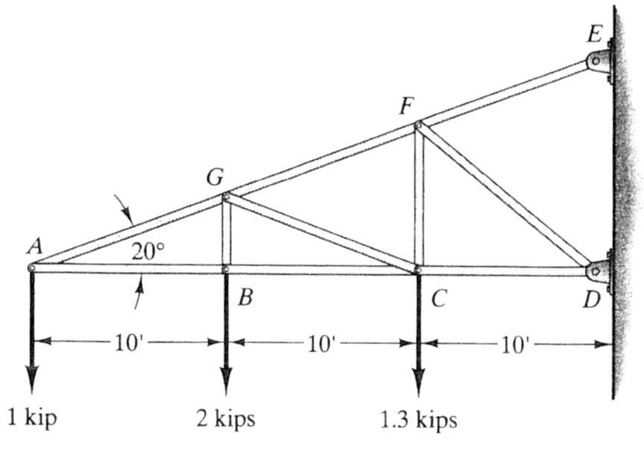

Figure P.18.32.

18.33. [18.8] Find the forces in the members. Take $E = 30 \times 10^6$ psi and $A = 724.1$ in.^{2} for each member (Fig. P.18.33). Use the total complementary energy principle.

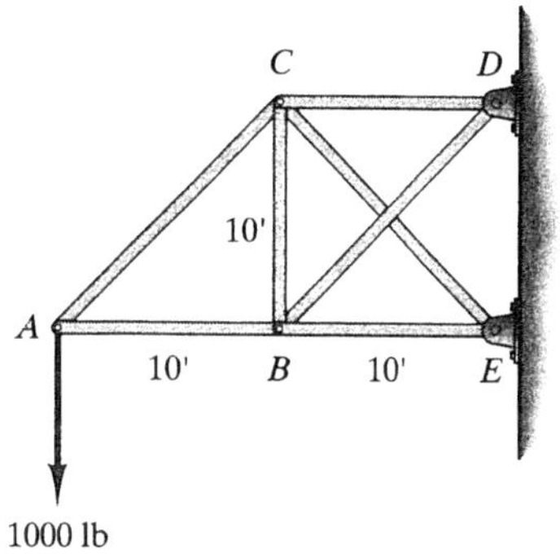

Figure P.18.33.

18.34. [18.8] Find the forces in all members in Fig. P.18.34. Take $E = 2 \times 10^{11}$ Pa an $A = 2 \times 10^4$ mm^2. Use the total complementary energy principle.

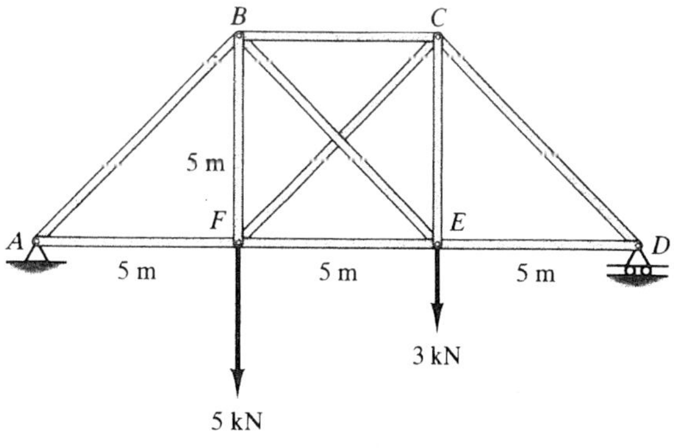

Figure P.18.34.

18.35. [18.8] Find the forces in all members in Fig. P.18.35. Take $E = 2 \times 10^{11}$ Pa and $A = 3 \times 10^4$ mm^2 for all members. The loads are

$$P_1 = 10 \text{ kN}$$
$$P_2 = 15 \text{ kN}$$
$$P_3 = 8 \text{ kN}$$

Use the total complementary energy principle.

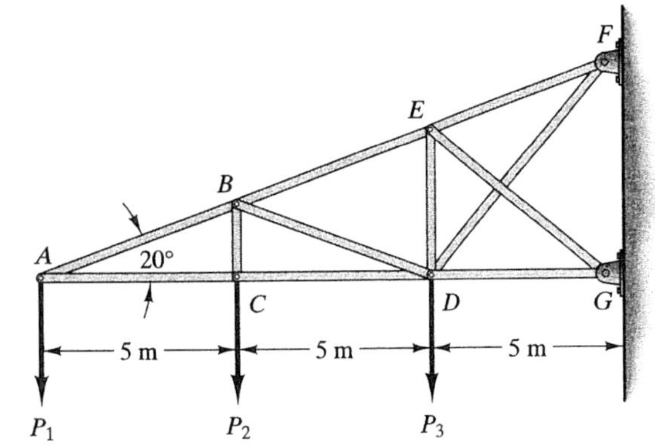

Figure P.18.35.

18.36. [18.8] Find the forces in all the members in Fig. P.18.36. Take $E = 30 \times 10^6$ psi and $A = 80$ in.^{2} for all members. The loads are

$$P_1 = 1000 \text{ lb}$$
$$P_2 = 3000 \text{ lb}$$
$$P_3 = 1500 \text{ lb}$$

Use the total complementary energy principle.

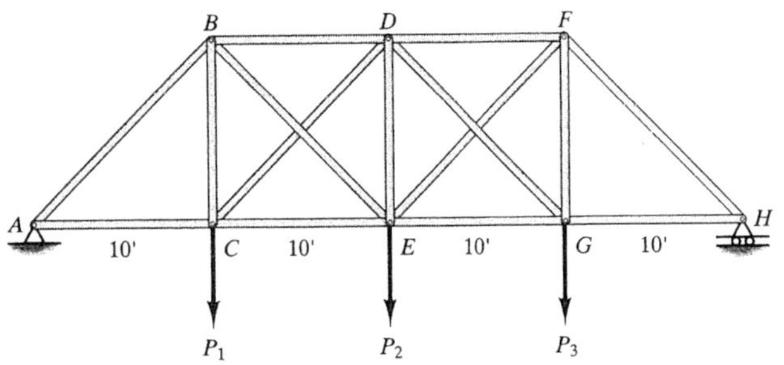

Figure P.18.36.

18.37. [18.9] Determine the force-displacement relation for the split-ring shown in Fig. P.18.37. Assume that the ring thickness t is small compared to the radius and use the second Castigliano theorem. (*Hint:* Use symmetry.)

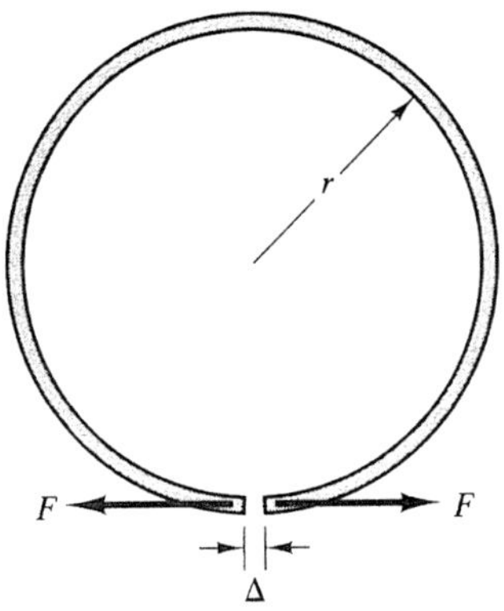

Figure P.18.37.

18.38. [18.9] What is the horizontal deflection of pin A for the truss shown in Fig. P.18.38? All members have the same AE. Use Castigliano's second theorem.

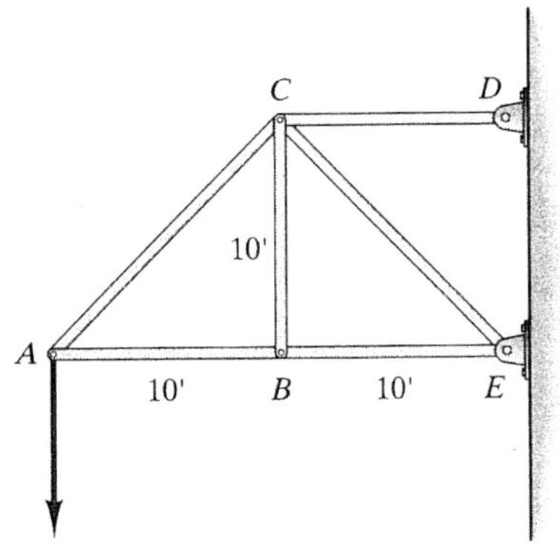

Figure P.18.38.

18.39. [18.9] Determine the horizontal deflection of pin *C* for the truss shown in Fig. P.18.39. All members have the same *AE*. Take $P_1 = 20$ kN, $P_2 = 30$ kN, and $P_3 = 15$ kN. Use the second Castigliano theorem.

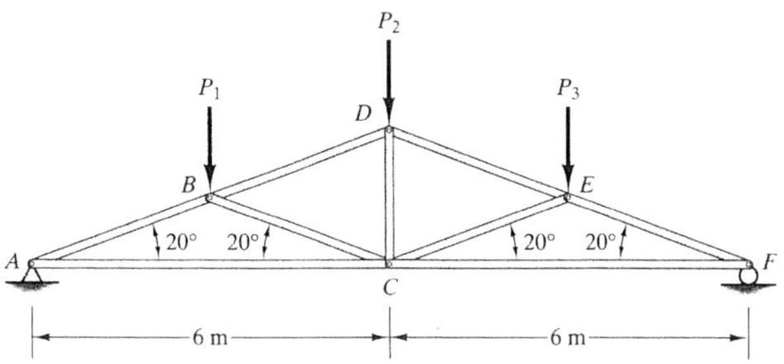

Figure P.18.39.

18.40. [18.9] A uniformly loaded beam with semirigid supports is modeled as shown in Fig. P.18.40. The rotational springs produce a resisting moment M_o. Determine M_o as a function of the end rotation θ. For a simple support, what is the end rotation? Use the second Castigliano theorem. *Hint*: form a cantilever beam using symmetry.

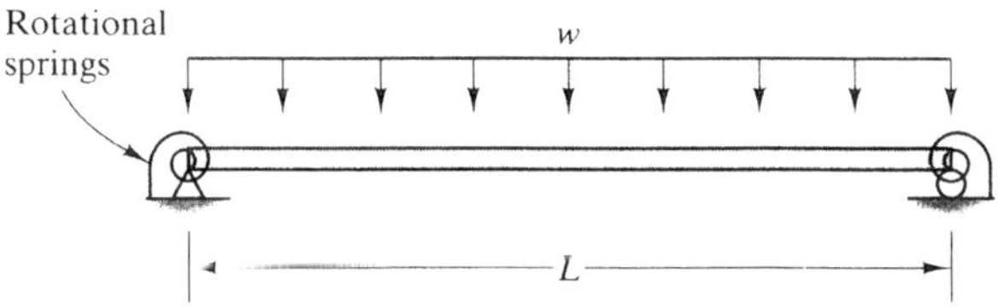

Figure P.18.40.

18.41. [18.9] For the thin arched frame shown in Fig. P.18.41, determine the redundant moment and axial force at point *A*. See hint in Problem 18.40.

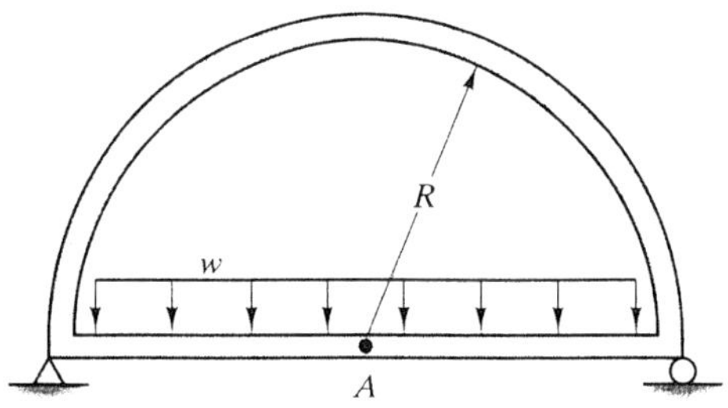

Figure P.18.41.

****18.42.** The semicircular cantilevered arch has a tip load in the vertical direction as shown in Fig. P.18.42. Determine the vertical displacement of the load point for an arbitrary arch segment between 0° and 90°. Plot the normalized vertical tip displacement $\dfrac{\Delta EI}{Pr^3}$ versus the arch angle θ. Use Castigliano's second theorem and consider bending energy only.

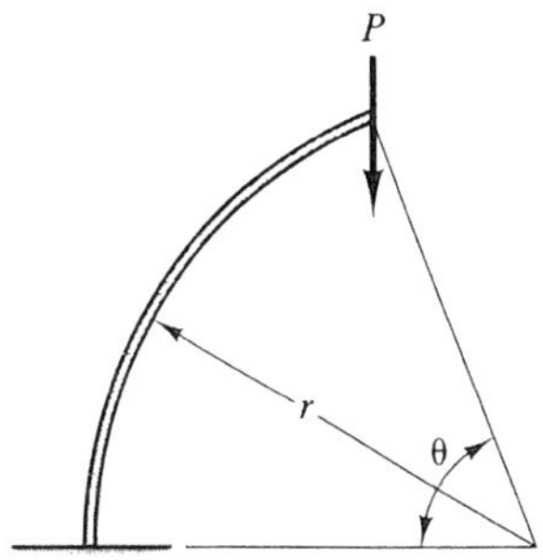

Figure P.18.42.

****18.43.** For the dog-leg frame loaded as shown in Fig. P.18.43, determine a general solution for the reactions at point *A*. For the case $L_1 = L_2 = L$, plot the reactions versus the ratio EI_1/EI_2 for the range from 0 to 10. What are the limiting reactions as $EI_1/EI_2 \to 0$? What is the physical meaning of this? Use the second Castigliano theorem.

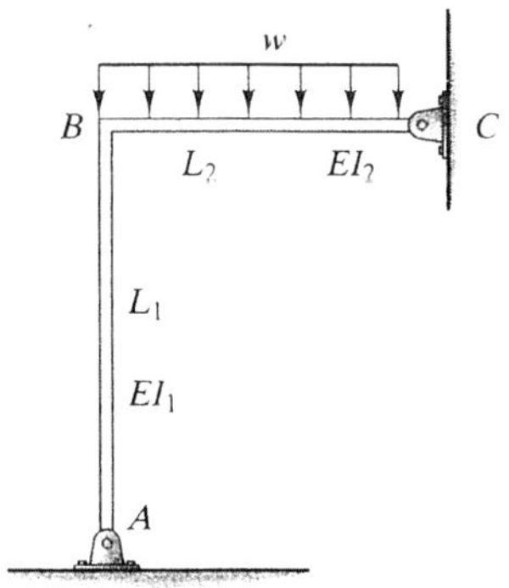

Figure P.18.43.

**18.44. Two beams each of total length of L_1 and L_2 respectively, are fastened to form an "X" as shown in Fig. P.18.44. Find the force transmitted between the two beams. Plot this reaction force R, normalized to the applied load, as (R/P) versus the beam stiffness ratio (EI_2/EI_1) for beam length ratios of $L_2/L_1 = 1/10, 1/5, 1/2, 1, 2, 5$, and 10. What happens to this force as one of the beams becomes much shorter than the other?

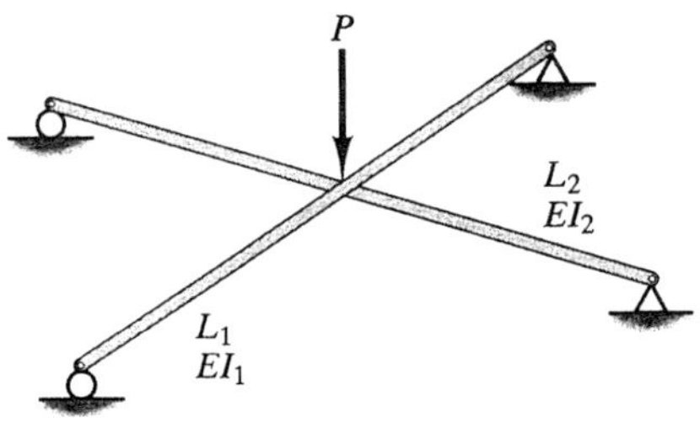

Figure P.18.44.

*CHAPTER 19

Introduction to Finite Elements[1]

19.1 A Comment

The advent of the computer has had a major impact on how we carry out analyses and design of complex structures. Beyond that, the methodologies of solving many kinds of boundary-value problems outside of solid mechanics have been similarly affected. Perhaps the most prominent new methodology is the method of *finite elements.* This method was developed in the 1950s in the aircraft industry when it became apparent that the usual methods of analysis were inadequate for design of modern high-speed aircraft.

The method of finite elements consists of decomposing the body under analysis into many small but finite contiguous elements. Certain points in the elements designated as *nodal points* are chosen for study in the case of solid mechanics (see Fig. 19.1). In heat transfer the nodal points would be called nodal temperatures. In general, they are nodal values and the numerical measures of these nodal values are what is sought in a finite element computation. Getting back to solid mechanics, we normally seek among other things the deflections of these points in the deformed geometry. Thus we have *discretized* the body. The deformation of the rest of the element is given in terms of the nodal displacements via carefully chosen functions called *interpolation functions* or *shape functions.* Choosing the proper nodes to use and the proper interpolation functions is a major part of the study of finite elements. Also, the continuity of deformation and its derivatives along contiguous sides of the neighboring elements is a vital consideration in the process.

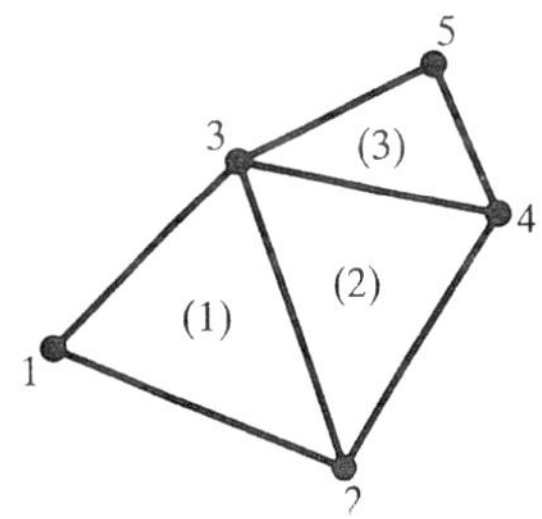

Elements (1), (2), (3)
Nodes 1, 2, 3, 4, 5

Figure 19.1. Triangular elements for plane stress.

[1]This chapter has been adapted from I. H. Shames and C. L. Dym, *Energy and Finite Element Methods in Structural Mechanics* (Hemisphere Publishing Corporation, New York, 1985), Chap. 9.

Restricting ourselves to solid mechanics, the general procedure is to extremize in each element the *total potential energy* (for elastic bodies) using as admissible deformations those that are *kinematically compatible* and that do not violate the constraints. When the total potential energy π has been expressed in terms of the nodal displacements, it degenerates from a functional[2] to a *function* of the nodal displacements a_i. When π has been extremized by the proper set of values for a_i, we have satisfied *equilibrium* for a deformation approximating that of the actual element. Since we have used only compatible deformation, we have already satisfied kinematic compatibility. Now if the continuity requirements between elements have been maintained as required by the theory, we can get more accurate results by making the elements ever smaller and more numerous. Alternatively, we could extremize the total complementary energy π^* by varying certain values related to stress at nodal points whereby, to be admissible, these values are such as always to satisfy equilibrium. When π^* has been extremized we will have then satisfied compatibility in the element. We shall consider only the extremization of π.

In Part A of the chapter we shall look at plane trusses, to introduce us partially to the finite element methodology in the most simple manner. In Part B we will discuss certain salient features of the finite element method that would be used for more complex problems, such as plane stress, torsion, plates, stability, and so on.

Part A: Finite Elements for Trusses

19.2 Introduction

The nodes for the plane truss will be the pins, and the elements will be the members of the truss. The nodal displacements must be made so as not to violate the constraints, and we must at all times satisfy *compatibility when varying the nodal displacements.* This is accomplished by simple trigonometric considerations, of exactly the type that you employed earlier in the book. Since the truss elements are contiguous *only* at the nodes, there is no need to be concerned with the *continuity of displacements* and their derivatives at the remaining boundaries between elements. This is of vital importance for more complex elements, such as those shown in Fig. 19.1, wherein we must be concerned with such questions along the contiguous boundaries *between* the elements.

We will *not need interpolation functions* to give the movement of any point on any member. Again, simple trigonometry is all that is needed to get this information from the nodal displacements. Thus each

[2]See *Looking Ahead* Section 6.11 for a comment on *functionals*.

member will undergo only uniform stretching or shrinking, and the centerline of each member will rotate—both effects of which are easily determined from the displacements of the end nodes of the member via elementary trigonometry.

Finally, instead of using the *total potential energy* for the truss, we will use the *first Castigliano theorem,* which is a special case of the total potential energy principle particularly suitable for trusses. Again we point out that for more complex elements such as those shown in Fig. 19.1, we must carefully choose interpolation functions to relate the displacement field for the whole element in terms of the nodal displacements. This must be done with great care and skill to achieve the aforementioned continuity requirements of displacement and its derivatives at contiguous boundaries between elements. If the interpolation functions do not give these continuity requirements properly, the process will not converge as the elements are made ever smaller and more numerous. You will learn more about this vital part of finite elements in finite element texts.

19.3 The Stiffness Matrix for an Element: Definition

We will now use the first Castigliano theorem to present the *stiffness method* of *linearly elastic,* statically indeterminate trusses. Recall from Section 19.2 that we could readily establish kinematically compatible deformation fields by giving those pins that are not fully constrained (we shall call them movable pins) virtual displacement components in the x and y directions. Suppose that we consider n such displacement components, denoting them as $\Delta_1, \Delta_2, \ldots, \Delta_n$. It will be understood that these deflections take the truss from an undeformed geometry to a kinematically compatible deformed geometry. The strain for any member, say the pth member, can be found by superposing strains developed from the separate displacements of the movable pins.[3] We thus may say that

$$\varepsilon_p = \sum_{s=1}^{n} \kappa_{ps}\Delta_s \tag{19.1}$$

where the κ_{ps} are constants, the first subscript of which refers to the member on which the strain is being considered and the second subscript refers to the virtual displacement. (Thus, κ_{12} is the strain in member 1 per unit deflection corresponding to Δ_2.) A strain energy expression can then be given in terms of the displacement Δ_p as follows for a truss with M members:

[3]It is because we are using here the *superposition principle* that we are restricted to *linear elastic behavior.*

$$U = \iiint_V \int_\varepsilon \tau \, d\varepsilon \, dv = \sum_{p=1}^{M} \int_\varepsilon (E_p\varepsilon_p)\, d\varepsilon_p (A_pL_p)$$
$$= \sum_{p=1}^{M} \frac{E_p\varepsilon_p^2}{2} A_pL_p = \sum_{p=1}^{M} \frac{E_p}{2}\left(\sum_{s=1}^{n} \kappa_{ps}\Delta_s\right)^2 A_pL_p \tag{19.2}$$

Now employ *Castigliano's first theorem.* We can then say that

$$\frac{\partial U}{\partial \Delta_r} = P_r = \sum_{p=1}^{M} E_p\left(\sum_{s=1}^{n} \kappa_{ps}\Delta_s\right)\kappa_{pr}A_pL_p = \sum_{p=1}^{M}\sum_{s=1}^{n} E_pA_pL_p\kappa_{ps}\kappa_{pr}\Delta_s$$
$$r = 1, 2, \ldots, n \tag{19.3}$$

We have thus a set of n simultaneous equations involving the forces on the members P_r, in the direction of the displacements Δ_r, and these displacements. These equations may be written as follows:

$$P_r = \sum_{s=1}^{n} K_{rs}\Delta_s \tag{19.4}$$

where

$$K_{rs} = \sum_{p=1}^{M} E_pA_pL_p\kappa_{pr}\kappa_{ps} \tag{19.5}$$

The constants K_{rs} are called *stiffness constants.* We may interpret K_{rq} as the force in the direction of P_r needed per unit deflection Δ_q. Now the forces P_r in Eq. (19.4) are those needed to maintain the deflections Δ_s. If we know K_{rs} and if we insert the *known values* of the external loads, we arrive at a system of simultaneous equations for directly determining the Δ_s, which clearly now must represent the *actual displacements* of the movable pins. The forces in the truss members may then be readily computed and the truss has been solved completely. We will pursue this procedure in detail.

19.4 Finite Elements and Trusses

Accordingly, consider the simple, plane truss in Fig. 19.2. The members of the truss are the *finite elements* of the system and are numbered separately from the joints. We shall call the joints *nodal points.* At each nodal point i (Fig. 19.3), we have indicated two displacement components (u_i, v_i). These are the *nodal displacement* components for the truss. Corresponding to each pair of nodal displacement components is a set of forces (U_i, V_i) that represent the total force components *on the members at joint i from node i.* A *double* arrow has been used, you will notice, for a nodal displacement component such as u_i and the corresponding nodal force component U_i, and so on.

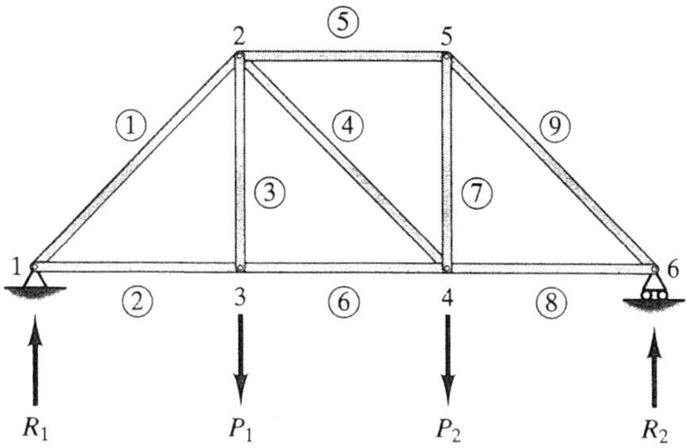

Figure 19.2. Truss with members and joints separately labeled.

In Eq. (19.4) we showed that the forces U_i, V_i are related to the nodal displacement component as follows:[4]

$$\begin{Bmatrix} U_1 \\ V_1 \\ \cdot \\ \cdot \\ \cdot \\ U_6 \\ V_6 \end{Bmatrix} = [K] \begin{Bmatrix} u_1 \\ v_1 \\ \cdot \\ \cdot \\ \cdot \\ u_6 \\ v_6 \end{Bmatrix} \tag{19.6}$$

where $[K]$ is the *stiffness* matrix for the *entire* truss or, as we say in finite elements, $[K]$ is the *global stiffness matrix.* Let us expand Eq. (19.6) to include the elements of $[K]$, which must be symmetric. Thus we have

$$\begin{Bmatrix} U_1 \\ V_1 \\ U_2 \\ V_2 \\ U_3 \\ V_3 \\ U_4 \\ V_4 \\ U_5 \\ V_5 \\ U_6 \\ V_6 \end{Bmatrix} = \begin{bmatrix}
K_{U_1u_1} & K_{U_1v_1} & K_{U_1u_2} & K_{U_1v_2} & K_{U_1u_3} & K_{U_1v_3} & K_{U_1u_4} & K_{U_1v_4} & K_{U_1u_5} & K_{U_1v_5} & K_{U_1u_6} & K_{U_1v_6} \\
K_{V_1u_1} & K_{V_1v_1} & K_{V_1u_2} & K_{V_1v_2} & K_{V_1u_3} & K_{V_1v_3} & K_{V_1u_4} & K_{V_1v_4} & K_{V_1u_5} & K_{V_1v_5} & K_{V_1u_6} & K_{V_1v_6} \\
K_{U_2u_1} & K_{U_2v_1} & K_{U_2u_2} & K_{U_2v_2} & K_{U_2u_3} & K_{U_2v_3} & K_{U_2u_4} & K_{U_2v_4} & K_{U_2u_5} & K_{U_2v_5} & K_{U_2u_6} & K_{U_2v_6} \\
K_{V_2u_1} & K_{V_2v_1} & K_{V_2u_2} & K_{V_2v_2} & K_{V_2u_3} & K_{V_2v_3} & K_{V_2u_4} & K_{V_2v_4} & K_{V_2u_5} & K_{V_2v_5} & K_{V_2u_6} & K_{V_2v_6} \\
K_{U_3u_1} & K_{U_3v_1} & K_{U_3u_2} & K_{U_3v_2} & K_{U_3u_3} & K_{U_3v_3} & K_{U_3u_4} & K_{U_3v_4} & K_{U_3u_5} & K_{U_3v_5} & K_{U_3u_6} & K_{U_3v_6} \\
K_{V_3u_1} & K_{V_3v_1} & K_{V_3u_2} & K_{V_3v_2} & K_{V_3u_3} & K_{V_3v_3} & K_{V_3u_4} & K_{V_3v_4} & K_{V_3u_5} & K_{V_3v_5} & K_{V_3u_6} & K_{V_3v_6} \\
K_{U_4u_1} & K_{U_4v_1} & K_{U_4u_2} & K_{U_4v_2} & K_{U_4u_3} & K_{U_4v_3} & K_{U_4u_4} & K_{U_4v_4} & K_{U_4u_5} & K_{U_4v_5} & K_{U_4u_6} & K_{U_4v_6} \\
K_{V_4u_1} & K_{V_4v_1} & K_{V_4u_2} & K_{V_4v_2} & K_{V_4u_3} & K_{V_4v_3} & K_{V_4u_4} & K_{V_4v_4} & K_{V_4u_5} & K_{V_4v_5} & K_{V_4u_6} & K_{V_4v_6} \\
K_{U_5u_1} & K_{U_5v_1} & K_{U_5u_2} & K_{U_5v_2} & K_{U_5u_3} & K_{U_5v_3} & K_{U_5u_4} & K_{U_5v_4} & K_{U_5u_5} & K_{U_5v_5} & K_{U_5u_6} & K_{U_5v_6} \\
K_{V_5u_1} & K_{V_5v_1} & K_{V_5u_2} & K_{V_5v_2} & K_{V_5u_3} & K_{V_5v_3} & K_{V_5u_4} & K_{V_5v_4} & K_{V_5u_5} & K_{V_5v_5} & K_{V_5u_6} & K_{V_5v_6} \\
K_{U_6u_1} & K_{U_6v_1} & K_{U_6u_2} & K_{U_6v_2} & K_{U_6u_3} & K_{U_6v_3} & K_{U_6u_4} & K_{U_6v_4} & K_{U_6u_5} & K_{U_6v_5} & K_{U_6u_6} & K_{U_6v_6} \\
K_{V_6u_1} & K_{V_6v_1} & K_{V_6u_2} & K_{V_6v_2} & K_{V_6u_3} & K_{V_6v_3} & K_{V_6u_4} & K_{V_6v_4} & K_{V_6u_5} & K_{V_6v_5} & K_{V_6u_6} & K_{V_6v_6}
\end{bmatrix} \begin{Bmatrix} u_1 \\ v_1 \\ u_2 \\ v_2 \\ u_3 \\ v_3 \\ u_4 \\ v_4 \\ u_5 \\ v_5 \\ u_6 \\ v_6 \end{Bmatrix}$$

We see from the preceding equation that $K_{U_iu_i}$ is the *horizontal* force *component* from pin i needed for a *unit horizontal* deflection of pin i

[4]We will use the convention that { } represents a *column* matrix while [ ] represents a *row* matrix.

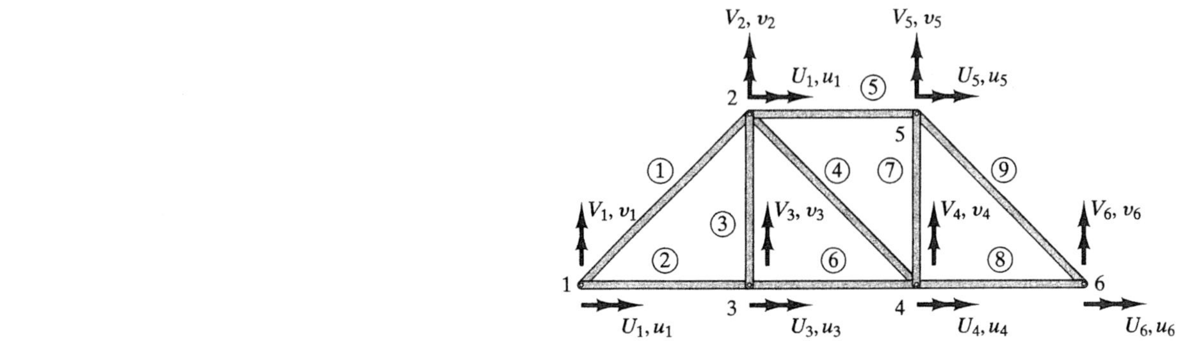

Figure 19.3. Truss showing nodal displacement components and forces from joints onto truss.

with all other nodal displacement components kept at *zero*. Similarly, $K_{V_p v_q}$ is the *vertical* force *component* from pin p needed for a unit *vertical* deflection of pin q with all other nodal displacement components held at *zero*.

To simplify the notation, we now define $\{q_i\}$ as the force vector from node i onto the members of joint i. That is,

$$\{q_i\} = \begin{Bmatrix} U_i \\ V_i \end{Bmatrix} \tag{19.7}$$

Also, we define $\{a_i\}$ as the *nodal displacement vector* at node i. That is,

$$\{a_i\} = \begin{Bmatrix} u_i \\ v_i \end{Bmatrix} \tag{19.8}$$

Then Eq. (19.6) can be given as follows, using *submatrices:*

$$\begin{Bmatrix} \{q_1\} \\ \cdot \\ \cdot \\ \cdot \\ \{q_6\} \end{Bmatrix} = \begin{bmatrix} [K_{11}] & \cdots & [K_{16}] \\ \cdot & & \cdot \\ \cdot & & \cdot \\ \cdot & & \cdot \\ [K_{61}] & \cdots & [K_{66}] \end{bmatrix} \begin{Bmatrix} \{a_1\} \\ \cdot \\ \cdot \\ \cdot \\ \{a_6\} \end{Bmatrix} \tag{19.9}$$

Clearly, we see that for a contribution to $\begin{Bmatrix} U_1 \\ V_1 \end{Bmatrix}$ we have, on considering $[K_{11}]\{a_1\}$,

$$[K_{11}] = \begin{bmatrix} K_{U_1u_1} & K_{U_1v_1} \\ K_{V_1u_1} & K_{V_1v_1} \end{bmatrix} \tag{19.10}$$

Also

$$[K_{ij}] = \begin{bmatrix} K_{U_iu_j} & K_{U_iv_j} \\ K_{V_iu_j} & K_{V_iv_j} \end{bmatrix} \tag{19.11}$$

Let us next consider an *element* of the truss, say the element 4. Then using Eq. (19.9) applied to this element, we can say for the relation between forces and displacements of nodes 2 and 4,

$$\begin{Bmatrix} \{q_2\} \\ \{q_4\} \end{Bmatrix} = \begin{bmatrix} [K_{22}] & [K_{24}] \\ [K_{42}] & [K_{44}] \end{bmatrix} \begin{Bmatrix} \{a_2\} \\ \{a_4\} \end{Bmatrix}$$

or

$$\{q\}^{e=4} = [K]^{e=4}\{a\}^{e=4} \tag{19.12}$$

where $[K]^{e=4}$ is the stiffness matrix for element 4. This formulation can be given as follows, using the format for all the nodal force vectors and nodal displacement vectors:

$$\begin{Bmatrix} \{q_1\} \\ \{q_2\} \\ \{q_3\} \\ \{q_4\} \\ \{q_5\} \\ \{q_6\} \end{Bmatrix} = \begin{bmatrix} 0 & 0 & 0 & 0 & 0 & 0 \\ 0 & [K_{22}] & 0 & [K_{24}] & 0 & 0 \\ 0 & 0 & 0 & 0 & 0 & 0 \\ 0 & [K_{42}] & 0 & [K_{44}] & 0 & 0 \\ 0 & 0 & 0 & 0 & 0 & 0 \\ 0 & 0 & 0 & 0 & 0 & 0 \end{bmatrix} \begin{Bmatrix} \{a_1\} \\ \{a_2\} \\ \{a_3\} \\ \{a_4\} \\ \{a_5\} \\ \{a_6\} \end{Bmatrix} \tag{19.13}$$

In expanded form, we can say for the preceding result:

$$\begin{Bmatrix} U_1 \\ V_1 \\ U_2 \\ V_2 \\ U_3 \\ V_3 \\ U_4 \\ V_4 \\ U_5 \\ V_5 \\ U_6 \\ V_6 \end{Bmatrix} = \begin{bmatrix} 0 & 0 & 0 & 0 & 0 & 0 & 0 & 0 & 0 & 0 & 0 & 0 \\ 0 & 0 & 0 & 0 & 0 & 0 & 0 & 0 & 0 & 0 & 0 & 0 \\ 0 & 0 & K_{U_2u_2} & K_{U_2v_2} & 0 & 0 & K_{U_2u_4} & K_{U_2v_4} & 0 & 0 & 0 & 0 \\ 0 & 0 & K_{V_2u_2} & K_{V_2v_2} & 0 & 0 & K_{V_2u_4} & K_{V_2v_4} & 0 & 0 & 0 & 0 \\ 0 & 0 & 0 & 0 & 0 & 0 & 0 & 0 & 0 & 0 & 0 & 0 \\ 0 & 0 & 0 & 0 & 0 & 0 & 0 & 0 & 0 & 0 & 0 & 0 \\ 0 & 0 & K_{U_4u_2} & K_{U_4v_2} & 0 & 0 & K_{U_4u_4} & K_{U_4v_4} & 0 & 0 & 0 & 0 \\ 0 & 0 & K_{V_4u_2} & K_{V_4v_2} & 0 & 0 & K_{V_4u_4} & K_{V_4v_4} & 0 & 0 & 0 & 0 \\ 0 & 0 & 0 & 0 & 0 & 0 & 0 & 0 & 0 & 0 & 0 & 0 \\ 0 & 0 & 0 & 0 & 0 & 0 & 0 & 0 & 0 & 0 & 0 & 0 \\ 0 & 0 & 0 & 0 & 0 & 0 & 0 & 0 & 0 & 0 & 0 & 0 \\ 0 & 0 & 0 & 0 & 0 & 0 & 0 & 0 & 0 & 0 & 0 & 0 \end{bmatrix} \begin{Bmatrix} u_1 \\ v_1 \\ u_2 \\ v_2 \\ u_3 \\ v_3 \\ u_4 \\ v_4 \\ u_5 \\ v_5 \\ u_6 \\ v_6 \end{Bmatrix} \tag{19.14}$$

By first evaluating each element stiffness matrix and then inserting the results into the above format, we can thereby get the global stiffness matrix for the whole truss.

We shall illustrate in the following section how we can easily formulate the stiffness matrix for any element of the truss.

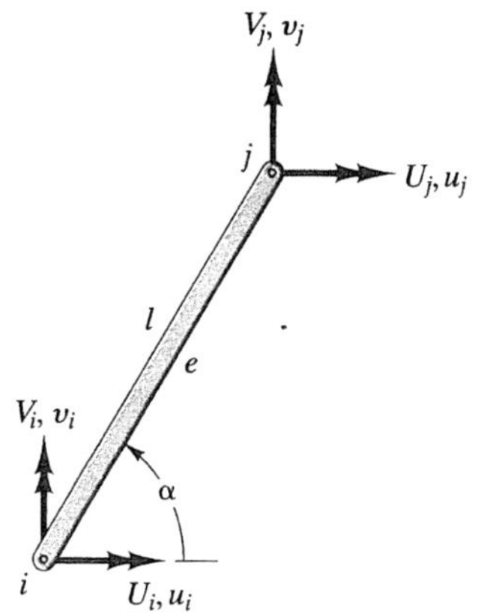

Figure 19.4. Element e from a truss.

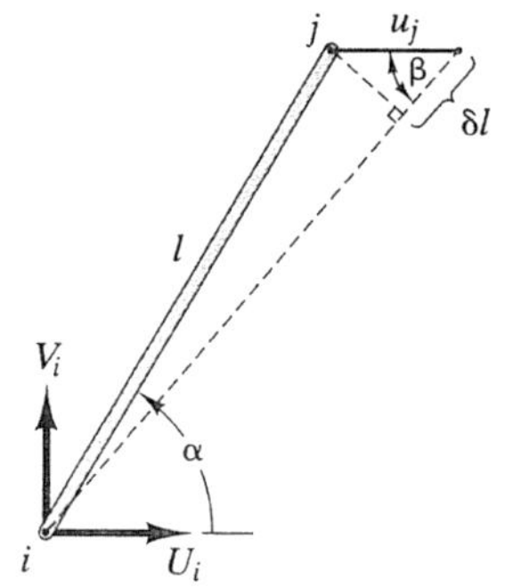

Figure 19.5. Element with displacement u_j.

19.5 Stiffness Matrix for an Element

We now find the stiffness matrix for an element e of the truss having nodes i and j as shown in Fig. 19.4. To get $[K]^e$, we form the following matrix equation for member ij:

$$\begin{Bmatrix} U_i \\ V_i \\ U_j \\ V_j \end{Bmatrix} = \begin{bmatrix} K_{U_iu_i} & K_{U_iv_i} & K_{U_iu_j} & K_{U_iv_j} \\ K_{V_iu_i} & K_{V_iv_i} & K_{V_iu_j} & K_{V_iv_j} \\ K_{U_ju_i} & K_{U_jv_i} & K_{U_ju_j} & K_{U_jv_j} \\ K_{V_ju_i} & K_{V_jv_i} & K_{V_ju_j} & K_{V_jv_j} \end{bmatrix} \begin{Bmatrix} u_i \\ v_i \\ u_j \\ v_j \end{Bmatrix} = [K]^e \begin{Bmatrix} \{a_i\} \\ \{a_j\} \end{Bmatrix} \tag{19.15}$$

In Fig. 19.5 we have instituted a small displacement u_j, keeping other nodal displacement components equal to zero. The elongation δl of the member is then

$$\delta l = u_j \cos \beta$$

For a small displacement u_j, we can take angles $\alpha \approx \beta$. Hence

$$\delta l \approx u_j \cos \alpha$$

With modulus of elasticity E, the force P in the member is easily determined as

$$P = \frac{\delta l}{l} EA = \frac{u_j \cos \alpha \, EA}{l} \qquad \text{(tension)} \tag{19.16}$$

where A is the cross-sectional area of the member. We can now give $K_{U_iu_j}$ as[5]

$$K_{U_iu_j} = -\frac{P \cos \alpha}{u_j} = -\frac{\cos^2\alpha \, EA}{l}$$

Similarly, for the vertical force from pin i stemming from a unit horizontal displacement of pin j, we have

$$K_{V_iu_j} = -\frac{P \sin \alpha}{u_j} = -\frac{\sin \alpha \cos \alpha \, EA}{l}$$

$$= -\frac{\sin 2\alpha \, EA}{2l}$$

Now if we consider the forces U_j and V_j from u_j at the other nodal point j, we get the same results as for U_i and V_i except for sign, as can readily be deduced from *equilibrium*. We can accordingly say for u_j that

[5]Recall that U_i is the horizontal force on the member from pin i to be associated with a horizontal unit nodal displacement component u_j at pin j. Since member ij will be in *tension* for u_j, U_i must be in the negative x direction, so that $K_{U_iu_j}$ is negative.

$$
\begin{aligned}
K_{U_i u_j} &= -\frac{\cos^2\alpha\, EA}{l} \\
K_{V_i u_j} &= -\frac{\sin 2\alpha\, EA}{2l} \\
K_{U_j u_j} &= +\frac{\cos^2\alpha\, EA}{l} \\
K_{V_j u_j} &= +\frac{\sin 2\alpha\, EA}{2l}
\end{aligned}
\qquad (19.17)
$$

If we now consider u_i (i.e., u at the other nodal point), we get compression of the member. It is easily seen that we get the preceding results with changed signs and with u_j replaced by u_i. Thus,

$$
\begin{aligned}
K_{U_i u_i} &= +\frac{\cos^2\alpha\, EA}{l} \\
K_{V_i u_i} &= +\frac{\sin 2\alpha\, EA}{2l} \\
K_{U_j u_i} &= -\frac{\cos^2\alpha\, EA}{l} \\
K_{V_j u_i} &= -\frac{\sin 2\alpha\, EA}{2l}
\end{aligned}
\qquad (19.18)
$$

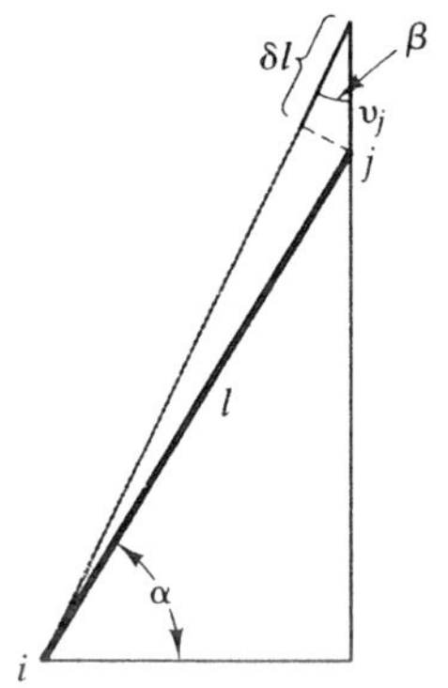

Figure 19.6. Element with displacement v_j.

Now consider v_j in Fig. 19.6. We can then say that for δl,

$$\delta l = v_j \cos\beta \approx v_j \sin\alpha$$

Hence, for the force P in the member, we have

$$P = \frac{v_j \sin\alpha}{l} EA \quad \text{(tension)}$$

We can now write for the horizontal force from pin i resulting from a unit vertical displacement of pin j:

$$K_{U_i v_j} = -\frac{P\cos\alpha}{v_j} = -\frac{\sin 2\alpha\, EA}{2l}$$

For the vertical force from pin i arising from the same unit vertical displacement of pin j:

$$K_{V_i v_j} = -\frac{P\sin\alpha}{v_j} = -\frac{\sin^2\alpha\, EA}{l}$$

For U_j and V_j, respectively (at pin j), we get the same results as above for v_j except for the sign (as a result of equilibrium), so that for v_j we have in total:

$$
\begin{aligned}
K_{U_i v_j} &= -\frac{\sin 2\alpha\, EA}{2l} \\
K_{V_i v_j} &= -\frac{\sin^2\alpha\, EA}{l} \\
K_{U_j v_j} &= +\frac{\sin 2\alpha\, EA}{2l} \\
K_{V_j v_j} &= +\frac{\sin^2\alpha\, EA}{l}
\end{aligned}
\tag{19.19}
$$

For v_i at the other end we get compression in the member. It is easily seen again that we get the same result as above but with opposite signs and with v_i replacing v_j. Thus,

$$
\begin{aligned}
K_{U_i v_i} &= +\frac{\sin 2\alpha\, EA}{2l} \\
K_{V_i v_i} &= +\frac{\sin^2\alpha\, EA}{l} \\
K_{U_j v_i} &= -\frac{\sin 2\alpha\, EA}{2l} \\
K_{V_j v_i} &= -\frac{\sin^2\alpha\, EA}{l}
\end{aligned}
\tag{19.20}
$$

The stiffness matrix for any element *ij*, with the angle α measured *counterclockwise* from the x axis to *ij*, can be given as follows:

$$
[K]^e = \frac{EA}{l}\begin{bmatrix}
\cos^2\alpha & \dfrac{\sin 2\alpha}{2} & -\cos^2\alpha & -\dfrac{\sin 2\alpha}{2} \\
\dfrac{\sin 2\alpha}{2} & \sin^2\alpha & -\dfrac{\sin 2\alpha}{2} & -\sin^2\alpha \\
-\cos^2\alpha & -\dfrac{\sin 2\alpha}{2} & \cos^2\alpha & \dfrac{\sin 2\alpha}{2} \\
-\dfrac{\sin 2\alpha}{2} & -\sin^2\alpha & \dfrac{\sin 2\alpha}{2} & \sin^2\alpha
\end{bmatrix}
\tag{19.21}
$$

We thus have available to us a simple stiffness matrix for an element that must be properly placed in a truss matrix [such as is shown in Eq. (19.14)] to get the global matrix.

19.6 The Global Stiffness Matrix

We now discuss the building of the global matrix. Examine for this purpose the 12 × 12 matrix in Eq. (19.14) for the truss in Fig. 19.2. The element stiffness matrix shown in the array corresponds to an element having *nodal points 2 and 4.* Hence, the array shown above for this element with $\alpha = -45°$ would be placed in the third and fourth rows (for nodal point 2) and the seventh and eighth rows (for nodal point 4). Furthermore, the columns also have the same numbering as the rows, namely, third and fourth and seventh and eighth. In addition, element 9 having nodal points 5 and 6 would occupy rows and columns 9 and 10 and rows and columns 11 and 12. The angle α would then be $-45°$.

We shall now illustrate this (as well as other considerations) in the following simple example.

Example 19.1

Find the stiffness matrix for the truss shown in Fig. 19.7.

There are five members and eight nodal displacement components. We will set up the format for the entire truss analogous to Eq. (19.14).

$$\begin{Bmatrix} U_1 \\ V_1 \\ U_2 \\ V_2 \\ U_3 \\ V_3 \\ U_4 \\ V_4 \end{Bmatrix} = \begin{bmatrix} K_{U_1u_1} & K_{U_1v_1} & K_{U_1u_2} & K_{U_1v_2} & K_{U_1u_3} & K_{U_1v_3} & K_{U_1u_4} & K_{U_1v_4} \\ K_{V_1u_1} & K_{V_1v_1} & K_{V_1u_2} & K_{V_1v_2} & K_{V_1u_3} & K_{V_1v_3} & K_{V_1u_4} & K_{V_1v_4} \\ K_{U_2u_1} & K_{U_2v_1} & K_{U_2u_2} & K_{U_2v_2} & K_{U_2u_3} & K_{U_2v_3} & K_{U_2u_4} & K_{U_2v_4} \\ K_{V_2u_1} & K_{V_2v_1} & K_{V_2u_2} & K_{V_2v_2} & K_{V_2u_3} & K_{V_2v_3} & K_{V_2u_4} & K_{V_2v_4} \\ K_{U_3u_1} & K_{U_3v_1} & K_{U_3u_2} & K_{U_3v_2} & K_{U_3u_3} & K_{U_3v_3} & K_{U_3u_4} & K_{U_3v_4} \\ K_{V_3u_1} & K_{V_3v_1} & K_{V_3u_2} & K_{V_3v_2} & K_{V_3u_3} & K_{V_3v_3} & K_{V_3u_3} & K_{V_3v_4} \\ K_{U_4u_1} & K_{U_4v_1} & K_{U_4u_2} & K_{U_4v_2} & K_{U_4u_3} & K_{U_4v_3} & K_{U_4u_4} & K_{U_4v_4} \\ K_{V_4u_1} & K_{V_4v_1} & K_{V_4u_2} & K_{V_4v_2} & K_{V_4u_3} & K_{V_4v_3} & K_{V_4u_4} & K_{U_4v_4} \end{bmatrix} \begin{Bmatrix} u_1 \\ v_1 \\ u_2 \\ v_2 \\ u_3 \\ v_3 \\ u_4 \\ v_4 \end{Bmatrix} \qquad \text{(a)}$$

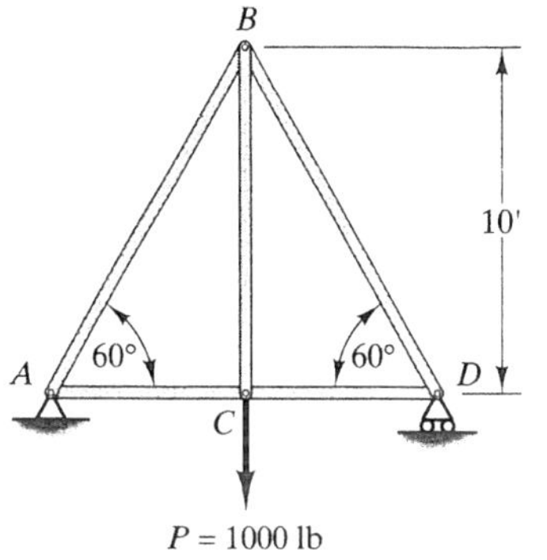

Figure 19.7. Simple truss.

As you may readily demonstrate, the stiffness matrix as indicated earlier will be symmetric, so only half the terms need be calculated.

The procedure is to get the stiffness matrix for each element (see Fig. 19.8) and then assemble the global stiffness matrix.

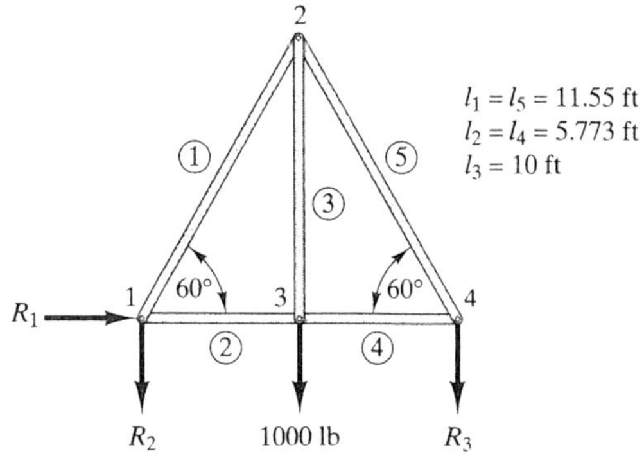

Figure 19.8. Truss with members and nodes numbered.

Example 19.1 (Continued)

Element ①, $\alpha = 60°$, $l = 11.55$ ft, nodal points 1–2:

$$[K]^1 = \frac{EA}{11.55}\begin{bmatrix} .250 & .433 & -.250 & -.433 \\ .433 & .750 & -.433 & -.750 \\ -.250 & -.433 & .250 & .433 \\ -.433 & -.750 & .433 & .750 \end{bmatrix} \quad \text{(b)}$$

Element ②, $\alpha = 0$, $l = 5.77$ ft, nodal points 1–3:

$$[K]^2 = \frac{EA}{5.77}\begin{bmatrix} 1 & 0 & -1 & 0 \\ 0 & 0 & 0 & 0 \\ -1 & 0 & 1 & 0 \\ 0 & 0 & 0 & 0 \end{bmatrix} \quad \text{(c)}$$

Element ③, $\alpha = 90°$, $l = 10$ ft, nodal points 3–2:

$$[K]^3 = \frac{EA}{10}\begin{bmatrix} 0 & 0 & 0 & 0 \\ 0 & 1 & 0 & -1 \\ 0 & 0 & 0 & 0 \\ 0 & -1 & 0 & 1 \end{bmatrix} \quad \text{(d)}$$

Element ④, $\alpha = 0$, $l = 5.77$ ft, nodal points 3–4:

$$[K]^4 = \frac{EA}{5.77}\begin{bmatrix} 1 & 0 & -1 & 0 \\ 0 & 0 & 0 & 0 \\ -1 & 0 & 1 & 0 \\ 0 & 0 & 0 & 0 \end{bmatrix} \quad \text{(e)}$$

Element ⑤, $\alpha = 120°$, $l = 11.55$ ft, nodal points 4–2:

$$[K]^5 = \frac{EA}{11.55}\begin{bmatrix} .250 & -.433 & -.250 & .433 \\ -.433 & .750 & .433 & -.750 \\ -.250 & .433 & .250 & -.433 \\ .433 & -.750 & -.433 & .750 \end{bmatrix} \quad \text{(f)}$$

We next assemble the global stiffness matrix by noting the nodal points and inserting terms to correspond to these nodal points. For instance, element ① involves nodal points 1 and 2. Merely lay matrix terms in Eq. (b), after multiplying by the coefficient 1/11.5, into appropriate locations in Eq. (a) to cover nodal points 1 and 2. Next

Example 19.1 (Continued)

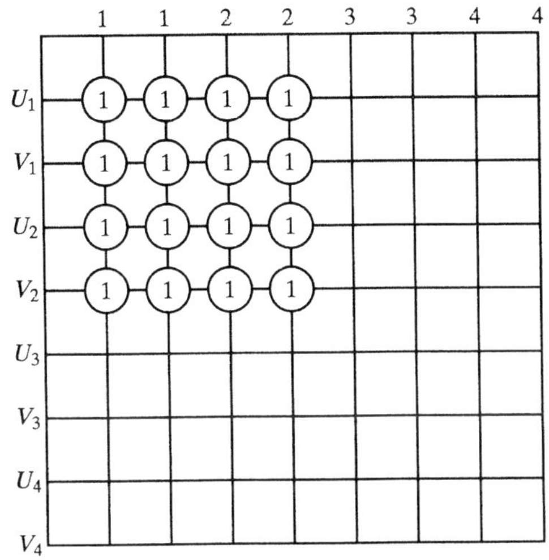

go to element ② involving nodal points 1 and 3. We insert Eq. (c), after multiplying by the coefficient 1/5.77, into the global array as shown below. For element ③ we simply place the matrix as it stands to occupy pairs of rows 2 and 3 and column sets 2 and 3 even though we went from pin 3 to pin 2 in determining α. We continue to insert the stiffness coefficients, *adding* those that appear at the same grid points. Of course, when we insert a stiffness coefficient, we use numerical values from Eqs. (b)–(f). The final result is presented next.

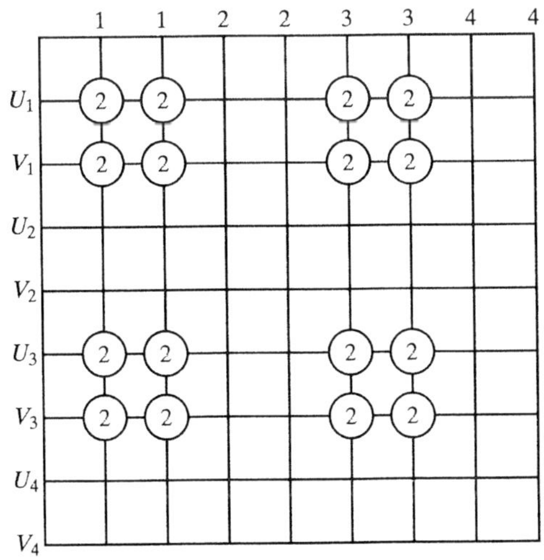

Example 19.1 (Continued)

$$[K] = EA\begin{bmatrix} .1950 & .03749 & -.02165 & -.03749 & -.1733 & 0 & 0 & 0 \\ .03749 & .06494 & -.03749 & -.06494 & 0 & 0 & 0 & 0 \\ -.02165 & -.03749 & .04330 & 0 & 0 & 0 & -.02165 & .03749 \\ -.03749 & -.06494 & 0 & .2299 & 0 & -.10 & .03749 & -.06494 \\ -.1733 & 0 & 0 & 0 & .3466 & 0 & -.1733 & 0 \\ 0 & 0 & 0 & -.10 & 0 & .10 & 0 & 0 \\ 0 & 0 & -.02165 & .03749 & -.1733 & 0 & .1950 & -.03749 \\ 0 & 0 & .03749 & -.06494 & 0 & 0 & -.03749 & .06494 \end{bmatrix} \quad \text{(g)}$$

19.7 Solution of a Truss Problem

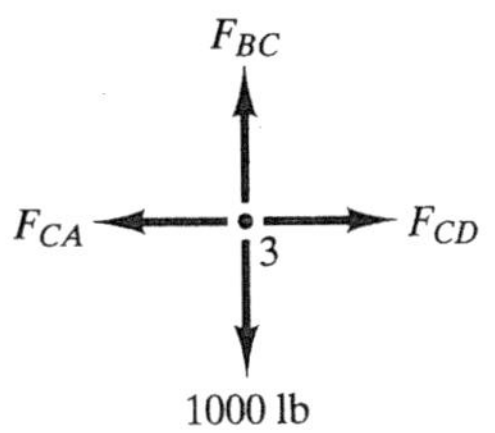

Figure 19.9. Pin 3 as a free body.

In Example 19.1 we formulated the global stiffness matrix for a simple truss. Now we shall use the data of this example to show how we can determine the nodal displacements and the forces in the members.

Recall that the forces U_i, V_i are forces *on the members from the pins or nodal points.* Accordingly, consider pin 3 as a free body in Fig. 19.9. The forces *on* pin 3 *from* members and external load are in **equilibrium,** so that we have:

$\underline{\Sigma F_x = 0}$:

$$F_{CD} - F_{CA} = 0$$

But $-(F_{CD} - F_{CA})$ is the horizontal force *from* the pin 3 *onto* the members at 3, and so $-(F_{CD} - F_{CA}) = U_3$. We can conclude from above that $U_3 = 0$. Now consider the vertical direction for equilibrium.

$\underline{\Sigma F_y = 0}$:

$$F_{BC} - 1000 = 0$$

Therefore,

$$F_{BC} = 1000 \text{ lb}$$

But $(-F_{BC})$ is the force in the vertical direction *from* pin 3 *onto* the members, and hence $(-F_{BC}) = V_3$. Using the equation above, we can readily see that

$$V_3 = -1000 \text{ lb}$$

From the consideration at nodal point 3, we conclude that we can replace U_3 and V_3, respectively, by the known *external loads* in the horizontal and vertical directions, using the regular signs for these loads. At nodal point 1, we replace U_1, V_1 by R_1, R_2 and at nodal point 4 we replace U_4, V_4 by R_3, 0. The supporting forces R_i are taken as unknown and are determined in the process. We can now write

$$\begin{Bmatrix} U_1 \\ V_1 \\ U_2 \\ V_2 \\ U_3 \\ V_3 \\ U_4 \\ V_4 \end{Bmatrix} = \begin{Bmatrix} R_1 \\ R_2 \\ 0 \\ 0 \\ 0 \\ -1000 \\ 0 \\ R_3 \end{Bmatrix}$$

We next go back to Eq. (19.6). Putting the force matrix on the right side of the equation as is customary, we have

$$\begin{bmatrix} & & \\ & K & \\ & & \end{bmatrix} \begin{Bmatrix} u_1 \\ v_1 \\ u_2 \\ v_2 \\ u_3 \\ v_3 \\ u_4 \\ v_4 \end{Bmatrix} = \begin{Bmatrix} R_1 \\ R_2 \\ 0 \\ 0 \\ 0 \\ -1000 \\ 0 \\ R_3 \end{Bmatrix} \tag{19.22}$$

The idea would now be to solve for $\{a\}$ in terms of the loads. We cannot do this yet because the stiffness matrix $[K]$ in Eq. (19.22) is *singular*; that is, the determinant of $[K]$ equals zero. This results from the fact that the boundary conditions have not as yet been applied; thus rigid-body movement has not yet been excluded.

To remedy this difficulty, we must insert the *boundary conditions.* Let us for a moment consider the following matrix equation:

$$\begin{bmatrix} K_{11} & K_{12} & K_{13} & K_{14} \\ K_{21} & K_{22} & K_{23} & K_{24} \\ K_{31} & K_{32} & K_{33} & K_{34} \\ K_{41} & K_{42} & K_{43} & K_{44} \end{bmatrix} \begin{Bmatrix} u_1 \\ v_1 \\ u_2 \\ v_2 \end{Bmatrix} = \begin{Bmatrix} U_1 \\ V_1 \\ U_2 \\ V_2 \end{Bmatrix} \tag{19.23}$$

where one of the nodal displacements, say u_2, is known to be zero. Equation (19.23) can be written as follows:

$$\begin{bmatrix} K_{11} & K_{12} & 0 & K_{14} \\ K_{21} & K_{22} & 0 & K_{24} \\ 0 & 0 & 0 & 0 \\ K_{41} & K_{42} & 0 & K_{44} \end{bmatrix} \begin{Bmatrix} u_1 \\ v_1 \\ 0 \\ v_2 \end{Bmatrix} = \begin{Bmatrix} U_1 \\ V_1 \\ 0 \\ V_2 \end{Bmatrix}$$

You may directly verify that when you multiply out the first, second, and fourth rows above you get the correct equations for Eq. (19.23) for these rows with $u_2 = 0$. If we strike out the third row and the third column, clearly we will not affect the results for U_1, V_1, and V_2. Thus, we can work with the following matrix equation:

$$\begin{bmatrix} K_{11} & K_{12} & K_{14} \\ K_{21} & K_{22} & K_{24} \\ K_{41} & K_{42} & K_{44} \end{bmatrix} \begin{Bmatrix} u_1 \\ v_1 \\ v_2 \end{Bmatrix} = \begin{Bmatrix} U_1 \\ V_1 \\ V_2 \end{Bmatrix}$$

The stiffness matrix so reduced is called the *reduced stiffness matrix.* If next both u_2 and v_2 are prescribed as equal to zero, we would delete the third and fourth rows and columns of Eq. (19.23) to form the reduced matrix. We would then have

$$\begin{bmatrix} K_{11} & K_{12} \\ K_{21} & K_{22} \end{bmatrix} \begin{Bmatrix} u_1 \\ v_1 \end{Bmatrix} = \begin{Bmatrix} U_1 \\ V_1 \end{Bmatrix}$$

As for the problem at hand, we form the reduced matrix by eliminating the first, second, and last rows and columns in Eq. (19.22) since $u_1 = v_1 = v_4 = 0$.

Suppose next that in Eq. (19.23) u_2 is specified as a *nonzero* value, a. Then we can rewrite Eq. (19.23) as follows:

$$\begin{bmatrix} K_{11} & K_{12} & 0 & K_{14} \\ K_{21} & K_{22} & 0 & K_{24} \\ 0 & 0 & 1 & 0 \\ K_{41} & K_{42} & 0 & K_{44} \end{bmatrix} \begin{Bmatrix} u_1 \\ v_1 \\ u_2 \\ v_2 \end{Bmatrix} = \begin{Bmatrix} (U_1 - K_{13}a) \\ (V_1 - K_{23}a) \\ a \\ (V_2 - K_{43}a) \end{Bmatrix} \tag{19.24}$$

When we multiply the third row out, we get $u_2 = a$ as required. When we multiply out the other rows, we get the correct results for Eq. (19.23) for these rows if we set $u_2 = a$. We can again delete row 3 and column 3 to form the following result:

$$\begin{bmatrix} K_{11} & K_{12} & K_{14} \\ K_{21} & K_{22} & K_{24} \\ K_{41} & K_{42} & K_{44} \end{bmatrix} \begin{Bmatrix} u_1 \\ v_1 \\ v_2 \end{Bmatrix} = \begin{Bmatrix} (U_1 - K_{13}a) \\ (V_1 - K_{23}a) \\ (V_2 - K_{43}a) \end{Bmatrix}$$

If $v_1 = a$ and $u_2 = b$, we extend the method as follows:

$$\begin{bmatrix} K_{11} & 0 & 0 & K_{14} \\ 0 & 1 & 0 & 0 \\ 0 & 0 & 1 & 0 \\ K_{41} & 0 & 0 & K_{44} \end{bmatrix} \begin{Bmatrix} u_1 \\ v_1 \\ u_2 \\ v_2 \end{Bmatrix} = \begin{Bmatrix} (U_1 - K_{12}a - K_{13}b) \\ a \\ b \\ (V_2 - K_{42}a - K_{43}b) \end{Bmatrix} \tag{19.25}$$

Clearly, if you evaluate the equations from Eq. (19.23) for the first and fourth rows and put in $v_1 = a$ and $u_2 = b$ after carrying out the steps, you will see that the above is a correct formulation. Now we can delete the second and third rows and columns above without invalidating the remaining formulation. Thus, we have

$$\begin{bmatrix} K_{11} & K_{14} \\ K_{41} & K_{44} \end{bmatrix} \begin{Bmatrix} u_1 \\ v_2 \end{Bmatrix} = \begin{Bmatrix} (U_1 - K_{12}a - K_{13}b) \\ (V_2 - K_{42}a - K_{43}b) \end{Bmatrix}$$

Note in Eq. (19.25) before reduction that for those rows for which the displacement components are *not* specified constants, we must subtract from the forces K-values corresponding to the position of the zeros in *these particular rows*. Each such K-value, furthermore, must be multiplied by the specific *known* displacement in vector $\{a\}$ that would normally be multiplied by the K-value when carrying out the matrix product $[K]\{a\}$. Thus, for the first row in Eq. (19.25), u_1 is unspecified. The zeros in this row correspond to K_{12} and K_{13}. Now K_{12} normally multiplies v_1, which is known to have value a, and K_{13} multiplies u_2, which is known to have value b. Thus we subtract from U_1 the value $K_{12}a$ and $K_{12}b$, as shown in the equation. It is thus a simple process to reduce the matrix equation. Clearly, the procedure described earlier with zero values of prescribed nodal displacements is but a special case of the procedure just described.

Example 19.2

Determine the nodal deflections and the supporting forces for the truss in Example 19.1.

On deleting the first two rows and columns as well as the last row and column, the basic matrix equation becomes

$$EA\begin{bmatrix} .04330 & 0 & 0 & 0 & -.02165 \\ 0 & .2299 & 0 & -.10 & .03749 \\ 0 & 0 & .3466 & 0 & -.1733 \\ 0 & -.10 & 0 & .10 & 0 \\ -.02165 & .03749 & -.1733 & 0 & .1950 \end{bmatrix} \begin{Bmatrix} u_2 \\ v_2 \\ u_3 \\ v_3 \\ u_4 \end{Bmatrix} = \begin{Bmatrix} 0 \\ 0 \\ 0 \\ -1000 \\ 0 \end{Bmatrix}$$

The reduced matrix is no longer singular, and we can now find the inverse of the reduced stiffness matrix. Inverting the matrix equation above using a computer we have

$$\begin{Bmatrix} u_2 \\ v_2 \\ u_3 \\ v_3 \\ u_4 \end{Bmatrix} = \frac{1}{EA}\begin{bmatrix} 25.98 & -1.667 & 2.885 & -1.667 & 5.770 \\ -1.667 & 8.661 & -1.667 & 8.661 & -3.331 \\ 2.885 & -1.667 & 5.770 & -1.667 & 5.770 \\ -1.667 & 8.661 & -1.667 & 18.661 & -3.331 \\ 5.770 & -3.331 & 5.770 & -3.331 & 11.540 \end{bmatrix} \begin{Bmatrix} 0 \\ 0 \\ 0 \\ -1000 \\ 0 \end{Bmatrix}$$

Example 19.2 (Continued)

We thus have

$$u_2 = \frac{1}{EA}1667 \text{ ft}$$
$$v_2 = -\frac{1}{EA}8861 \text{ ft}$$
$$u_3 = \frac{1}{EA}1667 \text{ ft}$$
$$v_3 = -\frac{1}{EA}18{,}661 \text{ ft}$$
$$u_4 = \frac{1}{EA}3331 \text{ ft}$$

where the units of EA are pounds force.

To get the supporting forces, go back to Eq. (19.22) and set up the equation for the R_1, R_2, and R_3 rows. Thus

$$EA\begin{bmatrix} .1950 & .0375 & -.02165 & -.0375 & -.1733 & 0 & 0 & 0 \\ .0375 & .0649 & -.0375 & -.0649 & 0 & 0 & 0 & 0 \\ 0 & 0 & .0375 & -.0649 & 0 & 0 & -.0375 & .0649 \end{bmatrix}\begin{Bmatrix} 0 \\ 0 \\ u_2 \\ v_2 \\ u_3 \\ v_3 \\ u_4 \\ 0 \end{Bmatrix} = \begin{Bmatrix} R_1 \\ R_2 \\ R_3 \end{Bmatrix}$$

Substituting the computed values for the nodal displacements, we now get the supporting forces from calculations involving eight-decimal-place accuracy:

$$R_1 = 0 \text{ lb}$$
$$R_2 = 500 \text{ lb}$$
$$R_3 = 500 \text{ lb}$$

These clearly are the expected results and act as a partial check on the work done up to this point. Finally, to get the forces in the members, we proceed as follows (see Fig. 19.10):

$$P_e = \frac{A_e E_e}{L_e}[(u_j \cos\phi_e + v_j \sin\phi_e) - (u_i \cos\phi_e + v_i \sin\phi_e)] \quad \text{(a)}$$

In matrix notation this becomes

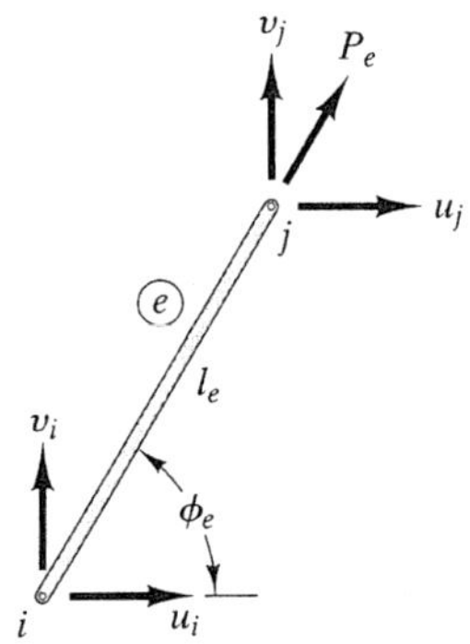

Figure 19.10. Element shown for finding force P_e in the element.

Example 19.2 (Continued)

$$P_e = \left(\frac{AE}{L}\right)_e [G]_e \begin{Bmatrix} \{a_i\} \\ \{a_j\} \end{Bmatrix}$$

where

$$[G] = [-\cos\phi_e, -\sin\phi_e, \cos\phi_e, \sin\phi_e] \tag{b}$$

Because we have used the correct relation between the nodal displacements and the displacement and deformation of the members themselves at all times, and furthermore, because we have maintained proper contiguity requirements between the members (needed only at the nodes), we get *exact* results even though the elements used are full-sized. Were these conditions not present such as in plane stress, torsion, and so on, we would only get approximations.

Part B: Some Preliminary General Considerations

19.8 Basic Considerations for Finite Elements

We will set up basic ideas and the accompanying notation for finite elements in this section. We will work with *plane stress* in formulating many of those ideas but will indicate as we go along when the results are more generally valid by shading in the pertinent results of general use. Accordingly, let us consider a triangular element (Fig. 19.11),

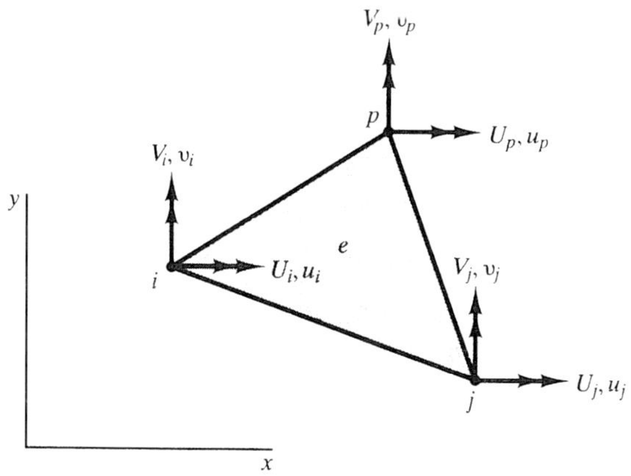

Figure 19.11. Triangular element.

which we denote as element e, with three nodes i, j, and p. The *nodal displacement components* for these three nodes are, respectively, (u_i, v_i) (u_j, v_j), and (u_p, v_p). The corresponding nodal forces on the element needed for this set of nodal displacements are (U_i, V_i), (U_j, V_j), and (U_p, V_p). We make the following additional definitions for the nodal displacement matrices for element e:

$$\begin{Bmatrix} u_i \\ v_i \end{Bmatrix} = \{a_i\}$$

$$\begin{Bmatrix} u_j \\ v_j \end{Bmatrix} = \{a_j\} \tag{19.26}$$

$$\begin{Bmatrix} u_p \\ v_p \end{Bmatrix} = \{a_p\}$$

$$\begin{Bmatrix} \{a_i\} \\ \{a_j\} \\ \{a_p\} \end{Bmatrix} = \{a\}^e \tag{19.27}$$

We next give the *displacement field* in terms of the nodal displacement vector $\{a\}^e$ for an element. The displacement function $\{\boldsymbol{u}(x,y)\}$ is defined as follows for plane strain:

$$\{\boldsymbol{u}(x, y)\} = \begin{Bmatrix} u(x,y) \\ v(x,y) \end{Bmatrix} \tag{19.28}$$

Note that this function gives the displacement components of any point in the element. Next we wish to relate the displacement field for an element to the nodal displacements for the element. For this purpose we introduce the *interpolation* functions or, as they are often called, the *shape* functions. Thus, we state for our triangular element

$$\begin{Bmatrix} u(x,y) \\ v(x,y) \end{Bmatrix} = [N] \begin{Bmatrix} a_1 \\ a_2 \\ a_3 \\ a_4 \\ a_5 \\ a_6 \end{Bmatrix} \tag{19.29}$$

where $[N]$ is the interpolation function matrix, which here must have 12 elements forming six columns and two rows. Each member of $[N]$ is a function of position. Unlike the truss, we have only an approximation of the actual displacement field of the element away from the nodal points. We point out yet again that we must be concerned that the interpolation functions employed have the proper continuity requirements at common borders other than at nodes. We may write Eq. (19.29) identifying the various terms with nodes as follows:

$$\{\boldsymbol{u}(x, y)\} = \lfloor [N_i][N_j][N_p] \rfloor \begin{Bmatrix} \{a_i\} \\ \{a_j\} \\ \{a_p\} \end{Bmatrix} \tag{19.30}$$

where for node i we have for the interpolation functions

$$[N_i] = \begin{bmatrix} (N_i)_{11} & (N_i)_{12} \\ (N_i)_{21} & (N_i)_{22} \end{bmatrix}$$

Associations with the other nodes may be formed similarly. Thus for our triangular element we have 12 interpolation functions. In the general case, we have

$$\{\boldsymbol{u}(x, y, z)\} = [N]\,\{a\} \tag{19.31}$$

Now the interpolation functions must be such that the displacement field $\{\boldsymbol{u}\}$ at the *position* of a nodal point gives the proper nodal *displacements* at that nodal point. Thus for nodal point i we have

$$\{\boldsymbol{u}(x_i, y_i)\} = [N_i\,(x_i, y_i)]\{a_i\} + [N_j\,(x_i, y_i)]\{a_j\} + [N_p\,(x_i, y_i)]\{a_p\} \tag{19.32}$$

and we require that $[N_j(x_i, y_i)] = 0$, $[N_p(x_i, y_i)] = 0$, and $[N_i(x_i, y_i)] = [I]$, where $[I]$ is the unitary matrix. Similarly,

$$\begin{cases} [N_i(x_j,y_j)] = 0 \\ [N_j(x_j,y_j)] = [I] \\ [N_p(x_j,y_j)] = 0 \end{cases} \quad \begin{cases} [N_i(x_p,y_p)] = 0 \\ [N_j(x_p,y_p)] = 0 \\ [N_p(x_p,y_p)] = [I] \end{cases} \tag{19.33}$$

Now we relate the *strain field* matrix $\{\varepsilon\}$ to the *displacement* field. Thus, for small strain we have in general

$$\{\varepsilon\} = [L]\,\{\boldsymbol{u}(x,y,z)\} \tag{19.34}$$

where $[L]$ is a suitable differential operator developed by geometric considerations. As an example, for *plane stress* we would have

$$\{\varepsilon\} \equiv \begin{Bmatrix} \varepsilon_{xx} \\ \varepsilon_{yy} \\ \varepsilon_{zz} \end{Bmatrix} = \begin{Bmatrix} \dfrac{\partial u}{\partial x} \\ \dfrac{\partial v}{\partial y} \\ \dfrac{1}{2}\left(\dfrac{\partial u}{\partial y} + \dfrac{\partial v}{\partial x}\right) \end{Bmatrix} = \begin{bmatrix} \dfrac{\partial}{\partial x} & 0 \\ 0 & \dfrac{\partial}{\partial y} \\ \dfrac{1}{2}\dfrac{\partial}{\partial y} & \dfrac{1}{2}\dfrac{\partial}{\partial x} \end{bmatrix} \begin{Bmatrix} u(x,y) \\ v(x,y) \end{Bmatrix} \tag{19.35}$$

where the operator $[L]$ acts on $\begin{Bmatrix} u(x,y) \\ v(x,y) \end{Bmatrix}$ in Eq. (19.35).

To relate $\{\varepsilon\}$ to the *nodal displacement vectors,* we introduce the *nodal displacement–strain* matrix $[B]$ by replacing $\{\boldsymbol{u}(x,y,z)\}$ in Eq. (19.34) using Eq. (19.31). We then get

$$\{\varepsilon\} = [L][N]\{a\} = [B]\{a\} \tag{19.36}$$

where

$$[B] = [L][N] \tag{19.37}$$

As a final step in this section, we consider the *constitutive law* matrix for linear elastic materials. Let $[\sigma]$ be the stress matrix. We then have the following relation:

$$[\sigma] = [D]([\varepsilon] - [\varepsilon_0]) + [\sigma_0] \tag{19.38}$$

where $[\varepsilon]$ is the total strain matrix, $[\varepsilon_0]$ represents the *initial strain* matrix, and $[\sigma_0]$ represents the *residual* stress matrix. The matrix $[D]$ represents the *constitutive law* matrix. In the case of plane stress

$$\{\sigma\} = \begin{Bmatrix} \tau_{xx} \\ \tau_{yy} \\ \tau_{xy} \end{Bmatrix} \tag{19.39}$$

and you may verify from the constitutive law for an isotropic, Hookean material that

$$[D] = \frac{E}{1 - \nu^2} \begin{bmatrix} 1 & \nu & 0 \\ \nu & 1 & 0 \\ 0 & 0 & \dfrac{1 - \nu}{2} \end{bmatrix} \tag{19.40}$$

We shall now consider the general theory for the *displacement method.*

19.9 General Theory for the Displacement Method

We will now employ the principle of total potential energy for an element to formulate the equation of equilibrium in terms of nodal displacements. That is, we shall extremize the total potential energy functional for an element with respect to the nodal displacement components of the element.[6] We shall consider that initial strain and resid-

[6]Recall that we used equilibrium equations at each joint to arrive at the stiffness matrix when we considered trusses in Section 19.3. In more general applications of the finite element approach, we shall use the procedure of extremizing an appropriate functional separately over each element domain.

ual stress are present, both indicated by the subscript zero. In the ensuing discussion, we will consider the nodes of an element as separate entities, leaving the element minus the nodes as a separate entity. This is very similar to the way in which we separately treated the pins of a truss and the members of the truss as discussed in Part A of this chapter.

Acting on an element from outside, we may have body force distributions b_i and traction force distributions T_i. We shall consider these forces to be replaced by an equivalent system of point forces acting only *on the nodes.* The forces *from* the nodes acting *on* the element are denoted as $\{q\}$. The method of total potential energy for the element then states that

$$\delta\pi = 0 = \iiint_V (\tau_{xx}\delta\varepsilon_{xx} + \cdots + \tau_{xy}\delta\varepsilon_{xy})\,dv - \lfloor \delta a \rfloor \{q\} \quad (19.41)$$

In matrix form we have[7]

$$\iiint_V [\delta\varepsilon]^T[\sigma]\,dv - \{\delta a\}^T\{q\} = 0 \quad (19.42)$$

Let us examine the first expression in Eq. (19.42). The integrand may be written as follows, using Eq. (19.38) for $[\sigma]$:

$$[\delta\varepsilon]^T[\sigma] = [\delta\varepsilon]^T([D]([\varepsilon] - [\varepsilon_0]) + [\sigma_0]) \quad (19.43)$$

Now use Eq. (19.36) for $[\delta\varepsilon]^T$ and $[\varepsilon]$ on the right side of Eq. (19.43). On separating the expressions, we get

$$[\delta\varepsilon]^T[\sigma] = \{\delta a\}^T[B]^T[D][B]\{a\} - \{\delta a\}^T[B]^T[D][\varepsilon_0] + \{\delta a\}^T[B]^I[\sigma_0] \quad (19.44)$$

Let us next evaluate the forces $\{q\}$, which, you will recall, apply *to* the element *from* the nodes. To do this we consider the nodes for the element as a system of free bodies. Acting on these nodes from the element we have the point force system $\{q'\}$, which is the *reaction* to $\{q\}$. Thus we have $\{q'\} = -\{q\}$. Also acting on the nodes is a loading system, which is statically equivalent to the external loadings on the element consisting for now of a body force distribution $b_i(x,y,z)$ and a traction force distribution $T_i(x,y,z)$. The static equivalent of these external force distributions must be in equilibrium with the force system $\{q'\}$.

[7]From the definition of a matrix product we have

$$A_{ij}B_{jk} \equiv [A][B] \quad \text{and} \quad A_{ji}B_{jk} = [A]^T[B]$$

where $[A]^T$ is the transport of $[A]$. Furthermore, note that the transpose of $[A][B]$ is equal to $[B]^T[A]^T$.

Hence, we may use the principle of virtual work for the nodal system of the element. Thus,

$$\{\delta a\}^T\{-q\} + \iiint_V \{\delta \mathbf{u}\}^T\{b\}\, dv + \oiint_A \{\delta \mathbf{u}\}^T\{T\}\, dA = 0 \quad (19.45)$$

Note that for the virtual work on the nodes, stemming from $\{b\}$ and $\{T\}$, we have used the virtual work of these loadings *directly on* the element itself. Replacing $\{\delta u\}^T$ by $\{\delta a\}^T[N]^T$, we have

$$\{\delta a\}^T\{q\} = \{\delta a\}^T\iiint_V \{N\}^T\{b\}\, dv + \{\delta a\}^T\oiint_A \{N\}^T\{T\}\, dA \quad (19.46)$$

where we have extracted $\{\delta a\}^T$ from the integrals. Solving for $\{q\}$ in Eq. (19.46), we have

$$\{q\} = \iiint_V \{N\}^T\{b\}\, dv + \oiint_A \{N\}^T\{T\}\, dA \quad (19.47)$$

Substitute from Eqs. (19.43) and (19.47) into Eq. (19.44) to replace $[\delta\varepsilon]^T[\sigma]$ and $\{q\}$, respectively. We get

$$\iiint_V [\{\delta a\}^T[B]^T[D][B]\{a\} - \{\delta a\}^T[B]^T[D][\varepsilon_0] + \{\delta a\}^T[B]^T[\sigma_0]]\, dv$$
$$- \{\delta a\}^T\iiint_V [N]^T\{b\}\, dv - \{\delta a\}^T\oiint_A [N]^T\{T\}\, dA = 0$$

Finally, extract $\{\delta a\}^T$ and $\{a\}$ from the integrals. We get

$$\{\delta a\}^T\left(\iiint_V [B]^T[D][B]\, dv\right)\{a\} - \{\delta a\}^T\iiint_V [B]^T[D][\varepsilon_0]\, dv +$$
$$\{\delta a\}^T\iiint_V [B]^T[\sigma_0]\, dv - \{\delta a\}^T\iiint_V [N]^T\{b\}\, dv \quad (19.48)$$
$$- \{\delta a\}^T\oiint_A [N]^T\{T\}\, dA = 0$$

We denote the first integral as $[K]^e$, and we will soon see that it represents the element stiffness matrix. That is,

$$[K]^e = \iiint_V [B]^T[D][B]\, dv \quad (19.49)$$

Let us denote the effects of initial strain and residual stress as $\{f\}^e$. That is,

$$\{f\}^e = \iiint_V [B]^T[\sigma_0]\, dv - \iiint_V [B]^T[D][\varepsilon_0]\, dv \tag{19.50}$$

With this notation and using Eq. (19.47), Eq. (19.48) becomes

$$[\delta a]^T[K]^e\{a\}^e + \{f\}^e - \{q\}^e = 0 \tag{19.51}$$

Since $\{\delta a\}^T$ is arbitrary, we conclude from Eq. (19.51) that for each element

$$[K]^e\{a\}^e = \{q\}^e - \{f\}^e \tag{19.52}$$

We thus arrive at the equation of **equilibrium** for the finite element methodology. It should be clear now why $[K]^e$ as defined by Eq. (19.49) must be the stiffness matrix for the element.

For the global equation, we assemble the stiffness matrix exactly as was done for the case of trusses. Thus, for 12 nodal displacement components, we form a 12 × 12 grid as shown below for the global stiffness matrix. Now consider an element having nodes 2, 3, and 5. The element stiffness matrix clearly will consist of 36 terms. We enter these 36 terms into the grid as shown. We do this for all the other elements, adding K-values as they superpose. Superposing all the inserted elemental stiffness elements, we can arrive at the global stiffness matrix.

1 u_1	2 v_1	3 u_2	4 v_2	5 u_3	6 v_3	7 u_4	8 v_4	9 u_5	10 v_5	11 u_6	12 v_6	
												U_1 1
												V_1 2
		K_{33}	K_{34}	K_{35}	K_{36}			K_{39}	$K_{3,10}$			U_2 3
		K_{43}	K_{44}	K_{45}	K_{46}			K_{49}	$K_{4,10}$			V_2 4
		K_{53}	K_{54}	K_{55}	K_{56}			K_{59}	$K_{5,10}$			U_3 5
		K_{63}	K_{64}	K_{65}	K_{66}			K_{69}	$K_{6,10}$			V_3 6
												U_4 7
												V_4 8
		K_{93}	K_{94}	K_{95}	K_{96}			K_{99}	$K_{9,10}$			U_5 9
		$K_{10,3}$	$K_{10,4}$	$K_{10,5}$	$K_{10,6}$			$K_{10,9}$	$K_{10,10}$			V_5 10
												U_6 11
												V_6 12

19.10 Closure

In the two-pronged approach of this chapter we hope that we have in Part A eased the reader into a nodding acquaintance with what the finite element method is about. In Part B, with this introduction behind us, we became more bold and looked into certain preliminary but vital generalizations vital for the method. We now summarize for the reader the key results of Part B.

$$\{\boldsymbol{u}\} = [N]\{a\} \tag{19.53a}$$

$$\{\varepsilon\} = [L]\,\{\boldsymbol{u}\} \tag{19.53b}$$

$$[B] = [L][N] \tag{19.53c}$$

$$\{\varepsilon\} = [B]\{a\} \tag{19.53d}$$

$$[\sigma] = [D]([\varepsilon] - [\varepsilon_0]) + [\sigma_0] \tag{19.53e}$$

$$[K] = \iiint_V [B]^T[D][B]\,dv \tag{19.53f}$$

$$[K]\{a\} = \{q\} - \{f\} \tag{19.53g}$$

In conclusion, we remind you that the finite element method is by no means restricted to solid mechanics. In other areas of study, such as heat transfer, fluid mechanics, electromagnetic theory, and many others, one can make good use of this powerful approach. Instead of extremizing the total potential energy functional[8] as we have done here, we must instead use other functionals. For further study in this field, you are advised to study this subject in conjunction with the calculus of variations.

[8]The functional became a function when we introduced the discretization process.

APPENDIX I

Deformation of Isotropic Materials

It will be valuable to prove the following statements for an *isotropic* material:[†]

1. Normal stresses can only generate normal strains.
2. A shear stress, say τ_{xy}, can only generate the corresponding engineering shear strain γ_{xy}

Consider statement 1 first. For this purpose, we have shown an element in Fig. I.1(a) under the normal stress τ_{zz}. We will assume that a shear strain γ_{xy} has resulted from the stress τ_{zz} as shown in the diagram. Now rotate the element 180° about the x axis to reach the configuration in Fig. I.1(b). We now have the same normal stress producing a shear strain that is different in sign. But for an isotropic material the relation between stress and the resulting strain should be independent of the orientation of an element relative to the axes. The only way to avoid the dilemma we find ourselves now in is to preclude the possibility of shear strain arising relative to a reference from normal stresses for that reference.

As for assertion 2, consider the element of Fig. I.1 now undergoing a pure shear stress τ_{xy}, as shown in Fig. I.2(a). We assume here that a normal strain ε_{yy} has resulted from τ_{xy}. By rotating the element 180° about an axis in the xy plane at an angle of 45° from the x axis, we arrive at the configuration shown by Fig. I.2(b). Here we have the same shear stress but a different normal strain. For an isotropic material, there should not be a change in material behavior arising from a change

[†]Although we are concerned in Chapter 4 with linear elastic materials, the geometric arguments we shall present in this appendix apply equally well to nonlinear and inelastic materials.

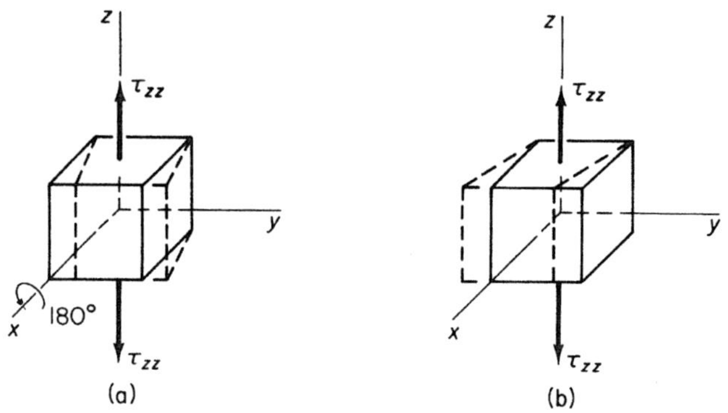

Figure I.1. Can τ_{zz} cause γ_{xy}?

in orientation of the element. To avoid this dilemma in which we find ourselves again, we must conclude that a shear stress on a face parallel to any pair of orthogonal axes cannot create a normal strain for these axes in an isotropic material Finally, to complete the substantiation of assertion 2, we assume that the shear stress τ_{xy} in Fig. I.2(a) causes a shear strain γ_{yz} as shown in Fig. I.3(a). This time we rotate the element 180° about the z axes to arrive at the configuration in Fig. I.3(b). For the same stress τ_{xy} we now have a strain γ_{yz} of opposite sign to that of Fig. I.3(a). This must not be permitted because of isotropy, so we conclude that a shear stress τ_{xy} can only produce the corresponding engineering shear strain γ_{xy}. A similar statement can, of course, be made for shear stresses τ_{yz} and τ_{zx}.

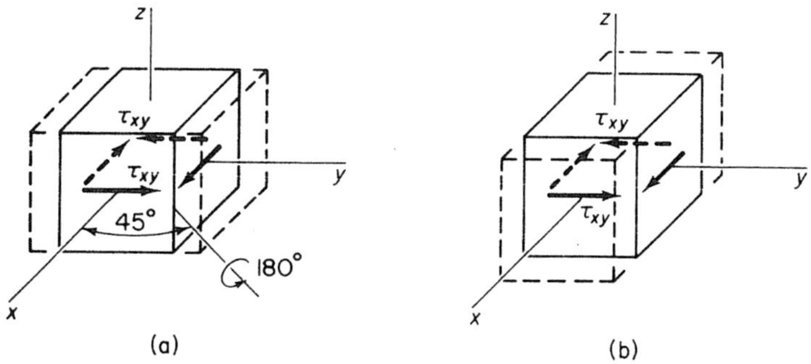

Figure I.2. Can τ_{xy} cause ε_{yy}?

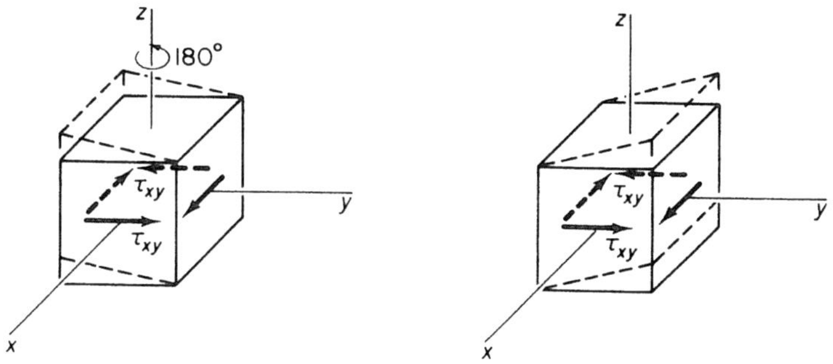

Figure I.3. Can τ_{xy} cause γ_{yz}?

APPENDIX II

Proof Using Tensor Notation That Strain is a Second-Order Tensor

Those readers that have studied the section on tensor notation may prove that strain is a second-order in a quick direct manner. Suppose that there are orthogonal axes rotated relative to xyz at a point. Then the strain-displacement relation for these new axes can be given as follows:

$$\varepsilon_{ns} = \frac{1}{2}\left(\frac{\partial u_n}{\partial s} + \frac{\partial u_s}{\partial n}\right) \tag{a}$$

where n and s are free indices representing the new axes. Thus for $n \equiv x'$ and $s \equiv x'$ we get

$$\varepsilon_{x'x'} = \frac{\partial u_{x'}}{\partial x'}$$

and for $n \equiv x'$ and $s \equiv y'$ we get

$$\varepsilon_{x'y'} = \frac{1}{2}\left(\frac{\partial u_{x'}}{\partial y'} + \frac{\partial u_{y'}}{\partial x'}\right)$$

The other strain displacements are reached by the same process. Next we can express u_n and u_s as follows by vector components [see Eq. (16.7)]:

$$u_n = u_i a_{ni}$$
$$u_s = u_j a_{sj}$$

We then get for $\boldsymbol{\varepsilon}_{ns}$:

$$\varepsilon_{ns} = \frac{1}{2}\left[\frac{\partial(u_i a_{ni})}{\partial s} + \frac{\partial(u_j a_{sj})}{\partial n}\right]$$
$$= \frac{1}{2}\left(a_{ni}\frac{\partial u_i}{\partial s} + a_{sj}\frac{\partial u_j}{\partial n}\right)$$

But $\partial u_i/\partial s$ and $\partial u_j/\partial n$ are directional derivatives and can be expressed as follows [see Eq. (16.9)]:

$$\frac{\partial u_i}{\partial s} = \frac{\partial u_i}{\partial x_j}\frac{dx_j}{ds}$$

$$\frac{\partial u_j}{\partial n} = \frac{\partial u_j}{\partial x_i}\frac{dx_i}{dn}$$

where the components of x_i are x, y, and z. Notice that i is a free index in the first equation with j as a dummy index, while in the second equation the roles are reversed. Hence, we have for $\boldsymbol{\varepsilon}_{ns}$:

$$\varepsilon_{ns} = \frac{1}{2}\left[a_{ni}\left(\frac{dx_j}{ds}\right)\frac{\partial u_i}{\partial x_j} + a_{sj}\left(\frac{dx_i}{dn}\right)\frac{\partial u_j}{\partial x_i}\right]$$

But from the Pythagorean theorem we have

$$\frac{dx_j}{ds} = a_{sj}$$

$$\frac{dx_i}{dn} = a_{ni}$$

Accordingly,

$$\varepsilon_{ns} = \frac{1}{2}\left(a_{ni}a_{sj}\frac{\partial u_i}{\partial x_j} + a_{sj}a_{ni}\frac{\partial u_j}{\partial x_i}\right)$$

$$= a_{ni}a_{sj}\left[\frac{1}{2}\left(\frac{\partial u_i}{\partial x_j} + \frac{\partial u_j}{\partial x_i}\right)\right]$$

Note that the expression in brackets on the right side of the equation is simply $\boldsymbol{\varepsilon}_{ij}$, which you may demonstrate as we did for Eq. (a). Hence we can say that

$$\boldsymbol{\varepsilon}_{ns} = a_{ni}a_{sj}\boldsymbol{\varepsilon}_{ij}$$

We thus have shown that the strain transforms as a second-order tensor and is accordingly a second-order tensor.

APPENDIX III

A Note on the Maxwell-Betti Theorem

We shall develop the Maxwell-Betti reciprocal theorem. For this purpose, consider a *linear*, elastic body constrained in space and acted on by point forces as shown in Fig. III.1. The total number of point forces both external and reactive will be taken as n.

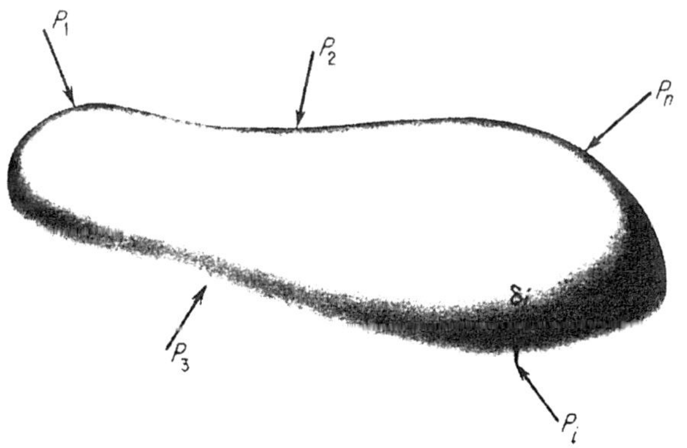

Figure III.1.

Now let $\boldsymbol{\delta}_i$ be the displacement of the point of application of load $\boldsymbol{P}_i$, remembering that this displacement is due to the total action of all the n forces and not due just to $\boldsymbol{P}_i$. Imagine for simplicity that all loads are applied in concert quasi-statically and always in the same proportion to each other as their final values. Thus, at any stage of the loading process, each load $\boldsymbol{P}_i$ can be given as $\lambda(\boldsymbol{P}_i)_f$, where λ is a fraction between zero and unity and $(\boldsymbol{P}_i)_f$ is the final value. Because of the linear, elastic behavior of the material, the deflections $\boldsymbol{\delta}_i$ at the aforementioned stage of loading are the same fraction λ of the final values $(\boldsymbol{\delta}_i)_f$.† Thus, we can say that $(\boldsymbol{\delta}_i) = \lambda(\boldsymbol{\delta}_i)_f$. Now we compute the work, W_k, done by the forces for this case‡:

$$W_k = \sum_{i=1}^{n} \int_0^{(\delta_i)_f} \boldsymbol{P}_i \cdot d\boldsymbol{\delta}_i = \sum_{i=1}^{n} \int_0^{(\delta_i)_f} (\lambda)(\boldsymbol{P}_i)_f \cdot d[\lambda(\boldsymbol{\delta}_i)_f]$$

$$= \int_0^1 \sum_{i=1}^{n} (\boldsymbol{P}_i)_f \cdot (\boldsymbol{\delta}_i)_f \,\lambda\, d\lambda$$

We can take $\sum_{i=1}^{n} (\boldsymbol{P}_i)_f \cdot (\boldsymbol{\delta}_i)_f$ as constant for the integration and so for W_k we have the desired result.

$$W_k = \sum_{i=1}^{n} (\boldsymbol{P}_i)_f \cdot (\boldsymbol{\delta}_i)_f \int_0^1 \lambda\, d\lambda = \sum_{i=1}^{n} \frac{(\boldsymbol{P}_i)_f \cdot (\boldsymbol{\delta}_i)_f}{2}$$

† We are tacitly assuming here that the superposition principle is satisfied.
‡ The order of loading makes no difference here in the computation of W_k. Our simple loading schedule permits us to arrive at Eq. (6.43) most easily.

APPENDIX IV

Tables

A. Wide-flange beams
B. Standard Channels
C. Standard Angles
D. Standard Pipes
E. Properties of Areas
F. Mechanical Properties of Materials

ACKNOWLEDGMENT. Data for Tables A–D are taken from *AISC Manual of Steel Construction* and are reproduced by permission of the American Institute of Steel Construction, Inc.

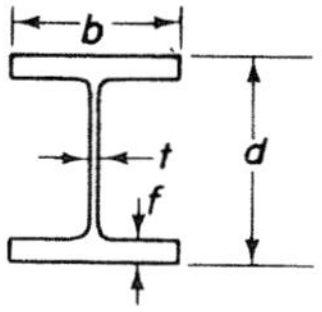

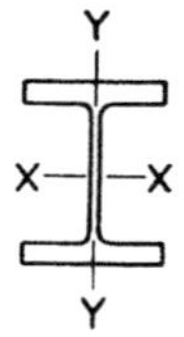

APPENDIX IV–A WIDE-FLANGE BEAMS

	Weight Per Foot (lb)	Cross-sectional Area (in.^{2})	d (in.)	b (in.)	f (in.)	t (in.)	X-X Axis			Y-Y Axis		
							I (in.^{4})	S (in.^{3})	r (in.)	I (in.^{4})	S (in.^{3})	r (in.)
36WF300	300	88.17	36.72	16.655	1.680	.945	20,290.2	1105.1	15.17	1225.2	147.1	3.73
36WF260	260	76.56	36.24	16.555	1.440	.845	17,233.8	951.1	15.00	1020.6	123.3	3.65
36WF230	230	67.73	35.88	16.475	1.260	.765	14,988.4	835.5	14.88	870.0	105.7	3.59
36WF170	170	49.98	36.16	12.027	1.100	.680	10,470.0	579.1	14.47	300.6	50.0	2.45
36WF150	150	44.16	35.84	11.972	.940	.625	9,012.1	502.9	14.29	250.4	41.8	2.38
33WF240	240	70.52	33.50	15.865	1.400	.830	13,585.1	811.1	13.88	874.3	110.2	3.52
33WF200	200	58.79	33.00	15.750	1.150	.715	11,048.2	669.6	13.71	691.7	87.8	3.43
33WF152	152	44.71	33.50	11.565	1.055	.635	8,147.6	486.4	13.50	256.1	44.3	2.39
33WF130	130	38.26	33.10	11.510	.855	.580	6,699.0	404.8	13.23	201.4	35.0	2.29
30WF210	210	61.78	30.38	15.105	1.315	.775	9,872.4	649.9	12.64	707.9	93.7	3.38
30WF190	190	55.90	30.12	15.040	1.185	.710	8,825.9	586.1	12.57	624.6	83.1	3.34
30WF132	132	38.83	30.30	10.551	1.000	.615	5,753.1	379.7	12.17	185.0	35.1	2.18
30WF108	108	31.77	29.82	10.484	.760	.548	4,461.0	299.2	11.85	135.1	25.8	2.06
27WF160	160	47.04	27.08	14.023	1.075	.658	6,018.6	444.5	11.31	458.0	65.3	3.12
27WF102	102	30.01	27.07	10.018	.827	.518	3,604.1	266.3	10.96	129.5	25.9	2.08
27WF94	94	27.65	26.91	9.990	.747	.490	3,266.7	242.8	10.87	115.1	23.0	2.04
24WF160	160	47.04	24.72	14.091	1.135	.656	5,110.3	413.5	10.42	492.6	69.9	3.23
24WF145	145	42.62	24.49	14.043	1.020	.608	4,561.0	372.5	10.34	434.3	61.8	3.19
24WF110	110	32.36	24.16	12.042	.855	.510	3,315.0	274.4	10.12	229.1	38.0	2.66
24WF76	76	22.37	23.91	8.985	.682	.440	2,096.4	175.4	9.68	76.5	17.0	1.85
21WF142	142	41.76	21.46	13.132	1.095	.659	3,403.1	317.2	9.03	385.9	58.8	3.04
21WF82	82	24.10	20.86	8.962	.795	.499	1,752.4	168.0	8.53	89.6	20.0	1.93
21WF62	62	18.23	20.99	8.240	.615	.400	1,326.8	126.4	8.53	53.1	12.9	1.71

APPENDIX IV–A (cont.)

	Weight Per Foot (lb)	Cross-sectional Area (in.^{2})	d (in.)	b (in.)	f (in.)	t (in.)	X-X Axis			Y-Y Axis		
							I (in.^{4})	S (in.^{3})	r (in.)	I (in.^{4})	S (in.^{3})	r (in.)
18WF114	114	33.51	18.48	11.833	.991	.595	2,033.8	220.1	7.79	255.6	43.2	2.76
18WF96	96	28.22	18.16	11.750	.831	.512	1,674.7	184.4	7.70	206.8	35.2	2.71
18WF85	85	24.97	18.32	8.838	.911	.526	1,429.9	156.1	7.57	99.4	22.5	2.00
18WF70	70	20.56	18.00	8.750	.751	.438	1,153.9	128.2	7.49	78.5	17.9	1.95
18WF64	64	18.80	17.87	8.715	.686	.403	1,045.8	117.0	7.45	70.3	16.1	1.93
18WF60	60	17.64	18.25	7.558	.695	.416	984.0	107.8	7.43	47.1	12.5	1.63
18WF55	55	16.19	18.12	7.532	.630	.390	889.9	98.2	7.41	42.0	11.1	1.61
18WF50	50	14.71	18.00	7.500	.570	.358	800.6	89.0	7.38	37.2	9.9	1.59
16WF96	96	28.22	16.32	11.533	.875	.535	1,355.1	166.1	6.93	207.2	35.9	2.71
16WF78	78	22.92	16.32	8.586	.875	.529	1,042.6	127.8	6.74	87.5	20.4	1.95
16WF71	71	20.86	16.16	8.543	.795	.486	936.9	115.9	6.70	77.9	18.2	1.93
16WF64	64	18.80	16.00	8.500	.715	.443	833.8	104.2	6.66	68.4	16.1	1.91
15WF58	58	17.04	15.86	8.464	.645	.407	746.4	94.1	6.62	60.5	14.3	1.88
16WF50	50	14.70	16.25	7.073	.628	.380	655.4	80.7	6.68	34.8	9.8	1.54
16WF45	45	13.24	16.12	7.039	.563	.346	583.3	72.4	6.64	30.5	8.7	1.52
16WF40	40	11.77	16.00	7.000	.503	.307	515.5	64.4	6.62	26.5	7.6	1.50
14WF426	426	125.25	18.69	16.695	3.033	1.875	6,610.3	707.4	7.26	2359.5	282.7	4.34
14WF370	370	108.78	17.94	16.458	2.658	1.655	5,454.2	608.1	7.08	1986.0	241.1	4.27
14WF246	246	72.33	16.25	15.945	1.813	1.125	3,228.9	397.4	6.68	1226.6	153.9	4.12
14WF184	184	54.07	15.38	15.660	1.378	.840	2,274.8	295.8	6.49	882.7	112.7	4.04
14WF142	142	41.85	14.75	15.500	1.063	.680	1,672.2	226.7	6.32	660.1	85.2	3.97
14WF127	127	37.33	14.62	14.690	.998	.610	1,476.6	202.0	6.29	527.6	71.8	3.76
14WF68	68	20.00	14.06	10.040	.718	.418	724.1	103.0	6.02	121.2	24.1	2.46
14WF48	48	14.11	13.81	8.031	.593	.339	484.9	70.2	5.86	51.3	12.8	1.91
14WF38	38	11.17	14.12	6.776	.513	.313	385.3	54.6	5.87	24.6	7.3	1.49

12WF190	190	55.86	14.38	12.670	1.736	1.060	1,892.5	263.2	5.82	589.7	93.1	3.25
12WF161	161	47.38	13.88	12.515	1.486	.905	1,541.8	222.2	5.70	486.2	77.7	3.20
12WF120	120	35.31	13.12	12.320	1.106	.710	1,071.7	163.4	5.51	345.1	56.0	3.13
12WF99	99	29.09	12.75	12.190	.921	.580	858.5	134.7	5.43	278.2	45.7	3.09
12WF92	92	27.06	12.62	12.155	.856	.545	788.9	125.0	5.40	256.4	42.2	3.08
12WF50	50	14.71	12.19	8.077	.641	.371	394.5	64.7	5.18	56.4	14.0	1.96
12WF45	45	13.24	12.06	8.042	.576	.336	350.8	58.2	5.15	50.0	12.4	1.94
12WF36	36	10.59	12.24	6.565	.540	.305	280.8	45.9	5.15	23.7	7.2	1.50
12WF31	31	9.12	12.09	6.525	.465	.265	238.4	39.4	5.11	19.8	6.1	1.47
10WF112	112	39.92	11.38	10.415	1.248	.755	718.7	126.3	4.67	235.4	45.2	2.67
10WF100	100	29.43	11.12	10.345	1.118	.685	625.0	112.4	4.61	206.6	39.9	2.65
10WF66	66	19.41	10.38	10.117	.748	.457	382.5	73.7	4.44	129.2	25.5	2.58
10WF45	45	13.24	10.12	8.022	.618	.350	248.6	49.1	4.33	53.2	13.3	2.00
10WF25	25	7.35	10.08	5.762	.430	.252	133.2	26.4	4.26	12.7	4.4	1.31
8WF67	67	19.70	9.00	8.287	.933	.575	271.8	60.4	3.71	88.6	21.4	2.12
8WF40	40	11.76	8.25	8.088	.558	.365	146.3	35.5	3.53	49.0	12.1	2.04
8WF28	28	8.23	8.06	6.540	.463	.285	97.8	24.3	3.45	21.6	6.6	1.62
8WF20	20	5.88	8.14	5.268	.378	.248	69.2	17.0	3.43	8.50	3.2	1.20

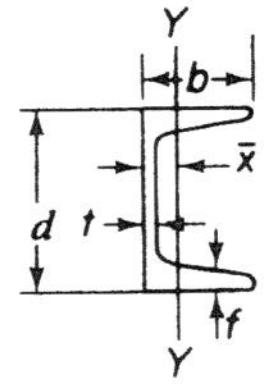

APPENDIX IV–B STANDARD CHANNELS

Depth, d (in.)	Weight Per Foot (lb)	Cross-sectional Area (in.^{2})	b (in.)	f (average) (in.)	t (in.)	Axis X-X I (in.^{4})	Axis X-X S (in.^{3})	Axis X-X r (in.)	Axis Y-Y I (in.^{4})	Axis Y-Y S (in.^{3})	Axis Y-Y r (in.)	$\bar{x}$ (in.)
18.00	58.0	16.98	4.200	.625	.700	670.7	74.5	6.29	18.5	5.6	1.04	.88
	45.8	13.38	4.000	.625	.500	573.5	63.7	6.55	15.8	5.1	1.09	.89
	42.7	12.48	3.950	.625	.450	549.2	61.0	6.64	15.0	4.9	1.10	.90
15.00	50.0	14.64	3.716	.650	.716	401.4	53.6	5.24	11.2	3.8	.87	.80
	40.0	11.70	3.520	.650	.520	346.3	46.2	5.44	9.3	3.4	.89	.78
	33.9	9.90	3.400	.650	.400	312.6	41.7	5.62	8.2	3.2	.91	.79
12.00	30.0	8.79	3.170	.501	.510	161.2	26.9	4.28	5.2	2.1	.77	.68
	25.0	7.32	3.047	.501	.387	143.5	23.9	4.43	4.5	1.9	.79	.68
10.00	30.0	8.80	3.033	.436	.673	103.0	20.6	3.42	4.0	1.7	.67	.65
	25.0	7.33	2.886	.436	.526	90.7	18.1	3.52	3.4	1.5	.68	.62
	20.0	5.86	2.739	.426	.379	78.5	15.7	3.66	2.8	1.3	.70	.61
9.00	20.0	5.86	2.648	.413	.448	60.6	13.5	3.22	2.4	1.2	.65	.59
	15.0	4.39	2.485	.413	.285	50.7	11.3	3.40	1.9	1.0	.67	.59
	13.4	3.89	2.430	.413	.230	47.3	10.5	3.49	1.8	.97	.67	.61
8.00	18.75	5.49	2.527	.390	.487	43.7	10.9	2.82	2.0	1.0	.60	.57
	13.75	4.02	2.343	.390	.303	35.8	9.0	2.99	1.5	.86	.62	.56
	11.5	3.36	2.260	.390	.220	32.3	8.1	3.10	1.3	.79	.63	.58
7.00	14.75	4.32	2.299	.366	.419	27.1	7.7	2.51	1.4	.79	.57	.53
	12.25	3.58	2.194	.366	.314	24.1	6.9	2.59	1.2	.71	.58	.53
6.00	13.0	3.81	2.157	.343	.437	17.3	5.8	2.13	1.1	.65	.53	.50
	10.5	3.07	2.034	.343	.314	15.1	5.0	2.22	.87	.57	.53	.50
5.00	9.0	2.63	1.885	.320	.325	8.8	3.5	1.83	.64	.45	.49	.49
	6.7	1.95	1.750	.320	.190	7.4	3.0	1.95	.48	.38	.50	.49
4.00	7.25	2.12	1.720	.296	.320	4.5	2.3	1.47	.44	.35	.46	.46
	5.4	1.56	1.580	.296	.180	3.8	1.9	1.56	.32	.29	.45	.46

APPENDIX IV–C STANDARD ANGLES

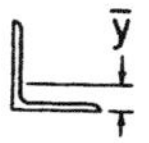

Size (in.)	Thickness (in.)	Weight Per Foot (lb)	Cross-sectional Area (in.²)	I (in.⁴)	S (in.³)	r (in.)	$\bar{y}$ (in.)
8 × 8	$1\frac{1}{8}$	56.9	16.73	98.0	17.5	2.42	2.41
	1	51.0	15.00	89.0	15.8	2.44	2.37
	$\frac{3}{4}$	38.9	11.44	69.7	12.2	2.47	2.28
	$\frac{1}{2}$	26.4	7.75	48.6	8.4	2.50	2.19
6 × 6	1	37.4	11.00	35.5	8.6	1.80	1.86
	$\frac{1}{2}$	19.6	5.75	19.9	4.6	1.86	1.68
5 × 5	$\frac{3}{4}$	23.6	6.94	15.7	4.5	1.51	1.52
	$\frac{1}{2}$	16.2	4.75	11.3	3.2	1.54	1.43
	$\frac{3}{8}$	12.3	3.61	8.7	2.4	1.56	1.39
4 × 4	$\frac{3}{4}$	18.5	5.44	7.7	2.8	1.19	1.27
	$\frac{1}{2}$	12.8	3.75	5.6	2.0	1.22	1.18
	$\frac{3}{8}$	9.8	2.86	4.4	1.5	1.23	1.14
	$\frac{1}{4}$	6.6	1.94	3.0	1.0	1.25	1.09
$3\frac{1}{2} \times 3\frac{1}{2}$	$\frac{1}{2}$	11.1	3.25	3.6	1.5	1.05	1.06
	$\frac{3}{8}$	8.5	2.48	2.9	1.2	1.07	1.01
	$\frac{1}{4}$	5.8	1.69	2.0	.79	1.09	.97
3 × 3	$\frac{1}{2}$	9.4	2.75	2.2	1.1	.90	.93
	$\frac{3}{8}$	7.2	2.11	1.8	.83	.91	.89
	$\frac{1}{4}$	4.9	1.44	1.2	.58	.93	.84
$2\frac{1}{2} \times 2\frac{1}{2}$	$\frac{1}{2}$	7.7	2.25	1.2	.73	.74	.81
	$\frac{1}{4}$	4.1	1.19	.70	.39	.77	.72
2 × 2	$\frac{3}{8}$	4.7	1.36	.48	.35	.59	.64
	$\frac{1}{4}$	3.19	.94	.35	.25	.61	.59
	$\frac{1}{8}$	1.65	.48	.19	.13	.63	.55
$1\frac{1}{2} \times 1\frac{1}{2}$	$\frac{1}{4}$	2.34	.69	.14	.13	.45	.47
	$\frac{1}{8}$	1.23	.36	.078	.072	.46	.42
1 × 1	$\frac{1}{4}$	1.49	.44	.037	.056	.29	.34
	$\frac{3}{16}$	1.16	.34	.030	.044	.30	.32
	$\frac{1}{8}$	.80	.23	.022	.031	.31	.30

APPENDIX IV–D STANDARD PIPES

Nominal Diameter (in.)	Weight Per Foot (lb)	Outside Diameter (in.)	Inside Diameter (in.)	Thickness (in.)	I (in.^{4})	A (in.^{2})	k (in.)
.50	.85	.840	.622	.109	.017	.250	.26
.75	1.13	1.050	.824	.113	.037	.333	.33
1.00	1.68	1.315	1.049	.133	.087	.494	.42
1.50	2.72	1.900	1.610	.145	.310	.799	.62
2.00	3.65	2.375	2.067	.154	.666	1.075	.79
2.50	5.79	2.875	2.469	.203	1.530	1.704	.95
3.00	7.58	3.500	3.068	.216	3.017	2.228	1.16
3.50	9.11	4.000	3.548	.226	4.788	2.680	1.34
4.00	10.79	4.500	4.026	.237	7.233	3.174	1.51
5.00	14.62	5.563	5.047	.258	15.16	4.300	1.88
6.00	18.97	6.625	6.065	.280	28.14	5.581	2.25
8.00	28.55	8.625	7.981	.322	72.49	8.399	2.94
10.00	40.48	10.750	10.020	.365	160.7	11.91	3.67
12.00	49.56	12.750	12.000	.375	279.3	14.58	4.38

Appendix IV -E Properties of Areas

Rectangle

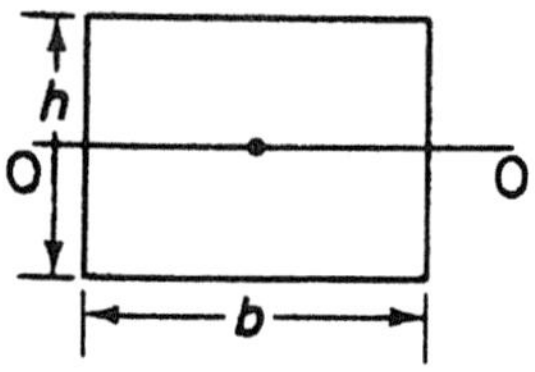

$A = bh$

$I_{OO} = \frac{1}{12} bh^3$

Triangle

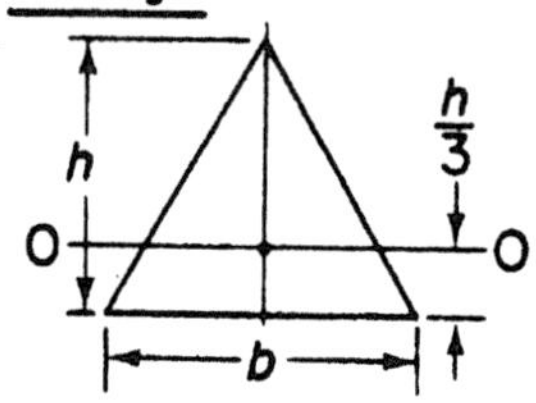

$A = bh/2$

$I_{OO} = bh^3/36$

Circle

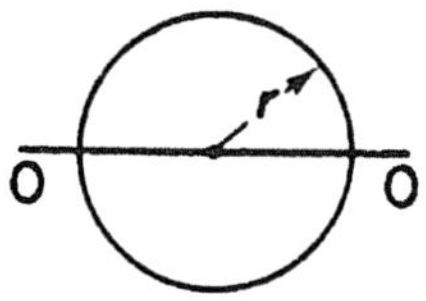

$A = \pi r^2$

$I_{OO} = \pi r^4/4$

$J = \pi r^4/2$

Thin Tube

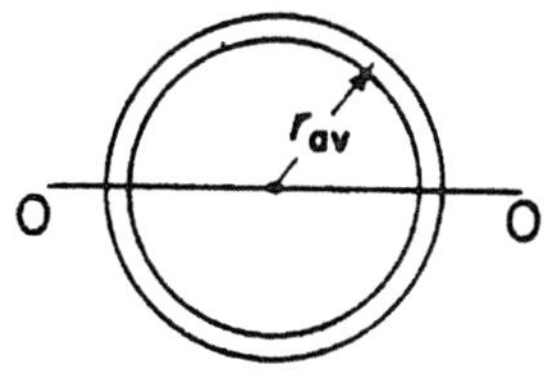

$A = 2\pi r_{av} t$

$I_{OO} \approx \pi r_{av}^3 t$

$J \approx 2\pi r_{av}^3 t$

Semicircle

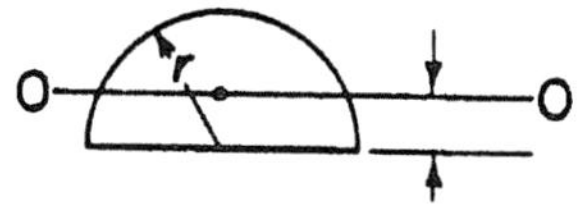

$\frac{4r}{3\pi} = .424r$

$A = \frac{\pi r^2}{2}$

$I_{OO} = .110 r^4$

Quadrant of a Circle

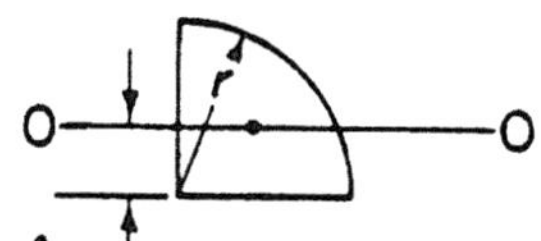

$\frac{4r}{3\pi}$

$A = \frac{\pi r^2}{4}$

$I_{OO} = \frac{1}{16} \pi r^4$

Right Triangle

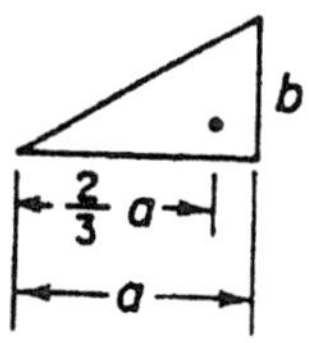

$A = \frac{1}{2} ab$

Ellipse

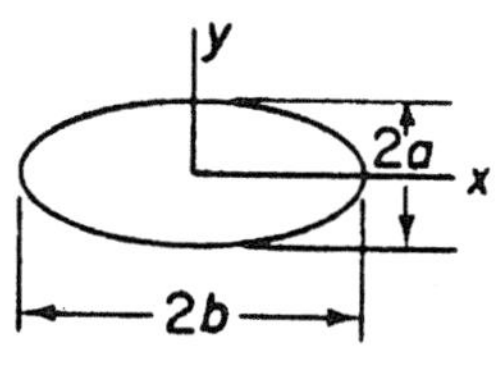

$A = ab\pi$

Appendix IV–F
Mechanical Properties of Materials

The following tables list the mechanical properties of selected engineering materials, Average values or a range of values are reported for materials at room temperature, loaded in tension. These values are provided for educational purposes only. For professional practice, consult the appropriate material manufacturer for information about a specific material. The following references were consulted during the preparation of these tables.

E. A. Avallone and T. Baumeister, *Marks' Standard Handbook for Mechanical Engineers*, 10th ed., McGraw-Hill, NY, NY, 1996.

M. Fogiel, *Handbook of Mathematical, Scientific, and Engineering Formulas*, Research and Education Assoc., Piscataway, NJ, 1984.

R. J. Roark and W.C. Young, *Formulas for Stress and Strain*, 5th ed, McGraw-Hill, NY, NY, 1975.

N.E. Dowling, *Mechanical Behavior of Materials*, Prentice-Hall, NJ, 1993.

Mechanical Properties of Common Materials at Room Temperature – (USCS)

Material	Density [slugs/ft^3]	Modulus of Elasticity [Mpsi]	Poisson's Ratio	Yield Stress [ksi]	Ultimate Stress[1] [ksi]	Thermal Expansion Coef. [10^{-6}/°F]
Structural Steel						
A36	15.2	29	0.28	36	58	6.7
A572 (50)	15.2	29	0.28	50	65	6.7
Spring Steel SAE 4068	15	30	0.27	175-240	200-270	8.3
Stainless Steel						
S40500	15	27.6	0.31	25	60	9.6
S17700	15	27.6	0.31	150	185	9.6
Aluminum Alloys						
2024-T3	5.3	10	0.33	50	70	12.6
6061-T6	5.3	10	0.33	40	45	13.5
7075-T6	5.3	10	0.33	73	83	13.7
Copper						
Soft (pure)	17.3	16	0.34	10	32	9.8
Hard (Be-Cu)	16.3	18	0.34	170	190	9.9
Magnesium Alloy AZ80A-T5	3.5	6.5	0.35	38	55	16
Cast Iron, Gray, No. 40	14	18.3	0.25	-	40 (130)	6.0
Wrought Iron	15	28	0.30	23-32	34-54	6.5
Glass	5.1	10	0.25	-	(10-125)	3-9
Titanium Alloy (5 Al, 2.5 Sn)	8.6	17	0.34	110	115	5.7
Nickel Steel	15	29	0.29	20-30	45-60	7.2
Wood (tension parallel to grain)						
Douglas Fir (coast)	0.9	1.3-1.9	-	-	0.2-1.0	-
Southern Pine	0.9	1.3-1.9	-	-	0.3-1.7	-
Hemlock	0.9	1.1-1.5	-	-	0.2-0.7	-
ABS/Nylon	2.1	0.29	0.40	-	5.3	56
PE-TFE	3.6	1.2	0.40	-	12	12
Tungsten	37.5	50	0.28	-	18-600	2.5
Granite	5.4	4-16	0.05-0.2	-	(13-55)	4.1
Sandstone	4.5	0.7-10	0.1-0.3	-	(5-20)	5.3
Brass, Hard	16	15	0.34	75	85	20

[1]Compression noted in parentheses

Mechanical Properties of Common Materials at Room Temperature – (SI)

Material	Density [kg/m³]	Modulus of Elasticity [GPa]	Poisson's Ratio	Yield Stress [MPa]	Ultimate Stress [MPa][2]	Thermal Expansion Coef. [10^{-6}/°C]
Structural Steel						
A36	7830	200	0.28	248	400	12.1
A572 (50)	7830	200	0.28	345	448	12.1
Spring Steel SAE 4068	7730	207	0.27	1210-1660	1380-1860	14.9
Stainless Steel						
S40500	7730	191	0.31	172	414	17.3
S17700	7730	191	0.31	1030	1280	17.3
Aluminum Alloys						
2024-T3	2730	69	0.33	345	483	22.7
6061-T6	2730	69	0.33	276	310	24.3
7075-T6	2730	69	0.33	503	572	24.7
Copper						
Soft (pure)	8920	110	0.34	69	220	17.6
Hard (Be-Cu)	8400	124	0.34	1170	1310	17.8
Magnesium Alloy AZ80A-T5	1800	45	0.35	262	379	28.8
Cast Iron, Gray, No. 40	7220	126	0.25	-	276 (896)	10.8
Wrought Iron	7730	193	0.30	159-220	234-372	11.7
Glass	2630	69	0.25	-	(69-862)	5.4-16.2
Titanium Alloy (5 Al, 2.5 Sn)	4430	117	0.34	758	793	10.3
Nickel Steel	7730	200	0.29	138-207	310-414	13.0
Wood (tension parallel to grain)						
Douglas Fir (coast)	464	9-13	-	-	1.4-7	-
Southern Pine	464	9-13	-	-	2-12	-
Hemlock	464	7.6-10	-	-	1.4-4.8	-
ABS/Nylon	1080	2.0	0.40	-	37	101
PE-TFE	1860	8.3	0.40	-	83	21.6
Tungsten	19300	345	0.28	-	124-4140	4.5
Granite	2780	27-110	0.05-0.2	-	(90-380)	7.4
Sandstone	2340	4.8-69	0.1-0.3	-	(34-138)	9.5
Brass, Hard	8250	103	0.34	517	586	36

[2]Compression noted in parentheses

ANSWERS

2.1. $P = 7069$ lb
2.2. $F = -8552$ N
2.3. $\tau = 1500$ psi
2.4. $\tau = 79.6$ MPa
2.5. $\tau = 36.3$ psi
2.6. 5.965×10^6 Pa
12.19×10^6 Pa
2.7. 1.479×10^5 Pa
2.8. 2.5 MPa
2.9. —
2.10. $\tau_{\text{Shear}} = 9.417 \times 10^5$ Pa
$\tau_{\text{Tens}} = 9.417 \times 10^6$ Pa
2.11. 924 psi
442 psi
2.12. $\tau_n = 3.03 \times 10^5$ Pa
$\tau_t = 1.75 \times 10^5$ Pa
2.13. $\tau_n = 6.578 \times 10^4$ Pa
$\tau_t = 10{,}018$ Pa
2.14. 1.334×10^6 Pa
2.669×10^6 Pa
2.15. $\tau_t = \dfrac{106.1}{r^2}$ Pa
2.16. 2.52×10^6 Pa
2.17. $\tau_t = 1.086 \times 10^8$ Pa
$\tau_n = 2.844 \times 10^7$ Pa
2.18. $d = 19.5$ mm
2.19. $\tau_t = 24{,}056$ Pa
$\tau_n = 41{,}666$ Pa
2.20. $\tau_n = 955$ Pa
$\tau_t = 551.4$ Pa
2.21. $\tau_{\max} = 8.842 \times 10^6$ Pa
$\tau_{\min} = 8.843 \times 10^4$ Pa
2.22. 1.201×10^8 Pa
2.23. $T = \dfrac{1}{2}\mu\pi d^2 Lp$
2.24. $\tau_{yy} = .0233(200y - 9779)$ psi
2.25. —
2.26. —
2.27. $\boldsymbol{F}_R = 6{,}280\hat{\mathbf{i}} - 8{,}380\hat{\mathbf{j}} + 28{,}820\hat{\mathbf{k}}$ N
2.28. —
2.29. —
2.30. $\boldsymbol{B} = -(12x + 3y)\hat{\mathbf{i}} - 10y\hat{\mathbf{k}}$ N/m^3
$\boldsymbol{B} = -42\hat{\mathbf{i}} - 20\hat{\mathbf{k}}$ N/m^3
2.31. —
2.32. —
2.33. 8.149×10^6 Pa
2.34. 1126 psi
2.35. 3,291 psi
2.36. dia = 0.4615 in.
2.37. 72,570 psi
2.38. 3/8″ Bolt is 2.24 times as strong as 1/4″ Bolt
2.39. 9424 N-m
2.40. 2.13 psi
1.598 psi
2.41. $\tau_n = \gamma(L - z)$
2.42. $(\tau_n)_{av.} = \dfrac{1}{3}(h - z)\gamma$
2.43. 497 psi
2.44. —
2.45. 5.57×10^5 Pa
2.46. $\tau_{zz} = 4.716[30^3 - z^3]$ psf
2.47. $\boldsymbol{F} = 2000\hat{\mathbf{i}} + 2167\hat{\mathbf{j}} + 917\hat{\mathbf{k}}$ kN
2.48. $\tau = 1.414\mu$
Drag/unit area = 2μ
2.49. $T = \dfrac{2\pi\mu\omega L r_0^3}{\epsilon}$
2.50. 13.59 g's
2.51. 17.08%
75%
2.52. 5090 psi
1143 psi
2.53. 1.970×10^9 Pa
2.54. 1.100×10^7 Pa
2.55. —
2.56. $P = 125$ lb
2.57. $T = \dfrac{2}{3}\mu F \dfrac{r_o^3 - r_i^3}{r_o^2 - r_i^2}$
2.58. $\alpha = 54.7°$
2.59. —

2.60. —
2.61. —
2.62. —
2.63. —
2.64. —
2.65. —

3.1. $13\hat{\mathbf{i}} + 2\hat{\mathbf{j}} + 9\hat{\mathbf{k}}$ m
3.2. 6.63 m
3.3. $\Delta d = 15.54$ m
3.4. Change in angle is 30°
3.5. $\Delta\theta = -18.4°$
3.6. $\varepsilon_{xx} = -.006$
$\varepsilon_{yy} = \varepsilon_{zz} = .003014$
3.7. $-.00260$
0.00417
3.8. $\Delta_{DIA} = \dfrac{2pr^2}{Et}$
3.9. $-.0987$
3.10. $\varepsilon_{xx} = \Delta/L$
$\varepsilon_{yy} = \delta/D$
$\varepsilon_{xy} = -\dfrac{D\theta\pi}{720L}$
3.11. $\varepsilon_{xx} = 9 \times 10^{-2}$
$\varepsilon_{xy} = 6 \times 10^{-2}$
3.12. 16×10^{-2}
24.68 m^3
3.13. —
3.14. $\varepsilon_{zz} = \dfrac{\gamma}{E}(L - z) + \dfrac{F}{E\left(\dfrac{\pi D^2}{4}\right)}$

$\Delta = \gamma L^2/2E + \dfrac{FL}{E\left(\dfrac{\pi D^2}{4}\right)}$
3.15. 5.79×10^{-7} ft
3.16. 1.365×10^{-6} m
3.17. 4.946×10^{-6} m
3.18. .004244 mm
3.19. $\varepsilon_{xx} = .75$
etc
3.20. $-.001454$
$-.001454$

3.21. $\dfrac{5}{6}$ sec
3.22. .510/sec
3.23. $\Delta = \dfrac{L^2}{4E}(\gamma_o + \gamma_A) + \dfrac{FL}{E\pi D^2/4} = \dfrac{\gamma L^2}{2E} + \dfrac{FL}{\dfrac{E\pi D^2}{4}}$
3.24. $\varepsilon_{yy} = -.01$
$\varepsilon_{yz} = .005$
etc
3.25. 2×10^{-2}
3×10^{-2}
3.26. $-.05\hat{\mathbf{i}} - .13\hat{\mathbf{j}} + .293\hat{\mathbf{k}}$
$\varepsilon_{xx} = .2$
$\varepsilon_{yz} = -.110$
etc
3.27. —
3.28. $\Delta = \dfrac{\gamma L^2}{2E} + \dfrac{FL}{\dfrac{E\pi D^2}{4}}$
3.29. $\Delta = \dfrac{L^2}{4E}(\gamma_o + \gamma_A)$
3.30. —
3.31. —
3.32. 3.77×10^{-6} in.
3.33. $\Delta = 1.678 \times 10^{-4}$ ft
3.34. $\Delta = 7.37 \times 10^{-6}$ ft
3.35. $\omega = 572$ rad/s
3.36. —
3.37. $\Delta ED = 2.5 \times 10^{-8}$ m
$\Delta BD = -2.5 \times 10^{-8}$ m
$\Delta DF = -5 \times 10^{-8}$ m
3.38. 7.91 m
3.39. 10.725 m
3.40. .01667
−.000873
3.41. 120 mm
3.42. .04324
−.026573
.02579
3.43. .04983
3.44. —
3.45. $\dfrac{P}{2\pi Gh}\ln\left(\dfrac{r_o}{r_i}\right)$

4.1. $E = 10.2 \times 10^6$ psi
$\nu = .2$

4.2. —

4.3. $K = 9 \times 10^4$ N/m
$E = 5.501 \times 10^8$ Pa

4.4. $\tau_{xx} = 17{,}080$ psi
Dia = 0.01931 in.
$K_{Av.} = 146.2$ lb/in.

4.5. 442 lb

4.6. $E = \dfrac{2\tau_0 r}{a - 2r}$

$\nu = -\dfrac{b - 2r}{a - 2r}$

4.7. $E = \dfrac{\tau_0}{\dfrac{a}{\sqrt{2}d} - 1}$

$\nu = -\dfrac{b - \sqrt{2}d}{a - \sqrt{2}d}$

4.8. —

4.9. For $\varepsilon = .001$
a) 4×10^8 Pa
b) $\tau = 2 \times 10^8$ Pa
c) $\tau = 2 \times 10^8$ Pa
etc

4.10. $\varepsilon_p = .001860$
etc

4.11. $F_{max} = .707 \times 10^6$ lb
$\Delta_{max} = .550$ in.
$F = .707 \times 10^6$ lb

4.12. $\Delta = .0283$ m

4.13. 7.275×10^{-4} in.

4.14. $P_{max} = 3.016 \times 10^6$ N
$\Delta = -.01804$ m

4.15. $\omega_{max} = 64.6$ rad/sec
$\Delta = .003555$ m

4.16. $\tau_{max} = 4.695 \times 10^8$ Pa
1.335 m
$\varepsilon_{max} = .006344$

4.17. 58 million cycles
$S = 29{,}000$ psi

4.18. 35×10^6 cycles
61.09 hours

4.19. 48×10^6 cycles
83.8 hours

4.20. $\tau_{max} = 3.20 \times 10^6$ Pa

4.21. $(\varepsilon_{zz})_A = 2.60 \times 10^{-4}$
$(\varepsilon_{zz})_B = 4.0 \times 10^{-4}$
$L_{new} = 22.00712'$
etc
$\Delta_A = 3.69 \times 10^{-3}$ ft
$\Delta_B = 6.12 \times 10^{-3}$ ft

4.22. 1.711×10^{-4}
$\Delta = 7.00 \times 10^{-4}$ m

4.23. $F = 22{,}619$ lb

4.24. $\tau = -1.988 \times 10^8$ Pa
$F = 9350$ N

5.1. $F = 5.154 \times 10^4$ N

5.2. $\Delta_D = 13.91 \times 10^{-5}$ m
$\Delta_L = 4.093 \times 10^{-5}$ m

5.3. Permanent set = .0365 in.

5.4. $\Delta = 7.617 \times 10^{-9}$ m

5.5. $\Delta = 5.046 \times 10^{-9}$ m

5.6. —

5.7. $\Delta_1 = 6.37 \times 10^{-4}$ in.
$\Delta_2 = 5.66 \times 10^{-4}$ in.

5.8. $E = 165$ MPa

5.9. $\Delta = 3.25$ mm

5.10. $E' = \dfrac{E_m A_m + E_f A_f}{A}$

5.11. $p = 2050$ psi

5.12. $L_{new} = 60\left(1 + \varepsilon_o \sinh \dfrac{318}{\tau_o}\right)$

5.13. $\Delta = 1.069 \times 10^{-5}$ in.

5.14. $\Delta = 7.607 \times 10^{-6}$ in.

5.15. 1.403 g's

5.16. $\Delta = 3.36 \times 10^{-4}$ ft

5.17. $\Delta = 66 \times 10^{-3}$ ft

5.18. $\tau_{Al} = 4.177 \times 10^8$ Pa
$\tau_{Stl} = 1.112 \times 10^9$ Pa

5.19. $p = 3.3 \times 10^7$ Pa

5.20. $\tau = 1.343 \times 10^9$ Pa
$\varepsilon_p = .0366$
$F = 3.142 \times 10^6$ N
$L = .6220$ m

5.21. $\Delta = .00995$ ft

5.22. $\Delta = 4.28 \times 10^{-6}$ m

$\omega = 8219$ RPM

5.23. $\omega = 7910$ RPM

5.24. $\Delta_{GA} = .004403$ mm

$\Delta_{CD} = 6.65 \times 10^{-6}$ m

5.25. $\Delta = 9.028 \times 10^{-5}$ in.

5.26. —

5.27. $u_F = -1.119 \times 10^{-4}$ m

$u_C = -1.119 \times 10^{-4}$ m

$v_F = -6.715 \times 10^{-4}$ m

$G_{EF} = 6.715 \times 10^{-4}$ m

5.28. $\Delta_A = 6.083 \times 10^{-5}$ mm

5.29. —

5.30. 196.3 sec

5.31. $F_{AB} = 0.395\,P$

5.32. $a = 9.4$ in.

5.33. $\delta = 5.151 \times 10^{-3}$ m

5.34. 6.64×10^{-6} ft

5.35. 694 g's

5.36. 1.907×10^{-6} m

5.37. $\theta = .001060$ rad

5.38. $a = 1.8918$ m

5.39. $\Delta_{CG} = .007542$ m

$\alpha = 2.637 \times 10^{-5}$ rad

5.40. $\Delta = 1.552 \times 10^{-5}$ m

5.41. .1822 mm

5.42. $\delta = 3.598$ mm

5.43. $\delta\theta = 1.639 \times 10^{-6}$ rad

5.44. 150,000 lb

$\Delta = 0.05$ in.

5.45. 6.136×10^5 N

1.227×10^5 N

5.46. 33,541 N

5.47. 8.518×10^{-5} rad

5.48. $\Delta = 5.627 \times 10^{-4}$ in.

5.49. $\Delta = 1.715 \times 10^{-5}$ m

5.50. $\Delta = 9.906$ mm

5.51. —

5.52. $A = 419.1\ \text{mm}^2$

$\Delta = 3.503$ mm

5.53. $-25{,}000$ psi

5.54. $w = 4.781 \times 10^6$ N

1.903×10^5 N

-3.296×10^5 N

$\Delta = .01400$ m

5.55. $D_{CD} = 20.26$ mm

$D_{AB} = 15.19$ mm

5.56. $D = 15.19$ mm

5.57. $L = 1.047$ ft

$D = .219$ in.

5.58. -5.382×10^7 Pa

-3.478×10^7 Pa

1.941×10^4 N

5.59. -2.692×10^7 Pa

-1.739×10^7 Pa

9.760×10^3 N

5.60. $\Delta L = 3.766 \times 10^{-4}$ m

5.61. $\Delta L = 2.511 \times 10^{-4}$ m

-1.147×10^7 Pa

-2.400×10^6 Pa

5.62. 9800 psi

5.63. $\tau_{zz} = 5.333\,(10 - z) + 945.1$ psi

$\Delta = 3.239 \times 10^{-4}$ ft

5.64. $\tau_{rr} = 3.071 \times 10^7\,(.36 - r^2)$ Pa

$\tau_{rr} = 3.071 \times 10^7\,(.16 - r^2) + 1.535 \times 10^6$ Pa

$\Delta_I = 3.276 \times 10^{-6}$ m

$\Delta_{II} = 3.575 \times 10^{-6}$ m

5.65. 67,600 lb

5.66. 63,500 lb

5.67. 16,875 N

28,125 N

$F_{max} = 9.896 \times 10^6$ N

5.68. $\delta = .01615$ m

$F_1 = 3.093 \times 10^6$ N

$F_2 = 2.855 \times 10^6$ N

5.69. $\Delta_{total} = 4.3496 \times 10^{-5}$ m

5.70. $\delta = 7.016 \times 10^{-5}$ m

5.71. $\Delta = 9.854 \times 10^{-4}$ ft

5.72. $\delta = (\alpha_{comp} - \alpha_{board})L\Delta T$

5.73. $(\tau_{xy})_{av} = 4{,}286$ psi

5.74. $\tau_{xx} = 7754$ psi

$(\tau_{xx})_{max} = -50{,}000$ psi

5.75. $H_x = -6.261 \times 10^5$ N

$H_y = 0$

5.76. $\Delta = .271$ mm

5.77. $T = 318°$ F

$\Delta L = .0243$ in.

5.78. $(\tau_{xx})_1 = -4.850 \times 10^7$ Pa

$(\tau_{xx})_2 = 4.200 \times 10^7$ Pa

5.79. 2.83×10^{-4} ft

5.80. 2263 RPM

5.81. $BE = 11{,}030$ N Tension

$BC = 6{,}620$ N Compression

5.82. $\Delta = 5.53 \times 10^{-3}$ ft

$F = 31{,}250$ lb

5.83. $F_{AB} = 4133$ lb

$F_{CB} = 9156$ lb

5.84. $(\tau_{ss})_{St} = -1680$ psi

$(\tau_{ss})_{Al} = 3360$ psi

5.85. $\Delta T = 200.3°$ C

5.86. $a = 8.18$ in.

5.87 3.01×10^{-3} Rev

5.88. 8.01×10^{-3} Rev

5.89. $D = 15.19$ mm

5.90. $L = 1.047$ ft

$F_{CD} = 7{,}035$ lb

$F_{AB} = 5{,}779$ lb

5.91. $F_{AD} = 1.53$ kN Tens

$F_{AC} = 1.085$ kN Tens

$F_{AB} = -2.08$ kN Comp

5.92. $F_1 = 1250$ lb

$F_2 = 6250$ lb

$.208 \times 10^{-3}$ ft to left

5.93. $\Delta = 8.033 \times 10^{-5}$ m

5.94. $A = 2.79 \times 10^{-3}$ m²

5.95. $\delta B = .04494$ mm

5.96. $\Delta = 87.3$ mm

5.97. $\tau_{xx} = 106$ MPa

5.98. (c) 3996.2 kg

$\Delta = -.1293$ mm

6.1. $\varepsilon_{yy} = 2.40 \times 10^{-6}$

$\varepsilon_{xy} = 1.2 \times 10^{-4}$

etc

6.2. $\varepsilon_{xx} = 5.5 \times 10^{-5}$

$\varepsilon_{yy} = 3.0 \times 10^{-5}$

$\varepsilon_{xy} = -.25 \times 10^{-3}$

etc

6.3. $\varepsilon_{xx} = -15 \times 10^{-6}$

$\varepsilon_{xy} = 4.34 \times 10^{-5}$

etc

6.4. $\varepsilon_{xx} = -8.00 \times 10^{-5}$

$\varepsilon_{yy} = 4.00 \times 10^{-4}$

$\varepsilon_{zz} = -8.00 \times 10^{-5}$

$\varepsilon_{xy} = 0$

etc

6.5. $\varepsilon_{xx} = 3.56 \times 10^{-6}$

$\varepsilon_{yy} = -1.067 \times 10^{-6}$

$\varepsilon_{xy} = 2.32 \times 10^{-6}$

6.6. $\varepsilon_{xx} = 2.54 \times 10^{-6}$

$\varepsilon_{yy} = -2.45 \times 10^{-6}$

$(\tau_{xx})_{TOTAL} = 33.20$ psi

6.7. $t = 0.22$ in.

6.8. $\tau_{xx} = 5.6784 \times 10^8$ Pa

$\tau_{zz} = 3.3376 \times 10^8$ Pa

$\tau_{xz} = 4.8716 \times 10^7$ Pa

6.9. $(\tau_{xx})_1 = 4.74 \times 10^4$ Pa

$(\tau_{xx})_2 = 2.09 \times 10^6$ Pa

$(\tau_{xz}) = -1.995 \times 10^5$ Pa

$\varepsilon_{xz} = 1.297 \times 10^{-6}$

etc

6.10. —

6.11. $\tau_{yy} = -\nu\tau_0$

6.12. $\tau_{xx} = -\nu p \qquad \tau_{yy} = -p \qquad \tau_{zz} = 0$

$$\varepsilon_{yy} = \frac{p}{E}(\nu^2 - 1)$$

$$\varepsilon_{zz} = \frac{\nu p}{E}(1 + \nu)$$

6.13. 56,590 Pa

-1.585×10^6 Pa

1.660×10^6 Pa

$\varepsilon_{xx} = -7.64 \times 10^{-6}$

etc

6.14. 1.317×10^6 Pa
-2.314×10^6 Pa
$\varepsilon_{xr} = 1.496 \times 10^{-6}$
$\varepsilon_{yx} = -5.731 \times 10^{-6}$
etc

6.15. $h = 7.5$ mm

6.16. —

6.17. $\varepsilon_{zz} = -1.250 \times 10^{-6}$
$\varepsilon_{rr} = -4.64 \times 10^{-6}$
$\varepsilon_{\theta\theta} = 7.55 \times 10^{-6}$
etc

6.18. —

6.19. $\tau_{xx} = \tau_{zz} = \dfrac{\nu p}{\nu - 1}$
$\tau_{yy} = -p$
etc

6.20. $\Delta T = 67°$ F

6.21. $\varepsilon_{xx} = 1 \times 10^{-3}$
$\tau_{xx} = -3500$ psi
etc

6.22. 0.19%

6.23. —

6.24. $\tau_{xx} = 10{,}000$ psi
$\tau_{xy} = 16{,}000$ psi
etc

6.25. $U = \dfrac{2.15 \times 10^5}{(Ct)^{2/3}}$ N-m

6.26. $\mathcal{U}_1 = 1.689 \times 10^{-3}$ lb/in^2.
$\mathcal{U}_2 = 54.0 \times 10^{-3}$ lb/in^2.
$U_1 = .637$ in.-lb
$U_2 = 5.09$ in.-lb

6.27. 1.357×10^7 in-lb

6.28. 116.6 in.-lb

6.29. $U = .821$ N-m
$\theta = .01255°$

6.30. $U = 5.22 \times 10^3$ N-m
$T = 5.98 \times 10^5$ N-m

6.31. .3540 in.-lb

6.32. $U = 1.907$ in.-lb

6.33. $\Delta = \dfrac{116{,}569}{AE}$ ft

6.34. $\Delta = \dfrac{3046}{AE}$ m

6.35. $\Delta = \dfrac{20.66}{AE} \times 10^{-3}$ m

6.36. $\Delta = \dfrac{4.964 \times 10^6}{AE}$ in.

6.37. $AB = -1000$
$BD = 625$
etc

6.38. $AF = 4.33$ kN
$BC = -4.83$ kN
$BE = .700$ kN
etc

6.39. $p = \dfrac{Et}{r\left(\dfrac{1}{2} - \nu\right)} \varepsilon_{\text{meas}}$

6.40. —

6.41. $\Delta = 2.26 \times 10^{-4}$ m

6.42. $\delta = \dfrac{\nu(1 + \nu)}{E} \gamma r^2$

6.43. $\Delta = 3.84 \times 10^{-2}$ in.

6.44. $\mathcal{U} = .2470$ in.-lb/in^3.

6.45. $\mathcal{U} = 3.55 \times 10^6$ N/m^2

6.46. 5% change

6.47. $\Delta = \dfrac{prL}{Et}\left(\dfrac{1}{2} - \nu\right)$

6.48. $\Delta D = .00265$ m

6.49. $\tau_{xx} = 4.845 \times 10^8$ Pa
$\tau_{yy} = -4.845 \times 10^8$ Pa
$\gamma_{xy} = 0$
$\gamma_{zx} = 0$

6.50. $\delta_{\text{end cap}} = \dfrac{pr^2}{2Et}(1 - \nu)$
$\delta_{\text{cyl}} \approx 2.4\delta_{\text{end-cap}}$

6.51. 103° F
$\tau_{xx} = -51{,}700$ psi
$\tau_{yy} = -44{,}000$ psi

6.52. $\Delta T = 142°$ F

7.1. $\tau_{nn} = -3.50 \times 10^5$ Pa
$\tau_{ns} = -5.25 \times 10^6$ Pa

7.2. $\tau_{x'x'} = 317$ psi
$\tau_{y'y'} = 683$ psi
$\tau_{x'y'} = -683$ psi

7.3. $\tau_{nn} = -4.20 \times 10^5$ Pa
$\tau_{x'y'} = -3.897 \times 10^5$ Pa

7.4. $\tau_{\text{normal}} = 3745$ psi
$\tau_{\text{parallel}} = 1755$ psi

7.5. $\tau_{nn} = 520$ psi
$\tau_{ns} = -436$ psi

7.6. $\theta = 45°$
$\tau_{nn} = 442.85$ psi

7.7. $\tau_{nn} = 7.02 \times 10^7$ Pa
$\tau_{ns} = 1.799 \times 10^7$ Pa

7.8. $2\theta = -46.33°; 133.7°$
$\tau_{\max} = 5.853 \times 10^6$ Pa
$\tau_{\min} = -9.353 \times 10^6$ Pa

7.9. $\tau_{\min} = -1118$ psi
$\theta = 121.72°$

7.10. $\tau_{x'x'} = 6.125 \times 10^6$ Pa
$\tau_{y'y'} = 4.375 \times 10^6$ Pa
$\tau_{x'y'} = 1.5155 \times 10^6$ Pa

7.11. $\tau_{\max} = 3.39$ psi
$\tau_{\min} = -12.77$ psi

7.12. $2\theta = 43.67°; 223.67°$
$(\tau_{xy})_{\text{ext}} = 7.603 \times 10^6$ Pa

7.13. -1.750×10^6 Pa

7.14. 625 psi
183.6 psi

7.15. $p = 677$ psi

7.16. $\tau_{rr} = 6.17 \times 10^7$ Pa
$\tau_{\theta\theta} = -4.03 \times 10^7$ Pa
$\tau_{r\theta} = 0$
$(\tau_{ns})_{\text{ext}} = -5.10 \times 10^7$ Pa

7.17.
$$\tau_{xx} = -\frac{P}{\pi}\frac{\cos\theta}{r}(1 + \cos 2\theta)$$
$$\tau_{yy} = -\frac{P}{\pi}\frac{\cos\theta}{r}(1 - \cos 2\theta)$$
$$\tau_{xy} = \frac{P}{\pi r}\cos\theta \sin 2\theta$$

7.18. —

7.19. $\tau_1 = 500$ kPa
$\tau_2 = 0$ kPa
$\tau_3 = -500$ kPa

7.20. $\tau_{nn} = 460$ psi
$\tau_{ns} = 400$ psi

7.21. $\tau_{x'x'} = 7 \times 10^6$ Pa
$\tau_{y'y'} = 0$ Pa
$\tau_1 = 14.57 \times 10^6$ Pa
$\tau_2 = -7.57 \times 10^6$ Pa

7.22. 20,900 psi

7.23. 15.35° c.c.

7.24. 2000 kPa

7.25. —

7.26. —

7.27. 44,142 psi
15,858 psi
0 psi
——
24,023 psi
35,977 psi
12,817 psi
——
1.3128×10^{-3}
8.718×10^{-5}
-6.00×10^{-4}

7.28. 2.132×10^8 Pa
-2.932×10^8 Pa
$-.3856 \times 10^8$ Pa

7.29. -3390 psi
4030 psi
15.35° c.c.

7.30. —

7.31. $(\tau_t)_E = 1.4713 \times 10^6$ Pa
$(\tau_{nn})_E = 0$
$(\tau_{nn})_G = -1.4713 \times 10^6$ Pa
$(\tau_{ns})_G = 0$

7.32. 4.0926×10^5 Pa
$\tau_{x'x'} = -1.2667 \times 10^6$ Pa
$\tau_{x'y'} = -2.0463 \times 10^5$ Pa

7.33. 197.24 psi
2198 psi
1802 psi
-1792 psi

7.34. $.446 \times 10^8$ Pa
$.334 \times 10^8$ Pa
1.886×10^8 Pa
——
2.287×10^8 Pa
-1.487×10^8 Pa
1.887×10^8 Pa

7.35. 24,136 psi
4136 psi

8.1. $21.141 + 23.28\,\Delta x - .990\,\Delta y + 3.141\,\Delta x^2 - .0706\,\Delta y^2 - 1.98\,\Delta x \Delta y$

8.2. $\frac{\partial \phi}{\partial n} = 16.496$

8.3. $\varepsilon_{x'x'} = -2.63 \times 10^{-4}$
$\varepsilon_{y'y'} = -5.37 \times 10^{-4}$
$\varepsilon_{x'y'} = -3.66 \times 10^{-5}$

8.4. $\varepsilon_1 = -.0002586$ at $\theta = 22.5°$
$\varepsilon_2 = -.0005414$ at $\theta = 112.5°$
$\varepsilon_{x'x'} = -.000437$
$\varepsilon_{y'y'} = -.0003634$
$\varepsilon_{x'y'} = .0001366$

8.5. $\varepsilon_{xx} = -.001616$
$\varepsilon_{yy} = .002616$

8.6. $\varepsilon_1 = .003191$
$\varepsilon_2 = -.002191$

8.7. $\varepsilon_{x'x'} = -1.533 \times 10^{-3}$
$\varepsilon_{y'y'} = 2.33 \times 10^{-3}$
$\varepsilon_{x'y'} = -5.985 \times 10^{-3}$

8.8. $\varepsilon_{xx} = 8.235 \times 10^{-6}$
$\varepsilon_{yy} = -7.293 \times 10^{-6}$
$\varepsilon_{nn} = 4.353 \times 10^{-6}$
$\varepsilon_{ns} = -6.724 \times 10^{-6}$

8.9. $\varepsilon_1 = .00514$
$\varepsilon_2 = -.00214$

8.10. $\varepsilon_{x'x'} = -2.616 \times 10^{-4}$
$\varepsilon_{y'y'} = 1.616 \times 10^{-4}$
$\varepsilon_{x'y'} = 1.665 \times 10^{-4}$
$\varepsilon_1 = 2.193 \times 10^{-4}$
$\varepsilon_2 = -3.193 \times 10^{-4}$

8.11. $\varepsilon_{max} = .00262$ @ $\theta = -22.5°$

8.12. $\varepsilon_{xy} = -.0035$
$\varepsilon_1 = .006036$
$\varepsilon_2 = -.001036$

8.13. $\tau_1 = 1.849 \times 10^9$ Pa
$\tau_2 = 1.503 \times 10^8$ Pa
$\tau_3 = 0$

8.14. .00358 @ $\theta = 41.7°$

8.15. —

8.16. $\varepsilon_1 = .005442$
$\varepsilon_2 = -.001942$
etc

8.17. $\Delta = -.15''$

8.18. .002799

8.19. $\varepsilon_{x'y'} = -.00634$
$\varepsilon_{x'x'} = .00410$
$\varepsilon_{y'y'} = -.00210$

8.20. $\varepsilon_1 = .00661$
$\varepsilon_2 = -.00061$

8.21. $\varepsilon_1 = .005472$
$\varepsilon_2 = -.003472$
$\varepsilon_{x'x'} = .005464$
$\varepsilon_{y'y'} = -.003464$
$\varepsilon_{x'y'} = .000268$
$\tau_1 = 9.725 \times 10^8$ Pa
$\tau_3 = -4.011 \times 10^8$ Pa

8.22. $\varepsilon_1 = .00250$
$\varepsilon_2 = -.00310$
$\tau_1 = 3.451 \times 10^8$ Pa
$\tau_2 = -5.165 \times 10^8$ Pa
$\varepsilon_3 = .000257$

8.23. —

8.24. $T = 1.724 \times 10^6$ N-m
3611 kW

8.25. $\varepsilon_1 = .00361$
$\varepsilon_2 = -.00361$

8.26. .002101
$-.008101$

8.27. For $p = 30$ Pa $\varepsilon = -.003416$
For $p = 90$ Pa $\varepsilon = .01025$

8.28. $\varepsilon_{yy} = -.0150$
$\varepsilon_{xy} = 9.815 \times 10^{-4}$

8.29. $\varepsilon_a = .003378$
$\varepsilon_b = -.004046$

9.1. —

9.2. Tresca -1.5×10^8 Pa
Mises -2.282×10^8 Pa

9.3. —

9.4. —

9.5. $t = .00406$ m

9.6. $d = 1.133$ m

9.7. $t_{min} = 1.893$ mm

9.8. Tresca $p = 13{,}333$ psi
Mises $p = 15{,}364$ psi

9.9. Mises $Y = 295$ MPa
Tresca $Y = 340$ MPa

9.10. $\delta = 1.0''$
Mises 1.25
Tresca 1.25

9.11. $\Delta T = 318.6°F$
$\tau_1 = -12{,}900$ psi
$\tau_2 = -36{,}000$ psi

9.12. —

9.13. Tresca 373.5 rad/sec
Mises 375.5 rad/sec

9.14. $P = 9185$ N

9.15. $p = 6.01 \times 10^7$ Pa

9.16. $t = .00497$ in.

9.17. $p = 1330$ psi

9.18. $\Delta T - 330°C$

10.1. a) $M = 3688$ ft-lb
$V = 737.5$ lb
b) $M = 6850$ ft-lb
$V = -262.5$ lb
c) $M = 1312.5$ ft-lb
$V = -262.5$ lb

10.2. $M_{max} = 30{,}000$ N-m
$V_{max} = -6000$ N

10.3. $\underline{0 < x < 10}$
$V = 453$ lb
$M = 453x$ ft-lb
etc

10.4. $\underline{10 < x < 25}$
$V = -200 - 30x$ lb
$M = -15x^2 - 200x + 1000$ ft-lb
etc

10.5. $\underline{16 < x < 20}$
$V = -1510$ N
$M = 4290x - 5000(x - 3) - 800(x - 9) + 8000$ N-m
etc

10.6. —

10.7. $\underline{8 \le x < 13}$
$V = 24.4$ lb
$M = -100x - 40(x - 4) + 164.4(x - 5)$ ft-lb
etc

10.8. $\underline{9 \le x < 12}$
$V = -650$ N
$M = 250x - 800(x - 4.5)$ N-m
etc

10.9. $\underline{3 \le x < 6}$

$$V = 750 - \frac{250}{3}(x - 3)^2 \text{ N}$$

$$M = 750x - 6750 - \frac{250}{9}(x - 3)^3 \text{ N-m}$$

etc

10.10. $\underline{14.14' \le s < 29.14'}$
$V = 52.5$ lb
$M = 52.5s - 218$ ft-lb
etc

10.11. $\underline{\frac{\pi}{4} < \theta < \frac{\pi}{2}}$
$V = -29.3 \sin\theta$ lb
$M = 176 \cos\theta$ ft-lb
etc

10.12. $M_y = \frac{1}{8}x^3 - 15x^2 + 400x$ N-m
etc

10.13. $M = -.85x^2 + 34x - 350$ ft-lb
etc

10.14. $\underline{10 \le x \le 12}$
$V_y = -1000$ N
$H = 0$
$M_x = -100$ N-m
$M_z = -1000x$ N-m
etc

10.15. $M_x = -5.68s^2 + 262s - 2194$ ft-lb
$M_y = 410$ ft-lb
etc

10.16. —

10.17. $\underline{22 < x < 30}$
$V = -46.7$ lb
$M = 453.3x - 800\ (x - 10) + 300\ (x - 22)$ ft-lb
etc

10.18. $\underline{10 \le x < 25}$
$V = -500 - 30\ (x - 10)$ lb
$$M = -500\ (x - 5) - \frac{30\ (x - 10)^2}{2} \text{ ft-lb}$$
etc

10.19. $\underline{13 \le x < 16}$
$V = -1510$ N
$M = 4290x - 5000\ (x - 3) - 800\ (x - 9)$ N-m
etc

10.20. —

10.21. $\underline{13 < x < 23}$
$V = 24.4$ lb
$M = -100x - 40\ (x - 4) + 164.4\ (x - 5) + 100$ ft-lb
etc

10.22. $\underline{12 < x < 15}$
$V = -1650$ N
$M = 250x - 900\ (x - 4.5) - 1000\ (x - 12)$ N-m
etc

10.23. $\underline{6 \le x < 8}$
$V = 0$
$M = -6750 + 750x - 750\ (x - 5)$ N-m
etc

10.24. $\underline{15 < x < 20}$
$V = -900$ lb
$M = 600x - 500\ (x - 5) - 1000\ (x - 15)$ ft-lb
etc

10.25. $\underline{6 < x \le 10}$
$V = 1000$ N
$M = 2000x - 23{,}000 - 1000\ (x - 3) + 5000$ N-m
etc

10.26. —

10.27. $\underline{10 < x < 15}$
$V = 50x + 250$
$M = 25x^2 + 250x - 9375$ ft-lb
etc

10.28. $\underline{13 < x < 17}$
$V = 0$ N
$M = -43{,}329 + 9333x + 10{,}000\ (x - 5) + 667\ (x - 13)$ N-m
etc

10.29. $\underline{30 < x < 40}$
$V = -262.5$ lb
$M = 737.5x - 1000\ (x - 10) + 500$ ft-lb
$M_{max} = 5439$ ft-lb
etc

10.30. $\underline{8 < x \le 10}$
$V = 112.5$ N
$M = -8900 + 1112.5x - 1000\ (x - 8)$ N-m
etc

10.31. $\underline{.8 \le x < 1.4}$
$V = 15 - 50\ (x - .8)$ N
$M = -1112 + 1015x - 1000\ (x - .4) + 700 - 25\ (x - .8)^2$ N-m
etc

10.32. —

10.33. —

10.34. $R_1 = 20$ K
$R_2 = 10$ K
etc

10.35. $\bar{x} = 13.6$ ft
$M(\bar{x}) = 43{,}000$ ft-lb

10.36. —

10.37. $M_{max} = \dfrac{400}{\pi^2}$ ft-lb

10.38. —

10.39. —

10.40. —

10.41. —

10.42. —

10.43. $M_{max} = 121.65$ kN-m

10.44. $b \approx 35.4$ ft

10.45. $a = 11.13$ ft

10.46. $\bar{x} = .707h$

10.47. $|M_{max}| = 500$ N-m

10.48. —

10.49. —

10.50. —

10.51. —

10.52. —

10.53. $\underline{10 \le x < 19}$

$V = -175.3$ lb

$M = 77.47 - 450\ (x - 7.5) - 500\ (x - 11.67)$ ft-lb

etc

10.54. $\underline{7 < x \le 13.2}$

$V = 2480 - 64.5\ (x - 7)^2$ N

$M = -32{,}911 - 3580x - 600\ (x - 3) - 500\ (x - 7) - 21.5\ (x - 7)^3$ N-m

etc

10.55. $\underline{5 < x < 10}$

$V = -930$ N

$M = 1070x - 1000\ (x - 2.5) + 1800 - 1000\ (x - 5)$ N-m

etc

10.56. —

10.57. $\underline{0 \le x \le 8}$

$H = -50x$ lb

etc

10.58. $\underline{90 < x < 120}$

$V = 0$

$M = -60{,}000$ ft-lb

10.59. $\underline{0 < x < 5}$

$V = -.596x^{3/2}$kN

$M = -.238x^{5/2}$kN-m

etc

10.60. $\underline{0 < x < 3}$

$V = 60 - 5x^2$ N

$M = 60x - \frac{5}{3}x^3$ N-m

etc

10.61. $\underline{6 < x < 11}$

$V = 1000$ N.

$M = 1000x - 11{,}000$ N-m.

etc.

10.62. —

10.63. —

10.64. —

10.65. —

11.1. $(\tau_{xx})_{max} = 5.76 \times 10^7$ Pa

$\tau_{xx} = -1.920 \times 10^7$ Pa

$R = 130.2$ m

11.2. $(\tau_{xx})_{max} = 6.48 \times 10^7$ Pa

$\varepsilon_{xx} = 3.24 \times 10^{-4}$

$\varepsilon_{yy} = -9.72 \times 10^{-5}$

11.3. $\tau_{xx} = -\dfrac{3My}{4\left[\dfrac{ad^2}{2} + c^3d\right]}$

11.4. $(\tau_{xx})_{max} = 3.673 \times 10^6$ Pa

11.5. Exact 697.3 in^4.

Approx. 574.5 in^4.

11.6. $(\tau_{xx})_{max} = 2.645 \times 10^6$ Pa

11.7. $\Delta = 5.598$ mm

11.8. $\pm$ 4500 psi

$\pm$ 11,700 psi

11.9. 3.73×10^6 Pa

7.466×10^6 Pa

11.10. $(\tau_{xy})_{max} = 3.24 \times 10^5$ Pa

$(\tau_{xx})_{max} = 1.296 \times 10^7$ Pa

11.11. $\tau_{xy} = 4.570$ psi

$\tau_{xx} = 274.5$ psi

11.12.

11.13. 2.250×10^4 Pa

0 Pa

11.14. 2.28×10^7 Pa

11.15. $\tau_{xx} = 3.645 \times 10^7$ Pa

$\tau_{xy} = 3.375 \times 10^5$ Pa

11.16. —

11.17. $\pm$188 MPa

11.18. $P = 31{,}800$ N

11.19. 23.44 psi

12.65 psi

11.20. 1.238×10^5 Pa

8.100×10^6 Pa

11.21. $(\tau_{xx})_{max} = -1.741 \times 10^7$ Pa

$|\bar{\tau}_{xy}| = 1.305 \times 10^5$ Pa

11.22. $|\overline{\tau}_{xy}| = 7.133 \times 10^5$ Pa
$\tau_{xx} = 7.012 \times 10^7$ Pa

11.23. $h_1 = .0578$ m
$h_2 = .0624$ m

11.24. $\frac{a}{b} = 0.1818$

11.25. —

11.26. $w_o = 192.6$ lb/ft
etc

11.27. 1.350×10^5 Pa

11.28. $P = 7360$ lb

11.29. 49.34 mm apart

11.30.
At A
$\tau_{xx} = 1.209 \times 10^8$ Pa
$|\overline{\tau}_{xy}| = 1.511 \times 10^6$ Pa
At B
$\tau_{xx} = 2.313 \times 10^8$ Pa
$|\overline{\tau}_{xy}| = 2.886 \times 10^5$ Pa

11.31. 1090 psi
3309 psi

11.32. 1335 Pa

11.33. $|\tau_{xx}| = 1.149 \times 10^7$ Pa
$|\overline{\tau}_{xy}| = 3.782 \times 10^4$ Pa.

11.34. 5.34×10^4 Pa

11.35. $\tau_{xx} = 10{,}794$ psi
$|\tau| = 84.03$ psi

11.36. $\tau_{xx} = 1.976 \times 10^8$ Pa
$|\tau| = 2.997 \times 10^6$ Pa

11.37. $|\overline{\tau}_{xy}| = 13.67$ psi
$\tau_{xx} = -229.2$ psi

11.38. $(\tau_{xx})_{max} = 9.52 \times 10^7$ Pa
$|\overline{\tau}_{xy}|_{max} = 1.735 \times 10^6$ Pa
$|\overline{\tau}_{xz}|_{max} = 6.948 \times 10^5$ Pa

11.39. 1590 psi

11.40. —

11.41. —

11.42. $P = 105.3$ lb

11.43. $(\tau_{xx})_{max} = 6.22 \times 10^8$ Pa

11.44. $(\tau_{xx})_{max} = 12{,}250$ psi

11.45. $[(\tau_{xx})_{max}]_{comp} = 1.011 \times 10^8$ Pa
$[(\tau_{xx})_{max}]_{tens} = 8.325 \times 10^7$ Pa

11.46. $\tau_1 = 1.931 \times 10^7$ Pa
$\tau_2 = 2.759 \times 10^7$ Pa

11.47. 1389 psi
4167 psi

11.48. $\tau_{xx} = 516.7$ psi
$|\overline{\tau}_{xz}| = 9.38$ psi

11.49. Formal
$\tau_{xx} = -2.003 \times 10^8$ Pa
Common sense
$\tau_{xx} = -2.003 \times 10^8$ Pa
$\tau_A = 7.847 \times 10^5$ Pa

11.50. $\tau_{xx} = 24{,}375$ psi
$\tau_{xx} = 24{,}375$ psi
$\tau = 325.6$ psi

11.51. $\tau_{xx} = 19{,}370$ psi
$\tau_{xx} = 19{,}367$ psi
$|\tau_{xy}| = 150$ psi

11.52. $(\tau_{xx})_{max} = 8.921 \times 10^4$ Pa
$|(\tau_{xy})|_{max} = 2.738 \times 10^3$ Pa
$|(\tau_{xz})|_{max} = 6.137 \times 10^5$ Pa

11.53. $|\tau_{sx}| = 9450s - 983s^2$
$|\tau_{sx}|_A = 1.810 \times 10^4$ psi

11.54. $|\tau_A| = 9840$ psi
$|\tau_B| = 106.1$ psi

11.55. $|\tau_{sx}| = |V_y[-1.859 \times 10^4 s + 2.867 \times 10^5 s^2]|$ Pa

11.56. $|\tau_{sx}| = |V_z[-4.869 \times 10^4 s + 4.262 \times 10^5 s^2]|$ Pa

11.57. $|\tau_{sx}| = |\frac{2V_y}{\pi Rt} \sin\theta|$

11.58. —

11.59. 1.2442″ from left edge

11.60. $e = -86.1$ mm

11.61. —

11.62. $d = .04331$ m
$R = 21.65$ m

11.63. $w_o = 133.3$ lb/ft
$R = 354$ in.

11.64. $P = 13{,}830$ N

11.65. Limit design
$P = 41{,}500$ N
Perfectly elastic
$P = 26{,}920$ N

11.66. 154.32 N/m
$w_o = 231.5$ N/m

11.67. 11,111 N/m
$w_o = 16{,}667$ N/m

11.68. $\begin{cases} -490.7 \text{ psi} \\ 708 \text{ psi} \end{cases}$

$\begin{cases} -555 \text{ psi} \\ 611 \text{ psi} \end{cases}$

11.69. 7.242×10^8 Pa

6.60×10^8 Pa

11.70. $\alpha = 52.4°$

11.71. $(\tau_{xx})_A = 1.169 \times 10^9$ Pa

$|\tau_{xy}| = 3.411 \times 10^6$ Pa

11.72. $P = 572$ lb

11.73. $\tau_{xx} = \pm 430$ MPa

11.74. $(\tau_{xx})_{max} = 1929$ psi

$|(\tau_{xy})_A| = 19.52$ psi

$(\tau_{ns})_{max} = 597.6$ psi

11.75. $\tau = 19{,}126$ psi

11.76. 400 rivets

11.77. 48.26 mm between nails

11.78. a) 4.089×10^7 Pa

b) 4.089×10^7 Pa

c) 1.678×10^5 Pa

11.79. $\tau_{xx} = 1.111 \times 10^6$ Pa

$\tau = 11.70 \times 10^4$ Pa

11.80. $t = \left[\frac{6Fr}{w\tau_{nn}}(1 - \cos\theta)\right]^{\frac{1}{2}}$

11.81. $\bar{y} = .02094$ m

$[(\tau_{xx})_1]_{max} = -7.716 \times 10^6$ Pa

$[(\tau_{xx})_2]_{max} = 8.953 \times 10^6$ Pa

11.82. $[(\tau_{xx})_1]_{max} = 3.577 \times 10^6$ Pa

$[(\tau_{xx})_2]_{max} = 1.073 \times 10^7$ Pa

11.83. $P = 15{,}000$ lb

11.84. a) 9.294×10^6 Pa

b) -7.311×10^8 Pa

11.85. —

11.86. 10,000 psi

11.87. $d = 1.00$ in.

11.88. 4.80×10^7 Pa

1.680×10^7 Pa

11.89. 17 screws per ft

11.90. $|\bar{\tau}_{xy}|_{max} = 4.567 \times 10^7$ Pa

$(\tau_{xy})_A = 4.629 \times 10^6$ Pa

$(\tau_{xz})_A = 8.641 \times 10^6$ Pa

$(\tau)_A = 9.803 \times 10^6$ Pa

11.91. 4.1 screws/foot

11.92. a) $\begin{cases} 33{,}750 \text{ psi} \\ 187.5 \text{ psi} \end{cases}$

b) 740 lb

11.93. $(\tau_{xx})_{max} = 1.729 \times 10^6$ Pa

$|\tau_{xy}|_{max} = 6000$ Pa

11.94. $\tau_{xx} = 1.744 \times 10^6$ Pa

$|(\tau_{xy})_{max}| = 4.80 \times 10^4$ Pa

11.95. $(\tau_{xx})_{max} = 17{,}200$ psi

$|\tau_{xy}|_{max} = 250$ psi

11.96. $\tau_{nn} = \frac{P}{wt}\left[6\left(\frac{r}{t}\right) + 1\right]$

11.97. a) -111.6 N-m

b) $\tau_{xx} = -4.3859 \times 10^5$ Pa

c) $|\tau_{xy}| = 2.388 \times 10^4$ Pa

d) $|\tau_{xz}| = 3.79 \times 10^5$ Pa

11.98. a) 4.912×10^{-6} m^4

b) -9.925×10^7 Pa

c) 1.323×10^5 Pa

11.99. —

11.100. $\frac{100t}{t + 2(R_m + a)}$

11.101. —

11.102. 13,900 psi

11.103. —

11.104. 501.4 psi

11.105. $R_n = 179.74$ mm

11.106. $P = 39.55$ lb

12.1. $v_{max} = .0776$ m

12.2. $v = \frac{1}{EI}[50x^3 - .4167x^5 - 1260x]$

12.3. $v = \frac{1}{EI}\left[-17{,}200\frac{x^2}{2} + 2000\frac{x^3}{6} - 1000\frac{(x-5)^3}{6} - 100\frac{(x-8)^4}{24} - 800\frac{x^2}{2} + C_7x + C_8\right]$

12.4. $\underline{0 < x < 6}$

$v = \frac{1}{EI}[6.25x^3 - .208x^5 + 45x]$

$\underline{6 < x < 9}$

$v = \frac{1}{EI}[-337x^2 + 3420x - 8370]$

12.5. $\Delta \mathbf{r} = 5.56\times10^{-6}\,\mathbf{i} - .01778\mathbf{j} + .0400\mathbf{k}$

12.6. $\underline{23 \le x \le 30}$

$$v = \frac{1}{EI}\left[-13.04x^3 + 100\frac{x^2}{2} - 100\frac{(x-15)^3}{6} + 213.04\frac{(x-23)^3}{6} + C_9x + C_{10}\right]$$

etc

12.7. $v(15) = \frac{1}{EI}[-6.41 \times 10^4]$

12.8. $x = 7.5$ m

$.01944/\text{m}^2$

12.9. $\underline{13 < x \le 28}$

$$v = \frac{1}{EI}\left[600.8\frac{x^2}{3} + 921.8\frac{(x-8.67)^3}{6} + 1000\frac{x^2}{2} + C_3x + C_4\right]$$

etc

12.10. $h_{alum} = 144$ mm

12.11. a) $M = 19{,}000$ N-m

$R = 3000$ N

b) $-.05595$ m

c) $-.2140$ m

12.12. $P = 446$ N

12.13. $EI\dfrac{d^4v}{dx^4} - kv = -\sin\dfrac{\pi x}{L}$

12.14. $\dfrac{d^2v}{dx^2} = -\dfrac{3}{2}\dfrac{PL^{\frac{3}{2}}}{Ea^3t\sqrt{x}}$

12.15. $v = \dfrac{-\gamma(4)}{3E\left(\dfrac{D}{L}\right)^2}\left[\dfrac{x^2}{2} - xL + \dfrac{L^2}{2}\right]$

12.16. —

12.17. 1.29′

12.18. .8632 m

12.19. 18.46 mm

12.20. $R_1 = 281$ lb

$R_2 = 1614$ lb

$R_3 = 605$ lb

12.21. $R_1 = 1818$ lb

$R_2 = 182.0$ lb

$M_2 = -603.3$ ft-lb

12.22. $R_1 = 111.45$ lb

$M_1 = -540$ ft-lb

-8.883×10^{-3} ft

12.23. $R_1 = 590.8$ N

$R_2 = -169.1$ N

$R_3 = 28.30$ N

12.24. $R_1 = 364$ lb

$R_2 = 1043$ lb

$R_3 = -407$ lb

$M = -675$ ft-lb

12.25. $\underline{15 \le x < 30}$

$$v = R_A\frac{x^3}{6} - \frac{25}{9}\frac{x^5}{20} - 1000\frac{(x-15)^2}{6} + C_3x + C_4$$

etc

12.26. $R_1 = 2522$ N

$R_2 = -722$ N

$M_2 = -1410$ N-m

12.27. $\underline{0 \le x \le 15}$

$$v = R_A\frac{x^3}{6} - \frac{25}{9}\frac{x^5}{20} + C_1x + C_2$$

etc

12.28. a) $R_1 = 2619$ N

b) $\Delta v = -\dfrac{1.860 \times 10^4}{EI}$ m

$\Delta x = \dfrac{7874}{EI}$ m

12.29. $\underline{10 \le x < 20}$

$$v = \frac{1}{EI}\left[-\frac{5}{120}x^5 + R_A\frac{(x - 5^3}{6} + (666.7 - 1.5\,R_A)\frac{(x-10)^3}{6} + C_5x + C_6\right]$$

etc

12.30. $(\tau_{xx})_{max} = 6.394\times10^6$ Pa

12.31. $\Delta_A = 6.87 \times 10^{-5}$ m

$\Delta_B = 1.758 \times 10^{-4}$ m

12.32. $\Delta_A = 1.299 \times 10^{-4}$ m

$\Delta_B = 1.147 \times 10^{-4}$ m

12.33. $P = 77.1$ lb

12.34. $\omega = 154$ RPM

12.35. $P = 6.791 \times 10^4$ N

12.36. $v = -\dfrac{6884}{EI}$ m

12.37. $v = -\dfrac{285{,}155}{EI}$

12.38. $R_1 = 2391$ N
$R_2 = 984.4$ N
$M = 420.8$ N-m

12.39. $R_1 = 2476$ N
$R_2 = 3409$ N
$R_3 = 1990$ N

12.40. $\dfrac{(v_S)_{\max}}{(v_B)_{\max}} = 10.83 \times 10^{-3}$
$\dfrac{(v_S)_{\max}}{(v_B)_{\max}} = .975$

12.41. $v_S = -2.411 \times 10^{-5}$ m
$v_B = -1.266 \times 10^{-4}$ m

12.42. $\underline{2.1 < x \leq 3.9}$
$v_3 = \dfrac{3}{2GA}[-1170x + 150x^2 - 1755]$

12.43. —

12.44. —

12.45. $\Delta = \dfrac{wL^4}{8EI}$

12.46. $\Delta = \dfrac{17wL^4}{384EI}$

12.47. $R_A = 112$ lb
$M_A = 540$ ft-lb
$R_B = 388$ lb
$M_B = -1024$ ft-lb

12.48. $R_1 = 2125$ lb
$R_2 = -125$ lb
$M_1 = 1250$ ft-lb

12.49. $R_1 = 4260$ lb
$R_2 = 16{,}600$ lb
$R_3 = 3110$ lb

12.50. $R_1 = 281$ lb
$R_2 = 1615$ lb
$R_3 = 604$ lb

12.51. $\dfrac{6.41 \times 10^4}{EI}$

12.52. $\Delta_Q = .0178\, Q$

12.53. $\Delta = \dfrac{\pi Q R^3}{2EI}$

12.54. $R_1 = R_3 = \dfrac{3}{16} wL$
$R_2 = \dfrac{5}{8} wL$

12.55. $\Delta = \dfrac{5wL^4}{384EI}$

12.56. $\delta = \dfrac{6PL^3}{bt^3E}$

12.57. $v(1) = (8.143 \times 10^{-3})x^3 - (3.353 \times 10^{-2})x$
$v(2) = (4.829 \times 10^{-3})x^3 - 2.415 \times 10^{-3}(x - 1.2)^3 - (5.312 \times 10^{-3})x - 2.262 \times 10^{-2}$

12.58. $v_A = -4.72$ in.

12.59. $v_{\max} = -.0327$ m
etc

12.60. $v = -5.75$ in.

12.61. $v = .07425$ m

12.62. $T = 1210$ lb
$\Delta = 0.33$ in

12.63. $R_1 = \dfrac{3}{8} wL$
$R_2 = \dfrac{5}{8} wL$
$M_1 = \dfrac{wL^2}{8}$

12.64. $\Delta = \dfrac{wL^4}{192EI}$

12.65. $\underline{5 \leq x \leq 10}$
$v = \dfrac{1}{EI}\left[R_A\dfrac{x^3}{6} - \dfrac{10}{6}\dfrac{x^5}{20} + (333.3 - 2R_A)\dfrac{(x-5)^3}{6} + C_3 + C_4\right]$
etc

12.66. 3700 lb

12.67. $R_1 = 1313$ N
$R_2 = 687$ N
$C = -3126$ N-m

12.68. $R_1 = 446$ lb
$R_2 = 54$ lb
$M_1 = 1983$ ft-lb

12.69. $\underline{5 \leq x \leq 13}$
$v = \dfrac{1}{EI}\left[(-13\, R_1 + 4.275 \times 10^4)\dfrac{x^2}{2} + R_1\dfrac{x^3}{6} - 5000\dfrac{(x-5)^3}{6} - 100\dfrac{(x-5)^4}{24} + C_3x + C_4\right]$

12.70. $0 < x \leq a$

$$v = \frac{1}{EI}\left\{\left[\frac{3P(L-a)^2}{L^2} - \frac{2P(L-a)^3}{L^3}\right]\frac{x^3}{6} - \left[\frac{P(L-a)^2}{L} - \frac{P(L-a)^3}{L^2}\right]\frac{x^2}{2}\right\}$$

etc

12.71. —

12.72. —

12.73. 0.527 in.

12.74. $\Delta_2 = 7.35 \times 10^{-10}\tau$ m

12.75. $40 < x \leq 50$

$$v = \frac{1}{EI}\left[374.1\frac{x^3}{6} - 200\frac{(x-10)^3}{6} - 500\frac{(x-30)^3}{6} - 707\frac{x^2}{2} + C_7x + C_8\right]$$

etc

12.76. $v = -1.439$ mm

etc

12.77. $v(10) = -.0704$ m

etc

12.78. $$v = -\frac{w_o}{EI}\left[\frac{L^4}{\pi^4}\sin\frac{\pi x}{L} + \frac{L}{\pi}\frac{x^3}{6} + \left(\frac{L^3}{\pi^3} - \frac{L^3}{2\pi}\right)x + \frac{L^4}{3\pi} - \frac{L^4}{\pi^3}\right]$$

12.79. $B_x = 100$ lb

$B_y = 66.7$ lb

$M_B = -1601$ ft-lb

12.80. $P = 7.167$ N

12.81. $$\Delta = \frac{P}{EI}\left[\frac{L_1^3}{3} + \frac{L_2^3}{3} - L_1L_2^2 + L_1^2L_2 + \pi RL_1^2 + 4L_1R^2 + \pi\frac{R^3}{2}\right]$$

12.82. 60% error

12.83. —

12.84. $P = 445$ N

12.85. $\Delta = \dfrac{4.32 \times 10^9}{EI}$ in.

12.86. $v_{\text{TIP}} = \dfrac{\alpha L^2}{2h}(T_b - T_t)$

12.87. $R_3 = -409$ lb

$R_2 = 1045$ lb

$R_1 = 364$ lb

$M_1 = -681$ ft-lb

12.88. $\Delta = \dfrac{wL^4}{6144EI}$

12.89. $\alpha = .427$

13.1. 0

200

300

13.2. $$v(x) = \frac{1}{EI}\left\{\frac{414}{6}x^3[u(x)] - \frac{250}{3}(x-10)^3[u(x-10)] - \frac{5}{6}(x-20)^4[u(x-20)] + \frac{5}{6}(x-30)^4[u(x-30)] - \frac{(286)(x-35)^3}{6}[u(x-35)] - 46{,}100x\right\}$$

$$(v)_{15} = -\frac{469{,}000}{EI}$$

13.3. $v(15) = -\dfrac{6.41 \times 10^4}{EI}$

$v'(20) = \dfrac{1.40 \times 10^4}{EI}$

13.4. $$v = \frac{1}{EI}\left\{-10\frac{x^4}{24}[u(x)] + \frac{10(x-10)^4}{24}[u(x-10)] + 100\frac{(x-5)^3}{6}[u(x-5)] + 404x - 1758\right\}$$

13.5. $R_1 = -300$ lb

$R_2 = 2800$ lb

$v(15) = \dfrac{616{,}670}{EI}$

13.6. $P = 446$ N

13.7. $(v)_{\max} = .866$ m

13.8. .07425 m

13.9. $R_1 = 374.1$ lb

$R_2 = 325.9$ lb

$v(30) = \dfrac{1}{EI}[-1.545 \times 10^6]f$

13.10. $R_1 = 111.5$ lb
$M_1 = 540.1$ ft-lb
$R_2 = 388.5$ lb
$M_2 = 1033$ ft-lb
$v = -8.884 \times 10^{-3}$ ft

13.11. $R_1 = 3804$ lb
$R_2 = 6196$ lb
$M_2 = 2.392 \times 10^4$ ft-lb

13.12. $v_{\max} = \dfrac{50{,}007}{EI}$ ft

13.13.
$$v = \frac{1}{EI}\left\{\frac{138.9}{6}x^3[u(x)] - \frac{(x-10)^6}{360}[u(x-10)] + \frac{(x-20)^6}{360}[u(x-20)] + \frac{(x-20)^5}{6}[u(x-20)] + \frac{50}{12}(x-20)^4[u(x-20)] + \frac{194.5}{6}(x-30)^3[u(x-30)] - 16{,}946x\right\}$$

13.14.
$$v = \frac{1}{EI}\left\{\left(\frac{700x^5}{6} - 5250x^2 - \frac{5}{6}x^4\right)[u(x)] + \frac{5}{6}(x-10)^4[u(x-10)] - \frac{250}{3}(x-15)^3[u(x-15)] + 1000(x-25)^2[u(x-25)\right\}$$
$$\left(\frac{d^3v}{dx^3}\right)_{17} = 0$$

13.15. $R_1 = 281$ lb
$R_2 = 1614$ lb
$R_3 = 605$ lb

13.16. $R_1 = 108.75$ N
$R_2 = 87.5$ N
$R_3 = 93.75$ N

13.17.
$$v = \frac{1}{EI}\left\{-M_1\frac{x^2}{2}[u(x)] + R_1\frac{x^3}{6}[u(x)] - P\frac{(x-a)^3}{6}[u(x-a)] + R_2\frac{(x-L)^3}{6}[u(x-L)] + M_2\frac{(x-L)^2}{2}[u(x-L)]\right\}$$

13.18. —

14.1. $\tau_{\max} = 8.15 \times 10^5$ Pa
$\phi = .00652$ rad
$\dfrac{M_x}{\phi} = 1.035 \times 10^5$ N-m
$\dfrac{M_x}{\tau_{\max}} = 8.25 \times 10^{-5}$ m^3

14.2. 11,600 N-m

14.3. $\tau_{\max} = 698$ psi
$\phi = .00397$ rad

14.4. .0296 rad

14.5. $L_{\text{eq}} = 1.825$ m

14.6. $\phi = .0264$ rad

14.7. $T_A = 655.5$ N-m
$(\phi)_B = .2450$ rad

14.8. $D_1 = .08471$ m
$D_2 = .3932$ m

14.9. $\omega = 77.2$ RPM
$L - 7.854$ m

14.10. 1.022×10^4 HP
21,800 psi

14.11. $D_B = .1191$ m

14.12. $\phi = .015$ rad
$\dfrac{d\phi}{dx} = .00333$ rad/m

14.13. $(\Delta\phi)_A = 5.861 \times 10^{-4}$ rad
$(\Delta\phi)_{D_2} = 1.189 \times 10^{-3}$ rad

14.14. $M_1 = -2387$ N-m
$M_2 = 4387$ N-m
$\theta = -.02387$ rad

14.15. 3712 HP
11,640 psi

14.16. $T_A = 483$ ft-lb

14.17. $\phi = .0512$ rad

14.18. $\phi_D = 1.577 \times 10^{-3}$ rad

14.19. $T_1 = 615.1$ N-m

$T_2 = 284.9$ N-m

$\phi_{max} = 4.358 \times 10^{-3}$ rad

14.20. $\phi_H = 2.085 \times 10^{-5}$ rad

14.21. $(\tau_1)_{max} = 1.367 \times 10^6$ Pa

$(\tau_2)_{max} = 3.907 \times 10^6$ Pa

14.22. $(\tau_{\theta x})_{Al} = 104.6$ psi

$(\tau_{\theta x})_{Stl} = 58.7$ psi

14.23. $(\tau_{max})_B = 28.33$ psi

$(\tau_{max})_A = 4.69$ psi

$\phi = 37.48 \times 10^{-6}$ rad

14.24. —

14.25. $K = 1.164 \times 10^4$ N/m

14.26. 1.776×10^8 Pa

1.881×10^8 Pa

5.58%

14.27. 38,600 psi

14.28. $T_1 = 0.941\, M$

$T_2 = 0.059\, M$

14.29. 1.206×10^{-6} rad/in.

48.2 psi

27.1 psi

14.30. $\phi = .00501$ rad

14.31. 17,175 psi

−305.8 psi

14.32. 3182 psi

14.33. 4074 psi

0 psi

−453 psi

14.34. −3.62 in.

14.35. $t = 0.758$ in.

14.36. $3\frac{1}{2}$ in.

14.37. $\theta = T\left[\dfrac{L_1}{12EI} + \dfrac{L_2}{GJ}\right]$

14.38. $\theta = T\left[\dfrac{L_1}{16EI} + \dfrac{L_2}{GJ}\right]$

14.39. 144.4 psi

$\dfrac{d\phi}{dx} = 8.26 \times 10^{-6}$ rad/in.

14.40. 5.5×10^4 N/m

$M_x = 3.644 \times 10^4$ N-m

$\dfrac{d\phi}{dx} = 5.263 \times 10^{-4} \dfrac{\text{rad}}{\text{m}}$

14.41. 3180 psi

$\dfrac{d\phi}{dx} = 1.007 \times 10^{-3}$ rad/in.

14.42. $a = .01733$ m

$T_{hinge} = 3600$ N-m

14.43. 6.795×10^4 N-m

14.44. $48.6 \times 10^{-9}\dfrac{G}{L}$

14.45. $3.98\dfrac{G}{L}$ (in.-lb)/rad

$3.15\dfrac{G}{L}$ (in.-lb)/rad

$3.55\dfrac{G}{L}$ (in.-lb)/rad

etc.

14.46. $3.98\,\dfrac{G}{L}\,\dfrac{\text{in.-lb}}{\text{rad}}$

$3.18\,\dfrac{G}{L}\,\dfrac{\text{in.-lb}}{\text{rad}}$

etc.

14.47. $.362\,\dfrac{G}{L}$

$.355\,\dfrac{G}{L}$

2.2% difference

14.48. 188.4 N-m

486.6 N-m

14.49. $\Delta = \pi F R^3\left[\dfrac{1}{EI} + \dfrac{3}{GK_T}\right]$

14.50. 496.2 ft-lb

3.848 ft-lb

14.51. —

14.52. $\phi_{AB} = .311°$

$\phi = .739°$

14.53. $\phi_A = .01321$ rad

$\phi_{1\text{-}2} = .000979$ rad

14.54. $.00207\,\dfrac{G}{L}\,\dfrac{\text{N-m}}{\text{rad}}$

$.00233\,\dfrac{G}{L}\,\dfrac{\text{N-m}}{\text{rad}}$

14.55. $T_A = 227.7$ N-m
$T_B = 572.3$ N-m
$\phi_A = .4554°$
$\tau_{max} = 2.33 \times 10^7$ Pa

14.56. $T_A = 978.6$ ft-lb

14.57. $M_A = 250$ N-m
$l = .721$ m

14.58. $\phi = 1.547 \times 10^{-4}$ rad

14.59. —

14.60. $\phi = 2.32 \times 10^{-4}$ rad

14.61. 33.75 N-m
67.55 N-m

14.62. —

14.63. 110 in.-lb

14.64. $t = .001733$ m

14.65. 5.371×10^7 Pa
3.007×10^7 Pa
$\phi_A = .07595$ rad

14.66. $\phi = .000538$ rad

14.67. 479 psi

14.68. 480 psi
$\phi = .00197$ rad

14.69. 304.3 ft-lb
495.7 ft-lb

14.70. 486.6 N-m
188.4 N-m

14.71. 452.2 N-m
347.8 N-m

14.72. $T_1 = -496.2$ ft-lb

14.73. 3761 psi
0 psi
−21,797 psi

14.74. $b = .577$ in.

14.75. Bredt $4.50\ ab^3$
Approx. $4.70\ ab^3$

14.76. $\Delta = 3.62$ in.

14.77. $M_1 = -495.7$ ft-lb
$M_2 = -304.3$ ft-lb

14.78. $\frac{d_1}{d_2} = 2.1$

15.1. $a_{nx} = .11$
$a_{ny} = .35$
-2.12×10^6 Pa

15.2. 199.8 psi

15.3. —

15.4. —

15.5. —

15.6. .543 μ-p

15.7. 4250 psi

15.8. —

15.9. 258 psi

15.10. $\tau_{x'y'} = 323$ psi

15.11. $\tau_{x'x'} = 3.5 \times 10^5$ Pa
$\tau_{x'y'} = 3.5 \times 10^5$ Pa
etc

15.12. $\begin{cases} \tau_1 = 1035 \text{ psi} \\ \tau_2 = 0 \text{ psi} \\ \tau_3 = -435 \text{ psi} \end{cases}$
$\begin{cases} a_{nx} = .750 \\ a_{ny} = .544 \\ a_{nz} = .375 \end{cases}$

15.13. $\begin{cases} \tau_1 = 1309 \text{ psi} \\ \tau_2 = 191 \text{ psi} \\ \tau_3 = 0 \text{ psi} \end{cases}$
$\begin{cases} a_{nx} = -.526 \\ a_{ny} = 0 \\ a_{nz} = .851 \end{cases}$

15.14. $\tau_1 = 6445$ psi
$\tau_2 = 2188$ psi
$\tau_3 = -1633$ psi

15.15. $\begin{cases} \tau_1 = 749 \text{ psi} \\ \tau_2 = 527 \text{ psi} \\ \tau_3 = -176 \text{ psi} \end{cases}$
$\begin{cases} a_{nx} = .90 \\ a_{ny} = -.266 \\ a_{nz} = -.348 \end{cases}$

15.16. —

15.17. 4676 psi

15.18. —

15.19. —

16.1. $.6198 \times 10^{-2}$

16.2. $\varepsilon_{nn} = .0400$
$\varepsilon_{ns} = .0318$

16.3. $\varepsilon_{x'x'} = .0100$
$\varepsilon_{x'y'} = .0354$

16.4. $\varepsilon_{x'x'} = .01375$
etc

16.5. —

16.6. 3.42×10^{-4}

16.7. —

16.8. $\gamma_{x'y'} = -1.732 \times 10^{-3}$

16.9. $\varepsilon_1 = 6 \times 10^{-6}$
$\varepsilon_2 = 3.621 \times 10^{-6}$
$\varepsilon_3 = -.6213 \times 10^{-6}$

16.10. $\varepsilon_1 = 3.828 \times 10^{-3}$
$\varepsilon_2 = 2 \times 10^{-3}$
$\varepsilon_3 = -1.828 \times 10^{-3}$

16.11. $\varepsilon_1 = 11.77 \times 10^{-4}$
$\varepsilon_2 = -7.597 \times 10^{-4}$
$\varepsilon_3 = -20.4 \times 10^{-4}$

16.12. $\varepsilon_1 = 3.998 \times 10^{-4}$
$\varepsilon_2 = -8.22 \times 10^{-5}$
$\varepsilon_3 = -3.812 \times 10^{-4}$

16.13. —

16.14. $\varepsilon_{x'x'} = 2.93 \times 10^{-2}$
$\varepsilon_{x'y'} = 1.116 \times 10^{-2}$

16.15. .0314
−.0414

16.16. —

16.17. —

16.18. —

16.19. —

17.1. $P_{cr} = \dfrac{k}{L}[(L-a)^2 + b^2]$

17.2. $P_{cr} = 12\,k$

17.3. $P_{cr} = K_T/L$

17.4. $P_{cr} = \dfrac{2K_T}{L}$

17.5. $P_{cr} = 20{,}439$ lb

17.6. —

17.7. $h = 1.457$ in.

17.8. $p = 4.509 \times 10^6$ Pa gage

17.9. $\Delta T_{cr} = \dfrac{4\pi^2 I}{\alpha A L^2}$

17.10. $W = 382{,}000$ lb

17.11. 6460 N

17.12. 112.4 g's

17.13. $P_{cr} = 702.8$ lb

17.14. $P_{max} = 31{,}810$ N

17.15. —

17.16. $\omega_1 = 19.49$ rad/sec
$\omega_2 = 38.98$ rad/sec

17.17. 3.432×10^4 lb

17.18. 2.796×10^3 lb
9.388×10^4 lb

17.19. —

17.20. $W_G = 163.9$ lb
$L = 24.8$ in.

17.21. $y = 19.92$ ft

17.22. clamped-free
$L_e = 2$ L
etc

17.23. 9791 N

17.24. 39,160 N

17.25. width = 9.62 in.

17.26. 3.238×10^5 lb

17.27. $h = .7644$ m

17.28. —

17.29. 2.667 in.

17.30. .03125 m

17.31. .249 in.

17.32. $P = 16{,}492$ lb

17.33. —

17.34. .03876 m

17.35. $F = 5126$ lb.

17.36. $w_o = 111$ N/m
$w_o = 999.3$ N/m

17.37. 37,111 N

17.38. $P_{max} = 51{,}404$ lb

17.39. $W = 990$ kN

17.40. $h = 1.014$ m
$W = 2356$ kN

17.41. $h = .984$ m

17.42. $w_o = 1.997 \times 10^4$ lb/ft

17.43. $P_{cr} = \dfrac{\pi^2 EI}{4L^2}$

17.44. $(P_{cr})_1 = 3.244 \times 10^5$ lb

$(P_{cr})_2 = 8.84 \times 10^5$ lb

17.45. $v = -\dfrac{w_o EI}{P^2 \sin\sqrt{\dfrac{P}{EI}}L} \sin\sqrt{\dfrac{P}{EI}}x +$

$\left(\dfrac{w_o L}{6P} + \dfrac{w_o EI}{LP^2}\right)x - \dfrac{w_o}{6LP}x^3$

17.46. $v = \dfrac{P_1 \sin\sqrt{\dfrac{P_2}{EI}}\dfrac{L}{2}}{2P_2\sqrt{\dfrac{P_2}{EI}}\cos\sqrt{\dfrac{P_2}{EI}}\dfrac{L}{2}} + \dfrac{P_1 L}{4P_2}$

17.47. $v_{max} = .0075$ in.

17.48. 183,948 lb

17.49. $P = 1730$ lb

17.50. 4.910×10^6 N

17.51. 857 lb

3418 lb

2140 lb

18.1. $\dfrac{939.7}{EA} = .057u + .00215v$

$-\dfrac{342}{EA} = .002148u + 1.100v$

18.2. $E\dfrac{u}{10} + .940\,E[.0883u - .0324v] +$

$.866\,E[.07499u + .0433v] = \dfrac{707}{A}$

18.3. $u_C = \dfrac{2.078 \times 10^5}{EA}$ in.

$v_C = \dfrac{6.369 \times 10^5}{EA}$ in.

18.4. $u = 0$ m

$v = \dfrac{4.35 \times 10^3}{AE}$ m

18.5. $u_C = -893\dfrac{L}{EA}$

$v_C = 442.6\dfrac{L}{EA}$

18.6. $u_B = \dfrac{2071}{EA}$ m

$v_B = 0.$

$v_C = \dfrac{7930}{EA}$ m

18.7. $u_A = -\dfrac{1045L}{EA}$

$u_B = \dfrac{1784L}{EA}$

18.8. $u_A = \dfrac{638L}{AE}$

$u_B = -\dfrac{1545L}{AE}$

18.9. $U = .0324$ ft-lb.

18.10. $U = .09714$ ft-lb.

18.11. $U_{\varepsilon_{nn}} = \dfrac{8 \times 10^4}{EI}$ ft-lb

18.12. $U_{\varepsilon_{nn}} = \dfrac{15{,}526}{EI_{zz}}$

18.13. $U_{\varepsilon_{nn}} = \dfrac{9.311 \times 10^6}{EI}$ ft-lb

18.14. $U = \dfrac{6667}{G}$ ft-lb

18.15. $U_{\varepsilon_{nn}} = \dfrac{1.102 \times 10^9}{EI}$ ft-lb

18.16. $U = .0544$ ft-lb

18.17. $U = \dfrac{1}{G}(1.046 \times 10^{12})$ N-m

18.18. $\pi = \dfrac{EA}{2}\left\{\dfrac{(u/5)^2}{5} + \dfrac{(.1500u + .0884v)^2}{5/.866} + \dfrac{(-.0884u + .0884v)^2}{8}\right\} - 866v - 500u$

18.19. $u = \dfrac{2.077 \times 10^5}{EA}$ in.

$v = \dfrac{6.368 \times 10^5}{EA}$ in.

18.20. $u = 0.$

$v = \dfrac{4345}{EA}$ m

18.21. $u = \dfrac{893L}{EA}$

$v = \dfrac{442L}{EA}$

18.22. $v_B = -\dfrac{7929}{EA}$ m

$u_B = \dfrac{2071}{EA}$ m

$u_C = 0.$

18.23. $u_B = \dfrac{1783L}{EA}$

$u_A = -\dfrac{1045L}{EA}$

18.24. $v = -\dfrac{6.370 \times 10^5}{EA}$ in.

$u = \dfrac{2.077 \times 10^5}{EA}$ in.

18.25. $u = 0$

$v = -\dfrac{4350}{EA}$ m

18.26. $u = -896\dfrac{L}{EA}$

$v = -442\dfrac{L}{EA}$

18.27. $u_B = \dfrac{2073}{EA}$

$v_B = -\dfrac{7928}{EA}$

18.28. $u_A = -\dfrac{1045L}{EA}$

$u_B = \dfrac{1784L}{EA}$

18.29. $\Delta = \dfrac{1}{AE}(116{,}569)$ ft

18.30. $\Delta_C = \dfrac{3046.02}{AE}$ m

18.31. $\dfrac{1}{EA}(61.5 \times 10^3)$ m

$\dfrac{1}{EA}(40.62 \times 10^3)$ m

18.32. $\dfrac{1}{EA}(4.126 \times 10^5)$ ft

18.33. $F_{AC} = 1414.2$ lb

$F_{BC} = -442.2$ lb

$F_{CD} = 1558$ lb

etc

18.34. $F_{AF} = 4.33$ K

$F_{BF} = 3.84$ K

$F_{CB} = -5.19$ K

etc

18.35. $F_{AB} = 29.24$ kN

$F_{AC} = -27.47$ kN

$F_{CD} = -27.47$ kN

$F_{BD} = -21.93$ kN

$F_{DG} = -50.14$ kN

etc

18.36. $F_{AB} = -3712$ lb

$F_{AC} = 2625$ lb

$F_{BD} = -3745$ lb

$F_{CD} = -715.7$ lb

$F_{DF} = -3912.7$ lb

etc

18.37. $\Delta = \dfrac{3\pi F r^3}{EI}$

18.38. $\Delta = \dfrac{116{,}569}{AE}$ ft

18.39. $\Delta = \dfrac{3046}{AE}$ m

18.40. $\theta = \dfrac{1}{24}\dfrac{wL^3}{EI}$

18.41. $M = -0.243\, wR^2$

$N = -0.327\, wR$

Index